SIPRI Yearbook 1993
World Armaments and Disarmament

**sipri**

**Stockholm International Peace Research Institute**

SIPRI is an independent international institute for research into problems of peace and conflict, especially those of arms control and disarmament. It was established in 1966 to commemorate Sweden's 150 years of unbroken peace.

The Institute is financed mainly by the Swedish Parliament. The staff, the Governing Board and the Scientific Council are international.

The Governing Board and the Scientific Council are not responsible for the views expressed in the publications of the Institute.


**Governing Board**

Professor Daniel Tarschys, MP, Chairman (Sweden)
Sir Brian Urquhart, Vice Chairman (United Kingdom)
Dr Oscar Arias Sánchez (Costa Rica)
Dr Gyula Horn (Hungary)
Professor Emma Rothschild (United Kingdom)
Dr Lothar Rühl (Germany)
The Director

**Director**

Dr Adam Daniel Rotfeld (Poland)



**sipri**

**Stockholm International Peace Research Institute**
Pipers väg 28, S-170 73 Solna, Sweden
Cable: SIPRI
Telephone: 46 8/655 97 00
Telefax: 46 8/655 97 33

# SIPRI Yearbook 1993

# World Armaments and Disarmament

**sipri**

Stockholm International Peace Research Institute

OXFORD UNIVERSITY PRESS
1993

*Oxford University Press, Walton Street, Oxford OX2 6DP*
*Oxford New York Toronto*
*Delhi Bombay Calcutta Madras Karachi*
*Petaling Jaya Singapore Hong Kong Tokyo*
*Nairobi Dar es Salaam Cape Town*
*Melbourne Auckland*
*and associated companies in*
*Berlin Ibadan*

*Oxford is a trade mark of Oxford University Press*

*Published in the United States*
*by Oxford University Press, New York*



*Yearbooks before 1987 published under title*
*'World Armaments and Disarmament:*
*SIPRI Yearbook [year of publication]'*



*British Library Cataloguing in Publication Data*
*Data available*
*ISSN 0953–0282*
*ISBN 0–19–829166–3*

*Library of Congress Cataloging in Publication Data*
*Data available*
*ISSN 0347–2205*
*ISBN 0–19–829166–3*

*Typeset and originated by Stockholm International Peace Research Institute*
*Printed and bound in Great Britain by*
*Biddles Ltd., Guildford and King's Lynn*

# Contents


**Preface** xv

**Glossary** xvi

**Introduction: parameters of change** 1
*Adam Daniel Rotfeld*

I. The new security environment 1
II. Arms control after the cold war 4
III. Consequences for the research agenda 7
IV. *Yearbook* findings 9

## Part I. The environment and security, 1992

**Chapter 1. The environment, security and development** 15
*Gro Harlem Brundtland*

I. Common security and the outlook for peace 15
II. A comprehensive concept of security: peace, the environment and development 17
III. Regional co-operation and global decision making 22

**Appendix 1A. Resource scarcity and environmental security** 27
*Richard H. Moss*

I. Introduction 27
II. The concept of environmental security 27
Resource scarcity, social mobilization and domestic political processes—Resources and international co-operation
III. Resource exploitation 29
Resource demands—Availability of resources
IV. Environmental indicators 32
Table 1A.1. Driving forces of resource demands for selected countries 33
Table 1A.2. Water resource use for selected countries 34
Table 1A.3. Land resource use for selected countries 35

**Appendix 1B. The United Nations Conference on Environment and Development** 37
The Rio Declaration—Agenda 21

## Part II. Global and regional security and conflicts, 1992

**Chapter 2. Preventive diplomacy, peace-making and peace-keeping** 45
*Roger Hill*

I. Introduction 45
II. The UN agenda for peace in 1992 46
*An Agenda for Peace*—Report of the work of the United Nations
III. The UN peace effort in 1992: main activities 49
Preventive diplomacy—Peace-making: mediation and negotiation—Peace-making: peace enforcement—Peace-keeping—Peace-building
IV. Regional and other contributions 58
V. Conclusion 60
Table 2.1. Operation Restore Hope—Somalia 51
Table 2.2. Composition of UN peace-keeping forces, end-November 1992 61

**Appendix 2A. An Agenda for Peace** 66

**Chapter 3. Major armed conflicts** 81
*Ramses Amer, Birger Heldt, Signe Landgren, Kjell Magnusson, Erik Melander, Kjell-Åke Nordquist, Thomas Ohlson and Peter Wallensteen*

I. Introduction 81
II. Changes in the list of conflicts for 1992 82
Conflicts recorded for 1981 which do not appear in the data for 1992—New conflicts in 1992
III. The level of conflict intensity and the Middle East peace process 83
Decreased intensity—Increased intensity—Regulation—The Israel/Palestine conflict and the Middle East peace process
IV. The end of the cold war: measuring the effects 86
V. Bosnia and Herzegovina 88
Events in 1992—The character of the war—The possibility of major-power intervention
VI. The post-Soviet armed conflicts 93
Azerbaijan: Nagorno-Karabakh—Georgia: three civil wars—Moldova: the Trans-Dniester region—Russia: the North Caucasus—Tajikistan
VII. Cambodia 107
Developments in 1978–91—Developments in 1992—Major power interest in Cambodia
VIII. Southern Africa 112
South Africa—Angola—Mozambique—War termination: going from war to peace

Table 3.1. Regional distribution of conflict locations with at least one major armed conflict, 1989–92 86
Table 3.2. Regional distribution, number and types of contested incompatibilities in major armed conflicts, 1989–92 87

**Appendix 3A. Major armed conflicts, 1992** 119
*Birger Heldt and Erik Melander*

Table 3A.1. Conflict locations with at least one major armed conflict in 1992 121

**Chapter 4. Post-Soviet conflict heritage and risks** 131
*Vladimir Baranovsky*

I. Introduction 131
II. Challenges of the transitional period 132
Uncertainties regarding the CIS—A political system in crisis—Frontiers and territorial integrity under question
III. Ethno-nationalism 140
Ethnocentrism in the political process—Minorities and relations between post-Soviet states—Disintegration trends
IV. The military dimension 148
Nuclear weapons—The transition from Soviet to national armed forces
V. Conclusion 155

**Appendix 4A. Chronology of conflict developments in the former USSR, 1992** 159
*Shannon Kile*

**Chapter 5. The CSCE: towards a security organization** 171
*Adam Daniel Rotfeld*

I. Introduction 171
II. New challenges, new tasks 172
III. Four meetings in Helsinki 175
IV. The Stockholm Decisions 180
The Forum for Security Co-operation

V. New participating states 184
VI. Concluding remarks 185

**Appendix 5A. Key CSCE documents in 1992** 190
Helsinki Document 1992—Decision on Peaceful Settlement of Disputes—Shaping a New Europe–The Role of the CSCE

# Part III. Weapons and technology proliferation, 1992

**Chapter 6. Nuclear weapon developments and proliferation** 221
*Dunbar Lockwood and Jon Brook Wolfsthal*

I. Introduction 221
II. US nuclear weapon programmes 222
Strategic nuclear weapons—Non-strategic nuclear weapons
III. Former Soviet and CIS nuclear weapon programmes 226
Strategic nuclear weapons—Non-strategic nuclear weapons
IV. French nuclear weapon programmes 228
Strategic nuclear weapons—Non-strategic nuclear weapons
V. British nuclear weapon programmes 230
Strategic nuclear weapons—Non-strategic nuclear weapons
VI. Chinese nuclear weapon programmes 232
Strategic nuclear weapons—Non-strategic nuclear weapons
VII. Summary of nuclear weapon developments 233
VIII. Nuclear weapon proliferation 240
Developments in the nuclear weapon non-proliferation regime—The Nuclear Suppliers Group
IX. Regional developments 244
Asia—South Asia—The Middle East—Africa

Table 6.1. US strategic nuclear forces, January 1993 234
Table 6.2. CIS strategic nuclear forces, January 1993 235
Table 6.3. French nuclear forces, January 1993 237
Table 6.4. British nuclear forces, January 1993 238
Table 6.5. Chinese nuclear forces, January 1993 239
Table 6.6. The Nuclear Suppliers Group list of nuclear-related dual-use equipment and materials and related technology 242
Figure 6.1. Projected US and Russian strategic nuclear forces after implementation of the START Treaty and the START II Treaty 236

**Appendix 6A. Nuclear explosions, 1945–92** 254
*Ragnhild Ferm*

Table 6A.1. Registered nuclear explosions in 1992 255
Table 6A.2. Estimated number of nuclear explosions 16 July 1945–5 August 1963 (the signing of the Partial Test Ban Treaty) 255
Table 6A.3. Estimated number of nuclear explosions 6 August 1963–31 December 1992 256
Table 6A.4. Estimated number of nuclear explosions 16 July 1945–31 December 1992 257

**Chapter 7. Chemical and biological weapons: developments and proliferation** 259
*Thomas Stock*

I. Introduction 259
II. Allegations of CW and BW use 260
The new republics—Mozambique—The former Yugoslavia—Iraq
III. Allegations of CW and BW possession 264
Non-lethal warfare
IV. CBW proliferation and measures to halt it 268

V. Destruction of chemical weapons 273
US–Russian bilateral agreements—The US CW destruction programme—The Russian CW destruction programme—Canadian CW destruction—Iraqi CW destruction
VI. Old CW ammunition and toxic armament wastes 282
VII. New developments in NBC protection 285
VIII. New BW developments 286
IX. Environmental implications of the Persian Gulf War 289
X. Conclusions 289
Table 7.1. US Army plans for destruction of the US chemical stockpile by 2004 275

**Appendix 7A. Benefits and threats of developments in biotechnology and genetic engineering** 293
*Tamas Bartfai, S. J. Lundin and Bo Rybeck*

I. Introduction 293
II. Advances in scientific knowledge and genome mapping 294
III. Biotechnology 295
Heterologous gene expression—Genetically engineered organisms—Protein engineering—Human monoclonal antibodies
IV. Medical and health improvement 298
V. Genomic diversity and DNA fingerprinting 299
The Genome Diversity Project—DNA fingerprinting
VI. The possible use of biotechnology for political and weapon purposes and countermeasures 300
Enhancement of bacterial and viral virulence—Heterologous gene expression and protein engineering of toxins—Natural variability—Use as weapons—Can 'genetic weapons' be developed?
VII. Summary and conclusions 305

**Chapter 8. Military technology and international security: the case of the USA** 307
*Eric H. Arnett and Richard Kokoski*

I. Introduction 307
II. R&D and US military supremacy 307
III. President Clinton and US aspirations 309
IV. Legacies of the cold war 311
V. Summary of US R&D programmes 313
DARPA—SDIO
VI. The new era in military space 321
Surveillance and intelligence—Navigation—Nuclear and space war-fighting
VII. Controlling the spread of technology 324
The role of normative arms control
VIII. Reconciling the aspirations and legacies 327
Reconstitution
IX. Conclusion 328

**Appendix 8A. Military satellites launched in 1992** 330
*John Pike*

# Part IV. Military expenditure, arms production and trade, 1992

**Chapter 9. World military expenditure** 337
*Saadet Deger*

I. Introduction 337
II. The United States 341
The budget—The past—The future

| | | |
|---|---|---|
| III. | Russia and the CIS<br>*Glasnost—Perestroika—Konversiya* | 351 |
| IV. | European NATO<br>NATO military expenditure—Procurement—The armed forces—The United Kingdom | 370 |
| V. | The European Community | 381 |
| VI. | China and Japan<br>China—Japan | 386 |
| VII. | The developing world | 393 |
| VIII. | Conclusion | 397 |
| Table 9.1. | US Department of Defense and 'national defense' expenditure, budget authority and outlay, FYs 1990–93 | 343 |
| Table 9.2. | Comparison of the presidential request for FY 1993 and Congress-approved final appropriations: procurement of major weapons | 344 |
| Table 9.3. | Comparison of the presidential request for military R&D programmes in FY 1993 and Congress-approved final appropriations | 345 |
| Table 9.4. | Allocation of US 'national defense' budget authority, FYs 1983–92 | 346 |
| Table 9.5. | Comparison of US defence-related manpower during the cold war (1975–90) and the projected figures for 1997 | 347 |
| Table 9.6. | Comparison of US military expenditure (budget authority) during the cold war (1975–90) and the projected figures for 1997 | 348 |
| Table 9.7. | US Department of Defense, TOA by service, FYs 1989–93 | 349 |
| Table 9.8. | Comparison of US force capabilities, 1990–97 (projected) | 350 |
| Table 9.9. | Comparison of US defence-related manpower, 1990–97 (projected) | 350 |
| Table 9.10. | Comparison of the military expenditure plans of Presidents Bush and Clinton, 1993–96 | 351 |
| Table 9.11. | Budget deficit of Russia, 1986–92 | 361 |
| Table 9.12. | Examples of collaborative ventures by defence industrial enterprises of the CIS countries | 364 |
| Table 9.13. | Foreign aid provided by the US Government for Central and Eastern Europe and the CIS countries (budget authority, FYs 1992 and 1993) | 367 |
| Table 9.14. | NATO military expenditure, in current price figures, 1983–92 | 368 |
| Table 9.15. | NATO military expenditure, in constant price figures, 1983–92 | 369 |
| Table 9.16. | NATO countries' military expenditure as a percentage of GDP, 1983–92 | 370 |
| Table 9.17. | NATO major weapon procurement expenditure, 1983–92 | 372 |
| Table 9.18. | NATO and EC major weapon procurement expenditure, 1983–92 | 373 |
| Table 9.19. | The rise and fall of European military procurement expenditures, 1983–92 | 375 |
| Table 9.20. | NATO armed forces, total military personnel, 1983–92 | 376 |
| Table 9.21. | NATO military and civilian personnel, as share of total labour force, 1983–92 | 377 |
| Table 9.22. | Applicants for EC membership: military expenditure, in current price figures, 1983–92 | 383 |
| Table 9.23. | Applicants for EC membership: military expenditure, in constant price figures, 1983–92 | 383 |
| Table 9.24. | Comparative economic and military indicators of the European Community countries, applicants for EC membership, the USA and Japan, 1991 | 384 |
| Table 9.25. | Official figures for China's military expenditure, 1983–92 | 388 |
| Table 9.26. | Allocations of the Japanese national defence budget, FYs 1990–92 | 391 |
| Table 9.27. | Japanese defence expenditure, FYs 1975–92 | 392 |
| Table 9.28. | Long-term debt and financial flows, growth rates per capita and military expenditure as a share of GDP/GNP of low- and middle-income economies, 1990–91 | 394 |

**Appendix 9A. Military expenditure in the Central and East European countries** 398
*Evamaria Loose-Weintraub*

I. Introduction 398
II. The Czech and Slovak Republics 401
The dissolution of the Federation—Economic development—Military postures and new military doctrine—Trends and military expenditure
III. Hungary 405
Economic development—Military postures and new military doctrine—Trends in military expenditure
IV. Poland 407
Economic development—Military postures and new military doctrine—Trends in military expenditure
V. Bulgaria 410
Economic development—Security, defence and the armed forces—Trends in military expenditure
VI. Romania 412
Economic development—Security, defence and the armed forces—Trends in military expenditure
Table 9A.1. Macroeconomic inicators in Central and East European countries, 1990 and 1991 399
Table 9A.2. Indicators of external indebtedness in Central and East European countries, 1990 400
Table 9A.3. Czechoslavakia's military expenditure allocation, official figures, 1989–92 404
Table 9A.4. Hungary's military expenditure allocation, official figures, 1989–92 407
Table 9A.5. Poland's military expenditure allocation, official figures, 1989–92 409
Table 9A.6. Bulgaria's military expenditure allocation, official figures, 1989–92 411
Table 9A.7. Romanian military expenditure allocation, official figures, 1990–92 413
Table 9A.8. Official estimates of Central and East European countries' CFE-limited weapons in Europe before and after the CFE Treaty implementation 414

**Chapter 10. Arms production and arms trade** 415
*Ian Anthony, Paul Claesson, Elisabeth Sköns and Siemon T. Wezeman*

I. Introduction 415
II. Important developments of 1992 418
The United Nations embargo on Yugoslavia—Fighter aircraft decisions in 1992
III. Arms production in the OECD and developing countries 427
The 'SIPRI 100': developments in the SIPRI list of companies—National restructuring—Internationalization
IV. The trade in major conventional weapons 443
The major arms exporters—Arms imports by Asian countries—Land-system retrofits
V. Arms transfer control initiatives 459
Multilateral export control processes in 1992—National arms transfer control initiatives in 1992
Table 10.1. EFA-related decisions taken in 1992 426
Table 10.2. Revised work-share for the EFA based on announced plans 427
Table 10.3. Regional/national shares of arms sales for the 100 largest arms-producing companies, 1991 compared to 1990 428
Table 10.4. Companies whose arms sales changed the most in 1991 429
Table 10.5. Employment changes among the largest arms-producing companies in OECD and developing countries, 1988–91 431
Table 10.6. Major acquisitions in the US arms industry announced in 1992 434

Table 10.7. International take-overs in the arms industry decided in 1992 440
Table 10.8. International joint ventures in the arms industry decided in 1992 441
Table 10.9. Teaming arrangements to bid for the US JPATS contract 442
Table 10.10. The leading exporters of major conventional weapons, 1988–92 444
Table 10.11. The leading importers of major conventional weapons, 1988–92 445
Table 10.12. Official arms export data 446
Table 10.13. Value of US Foreign Military Sales agreements and deliveries, 1982–91 447
Table 10.14. Regional distribution of deliveries of arms and military equipment by the former Soviet Union in 1991 449
Table 10.15. Distribution by weapon category of deliveries of arms and military equipment by the former Soviet Union in 1991 450
Table 10.16. Selected international retrofits of armoured vehicles in progress in 1992 458
Figure 10.1. The downward trend in the aggregate value of deliveries of major conventional weapon systems, 1988–92 417

**Appendix 10A. The 100 largest arms-producing companies, 1991** 469
*Ian Anthony, Paul Claesson, Gerd Hagmeyer-Gaverus, Elisabeth Sköns and Siemon T. Wezeman*

Table 10A. The 100 largest arms-producing companies in the OECD and the developing countries, 1991 470

**Appendix 10B. Tables of the value of the trade in major conventional weapons** 476
*Ian Anthony, Paul Claesson, Gerd Hagmeyer-Gaverus, Elisabeth Sköns and Siemon T. Wezeman*

Table 10B.1. Values of imports of major conventional weapons, 1983–92 476
Table 10B.2. Values of exports of major conventional weapons, 1983–92 477
Table 10B.3. World trade in major conventional weapon systems, 1988–92 479

**Appendix 10C. Register of the trade in and licensed production of major conventional weapons in industrialized and developing countries, 1992** 483
*Ian Anthony, Paul Claesson, Gerd Hagmeyer-Gaverus, Elisabeth Sköns and Siemon T. Wezeman*

**Appendix 10D. Sources and methods** 519
I. The SIPRI sources 519
II. Selection criteria 519
III. The value of the arms trade 520
IV. Conventions 520

**Appendix 10E. An overview of the arms industry modernization programme in Turkey** 521
*Gülay Günlük-Senesen*

I. Introduction 521
II. The strategic context 522
III. The modernization programme 523
Background—The administrative basis—Financing—Domestic spill-over expectations—The modernization programme in practice—The aerospace industry
IV. Conclusions 532
Table 10E.1. Administration of the Turkish military modernization programme, in order of decision and implementation, since 1989 525
Table 10E.2. Leading arms producers in Turkey, on the basis of SASAD membership 526
Table 10E.3. Military projects initiated in Turkey as of mid-1992 529
Table 10E.4. Military projects under negotiation as of mid-1992 529

**Appendix 10F. The United Nations Register of Conventional Arms** 533
*Herbert Wulf*

I. Introduction 533
II. The history of the Register and transparency in the armaments process 534
III. The 1991 UN General Assembly decision 535
IV. Technical procedures 537
The mandate of the Panel—Adjustment and definitions of weapon categories—Designations of weapon systems—Modalities for expansion of the Register
V. The objectives of the Register and the prospects for implementation 542
Figure 10F.1. Standardized forms for reporting international transfers of conventional arms 541

**Appendix 10G. Documents on arms export control, 1992** 545
Declaration of the CSCE Council on Non-Proliferation and Arms Transfers—Interim Guidelines related to Weapons of Mass Destruction

# Part V. Arms control and disarmament, 1992

**Chapter 11. Nuclear arms control** 549
*Dunbar Lockwood*

I. Introduction 549
II. The 1991 START Treaty 549
III. The 1993 START II Treaty 554
SS-18 missile silos—SS-19 downloading—Bomber issues—The START II Treaty benefits both countries—Savings from the START II Treaty—The future of US–Russian strategic arms control
IV. The Non-Proliferation Treaty 560
The IAEA
V. A comprehensive test ban 561
The United States—Russia—France—China—The United Kingdom
VI. The Nuclear Suppliers Group 565
VII. The Safety, Security and Dismantlement Talks 566
Russia—Belarus—Ukraine—Kazakhstan—The 1993 US–Russian HEU agreement
VIII. Fissile material production cut-off 571
IX. The Treaty of Tlatelolco 571
X. Conclusion 572

**Appendix 11A. Documents on nuclear arms control** 574
State of the Union Address by President George Bush—Protocol to Facilitate the Implementation of the START Treaty (Lisbon Protocol)—Joint US–Russian Understanding on Further Reductions in Strategic Offensive Arms (De-MIRVing Agreement)—Treaty between the United States of America and the Russian Federation on Further Reduction and Limitation of Strategic Offensive Arms (START II Treaty)

**Chapter 12. Conventional arms control in Europe** 591
*Jane M. O. Sharp*

I. Introduction 591
II. Ratification of the CFE Treaty 592
The Baltic states opt out of the CFE Treaty—Former Soviet republics become parties to the CFE Treaty—Attitudes to the CFE Treaty among the former Soviet republics—Russian problems with the CFE Treaty—Peace-keeping or military intervention?—Russian troop withdrawals from the Baltic States—Russian troops withdrawals from the Transcaucasus and Moldova

III. Implementation of the CFE Treaty 606
Exchange of data—Inspections
IV. The CFE-1A Agreement 613
Issues in the negotiations
V. Conclusion 615
Table 12.1 Ratification of the CFE Treaty 592
Table 12.2. Allocation of treaty-limited equipment entitlements among the former Soviet republics 597
Table 12.3. Treaty-limited equipment in the former Soviet republics compared with Soviet holdings in 1988 602
Table 12.4. NATO TLE holdings in 1990–92 and 1995 CFE Treaty ceilings 608
Table 12.5. Former WTO TLE holdings in 1990–92 and 1995 CFE Treaty ceilings 609
Table 12.6. Estimated passive and challenge inspections, June 1992 611
Table 12.7. CFE Treaty declared sites and objects of verification 612
Table 12.8 CFE-1A manpower limitations 614
Figure 12.1. Implementation and verification of the CFE Treaty and the CFE-1A Agreement 610

**Appendix 12A. The Vienna confidence- and security-building measures in 1992** 618
*Zdzislaw Lachowski*

I. Introduction 618
II. Implementation 620
Military activities—Annual exchange of military information—Risk reduction—Military contacts—Notification and observation—Constraining provisions—Compliance and verification—Communications—Annual assessment of implementation—New participants—The data bank and public access to the data
III. Assessment and outlook 630
Table 12A.1. Annual numbers of military exercises conducted by NATO, the WTO/ former WTO, and the neutral and non-aligned countries in 1989–92 and forecast for 1993 621
Table 12A.2. Notifiable military activities which were scaled down in 1992 621
Table 12A.3. Calendar of planned notifiable military activities in 1993, as required by the Vienna Document 1992 623
Table 12A.4. CBM/CSBM notification and observation thresholds, 1975–92 626

**Appendix 12B. The Treaty on Open Skies** 632
*Richard Kokoski*

I. Introduction 632
II. The Consultative Commission 632
III. Trial overflights in 1992 633
IV. Status of the Treaty and conclusion 634

**Appendix 12C. Documents on conventional arms control in Europe, 1992** 635
Vienna Document 1992—Treaty on Open Skies—Tashkent Document—Final Document of the Extraordinary Conference of the States Parties to the CFE Treaty (Oslo Document)—Provisional Appplication of the Treaty on Conventional Armed Forces in Europe of 19 November 1990—Concluding Act of the Negotiation on Personnel Strength of Conventional Armed Forces in Europe (CFE-1A Agreement)

**Chapter 13. The United Nations Special Commission on Iraq: activities in 1992** 691
*Rolf Ekéus*

I. Introduction 691
II. Iraqi non-compliance 692
III. UNSCOM surveillance activities 694

IV. UNSCOM inspection activities 697
Nuclear weapons—Ballistic missiles—Chemical weapons—Biological weapons—Inspection developments
V. Conclusion 702
Table 13.1. The UNSCOM inspection schedule, as of 31 December 1992 698

**Chapter 14. The Chemical Weapons Convention: the success of chemical disarmament negotiations** 705
*J. P. Perry Robinson, Thomas Stock and Ronald G. Sutherland*

I. Introduction 705
II. Historical overview of the CWC negotiation 708
Introduction—Security aspects of toxic weapons—Entry of CBW into the agenda of the Geneva disarmament conference—The exploratory talks on chemical weapons—The start of negotiations on chemical weapons—The end of the negotiations
III. The National Implementation provision 719
IV. The international organization 721
The Conference of the States Parties—The Executive Council—The Technical Secretariat—The Inspectorate
V. Destruction requirements under the CWC 724
VI. The verification regime under the CWC 726
Verification of activities not prohibited by the Convention—Challenge inspection—Investigation of alleged use
VII. The Preparatory Commission 730
VIII. Entry into force 732
IX. The CWC: some conclusions 732
Table 14.1. Signatory status of the Chemical Weapons Convention as of 8 February 1993 709

**Appendix 14A. The Convention on the Prohibition of the Development, Production, Stockpiling and Use of Chemical Weapons and on their Destruction** 735

# Annexes

**Annexe A. Major multilateral arms control agreements** 759
*Ragnhild Ferm*

I. Summaries of the agreements 759
II. Status of the implementation of the major multilateral arms control agreements, as of 31 December 1992 763
III. UN member states and year of membership, as of 31 December 1992 790

**Annexe B. Chronology 1992** 793
*Ragnhild Ferm*

**About the contributors** 805

**Abstracts** 812

**Errata** 818

**Index** 819

# Preface

The structure of this twenty-fourth edition of the *SIPRI Yearbook* differs from that of previous editions. The first two parts of the volume deal with the non-military aspects of security—activities connected with preventive diplomacy, peace-making and peace-keeping in the context of post-cold war conflicts and new security institutions. The volume contains extensive documentation. Agreements reached in 1992 closed an important chapter in the history of arms control and disarmament and many of them are reproduced here. It is hoped that, along with the analyses contained in this volume, the inclusion of texts of documents which are often not easily accessible will facilitate the work of experts, negotiators and journalists.

The intention of the authors was to publish material that, while chiefly documenting and analysing events of the past year, will be of significance for the future. Looking at the last League of Nations yearbook on armaments I noted that, although Europe was caught up in the greatest war in world history at the time of its publication, its authors stressed that 'it does show the world in arms organized on a peace footing as it was on the eve of the hostilities which broke out in Europe in September 1939' (*Armament Year-book 1939–40*, Geneva, 1940, p. 4). Our intention and assumption are quite different. The aim of the *SIPRI Yearbooks* is not only to report on current developments but also to encourage thinking that could help stave off hostilities and conflicts.

Apart from the chapters prepared by the in-house research staff, who continuously monitor the data and events, this *Yearbook* also contains analyses by prominent outside experts. We are proud to publish a chapter on the environment, security and development by Norwegian Prime Minister Gro Harlem Brundtland and a report by Ambassador Rolf Ekéus in his capacity as Executive Chairman of the UN Special Commission on Iraq, as well as several other outstanding contributions. We are also grateful to those who provided us with valuable suggestions in their role as external referees of the material.

Given the accelerated pace of recent international change, the editors deserve special mention for their perseverance in preparing final camera-ready copy of the texts. All the editors—Billie Bielckus, Jetta Gilligan Borg, Eve Johansson and Don Odom—have done a superb job under the experienced editorial leadership of Connie Wall. Their enthusiasm, devotion and skill made it possible to publish this volume in a short time. My thanks also go to Ragnhild Ferm and Shannon Kile, whose close co-operation with the editors and careful reading contributed to the accuracy of the *Yearbook.* I would like to thank Gerd Hagmeyer-Gaverus for programming and other computer support. I wish to express my appreciation to the secretaries—Cynthia Loo, Marianne Lyons, Miyoko Suzuki and Catherine Walsh Söderquist—for their assistance in preparing this *Yearbook.* The index was prepared by Peter Rea, UK.

Dr Adam Daniel Rotfeld
Director
May 1993

# GLOSSARY

## Acronyms

| | |
|---|---|
| ABM | Anti-ballistic missile |
| ABACC | Argentinian–Brazilian Agency for Accounting and Control of Nuclear Materials |
| ACE | Allied Command Europe (NATO) |
| ACM | Advanced cruise missile |
| ACV | Armoured combat vehicle |
| ADM | Atomic demolition munition |
| AFAP | Artillery-fired atomic projectile |
| AIFV | Armoured infantry fighting vehicle |
| ALCM | Air-launched cruise missile |
| AMF | Allied Mobile Force |
| ANC | African National Congress |
| ASAT | Anti-satellite |
| ASEAN | Association of South-East Asian Nations |
| ASLCM | Advanced sea-launched cruise missile |
| ASM | Air-to-surface missile |
| ASUW | Anti-surface warfare |
| ASW | Anti-submarine warfare |
| ATBM | Anti-tactical ballistic missile |
| ATC | Armoured troop carrier |
| ATTU | Atlantic-to-the-Urals (zone) |
| AWACS | Airborne warning and control system |
| BCC | Bilateral Consultation Commission |
| BMD | Ballistic missile defence |
| BW | Biological warfare/weapons |
| BWC | Biological Weapons Convention |
| CARICOM | Caribbean Common Market |
| CAS | Committee on Assurances of Supply |
| CBM | Confidence-building measure |
| CBW | Chemical and biological warfare/weapons |
| CD | Conference on Disarmament |
| CEP | Circular error probable |
| CFE | Conventional Armed Forces in Europe |
| $C^3I$ | Command, control, communications and intelligence |
| CIS | Commonwealth of Independent States |
| CMEA | Council for Mutual Economic Assistance (as COMECON) |
| COCOM | Coordinating Committee (on Multilateral Export Controls) |
| COMECON | Council for Mutual Economic Assistance (as CMEA) |
| CORRTEX | Continuous reflectometry for radius versus time experiments |
| CPC | Conflict Prevention Centre |
| CPI | Consumer price index |
| CSBM | Confidence- and security-building measure |
| CSCE | Conference on Security and Co-operation in Europe |
| CSO | Committee of Senior Officials |
| CTB(T) | Comprehensive test ban (treaty) |
| CTOL | Conventional take-off and landing |
| CW | Chemical warfare/weapons |
| CWC | Chemical Weapons Convention |

| | |
|---|---|
| CWFZ | Chemical weapon-free zone |
| DEW | Directed-energy weapon |
| DOD | Department of Defense (US) |
| DST | Defence and Space Talks |
| EC | European Community |
| ECOWAS | Economic Community of West African States |
| ECU | European Currency Unit |
| EEA | European Economic Area |
| EFA | European Fighter Aircraft |
| EFTA | European Free Trade Area |
| ELINT | Electronic intelligence |
| ELV | Expendable launch vehicle |
| EMP | Electromagnetic pulse |
| EMU | Economic and Monetary Union |
| Enmod | Environmental modification |
| EPU | European Political Union |
| ERW | Enhanced radiation (neutron) weapon |
| EU | European Union |
| EUCLID | European Cooperative Long-term Initiative on Defence |
| FAO | Food and Agriculture Organization |
| FBS | Forward-based system |
| FOC | Full operational capability |
| FOST | Force Océanique Stratégique |
| FOTL | Follow-on to Lance |
| FROD | Functionally related observable difference |
| FROG | Free-rocket-over-ground |
| FY | Fiscal year |
| GATT | General Agreement on Tariffs and Trade |
| GBR | Ground-based radar |
| GCC | Gulf Co-operation Council |
| GDP | Gross domestic product |
| GLCM | Ground-launched cruise missile |
| GNP | Gross national product |
| GPALS | Global Protection Against Limited Strikes |
| GPS | Global Protection System |
| HACV | Heavy armoured combat vehicle |
| HEU | Highly enriched uranium |
| HLTF | High Level Task Force |
| HLWG | High Level Working Group |
| IAEA | International Atomic Energy Agency |
| ICBM | Intercontinental ballistic missile |
| IEPG | Independent European Programme Group |
| IFV | Infantry fighting vehicle |
| IMF | International Monetary Fund |
| INF | Intermediate-range nuclear forces |
| IOC | Initial operational capability |
| IRBM | Intermediate-range ballistic missile |
| JCC | Joint Consultative Commission |
| JCG | Joint Consultative Group |
| JCIC | Joint Compliance and Inspection Commission |
| JSG | Joint Strategy Group |
| LDC | Less developed country |
| LDDI | Less developed defence industry |
| MAD | Mutual assured destruction |
| MARV | Manœuvrable re-entry vehicle |
| MD | Military District |
| MIC | Military–industrial complex |
| MIRV | Multiple independently targetable re-entry vehicle |
| MLRS | Multiple launcher rocket system |

| | |
|---|---|
| MOU | Memorandum of Understanding |
| MRV | Multiple re-entry vehicle |
| MSC | Military Staff Committee |
| MTCR | Missile Technology Control Regime |
| MTM | Multinational technical means (of verification) |
| NACC | North Atlantic Cooperation Council |
| NATO | North Atlantic Treaty Organization |
| NBC | Nuclear, biological and chemical (weapons) |
| NMP | Net material product |
| NNA | Neutral and non-aligned (states) |
| NPG | Nuclear Planning Group |
| NPT | Non-Proliferation Treaty |
| NRRC | Nuclear Risk Reduction Centre |
| NSG | Nuclear Suppliers Group |
| NST | Nuclear and Space Talks |
| NSWTO | Non-Soviet WTO |
| NTI | National trial inspection |
| NTM | National technical means (of verification) |
| NTS | Nevada test site |
| NWFZ | Nuclear weapon-free zone |
| OAS | Organization of American States |
| ODA | Official development assistance |
| OAU | Organization for African Unity |
| OECD | Organization for Economic Co-operation and Development |
| OMG | Operational Manœuvre Group |
| O&M | Operation and maintenance |
| OOV | Object of verification |
| OPANAL | Agency for the Prohibition of Nuclear Weapons in Latin America |
| OSI | On-site inspection |
| OSIA | On-Site Inspection Agency |
| PLO | Palestine Liberation Organization |
| PNE(T) | Peaceful Nuclear Explosions (Treaty) |
| POMCUS | Prepositioned Organizational Material Configured to Unit Sets |
| PTB(T) | Partial Test Ban (Treaty) |
| R&D | Research and development |
| RDT&E | Research, development, testing and evaluation |
| RMA | Restricted Military Area |
| RPV | Remotely piloted vehicle |
| RV | Re-entry vehicle |
| SACEUR | Supreme Allied Commander, Europe |
| SALT | Strategic Arms Limitation Talks |
| SAM | Surface-to-air missile |
| SCC | Standing Consultative Commission |
| SDI | Strategic Defense Initiative |
| SDIO | SDI Organization |
| SICBM | Small ICBM |
| SLBM | Submarine-launched ballistic missile |
| SLCM | Sea-launched cruise missile |
| SLV | Space launch vehicle |
| SNDV | Strategic nuclear delivery vehicle |
| SNF | Short-range nuclear forces |
| SSD | Safety, Security and Dismantlement (Talks) |
| SS(M) | Surface-to-surface (missile) |
| SRAM | Short-range attack missile |
| SRBM | Short-range ballistic missile |

| | |
|---|---|
| SSBN | Nuclear-powered, ballistic-missile submarine |
| SSGN | Nuclear-powered, guided-missile submarine |
| SSN | Nuclear-powered attack submarine |
| START | Strategic Arms Reduction Talks |
| SVC | Special Verification Commission |
| SWS | Strategic weapon system |
| TASM | Tactical air-to-surface missile |
| TEL | Transporter–erector–launcher |
| TLE | Treaty-limited equipment |
| TNF | Theatre nuclear forces |
| TTB(T) | Threshold Test Ban (Treaty) |
| UNCED | United Nations Conference on Environment and Development |
| UNIKOM | United Nations Iraq–Kuwait Observation Mission |
| UNOSOM | United Nations Operation in Somalia |
| UNPROFOR | United Nations Protection Force |
| UNSCOM | United Nations Special Commission on Iraq |
| UNTAC | United Nations Transitional Authority in Cambodia |
| UNTAG | United Nations Transition Assistance Group |
| UNTEA | United Nations Temporary Executive Authority |
| VCC | Verification Co-ordinating Committee |
| V/STOL | Vertical/short take-off and landing |
| WEU | Western European Union |
| WTO | Warsaw Treaty Organization (Warsaw Pact) |

# Glossary

| | |
|---|---|
| Anti-ballistic missile (ABM) system | Weapon system designed to defend against a ballistic missile attack by intercepting and destroying ballistic missiles and their warheads in flight. |
| Anti-Ballistic Missile (ABM) Treaty | Treaty signed by the USSR and the USA in 1972 in the SALT I process which prohibits the development, testing and deployment of sea-, air-, space- or mobile land-based ABM systems. |
| ATTU zone | The Atlantic-to-the-Urals zone of the 1990 Treaty on Conventional Armed Forces in Europe (CFE). The zone, stretching from the Atlantic Ocean to the Ural Mountains, comprises the entire land territory of the European NATO states, former WTO states, and European former Soviet republics (Armenia, Azerbaijan, Belarus, Georgia, Kazakhstan, Moldova, Russia and Ukraine). *See also* CFE-1A Agreement, Conventional Armed Forces in Europe (CFE) Treaty, Treaty-limited equipment (TLE). |
| Ballistic missile | A missile which follows a ballistic trajectory (part of which may be outside the earth's atmosphere) when thrust is terminated. |
| Binary chemical weapon | A shell or other device filled with two chemicals of relatively low toxicity which mix and react while the device is being delivered to the target, the reaction product being a super-toxic chemical warfare agent, such as nerve gas. |
| Biological weapon (BW) | A weapon containing living organisms, whatever their nature, or infective material derived from them, which are intended for use in warfare to cause disease or death in man, animals or plants, and which for their effect depend on their ability to multiply in the person, animal or plant attacked, as well as the means of their delivery. |
| CFE-1A Agreement | The Concluding Act of the Negotiation on Personnel Strength of Conventional Armed Forces in Europe (the CFE-1A Agreement) was signed in Helsinki on 10 July 1992 by the NATO states, Armenia, Azerbaijan, Belarus, Bulgaria, Czechoslovakia, Georgia, Hungary, Kazakhstan, Moldova, Poland, Romania, Russia and Ukraine, and entered into force on 17 July 1992. It sets limits on the number of military personnel permitted in the ATTU zone. *See also* ATTU zone, Treaty-limited equipment (TLE). |
| Charter of Paris for a New Europe | *See* Paris Documents. |
| Chemical weapon (CW) | Chemical substances—whether gaseous, liquid or solid—which might be employed as weapons in combat because of their direct toxic effects on man, animals or plants, and the means of their delivery. |

| | |
|---|---|
| Chemical Weapons Convention (CWC) | The multilateral Convention on the Prohibition of the Development, Production, Stockpiling and Use of Chemical Weapons and on their Destruction was opened for signature on 13 January 1993. It bans all chemical weapons world-wide, imposes a wide spectrum of inspections to verify the ban, outlaws any use of these weapons and imposes a strict ban on all activities to develop new chemical weapons. |
| Commonwealth of Independent States (CIS) | Established by Belarus, Russia and Ukraine in the Agreement on the Commonwealth of Independent States signed in Minsk on 8 December 1991 and joined by eight additional former Soviet republics in Alma-Ata on 21 December 1991. |
| Comprehensive test ban (CTB) | A proposed ban on all nuclear weapon tests in all environments. |
| Conference on Disarmament (CD) | Multilateral arms control negotiating body, based in Geneva, composed of states representing all the regions of the world and including all the permanent members of the UN Security Council. The CD reports to the UN General Assembly. |
| Conference on Confidence- and Security-Building Measures and Disarmament in Europe | The so-called Stockholm Conference, part of the CSCE process, was held in 1984–86. The Stockholm Document, in which the confidence-building measures adopted at Helsinki in 1975 were improved and expanded, was signed in 1986. *See also* Vienna Documents 1990 and 1992 on CSBMs. |
| Conference on Security and Co-operation in Europe (CSCE) | A conference which began in 1973 with the participation of all the European states except Albania plus the USA and Canada, and in 1975 adopted a Final Act, containing, among others, a Document on confidence-building measures and certain aspects of security and disarmament. Follow-up meetings were held in Belgrade (1977–78), Madrid (1980–83), Vienna (1986–89) and Helsinki (1992). The main new CSCE institutions are: the Council of Foreign Ministers, the Committee of Senior Officials (CSO), the Secretariat, the Conflict Prevention Centre (CPC), the Office for Democratic Institutions and Human Rights, the Parliamentary Assembly, the Forum for Security Co-operation (FSC), the Chairman-in-Office (CIO), the High Commissioner on National Minorities (HCNM) and the Court (on Conciliation and Arbitration). *See also* Conventional Armed Forces in Europe (CFE) Treaty, Paris Documents, Vienna Documents 1990 and 1992 on CSBMs. |

| | |
|---|---|
| Conventional Armed Forces in Europe (CFE) Treaty | The CFE Treaty, included in the set of Paris Documents, was signed in 1990 by 22 original NATO and WTO signatories and entered into force on 9 November 1992. It sets ceilings on treaty-limited equipment (TLE) in the ATTU zone. On 15 May 1992, the former Soviet republics with territory in the ATTU zone signed, in Tashkent, the Agreement on the Principles and Procedures of Implementation of the CFE Treaty, with four protocols, confirming the allocation of CFE-limited weapons on their territories. On 5 June the NATO states and Armenia, Azerbaijan, Belarus, Bulgaria, Czechoslovakia, Georgia, Hungary, Kazakhstan, Moldova, Poland, Romania, Russia and Ukraine signed the Final Document of the Extraordinary Conference of the States Parties to the CFE Treaty (Oslo Document), making these states parties to the CFE Treaty. |
| Conventional weapon | Weapon not having mass destruction effects. *See also* Weapon of mass destruction. |
| De-MIRVing Agreement | The Joint US–Russian Understanding on Further Reductions in Strategic Offensive Arms, reached on 17 June 1992, in which both countries pledge to eliminate their MIRVed ICBMs. This understanding was codified in the 1993 START II Treaty. |
| European Community (EC) | The EC was created in the 1950s by six governments—Belgium, France, the Federal Republic of Germany, Italy, the Netherlands and Luxembourg. In 1991 the texts of draft treaties on an Economic and Monetary Union and a European Political Union were agreed at the EC summit meeting in Maastricht, the Netherlands, and were signed on 7 February 1992. The Maastricht Treaty is to be ratified by the parliaments of the EC states. Today there are 12 EC member states. |
| First-strike capability | Theoretical capability to launch a pre-emptive attack on an adversary's strategic nuclear forces that eliminates the retaliatory, second-strike capability of the adversary. |
| Global Protection Against Limited Strikes (GPALS) | *See* Strategic Defense Initiative (SDI). |
| Intercontinental ballistic missile (ICBM) | Ground-launched ballistic missile capable of delivering a warhead to a target at ranges in excess of 5500 km. |
| Intermediate-range nuclear forces (INF) | Theatre nuclear forces with a range of from 1000 km up to and including 5500 km. *See also* Theatre nuclear forces. |
| Intermediate-range nuclear forces (INF) Treaty | The 1987 US–Soviet Treaty on the Elimination of Intermediate-Range and Shorter-Range Missiles obliged the USA and the USSR to destroy all land-based missiles with a range of 500–5500 km (intermediate-range, 1000–5500 km, and shorter-range, 500–1000 km) and their launchers by 1 June 1991. *See also* Theatre nuclear forces. |

| | |
|---|---|
| International Atomic Energy Agency (IAEA) | With headquarters in Vienna, the IAEA is endowed by its Statute, which entered into force in 1957, with the twin purposes of promoting the peaceful uses of atomic energy and ensuring that nuclear activities are not used to further any military purpose. *See also* Non-Proliferation Treaty (NPT). |
| Joint Consultative Group (JCG) | Established by the CFE Treaty to reconcile ambiguities of interpretation and implementation of the CFE Treaty. |
| Kiloton (kt) | Measure of the explosive yield of a nuclear weapon equivalent to 1000 tons of trinitrotoluene (TNT) high explosive. (The bomb detonated at Hiroshima in World War II had a yield of about 12–15 kilotons.) |
| Launcher | Equipment which launches a missile. ICBM launchers are land-based launchers which can be either fixed or mobile. SLBM launchers are missile tubes on submarines. |
| Lisbon Protocol | *See* START Treaty. |
| Megaton (Mt) | Measure of the explosive yield of a nuclear weapon equivalent to 1 million tons of trinitrotoluene (TNT) high explosive. |
| Multiple independently targetable re-entry vehicle (MIRV) | Re-entry vehicle, carried by a nuclear missile, which can be directed to separate targets along separate trajectories (as distinct from MRVs). A missile can carry one or several RVs. |
| Multiple re-entry vehicle (MRV) | Re-entry vehicle, carried by a nuclear missile, directed to the same target as the missile's other RVs. |
| Mutual assured destruction (MAD) | Concept of reciprocal deterrence which rests on the ability of the nuclear weapon powers to inflict intolerable damage on one another after receiving a nuclear attack. *See also* Second-strike capability. |
| National technical means of verification (NTM) | The technical intelligence means used to monitor compliance with treaty provisions which are under the national control of individual signatories to an arms control agreement. |
| Non-Proliferation Treaty (NPT) | A multilateral treaty opened for signature in 1968 which established a regime to prevent the proliferation of nuclear weapons while guaranteeing the peaceful uses of nuclear energy. Under the NPT, non-nuclear weapon states undertake to conclude safeguards agreements with the IAEA to prevent the diversion of nuclear energy from peaceful to weapon use. |
| North Atlantic Cooperation Council (NACC) | Created in 1991 as a NATO institution for consultation and cooperation on political and security issues between NATO and the former WTO states and European former Soviet republics. |
| North Atlantic Treaty Organization (NATO) | A security alliance established in 1949 by the North Atlantic Pact concluded between the USA, Canada and 10 West European states. Since 1966, NATO Headquarters are in Brussels. Today there are 16 member states. |

| | |
|---|---|
| Nuclear Risk Reduction Centres (NRRC) | Established by the 1987 US–Soviet NRRC Agreement. The two centres, which opened in Washington and Moscow in 1988, exchange information by direct satellite link in order to minimize misunderstandings which might carry a risk of nuclear war. Notifications concerning exchange of information about nuclear explosions under the 1974 Threshold Test Ban Treaty, the 1976 Peaceful Nuclear Explosions Treaty and the 1990 Protocols to the two treaties shall also be submitted through the two NRRCs. |
| Open Skies Treaty | An agreement signed by 25 CSCE states in 1992, permitting flights by unarmed military or civilian surveillance aircraft over the territory of the signatory states. Negotiations opened in 1990 between the NATO and the then WTO states. |
| Organization for Economic Co-operation and Development (OECD) | Established in 1961 to replace the Organization for European Economic Co-operation (OEEC). With the accession of Canada and the USA, it ceased to be a purely European body. OECD objectives are to promote economic and social welfare by co-ordinating policies. The 24 members as of 1 May 1993 are Australia, Austria, Belgium, Canada, Denmark, Finland, France, Germany, Greece, Iceland, Ireland, Italy, Japan, Luxembourg, the Netherlands, New Zealand, Norway, Portugal, Spain, Sweden, Switzerland, Turkey, the UK and the USA. |
| Oslo Document | *See* Conventional Armed Forces in Europe (CFE) Treaty. |
| Paris Documents | A set of five documents adopted at the 1990 Paris CSCE summit meeting. They include the CFE Treaty, the Joint Declaration of Twenty-Two States, the Charter of Paris for a New Europe, the Supplementary Document to give new effect to certain provisions contained in the Charter, and the Vienna Document 1990. Several new CSCE institutions were set up in the Paris Documents. |
| Peaceful nuclear explosion (PNE) | Application of a nuclear explosion for non-military purposes such as digging canals or harbours or creating underground cavities. |
| Re-entry vehicle (RV) | That part of a ballistic missile which carries a nuclear warhead and penetration aids to the target, re-enters the earth's atmosphere and is destroyed in the terminal phase of the missile's trajectory. A missile can have one or several RVs; each RV contains a warhead. |
| Safety, Security and Dismantlement (SSD) Talks | A nuclear arms control forum established in 1992 to institutionalize continuous co-operation between the USA and the former Soviet republics in the safe and environmentally responsible storage, transportation, dismantlement and destruction of former Soviet nuclear weapons. Talks have resulted in bilateral agreements between the USA and some of these states for US funding to assist these countries in the destruction of their nuclear weapons. |

| | |
|---|---|
| Second-strike capability | Ability to receive a nuclear attack and launch a retaliatory blow large enough to inflict intolerable damage on the opponent. *See also* Mutual assured destruction. |
| Short-range nuclear forces (SNF) | Nuclear weapons, including gravity bombs on aircraft, artillery, mines, etc., with ranges up to 500 km. *See also* Theatre nuclear forces. |
| START Treaty | The US–Soviet Treaty on the Reduction and Limitation of Strategic Offensive Arms (referred to as the START I Treaty), signed on 31 July 1991, which reduces US and Soviet offensive strategic nuclear weapons to equal aggregate levels over a seven-year period. It sets numerical limits on deployed strategic nuclear delivery vehicles (SNDVs)—ICBMs, SLBMs and heavy bombers—and the nuclear warheads they carry. In the Protocol to Facilitate the Implementation of the START Treaty (the Lisbon Protocol), signed on 23 May 1992 by the USA, Russia, Belarus, Kazakhstan and Ukraine, the latter three states pledge to accede to the START Treaty, to eliminate all strategic weapons on their territories within the seven-year START Treaty reduction period and to join the Non-Proliferation Treaty as non-nuclear weapon states in the shortest possible time. Not in force on 1 May 1993. |
| START II Treaty | The US–Russian Treaty on Further Reduction and Limitation of Strategic Offensive Arms, signed on 3 January 1993, which requires the USA and Russia to eliminate their MIRVed ICBMs and sharply reduce their strategic nuclear warheads to no more than 3500 each (of which no more than 1750 may be deployed on SLBMs) by 1 January 2003 or no later than 31 December 2000 if the USA and Russia reach a formal agreement committing the USA to help finance the elimination of strategic nuclear weapons in Russia. It will not enter into force until the 1991 START Treaty enters into force. |
| Stockholm Conference | *See* Conference on Confidence- and Security-Building Measures and Disarmament in Europe. |
| Strategic Arms Limitation Talks (SALT) | Negotiations between the USSR and the USA which opened in 1969 and sought to limit the strategic nuclear forces, both offensive and defensive, of both sides. The SALT I Interim Agreement and the ABM Treaty were signed in 1972. The negotiations were terminated in 1979 when the SALT II Treaty was signed (it never entered into force). *See also* START Treaty, START II Treaty. |
| Strategic Defense Initiative (SDI) | The programme announced by President Reagan in his 1983 'Star Wars' speech for research and development of systems capable of intercepting and destroying nuclear weapons in flight and rendering the USA safe from the threat of a nuclear strike by another state. The GPALS (Global Protection Against Limited Strikes) programme was initiated in 1990 and accelerated in 1991 to test and deploy ground- and space-based ABM systems for territorial defence of the USA against limited ballistic missile attack, whatever the source. |

| | |
|---|---|
| Strategic nuclear weapons | ICBMs, SLBMs and bomber aircraft carrying nuclear weapons of intercontinental range (over 5500 km). |
| Submarine-launched ballistic missile (SLBM) | A ballistic missile launched from a submarine usually with a range in excess of 5500 km. |
| Tactical nuclear weapon | A nuclear weapon usually with a range less than 500 km which is deployed with general-purpose forces along with conventional weapons. *See also* Theatre nuclear forces. |
| Theatre nuclear forces (TNF) | Nuclear weapons with ranges of up to and including 5500 km. In the 1987 INF Treaty, nuclear missiles are divided into intermediate-range (1000–5500 km) and shorter-range (500–1000 km). Also called non-strategic nuclear forces. Nuclear weapons with ranges up to 500 km are called short-range nuclear forces. |
| Throw-weight | The sum of the weight of a ballistic missile's re-entry vehicle(s), dispensing mechanisms, penetration aids, and targeting and separation devices. |
| Toxins | Poisonous substances which are products of organisms but are inanimate and incapable of reproducing themselves as well as chemically induced variants of such substances. Some toxins may also be produced by chemical synthesis. |
| Treaty-limited equipment (TLE) | The five categories of equipment on which numerical limits are established in the 1990 CFE Treaty: battle tanks, armoured combat vehicles, artillery, combat aircraft and attack helicopters. |
| Vienna Documents 1990 and 1992 on CSBMs | The Vienna Document 1990 on CSBMs, included in the set of Paris Documents, repeats many of the provisions in the 1986 Stockholm Document and expands several others. It established a communications network and a risk reduction mechanism. The Vienna Document 1992 on CSBMs builds on the Vienna Document 1990 and supplements its provisions with new mechanisms and constraining provisions. *See also* Paris Documents. |
| Warhead | That part of a weapon which contains the explosive or other material intended to inflict damage. |
| Warsaw Treaty Organization (WTO) | The WTO, or Warsaw Pact, was established in 1955 by a Treaty of friendship, co-operation and mutual assistance between seven East European countries and the USSR. On 31 March 1991 the military organs and structures of the WTO were dismantled, and on 1 July 1991 it was dissolved. |
| Weapon of mass destruction | Nuclear weapon and any other weapon which may produce comparable effects, such as chemical and biological weapons. |

| | |
|---|---|
| Western European Union (WEU) | Established in the 1954 Protocol to the 1948 Treaty of Brussels of Collaboration and Collective Self-Defence among Western European States. In 1950 its defence organization functions were transferred to the NATO command. The WEU Consultative Forum was established in 1992. In addition to the members of the WEU, the Forum includes Bulgaria, the Czech Republic, Estonia, Hungary, Latvia, Lithuania, Poland, Romania and the Slovak Republic. Today there are 10 full member states. |
| Yield | Released nuclear explosive energy expressed as the equivalent of the energy produced by a given number of tons of trinitrotoluene (TNT) high explosive. *See also* Kiloton, Megaton. |

# Conventions

| | |
|---|---|
| . . | Data not available or not applicable |
| — | Nil or a negligible figure |
| ( ) | Uncertain data |
| m. | million |
| b. | billion (thousand million) |
| $ | US $, unless otherwise indicated |

## Membership of international organizations with security functions, as of 1 May 1993

| Country | CSCE 1973 | NATO 1949 | NACC 1991 | WEU 1954 | EC 1957 | CIS 1991 |
|---|---|---|---|---|---|---|
| Albania | 1991 | | 1992 | | | |
| Armenia | 1992 | | 1992 | | | • |
| Austria | • | | | | | |
| Azerbaijan | 1992 | | 1992 | | | •[a] |
| Belarus | 1992 | | 1992 | | | • |
| Belgium | • | • | • | • | • | |
| Bosnia and Herzegovina | 1992 | | | | | |
| Bulgaria | • | | • | | | |
| Canada | • | • | • | | | |
| Croatia | 1992 | | | | | |
| Cyprus | • | | | | | |
| Czech Republic | 1993[b] | | 1993[b] | | | |
| Denmark | • | • | • | [c] | 1973 | |
| Estonia | 1991 | | • | | | |
| Finland | • | | [d] | | | |
| France | • | •[e] | • | • | • | |
| Georgia | 1992 | | 1992 | | | |
| Germany | •[f] | 1955[g] | • | •[g] | •[g] | |
| Greece | • | 1952 | • | 1992 | 1981 | |
| Holy See | • | | | | | |
| Hungary | • | | • | | | |
| Iceland | • | • | • | [c] | | |
| Ireland | • | | | [c] | 1973 | |
| Italy | • | • | • | • | • | |
| Kazakhstan | 1992 | | 1992 | | | • |
| Kyrgyzstan | 1992 | | 1992 | | | • |
| Latvia | 1991 | | • | | | |
| Liechtenstein | • | | | | | |
| Lithuania | 1991 | | • | | | |
| Luxembourg | • | • | • | • | • | |
| Malta | • | | | | | |
| Moldova | 1992 | | 1992 | | | •[a] |
| Monaco | • | | | | | |
| Netherlands | • | • | • | • | • | |
| Norway | • | • | • | [c] | | |
| Poland | • | | • | | | |
| Portugal | • | • | • | 1988 | 1986 | |
| Romania | • | | • | | | |
| Russia | •[h] | | • | | | • |
| San Marino | • | | | | | |
| Slovak Republic | 1993[b] | | 1993[b] | | | |
| Slovenia | 1992 | | | | | |
| Spain | • | 1982[c] | • | 1988 | 1986 | |
| Sweden | • | | | | | |
| Switzerland | • | | | | | |
| Tajikistan | 1992 | | 1992 | | | • |
| Turkey | • | 1952 | • | [c] | | |
| Turkmenistan | • | | 1992 | | | • |
| UK | • | • | • | • | 1973 | |
| Ukraine | 1992 | | 1992 | | | • |
| USA | • | • | • | | | |
| Uzbekistan | 1992 | | 1992 | | | • |
| Yugoslavia | •[i] | | | | | |

| | |
|---|---|
| CSCE | Conference on Security and Co-operation in Europe |
| NATO | North Atlantic Treaty Organization |
| NACC | North Atlantic Cooperation Council |
| WEU | Western European Union |
| EC | European Community |
| CIS | Commonwealth of Independent States |

A • in the column for membership in an organization indicates that the country is one of the original members, that is, since the date given in the column heading for establishment of the organization. A year in the column indicates the year in which a country that is not an original member joined the organization.

[a] As of 1 May 1993, Azerbaijan and Moldova had not ratified the CIS Agreement.
[b] The former state of Czechoslovakia was an original member of the CSCE and NACC.
[c] Iceland, Norway and Turkey are associate members of the WEU. Denmark and Ireland are observers.
[d] Finland has observer status at NACC.
[e] France and Spain are not in the integrated military structures of NATO.
[f] The original members of the CSCE were the former Federal Republic of Germany (West Germany) and the German Democratic Republic (East Germany). After unification in 1991, Germany assumed the membership of this organization.
[g] The original member of NATO, the WEU and the EC was the former Federal Republic of Germany (West Germany). After unification of West Germany and East Germany in 1991, Germany assumed the membership of these organizations.
[h] Russia is a member of the CSCE by virtue of its status as successor to the USSR.
[i] As from 7 July 1992, Yugoslavia is suspended from the CSCE.

# Introduction: parameters of change

ADAM DANIEL ROTFELD

## I. The new security environment

A number of arms control negotiations, some of which had been conducted for over 20 years, reached conclusion in 1992. Fundamental and previously unimaginable agreements on arms control and disarmament were achieved.

In the bipolar world, relations between the chief partners were characterized and determined by confrontation, tension and distrust. Demands to limit armaments radically and eliminate weapons of mass destruction were generally seen as a reflection of idealism and pacifism, far removed from the harsh realities of the political situation. The dramatic end of the cold war at the end of 1991 was accompanied by important decisions in the field of arms limitation. The postulates put forward by SIPRI and some other research centres for many years, once thought idealistic, were not only proven feasible but have also become part of the security policies of states. Proposals for an international co-operative security system, until recently regarded as Utopian, are now seen to be realistic and workable.

Not long ago, each of these recently concluded agreements would have been proclaimed a landmark in history. Today, fascination with the enormous changes that have taken place and preoccupation with new conflicts have led to underestimating the historic turning-point which the agreements represent. Signed on 3 January 1993, the second Strategic Arms Reduction Treaty (START II)[1] will reduce US and Russian strategic nuclear warhead stockpiles to their lowest levels in decades—almost one-third the number of warheads allowed under the 1991 START Treaty ceilings. Moreover, the reductions apply to those weapon systems which had caused the greatest anxiety: on the US side to nuclear warheads carried on highly accurate submarine-launched ballistic missiles (SLBMs); and on the Russian side to nuclear warheads carried on land-based heavy intercontinental ballistic missiles (ICBMs).

Another agreement of crucial importance was signed in Paris on 13 January 1993, after 23 years of negotiation: the Convention on the Prohibition of the Development, Production, Stockpiling and Use of Chemical Weapons and Their Destruction (the CWC).[2] At the regional level, a document of great significance is the Helsinki Document 1992, putting forward the decisions adopted in July at the summit meeting of the leaders of the participating states of the Conference on Security and Co-operation in Europe (CSCE) concerning

[1] For an analysis and the text of the Treaty between the USA and the Russian Federation on the Further Reduction and Limitation of Strategic Offensive Arms, see chapter 11 and appendix 11A in this volume.

[2] For an analysis and the text of the CWC, see chapter 14 and appendix 14A in this volume.

the strengthening of existing, and the calling into being of new, CSCE institutions and structures. Those decisions lay the foundations of a security organization to embrace all states in the zone from Vancouver to Vladivostok.[3] Important agreements on conventional arms reductions in Europe and on new confidence- and security-building measures have also entered into force.[4]

Responses to the new challenges on the global scale could be seen in the decisions of the 1992 World Summit, held in Rio de Janeiro under the auspices of the UN Conference on Environment and Development,[5] and in the momentous report of the UN Secretary-General, *An Agenda for Peace,* setting forth a programme for preventive diplomacy, peace-making and peace-keeping.[6] Both the Rio Declaration and *An Agenda for Peace* herald new priorities in international security policy and bring home the magnitude of the tasks facing the international community. That traditional institutions and procedures are not adequate to the new political circumstances may best be illustrated by the number of major armed conflicts around the world in 1992; in all but one of the 30 conflict locations (India–Pakistan) the conflicts were domestic, intra-state conflicts.[7] Hitherto, the international system and the means available to international security organizations have been tailored to resolving conflicts *between* states, not *within* them.

The profound changes in the politico-military environment mean that states must reconsider how to ensure their security. The Warsaw Treaty Organization (WTO) and the USSR collapsed in 1991; Central and East European peoples shook off the totalitarian system and embarked on the road towards a democratic system and market economy; Russian troops are being withdrawn from the central parts of Europe; the US military presence in Europe and elsewhere has been radically reduced, and essential cuts in nuclear, chemical and conventional weapons are under way. Despite the end of the cold war and the corresponding dissolution of the bipolar division old security structures, some universal and some European, live on. The place, role and operation of the United Nations, the Atlantic Alliance, the Western European Union (WEU), the European Community (EC) and many other multilateral institutions have all been affected by recent events. It would be a simplification to say that some of these organizations have outlived their usefulness and ought to be dissolved like the WTO and the Council for Mutual Economic Assistance (CMEA). A strong, heavily militarized Soviet Russia, armed with the most sophisticated weapons of mass destruction, once posed a real threat of sudden and unexpected attack. Such a menace to the West no longer exists. There is no longer a military threat *from* the East, but there is an increasing threat of war *in* the East. The paradox, however, is that today the greatest risk stems from Russia's domestic weakness. Should the deep structural crisis precipitate

[3] For details of developments towards a CSCE security organization, see chapter 5 and the background documents in appendix 5A in this volume.

[4] Developments in conventional arms control in Europe in 1992 are described in chapter 12, with corresponding documents and analysis in appendices 12A, 12B and 12C in this volume.

[5] See chapter 1 and appendices 1A and 1B in this volume.

[6] See chapter 2 and appendix 2A in this volume.

[7] See chapter 3 and appendix 3A in this volume.

the disintegration of the Russian Federation, this may cause a loss of control over the gigantic military potential inherited from the Soviet empire. In other words, the greatest threat today is chaos and disorder, accompanied by numerous ethnic and other local conflicts and wars. If the programme of economic reform and building a democratic state of law fails, then the vacuum left by the discredited communist ideology may be filled by nationalism and religious and political fundamentalism. This could be averted much more effectively by swift, decisive action than by a time-consuming shaping of new institutions. NATO military forces need to be redesigned for preventive diplomacy and pre-emptive intervention to deter regional crises and provide the conditions for ending conflicts.[8]

In the post-cold war period security is not only the absence of conflict. Formerly, any conflict in Europe was seen as a potential clash of superpower interests. Fearing the spill-over of local conflicts into a global confrontation, Western states stood by as the USSR quelled liberation and democratic movements in Central and Eastern Europe. The Soviet Army 'restored order' in East Berlin (1953), Budapest (1956) and Prague (1968). New world order amounts to the restoration of common democratic values; it cannot be based on the law of power but rather on the power of law.[9] An international system of security based on the rule of law will not be the outcome of a theoretical concept or a comprehensive proposal negotiated at an international conference. It will evolve gradually, responding to the requirements of the situation.

In forming security systems a catalytic role is usually played by wars and crises. The 1991 Persian Gulf War and the war in Yugoslavia are more decisive in this respect than political rhetoric, blueprints and theoretical debates. Stephen R. Rock, one of the theoreticians of international relations who presaged the end of the cold war, wrote: 'Even if the exercise of power, economic activities, and societal attributes favour pacific relations, some catalytic event may be required to set the process of reconciliation in motion. The most probable candidate for this role is an acute crisis between two states.'[10]

This is even more true in a crisis situation involving a larger group of states on a regional or global scale. John Lewis Gaddis rightly asserts that no major theoretical school (realist, behaviourist or structuralist) envisaged the end of the cold war or the development of international relations after the collapse of the bipolar system.[11] The dissolution of the bipolar system resulted from an implosion (the exhaustion of the domestic driving forces of the social and economic development of the communist states). While the impact of external forces and factors on the withering totalitarian systems should not be under-

[8] See the views presented by German Defence Minister Volker Rühe in his lecture at the International Institute of Strategic Studies in London. These are referred to in Asmus, R. D., 'An outward-looking new NATO or no NATO at all', *International Herald Tribune*, 13 Apr. 1993, p. 6.

[9] Engholm, B., *Eine neue Weltordnung—Vor welchen Herausforderungen stehen Deutschland und die atlantische Allianz?,* Statement by the Chairman of the Social Democratic Party of Germany (Forum für Deutschland: Berlin, 10 Mar. 1993).

[10] Rock, R. S., *Why Peace Breaks Out: Great Power Rapprochement in Historical Perspective* (University of North Carolina Press: Chapel Hills, N.C., 1989), pp. 12–18.

[11] Gaddis, J. L., 'International relations: theory and the end of the cold war', *International Security*, no. 3 (winter 1992/93), pp. 53–58.

estimated, the predominant role was played by domestic shortcomings. Analysis of internal relations is of crucial importance to an understanding of the causes of the collapse of the communist system. To a great extent the effects of the breakup of the USSR and the repudiation of communist ideology have determined the new political and military international security environment. Arms control and disarmament play one role in a security system in which Russia sides with the USA and the system of values represented by the West, but played another when the USSR was an adversary of the USA.

## II. Arms control after the cold war

Justifying the need to strengthen NATO, British Foreign Secretary Douglas Hurd voiced fears, widespread in the West, that familiar threats might recur: 'Russia still bristles with nuclear weapons, a deadly inheritance from the Communists. The authorities there and in Ukraine, Belarus and Kazakhstan are committed by treaty to dismantling most of this arsenal. But the risks of proliferation abound when there is disorder.'[12] Confronted with this new type of threat, some analysts argue that the search for a new collective or co-operative security system and arms control, contrary to the conventional wisdom, 'may create not the conditions for peace but those for war'.[13] Are such pessimistic forecasts warranted, and if so, to what degree?

There is no sphere in international relations in which the end of the cold war brought so many favourable changes as in arms control and disarmament. The scepticism and disappointment of many authors seem to stem from the fact that the arms race was itself seen as a source of evil and its ending as tantamount to removing the causes of tension, confrontation and destabilization. This reasoning mixes cause and effect. While amassing huge weapon arsenals constituted a menace *per se*, the arms race was not the cause but the effect of the former situation. Decisions concerning the elimination of certain weapon systems and the limitation of armaments resulted in a slightly safer, if somewhat more complex, world. The simple divisions disappeared; a Manichean world view in which everything was seen as a dichotomy—'the children of light *versus* the children of darkness'—has been abandoned.

At the same time, however, new and formerly unknown threats and challenges have emerged. It is evident that many of the tasks of and methods applied by the international community to arms control should undergo scrutiny and reform.[14] One must agree with the UN Secretary-General who rejects the view shared by many experts that disarmament 'is no longer centrally relevant to international security needs'.[15] Three concepts—integration, globalization and revitalization—are presented in the UN report as the future-

[12] 'The new disorder', Speech by British Foreign Secretary Douglas Hurd to Chatham House, London, 27 Jan. 1993.

[13] Betts, R. K., 'System for peace or causes for war? Collective security, arms control and the new Europe', *International Security*, vol. 17, no. 1 (summer 1992), pp. 3 and 40–42.

[14] United Nations, *New Dimensions of Arms Regulation and Disarmament in the Post-Cold-War Era,* Report of the Secretary-General, UN document A/C.1/47/7 (United Nations: New York, 23 Oct. 1992).

[15] United Nations (note 14), p. 1.

oriented programme of arms control after the cold war. Boutros Boutros-Ghali asserts that 'the time has come for the practical integration of disarmament and arms regulation issues into the broader structure of the international peace and security agenda'. Moreover, there is a need for globalization of the process. Not only the great powers but all states have to be engaged in the process of disarmament. This would imply 'an all-inclusive, multi-dimensional, non-compartmentalized approach.'[16]

In the bipolar world, arms control and disarmament were seen to have the highest priority. Negotiation in this field was the main channel of dialogue between East and West and the agreements achieved were a unique barometer of tension versus *détente* in the international climate. This substitute function of negotiations has gradually faded away and arms control developments have assumed their real dimension and place. Although armaments and armed forces, the proliferation of weapon systems, military expenditures and arms transfers continue to play a substantial role in international relations, the nature of the threats has changed dramatically. The probability of the outbreak of a third world war in the near future has reached the zero mark. The end of the cold war enabled the ending of several conflicts in Latin America, Africa and Asia; a peace process is under way in the Middle East; and activity aimed at conflict resolution, preventive diplomacy, peace-making and peace-keeping has gained in importance.[17]

The new scope and forms of this activity are directly and closely related to the new character and scale of conflicts. Similarly, the geographical pattern of conflicts has changed. Formerly conflicts were waged far from the line demarcating the two opposing blocs; they took place on the rim of the potential area of confrontation of the main adversaries leading the antagonistic military alliances. Totalitarian regimes prevented domestic conflicts by means of terror and repression, and conflicts in the former Yugoslavia and the former USSR were nipped in the bud, although the underlying problems and tensions remained unresolved. The overthrow of the totalitarian systems laid them bare, especially where ethnic, national and religious conflicts were often treated as a tool in the power struggle. As a result, local conflicts in and outside Europe have intensified and slipped out of the control of the major powers. The wars in the former Yugoslavia, Nagorno-Karabakh and Tajikistan are not being waged by proxy. Their origins, causes and motives are rooted in history and evidence above all the conflict of interests among various social groups and political cliques striving for power and domination.

This poses the question of whether and to what extent arms control agreements and on-going negotiations prevent or promote the resolution of existing and potential conflicts. Recent major agreements—the 1991 and 1993 START treaties, the 1990 CFE Treaty and the 1992 CFE-1A Agreement, the 1992 Open Skies Treaty and the 1993 Chemical Weapons Convention—closed the negotiations carried out in the cold war years. Their significance can hardly be

[16] United Nations (note 14), p. 4.
[17] See chapter 2 in this volume.

over-estimated. Irrespective of operational limitations, they introduced specific procedures and rules of conduct for relations between states. However, their goals were essentially to halt the arms race in a bipolar world (making specific weapon systems subject to limitation, reduction and elimination) to lessen the danger of the outbreak of a major war between the antagonistic blocs and to reduce the threat of the use of weapons of mass destruction. They are highly effective in promoting the new co-operative security system but their functions are not addressed to helping resolve local conflicts.

In the new political circumstances, the most crucial challenge for these agreements is effectively to prevent the proliferation of weapons and their means of delivery. This applies equally to conventional, nuclear, chemical and biological weaponry. The side-effect of the breakup of the USSR is that nuclear weapons and their means of delivery are possessed not only by Russia but also by Ukraine, Belarus and Kazakhstan. This situation should be resolved urgently, on a bilateral and multilateral basis, not only with Russia but also with the other states directly concerned. Implementation of the agreements and the resultant transfer and scrapping of weapons and their means of delivery have fostered close co-operation between the USA and Russia.[18] None the less, it is clear that the political philosophy underlying the START and CFE treaties was past- rather than future-oriented, aiming to match and balance nuclear and conventional potentials within a bipolar system that no longer exists. More important, neither a uni- nor a multi-polar system emerged from the process of change. It is hard to predict what kind of order will evolve from the present-day chaos. One thing is for certain: the implementation of the arms control accords and the work of the institutions that they called into being will be carried out under new political and military premises. While they must address a different reality it would be a mistake to underestimate their significance. They constitute an important part of the new politico-military environment. As stated by the UN Secretary-General in his report, the legacy of 11 global multilateral agreements, 4 major regional multilateral agreements and 16 bilateral agreements between the USA and the Russian Federation provides a basis for the disarmament and arms control process today and in the immediate future and establishes some procedures and rules of conduct in the search for a co-operative security system.[19] Their very existence together with that of institutions and security organizations which, like the UN, NATO and the CSCE, are quickly adapting to the new requirements and challenges, furthers and facilitates the process of change and restructuring.[20]

[18] At the US–Russian summit (Vancouver, 4 Apr. 1993) the new US Administration reiterated its readiness to provide, within the framework of the Safety, Security and Dismantlement (SSD) talks with Russia, assistance to dismantle nuclear, chemical and other weapons and to establish safeguards against their proliferation. See the Joint Statement issued by the US and Russian presidents, Vancouver Declaration, 4 Apr. 1993, Special Wireless File, United States Information Service, US Embassy, Stockholm, 3–4 Apr. 1993, pp. 10–11.

[19] United Nations (note 14), p. 1.

[20] See chapter 5 in this volume.

The close interrelationship between arms control, preventive diplomacy and peace-making is illustrated by the mechanisms for weapon inspection and monitoring troop withdrawals and the establishment of disengagement or demilitarized zones. However, it would be wrong to compare peace enforcement measures (like the compulsory measures applied by the UN in Iraq) with disarmament measures.[21]

The 1968 Non-Proliferation Treaty (NPT) is an example of a future-oriented agreement. Its significance is growing with time. For many reasons its verification and safeguards arrangements need to be strengthened: 'it should be extended indefinitely and unconditionally at the NPT Extension Conference in 1995'.[22]

In the post-cold war period, one of the key issues to be resolved is that of arms transfers. The flow of arms is an offshoot of disarmament agreements and arms reductions, stemming from the collapse of the economies of the socialist states—primarily Russia, Ukraine and other former Soviet republics. For Central and East European states, entering new arms markets can be seen as one of the ways to solve their domestic economic problems. While the volume of arms trade in recent years has gone down considerably, 'the problems related to excessive arms transfers are daunting,[23] as the UN Secretary-General stated in his report.

Increased transparency concerning armaments can build confidence among states and facilitate non-proliferation efforts. The recently created UN Conventional Arms Register paved the way for a policy of institutional openness and transparency in this field. However, it is alarming that not all information exchanged among governments is not available to independent research centres. The monitoring of the arms control process by public opinion and the community of independent experts cannot be considered to pose a major threat so data on armaments and arms control should be made accessible.

## III. Consequences for the research agenda

The collapse of the old order was a starting-point for proposals for a new security system. Much publicity was given to the idea of a new world order.[24] Perhaps the rhetoric promised too much—the higher the expectation the greater the disappointment. Based on the changing political circumstances President George Bush defined the new world order as embracing the follow-

[21] The UN Secretary-General rightly noted in his report that '[t]he use of disarmament measures within the framework of peace enforcement is quite distinct from the process of disarmament through negotiation, which several States and elements of the international community have been pursuing for years. The two should never be confused, even if there may be some conceptual overlap in terms of the mechanics of weapons inspection and disposal.' See United Nations (note 14), p. 5.

[22] United Nations (note 14), p. 7.

[23] United Nations (note 14), p. 9.

[24] The beginning of the concept is generally linked to the formula used by President George Bush in autumn 1990 and developed later in a number of his speeches. For the sake of precision, however, it should be noted that the phrase was previously used in Mikhail Gorbachev's famous address before the United Nations General Assembly on 7 Dec. 1988: 'further world progress is only possible through a search for universal human consensus as we move forward to a new world order.'

ing principles: the rule of law and peaceful settlement of disputes; the strong force of democracy and solidarity against aggression; reduced and controlled arsenals; strengthening the United Nations; and just treatment of all peoples.[25] In the former world polarization made it impossible to build regional security systems, and even encouraged potential aggressors (e.g., Iraq) to take advantage of the situation and use the controversies to play off the world powers against each other. The course of events in Europe in 1989–90 and the response to the Iraqi aggression in Kuwait in 1991 showed the logic of the cold war to be a thing of the past. The 1990 Charter of Paris for a New Europe ushered in the adoption of a common system of values by states belonging to opposing blocs, and the response to the Iraqi aggression by the international coalition heralded the possibility of a new, effective co-operative security system.[26]

Developments, and particularly domestic conflicts requiring international intervention, showed that the expectations connected with the concept of a new world order were hard to fulfil. Not only was the concept not sufficiently considered but, more important, most states are not prepared to incur material, financial and human costs in defence of declared values when their vital national interests are not directly endangered. Robert S. McNamara found that the new world order should accomplish six objectives:

1. Provide to all states guarantees against external aggression;
2. Codify, and provide means of protecting, the rights of minorities and ethnic groups within states;
3. Establish a mechanism for resolution of regional conflicts without unilateral action by the Great Powers;
4. Commit the Great Powers to termination of military support of conflicts between other nations and conflicts between opposed political parties within those nations;
5. Increase the flow of both technical and financial assistance to the developing countries to help them accelerate their rates of social and economic advance;
6. Assure preservation of the global environment as a basis of sustainable development for all.[27]

From the point of view of research, it is not significant whether this catalogue of objectives is comprehensive as regards the list of expectations connected with the new system. It is more important to ask why states postulate the creation of a new international order, but in practice go back to the traditional concepts and ways of ensuring their national security. In his address to Chatham House, British Foreign Secretary Douglas Hurd began as follows: 'British foreign policy exists to protect and promote British interests. Despite

[25] See more on this in Sloan, R. S., 'The US role in a new world order: prospects for George Bush's global vision', CRS Report 91—294 RCO (Congressional Research Service, Washington, DC, 28 Mar. 1991).

[26] See Goodby, J. E., 'Introduction', eds J. E. Goodby and B. Morel, SIPRI, *The Limited Partnership: Building a Russian–US Security Community* (Oxford University Press: Oxford, 1993), pp. 1–2.

[27] McNamara, R. S., *The Changing Nature of Global Security and its Impact on South Asia*, Address to the Indian Defense Policy Forum, New Delhi, 20 Nov. 1992 (Washington Council on Non-Proliferation: Washington, DC, Dec. 1992), pp. 3–4.

all the changes in the world that underlying truth has not changed.'[28] Removing ideology from foreign policy is often seen as a return to the concept of security in which the only motives are national security interests. 'Policymakers are guided, consciously or unconsciously, by concepts and axioms as they address particular strategic challenges.'[29]

The task facing scholars is to suggest, on the basis of analysis of the new problems and challenges, a future-oriented conceptual tool that might be applied to manage the change. The main difficulty seems to lie in the fact that reality changes faster than the capability of researchers to perceive, analyse and conceptualize those changes, and that of politicians and decision makers to take the necessary action. Numerous firm notions and principles guiding mutual relations among states should be reassessed and reinterpreted. This applies in particular to the essence of the right of nations to self-determination, commonly identified with the right to secede. What are the limits of the exercise of this right in the context of respect for the principle of the territorial integrity of states? The Iraqi aggression against Kuwait, the war in Yugoslavia and the post-Soviet armed conflicts bring home to the global community the need for legitimization of international intervention to protect basic universal values. This calls for reassessment of the principle of non-intervention in internal affairs. The chances for shaping an international security system based on the rule of law are greater than in the past. John Chipman is correct in stating that 'there is now an opportunity to re-establish "rules of the game" that diminish certain risks of insecurity and correspondingly heighten the possibilities of moral interaction among states'.[30]

## IV. *Yearbook* findings

The materials and documents presented in this volume not only embrace a description and analysis of the most important developments in the field of arms control and disarmament, but also draw attention to new problems and priorities in the sphere of international security. A *signum temporis* showing the change in SIPRI's research agenda is the chapter devoted to major armed conflicts. The war in Bosnia and Herzegovina, the post-Soviet armed conflicts—in Nagorno-Karabakh, Georgia, Northern Caucasus, Moldova and Tajikistan—and the conflicts in Southern Africa and Cambodia are the subject of meticulous analysis.

In contradiction to popular belief the finding is that the post-Soviet conflicts are primarily political and ideological struggles between the defenders of the old system and the forces of nation-building or national revival. Ethnic and national motives are used as instruments in the power struggles. The list of conflict-generating problems connected with the post-Soviet heritage is a long one: ill-prepared separation of economic and military assets; questionable

[28] Note 12.

[29] Chipman, J., 'The future of strategic studies: beyond even grand strategy', *Survival*, vol. 34, no. 1 (spring 1992), p. 124.

[30] Chipman (note 29), p. 130.

legitimacy of the borders between the new states and tendencies towards further disintegration; politically sensitive disputes over citizenship and human rights; weak democratic traditions and ineffectiveness of political power; increasing nationalism and the search by numerous ethnic groups for self-identification; accelerating migration and the worsening refugee problem. The issue of nuclear weapons and armed conflicts over a number of contested territories is of special importance.[31]

An important conclusion of the chapter on nuclear weapon developments and proliferation is that in the former USSR, Iraq and the developing nations in 1992 the end of the East–West military confrontation increased rather than decreased the dangers of and incentives for the proliferation of nuclear weapons and ballistic missiles.[32] In this connection one should note that the authors of the analysis of US military technology and international security stated that there is a new Western consensus, especially pronounced in the USA, that military research and development (R&D) must be maintained at its cold war level, if not accelerated, even as funding for personnel and procurement programmes are cut.[33] This consensus ensures that the USA will maintain its technological supremacy for the foreseeable future but does not rule out deeper cuts in defence expenditures.

The chapter on the Chemical Weapons Convention[34] presents SIPRI's first attempt to analyse the text of the agreement and provide information about the destruction obligations, including the verification provisions, and to illustrate the interplay between national implementation obligations and international compliance measures. Although the conclusion of the CWC and the START and CFE treaties is an achievement of historic proportions, many problems related to nuclear, biological and chemical (NBC) and conventional warfare remain.[35] The main question for the near future will be whether or not the signing and implementation of these agreements will have a significant impact on: (*a*) future proliferation of chemical and biological weapons; (*b*) CW stockpile destruction with respect to ecological, health and safety concerns; (*c*) future research in the area of new toxic agents; and (*d*) NBC defence research and development. The 'new environment' will need to be monitored closely to identify any changes which may result from the CWC. An important conclusion of one of the analyses is that developments in molecular genetics and biotechnology (genetic engineering) have proceeded more rapidly than expected, particularly as regards the mapping of the human genome.[36]

What impact have achievements in disarmament had on military expenditures? In 1992, for the first time since the end of the cold war, global military spending fell steeply by about 15 per cent. The Organization for Economic Co-operation and Development (OECD) countries are reducing military

[31] See chapter 4 in this volume.
[32] See chapter 6 in this volume.
[33] See chapter 8 in this volume.
[34] See chapter 14 in this volume.
[35] See chapter 7 in this volume.
[36] See appendix 7A in this volume.

spending cautiously. The decline in spending stems partly from uncertainty in international relations; partly from the fact that the cost of restructuring has been higher than anticipated; and partly from the new roles being created (e.g., peace-keeping). The developing world and Central and East European countries have also cut their defence spending. The dramatic fall in world military expenditures can be chiefly attributed to the collapse of military spending in the former USSR. Russia, Ukraine and other CIS states halved their defence spending in one year in response to the economic crises and the improved security environment. All this suggests a growing trend of demilitarization throughout the world.

Within the OECD and developing countries arms sales and arms industry employment are falling. The combined value of arms sales by the 100 largest arms-producing companies in these countries decreased from $183.7 billion in 1990 to $178.8 billion in 1991, representing a fall of 7 per cent in real terms. Almost 80 per cent of those companies among the 100 largest arms-producing companies and their largest subsidiaries that depended on arms sales for more than half of total sales in 1988 had reduced their employment levels by 1991. The global value of foreign deliveries of major conventional weapons in 1992 is estimated by SIPRI to have been $18.4 billion (in 1990 US dollars).[37]

***

The context and premises in which decisions are taken with regard to arms control and regulation as well as disarmament have changed dramatically. Tools for conflict resolution and ways of organizing the international security system are also undergoing a change. It would be a mistake to think there is only one way or one model for solving problems and eliminating the threats facing the world. The transformations we are facing today are unprecedented. This applies both to the scale of the problems, the volatility and unpredictability of the situation, and the need for simultaneous decisions on a host of issues. Decision-makers expect researchers to ask essential questions and set forth a hierarchy of matters for consideration.[38] The key questions are (*a*) whether nations and states can rely on international instruments for their security or whether the decisive role will continue to be played by national means of security; (*b*) whether centripetal or centrifugal forces, integration or disintegration forces, national or international security interests will dominate; and (*c*) whether the international community will be able to manage the international politics of parochialism.[39] The costs and difficulties of shaping a new co-operative security system should not be underestimated. There should be a

[37] See chapter 10 in this volume. Since the SIPRI arms trade statistics do not reflect purchase prices, they are not comparable with economic statistics such as national accounts of foreign trade statistics, nor with the arms sales data reported in the sections of chapter 10 dealing with arms production. The methods used for the valuation of SIPRI arms trade statistics are described in appendix 10D.

[38] In a debate on co-operative security, Stephen van Evera rightly noted: 'The international community cannot address every issue at once, hence key issues should be addressed first'; *Boston Review*, vol. 17, no. 6 (Nov.–Dec. 1992), p. 10.

[39] See Chipman, J., 'Managing the politics of parochialism', *Survival*, vol. 35, no. 1 (spring 1993), p. 168.

focus on strengthening existing bilateral and multilateral arms control regimes, 'both to bolster the habit of security co-operation and to contribute to force restructurings'.[40] A number of important decisions in this regard were made in 1992.

Shaping a new, effective system of international security will be a long and difficult process. In making decisions that are of significance for ensuring security today and tomorrow, the crucial thing is to understand what is substantial in the on-going processes. The documents, analyses and conclusions contained in this volume are intended to facilitate this understanding.

[40] Miller, S., 'Dilemmas of cooperative security', *Boston Review*, vol. 17, no. 6 (Nov./Dec. 1992), pp. 15–16.

# Part I. The environment and security, 1992

## Chapter 1. The environment, security and development

# 1. The environment, security and development*

GRO HARLEM BRUNDTLAND

## I. Common security and the outlook for peace

When the Palme Commission, the Independent Commission on Disarmament and Security Issues, was launched in 1980, there was very little discussion about the prospects of ending the arms race, let alone achieving real disarmament. While the Commission was working, relations between the United States and the Soviet Union deteriorated sharply. Conflicts in Afghanistan and Poland, between Iran and Iraq, and elsewhere in the Third World contributed to the escalating arms race. The policies of the major powers offered few alternatives. However, at this low ebb in international co-operation, many of us were convinced that policies would have to change, that we would have to forge a new concept of security rather than continue the competition for military supremacy and the development of ever more effective means of destruction.

The report of the Palme Commission[1] was unique because it was the first political document in which representatives from the North Atlantic Treaty Organization (NATO), the Warsaw Pact and neutral countries alike were able to agree on a common analysis of the dangers to peace and security and on a broad programme of action to break out of the apparent deadlock in world affairs. We agreed that a nuclear war could never be won and must never be fought. We offered an alternative concept to mutual deterrence, that of common security.

[1] Independent Commission on Disarmament and Security Issues, *Common Security: A Programme for Disarmament* (Pan Books: London, 1982), p. vii.

* This chapter is the text of the sixth SIPRI Olof Palme Memorial Lecture, given by Gro Harlem Brundtland, Prime Minister of Norway, on 18 Nov. 1992. In Oct. 1986, SIPRI's Governing Board decided to arrange an annual public lecture, named after the late Swedish Prime Minister Olof Palme. The lecture is to be delivered in Stockholm by a political leader of international stature or an eminent scholar in order to highlight the need for, and problems of, peace and security, in particular of arms control and disarmament. The lecture is also intended to draw attention to SIPRI's commitment to a future with fewer arms and more freedom. The first annual Olof Palme Memorial Lecture was given in 1987 by the late Willy Brandt, former Chancellor of the Federal Republic of Germany, and subsequent lectures by the late Sergey F. Akhromeyev, Chief of the General Staff, First Deputy Minister of Defence and Marshal of the Soviet Union (1988); Victor F. Weisskopf, Professor Emeritus at the Massachusetts Institute of Technology, USA (1989); Oscar Arias Sánchez, former President of the Republic of Costa Rica and Nobel Peace Prize Laureate (1990); and Sir Shridath Ramphal, former Secretary-General of the Commonwealth (1991).

The essence of this concept is that countries can never achieve security against their adversaries. Common security can only be achieved if countries work together, defining their common interests in arms reductions and recognizing that co-operation would have to replace confrontation as the basis for a programme for joint survival. Although the report of the Commission was controversial for many years, important parts of its analysis and conclusions are, in fact, widely accepted and valued today.

Since the end of World War II the world has spent about $20 trillion for military purposes. Industrialized countries have doubled their defence spending since 1960 and developing countries increased their expenditures more than sixfold.

Nevertheless, we are living in a period of genuine disarmament. SIPRI figures show that in 1990 world military expenditure declined by an estimated 6 per cent to about $950 billion[2] as compared with the all-time high of more than $1 trillion in 1987.

Many of the specific recommendations of the Commission are now being implemented. The demise of totalitarian regimes in Eastern Europe and the progress made by the ideas of self-determination, democracy, the rule of law and freedom of expression have profoundly changed the outlook for peace and co-operation.

These trends are global and not confined to the European continent. In a security policy context, Europe has been extremely successful in dealing with the revolutionary changes we have witnessed. The Soviet leadership, breaking with earlier practice, respected the right of self-determination throughout Eastern Europe and supported the process of peaceful change.

NATO's response to the dissolution of the Warsaw Pact and the fundamental changes in Moscow has been to adopt a co-operative approach to security. NATO was able to redefine its concept of security. This would not have been possible without the Alliance's open and non-exclusive approach.

The arms control process has produced an impressive set of agreements which not only prescribe reduction in armaments, but also stipulate co-operative ways of following them up.

The most recent agreements between the United States and Russia in the field of nuclear disarmament represent a quantitative breakthrough. For certain categories of weapons, there will be reductions of up to 80 per cent compared with the early 1980s, and implementation is well under way. The agreements will require very close co-operation of a qualitatively new kind, involving everything from verification and control to destruction technologies and financing.

The issue of a global, verifiable test ban, as advocated by the Palme Commission, has been on the international agenda for a long time. Today, a window of opportunity has opened. Now that nuclear weapons are being destroyed on a large scale, it can no longer be argued that nuclear tests are

[2] SIPRI, *SIPRI Yearbook 1991: World Armaments and Disarmament* (Oxford University Press: Oxford, 1991), p. xxxvii.

necessary for security. They were part of the old, competitive and confrontational order. Several nuclear powers have already announced short-term unilateral moratoria. A lasting moratorium must be global. The opportunity should now be seized to agree on a lasting, global nuclear test ban as a matter of first priority.

Another area of priority today must be to counter the threat of a spread of nuclear technology. To achieve this, we must be prepared to assist, both practically and financially, by co-operating to detect clandestine nuclear activities and to find alternative uses for the huge nuclear establishments. The initiative to establish international research centres in Moscow and Kiev is a commendable contribution.

Considerable progress has been made with respect to conventional arms reductions. Under the 1990 Treaty on Conventional Armed Forces in Europe (CFE), meticulously elaborated procedures for the exchange of information and verification are being implemented.[3] A network of military contacts and confidence-building channels is thus being established throughout the European continent. The provisions of the treaty are practical in nature and may even seem prosaic. Nevertheless, this qualitatively new way of building security may well prove to be one of the most outstanding achievements of diplomacy in our time. The CFE Treaty has aptly been described as the first solid corner-stone of the new European security architecture. In my view, its soundness lies in the integration of the intentions of the treaty with the practical steps that must be taken to ensure their realization at a national or local level.

At the global level, all states are rallying behind the United Nations in its role as our common global peace organization. Never has the UN been involved in so many peace-keeping operations. The Secretary-General's Agenda for Peace is widely recognized as a starting-point for a serious discussion on how to revitalize the collective security system envisaged in the UN Charter.[4]

## II. A comprehensive concept of security: peace, the environment and development

Still more time will be needed to establish lasting new patterns of global stability. Our challenge is to move on from the cold war era, through policies of common security among states towards a wider comprehensive security concept which must include social, economic and environmental progress. Our new comprehensive security concept must fully include the anticipation

[3] SIPRI, *SIPRI Yearbook 1991: World Armaments and Disarmament* (Oxford University Press: Oxford, 1991), pp. 471–73 for Articles XIII–XVII. See also *Treaty on Conventional Armed Forces in Europe* (United State Information Agency: Paris, 1990), pp. 17–20 for Articles XIII–XVII and pp. 58–107 for the Protocols on Notification and Exchange of Information and on Inspection.

[4] United Nations, General Assembly/Security Council, *An Agenda for Peace: Preventive Diplomacy, Peacemaking and Peacekeeping*. UN documents A/47/277 and S/24111 (United Nations: Geneva, 17 June 1992).

and prevention of the underlying political, social and environmental causes of tension, preventing war by building and managing peace.

When I was called upon in 1983 to establish and chair an independent commission it was clear that the international community was unable to deal effectively with the vital issues confronting us. Throughout the 1970s, the United Nations had dealt with important areas such as population, housing, safe water, and new and renewable energy sources by holding major conferences. This offered hope, but the United Nations system was on the whole too weak and fragmented to deal with human needs in an integrated way.

The World Commission on Environment and Development was fortunate to be able to build on the reports of the Brandt Commission and the Palme Commission. It was clear to me that after Brandt's *Common Crisis*[5] and Palme's *Common Security*,[6] *Our Common Future*[7] would have to be the next step in a major effort to persuade countries to return to multilateralism in an integrated effort to address the issues of peace, the environment and development.

In the early 1970s, the Club of Rome had presented for the first time the ways in which limited resources could set limits to growth.[8] The ecological movement and many scientists had since the late 1960s become increasingly aware that we were approaching limits to the burdens that we could load upon nature's capacity to absorb the effects of human activities. The Stockholm Conference in 1972 was the first major international effort to address these new threats.[9]

The increasing knowledge which we acquired throughout the 1970s was new to our generation. Never before in human history had we had the capacity to destroy the environment and to reduce the options of future generations. Our generation was the first which had to be cognizant of its responsibility for the environment on behalf of generations yet unborn.

The South was sceptical of the new environmental awareness of the North, seeing it as a threat to their development ambitions. The North had been developing for decades without showing much concern for environmental degradation and destruction. The developing countries were facing completely different challenges. They were caught in a downward spiral of increasing poverty, crushing debt burdens, deteriorating terms of trade and inadequate access to world markets. They felt unable to afford the apparent luxury of protecting their own resource base.

*Our Common Future* played perhaps its most important role in establishing the link between the environment and development. These were formerly

[5] Independent Commission on International Development Issues, *Report of the Independent Commission on International Development Issues* (ICIDI Secretariat: Bonn, 1980).

[6] See Independent Commission on Disarmament and Security Issues (note 1).

[7] Brundtland, G. H. *et al.*, *Our Common Future: World Commission on Environment and Development*, rev. edn (Oxford University Press: Oxford, 1987).

[8] Club of Rome, *The Limits to Growth. A Report for the Club of Rome's Project on the Predicament of Mankind* (Friends of the Earth: London, 1972).

[9] *The Stockholm Conference: Only One Earth. An Introduction to the Politics of Survival* (Friends of the Earth: London, 1972).

viewed as separate issues and were dealt with by different institutions internationally and different ministries at the national level.

The World Commission managed to forge the basis for a global consensus because we made it explicit that it was only by solving social and economic problems that we could hope to solve the threats to the environment. We firmly believed that we could not protect the global environment without establishing a more just international economic order, nor provide the basis for a more just and equitable future for all, if global trends that threaten the resource base were allowed to continue.

We developed the concept of sustainable development, which means that we must meet the needs of the present generation without compromising the ability of future generations to meet their needs. Sustainable development is a political concept for human social, economic and environmental progress. It would require a new era of international co-operation and greater participation by people themselves. They must become more actively involved in political life so that they can have a say in decisions of importance to their own lives and futures. Thus, democracy, human rights and practical solidarity had to become the basis of all effective policies for the environment and development.

Like the earlier commissions, our analysis led to the call for a strengthening of international co-operation. Only by working together, not against each other, can we have a vision of a better managed world, better governance and global adherence to the fundamental principles of democracy and to the principle that economic and social development must be sustainable. Peace, democracy, the environment and development would have to be the core issues of our common agenda for the 21st century.

We must recognize how interdependent we have all become. World population trends indicate a doubling or trebling of our numbers some time in the next century. Ninety per cent of the increase will take place in the developing countries, and unless corrective action is taken this will aggravate the vicious circle of poverty and environmental degradation in which they are already caught. Combined with unsustainable patterns of production and consumption, especially in the North, these trends will place intolerable strains on finite natural resources.

We in the North must also recognize that no one, not even the richest of us, can escape these global trends. There will be no sanctuaries where some people can escape the harsh realities. We will all suffer from the radiation if the ozone layer is further damaged. Climate change can cause drought, floods and disruption of agricultural patterns both in the North and in the South.

Hundreds of millions of people are living in areas that will be affected by a rise in the sea level. Toxic substances are travelling with winds and currents, and everyone has to breathe. Pollutants originating in the temperate zone are already to be found in the food chain in the Arctic. Clearly we need fundamental changes in the way we use the earth's crust and in the way we distribute the benefits of economic growth.

Our security depends at least as much on economic well-being, social justice and ecological stability as it does on military security. Throughout human history, struggles over access to and control over natural resources have been one of the root causes of tension and armed conflict. We risk a proliferation of such disputes if the rapid deterioration of environmental quality is allowed to continue. Our common future depends on our collective ability to change. We must address issues of peace in a precautionary, integrated manner, and we must deal decisively with all the underlying causes of human conflict and distress.

Above all we must be uncompromising in our determination to alleviate poverty. Poverty is a major cause of environmental degradation in the developing world. Poor people will concentrate on their daily survival. They will be forced to cut down trees, over-graze pastures and over-use farmland in order to stay alive. Poor countries, too, will have to over-exploit their natural resources in order to produce the export goods needed to pay for necessary imports. When prices go down, they will have to produce more and more basic commodities and extract more and more of their natural resources to pay for goods that they do not produce themselves.

Poverty is also in itself intolerable and cannot be reconciled with human dignity. We must adamantly oppose any tendency to ignore the fundamental challenges of the continuing North–South divide. Otherwise the very future of our planet is in danger.

In spite of remarkable economic and social progress in many developing countries, the inequalities persist. According to the United Nations Development Programme's latest *Human Development Report,* the richest 20 per cent of the world's population receive 83 per cent of total world income, whereas the poorest 20 per cent have 1.4 per cent.[10] We cannot allow this to continue. The African continent has been particularly hard hit by economic decline, and a concerted international effort must be mounted to reverse this unfortunate situation. Look at Sub-Saharan Africa where the vast majority of the population is being robbed of any hope of a decent future. Just look at the terrible gaps between the opulent wealthy and the most miserable poor. It is appalling that hundreds of millions of people are forced to live on less than a dollar a day. How can we live with a situation where 40 000 children die each day of malnutrition and disease?

To break out of the present situation of uneven, unsustainable development, we will have to improve both the way the world economy can generate more benefits and the way we distribute the benefits of growth within and among countries. A broad set of co-ordinated measures will have to be applied.

Debt relief is necessary. How can developing countries make the investments needed to provide health, education and basic amenities for such growing populations when today they are suffocating under crushing burdens of debt and when financial flows are going from the poor countries to the rich?

[10] United Nations Development Programme (UNDP), *Human Development Report 1992* (Oxford University Press: Oxford and New York, 1992), p. 34.

We must create economic growth in the developing countries. This is essential and the concept of growth must be adjusted to the real requirements of sustainable development.

However, the slow rate of economic growth and high level of unemployment in most of the countries in the Organization for Economic Co-operation and Development (OECD), including Sweden and my own country, Norway, limit demands for products from the developing countries. The current financial and monetary instability poses great risks to countries and individuals. We need to improve stability and prevent speculation from throwing national economies into peril. There is no alternative to effective co-ordination of financial and monetary policies, but it can only succeed when coupled with social purpose, a fair distribution and public efforts to create employment. In fact most economic problems that we are faced with are linked to a lack of co-ordination, to rivalry and to *laissez-faire* attitudes among industrialized countries.

While economic policies are important, all evidence supports the assertion that social development depends on democracy and pluralism. Even the best economic policies alone will not suffice unless the human potential of a healthy and educated population can be unleashed and unless people can participate in political life without fear.

To get out of the crisis we also need to improve the developing countries' access to world markets. The Uruguay Round of negotiations of the General Agreement on Tariffs and Trade (GATT) is vital and must be successfully concluded. World trade must be governed by common rules. GATT must be the stronghold of trade discipline. Enforcing mechanisms are important particularly for weaker parties.

A striking illustration has been offered by the World Bank. Developing countries would benefit by some $50 billion if they were granted unrestricted access to the markets of industrialized countries. This sum is equivalent to what they now, in sum, receive in aid.[11]

Aid will remain important, in particular for Africa. Many donor countries can increase the quantity of their aid and improve its quality. I feel I can point to this since Norway maintains its development assistance in excess of 1 per cent of its gross domestic product (GDP), the highest in the world. Yet aid alone can never solve the poverty problem. Aid must be designed to help in building sound national economies and in implementing policies of social reform.

If we should fail, our predicament can be variously described. Steady deterioration of the quality of life, traumatic for the rich, catastrophic for the poor, is perhaps the least dramatic way of describing humanity's future. Still, there are signs that international co-operation is experiencing a period of maturing.

[11] World Bank, *Annual Report 1992* (World Bank: Washington, DC, [1992]), p. 44.

## III. Regional co-operation and global decision making

The new spirit of regional co-operation bodes well as a means of overcoming the impediments to economic growth and social progress. Groups of states are in fact pooling their efforts and building down barriers between them to ensure the free flow of people, ideas, capital investment and goods and are also including sustainable development as an overriding objective.

The Nordic countries are strengthening their long-standing co-operation in facing and adjusting to the profound changes taking place in Europe. We are able to deal with foreign and security policy issues in a new way. We are actively facing the prospects of European co-operation, strengthening our ties to the European Community (EC) and European Free Trade Area (EFTA) countries through the European Economic Area (EEA) agreement and through applications for membership of the EC. Our co-operation also includes the environmental aspects of our security in Northern Europe through our assistance to Russia in dealing with nuclear and other environmental dangers which concern all our people.

The European Community is also inspiring action on other continents, but so far it has only been able to adopt the necessary institutional means that are available when countries decide to exercise some of their sovereignty jointly. The European Coal and Steel Community established an international authority in charge of the resources and industrial equipment necessary for waging war in Western Europe. Today, the Community's potential as a peace organization is clearer than ever, given its success in binding the European powers closer together and its potential as an engine of renewal on a pan-European basis. The extension of EC responsibility to the spheres of foreign policy, environmental co-operation and monetary stability is taking place at a time of great opportunity. In Russia, a new way of thinking and a new approach to the world are dominant. Soon, a new generation will be at the helm in the United States. All in all, I take an optimistic perspective, notwithstanding the problems which are evident in the world today in many spheres.

The Palme Commission as early as 1982 proposed a strengthening of regional co-operation and the establishment of links to the United Nations. It is not a question of choosing either regional or global co-operation, although some political parties in my country seem to think so. Regional and global organizations must be mutually reinforcing, each organization doing what it can do best. Our goal must be to create an appropriate division of responsibility between regional organizations and the United Nations system. In other parts of the world, regional organizations such as the Organization of American States (OAS) and the Organization for African Unity (OAU) and others could define and sustain their contributions to regional peace, stability and integration. The UN's Regional Commissions should also play a growing role in this respect.

The way in which the UN, the Conference on Security and Co-operation in Europe (CSCE), the countries in NATO, the Council of Europe and the EC are co-operating in addressing the difficult situation in the former Yugoslavia shows how we can reap the benefits of each organization's specialities. The people behind the 'ethnic cleansing' and the war for territorial gains there represent the negation of everything that has been achieved in this century. Their ill-conceived policy and cowardly, savage attacks on innocent people are at the same time a brutal onslaught on the very foundations for any functioning international order.

It must be made clear that basic international standards as they have evolved during this century apply without exception. It is particularly important to demonstrate this here in Europe. Norway has proposed that an international tribunal should be set up to punish all those responsible for the war crimes that are now being committed, and the Nordic countries have supported that proposal. Steps to explore our judicial possibilities are also being taken by the CSCE and the Council of Europe. The inviolability of human rights must permeate and guide all our efforts in a world community based on the rule of law.

Our efforts regionally and globally to deal with concrete environmental problems have so far had uneven success. Despite the progress in some environmental fields such as ozone depletion, the international community is still seriously lacking adequate means of arriving at effective decisions. While the whole range of regulatory and economic measures is available to us on the national level, we lack corresponding measures of governance on a global basis.

Most of the critical decisions that shape the world today are still taken within the confines of national polities. This will have to change. Decisions intended to shape the world of tomorrow must be taken at the level where the problems occur. Just as we are used to applying the necessary political measures to deal with domestic problems, we will have to develop the necessary international measures to deal with international problems.

This is the essence of what we must deal with as we open a new chapter in the development of democracy itself. We must not only spread the benefits of democracy to all corners of the globe, but also ensure that political decision making can be made to work at the same level as other phenomena that influence our daily lives.

At the United Nations Conference on Environment and Development (UNCED) in Rio de Janeiro this summer, it was made clear that we are heading towards a crisis of uncontrollable dimensions unless we change course. Assessments of the conference have varied widely. I believe that the most important outcome was the recognition of the shared but differentiated responsibility of countries. What this means is that the industrialized countries, which are still the major polluters, will have to shoulder the greatest commitments to reducing the strain they are placing on the global environment. They must allow the South sufficient environmental space for their development. They cannot say to developing countries 'Sorry, we have filled the wastebasket, there is no room left for you'.

We also agreed in principle that the North must provide new and additional resources to be transferred to the South to enable them to fulfil their obligations. The South is facing enormous problems of the environment and development which are mostly of a regional or national nature. In order to allow these countries to take part in efforts to counter the truly global environmental threats, such as climate change and loss of biodiversity, they need assistance to be able to cover the additional costs of implementation.

Implementing what is called Agenda 21, the action programme adopted by the conference, will require an extremely high level of activity. It has been estimated that the cost of implementation to developing countries alone would amount to $600 billion a year between now and the year 2000. Four-fifths of this would have to be covered by the developing countries themselves. The remaining $125 billion a year would have to come as financial transfers from the industrialized countries. While this may appear to be a large sum, it is in fact equivalent to the funds that would be available if all the OECD countries raised their level of development assistance to that of Norway and the Netherlands and designed that aid to support sustainable development.

The Nordic countries were forthcoming towards the demands of the South at the conference by making concrete proposals regarding the financial issues. For us it was hard to accept that even reaching the old target of 0.7 per cent of GDP as development assistance by the year 2000 should be denounced as unrealistic by other industrialized countries.

There are many reasons why we must delay making our final judgement on the overall success of UNCED. Although progress was made in some fields, many of the conclusions of the conference were inadequate and much will depend on the road from Rio. Many of the decisions which could not be taken at UNCED will remain with us in urgent need of attention. However, I believe that one fundamental conclusion can be drawn.

Traditional international conferences run by consensus can only advance at the pace of the most reluctant mover in each field. The future requires stronger decision-making procedures. The suspicion that some countries are jockeying for advantage at the expense of others is a crucial problem. It threatens to blur the recognition of mutual dependence. Everyone seems to be waiting for someone else to make a move.

Faced with these challenges and ever-dwindling natural resources, I see UNCED as a step on to the staircase leading to what will inevitably have to come—a better-organized world community in which we pool resources and formal sovereignty in order to obtain more real sovereignty and wider choices for the future, without depriving future generations of their freedom of choice.

However, at present we do not have global institutions that are strong enough to determine new directions or implement effective global policies. It is difficult to see how decision making in international institutions can be made effective unless we introduce new elements of supranationality or make more frequent use of majority vote. This must be the next chapter in developing a system of global governance that can serve our real interests, across national barriers and across generations.

If international co-operation is to become more effective, countries must co-ordinate their representation in different international organizations much more closely. We cannot move forward if a representative of a country's ministry of agriculture pursues one policy in the Food and Agriculture Organization (FAO) while its foreign ministry's representative says something quite different in the UN General Assembly.

Furthermore, it is essential that the various organs of the United Nations system improve the co-ordination of their policies and activities. We need a more unified approach and stronger direction within the UN system. Over the years the number of agencies, programmes and councils has mushroomed, with too little inter-agency co-ordination. This cannot continue. We must streamline our organization to avoid duplication and improve effectiveness. The Secretary-General himself must be supported by all countries to enable him to perform his co-ordinating role.

One of the most important prerequisites for change, which is obviously needed, will be building on the public commitment generated during the years of preparation for UNCED, managing to harness the broad-based dialogue among concerned people, and developing shared perspectives and experiences. Such support and commitment are crucial if we are to be able to take far-reaching decisions, particularly when the necessary measures seem costly in a short-term perspective.

The industrialized democracies will again have to take the lead. We must constantly improve the way our own democracies work and at all costs prevent ignorance and apathy from being allowed to gain a foothold among the millions who, for instance, now have fallen victim to unemployment also in the industrialized world. Our challenge is to help them retain their faith in the future of democracy even in a period when their own most fundamental needs are being inadequately met.

The political system in our countries is often judged in terms of its ability to produce results that are in reality beyond its power. This can lead to a feeling of alienation from the political system. To overcome this danger, of which we see daily examples, politicians must take great care to explain what can be done nationally and what can only be done when countries work together.

The concept of the nation-state—which has been the building-block in our system of international organization during this century—is today very much part of the process of global change. The political leaders of our time have had their basis in the nation-state. They will increasingly have to carry out their work at the international level. They will be dependent upon attitudes and perspectives which can be shared in a democratic sense with other countries and other nations. Those who advocate democracy locally and nationally must also be its champions internationally, for its values and principles are indivisible.

The new Independent Commission chaired by former Swedish Prime Minister Ingvar Carlsson and Sir Shridath Ramphal, who delivered the Olof Palme Memorial Lecture in 1991, will deal with all these pressing issues, benefitting from the work of the earlier commissions, including the South Commission,[12] and basing its outlook on the 1991 Swedish Initiative.[13]

Global governance will depend upon our ability to develop international policies of legislation, redistribution, and a system of caring and sharing with those who risk marginalization.

Burden-sharing will remain essential. There are several bills that need to be covered in a turbulent, troubled world—bills for peace-keeping, refugee relief, famine and natural disasters. Environmental threats and poverty, however, are cross-cutting, long-term, predictable and unavoidable unless we establish a world order of burden-sharing, common perceptions and common responsibility.

From my experience of working with the previous independent commissions, following the Brandt Commission at close range, being a member of the Palme Commission and later serving as chairman of the World Commission on Environment and Development, I am convinced that the international community will have to move from one earth to one world.

There is no such thing as a separate Swedish way to monetary stability or an independent Norwegian way to full employment. There is no such thing as an Ethiopian solution to the drought problem, and Bangladesh alone cannot free itself from the threat of floods.

It has been said that we are the first generation that has the ability really radically to change the course of world development and that we may be the last to have the possibility of doing it. That is precisely why our generation has a unique responsibility and opportunity to manage global change, and to do it in time.

[12] South Commission, *The Challenge to the South, an Overview and Summary of the South Commission Report* (South Commission: [Geneva], 1990).

[13] Sweden, Prime Minister's Office, *Common Responsibility in the 1990s: the Stockholm Initiative on Global Security and Governance, 22 April 1991* (Prime Minister's Office: Stockholm, 1991).

# **Appendix 1A.** Resource scarcity and environmental security

RICHARD H. MOSS

## I. Introduction

The challenge of raising living standards for a growing human population while simultaneously protecting the long-term ability of environmental systems to renew themselves will increase the scarcity of environmental resources. In some locations, resource degradation and shortages will interact with increased demands, seriously straining both environmental and human systems. Recent analyses predict that conflicts over renewable resources will become more common and severe during the coming decades. Conflict is only one possible outcome of this situation, however. Environmental problems may increase the likelihood of co-operation by altering political processes, incentives to co-operate or the potential for finding compromises that serve mutual interests. This appendix points to some of the environmental and social factors that can affect the likelihood of conflict or co-operation and examines trends in the use and availability of two vital resources, land and water. It suggests that quantitative indicators of environmental security could help in assessing the likelihood of conflict and managing scarce renewable resources.

## II. The concept of environmental security

Environmental security is defined as the condition which exists when governments are able to mitigate the social and political impacts of environmental scarcity of resources, drawing on their own capabilities as well as the capabilities of intergovernmental organizations and non-governmental organizations. Environmental resources include not only (*a*) non-renewable resources such as oil and minerals and (*b*) renewable resources such as fisheries products, biomass and fresh water, but also (*c*) environmental services such as waste assimilation, nutrient recycling, generation of soils, regulation of atmospheric conditions and climate, and the creation and maintenance of genetic diversity. Environmental security is thus a function of three sets of factors: (*a*) current and projected levels of resource exploitation; (*b*) the social and political impacts of scarcity, and (*c*) the response capabilities that are available to mitigate the effects of scarcity.

### Resource scarcity, social mobilization and domestic political processes

Resource scarcity has long been postulated to be a cause of violent inter-group and international conflict. Current interpretations are based on a narrow view of the consequences of resource scarcity, namely that it produces social dislocation, which

The author acknowledges advice and comments from Professor Harold Jacobson, Dr John Marks, Dr Richard Rockwell and Dr Phillip Williamson and also wishes to thank Mr Krister Svärd, Librarian of the Stockholm Environment Institute.

then leads to flows of environmental refugees, increased support for insurgencies, heightened ethnic or religious tensions among marginalized groups, and other social processes that lead to conflict.[1] The internal social and political consequences of scarcity are likely to be much broader and to include shifts in residence, occupation, means of subsistence, standards of living, social associations, political activity, expectations, habits and needs. In countries where the impacts of resource scarcity are great, these changes would happen to large numbers of people. Singly, and even more in their cumulative impact, these changes would tend to influence and sometimes transform political behaviour in a process that could be referred to as 'environmentally-driven social mobilization'.[2]

The consequences of this process of environmentally driven social mobilization appear to depend to a large extent on internal social, political and economic structures and conditions. For example, in India, resource degradation in the highlands has resulted in the allocation of increasingly scarce natural resources to the market sector and the displacement of individuals involved in subsistence activities. In some cases, these displaced individuals have used non-violent means of political and social action to reclaim access to the resource base and obtain support for the development of common property resources.[3] In the former Soviet republics and the states of Eastern Europe, resource degradation and associated health problems contributed to growth of anti-Russian sentiments and 'served as a catalyst for the nationalist struggles' which contributed to the breakup of the Soviet empire.[4] Internal social, political, and economic factors affecting the consequences of resource degradation and scarcity include: (*a*) level of development, as measured, for example, by the United Nations Development Programme (UNDP) Human Development Index (based on gross domestic product, GDP, per capita, literacy and life expectancy); (*b*) the country's dependence on activities based on primary renewable resources that were becoming scarce; (*c*) the percentage of the population engaged in subsistence; (*d*) the distribution and size of land holdings; (*e*) the percentage of energy derived from renewable sources, particularly biomass; (*f*) the degree of ethnic homogeneity of the society; and (*g*) the extent to which political and legal institutions are well-functioning and have legitimacy.

Governmental and societal responses will also affect the consequences of scarcity. Fortunately, measures already exist to adapt to shifts in the availability of renewable resources such as water due to the intrinsic variability of climate. By providing incentives to apply existing measures to improve management of the resource base, such as improving collection of precipitation runoff or soil conservation practices, governments can increase the resilience and supply of resources. Many factors will

[1] This appendix is not intended as a review of the literature linking renewable resource scarcity and conflict, but interested readers may wish to consult a number of recent works including: Homer-Dixon, T., 'On the threshold: environmental changes as causes of acute conflict,' *International Security*, vol. 16, no. 2 (1991), pp. 76–116; Homer-Dixon, T., Boutwell, J. H. and Rathjens, G. W., 'Environmental change and violent conflict', *Scientific American*, vol. 268, no. 3 (Mar. 1993); Westing, A. H. (ed.), SIPRI, *Global Resources and International Conflict: Environmental Factors in Strategic Policy and Action* (Oxford University Press: Oxford, 1986); 'Environment and security: the case of Africa', *Ambio*, Special Issue, vol. 20, no. 5 (Aug. 1991).

[2] Deutsch, K. W., 'Social mobilization and political development', *American Political Science Review*, vol. 55, no. 3 (1961), pp. 493–511 defines a similar process of 'social mobilization' caused by the process of modernization.

[3] Shiva, V., *Ecology and the Politics of Survival* (Sage Publications: New Delhi and London, 1991).

[4] Goldman, M. I., 'Environmentalism and nationalism: an unlikely twist in an unlikely direction,' ed. J. M. Stewart, *The Soviet Environment: Problems, Policies, and Politics* (Cambridge University Press: Cambridge, 1991).

affect the ability of governments to cope with scarcity, including: (*a*) availability of capital and access to investment funds; (*b*) access to markets and trade, both domestic and international; (*c*) the existence of national programmes to encourage fomation of locally managed development schemes; (*d*) the existence of functioning regimes (such as regional land-use commissions and watershed management commissions) for managing resource scarcities; (*e*) mechanisms to promote sharing and diffusion of technologies and management practices that have already been successful in adapting to resource scarcity in some locations; and (*f*) training and education programmes to enable affected individuals to develop new livelihoods.

**Resources and international co-operation**

Hostilities resulting from spontaneous public violence or failed negotiations are more dramatic and visible than successful negotiations that lead to the avoidance of hostilities. There are, however, many instances in which governments have successfully negotiated over resources. A principal incentive for negotiation and co-operation is that joint management of renewable resources can do more than produce agreement over how to 'divide the pie'—it can actually increase the productivity of the resource base and create a bigger pie. A few examples drawn from a data base of successful river basin negotiations compiled by the International Institute for Applied Systems Analysis include: (*a*) the 1960 Indus Water Treaty between India and Pakistan, which successfully avoided an outbreak of hostilities over use of the river; (*b*) the 1972 agreement between Mali, Mauritania and Senegal to form the Organisation pour la mise en valeur du fleuve Sénégal (OMVS) to provide for integrated development of the Senegal River; (*c*) the 1973 De La Plata River agreement between Argentina and Uruguay, which covered control over river islands, access to navigation channels, and joint development of oil resources; and (*d*) the 1980 agreement among the nine basin states of the Niger River (Benin, Cameroon, Chad, Guinea, Ivory Coast, Mali, Niger, Nigeria and Upper Volta, now Burkina Faso) to establish the Niger River Basin Authority because of the perceived dangers of unco-ordinated national projects.[5]

## III. Resource exploitation

Scarcity of renewable resources is a function of demands for resources and current levels of exploitation in relation to natural renewal rates. Tables 1A.1–1A.3 extrapolate from available data concerning demands for and availability of land and fresh water. These resources are vital for the continued production of most renewable resources and environmental services and are likely to present the greatest management problems for governments in the future. The tables cover developing countries (with comparison to two contrasting industrialized countries) because increased demands and resource degradation will pose the most serious challenges to these countries, many of which are already vulnerable to intrinsic variability in environmental conditions. The countries selected comprise nearly 60 per cent of the world population and represent a variety of stages of demographic transition and economic and social development.

[5] McDonald, A., *International River Basin Negotiations: Building a Database of Illustrative Successes* (International Institute for Applied Systems Analysis: Laxenburg, Austria, 1988).

## Resource demands

Increasing numbers of humans, consuming larger amounts of energy and materials, will inevitably increase demands on available flows of environmental resources. Mid-range UN projections are for the world population to increase to 8.5 billion by 2020. While fertility rates in many developing countries are declining, the growing number of women of child-bearing age in these countries will by 2025 result in 85 per cent of the world's population living in countries now classified as 'developing'.[6] Table 1A.1 dramatically illustrates the momentum of population growth in all African and many Asian nations. In contrast, most Latin American countries are expected to attain replacement levels of fertility by the end of the century.

The implications of population trends for resource use cannot be appreciated without reference to data on living standards in developing countries that illustrate the need for human development and economic growth. Over 1.1 billion people (one in five of the world's population) have annual incomes lower than $370, and an estimated 630 million of these are 'extremely poor', with incomes lower than $275 per annum. Much of this poverty is concentrated in developing countries: in South Asia and sub-Saharan Africa, close to 50 per cent of the population have annual incomes lower than $370.[7] The data in table 1A.1 illustrate the variation in per capita gross national product (GNP) and the annual percentage change in real GNP, which is in some cases inadequate to keep pace with increases in population. Because they are averages, these data mask the pronounced concentration of wealth and skewness in the distribution of income in most developing countries, factors which also have important implications for resource demands. The extreme poverty and high population growth rates of many developing countries affect resource demand and degradation in two ways: (*a*) poverty itself can slow demographic transition and contribute to resource degradation;[8] and (*b*) the increases in economic activity needed to narrow the income gap between industrialized and developing countries will increase stress on environmental resources that are already over-stretched in many countries and regions.

Technology used in production and consumption is also an important factor in determining demands for resources. Table 1A.1 includes information on energy intensity of economic production and dependence on traditional biomass fuels as surrogate measures for the environmental implications of economic activity. The best prospect for increasing energy supplies in many developing countries is through greater reliance on fuels derived from biomass production. However, current biomass stocks are already committed to production of food, fodder, building materials, and raw industrial materials and will thus prove inadequate unless significantly more efficient technologies of energy production are adopted. In cases where current GNP per capita is low, but economic growth is increasing and reliance on traditional fuels is high, the expected increase in population can be expected to create particularly taxing demands on biomass resources. Even in countries which have relatively modern energy infrastructures and low dependence on traditional fuels, drops in bio-

[6] World Bank, *World Development Report 1992: Development and the Environment* (Oxford University Press: Oxford, 1992), p. 30.

[7] World Bank (note 6).

[8] See, for example, Dasgupta, P., 'Population, resources, and poverty', *Ambio,* vol. 21, no. 1 (1992), pp. 95–101.

mass production due to drought conditions have led to reduced production of renewable fuels, which has serious economic consequences.

Data on population, consumption and technology provide crude indicators of demand for renewable resources. It would be simplistic, however, to view demands for resources simply as a function of these factors. Other factors, such as attitudes and values, determine basic preferences for the types and quantities of goods and services that are produced. These preferences, as well as production possibilities, are also affected by such factors as trade flows and access to capital. Additional social factors, such as property rights and decision-making structures, influence access to or control over resources, and hence the way in which they will be used.

## Availability of resources

Given demand trends and current consumption and production patterns, land and water resources will be inadequate in some countries to support growing populations even at the currently inadequate standards of living.[9]

Reliable supplies of fresh water for domestic use are already beyond the reach of many in developing countries and are increasingly threatened in some developed countries. For example, consumption outstrips supply in northern China, and shortages could reach crisis proportions in the Middle East and North Africa, where political tensions exacerbate shortages.[10] As shown in table 1A.2, the distribution of fresh water resources is uneven. Most countries have plentiful resources and, in fact, use only a small percentage of their available resources due to lack of institutional capacity and infrastructure. Globally, however, some 33 nations (including a number of developed countries) are projected by 2025 to fall below the minimum per capita water requirement for an industrialized nation of 1000 $m^3$ per person per year.[11] The data on water use by sector illustrate the generally high percentage of water used in agriculture in most developing countries, something which will make these countries particularly vulnerable to shortages. Another measure of vulnerability is the percentage of water resources flowing in from other countries. Niger, for example, obtains nearly three-fourths of its water from external sources.

Land is a key resource for most human activities including agriculture, industry, forestry, energy production, settlement and recreation. The two human activities that use the most land are arable cultivation and livestock production. Approximately 14–15 million $km^2$, an area about the size of South America, is in some form of cultivation. An additional 70 million $km^2$ is used for some form of livestock production. Most of the prime agricultural lands of the world are already cultivated, and thus major increases in food production are likely to come from greater use of fertilizers, pesticides and irrigation.[12]

[9] In 1984 the UN Food and Agriculture Organization (FAO) identified 64 countries (of 117 studied) which would be unable to feed their populations in 2000 using low levels of agricultural inputs even if they used all of their available land for food production. With intermediate use of inputs the critical list includes only 36, and with high input levels (an unrealistic assumption in some cases), 18 remain critical. See Norse, D., 'A new strategy for feeding a crowded planet', *Environment,* vol. 34, no. 5 (June 1992), pp. 10–11.

[10] World Resources Institute (WRI), *World Resources 1992–93* (Oxford University Press: New York, 1992), p. 159.

[11] Gleick, P. H., *Water and Conflict* (American Academy of Arts and Sciences: Boston, Md., 1992), Occasional Paper of the Project on Environmental Change and Acute Conflict, p. 17.

[12] Turner II, B. L., Moss, R. H. and Skole, D. S., *Relating Land Use and Global Land-Cover Change* (International Geosphere–Biosphere Programme (IGBP) and Human Dimensions of Global

Data in table 1A.3 on percentage changes in different categories of land cover illustrate the rapid recent changes in land cover, primarily due to human activities. The data on population density, population growth rates and cropland per capita indicate that pressures on land resources will increase dramatically, particularly in Asia. If the UN mid-range estimate of population growth proves to be accurate, the global average of 0.28 ha. of cropland per capita is expected to decline to 0.17 ha. in 2025. In Asia, cropland per capita is expected to decline to an average of 0.09 ha., although some countries such as Bangladesh and the Republic of Korea already experience lower per capita levels. Expanding trade and creating institutions to promote technology co-operation, particularly in the agricultural and energy sectors, will become increasingly important in managing the consequences of land scarcity.

# IV. Environmental indicators

Environmental resources will come under growing stress in the coming decades. These stresses and resulting scarcities will have social and political effects, including but not limited to outbreaks of violent conflict. In examining the international security implications of these trends, research needs to focus not only on the ways in which scarcity can lead to conflict, but also on the factors that affect the probability of co-operation.

One way to lay the basis for effective inter-governmental co-operation and policy co-ordination is to establish mechanisms for sharing information about national plans for resource use. Governments agreed at the UN Conference on Environment and Development (UNCED) to make periodic reports on the implementation of their commitments to environmentally sustainable economic development. *Inter alia*, these reports are expected to assess 'national population carrying capacity in the context of satisfaction of human needs and sustainable development', with special attention given to 'critical resources such as water and land, and environmental factors, such as ecosystem health and biodiversity'.[13]

The development and adoption of a series of quantitative indicators which can be used to pinpoint potential resource conflicts would facilitate effective policy co-ordination. 'Environmental indicators' could be used 'to inform the ongoing process of policy dialogue among countries and to lay the basis for international co-operation and agreements . . [thus to] parallel the role of economic indicators used in economic policy co-ordination'.[14] The indicators should include data on current and projected levels of resource exploitation; the social and political impacts of scarcities; and the response capabilities that are available to mitigate the effects of scarcity—those factors affecting environmental security as defined in this appendix. They should focus on specific environmental issues and economic sectors as well as include a composite indicator that would help to summarize where and how plans for development may outstrip resource capabilities and create conflicts.

Environmental Change Programme (HDP): Stockholm, 1993). HDP Report Series, no. 5; IGBP Report Series no. 24.

[13] United Nations, *Agenda 21*, UN document A/CONF.151.L.3/Add. 38.

[14] Organization for Economic Co-operation and Development (OECD), *Environmental Indicators* (OECD: Paris, 1992).

**Table 1A.1.** Driving forces of resource demands for selected countries

| | | Population (m.) | | Average population change (% per year) | | Total fertility (average no. of births per woman) | | GNP (US $ per cap.) | Average annual change in real GNP (%) | | Energy intensity | | Traditional fuels[c] |
|---|---|---|---|---|---|---|---|---|---|---|---|---|---|
| | 1990 | Projected 1995 | Projected 2025 | Actual 1985–90 | Projected 1995–2000 | 1970–75 | 1990–95 | 1989 | 1969–79 | 1979–89 | 1989[a] | 1989[b] | 1989 |
| World | 5 292.2 | 5 770.29 | 8 504.22 | 1.74 | 1.63 | 4.5 | 3.3 | . . | . . | . . | . . | . . | 6 |
| Kenya | 24.03 | 28.98 | 79.11 | 3.58 | 3.81 | 8.1 | 6.8 | 380 | 7.1 | 4.2 | 52 | – 9 | 79 |
| Niger | 7.73 | 9.10 | 21.48 | 3.14 | 3.33 | 7.1 | 7.1 | 290 | 1.5 | – 0.6 | 26 | 56 | 73 |
| Nigeria | 108.54 | 127.69 | 280.89 | 3.30 | 3.17 | 6.9 | 6.6 | 250 | 6.3 | – 0.1 | 63 | 64 | 62 |
| Senegal | 7.33 | 8.42 | 16.99 | 2.78 | 2.86 | 7.0 | 6.2 | 650 | 2.8 | 2.2 | 18 | – 6 | 51 |
| Zimbabwe | 9.71 | 11.34 | 22.62 | 3.16 | 2.92 | 7.2 | 5.3 | 640 | 4.9 | 4.3 | 51 | –15 | 25 |
| Bangladesh | 115.59 | 132.22 | 234.99 | 2.67 | 2.60 | 7.0 | 5.1 | 180 | 2.4 | 3.5 | 27 | – 2 | 54 |
| China | 1 139.06 | 1 222.56 | 1 512.59 | 1.45 | 1.22 | 4.8 | 2.3 | 360 | 7.4 | 8.9 | 84 | – 32 | 6 |
| India | 853.09 | 946.72 | 1 442.39 | 2.07 | 1.91 | 5.0 | 4.1 | 350 | 3.0 | 5.6 | 37 | – 1 | 25 |
| Korea (South) | 42.79 | 44.66 | 51.63 | 0.95 | 0.77 | 4.10 | 1.7 | 4 400 | 9.6 | 8.1 | 21 | – 12 | 1 |
| Pakistan | 122.63 | 141.52 | 267.11 | 3.44 | 2.75 | 7.0 | 5.9 | 370 | 4.7 | 6.8 | 36 | 5 | 21 |
| Philippines | 62.41 | 69.94 | 111.51 | 2.49 | 2.05 | 5.3 | 3.9 | 700 | 6.3 | 1.8 | 25 | 1 | 38 |
| Thailand | 55.7 | 59.61 | 80.91 | 1.53 | 1.32 | 5.0 | 2.2 | 1 170 | 7.0 | 7.3 | 27 | – 6 | 34 |
| Argentina | 32.32 | 34.26 | 45.51 | 1.27 | 1.12 | 3.2 | 2.8 | 2 160 | 2.9 | – 1.6 | 31 | 47 | 5 |
| Bolivia | 7.31 | 8.42 | 18.29 | 2.76 | 2.88 | 6.5 | 5.8 | 600 | 3.9 | – 0.4 | 24 | 25 | 16 |
| Brazil | 150.37 | 165.08 | 245.81 | 2.07 | 1.67 | 4.7 | 3.2 | 2 550 | 8.4 | 2.7 | 25 | 2 | 30 |
| Costa Rica | 3.02 | 3.37 | 5.25 | 2.64 | 1.90 | 4.3 | 3.0 | 1 790 | 6.1 | 1.8 | 22 | 9 | 33 |
| Honduras | 5.14 | 5.97 | 11.51 | 3.18 | 2.75 | 7.4 | 4.9 | 900 | 5.6 | 2.1 | 21 | 1 | 62 |
| Mexico | 88.60 | 97.97 | 150.06 | 2.20 | 1.81 | 6.4 | 3.1 | 1 990 | 9.2 | 2.0 | 33 | 14 | 5 |
| Sweden | 8.44 | 8.51 | 8.58 | 0.22 | 0.12 | 1.9 | 1.9 | 21 710 | 2.4 | 1.8 | 14 | – 9 | 5 |
| USA | 249.22 | 258.16 | 299.88 | 0.81 | 0.600 | 2.0 | 1.9 | 21 100 | 2.8 | 2.6 | 17 | – 18 | 2 |

**Table 1A.2.** Water resource use for selected countries

| | Annual internal renewable water resources ($km^3$) | Annual river flows from other countries ($km^3$) | Total annual water resources ($km^3$) | Resources from other countries as % of total resources | Internal resources per capita 1990 ($1000\ m^3$) | Percentage resources withdrawn 1987 | Percentage used in agriculture 1987 | Internal resources per capita 2025 (projected) ($1000\ m^3$) |
|---|---|---|---|---|---|---|---|---|
| World | . . | . . | 40 673 | . . | 7.69 | *8* | *69* | 4.78 |
| Kenya | 15 | . . | . . | . . | 0.59 | *7* | *62* | 0.19 |
| Niger | 14 | 30 | 44 | *68* | 1.81 | *1* | *74* | 0.65 |
| Nigeria | 261 | 47 | 308 | *15* | 2.31 | *1* | *54* | 0.93 |
| Senegal | 23 | 12 | 35 | *34* | 3.15 | *4* | *92* | 1.37 |
| Zimbabwe | 23 | . . | . . | . . | 2.37 | *5* | *79* | 1.02 |
| Bangladesh | 1 357 | 1 000 | 2 357 | *42* | 11.74 | *1* | *96* | 5.77 |
| China | 2 800 | – | 2 800 | – | 2.47 | *16* | *87* | 1.85 |
| India | 1 850 | 235 | 2 085 | *11* | 2.17 | *18* | *93* | 1.28 |
| Korea (South) | 63 | . . | . . | . . | 1.45 | *17* | *75* | 1.22 |
| Pakistan | 298 | 170 | 468 | *36* | 2.43 | *33* | *98* | 1.12 |
| Philippines | 323 | – | 323 | – | 5.18 | *9* | *61* | 2.90 |
| Thailand | 110 | 69 | 179 | *39* | 1.97 | *18* | *90* | 1.36 |
| Argentina | 694 | 300 | 994 | *30* | 21.47 | *3* | *73* | 15.25 |
| Bolivia | 300 | . . | . . | . . | 41.02 | – | *85* | 16.40 |
| Brazil | 5 190 | 1 760 | 6 950 | *25* | 34.52 | *1* | *40* | 21.11 |
| Costa Rica | 95 | . . | . . | . . | 31.51 | *1* | *89* | 18.10 |
| Honduras | 102 | . . | . . | . . | 19.85 | *1* | *91* | 8.86 |
| Mexico | 36 | . . | . . | . . | 4.03 | *15* | *86* | 0.24 |
| Sweden | 176 | 4 | 180 | 2 | 21.11 | 2 | *9* | 20.51 |
| USA | 2 478 | . . | . . | . . | 9.94 | *19* | *42* | 8.26 |

**Table 1A.3.** Land resource use for selected countries

| | Population density 1990 (per 1000 ha.) | Population growth rate 1995–2000 (projected, %) | Cropland, ha. per capita 1990 | Cropland, % change since 1977–79 | Pasture, % change since 1977–79 | Forest, % change since 1977–79 | Degraded area as % of all vegetated land, late 1980s |
|---|---|---|---|---|---|---|---|
| World | 403 | *1.63* | 0.28 | *2.2* | *0.1* | *– 1.8* | *17* |
| *Africa* | | | | | | | *22* |
| Kenya | 422 | *3.81* | 0.10 | *6.8* | – | *– 7.8* | |
| Niger | 61 | *3.33* | 0.47 | *13.0* | *– 4.0* | *– 22.1* | |
| Nigeria | 1 192 | *3.17* | 0.29 | *3.5* | – | *– 19.4* | |
| Senegal | 381 | *2.86* | 0.71 | *1.5* | – | *– 2.6* | |
| Zimbabwe | 251 | *2.92* | 0.29 | *10.3* | – | *– 4.0* | |
| *Asia* | | | | | | | *20* |
| Bangladesh | 8 880 | *2.60* | 0.08 | *1.5* | – | *10.4* | |
| China | 1 221 | *1.22* | 0.08 | *– 3.9* | – | *– 7.7* | |
| India | 2 869 | *1.91* | 0.20 | *0.5* | *– 3.4* | *– 0.7* | |
| Korea (South) | 4 334 | *0.77* | 0.05 | *– 3.8* | *116.4* | *– 1.4* | |
| Pakistan | 1 591 | *2.75* | 0.17 | *3.3* | – | *17.3* | |
| Philippines | 2 093 | *2.05* | 0.13 | *4.4* | *23.1* | *– 16.4* | |
| Thailand | 1 090 | *1.32* | 0.40 | *21.8* | *26.8* | *– 15.6* | |
| *South America* | | | | | | | *14* |
| Argentina | 118 | *1.12* | 1.11 | *1.9* | *– 0.7* | *– 1.3* | |
| Bolivia | 67 | *2.88* | 0.47 | *3.7* | *– 1.5* | *– 1.1* | |
| Brazil | 178 | *1.67* | 0.52 | *17.1* | *6.3* | *– 4.2* | |
| Costa Rica | 590 | *1.90* | 0.18 | *5.5* | *24.0* | *– 17.0* | |
| Honduras | 459 | *2.75* | 0.35 | *2.3* | *7.2* | *– 18.8* | |
| Mexico | 464 | *1.81* | 0.28 | *1.9* | – | *– 12.0* | |
| *Europe* | | | | | | | *23* |
| Sweden | 205 | *0.12* | 0.33 | *– 4.1* | *– 22.8* | *0.5* | |
| *North and Central America* | | | | | | | *8* |
| USA | 272 | *0.60* | 0.76 | – | *1.0* | *– 1.1* | |

**Notes to tables 1A.1–1A.3**

All data are based on World Resources Institute (WRI), *World Resources 1992–93* (Oxford University Press: New York, 1992). Additional information about sources and technical issues accompany the WRI tables in the source. The data are compiled by WRI from a variety of sources and models, and as a result caution should be used in comparing countries.

Figures in italics are percentages.

*Table 1A.1*

[a] Energy intensity (in conventional fuel equivalent) in megajoules per 1987 US $ of GNP. An estimate of the amount of energy from all sources (including traditional fuels such as fuelwood, charcoal, etc.) per unit production of GNP in 1989.

[b] Percentage change in energy intensity since 1989.

[c] Traditional fuels as a percentage of total requirements, 1989. Estimated by WRI, relying on data from a variety of sources. The estimates are not regarded as precise.

*Sources:* Data from *World Resources 1992–93*: table 15.1 ('Gross National Product and Official Development Assistance'); table 16.1 ('Size and Growth of Population and Labour Force, 1950–2025'); table 16.2 ('Trends in Births, Life Expectancy, Fertility, and Age Structure, 1970–95'); and table 21.2 ('Energy Consumption and Requirements, 1979–89').

*Table 1A.2*

*Sources:* Data from (or derived from) *World Resources 1992–93* : table 22.1 ('Freshwater Resources and Withdrawals') and table 16.1 ('Size and Growth of Population and Labour Force, 1950–2025). The columns 'Total annual water resources', 'Resources from other countries as % of total resources', and 'Internal resource per capita projected 2025' are calculated from WRI data by the author. Calculations are approximate due to rounding. Some of the data in the columns 'Percentage Resources Withdrawn' and 'Percentage Used in Agriculture' are from years other than 1987. It should also be noted that these annual averages mask large variations of precipitation and flow between different years.

*Table 1A.3*

*Sources:* Data from *World Resources 1992–93*: table 16.1 ('Size and Growth of Population and Labour Force, 1950–2025'); table 17.1 ('Land Area and Use, 1977–89'); table 18.2 ('Agricultural Inputs, 1975–89'); and table 19.3 ('Human-Induced Soil Degradation, 1945 to late 1980s'). Data on land use are reported to the UN Food and Agriculture Organization (FAO) by national governments, whose classifications of land uses often differ. The data on percentage changes in different types of land use reflect not only trends in resource use, but differences in reporting and changes in classification, including by FAO. For this reason, trends must be interpreted with caution. Data on land degradation are derived from digitized maps of degradation developed by an expert group convened by the UN Environment Programme in 1990 and are only available at the regional level.

# **Appendix 1B.** The United Nations Conference on Environment and Development

## THE RIO DECLARATION

The United Nations Conference on Environment and Development,

Having met at Rio de Janeiro from 3 to 14 June 1992,

Reaffirming the Declaration of the United Nations Conference on the Human Environment, adopted at Stockholm on 16 June 1972, and seeking to build upon it,

With the goal of establishing a new and equitable global partnership through the creation of new levels of cooperation among States, key sectors of societies and people,

Working towards international agreements which respect the interests of all and protect the integrity of the global environmental and developmental system,

Recognizing the integral and interdependent nature of the Earth, our home,

Proclaims that:

### Principle 1

Human beings are at the centre of concerns for sustainable development. They are entitled to a healthy and productive life in harmony with nature.

### Principle 2

States have, in accordance with the Charter of the United Nations and the principles of international law, the sovereign right to exploit their own resources pursuant to their own environmental and developmental policies, and the responsibility to ensure that activities within their jurisdiction or control do not cause damage to the environment of other States or of areas beyond the limits of national jurisdiction.

### Principle 3

The right to development must be fulfilled so as to equitably meet developmental and environmental needs of present and future generations.

### Principle 4

In order to achieve sustainable development, environmental protection shall constitute an integral part of the development process and cannot be considered in isolation from it.

### Principle 5

All States and all people shall cooperate in the essential task of eradicating poverty as an indispensable requirement for sustainable development, in order to decrease the disparities in standards of living and better meet the needs of the majority of the people of the world.

### Principle 6

The special situation and needs of developing countries, particularly the least developed and those most environmentally vulnerable, shall be given special priority. International actions in the field of environment and development should also address the interests and needs of all countries.

### Principle 7

States shall cooperate in a spirit of global partnership to conserve, protect and restore the health and integrity of the Earth's ecosystem. In view of the different contributions to global environmental degradation, States have common but differentiated responsibilities. The developed countries acknowledge the responsibility that they bear in the international pursuit of sustainable development in view of the pressures their societies place on the global environment and of the technologies and financial resources they command.

### Principle 8

To achieve sustainable development and a higher quality of life for all people, States should reduce and eliminate unsustainable patterns of production and consumption and promote appropriate demographic policies.

### Principle 9

States should cooperate to strengthen endogenous capacity-building for sustainable development by improving scientific understanding through exchanges of scientific and technological knowledge, and by enhancing the development, adaptation, diffusion and transfer of technologies, including new and innovative technologies.

**Principle 10**

Environmental issues are best handled with the participation of all concerned citizens, at the relevant level. At the national level, each individual shall have appropriate access to information concerning the environment that is held by public authorities, including information on hazardous materials and activities in their communities, and the opportunity to participate in decision-making processes. States shall facilitate and encourage public awareness and participation by making information widely available. Effective access to judicial and administrative proceedings, including redress and remedy, shall be provided.

**Principle 11**

States shall enact effective environmental legislation. Environmental standards, management objectives and priorities should reflect the environmental and developmental context to which they apply. Standards applied by some countries may be inappropriate and of unwarranted economic and social cost to other countries, in particular developing countries.

**Principle 12**

States should cooperate to promote a supportive and open international economic system that would lead to economic growth and sustainable development in all countries, to better address the problems of environmental degradation. Trade policy measures for environmental purposes should not constitute a means of arbitrary or unjustifiable discrimination or a disguised restriction on sinternational trade. Unilateral actions to deal with environmental challenges outside the jurisdiction of the importing country should be avoided. Environmental measures addressing transboundary or global environmental problems should, as far as possible, be based on an international consensus.

**Principle 13**

States shall develop national law regarding liability and compensation for the victims of pollution and other environmental damage. States shall also cooperate in an expeditious and more determined manner to develop further international law regarding liability and compensation for adverse effects of environmental damage caused by activities within their jurisdiction or control to areas beyond their jurisdiction.

**Principle 14**

States should effectively cooperate to discourage or prevent the relocation and transfer to other States of any activities and substances that cause severe environmental degradation or are found to be harmful to human health.

**Principle 15**

In order to protect the environment, the precautionary approach shall be widely applied by States according to their capabilities. Where there are threats of serious or irreversible damage, lack of full scientific certainty shall not be used as a reason for postponing cost-effective measures to prevent environmental degradation.

**Principle 16**

National authorities should endeavour to promote the internalization of environmental costs and the use of economic instruments, taking into account the approach that the polluter should, in principle, bear the cost of pollution, with due regard to the public interest and without distorting international trade and investment.

**Principle 17**

Environmental impact assessment, as a national instrument, shall be undertaken for proposed activities that are likely to have a significant adverse impact on the environment and are subject to a decision of a competent national authority.

**Principle 18**

States shall immediately notify other States of any natural disasters or other emergencies that are likely to produce sudden harmful effects on the environment of those States. Every effort shall be made by the international community to help States so afflicted.

**Principle 19**

States shall provide prior and timely notification and relevant information to potentially affected States on activities that may have a significant adverse transboundary environmental effect and shall consult with those States at an early stage and in good faith.

**Principle 20**

Women have a vital role in environmental management and development. Their full participation is therefore essential to achieve sustainable development.

**Principle 21**

The creativity, ideals and courage of the youth of the world should be mobilized to forge a global partnership in order to achieve sustainable development and ensure a better future for all.

**Principle 22**

Indigenous people and their communities, and other local communities, have a vital role in environmental management and development because of their knowledge and traditional practices. States should recognize and duly support their identity, culture and interests and enable their effective participation in the achievement of sustainable development.

**Principle 23**

The environment and natural resources of people under oppression, domination and occupation shall be protected.

**Principle 24**

Warfare is inherently destructive of sustainable development. States shall therefore respect international law providing protection for the environment in times of armed conflict and cooperate in its further development, as necessary.

**Principle 25**

Peace, development and environmental protection are interdependent and indivisible.

**Principle 26**

States shall resolve all their environmental disputes peacefully and by appropriate means in accordance with the Charter of the United Nations.

**Principle 27**

States and people shall cooperate in good faith and in a spirit of partnership in the fulfilment of the principles embodied in this Declaration and in the further development of international law in the field of sustainable development.

---

*Source:* United Nations document A/CONF.151/5/REV.1, 13 June 1992.

---

## AGENDA 21

*Adopted on 14 June 1992*
*Excerpt*

**Chapter 39. International legal instruments and mechanisms***

39.1. The recognition that the following vital aspects of the universal, multilateral and bilateral treaty-making process should be taken into account:

(a) The further development of international law on sustainable development, giving special attention to the delicate balance between environmental and developmental concerns;

(b) The need to clarify and strengthen the relationship between existing international instruments or agreements in the field of environment and relevant social and economic agreements or instruments, taking into account the special needs of developing countries;

(c) At the global level, the essential importance of the participation in and the contribution of all countries, including the developing countries, to treaty-making in the field of international law on sustainable development. Many of the existing international legal instruments and agreements in the field of environment have been developed without adequate participation and contribution of developing countries, and thus may require review in order to reflect the concerns and interests of developing countries and to ensure a balanced governance of such instruments and agreements;

(d) Developing countries should also be provided with technical assistance in their attempts to enhance their national legislative capabilities in the field of sustainable development;

(e) Future codification projects for the progressive development and codification of international law on sustainable development should take into account the ongoing work of the International Law Commission; and

(f) Any negotiations for the progressive development and codification of international law concerning sustainable development should, in general, be conducted on a universal basis, taking into account special circumstances in the various regions.

---

* This is a final, advance version of a chapter of Agenda 21. This document will be further edited, translated into the official languages, and published by the United Nations for the General Assembly in the autumn of 1993.

*Objectives*

39.2. The overall objective of the review and development of international environmental law should be to evaluate and to promote the efficacy of that law and to promote the integration of environment and development policies through effective international agreements or instruments, taking into account both universal principles and the particular and differentiated needs and concerns of all countries.

39.3. Specific objectives are:

(a) To identify and address difficulties which prevent some States, in particular developing countries, from participating in or duly implementing international agreements or instruments and, where appropriate, to review or revise them with the purposes of integrating environmental and developmental concerns and laying down a sound basis for the implementation of these agreements or instruments;

(b) To set priorities for future international law- making on sustainable development at the global, regional or sub-regional level, with a view to enhancing the efficacy of international law in this field through, in particular, the integration of environmental and developmental concerns;

(c) To promote and support the effective participation of all countries concerned, in particular developing countries in the negotiation, implementation, review and governance of international agreements or instruments, including appropriate provision of technical and financial assistance and other available mechanisms for this purpose, as well as the use of differential obligations where appropriate;

(d) To promote, through the gradual development of universally and multilaterally negotiated agreements or instruments, international standards for the protection of the environment that take into account the different situations and capabilities of countries. States recognize that environmental policies should deal with the root causes of environmental degradation, thus preventing environmental measures from resulting in unnecessary restrictions to trade. Trade policy measures for environmental purposes should not constitute a means of arbitrary or unjustifiable discrimination or a disguised restriction on international trade. Unilateral actions to deal with environmental challenges outside the jurisdiction of the importing country should be avoided. Environmental measures addressing international environmental problems should, as far as possible, be based on an international consensus. Domestic measures targeted to achieve certain environmental objectives may need trade measures to render them effective. Should trade policy measures be found necessary for the enforcement of environmental policies, certain principles and rules should apply. These could include, inter alia, the principle of non-discrimination; the principle that the trade measure chosen should be the least trade-restrictive necessary to achieve the objectives; an obligation to ensure transparency in the use of trade measures related to the environment and to provide adequate notification of national regulations; and the need to give consideration to the special conditions and development requirements of developing countries as they move towards internationally agreed environmental objectives.

(e) To ensure the effective, full and prompt implementation of legally binding instruments, and to facilitate timely review and adjustment of agreements or instruments by the parties concerned, taking into account the special needs and concerns of all countries, in particular developing countries;

(f) To improve the effectiveness of institutions, mechanisms and procedures for the administration of agreements and instruments;

(g) To identify and prevent actual or potential conflicts, particularly between environmental and social/economic agreements or instruments, with a view to ensuring that such agreements or instruments are consistent. Where conflicts arise, they should be appropriately resolved;

(h) To study and consider the broadening and strengthening of the capacity of mechanisms, inter alia in the United Nations system, to facilitate, where appropriate and agreed by the parties concerned, the identification, avoidance and settlement of international disputes in the field of sustainable development, duly taking into account existing bilateral and multilateral agreements for the settlement of such disputes.

*Activities*

39.4. Activities and means of implementation should be considered in the light of the above Basis for Action and Objectives without prejudice to the right of every State to put forward suggestions in this regard in the General Assembly of the United Nations. These suggestions could be reproduced in a separate compilation on sustainable development.

*A. Review, assessment and fields of action in international law for sustainable development*

39.5. While ensuring the effective participation of all countries concerned, Parties should at periodic intervals review and assess both the past performance and effectiveness of existing international agreements or instruments as well as the priorities for future law-making on sustainable development. This may include an examination of the feasibility of elaborating general rights and obligations of States, as appropriate, in the field of sustainable development, as provided by General Assembly resolution 44/228. In certain cases, attention should be given to the possibility of taking into account varying circumstances through differential obligations or gradual application. As an option for carrying out this task, earlier UNEP practice may be followed whereby legal experts designated by governments could meet at suitable intervals to be decided later with a broader environmental and developmental perspective.

39.6. (a) Measures in accordance with international law should be considered to address, in times of armed conflict, large-scale destruction of the environment that cannot be justified under international law. The General Assembly and the Sixth Committee are the appropriate fora to deal with this subject. The specific competence and role of the International Committee of the Red Cross should be taken into account.

(b) In view of the vital necessity to ensure safe and environmentally sound nuclear power, and in order to strengthen international cooperation in this field, efforts should be made to conclude the ongoing negotiations for a nuclear safety convention in the framework of the International Atomic Energy Agency.

*B. Implementation mechanisms*

39.7. The parties to international agreements should consider procedures and mechanisms to promote and review their effective, full and prompt implementation. To that effect, States could, inter alia:

(a) Establish efficient and practical reporting systems on the effective, full and prompt implementation of international legal instruments;

(b) Consider appropriate ways in which relevant international bodies, such as UNEP, might contribute towards the further development of such mechanisms.

*C. Effective participation in international law-making*

39.8. In all these activities and others that may be pursued in the future, based on the above Basis for Action and Objectives, the effective participation of all countries, in particular developing countries, should be ensured through appropriate provision of technical assistance and/or financial assistance. Developing countries should be given 'headstart' support not only in their national efforts to implement international agreements or instruments, but also to participate effectively in the negotiation of new or revised agreements or instruments and in the actual international operation of such agreements or instruments. Support should include assistance in building up expertise in international law particularly in relation to sustainable development, and in assuring access to the necessary reference information and scientific/technical expertise.

*D. Disputes in the field of sustainable development*

39.9. In the area of avoidance and settlement of disputes, States should further study and consider methods to broaden and make more effective the range of techniques available at present, taking into account, among others, relevant experience under existing international agreements, instruments or institutions and, where appropriate, their implementing mechanisms such as modalities for dispute avoidance and settlement. This may include mechanisms and procedures for the exchange of data and information, notification and consultation regarding situations that might lead to disputes with other States in the field of sustainable development and for effective peaceful means of dispute settlement in accordance with the Charter of the United Nations including, where appropriate, recourse to the International Court of Justice, and their inclusion in treaties relating to sustainable development.

---

*Source:* United Nations document A/21/39, 14 June 1993.

# Part II. Global and regional security and conflicts, 1992

Chapter 2. Preventive diplomacy, peace-making and peace-keeping

Chapter 3. Major armed conflicts

Chapter 4. Post-Soviet conflict heritage and risks

Chapter 5. The CSCE: towards a security organization

# 2. Preventive diplomacy, peace-making and peace-keeping

ROGER HILL*

## I. Introduction

During 1992 efforts to control regional and intra-state conflicts assumed a place of central importance on the international agenda.

In September Boutros Boutros-Ghali, Secretary-General of the United Nations, declared that 'A widely varying array of resentments, ambitions, rivalries and hatreds masked for decades have come to the fore to threaten international harmony and shared purpose'.[1] World opinion was shocked by the barbarism displayed so openly in the former Yugoslavia, the scenes of mass starvation in Somalia and similar instances of inhumanity elsewhere. The urgency of effective international action to prevent, control or resolve such problems was widely acknowledged.

At the same time the end of the cold war appeared to have opened up a great new opportunity to use the United Nations to deal with critical conflict problems. This prospect was set out most graphically in Boutros-Ghali's *An Agenda for Peace*, released in June 1992.[2] Stressing the central role that ought to be played by the United Nations, it reviewed the various methods of preventing, controlling or resolving disputes, and briefly discussed the UN's work in each area. It now constitutes a focal point for international thinking about preventive diplomacy, peace-making, peace-keeping and related peace efforts.

This chapter examines key UN and related developments in the field of conflict resolution, peace-keeping and other peace activities. It also considers the requirements of effective UN action, and the need for new initiatives at the global level.

[1] United Nations General Assembly, *Report of the Secretary-General on the Work of the Organization, September 1992,* UN document A/47/1 (United Nations: New York, 11 Sep. 1992), p. 43 (hereafter referred to as the *UN September 1992 Report.*)

[2] United Nations Security Council, *An Agenda for Peace: Preventive Diplomacy, Peacemaking and Peace-keeping*, Report of the Secretary-General pursuant to the statement adopted by the Summit Meeting of the Security Council on 31 January 1992, UN document A/47/277 (S/24111) (United Nations: New York, 17 June 1992), paras 1–15 (hereafter referred to as *An Agenda for Peace* and reproduced as appendix 2A).

* The Department of External Affairs and International Trade Canada, the Department of National Defence in Ottawa, the United Nations Association in Canada, the Canadian Centre for Global Security and the United Nations Organization in New York provided information for this chapter.

# II. The UN agenda for peace in 1992

The arrival in office in January 1992 of a new Secretary-General, Boutros-Ghali, and the convening of a special summit meeting of the United Nations Security Council in New York heralded a fresh attempt by the United Nations to tackle the problems of international conflict prevention and resolution. Held at heads of state and government level on 31 January, the Security Council meeting focused on the primary responsibility of the Council for the maintenance of international peace and security. The Council expressed a renewed commitment to working collectively for improvements in such fields as peace-making and peace-keeping as well as arms control and disarmament. Participants asked the Secretary-General to analyse the situation and recommend ways of making the organization stronger and more efficient in these areas.[3]

### *An Agenda for Peace*

Boutros-Ghali responded to the Security Council in June with *An Agenda for Peace*, a report in which he foresaw the United Nations playing a central role in world peace efforts. He expected the organization to be active in such areas as fact-finding, early warning and mediation. When necessary, the collective security mechanism of the Security Council would come into play, including the use of national forces authorized by the UN or, he suggested, standing forces made available on a permanent basis to the Security Council. 'Peace-enforcement' units, including heavily armed elements, could, the Secretary-General proposed, be used in some circumstances, for example, in enforcing cease-fires. Peace-keeping forces would be used to help preserve peace where fighting had been halted and to assist in implementing agreements reached by peace-makers.[4] The United Nations was to be equipped with a powerful and diverse assemblage of military strength that would deter almost all but the largest countries and far exceed anything that could have been contemplated during the long and divisive years of the cold war.

An important contribution of this document was the attempt to spell out the different types of activity the UN might pursue:

1. *Preventive diplomacy* was described as action to prevent disputes arising between parties, to prevent existing disputes from escalating into conflicts, or

[3] United Nations Security Council, *Note by the President of the Security Council*, UN document S/23500 (United Nations: New York, 31 Jan. 1992).

[4] See *An Agenda for Peace* (note 2), paras 8–54; and *The Charter of the United Nations*, Chapter VI, 'Pacific Settlement of Disputes', and Chapter VII, 'Action with Respect to Threats to the Peace, Breaches of the Peace, and Acts of Aggression'. In *An Agenda for Peace*, the Secretary-General evidently envisaged a range of military forces to uphold collective security: (*a*) national forces authorized to act on behalf of the Security Council to respond to major threats such as those that might be posed by countries with large armies equipped with sophisticated weapons, e.g., as in the 1991 Persian Gulf Crisis (see paras 42–43 of *An Agenda for Peace*); (*b*) forces made available to the Security Council on a permanent basis, to meet 'any threat posed by a military force of a lesser order' (para. 43); (*c*) 'peace-enforcement' forces, under the command of the Security Council and the Secretary-General, to restore and maintain cease-fires (para. 44); and (*d*) peace-keeping forces.

to limit the spread of conflicts when they occur. Such action includes measures to build confidence, such as systematic exchanges of military missions; fact-finding, to obtain information and clarify confused situations; early warning, for example of famines or mass movements of populations; preventive deployment of UN and other forces; and the creation of demilitarized zones.[5]

2. *Peace-making* was described as action to bring hostile parties to agreement, including such traditional peace-making activities as mediation and negotiation, and the stronger concept of 'peace enforcement'. The section on peace-making also contained discussion of two issues normally considered under the rubric of 'Collective security', namely, 'Sanctions and special economic problems' and the 'Use of military force'. These are both measures that could be taken under Article VII of the United Nations Charter to maintain or restore international peace and security in the face of a threat to the peace, breach of the peace or act of aggression.[6]

3. *Peace-keeping* was presented as the deployment of UN or similar forces to bring stability to areas of tension, help implement agreements among parties to a dispute and contribute to settlements. The mandates of peace-keeping forces have been broadened in recent years, the report indicated, as the UN has responded to new demands.

The report also noted that the deployment of a UN peace-keeping force has hitherto required the consent of all the main parties to a dispute. The basic conditions for the success of such operations, the report said, have remained unchanged over the years and include: a clear and practicable mandate; the co-operation of the parties in implementing that mandate; the continuing support of the Security Council; the readiness of member states to contribute the military, police and civilian personnel required; effective UN command at headquarters and in the field; and adequate financial and logistic support. The report noted that 13 peace-keeping missions had been established between 1945 and 1987, and another 13 between then and 1992. More than half a million personnel have served in these operations; and over 800, from 43 countries, have died in the performance of their duties.[7]

4. *Peace-building* was described as action designed to identify and support structures which will consolidate peace, promote confidence, strengthen links among nations formerly at war and help to develop democratic societies. The report also stressed the importance of tackling the *deepest* causes of conflict, such as political oppression, social injustice and economic despair.[8]

*An Agenda for Peace* welcomed the contributions of regional organizations to the conflict resolution effort, mentioning in particular such bodies as the Organization for African Unity (OAU), the League of Arab States, the Organization of American States (OAS), the European Community (EC) and the

[5] See *An Agenda for Peace* (note 2), paras 20 and 23–33.

[6] See *An Agenda for Peace* (note 2), paras 20 and 34–45. Note also the comments on the World Court and amelioration through assistance.

[7] See *An Agenda for Peace* (note 2), paras 20 and 46–53.

[8] See *An Agenda for Peace* (note 2), paras 15 and 55–59.

Conference on Security and Co-operation in Europe (CSCE). It also discussed the safety of personnel involved in UN missions, the growing problem of financing UN operations in a time of tardy contributions and rising demands for the UN's services, and renewal and strengthening of the United Nations by its 50th anniversary in 1995.[9]

The UN membership welcomed *An Agenda for Peace*, but expressed varying views on particular parts of it. The idea of giving the UN a central role in peace enforcement was regarded as requiring a great deal of further consideration. There were also worries—especially among the non-aligned countries—about the report's implications for the sanctity of national sovereignty and territorial integrity as enshrined in Article 2 of the UN Charter.[10]

## Report on the work of the United Nations

A further report from the United Nations in the latter part of 1992 filled out the conceptual framework for the treatment of conflict issues and described trends in UN peace activities. *The Report on the Work of the Organization from the Forty-sixth to the Forty-seventh Session of the General Assembly*, issued by Boutros-Ghali in September (hereafter referred to as the *UN September 1992 Report*) reiterated some of his earlier points about the challenge and the opportunity now before the world body, and discussed a number of critical practical issues: the improvement and streamlining of the UN organization; the requirements of effective UN action in the area of economic development; and the UN's work in the field of 'peace endeavours'. It also took a closer look at the importance of humanitarian assistance in conflict situations, especially in such man-made crises as those in the former Yugoslavia and in Somalia.[11]

The report provides remarkable figures on the greatly expanding peace activities of the United Nations, including statistics on the numbers of official meetings of the Security Council, on fact-finding and similar missions, and on peace-keeping operations.

From 1991 to 1992 the number of official meetings of the Security Council increased by more than one-third. After averaging around 60 a year from 1987 to 1990 and declining slightly in 1991, they increased to more than 80 in 1992. This is one indicator of the increasing consultation, preparation of resolutions and other diplomatic support for conflict resolution that occurred in UN headquarters during the year.[12]

The number of requests from the General Assembly to the Secretary-General for reports on conflict situations also increased markedly. In 1987, 87

[9] See *An Agenda for Peace* (note 2), paras 60–86.

[10] On the international response to *An Agenda for Peace* and Boutros-Ghali's efforts to strengthen the UN's peace-making capabilities, see, for example, The Permanent Mission of Canada to the United Nations (New York), Statement by the Hon. Barbara McDougall, Secretary of State for External Affairs, before the United Nations General Assembly (24 Sep. 1992); and Serrill, M. S., 'Under fire', *Time*, 18 Jan. 1993, pp. 14–16.

[11] See the *UN September 1992 Report* (note 1).

[12] See the *UN September 1992 Report* (note 1), p. 21, figure 1.

such requests were received, but by 1992 the number of responses had increased to 189. In the first eight months of 1992, the Secretary-General authorized 75 fact-finding, representational and good-offices missions.[13]

The most striking figures are in the area of peace-keeping. After stable levels of 10 000–15 000 throughout the period 1987–91, the number of military personnel involved in peace-keeping increased rapidly to 40 000 in August 1992. Police personnel increased from 35 in 1987 to over 3600 in August 1992, and civilian personnel increased from 877 to over 9400 in the same period.[14] By the end of November 1992, the numbers had grown to about 48 000 military personnel, 4400 police and 10 000 civilians, a total of more than 62 000.[15] The number of peace-keeping operations increased from 8 in 1991 to 12 in August 1992, while annual costs were quadrupled.[16] The actual cost of peace-keeping operations in the calendar year 1992 stood at $1.4 billion.[17]

## III. The UN peace effort in 1992: main activities

### Preventive diplomacy

As conflicts related to decolonization and the cold war have disappeared the focus has shifted to disputes arising from the collapse of the Soviet Union as well as from ethnic divisions and troubled political, social and economic conditions in many countries. A number of protracted conflicts of national interests, and disputes over borders, territories or resources, add to the list of concerns.[18]

In these circumstances, active preventive diplomacy designed to ease tensions before they result in conflict is essential. Preventive deployment of military forces may also be necessary at times to dampen dangerous situations or contain conflicts which threaten to spread.

The significance of such activity was demonstrated in a number of instances in 1992. In July, for example, Boutros-Ghali named Cyrus Vance as his Special Representative to report on continuing troubles in South Africa, and the following month the UN sent a group of 10 observers to that country to witness planned mass demonstrations and political rallies. In the same period, the Secretary-General sent an observer to Haiti with a fact-finding mission dispatched by the OAS, and stated his readiness to help in any other way he could to solve the Haitian crisis. Between March and August the UN sent three fact-finding missions to Armenia and Azerbaijan to investigate the

[13] See the *UN September 1992 Report* (note 1), p. 8.

[14] See the *UN September 1992 Report* (note 1), pp. 8 and 65, figure 4. Updated figures for the end of Aug. 1992 provided by the United Nations Organization, New York.

[15] Figures provided by the United Nations Organization, New York.

[16] *UN September 1992 Report* (note 1), p. 65, figure 5.

[17] *Financing an Effective United Nations: A Report of the Independent Advisory Group on UN Financing* (The Ford Foundation: New York, Feb. 1993).

[18] Analysis of the types of conflict and their changing patterns on the world scene is a major field of academic enquiry. See, for example, Bloomfield, L. P., 'Coping with conflict in the late twentieth century', *International Journal* (Toronto), vol. 44, no. 4 (autumn 1989), pp. 772–802.

Nagorno-Karabakh dispute; and later it sent two fact-finding missions to Moldova to help establish and maintain a truce in the Trans-Dniester region. In these last two cases the UN operated in support of local or regional peace initiatives, including those of national authorities, the Commonwealth of Independent States (CIS) and the CSCE.[19]

### Peace-making: mediation and negotiation

The United Nations continued to play an active role in 1992 in the effort to mediate disputes and negotiate peaceful settlements. Highlights included a further, frustrating attempt to settle the Cyprus problem and new efforts to break a deadlock holding up a referendum on the future of Western Sahara. The situations in Angola, Mozambique and the Middle East also involved the UN in a variety of negotiations.[20]

### Peace-making: peace enforcement

While in 1992 the United Nations did not need to authorize any actions similar to the 1991 Persian Gulf War operation, Coalition aircraft did enforce two no-fly zones in Iraq during the year, one in the mainly Kurdish northern part of the country and the other in the partly Shiite south. They shot down one Iraqi fighter aircraft on 27 December, as well as threatening punitive action on a number of occasions. This did not deter Saddam Hussein from sending military raiding parties into demilitarized Kuwaiti territory early in 1993. As a result allied air raids were launched against military facilities in Iraq as a warning to Baghdad not to continue its cease-fire violations. There was some controversy over whether or not the allied action was consistent with existing Security Council resolutions.

The possibilities of peace-enforcement in other regions also received a good deal of attention as 1992 progressed. A no-fly zone was declared over Bosnia in October, followed by speculation about armed enforcement of the ban or even a major, US-led military strike against Serbia.[21]

International action in Somalia moved from peace-keeping to a form of peace-enforcement in December, when the United States sent in a powerful task force and other countries moved quickly to provide additional forces. During the following month the numbers of military personnel deployed with Operation Restore Hope rose rapidly (see table 2.1). The mandate of this force required it to 'create a secure environment for humanitarian relief operations

[19] See the *UN September 1992 Report* (note 1), pp. 43–50. See also section IV below, on CIS, CSCE and national peace efforts.

[20] See the *UN September 1992 Report* (note 1), pp. 45–50. On the failure to achieve a breakthrough on the Cyprus question, see, for example, Ferguson, A., 'Cyprus solution still eludes UN after 28 years', *Toronto Star*, 1 Nov. 1992, p. F7.

[21] See, for example, Adams, J., 'US draws up plans to fight war in Balkans', *Sunday Times*, 29 Nov. 1992, p. 24.

**Table 2.1.** Operation Restore Hope—Somalia

Figures are approximate and include army and air force personnel in Somalia and naval personnel in the country or off-shore in mid-January 1993.

| Country | Troops deployed |
|---|---|
| United States | 21 000 |
| Italy | 3 540 |
| France | 2 370 |
| Canada | 1 360 |
| Morocco | 1 240 |
| Australia | 1 140 |
| Belgium | 620 |
| Botswana | 300 |
| India | 280 |
| Egypt | 240 |

*Notes:*

1. Smaller contingents or other support were provided by a number of other countries, including Britain, Germany, Japan, Jordan, Kuwait, New Zealand, Saudi Arabia and Turkey.

2. Additional major contingents were reportedly offered by various countries, including Mauritania, Nigeria, Saudi Arabia, Tunisia, Turkey and the United Arab Emirates.

3. Note also the Pakistani battalion and 62 other military personnel from 12 other countries originally deployed in Somalia as a peace-keeping force. See text and table 2.2, under United Nations Operation in Somalia (UNOSOM).

*Sources:* Figures from government sources. On the buildup during Dec. 1992, see Associated Press report, 'Restore Hope troops', *Ottawa Citizen*, 17 Dec. 1992, p. C1; Associated Press and Reuters, 'Countries contribute to military mission', *Globe and Mail* (Toronto), 7 Dec. 1992, p. A8; 'The mission: week two', *Newsweek*, 28 Dec. 1992, p. 39; and *Facts on File, 1992*, vol. 52, no. 2718 (31 Dec. 1992), p. 966.

in Somalia', by force when necessary. Boutros-Ghali evidently hoped that the force would also disarm rival gangs interfering with food deliveries and pacify the country before it handed over its responsibilities to a peace-keeping operation at some time in 1993. The US Government was reluctant to accept such broader responsibilities.[22] During December the force established itself in the Mogadishu area and began to move out to other key centres such as Baidoa. There were a number of military actions against armed Somali groups as the UN force went about its main task of ensuring the safe distribution of relief supplies.

In general the experience of peace-enforcement in 1992 was mixed. Action taken in Somalia and inaction in the former Yugoslavia may well have been correct responses in the circumstances, but they did raise difficult questions. Would peace-enforcement be used only in countries which did not have powerful armed forces, and what did this imply about equality of status and treatment for members of the international community? What did it imply for the sovereignty and territorial integrity of the smaller, weaker nations? How extensive would its use be for tackling civil wars? Was the UN about to

[22] See Lewis, P., 'UN chief says letter to Bush outlines US commitment to disarm Somali gangs', *New York Times* (International Section), 13 Dec. 1992, p. 14.

embark upon a path which might bring serious dangers for the organization itself as well as for some of its members? At the end of the year, there was an obvious need to give further thought to the whole question of the part that peace-enforcement might play in global collective security.[23] The views of the new US Administration would be vital to this debate.

## Peace-keeping

At the outset of 1992, the UN was still engaged in long-standing peace-keeping missions in Cyprus, the Middle East and the Indian sub-continent. Another set of ongoing missions had been established in the recent past: the UN Angola Verification Mission (UNAVEM) in 1989 and three operations in 1991—the UN Iraq–Kuwait Observer Mission (UNIKOM), the UN Mission for the Referendum in Western Sahara (MINURSO), and the UN Observer Mission in El Salvador (ONUSAL). Three other peace-keeping missions had also been established in 1988–92 but had completed their activities: the UN Iran–Iraq Military Observer Group (UNIIMOG), the UN Observer Group in Central America (ONUCA) and the UN Transition Assistance Group in Namibia (UNTAG).[24] The composition of the main peace-keeping operations in late 1992, is shown in table 2.2.

During 1992 several new peace-keeping missions were launched. ONUSAL in El Salvador was essentially transformed into a new peace-keeping force when its mandate was greatly broadened. Additional peace-keeping missions were authorized for Cambodia, the former Yugoslavia, Somalia and Mozambique.

Peace-keeping has evolved markedly over recent years. Starting with observer missions in Greece, Palestine and elsewhere in the immediate post-war period, it acquired a completely new dimension in the 1950s and 1960s with the establishment of three major emergency forces in Egypt (United Nations Emergency Forces—UNEF), Cyprus (United Nations Peace-keeping Force in Cyprus—UNFICYP) and the Congo (United Nations Operation in the Congo—ONUC)—each designed not only to observe and report on cease-fires and similar truce arrangements, but also to neutralize dangerous conflicts by interpositioning UN forces between antagonists. The UN force in the Congo also had the task of helping to maintain the unity of that country in the face of internal chaos and secessionist bids from Katanga and elsewhere, while that in

[23] Some of the key issues that will need to be discussed in any future debate on global collective security are included in Cox, D. (ed.), *The Use of Force by the Security Council for Enforcement and Deterrent Purposes: A Conference Report* (The Canadian Centre for Arms Control and Disarmament: Ottawa, 1990). This collection of very useful papers includes contributions by D. Cox, J. Boulden, B. M. Russett and J. S. Sutterlin, J. Dedring and B. S. Krasulin.

[24] See the *UN September 1992 Report* (note 1), pp. 4–64. See also United Nations, *The Blue Helmets: A Review of United Nations Peace-keeping*, 2nd edn (United Nations: New York, 1990); and United Nations, *United Nations Peace-keeping Operations: Information Notes* (UN Department of Public Information, Communications and Project Management Division: New York, Sep. 1992).

Cyprus also had policing and emergency assistance functions.[25] In 1962, the UN established a third kind of peace-keeping mission, when it dispatched a security force to West Irian as part of the United Nations Temporary Executive Authority (UNTEA), charged with administering that territory while sovereignty was being transferred from the Netherlands to Indonesia.[26]

The great upsurge in peace-keeping activity that occurred after the mid-1980s led to the establishment of several UN forces with extensive mandates, in what was often described as the new era or wave of peace-keeping. The successful UN mission in Namibia especially, between February 1989 and March 1990, provided a good deal of the inspiration for this more activist approach and led to the development of some tendency to believe that broad-based flexible peace-keeping missions might be the answer to many—if not all—of the world's local conflict problems. Set up to help bring independence to Namibia, UNTAG carried out a wide variety of tasks to create conditions for free and fair elections. Its duties included monitoring cease-fires and force reductions, serving as a watch-dog over the country's security forces, promoting confidence in the electoral process, helping to register voters, and ensuring that voting took place smoothly and peacefully on polling day. At its height, UNTAG comprised some 4500 military personnel, over 1000 police and 2000 civilians, from 120 countries. It resembled the UN Security Force in West Irian more than the classic, interpositional models of UNEF and UNFICYP.[27]

However, the experience of 1992 suggests that a more careful appraisal of requirements and possibilities for peace-keeping operations is necessary. This is amply demonstrated by a comparison of the five new peace-keeping missions established during 1992 in El Salvador, Cambodia, the former Yugoslavia, Somalia and Mozambique.

The mission begun in El Salvador at the outset of 1992 was similar in many ways to that in Namibia. In El Salvador, ONUSAL was greatly expanded shortly after a peace agreement between the national government and the Farabundo Martí Front for National Liberation (FMLN) was signed on 16 January 1992. It was also given a much broader mandate aimed at helping that country to transform itself into a new nation with reformed institutions. One of its main responsibilities was to assist El Salvador in a transitional

[25] The functions of UNFICYP and its displacements changed following the Turkish intervention in 1974. Afterwards, UNFICYP concentrated its units along the new dividing line between the Greek- and Turkish-controlled areas of the island, and assumed more of an observer role than a true interpositioning one. It has continued to carry out these modified duties for almost two decades. However, in Dec. 1992 Denmark pulled the last of its troops out of UNFICYP, and Canada announced that its contingent would be withdrawn by Sep. 1993. At the end of 1992, Britain and Austria were also reducing their troop levels, and the whole future of UNFICYP was increasingly in doubt.

[26] The experience of the 1950s and 1960s gave rise to some important studies on peace-keeping, notably the following: James, A., *The Politics of Peacekeeping* (Chatto & Windus: London, 1969); Wainhouse, D. W., *International Peace Observation* (Johns Hopkins Press: Baltimore, Md., 1966); and Bowett, D. W., *United Nations Forces* (Praeger: New York, 1964). See also Hill, R. J., *Command and Control Problems of UN and Similar Peacekeeping Forces*, ORD Report No. 68/R5 (Department of National Defence, Operational Research Division: Ottawa, Apr. 1968).

[27] See United Nations, *The Blue Helmets: A Review of United Nations Peace-keeping* (2nd edn) (note 24), pp. 341–88; and Gardam, J., *The Canadian Peacekeeper* (General Store Publishing House: Burnstown, Ontario, 1992), pp. 56–57.

period leading to new national elections in early 1994. The new, wider duties of ONUSAL included verifying the cease-fire agreement by observing the concentration of combatants, helping FMLN members to re-integrate into society, and monitoring the reform of the armed forces and the police. In addition, the UN was charged with overseeing various activities relating to land and other economic and social issues. With about 230 military personnel and 300 police in its ranks, ONUSAL seemed well on the way to success in the latter part of 1992: the cease-fire was holding, the reform process was taking hold—despite difficulties with various aspects—and the movement towards national elections was under way. It has been suggested that ONUSAL's mandate should be extended until the national elections take place in 1994.[28]

Another exercise in social construction and political transition was launched in Cambodia in accordance with the Agreements on a Comprehensive Political Settlement signed at the Paris Conference in October 1991 and with UN Security Council Resolution 745 of 28 February 1992. The latter set up the United Nations Transitional Authority in Cambodia (UNTAC), with the task of administering the country and bringing it through democratic national elections. As the *UN September 1992 Report* indicates, the mission in Cambodia is one of the most ambitious and complex peace-keeping operations in the UN's history and its mandate unprecedented. 'On the military side', the *UN September 1992 Report* comments,

> the operation involves performing the difficult tasks of supervision, monitoring and verification of the cease-fire; the withdrawal of foreign troops; and the regrouping, cantonment, disarming and subsequent demobilization of the armed forces of the four Cambodian factions. On the civilian side, it includes innovative responsibilities such as the control and supervision of the activities of existing administrative structures and police forces, as well as measures to promote respect for human rights and fundamental freedoms, including the investigation and redress of human rights violations.

For the first time the UN was also given responsibility for organizing and conducting free and fair elections, scheduled for May 1993; for repatriating some 360 000 refugees and displaced persons; and for co-ordinating rehabilitation assistance.[29]

By the end of 1992, UNTAC had about 15 500 military personnel, about 3300 police and over 1000 civilians deployed in Cambodia. More than half of the refugees and displaced persons had been repatriated, and the UN's efforts were having an impact on the whole country. However, the Khmer Rouge were continuing to prove reluctant to comply with the Paris Agreements, and

[28] See United Nations Security Council, Resolution 729 (1992), Document S/RES/729 (1992) (United Nations: New York, 14 Jan. 1992); the *UN September 1992 Report* (note 1), pp. 59–60; and *Economist* Editorial, 'Two steps the world can take in El Salvador', *Globe and Mail* (Toronto), 21 Dec. 1992, p. A21. Figures on ONUSAL military and police personnel were provided by the United Nations Organization, New York.

[29] See United Nations, *Agreements on a Comprehensive Political Settlement of the Cambodian Conflict* (United Nations: Paris, 23 Oct. 1991); United Nations Security Council, Resolution 745 (1992), Document S/RES/745 (1992) (United Nations: New York, 28 Feb. 1992); and the *UN September 1992 Report* (note 1), pp. 52–55.

political violence around the country was adversely affecting preparations for democratic elections. There were continuing doubts about the long-term future of Cambodia: Would it gradually resume life as a peaceful, unified country or would internal rivalries born of two decades of ideological battles, local strife, outside intervention and genocide give way to new horrors or the possible disintegration of this nation?[30]

The peace-keeping mission established in the former Yugoslavia in 1992 has had an even more difficult situation to face. The United Nations Protection Force (UNPROFOR) has had to operate within a bewildering diplomatic context and in the midst of conflicting claims of antagonistic and warring nationalities. Its areas of operation have sprawled over parts of Croatia and Bosnia and Herzegovina, with a recent extension into Macedonia. By the end of November 1992, its numbers included more than 21 400 military personnel and 600 police, but it had no mandate to effect a general settlement or enforce local cease-fires. Its members do have the power to defend themselves when threatened, but have sometimes been unable to protect those looking to them for safety. Economic sanctions, naval patrols in the Adriatic and the establishment of a no-fly zone over Bosnia have not helped significantly to stabilize the situation or reduce the fighting.[31]

The main tasks of UNPROFOR have varied according to its area of operation, but resemble the interpositioning and emergency assistance roles of UNFICYP rather than the transitional authority functions of the current missions in El Salvador or Cambodia. In Bosnia and Herzegovina the cease-fires the force should be monitoring are broken as soon as they are agreed upon. In the circumstances, UNPROFOR simply does what it can to contain the situation, while carrying out mine-clearance, keeping Sarajevo airport open, delivering humanitarian assistance, transporting supplies to beleaguered centres of population, and aiding refugees and others in distress. In December and January 1993 it also sent Canadian troops into Macedonia to deter the spread of the conflict.[32]

[30] Figures on military personnel and police in Cambodia were provided by the United Nations Organization, New York. On the situation in Cambodia, see, for example: Swain, J., 'Blood runs again in Cambodia's killing fields', *Sunday Times*, 29 Nov. 1992, p. 19; and 'UN measures stepped up', *Globe and Mail* (Toronto), 8 Jan. 1993, p. A7.

[31] An advance mission of UN military, police and Secretariat officers was sent to Yugoslavia at the end of Dec. 1991 to look into the implementation of the plans for a peace-keeping force, but UNPROFOR itself was not established until 21 Feb. 1992, by Security Council Resolution 743. The EC and the CSCE, among others, have also been actively involved in efforts to resolve this crisis, but the international community has not been able, in the circumstances, to establish a unified approach to bringing about a settlement. Figures on UNPROFOR personnel were provided by the United Nations Organization in New York. See also the *UN September 1992 Report* (note 1), pp. 55–56; United Nations, *United Nations Focus, United Nations Protection Force,* Document DP/1257 (United Nations Department of Public Information: New York, May 1992); and United Nations, *United Nations Peace-keeping Operations: Information Notes* (note 24), pp. 29–32.

[32] A Jan. 1993 report on the work of UNPROFOR notes that, of its 23 000 troops, some 16 000 'are assigned to keep Serbs and Croats apart in Croatia; the remainder are occupied with ferrying food and supplies to Sarajevo and other beleaguered towns in Bosnia. So far, the UN presence in Bosnia has done nothing to stop the fighting and little to relieve the suffering of Bosnians, who are still dying from shelling, sniper fire, hunger and intense cold'. See Serrill (note 10). On the dispatch of Canadian troops to Macedonia, see 'Canadians take up preventive positions', *Globe and Mail* (Toronto), 8 Jan. 1993, p. A6.

At the end of the year attention was focused on new peace talks on Bosnia and Herzegovina in Geneva, but further international moves were anticipated in the event of their failure. A good deal was likely to depend on the policies of the new Clinton Administration in the United States. Peace-enforcement or other forms of collective security action appeared to be one possibility, although the political, military and financial costs of intervention against well-armed Serbian forces were not being taken lightly.

The experience in Somalia in 1992 was almost as frustrating as that in the former Yugoslavia. The central problem in this case was that civil society and government had largely broken down, and rival armed gangs were terrorizing the population. Aid workers were also being threatened, and hundreds of thousands of Somalis were dying of starvation. On 24 April the Security Council adopted Resolution 751, calling for an immediate cease-fire among warring factions throughout the country, and establishing UNOSOM. This led to the creation of a peace-keeping force charged with the supply of humanitarian relief, consolidating cease-fires, reducing violence and promoting national reconciliation. The task in Somalia, as the Secretary-General noted in September, was nothing less than the reconstruction of an entire society and nation.[33]

By December, however, the UN effort in Somalia was facing the prospect of failure. Although 300 000 people had already died of famine and warfare since the beginning of the crisis, and another 1000 were dying each day, the 500 Pakistani peace-keepers sent to Mogadishu to monitor a local cease-fire were being obstructed by armed gangs. Consequently, the UN authorized the USA and other member countries to send in a peace-enforcement force.

A further UN peace-keeping operation was launched in Mozambique in mid-December 1992, when the Security Council authorized the dispatch of 7500 troops and civilians to oversee a peace accord signed by the government and (Renamo) the Mozambican National Resistance. A first contingent of Italian troops was due to fly in to the country in January 1993 to begin policing the cease-fire and observing rival forces disarming and demobilizing. Multi-party elections will also be held under international supervision. The mission of this force is essentially a transitional one similar to that of ONUSAL, although in even more difficult circumstances.[34]

Looking at 1992 overall, peace-keeping won some new laurels in El Salvador and Cambodia, and the peace-keepers themselves performed well and often bravely in all missions, especially those in the former Yugoslavia and in Somalia. However, the lessons of the year must surely include the recognition that peace-keeping is not a panacea for all the world's conflicts and should not be applied indiscriminately in every dispute. The world community has to meet the challenge launched by the UN Secretary-General's *Agenda for Peace*, and work out viable new approaches to world order. These must include effective collective security and realistic conceptions of peace-

[33] See United Nations Security Council, Resolution 751 (1992), Document S/RES/751 (1992), (United Nations: New York, 24 Apr. 1992). See also the *UN September 1992 Report* (note 1), pp. 56–57.

[34] See 'Mozambique: UN peace forces authorized', *Ottawa Citizen*, 17 Dec. 1992, p. A7; and Bierman, J., 'UN troops moving into Mozambique', *Toronto Star*, 3 Jan. 1993, p. F2.

keeping, rather than reliance on *ad hoc* approaches. The different types of peace-keeping operation have to be clarified, and each force must be given the diplomatic backing, mandate, resources, structures, command and control systems, and, in most cases, local consent, that will give it a solid chance to succeed.[35]

## Peace-building

The large number of international and local civilian personnel involved in peace-keeping missions in 1992 is one indicator of increased UN peace-building activity. Their expertise is generally in such areas as electoral procedures, health, finance, engineering and administration. Numbers increased almost tenfold between 1991 and the end of 1992, from around 1000 to over 10 000.[36]

Military personnel attached to peace-keeping missions also frequently assist in peace-building activities. For example, the mine-clearance operations conducted by UNIKOM forces in the Iraq–Kuwait border area in 1992 are an important instance of peace-building. Military support for repatriation of refugees and displaced persons in Cambodia is another.[37]

However, most peace-building activities authorized or supported by the UN occur outside the framework of peace-keeping operations. For example, in 1992, the Secretary-General dispatched a number of special missions to Eritrea, Ethiopia and Mozambique to provide technical or other assistance for the organization of elections or referenda.[38] The specialized agencies of the UN, such as the Office of the United Nations High Commissioner for Refugees and the United Nations Children's Fund (UNICEF), continued to assume broad and demanding responsibilities for humanitarian assistance in such strife-torn areas as the Middle East and the Horn of Africa.[39] Member countries of the United Nations, together with private foundations and such bodies as the Inter-Parliamentary Union, meanwhile continued to provide specialized expertise on the functioning of democratic institutions and other aspects of peace-building.

[35] The requirements of effective peace-keeping and an effective United Nations have been debated for many years. For two recent, important contributions, see Rikhye, I. J. and Skjelsbaek, K. (eds), *The United Nations and Peacekeeping* (Macmillan, in Association with the International Peace Academy: Basingstoke, 1990); and Urquhart, B. and Childers, E., *Towards a More Effective United Nations* (Dag Hammarskjold Foundation: Uppsala, Sweden, 1992).

[36] The exact total numbers are not certain, partly because a variety of civilian personnel are involved, including UN Secretariat staff, the staff of other UN agencies, international personnel on contract and locally engaged people. Present numbers are estimated to be over 10 000.

[37] See *United Nations Peace-keeping Operations: Information Notes* (note 24), pp. 15–17 and 33–37. See also *An Agenda for Peace* (note 2), para. 58, on the importance of mine-clearance in peace-building operations.

[38] See the *UN September 1992 Report* (note 1), pp. 45–50.

[39] See the *UN September 1992 Report* (note 1), pp. 60–64.

# IV. Regional and other contributions

Regional organizations made significant contributions to international peace efforts in 1992. In some parts of the world they continued to play a more direct role in the maintenance of international peace and security than the United Nations.

The EC, the Western European Union (WEU) and NATO continued to provide guarantees for the security of Western Europe and also took an interest in security and conflict situations across the continent. However, their roles in efforts to resolve the conflict in the former Yugoslavia were disappointing. The EC has been involved in mediation, cease-fire monitoring and emergency assistance there since 1991, while the WEU and NATO became directly involved only in 1992, notably by maintaining flotillas in the Adriatic Sea to enforce the economic sanctions on Serbia and Montenegro.[40] All were growing increasingly frustrated by the end of the year and were beginning to consider stronger military action against Serbia if the conflict continued.

An increasingly promising mechanism for conflict resolution is the CSCE, a forum that began in the early 1970s as a vehicle for managing East–West relations and promoting human rights throughout Europe. In recent years it has acquired a Secretariat in Prague and a Conflict Prevention Centre in Vienna, initially to handle disputes over confidence- and security-building measures (CSBMs), but potentially to work on more general conflict issues. During 1992—as mentioned above—the CSCE played an active role in efforts to mediate the Trans-Dniester and Nagorno-Karabakh disputes.[41]

In the territories of the former USSR, the CIS created a multilateral force in 1992 to contain conflicts among or within its members, and afterwards established peace-keeping operations in South Ossetia and the Trans-Dniester region. Russian troops were also employed to help contain conflict in Tajikistan, and between North Ossetians and Ingush in the northern Caucasus. Russian negotiating efforts have been directed at the Nagorno-Karabakh and Georgia–Abkhazia conflicts. The aim in each case was to calm ethnic tensions and inter-group warfare so that territorial disputes and other contentious issues might be resolved. These efforts have not always been successful, but at least may have helped to contain some of the worst excesses in a number of volatile situations.[42]

[40] See Reuters dispatch, 'La France pense à une intervention militaire en Bosnie: l'OTAN en réunion pour discuter de la Yugoslavie et de l'armée européenne', *Le Devoir* (Montreal), 10 Dec. 1992, p. B5.

[41] The CSCE also takes a keen interest in such developments as relations between the Baltic states and Russia, the separation of Czechoslovakia into two independent states, and the continuing strife in Georgia and the former Yugoslavia. For example, it sent fact-finding missions to Kosovo, Krajina and Macedonia. The Geneva conferences on the former Yugoslavia are a joint UN–EC exercise, however.

[42] 1500 Russian, Georgian and Ossetian troops were deployed as a CIS peace-keeping force in South Ossetia in July 1992 (see *Moscow News Report*, 30 Nov. 1992, p. 4). Six battalions of Russian troops were also deployed in Trans-Dniester in July, as the first contingent of a CIS peace-keeping force agreed by Russia, Moldova, Romania and Ukraine (see *The Guardian*, 30 July 1992, p. 4). In Sep., Russian troops in Tajikistan were increased by request to 10 000 for civil protection (see *International Herald Tribune*, 29 Sep. 1992, p. 2). Russia and Ukraine also have contingents with the UN peace-keeping forces in former Yugoslavia (see table 2.2); and Crow, S., 'Russian peacekeeping: defense, diplomacy or

In the Western hemisphere, the OAS sees itself as a regional collective security mechanism under the United Nations. It has gained strength in recent years with the restoration of democracy in several major countries of Latin America, the admission of Canada to the organization, and the decline of ideological confrontation and war in Central America. In 1992, the OAS continued to press for a solution to the Haitian crisis by means of a trade embargo, observation missions and mediation.[43] Various sub-regional bodies, such as the Caribbean Common Market (CARICOM), also continued to be helpful in maintaining common purposes and promoting peace in this hemisphere.

Elsewhere, the main regional organizations such as the OAU, the League of Arab States and the Association of South East Asian Nations (ASEAN) have promoted common purposes among their members and on occasions attempted to contain the dangers of disputes over borders, territories, economic resources or other interests.[44] There have also been attempts at conflict resolution and peace-enforcement, as for example by the OAU in Somalia[45] and by the Economic Community of West African States in Liberia.[46] However, in these areas of the world, countries generally tend to look to the United Nations rather than regional security mechanisms when there are major conflicts with international dimensions to resolve or manage.

The role of national states in international security should not be forgotten. Sometimes they act independently as third parties in conflict situations, for example, as mediators between two other countries, or as power-brokers between two or more antagonistic groups within a state or region. The continuing US effort to promote a settlement between Arabs and Israelis is a case in point.

Private institutes and university researchers are also working actively in the conflict resolution field. For example, the Carter Center established by former US President Jimmy Carter in Atlanta, Georgia, has contributed to the search for peaceful outcomes in the Middle East and the Horn of Africa. The International Peace Academy in New York continues to work on peace-keeping requirements.

imperialism?', *RFE/RL Research Report*, vol. 1, no. 39 (18 Sep. 1992), pp. 37–40; Sheehy, A., 'The CIS: a progress report', *RFE/RL Research Report,* vol. 1, no. 38 (25 Sep. 1992), pp. 1–6; '140,000 Tajik refugees stranded', *International Herald Tribune*, 7 Dec. 1992, p. 2; and 'Russia: troops sent to latest ethnic tinderbox', *Ottawa Citizen*, 2 Nov. 1992, p. A8.

[43] See, for example, 'Haiti: Canada asks UN to enforce trade embargo', *Ottawa Citizen,* 14 Dec. 1992, p. A10; and the *UN September 1992 Report* (note 1), pp. 45–46.

[44] For a very useful, short survey of peace-keeping activities carried out by the League of Arab States, the OAU, the OAS, and other groups, see Congressional Research Service, *Middle East Arms Control and Related Issues*, CRS Report for Congress (Library of Congress: Washington, DC, 1 May 1991).

[45] See the *UN September 1992 Report* (note 1), p. 57.

[46] The seven-nation, 15 000-strong multinational force in Liberia was originally described as a peace-keeping force, but is now engaged in major military actions against rebel leader Charles Taylor and his forces. Among other units, it includes some strongly armed Nigerian and Guinean army and air force elements. Its basic mandate is to enforce a cease-fire while disarming guerrilla forces and bringing them into camps. See 'Coalition set to fight rebel forces in Liberia', *Toronto Star*, 15 Nov. 1992, p. A16; and *UN September 1992 Report* (note 1), p. 46.

# V. Conclusion

In 1992 the effort to promote world peace by preventing, containing or resolving regional and intra-state conflicts featured prominently on the international scene. The United Nations, regional organizations and others were actively involved in a greatly expanded number of peace operations and initiatives, ranging from fact finding and other forms of preventive diplomacy to mediation, negotiation, peace-enforcement, peace-keeping and peace-building.

However, at the end of the year, the approach remained a piecemeal one, and 1992 witnessed disappointments and frustrations as well as achievements. UN members had still to address the challenge issued by the Secretary-General in *An Agenda for Peace*, and to work out a new, comprehensive set of policies for enhancing international peace and security.

There is still an urgent need to define the concrete requirements of international security for the period ahead, and to address such key issues as the required balance between global power on the one hand and national sovereignty on the other. A greater understanding of the fundamentals of security in the post-cold war world is essential, together with a better grasp of its changing dimensions and more conviction about long-term objectives. The world community must still decide how much reliance it intends to place in the collective security mechanism of the United Nations Security Council, and how central a role that body should be given in such areas as fact finding and peace-enforcement. If the role is to be significantly enhanced, then the institutions of the United Nations will need to be greatly strengthened, much greater financial and other resources will need to be devoted to their operations, and more effective processes and procedures will have to be put in place for tackling the world's many conflict problems. More clarity about terminology and types of peace-keeping and other conflict measures will be necessary; and related regional organizations and regimes will need to be significantly strengthened wherever possible.

It is also clear that new acts of political will and a renewed commitment to shared purposes will be an essential precondition for any new drive towards true international peace and security. A renewed debate on the issues raised by the Secretary-General in his *Agenda for Peace* is urgently required in order to galvanize international opinion and persuade the member states to seize the new opportunities for world order and conflict resolution now before them.

**Table 2.2.** Composition of UN peace-keeping forces, end-November 1992

| Country of origin | UNTSO | UNMOGIP | UNFICYP | UNDOF | UNIFIL | UNAVEM | UNIKOM | MINURSO | ONUSAL | UNTAC | UNPROFOR | UNOSOM |
|---|---|---|---|---|---|---|---|---|---|---|---|---|
| Algeria | | | | | | 5 | | | | 18 | | |
| | | | | | | | | | | *155 | | |
| Argentina | 6 | | | | | 7 | 7 | 7 | 7 | 2 | 899 | |
| | | | | | | *10 | | | | | *30 | |
| Australia | 13 | | | | | | | 45 | | 499 | 1 | 11 |
| | | | *20 | | | | | | | *10 | | |
| Austria | 13 | | 409 | 452 | | | 7 | 1 | | 17 | | 5 |
| | | | | | | | | | *3 | *19 | | |
| Bangladesh | | | | | | | 7 | 1 | | 931 | 24 | 5 |
| | | | | | | | | | | *225 | *45 | |
| Belgium | 7 | 2 | | | | | | 1 | | 2 | 781 | |
| Brazil | | | | | | 14 | | | 45 | | 15 | |
| | | | | | | *10 | | | | | | |
| Bulgaria | | | | | | | | | | 739 | | |
| | | | | | | | | | | *75 | | |
| Brunei | | | | | | | | | | 3 | | |
| Cameroon | | | | | | | | | | 14 | | |
| | | | | | | | | | | *75 | | |
| Canada | 13 | | 581 | 178 | | 15 | 42 | 32 | 10 | 212 | 2 087 | 1 |
| | | | | | | | | | | | *45 | |
| Chile | 4 | 3 | | | | | | | | 52 | | |
| | | | | | | | | | *25 | | | |
| China | 5 | | | | | | 15 | 20 | | 448 | | |
| Colombia | | | | | | | | | 6 | | 4 | |
| | | | | | | *3 | | | | *150 | *46 | |

| Country of origin | UNTSO | UNMOGIP | UNFICYP | UNDOF | UNIFIL | UNAVEM | UNIKOM | MINURSO | ONUSAL | UNTAC | UNPROFOR | UNOSOM |
|---|---|---|---|---|---|---|---|---|---|---|---|---|
| Congo | | | | | | 14 | | | | | | |
| Czechoslovakia | | | | | | 7 | | | | | 515 | 5 |
| Denmark | 11 | 6 | 352 | | | | 51 | | | | 1 048 | |
| | | | | | | | | | | | *40 | |
| Ecuador | | | | | | | | | 3 | | | |
| Egypt | | | | | | 14 | | 9 | | | 415 | 5 |
| | | | | | | | | | | *100 | *17 | |
| Fiji | | | | | 724 | | 6 | | | | | 5 |
| | | | | | | | | | | *50 | | |
| Finland | 18 | 5 | 8 | 355 | 542 | | 6 | | | | 303 | 5 |
| France | 17 | | | | 491 | | 15 | 29 | | 1 325 | 4 133 | |
| | | | | | | | | | *28 | *101 | *36 | |
| Germany | | | | | | | | | | 141 | | |
| | | | | | | | | | | *76 | | |
| Ghana | | | | | 819 | | 6 | 1 | | 901 | 13 | |
| | | | | | | | | | | *224 | | |
| Greece | | | | | | | 6 | 1 | | | | |
| Guyana | | | | | | | | | *10 | | | |
| Guinea | | | | | | | | 1 | | | | |
| Guinea-Bissau | | | | | | 15 | | | | | | |
| Honduras | | | | | | | | 14 | | | | |
| Hungary | | | | | | 7 | 6 | | | | | |
| | | | | | | | | | | *100 | | |
| India | | | | | | 10 | 6 | | 2 | 1 363 | 2 | |
| | | | | | | | | | | *366 | | |

| | | | | | | | | | | | |
|---|---|---|---|---|---|---|---|---|---|---|---|
| Indonesia | | | | | | 6 | | | 1 752 | | 5 |
| | | | | | | | | | *225 | | |
| Ireland | 20 | | 8 | 740 | 2 | 6 | 6 | 2 | 1 | 7 | |
| | | | | | | | | | | *20 | |
| Italy | 8 | 5 | | 56 | | 6 | 6 | | | | |
| | | | | | | | | *10 | *75 | | |
| Japan | | | | | | | | | 608 | | |
| | | | | | | | | | *75 | | |
| Jordan | | | | | 7 | | | | | 892 | 5 |
| | | | | | | | | | *84 | *45 | |
| Kenya | | | | | | 6 | 10 | | | 918 | |
| | | | | | | | | | *100 | *45 | |
| Luxembourg | | | | | | | | | | 35 | |
| Malaysia | | | | | 8 | 6 | 1 | | 908 | | |
| | | | | | *12 | | | | *150 | | |
| Mexico | | | | | | | | *111 | | | |
| Morocco | | | | | 5 | | | | | | 5 |
| | | | | | *6 | | | | *98 | | |
| Nepal | | | | 806 | | | | | | 886 | |
| | | | | | | | | | *85 | *45 | |
| Netherlands | 15 | | | | 12 | | | | 847 | 1 131 | |
| | | | | | *9 | | | | *2 | | |
| New Zealand | 5 | | | | 6 | | | | 92 | 9 | |
| Nigeria | | | | | 15 | 7 | 1 | | | 893 | |
| | | | | | *12 | | | | *150 | *30 | |
| Norway | 15 | 8 | | 874 | 4 | 27 | | | | 139 | |
| | | | | | | | | *3 | *20 | *30 | |
| Pakistan | | | | | | 7 | 1 | | 1 109 | | 505 |
| | | | | | | | | | *190 | | |

| Country of origin | UNTSO | UNMOGIP | UNFICYP | UNDOF | UNIFIL | UNAVEM | UNIKOM | MINURSO | ONUSAL | UNTAC | UNPROFOR | UNOSOM |
|---|---|---|---|---|---|---|---|---|---|---|---|---|
| Philippines | | | | | | | | | | 127 | | |
| | | | | | | | | | | *225 | | |
| Poland | | | | 135 | 87 | | 6 | 2 | | 688 | 919 | |
| | | | | | | | | | | | *30 | |
| Portugal | | | | | | | | | | | 12 | |
| | | | | | | | | | | | *35 | |
| Romania | | | | | | | 6 | | | | | |
| Russian Federation | 33 | | | | | | 14 | 26 | | 54 | 917 | |
| | | | | | | | | | | | *36 | |
| Senegal | | | | | | 15 | 7 | | | 3 | | |
| Singapore | | | | | | 6 | 7 | | | | | |
| | | | | | | | | | | *75 | | |
| Spain | | | | | | 5 | | | 120 | | 764 | |
| | | | | | | | | | *109 | | | |
| Sweden | 31 | 8 | 8 | | 504 | 5 | 6 | | 3 | | 109 | |
| | | | *18 | | | *4 | | | *4 | *40 | *30 | |
| Switzerland | 5 | | | | | | | 54 | | | 6 | |
| Thailand | | | | | | | 7 | | | 720 | | |
| Tunisia | | | | | | | | 9 | | 884 | | |
| | | | | | | | | | | *32 | *12 | |
| Turkey | | | | | | | 6 | | | | | |
| Ukraine | | | | | | | | | | | 387 | |
| United Kingdom | | | 793 | | | | 15 | 13 | | 113 | 2 822 | |
| United States | 20 | | | | | | 15 | 28 | | 44 | 354 | |
| Uruguay | | 2 | | | | | 7 | | | 932 | | |
| Venezuela | | | | | | | 6 | 13 | 29 | | 6 | |

| | | | | | | | | | | | | |
|---|---|---|---|---|---|---|---|---|---|---|---|---|
| Yugoslavia | | | | | | 8 | | | | | | |
| Zimbabwe | | | | | | 15<br>*11 | | | | | | 5 |
| **Totals** | | | | | | | | | | | | |
| **Military** | **259** | **39** | **2 159** | **1 120** | **5 643** | **221** | **353** | **332** | **227** | **15 549** | **21 446** | **567** |
| **Police** | | | ***38** | | | ***77** | | | ***303** | ***3 352** | ***617** | |

*Notes*:

1. Figures preceded by an asterisk (*) are for police. The others are for military personnel, including observers and troops (i.e., infantry, logistics, engineering, air, naval, staff, etc.). Figures are provided by the United Nations Organization, New York.

2. Figures are for the end of November 1992, and therefore do not include those for the new mission sent to Mozambique at the end of the year. Figures for Somalia are for the peace-keeping mission (UNOSOM), not for the peace-enforcement mission created in December (see table 2.1 for these figures).

3. Figures are for UN peace-keeping missions only. See the text and footnotes for some figures on CIS and other 'non-UN' peace-keeping operations.

4. The UN peace-keeping missions listed in the table are as follows:

UNTSO United Nations Truce Supervision Organization (in the Middle East since 1948; estimated cost to UN in 1992: $25 m.);
UNMOGIP United Nations Military Observer Group in India and Pakistan (since 1949; estimated cost to UN in 1992: $6 m.);
UNFICYP United Nations Peace-keeping Force in Cyprus (since 1964; estimated cost to UN in 1992: $31 m.);
UNDOF United Nations Disengagement Observer Force (on Israeli–Syrian border since 1974; estimated cost to UN in 1992: $39 m.);
UNIFIL United Nations Interim Force in Lebanon (since 1978; estimated cost to UN in 1992: $153 m.);
UNAVEM United Nations Angola Verification Mission (present mission, UNAVEM II, in operation since 1991; estimated cost to UN in 1992: $67 m.);
UNIKOM United Nations Iraq–Kuwait Observer Mission (since 1991; estimated cost to UN in 1992: $68 m.);
MINURSO United Nations Mission for the Referendum in Western Sahara (since 1991; estimated cost to UN in 1992: $18 m.);
ONUSAL United Nations Observer Mission in El Salvador (in operation since 1991, mandate broadened in 1992; estimated cost to UN in 1992: $35 m.);
UNTAC United Nations Transitional Authority in Cambodia (since March 1992; estimated cost to UN in 1992: $637 m.);
UNPROFOR United Nations Protection Force (in former Yugoslavia since March 1992; estimated cost to UN in 1992: $222 m.); and
UNOSOM United Nations Operation in Somalia (from April 1992; estimated cost to UN in 1992: $39 m.).

*Sources:* United Nations, *Current Peace-keeping Operations*, Note PS/DPI/15/Rev. 2–Sep. 1992 (United Nations Department of Public Information, Communications and Project Management Division: New York, Sep. 1992); estimated costs are from *Financing an Effective United Nations: A Report of the Independent Advisory Group on UN Financing* (The Ford Foundation: New York, Feb. 1993).

# **Appendix 2A.** An Agenda for Peace

**AN AGENDA FOR PEACE**

**Preventive diplomacy, peacemaking and peace-keeping**

**Report of the Secretary-General 17 June 1992**

**Introduction**

1. In its statement of 31 January 1992, adopted at the conclusion of the first meeting held by the Security Council at the level of Heads of State and Government, I was invited to prepare, for circulation to the Members of the United Nations by 1 July 1992, an 'analysis and recommendations on ways of strengthening and making more efficient within the framework and provisions of the Charter the capacity of the United Nations for preventive diplomacy, for peacemaking and for peace-keeping'.[1]

2. The United Nations is a gathering of sovereign States and what it can do depends on the common ground that they create between them. The adversarial decades of the cold war made the original promise of the Organization impossible to fulfil. The January 1992 Summit therefore represented an unprecedented recommitment, at the highest political level, to the Purposes and Principles of the Charter.

3. In these past months a conviction has grown, among nations large and small, that an opportunity has been regained to achieve the great objectives of the Charter—a United Nations capable of maintaining international peace and security, of securing justice and human rights and of promoting, in the words of the Charter, 'social progress and better standards of life in larger freedom'. This opportunity must not be squandered. The Organization must never again be crippled as it was in the era that has now passed.

4. I welcome the invitation of the Security Council, early in my tenure as Secretary-General, to prepare this report. It draws upon ideas and proposals transmitted to me by Governments, regional agencies, non-governmental organizations, and institutions and individuals from many countries. I am grateful for these, even as I emphasize that the responsibility for this report is my own.

5. The sources of conflict and war are pervasive and deep. To reach them will require our utmost effort to enhance respect for human rights and fundamental freedoms, to promote sustainable economic and social development for wider prosperity, to alleviate distress and to curtail the existence and use of massively destructive weapons. The United Nations Conference on Environment and Development, the largest summit ever held, has just met at Rio de Janeiro. Next year will see the second World Conference on Human Rights. In 1994 Population and Development will be addressed. In 1995 the World Conference on Women will take place, and a World Summit for Social Development has been proposed. Throughout my term as Secretary-General I shall be addressing all these great issues. I bear them all in mind as, in the present report, I turn to the problems that the Council has specifically requested I consider: preventive diplomacy, peacemaking and peace-keeping—to which I have added a closely related concept, post-conflict peace-building.

6. The manifest desire of the membership to work together is a new source of strength in our common endeavour. Success is far from certain, however. While my report deals with ways to improve the Organization's capacity to pursue and preserve peace, it is crucial for all Member States to bear in mind that the search for improved mechanisms and techniques will be of little significance unless this new spirit of commonality is propelled by the will to take the hard decisions demanded by this time of opportunity.

7. It is therefore with a sense of moment, and with gratitude, that I present this report to the Members of the United Nations.

**I. The Changing Context**

8. In the course of the past few years the immense ideological barrier that for decades gave rise to distrust and hostility—and the terrible tools of destruction that were their inseparable companions—has collapsed. Even as the issues between States north and south grow more acute, and call for attention at the highest levels of government, the improvement in relations between States east and west affords new possibilities, some already realized, to meet successfully threats to common security.

9. Authoritarian regimes have given way

to more democratic forces and responsive Governments. The form, scope and intensity of these processes differ from Latin America to Africa to Europe to Asia, but they are sufficiently similar to indicate a global phenomenon. Parallel to these political changes, many States are seeking more open forms of economic policy, creating a world-wide sense of dynamism and movement.

10. To the hundreds of millions who gained their independence in the surge of decolonization following the creation of the United Nations, have been added millions more who have recently gained freedom. Once again new States are taking their seats in the General Assembly. Their arrival reconfirms the importance and indispensability of the sovereign State as the fundamental entity of the international community.

11. We have entered a time of global transition marked by uniquely contradictory trends. Regional and continental associations of States are evolving ways to deepen cooperation and ease some of the contentious characteristics of sovereign and nationalistic rivalries. National boundaries are blurred by advanced communications and global commerce, and by the decisions of States to yield some sovereign prerogatives to larger, common political associations. At the same time, however, fierce new assertions of nationalism and sovereignty spring up, and the cohesion of States is threatened by brutal ethnic, religious, social, cultural or linguistic strife. Social peace is challenged on the one hand by new assertions of discrimination and exclusion and, on the other, by acts of terrorism seeking to undermine evolution and change through democratic means.

12. The concept of peace is easy to grasp; that of international security is more complex, for a pattern of contradictions has arisen here as well. As major nuclear Powers have begun to negotiate arms reduction agreements, the proliferation of weapons of mass destruction threatens to increase and conventional arms continue to be amassed in many parts of the world. As racism becomes recognized for the destructive force it is and as apartheid is being dismantled, new racial tensions are rising and finding expression in violence. Technological advances are altering the nature and the expectation of life all over the globe. The revolution in communications has united the world in awareness, in aspiration and in greater solidarity against injustice. But progress also brings new risks for stability: ecological damage, disruption of family and community life, greater intrusion into the lives and rights of individuals.

13. This new dimension of insecurity must not be allowed to obscure the continuing and devastating problems of unchecked population growth, crushing debt burdens, barriers to trade, drugs and the growing disparity between rich and poor. Poverty, disease, famine, oppression and despair abound, joining to produce 17 million refugees, 20 million displaced persons and massive migrations of peoples within and beyond national borders. These are both sources and consequences of conflict that require the ceaseless attention and the highest priority in the efforts of the United Nations. A porous ozone shield could pose a greater threat to an exposed population than a hostile army. Drought and disease can decimate no less mercilessly than the weapons of war. So at this moment of renewed opportunity, the efforts of the organization to build peace, stability and security must encompass matters beyond military threats in order to break the fetters of strife and warfare that have characterized the past. But armed conflicts today, as they have throughout history, continue to bring fear and horror to humanity, requiring our urgent involvement to try to prevent, contain and bring them to an end.

14. Since the creation of the United Nations in 1945, over 100 major conflicts around the world have left some 20 million dead. The United Nations was rendered powerless to deal with many of these crises because of the vetoes—279 of them—cast in the Security Council, which were a vivid expression of the divisions of that period.

15. With the end of the cold war there have been no such vetoes since 31 May 1990, and demands on the United Nations have surged. Its security arm, once disabled by circumstances it was not created or equipped to control, has emerged as a central instrument for the prevention and resolution of conflicts and for the preservation of peace. Our aims must be:

To seek to identify at the earliest possible stage situations that could produce conflict, and to try through diplomacy to remove the sources of danger before violence results;

Where conflict erupts, to engage in peace-making aimed at resolving the issues that have led to conflict;

Through peace-keeping, to work to preserve peace, however fragile, where fighting has been halted and to assist in implementing agreements achieved by the peacemakers;

To stand ready to assist in peace-building in its differing contexts: rebuilding the institutions and infrastructures of nations torn by civil war and strife; and building bonds of peaceful mutual benefit among nations formerly at war;

And in the largest sense, to address the deepest causes of conflict: economic despair, social injustice and political oppression. It is possible to discern an increasingly common moral perception that spans the world's nations and peoples, and which is finding expression in international laws, many owing their genesis to the work of this Organization.

16. This wider mission for the world Organization will demand the concerted attention and effort of individual States, of regional and non-governmental organizations and of all of the United Nations system, with each of the principal organs functioning in the balance and harmony that the Charter requires. The Security Council has been assigned by all Member States the primary responsibility for the maintenance of international peace and security under the Charter. In its broadest sense this responsibility must be shared by the General Assembly and by all the functional elements of the world Organization. Each has a special and indispensable role to play in an integrated approach to human security. The Secretary-General's contribution rests on the pattern of trust and cooperation established between him and the deliberative organs of the United Nations.

17. The foundation-stone of this work is and must remain the State. Respect for its fundamental sovereignty and integrity are crucial to any common international progress. The time of absolute and exclusive sovereignty, however, has passed; its theory was never matched by reality. It is the task of leaders of States today to understand this and to find a balance between the needs of good internal governance and the requirements of an ever more interdependent world. Commerce, communications and environmental matters transcend administrative borders; but inside those borders is where individuals carry out the first order of their economic, political and social lives. The United Nations has not closed its door. Yet if every ethnic, religious or linguistic group claimed statehood, there would be no limit to fragmentation, and peace, security and economic well-being for all would become ever more difficult to achieve.

18. One requirement for solutions to these problems lies in commitment to human rights with a special sensitivity to those of minorities, whether ethnic, religious, social or linguistic. The League of Nations provided a machinery for the international protection of minorities. The General Assembly soon will have before it a declaration on the rights of minorities. That instrument, together with the increasingly effective machinery of the United Nations dealing with human rights, should enhance the situation of minorities as well as the stability of States.

19. Globalism and nationalism need not be viewed as opposing trends, doomed to spur each other on to extremes of reaction. The healthy globalization of contemporary life requires in the first instance solid identities and fundamental freedoms. The sovereignty, territorial integrity and independence of States within the established international system, and the principle of self-determination for peoples, both of great value and importance, must not be permitted to work against each other in the period ahead. Respect for democratic principles at all levels of social existence is crucial: in communities, within States and within the community of States. Our constant duty should be to maintain the integrity of each while finding a balanced design for all.

## II. Definitions

20. The terms preventive diplomacy, peacemaking and peace-keeping are integrally related and as used in this report are defined as follows:

*Preventive diplomacy* is action to prevent disputes from arising between parties, to prevent existing disputes from escalating into conflicts and to limit the spread of the latter when they occur.

*Peacemaking* is action to bring hostile parties to agreement, essentially through such peaceful means as those foreseen in Chapter VI of the Charter of the United Nations.

*Peace-keeping* is the deployment of a United Nations presence in the field, hitherto with the consent of all the parties concerned, normally involving United Nations military and/or police personnel and frequently civilians as well. Peace-keeping is a technique that expands the possibilities for both the prevention of conflict and the making of peace.

21. The present report in addition will address the critically related concept of post-conflict *peace-building*—action to identify and support structures which will tend to strengthen and solidify peace in order to avoid a relapse into conflict. Preventive diplomacy seeks to resolve disputes before violence breaks out; peacemaking and peace-keeping are required to halt conflicts and preserve peace once it is attained. If successful, they strengthen the opportunity for post-conflict peace-building, which can prevent the recurrence of violence among nations and peoples.

22. These four areas for action, taken together, and carried out with the backing of all Members, offer a coherent contribution towards securing peace in the spirit of the Charter. The United Nations has extensive experience not only in these fields, but in the wider realm of work for peace in which these four fields are set. Initiatives on decolonization, on the environment and sustainable development, on population, on the eradication of disease, on disarmament and on the growth of international law—these and many others have contributed immeasurably to the foundations for a peaceful world. The world has often been rent by conflict and plagued by massive human suffering and deprivation. Yet it would have been far more so without the continuing efforts of the United Nations. This wide experience must be taken into account in assessing the potential of the United Nations in maintaining international security not only in its traditional sense, but in the new dimensions presented by the era ahead.

## III. Preventive Diplomacy

23. The most desirable and efficient employment of diplomacy is to ease tensions before they result in conflict—or, if conflict breaks out, to act swiftly to contain it and resolve its underlying causes. Preventive diplomacy may be performed by the Secretary-General personally or through senior staff or specialized agencies and programmes, by the Security Council or the General Assembly, and by regional organizations in cooperation with the United Nations. Preventive diplomacy requires measures to create confidence; it needs early warning based on information gathering and informal or formal fact-finding; it may also involve preventive deployment and, in some situations, demilitarized zones.

### Measures to build confidence

24. Mutual confidence and good faith are essential to reducing the likelihood of conflict between States. Many such measures are available to Governments that have the will to employ them. Systematic exchange of military missions, formation of regional or subregional risk reduction centres, arrangements for the free flow of information, including the monitoring of regional arms agreements, are examples. I ask all regional organizations to consider what further confidence-building measures might be applied in their areas and to inform the United Nations of the results. I will undertake periodic consultations on confidence-building measures with parties to potential, current or past disputes and with regional organizations, offering such advisory assistance as the Secretariat can provide.

### Fact-finding

25. Preventive steps must be based upon timely and accurate knowledge of the facts. Beyond this, an understanding of developments and global trends, based on sound analysis, is required. And the willingness to take appropriate preventive action is essential. Given the economic and social roots of many potential conflicts, the information needed by the United Nations now must encompass economic and social trends as well as political developments that may lead to dangerous tensions.

(a) An increased resort to fact-finding is needed, in accordance with the Charter, initiated either by the Secretary-General, to enable him to meet his responsibilities under the Charter, including Article 99, or by the Security Council or the General Assembly. Various forms may be employed selectively as the situation requires. A request by a State for the sending of a United Nations fact-finding mission to its territory should be considered without undue delay.

(b) Contacts with the Governments of Member States can provide the Secretary-General with detailed information on issues of concern. I ask that all Member States be ready to provide the information needed for effective preventive diplomacy. I will supplement my own contacts by regularly sending senior officials on missions for consultations in capitals or other locations. Such contacts are essential to gain insight into a situation and to assess its potential ramifications.

(c) Formal fact-finding can be mandated by the Security Council or by the General Assembly, either of which may elect to send a mission under its immediate authority or may invite the Secretary-General to take the necessary steps, including the designation of a special envoy. In addition to collecting information on which a decision for further action can be taken, such a mission can in some instances help to defuse a dispute by its presence, indicating to the parties that the Organization, and in particular the Security Council, is actively seized of the matter as a present or potential threat to international security.

(d) In exceptional circumstances the Council may meet away from Headquarters as the Charter provides, in order not only to inform itself directly, but also to bring the authority of the Organization to bear on a given situation.

### Early warning

26. In recent years the United Nations system has been developing a valuable network of early warning systems concerning environmental threats, the risk of nuclear accident, natural disasters, mass movements of populations, the threat of famine and the spread of disease. There is a need, however, to strengthen arrangements in such a manner that information from these sources can be synthesized with political indicators to assess whether a threat to peace exists and to analyse what action might be taken by the United Nations to alleviate it. This is a process that will continue to require the close cooperation of the various specialized agencies and functional offices of the United Nations. The analyses and recommendations for preventive action that emerge will be made available by me, as appropriate, to the Security Council and other United Nations organs. I recommend in addition that the Security Council invite a reinvigorated and restructured Economic and Social Council to provide reports, in accordance with Article 65 of the Charter, on those economic and social developments that may, unless mitigated, threaten international peace and security.

27. Regional arrangements and organizations have an important role in early warning. I ask regional organizations that have not yet sought observer status at the United Nations to do so and to be linked, through appropriate arrangements, with the security mechanisms of this Organization.

### Preventive deployment

28. United Nations operations in areas of crisis have generally been established after conflict has occurred. The time has come to plan for circumstances warranting preventive deployment, which could take place in a variety of instances and ways. For example, in conditions of national crisis there could be preventive deployment at the request of the Government or all parties concerned, or with their consent; in inter-State disputes such deployment could take place when two countries feel that a United Nations presence on both sides of their border can discourage hostilities; furthermore, preventive deployment could take place when a country feels threatened and requests the deployment of an appropriate United Nations presence along its side of the border alone. In each situation, the mandate and composition of the United Nations presence would need to be carefully devised and be clear to all.

29. In conditions of crisis within a country, when the Government requests or all parties consent, preventive deployment could help in a number of ways to alleviate suffering and to limit or control violence. Humanitarian assistance, impartially provided, could be of critical importance; assistance in maintaining security, whether through military, police or civilian personnel, could save lives and develop conditions of safety in which negotiations can be held; the United Nations could also help in conciliation efforts if this should be the wish of the parties. In certain circumstances, the United Nations may well need to draw upon the specialized skills and resources of various parts of the United Nations system; such operations may also on occasion require the participation of non-governmental organizations.

30. In these situations of internal crisis the United Nations will need to respect the sovereignty of the State; to do otherwise would not be in accordance with the understanding of Member States in accepting the principles of the Charter. The Organization must remain mindful of the carefully negotiated balance of the guiding principles annexed to General Assembly resolution 46/182 of 19 December 1991. Those guidelines stressed, *inter alia*, that humanitarian assistance must be provided in accordance

with the principles of humanity, neutrality and impartiality; that the sovereignty, territorial integrity and national unity of States must be fully respected in accordance with the Charter of the United Nations; and that, in this context, humanitarian assistance should be provided with the consent of the affected country and, in principle, on the basis of an appeal by that country. The guidelines also stressed the responsibility of States to take care of the victims of emergencies occurring on their territory and the need for access to those requiring humanitarian assistance. In the light of these guidelines, a Government's request for United Nations involvement, or consent to it, would not be an infringement of that State's sovereignty or be contrary to Article 2, paragraph 7, of the Charter which refers to matters essentially within the domestic jurisdiction of any State.

31. In inter-State disputes, when both parties agree, I recommend that if the Security Council concludes that the likelihood of hostilities between neighbouring countries could be removed by the preventive deployment of a United Nations presence on the territory of each State, such action should be taken. The nature of the tasks to be performed would determine the composition of the United Nations presence.

32. In cases where one nation fears a cross-border attack, if the Security Council concludes that a United Nations presence on one side of the border, with the consent only of the requesting country, would serve to deter conflict, I recommend that preventive deployment take place. Here again, the specific nature of the situation would determine the mandate and the personnel required to fulfil it.

### Demilitarized zones

33. In the past, demilitarized zones have been established by agreement of the parties at the conclusion of a conflict. In addition to the deployment of United Nations personnel in such zones as part of peace-keeping operations, consideration should now be given to the usefulness of such zones as a form of preventive deployment, on both sides of a border, with the agreement of the two parties, as a means of separating potential belligerents, or on one side of the line, at the request of one party, for the purpose of removing any pretext for attack. Demilitarized zones would serve as symbols of the international community's concern that conflict be prevented.

## IV. Peacemaking

34. Between the tasks of seeking to prevent conflict and keeping the peace lies the responsibility to try to bring hostile parties to agreement by peaceful means. Chapter VI of the Charter sets forth a comprehensive list of such means for the resolution of conflict. These have been amplified in various declarations adopted by the General Assembly, including the Manila Declaration of 1982 on the Peaceful Settlement of International Disputes[2] and the 1988 Declaration on the Prevention and Removal of Disputes and Situations Which May Threaten International Peace and Security and on the Role of the United Nations in this Field.[3] They have also been the subject of various resolutions of the General Assembly, including resolution 44/21 of 15 November 1989 on enhancing international peace, security and international cooperation in all its aspects in accordance with the Charter of the United Nations. The United Nations has had wide experience in the application of these peaceful means. If conflicts have gone unresolved, it is not because techniques for peaceful settlement were unknown or inadequate. The fault lies first in the lack of political will of parties to seek a solution to their differences through such means as are suggested in Chapter VI of the Charter, and second, in the lack of leverage at the disposal of a third party if this is the procedure chosen. The indifference of the international community to a problem, or the marginalization of it, can also thwart the possibilities of solution. We must look primarily to these areas if we hope to enhance the capacity of the Organization for achieving peaceful settlements.

35. The present determination in the Security Council to resolve international disputes in the manner foreseen in the Charter has opened the way for a more active Council role. With greater unity has come leverage and persuasive power to lead hostile parties towards negotiations. I urge the Council to take full advantage of the provisions of the Charter under which it may recommend appropriate procedures or methods for dispute settlement and, if all the parties to a dispute so request, make recommendations to the parties for a pacific settlement of the dispute.

36. The General Assembly, like the Security Council and the Secretary-General, also has an important role assigned to it under the Charter for the maintenance of international peace and security. As a universal forum, its capacity to consider and recommend appropriate action must be recognized. To that end it is essential to promote its utilization by all Member States so as to bring greater influence to bear in pre-empting or containing situations which are likely to threaten international peace and security.

37. Mediation and negotiation can be undertaken by an individual designated by the Security Council, by the General Assembly or by the Secretary-General. There is a long history of the utilization by the United Nations of distinguished statesmen to facilitate the processes of peace. They can bring a personal prestige that, in addition to their experience, can encourage the parties to enter serious negotiations. There is a wide willingness to serve in this capacity, from which I shall continue to benefit as the need arises. Frequently it is the Secretary-General himself who undertakes the task. While the mediator's effectiveness is enhanced by strong and evident support from the Council, the General Assembly and the relevant Member States acting in their national capacity, the good offices of the Secretary-General may at times be employed most effectively when conducted independently of the deliberative bodies. Close and continuous consultation between the Secretary-General and the Security Council is, however, essential to ensure full awareness of how the Council's influence can best be applied and to develop a common strategy for the peaceful settlement of specific disputes.

### The World Court

38. The docket of the International Court of Justice has grown fuller but it remains an under-used resource for the peaceful adjudication of disputes. Greater reliance on the Court would be an important contribution to United Nations peacemaking. In this connection, I call attention to the power of the Security Council under Articles 36 and 37 of the Charter to recommend to Member States the submission of a dispute to the International Court of Justice, arbitration or other dispute-settlement mechanisms. I recommend that the Secretary-General be authorized, pursuant to Article 96, paragraph 2, of the Charter, to take advantage of the advisory competence of the Court and that other United Nations organs that already enjoy such authorization turn to the Court more frequently for advisory opinions.

39. I recommend the following steps to reinforce the role of the International Court of Justice:

(a) All Member States should accept the general jurisdiction of the International Court under Article 36 of its Statute, without any reservation, before the end of the United Nations Decade of International Law in the year 2000. In instances where domestic structures prevent this, States should agree bilaterally or multilaterally to a comprehensive list of matters they are willing to submit to the Court and should withdraw their reservations to its jurisdiction in the dispute settlement clauses of multilateral treaties;

(b) When submission of a dispute to the full Court is not practical, the Chambers jurisdiction should be used;

(c) States should support the Trust Fund established to assist countries unable to afford the cost involved in bringing a dispute to the Court, and such countries should take full advantage of the Fund in order to resolve their disputes.

### Amelioration through assistance

40. Peacemaking is at times facilitated by international action to ameliorate circumstances that have contributed to the dispute or conflict. If, for instance, assistance to displaced persons within a society is essential to a solution, then the United Nations should be able to draw upon the resources of all agencies and programmes concerned. At present, there is no adequate mechanism in the United Nations through which the Security Council, the General Assembly or the Secretary-General can mobilize the resources needed for such positive leverage and engage the collective efforts of the United Nations system for the peaceful resolution of a conflict. I have raised this concept in the Administrative Committee on Coordination, which brings together the executive heads of United Nations agencies and programmes; we are exploring methods by which the inter-agency system can improve its contribution to the peaceful resolution of disputes.

### Sanctions and special economic problems

41. In circumstances when peacemaking requires the imposition of sanctions under

Article 41 of the Charter, it is important that States confronted with special economic problems not only have the right to consult the Security Council regarding such problems, as Article 50 provides, but also have a realistic possibility of having their difficulties addressed. I recommend that the Security Council devise a set of measures involving the financial institutions and other components of the United Nations system that can be put in place to insulate States from such difficulties. Such measures would be a matter of equity and a means of encouraging States to cooperate with decisions of the Council.

**Use of military force**

42. It is the essence of the concept of collective security as contained in the Charter that if peaceful means fail, the measures provided in Chapter VII should be used, on the decision of the Security Council, to maintain or restore international peace and security in the face of a 'threat to the peace, breach of the peace, or act of aggression'. The Security Council has not so far made use of the most coercive of these measures—the action by military force foreseen in Article 42. In the situation between Iraq and Kuwait, the Council chose to authorize Member States to take measures on its behalf. The Charter, however, provides a detailed approach which now merits the attention of all Member States.

43. Under Article 42 of the Charter, the Security Council has the authority to take military action to maintain or restore international peace and security. While such action should only be taken when all peaceful means have failed, the option of taking it is essential to the credibility of the United Nations as a guarantor of international security. This will require bringing into being, through negotiations, the special agreements foreseen in Article 43 of the Charter, whereby Member States undertake to make armed forces, assistance and facilities available to the Security Council for the purposes stated in Article 42, not only on an ad hoc basis but on a permanent basis. Under the political circumstances that now exist for the first time since the Charter was adopted, the long-standing obstacles to the conclusion of such special agreements should no longer prevail. The ready availability of armed forces on call could serve, in itself, as a means of deterring breaches of the peace since a potential aggressor would know that the Council had at its disposal a means of response. Forces under Article 43 may perhaps never be sufficiently large or well enough equipped to deal with a threat from a major army equipped with sophisticated weapons. They would be useful, however, in meeting any threat posed by a military force of a lesser order. I recommend that the Security Council initiate negotiations in accordance with Article 43, supported by the Military Staff Committee, which may be augmented if necessary by others in accordance with Article 47, paragraph 2, of the Charter. It is my view that the role of the Military Staff Committee should be seen in the context of Chapter VII, and not that of the planning or conduct of peace-keeping operations.

**Peace-enforcement units**

44. The mission of forces under Article 43 would be to respond to outright aggression, imminent or actual. Such forces are not likely to be available for some time to come. Cease-fires have often been agreed to but not complied with, and the United Nations has sometimes been called upon to send forces to restore and maintain the cease-fire. This task can on occasion exceed the mission of peace-keeping forces and the expectations of peace-keeping force contributors. I recommend that the Council consider the utilization of peace-enforcement units in clearly defined circumstances and with their terms of reference specified in advance. Such units from Member States would be available on call and would consist of troops that have volunteered for such service. They would have to be more heavily armed than peace-keeping forces and would need to undergo extensive preparatory training within their national forces. Deployment and operation of such forces would be under the authorization of the Security Council and would, as in the case of peace-keeping forces, be under the command of the Secretary-General. I consider such peace-enforcement units to be warranted as a provisional measure under Article 40 of the Charter. Such peace-enforcement units should not be confused with the forces that may eventually be constituted under Article 43 to deal with acts of aggression or with the military personnel which Governments may agree to keep on stand-by for possible contribution to peace-keeping operations.

45. Just as diplomacy will continue across the span of all the activities dealt with in the present report, so there may not be a dividing line between peacemaking and peace-keeping. Peacemaking is often a prelude to peace-keeping—just as the deployment of a United Nations presence in the field may expand possibilities for the prevention of conflict, facilitate the work of peacemaking and in many cases serve as a prerequisite for peace-building.

## V. Peace-keeping

46. Peace-keeping can rightly be called the invention of the United Nations. It has brought a degree of stability to numerous areas of tension around the world.

### Increasing demands

47. Thirteen peace-keeping operations were established between the years 1945 and 1987; 13 others since then. An estimated 528,000 military, police and civilian personnel had served under the flag of the United Nations until January 1992. Over 800 of them from 43 countries have died in the service of the Organization. The costs of these operations have aggregated some $8.3 billion till 1992. The unpaid arrears towards them stand at over $800 million, which represent a debt owed by the Organization to the troop-contributing countries. Peace-keeping operations approved at present are estimated to cost close to $3 billion in the current 12-month period, while patterns of payment are unacceptably slow. Against this, global defence expenditures at the end of the last decade had approached $1 trillion a year, or $2 million per minute.

48. The contrast between the costs of United Nations peace-keeping and the costs of the alternative, war—between the demands of the Organization and the means provided to meet them—would be farcical were the consequences not so damaging to global stability and to the credibility of the Organization. At a time when nations and peoples increasingly are looking to the United Nations for assistance in keeping the peace—and holding it responsible when this cannot be so—fundamental decisions must be taken to enhance the capacity of the Organization in this innovative and productive exercise of its function. I am conscious that the present volume and unpredictability of peace-keeping assessments poses real problems for some Member States. For this reason, I strongly support proposals in some Member States for their peace-keeping contributions to be financed from defence, rather than foreign affairs, budgets and I recommend such action to others. I urge the General Assembly to encourage this approach.

49. The demands on the United Nations for peace-keeping, and peace-building, operations will in the coming years continue to challenge the capacity, the political and financial will and the creativity of the Secretariat and Member States. Like the Security Council, I welcome the increase and broadening of the tasks of peace-keeping operations.

### New departures in peace-keeping

50. The nature of peace-keeping operations has evolved rapidly in recent years. The established principles and practices of peace-keeping have responded flexibly to new demands of recent years, and the basic conditions for success remain unchanged: a clear and practicable mandate; the cooperation of the parties in implementing that mandate; the continuing support of the Security Council; the readiness of Member States to contribute the military, police and civilian personnel, including specialists, required; effective United Nations command at Headquarters and in the field; and adequate financial and logistic support. As the international climate has changed and peace-keeping operations are increasingly fielded to help implement settlements that have been negotiated by peacemakers, a new array of demands and problems has emerged regarding logistics, equipment, personnel and finance, all of which could be corrected if Member States so wished and were ready to make the necessary resources available.

### Personnel

51. Member States are keen to participate in peace-keeping operations. Military observers and infantry are invariably available in the required numbers, but logistic units present a greater problem, as few armies can afford to spare such units for an extended period. Member States were requested in 1990 to state what military personnel they were in principle prepared to make available; few replied. I reiterate the request to all Member States to reply frankly and promptly. Stand-by arrangements should be confirmed, as appropriate, through exchanges of letters between the Secretariat

and Member States concerning the kind and number of skilled personnel they will be prepared to offer the United Nations as the needs of new operations arise.

52. Increasingly, peace-keeping requires that civilian political officers, human rights monitors, electoral officials, refugee and humanitarian aid specialists and police play as central a role as the military. Police personnel have proved increasingly difficult to obtain in the numbers required. I recommend that arrangements be reviewed and improved for training peace-keeping personnel—civilian, police, or military—using the varied capabilities of Member State Governments, of non-governmental organizations and the facilities of the Secretariat. As efforts go forward to include additional States as contributors, some States with considerable potential should focus on language training for police contingents which may serve with the Organization. As for the United Nations itself, special personnel procedures, including incentives, should be instituted to permit the rapid transfer of Secretariat staff members to service with peace-keeping operations. The strength and capability of military staff serving in the Secretariat should be augmented to meet new and heavier requirements.

### Logistics

53. Not all Governments can provide their battalions with the equipment they need for service abroad. While some equipment is provided by troop-contributing countries, a great deal has to come from the United Nations, including equipment to fill gaps in under-equipped national units. The United Nations has no standing stock of such equipment. Orders must be placed with manufacturers, which creates a number of difficulties. A pre-positioned stock of basic peace-keeping equipment should be established, so that at least some vehicles, communications equipment, generators, etc., would be immediately available at the start of an operation. Alternatively, Governments should commit themselves to keeping certain equipment, specified by the Secretary-General, on stand-by for immediate sale, loan or donation to the United Nations when required.

54. Member States in a position to do so should make air- and sea-lift capacity available to the United Nations free of cost or at lower than commercial rates, as was the practice until recently.

## VI. Post-conflict Peace-building

55. Peacemaking and peace-keeping operations, to be truly successful, must come to include comprehensive efforts to identify and support structures which will tend to consolidate peace and advance a sense of confidence and well-being among people. Through agreements ending civil strife, these may include disarming the previously warring parties and the restoration of order, the custody and possible destruction of weapons, repatriating refugees, advisory and training support for security personnel, monitoring elections, advancing efforts to protect human rights, reforming or strengthening governmental institutions and promoting formal and informal processes of political participation.

56. In the aftermath of international war, post-conflict peace-building may take the form of concrete cooperative projects which link two or more countries in a mutually beneficial undertaking that can not only contribute to economic and social development but also enhance the confidence that is so fundamental to peace. I have in mind, for example, projects that bring States together to develop agriculture, improve transportation or utilize resources such as water or electricity that they need to share, or joint programmes through which barriers between nations are brought down by means of freer travel, cultural exchanges and mutually beneficial youth and educational projects. Reducing hostile perceptions through educational exchanges and curriculum reform may be essential to forestall a re-emergence of cultural and national tensions which could spark renewed hostilities.

57. In surveying the range of efforts for peace, the concept of peace-building as the construction of a new environment should be viewed as the counterpart of preventive diplomacy, which seeks to avoid the breakdown of peaceful conditions. When conflict breaks out, mutually reinforcing efforts at peacemaking and peace-keeping come into play. Once these have achieved their objectives, only sustained, cooperative work to deal with underlying economic, social, cultural and humanitarian problems can place an achieved peace on a durable foundation. Preventive diplomacy is to avoid a crisis; post-conflict peace-building is to prevent a recurrence.

58. Increasingly it is evident that peace-building after civil or international strife

must address the serious problem of land mines, many tens of millions of which remain scattered in present or former combat zones. De-mining should be emphasized in the terms of reference of peace-keeping operations and is crucially important in the restoration of activity when peace-building is under way: agriculture cannot be revived without de-mining and the restoration of transport may require the laying of hard surface roads to prevent re-mining. In such instances, the link becomes evident between peace-keeping and peace-building. Just as demilitarized zones may serve the cause of preventive diplomacy and preventive deployment to avoid conflict, so may demilitarization assist in keeping the peace or in post-conflict peace-building, as a measure for heightening the sense of security and encouraging the parties to turn their energies to the work of peaceful restoration of their societies.

59. There is a new requirement for technical assistance which the United Nations has an obligation to develop and provide when requested: support for the transformation of deficient national structures and capabilities, and for the strengthening of new democratic institutions. The authority of the United Nations system to act in this field would rest on the consensus that social peace is as important as strategic or political peace. There is an obvious connection between democratic practices—such as the rule of law and transparency in decision-making—and the achievement of true peace and security in any new and stable political order. These elements of good governance need to be promoted at all levels of international and national political communities.

## VII. Cooperation with Regional Arrangements and Organizations

60. The Covenant of the League of Nations, in its Article 21, noted the validity of regional understandings for securing the maintenance of peace. The Charter devotes Chapter VIII to regional arrangements or agencies for dealing with such matters relating to the maintenance of international peace and security as are appropriate for regional action and consistent with the Purposes and Principles of the United Nations. The cold war impaired the proper use of Chapter VIII and indeed, in that era, regional arrangements worked on occasion against resolving disputes in the manner foreseen in the Charter.

61. The Charter deliberately provides no precise definition of regional arrangements and agencies, thus allowing useful flexibility for undertakings by a group of States to deal with a matter appropriate for regional action which also could contribute to the maintenance of international peace and security. Such associations or entities could include treaty-based organizations, whether created before or after the founding of the United Nations, regional organizations for mutual security and defence, organizations for general regional development or for cooperation on a particular economic topic or function, and groups created to deal with a specific political, economic or social issue of current concern.

62. In this regard, the United Nations has recently encouraged a rich variety of complementary efforts. Just as no two regions or situations are the same, so the design of cooperative work and its division of labour must adapt to the realities of each case with flexibility and creativity. In Africa, three different regional groups—the Organization of African Unity, the League of Arab States and the Organization of the Islamic Conference—joined efforts with the United Nations regarding Somalia. In the Asian context, the Association of South-East Asian Nations and individual States from several regions were brought together with the parties to the Cambodian conflict at an international conference in Paris, to work with the United Nations. For El Salvador, a unique arrangement—'The Friends of the Secretary-General'—contributed to agreements reached through the mediation of the Secretary-General. The end of the war in Nicaragua involved a highly complex effort which was initiated by leaders of the region and conducted by individual States, groups of States and the Organization of American States. Efforts undertaken by the European Community and its member States, with the support of States participating in the Conference on Security and Cooperation in Europe, have been of central importance in dealing with the crisis in the Balkans and neighbouring areas.

63. In the past, regional arrangements often were created because of the absence of a universal system for collective security; thus their activities could on occasion work at cross-purposes with the sense of solidarity

required for the effectiveness of the world Organization. But in this new era of opportunity, regional arrangements or agencies can render great service if their activities are undertaken in a manner consistent with the Purposes and Principles of the Charter, and if their relationship with the United Nations, and particularly the Security Council, is governed by Chapter VIII.

64. It is not the purpose of the present report to set forth any formal pattern of relationship between regional organizations and the United Nations, or to call for any specific division of labour. What is clear, however, is that regional arrangements or agencies in many cases possess a potential that should be utilized in serving the functions covered in this report: preventive diplomacy, peace-keeping, peacemaking and post-conflict peace-building. Under the Charter, the Security Council has and will continue to have primary responsibility for maintaining international peace and security, but regional action as a matter of decentralization, delegation and cooperation with United Nations efforts could not only lighten the burden of the Council but also contribute to a deeper sense of participation, consensus and democratization in international affairs.

65. Regional arrangements and agencies have not in recent decades been considered in this light, even when originally designed in part for a role in maintaining or restoring peace within their regions of the world. Today a new sense exists that they have contributions to make. Consultations between the United Nations and regional arrangements or agencies could do much to build international consensus on the nature of a problem and the measures required to address it. Regional organizations participating in complementary efforts with the United Nations in joint undertakings would encourage States outside the region to act supportively. And should the Security Council choose specifically to authorize a regional arrangement or organization to take the lead in addressing a crisis within its region, it could serve to lend the weight of the United Nations to the validity of the regional effort. Carried forward in the spirit of the Charter, and as envisioned in Chapter VIII, the approach outlined here could strengthen a general sense that democratization is being encouraged at all levels in the task of maintaining international peace and security, it being essential to continue to recognize that the primary responsibility will continue to reside in the Security Council.

**VIII. Safety of Personnel**

66. When United Nations personnel are deployed in conditions of strife, whether for preventive diplomacy, peacemaking, peace-keeping, peace-building or humanitarian purposes, the need arises to ensure their safety. There has been an unconscionable increase in the number of fatalities. Following the conclusion of a cease-fire and in order to prevent further outbreaks of violence, United Nations guards were called upon to assist in volatile conditions in Iraq. Their presence afforded a measure of security to United Nations personnel and supplies and, in addition, introduced an element of reassurance and stability that helped to prevent renewed conflict. Depending upon the nature of the situation, different configurations and compositions of security deployments will need to be considered. As the variety and scale of threat widens, innovative measures will be required to deal with the dangers facing United Nations personnel.

67. Experience has demonstrated that the presence of a United Nations operation has not always been sufficient to deter hostile action. Duty in areas of danger can never be risk-free; United Nations personnel must expect to go in harm's way at times. The courage, commitment and idealism shown by United Nations personnel should be respected by the entire international community. These men and women deserve to be properly recognized and rewarded for the perilous tasks they undertake. Their interests and those of their families must be given due regard and protected.

68. Given the pressing need to afford adequate protection to United Nations personnel engaged in life-endangering circumstances, I recommend that the Security Council, unless it elects immediately to withdraw the United Nations presence in order to preserve the credibility of the Organization, gravely consider what action should be taken towards those who put United Nations personnel in danger. Before deployment takes place, the Council should keep open the option of considering in advance collective measures, possibly including those under Chapter VII when a threat to international peace and security is also involved, to come into effect should the purpose of the United

Nations operation systematically be frustrated and hostilities occur.

## IX. Financing

69. A chasm has developed between the tasks entrusted to this Organization and the financial means provided to it. The truth of the matter is that our vision cannot really extend to the prospect opening before us as long as our financing remains myopic. There are two main areas of concern: the ability of the Organization to function over the longer term; and immediate requirements to respond to a crisis.

70. To remedy the financial situation of the United Nations in all its aspects, my distinguished predecessor repeatedly drew the attention of Member States to the increasingly impossible situation that has arisen and, during the forty-sixth session of the General Assembly, made a number of proposals. Those proposals which remain before the Assembly, and with which I am in broad agreement, are the following:

*Proposal one*. This suggested the adoption of a set of measures to deal with the cash flow problems caused by the exceptionally high level of unpaid contributions as well as with the problem of inadequate working capital reserves:

(a) Charging interest on the amounts of assessed contributions that are not paid on time;

(b) Suspending certain financial regulations of the United Nations to permit the retention of budgetary surpluses;

(c) Increasing the Working Capital Fund to a level of $250 million and endorsing the principle that the level of the Fund should be approximately 25 per cent of the annual assessment under the regular budget;

(d) Establishment of a temporary Peacekeeping Reserve Fund, at a level of $50 million, to meet initial expenses of peacekeeping operations pending receipt of assessed contributions;

(e) Authorization to the Secretary-General to borrow commercially, should other sources of cash be inadequate.

*Proposal two*. This suggested the creation of a Humanitarian Revolving Fund in the order of $50 million, to be used in emergency humanitarian situations. The proposal has since been implemented.

*Proposal three*. This suggested the establishment of a United Nations Peace Endowment Fund, with an initial target of $1 billion. The Fund would be created by a combination of assessed and voluntary contributions, with the latter being sought from Governments, the private sector as well as individuals. Once the Fund reached its target level, the proceeds from the investment of its principal would be used to finance the initial costs of authorized peace-keeping operations, other conflict resolution measures and related activities.

71. In addition to these proposals, others have been added in recent months in the course of public discussion. These ideas include: a levy on arms sales that could be related to maintaining an Arms Register by the United Nations; a levy on international air travel, which is dependent on the maintenance of peace; authorization for the United Nations to borrow from the World Bank and the International Monetary Fund—for peace and development are interdependent; general tax exemption for contributions made to the United Nations by foundations, businesses and individuals; and changes in the formula for calculating the scale of assessments for peace-keeping operations.

72. As such ideas are debated, a stark fact remains: the financial foundations of the Organization daily grow weaker, debilitating its political will and practical capacity to undertake new and essential activities. This state of affairs must not continue. Whatever decisions are taken on financing the Organization, there is one inescapable necessity: Member States must pay their assessed contributions in full and on time. Failure to do so puts them in breach of their obligations under the Charter.

73. In these circumstances and on the assumption that Member States will be ready to finance operations for peace in a manner commensurate with their present, and welcome, readiness to establish them, I recommend the following:

(a) Immediate establishment of a revolving peace-keeping reserve fund of $50 million;

(b) Agreement that one third of the estimated cost of each new peace-keeping operation be appropriated by the General Assembly as soon as the Security Council decides to establish the operation; this would give the Secretary-General the necessary commitment authority and assure an adequate cash flow; the balance of the costs would be appropriated after the General Assembly approved the operation's budget;

(c) Acknowledgement by Member States

that, under exceptional circumstances, political and operational considerations may make it necessary for the Secretary-General to employ his authority to place contracts without competitive bidding.

74. Member States wish the Organization to be managed with the utmost efficiency and care. I am in full accord. I have taken important steps to streamline the Secretariat in order to avoid duplication and overlap while increasing its productivity. Additional changes and improvements will take place. As regards the United Nations system more widely, I continue to review the situation in consultation with my colleagues in the Administrative Committee on Coordination. The question of assuring financial security to the Organization over the long term is of such importance and complexity that public awareness and support must be heightened. I have therefore asked a select group of qualified persons of high international repute to examine this entire subject and to report to me. I intend to present their advice, together with my comments, for the consideration of the General Assembly, in full recognition of the special responsibility that the Assembly has, under the Charter, for financial and budgetary matters.

**X. An Agenda for Peace**

75. The nations and peoples of the United Nations are fortunate in a way that those of the League of Nations were not. We have been given a second chance to create the world of our Charter that they were denied. With the cold war ended we have drawn back from the brink of a confrontation that threatened the world and, too often, paralysed our Organization.

76. Even as we celebrate our restored possibilities, there is a need to ensure that the lessons of the past four decades are learned and that the errors, or variations of them, are not repeated. For there may not be a third opportunity for our planet which, now for different reasons, remains endangered.

77. The tasks ahead must engage the energy and attention of all components of the United Nations system—the General Assembly and other principal organs, the agencies and programmes. Each has, in a balanced scheme of things, a role and a responsibility.

78. Never again must the Security Council lose the collegiality that is essential to its proper functioning, an attribute that it has gained after Such trial. A genuine sense of consensus deriving from shared interests must govern its work, not the threat of the veto or the power of any group of nations. And it follows that agreement among the permanent members must have the deeper support of the other members of the Council, and the membership more widely, if the Council's decisions are to be effective and endure.

79. The Summit Meeting of the Security Council of 31 January 1992 provided a unique forum for exchanging views and strengthening cooperation. I recommend that the Heads of State and Government of the members of the Council meet in alternate years, just before the general debate commences in the General Assembly. Such sessions would permit exchanges on the challenges and dangers of the moment and stimulate ideas on how the United Nations may best serve to steer change into peaceful courses. I propose in addition that the Security Council continue to meet at the Foreign Minister level, as it has effectively done in recent years, whenever the situation warrants such meetings.

80. Power brings special responsibilities, and temptations. The powerful must resist the dual but opposite calls of unilateralism and isolationism if the United Nations is to succeed. For just as unilateralism at the global or regional level can shake the confidence of others, so can isolationism, whether it results from political choice or constitutional circumstance, enfeeble the global undertaking. Peace at home and the urgency of rebuilding and strengthening our individual societies necessitates peace abroad and cooperation among nations. The endeavours of the United Nations will require the fullest engagement of all of its Members, large and small, if the present renewed opportunity is to be seized.

81. Democracy within nations requires respect for human rights and fundamental freedoms, as set forth in the Charter. It requires as well a deeper understanding and respect for the rights of minorities and respect for the needs of the more vulnerable groups of society, especially women and children. This is not only a political matter. The social stability needed for productive growth is nurtured by conditions in which people can readily express their will. For this, strong domestic institutions of participation are essential. Promoting such institutions means

promoting the empowerment of the unorganized, the poor, the marginalized. To this end, the focus of the United Nations should be on the 'field', the locations where economic, social and political decisions take effect. In furtherance of this I am taking steps to rationalize and in certain cases integrate the various programmes and agencies of the United Nations within specific countries. The senior United Nations official in each country should be prepared to serve, when needed, and with the consent of the host authorities, as my Representative on matters of particular concern.

82. Democracy within the family of nations means the application of its principles within the world Organization itself. This requires the fullest consultation, participation and engagement of all States, large and small, in the work of the Organization. All organs of the United Nations must be accorded, and play, their full and proper role so that the trust of all nations and peoples will be retained and deserved. The principles of the Charter must be applied consistently, not selectively, for if the perception should be of the latter, trust will wane and with it the moral authority which is the greatest and most unique quality of that instrument. Democracy at all levels is essential to attain peace for a new era of prosperity and justice.

83. Trust also requires a sense of confidence that the world Organization will react swiftly, surely and impartially and that it will not be debilitated by political opportunism or by administrative or financial inadequacy. This presupposes a strong, efficient and independent international civil service whose integrity is beyond question and an assured financial basis that lifts the Organization, once and for all, out of its present mendicancy.

84. Just as it is vital that each of the organs of the United Nations employ its capabilities in the balanced and harmonious fashion envisioned in the Charter, peace in the largest sense cannot be accomplished by the United Nations system or by Governments alone. Non-governmental organizations, academic institutions, parliamentarians, business and professional communities, the media and the public at large must all be involved. This will strengthen the world Organization's ability to reflect the concerns and interests of its widest constituency, and those who become more involved can carry the word of United Nations initiatives and build a deeper understanding of its work.

85. Reform is a continuing process, and improvement can have no limit. Yet there is an expectation, which I wish to see fulfilled, that the present phase in the renewal of this Organization should be complete by 1995, its fiftieth anniversary. The pace set must therefore be increased if the United Nations is to keep ahead of the acceleration of history that characterizes this age. We must be guided not by precedents alone, however wise these may be, but by the needs of the future and by the shape and content that we wish to give it.

86. I am committed to broad dialogue between the Member States and the Secretary-General. And I am committed to fostering a full and open interplay between all institutions and elements of the Organization so that the Charter's objectives may not only be better served, but that this Organization may emerge as greater than the sum of its parts. The United Nations was created with a great and courageous vision. Now is the time, for its nations and peoples, and the men and women who serve it, to seize the moment for the sake of the future.

---

*Notes:*

[1] See S/23500, statement by the President of the Council, section entitled 'Peacemaking and peace-keeping'.

[2] General Assembly resolution 37/10, annex.

[3] General Assembly resolution 43/51, annex.

*Source:* UN document A/47/277 (S/24111), 17 June 1992.

# 3. Major armed conflicts

RAMSES AMER, BIRGER HELDT, SIGNE LANDGREN, KJELL MAGNUSSON, ERIK MELANDER, KJELL-ÅKE NORDQUIST, THOMAS OHLSON and PETER WALLENSTEEN*

## I. Introduction

In 1992 major armed conflicts were waged in 30 locations around the world. All of these conflicts except one (India–Pakistan) were intra-state.[1] A major armed conflict is characterized by prolonged combat between the military forces of two or more governments or of one government and at least one organized armed group, and incurring the battle-related deaths of at least 1000 persons during the entire conflict.[2] As some countries are the location of several major armed conflicts, the actual number of conflicts is higher than the number of locations. The locations, contested incompatibilities, warring parties, and figures for active armed forces and deaths incurred in the conflicts are presented in appendix 3A.

This chapter provides an overview of 1992. It comments upon the changes from 1991 in the table of conflicts and developments in the new conflicts recorded for 1992 as well as the Israel/Palestine peace process.

The impact of the end of the cold war on the number and nature of the armed conflicts is also assessed. As the cold war was winding down in the late 1980s, there were contradictory expectations regarding the outcome for world politics—ranging from scenarios of a 'peace dividend' of some magnitude, to the erosion of order and stability with the collapse of the East–West divide and the nuclear threat. The data on major armed conflicts do not support the expectation that the end of the cold war would result in increased global disorder but rather show a very gradual decrease in the annual total number of conflict locations since 1989 (see table 3.1 below).

[1] The relationship between the number of inter-state and intra-state major armed conflicts has not changed appreciably since the mid-1980s. A clear trend towards a greater number of intra-state wars can be observed for the period from 1945 until the late 1980s. The annual number of inter-state wars in this period did not exceed 8, with a peak in 1965–72. Lindgren, K., *Världens Krig* (Swedish Institute of International Affairs: Stockholm, 1990).

[2] See Heldt, B. (ed.), *States in Armed Conflict 1990–91* (Department of Peace and Conflict Research, Uppsala University: Uppsala, 1992), chapter 3, for the full definition.

* R. Amer (section VII), B. Heldt, E. Melander, K-Å. Nordquist, T. Ohlson (section VIII) and P. Wallensteen, the Department of Peace and Conflict Research, Uppsala University, Sweden; K. Magnusson (section V), Department of Soviet and East European Studies, Uppsala University, Sweden; S. Landgren (section VI), SIPRI.

# II. Changes in the list of conflicts for 1992

## Conflicts recorded for 1991 which do not appear in the data for 1992

In five of the 30 conflict locations recorded for 1991 there was no major armed conflict in 1992: El Salvador, Ethiopia, Iraq–Kuwait, Morocco/Western Sahara and Uganda. (Furthermore, the conflicts in the Indonesian region of Aceh and the Iranian region of Kurdistan showed no military activity in 1992.)

Regulation[3] of conflicts took place in three conflict locations: El Salvador, Iraq–Kuwait and Morocco/Western Sahara. In El Salvador the parties had in December 1991 agreed on a peace accord, mediated with the help of the United Nations. The Iraq–Kuwait conflict concerning the status of Kuwait was regulated by United Nations Security Council (UNSC) Resolution 687 of 3 April 1991. The conflict was not solved, however, since Iraq continued to state its claim to sovereignty over Kuwait. The conflict in Morocco/Western Sahara concerning the status of Western Sahara was regulated by a plan that was drawn up and authorized by the UNSC. A UN-monitored cease-fire came into effect at the end of 1991. A referendum giving the inhabitants of Western Sahara a choice between independence and integration with Morocco was scheduled for January 1992. However, because of disagreement between the previously warring parties regarding the list of eligible voters, the referendum was postponed. Breaches of the cease-fire related to the confinement of troops were reported.

The conflicts in Ethiopia were terminated in 1991 as one of the warring parties ceased to exist. Ethiopia's leader Mengistu Haile Mariam, facing a military defeat, fled the country in May 1991. The conflict concerning government thus disappeared, a new government was installed, and the final decision on the conflict concerning the status of Eritrea was left to a referendum scheduled for 1993. Despite tension and skirmishes between the new government and Oromo groups, no major armed conflict in Ethiopia was recorded for 1992.

Conflicts in three locations reported for 1991—Indonesia, Iran and Uganda—were neither solved nor regulated in 1992. In the case of Iran, the reason for the absence of military activity in the conflict over the status of the Kurdish region in Iran between the Kurdish Democratic Party of Iran (KDPI) and the government was unclear.[4] This was not, however, a significant change compared to 1991, since there were only minor skirmishes that year. In Indonesia, the Government subdued the secessionist Aceh Merdeka movement in the Aceh province at the northern tip of Sumatra. In Uganda, offensives were launched in 1991 against already seriously weakened opposition groups. By 1992 the groups had reportedly ceased to exist and approximately 100 of their members were scattered into small groups, trying to evade arrest.

[3] Regulation is defined as 'an agreement to contest the incompatibility without the use of armed force'. See section IV for the definition of 'incompatibility'.

[4] An 'on–off' pattern of military activity could be seen in both large-scale and small-scale armed conflicts during 1989–91; see Heldt, B. (ed.), *States in Armed Conflict 1990–91* (Department of Peace and Conflict Research, Uppsala University: Uppsala, 1992), chapters 1 and 3.

### New conflicts in 1992

Major armed conflicts were waged in five new locations in 1992 compared to 1991: Azerbaijan, India–Pakistan, Laos, Tajikistan, and Bosnia and Herzegovina.[5] Two of these, Laos and Tajikistan, concerned government and the remaining three concerned territory. In 1992 Tajikistan and Bosnia and Herzegovina were probably the two most devastating conflicts in terms of deaths.[6] (For a case study of the conflict in Bosnia and Herzegovina, see section V in this chapter, and for the conflicts in the newly independent states of the former Soviet Union, see section VI.)

The conflict between India and Pakistan concerning Kashmir showed some heightened military activity in 1992. An attempt by Pakistan to occupy territory in the contested area by building new military installations resulted in heavy artillery exchanges and reportedly heavy casualties. In Laos, the conflict concerning government re-emerged, with fighting in January and June 1992. Both India–Pakistan and Laos were recorded as major armed conflict locations in 1990, but neither was listed for 1991 because the fighting there continued to be sporadic.

## III. The level of conflict intensity and the Middle East peace process

### Decreased intensity

Four locations with major armed conflicts in 1992 showed a sharp decrease in the number of battle-related deaths compared to 1991: Angola, Croatia, Iraq (the Kurdish conflict in the north) and Somalia.[7]

In Angola, fighting resumed in October 1992 as UNITA refused to accept defeat in the elections (for descriptions of the conflicts in Southern Africa, see section VIII). Although this meant that the 1991 peace accord collapsed, the number of battle-related deaths in 1992 was significantly lower than in 1991. In Croatia, only small-scale fighting took place. The deaths incurred are estimated at fewer than 100, as compared to 6000–10 000 in 1991. The Kurdish conflict in Iraq involved minor skirmishes, in contrast to the heavy fighting in 1991 following unco-ordinated Kurdish offensives against the Iraqi Govern-

[5] Several additional conflicts came close to satisfying the criteria for inclusion among the major armed conflict locations for 1992. Among these are Georgia and Moldova. In the case of *Georgia*, none of the three conflicts within its borders incurred 1000 battle-related deaths. In *Moldova*, the death figure was 700–800. A description of these two conflicts and those in the North Caucasus region of Russia is presented in section VI. Another example is *Niger*, where an armed conflict between the Government and the Front de Libération de l'Air et l'Azawad (Air and Azawad Liberation Front, FLAA) has been active since 1990; the death toll was probably approximately 600 up to Dec. 1992. There are no reliable estimates of the number of deaths in the conflict between the Government of *Sierra Leone* and the Revolutionary United Front (RUF).

[6] This assumes that the figure for deaths incurred in the conflict in Tajikistan was 20 000–30 000. Note that there are no reliable estimates for the numbers of battle-related deaths in Afghanistan and Sudan. The numbers might be higher than those for the conflict in Bosnia and Herzegovina.

[7] For the level of intensity of each conflict recorded for 1992, see the column 'change from 1991' in the table, appendix 3A in this volume.

ment which had just been defeated in the Persian Gulf War. Conflict in Somalia between the two factions of the United Somali Congress (USC) decreased in 1992, with most of the fighting taking place during the first six months of 1992.

### Increased intensity

Two locations with major armed conflicts showed a sharp increase in battle-related deaths compared to 1991: Liberia and Turkey.[8] In 1991 there were skirmishes between the Liberian Interim Government of National Unity (IGNU), supported by the Economic Community of West African States' Monitoring Group (ECOMOG), controlling the capital Monrovia, and the National Patriotic Forces of Liberia (NPFL), controlling nearly all of the rest of the country and having established a parallel government. Several negotiated peace agreements were not implemented. Heavy fighting erupted in October 1992 as the NPFL launched attacks on Monrovia. The ensuing fighting was the heaviest since 1990. In Turkey, an offensive by the Kurdish Worker's Party (PKK) led to sharply escalated fighting, spilling over into Iraq. The PKK suffered severe losses and, by the end of 1992, the number of deaths in this nine-year conflict had nearly doubled, from 3200 to about 6000.

### Regulation

In 1992 only one major armed conflict was clearly regulated:[9] that in Mozambique. A comprehensive peace agreement was signed in October 1992 (see section VIII for a case study of the conflicts in Southern Africa).

### The Israel/Palestine conflict and the Middle East peace process

In the Israel/Palestine conflict, the talks between Israel and delegations from neighbouring countries continued in 1992 as a parallel process of multilateral and bilateral talks. Three events during the year were regarded as particularly important for this process: the Israeli elections in June, the US presidential elections in November and the Israeli deportation in December of 415 alleged Hamas activists. In the Israeli elections, the Likud coalition government was replaced by a Labour coalition government headed by Prime Minister Itzak Rabin, a change which many parties in the peace process considered positive for the process. In the US presidential elections, the defeat of President George Bush, whose Administration together with the Soviet Union orchestrated the Madrid peace talks, created uncertainty regarding future US

[8] Two additional locations with conflicts that have have intensified in 1992 compared to 1991 are Sudan and Afghanistan. In Sudan, the government took advantage of the in-fighting between factions of the Sudanese People's Liberation Army (SPLA) and launched several offensives. Heavy casualties were reported, but no exact figures have been found. In Afghanistan, fighting was heavy following the fall of the Najibullah Government.

[9] See note 3.

policy *vis-à-vis* the Middle East. The Israeli deportation of Hamas activists to the Israeli–Lebanese border following the killing of Israeli border police and military personnel blocked the peace process at the end of the year.

Five rounds of bilateral talks were held in 1992. The stage of multilateral meetings in the Madrid process was initiated by talks held in Moscow on 28–29 January. In May, separate multilateral talks were held on each of the following themes: economic co-operation (held in Brussels), arms control (Washington), refugees (Ottawa), sharing of water resources (Vienna) and environmental protection (Tokyo). While the bilateral talks involved delegations from Israel, Jordan (including a Palestinian delegation), Lebanon and Syria, the multilateral talks also included representations from *inter alia* Algeria, Bahrain, Canada, China, the European Community (EC), India, Japan, Kuwait, Mauritania, Morocco, Oman, Qatar, Russia, Saudi Arabia, Tunisia, Turkey, the United Arab Emirates, the USA and Yemen. Syria and Lebanon did not attend any of the multilateral talks, stating that progress in the bilateral talks had to be made before multilateral talks would be meaningful.

In a bilateral meeting in January 1992, Israel proposed an 'interim self-government authority' for Palestinians, covering 15 spheres of civil administration but excluding security, foreign relations, the status of Israelis living in the occupied territories and 'vital Israeli needs'. A Palestinian counter-proposal also included provisions for Arab self-rule in Jerusalem and the election of an organ assuming authority over the land, people and resources in the occupied territories, pending an agreement on the final status of the territories. In talks in April, Israel proposed 'municipal' elections, which later was changed to 'general' elections by the new Labour Government. Also in 1992, Israel agreed to talk to a separate Palestinian representation (reducing the Jordanian delegation to a formality) about the nature of Palestinian authority and the fate of the Jewish settlements in the Occupied Territories.

Other topics discussed in 1992 were Israeli–Syrian positions in the dispute over the Golan Heights and a possible Israeli–Jordanian agreement on a 'full agenda' for their bilateral peace agreement negotiations. The latter pre-negotiations were reported to be close to finalization in October, but proved to need more time when the seventh round of bilateral talks was held in November. The October agenda was reported to deal with a peace treaty, water and land issues, Palestinian refugees and arms control.

In terms of militarization, the conflict showed no sign of decline: the number of Israeli Defence Forces shootings related to the *intifada* was 340 in 1992. The Hezbollah guerrilla movement, Israel and the Israel-backed South Lebanese Army clashed almost routinely in southern Lebanon in early 1992. The Israeli killing of Hezbollah Secretary-General Musawi on 16 February fuelled the violent spiral of measures and counter-measures with artillery and rocket firing into northern Israel, causing Israeli incursion into United Nations Interim Force in Lebanon (UNIFIL)-patrolled areas of Lebanon. Following Israeli withdrawal from these areas, bombardments continued for some time from both sides. Apart from Hezbollah, a number of other groups or organizations also used armed force against the Government of Israel.

## IV. The end of the cold war: measuring the effects

As the cold war was winding down during the latter half of the 1980s, there was a gradual decline in the number of locations where major armed conflicts were being fought.

If the figures in table 3.1 are seen as a measure of a 'peace dividend', then the five world regions have been unevenly affected: Central and South America was the only clear beneficiary throughout the period 1989–92. The figures for Africa showed a decline only in 1992. In Asia the number of conflicts declined from 1989 to 1991 but increased in 1992. Europe was also affected, with an increase in the number of conflicts since 1990. The figure for conflicts in the Middle East was stable until 1992, when it decreased slightly.

**Table 3.1.** Regional distribution of conflict locations with at least one major armed conflict, 1989–92

| Region | 1989 | 1990 | 1991 | 1992 |
|---|---|---|---|---|
| Africa | 9 | 10 | 11 | 8 |
| Asia | 11 | 10 | 8 | 11 |
| Central and South America | 5 | 5 | 4 | 3 |
| Europe | 2 | 1 | 2 | 4 |
| Middle East | 5 | 5 | 5 | 4 |
| **Annual total** | **32** | **31** | **30** | **30** |

*Source*: Uppsala Conflict Data Project.

A drawback of the method of counting the locations with at least one major armed conflict is that it is a crude measure of change: it does not assess the change in actual number of conflicts, since a country may be the location of several conflicts.

Another method of measuring change in terms of conflicts is to establish the number of dyads, or pairs, of warring parties. However, it is difficult to establish a reliable figure for the number of dyads of warring parties, of which at least one is the government of a state, that have incurred at least 1000 battle-related deaths during the course of the conflict. For instance, approximately 180 armed groups reportedly existed in the Indian part of Kashmir in 1992. The precise number of Sikh groups in Punjab, Kurdish and Shiite groups in Iraq, and Palestinian groups in Israel is also unknown and seems to vary over time. In Afghanistan, there may have been 20 active dyads (the government versus 20 opposition groups) that have incurred at least 1000 battle-related deaths in the conflict in 1992. Another problem is determining what is to be considered as a party or organization—for instance, how umbrella organizations should be treated.

**Table 3.2.** Regional distribution, number and types of contested incompatibilities in major armed conflicts, 1989–92[a]

| | 1989 | | 1990 | | 1991 | | 1992 | |
|---|---|---|---|---|---|---|---|---|
| Region | Govt | Terr. | Govt | Terr. | Govt | Terr. | Govt | Terr. |
| Africa | 7 | 3 | 8 | 3 | 9 | 3 | 7 | 1 |
| Asia | 6 | 8 | 5 | 10 | 3 | 8 | 5 | 9 |
| Central and South America | 5 | – | 5 | – | 4 | – | 3 | – |
| Europe | 1 | 1 | – | 1 | – | 2 | – | 4 |
| Middle East | 1 | 4 | 1 | 4 | 2 | 5 | 2 | 2 |
| *Total* | *20* | *16* | *19* | *18* | *18* | *18* | *17* | *16* |
| **Annual total** | **36** | | **37** | | **36** | | **33** | |

[a] The total annual number of conflicts does not necessarily correspond to the number of conflict locations in table 3A.1, appendix 3A, since there may be more than one armed conflict in each location.

*Source*: Uppsala Conflict Data Project.

A more promising method of measuring change in conflicts is to count the contested incompatibilities in which at least one active dyad of warring parties, of which at least one is the government of a state, has incurred at least 1000 battle-related deaths. This method offers more reliable and comparable figures. The data for incompatibilities in this chapter and appendix 3A are for contested incompatabilities over government and territory.[10]

A contested incompatibility is defined as the stated general goals that the warring parties seek to realize through the use of armed force. An incompatibility concerning government is at hand when the warring parties have a stated general incompatible position concerning the type of political system or change of the central government or its composition. An incompatibility concerning territory is at hand when the warring parties have a stated general incompatible position concerning control of territory (inter-state conflict), secession or autonomy (intra-state conflict).[11] Table 3.2 shows the regional distribution of major armed conflicts with at least one active dyad of warring parties.

The annual number of contested incompatibilities fell from 37 in 1990 to 33 in 1992. The total number of incompatibilities concerning government has consistently declined, while the number of incompatibilities concerning territory varied. Central and South America, a region with incompatibilities only concerning government, had an unbroken downward trend each year from

[10] Incompatibilities concerning government or territory have played central roles in all large-scale armed conflicts ('wars') since 1648. See Holsti, K. J., *Armed Conflicts and International Order 1648–1989* (Cambridge University Press: Cambridge, 1991).

[11] The focus on incompatibilities as defined above means that in a single country there can only be one incompatibility over government, while there may be several incompatibilities regarding territory. This is because each country has only one government but sometimes several distinct and contested regions.

1990. The annual figures for Africa showed a decline only in 1992. Asia and the Middle East did not show any clear trends—the annual totals fluctuated from one year to another. Of all the regions, Europe seems to show the worst development, with increasing annual totals since 1990; the increase concerned incompatibilities concerning only territory.

The contested incompatibilities in Africa mostly concern government, and the region with the most incompatibilities concerning territory was Asia. In many of the contested incompatibilities concerning government in Africa, as in other regions, the non-governmental groups could be ethnically identified, as in Angola, Chad, Liberia and Rwanda. This indicates that ethnicity is an important element of most armed conflicts.

An effect of the end of the cold war was the increased resort to the UN. In 1992 the UN was engaged in promoting solutions, observing developments, implementing decisions or peace-keeping operations in a number of locations with at least one major armed conflict: Afghanistan, Angola, Azerbaijan, Cambodia, Croatia, India–Pakistan, Iraq, Israel/Palestine, Mozambique, Somalia, South Africa, and Bosnia and Herzegovina.[12] In reality, however, the amount of UN activity related to conflicts was greater. For instance, supervision of a functioning cease-fire meant that the conflicts were no longer recorded as active. Such cases are El Salvador, Iraq–Kuwait and Morocco/ Western Sahara.

## V. Bosnia and Herzegovina[13]

A fundamental problem underlying the conflicts in the former Yugoslavia was how the principle of national self-determination should be implemented. Did the right to self-determination apply to the Yugoslav republics, as Slovenia and Croatia maintained, or was it to be exercised by the Yugoslav peoples, as Serbia argued?

Bosnia and Herzegovina was the only Yugoslav republic without a titular nation. No ethnic group could claim Bosnia as its state in the way that Serbia or Croatia were the national states of the Serbs and Croats. The Serbo-Croatian-speaking Muslims, forming the largest ethnic group, were recognized as one of the Yugoslav nations in 1971:[14] in 1991 Muslims constituted

[12] Four new UN missions were established in 1992. See chapter 2 in this volume for an overview of the international efforts at preventive diplomacy, peace-making and peace-keeping in 1992.

[13] For a general background, see Garde, P., *Vie et mort de la Yougoslavie* (Fayard: Paris, 1992); Glenny, M., *The Fall of Yugoslavia: The Third Balkan War* (Penguin: London, 1992). For developments in 1991–92, see issues of *Radio Free Europe/Radio Liberty Research Report*, *Yugofax/War Report*, *East European Reporter*, *Südosteuropa* and *Keesing's Contemporary Archives*.

[14] In Yugoslav constitutional and political terminology, there was a distinction between, on the one hand, the South Slavic nations (Serbs, Croats, Muslims, Slovenes, Macedonians and Montenegrines) forming the Yugoslav Federation and, on the other hand, the nationalities (ethnic minorities) such as Albanians, Hungarians, Turks, etc.

43.7 per cent of the population, the remainder being Serbs (31.3 per cent), Croats (17.3 per cent), 'Yugoslavs' (5.5 per cent) and others (2.2 per cent).[15]

The very idea of Bosnia and Herzegovina as an independent political entity was, in fact, directly linked to the existence of the Yugoslav federation. If Yugoslavia were to disintegrate, Bosnia would immediately find itself in a precarious situation. Both Serbs and Croats had traditional claims on all or parts of Bosnian territory, and it was doubtful whether they had accepted the idea of a distinct Muslim nationality.

In the 1990–91 negotiations on the future of Yugoslavia, it was natural that Bosnia and Herzegovina—as well as Macedonia—would try to find a compromise which would guarantee the continued existence of a Yugoslav community.

In contrast to both the model of a Yugoslav Confederation suggested by Slovenia and Croatia and a Federal Republic of Yugoslavia put forward by Serbia and Montenegro, Bosnia and Herzegovina and Macedonia advocated a solution in which the constituent republics of a new Yugoslav community would be regarded as independent states, at the same time as they would relinquish part of their sovereignty to a central government. The latter idea was similar to important features of the Serbian model, but on the crucial issue of self-determination the Bosnian–Macedonian proposal was closer to that of Slovenia and Croatia, since an endorsement of the Serbian position would have meant that both Bosnia and Herzegovina and Macedonia would be partitioned according to ethnic principles. The explicit goal of Bosnian President Alija Izetbegovic was therefore some type of Yugoslav association where Serbia *and* Croatia would be members. If that were to prove impossible, Bosnia and Herzegovina should become a sovereign and unified state.

The Bosnian Serbs had made it clear that they would not accept either minority status in a Muslim state or a confederation with Croatia. They wanted to remain part of Yugoslavia but were ready to discuss a 'cantonization' of Bosnia. The Croats wavered. On the one hand they recognized Bosnian sovereignity, but on the other, like the Serbs, they seemed to prefer a division along ethnic lines. However, such a division was impossible to achieve without large-scale migration. According to estimates made before the war,[16] over 1.5 million people would have to leave their homes if Bosnia and Herzegovina was to be transformed into ethnically homogeneous areas.[17] Nevertheless, during 1991 both Serbs and Croats formed autonomous areas which might be a first step towards an eventual partition of Bosnia and Herzegovina.

[15] Andrejevich, M., 'Bosnia and Herzegovina: a precarious peace', *RFE/RL Research Report*, 28 Feb. 1992, p. 7; Büschenfeld, H., 'Ergebnisse der Volkszählung 1991 in Jugoslawien', *Osteuropa*, vol. 42, no. 12 (1992), p. 1101.

[16] *Borba*, 27 Jan. 1992.

[17] According to the 1981 census there was a Muslim majority in 35 of Bosnia's 109 communes; the Serbs constituted a majority in 32 and the Croats in 14 communes. These 81 communes had a total population of 2 700 000 inhabitants. In the remaining 28 communes no group formed an absolute majority; that is, 1 420 000 people, or 35% of Bosnia's population, lived in these areas.

**Events in 1992**

In a referendum on 29 February–1 March 1992, over 99 per cent of the voters voted in favour of Bosnian independence, and the government immediately asked for international recognition. However, the turn-out was only 64 per cent, as the Serbs boycotted the referendum, arguing that any radical change of Bosnia's status must be based on consensus between the three constituent nations. Immediately afterwards, violence erupted throughout the country.

The international conference on Bosnia and Herzegovina, which met in Sarajevo on 17–18 March 1992 in order to reconcile the conflicting positions, ended in a compromise: Bosnia and Herzegovina was to be a unified state within existing borders, but would at the same time consist of three national units. According to a map presented at the conference, Muslims and Serbs would each get 44 per cent of the territory and the Croats 12 per cent. As this solution was contrary to the official position of the Bosnian Government, it came as no surprise when the Muslims after a few days rejected the idea of 'three Bosnias in one'.

Tension increased and armed incidents occurred in several parts of Bosnia and Herzegovina. In Bosanski Brod, Serbian paramilitary forces fought Croat and Muslim militias. In Neum, the only Bosnian outlet to the Adriatic Sea, mainly populated by Croats, fighting broke out on 23 March between the Yugoslav Army and the Croatian Armed Forces (Hrvatske Oruzane Snage, HOS), a military branch of the nationalist Croatian Party of the Right (Hrvatska Stranka Prava, HSP).

Bosnian leaders met again in Brussels on 30 March, without reaching an agreement on the future of the republic. On 4 April the state presidency of Bosnia and Herzegovina ordered a general mobilization of the territorial defence.

On 6–7 April 1992, Bosnia and Herzegovina was recognized by the EC and the USA, after which fighting in Bosnia and Herzegovina escalated as Croat and Serb forces tried to consolidate their control of regions where they were the dominant ethnic population. On 27 April Serbia and Montenegro announced the formation of a new state, the Federal Republic of Yugoslavia.

On 14 May the Yugoslav Army formally withdrew from Bosnia and Herzegovina. However, Serbian officers and soldiers of Bosnian origin—about 80 per cent—remained and were incorporated in the Serb Army in Bosnia. On 22 May Bosnia and Herzegovina, as well as Croatia and Slovenia, were granted membership of the UN. On the same day the Security Council declared an oil embargo against Serbia. On 30 May, harsh sanctions were imposed on Serbia and Montenegro. In addition, the international community provided humanitarian aid that brought some relief to the citizens of Bosnia and Herzegovina. Fighting did not stop, however, and it has been extremely difficult to aid the large number of people living under siege in provincial towns and villages.

Efforts at mediation have been led by the EC and the UN. Representatives of the contending parties met in London on 26–28 August 1992. The conference formulated a set of general preconditions for peace: respect of the rights of both individuals and minorities, rejection of violence as a means of resolving conflicts, a programme involving an immediate cease-fire and international control of heavy weapons, as well as a general framework for negotiations. Under the leadership of Cyrus Vance and Lord Owen, six subcommittees were to deal with specific issues.

By the end of 1992 the Serb forces in Bosnia and Herzegovina numbered some 70 000–80 000 troops, of which 35 000 were irregulars. On the Croatian side there were 50 000 troops, 15 000 irregulars and 35 000 belonging to the Croatian Defence Council (Hrvatsko Vijece Obrane, HVO). The Muslims controlled 80 000 men, of which 30 000 were volunteers.[18]

## The character of the war

The number of civilian deaths in the war in Bosnia and Herzegovina is appalling. Instead of being indirect casualties of military operations, the civilian population of Bosnia—especially the Muslims—has been a major target. The purpose has obviously been to create ethnically homogeneous areas by expelling those with the 'wrong' identity. During these actions people have been harassed, molested and killed, and their property confiscated or devastated. While all sides are guilty of acts of cruelty, according to unanimous evidence 'ethnic cleansing' has been practised most often by Serbian forces.

During the summer of 1992 there were persistent rumours about concentration camps, and in August pictures of starving prisoners detained by the Serbs shocked the world. Muslim and Croat sources claimed that 130 000 prisoners were kept in 94 camps, while Serbian representatives refered to 42 000 prisoners held by Muslims and Croats in 45 camps.[19] According to figures of the International Committee of the Red Cross (ICRC), in September 1992 there was a total of 8485 internees: 6718 held by Serbs, 854 by Muslims and 913 by Croats.[20]

In the autumn an increasing number of reports referred to serious and large-scale violations of human rights: torture, murder and gang rapes. Especially alarming was the frequent and systematic abuse of Muslim women and children.[21]

[18] Figures given by Major General Jörn Beckman, at a Public Hearing on the situation in former Yugoslavia organized by the Foreign Affairs Committee of the Swedish Parliament on 12 Nov. 1992. According to Jens Reuter, referring to NATO sources, there are approximately 100 000 Serb troops (60 000 belonging to the Serb Army of Bosnia and 35 000 irregulars), 35 000 troops belonging to the Croatian Territorial Defence, and approximately 100 000 Bosnian (mostly Muslim) troops (30 000 belonging to the Bosnian Army formed on 20 May 1992, 30 000 volunteers and 20 000 Bosnian police forces). See Reuter, J., 'Die politische Entwicklung in Bosnien-Hercegowina', *Südosteuropa*, vol. 41, no. 11–12 (1992), pp. 675–76.

[19] Moore, P., Ethnic cleansing in Bosnia: outrage but little action', *RFE/RL Research Report*, vol. 1, no. 23 (28 Aug. 1992), p. 2.

[20] *Bosnia-Herzegovina: Gross Abuses of Basic Human Rights* (Amnesty International: London, Oct. 1992), p. 11.

[21] *War Crimes in Bosnia-Herzegovina*, A Helsinki Watch Report (Human Rights Watch: New York, Washington, Los Angeles, London, Aug. 1992).

There are no reliable figures on casualties in the Bosnian war. According to a US Senate Committee, 35 000 people were killed between March and August. On 1 October medical authorities in Sarajevo estimated the number of deaths between April and September at 14 363, of whom 1447 were children.[22] If Serbian losses and missing persons (53 200) are added, the total number is much higher, and it is often assumed that over 100 000 people have died.

The war in the former Yugoslavia has resulted in the most serious refugee problem in Europe since World War II. According to UN estimates, the total number of refugees and displaced persons in the autumn of 1992 was 2.7 million; in Bosnia and Herzegovina, 1 350 000; Croatia, 640 000; Serbia, 500 000; in UN-controlled areas of Croatia, 70 000; Slovenia, 75 000; Montenegro, 85 000; and Macedonia, 60 000.[23] By the end of 1992 approximately 1.6 million Bosnians had been forced to leave their homes.

## The possibility of major-power intervention

It seems unlikely at the time of writing that the war in Bosnia and Herzegovina will lead to unilateral or joint military intervention by the major powers. In spite of strong pressure from a public opinion outraged by serious human rights violations, the measures adopted have been restricted to the protection of humanitarian aid convoys and the enforcement of sanctions. Moreover, the permanent members of the UN Security Council have prefered to co-ordinate their activities, and in general military intervention has been regarded as too costly. One factor inhibiting military action has been the lack of a clear political objective, or different views on the character of the conflict. It is known, for example, that Russia has argued against measures exclusively directed against the Serbs.

However, intervention cannot be excluded. If the enforcement of 'no-fly' zones were to result in causalties, or if the safety of foreign soldiers were jeopardized, one would expect retaliations, such as air-strikes on military targets. Most probably this would be a common undertaking.

Another factor is the Islamic dimension. The inability of the international community to help the Bosnian Muslims is causing widespread bitterness in the Muslim world, and whether this represents an immediate military threat or not, the long-term implications are serious. If the war continues, it will be increasingly difficult to resist Muslim demands for military action.

There is general concern that the war in Bosnia and Herzegovina might spread and lead to a major conflict with unforeseen consequences. In many respects the present situation in the Balkans is reminiscent of the period before the Balkan Wars (1912–13). The breakup of Yugoslavia has again made salient a number of unresolved questions: those of Macedonia, Albania, Serbia and Bulgaria.

[22] Human Rights Watch (note 20), p. 9.

[23] *United Nations Consolidated Inter-Agency Programme of Action and Appeal for Former Yugoslavia*, 4 Sep. 1992, pp. 5–6.

In this perspective a variety of more or less speculative scenarios has been formulated. According to one, hostilities might unintentionally spread to the Sandzak, with its large Muslim population, and then affect Kosovo and Macedonia. According to another scenario, Serbia would deliberately open a southern front in Kosovo and use any pretext to invade Macedonia, perhaps together with Greece. Should that happen, Bulgaria—which has recognized the Macedonian state but not its people—will have to defend its interests. Similarly, if war erupts in Kosovo, it will most likely involve the Albanians of Macedonia, and in that situation, Albania, although military weak, would have to react.

It is doubtful, however, whether Serbia is capable of waging war on several fronts. The Serbs already have difficulties controlling the occupied areas in Bosnia. Moreover, the population in Serbia proper is tired of war. It should also be pointed that Croatia strengthened its military capacity during 1992. In general, all parties concerned seem to be acting in a cautious manner.

A more probable, but equally dangerous, scenario might be that Serbian right-wing parties and paramilitary groups provoke unrest in Kosovo and Macedonia, which would be impossible to contain. Should this happen, the West would probably have to intervene. In a letter of 25 December, President Bush warned Serbia that the USA would not tolerate 'ethnic cleansing' in Kosovo and would, if necessary, intervene unilaterally. The question is how Russia would react. It is generally assumed that Russia at the moment is unable to undertake complex military operations abroad. However, sympathies with the Serbs are widespread among the Russian population, and the instability of Russia itself makes any prognosis hazardous.

## VI. The post-Soviet armed conflicts[24]

The year 1992 saw both an intensification of armed conflicts on the territory of the former USSR and the outbreak of new wars, testifying to the tragic heritage left behind by the superpower for the newly independent states to cope with as best they can. Attempts to generalize the causes and determinants of these conflicts run the risk of over-simplifying highly complex issues, but certain common denominators can be distinguished. The locations of conflicts described in this section—Azerbaijan, Georgia, Moldova, the North Caucasus region of Russia, and Tajikistan—were all previously constituent parts of the Soviet 'federative' state system. The newly independent states present the same picture of: (*a*) a total lack of experience and structures for independent government; (*b*) a total lack of experience in military decision making, coupled with a lack of national armies, leading to a situation where wars are fought by newly set up irregular armed forces, militia and paramilitary police with local rather than nation-wide loyalties; (*c*) unsolved territorial and

[24] Of the case studies of 5 conflict locations in this section, 2—Azerbaijan and Tajikistan—satisfied the criteria for inclusion in the table of major armed conflicts in 1992 in appendix 3A.

regional disputes;[25] and (*d*) the presence of former Soviet, now Russian, armed forces on their territories. In addition, any attempts at conflict solution are influenced by the dominance of Russia and the future political and strategic interests of Russia. Coupled to this 'Russia factor' is the further complication of the fate of the Russian minority populations. Finally, all the post-Soviet states and sub-state entities share the plight of economic disarray, which further endangers peaceful solutions.

All the above represent negative factors for peace-building. It might be noted on the positive side that all parties involved in the 1992 wars demanded outside aid to solve the conflicts, to prevent war and to help find solutions. Such appeals were constantly made to Russia, to the Commonwealth of Independent States (CIS), to the UN and the Conference on Security and Co-operation in Europe (CSCE), and to individual foreign countries such as Iran, Turkey and the USA. In retrospect, these hopes of outside aid rather add to the tragedy of the wars that none the less took place.

These conflicts are here held to be primarily political conflicts between the defenders of the old system and the forces of a new order accompanying the process of nation-building or national revival.[26] In the words of one analyst, the massive expressions of independence first in the Baltic states but nearly simultaneously from 1988 in Georgia, Nagorno-Karabakh and Moldova 'represented the emergence of civil society and were far more the product of Soviet history than of a primordial ethnicity'.[27]

The armed conflict in Moldova cannot be explained in terms of ethnicity as far as the status of the Trans-Dniester region is concerned, even if the Russian leadership of this region played a dominant role in the rebellion against the Moldovan Government.[28] The civil war in Tajikistan entirely lacked an 'ethnic' factor since the warring sides represented different Tajik political sides. The armed confrontation between the small republics within Russia in North Caucasus—North Ossetia and the Ingush Republic—concerned territory rather than ethnicity, and the same can be said for the war between Georgia and its sub-regions of South Ossetia and Abkhazia. The enclave of Nagorno-Karabakh insists on either being transferred from Azerbaijani to Armenian jurisdiction or becoming an independent state, both unacceptable to Azerbaijan as a sovereign state.

In addition to this political background, the nations at war in 1992 exhibit other common factors at work. Among these is the fear of sub-republic

[25] The borders within the former USSR were arbitrarily drawn and redrawn over the decades and were never intended to be more than administrative delineations. Now, however, they have become international borders subject to many objections and protests, even if the border issue has not led to armed conflicts in the majority of cases.

[26] In the last years of the existence of the USSR, Soviet authorities exclusively used the concepts of 'ethnic conflict', 'nationality conflicts', etc. to explain away all resistance to Soviet central power.

[27] Suny, R., 'State, civil society and ethnic cultural consolidation in the USSR—roots of the national question', ed. G. Lapidus, *From Union to Commonwealth—Nationalism and Separatism in the Soviet Republics* (Cambridge University Press: Cambridge, 1992), chapter 2, p. 35.

[28] See the Interim Report by the personal representative of the Chairman-in-Office of the CSCE Council, Adam Daniel Rotfeld, on the conflict in the Left Bank Dniester Areas, CSCE Communication no. 281 (mimeo), Prague, 16 Sep. 1992.

minority populations for their survival as nations within the newly independent states after the dissolution of the Soviet Union. However, the political factor is present also in these cases where autonomous regions are rebelling against post-communist governments—the Trans-Dniester region and the Gagauz region in Moldova from 1988 objected to Moldova's struggle for independence from the Soviet Union, voted in favour of Gorbachev's proposal for a new Union treaty and supported the August 1991 coup attempt in Moscow. In Georgia, the South Ossetian and Abkhazian autonomies took the same stance. In Tajikistan, the republic government supported the coup attempt and suppressed the opposition, demanding a multi-party system.

The role of religion varies between the conflict areas and is inconclusive—in North Caucasus, the most anti-Russian new republic is Muslim Chechnia; the Ingush are Muslims; while the Ossetians are Orthodox. Georgia and Armenia represent ancient Christianity on the border to the Muslim world, here represented by Azerbaijan. This has brought accusations against Russia of anti-Muslim policies, but there has not been enough concrete action—such as a military intervention, for example—to support such accusations. In Tajikistan, Muslims fought against Muslims in 1992, but the allied opposition rallying together democratic and religious forces was in the end denounced as fundamentalists intending to set up an Islamic state.

## Azerbaijan: Nagorno-Karabakh

The conflict in Nagorno-Karabakh, an Armenian-populated enclave in Azerbaijan was known in the former Soviet Union as 'the touchstone of perestroika',[29] standing out as the living example of the failure of the Soviet leadership to solve an armed conflict within its territory.[30] By early 1992 the situation was often likened to that of Lebanon and thus the prospect of an unending war. In 1992 this war developed into a series of Nagorno-Karabakh victories—officially involving only the Nagorno-Karabakh Armenian 'self-defence forces' against the Azerbaijani national army—both within the enclave where they conquered several previously Azeri-controlled population centres and villages and on Azerbaijani territory.

On 27 February 1992, the Azerbaijani town of Khodzhaly was burnt down by Armenian forces and the massacred civilians were filmed in a documentary shown in Moscow on 4 March.[31] The Azerbaijani side claimed that over 1000[32] were killed in Khodzhaly alone, as against Armenian reports of 30–40 Azeri troops killed in the battle. In February Nagorno-Karabakh units also

[29] The expression was first formulated by Andrey Sakharov, as quoted in Bonner, Y., 'Karabakh is perestroika's touchstone', *Moscow News*, no. 7 (1992), p. 6.

[30] The revival of Armenian statehood, including the Karabakh demand that a historical promise of reunification with Armenia made by the Soviet central power of the 1920s be fulfilled, collided with the revival of Azerbaijani statehood, escalating in 1988 into armed clashes.

[31] *Izvestia*, 4 Mar. 1992, pp. 1–2.

[32] According to the Azerbaijani representative in Moscow who presented a document to the CIS heads of state, the entire town, of 10 000 inhabitants, was destroyed and over 1000 were killed, while 1500 were listed as missing; as reported in *Izvestia*, 3 Mar. 1992, p. 1.

opened heavy grenade fire against the 366th military garrison, then under CIS command in Stepanakert, the Nagorno-Karabakh capital. The Russian military command of the Transcaucasus Military District first ordered the troops to fight back, but later in February Marshal Yevgeniy Shaposhnikov, Commander-in-Chief of the CIS Joint Forces, ordered the troops to leave Nagorno-Karabakh. The former Soviet armed forces located in Nagorno-Karabakh, Armenia and Azerbaijan experienced much the same fate as the 14th Army in Trans-Dniester, Moldova, becoming a target for both of the warring sides while individual units decided to join one or the other side. In February fear that the war over Nagorno-Karabakh would cross international borders mounted as the Turkish 3rd Field Army began conducting large manœuvres close to the border of the Azerbaijani region of Nakhichevan. The prospect of the former Soviet 4th Army in Armenia or the former Soviet 7th Army in Azerbaijan coming into direct contact with Turkish armed forces, that is, the prospect of a NATO country becoming more closely involved, did not seem too far-fetched since Nakhichevan had previously been subjected to military raids by Armenia.

In May 1992, after a heavy battle, Armenian forces managed to take the Azerbaijani city of Lachin, thereby opening a corridor for supplies from Armenia into besieged Nagorno-Karabakh. With the conquest of the Azerbaijani town of Shusha, Armenia came close to taking all the land between Nagorno-Karabakh and Armenia and to an inter-state war with Azerbaijan. Turkey, intervening on the side of Azerbaijan, demanded US intervention.

### *The search for peace*

Simultaneously, efforts to find a solution continued—on 4 February 1992 both Armenia and Azerbaijan accepted an invitation to open peace talks in Moscow which had been extended by Russian Foreign Minister Andrey Kozyrev at the Istanbul meeting on the Black Sea economic zone in December 1991.

International attention, for a long time focused on the Nagorno-Karabakh conflict, was expressed, for example, by US Secretary of State James Baker during his visit to Azerbaijan and Armenia in early February, when he called for a peaceful solution. Azerbaijani President Ayaz Mutalibov, who had earlier resisted calling for UN peace-keeping troops, expressed his support for such a measure. In addition, both Armenia and Azerbaijan asked for CIS help in the conflict. The Nagorno-Karabakh leadership also turned to the CIS, with a request for help to lift the blockade of the enclave in a statement which also claimed that Azerbaijan's blockade as well as its military attacks had intensified after the dissolution of Soviet power and the establishment of the CIS.

On 13 February CSCE representatives arrived in Baku on a fact-finding mission to Nagorno-Karabakh, and on 20 February peace talks between Armenia and Azerbaijan in Moscow took place, although still without the participation of Nagorno-Karabakh representatives—which had rendered pointless all previous attempts at reaching a solution. On this occasion, Armenian President Levon Ter-Petrosyan expressed support for the idea of accepting a CIS peace-keeping force and said that, if this should fail, Nagorno-Karabakh should be put under international protection and UN troops should be engaged. Azerbaijani President Mutalibov expressed a preference for Iran as a mediator and stated that he was prepared for a dialogue also with the new leadership in Nagorno-Karabakh and that any mediator must be tried, mentioning in addition to Iran the efforts undertaken by Turkey and Kazakhstan to propose solutions.[33]

At the meeting in Moscow on 20 February, agreement was reached in principle on a peace process for Nagorno-Karabakh. A joint communiqué listed the measures needed as: (*a*) a de-blockade of Nagorno-Karabakh and of Armenia, (*b*) a cease-fire, (*c*) humanitarian aid, (*d*) the involvement of CSCE and UN peace-keeping forces, (*e*) the commencement of the negotiating process, and (*f*) the establishment of a three-party working group.

In early March President Nursultan Nazarbayev of Kazakhstan proposed that the Council of CIS Heads of State demand a cease-fire in Nagorno-Karabakh, to be followed by a temporary halt in the buildup of national armed forces in all the CIS member nations. Instead, all efforts should be devoted to setting up a CIS peace-keeping force.[34] Later the same month UN Secretary-General Boutros Boutros-Ghali decided to send Cyrus Vance as his personal representative to the area, and the CSCE proposed an international peace conference for Nagorno-Karabakh as a result of the work of the CSCE Committee of Senior Officials, led by Czechoslovak Foreign Minister Jiri Dienstbier. On 15 March a joint NATO–CIS plan to start a peace process was adopted at a meeting of the North Atlantic Cooperation Council (NACC) and a CSCE mediatory mission was set up, led by Dienstbier.

After an unsuccessful attempt to stop the war in Nagorno-Karabakh at a CSCE Preparatory Conference of 11 states in Minsk,[35] a conference under the auspices of the CSCE opened in Rome on 15 June 1992, with the participation of seven nations. For the rest of the year both the peace talks and the war continued. The end-result was that neither side could win this war—Armenia was exhausted economically because of the blockade, and Azerbaijan suffered heavy military losses to Nagorno-Karabakh units. A US Government report held that by December at least 4000 Azeris and 3500 Armenians had been killed in the 11 months since Nagorno-Karabakh had declared independence.[36]

With the re-opening of the CSCE conference in Rome on 25 February 1993, the chances for an end to the conflict for the first time seemed to be at hand, if only because of mutual exhaustion and military deadlock. New Armenian offensives during March and April put an end to the optimism, however.

[33] *Izvestia,* 21 Feb. 1992, p. 1.

[34] *Komsomol'skaya Pravda*, 5 Mar. 1992, p. 2.

[35] The participating states were Armenia, Azerbaijan, Belarus, the Czech and Slovak Republic, France, Germany, Italy, the Russian Federation, Sweden, Turkey and the USA.

[36] *International Herald Tribune,* 21 Dec. 1992, p. 5. Other reports held that at least 2500 persons had been killed in this war by the end of Jan. 1993; *Svenska Dagbladet* (Stockholm, Sweden), 6 Feb. 1993, quoting TT–Reuter, Moscow.

Rather, the war seemed to take a direction towards an outright inter-state war between Armenia and Azerbaijan.

## Georgia: three civil wars

The year 1992 did not bring peace to war-torn Georgia[37] in spite of the hopes connected with the country's new leader, Eduard Shevardnadze, who was called back to Georgia from Moscow in March to take up the position as Chairman of the State Council. The immediate task of the new leadership was to defeat the armed opposition in Mingrelia in western Georgia, which supported the dismissed President Zviad Gamsakhurdia.[38] In addition to this civil war, it also had to bring the armed conflict between the government forces and the autonomous region of South Ossetia to an end.

### *South Ossetia*

On 18 March a cease-fire was achieved with the Gamsakhurdia units in the town of Zugdidi. Throughout the spring of 1992, the peace process in South Ossetia slowly progressed, although against the background of armed clashes in particular in the capital Tskhinvali. With Shevardnadze in power, international support was enlisted and the CSCE placed the Ossetian conflict on its agenda. The presence of former Soviet armed forces in Ossetia created the same problems as in Moldova and Nagorno-Karabakh—they came under attack from both Georgian and Ossetian armed units who raided weapons stores, and they were accused of interference in a local conflict. As in Moldova and Nagorno-Karabakh, it became clear that neither government military forces, the opposition nor Russian troops were at all times under control of the central commands. Georgian armed forces were chiefly made up of the National Guard units under the personal command of Defence Minister Tenguiz Kitovani, and the Mkhedrioni (Horsemen) forces under the personal command of Dzhaba Yoselyani; the task of setting up a national army under government and parliamentary control remained to be achieved by the Shevardnadze leadership.

On 3 June President Yeltsin and the Georgian leadership agreed to send a Russian parachute battalion to South Ossetia to act in the role of a peacekeeping force. On 24 June an agreement was reached between Russia and

[37] Georgia, which together with Armenia represents ancient Christianity in the Caucasus region and independent statehood preceding the existence of the Russian state itself, has a long history of manifestation of its separate culture against central rule from Moscow. Emphasizing Georgian national distinctiveness historically included suppression of the minority rights of the Abkhazians and Ossetians, both belonging to the Muslim Caucasian peoples who were victimized during the establishment of the Soviet system in the 1920s and 1930s.

[38] Zviad Gamsakhurdia, a former dissident imprisoned by the communist leadership of Georgia, was elected President in 1990. He introduced a severe anti-minority policy intended to end the autonomous status of South Ossetia and Abkhazia. In 1989 South Ossetia had declared its intention to re-unite with North Ossetia, thus becoming a republic of the Russian Federation rather than Georgia. This signalled the beginning of the war in Ossetia, which reached its peak in 1991. According to the 1989 USSR census, the population comprised: Georgians, 69%; Armenians, 9%; Russians, 7%; Ossetians, 3%; and Abkhazians, 1%.

Georgia to solve the conflict in accordance with the CSCE principles on the rights of minorities and the territorial integrity of Georgia. A trilateral peace-keeping force was set up and began to function from July—Georgian military units took charge of Georgian villages in the region, and South Ossetian forces were posted in Ossetian villages. A small CSCE team was set up in Tbilisi to monitor the peace process. Like the developments in Moldova, the South Ossetian conflict at this stage was presented as an example of progress for the CSCE principles, although its local representatives admitted that the peace achieved was precarious and the conflict might be renewed. Towards the end of 1992 it was furthermore generally expected that the South Ossetian opposition would renew its attacks again, inspired by the Abkhazian stance.

### *Abkhazia*

In July 1992 the Autonomous Republic of Abkhazia[39] declared independence, which signalled the beginning of the third civil war in Georgia which throughout the autumn devastated the former Soviet sea resorts along the Black Sea coast and eventually turned the entire region into a battleground. The Georgian National Guard immediately entered Abkhazia and took the capital Sukhumi, where a provisional regional government was set up, consisting of the Georgian part of the former Abkhazian Supreme Soviet. Georgian forces were in control of most of the region by the end of August. This first campaign reportedly caused 100 deaths,[40] but Abkhazia mobilized and fought back, aided by voluntary contingents from the Confederation of Mountainous Peoples of the Caucausus, which was interpreted in Georgia as evidence of Russian aid to the secessionists.

Attempts to find a solution were seriously jeopardized by the declaration taken on 25 September by the Russian Supreme Soviet, headed by Ruslan Khasbulatov, who demanded that not only the Confederation of Mountainous Peoples of the Caucausus armed units but also the Georgian forces must leave Abkhazia and hand over their armaments to the Russian troops.[41]

On 11 October Shevardnadze won an overwhelming victory in the general elections which strengthened Georgian central state power and legitimized Shevardnadze's position as the leader of the nation. Emphasizing the political nature of the conflict he said: 'This is not an ethnic war, not even a civil war'.[42] The Georgian leadership also pointed to the strategic interests of Russia in achieving control over an independent Abkhazia—with a large Russian population—connecting the Russian North Caucasus with the Black Sea coast, to which Russian access had diminished after Ukraine's and Georgia's independence. It was also asserted that economic factors played a

[39] Abkhazia had tried in vain since the 1970s to be transferred from Georgian jurisdiction to Russia. The Abkhazian population was forcefully incorporated in the Georgian Soviet Republic in the 1930s and was deprived of its Arabic script. The population according to the 1989 census was: Georgians, 43.9%; Abkhazians, 17.1%; and Russians, 16.4%.

[40] *Komsomol'skaya Pravda,* 29 Sep. 1992, p. 1.

[41] ITAR-TASS, *Krasnaya Zvezda,* 29 Sep. 1992, p. 3.

[42] *Izvestia,* 27 Oct. 1992, p. 3.

role since Abkhazia was a major exporter of citrus fruits and local financial interests would profit from a separation from Georgia—hence the epithet of the 'mandarin war'.

By the end of November the Abkhazian Defence Ministry stated that during the four months of warfare Abkhazia had lost 400 men, 277 of them conscripts. Georgia had lost 1100 and some 3000 were wounded, and after the fall of Gagra into rebel hands the war had developed into a battle for positions where neither side could expect a rapid military victory. Georgia remained in control of only 50 per cent of Abkhazia.[43] Political positions were locked, too—Abkhazia declared that it would continue the war until all Georgian troops had left; Defence Minister Kitovani declared that Abkhazia will never achieve independence or even autonomy; and Shevardnadze demanded that Russian troops leave both Abkhazia and Georgia.

### *The search for peace*

The first attempts at a cease-fire in late August led to no result, but the efforts to find a solution continued—Shevardnadze invited a NATO delegation on 16 September to visit Georgia, and he turned to both the UN and the USA for support, which he received in principle. Both Secretary-General Boutros-Ghali and President Bush expressed their support for Georgia's territorial integrity. At a late September meeting in Sukhumi of representatives of the Georgian Government, Russia and Chechnia's military commander in Abkhazia, and a government representative from Chechnia, it was agreed that the Chechnian fighters be withdrawn from Abkhazia.

The Abkhazian leader Vladislav Ardzinba proposed a federation with Georgia, and talks were held in Gudauta from 9 December. On 15 December Georgia and Abkhazia signed five agreements on ending military action and a withdrawal of all heavy arms and troops from the front by 18 December.

After a short period of relative calm, new fighting broke out in 1993, however, as the Abkhazians launched a new offensive in the direction of Sukhumi during which some 40 troops on both sides were killed.[44] Shevardnadze asked the UN for peace-keeping troops. In February 1993 Russian military forces became ever more involved in the fighting as Georgian troops kept attacking the former Soviet military laboratory in the village of Nizhnye Eshery, which led to Russian air bombings of Sukhumi.

March 1993 brought a dangerous escalation of the war instead of a peace settlement. A settlement thus remains to be achieved in Georgia, and the consequences of failure may, in the words of a Russian political expert, be even more ominous for the Russian Federation than for Georgia.[45]

[43] *Krasnaya Zvezda,* 24 Nov. 1992, p. 1.

[44] *Izvestia,* 5 Jan. 1993, p. 4.

[45] 'The aftermath of Abkhazia may be that a precedent has been set for voluntary military forces that move from territory to territory to help their ethnical or political kin. This could paralyze all interstate relations and structures in the whole post-Soviet sphere.' Pain, E., 'A Russian echo of the Caucasus War', *Izvestia,* 9 Oct. 1992, p. 3.

## Moldova: the Trans-Dniester region

### *Developments in 1992*

At the beginning of 1992, positions were locked between the Moldovan Government in Cisinau (former Kishinev) and the Trans-Dniester leadership in Tiraspol.[46] The situation deteriorated into what was described as threatening to become a 'second Karabakh'—a civil war with little prospect for political solution.

In early March armed fighting broke out in Trans-Dniester, and President Mircea Snegur declared a state of emergency and issued a call for general mobilization. In an address to the people of Moldova, he also accused the former Soviet, now Russian, media of depicting the conflict in Moldova as 'ethnic' when, according to Snegur, it was rather a political conflict between the central government of Moldova and the breakaway region which was a 'militant Communist pseudo-republic trying to attach forces of Russian national patriots'.[47]

The number of armed incidents increased on the left bank of the Dniester River, involving also Cossack armed patrols incorporated in the so-called 'Dniester Republican Guard'. In Cisinau, this led to mass demonstrations which blocked the Foreign Ministry building, demanding the hand-out of arms to all citizens for participation in what then was described as a war of liberation against the Russian-led Cossack threat. Tension rose further when Russian Vice-President Alexander Rutskoy, during a visit to Tiraspol, promised support for the breakaway 'republic' and when officers of the Russian 14th Army in the region demanded recognition of Trans-Dniester. Heavy battles took place in April around the town of Bendery, belonging administratively to Trans-Dniester but situated on the right bank of the river.

On 19 May the 14th Army Command issued an ultimatum that its troops would open fire if the persistent raids on arms stores and centres did not stop. The same night, individual troops left their posts and intervened with tanks in the fighting in the Dubossary region, where they opened fire on Moldovan Government units.

The worst fighting during the civil war in Moldova took place in June in the Bendery region and in the city of Bendery. On 19 June a three-day battle began between the Moldovan paramilitary police and the Trans-Dniesterian Republican Guard, with heavy artillery and the use of tanks on both sides. The

[46] The Soviet republic of Moldavia was created by Stalin after the acquisition of Romanian Bessarabia under the Molotov–Ribbentrop Pact of 1939. In 1940 the former Ukrainian Trans-Dniester region was attached to the new Soviet entity. Ukraine in turn received as compensation the former Romanian Northern Bukovina and Bugeac regions between Bessarabia and the Black Sea. Together with Bessarabia, the enclave populated by the 150 000 Orthodox Turcic Gagauz people also became part of the Soviet Union as an autonomous area within Moldavia. The population of Moldova in 1991 was 4.3 million, of which Moldovans, 64.5% (Romanian origin); Ukrainians, 13.8%; Russians, 13%; Gagauz, 3.5%; and Bulgarians, 2%. The total number of nationalities is no less than 96, according to information made available to the CSCE representative in Sep. 1992. In Trans-Dniester the 1991 population was 780 000, of which Moldovans, 40%; Ukrainians, 28.3%; and Russians, 27%. CSCE Communication no. 281 (mimeo), Prague, 16 Sep. 1992, p. 14.

[47] *Krasnaya Zvezda,* 6 Mar. 1992, p. 1.

government forces were driven out of Bendery on 21 June after heavy casualties were inflicted on both sides. However, a cease-fire was established the same day by the local military commanders and remained in force in principle for the rest of the year, suddenly breaking the spiral of escalating hostilities.

According to data released to the CSCE, the total casualties in the armed conflict were 231 persons killed and 845 wounded on the Moldovan Government side, and 600–700 killed and approximately 3500 wounded on the Trans-Dniestrian side. The civil war led both to a stream of approximately 100 000 refugees from the conflict zone into Moldova west of Dniester and into the Odessa district of Ukraine and to extensive material destruction and damage to the economy.[48]

### *Peace-building*

Russia began to act from April 1992, and a number of foreign ministers' meetings took place, involving also Ukraine and Romania. A plan for a CIS peace-keeping force in the conflict zone was first proposed by Marshal Shaposhnikov.

International attention was abruptly focused on Moldova with the fear of a Russian intervention on the Trans-Dniestrian side, and on 9 April the USA issued a statement in principle on the need to negotiate a cease-fire. On 11 April President Snegur, in a letter to all the CSCE member states, outlined Moldova's demands for a cease-fire, emphasizing as a first condition the withdrawal of the Russian 14th Army from the territory and that Moldova's territory must be kept intact. On 18 May Romanian President Ion Iliescu arrived for a two-day visit to Cisinau, during which the 'Snegur–Iliescu' doctrine was established of two independent Romanian states in Europe, thereby officially closing the option of re-unification.

On 21 July Presidents Yeltsin and Snegur signed an agreement in Moscow where the basic principles for a peaceful solution were set out, most importantly stating the inviolability of present borders and a guarantee of the observance of human rights and the rights of minorities in the country in accordance with the CSCE principles. The role of the CSCE was enhanced by its decision to invite a special representative to the conflict zone. Trans-Dniester and Gagauz were guaranteed a special status, to be negotiated. A trilateral peace-keeping force made up of Moldovan, Trans-Dniestrian and Russian contingents was to be set up to guarantee the cease-fire and return to peaceful conditions. Negotiations between Moldova and Russia on the final withdrawal of the 14th Army began during the year. The CSCE representative conducted talks with all parties involved, including the Trans-Dniester leadership. After his consultations—conducted in Bucharest, Kiev and Moscow with the governments of Romania, Russia and Ukraine[49]—the CSCE Committee of

[48] Data released to the CSCE representative, as reported in CSCE Communication no. 281 (mimeo), Prague, 16 Sep. 1992, p. 7.

[49] See all the reports in CSCE Communication no. 43 (mimeo), Prague, 2 Feb. 1993.

Senior Officials decided to establish a long-duration mission in Moldova to assist in peace-building.[50]

## Russia: the North Caucasus

### *Ingushetia and North Ossetia*

If the Nagorno-Karabakh conflict became the touchstone of *perestroika*, North Caucasus[51] presents the same kind of challenge to democratic Russia. In 1992 war came to this region of the Russian Federation when the formally re-established Ingush Republic demanded parts of its original lands back from North Ossetia—the Prigorodny district, including part of the North Ossetian capital Vladikavkaz. This led to armed fighting at the end of October/early November between the two small nations, resulting in a large number of casualties, material destruction and refugees. After six days of fighting, Russian parachute divisions were flown in and intervened directly to stop the war. On 11 November Chechnia mobilized and threatened to attack the Russian troops dispatched to Ingushetia, claiming they were located on Chechnian territory. The troops were eventually drawn back and an escalation of the war was thereby averted.

Subsequently, Russian troops received the task of peace-keeping in the area, but a political solution proved difficult to achieve. Throughout 1992 Russia was unable to present a comprehensive policy for the area against the background of the power struggle between the President and the Government, on the one side, and the Supreme Soviet, on the other. The weakness of democratic Russia's central power aggravated the Ingush conflict and fomented a dangerous growth of conspiracy theories ostensibly claiming that Russia supported the Orthodox Ossetians against the Sufi Muslim Ingush by deliberate disinformation in the mass media on the armed conflict. Be that as it may, the situation by the end of 1992 was that some 60 000 Ingush—the entire Ingush population in the Prigorodny district—had fled and no political measures were taken to relocate them, which led to Ingush accusations of a deliberate 'ethnic cleansing' in favour of the Ossetians.

[50] The decision on the CSCE mission to Moldova was taken by the CSO Vienna Group; see Journal no. 7 (mimeo), Annex 1, Vienna, 12 Mar. 1993.

[51] The total population is 5 million, of some 60 nationalities, mostly Muslims but also Christians (e.g., Orthodox North Ossetians). These nationalities are the descendants of those who were defeated in the first Caucasus War which began in 1817 and involved the large Abkhazian Army that was defeated in 1864 by the Russian empire which then could incorporate the Caucasus region. Sovietization in the 1920s deliberately ignored ethnic territorial boundaries and instead separated several of the nationalities into 'autonomous areas'. Resistance to Soviet power was never extinguished, which in 1944 led to Stalin's deportation of entire populations (Chechens, Ingush, Kabardians and Balkarians) to Central Asia, ostensibly charged with treason for 'collaboration' with the Axis powers. Rehabilitation by Khrushchev and resettlement into the region did not mean resettlement into original territories—the Russian Government-appointed head of the provisional administration in North Ossetia and Ingushetia, Sergey Shakhray, in 1992 remarked that during the Soviet era, the internal borders in North Caucasus had been changed over 30 times.

The armed conflict on North Ossetian territory reportedly resulted in 340 dead and 753 wounded.[52] The Russian military forces that had been strengthened in North Ossetia by mid-1992 because of the conflict in South Ossetia lost 12 and 32 were wounded; 3500 houses were destroyed, including schools and hospitals; and the cost of accountable losses according to the North Ossetian reporting committee which was set up reached 11 billion roubles. North Ossetia registered over 10 000 refugees on its territory and the Ingush republic some 55 000 from Ossetia.[53]

The political picture of the North Caucasus region is contradictory and presents several parallel and even mutually exclusive currents. The small North Caucasian republics of the Russian Federation remain governed by their former Communist Party élites with the exception of Chechnia. Their presidents, however, all signed the federal agreement with Russia proposed by President Yeltsin in the spring of 1992, again with the exception of Chechnia which declared itself an independent republic in November 1991 and since then has been recognized by four countries (Estonia, Iran, Lithuania and Turkey). The nationalist tide is threatening the power establishments in the other republics—Dagestan, North Ossetia, Kabardino-Balkaria, Karachay-Circassia and Adygey.

In early 1993 Sergey Shakhrai became the Russian administrator of both North Ossetia and Ingushetia in his capacity as the Chairman of the Russian Federation State Committee for Nationalities, with the task of organizing a political solution to the territorial issue. A state of emergency was proclaimed for both these small states until 31 March 1993. His plan included a reorganization of the Russian military presence in the region through the formation of entirely new units under the Ministry of Interior, and he emphasized that North Caucasus remains a Russian sphere of interest, all the republics there being part of the Russian Federation.[54]

## Tajikistan

The large-scale civil war that spread to encompass the entire Republic of Tajikistan[55] during the autumn of 1992 fulfilled all the worst scenarios that were discussed as early as the 1920s, when Central Asia was divided up into new nations. It resulted in an estimated 20 000 deaths—more than during all armed conflicts on Soviet territory in 1987–91—and half a million refugees. The war was brought to a halt because of the near extinction of the nation.[56] The traditional incapacity of communist regimes to cope with organized

[52] *Krasnaya Zvezda,* 3 Feb. 1993, p. 1.

[53] *Krasnaya Zvezda,* 16 Dec. 1992, p. 2.

[54] On the seemingly insoluble territorial issue in a region with far more nationalities than land, Shakhrai said: 'What can we do, no one will take off to the moon from here, all peoples who have lived here for centuries will continue to live here'. *Izvestia,* 10 Jan. 1993.

[55] Tajikistan was set up as a state by Soviet power in 1924 for the 'Tajik' nationality, a people of Iranian origin with a language close to Farsi, and of the Sunni Muslim religion. According to the 1989 census, the population was 5.1 million, of which Tajiks, 59%, and Uzbeks, 24%.

[56] Swedish radio report of 1 Nov. 1992, quoting ITAR-TASS; *Rossiyskye Vesti,* 30 Dec. 1992, p. 1.

opposition was tragically evidenced in Tajikistan where the societal forces released by liberalization during the years of *perestroika* were suppressed and driven underground. By December 1991, opposition to the hard-line traditional government was uniting all the anti-communist forces in the country, including the Democratic Party and Islamic Renaissance Party (IRP), the Islamic National Front, and other groupings representing the districts and regions in the south of the country under different religious leaders.

During the spring of 1992 this massive opposition conducted a political campaign where at times direct confrontation with interior troops was close. In May, after one and a half months of mass demonstrations in the capital, Dushanbe, armed members of the opposition took over the television station and the presidential palace and blocked all roads to the city, effectively signalling the breakdown of governmental power. Armed clashes broke out with forces loyal to President Nabiyev and led to the establishment of an interim government that proved to be too weak to prevent the escalation of tension.[57] The opposition declared their aim as to oust the current pro-communist regime from power, and the Russian population was told that the struggle in Tajikistan was not 'nationalistic'.

The former Soviet, former CIS and finally designated Russian armed forces remaining in Tajikistan found themselves caught between opposing armies in a civil war, as elsewhere. In the south of the country, Russian border troops guarded the 2000-km border with Afghanistan according to a January 1992 agreement between the CIS High Command and Tajikistan. These troops were increasingly accused of handing over or selling weapons to the combatants, and they were increasingly subjected to raids on their weapon stores. The 201st Motorized Rifle Division was requested by what remained of the central government to guard vital objects in the country—the Nuryek power station, the chemical factory in Yavan and the Vakhsh industrial plant.[58] As elsewhere on former Soviet territory, these troops—95 per cent of whom were made up of local conscripts—were ordered to observe a strict neutrality.

In July a peace agreement was reached in principle, where all present agreed that a civil war must be prevented. At the end of August, Marshal Shaposhnikov and representatives from Russia's Defence Ministry and Kyrgyzstan and Kazakhstan met in Dushanbe with the Tajik authorities to discuss collective peace-keeping under CIS command. It was decided to organize a strengthening of the border troops to stop the arms traffic from Afghanistan to the opposition forces in the south. Presidents Yeltsin, Nazarbayev of Kazakhstan, Akayev of Kyrgyzstan and Karimov of Uzbekistan issued a joint declaration stating that it was impossible to allow any escalation of the civil war in Tajikistan, as this development also threatened other states and regional stability in all of Central Asia.[59]

[57] *Izvestia,* 6 May 1992, p. 1.

[58] The task of these troops was to divert attacks that could bring ecological disaster not limited to Tajikistan but also affecting Uzbekistan, Kazakhstan and Turkmenistan, since over 500 tonnes of ammonia and 100 tonnes of formalin were stored at the Vakhsh factory situated only 12 km from Kurgan-Tyube.

[59] *Krasnaya Zvezda,* 29 Aug. 1992, p. 1.

By early September armed fighting in the Kurgan-Tyube area escalated, however, as opposition forces clashed with Nabiyev supporters, reportedly causing some 1500 deaths in the Vakhsh valley and a stream of 90 000 refugees into the Kulyab district.[60]

On 24 October, the Russian Defence Ministry called on the CIS, the UN and other international organizations to help create peace in Tajikistan as war approached the capital Dushanbe, where a large battle took place in late October, with at least 600 deaths on each side.[61]

The military aid extended from Uzbekistan to the Tajik Government forces finally proved decisive and led to a military victory. On 11 November a cease-fire agreement was reached and on 16 November the Tajik Supreme Soviet managed to open its session in Hojend, with the statement that the nation's very existence was at stake, at the same time as the Hojend rulers decided to hold the requested national Mejlis meeting to stop the war. CIS observers from Central Asia and Russia were present. State Counsellor Khudonazarov declared that a compromise solution was needed unless Tajikistan was to run the risk of developing into not only a 'Lebanon' in Central Asia but also a 'Somalia', with its population starving against the background of an unending civil war.

In November the Supreme Soviet formally dismissed President Nabiyev and confirmed the newly elected Prime Minister Abdulajanov and the former rebel leader from southern Kulyab Imamali Rakhmonov as the new Chairman ('president') of the Supreme Soviet. Defence Minister Emran Shah Goldas stated during this parliamentary session that 50 000 had died in the civil war since July.[62] On 25 November some 22 military commanders agreed to sign a traditional peace agreement in Hojend and 27 November was designated as the first day of peace.

The presidency was formally abolished and Tajikistan was declared a parliamentary republic. A coalition government was formed and a district reorganization was carried out, but fighting continued between military units in Kulyab and along the Afghan border well through December. The Red Cross mission in Dushanbe stated that up to 500 000 persons, or 10 per cent of the population, had been displaced by the civil war and were in urgent need of aid.[63] Along the Pianj River constituting the border with Afghanistan, 150 000 refugees were stranded in camps; reports of atrocities from the area included attacks on the refugees, starvation and drowning as they tried to cross the border river. The formally organized CIS peace-keeping troops began their work in December. On 11 December the city of Dushanbe was taken by force from the opposition, and the new authorities could enter the capital. In his summary of the 1992 war, President of the Supreme Soviet Rakhmonov claimed that individual political parties and groupings were responsible for the war and reiterated that the number of deaths during the six months of conflict was

[60] Lugovskaya, A., 'The political crisis in Tajikistan was inevitable', *Izvestia,* 4 Sep. 1992, p. 3.
[61] AFP Dushanbe report, quoted in *Dagens Nyheter* (Stockholm, Sweden), 27 Oct. 1992, p. A12.
[62] AFP Hojend, in *Dagens Nyheter* (Stockholm, Sweden), 25 Nov. 1992, p. A10.
[63] *Financial Times,* 7 Dec. 1992, p. 3.

close to 20 000.[64] Other observers considered that the main responsibility for the war in Tajikistan must be placed on the communist heritage.[65]

In early 1993 the government forces, aided by Russian troops, began cleaning-up operations, but fierce resistance continued in some border areas, in particular along the Afghanistan border. It remains to be seen whether the opposition driven into exile was really extinguished or whether it will stage a come-back with repercussions for all of former Soviet Central Asia.

At the end of the year 1992 it was still too early to discern future trends that would in any way facilitate a prediction of future events on the former Soviet territory. Of the wars in 1992, three may have been stopped or pacified and brought into a peace process—Moldova, North Caucasus and Tajikistan—while two remained unsolved—Georgia and Nagorno-Karabakh. The conflict over the Nagorno-Karabakh territory even seemed to be developing into a full-scale war between Armenia and Azerbaijan.

Ultimately, the fate of Russia's new neighbouring states will to a large extent depend on the fate of Russia itself.

## VII. Cambodia[66]

### Developments in 1978–91

Viet Nam's military intervention in December 1978 toppled the Cambodian Government. The country was renamed the People's Republic of Kampuchea (PRK); in April 1989 the PRK changed its official name to the State of Cambodia (SOC).

The armed opposition against the PRK was not only made up of the Party of Democratic Kampuchea (PDK), that is, the overthrown government, but also of two smaller non-communist groups—the Khmer People's National Liberation Front (KPNLF) led by Son Sann, and the Front uni national pour un Cambodge indépendant, neutre, pacifique et coopératif (FUNCINPEC) led by Prince Sihanouk. These three groups formed the Coalition Government of Democratic Kampuchea (CGDK) in June 1982.

Efforts in the 1980s to bringing about a settlement of the Cambodia conflict were not successful. Despite the impasse on the diplomatic front, Viet Nam withdrew the last of its troops from Cambodia in September 1989. During 1990 the five permanent members (the 'P5') of the UN Security Council reached an common understanding on 28 August and presented a document entitled the Framework for a Comprehensive Political Settlement of the Cambodia Conflict which, among other things, included provisions for the creation of a UN Transitional Authority in Cambodia (UNTAC). The P5

[64] *Rossiyskye Vesti,* 30 Dec. 1992, p. 1.

[65] Bonner, Y., 'The communist heritage led to the civil war in Tajikistan', *Izvestia,* 15 Jan. 1993, p. 8.

[66] The period up to the end of 1991 is based on Amer, R., 'The United Nations' peace plan for Cambodia: from confrontation to consensus', *Interdisciplinary Peace Research*, vol. 3, no. 2 (Oct./Nov. 1991), pp. 3–27. For 1992 the following sources were consulted: *Far Eastern Economic Review* (Hong Kong), *Keesing's Contemporary Archives* (London), *The Economist* (London), *New Strait Times* (Kuala Lumpur), *The Star* (Kuala Lumpur), *The Times* (London), and UN documents.

urged the Cambodian parties to create a Supreme National Council (SNC) to act as the legitimate body and source of authority in Cambodia and represent Cambodia in international organizations.[67] The Cambodian parties formed the SNC at a meeting held in Jakarta on 9–10 September.

By September 1991 the Cambodian parties had reached agreements on the major issues of dispute. Following these agreements the Paris Conference on Cambodia re-convened in October 1991. Two agreements were signed on 23 October: the Agreement on a Comprehensive Political Settlement of the Cambodia Conflict and the Agreement Concerning the Sovereignty, Independence, Territorial Integrity and Inviolability, Neutrality and National Unity of Cambodia.[68] The Paris Agreements gave UNTAC extensive powers in the field of 'civil administration', including direct control of all administrative units acting in the fields of foreign affairs, national defence, finance, public security and information. UNTAC was also to supervise the cease-fire, verify the withdrawal of foreign forces, and supervise the cessation of foreign military assistance and the demobilization and cantonment of the military forces of the Cambodian parties. Finally, UNTAC was to organize and conduct general elections to be held in Cambodia.

## Developments in 1992

The formal decision to set up UNTAC was taken by a unanimous UNSC in Resolution 745 on 28 February 1992. UNTAC was officially established on 15 March.

The military situation in Cambodia deteriorated in January 1992, with fighting erupting in the province of Kompong Thom. The fighting was reported to have involved the PDK and the SOC. More fighting was reported in March.

On 9 April 1992 the UN Secretary-General's Personal Representative in Cambodia, Yasushi Akashi, condemned the PDK for not co-operating fully with the UN. Other UN officials complained about the PDK's refusal to allow UNTAC officials access to territory under its control. Access was granted on 20 April after UNTAC had opened three checkpoints on the border between Cambodia and Viet Nam. The deployment of UNTAC personnel along that border was a prerequisite set up by the PDK. By mid-May all but one of the targeted nine checkpoints had been established along the border between Cambodia and Viet Nam.

Fighting continued in Kompong Thom province during the month of May, with the PDK temporarily cutting off Highway 12 close to the provincial capital. The PDK claimed that it continued to fight because thousands of Vietnamese troops remained in Cambodia. However, UN military observers had found no evidence to support these claims.

[67] UN documents A/45/472 and S/21689, 31 Aug. 1990.
[68] UN documents A/46/608 and S/23177, 30 Oct. 1991.

Phase two of the military provisions of the Paris Agreements was launched on 13 June and involved the demobilization of 70 per cent of the armed forces of the four parties (estimated at some 200 000 troops and 250 000 militiamen), while the remaining 30 per cent were to enter into cantonment. However, the PDK refused to join in this process, stating two main preconditions: UNTAC should thoroughly investigate whether Vietnamese forces had left Cambodia and ensure that any remaining troops be withdrawn; and the SOC administrative structure should be dismantled and its powers transferred to the SNC. The PDK also accused the UN of propping up the SOC and thus not abiding by the Paris Agreements. The PDK was particularly upset about a proposal to grant $110 million to enable the SOC to run the country up to the elections scheduled for May 1993. The refusal by the PDK to join in phase two was criticized by several nations and indirectly criticized by a unanimous UNSC in Resolution 766 on 21 July.

In late April and again in May 1992, attacks on ethnic Vietnamese, causing several deaths, were reported to have taken place in the province of Kompong Chhang. From early July the PDK began using the presence of ethnic Vietnamese in Cambodia in its attacks against the UN, claiming that the UN was neglecting an alleged massive illegal immigration of Vietnamese to Cambodia and that 700 000 Vietnamese had obtained Cambodian identity cards. In July anti-Vietnamese sentiments seemed to be gaining momentum as other political groups voiced concerns similar to those of the PDK. There seems to have been unity among the Cambodian parties, except the SOC, to put pressure on UNTAC to take action and solve their 'Vietnamese problem'. Anti-Vietnamese feelings were further reinforced by the influx of Vietnamese migrants attracted by the economic liberalization in Cambodia and by the arrival of thousands of UNTAC personnel and other foreigners.

In late August the SNC, despite PDK opposition, adopted an electoral law drafted by UNTAC which enfranchises any 18-year-old whose parents or, in the case of those born overseas, grandparents were born in Cambodia. This constitutes a revision of the Paris Agreements which stated that any 18-year-old born in Cambodia or the child of a person born in Cambodia would be eligible to vote.[69] The intention of the law was to disenfranchise new Vietnamese settlers but not ethnic Vietnamese who lived in the country in the pre-1970 period. It is noteworthy that the law disqualifies a number of senior politicians from opposition parties who originate from southern Viet Nam. The PDK opposed the law because it would allow ethnic Vietnamese in the country to vote.

On 13 October the UNSC unanimously adopted Resolution 783, which demanded that the PDK 'fulfil immediately its obligations' under the Paris Agreements. The resolution also included a warning that if the 'present difficulties' were not overcome the UNSC would consider 'what further steps' would be necessary and appropriate to ensure the realization of the 'fundamental' objectives of the Paris Agreements. The UN Secretary-General

[69] UN documents A/46/608, p. 40, and S/23177, p. 40 (note 68).

was requested to report back at the latest on 15 November on the implementation of the resolution.

In his November report, the Secretary-General presented a broad description of the achievements made by UNTAC. It can be noted that about 1 million of an estimated 4.5 million potential voters had been registered, that over 170 000 Cambodians had been repatriated from camps in Thailand and that 55 000 troops had entered cantonment and handed over their weapons. However, the cantonment process had not been completed because of the PDK's refusal to take part in it. The Secretary-General could only confirm that the PDK did not comply with its obligations under the Paris Agreements. Nevertheless, he did not recommend an approach imvolving 'specific measures' to get the PDK to comply. Thus, the Secretary-General urged continued diplomatic efforts in dealing with the PDK and not any sanctions.[70]

The UNSC adopted Resolution 792 on 30 November, with 14 votes in favour and China abstaining in the vote. The resolution took note of the report by the Secretary-General. It also determined that UNTAC should proceed with preparations for general elections to be held in April or May 1993 in all areas of Cambodia to which UNTAC would have access by 31 January 1993. There was also a demand that the PDK 'fulfil immediately' its obligations under the Paris Agreements. The resolution called on those 'concerned' to prevent the supply of petroleum products to areas controlled by 'any' Cambodian party that does not comply with the military provisions of the Paris Agreements. Furthermore, support was expressed for the decision on a moratorium on the export of logs from Cambodia taken by the SNC on 22 September 1991 and UNTAC was requested to take 'appropriate measures' to secure its implementation. Finally, a stern warning was addressed to the PDK that if it would 'obstruct' the implementation of the 'peace plan' the UNSC would 'consider appropriate measure to be implemented' and an example was given: namely, to freeze PDK's assets outside of Cambodia. UNTAC began enforcing the ban on the export of logs from Cambodia on 31 December 1992. The ban was directed at all Cambodian parties and not only at the PDK.

The PDK's response to these moves by the UN was to step up its activities directed at the UN presence in Cambodia. In December incidents were reported involving the PDK taking UNTAC military personnel as hostages and holding them for one or a few days before their release. An interesting move by the PDK was to announce the formation of a new political party, the National Unity of Cambodia Party (NUCP), on 30 November. The NUCP was to compete in the forthcoming general elections, but only if the PDK conditions were met.

### *Remarks on the developments*

The UN intervention in Cambodia was plagued with problems in 1992. The differences between UNTAC and the PDK have led to a situation in which the

[70] UN document S/24800, 15 Nov. 1992.

PDK is in practice not taking part in the peace process except as a party represented in the SNC. UNTAC has met difficulties in ensuring that opposition parties are not subject to harassment in SOC-controlled areas. On a more positive note the repatriation of Cambodians from Thailand is expected to be completed prior to the general elections. However, the registration of voters has not been carried out in PDK-controlled areas.

If the situation of late 1992 prevails and general elections are held in Cambodia in May 1993, the PDK will not take part and the population living in areas under PDK control will not have the opportunity to vote in the elections. An alternative scenario is that the PDK decides to participate in the elections but at such a late date as to make UNTAC registration of voters difficult as well as making supervision of PDK-controlled areas during the elections less stringent compared with other parts of the country. Such a state of affairs would undoubtedly benefit the PDK. Whether or not the PDK takes part in the elections, it could still be included in a post-election government.

Since the demobilization and cantonment process has been halted, some of the participating parties will still have armed forces at their disposal after the elections, as will the PDK. This situation does not augur well for a peaceful evolution in the post-election period. On the other hand, if all parties except the PDK were to demobilize their forces there would be no party left in the country to resist militarily an attempt by the PDK to gain power through the use of force.

## Major-power interest in Cambodia

The impact of major-power involvement in Cambodia since the late 1970s has been fundamentally altered. Prior to 1990 the major powers supported different parties in the Cambodian conflict and thus contributed to deepening the conflict. Since 1990 they have been actively involved in the Cambodian peace process which led to the Paris Agreements. At present, major-power involvement is one of promoting direct intervention by the UN in Cambodia. Any attempt at unilateral military involvement by any of the major powers is inconceivable as long as the Paris Agreements are being implemented and as long as there is a strong UN presence and involvement in Cambodia.

Due to the uncertainty about the future internal political evolution in Cambodia it is extremely difficult to assess how Cambodia's relations with its neighbours will evolve. However, a few observations can be made. First, the treatment of the Vietnamese minority in Cambodia could potentially lead to a conflict with Viet Nam. Second, the borders with Thailand and Viet Nam have to be regulated, or the contested areas could lead to conflicts. Despite these potentially troublesome issues neither Thailand nor Viet Nam is likely to intervene militarily in Cambodia. Thailand will most likely pursue the policy of expanding its economic influence in Cambodia and military confrontation would jeopardize such ambitions. Viet Nam had to endure widespread international isolation following its military intervention in Cambodia in late 1978 and would not risk its improved relations with the member states of the Association of South-East Asian Nations (ASEAN), China, Japan and Western nations such as France and Australia by intervening again in Cambodia. Both Thailand and Viet Nam can be expected to bring conflicts arising with Cambodia which threaten to get out of hand to the attention of the UN in order to settle the differences.

In the event of renewed civil war in Cambodia, after the planned general elections and the withdrawal of UN troops, none of the neighbouring countries is likely to act unless they face a direct security threat, and in such circumstances they would seek a multilateral response sanctioned by the UN. No major power has indicated any willingness to carry out a unilateral military intervention if renewed civil war breaks out after the elections and a UN withdrawal. This is partly because none of the major powers would perceive their vital interests as threatened by civil war in Cambodia. Only a foreign military intervention in Cambodia could create such a response by a major power. It can therefore be argued that as long as Thailand and Viet Nam do not intervene militarily in Cambodia, unilaterally and without international sanction, it is inconceivable that any major power will do so.

## VIII. Southern Africa[71]

For three decades a conflict has been waged in Southern Africa between white minority rule and African liberation. Hopes for peace and stability were raised in the early 1990s, with the legalization of the African National Congress (ANC) and other anti-apartheid organizations, the release from prison of Nelson Mandela, the elimination of apartheid legislation, Namibian independence, and the peace accords in Angola and Mozambique.

Much of this optimism had dissipated by late 1992. The parties to the conflict in South Africa had not agreed on a settlement or managed to curb political violence. The war was renewed in Angola as the União Nacional Para a Independência Total de Angola (National Union for the Total Independence of Angola, UNITA) refused to accept defeat in the UN-supervised elections. While the Mozambican cease-fire held throughout 1992, subsequent steps in the peace process did not get under way.

The focus in the sections below is on South Africa, Angola and Mozambique. In all three conflict locations, the conflict reflects internal incompatibilities over the issue of government. There are also external dimensions: the cold war has played a role and the apartheid government of South Africa has trained and supported UNITA and RENAMO as part of their strategy to maintain white supremacy at home.

[71] Southern Africa is defined as the member states of the Southern African Development Community (SADC) and the Republic of South Africa. The principal sources for information for 1992 are: *Facts and Reports* (Amsterdam), *SouthScan* (London), *SouthScan Monthly Regional Bulletin* (London), *AIM Report* (London), *Mozambique Information Office News Review* (London), *ANGOP Newsletter* (Luanda), *SA Barometer* (Johannesburg), *Africa Confidential* (London) and press reports from international media. The texts of the Bicesse Accord and the Rome Agreement and related protocols have been consulted.

## South Africa

The declared, central objective of the South African Government has been consistent since the National Party (NP) came to power in 1948: to preserve the political, economic and military domination of the white minority in South Africa. The declared goal of the ANC since its establishment in 1912 has been the creation of a non-racial, democratic and unitary South Africa. When the crisis of apartheid deepened in the late 1980s, a new approach towards the ANC was initiated. The de Klerk Government openly recognized that negotiations with representatives of the majority over the future of the country were necessary.

Following bi- and multilateral talks, 31 organizations—not including the Conservative Party (CP)—signed a National Peace Accord in September 1991. The Accord sought to end violence and establish the background for all-party constitutional talks. In December 1991 the CODESA (Convention for a Democratic South Africa) all-party conference on a new constitutional dispensation held its first session. Eighteen political organizations, including the government and the ANC, participated. The Declaration of Intent signed after CODESA-I pledged allegiance to the notion of a democratic, non-racial South Africa and gave CODESA decisions the status of law. The five CODESA working committees resumed work in February 1992 to prepare for a second session. In March 1992 the government called a whites-only referendum on whether de Klerk should continue his reforms 'aimed at a new constitution through negotiation'. After an election campaign during which nearly 300 were killed, white voters turned out in massive numbers (86 per cent) and voted in favour (68.7 per cent). Shortly afterwards the NP aborted tacit agreements made on the political transition in CODESA-I and in subsequent talks, and took a no-concession stance on the important issue of the percentage of votes that would be necessary to pass legislation in a future constituent assembly/interim government. The NP stance weakened those in favour of compromise within the ANC, destroyed the momentum of trust between the parties, and contributed to the deterioration of the relationship between the two main parties.

CODESA-II began on 15 May and ended the following day since no agreement was reached on key constitutional issues, such as the voting procedures for a constitution-making body (with the government demanding a three-fourths majority and the ANC first suggesting two-thirds and then proposing a compromise of 70 per cent). The disagreement over these percentages reflected the government's stated objective of seeking constitutionally enshrined minority rights. The impasse led the ANC to announce that a mass action campaign would follow if the government and the ANC could not come to an agreement. Various bilateral meetings failed to break the deadlock and the campaign started on 16 June. Following a massacre in the Boipatong township in Vaal, the ANC broke off bilateral talks with the government, withdrew from CODESA (whose working groups were still active) and demanded UN Security Council intervention in the negotiation impasse.

In July the UNSC unanimously adopted Resolution 765 condemning the escalation of violence in South Africa, demanding measures from the government to bring the violence to an end, and agreeing to send a special envoy, Cyrus Vance, on a fact-finding mission to South Africa. Vance proposed a modest presence of UN monitors in South Africa. UNSC Resolution 772 established the UN Observer Mission in South Africa (UNOMSA) and by November there were some 50 UN monitors in South Africa, alongside some from the Commonwealth, the EC and different independent observer missions.

The physical presence of international monitors at rallies, marches, etc. had a pacifying impact. However, the limited number present—fewer than 100 by late 1992—could not satisfactorily fulfil their monitoring tasks. For example, in September 28 people were killed and about 200 were wounded in Bisho, Ciskei, as Ciskei homeland troops fired indiscriminately on ANC marchers. In other cases, such acts were replaced by targeted assassinations of political and community leaders. The end result—raising the overall level of violence—remained the same.[72] Demands for more monitors and for a switch from monitoring to peace-keeping/peace-making were made by monitoring organizations, the ANC and other political actors.

In late September a Record of Understanding between the government and the ANC to resume the negotiation process was signed. The outlines of a bilateral agreement on how to restart the transition process emerged early in 1993.

## Angola

On 31 May 1991 the Angolan Government and UNITA signed the Bicesse Peace Accord, stipulating a complete cease-fire, the creation of a unified 50 000-strong defence force, and the holding of multi-party elections in the autumn of 1992. The task of overseeing the political and cease-fire process was given to a Joint Politico-Military Commission (CCPM), including various sub-groups and comprising members of the government, UNITA, and representatives of Portugal, the USA and the USSR (now Russia). A UN verification operation (UNAVEM-2, the Second UN Angola Verification Mission) comprising 440 military and police observers was deployed. Several problems were encountered. Most importantly, the regrouping of 150 000 soldiers to assembly points and their disarmament and demobilization proved complicated. In effect demobilization did not take place and the unified army was not established. The planned elections nevertheless went ahead in September 1992, following a peaceful election campaign.

The governing MPLA (Popular Movement for the Liberation of Angola) won 53.7 per cent of the votes in the parliamentary elections, giving it 129 of 220 seats in the legislature, while UNITA won 34.1 per cent, giving it 70 seats. In the presidential election President José Eduardo dos Santos won

[72] According to the South African Human Rights Commission, approximately 3500 people were killed in political violence in South Africa in 1992. Of these, 1822 were killed in the Pretoria–Witwatersrand–Vaal area and 1430 in Natal; *SouthScan*, vol. 8, no. 3 (22 Jan. 1993), p. 19.

49.57 per cent while UNITA leader Jonas Savimbi received 40.07 per cent. The elections were judged free and fair by UN and official observers. However, UNITA threatened to take up arms if the MPLA were officially declared to have won. Intervention by South African Foreign Minister Pik Botha to arrange a compromise, in which a second round run-off for the presidency would be held and UNITA guaranteed a prominent position in a power-sharing arrangement regardless of the result of the run-off, failed to avert violence. UNITA forces went on the offensive throughout the country. They were driven out of the capital Luanda, where over 1000 people were killed, but made major advances elsewhere in the country. Large-scale warfare recommenced. UNITA was believed to control at least 50 per cent of Angolan territory by the end of the year and by early 1993 UN figures suggested that over 16 000 people had been killed since the election.[73]

## Mozambique

RENAMO (Resistência Nacional Moçambicana, Mozambique National Resistance or MNR) started to fight the Mozambican Government in 1976. While originally created by Rhodesia as a strike force against the Zimbabwean African National Union (ZANU) guerrillas based in Mozambique, the arming, training and control of RENAMO was taken over by the South African Defence Force (SADF) at Zimbabwean independence in 1980. The first direct peace talks were held in Rome in July 1990, mediated by the Italian Government, church representatives and the Archbishop of Beira.[74]

The difficult peace process was speeded up by three external factors as of mid-1992: strong pressure from external observers to the talks, most notably exerted on RENAMO by the United States, Britain and Portugal; the gradual siphoning off of South African support to RENAMO and an ensuing breakdown of RENAMO command structures; and the drought that was then being seriously felt throughout southern Mozambique.

The October 1992 Rome Agreement consists of a general peace agreement and seven protocols covering different issues linked to termination of the war, future elections, international supervision, and so on. The key stipulations are: all armed forces are to report to 49 assembly points were they will be demobilised under UN supervision; all weapons are to be 'disposed of'; all prisoners of war are to be released; all demobilized soldiers will be assisted in terms of re-integration in their home areas; freedom of movement throughout the country is guaranteed for everyone; a UN-led commission for the control and supervision of the accord is to be formed, along with seven sub-commissions for specific issues; and a unitary 30 000-strong military force is to be set up. The major responsibility for carrying out the Agreement rests with—apart

[73] Brittain, V., 'The worst is yet to come in Angola,' *Weekly Mail*, vol. 9, no. 5 (5 Feb. 1993), p. 15, quoting UN figures.

[74] A 'stop–go' negotiation process evolved slowly until the spring of 1992. For details, see *SIPRI Yearbook 1992: World Armaments and Disarmament* (Oxford University Press: Oxford, 1992), chapter 11.

from FRELIMO (Frente de Libertaçao de Moçambique) and RENAMO—the international community, through the UN.

On 16 December, the UN Security Council approved a $331 million operation, called ONUMOZ, covering a troop contingent of at least 4000, possibly up to 8000 UN troops and 354 military observers. An advance team of 20 military observers and a UN Special Representative arrived shortly after the signing of the peace accord and more arrived in early 1993. The UN is to guarantee the implementation of the accord and it chairs the three key commissions set up. It is also to control humanitarian assistance.

The cease-fire has largely held but the implementation of the accord has been delayed. Under the Rome Agreement, all troops were to be confined at the assembly points by mid-November, supervised by at least five UN observers at each assembly point. By late February 1993 there was not even an agreed list of assembly points. Since this is the first step in the implementation process, subsequent steps are also delayed. Further factors contributing to the delays were the refusal of the RENAMO leader Dhlakama to come to Maputo, and RENAMO's refusal to take its position in some of the commissions agreed in the accord. Elections are now being re-scheduled for mid-1994.

### War termination: going from war to peace

Neither Angola nor Mozambique have seen peace since the early 1960s. The basic formula for peace in these two countries, mapped out in a long and complicated process involving both the international community and key regional actors, was essentially the same in both countries. The first step was negotiations aimed at reducing the levels of external involvement in the armed conflicts. The most successful result was the 1988 New York Accord on Angola and Namibia. The second step was the peace negotiations between the warring parties, completed with the 1991 and 1992 accords. The third step was the holding of multi-party elections, which in the case of Mozambique is to take place within a year of the signing of the accord. The fourth step will be a social and economic reconstruction programme to facilitate national reconciliation and peace-building.

In 1992 South African politics was marked by two competing trends: (*a*) problem-solving and bargaining, and (*b*) electioneering, positioning and a struggle for ascendancy. These trends translated into a double agenda for the NP and the government. On the one hand, the government portrayed itself as a responsible political actor, trying to find solutions to problems through dialogue and negotiation, and preparing for a democratic and transparent political competition with the ANC over South Africa's future. On the other hand, elements within the government in collusion with other forces sought either to undermine the ANC as a credible political force prior to a settlement or to derail or block the entire transition process.[75]

[75] Defections from the security apparatus were an almost daily occurrence in South Africa by 1992. The defectors often describe to the media the activities they have taken part in. Many such claims have been substantiated in the reports of the Goldstone Commission, which investigates acts of political

There is—with important exceptions, such as the highly militarized white right wing—general agreement that the conflict over the lack of formalized political participation of the majority in the country has to be resolved. However, beyond that, agreement fragments into competing programmes for structuring participation, with ethnic groups putting forward constitutional proposals for enshrining group rights and group representation, in contrast to other constitutional schemes that would minimize group-based politics.

While it has been suggested that 'the NP's conversion to constitutional democracy is highly situational, and that its present proposals are designed to "non-racially" entrench the existing disparities of property, wealth and power,' there are also positive signs.[76] Towards the end of 1992 there was a visible shift within the NP towards making agreements with the ANC rather than with the Zulu-based Inkhata Freedom Party (IFP) and the CP. Many members of government moved towards the ANC, while many of those opposed were moved to the background or opted out of politics as a result of fatigue and attrition in the NP upper ranks.

The early 1993 the government–ANC bilateral agreement seeks to establish an acceptable platform for a resumption of a multi-party process. The following has tentatively been agreed by the two parties:

1. There will be elections in late 1993 or early 1994 to a single-chamber, sovereign constitution-making body (CMB) which will be tasked with drawing up a new constitution. The CMB will be based on proportional representation with a 5 per cent threshold. Decisions will be taken by a two-thirds majority.
2. An interim government of national unity, based in the CMB and reflecting its composition, will rule the country until new elections on the basis of the new constitution are held. The interim period is expected to be five years.

Many political problems remain unsettled, such as the crucial issues of federalism and the autonomy of local government. Another problem is whether the constituencies of the two main parties, that of the ANC in particular, will accept an élite-level agreement that may be seen to go against their expectations. The IFP and the Pan African Congress (PAC) have also declared that they will not accept a political deal which has the character of a pact that excludes other actors.[77]

The political power struggle aside, South Africa must urgently address three other interrelated sets of problems in order to avoid a deterioration of the situation: (*a*) violence; (*b*) race, ethnicity and identity; and (*c*) economic growth,

violence. See, for example, the Goldstone Commission reports cited in *Business Day* and *The Star,* 17–18 Nov. 1992.

[76] Southall, R., 'The contradictory state! The proposals of the National Party for a new Constitution', *Monitor*, Oct. 1991, pp. 90–92.

[77] In late 1992 IFP leader Buthelezi launched COSAG (Concerned South Africans Group) as a signal that bilateral deals between the Government and the ANC would not resolve the South African problem. Other members of COSAG include the CP and the leaders of the Ciskei and Bophuthatswana homelands. Buthelezi also tabled a federalist constitutional blueprint—drafted by a US academic, Professor Albert Blaustein—for a Natal/KwaZulu secession.

redistribution and transformation. Reducing the high level of political violence is particularly crucial to a successful democratic transition.

Angola and Mozambique face virtually identical problems of war termination:

1. The first problem concerns the behaviour of external parties, historically involved in arming, training and leading UNITA and RENAMO. Complex networks and long-standing friendships and interdependencies are involved. Logistical and other support continue to be provided to both UNITA and RENAMO from South African territory. RENAMO is a major supplier of arms to Inkhata. Furthermore, while the trend is towards reducing such involvement, there is no guarantee that there will not be a resumption of previous patterns of involvement. With so much covert external involvement, it becomes more difficult to go from war to peace even if the main parties act in good faith.

2. They face the problem of avoiding the spill-over effects of violence after a formal cease-fire. Violence has become institutionalized in the countryside and the two societies are highly militarized in socio-psychological terms. Many have concluded that living by the gun is easier than without it. Unless aid can be directed to those most in need—the most marginalized of the rural poor—local warlords may find it easy to recruit people for social banditry or for purposes linked to ethnic or other sectarian or secessionist objectives.

3. The third problem concerns the technicalities of carrying out the terms of the peace agreements. The lessons from Angola underline the risks involved in having too optimistic time schedules for, for example, troop assembly, weapon collection and demobilization. Apart from the practical military problems, the demobilization in Mozambique will leave over 100 000 former soldiers unemployed. The format, tasks and efficiency of the international supervisory component is another problem.

4. Finally, they do not want a peace accord based on military realities but with major political issues unresolved. There is no guarantee that the choice of the electorates will be respected. Post-election developments in Angola illustrate the danger of holding elections before a unified army has been set up and institutionalized. As UNITA reverted to large-scale warfare, the behaviour of the UN and the international community began to be questioned. They refused support to the democratically elected government and instead urged both sides to make concessions to restore peace. This signals to RENAMO in Mozambique, whose election defeat is a forgone conclusion, that resumption of warfare is an option.

Looking beyond the immediate problems related to war termination and reconciliation, there are many other problems and conflict issues of a structural nature having to do with nation building, state formation, the states' lack of popular legitimacy and economic distress. One way of achieving legitimacy is to satisfy peoples' material needs, but the capacity to do so is limited. The prospects for peace and stability in Southern Africa remain grim.

# Appendix 3A. Major armed conflicts, 1992

BIRGER HELDT and ERIK MELANDER*

The following notes and sources apply to the locations listed in table 3A.1:[1]

[a] The stated general incompatible positions. 'Govt' and 'Territory' refer to contested incompatibilities concerning government (type of political system, a change of central government or in its composition) and territory (control of territory [inter-state conflict], secession or autonomy), respectively.

[b] 'Year formed' is the year in which the incompatibility was stated. 'Year joined' is the year in which use of armed force began or recommenced.

[c] The non-governmental warring parties are listed by the name of the parties using armed force. Only those parties which were active during 1992 are listed in this column.

[d] The figures for 'No. of troops in 1992' are for total armed forces (rather than for army forces, as in the *SIPRI Yearbooks 1988–1990*), unless otherwise indicated by a note (*).

[e] The figures for deaths refer to total battle-related deaths during the conflict. *'Mil.'* and *'civ.'* refer, where figures are available, to *military* and *civilian* deaths; where there is no such indication, the figure refers to total military and civilian battle-related deaths in the period or year given. Information which covers a calendar year is by necessity more tentative for the last months of the year. Experience has also shown that the reliability of figures is improved over time; they are therefore revised each year.

[f] The 'change from 1991' is measured as the increase or decrease in battle-related deaths in 1992 compared with battle-related deaths in 1991. Although based on data that cannot be considered totally reliable, the symbols represent the following changes:

| | |
|---|---|
| + + | increase in battle deaths of > 50% |
| + | increase in battle deaths of 10–50% |
| 0 | stable rate of battle deaths (+ or – 10%) |
| – | decrease in battle deaths of > 10% to < 50% |
| – – | decrease in battle deaths of > 50% |
| n.a. | not applicable, since the major armed conflict was not recorded for 1991. |

[1] Note that there is at least one major armed conflict for each location. All the conflicts, that is, not only major armed conflicts, are presented for each location. Compare the tables of conflict locations with major armed conflicts in previous *SIPRI Yearbooks*. Reference to these tables is given in the list of sources. It should be noted that this year's table does not include comments describing events in 1992; significant developments are discussed in chapter 3.

* Erik Melander was responsible for the data for the conflict locations of Azerbaijan, Bangladesh, Cambodia, Croatia, Iran, Tajikistan, and Bosnia and Herzegovina. Björn Holmberg and Christer Ahlström were responsible for Israel/Palestine and India, respectively. Birger Heldt was responsible for the remaining conflict locations. Ylva Nordlander, Carl Åsberg, Susane El-Sarraj and Goshka Wojtasik provided assistance in the data collection.

*Sources*: For additional information on these conflicts, see chapters in previous editions of the *SIPRI Yearbook*: Heldt, B., Wallensteen, P. and Nordquist, K-Å., 'Major armed conflicts in 1991', *SIPRI Yearbook 1992: World Armaments and Disarmament* (Oxford University Press: Oxford, 1992), chapter 11; Lindgren, K., Heldt, B., Nordquist, K-Å. and Wallensteen, P., 'Major armed conflicts in 1990', *SIPRI Yearbook 1991: World Armaments and Disarmament* (Oxford University Press: Oxford, 1991), chapter 10; Lindgren, K., Wilson, G. K., Wallensteen, P. and Nordquist, K.-Å., 'Major armed conflicts in 1989', *SIPRI Yearbook 1990* (OUP: Oxford, 1990), chapter 10; Lindgren, K., Wilson, G. K. and Wallensteen, P., 'Major armed conflicts in 1988', *SIPRI Yearbook 1989* (OUP: Oxford, 1989), chapter 9; Wilson, G. K. and Wallensteen, P., 'Major armed conflicts in 1987', *SIPRI Yearbook 1988* (OUP: Oxford, 1988), chapter 9; and Goose, S., 'Armed conflicts in 1986, and the Iraq–Iran War', *SIPRI Yearbook 1987* (OUP: Oxford, 1987), chapter 8.

The following journals, newspapers and news agencies were consulted: *Africa Confidential* (London); *Africa Events* (London); *Africa News* (Durham); *Africa Research Bulletin* (Oxford); *Africa Reporter* (New York); *AIM Newsletter* (London); *Asian Defence Journal* (Kuala Lumpur); *Conflict International* (Edgware); *Dagens Nyheter* (Stockholm); Dialog Information Services Inc. (Palo Alto); *The Economist* (London); *Facts and Reports* (Amsterdam); *Far Eastern Economic Review* (Hong Kong); *The Guardian* (London); *Horn of Africa Bulletin* (Uppsala); *Jane's Defence Weekly* (Coulsdon, Surrey); *The Independent* (London); *International Herald Tribune* (Paris); *Kayhan International* (Teheran); *Keesing's Contemporary Archives* (Harlow, Essex); *Latin America Weekly Report* (London); *Mexico and Central America Report* (London); *The Middle East* (London); *MIO Mozambique News Review* (London); *Moscow News* (Moscow); *Newsweek* (New York); *New Times* (Moscow); *New York Times* (New York); *RFE/RL (Radio Free Europe/Radio Liberty) Research Report* (Munich); *Pacific Report* (Canberra); *Pacific Research* (Canberra); *S.A. Barometer* (Johannesburg); *Selections from Regional Press* (Institute of Regional Studies: Islamabad); *SouthScan* (London); *Sri Lanka Monitor* (London); *The Statesman* (Calcutta); *Svenska Dagbladet* (Stockholm); *Teheran Times* (Teheran); *The Times* (London).

**Table 3A.1.** Conflict locations with at least one major armed conflict in 1992

| Location | Incompat-ibility[a] | Year formed/ year joined[b] | Warring parties[c] | No. of troops in 1992[d] | Total deaths[e] (incl. 1992) | Deaths in 1992 | Change from 1991[f] |
|---|---|---|---|---|---|---|---|
| **Europe** | | | | | | | |
| Azerbaijan | Territory | 1988/1990 | Govt of Azerbaijan<br>*vs.* Republic of Nagorno-Karabakh,<br>Govt of Armenia | 20 000–25 000<br>1 500–7 000<br>30 000–50 000 | 2 000 | 1 200 | n.a. |
| Bosnia and Herzegovina* | Territory | 1992/1992 | Serbian Rep. of Bosnia-Herzegovina,<br>Govt of Yugoslavia,<br>Serbian irregulars<br>*vs.* Rep. of Bosnia-Herzegovina,<br>Govt of Croatia,<br>Muslim irregulars,<br>Croatian irregulars | 40 000–48 000<br>135 000<br>. .<br>20 000<br>70 000–105 000<br>. .<br>10 000–16 000 | 10 000–20 000 | 10 000–20 000 | n.a. |

* It has not been possible to determine if all reported irregular groups were allied as described. The parties listed include all those involved in the conflict during 1992. In chapter 3 this conflict is classified as an intra-state conflict, as it began in 1992. Bosnia and Herzegovina became an independent state in late Apr. 1992.

| Location | Incompat-ibility | Year formed/ year joined | Warring parties | No. of troops in 1992 | Total deaths (incl. 1992) | Deaths in 1992 | Change from 1991 |
|---|---|---|---|---|---|---|---|
| Croatia | Territory | 1990/1990 | Govt of Croatia<br>*vs.* Serbian irregulars* | 70 000–105 000<br>16 000 | 6 000–10 000** | < 100 | – – |

* Only the irregulars commanded by the Krajina Serbian Republic were reported as active.
** This figure includes the fighting during 1991 where not only the two parties participated (see SIPRI *Yearbook 1992*, chapter 11).

| Location | Incompat-ibility[a] | Year formed/ year joined[b] | Warring parties[c] | No. of troops in 1992[d] | Total deaths[e] (incl. 1992) | Deaths in 1992 | Change from 1991[f] |
|---|---|---|---|---|---|---|---|
| United Kingdom | | | Govt of UK | 293 500 | 3 000* | < 100 | 0 |
| | Territory | 1969/1969 | *vs.* PIRA | 200–400 | | | |
| | | 1975/1992 | *vs.* INLA | . . | | | |

PIRA: Provisional Irish Republican Army.
INLA: Irish National Liberation Army.
* Approximately half of these deaths were related to the conflict between the Govt and PIRA. The other half was almost totally caused by sectarian violence by para-military organizations.

**Middle East**

| Location | Incompat-ibility[a] | Year formed/ year joined[b] | Warring parties[c] | No. of troops in 1992[d] | Total deaths[e] (incl. 1992) | Deaths in 1992 | Change from 1991[f] |
|---|---|---|---|---|---|---|---|
| Iran | | | Govt of Iran | 528 000* | . . | <100 | . . |
| | Govt | 1970/1991 | *vs.* Mujahideen Khalq | 4 500 | | | |

* Including the Revolutionary Guard.

| Location | Incompat-ibility[a] | Year formed/ year joined[b] | Warring parties[c] | No. of troops in 1992[d] | Total deaths[e] (incl. 1992) | Deaths in 1992 | Change from 1991[f] |
|---|---|---|---|---|---|---|---|
| Iraq | | | Govt of Iraq | 350 000–450 000 | . . | 300–500**** | – – |
| | Govt | 1980/1991 | *vs.* SAIRI* | 10 000*** | | | |
| | Territory | 1987/1987 | *vs.* Kurdistan Front** | 60 000–100 000 | | | |

SAIRI: Supreme Assembly for the Islamic Revolution in Iraq.
* This is the largest of reportedly 10 groups active in southern Iraq.
** Consists of 8 groups, of which the largest are the Democratic Party of Kurdistan (DPK) and the Patriotic Union of Kurdistan (PUK).
*** The figure refers to the total strength of all groups.
**** Only the conflict between the Govt and Kurdish groups. A plot, or a coup attempt, by officers in the Republican Guard reportedly took place in June. This case is excluded, as information is scarce.

| | | | | | | | |
|---|---|---|---|---|---|---|---|
| Israel/Palestine | | | Govt of Israel | 175 000 | 1948–: | < 250 | . . |
| | Territory | 1964/1964 | *vs.* PLO* | . . | > 12 300 | | |
| | | . . / . . | *vs.* Non-PLO groups** | . . | | | |

* The Palestine Liberation Organization (PLO) is an umbrella organization; armed action is carried out by member organizations. The main groups represented on the Executive Committee are Al-Fatah, PFLP (Popular Front for the Liberation of Palestine; George Habash), DFLP (Democratic Front for the Liberation of Palestine; Branch of Nayef Hawatmeh), DFLP (Democratic Front for the Liberation of Palestine; Branch of Yassar Abed Rabbo), ALF (Arab Liberation Front), PPSF (Palestine Popular Struggle Front; Samir Ghosheh), PLP (Palestinian Liberation Front; Mahmoud Abul Abbas) and PPP (Palestinian People's Party, formerly PCP Palestinian Communist Party). Apart from these groups, 10 other members of the Executive Committee are not affiliated with any particular political party, ideology or organization.
** Examples of these groups are Hamas and PFLP–GC (Popular Front for the Liberation of Palestine–General Command).

| | | | | | | | |
|---|---|---|---|---|---|---|---|
| Turkey* | | | Govt of Turkey | 560 300 | 6 200** | 3 000 | + + |
| | Govt | 1978/1978 | *vs.* Devrimci Sol | 100 | | | |
| | Territory | 1974/1984 | *vs.* PKK | 8 000–10 000 | | | |

Devrimci Sol: Revolutionary Left.
PKK: or Apocus, Kurdish Worker's Party.
* During 1992 the Turkish People's Liberation Front Party claimed responsibility for an assault. The 'Red Army' was also reportedly responsible for an assault. Since the information in these two cases is scarce, they are excluded from the table.
** Only the conflict between the Govt and the PKK.

## Asia

| | | | | | | | |
|---|---|---|---|---|---|---|---|
| Afghanistan* | | | Govt of Afghanistan | 45 000 | 1978–90 | . . | . . |
| | Govt | 1978/1978 | Mujahideen based in | | 1 000 000 | | |
| | | | Afghanistan, | . . | (estimated direct | | |
| | | | Iran, | 116 000 | and indirect deaths) | | |
| | | | Pakistan | 40 000 | | | |
| | | 1992/1992 | Military/militia factions | . . | | | |

* No general '*vs.*' or Govt can be distinguished for the entire year 1992.

| Location | Incompat-ibility[a] | Year formed/ year joined[b] | Warring parties[c] | No. of troops in 1992[d] | Total deaths[e] (incl. 1992) | Deaths in 1992 | Change from 1991[f] |
|---|---|---|---|---|---|---|---|
| Bangladesh | | | Govt of Bangladesh | 107 000 | 1975–: | < 100 | 0 |
| | Territory | 1971/1982 | *vs.* JSS/SB | 5 000* | > 2 000 | | |

JSS/SB: Parbatya Chattagram Jana Sanghati Samiti (Chittagong Hill Tracts People's Co-ordination Association)/Shanti Bahini (Peace Force).
* Figure for 1991.

| Location | Incompat-ibility[a] | Year formed/ year joined[b] | Warring parties[c] | No. of troops in 1992[d] | Total deaths[e] (incl. 1992) | Deaths in 1992 | Change from 1991[f] |
|---|---|---|---|---|---|---|---|
| Cambodia | | | Govt of Cambodia | 100 000 | >25 300* | < 200 | – – |
| | Govt | 1979/1979 | *vs.* PDK | 27 000–35 000 | | | |

PDK: Party of Democratic Kampuchea (Khmer Rouge).
* For figures for battle-related deaths in this conflict before 1979, see *SIPRI Yearbook 1990*, p. 405, and note *p,* p. 418. Regarding battle-related deaths in 1979–89, that is, not only involving the Govt and PDK, the only figure available is from official Vietnamese sources, indicating that 25 300 Vietnamese soldiers died in Cambodia. An estimated figure for the period 1979–89, based on various sources, is > 50 000, and for 1989 > 1000. The figures for 1990 and 1991 were lower.

| Location | Incompat-ibility[a] | Year formed/ year joined[b] | Warring parties[c] | No. of troops in 1992[d] | Total deaths[e] (incl. 1992) | Deaths in 1992 | Change from 1991[f] |
|---|---|---|---|---|---|---|---|
| India | | | Govt of India | 1 265 000 | >30 000*** | 5 600**** | – |
| | Territory | . . / . . | *vs.* Kashmir insurgents* | . . | | | |
| | Territory | . . /1981 | *vs.* Sihk insurgents** | . . | | | |
| | Territory | . . /1992 | *vs.* ATTF | . . | | | |
| | | . . /1992 | *vs.* BSF | . . | | | |
| | | 1978/ . . | *vs.* NSCN | . . | | | |
| | | . . /1991 | *vs.* PLA | . . | | | |
| | | 1992/1992 | *vs.* ULFA-faction | . . | | | |
| | Territory | . . /1992 | *vs.* JMM | . . | | | |
| | | . . /1992 | *vs.* MCC | . . | | | |
| | Territory | 1967/1967 | *vs.* PWG | . . | | | |

ATTF: All Tripura Tribal Force.
BSF: Bodo Security Force.
NSCN: National Socialist Council of Nagaland.
PLA: People's Liberation Army.
ULFA: United Liberation Front of Assam.
JMM: Jharkand Multi Morcha.
MCC: Maoist Communist Centre.
PWG: People's War Group.
* According to the Govt, approximately 180 groups exist. The figure was reportedly 140 and 60 for 1991 and 1990, respectively.
** There reportedly exist over 24 organizations and splinter groups.
*** The Kashmir and Punjab conflicts only. Of these deaths, approximately 25 000 were killed in the Sikh conflict.
**** Approximately 2000 were killed in Kashmir. 3600 people were killed in Punjab. Excluding all other conflicts between parties listed above.

| | | | | | | | |
|---|---|---|---|---|---|---|---|
| India–Pakistan | | | Govt of India | 1 265 000 | 1971: 11 000 | . . | n.a. |
| | Territory | 1947/1992 | *vs.* Govt of Pakistan | 580 000 | 1982–90: | | |
| | | | | | < 700 (mil.) | | |

| | | | | | | | |
|---|---|---|---|---|---|---|---|
| Indonesia | | | Govt of Indonesia | 283 000 | 15 000– | < 100 | . . |
| | Territory | 1975/1975 | *vs.* Fretilin | 100 | 16 000 (mil.)* | | |
| | Territory | 1963/1984 | *vs.* OPM | . . | | | |

Fretilin: Frente Revolucionára Timorense de Libertação e Independência (Revolutionary Front for an Independent East Timor).
OPM: Organisasi Papua Merdeka (Free Papua Movement).
* Only the conflict between the Govt and Fretilin.

| | | | | | | | |
|---|---|---|---|---|---|---|---|
| Laos | | | Govt of Laos | 37 000 | . . | . . | n.a. |
| | Govt | 1975/1975 | *vs.* Opposition group* | 2 000 | | | |
| | | . . /1992 | *vs.* Free Democratic Lao National Salvation Force | . . | | | |

* Probably ULNLF: United Lao National Liberation Front.

| Location | Incompat- ibility[a] | Year formed/ year joined[b] | Warring parties[c] | No. of troops in 1992[d] | Total deaths[e] (incl. 1992) | Deaths in 1992 | Change from 1991[f] |
|---|---|---|---|---|---|---|---|
| Myanmar | | | Govt of Myanmar | 286 000 | 1948–49: 3 000 | < 2 000 | + |
| | Govt | 1988/1991 | vs. ABSDF | 2 000 | 1950: 5 000 | | |
| | Territory | ../1991 | vs. Arakan insurgents* | 1 000–6 000 | 1981–84: 400–600 yearly | | |
| | Territory | 1961/.. | vs. KIO/KIA | 8 000 | 1985–87: 1 000 yearly | | |
| | Territory | 1950s/1992 | vs. KNPP | 800–1 000 | 1988: 500–3 000 | | |
| | Territory | 1948/1948 | vs. KNU | 4 000–6 000 | | | |
| | Territory | 1979/1992 | vs. Naga insurgents** | .. | | | |

ABSDF: All-Burma Students' Democratic Front.
KIO/KIA: Kachin Independence Organization/Army.
KNPP: Karenni National Progressive Party.
KNU/KNLA: Karen National Union/Karen National Liberation Army.
\* At least 6 groups are reported to exist. The largest of these is the Rohingya Solidarity Organization (RSO).
\*\* Probably the National Socialist Council of Nagaland (NSCN).

| The Philippines* | Govt | 1968/1986 | Govt of the Philippines vs. NPA | 106 500 13 500–15 000 | 21 000– 25 000** | .. | .. |

NPA: New People's Army.
\* A number of violent incidents took place on Mindanao. However, none of them seems to have been between secessionist groups and the Govt. They were related to the conflict concerning the territorial status of Mindanao.
\*\* Official military sources claim that 6500 civilians were killed during 1985–91.

| Sri Lanka | Territory | 1976/1983 | Govt of Sri Lanka vs. LTTE | 125 000 7 000 | 24 000* | 4 000** | + |

LTTE: Liberation Tigers of Tamil Eelam.
\* Includes fighting involving all groups since 1983, that is, not only the fighting between the Govt and LTTE.
\*\* Excluding civilian deaths.

| | | | | | | | |
|---|---|---|---|---|---|---|---|
| Tajikistan | | | Govt of Tajikistan | . . | 4 000– 30 000 | 4 000– 30 000 | n.a. |
| | Govt | 1991/1992 | *vs.* Popular Democratic Army | 16 000 | | | |

**Africa**

| | | | | | | | |
|---|---|---|---|---|---|---|---|
| Angola | | | Govt of Angola | 30 000–50 000 | > 32 000 (mil.)* | 3 000* | – – |
| | Govt | 1975/1992 | *vs.* UNITA | 30 000–50 000 | > 70 000 (civ.)* | | |
| | Territory | 1975/1975 | *vs.* FLEC | . . | | | |

UNITA: União Nacional Para a Independência Total de Angola (National Union for the Total Independence of Angola).
FLEC: Frente da Libertação do Enclave de Cabinda (Front for the Liberation of the Enclave of Cabinda).
* Only the conflict between the Govt and UNITA.

| | | | | | | | |
|---|---|---|---|---|---|---|---|
| Chad | | | Govt of Chad | 50 000 | . . | 300–600 | + |
| | Govt | . . /1992 | *vs.* CSNPD | 5 000 | | | |
| | | 1992/1992 | *vs.* Forces of Koti | 100 | | | |
| | | . . /1992 | *vs.* FNT | . . | | | |
| | | 1989/1989 | *vs.* MDD (-FANT) | 500–1 000 | | | |

MDD: Mouvement pour la Démocratie et le Développement (Movement for Democracy and Development). A coalition of former President Habré loyalists and remnants of
CSNPD: Committee of National Revival for Peace and Democracy.
FANT: Forces Armées Nationales du Tchad (Chad National Armed Forces).
FAO: Forces Armées Occidentales (Western Armed Forces).
FNT: Front National Chadienne (Chadian National Front).

| | | | | | | | |
|---|---|---|---|---|---|---|---|
| Liberia | | | Govt of Liberia | 250 | 20 000* | 4 000–5 000 | + + |
| | | | ECOMOG | 8 000–12 000 | | | |
| | Govt | 1989/1989 | *vs.* NPFL | 10 000 | | | |

| Location | Incompat-ibility[a] | Year formed/ year joined[b] | Warring parties[c] | No. of troops in 1992[d] | Total deaths[e] (incl. 1992) | Deaths in 1992 | Change from 1991[f] |
|---|---|---|---|---|---|---|---|
| ECOMOG: | Economic Community of West African States Monitoring Group. | | | | | | |
| NPFL: | National Patriotic Forces of Liberia. | | | | | | |
| * | Note that this figure includes the fighting in 1990–91 (incurring 15 000 deaths) where other than only the two parties participated. | | | | | | |
| Mozambique | | | Govt of Mozambique, | 62 000 | 10 000– | . . | . . |
| | | | Govt of Zimbabwe | 5 000 | 12 000 (mil.)* | | |
| | Govt | 1975/1976 | *vs.* RENAMO | 21 000 | 110 000 (civ.)* | | |
| RENAMO: | Resistência Nacional Moçambicana: Mozambican National Resistance, MNR. | | | | | | |
| * | Figures for 1991. | | | | | | |
| Rwanda | | | Govt of Rwanda | 40 000 | 5 000 | . . | . . |
| | Govt | 1987/1990 | *vs.* FPR | 3 000–5 000 | | | |
| FPR: | Front Patriotique Rwandais (Rwandan Patriotic Front). | | | | | | |
| Somalia | | | Govt of Somalia | 2 000–6 000 | . . | 3 000–4 000 | – – |
| | Govt | 1991/1991 | *vs.* USC (Aydeed) faction | 2 000–6 000 | | | |
| USC: | United Somali Congress. | | | | | | |
| South Africa | | | Govt of South Africa | 72 400 | 1984–92: | 3 500* | + |
| | Govt | 1948/1961 | *vs.* ANC | 6 000–10 000 | 14 500* | | |
| | | 1963/1992 | *vs.* PAC | 350 | | | |
| | | 1977/1992 | *vs.* AZANLA | . . | | | |

ANC: African National Congress.
PAC: Pan Africanist Congress.
AZANLA: Azanian National Liberation Army.
* Including victims of 'political violence', that is, not only between the Govt and the ANC. Excluding the conflict with PAC and AZANLA.

| | | | | | | | |
|---|---|---|---|---|---|---|---|
| Sudan | | | Govt of Sudan | 82 500 | 37 000– | . . | . . |
| | Territory | 1980/1983 | *vs.* SPLA (Torit faction) | . .* | 40 000 (mil.)** | | |
| | | 1991/1991 | *vs.* SPLA (Nazir faction) | . .* | | | |

SPLA: Sudanese People's Liberation Army.
* The combined forces of the factions are estimated at 40 000–55 000.
** Figure for 1991.

## Central and South America

| | | | | | | | |
|---|---|---|---|---|---|---|---|
| Colombia | | | Govt of Colombia | 139 000 | 1980–92: | 1 600 (mil.)* | . . |
| | Govt | 1949/1978 | *vs.* FARC | 5 000–7 000 | > 11 000 | | |
| | | 1965/1978 | *vs.* ELN | 1 500–4 000 | | | |

FARC: Fuerzas Armados Revolucionarias Colombianas (Revolutionary Armed Forces of Colombia).
ELN: Ejército de Liberación Nacional (National Liberation Army).
Some activity from an organization composed of dissidents from EPL (Ejército Popular de Liberación: Popular Liberation Army), Quintin Lame, M19 (Movimiento 19 de Abril) and the Corriente de Renovación Socialista faction of ELN was reported. Since it is unclear whether armed force was used and the type of incompatibility is unknown, this organization is excluded from the table.
* The figure is 3600 (mil. and civ.) if activities of death squads and paramilitary groups are included. This figure was 3700 in 1991.

| | | | | | | | |
|---|---|---|---|---|---|---|---|
| Guatemala | | | Govt of Guatemala | 44 600 | < 2 800 (mil.) | 680* | . . |
| | Govt | 1967/1968 | *vs.* URNG | 1 000 | < 43 500 (civ.) | | |

URNG: Unidad Revolucionaria Nacional Guatemalteca (Guatemalan National Revolutionary Unity). Consists of EGP (Ejército Guerrilleros de los Pobres: Guerrilla Army of the Poor); PGT (Partido Guatemalteco del Trabajo: Guatemalan Worker's Party); FAR (Fuerzas Armadas Rebeldes: Rebel Armed Forces) and ORPA (Organización del Pueblo en Armas: Organization of Armed People).
* Only for the first 6 months of 1992.

| Location | Incompat- ibility[a] | Year formed/ year joined[b] | Warring parties[c] | No. of troops in 1992[d] | Total deaths[e] (incl. 1992) | Deaths in 1992 | Change from 1991[f] |
|---|---|---|---|---|---|---|---|
| Peru | | | Govt of Peru | 112 000 | > 27 000* | 3 100 | 0 |
| | Govt | 1980/1981 | *vs.* Sendero Luminoso | 5 000 | | | |
| | | 1984/1986 | *vs.* MRTA | 200–500 | | | |

Sendero Luminoso: Shining Path.
MRTA: Movimiento Revolucionario Tupac Amaru (Tupac Amaru Revolutionary Movement).
* Only the conflict between the Govt and Sendero Luminoso.

# 4. Post-Soviet conflict heritage and risks

VLADIMIR BARANOVSKY

## I. Introduction

The collapse of the Soviet Union in December 1991 could have resulted in a large-scale armed conflict between the newly independent states. Fortunately, this 'worst case scenario' was avoided in 1992, in part (paradoxically) as a result of the failed *coup d'état* of August 1991. The coup, which was rationalized as an attempt to preserve the union, had the effect of discrediting union supporters. The breakup of the USSR into newly independent states followed peacefully, and a scenario such as that being played out in the former Yugoslavia was avoided.

The breakup itself was a rapid process which was hardly prepared from a political or an economic standpoint. There was no legal basis for the process, and the prerequisites for a 'divorce', as in the case of the former Czechoslovakia, were not met. The ill- or non-prepared separation, in fact, was probably the most important source of conflict in the former Soviet Union in 1992. Numerous conflicts developed, many of them with serious implications for international stability.

The global trends in the economic, political, ideological and cultural evolution of the country which facilitated the breakup of the Soviet Union provide a common background for these conflicts. Nevertheless, each conflict has its own sources and specific logic which can hardly be reduced to a certain common denominator. Complicating the analysis even further, many conflicts are closely interrelated and intensify each other in different ways. The analytic scheme suggested in this volume, which differentiates between internal, interstate and state-formation conflicts,[1] is helpful for classifying conflicts in general, but in the former Soviet Union there is such a striking overlap of all three categories that none of the conflicts could be classified as strictly domestic or as strictly external in nature.

Furthermore, it is not always possible to draw a clear line separating actual conflicts from potential ones. There is certainly a difference between large-scale armed violence and low-profile political disputes with the potential to escalate to armed violence, but the distinction could at any time become irrelevant in light of the unstable domestic situation in the new states and the unpredictable turns of their foreign policies.

This chapter examines three major groups of conflict-generating trends at work in the former Soviet Union. First, there is the extremely difficult transition from a totalitarian regime and empire to a new type of society wherein

[1] See chapter 3 in this volume.

newly independent states develop new economic and political structures and patterns of relations. Second, there is the phenomenon of ethno-nationalism which is perhaps one of the most destabilizing conflict-generating factors. Third, there is the military legacy of the former superpower, which provides both the means to make conflicts more explosive and the capacity to contain them.

This chapter concentrates on those factors which can be clearly linked to the development of conflicts. It does not pretend to analyse the post-Soviet situation *per se*. Such broad important issues as overall economic performance, the transition towards market economies and the international implications of the post-Soviet conflict heritage deserve much more attention and analysis than is possible in this chapter.

## II. Challenges of the transitional period

The post-Soviet societies in the newly independent states are in a state of profound transition. Previous values, beliefs, structures, institutions, links, economic mechanisms and behavioural patterns have either been destroyed or discredited, while new ones are either non-existent or just beginning to emerge. The social fabric of the former republics is extremely weak and receptive to conflict.

The transitional period presents numerous challenges to the new states. First, there is the uncertainty concerning the role and status of the Commonwealth of Independent States (CIS), given the fact that the patterns of relations among the states are just now taking shape. Second, the process of state formation in the new states is still far from ensuring domestic stability. Third, there is the complex issue of frontier and territorial claims as the states develop and express their independence and sovereignty. All of these factors have the potential to generate conflicts while at the same time undermining the possibilities for solving them.

### Uncertainties regarding the CIS

The CIS proved to be a helpful mechanism for containing the effects of the breakup of the Soviet Union in 1992, but it failed to emerge as a viable structure to replace it. The road 'from Minsk to Minsk' (that is, from the first meeting to establish the CIS to the summit meeting at which the CIS Charter was adopted) showed that the requirements of nation-building were driving the states apart rather than bringing them together.[2]

It is quite clear that the future place, role and status of the newly independent states in the international arena will be substantially affected by political and other assets they can acquire now, when the old international

[2] Belarus, Russia and Ukraine proclaimed the establishment of the CIS on 8 Dec. 1991 in Minsk. At the eighth summit meeting of 10 new independent states on 22 Jan. 1993, the draft of the CIS Charter was adopted by seven participants.

system has collapsed and a new one is in the making. This provides a rationale for more activism and less readiness to compromise, and created a rather unfavourable background for the CIS during its first full year of operation.

Efforts to make the CIS effective have been blocked by different approaches towards its role and status. Azerbaijan, Moldova, Turkmenistan and Ukraine, fearing that the CIS would operate as a type of superstructure which would undermine their independence, consistently attempted to downgrade the CIS into the loosest of alliances.[3] The other Central Asian republics and especially Kazakhstan called for strengthening the CIS into a kind of confederation; Belarus combined a spirit of co-operation with reluctance to damage its freshly proclaimed neutrality; Russia was hesitant in facing a well-known European integration-type dilemma, that of either 'widening or deepening'.

In a sense, the CIS itself has become a matter of conflict between its participants, its main (and unavoidable) weakness being its highly unbalanced composition, with Russia practically doomed to play a hegemonic role even if only for economic reasons and whatever the political choice of Moscow might be. Moreover, there were serious (and not unjustified) suspicions on the part of some CIS members that the organization was conceived by Russia as an instrument of a 'special relationship' with the other post-Soviet states.[4]

Nevertheless, the CIS has survived as a framework for interaction between the new states which will most probably operate on the basis of the 'variable geometry' pattern. This will neither eliminate nor marginalize potential conflicts between the participants, but will hopefully provide the parties with a certain spirit of co-operation in addressing issues of disagreement. Three aspects of the emerging CIS were of special relevance to post-Soviet conflicts in 1992.

1. The CIS has contributed to make the disintegration of the former Soviet armed forces less dramatic than it might have been. Early on, the military establishment argued that the armed forces should be largely preserved as a CIS entity. This maximalist position proved to be extremely counter-productive. More effective was the focus on the practical issues of organizing the functioning of the military infrastructure under circumstances of political uncertainty. During 1992, over 100 agreements on military issues were concluded, and although the CIS proved to be much more productive in producing documents than in implementing their provisions, the very fact of this activity was undoubtedly helpful, both in preventing chaos and in neutralizing the military establishment's growing frustration.

[3] The CIS agreement has never been ratified by Azerbaijan and Moldova. In fact, Azerbaijan has left the CIS to be in the same position as Georgia which had declined to participate from the very beginning. However, both Azerbaijan and Moldova (contrary to the Baltic states) do not exclude a certain level of partnership with the CIS.

[4] Significantly, in Russian political parlance, the term 'near abroad' (that is, not actually abroad) was invented to refer to the other post-Soviet states. Some prominent Russian experts include them in the 'first circle' of Moscow's security policy interests. See Goodby, J. E. and Morel, B. (eds), SIPRI, *The Limited Partnership: Building a Russian–US Security Community* (Oxford University Press: Oxford, 1993), p. 76.

2. On 15 May 1992, the 'core area' participants to the CIS (Armenia, Kazakhstan, Kyrgyzstan, Russia, Tajikistan and Uzbekistan) signed the Tashkent Treaty on Collective Security which provides for mutual military assistance in the event of outside aggression. For the moment it is hardly possible to speak about a viable defensive alliance in the traditional sense. Armenia, for example, had hoped that the Treaty would ensure assistance in the war against Azerbaijan (which was the main rationale for Yerevan to participate), but this proved to be illusory. Still, the legal structure exists and in principle could be brought to bear in the future, although any speculation concerning how this could affect conflict development is premature.

3. The CIS has addressed the issue of post-Soviet conflicts and has tried to develop some approaches towards peace-making. A mechanism has been agreed which provides for military observers and collective CIS peace-keeping forces.[5] The first multilateral peace-keeping forces on former Soviet territory began operation on 14 July 1992 in the Tskhinvali area in South Ossetia.[6] Two weeks later, trilateral peace-keeping forces began operations in the Trans-Dniester region.[7] Collective peace-making efforts were also decided upon with respect to the civil war in Tajikistan.

Whether the CIS conflict-management system will develop into a viable structure or whether the participants will prefer to act bilaterally on an *ad hoc* basis remains unclear. The former option has the advantage of providing procedures for conflict-management and facilitating interaction with existing international bodies such as the United Nations (UN) and the Conference on Security and Co-operation in Europe (CSCE). This option, however, can only be effective if the CIS itself becomes operational and acquires a certain legitimacy. Furthermore, conflicts outside the CIS area are hardly eligible for being dealt with by such a mechanism. However, it is ironic that almost all of the above-mentioned conflicts addressed by the CIS have been outside its area.

The latter option seems to be more practical in terms of rapid reaction to problems in possible 'hot spots', but it is conducive to political complications. Questions will inevitably arise concerning the role of the forces operating under national control, the influence of the involved parties and so on,[8] issues which could be avoided if the multilateral option is pursued.

[5] Decisions were taken at the CIS summit meeting in Kiev (20 Mar. 1992) and at a meeting of foreign and defence ministers in Tashkent (16 July 1992).

[6] This peace-keeping force is comprised of 1500 Georgian, Ossetian and Russian troops.

[7] The forces were comprised of 3800 Russians, 1200 Moldovans and 1200 members of the Trans-Dniester national guard.

[8] When negotiating the issue of Abkhazia in Sep. 1992, Georgia agreed that Russian armed forces could operate as peace-keepers—but later questioned their impartiality. See *Krasnaya Zvezda*, 12 Dec. 1992, p. 1. While discussing the issue of peace-keeping operations in Tajikistan, the countries involved had serious problems in reaching agreement on the national composition of the forces to be used. See *Le Monde*, 10 Oct. 1992, p. 3; *Krasnaya Zvezda*, 9 Dec. 1992, p. 3.

## A political system in crisis

In terms of domestic conflict-generating phenomena, the most serious problem seems to be the lack of functioning political structures. Although the bureaucratically over-centralized Soviet state was increasingly deteriorating prior to the breakup, there existed developed institutional mechanisms for exercising central control and to prevent or minimize potential conflicts. After the top level of the political mechanism disappeared, however, lower elements became either disorganized or dysfunctional. The elimination of the 'centre' created a strong and possibly decisive incentive for the breakdown and atomization of the entire political structure, which in itself became a powerful conflict-generating phenomenon.

Such developments may have been inevitable, due to the fact that a civil society based on democratic values was non-existent, whereas the political system was based on the predominance of the highly centralized Communist Party of the Soviet Union (CPSU) *nomenklatura*. The latter, even if remaining in many cases powerful (unlike in other European post-socialist states) after the elimination of the CPSU, cannot serve as a consolidating element. What was once the corner-stone of the totalitarian state and society no longer exists.

The elements of a new political structure based on a multi-party system, separation of powers, respect for fundamental freedoms, private ownership, market economy and the rule of law is emerging very slowly. The absence of appropriate legislation and decreasing political participation is compounded by the lack of mature democratic traditions, the inherited patterns of political behaviour and the populist inclinations of new leaders which fertilize the ground for conflicts. The lack of legitimacy and efficiency tends to generate conflict developments at two levels.

1. Almost nowhere in the former Soviet Union (with the possible exception of the Baltic states—Estonia, Latvia and Lithuania—and, to some extent, Armenia) do competing political forces operate in an established framework approaching a 'normal' political process. Instead, consolidation of power positions is sought at whatever price, and open conflict with opponents is not uncommon. Taken to an extreme, such developments could result in either a *coup d'état* or an outright civil war and armed conflict, such as that which occurred in Georgia and Tajikistan in 1992. Russia was also approaching a similar situation by year's end, when an open collision between the Russian President and the Parliament threatened to destroy the existing minimal political stability, with dangerous and unpredictable consequences.

The potential for conflict is aggravated by the fact that key elements of the political system often represent vestiges of the past that must be eliminated before a transition to a civil order based on democratic values can succeed.[9] Greatly complicating this transition, however, is the fact that it must occur

[9] For example, the composition and norms of the Russian and Ukrainian Parliaments are a reflection of the CPSU 'leading role' period; in Turkmenistan, the authorities exercise total control over the media; and in Uzbekistan, the opposition is persecuted.

contemporaneously with the collapse of the previous legal and constitutional order, a process which in itself has the potential to provoke serious conflicts. To avoid or at least to minimize the effects of this process while breaking this vicious circle is possible only with exceptional effort by the society at large. Indeed, massive political participation appears to be a *sine qua non*.

The Baltic states were relatively successful in making this transition, not least because a certain legitimacy was implicitly provided by the restoration of the pre-war statehood. This factor is absent in the other former Soviet republics. Many of them have never existed as independent states and continue the constitutional order and the political culture of the previous regime in flagrant contradiction with the requirements of reforms. In Russia, an opportunity for a large-scale and relatively painless political transformation was missed during the first months after the failed *coup d'état*. This provided the basis for an escalating political confrontation which culminated in the spring of 1993 in the form of a constitutional crisis. In Central Asia the lack of serious political changes results in an extremely uncertain situation and runs the risk of a major explosion of unpredictable scope and configuration, such as with the war in Tajikistan. In fact, the residual conflict potential within existing political structures which were inherited from the past could be released at any moment.

2. While old institutions and structures are being dismantled (and new ones are taking shape), political administration becomes especially inefficient and difficult. The bureaucracy is often demoralized and frustrated, and there is resentment at the ascendance of numerous inexperienced 'newcomers' to top positions. Corruption is also a problem.[10] The inability to implement decisions[11] can degenerate very rapidly into lawlessness, further undermining legitimacy and creating a favourable environment for criminality.[12] One of the consequences of this trend (and probably the most destabilizing) is the increasing illegal arms trade and the emergence of numerous armed criminal groupings.[13] The role played by these groups in causing and exacerbating domestic conflicts can be serious indeed, as evidenced in Georgia and Tajikistan, as well as in some areas of Northern Caucasus in Russia.

The confusion and disorder are only increased by the activities of some new bodies which either challenge or disregard state institutions. It is true that some of these organizations emerged 'from below' as forms of legitimate claims to correct the injustices of the Soviet period and to protest against the predominance of the old *nomenklatura* system—as in the case of the Cossack

[10] In Russia, 3331 cases of bribes were officially registered during 1992. This is estimated to be only about 5 to 6% of the total number. See *Literaturnaya Gazeta*, 10 Mar. 1993, p. 10.

[11] According to unofficial estimates, in Russia only 7% of presidential decrees are being implemented. See *Literaturnaya Gazeta*, 27 Jan. 1993, p. 10.

[12] In Russia, 2.8 million crimes were officially registered in 1992, which is the highest level ever for the country. See *Komsomol'skaya Pravda*, 24 Mar. 1993, p. 4.

[13] According to the Russian Interior Ministry, about 3000 criminal organizations with tens of thousand of members operate in Russia (*Financial Times*, 30 Dec. 1992, p. 2). Some experts estimate the number of illegal armed groupings in the former Soviet Union to be as many as 500, with hundreds of thousands of weapons at their disposal. See *Krasnaya Zvezda*, 13 Aug. 1992, p. 2.

movement. However, the attempts to integrate these movements into a new political mechanism have not always been successful: in some areas, the Cossack organizations pretend to be an alternative to local legislative, executive and judicial bodies; in Tatarstan, the *melli-mejlis* (an organization pretending to represent all Tatars) denies to the local parliament its status as a representative body; the Confederation of the Mountainous Peoples of the Caucasus claims to operate on behalf of all the major ethnic groups in the region. Moreover, the 'shadow' political power of mafia-type structures is becoming a matter of serious concern throughout the former Soviet Union.[14] There is suspicion that some of the conflicts may even be artificially stimulated and maintained in order to ensure continuation of disorder that paves the way for illegal economic activities.[15]

The elaboration and implementation of coherent policies for dealing with conflict-generating problems is made even more difficult because of chaos in the decision-making apparatus. The incompetence of many new administrators exacerbates this difficulty, as does the lack of a diplomatic corps in almost all the post-Soviet states. It is next to impossible to figure out what decisions are being made and by whom. More often than not, the signals emanating from different elements are contradictory even on basic points.

Even if deliberate efforts to minimize the effects of emerging conflicts are made, their results cannot be guaranteed in the absence of a viable political system. Often, political leaders are unable to control the situation or even simply to fulfil the obligations taken under the agreements with other parties to the conflicts. There have been numerous examples of the effects of this political vacuum: in Abkhazia and Nagorno-Karabakh cease-fire agreements remained dead letter not because of the intention of a party to violate them but due to an inability to control those who had participated directly in military actions.[16] On the contrary, the 'central' political leadership could easily become involved in belligerent activities or removed from the political scene. In Tajikistan, stopping the civil war required negotiations with 'field commanders' rather than between politicians.

In some cases, competing political forces were interested in organizing or continuing fighting in order both to increase their own role and to discredit

[14] As one of the popular newspapers in Moscow put it: 'In Russia the era of merciless gangsterism starts—that of bloody fighting for large sums of money, for spheres of influence in commerce, in economics and, apparently, in politics and in power structures of society. Mafia becomes a part of our life'. See *Komsomol'skaya Pravda*, 12 Feb. 1993, p. 1.

[15] A so-called 'third force' in the Trans-Dniester region has been often mentioned in this respect. The conflict in Abkhazia was described as having been initiated by competing clans seeking to control the extremely profitable 'mandarin business'. The head of the interim administration in North Ossetia and Ingushetia, Sergey Shakray, mentions the mafia activities as one of the main reasons for the inter-ethnic explosion in the area. See *Ogon'ok*, no. 2 (Jan. 1993), p. 24.

[16] The three-party agreement on Abkhazia was signed in Moscow on 3 Sep. 1992, but military actions continued. See *Krasnaya Zvezda*, 15 Sep. 1992, pp. 1 and 3. This may have been because 'the Russians and the Georgians engaged themselves to make a cease-fire respected in an area which neither of them actually controls'. See *Le Monde*, 10 Feb. 1993, p. 5.

their political opponents.[17] This argument has been comprehensively developed by Russian Foreign Minister Andrey V. Kozyrev in the article, 'The party of war is on the offensive everywhere—in Moldova, in Georgia and in Russia'.[18] Although his critics reproached him for overdramatizing the problem and emphasizing its superficial part, it is striking that the most 'primitive' schemes aimed at inflaming external conflicts for gain in domestic political battles have surfaced more than once in the former Soviet Union.[19]

## Frontiers and territorial integrity under question

The transition from the USSR as a single political entity to 15 newly independent states has highlighted the necessity of defining their relations in terms of territorial sovereignty. Since territorial inviolability is one of the most important attributes of independence and statehood, it should come as no surprise that the new states are extremely sensitive to the issue of frontiers. Any actual or potential territorial claims will inevitably become a source of conflict—in fact, the most serious one on the level of relations between the new states.

The 'internal' frontiers in the former Soviet Union were administrative in nature and politically insignificant. This of course changed dramatically when these frontiers were suddenly upgraded to the status of inter-state borders. The fact that many of these borders are not perceived as legitimate[20] represents a serious challenge to relations between the states. Mutual territorial claims could open up a long list of conflicts virtually among all of the new states.

The only pragmatic option for the states was to recognize officially the existing frontiers and territorial integrity of each other. However, this recognition alone will most likely neither prevent open conflicts nor attempts by some states to use territorial problems for political advantage. As the record of 1992 illustrates, territorial disputes have the potential to become very explosive.

The most violent conflicts—in fact, war-fighting—occurred in Nagorno-Karabakh (in Azerbaijan), South Ossetia and Abkhazia (both in Georgia) and the Trans-Dniester region (in Moldova). Strong separatist movements

[17] One of the most characteristic examples took place in Azerbaijan where President Ayaz Mutalibov was removed under strong accusations of not being active enough in resisting the 'Armenian aggression'.

[18] See *Izvestia*, 30 June 1992, p. 3. See also *RFE/RL Research Report*, vol. 1, no. 32 (14 Aug. 1992), p. 22.

[19] Commenting on the attacks against the Russian units, organized since the beginning of the conflict in Abkhazia by Georgian Minister of Defence Tenguiz Kitovani, numerous observers pointed that these provocations were oriented not so much against Russia, but against Eduard Shevardnadze. See *Izvestia*, 12 Jan. 1993, p. 1. It is striking that Shevardnadze expressed similar suppositions about the reasons of the active participation of the Russian troops in the Abkhazian offensive against the city of Sukhumi in Georgia in Mar. 1993: 'I believe that the conservative forces having gained the upper hand during the Congress in Russia, are attempting to shift the center of chaos towards Georgia thus releasing the steam from the boiler'. See *Izvestia*, 17 Mar. 1993, p. 1.

[20] By the beginning of 1992, 180 border disputes had been reported—in regions covering one-third of the former Soviet territory with a total population of about 30 million people. Potential transborder conflicts are unlikely only in two cases: between Latvia and Lithuania, and between Russia and Belarus. Less than 20% of the CIS 'internal' borders are considered to be relatively legitimate. See *Voyennaya Mysl'*, 1992, no. 10, p. 8.

(multiplied in the last case by a certain nostalgia towards the old Soviet-type order) exist in these areas, which the central authorities of the new states seek to suppress.[21] Armenia in the first case and Russia in three others have been directly or indirectly involved, which actually makes them a party to the respective conflicts. Neither Yerevan nor Moscow have made territorial claims in these cases, but the mere possibility that the frontiers will be changed contributes to the conflict development. This is perhaps the most dangerous dimension of post-Soviet conflicts paving the way to war between former Soviet republics, such as between Armenia and Azerbaijan. Fortunately, Russia and Moldova have avoided armed conflict, but it remains highly probable between Russia and Georgia.

The dispute between Russia and Ukraine over Crimea did not lead to armed conflict in 1992, but it was exceptional in two respects: (*a*) Russia openly raised the question of territorial changes; and (*b*) the problem provoked extreme emotional and political tensions in the two most powerful of the post-Soviet states. The issue will likely remain a real source of discord because of the rather strong support in Russia for a revisionist line—actively endorsed, among others, by the Russian Parliament. This reflects the reluctance of the political leadership and the public to accept 'the loss' of Ukraine, as well as strategic considerations of the military establishment (for example, concerning access to the Black Sea). The overall suspicions of Moscow with respect to Kiev will probably contribute to this trend.[22]

There is also a hypothetical possibility that the problem of frontiers between Russia and Ukraine could be raised because of the preponderant Russian population in the eastern part of Ukraine. So could questions with respect to the frontiers between Russia and Kazakhstan, taking into account the demographic composition of Kazakhstan's northern and north-eastern regions and the doubtful historical legitimacy of the borders separating these areas from Russia. In both cases, however, raising the question of borders to the point of serious conflict seems almost unthinkable, unless relations between Russia and the two other states dramatically deteriorate. Moreover, the very prospect of Russian irredentism will likely make Kiev and Alma-Ata more open to compromise in their policies towards Moscow (whereas Moscow, on the contrary, could be tempted to use the the threat of territorial claims in case of a crisis).

In Central Asia, the situation could deteriorate very rapidly. There, the ethnic composition is complex, much of the area is overpopulated, and during the Soviet period borders were arbitrarily changed on a number of occasions.

In the Baltic region, some territorial disputes (Estonia–Russia and Lithuania–Belarus) could become politically relevant. The arguments for returning to the borders which existed before World War II are questionable

[21] These conflicts are analysed in chapter 3 of this volume.

[22] By the end of 1992, the issue of Crimea belonging to Ukraine was played down, but in Dec. the Russian Parliament re-initiated the problem by deciding to review the status of the port city of Sevastopol on the Crimean Black Sea coast. See *Moscow News*, no. 51 (20–27 Dec. 1992), p. 2.

and would hardly be supported by the international community. It is still unclear, however, whether the issue could become a top priority for the parties involved or will remain on the level of diplomatic routine not going beyond the declaration of official positions.[23]

A special category of frontier and territorial problems is that which concerns (or could concern) post-Soviet states and other countries. Japan's claim over the southern Kurile Islands represents just one of these problems inherited by Russia. Although Tokyo had some basis in believing that Moscow would be ready to renounce this territory after the breakup of the Soviet Union, the issue of the Kurile Islands became a subject of extremely emotional domestic debate in Russia, which prevented the parties from breaking the deadlock.

Hypothethical territorial claims include those which could arise from the changes of the western border of the Soviet Union on the eve of or immediately after World War II. These could involve pretensions on the territories of Russia (by Finland), Belarus (by Poland) and Ukraine (by Poland, Romania and Slovakia). Even more hypothetical are the potential claims of China towards Russia and Kazakhstan, and those concerning the Kaliningrad region, which has become an exclave of the Russian Federation. The political salience of these issues is today very low, but it is clear that with respect to irredentist claims the new states are much more vulnerable than the former superpower.

At the same time there is uncertainty concerning the status of 'external' frontiers, both in Central Asia and to a lesser extent in Transcaucasus. Ethnic and religious closeness to 'outsiders', as in the cases of Tajikistan (to Afghanistan), and Azerbaijan (to Turkey), coupled with the collapse of viable political structures, could result in a startling *rapprochement* between these states.

## III. Ethno-nationalism

One of the main causes of the collapse of the Soviet Union was the inability to find appropriate responses to the challenge of national self-identification generated by numerous ethnic groups. With the emergence of 15 new states, the 'top level' of the problem was eliminated, and the old centre which suppressed the search for independence disappeared. In terms of inter-ethnic relations, however, the imperial legacy proved to be explosive in 1992.

[23] The Constitution of Estonia, introduced on 17 July 1992, contains a reference to the 'legal frontiers' of the country, which provoked a protest from the Russian Foreign Minister and a response of his Estonian counterpart. See *Izvestia*, 20 Aug. 1992, p. 5. The issue has certainly created additional problems with the negotiations between Russia and Estonia, the latter insisting on re-establishing the validity of the 1920 Tartu Peace Treaty. See *Krasnaya Zvezda*, 6 Mar. 1993, p. 2.

## Ethnocentrism in the political process

Political development in the new states is gravely affected by ethno-nationalism. The movements towards independence in the former republics were deeply coloured by aspirations for nationhood.[24] This was undoubtedly helpful in terms of mobilizing popular support, but turned out to be extremely dangerous from the point of view of further political development. The search for national identity, if not carefully channelled and correlated with basic values of democracy and human rights, can easily degenerate into primitive nationalism with all of the associated extremity and intolerance. Because democratic traditions in the former USSR were almost non-existent, this has become a major problem, even in the Baltic states with their relatively more developed political cultures.

In many parts of the former Soviet Union, ethno-nationalism has begun to flourish as a kind of official ideology substituting for communism. The values of this politico-ideological stream are manifested across a broad spectrum: from primitive slogans (such as 'Georgia for Georgians' of Zviad Gamsakhurdia) and attempts to preserve national identity through discrimination (such as that practiced in Latvia against non-Latvians) to much more sophisticated ideas of the 'national interest' (such as those expressed in Russia).

What seems significant is the fact that not only has ethno-nationalism replaced the communist ideology, it has also corrupted the new democratic credo. Furthermore, conflict developments have not been mitigated by official proclamations of democratic values. The victory of the democrats in Moscow was crucial for the independence of the Baltic states, but this did not prevent or even soften the later conflicts with Russia. Conversely, Russia, all of its anti-communist rhetoric notwithstanding, prefers to accept neo-communist regimes in some of the former Central Asian republics as factors of stability, rather than to support vigorously the development of democracy in these countries.[25]

The backgrounds of the new political élites in the former Soviet Union certainly affect the attitudes expressed towards potential or emerging conflicts. These attitudes, however, are not necessarily explained by a superficial dichotomy of those recruited from the active opposition to the 'old regime'[26] and those from the former *nomenklatura* who had the opportunity, the time,

[24] Russia was probably the only exception because of its *de facto* predominance in the former Soviet Union, whereas in the other republics the Moscow-based central power was perceived not only as the bearer of the administrative dictate but also of the national one.

[25] The old *nomenklatura* in Central Asia has actively capitalized on the so-called 'threat of fundamentalism' in order to prevent democratization and to gain support from Russia. As early as May 1992, President of Uzbekistan Islam Karimov appealed to the Russian forces deployed in Central Asia 'to guarantee the stability' in the area. See *Le Monde*, 30 Jan. 1993, p. 6. This was actually done later in Tajikistan.

[26] Here note Ebulfez Elcibey of Azerbaijan, Zviad Gamsakhurdia of Georgia, Vytautas Landsbergis of Lithuania and Levon Ter-Petrosyan of Armenia.

the intuition and, in some cases, the courage to distance themselves from traditional institutions and approaches.[27]

The former group is characterized by a higher level of political radicalism and a general lack of practical experience. When new leaders assume power as resolute, firm and uncompromising challengers to communist rule (or to the 'imperial dictate' of Moscow), they easily (even if temporarily) can become hostages of their own reputations and the expectations raised by their struggle. 'Newcomers' tend to proceed from simplified assumptions about the nature of the conflicts and the methods that may be used to resolve them. In many cases, they ignore international realities, lack knowledge about the existing patterns of state behaviour and have a distorted perception of the ability and readiness of other actors to get involved in their problems. Although the political learning curve can be rather quick, it still does not necessarily eliminate the errors and false steps which can trigger or aggravate conflicts.

The 'converted' representatives of the previous political system are more cautious and less inclined to excessive and misleading romanticism. Their pragmatism does not exclude cynical calculations but proceeds from an assessment of existing realities. Furthermore, since they have had some practical experience in leadership roles in the unified USSR, they can find a certain 'common language' among themselves. On the other hand, these ex-*apparatchiks* often tend to be 'more royal than the king' and are extremely vulnerable to pressures initiated by the extremist part of the political spectrum. They are prime targets for accusations of betraying national interests and very often attempt to neutralize or to prevent such accusations through more 'resolute' policies (such as with Shevardnadze in Georgia and Kravchuk in Ukraine).

The differences between both groups notwithstanding, it is alarming that many new political leaders have come to power under slogans of national self-identification. Often, they are subject to the psychological inertia of the recent past, as well as to the populist beliefs to which they have actively contributed. In some cases, there is good reason to suspect that capitalizing on nationalist sentiment is a deliberate policy. Indeed, the previous CPSU 'partocracy' based on the *nomenklatura* is widely believed to have been replaced by (or transformed itself into) the new 'ethnocracy'.[28]

Policies driven by ethno-nationalism are extremely destabilizing and conflict-generating both at the domestic and inter-state levels. This applies to all of the post-Soviet states, but the case of Russia deserves particular attention. Although nationalist right-wing extremism is for the moment only a marginal phenomenon in the country,[29] its arguments capitalizing on the

[27] Here note Anatolijs Gorbunovs of Latvia, Leonid Kravchuk of Ukraine, Nursultan Nazarbayev of Kazakhstan, Eduard Shevardnadze of Georgia and Boris Yeltsin of Russia. Some of the new leaders (Stanislav Shushkevich of Belarus and Askar Akayev of Kyrgyzstan) do not fit in either category. Belarus and Kyrgyzstan, incidentally, are much less associated with the post-Soviet conflicts.

[28] With respect to the post-Soviet states the term 'ethnocracy' refers both to the ethnic roots of the new élites and to their orientation towards defending 'national values'.

[29] However, right-wing extremist organizations have been reportedly established in 200 towns in the former Soviet Union. See *Komsomol'skaya Pravda*, 25 Mar. 1993, p. 1.

alleged humiliation of what used to be 'a great Russia' could affect considerably a public conciousness frustrated and disoriented by the constraints of the transitional period. At the same time the increasing support for the search for a 'Russian idea' and for the policies of 'strong statehood' actively endorsed by some prominent democrats in Moscow cannot be ignored.

## Minorities and relations between post-Soviet states

With the breakup of the USSR, the status and treatment of national minorities have become serious sources of tension between the new states. When the Soviet flag was lowered over the Kremlin, over 70 million people found themselves outside the territories of the states they considered 'theirs' on the basis of national criteria. This has resulted in serious psychological difficulties for many people in adjusting to a new situation, and a number of grave political problems have emerged.

In the Baltic states, a number of legislative acts have been enacted either denying or impeding rights of citizenship to the non-indigenous population. This has resulted in increased domestic tensions and aggravated relations with Russia, with the effect of directly contributing to the development of nationalist trends in Russia. In Armenia and Azerbaijan, which are at war, the remaining nationals from each side face increasing intolerance, in many cases with official blessing. In Central Asia, acts of violence and pressure against 'aliens' have become elements of everyday life, with the potential for escalating to a policy of 'ethnic cleansing' and open inter-state conflicts.

In other areas of the former Soviet Union, the picture is not quite as bleak. In Belarus and (somewhat unexpectedly) in Ukraine, the record seems perfect. This can also be said (with some qualification) about Kazakhstan. However, even in Russia with its relatively low level of nationalism there are alarming reports of 'persons of Caucasian nationality' becoming objects of intolerance and discrimination, sometimes openly or discretely supported by local officials.

The Trans-Dniester region represents a special case in which the political activism of the separatist movement is based on fears of Moldova's unification with Romania, rather than on nationalistic feelings. But as the appeals for support are addressed in the first instance to Russia, influential political forces there are inclined to consider this case as part of the more general problem—that of Russians and Russian-speaking people living in the 'near abroad'.[30]

After the breakup of the Soviet Union and the emergence of 15 new states, 25 million ethnic Russians and over 11 million people of other ethnic groups considering Russian as their native language found themselves outside of the Russian Federation.[31] Many of them are in an extremely difficult position, especially when there are serious grounds to believe that the situation could be further aggravated (such as in Central Asia). On the one hand, concerns about

[30] See note 4.

[31] See *Svobodnaya Mysl'*, no. 2 (Jan. 1993), pp. 67–68.

the linguistic and cultural environment, and increasing insecurity over guarantees by the authorities prevent them from making a definite decision to stay. On the other hand, a general lack of funds and organizational support from Russia make repatriation to the 'historical motherland' extremely difficult. Funding is no doubt a serious problem for Moscow, as the stimulation of a massive influx could be extremely costly.[32] But such a course may also aggravate the position of Russians who prefer to remain behind: being eligible for support from Moscow they could become the objects of even stronger discrimination intended to force them out.

During 1992, the problem of Russians in the 'near abroad' has become politically much more sensitive for Moscow. Arguments in favour of a policy envisaging the use of force to protect Russians are met with increasing sympathy. Serious clashes involving these people within some of the newly independent states could be very serious and could result in increased stridency on the part of Russian politicians in their calls for action. It is not clear how the security of millions of Russians who could become actual or potential victims of conflicts in the 'near abroad' will be ensured by regular armed forces, but it seems quite obvious that endorsing such a policy would be extremely damaging both to the international prestige of Russia and to the course of post-Soviet developments.

## Disintegration trends

Managing inter-ethnic relations between the newly independent states is only one part of the problem. Another part concerns the multi-ethnic composition of these states which in a number of cases creates a serious incentive for disintegration and related conflict. Put simply, the basic question is: Can the trend that resulted in the breakup of the Soviet Union be stopped, or will it lead to the disintegration of the newly independent states as well?

Disintegration is generated not only by ethnic diversity, but also by the insufficient viability of political systems.[33] Poor economic performance, especially in light of the urgently needed transition to market economies, also plays a substantial role.[34] But ethnic factors, as in the case of the USSR, could dramatically accelerate the trend towards disintegration in the new states. The existence of political and administrative structures which represent certain

[32] By Dec. 1992, according to official Russian statistics, 470 000 refugees had fled from conflict zones and 800 000 'constrained migrants' had left other former Soviet republics fearing violence or discrimination. The total number of people accepted by Russia from the 'near abroad' was expected to reach 2 million by the end of 1992. See *Rossiyskie Vesti*, 10 Dec. 1992, p. 2.

[33] Federal law is reported to be neglected almost everywhere in the Russian provinces. According to Russian Procurator-General Valentin Stepankov, his office has uncovered in 1992 over 13 000 new local laws which contradict either the Federal Constitution or specific federal laws. See *Moscow News*, 20–27 Sep. 1992, p. 7.

[34] The central government is often blamed for regional disparities. For example, even in the relatively developed city of Novosibirsk, per capita income was reportedly two times lower than that in Moscow. See *Rossiyskie Vesti*, 24 Nov. 1992, p. 1. Grigoriy Yavlinsky, a prominent economist, considers that the most serious challenge that Russia faces is generated by economic incentives for disintegration. See *Moscow News*, 20–27 Sep. 1992, pp. 8–9.

national minorities but by and large had only symbolic character in the past becomes a very important point of reference for a new separatism.[35]

Actually all of the former Soviet republics that included some so-called 'national autonomies' (in the form of 'autonomous republics', 'autonomous regions' or 'national districts') have been affected by this development. In Azerbaijan, this refers not only to Nagorno-Karabakh (a former 'autonomous region') fighting for independence but also to Nakhichevan (a former 'autonomous republic') which has *de facto* escaped the control of Baku. The case of Georgia with respect to Abkhazia and South Ossetia was mentioned above. Significantly, the first civil war in Central Asia broke out in Tajikistan which includes the Gorno-Badakhshan 'autonomous region' with strong separatist inclinations. Only Uzbekistan (which has the Kara-Kalpak 'autonomous republic') seems still relatively quiet, but this may be explained by the slow pace of post-Soviet development, with the main features of the old regime being preserved. This does not, however, guarantee that Uzbekistan will not experience ethnic violence.

Among the newly independent states, the Russian Federation is undoubtedly the most vulnerable to the dangers of ethnic separatism.[36] Russia has 21 'autonomous republics' and 10 'autonomous districts'. Of the autonomous republics, 16 date from the Soviet era; 1 was 'restored' by decision of the Russian Parliament;[37] and 4 were upgraded from former 'autonomous regions'.[38] This patchwork of autonomous republics and districts represents a strong element of decentralization in the political system. In some of these entities, separatist movements are actively initiated and supported by local political élites.[39]

Here again, the situation is not the same in all regions. In some cases, the ethnic group after which the administrative entity is named represents only a small part of the total population (which is predominantly Russian) and does not have a tradition of political representation (apart from the symbolic one in the framework of the former Soviet Union). The search for greater autonomy from Moscow usually does not go beyond demands for more substantial (but not necessarily exclusive) powers in some fields, such as to define and carry out economic, environmental and cultural policies, and to organize local

[35] However, a process of self-identification of some smaller ethnic groups which do not have their 'own' administrative entities is also under way and could result in conflicts. For example, the 'Association of ethnic Koreans' insists on creation of an independent Korean economic zone in the Far East (significantly, as a constituent part of the CIS, not Russia) and appeals to Korean families banished by Stalin from their homeland to Central Asia during the 1930s to return. This immediately provoked tension within the local population: the Ussuri Cossacks threatened to use weapons for defending their lands. See *Moscow News*, 20–27 Sep. 1992, p. 5.

[36] See Salmin, A., 'Dezintegratsiya Rossii?', *Nezavisimaya Gazeta*, 12 Dec. 1992, p. 5.

[37] The Ingush Republic.

[38] The only 'autonomous region' preserving the old status is a tiny Jewish autonomous region in the Far East which has always been an artificial administrative entity without any sizable Jewish population; hence there are no signs of a 'grass root' search for national identity.

[39] Out of 18 constitutions and draft constitutions at work in the Russian Federation, one proclaims independence (Chechnia), one—the status of a state associated with Russia (Tatarstan), one—the right to free secession (Bashkortostan) and six—the primacy of local laws over the all-Russian ones. See *Moscow News*, no. 6 (4 Feb. 1993), p. 3.

administration. This is the case, for example, in Yakut-Sakha (in Siberia), Komi (in the North European part of Russia) and Udmurt (in the Volga river basin) republics, and in all of the 'autonomous districts'.[40] Separatist movements based purely on ethnic grounds in these areas are politically irrelevant today, but hope for economic self-sufficiency from control of natural resources could generate demands for increasing autonomy, as one could well anticipate on the part of the Yakut-Sakha republic. Moreover, those entities, even if not initiating the process, will hardly refrain from 'solidarity' with the other republics pursuing a more active policy line towards Moscow.

In some other republics, separatist trends are much more developed, especially in Tatarstan (the most populated and industrially developed republic in Russia, located in the Volga river basin) and in Chechnia (in Northern Caucasus). Both have refused to sign the Federal Treaty and pretend to deal with Russia on a bilateral basis, as *de facto* or even *de jure* independent states, without, however, leaving the 'common economic, currency and military space'. Tatarstan's political élite has moved cautiously to marginalize politically extremist trends,[41] as it is apparently more confident of the success of independence considering the relative economic might of the republic, the demographic weight of Tatars in Russia (who are the second-largest national group) and the 'Muslim factor' (which Russia must consider). In Chechnia, on the other hand, Chechnian leader Dzhokhar Dudayev has capitalized on anti-Russian nationalism, threatening to initiate 'the second Caucasian war' and even to start a terrorist campaign against Moscow.[42] However, the confrontation with Moscow could also be an important instrument in domestic political struggles—as it was during the time of the Soviet Union.[43]

Ethnic problems in Russia are not limited to the confrontation between the centre and the republics searching for greater autonomy. The post-Soviet legacy also includes a number of unresolved problems between autonomous republics and between ethnic groups. In Northern Caucasus, for example, numerous administrative changes and mass deportations affected the lives of hundreds of thousands of people during the Soviet era.

The explosiveness of these issues was clearly demonstrated in 1992, when the Russian Parliament approved a law re-establishing the 'Ingush Republic'

[40] The 'autonomous districts' became full-fledged participants to the Federal Treaty, which contradicts the principle of administrative subordination to the 'territories', and opens up one more field of political conflict between the constituent parts of the Russian Federation. There are a number of reports of attempts in 'autonomous districts' to change the existing administrative subordination, to eliminate it completely, to reconsider the territorial composition and so on. See *Izvestia*, 5 Jan. 1993, p. 2.

[41] Representative, for example, is the movement 'Sovereignty' in Tatarstan. Among its goals are: the proclamation of Tatarstan as a neutral demilitarized state; the withdrawal of Russian military units from the area; and the establishment of independent security forces . See *Izvestia*, 10 Feb. 1993, p. 2.

[42] Dzhokhar Dudayev stated flatly in an interview that 'we shall not fight on our territory'. See *Literaturnaya Gazeta*, 12 Aug. 1992, p. 12. He was quite clear in pointing out that Chechens living outside the breakaway republic (300 000 according to some estimates; see *Komsomol'skaya Pravda*, 21 Aug. 1992, p. 2) could be used in terrorist activity.

[43] One of the opponents of Dzhokhar Dudayev noticed that each time he loses political points domestically, the tension in relations between Chechnia and Russia increases. See *Komsomol'skaya Pravda*, 20 Feb. 1993, p. 1.

without, however, defining its borders or specifying the ways to implement the decision. As the territory of the would-be republic (already inhabited by other nationals) could only be delimited from neighbouring ones, this inevitably resulted in serious tensions in the region. Ingushi activism, fuelled by frustration over the absence of any practical moves from Moscow to implement the decision, exacerbated these tensions, resulting in a bloody interethnic explosion involving hundreds of casualties, hostage-taking and thousands of refugees.[44] To restore peace and order, Moscow introduced martial law and moved armed forces into the area[45] and by doing so essentially recognized the outbreak of the first war on Russian territory of the post-Soviet era.[46]

Northern Caucasus is perhaps the most unstable area of the Russian Federation. Several options for this region have been discussed which, in itself, contributes to political confusion and uncertainty. For example, it has been suggested to divide existing 'autonomous republics' (as in the case of Kabardino-Balkaria), to amalgamate others and to establish new ones (for example, the attempts to unify the various Adygo-Cherkess ethnic groups).[47] The developments in Northern Caucasus are of crucial concern for Moscow. The territorial integrity of the country is especially vulnerable in light of secessionist trends generated by the influential Confederation of the Mountainous Peoples of the Caucasus (established in 1990 and pretending to represent all major ethnic groups in the area). The Russian-speaking peoples are being quickly radicalized as well, insisting that the central government take strong measures to protect their interests. However, political indecisiveness and resultant frustration have led to the development of populist-oriented Cossack movements.[48] The local ethnic extremism is also generated by an increasing flow of refugees from Transcaucasus.[49] Ethnic problems and instability in Northern Caucasus threaten to involve Russia in serious conflicts with Georgia (which in fact occurred in 1992 over the war in Abkhazia) and Azerbaijan (which is increasingly anxious about developing

[44] It was reported that by mid-Dec. 1992 there were 343 killed, 753 wounded, over 2000 hostages and 65 000 refugees as a result of clashes. See *Krasnaya Zvezda*, 16 Dec. 1992, p. 2. Of the Ingushi minority living in Northern Ossetia, 90% fled. See *Le Monde*, 1 Dec. 1992, p. 3.

[45] Up to 20 000 Russian interior and airborne troops were deployed in North Caucasus. See *Rossiyskie Vesti*, 21 Nov. 1992.

[46] Two other attempts 'to restore the rights of peoples subject to repressions' should be mentioned: the case of the Crimean Tatars and that of the Volga Germans (these ethnic groups had their own 'autonomous republics' until World War II). Without going into detail, it should be noted that in both cases the intentions and political decisions were not backed by concrete measures; the most 'visible' result was increasing tension in the areas where the deportees were expected to return.

[47] See *Rossiyskie Vesti*, 16 Dec. 1992, p. 2

[48] In the Kuban area alone, Cossack armed formations total about 140 000 people. See *Ogon'ok*, no. 1 (Jan. 1993), p. 7.

[49] In some areas, the share of refugees in the local population has surpassed 15%. See *Komsomol'skaya Pravda*, 4 Feb. 1993, p. 2.

transborder activities of Lezghians who also inhabit Dagestan, a neighbouring Russian 'autonomous republic').[50]

## IV. The military dimension

The military legacy of the former Soviet Union has resulted in a number of conflicts. The conflicts mostly concern the disposition and control of nuclear weapons and the transition from Soviet to national armed forces. Not all of the conflicts over these issues were resolved in 1992, although developments were more positive than one might have expected. This section focuses on those aspects of the Soviet military legacy which directly affect relations between the post-Soviet states. The military implications of the dissolution of the USSR for the international community, as well as the military aspect of the security interests of the newly independent states deserve much more attention than is possible in this chapter.

### Nuclear weapons

The most urgent issue with serious international implications concerns the disposition of nuclear weapons inherited from the former Soviet Union.[51] The vast strategic arsenal of the former USSR is deployed on the territories of four former republics—Belarus, Kazakhstan, Russia and Ukraine. Although Russia is regarded as the nuclear weapon successor state *par excellence*, the others have been tempted to capitalize on *de facto* (even if transitional) nuclear weapon status.

Only Belarus opted immediately to unconditionally relinquish its nuclear weapons. This decision pre-empted conflict over the issue and only required negotiations of a technical nature with Russia. Kazakhstan followed suit, but not before making some initially ambiguous statements with the apparent intent of highlighting its international status. The approach taken by Ukraine, however, raised Russian concerns and resulted in a number of disputes. There were uncertainties with respect to the schedule for the withdrawal of tactical nuclear weapons, pretensions concerning Ukrainian 'administrative control' over units of the CIS strategic forces, insistence on participating in the process of dismantlement of nuclear weapons, demands for 'special guarantees' against nuclear attack, requests for funds to cover the expenses associated with nuclear disarmament and claims of ownership of fissionable materials extracted from nuclear warheads.[52]

[50] The idea of introducing Lezghistan into Russia has recently become a matter of discussion. See *Rossiyskie Vesti,* 23 Dec. 1992, p. 2. See also Achundova, E., 'Vtoroy Karabakh na beregakh Samura?', *Literaturnaya Gazeta*, 23 Dec. 1992, p. 12.

[51] The issue is analysed in more detail in chapter 6 of this volume.

[52] Ukraine estimated the cost of transferring strategic nuclear weapons to Russia to be $1.2–1.5 billion US (compared to $175 million suggested by the USA). See *Rossiyskie Vesti*, 30 Dec. 1992, p. 1; and *Izvestia* 10 Jan. 1993, p. 3

Despite international pressure, by the beginning of 1993 the parties were unable to reach a consensus concerning the status of strategic nuclear forces deployed outside Russia.[53] The latter insisted on their full transfer under its jurisdiction. Belarus agreed with this approach whereas Kazakhstan stressed the necessity to have the weapons under the control of the CIS, not Russia alone. Ukraine proposed placing strategic nuclear forces deployed on its territory under a 'joint command' and to keep them under Ukrainian jurisdiction, as well as to preserve the technical possibility of averting the launch of the missiles.

The military and political élites in Moscow apparently have serious suspicions that Ukraine could go nuclear, thus creating grave political and security problems for Russia. These suspicions are exacerbated by the ambiguous statements of Ukrainian officials, as well as by the apparent consideration being given to the idea of a small nuclear deterrent by political circles in Kiev, including the Parliament. Moreover, it is widely believed that although Ukraine neither has operational control over nuclear weapons nor the command system they require, it could eventually overcome these hurdles. Furthermore, Ukraine may even be able to develop an indigenous nuclear weapon-production capability.

The political line of Russia (backed by the international community) consists of emphasizing the most negative implications for the non-proliferation regime if there remains any ambiguity about the non-nuclear status of the other successors to the Soviet Union. Officials in Moscow have also expressed concern over the safety of nuclear missiles deployed in Ukraine, since Ukraine refuses to give Russia access to the weapons for maintenance purposes.[54] Russia, however, had certainly played into the hands of the pro-nuclear lobby in Ukraine through its unwise belligerence on such a sensitive issue as territorial claims. Moreover, it is an open question whether ignoring Belarus, Kazakhstan and Ukraine on the eve of the signature of the START II Treaty was a better tactic than non-binding consultations with the three states that might have been helpful to minimize both their concerns and their feelings of 'alienation'.

The scope of an eventual open conflict over the issue remains unclear; however, to prevent its overdramatization and to find appropriate solutions, Russia will most probably have to exercise its diplomatic art and political skill to their fullest extent. At stake are not only the future of the non-proliferation regime or relations between the post-Soviet states, but also the fate of the Russian nuclear arsenal in the event Russian Federation should break up.

[53] As of Apr. 1993, Ukraine had yet to ratify the 1991 START Treaty and the 1968 Non-Proliferation Treaty (NPT).

[54] Head of the General Staff of the Russian Armed Forces Mikhail Kolesnikov stated that radiation levels from some sites in Ukraine exceed permissible levels by thousands of times, an allegation which was vigorously denied by Ukrainian officials. See *Financial Times*, 3 Mar. 1993, p. 2.

## The transition from Soviet to national armed forces

During 1992, the process of transition from Soviet to national armed forces had two closely interrelated aspects with the potential for generating conflict: the sharing of military assets, on the one hand; and the withdrawal of military units, on the other.

Initially, the reluctance of the Russian political leadership to establish national armed forces was officially rationalized as preventing a collapse of the existing military potential. Other participants in the CIS, however, regarded it as the desire to retain a unified military establishment controlled by Moscow. Thus, the debates in 1992 concerning 'unified', 'joint' or 'common' armed forces were largely guided by suspicions about Russian political motives, rather than by strategic considerations. Even the idea of 'collective security', which came some time after, became tainted by these suspicions.

However, any forcible attempts to impose the preservation of the 'common military–strategic space' on the former Soviet republics would have had dangerous implications. Fortunately, this line was abandoned, which served to prevent serious conflicts and to promote the acceptance of each newly independent state's right to have its own armed forces.

External factors also played a role in this process. The new states were under pressure from their Western partners to implement the 1990 Conventional Armed Forces in Europe (CFE) Treaty. This was possible only if the new states could agree upon the quotas of treaty-limited equipment allocated to the former Soviet Union. An agreement was reached in Tashkent on 15 May 1992 by all eight former republics sharing part of the 'Atlantic to the Urals' (ATTU) zone.[55] This agreement not only gave some measure of international endorsement to the idea of separate armed forces but also defined certain guidelines for their practical realization. In the absence of the concrete figures contained in the CFE Treaty, the issue of sharing military equipment would likely have been much more difficult to resolve. It is interesting that the CFE Treaty, a traditional (albeit unprecedented in scope) arms control agreement conceived during and inherited from the epoch of bipolar confrontation, has played an extremely important role in the post-Soviet period.

Nevertheless, the reorganization of the military assets of the former Soviet Union has been extremely controversial and a source of conflict.[56] All of the new states have had to deal with Russia over this reorganization, since Russia assumed responsibility for the former Soviet armed forces deployed outside its territory. However, the patterns of conflict among the parties concerning this

[55] The Baltic states opted for non-participation in the 1990 CFE Treaty. See also chapter 12 in this volume.

[56] According to the Centre for Military and Strategic Research of the General Staff of the Russian Armed Forces, about 30 conflicts and over 70 disputes have emerged on the territory of the former Soviet Union during 1991–92 because of the problems associated with sharing the assets, military and otherwise, of the dissolved state. See *Rossiyskie Vesti*, 12 Dec. 1992, p. 2. See also *Le Monde diplomatique*, Apr. 1993, p. 17.

issue have been different, reflecting the specific characteristics of each situation and the priorities of the states involved.

The Baltic states insisted on rapid withdrawal of the former Soviet armed forces from their territories. Russia agreed, but has since tried to drag out the process.[57] By the end of 1992, Russia actually linked the problem of the withdrawal to the rights of the Russian-speaking population, thus responding to Russian domestic demands and attempting to exert pressure on Latvia and Estonia.[58] Diplomatic conflict over this issue has manifested itself in all of the main international institutions, including the UN and the CSCE.

Belarus and Ukraine assumed control of the armed forces deployed on their territories. Moscow and Minsk had actually no serious problems in agreeing upon all the main issues related to the 'nationalization' of the former Soviet arsenal,[59] whereas relations between Russia and Ukraine were heavily damaged not only by the fact that Kiev was perceived as bearing the main responsibility for failure of the CIS common forces approach,[60] but also by the dispute over the Black Sea fleet.[61] After protracted discussions, both sides decided to keep the fleet under joint command until 1995, which apparently adjourned the conflict rather than solved it.[62] In fact, despite the agreement on joint command, attempts have been made to appropriate unilaterally some elements of the fleet. During 1992, Ukraine twice 'nationalized' warships, and on both occasions the danger for an escalation to a military confrontation existed. Some warships under construction in the Black Sea shipyards were also placed unilaterally by Ukraine under its jurisdiction.[63]

[57] The negotiations started in Feb. 1992. During 1992 about half of the Russian troops deployed in the Baltic states were withdrawn; the Russian Ministry of Defence estimates the total Russian military presence in the three Baltic states to be about 75 000 by late 1992. See *Krasnaya Zvezda,* 25 Sep. 1992.

[58] On 29 Oct. 1992 Russia announced a suspension of the withdrawal—but, significantly, officially referred not to the violation of human rights in the Baltic states but to the urgent social needs of the military and the housing problems for the withdrawing troops. See *Komsomol'skaya Pravda*, 11 Nov. 1992, p. 2. In fact, according to numerous reports, the withdrawal continued even after this statement. The withdrawal from Lithuania must be completed by 31 Aug. 1993. See *Izvestia*, 30 Dec. 1992, p. 1.

[59] On 20 July 1992, Russia and Belarus signed a package of 24 agreements; 5 of them specified interaction of the two countries with respect to military assets. See *Vestnik Voyennoy Informatsii*, no. 9 (Sep. 1992), part I, p. 1–8.

[60] Ukraine was the first to announce national control over the armed forces deployed on its territory: in fact, immediately after the failed coup in Moscow (on 24 Aug. 1991). Kiev began to restructure military formations as early as 3 Jan. 1992; it also initiated oaths of allegiance for military personnel on Ukrainian territory, thus sowing confusion in the armed forces and irritating military and political establishments in Moscow. See *Krasnaya Zvezda*, 13 Jan. 1993, p. 2.

[61] Ukraine used two arguments in support of its claim: territorial basing of the fleet in the Crimea (Sevastopol); and the fact that all the other naval fleets of the former Soviet Union had been unilaterally appropriated by Russia. Moscow referred to historical considerations and to a lesser degree to the strategic ones.

[62] Russia and Ukraine are the main parties to the conflict over the Black Sea fleet but not the only ones. Belarus expressed its dissatisfaction because of the strictly bilateral character of the Russian–Ukrainian deal on what had been officially proclaimed a part of the joint CIS armed forces. See *Nezavisimaya Gazeta*, 12 Aug. 1992, p. 1. Moldova put under its jurisdiction the naval aircraft fighters (MiG-29) regiment deployed in Markulesty. See *Krasnaya Zvezda,* 24 Dec. 1992, p. 2. Georgia claimed its part after the beginning of the conflict in Abkhazia and tried to organize the blockade of the naval base in Poti. See *Krasnaya Zvezda*, 24 Nov. 1992, p. 2.

[63] *Krasnaya Zvezda*, 24 Dec. 1992, p. 2

In Moldova, the most controversial aspect of the issue concerns the Russian 14th Army deployed in the Trans-Dniester region which has alledgedly been actively involved in the conflict on the side of the separatists. The issue exploded in the summer of 1992 when armed conflict resulted in numerous casualties. Although Moscow accepted the principle of withdrawal,[64] the negotiations to implement it have so far been fruitless.[65] Apparently, two reasons exist for this reluctance. First, there are increasing domestic pressures in Russia to retain the instrument of military intervention to protect the Russian-speaking population (and to influence Moldova). Second, it remains unclear how to withdraw an army which is composed mainly of local inhabitants. It should be noted that Moldova (unlike the Baltic states) is reluctant to dramatize the problem, perhaps not least because the Russian 14th Army could be helpful in neutralizing the extremist tendencies in the self-proclaimed 'Trans-Dniester Republic'.[66]

The case of the Russian 14th Army revealed another aspect of the Soviet military legacy—that is, the danger of a high level of autonomy in the activities of organized armed forces when central control is weak and the political leadership is unclear. A scenario involving a 'frustrated and independent military' has often been described by analysts as potentially one of the most dangerous aspects of the Soviet military legacy. In Moldova, however, developments in this regard have been relatively quiet.[67]

In Transcaucasus, the process was much more dramatic. Transcaucasus is one of the most heavily militarized areas in the world, with intense on-going fighting.[68] Armenia and Azerbaijan are at war, and Georgia is actively using military means in domestic conflicts. On the basis of developments in 1992, the three states seem to share similar interests with respect to the former Soviet armed forces deployed in the region: (*a*) to have them on its side in war-fighting or at least to prevent them from helping the opponent; (*b*) to employ their weapons, equipment and so on for increasing indigenous military might; and (*c*) to use them as a vehicle to articulate relations (either positive or negative) with Russia. Moscow, for its part, has been hesitant concerning whether to keep its military forces in place or to withdraw. By staying, Moscow hopes somehow to prevent further destabilization and to influence developments. By withdrawing, Moscow loses the military lever and faces difficult problems of transferring weapons and military equipment.

[64] The decision to withdraw the 14th Army was announced in May 1992. See *RFE/RL Research Report*, vol. 1, no. 37 (18 Sep. 1992), p. 73.

[65] Talks on the terms of the eventual withdrawal and status of the troops began on 12 Aug. 1992. See *RFE/RL Research Report*, vol. 1, no. 34 (28 Aug. 1992), p. 76.

[66] By the end of 1992 tensions on the issue of withdrawal of the 14th Army have been substantially reduced, alongside the successful development of political dialogue between Russia and Moldova. See *Izvestia*, 11 Feb. 1993, p. 1–2.

[67] Earlier the former Soviet detachments deployed in Nagorno-Karabakh had found themselves in a similar situation.

[68] Military assets include hardware, infrastructure and depots of the Transcaucasian MD, the Transcaucasian Borderguard District, the 19th Air Defence Army, the 34th Air Force Army, the Caspian Flotilla and the Black Sea Fleet deployed on a strip 300 km wide and 700 km long. See *Moscow News*, no. 51 (20–27 Dec. 1992), p. 6.

Apart from contradictory political signals from Moscow which have created an ambiguous situation, Russian armed forces have had to deal increasingly with local pressures and provocations. This has resulted in demoralization of the forces, undermining command and control and reducing force reliability.[69] Numerous weapons (including armoured combat vehicles, heavy guns and even aircraft) have been seized from the military.[70] This not only exacerbates the on-going violence, but also could result in dire consequences in the event the military resorts to arms to defend its assets.[71] At the same time deteriorating morale and poor maintenance have pushed some military personnel and units to become involved in the illegal arms trade. Certainly, any weapon transfers (either deliberate or not) in the areas of armed conflicts couldn't help but contribute to their development.[72]

The agreement of May 1992 to transfer partially the military equipment of the Transcaucasian Military District (MD) to Armenia, Azerbaijan and Georgia 'on the basis of parity'[73] and the decision to begin the withdrawal of the 7th Army from Armenia and the 4th Army from Azerbaijan was helpful in defining the political perspective. It was envisaged that the Transcaucasian MD would be dissolved while Russia would keep an airborne division and an air-defence unit in Azerbaijan (until 1994) and three divisions in Armenia and Georgia. Georgia had not raised the issue of withdrawal as a matter of immediate priority, but the conflict in Abkhazia (in August 1992) changed the situation sharply: Tbilisi accused Moscow of interference,[74] suspended nego-

[69] In 1992 there were about 600 attacks against Russian military personnel in the 'near abroad'. See *Krasnaya Zvezda*, 30 Dec. 1992, p. 1. Up to 80% of all the attacks took place in Georgia and Azerbaijan. See *Rossiyskie Vesti*, 30 Dec. 1992, p. 1.

[70] During the first half of 1992 alone over 200 incidents were reported in Transcaucasus. Over 100 tanks and armoured vehicles, about 100 each of guns and mortars, and over 2000 sub-machine guns were captured. See *Krasnaya Zvezda,* 5 Jan. 1993, p. 2. As for the aircraft, the first large-scale seizure took place in Feb. 1992 when the Azeri military began a blockade of separate helicopter squadrons deployed in Sangachaly and announced that all its hardware had been nationalized, including 20 helicopters Mi-4 (considered as the world best fire support helicopters up to the end of 1970s) and Mi-8 which were immediately used in the conflict with Armenia. In Apr. 1992 an Su-25 attack aircraft was seized as well as 70 Czechoslovakian-made training aircraft L-29 which can be used in military actions. In June 16 MiG-25 and Su-24 aircraft were seized at the airfield in Dallar. There was also unconfirmed information on seizure of a number of MiG-21. In Armenia a squadron of helicopters Mi-24 and Mi-8 took active part in the fighting. See *Krasnaya Zvezda*, 12 Dec. 1992, p. 3.

[71] On 20 June 1992 the Russian Government took a decision giving the armed forces deployed in 'near abroad' the right 'to defend their honour, dignity and life' by any means, including military. See *Krasnaya Zvezda*, 30 Dec. 1992, p. 1.

[72] On 20 June 1992 the Russian Government adopted a decision granting the armed forces deployed in the 'near abroad' the right 'to defend their honour, dignity and life' by any means, including military ones. See *Krasnaya Zvezda*, 30 Dec. 1992, p. 1.

[73] The CFE ceilings agreed in Tashkent gave the same quotas in tanks (220), armoured combat vehicles (220), artillery (285), aircraft (100) and helicopters (50) to each of the three states.

[74] Characteristically, at the very initial stage of the conflict, Russiá was accused by the opponents of Tbilisi as well for having handed over to Georgia weapons and equipment sufficient to arm a whole division, which allegedly made the invasion in Abkhazia possible. See *Moscow News*, no. 35 (31 Aug.–6 Sep. 1992), p. 4. Moscow, according to this logic, gave its 'blessing' to the Georgians to use force against separatists in order to reinforce the principle of the state's integrity and to prevent North Caucasus from being 'contaminated' by the Abkhaz example. See *Komsomol'skaya Pravda*, 17 Sep. 1992, p. 2. Following the Tashkent agreement of 15 May 1992 the 10th Motorized Rifle Division of the Transcaucasian MD transferred to Georgian armed forces 108 tanks and 129 combat vehicles—under

tiations on the bilateral treaty, threatened to request the immediate withdrawal of the troops[75] and attempted to take over military property.[76] Negotiations were resumed only after the political tension diminished by the end of 1992;[77] however, the key factor for resolving the issue will undoubtedly be the general context of Russian–Georgian relations which are overburdened by the Abkhazian problem.

Compared to the European part of the former Soviet Union, the transition to a new military pattern in Kazakhstan and the rest of Central Asia seems to have been much less painful. This could be attributed to three factors: (*a*) the area was only of secondary importance for the former Soviet armed forces—both in terms of deployment and in terms of military infrastructure; (*b*) all the states in the area are deeply dependent on Russia economically and consider co-operative relations with Moscow as imperative; and (*c*) so far there have not been any major conflicts between them. Their decision to place the armed forces under national jurisdiction was more than balanced by their predominantly 'loyal' participation in the CIS; moreover, Russia promised assistance in formation of national armed forces and in jointly organizing the external border controls.[78]

Because of heavy financial constraints, the newly independent states in Central Asia will hardly be able to allocate any substantial resources to developing military potential.[79] Whether they can continue a low-profile policy will mainly depend upon the impact of the global political situation on the area; domestic and inter-state conflicts will certainly create serious incentives for both the arms race and political realignments. Apart from that, the difficult challenge faced by residual Russian armed forces was highlighted by the war in Tajikistan, where the 201st Motorized Rifle Division found itself in a dramatic position of being both a hostage in the violence which had broken out and the only viable force which could restore order.[80]

By and large, post-Soviet political developments in 1992 were significantly affected by the conflicts over the military heritage of the former superpower. Although these conflicts undoubtedly contributed to overall disorder and

the condition, however, that they would not be used in domestic conflicts. See *Krasnaya Zvezda*, 24 Nov. 1992, p. 2.

[75] The total number of Russian troops deployed in Georgia has been reduced from 150 000 (in 1991) to 18 700 by Apr. 1993. See *Izvestia*, 7 Apr. 1993, p. 1.

[76] *Izvestia*, 28 Sep. 1992, p. 2; *Izvestia*, 18 Dec. 1992, p. 1.

[77] As of Feb. 1993, the draft agreement envisaged that Russian armed forces would remain until the end of 1995 and border troops until 1996. See *Krasnaya Zvezda,* 6 Feb. 1993, p. 2.

[78] Turkmenistan even opted for having joint operational command over the emerging national armed forces (during the transitional 5-year period). See *Izvestia*, 28 July 1992, p. 2; *Izvestia*, 2 Apr. 1993, p. 2. Both countries are interested in protecting the 2000-km frontier with Afghanistan. This decision also seems quite significant in light of the fact that Turkmenistan is the only one of the non-European newly independent states that has reacted coolly towards the CIS and refused to take part in the Tashkent Treaty on Collective Security.

[79] President of Kyrgyzstan Askar Akayev stated that 'we cannot afford a powerful army, whereas there is no sense to have a small one only as a symbol', which explains why the country will set up only a national guard (at 1000 strength) and border troops. See interview in *Ogon'ok*, no. 47–49 (Nov. 1992), p. 7.

[80] Significantly, the new political leadership in Tajikistan insists on the continued presence of Russian armed forces 'as stabilizing factor'. See *Komsomol'skaya Pravda*, 12 Jan. 1993, p. 1.

tension, the dramatic scenarios describing the process of the disintegration of the Soviet armed forces were not realized. The conflicts were affected more by the general political context than vice versa.

Disagreements over military assets could still have extremely destabilizing consequences. The newly independent states will likely remain highly sensitive towards what is considered both one of the most important elements of their independence and an effective instrument enabling them to defend their interests—namely, armed forces. This is especially the case with 'hot spots' in which even minor episodes can rapidly escalate up to large-scale clashes involving substantial civilian and military resources.

The process of 'nationalization' of the armed forces is far from over, and a new balance between the actors operating in the former USSR has not been established. Meanwhile, the process of state-building within the new states could highlight the importance of the military dimension of their status even further[81]—but also hopefully the necessity of co-operative approaches towards defining their security requirements.

## V. Conclusion

The initial phase of developments of the post-Soviet era provides clear evidence that a highly unstable grouping of states has emerged in place of the former superpower. Some specialists anticipate a long period of conflicts and wars for at least the next 30–40 years.[82] Such alarming forecasts may be exaggerated, but the experience of 1992 clearly illustrates that both domestic developments within the new states and their external interactions are fraught with the danger of numerous conflicts.

The year 1992 began with the civil war in Georgia and ended with the civil war in Tajikistan. It seems significant that these were the only cases in which domestic disputes over power and basic political orientation escalated into open armed conflict. In Tajikistan, the violence was unprecedented[83] and resulted in a new phenomenon—the first massive outflow of refugees from former Soviet territory (in this case, to Afghanistan).[84] Another new

[81] It seems quite revealing that Askar Akayev in the above-mentioned interview, arguing for neutrality and full dissolution of the armed forces, mentioned Switzerland as a model to follow. See *Ogon'ok*, no. 47–49 (Nov. 1992), p. 7. Apparently he did not consider the military dimension of Swiss neutrality nor Switzerland's non-participation in alliances in his analysis. This is certainly not the case with Kyrgyzstan which is a participant in the Tashkent Treaty on Collective Security which provides *inter alia* for mutual military assistance.

[82] See *Izvestia*, 18 Dec. 1992, p. 5.

[83] The total number killed during the conflict has been estimated at 20 000–50 000. The official estimate is 25 000. See *Le Monde*, 14 Jan. 1993, p. 4; *Le Monde*, 27 Jan. 1993, p. 4. The total number of refugees may have reached 500 000 (an extremely high figure for a population of 5 million). See *Izvestia*, 11 Jan. 1993, p. 5; and *Izvestia*, 24 Feb. 1993, p. 5.

[84] Estimates range from 30 000–50 000. See *Izvestia*, 28 Dec. 1992, p. 2; and *Krasnaya Zvezda*, 15 Jan. 1993, p. 3. The office of the UN High Commissioner for Refugees put the figure at 60 000–85 000. See *Le Monde*, 6 Feb. 1993, p. 1; and *Komsomol'skaya Pravda*, 22 Apr. 1993, p. 2. Even 120 000–130 000 have been estimated. See *Izvestia*, 24 Feb. 1993, p. 5.

phenomenon is an open and decisive external military involvment in domestic developments.[85]

The 'old conflicts' in Nagorno-Karabakh, South Ossetia and the Trans-Dniester region continued, but developed along different lines. In Nagorno-Karabakh, the war-fighting escalated until almost all kinds of weapons were in use with full-size armed forces involved. This has exhausted the resources of the participants to the utmost and resulted in a deadlock in which no side could win. In South Ossetia and Trans-Dniester, tensions had been significantly reduced by the end of the year (although aerial bombardment of Bendery by Moldova in July 1992 resulted in a genuine crisis).

Apart from Tajikistan (see above), two more 'hot spots' appeared on the map of the former Soviet Union: Abkhazia and Ossetia/Ingushetia. Both could be considered as models of possible future conflicts: generated by controversies between territorial integrity versus autonomy and separatism in the first case and by ethnic clashes and territorial redistributions or claims between state's constituent entities in the second. The fact that Russia found itself directly involved in Abkhazia[86] and had to use force in order to stabilize the situation in North Caucasus might also be relevant with respect to the future post-Soviet conflicts. The same could be said about the role of the Russian military in Tajikistan which is surprisingly reminiscent of the beginning of the Soviet adventure in Afghanistan 13 years earlier.[87]

Unlike the intervention in Afghanistan, however, Moscow insists on a kind of international mandate for Russia to act freely on former Soviet territory—which could be rationalized, among other factors, by the lesson of Yugoslavia. In February 1993, President Yeltsin formulated what could become the 'Yeltsin doctrine': 'The moment has come when responsible international institutions, including the United Nations, should grant Russia special powers as guarantor of peace and stability in the region of the former [Soviet] Union'.[88] There are also some indications that ensuring stability in the immediate vicinity could be only one rationale—the others probably being related to traditional considerations of power projection.[89]

The potential for conflict in the post-Soviet era is exacerbated by political instability, the increasing role of ethno-nationalism and controversies over military assets. The insufficient experience of the political élites and the feverish search by the new states for political self-identification and higher international status have also contributed to conflicts.

[85] Tanks and aircraft (including helicopters and Su-25 attack bombers) from Uzbekistan were reported to participate in fighting against so-called 'islamo-democratic' forces retaining power in the second half of 1992. See *Le Monde*, 27 Jan. 1993, p. 4.

[86] Despite official statements from Moscow concerning the 'strict neutrality' of Russian troops deployed in the conflict area, by mid-March 1993 there have appeared numerous reports of their direct participation in combat as well as of the supply of quantities of weapons and military equipment to Abkhazian irregulars. See *Izvestia*, 17 Mar. 1993, p. 1–2.

[87] Makartsev, V., 'Afghanskiy sindrom stanovits'a real'nostyu', *Izvestia*, 24 Feb. 1993, p. 5.

[88] See *Financial Times*, 1 Mar. 1993, p. 1.

[89] Defence Minister Pavel Grachev stated flatly that Russian troops would not be taken out of Abkhazia because 'otherwise we would lose the Black Sea'. See *Financial Times*, 25 Feb. 1993, p. 2.

However, the post-Soviet heritage is not exclusively a conflict-generating factor. The new states in many respects are so interdependent that they are linked in ways which were not and cannot be eliminated by a political act. Their viability will substantially depend on their ability to maintain co-operative interaction with each other, whatever disagreements and disputes may arise.

Economic interdependence among the former republics failed to prevent political disintegration. As it turned out, the belief that it would was not only illusory but also even misleading. Conversely, other conditions being equal, economic ties could be helpful in upgrading political interaction—such as in the case of Russia and Kazakhstan. Still, economic ties become almost irrelevant (at least in the short run) in light of non-economic conflicts over territories, ethnic issues, political status and so on (such as in the case of Russia and Ukraine, or Armenia and Azerbaijan). During 1992 'high politics' certainly prevailed over 'low politics', and the behaviour of the newly independent states was much more affected by the emerging realities (and myths) of national self-identification than by pragmatic economic considerations.

But this could (and most probably will) change over the long run—hopefully before the total collapse of economy occurs. Although economic interdependence did not generate incentives sufficient to reduce conflict potential in 1992, some positive signs were evident by the end of the year. If the post-independence euphoria is still alive in some of the new states, the initial enthusiasm seems gradually to be giving way to more sober assessments of the catastrophic consequences that could result from breaking economic links. Significantly, this refers both to the Baltic states (despite strong political momentum for a complete disengagement) and to Ukraine (despite its potential for self-sufficiency which is greater than any other former Soviet republic except Russia).[90] When political leaders persist in ignoring this reality, they face the prospect of losing power, whatever their 'merits' in achieving independence might have been.[91]

There is, however, the opposite side of the coin as well. The logic of conflicts pushs the participants towards using 'economic weapons' for achieving their goals—such as in the Trans-Dniester conflict, the war between Armenia and Azerbaijan and the case of the Chechnian breakaway republic. In 1992 there were hardly any visible results in terms of conflict resolution—which, however, does not mean that such attempts will be abandoned in the future. On the contrary, there are clear indications that the linkage approach will be implicitly or explicitly used on an even greater scale—especially by economi-

[90] The new Prime Minister of Ukraine, Leonid Kuchma, stated in Parliament that during the first year of independence Ukraine 'had gone through the economic war—and lost it'. See *Komsomol'skaya Pravda*, 5 Jan 1993, p. 1.

[91] The overthrow of Zviad Gamsakhurdia in Georgia was probably accelerated by the dramatic situation in the national economy. There are numerous reports on increasing opposition to Dzhokhar Dudayev in Chechnia experiencing serious difficulties because of isolation from Russia. As for the 'normal' democratic patterns, the defeat of Vytautas Landsbergis in the parliamentary elections in Lithuania also seems more than symptomatic.

cally more powerful post-Soviet actors (which, certainly, applies in the first instance to Russia).[92]

Thus a more pragmatic and economically focused agenda will not necessarily be less fraught with the danger of conflicts between the former Soviet republics. However, hopefully it will create incentives for depoliticizing the whole process of mutual accommodation and making it less dramatic as compared with the very initial stage of 'divorce' between the post-Soviet states. The confrontational orientation (reminiscent, in a sense, of the old spirit of East–West relations) could become gradually (although painfully) mitigated by elements of co-operation similar to those prevailing among OECD (Organization for Economic Co-operation and Development) countries. This will certainly not eliminate the conflicts (such as those existing in trans-Atlantic or in Japanese–US relations), but it could change their character and make them more manageable.

The current attempts to find solutions to (or at least to prevent the aggravation of) conflicts inherited from the USSR and initiated by its breakup represent only the first steps. Even if the results of conflict-management—bilateral or multilateral on the level of the CIS—have not been particularly impressive in 1992, the new states have also acquired a certain positive experience. Still there will certainly be a need for more consistent efforts in guiding future post-Soviet developments.

[92] In Feb. 1993, Moscow announced an increase of the price for natural gas delivered to Ukraine to world levels—an increase by 2500%. This move was in accordance both with the logic of the free market as well as the status of relations between Russia and Ukraine as independent states. Significantly, however, Russian Deputy Prime Minister Viktor Shokhin openly stated that Ukraine could have subsidized energy only if it made concessions over the Black Sea fleet and allowed Russian military bases to be established in Ukraine. See *Financial Times*, 19 Feb. 1993, p. 2.

# **Appendix 4A.** Chronology of conflict developments in the former USSR, 1992

SHANNON KILE

For the convenience of the reader, key words are indicated in the right-hand column, opposite each entry. They refer to the subject-areas covered in the entry. Definitions of the acronyms can be found in the glossary.

| | | |
|---|---|---|
| *2 Jan.* | The Ukrainian Ministry for Defence announces that effective 3 Jan. all non-strategic former Soviet military forces stationed on Ukrainian territory are placed under the command authority of the President; the strategic forces based in Ukraine are to remain under joint CIS (Commonwealth of Independent States) command. The announcement emphasizes that the Black Sea Fleet is not considered to be part of the strategic forces, although it does have a nuclear capability. | Ukraine |
| *2 Jan.* | In Georgia, opposition leaders form a ruling Military Council and announce the suspension of Parliament and the deposing of President Zviad Gamsakhurdia. | Transcaucasus |
| *6 Jan.* | Ousted Georgian President Gamsakhurdia flees Tbilisi for temporary exile in Armenia. Growing unrest among Gamsakhurdia loyalists is reported in the major towns in western Georgia. | Transcaucasus |
| *11 Jan.* | Parliament votes to subordinate all former Soviet troops in the republic to Belarussian jurisdiction and to create a national defence ministry. The strategic forces are to remain under joint CIS command. | Belarus |
| *19 Jan.* | A referendum held in South Ossetia in Georgia produces a heavy majority in favour of integration into the Russian Federation. | Transcaucasus |
| *29 Jan.* | In response to US President George Bush's announcement of unilateral defence cuts in his State of the Union Address, Russian President Boris Yeltsin unveils a set of deep military cuts, including a proposal to reduce Russian and US strategic nuclear arsenals to 2000–2500 warheads. Ukrainian President Leonid Kravchuk and Kazakh President Nursultan Nazarbayev complain that they had not been consulted by Yeltsin. | Russia/ International |

| | | |
|---|---|---|
| *30 Jan.* | Armenia, Azerbaijan, Belarus, Kazakhstan, Kyrgyzstan, Moldova, Tajikistan, Turkmenistan, Ukraine and Uzbekistan are admitted as CSCE (Conference on Security and Co-operation in Europe) member states. Georgia had not applied for membership; Russia and the Baltic states were already members. | CIS/CSCE |
| *5 Feb.* | Azerbaijani President Ayaz Mutalibov rejects Armenian calls for the involvement of UN peace-keeping forces in Nagorno-Karabakh. | Transcaucasus |
| *12 Feb.* | A CSCE-mandated delegation begins a fact-finding mission to Armenia and Azerbaijan. | Transcaucasus/ CSCE |
| *14 Feb.* | At a CIS summit meeting in Minsk, 10 member states (all except Moldova and with the qualified agreement of Kazakhstan and Ukraine) sign an accord broadly defining the status of the former Soviet strategic forces and stipulating that their commander is subordinate to the CIS Council of Heads of State. In addition, Russia and seven other republics agree to central command over conventional forces for an interim period of at least two years; Azerbaijan, Moldova and Ukraine insist on creating their own independent national armed forces. Qualified agreement is also reached on drafting a single CIS defence budget.<br>Former Soviet Defence Minister Yevgeniy Shaposhnikov is formally confirmed in his appointment as Commander-in-Chief of the CIS Joint Forces for a 2–3 year period. | CIS |
| *16 Feb.* | President Nazarbayev states that while Kazakhstan aspires to non-nuclear weapon status, it is a nuclear weapon state as defined under the terms of the 1968 Non-Proliferation Treaty (NPT). | Kazakhstan |
| *20 Feb.* | Meeting in Moscow, the foreign ministers of Armenia, Azerbaijan and Russia sign a joint communiqué agreeing on the need for an immediate cease-fire and further negotiations on a settlement of the Nagorno-Karabakh conflict. Armenia and Azerbaijan continue to disagree over the participation of representatives from the disputed enclave and over the possible deployment of CSCE or UN peace-keeping troops there. | Transcaucasus |
| *1–3 Mar.* | Fighting erupts on the left bank of the Dniester River when armed 'Trans-Dniester Guard' units attack Moldovan police stations. | Moldova |
| *10 Mar.* | Georgia's ruling Military Council dissolves itself and is replaced by a temporary State Council, combining both executive and legislative powers, to be headed by former Soviet Foreign Minister Eduard Shevardnadze. | Transcaucasus |

| | | |
|---|---|---|
| *12 Mar.* | President Kravchuk halts the transfer of former Soviet tactical nuclear weapons from Ukraine, citing concern that the weapons might not be destroyed by Russia as promised or might 'fall into the wrong hands'. | Ukraine |
| *16 Mar.* | President Yeltsin issues a decree creating a Russian Federation Defence Ministry to be headed temporarily by himself. | Russia |
| *16 Mar.* | While insisting that Kazakhstan wants to remain part of the CIS Joint Forces, President Nazarbayev issues a decree establishing an independent national guard. | Kazakhstan |
| *20 Mar.* | CIS leaders meet in Kiev to discuss the division of former Soviet military assets and arrangements for a common defence framework. Seven CIS member states (all except Azerbaijan, Moldova, Turkmenistan and Ukraine) sign an agreement designating the Council of Heads of State as the highest body in defence matters. The seven also sign a document defining the status of the High Command of the Joint CIS Armed Forces; however, only four states—Armenia, Kazakhstan, Russia and Tajikistan—sign an agreement defining the status of the CIS general purpose forces. All the member states (except Turkmenistan) agree to the creation of a voluntary peace-keeping force which can be deployed to implement cease-fire arrangements upon the request of all parties involved in a conflict.<br>Many other key issues, such as the implementation of the Treaty between the United States and the Soviet Union on the reduction and limitation of strategic offensive arms (the START Treaty), conscription and the financing of the armed forces, are unresolved. | CIS |
| *20 Mar.* | The Belarussian Supreme Soviet votes to set up national armed forces to consist initially of 90 000–100 000 men. | Belarus |
| *21 Mar.* | Voters in Tatarstan approve a referendum calling for the republic to be granted sovereign status within the Russian Federation. | Russia |
| *31 Mar.* | Delegates from 18 of the 20 autonomous republics constituting the Russian Federation sign a Federal Treaty defining the powers of the central government and the regions. Chechnia and Tatarstan reject the Treaty. | Russia |
| *6 Apr.* | A quadripartite meeting in Chisinau between the foreign ministers of Moldova, Romania, Russia and Ukraine results in the signing of a cease-fire agreement for Trans-Dniester, effective 7 Apr. | Moldova |

| | | |
|---|---|---|
| *11 Apr.* | At a meeting in Moscow, defence and foreign ministry officials from Belarus, Kazakhstan, Ukraine and Russia fail to reach agreement on how to implement the START Treaty; Russia reportedly wants the Treaty with the USA to remain a bilateral one, while Ukraine insists that it be made a formal party to the Treaty. | CIS |
| *16 Apr.* | The suspended withdrawal of former Soviet tactical nuclear weapons from Ukrainian territory resumes as Presidents Yeltsin and Kravchuk agree to form a joint commission to monitor the transfer of the weapons to Russia and to oversee their subsequent dismantlement. | Ukraine/Russia |
| *28 Apr.* | The Belarussian Defence Ministry announces that the transfer of former Soviet tactical nuclear weapons from Belarus to Russia is complete. | Belarus/Russia |
| *5 May* | Tajik President Rakhmon Nabiyev declares a temporary state of emergency in the capital, Dushanbe. Facing mounting anti-government protests spearheaded by the Islamic Renaissance Party, government and opposition representatives agree on 11 May to form a coalition cabinet, with Nabiyev remaining as President. | Central Asia |
| *5 May* | The Crimean Supreme Soviet proclaims its intention to declare independence from Ukraine, pending the outcome of negotiations with Kiev and a referendum scheduled for 2 Aug. | Ukraine |
| *7 May* | President Kravchuk announces the completion of the transfer to Russia of the remaining former Soviet tactical nuclear weapons stockpiled on Ukrainian territory. | Ukraine/Russia |
| *6 May* | President Yeltsin issues a decree creating independent Russian Federation armed forces and installing himself as Commander-in-Chief. The new armed forces will include all former Soviet soldiers and military installations on Russian territory, as well as troops and naval forces under Russian jurisdiction based outside the republic. Strategic nuclear weapons will remain under the command authority of the CIS Joint Forces. | Russia/CIS |
| *14 May* | Meeting in emergency session, the Azerbaijani Parliament reinstates President Mutalibov, who had been ousted on 6 Mar. after a series of military reversals, and cancels the presidential elections set for 7 June.<br>On 15 May leaders of the Azerbaijani Popular Front, which had been expected to win the June poll, denounce Mutalibov's return as a 'coup'; with the tacit support of the army, Popular Front supporters force the government to flee after violent street protests. The formation of a new government and suspension of the day-old state of emergency are announced. | Transcaucasus |

| | | |
|---|---|---|
| *15 May* | At a CIS summit meeting in Tashkent (Uzbekistan), Armenia, Kazakhstan, Kyrgyzstan, Russia, Tajikistan and Uzbekistan sign a five-year collective security treaty providing for mutual military aid in case of aggression against any of the signatories.<br>The meeting also produces an agreement signed by Armenia, Azerbaijan, Belarus, Georgia, Kazakhstan, Moldova, Russia and Ukraine confirming the allocation of CFE Treaty-limited equipment on their territories. | CIS |
| *21 May* | On the last day of the deadline set by Kiev, the Crimean Supreme Soviet votes to rescind its earlier independence proclamation and to cancel a planned referendum on the issue. | Ukraine |
| *21 May* | The Russian Supreme Soviet votes in closed session to rescind the 1954 decree ceding Crimea from Russia to Ukraine. | Russia/Ukraine |
| *22 May* | President Yeltsin and Polish President Lech Walesa sign a treaty of friendship and co-operation in Moscow. | Russia/Poland/ International |
| *23 May* | Officials from Belarus, Kazakhstan, Russia and Ukraine confer in Lisbon, Portugal, and sign a protocol with the United States committing themselves to adhere to the former Soviet Union's START obligations. In return for becoming formal parties to the accord, Belarus, Kazakhstan and Ukraine pledge to accede to the 1968 NPT as non-nuclear weapon states. | CIS/ International |
| *25 May* | Russia and Kazakhstan sign a bilateral treaty on friendship, co-operation and mutual assistance. | Kazakhstan/ Russia |
| *4 June* | The Ukrainian Parliament approves a resolution rejecting the 21 May vote by the Russian Supreme Soviet to annul the 1954 decree transferring Crimea from Russia to the Ukraine. | Ukraine |
| *4 June* | The Russian Supreme Soviet approves legislation making Ingushetia, formerly part of the Chechen-Ingushetia Autonomous Republic, a new republic within the Russian Federation. The split had been forced by Chechnia's earlier secession declaration and refusal to sign the Russian Federal Treaty. | Russia |
| *16 June* | The separatist 'Trans-Dniester Republic Parliament' rejects the cease-fire proposals approved by the Moldovan Parliament on 9 June. | Moldova |
| *21 June* | The constituent congress establishing the Civic Union is held in Moscow, bringing together into a 'constructive' opposition coalition a number of influential political groupings calling for a more gradual transition to a market economy and a greater emphasis on halting the slide of industrial production in the country. | Russia |

| | | |
|---|---|---|
| *21–22 June* | The town of Bendery, which had been seized by Moldovan forces on 20 June, is recaptured by Trans-Dniestrian insurgents after heavy fighting. Senior Russian army officers acknowledge that units of the 14th Army participated in the counter-attack to dislodge the Moldovan forces from the city following a reported Moldovan air raid. | Moldova |
| *22 June* | Marking a change in the Ukrainian position on the conflict in neighbouring Moldova, President Kravchuk urges the Moldovan leadership to grant the disputed Trans-Dniester region the status of an autonomous republic within the country. | Ukraine/ Moldova |
| *23 June* | Meeting in the Black Sea town of Dagomys, Presidents Yeltsin and Kravchuk agree to draw up a wide-ranging political treaty aimed at improving bilateral relations between the two republics. Among the points to be included in the agreement are decisions on the division of the assets the Black Sea Fleet and joint command arrangements for the CIS strategic forces. The issue of Crimea is excluded from the agenda. | Russia/Ukraine |
| *28 June* | In a nation-wide referendum drawing a heavy turnout, 91.37% of Estonians vote in favour of the new post-Soviet constitution. | Baltic states |
| *1 July* | The Ukrainian Parliament ratifies the 1990 CFE Treaty. | Ukraine |
| *2–3 July* | CIS defence and foreign ministers hold talks in Moscow but are unable to resolve differences arising from Ukraine's insistence on retaining 'administrative control' over the strategic nuclear weapons based on its territory. | CIS/Ukraine |
| *2 July* | The Kazakh Supreme Soviet ratifies the 1991 START Treaty. | Central Asia |
| *6 July* | At a summit meeting in Moscow (Azerbaijan does not participate), CIS leaders agree in principle to establish joint peace-keeping forces to intervene in ethnic conflicts within the CIS; the first deployment of the peace-keepers is planned for eastern Moldova. In addition, the leaders reach agreements on financing the CIS Joint Forces and on establishing a council of defence ministers. They also sign protocols concerning missile early warning systems, space flight mission control and air defence arrangements. | CIS |
| *8 July* | The Russian Parliament ratifies the 1990 CFE Treaty. | Russia/CFE |
| *9 July* | In a speech delivered at the CSCE summit meeting in Helsinki, Moldovan President Mircea Snegur calls for the employment of CSCE peace-keeping mechanisms in the ongoing Trans-Dniester conflict. | Moldova/CSCE |

| | | |
|---|---|---|
| *14 July* | Pursuant to the 24 June agreement between Presidents Yeltsin and Shevardnadze, a peace-keeping force comprised of Russian, Georgian and Ossetian troops deploys in the Tskhinvali region of South Ossetia to create a buffer zone between the warring parties; the deployment marks the first use of multilateral peace-keeping forces on the territory of the former USSR. | Transcaucasus/ Russia |
| *15 July* | Meeting in Tashkent, CIS foreign and defence ministers (Moldovan and Turkmen ministers do not attend) approve a protocol on deploying peace-keeping forces to areas of ethnic conflict; the deployment of such forces would require the consent of all parties involved in the conflict. Belarussian representatives refuse to sign the accord. | CIS |
| *17 July* | The Russian Parliament passes a resolution threatening to impose unspecified economic sanctions against Estonia in response to alleged legislative discrimination against ethnic Russians and other 'human rights violations' in the Baltic republic. | Russia/Baltic states |
| *20 July* | Belarussian and Russian officials sign a bilateral military co-ordination agreement. Strategic forces units are to remain in Belarus under Moscow's command. | Belarus/Russia |
| *21 July* | Presidents Yeltsin and Snegur sign an accord in Moscow on ending the fighting in eastern Moldova. The agreement envisages granting special status to the Trans-Dniester area and a right to self-determination in case of Moldova's future reunification with Romania. It also reaffirms the inviolability of present borders and the commitment of both parties to observe human rights in accordance with CSCE principles. | Moldova/Russia |
| *23 July* | The Georgian Autonomous Republic of Abkhazia issues a declaration of sovereignty; the declaration is rejected by the Georgian State Council two days later. | Transcaucasus |
| *29 July* | The trilateral Moldovan, Russian and 'Trans-Dniestrian' peace-keeping force begins to operate in Moldova. | Moldova |
| *31 July* | Moldovan President Snegur appeals for the sending of UN observers to oversee the implementation of the Russian–Moldovan accord for resolving the conflict in Trans-Dniester. | Moldova |
| *3 Aug.* | Meeting in Yalta, Presidents Kravchuk and Yeltsin conclude an interim agreement for resolving the dispute over the control of the Black Sea Fleet. The Fleet is to be removed from the command authority of the CIS Joint Forces and temporarily placed directly under the joint jurisdiction of Kiev and Moscow. Both states are to be allowed equal access to the Fleet's bases and facilities dur- | Russia/Ukraine |

| | | |
|---|---|---|
| | ing a transition period extending to 1995, at which time the assets of the fleet will be divided. | |
| *9 Aug.* | With Azerbaijani military pressure intensifying, Armenian President Levon Ter-Petrosyan issues an appeal to the other CIS states party to the Tashkent collective security agreement urging them to 'fulfil their obligations to the Republic of Armenia'. | CIS/Trans-caucasus |
| *14 Aug.* | The ruling State Council orders Georgian army units into Abkhazia to secure the release of a group of Georgian officials abducted in Sukhumi; the units are reportedly involved in a series of skirmishes with Abkhaz interior ministry troops. The Abkhaz Parliament denounces the Georgian incursion as 'an occupation'. | Transcaucasus |
| *25 Aug.* | Following stormy debates on the future of the Kurile Islands, Russian parliamentary leaders caution President Yeltsin against handing over sovereignty of the islands to Japan. Yeltsin's visit to Tokyo is subsequently cancelled. | Russia/ International |
| *27 Aug.* | Armenian President Levon Ter-Petrosyan announces that he and his Azerbaijani counterpart, Ebulfez Elcibey, accept a CSCE-sponsored proposal for a 60-day cease-fire. Representatives of the two republics sign a cease-fire agreement on 28 Aug:, but fighting erupts before the agreement takes effect. | Transcaucasus |
| *3 Sep.* | Following several weeks of heavy fighting in western Georgia, tripartite negotiations between Abkhazia, Georgia and Russia result in the signing of a cease-fire agreement to go into effect on 3 Sep. A control commission is established to oversee implementation of the cease-fire, which is violated almost immediately. | Transcaucasus |
| *3–4 Sep.* | CIS defence ministers meet in Moscow to conclude an agreement covering, among other issues, the transfer of personnel from the former Soviet armed forces to the armed forces of individual CIS member states and the legal status of the CIS Joint Forces High Command. No agreement is reached on the status of the strategic nuclear forces. | CIS |
| *3 Sep.* | The presidents of Kazakhstan, Kyrgyzstan, Russia and Uzbekistan issue a joint communiqué warning government and opposition factions inside Tajikistan that the widening civil conflict there poses a danger to the CIS that could provoke a CIS intervention. | Central Asia/ CIS |
| *7 Sep.* | Tajik President Nabiyev announces his resignation following his detention by armed opponents at the Dushanbe airport. On 3 Sep. the Presidium of the Tajik Supreme Soviet had issued a call for Nabiyev to be ousted. | Central Asia |

| | | |
|---|---|---|
| *8 Sep.* | The defence ministers of Lithuania and Russia agree to a schedule that provides for the complete withdrawal of former Soviet troops stationed on Lithuanian territory by 31 Aug. 1993. | Baltic states |
| *19 Sep.* | The defence ministers of Armenia, Azerbaijan, Georgia and Russia sign a cease-fire agreement to halt the fighting spreading along the Armenian–Azerbaijani border and in the disputed enclave of Nagorno-Karabakh. The agreement includes a protocol on stationing cease-fire observers from other CIS member states. Clashes continue to be reported after the cease-fire goes into effect on 25 Sep. | CIS |
| *20 Sep.* | In Estonia's first elections as an independent state since Soviet occupation began in 1940, the nationalist Fatherland coalition (Isamaa) win a majority of the seats in the Estonian Parliament. The vote, conducted under laws barring most ethnic Russians from the polls as non-citizens, draws complaints from Russia and complicates negotiations over the withdrawal of former Soviet troops from Estonia. | Baltic states |
| *25 Sep.* | The Russian Supreme Soviet issues a declaration denouncing the Georgian Government for resorting to military force to solve inter-ethnic problems in Abkhazia and demanding that it withdraw its armed militia units from the region. | Transcaucasus/ Russia |
| *27 Sep.* | A 60-day state of emergency is imposed in Nalchik, capital of the Russian autonomous republic of Kabardino-Balkaria, following violent demonstrations protesting the detention by Russian Interior Ministry troops of the leader of the Confederation of Mountainous Peoples, Musa Shanibov, for his alleged involvement in the unrest in Abkhazia. | Transcaucasus |
| *7 Oct.* | The Azerbaijani National Council, the country's interim supreme legislative body, votes against membership in the CIS by refusing to ratify the Dec. 1991 Alma-Ata agreement. | Transcaucasus/ CIS |
| *9 Oct.* | A CIS summit meeting in Bishkek results in the signing of 15 documents relating to military and economic matters. The meeting confirms the emergence of a 'dual-track' approach to the CIS, with Moldova and Ukraine resisting calls for a larger CIS role in economic and defence matters. The contentious issue of future command and control arrangements for the strategic nuclear forces is unresolved as Ukraine rejects proposals to transfer control of the former Soviet nuclear arsenal to Russia. The leaders also fail to agree on dispatching CIS peace-keeping troops to curb an upsurge in the fighting in Tajikistan. | CIS |

| | | |
|---|---|---|
| *11 Oct.* | In Georgia's legislative elections, Eduard Shevardnadze is overwhelmingly elected to be chairman of the new parliament, in effect making him president and head of state. | Transcaucasus |
| *12 Oct.* | President Yeltsin and Azerbaijani President Ebulfez Elcibey sign an economic co-operation and mutual security agreement in Moscow. | Russia/Transcaucasus |
| *23 Oct.* | Moldovan President Snegur and Ukrainian President Kravchuk sign a Treaty of Friendship and Co-operation. The Treaty prohibits armed groups hostile to one of the signatories from transit across the territory of the other (a provision added to prevent Russian Cossack volunteers from crossing Ukraine to reach the disputed Trans-Dniester region). | Moldova/Ukraine |
| *25 Oct.* | In the first round of Lithuania's first post-Soviet era parliamentary elections, the Democratic Labour Party led by former Lithuanian Communist Party General Secretary Algirdas Brazauskas wins a surprise victory over the ruling nationalist Sajudis coalition and its chairman and head of state, Vytautas Landsbergis. | Baltic states |
| *29 Oct.* | President Yeltsin orders the suspension of the withdrawal of Russian troops from the Baltic states, citing a lack of facilities in Russia to house the returning soldiers and the need for the Baltic governments to provide 'social guarantees' for the troops. | Baltic states/Russia |
| *2–3 Nov.* | A one-month state of emergency is imposed along the borders of North Ossetia and Ingushetia following efforts by armed Ingush militias to recover territory from which Ingush residents had been deported in 1944. | Russia |
| *3 Nov.* | The Georgian Defence Ministry accuses Russia of having launched air and artillery attacks against Georgian militia positions in Abkhazia. Russian Defence Minister Pavel Grachev warns that he will order 'direct military action' if Georgian forces do not return the arms and ammunition they allegedly seized from a Russian base the previous day. | Transcaucasus/Russia |
| *4 Nov.* | Meeting in Moscow, the CIS Council of Defence Ministers reach a compromise agreement on the composition of the CIS strategic forces. Ukraine continues to reject CIS Commander-in-Chief Shaposhnikov's view that the strategic forces should be subordinated to Russian command. The ministers agree to set up a 'Committee on Nuclear Policy' to promote a common approach to nuclear issues. | CIS |
| *4 Nov.* | The Russian Parliament ratifies the 1991 START Treaty. | Russia/International |

| | | |
|---|---|---|
| *4 Nov.* | Meeting in Alma-Ata to discuss ways to halt the widening civil war in Tajikistan, the leaders of Kazakhstan, Kyrgyzstan, Russia and Uzbekistan agree on a five-point plan calling for, among other things, the continued presence of Russian troops and the formation of an all-party state council. | Central Asia |
| *6 Nov.* | The Tatar Parliament approves a controversial new constitution for the 'sovereign state' of Tatarstan which, while recognizing that the republic is 'associated with the Russian Federation', does not acknowledge that it is part of Russia. | Russia |
| *6 Nov.* | Russian President Yeltsin appeals to the UN Secretary-General to 'take all necessary measures' to halt the alleged abuses of the human rights of ethnic Russians living in the three Baltic states. | Baltic states/ Russia/UN |
| *10 Nov.* | Tajik President Akbarsho Iskandarov and his coalition government are forced to resign as militia forces backing the former leadership, having gained control of the southern provinces of the country, lay siege to the capital, Dushanbe. | Central Asia |
| *16 Nov.* | Russian and Chechnian officials reach an agreement to pull back their forces along the disputed Checheno-Ingushetian border. | Russia |
| *19 Nov.* | Meeting in the town of Hojend, the Tajik Supreme Soviet elects Imamali Rakhmonov as its new chairman and *de facto* head of state. | Central Asia |
| *25 Nov.* | The UN General Assembly adopts a resolution supporting an expeditious withdrawal of Russian troops from Estonia, Latvia and Lithuania. | Baltic states/ Russia/UN |
| *27 Nov.* | The communist-dominated Tajik Supreme Soviet declares the establishment of a parliamentary republic and votes to abolish presidential rule. | Central Asia |
| *30 Nov.* | The defence ministers of Kazakhstan, Kyrgyzstan, Russia and Uzbekistan reach an agreement to establish a joint peace-keeping force to be deployed in Tajikistan. The force will consist of units drawn from each of the republics. | Central Asia/ Russia |
| *1–14 Dec.* | Delegates attending the Congress of People's Deputies vote by a narrow margin to reject Yeltsin's nomination of Yegor Gaidar to the post of prime minister. Denouncing the Congress as a 'stronghold of conservative forces and reaction', Yeltsin threatens to appeal directly to Russian voters to launch a referendum drive to dissolve the Congress and the Supreme Soviet. | Russia |

The Congress adopts several laws to limit the executive powers of the presidency and to block what it calls Yeltsin's 'anti-constitutional activities'.

Following lengthy negotiations between President Yeltsin and conservative opposition leaders, the Congress approves Yeltsin's compromise nomination of a centrist candidate, Viktor Chernomyrdin, as prime minister. A referendum to resolve the constitutional dispute over the powers of the presidency and the legislature is scheduled to be held in April 1993.

*10 Dec.* Following a week of heavy fighting, pro-government forces led by Tajik Supreme Soviet Chairman Imamali Rakhmonov retake control of the capital, Dushanbe. Despite the commencement of peace talks with Islamic opposition forces, fierce fighting continues to be reported in the southern provinces of the country and around Dushanbe. Central Asia

*17 Dec.* Speaking before Parliament, Georgian Prime Minister Tengiz Sigua describes relations between his country and Russia as being in a 'crisis' following the alleged downing by Georgian forces of a Russian helicopter; he denies that the countries are at war. Russian Defence Minister Pavel Grachev threatens to take 'military actions of a decisive nature'. Transcaucasus/ Russia

# 5. The CSCE: towards a security organization

ADAM DANIEL ROTFELD

## I. Introduction

The end of the cold war brought about a revolutionary change in the nature of the threats to security in Europe. With the threat of a sudden and unexpected attack from outside Europe gone, the risk that domestic conflicts might escalate into international conflicts became part and parcel of European reality. The shortcomings of existing security structures, their relatively low efficiency and their inadequacy in the face of new requirements and challenges were laid particularly bare by the events and conflicts in 1992.

However, the multilateral institutions established by the Conference on Security and Co-operation in Europe (CSCE) to further European security underwent a progressive transformation, and in 1992 considerable progress was made towards the construction of a security system across Europe, North America and the vast expanse of Asia. Decisions adopted at the summit meeting of the heads of state or government of the CSCE participating states (held in Helsinki, 9–10 July 1992)[1] and during the meeting of the CSCE Council of Foreign Ministers (Stockholm, 14–15 December 1992)[2] further developed the new political strategy and character of CSCE activities, agreed upon at the Paris summit meeting (19–21 November 1990) and the subsequent Berlin (19–20 June 1991) and Prague meetings (30–31 January 1992) of the CSCE Council of Foreign Ministers.[3]

In the wake of the revolutionary changes in Europe in 1989–90, the decisions of the 1990 Paris summit meeting laid common democratic foundations for the CSCE process, applicable from Vancouver to Vladivostok, and numerous institutions and mechanisms were established for co-operation among states in the security field. Most important, the documents adopted in Paris epitomized an accepted common system of values: democracy and the rule of law, pluralism and market economics, and respect for human rights and fundamental freedoms, including the rights of national minorities.[4] Originally, the

[1] Decisions of the summit meeting are published in CSCE *Helsinki Document 1992: The Challenges of Change*, Helsinki summit meeting, Helsinki, 10 July 1992. See appendix 5A for excerpts. Since July 1992 the participation of Yugoslavia (Serbia and Montenegro) in the CSCE has been suspended. At the time of writing there are 53 CSCE states (on 1 Jan. 1993, Czechoslovakia split into two states, the Czech Republic and the Slovak Republic). For a list of members see p. xxviii.

[2] Excerpts from the Summary of Conclusions of the Stockholm Council Meeting, Third Meeting of the CSCE Council, Stockholm, 15 Dec. 1992 are reproduced in appendix 5A.

[3] For more information see Rotfeld, A. D., 'European structures in transition', *SIPRI Yearbook 1992: World Armaments and Disarmament* (Oxford University Press: Oxford, 1992), pp. 563–82, and the relevant documents in appendix 15A of the same volume, pp. 583–94.

[4] See 'The Charter of Paris for a New Europe', in *SIPRI Yearbook 1991: World Armaments and Disarmament* (Oxford University Press: Oxford, 1991), appendix 17B, pp. 603–10.

CSCE principles were intended to guide relations between states. A new phenomenon which has arisen since the 1990 CSCE Paris summit meeting is that the system agreed by the participating states is now addressed chiefly, if not exclusively, to the sphere of domestic rather than international relations. Historically, the political and social system of a state has been considered to be entirely an internal matter. Since the adoption of the 1945 UN Charter and the 1948 Declaration of Universal Human Rights, however, human rights have been increasingly subject to obligations that have constrained the discretionary power of states. The principles and norms of the CSCE have not only given teeth to these commitments, but they have also been included in regulations for ensuring international peace and security. The 1975 Helsinki Final Act made it more difficult for the participating states to invoke the principle of non-intervention in internal affairs in response to allegations of violations of human rights and fundamental freedoms. None the less it was a principle which was intended to and did create a kind of umbrella protecting states that did not want to be subject to international assessment of their domestic practices and regulations.

The developments and changes in Central and Eastern Europe in 1989–92 made it possible to widen the area of application of CSCE agreements. In place of antagonistic blocs, a common system of values that includes the foundations of a democratic order now prevails across the whole CSCE area. The adoption of those principles, however, is not tantamount to their practical application in the newly founded states. The key issue now is to make the newly adopted principles and norms that determine the common system of values more operational. In 1992, with this aim in mind, new procedures were agreed upon, and new institutions and tools for monitoring the implementation of decisions were established.

## II. New challenges, new tasks

In post-cold war Europe it is not relations between states but domestic developments, and particularly mass-scale violations of human rights and the collective rights of national, ethnic and religious groups, that constitute the main sources of conflict.[5] The structures of the CSCE seem more appropriate than those of NATO or the Western European Union (WEU) for solving such conflicts.

The inclusion in the Helsinki Final Act of Basket III issues, 'Co-operation in Humanitarian and other Fields' (human contacts, information, culture and education), combined with security principles, determined the uniqueness of the CSCE process in 1975–90 and helped overcome the existing East–West divisions. The end of the cold war, corroborated by the 1990 Paris summit

[5] See chapter 3 in this volume.

decisions,[6] thrust new tasks upon the CSCE, and new structures[7] and institutions[8] were created in order to execute them.

A first test of the workability and effectiveness of the new institutions was the outbreak of armed conflict in Yugoslavia.[9] The first meeting of the CSCE Council of Foreign Ministers in Berlin (19–20 June 1991) issued an agreed statement on the situation in Yugoslavia; it also established procedures for consultation and co-operation in emergency situations.[10] Looking back on two years of conflict in Yugoslavia, it is difficult to estimate the extent to which the CSCE could have averted a situation in which giving effect to the right of self-determination of the nations constituting the Yugoslav Federation led to armed conflict. What is clear is that general messages and appeals drawn up in the CSCE framework have virtually remained a dead letter.[11] The confidence- and security-building measures (CSBMs) and emergency mechanisms agreed within the CSCE were adequate to the new state of trust and mutual understanding developing among the negotiators, but fared poorly when faced with the real world. Admittedly, the agreed procedures and mechanisms were addressed to states; their aim was to ward off situations in which conflict of interests between states could escalate into armed conflict. It was assumed that obtaining information and convening a CSCE meeting within 48 hours to consider an emergency situation would in itself encourage restraint among the parties and prevent conflicts. In reality, this is not the case. The prevention and solution of conflict situations require immediate and determined responses rather than meetings, debates and resolutions. This implies a radical change in the method of operation of the CSCE: the process initiated in Helsinki should be given the character of a security organization, with organs not bound by the consensus rule. Such bodies would also need to be equipped to ensure implementation.

The decisions of the meeting of the Council of Foreign Ministers in Berlin on emergency situations marked a departure from the consensus rule, establishing that the agreed procedures could be set in motion at the request of one state supported by at least 12 other participating states.[12] Under this formula,

[6] The Charter of Paris for a New Europe (note 4).

[7] Rotfeld, A. D., 'New security structures', *SIPRI Yearbook 1991* (note 4), pp. 585–86.

[8] Rotfeld (note 3), p. 577.

[9] The background of the conflict in Yugoslavia has been analysed by Nelson, N. D., *Balkan Imbroglio: Politics and Security in Southeastern Europe* (Westview Press, Boulder, Colo., 1991); and Vukadinovic, R, 'Yugoslavia and the East: from non-alignment to disintegration', *Yearbook of European Studies*, vol. 5 (Rodopi: Amsterdam 1992), pp. 147–73.

[10] Summary of Conclusions, Annex 2, Berlin Meeting of the CSCE Council, Berlin, 19–20 June 1991, pp. 12–13; reproduced in *SIPRI Yearbook 1992* (note 3), appendix 15A.

[11] 'Ministers stressed that it is only for the peoples of Yugoslavia themselves to decide on the country's future . . . They urged all parties concerned to redouble their efforts to resolve their differences peacefully through negotiations.' Statement on the situation in Yugoslavia, Berlin, 19–20 June 1991, p. 16. The Committee of Senior Officials of the CSCE in Prague (10 Oct. 1991) adopted a document about the situation in Yugoslavia, the terminology of which is reminiscent of many UN General Assembly resolutions: 'resolves', 'supports', 'expresses', 'takes note of', 'urges', 'insists', 'considers', 'underlines', 'will examine', etc. This language reflects the ineffectiveness of the CSCE institutions when confronted with military conflict.

[12] The first meeting of the Council of Foreign Ministers decided that 'As soon as 12 or more participating States have seconded the request within a maximum period of 48 hours by addressing their support to the Chairman, he will immediately notify all participating States of the date and time of the

new political developments, such as the adoption of common positions by the 12 European Community (EC) states, could also have an impact.

Further progress was made at the second meeting of the CSCE Council of Foreign Ministers, in Prague (January 1992).[13] It was decided to enhance the competence of the Committee of Senior Officials (CSO) (to include 'overview, management and co-ordination' and taking appropriate decisions as the Council's agent between Council meetings) and to increase its effectiveness by holding regular meetings and delegating its tasks to other CSCE institutions or to *ad hoc* groups of participating states.[14] To strengthen the political consultation process, it was also recommended that the CSO devote certain sessions, or parts of sessions, to previously agreed specific issues.

An innovation of the Prague meeting was the decision that the CSCE Council or the CSO should take appropriate action '*if necessary, in the absence of the consent of the State concerned*' in cases of clear, gross and uncorrected violations of relevant CSCE commitments to safeguard human rights, democracy and the rule of law. At this stage 'action' was considered to mean political declarations or other steps of a political nature applicable outside the territory of the state concerned.[15]

Decisions were also taken to extend practical co-operation in the human dimension.[16] The most important of these incorporated recommendations to the Helsinki follow-up meeting to improve CSCE capabilities in crisis management and conflict prevention and resolution. In this context, the following instruments were mentioned: fact-finding and rapporteur missions; monitoring missions; good offices; counselling and conciliation; and dispute settlement. The Council also committed the Helsinki follow-up meeting to consider possibilities for direct and indirect involvement of the CSCE in peace-keeping activities.[17]

Finally, the Prague meeting decided to increase the role of the Consultative Committee of the Conflict Prevention Centre (CPC). The tasks of the Consultative Committee were defined as follows:

1. To serve as a forum in the security field wherein the CSCE participating states will conduct comprehensive and regular consultations on security issues with politico-military implications and 'as a forum for consultation and co-operation in conflict prevention and for co-operation in the implementation of

meeting, which will be held at the earliest 48 hours and at the latest three days after this notification.' See note 10, p. 12.

[13] Second Meeting of the Council of Foreign Ministers, Summary of Conclusions, Prague Document on Further Development of CSCE Institutions and Structures, Prague, 1992; for excerpts see *SIPRI Yearbook 1992* (note 3), appendix 15A

[14] Prague Document (note 13), p. 12.

[15] Prague Document (note 13), p. 15 (emphasis added).

[16] 'The Ministers agreed that monitoring and promoting progress in the human dimension remain a key function of the CSCE'. They decided to assign additional functions to the Office for Free Elections and transform it into the Office for Democratic Institutions and Human Rights (ODIHR). See Prague Document (note 13), p. 13.

[17] Prague Document (note 13), p. 16.

decisions on crisis management taken by the Council or by the CSO acting as its agent';[18]

2. To initiate and, with the assistance of the CPC Secretariat, execute fact-finding and monitoring missions;[19]

3. To execute any additional tasks assigned to it by the CSCE Council, or by the CSO acting as its agent.

It was decided that meetings of the Consultative Committee be convened regularly—as a rule at least once a month.[20]

The many decisions adopted in Berlin and Prague signalled that the CSCE urgently needed a new institutional character. These decisions were a response to the developments and expectations arising from the new CSCE tasks, on the one hand, and expressed ambitious plans to impart a new character to the CSCE—that of an organization for security in Europe—on the other.[21] The Helsinki follow-up meeting accordingly discussed and elaborated a new institutional concept, which was decided upon at the summit meeting in July 1992.

## III. Four meetings in Helsinki

A strategy and *modus operandi* for institutionalizing the CSCE were mapped out at four subsequent meetings in Helsinki: (*a*) the preparatory meeting (10–20 March 1992); (*b*) the additional meeting of the CSCE Council of Foreign Ministers (28 March 1992); (*c*) the follow-up meeting (24 March–8 July 1992); and (*d*) the summit meeting (9–10 July 1992).

The tasks of the follow-up meeting were set out by the preparatory meeting and covered four elements:[22]

1. Exchange of views on the implementation of all CSCE commitments;

2. Examination of the results of the negotiations, conferences and expert meetings of the CSCE since the 1986–89 Vienna follow-up meeting;

3. Consideration of proposals to give new impetus to a balanced and comprehensive development of the CSCE process; and

4. Preparation of a document to be adopted at the Helsinki summit meeting.

[18] Prague Document (note 13), p. 17.

[19] This should be done in accordance with para. 17 of the Vienna Document 1990 of the Negotiations on CSBMs concerning the mechanism for consultation and co-operation as regards unusual military activities. The Vienna Document 1990 is reproduced in *SIPRI Yearbook 1991*, pp. 475–88.

[20] The Consultative Committee may establish subsidiary working bodies, including open-ended *ad hoc* groups entrusted with specific tasks. The representatives of the Consultative Committee should attend meetings of the CSO relevant to the tasks of the CPC.

[21] See also Gärtner, H., 'The future of institutionalization: the CSCE example', ed. I. M. Cuthbertson, *Redefining the CSCE: Challenges and Opportunities in the New Europe* (Finnish Institute of International Affairs (IEWSS): New York, 1992), pp. 233ff; Ghebali, V.-Y., 'The institutionalization of the CSCE process: towards an instrument for the "Greater Europe"?', and Antola, E., 'Hegemony versus institutionalization: the erosion of the postwar order', eds K. Holder, R. E. Hunter and P. Lipponen, *Conference on Security and Co-operation in Europe. The Next Phase: New Security Arrangements in Europe,* Significant Issues Series (Center for Strategic and International Studies: Washington, DC, 1992).

[22] Decisions of the Preparatory Meeting of the Helsinki follow-up meeting of the CSCE, Helsinki, 1992.

Before the opening of the follow-up meeting, the CSCE Council of Foreign Ministers held its first additional meeting. The ministers reviewed and endorsed the decisions taken by the CSO for an immediate and effective cease-fire in and around Nagorno-Karabakh and praised the efforts made in this respect by the EC and its members, by the member states of the Commonwealth of Independent States (CIS), by the North Atlantic Cooperation Council (NACC) and by the UN Secretary-General. They mandated the CSCE Chairman-in-Office, Jiri Dienstbier, to contribute to the establishment of a cease-fire and an overall peaceful solution and to convene a conference on Nagorno-Karabakh under CSCE auspices.[23]

In order to accomplish the tasks listed above, the follow-up meeting was divided into four working groups. The first dealt with questions relating to the further development of CSCE institutions and structures and addressed the political consultation and decision-making processes, and instruments for crisis management, conflict prevention and peaceful settlement of disputes. In addition, legal, financial and administrative arrangements and relations with international organizations, Mediterranean non-participating states and NGOs were worked out by this group. The second group worked on the mandate for the planned CSCE Forum for Security Co-operation (FSC). The third group dealt with matters relating to the human dimension, while the fourth addressed problems of economic, scientific, technological and environmental co-operation, as well as those accompanying the transition to a free market economy and related social problems.

The decisions of the July 1992 Helsinki summit meeting were of crucial importance for institutionalizing the CSCE process and mapping out a strategy for mutually reinforcing institutions for security in Europe. In Berlin, the foreign ministers had encouraged the exchange of information and relevant documents among CSCE and other main European and transatlantic institutions.[24] In Prague, the list of CSCE relationships with international organizations had been expanded to embrace the Council of Europe, the UN Economic Commission for Europe (ECE), NATO, the WEU, the Organization for Economic Co-operation and Development (OECD), the European Bank for Reconstruction and Development (EBRD), the European Investment Bank (EIB) 'and other European and transatlantic organizations which may be agreed' with the aim of inviting them to make contributions to specialized CSCE meetings for which they have relevant expertise.[25]

At the summit meeting, the leaders of the participating states welcomed the rapid adaptation of European and transatlantic institutions which were

[23] The Conference was convened in Minsk with the participation of Armenia, Azerbaijan, Belarus, the Czech and Slovak Federal Republic, France, Germany, Italy, the Russian Federation, Sweden, Turkey and the USA. See First Additional Meeting of the CSCE Council, Summary of Conclusions, Helsinki, 1992, pp. 14–15.

[24] In the Summary of Conclusions of the Berlin Meeting of the CSCE Council in June 1991 the following organizations were mentioned: the EC, the Council of Europe, the ECE, NATO and the WEU.

[25] In the Prague Document, the Ministers requested these organizations to inform the CSCE Secretariat annually of their current work programme and of the facilities available for work relevant to the CSCE.

'increasingly working together to face up to the challenges before us and to provide a solid foundation for peace and prosperity'.[26] The meeting laid down guidelines for CSCE co-operation with individual organizations. The Helsinki Document stated that the European Community, 'fulfilling its important role in the political and economic development in Europe . . . is closely involved in CSCE activities'. NATO, through NACC, 'has established patterns of co-operation with new partners in harmony with the process of the CSCE. It has also offered practical support for the work of the CSCE'.[27] The WEU, states the Helsinki Document, as an integral part of the development of the European Union, is 'opening itself to additional co-operation with new partners and has offered to provide resources in support of the CSCE'.[28] A framework of co-operation was also established linking the CSCE with the Council of Europe, the Group of Seven (G7) and the Group of Twenty-Four as well as with the OECD, the ECE and the EBRD.

The Helsinki Document also indicates possibilities for such regional and sub-regional organizations as the Council of Baltic States, the Visegrad Triangle, the Black Sea Economic Co-operation, the Central European Initiative and the Commonwealth of Independent States to co-operate with and assist the CSCE. This list of diverse organizations reflects the excessive bureaucratization of multilateral relations among European, North American and Central Asian states; the doubling of the functions and tasks of these institutions and structures brings with it the risk that they will be more competitive than compatible in mutual relations, more 'inter-blocking' than interlocking and more likely to weaken than reinforce each other.

Finally the leaders of the participating states declared their understanding that 'the CSCE is a regional arrangement in the sense of chapter VIII of the Charter of the United Nations'.[29] No enforcement action shall be taken under

[26] *Helsinki Document 1992* (note 1), Helsinki Declaration, para. 10.

[27] *Helsinki Document 1992* (note 1). Proposed by the NATO Rome summit meeting on 7–8 Nov. 1991, NACC was called into being on 20 Dec. 1991 to establish 'liaison' between the Alliance and the new democracies of Central and Eastern Europe (CEE). Its declared goal is consultation and co-operation (but not guarantees) on security and related issues, such as defence planning, conceptual approaches to arms control, democratic concepts of civilian–military relations, civilian–military co-ordination of air traffic management and the conversion of defence production to civilian purposes. Apart from the institutional structure (meetings at foreign minister, ambassadorial and other levels), an informal High-Level Working Group was established to redistribute among the CIS states the TLE ceilings in relation to the CFE Treaty, which contributed to its successful conclusion (see chapter 12 in this volume). On 1 Apr. 1992 the first meeting of NACC defence ministers took place; it agreed on a further co-operation programme in such defence-related matters as military strategies, defence management, the legal framework for military forces, harmonization of defence planning and arms control, exercises and training, defence education, reserve forces, environmental protection, air traffic control, search and rescue, military contribution to humanitarian aid and military medicine. As of 31 Dec. 1992 there were 37 NACC member states (16 NATO, 5 CEE, 15 former Soviet republics plus Albania—see p xxviii). The division of the Czech and Slovak Federal Republic brought the number of member states to 38 on 1 Jan. 1993. Finland attended the Oslo NACC meeting on 5 June 1992 as an observer.

[28] *Helsinki Document 1992* (note 1). See also the Petersberg Declaration (19 June 1992) adopted at the WEU Council of Ministers Meeting. The Petersberg Declaration structures the WEU–Central European states' dialogue, consultations and co-operation with regard to the European security architecture and stability.

[29] Chapter VIII of the UN Charter deals with regional arrangements (articles 52, 53 and 54). Article 52, para. 2, reads as follows: 'The members of the United Nations entering into such arrangements or constituting such agencies shall value every effort to achieve pacific settlement of local dis-

regional arrangements without the authorization of the UN Security Council.[30] The Helsinki Document reaffirmed that 'The rights and responsibilities of the Security Council remain unaffected in their entirety'. For the first time this established an important link between the CSCE and the United Nations or, more broadly, between European and global security.

The Helsinki summit meeting also strengthened the CSCE institutions and structures set forth in the Charter of Paris and the Prague Document. It was decided that:

1. Meetings of heads of state or government would be convened, as a rule, every two years following review conferences (former 'follow-up meetings'). Summit meetings would set priorities and provide orientation at the highest political level.

2. Follow-up meetings would thus be replaced by review conferences. The traditional mandate of CSCE follow-up meetings, as defined in the Helsinki Final Act, covered a 'thorough exchange of views', which in fact made the general political debate and ideological polemics between the cold war antagonists excessively prolonged and more important than negotiations. Review conferences were to be operational and of short duration.[31]

3. The CSCE Council constitutes the central decision-making and governing body of the CSCE.[32]

4. The CSO will be responsible—between CSCE Council meetings—for overview, management and co-ordination and 'will act as the Council's agent in taking appropriate decisions'. The new functions, tasks and primary responsibilities are attributed to the CSO regarding early warning, preventive actions and crisis management. It was also recommended to make greater use of the communications network and the points of contact to manage the flow of information more efficiently.[33]

5. The CSCE Chairman-in-Office (CIO) is now responsible on behalf of the Council/CSO for co-ordination of and consultation on current activities. The CIO may be assisted by the preceding and succeeding Chairmen (the 'Troika'), *ad hoc* steering groups and personal representatives.

putes through such regional arrangements or by such regional agencies before referring them to the Security Council.' Charter of the United Nations and Statute of the International Court of Justice (UN Office of Public Information: New York, 1963), p. 28.

[30] An exception to this rule is the out-of-date provision of the UN Charter which provided (articles 53 and 107) that measures against 'any enemy state' could be taken without authorization of the Security Council. The intention was to prevent 'renewal of aggressive policy on the part of any such state'. The UN Charter term 'enemy state' applies to any state which during World War II was an enemy of any signatory of that Charter. The UN membership of the former 'enemy states' (Bulgaria, Finland, Germany, Hungary, Italy, Japan, Romania) made these provisions of the UN Charter irrelevant, since in accordance with article 4, para. 1, all the UN members were recognized as 'peace-loving states which accept the obligations contained in the present Charter and, in the judgement of the organization, are able and willing to carry out these obligations.', Charter of the United Nations (note 29), p. 6.

[31] The 'thorough review of the implementation of CSCE commitments', would now be co-operative in nature, comprehensive in scope and, at the same time, 'able to address specific issues.' See *Helsinki Document 1992* (note 1), Helsinki Decisions, chapter I, para. 28.

[32] *Helsinki Document 1992* (note 1), Helsinki Decisions, chapter I, para. 6.

[33] *Helsinki Document 1992* (note 1), Helsinki Decisions, chapter I, paras 9 and 10.

6. A new institution—the High Commissioner on National Minorities (HCNM)—was established with the aim of providing early warning and early action to prevent tensions involving national minority issues developing into a conflict.[34] The essence of the mandate of the HCNM is that it should act under the aegis of the CSO and thus be 'an instrument of conflict prevention at the earliest possible stage.'[35]

The decisions concerning a system of early warning, conflict prevention and crisis management should be seen as a qualitatively new stage of the CSCE process. Fact-finding and rapporteur missions as well as CSCE peace-keeping are considered as instruments of conflict prevention and crisis management. Early warning and preventive action provisions envisage that crisis or conflict situations should be brought to the attention of the CSO not only by a state directly involved in a dispute, but also by 11 states that are not involved in it. The HCNM and the Consultative Committee of the CPC also have the right to do this. An action may also be taken through the Human Dimension Mechanism and the Procedure for Peaceful Settlement of Disputes. The CSO has the right to set up a framework for a negotiated settlement or to dispatch a rapporteur or fact-finding mission. The list of such missions sent in 1992 is impressive;[36] however, their tangible results are fairly modest and have by and large fallen short of expectations.

The exercise of good offices, mediation or conciliation may also be initiated by the CSO. The Helsinki Decisions envisaged new types of CSCE instrument for conflict prevention, crisis management and peace-keeping (specifying a precise mandate, the chain of command, reporting and financial arrangements) as well as co-operation with regional and transatlantic organizations.[37]

[34] The Stockholm Meeting of the CSCE Council appointed Max van der Stoel, former Netherlands Minister for Foreign Affairs, as the first HCNM. Summary of the Conclusions of the Stockholm Council Meeting (note 2), Decision 3.

[35] *Helsinki Document 1992* (note 1), Helsinki Decisions, chapter II, para. 2.

[36] The first CSO Emergency Meeting was convened on Yugoslavia (Prague, 3–4 July 1991) and resulted in the sending of a monitoring mission and offering a CSCE good-offices mission as well as assistance to the negotiations on the future of former Yugoslavia. The first rapporteur mission was dispatched to Albania by the Consultative Committee of the CPC (Vienna, 16–19 Sep. 1991). Since then, in 1992 rapporteur missions have been carried out in the following countries: Armenia and Azerbaijan (12–18 Feb.); Ukraine, Moldova and Belarus (8–16 Mar.); Turkmenistan, Uzbekistan and Tajikistan (10–19 Mar.); Kazakhstan and Kyrgyzstan (31 Mar.–6 Apr.); Georgia (5–22 May), and Bosnia and Herzegovina (29 Aug.–4 Sep.).

Fact-finding missions were sent to the following areas: Kosovo (18–21 May); the region of the Georgian–Ossetian conflict (25–30 July); and Georgia (13–22 Oct.). An exploratory mission was embarked on in Kosovo, Vojvodina and Sandjak (2–8 Aug.). A mission was carried out in Nagorno-Karabakh (19–22 Aug.). A spill-over monitoring mission visited Macedonia (Skopje, 10–13 Sep.); human rights rapporteur missions went to Yugoslavia (12 Dec. 1991 and 7–10 Jan. 1992) and a mission under the Human Rights Mechanism was sent to Croatia (30 Sep.–5 Oct.). The CIO of the CSCE Council visited Azerbaijan and Armenia (30 Mar.–3 Apr.) and Yugoslavia (19–20 Aug.). The personal representative of the CIO paid visits to Moldova (8–12 Sep.), Romania (21–26 Oct.), Ukraine (26–27 Nov.) and Russia (17–21 Jan. 1993). (List compiled from the summary of missions prepared by the CSCE Secretariat, Prague.)

[37] The Helsinki Decision in this respect reads as follows: 'The CSCE may benefit from resources and possible experience and expertise of existing organizations such as the EC, NATO, and the WEU, and could therefore request them to make their resources available in order to support it in carrying out peace-keeping activities.' *Helsinki Document 1992* (note 1), Helsinki Decisions, chapter III, para. 52.

As regards CSCE peace-keeping endeavours, the CSCE will either create its own forces or employ forces assigned to it by its participating states or other international organizations. For the time being the CSCE is equipped to make use of the peace-keeping forces of other multilateral organizations rather than deploy its own force. It is possible that international military deployments will soon be carried out under the auspices of the CSCE. However, experience indicates that the CSCE is rather destined to develop other, non-military ways of resolving crisis situations—the new instruments of peaceful settlement of disputes that have been agreed.

This relatively modest progress was supplemented with the decisions of the third meeting of the CSCE Council in Stockholm,[38] based on the recommendations made by the CSCE Meeting on Peaceful Settlement of Disputes (Geneva, 12–23 October 1992). The most significant resolution was to adopt a Convention on Conciliation and Arbitration within the CSCE.

## IV. The Stockholm Decisions

In accordance with the strategy adopted, the Stockholm Council Meeting (14–15 December 1992) sought chiefly to consolidate 'the CSCE's operational capabilities through structural reforms and the appointment of a Secretary General'.[39] That decision was an important step in turning the CSCE process into an organization, and its negotiating structures into operational bodies.

The meeting devoted much time and attention to the peaceful settlement of disputes, a key issue since the first days of drafting the Helsinki Final Act (in 1973).[40] It was, however, the decisions of the 1986–89 Vienna follow-up meeting that promised an obligatory settlement of disputes with the participation of a third party. That evolution drew to a close with the resolutions of the 1990 Charter of Paris, the 1991 Valletta Experts' Meeting, the 1991 CSCE Berlin Council meeting and the 1992 Helsinki summit meeting.[41] In Geneva (12–23 October 1992), experts prepared a comprehensive and coherent set of measures that were formally adopted during the Stockholm meeting of the CSCE Council.

In the Stockholm Decision on Peaceful Settlement of Disputes the foreign ministers agreed on (*a*) measures to enhance the Valletta provisions through modification of the procedure for selecting Dispute Settlement Mechanisms; (*b* ) the text of a Convention on Conciliation and Arbitration; (*c*) a conciliation procedure as an option available to participating states, and (*d*) a decision that the Council or CSO 'may direct any two participating States to seek

[38] Summary of Conclusions of the Stockholm Council Meeting (note 2).

[39] Summary of Conclusions of the Stockholm Council Meeting (note 2).

[40] It was Professor Rudolf Bindschedler (Switzerland) who presented the first draft convention on European system of peaceful settlement of disputes at the stage of drafting the Helsinki Final Act, CSCE document CSCE/B/1, Geneva, 18 Sep. 1973. See more on this in Rotfeld, A. D., *Europejski system bezpieczenstwa in statu nascendi* [The European Security System in *Statu Nascendi*], (Polish Institute of International Affairs: Warszaw, 1990), pp. 67–91.

[41] This matter was debated during four meetings of CSCE experts in Montreux (1978), Athens (1984), Valletta (1991) and Geneva (12–23 Oct. 1992).

conciliation to assist them in resolving a dispute that they have not been able to settle within a reasonable period of time and adopted provisions related thereto'.[42]

The Stockholm Convention on Conciliation and Arbitration was signed by 29 CSCE participating states on 14 December 1992. It provides for pre-appointed conciliators and arbitrators that can be called in to perform their duties. Although the new body is called 'the Court', in fact it is not a permanent court; conciliation commissions and arbitration tribunals will be set up on an *ad hoc* basis. A conciliation proposal will not be binding on the parties unless they have made a specific commitment to be bound by it, but the verdicts of an arbitration tribunal will be binding. However, a tribunal can only commence its work if the parties have agreed to this, whether for general purposes or for a specific case.

The Stockholm Convention can be seen as a major achievement, the negotiators having drawn up a very complex mechanism.[43] The degree to which this legal masterpiece will prove workable remains an open question. The apparant complexity and cost of the proposed instruments may deter states from making use of the system.[44] Some analogies can be drawn with the International Court of Justice (ICJ). The Permanent International Court of International Justice at the League of Nations played a greater role in the inter-war period than its successor does as a judicial organ of the UN system, basically because the compulsory jurisdiction of the ICJ, based on Article 36 of its Statute, was accepted only by some small- and medium-sized countries. The USA and other great powers do not recognize the jurisdiction of the ICJ as compulsory and exclude from its jurisdiction any matter falling essentially within their domestic jurisdiction as determined by themselves.[45]

Similar factors may influence the effectiveness of the new instruments. The heart of the matter lies in the degree to which states are ready to use the new institutions and arrangements. The measure of readiness of a state is the means it puts at others' disposal in order to set in motion new mechanisms, and the extent of its submission to the decisions agreed within the new structures and institutions. Analysis of the activities of the CSCE institutions called into being in the wake of the Paris summit meeting shows that, for political and financial reasons, they do not measure up to their tasks (see, e.g., the CPC). A gap is widening between the declared will to build up various CSCE institutions and a remarkable degree of restraint in embarking on this path.

[42] The Stockholm Decision on Peaceful Settlement of Disputes, Stockholm, 14 Dec. 1992, para. 5(d) (for excerpts see appendix 5A); see also Modification to Section V of the Valletta Provisions—Annex 1 and Convention on Conciliation and Arbitration within the CSCE—Annex 2.

[43] The Stockholm Convention was prepared at an expert meeting in Geneva (12–23 Oct. 1992) under the chairmanship of Swedish diplomat Hans Corell, the Under-Secretary for Legal and Consular Affairs of the Swedish Ministry for Foreign Affairs. For excerpts from the Convention see the Stockholm Decision on Peaceful Settlement of Disputes, appendix 5A.

[44] The Financial Protocol to the Stockholm Convention was negotiated during the special meetings in Vienna on 1–2 Dec. 1992, 19–20 Jan. 1993 and 8–9 Mar. 1993, under Swedish chairmanship (Ambassador Hans Corell).

[45] The US reservation has come to be known as the Connolly Amendment. See more on this in Lauterpacht, H., *The Development of International Law by the International Court* (Stevens: London, 1958); Rosenne, S., *The Law and Practice of the International Court* (Sijthoff: Leiden, 1965).

The disproportion between the expectations and their realization is the main cause of scepticism over the establishment of further CSCE institutions.

The chief issue on the agenda of shaping a new security system is not the need for new structures or institutions, but rather the question of how to make the existing ones more efficient. The reason CSCE participating states keep establishing new organs without making use of them, and why they link their national security more to NATO or the WEU than to the CSCE or NACC, is mainly because NATO and the WEU have proven military structures at their command. It is an open question whether the CSCE and NACC should have such resources themselves or whether should take advantage of co-operation with existing military structures. The CSCE has the experience and capability to operate in the field of preventive diplomacy and to defuse crises before they escalate into open armed conflict.[46] It seems that in cases where armed intervention cannot be avoided NACC could play some role in the field of military operations.[47] In the foreseeable future the military security of states will be based on the guarantees which NATO accords to its member states and on national forces. At the same time, the North Atlantic Alliance, the transformation of which has already started, will continue to change, not only because of the new political circumstances and tasks but also because of the increased membership and geographical scope expected in the future. It can be assumed that the inclusion of Central and East European states in NATO will be a gradual process, differentiated both in terms of the links established and the scope of guarantees given and commitments undertaken. NACC and the FSC in Vienna could play an essential role in this process.

## The Forum for Security Co-operation

The FSC opened in Vienna on 22 September 1992, in keeping with the provisions of the Helsinki summit meeting, to meet the need for a new permanent negotiating body on disarmament and CSBMs for all participating states. The Negotiation on Conventional Armed Forces in Europe (CFE) was originally confined to the group of 23 (16 NATO and then 7 Warsaw Treaty Organization—WTO) states. The conclusion of the CFE and Open Skies treaties, the Concluding Act of the Negotiation on Personnel Strength of Conventional Armed Forces in Europe (CFE-1A), and, most importantly, the breakup of the USSR and Yugoslavia and the formation of many new states, as well as the need to include the former neutral and non-aligned states in new agreements, indicated the need for the establishment of this type of framework as an integral part of the new CSCE structure.

[46] NATO has acknowledged that role. The Final Communique of the Ministerial meeting of the North Atlantic Council stated: 'The CSCE has an essential role to play in developing a co-operative approach to security and in conflict prevention and crisis management.' Press Communique M-NAC-(92) 106, Brussels, 17 Dec. 1992.

[47] The Ministerial Meeting of NACC (18 Dec. 1992) decided to initiate consultations in peace-keeping activities. Within the NACC framework an *Ad Hoc* Group on Co-operation in Peace-keeping was established which met several times in Brussels in Feb. and Mar. 1993, and elaborated specific recommendations for an initial programme of co-operation in this field.

Three goals were set for the Forum at Helsinki:

1. To strengthen security and stability through the negotiation of concrete measures 'aimed at keeping or achieving the levels of armed forces to a minimum commensurate with common or individual legitimate security needs within Europe and beyond';
2. To harmonize obligations agreed among CSCE states under the various existing instruments of arms control, disarmament and CSBMs because not all the CSCE states are parties to all those agreements (e.g., the former neutral and non-aligned states did not participate in the CFE Negotiation and consequently are not parties to that Treaty);
3. To work out new stabilizing measures with respect to military forces and new CSBMs designed to ensure greater transparency.

A number of other specific tasks were set up and a Programme for Immediate Action[48] was determined, embracing harmonization, the development of the Vienna Document 1992, the global exchange of military information, and co-operation on non-proliferation and regional measures. In the domain of security enhancement and military co-operation it was decided to work towards decisions aimed at providing transparency on force planning (the size, structure and equipment of the armed forces as well as defence policy, doctrines and budgets), co-operation in defence conversion and the strengthening of non-proliferation regimes for the transfer of sensitive expertise as well as the establishment of a responsible approach to international armaments transfers. Other activities include the prevention of conflicts in co-operation with the CPC; training, exchanges and participation in evaluation and inspection teams; and consolidation of the verification regime.

The FSC has conducted intensive negotiations during its first few months, primarily on harmonization, information exchange on defence planning, regional measures (including regional 'tables'?), restriction of arms transfers and the code of conduct for security relations among CSCE states. In Stockholm the Council of Foreign Ministers decided that the Programme for Immediate Action should be carried out by the time of their next meeting.[49] Non-proliferation was found to be of special importance at Stockholm.[50] Consequently, the CSCE states which were not parties to the 1968 Non-Proliferation Treaty undertook to accede to that Treaty as non-nuclear weapon states 'in the shortest time possible'.[51] Analysis of the provisions adopted in Stockholm shows that the newly established structures and institutions have been actively getting on with their job. Nevertheless, it seems that the activity

[48] Annexed to the Stockholm Document 1992, see appendix 5A.
[49] The next meeting of the CSCE Council will be held in Rome in Nov.–Dec. 1993.
[50] In this context, it was decided that CSCE states will become original signatories of the Convention on the Prohibition of the Development, Production, Stockpiling and Use of Chemical Weapons and on Their Destruction. The Convention was opened for signature in Paris on 13 Jan. 1993. See also chapter 14 in this volume.
[51] See note 2.

of the FSC still falls short of its capabilities. The role of the Forum can and should be much more influential.

## V. New participating states

The Prague Council meeting of January 1992 accorded the status of 'participating state' to all former Soviet republics.[52] Enlarging the number of participants was accompanied by a considerable expansion of tasks, and the admission of all the former Soviet republics gives institutional scope to a new security area from Vancouver to Vladivostok. CSCE decisions already apply not only to Europe ('from the Atlantic to the Urals') and North America (the United States and Canada) but also to states of Central Asia and the Far East. Such a significant expansion of geographic scope and the inclusion of new participating states necessitate a differentiation of tasks and expectations connected with the implementation of the provisions already adopted and those yet to be negotiated. States that have emerged as a result of the collapse of the Soviet Union are at a cross-roads and face difficult choices about how to proceed with their development. Their acceptance as participating states in the CSCE process was contingent upon an undertaking by each of them to accept 'in their entirety all commitments and responsibilities' contained in the CSCE documents.[53] Indeed, they declared their determination to act in accordance with these provisions. Specific commitments were made regarding the Vienna Documents 1990 and 1992 requirements on CSBMs and the prompt ratification of the 1990 CFE Treaty.

To implement these commitments, it was agreed in Prague that the governments of the newly admitted states will invite a rapporteur mission (arranged by the Chairman of the CSCE Council of Foreign Ministers) to visit, and will fully facilitate its activities.[54] This mission reported back to the CSCE on progress towards full implementation of CSCE commitments in those states and provided assistance towards that objective. The procedures adopted within the CSCE, and the established institutions and structures, ought, on the one hand, to facilitate a stabilization of democracy and the rule of law in the post-totalitarian states, and on the other hand to help prevent Central Asian states from sliding into political and religious Islamic fundamentalism. It was also envisaged that informal consultations under the direction of the CSO Chairman should take place at Helsinki during the follow-up meeting in order to establish the modalities for a programme of co-ordinated support to recently

[52] Albania joined the CSCE during the Berlin Meeting (June 1991). The Baltic states (Estonia, Latvia and Lithuania) had been accepted in Sep. 1991 before the Prague Meeting and Georgia joined the CSCE on 24 Mar. 1992 during the follow-up meeting in Helsinki. On the same day Croatia and Slovenia were admitted as participating states. The letters of admission are annexed to the Summary of the Conclusions of the First Additional Meeting of the CSCE Council (24 Mar. 1992). The latest newcomer is Bosnia and Herzegovina, accepted on 30 Apr. 1992.

[53] Prague Document (note 13).

[54] A relevant identical formula is contained in the letters of all foreign ministers of the newly admitted states addressed to the Chairman-in-Office of the CSCE Council of Foreign Ministers, Jiri Dienstbier, Foreign Minister of the Czech and Slovak Federal Republic. See Prague Document (note 13).

admitted states, through which appropriate diplomatic, academic, legal and administrative expertise and advice on CSCE matters could be made available.[55]

The leaders gathered at the Helsinki summit meeting welcomed the commitment by all participating states 'to our shared values'.[56] In essence, the unique character of the CSCE lies in unifying and organizing all the states of Europe, North America and Central Asia as well as a considerable part of the Far East. The Helsinki summit meeting took the decision to invite Japan, as a non-participating state, to attend CSCE meetings, including those of heads of state or government, the CSCE Council, the CSO and 'other appropriate CSCE bodies which consider specific topics of expanded consultation and co-operation.'[57] In practice, this is a considerably broader scope of participation and co-operation than that envisaged by the Helsinki Final Act for non-participating Mediterranean states, whose 'contributions' are purely formal and intermittent in character. Although the expansion of CSCE membership is desirable, at the same time it is one of the causes of the operational difficulties of the new institutions.

## VI. Concluding remarks

The collapse of the bipolar world has led to a need not only to adapt the structures and institutions shaped in the cold war period but also to tailor the principles and norms by which states are guided in their mutual relations to the qualitatively new political circumstances. A completely new pattern of relationships is emerging.[58] What are the practical consequences?

First, there is a need to redefine the old principles. If the main threats are domestic, then the principle of non-intervention in internal affairs (Principle VI of the Helsinki Final Act) cannot constitute an excuse for staving off or an obstacle to carrying out an urgent and just international intervention.[59] The situation in Yugoslavia is a glaring example of human rights and those of national minorities being violated on a massive scale; genocide is being perpetrated in full view of the world public. A new principle of legitimized international interventionism is needed.

There is also a need to re-interpret the existing CSCE principles on the right of peoples to self-determination and of territorial integrity. With the end of the process of de-colonization new states may emerge on the territories of existing states as a result of division, secession or unification with neighbouring states.

[55] Prague Document (note 13), para. 19, p. 8.

[56] The CSCE summit stated: 'Adherence to our commitments provides the basis for participation and co-operation in the CSCE and a cornerstone for further development of our societies', *Helsinki Document 1992* (note 1).

[57] *Helsinki Document 1992* (note 1), Helsinki Decisions, chapter IV, para. 11 reads: 'Representatives of Japan may contribute to such meetings, without participating in the preparation and adoption of decisions, on subjects in which Japan has a direct interest and/or wishes to co-operate actively with the CSCE.'

[58] See Europe in Transition: the role of the CSCE, Statement by Krzysztof Skubiszewski, Minister of Foreign Affairs of Poland, at the Stockholm Meeting of the CSCE Council (14 Dec. 1992).

[59] For a discussion of this issue see Rotfeld (note 3).

The international developments in Europe in 1990–92 provided examples of unification (FR Germany and the German Democratic Republic), division (Czechoslovakia), breakup (the USSR and Yugoslavia), and secession (proclamation of new states on territories of the Russian Federation, Georgia and other newly founded states). For many reasons these processes are difficult, complex and conflict-generating. The search for peaceful solutions always requires an analysis of a given situation, and not a mechanistic or formalistic application of general rules. Equal rights and self-determination of peoples means that all the other purposes and principles of the UN Charter and the relevant norms of international law ('including those relating to territorial integrity of States'[60]) should be respected. In essence the principle of self-determination is closely related to democratization of domestic relations.[61] It should by no means be identified with the right of part of a territory to secede. Given the situation in Europe at the end of 1992 the principle of self-determination should be reformulated in such a way that the right to secession be treated both as an exception and as a last resort.

Second, the states that may be called a Euro-Atlantic community have at their command a relevant set of instruments and capabilities that enable them quickly to agree on new decisions. A review of the resolutions adopted within NATO, the WEU, the CSCE, the Council of Europe and other sub-regional groupings (the Visegrad Group, the Central European Initiative, the Nordic Council, the Black Sea Economic Co-operation, etc.) indicates that new decisions are considered to have a value *per se,* and their number remains in blatant disproportion to examples of their effectiveness. New principles and norms, a new code of conduct in the field of security and other new security arrangements and institutions are clearly necessary. At the same time, the most important thing is to have the adopted resolutions put into effect in order to prevent conflicts and help resolve crisis situations.

Third, the new structures evolving within the CSCE constitute a framework for a pan-European security organization. Loose structures have taken on organizational forms: the Chairman-in-Office, the Committee of Senior Officials and the Vienna group permanently operating between sessions of the CSO, the new institutions of Secretary-General of the CSCE and the High Commissioner for National Minorities, different instruments of the European system for peaceful settlement of disputes and—last but not least—the Forum for Security Co-operation. The latter framework is not only a convenient platform for conducting negotiations, but also provides a common basis for operational activity and collective action. It must be considered how these and other bodies, structures and institutions could help expedite the decision-making process, the efficiency of which is contingent upon speed, the scope and type of means available, and the accountability of those who carry out the

[60] See Helsinki Final Act (note 40), Principle VIII.

[61] The Helsinki Final Act 1975 states: 'By virtue of the principle of equal rights and self-determination of peoples, all peoples always have the right, in full freedom, to determine, when and as they will, their internal and external political status, without external interference, and to pursue as they wish their political, economic, social and cultural development.'

decisions. Excessive bureaucratization should be avoided, but, on the other hand, the CSCE process should acquire the character of a European security organization—either through transformation of existing structures and institutions or by establishing an organization that would act upon predetermined assumptions.[62] It seems that in this respect the political will of states will be more decisive than theoretical concepts or the intellectual advantages of the proposed solutions. The usefulness of new institutions is determined by whether major states, and great powers in particular, are ready to make use of them—whether they are prepared to make their security interests dependent on the effectiveness of multilateral organizations. An example is the role played by such organizations as NATO, the WEU and the EC in the security policy of Western states.

Until recently security was generally seen as tantamount to arms control. Now it seems that grand agreements and treaties (such as the START and CFE treaties) are a thing of the past. In years to come the most important security issue will be conflict prevention and crisis management. Accordingly, the main role will fall not to arms control but rather to a new organization and new principles for a peaceful order. The tasks entrusted to the CSCE 20 years ago have been partly fulfilled or have become irrelevant. The new European security regime will probably combine the concept of an Executive Committee[63] and an appropriate role for NATO, the EC and the WEU. To give the CSCE the character of an operational security organization, it is necessary to streamline its decision making: operational decisions cannot be taken in a group of some 50 states, particularly by consensus. There is now an urgent need to create a CSCE Executive Committee, perhaps composed of the permanent CSCE members of the UN Security Council (the USA, the UK, France, Russia) and Germany. Other states might be represented by representatives of sub-regional organizations or groupings (the Visegrad Group, the Central European Initiative, Baltic and Balkan states, the CIS, etc.), delegated on the basis of rotation. EC, NATO or WEU countries could be represented on the same basis. Such a European Executive Committee, with permanent and non-permanent members, should not exceed 11 states. It is high time that such a decision be made, and the details could well be drawn up in the framework of preparations for the next CSCE summit meeting.

Clearly the new post-cold war conflicts will not just disappear. They are rather becoming more numerous, even if they do not pose a direct threat of world war. The new peace order will be based to a greater degree on political and legal instruments than on military deterrence. In future that order is likely

[62] The Institute for Peace Research and Security Policy at the University of Hamburg (IFSH) proposed a new organization—a European Security Community comprising the following bodies: a European Security Council, Permanent Representatives, a Standing Commission, a Secretary-General, a Military Staff and a Court of Justice. Except for the Security Council, all these bodies belong to the CSCE. The Institute's report was discussed at IFSH on 1–2 Feb. 1993. See 'Vom Recht des Stärkeren zur Stärke des Rechts', *Hamburger Beiträge zur Friedensforschung und Sicherheitspolitik*, no. 75 (Apr. 1993).

[63] Goodby, J., 'Commonwealth and concert: organizing principles of post-containment order in Europe', *Washington Quarterly*, summer 1991.

to include a comprehensive ban on weapons of mass destruction combined with a non-proliferation regime for conventional weapons and an efficient system of peace-keeping activities.[64] New dimensions of the CSCE process can and should promote the shifting of the centre of gravity away from national security towards a co-operative pan-European and global security system.

Change, by its very nature, is inevitably accompanied by instability. The transitional period of the early 1990s may last for a relatively long time. Old threats have faded away but at the same time the sense of cold war stability has also disappeared. There is a rather widespread conviction that the bipolar system created in the wake of World War II could be replaced by an equally stable and less costly international structure. This is wishful thinking reflecting needs and hopes as opposed to a clear political programme. In effect, a gap has appeared between the sought-after 'world order' or 'new security order in Europe', and reality.

There is also a gap between the expectations connected with the CSCE and its capabilities. In the past, this process was overestimated in the East and underrated in the West. Now the situation is reversed. Nevertheless, in the light of the negotiations and actions undertaken so far within the framework of the CSCE process, some general conclusions can be drawn:

1. The norms and procedures agreed upon 20 years ago are only of limited application to the new situation in Europe. This applies both to the concrete recommendations that focused on Basket III and human rights issues in the past, and to some of the principles and decision-making procedures.

2. The main features of the CSCE have changed and its drawbacks have been altered to a considerable degree; none the less the process does not meet the new expectations. For instance, within its framework treaties are being negotiated that are binding on the basis of international law; consensus is not always observed and the decisions adopted are obligatory in character; an increasingly important role is played by military questions; a number of institutions and structures have been established to ensure proper monitoring and the effectiveness of the decisions. All in all, however, the CSCE is neither an alliance nor a security organization.

3. The role of the Helsinki process is recognized as that of a forum for ongoing and future negotiations concerning not only human dimension issues but also to an increasing degree, if not chiefly, arms control and the dialogue on military security. However, the need for negotiating grand treaties on arms control is no longer of primary significance. What is essential is to put the provisions adopted into effect and abide by the limitations set.

4. The role of the CSCE in ensuring security will be determined by whether NATO and its structures will make use of the CSCE in peace-keeping activities and in solving conflict and crisis situations. The question of accepting Central and East European states to NATO is now on the agenda. Such a deci-

[64] Burns, H. W., 'Law and alternative security: towards a just world peace', *Alternative Security: Living without Nuclear Deterrence* (Westview Press: Boulder, Colo., 1991), pp. 78–107.

sion would impart to the Atlantic Alliance a new dimension in the area of security.[65] Collaboration within NACC might, at an intermediate stage, have a significant role to play.

In sum, the significance of the CSCE process will be largely determined not by the establishment of new structures and institutions but rather by how efficiently this existing, well-functioning, trusted organization, to which the participating states have delegated some of their competencies, uses and adapts to the new circumstances.[66] Indeed, the role of structures and institutions dealing with security problems is determined not by the bodies and tasks entrusted to them but rather by the states' readiness to make use of those bodies. CSCE institutions are no exception to this rule.

[65] In this context an interesting debate on co-operative security was published: Forsberg, R. and Van Evera, S., 'After the cold war: a debate on cooperative security', *Boston Review*, vol. 17, no. 6 (Nov./Dec. 1992); with responses by H. R. Alker, J. Dean, K. Kaysen, J. Landy, S. Miller and J. M. O. Sharp. See also a discussion document on European Security by London-based European Security Working Group, The British American Security Information Council, Nov. 1992.

[66] See also Mandelbaum, M., 'Reconstructing the European security order', *Critical Issues, 1990–91* (Council on Foreign Relations: New York, 1990), pp. 12–21; Baumann, C. E., 'Europe emergent: a web of institutions', ed. R. J. Jackson, *Europe in Transition: The Management of Security After the Cold War* (Adamantine Press: London, 1992), pp. 156–68; Ropers, N. and Schlotter, P., 'The CSCE. Multilateral conflict management in a transforming world order: future perspectives and new impulses for regional peace strategies', *Interdependence*, no. 14 (Foundation Development and Peace: Bonn, 1993).

# Appendix 5A. Key CSCE documents in 1992

## HELSINKI DOCUMENT 1992

## THE CHALLENGES OF CHANGE

*Helsinki, 10 July 1992*

### The Helsinki Summit Declaration
### Promises and Problems of Change

1. We, the Heads of State or Government of the States participating in the Conference on Security and Co-operation in Europe, have returned to the birthplace of the Helsinki process, to give new impetus to our common endeavour.

2. The Charter of Paris for a New Europe, signed at the last Summit, defined a common democratic foundation, established institutions for co-operation and set forth guidelines for realization of a community of free and democratic States from Vancouver to Vladivostok.

3. We have witnessed the end of the cold war, the fall of totalitarian regimes and the demise of the ideology on which they were based. All our countries now take democracy as the basis for their political, social and economic life. The CSCE has played a key role in these positive changes. Still, the legacy of the past remains strong. We are faced with challenges and opportunities, but also with serious difficulties and disappointments.

4. We have met here to review the recent developments, to consolidate the achievements of the CSCE and to set its future direction. To meet new challenges we have approved here today a programme to enhance our capabilities for concerted action and to intensify our co-operation for democracy, prosperity and equal rights of security.

5. The aspirations of peoples freely to determine their internal and external political status have led to the emergence of new sovereign States. Their full participation brings a new dimension to the CSCE.

6. We welcome the commitment of all participating States to our shared values. Respect for human rights and fundamental freedoms, including the rights of persons belonging to national minorities, democracy, the rule of law, economic liberty, social justice and environmental responsibility are our common aims. They are immutable. Adherence to our commitments provides the basis for participation and co-operation in the CSCE and a cornerstone for further development of our societies.

7. We reaffirm the validity of the guiding principles and common values of the Helsinki Final Act and the Charter of Paris, embodying the responsibilities of States towards each other and of governments towards their people. They are the collective conscience of our community. We recognise our accountability to each other for complying with them. We underline the democratic rights of citizens to demand from their governments respect for these values and standards.

8. We emphasize that the commitments undertaken in the field of the human dimension of the CSCE are matters of direct and legitimate concern to all participating States and do not belong exclusively to the internal affairs of the State concerned. The protection and promotion of the human rights and fundamental freedoms and the strengthening of democratic institutions continue to be a vital basis for our comprehensive security.

9. The transition to and development of democracy and market economy by the new democracies is being carried forward with determination amidst difficulties and varying conditions. We offer our support and solidarity to participating States undergoing transformation to democracy and market economy. We welcome their efforts to become fully integrated into the wider community of States. Making this transition irreversible will ensure the security and prosperity of us all.

10. Encouragement of this sense of wider community remains one of our fundamental goals. We welcome in this connection the rapid adaptation of European and transatlantic institutions and organizations which are increasingly working together to face up to the challenges before us and to provide a solid foundation for peace and prosperity.

The European Community (EC) fulfilling its important role in the political and economic development of Europe, is moving towards a union and has decided to broaden its membership. It is closely involved in CSCE's activities.

The North Atlantic Treaty Organization (NATO), one of the essential transatlantic links, has adopted a new strategic concept and strengthened its role as an integral aspect for security in Europe. Through establishment of the North Atlantic Co-operation Council (NACC) it has established patterns of co-operation with new partners in harmony with the goals of the CSCE. It has also offered practical support for the work of the CSCE.

The Western European Union (WEU) is an integral part of the development of the European Union; it is also the means to strengthen the European pillar of the Atlantic Alliance; it is developing an operational capacity; it is opening itself to additional co-operation with new partners and has offered to provide resources in support of the CSCE.

The Council of Europe is elaborating its own programmes for new democracies, opening up to new members and is co-operating with the CSCE in the human dimension.

The Group of Seven and the Group of Twenty-Four are deeply engaged in assistance to countries in transition.

The Organization for Economic Co-operation and Development (OECD), the United Nations Economic Commission for Europe (ECE) and the European Bank for Reconstruction and Development (EBRD) have a key role to play in the construction of a new Europe.

The Commonwealth of Independent States (CIS) has stated its readiness to assist the CSCE in pursuit of its objectives.

These and other forms of regional and sub-regional co-operation which continue to develop, such as the Council of Baltic Sea States, the Visegrad Triangle, the Black Sea Economic Co-operation and the Central European Initiative, multiply the links uniting CSCE participating States.

11. We welcome the adoption of the Vienna 1992 Document on Confidence- and Security-Building Measures and the signature of the Treaty on Open Skies, with the adoption of the Declaration on the Treaty on Open Skies. We also welcome the imminent entry into force of the Treaty on Conventional Armed Forces in Europe (CFE) and the Concluding Act of the Negotiation on Personnel Strength of Conventional Armed Forces in Europe. These agreements provide a solid foundation for our further security co-operation. We welcome the recent United States–Russian joint understanding on Strategic Offensive Arms. We reaffirm our commitment to become original signatories to the forthcoming convention on the prohibition of the development, production, stockpiling and use of chemical weapons and on their destruction, and urge other States to do so.

12. This is a time of promise but also a time of instability and insecurity. Economic decline, social tension, aggressive nationalism, intolerance, xenophobia and ethnic conflicts threaten stability in the CSCE area. Gross violations of CSCE commitments in the field of human rights and fundamental freedoms, including those related to national minorities, pose a special threat to the peaceful development of society, in particular in new democracies.

There is still much work to be done in building democratic, pluralistic societies, where diversity is fully protected and respected in practice. Consequently, we reject racial, ethnic and religious discrimination in any form. Freedom and tolerance must be taught and practised.

13. For the first time in decades we are facing warfare in the CSCE region. New armed conflicts and massive use of force to achieve hegemony and territorial expansion continue to occur. The loss of life, human misery, involving huge numbers of refugees have been the worst since the Second World War. Damage to our cultural heritage and the destruction of property have been appalling.

Our community is deeply concerned by these developments. Individually and jointly within the CSCE, the United Nations and other international organisations we have sought to alleviate suffering and seek long term solutions to the crises which have arisen.

With the Helsinki decisions we have put in place a comprehensive programme of co-ordinated action which will provide additional tools for the CSCE to address tensions before violence erupts and to manage crises which may regrettably develop. The Council and Committee of Senior Officials have already established for the CSCE an important role in dealing with crises that have developed within our area.

No international effort can be successful if those engaged in conflicts do not reaffirm their will to seek peaceful solutions to their differences. We stress our determination to hold parties to conflicts accountable for their actions.

14. In times of conflict, the fulfilment of basic human needs is most at risk. We will exert every effort to ensure that they are met

and that humanitarian commitments are respected. We will strive to relieve human suffering by humanitarian ceasefires and to facilitate the delivery of assistance under international supervision, including its safe passage. We recognize that the refugee problems resulting from these conflicts require the co-operation of all of us. We express our support for, and solidarity with, those countries which bear the brunt of these refugee problems resulting from these conflicts. In this context, we recognise the need for co-operation and concerted action.

15. Even where violence has been contained, the sovereignty and independence of some States still needs to be upheld. The participating States express support for efforts by CSCE participating states to remove, in a peaceful manner and through negotiations, the problems that remain from the past, like the stationing of foreign armed forces on the territories of the Baltic States without the required consent of those countries.

Therefore, in line with basic principles of international law and in order to prevent any possible conflict, we call on the participating States concerned to conclude, without delay, appropriate bilateral agreements, including timetables, for the early, orderly and complete withdrawal of such foreign troops from the territories of the Baltic States.

16. The degradation of the environment over many years threatens us all. The danger of nuclear accidents is a pressing concern. So are, in several parts of the CSCE area, defence-related hazards for the environment.

17. The present proliferation of weapons increases the danger of conflict and is an urgent challenge. Effective export controls on nuclear materials, conventional weapons and other sensitive goods and technologies are a pressing need.

**The CSCE and the Management of Change**

18. The CSCE has been instrumental in promoting changes; now it must adapt to the task of managing them. Our decisions in Helsinki are making the CSCE more operational and effective. We are determined to fully use consultations and concerted action to enable a common response to the challenges facing us.

19. In approaching these tasks, we emphasize the central role of the CSCE in fostering and managing change in our region. In this era of transition, the CSCE is crucial to our efforts to forestall aggression and violence by addressing the root causes of problems and to prevent, manage and settle conflicts peacefully by appropriate means.

20. To this end, we have further developed structures to ensure political management of crises and created new instruments of conflict prevention and crisis management. We have strengthened the Council and the Committee of Senior Officials (CSO) and devised means to assist them. The CSCE capacities in the field of early warning will be strengthened in particular by the activities of the newly established High Commissioner on National Minorities.

We have provided for CSCE peacekeeping according to agreed modalities. CSCE peacekeeping activities may be undertaken in cases of conflict within or among participating States to help maintain peace and stability in support of an ongoing effort at a political solution. In this respect, we are also prepared to seek, on a case-by-case basis, the support of international institutions and organizations, such as the EC, NATO and WEU. We welcome their readiness to support peacekeeping activities under the responsibility of the CSCE, including by making available their resources.

We are further developing our possibilities for peaceful settlement of disputes.

21. Our approach is based on our comprehensive concept of security as initiated in the Final Act. This concept relates the maintenance of peace to the respect for human rights and fundamental freedoms. It links economic and environmental solidarity and co-operation with peaceful inter-State relations. This is equally valid in managing change as it was necessary in mitigating confrontation.

22. The CSCE is a forum for dialogue, negotiation and co-operation providing direction and giving impulse to the shaping of the new Europe. We are determined to use it to give new impetus to the process of arms control, disarmament and confidence and security-building, to the enhancement of consultation and co-operation on security matters and to furthering the process of reducing the risk of conflict. In this context, we will also consider new steps to further strengthen norms of behaviour on politico-military aspects of security. We will ensure that our efforts in these fields are coherent, interrelated and complementary.

23. We remain convinced that security is indivisible. No State in our CSCE commu-

nity will strengthen its security at the expense of the security of other States. This is our resolute message to States which resort to the threat or use of force to achieve their objectives in flagrant violation of CSCE commitments.

24. Essential to the success of our efforts to foster democratic change will be increased co-operation with other European and transatlantic organizations and institutions. Therefore, we are convinced that a lasting peaceful order for our community of states will be built on mutually reinforcing institutions, each with its own area of action and responsibility.

25. Reaffirming the commitments to the Charter of the United Nations as subscribed by our States, we declare our understanding that the CSCE is a regional arrangement in the sense of Chapter VIII of the Charter of the United Nations. As such, it provides an important link between European and global security. The rights and responsibilities of the Security Council remain unaffected in their entirety. The CSCE will work together closely with the United Nations, especially in preventing and settling conflicts.

26. We restate our unreserved condemnation of all acts, methods and practices of terrorism. We are determined to enhance our co-operation to eliminate this threat to security, democracy and human rights. To this end, we will take measures to prevent in our territories criminal activities that support acts of terrorism in other States. We will encourage exchange of information concerning terrorist activities. We will seek further effective avenues for co-operation as appropriate. We will also take the necessary steps at a national level to fulfil our international obligations in this field.

27. Illicit trafficking in drugs represents a danger to the stability of our societies and democratic institutions. We will act together to strengthen all forms of bilateral and multilateral co-operation in the fight against illicit trafficking in drugs and other forms of international organized crime.

28. We will work to reinforce the close link which exists between political pluralism and the operation of a market economy. Enhanced co-operation in the field of economy, science and technology has a crucial role to play in strengthening security and stability in the CSCE region.

29. Economic co-operation remains an essential element of the CSCE. We will continue to support the transformations under way to introduce market economies as the means to enhance economic performance and increased integration into the international economic and financial systems.

30. We will also facilitate expanded economic co-operation which must take account of the prevailing political and economic conditions. We welcome the contribution of economic, financial and technical assistance programmes of the Group of Seven and the Group of Twenty-Four to the transition process. In the framework of our co-operation we fully support the further development of the European Energy Charter which is of particular importance in the period of transition.

31. We will work together to facilitate means of transportation and communication in order to deepen co-operation among us.

32. We renew our commitment to co-operate in protecting and improving the environment for present and future generations. We stress in particular the importance of co-operation to effectively ensure the safety of nuclear installations and to bring defence-related hazards for the environment under control.

We emphasize the need for greater public awareness and understanding of environmental issues and for public involvement in the planning and decisionmaking process.

We welcome the important outcome of the United Nations Conference on Environment and Development (UNCED) held in Rio de Janeiro in June 1992. We emphasize the need for effective and sustained implementation of the UNCED decisions.

33. Further steps must be taken to stop the proliferation of weapons. It remains vital to ensure non-proliferation of nuclear weapons and the relevant technology and expertise. We urge all States which have not acceded to the Treaty on Non-proliferation of Nuclear Weapons to do so as non-nuclear weapons States and to conclude safeguards agreements with the International Atomic Energy Agency (IAEA). We commit ourselves to intensify our co-operation in the field of effective export controls applicable to nuclear materials, conventional weapons and other sensitive goods and technologies.

34. We welcome the development of regional co-operation among CSCE participating States as a valuable means of promoting pluralistic structures of stability. Based on the CSCE principles and commitments,

regional co-operative activities serve the purpose of uniting us and promoting comprehensive security.

35. We encourage wide-ranging transfrontier co-operation, including human contacts, involving local and regional communities and authorities. This co-operation contributes to overcoming economic and social inequalities and enhancing ethnic understanding, fostering good-neighbourly relations among States and peoples.

36. In order to ensure full participation and co-operation by recently admitted participating States we are initiating a programme of co-ordinated support.

37. We reaffirm our conviction that strengthening security and co-operation in the Mediterranean is important for stability in the CSCE region. We recognize that the changes which have taken place in Europe are relevant to the Mediterranean region and that, conversely, economic, social, political and security developments in that region have a direct bearing on Europe.

38. We will therefore widen our co-operation and enlarge our dialogue with the non-participating Mediterranean States as a means to promote social and economic development, in order to narrow the prosperity gap between Europe and its Mediterranean neighbours and protect the Mediterranean ecosystems. We stress the importance of intra-Mediterranean relations and the need for increased co-operation within the region.

39. We welcome and encourage the continuation of initiatives and negotiations aimed at finding just, lasting and viable solutions, through peaceful means, to the outstanding crucial problems of the Mediterranean region.

40. We have expanded dialogue with non-participating States, allowing them to take part in our activities on a selective basis when they can make a contribution.

41. We welcome the establishment of the CSCE Parliamentary Assembly which held its first meeting in Budapest on 3 to 5 July and look forward to the active participation of parliamentarians in the CSCE process.

42. We attach particular importance to the active participation of our publics in CSCE. We will expand the opportunities for contributions by and co-operation with individuals and non-governmental organizations in our work.

43. In order to foster our partnership, and to better manage change, we have today in Helsinki adopted an agenda for a strengthened and effective CSCE through the Helsinki Decisions. These decisions will be implemented fully and in good faith.

44. We entrust the Council with the further steps which may be required to implement them. The Council may adopt any amendment to the decisions which it may deem appropriate.

45. The full text of the Helsinki Document will be published in each participating State, which will make it known as widely as possible.

46. The Government of Finland is requested to transmit to the Secretary-General of the United Nations the text of the Helsinki Document, which is not eligible for registration under Article 102 of the Charter of the United Nations, with a view to its circulation to all the members of the Organization as an official document of the United Nations.

47. The next review conference will be held in Budapest in 1994 on the basis of modalities of the Helsinki Follow-up Meeting, *mutatis mutandis,* to be further specified by the CSO which may decide to organize a special preparatory meeting.

## Helsinki Decisions

*Excerpts*

### I. Strengthening CSCE Institutions and Structures

(1) In order to enhance the coherence of their consultations and the efficiency of their concerted action based on their joint political will, as well as to further develop the practical aspects of co-operation among them, the participating States have decided to reaffirm and develop the decisions on CSCE structures and institutions set forth in the Charter of Paris and the Prague Document on Further Development of CSCE Institutions and Structures.

To this end, they have agreed as follows:

#### Meetings of Heads of State or Government

(2) Meetings of Heads of State or Government, as laid down in the Charter of Paris, will take place, as a rule, every two years on the occasion of review conferences.

(3) They will set priorities and provide

orientation at the highest political level.

**Review Conferences**

(4) Review Conferences will precede the meetings of Heads of State or Government. They will be operational and of short duration. They will:

– review the entire range of activities within the CSCE, including a thorough implementation debate, and consider further steps to strengthen the CSCE process;

– prepare a decision-oriented document to be adopted at the meeting.

(5) Preparation of review conferences, including the agenda and modalities, will be carried out by the Committee of Senior Officials (CSO), which may decide to organize a special preparatory meeting.

**CSCE Council**

(6) The Council constitutes the central decision-making and governing body of the CSCE.

(7) The Council will ensure that the various CSCE activities relate closely to the central political goals of the CSCE.

(8) The participating States have agreed to enhance the working methods of the Council and promote effective consultations at its meetings.

**Committee of Senior Officials**

(9) Further to the decisions contained in the Charter of Paris and as set forth in the Prague Document, the CSO, between the meetings of the CSCE Council, will be responsible for overview, management and co-ordination and will act as the Council's agent in taking appropriate decisions. Additional responsibilities are described in Chapter III of this document.

(10) Greater use will be made of the points of contact and communications network in order to manage the flow of information more efficiently.

(11) The functions of the CSO convening as the Economic Forum are set out in Chapter VII of this document.

**Chairman-in-Office**

(12) The Chairman-in-Office will be responsible on behalf of the Council/CSO for the co-ordination of and consultation on current CSCE business.

(13) The Chairman-in-Office will be requested to communicate Council and CSO decisions to the CSCE institutions and to give them such advice regarding those decisions as may be required.

(14) In carrying out entrusted tasks, the Chairman-in-Office may be assisted, *inter alia*, by:

– the preceding and succeeding Chairmen, operating together as a Troika;

– *ad hoc* steering groups;

– personal representatives, if necessary.

**Assistance to the Chairman-in-Office**

*Troika*

(15) The Chairman-in-Office may be assisted by the preceding and succeeding Chairmen, operating together as a Troika, in carrying out entrusted tasks. The Chairman-in-Office will retain the responsibility for such tasks and for reporting on Troika activities to the Council/CSO.

**Ad Hoc *Steering Groups***

(16) *Ad hoc* steering groups may be established on a case-by-case basis in order to further assist the Chairman-in-Office, in particular in the field of conflict prevention, crisis management and dispute resolution.

(17) The decision of the Council/CSO to establish an *ad hoc* steering group will, in principle, be taken upon recommendation of the Chairman-in-Office and will include a description of its composition and mandate which will set out the specific tasks and objectives and specify the duration.

(18) If the matter is urgent, the Chairman-in-Office may consult the participating States to propose the establishment of an *ad hoc* steering group under a silence procedure. If objections to the proposal are voiced within five days and if further consultations by the Chairman-in-Office have not led to consensus, the CSO must address the question.

(19) In order to ensure efficiency, an *ad hoc* steering group will be composed of a restricted number of participating States which will include the Troika. Its composition and size will be decided taking into account the need for impartiality and efficiency.

(20) The Council/CSO may decide to terminate or extend for a specific period of time the term of the activities of an *ad hoc* steering group as well as to amend the mandate, composition and instructions given to a steering group.

(21) The Chairman-in-Office will report comprehensively and on a regular basis to

the CSO the activities of the *ad hoc* steering group and on related developments.

***Personal Representatives***

(22) When dealing with a crisis or a conflict, the Chairman-in-Office may, on his/her own responsibility, designate a personal representative with a clear and precise mandate in order to provide support. The Chairman-in-Office will inform the CSO of the intention to appoint a personal representative and of the mandate. In reports to the Council/CSO, the Chairman-in-Office will include information on the activities of the personal representative as well as any observations or advice submitted by the latter.

**High Commissioner on National Minorities**

(23) The Council will appoint a High Commissioner on National Minorities. The High Commissioner provides "early warning" and, as appropriate, "early action", at the earliest possible stage in regard to tensions involving national minority issues that have the potential to develop into a conflict within the CSCE area, affecting peace, stability, or relations between participating States. The High Commissioner will draw upon the facilities of the Office for Democratic Institutions and Human Rights (ODIHR) in Warsaw.

**Other Institutions and Structures**

(24) Further to the Charter of Paris and the Prague Document, additional functions of the other CSCE institutions and structures are described in Chapters II, III, IV, V and VI of this document.

(25) The particular States mandate the CSO to study ways and means which would enable the three CSCE institutional arrangements to better accomplish their functions. In this regard they will consider the relevance of an agreement granting a internationally recognized status to the CSCE Secretariat, the Conflict Prevention Centre (CPC) and the ODIHR.

**Implementation Reviews**

(26) Thorough review of the implementation of CSCE commitments will continue to play a prominent role in CSCE activities, thus enhancing co-operation among participating States.

(27) Reviews of implementation will be held regularly at review conferences as well as at special meetings convened for this purpose at the ODIHR and the CPC, and when the CSO convenes as the Economic Forum as provided for in the relevant CSCE documents.

(28) These reviews of implementation will be of a co-operative nature, comprehensive in scope and at the same time able to address specific issues.

(29) The particular States will be invited to offer contributions on their implementation experience, with particular reference to difficulties encountered, and to provide their views of implementation throughout the CSCE area. Participating States are encouraged to circulate descriptions of contributions in advance of the meeting.

(30) Reviews should offer the opportunity to identify action which may be required to address problems. Meetings at which reviews of implementation take place may draw to the attention of the CSO any suggestions for measures to improve implementation which they deem advisable.

**Communications**

(31) The CSCE communications network is an important instrument for the implementation of the Vienna Document 1992 and other documents and agreements. As the CSCE's capacity to deal with emergency situations is being developed, the network is assuring a new and vital role in providing the participating States with up-to-date means for urgent communications. In this respect, it is essential that all participating States be connected to the system. The Consultative Committee of the CPC will monitor progress and, if necessary, recommend solutions for technical problems.

**II. CSCE High Commissioner on National Minorities**

(1) The participating States decide to establish a High Commissioner on National Minorities.

**Mandate**

. . .

## III. Early Warning, Conflict Prevention and Crisis Management (including Fact-Finding and Rapporteur Missions and CSCE Peacekeeping), Peaceful Settlement of Disputes

### Early Warning, Conflict Prevention and Crisis Management (including Fact-Finding and Rapporteur Missions and CSCE Peacekeeping)

(1) The participating States have decided to strengthen the structure of their political consultations and increase their frequency, and to provide for more flexible and active dialogue and better early warning and dispute settlement, resulting in a more effective role in conflict prevention and resolution, complemented, when necessary, by peacekeeping operations.

(2) The participating States have decided to enhance their capability to identify the root causes of tensions through a more rigorous review of implementation to be conducted both through the ODIHR and the CPC. They have also decided to improve their capability to gather information and to monitor developments, as well as their ability to implement decisions about further steps. They have recommitted themselves to co-operating constructively in using the full range of possibilities within the CSCE to prevent and resolve conflicts.

### Early Warning and Preventive Action

(3) In order to have early warning of situations within the CSCE areas which have the potential to develop into crises, including armed conflicts, the participating States will make intensive use of regular, in-depth political consultations, within the structures and institutions of the CSCE, including implementation review meetings.

(4) The CSO, acting as the Council's agent, will have primary responsibility in this regard.

(5) Without prejudice to the right of any State to raise any issue, the attention of the CSO may be drawn to such situations through the Chairman-in-Office, *inter alia,* by:

– any State directly involved in a dispute;

– a group of 11 States not directly involved in the dispute;

– the High Commissioner on National Minorities in situations he/she deems escalating into a conflict or exceeding the scope of his action;

– the Consultative Committee of the CPC in accordance with paragraph 33 of the Prague Document;

– the Consultative Committee of the CPC following the use of the mechanism for consultations and co-operation as regards unusual military activities;

– the use of the Human Dimension Mechanism or the Valletta Principles for Dispute Settlement and Provisions for a CSCE Procedure for Peaceful Settlement of Disputes.

### Political Management of Crisis

(6) The CSO will promote steps by the State or States concerned to avoid any action which could aggravate the situation and, if appropriate, recommend other procedures and mechanisms to resolve the dispute peacefully.

(7) In order to facilitate its consideration of the situation, it may seek independent advice and counsel from relevant experts, institutions and international organizations.

(8) If the CSO concludes that concerted CSCE action is required, it will determine the procedure to be employed in the light of the nature of the situation. It will have, acting on behalf of the Council, overall CSCE responsibility for managing the crisis with a view to its resolution. It may, *inter alia,* decide to set up a framework for a negotiated settlement, or to dispatch a rapporteur or fact-finding mission. The CSO may also initiate or promote the exercise of good offices, mediation or conciliation.

(9) In this context the CSO may delegate tasks to:

– the Chairman-in-Office, who may designate a personal representative to carry out certain tasks, as defined in paragraph (22) of Chapter I of this document;

– the Chairman-in-Office, assisted by the preceding and succeeding Chairmen-in-Office operating together as a Troika, as defined in paragraph (15) of Chapter I of this document;

– an *ad hoc* steering group of participating States, as defined in paragraphs (16) to (21) of Chapter I of this document;

– the Consultative Committee of the CPC, or other CSCE institutions.

(10) Once the CSO has determined the procedure to be applied, it will establish a precise mandate for action, including provisions for reporting back within an agreed

period. Within the limits of that mandate, those to whom the CSO has delegated tasks under the preceding paragraph will retain the freedom to determine how to proceed, with whom to consult, and the nature of any recommendations to be made.

(11) All participating States concerned in the situation will fully co-operate with the CSO and the agents it has designated.

**Instruments of Conflict Prevention and Crisis Management**

***Fact-finding and Rapporteur Missions***

(12) Fact-finding and rapporteur missions can be used as an instrument of conflict prevention and crisis management.

(13) Without prejudice to the provisions of paragraph 13 of the Moscow Document in respect of Human Dimension issues, and paragraph 29 of the Prague Document in respect of Unusual Military Activities, the Committee of Senior Officials or the Consultative Committee of the Conflict Prevention Centre may decide, by consensus, to establish such missions. Such decisions will in every case contain a clear mandate.

(14) The participating State(s) will co-operate fully with the mission on its territory in pursuance of the mandate and facilitate its work.

(15) Reports of fact-finding and rapporteur missions will be submitted for discussion to the Committee of Senior Officials or the Consultative Committee of the Conflict Prevention Centre as applicable. Such reports and any observations submitted by the State(s) visited will remain confidential until they are discussed. The reports will normally be made public. If, however. the mission or the participating State(s) visited request that they should be kept confidential, they will not be made public, unless otherwise decided by the participating States.

(16) Except where provided on a voluntary basis, the expenses of fact-finding and rapporteur missions will be borne by all participating States in accordance with the scale of distribution.

**CSCE Peacekeeping**

(17) Peacekeeping constitutes an important operational element of the overall capability of the CSCE for conflict prevention and crisis management intended to complement the political process of dispute resolution. CSCE peacekeeping activities may be undertaken in cases of conflict within or among participating States to help maintain peace and stability in support of an ongoing effort at a political solution.

(18) A CSCE peacekeeping operation, according to its mandate, will involve civilian and/or military personnel, may range from small-scale to large-scale, and may assume a variety of forms including observer and monitor missions and larger deployments of forces. Peacekeeping activities could be used, *inter alia,* to supervise and help maintain cease-fires, to monitor troop withdrawals, to support the maintenance of law and order, to provide humanitarian and medical aid and to assist refugees.

(19) CSCE peacekeeping will be undertaken with due regard to the responsibilities of the United Nations in this field and will at all times be carried out in conformity with the Purposes and Principles of the Charter of the United Nations. CSCE peacekeeping will take place in particular within the framework of Chapter VIII of the Charter of the United Nations. The CSCE, in planning and carrying out peacekeeping operations, may draw upon the experience and expertise of the United Nations.

(20) The Chairman-in-Office will keep the United Nations Security Council fully informed of CSCE peacekeeping activities.

(21) The Council, or the CSO acting as its agent, may conclude because of the specific character of an operation and its envisaged size that the matter should be referred by the participating States to the United Nations Security Council.

(22) CSCE peacekeeping operations will not entail enforcement action.

(23) Peacekeeping operations require the consent of the parties directly concerned.

(24) Peacekeeping operations will be conducted impartially.

(25) Peacekeeping operations cannot be considered a substitute for a negotiated settlement and therefore must be understood to be limited in time.

(26) Requests to initiate peacekeeping operations by the CSCE may be addressed by one or more participating States to the CSO through the Chairman-in-Office.

(27) The CSO may request the Consultative Committee of the CPC to consider which peacekeeping activities might be most appropriate to the situation and to submit its recommendations to the CSO for decision.

(28) The CSO will exercise overall politi-

cal control and guidance of a peacekeeping operation.

(29) Decisions to initiate and dispatch peacekeeping operations will be taken by consensus by the Council or the CSO acting as its agent.

(30) The Council/CSO will only take such decisions when all parties concerned have demonstrated their commitment to creating favourable conditions for the execution of the operation, *inter alia,* through a process of peaceful settlement and their willingness to co-operate. Before the decision to dispatch a mission is taken, the following conditions must be fulfilled:

– establishment of an effective and durable cease-fire;

– agreement on the necessary Memoranda of Understanding with the parties concerned, and

– provision of guarantees for the safety at all times of personnel involved.

(31) Missions will be dispatched as soon as possible following such a decision.

(32) Decisions by the CSO to establish a peacekeeping operation will include the adoption of a clear and precise mandate.

(33) When establishing a mission, the CSO will take into account the financial implications involved.

(34) The terms of reference of a peacekeeping operation will define practical modalities and determine requirements for personnel and other resources. Preparation of the terms of reference will be carried out, as appropriate, by the Consultative Committee of the CPC. They will be adopted by the CSO unless it has agreed otherwise.

(35) All participating States are eligible to take part in CSCE peacekeeping operations. Appropriate consultations by the Chairman-in-Office will take place. Participating States will be invited by the Chairman-in-Office of the CSO to contribute, on an individual basis, to an operation case by case.

(36) Personnel will be provided by individual participating States.

(37) Parties concerned will be consulted about which participating States will contribute personnel to the operation.

(38) The Council/CSO will regularly review an operation and make any necessary decision related to its conduct, taking into account political developments and developments in the field.

***Chain of Command***

(39) The Council/CSO will assign overall operational guidance of an operation to the Chairman-in-Office assisted by an *ad hoc* group established at the CPC. The Chairman-in-Office will chair the *ad hoc* group and, in this capacity, be accountable to it, and will receive, on behalf of the *ad hoc* group, the reports of the Head of Mission. The *ad hoc* group will, as a rule, consist of representatives of the preceding and succeeding Chairmen-in-Office, of the participating States providing personnel for the mission and of participating States making other significant practical contributions to the operation.

(40) The *ad hoc* group will provide overall operational support for the mission and will monitor it. It will act as a 24-hour point of contact for the Head of Mission and assist the Head of Mission as required.

(41) Continuous liaison between the operation and all participating States will be ensured by the Consultative Committee of the CPC through the regular provision of information to it by the *ad hoc* group.

(42) In all cases where the CSO assigns tasks related to peacekeeping to the CPC, the Consultative Committee of the CPC will be responsible to the CSO for the execution of those tasks.

***Head of Mission***

(43) The Chairman-in-Office, after appropriate consultations, will nominate a Head of Mission for endorsement by the CSO.

(44) The Head of Mission will be responsible to the Chairman-in-Office. The Head of Mission will consult and be guided by the *ad hoc* group.

(45) The Head of Mission will have operational command in the mission area.

***Financial Arrangements***

(46) Peacekeeping operations require a sound financial basis and must be planned with maximum efficiency and cost-effectiveness on the basis of clear cost projections.

(47) Costs of CSCE peacekeeping activities will be borne by all CSCE participating States. At the beginning of each calendar year, the CSO will establish a reasonable ceiling for the cost of peacekeeping operations to which the CSCE scale of distribution will be applied. Beyond that limit, other spe-

cial arrangements will be negotiated and agreed to by consensus. Full and timely payments will be required.

(48) Additional contributions could be provided by participating States on a voluntary basis.

(49) Financial accountability will be ensured by the Chairman-in-Office through regular reports to the participating States.

(50) A start-up fund will, if appropriate, be established to cover the initial costs of an operation. Contributions by a participating State to the start-up fund will be deducted from that State's regular assessed share of the costs relating to the operation.

(51) The Consultative Committee of the CPC is charged to submit to the CSO by the end of 1992 a recommendation with regard to financial modalities of CSCE peacekeeping operations, specifying, *inter alia,* the costs to be shared among participating States in accordance with the preceding paragraphs.

***Co-operation with regional and transatlantic organizations***

(52) The CSCE may benefit from resources and possible experience and expertise of existing organizations such as the EC, NATO and the WEU, and could therefore request them to make their resources available in order to support it in carrying out peacekeeping activities. Other institutions and mechanisms, including the peacekeeping mechanism of the Commonwealth of Independent States (CIS), may also be asked by the CSCE to support peacekeeping in the CSCE region.

(53) Decisions by the CSCE to seek the support of any such organization will be made on a case-by-case basis, having allowed for prior consultations with the participating States which belong to the organization concerned. The CSCE participating States will also take into account the consultations by the Chairman-in-Office regarding prospective participation in the mission, in light of the envisaged size of the operation and the specific character of the conflict.

(54) Contributions by such organizations will not affect the procedures for the establishment, conduct and command of CSCE peacekeeping operations as set out in paragraphs (17) to (51) above, nor does the involvement of any such organization affect the principle that all participating States are eligible to take part in CSCE peacekeeping operations as set out in paragraph (35) above.

(55) Organizations contributing to CSCE peacekeeping would carry out defined and mutually agreed tasks in connection with the practical implementation of a CSCE mandate.

(56) The *ad hoc* group will establish and maintain effective communication with any organization whose resources may be drawn upon in connection with CSCE peacekeeping activities.

***Peaceful Settlement of Disputes***

(57) The participating States consider their commitment to settle disputes among themselves by peaceful means to form a cornerstone of the CSCE process. In their view, the peaceful settlement of disputes is an essential component of the CSCE's overall ability to manage change effectively and to contribute to the maintenance of international peace and security.

(58) The participating States welcome the work done to this end by the Helsinki Follow-up Meeting. In particular they were encouraged by significant progress made on issues relating to creating a conciliation and arbitration court within the CSCE, enhancing the Valletta mechanism and establishing a CSCE procedure for conciliation including directed conciliation, for which proposals were submitted.

(59) In the light of the important subject matter and of the discussions held here in Helsinki, they have decided to continue to develop a comprehensive set of measures to expand the options available within the CSCE to assist States to resolve their disputes peacefully.

(60) In this respect, the Council of Ministers and the CSO could play an important role, in particular by encouraging wider use of conciliation.

(61) Accordingly, intending to reach early results, they have decided to convene a CSCE meeting in Geneva, with a first round from 12 to 23 October 1992, to negotiate a comprehensive and coherent set of measures as mentioned above. They will take into account the ideas expressed regarding procedures for a compulsory element in conciliation, setting up of a court of conciliation and arbitration within the CSCE, and other means.

(62) The results of the meeting will be submitted to the Council of Ministers at the Stockholm Meeting on 14 and 15 December

1992 for approval and, as appropriate, opening for signature.

## IV. Relations with International Organizations, Relations with Non-Participating States, Role of Non-Governmental Organizations (NGOs)

(1) The new tasks before the CSCE require clearer relations and closer contacts with international organizations, in particular with the United Nations, and non-participating States. The CSCE remains at the same time a process whose activities go far beyond formal relations among governments to involve citizens and societies of the participating States. Successful efforts to build a lasting peaceful and democratic order and to manage the process of change require more structured and substantive input from groups, individuals, States and organizations outside the CSCE process.

To this end, the participating States have decided as follows:

### Relations with International Organizations

(2) The participating States, reaffirming their commitments to the Charter of the United Nations as subscribed to by them, declare their understanding that the CSCE is a regional arrangement in the sense of Chapter VIII of the Charter of the United Nations and as such provides an important link between European and global security. The rights and responsibilities of the United Nations Security Council remain unaffected in their entirety.

(3) Recalling the relevant decisions of the Prague Document, the participating States will improve contact and practical co-operation with appropriate international organizations.

(4) They may accordingly agree to invite presentations by those international organizations and institutions mentioned in the Prague Document and others, as appropriate.

(5) Those organizations, institutions and others as agreed may be invited to attend CSCE meetings and seminars as guests of honour with appropriate name-plates.

(6) They will make full use of the information exchange under paragraph 44 of the Prague Document.

### Relations with Non-Participating Mediterranean States

(7) Recalling the provisions of the Final Act and other CSCE relevant documents and consistent with established practice, the non-participating Mediterranean States will continue to be invited to contribute to CSCE activities.

(8) Measures to widen the scope of co-operation with non-participating Mediterranean States are set forth in Chapter X.

### Relations with Non-Participating States

(9) In accordance with paragraph 45 of the Prague Document, the participating States intend to deepen their co-operation and develop a substantial relationship with non-participating States, such as Japan, which display an interest in the CSCE, share its principles and objectives, and are actively engaged in European co-operation through relevant organizations.

(10) To this end, Japan, will be invited to attend CSCE meetings, including those of Heads of State and Government, the CSCE Council, the Committee of Senior Officials and other appropriate CSCE bodies which consider specific topics of expanded consultation and co-operation.

(11) Representatives of Japan may contribute to such meetings, without participating in the preparation and adoption of decisions, on subjects in which Japan has a direct interest and/or wishes to co-operate actively with the CSCE.

### Increasing Openness of CSCE Activities, Promoting Understanding of the CSCE, Expanding the Role of NGOs

(12) The participating States will increase the openness of the CSCE institutions and structures and ensure wide dissemination of information on the CSCE.

(13) To this end:

– the Chairman-in-Office assisted by the CSCE Secretariat will arrange briefings on the political consultation process;

– the CSCE institutions will, within existing budgets, provide information to the public and organize public briefings on their activities;

– the Secretariat will facilitate the flow of information to and contacts with the media, bearing in mind that CSCE policy issues remain the responsibility of participating States.

(14) The participating States will provide opportunities for the increased involvement of non-governmental organizations in CSCE activities.

(15) They will, accordingly:

– apply to all CSCE meetings the guidelines previously agreed for NGO access to certain CSCE meetings;

– make open to NGOs all plenary meetings of review conferences, ODIHR seminars, workshops and meetings, the CSO when meeting as the Economic Forum, and human rights implementation meetings, as well as other expert meetings. In addition each meeting may decide to open some other sessions to attendance by NGOs;

– instruct Directors of CSCE institutions and Executive Secretaries of CSCE meetings to designate an 'NGO liaison person' from among their staff;

– designate, as appropriate, one member of their Foreign Ministries and a member of their delegations to CSCE meetings to be responsible for NGO liaison;

– promote contacts and exchanges of views between NGOs and relevant national authorities and governmental institutions between CSCE meetings;

– facilitate during CSCE meetings informal discussion meetings between representatives of participating States and of NGOs;

– encourage written presentations by NGOs to CSCE institutions and meetings, titles of which may be kept and provided to the participating States upon request;

– provide encouragement to NGOs organizing seminars on CSCE-related issues;

– notify NGOs through the CSCE institutions of the dates of future CSCE meetings, together with an indication, when possible, of the subjects to be addressed, as well as, upon request, the activations of CSCE mechanisms which have been made known to all participating States.

(16) The above provisions will not be applied to persons or organizations which resort to the use of violence or publicly condone terrorism or the use of violence.

(17) The participating States will use all appropriate means to disseminate as widely as possible within their societies knowledge of the CSCE, its principles, commitments and activities.

(18) The concept of a CSCE Prize will be considered.

## V. CSCE Forum for Security Co-operation

The participating States of the Conference on Security and Co-operation in Europe.

(1) Reaffirming their commitments undertaken in the Charter of Paris for a New Europe and, in particular, their determination to establish new negotiations on disarmament and confidence- and security-building open to all participating States,

(2) Encouraged by the opportunities for new co-operative approaches to strengthening security offered by the historic changes and by the process of consolidation of democracy in the CSCE community of States,

(3) Welcoming the adoption of the Vienna Document 1992 on Confidence and Security-Building Measures, the conclusion of the Treaty on Open Skies and the adoption of the CSCE Declaration on the Treaty on Open Skies and the Concluding Act of the Negotiation on Personnel Strength of Conventional Armed Forces in Europe as well as the imminent entry into force of the Treaty on Conventional Armed Forces in Europe (CFE),

(4) Determined to build upon those important achievements and to give a new impetus to arms control, disarmament and confidence- and security-building, security co-operation and conflict prevention in order to better contribute to the strengthening of security and stability and the establishment of a just and lasting peace within the CSCE community of States,

(5) Underlining the equality of rights and the equal respect for the security interests of all CSCE participating States,

(6) Reaffirming their right to choose their own security arrangements,

(7) Recognizing that security is indivisible and that the security of every participating State is inseparably linked to that of all others,

(8) Have decided:

– to start a new negotiation on arms control, disarmament and confidence- and security-building,

– to enhance regular consultation and to intensify co-operation among them on matters related to security, and

– to further the process of reducing the risk of conflict.

(9) To carry out these tasks the participating States have decided to establish a new

CSCE Forum for Security Co-operation, with a strengthened Conflict Prevention Centre, as an integral part of the CSCE.

(10) The participating States will ensure that their efforts in the Forum towards arms control, disarmament and confidence- and security-building, security co-operation and conflict prevention are coherent, interrelated and complementary.

**Objectives**

(11) The participating States will strengthen security and stability through the negotiation of concrete measures aimed at keeping or achieving the levels of armed forces to a minimum commensurate with common or individual legitimate security needs within Europe and beyond. These new measures may entail reductions of and limitations on conventional armed forces and may, as appropriate, include measures of a regional character.

(12) They will address the question of the harmonization of obligations agreed among participating States under the various existing instruments concerning arms control, disarmament and confidence- and security-building.

(13) They will develop the Vienna Document 1992 on the basis of a review of its implementation.

(14) They will negotiate new stabilizing measures in respect of military forces and new confidence- and security-building measures designed to ensure greater transparency in the military field. Such measures may be of a regional character and/or may apply in relation to certain border areas.

* * *

(15) The participating States will aim at establishing among themselves new security relations based upon co-operative and common approaches to security. To this end, they will develop consultation, goal-oriented continuing dialogue and co-operation in the field of security.

(16) They will promote increased predictability about their military plans, programmes and capabilities, including the introduction of major new weapons systems.

(17) They will support and enhance regimes on non-proliferation and arms transfers.

(18) They will enhance contacts, liaison, exchanges and co-operation between their armed forces.

(19) They will promote consultation and co-operation in respect of challenges to their security from outside their territories.

(20) They will also consider other measures to foster security among the participating States in order to contribute to a just and lasting peace among them, including the possibility of further strengthening the norms of behaviour among them through the elaboration of additional security instruments.

* * *

(21) They will make every effort to prevent conflict and give full effect to relevant provisions.

(22) They will further enhance the capability of the CPC to reduce the risks of such conflicts through relevant conflict prevention techniques.

(23) They will foster their co-operation in the field of the implementation and verification of existing and future arms control, disarmament and confidence- and security-building agreements.

* * *

(24) The negotiations on new measures of arms control, disarmament and confidence- and security-building will proceed in distinct phases, taking into account progress made in the implementation of existing arms control agreements. They will also take into consideration ongoing reduction, restructuring and re-deployment processes regarding armed forces as well as further relevant political and military developments. Such new measures will build upon the achievements of existing agreements and will be effective, concrete and militarily significant.

(25) All measures negotiated in the Forum will be developed in a way which precludes circumvention.

**Programme for Immediate Action**

(26) A Programme for Immediate Action is set out in the Annex. It can be amended, supplemented or extended by consensus. It will be reviewed, together with the progress and results obtained, at the review conference preceding the next meeting of CSCE Heads of State or Government.

(27) Additional proposals can be tabled and discussed at any time.

**Area of Application**

(28) Each measure to be negotiated in the Forum will have an area of application according to its nature. The areas of application for negotiations under the Programme for Immediate Action are set out therein in relation to its relevant elements. This is without prejudice to subsequent negotiations on arms control, disarmament and confidence- and security-building or security co-operation in the Forum. Consideration of decisions concerning the area of application will take into account existing agreements and the need for greater transparency.

**Constitution and Organization of the Forum**

(29) The arrangements for the Forum will be as follows:

(30) *The Special Committee* meeting either:

(a) for negotiations on arms control, disarmament and confidence- and security-building, or

(b) for consideration of, goal-oriented dialogue on and, as appropriate, elaboration or negotiation of proposals for security enhancement and co-operation.

(31) *The Consultative Committee* in respect of the existing and future tasks of the CPC.

(32) In order to ensure coherence the representation of the participating States on the Special Committee and the Consultative Committee will in principle be assured by the same delegation. Appropriate meetings will be held as necessary for organizational purposes.

**Procedures**

(33) The Forum will, unless otherwise agreed below, work according to the CSCE procedures.

1. THE SPECIAL COMMITTEE

(34) The Special Committee may establish under its authority subsidiary working bodies open to all participating States. They will work on an *ad referendum* basis and report regularly to the Special Committee. Any question under consideration by such subsidiary working bodies may at any time be raised before the Special Committee.

(35) Consideration and negotiation of regional measures undertaken within the CSCE framework will form an integral part of the activity of the Forum.

(36) They will be dealt with in open-ended working groups established by the Special Committee.

(37) Alternatively, the Special Committee may decide, on the initiative of a limited number of participating States, and on the basis of information provided by them on the nature and the scope of the measures envisaged, that these States form a working group in order to consider, negotiate or develop among themselves certain regional measures. Such working groups will on a regular basis provide appropriate information on their activities to the Special Committee and will submit to it the results.

(38) Any question under consideration by such working groups may at any time be raised before the Special Committee.

(39) This is without prejudice to the right of States to consider, negotiate or develop measures among themselves outside the framework of the CSCE. In such cases they are invited to inform the Forum about progress and results of their work.

2. THE CONSULTATIVE COMMITTEE

(40) The procedures of the Consultative Committee will be based on the relevant decisions of the CSCE Council.

**Form of Commitments**

(41) The results of the negotiations of the Forum will be expressed in international commitments. The nature of obligations will be determined by the character of the measures agreed. They will enter into force in the forms and according to the procedures to be agreed by the negotiations.

**Verification**

(42) Measures will, if appropriate, be provided with suitable forms of verification according to their nature.

**Conference Services**

(43) Common conference services for the Special Committee and the Consultative Committee as well as for all their subsidiary bodies (including seminars), will be provided by an Executive Secretary to be nominated by the host country. The Executive Secretary may also, if so decided by those concerned, provide conference services for meetings of the CFE Joint Consultative Group and the Open Skies Consultative Commission. The Executive Secretary will assume full respons-

ibility for the organization of all the relevant meetings as well as for all related administrative and budgeting arrangements, for which he will be accountable to the participating States according to procedures to be agreed.

(44) The Special Committee and the Consultative Committee will use the same premises.

(45) The new CSCE Forum for Security Co-operation shall commence in Vienna on 22 September 1992.

## ANNEX

### Programme for Immediate Action

(46) The participating States have decided to give early attention to the following:

### *A. Arms Control, Disarmament and Confidence- and Security-Building*

Measures to be negotiated under paragraphs 1–3 will apply to the territory of the participating States in Europe or in Asia as defined below in relation to the area of application of each measure. Measures to be negotiated under paragraphs 4 and 5 will apply to the conventional armed forces and facilities of the participating States both on the territory of all the participating States and beyond. Measures to be negotiated under paragraph 6 will apply to the territory or part thereof of the participating States involved in the measures. Exceptions to these rules on the area of application may be agreed by consensus.

#### 1. HARMONIZATION OF OBLIGATIONS CONCERNING ARMS CONTROL, DISARMAMENT AND CONFIDENCE- AND SECURITY-BUILDING

An appropriate harmonization of the obligations of participating States under existing international instruments applicable to conventional armed forces in Europe, in particular of those concerning the exchange of information, verification and force levels. The harmonization of obligations concerning arms control, disarmament and confidence- and security-building will apply to the areas of application in respect of which the obligations have been undertaken.

#### 2. DEVELOPMENT OF THE VIENNA DOCUMENT 1992

Improvement and further development of confidence- and security-building measures contained in this document. The area of application will be as set out in the Vienna Document 1992.

#### 3. THE FURTHER ENHANCEMENT OF STABILITY AND CONFIDENCE

The negotiation of new stabilizing measures and confidence-building measures related to conventional armed forces, including, with due regard to the specific characteristics of the armed forces of individual participating States, measures to address force generation capabilities of active and non-active forces. These measures may be of a constraining kind. They will apply within the area of application set out in the Vienna Document 1992. This is without prejudice to the possibility that participating States may, if they so choose, decide to offer certain assurances in respect of their conventional armed forces in parts of their territory adjacent to this area of application if they consider such forces relevant to the security of other CSCE participating States.

#### 4. GLOBAL EXCHANGE OF MILITARY INFORMATION

The negotiation of further transparency by means of a global annual appropriately aggregated or disaggregated exchange of information encompassing armaments and equipment, including information on armaments and equipment categories limited by the CFE Treaty, and personnel in the conventional armed forces of the participating States. The regime will also include information on the production of military equipment. The regime will be separate from other information exchange regimes and, because of its special nature, will not involve limitations, constraints or verification.

#### 5. CO-OPERATION IN RESPECT OF NON-PROLIFERATION

Co-operation in respect of the strengthening of multilateral non-proliferation regimes, including the transfer of sensitive expertise, and the establishment of a responsible approach to international armaments transfers.

6. REGIONAL MEASURES

The negotiation by the participating States of suitable measures, including, where appropriate, reductions or limitations in accordance with the objectives set out above, for example in relation to certain regions or border areas. The area of application will be the territory or part thereof of the participating States' territories involved in a regional measure.

**B. Security Enhancement and Co-operation**

Proposals for and dialogue on measures and activities under paragraphs 7-12 will apply to all participating States, unless otherwise agreed or specified below.

7. FORCE PLANNING

The elaboration of provisions to provide transparency about each CSCE participating State's intentions in the medium to long term as regards the size, structure, and equipment of its armed forces, as well as defence policy, doctrines and budgets related thereto. Such a system should be based on each participating State's national practice, and should provide the background for a dialogue among the participating States.

8. CO-OPERATION IN DEFENCE CONVERSION

The development of a programme of exchanges, co-operation and the sharing of expertise in the field of defence conversion throughout all the territory of the participating States.

9. CO-OPERATION IN RESPECT OF NON-PROLIFERATION

Co-operation in respect of the strengthening of multilateral non-proliferation regimes, including the transfer of sensitive expertise, and the establishment of a responsible approach to international armaments transfers.

10. DEVELOPMENT OF PROVISIONS ON MILITARY CO-OPERATION AND CONTACTS

The development of a programme of military contacts, liaison arrangements, co-operation and exchanges, particularly in the fields of the training and organization of armed forces. Participation in this programme will be open to all CSCE participating States in respect of all their armed forces and territory.

11. REGIONAL SECURITY ISSUES

Discussion and clarification of regional security issues or specific security problems for example in relation to border areas.

12. SECURITY ENHANCEMENT CONSULTATIONS

Goal-oriented dialogue and consultations aimed at enhancing security co-operation, including through the further encouragement of responsible and co-operative norms of behaviour on politico-military aspects of security. The participating States will undertake consultations with a view to strengthening the role of the CSCE, by establishing a code of conduct governing their mutual relations in the field of security.

***Conflict Prevention***

Consistent with and further to the decisions taken in Paris, Prague and Helsinki about the tasks of the CPC, the following parts of this work programme will be undertaken in the CPC.

13. RELEVANT TECHNIQUES

Without prejudice to other tasks of the CPC or to the competence of the Committee of Senior Officials in the field of conflict prevention and crisis management, the Consultative Committee will, particularly in the light of experience gained in the execution of its own tasks, maintain under consideration the need for improvements in relevant techniques.

14. CO-OPERATION IN THE FIELD OF VERIFICATION

The encouragement of practical co-operation, through training, exchanges and participation in evaluation and inspection teams, in the implementation of the verification provisions of arms control, disarmament and confidence- and security-building agreements among CSCE participants who are parties to such agreements. The area of application will correspond to that of the relevant agreements.

**VI. The Human Dimension**

(1) The participating States conducted a useful review of implementation of CSCE commitments in the Human Dimension. They based their discussion on the new community of values established among them, as set forth by the Charter of Paris for a New

Europe and developed by the new standards created within the CSCE in recent years. They noted major progress in complying with Human Dimension commitments, but recognized developments of serious concern and thus the need for further improvement.

(2) The participating States express their strong determination to ensure full respect for human rights and fundamental freedoms, to abide by the rule of law, to promote the principles of democracy and, in this regard, to build, strengthen and protect democratic institutions, as well as to promote tolerance throughout society. To these ends, they will broaden the operational framework of the CSCE, including by further enhancing the ODIHR, so that information, ideas, and concerns can be exchanged in a more concrete and meaningful way, including as an early warning of tension and potential conflict. In doing so, they will focus their attention on topics in the Human Dimension of particular importance. They therefore keep the strengthening of the Human Dimension under constant consideration, especially in a time of change.

(3) In this regard, the participating States adopt the following:

**Framework for Monitoring Compliance with CSCE Commitments and for Promoting Co-operation in the Human Dimension**

(4) In order to strengthen and monitor compliance with CSCE commitments as well as to promote progress in the Human Dimension, the participating States agree to enhance the framework of their co-operation and to this end decide the following:

***Enhanced Role of the ODIHR***

(5) Under the general guidance of the CSO and in addition to its existing tasks as set out in the Charter of Paris for a New Europe and in the Prague Document on Further Development of CSCE Institutions and Structures, the ODIHR will, as the main institution of the Human Dimension:

(a) assist the monitoring of implementation of commitments in the Human Dimension

. . .

(b) act as a clearing-house for information

. . .

(c) assist other activities in the field of the Human Dimension, including the building of democratic institutions

. . .

(6) The activities on Human Dimension issues undertaken by the ODIHR may, *inter alia,* contribute to early warning in the prevention of conflicts.

**Human Dimension Mechanism**

(7) In order to align the Human Dimension Mechanism with present CSCE structures and institutions the participating States decide that:

Any participating State which deems it necessary may provide information on situations and cases which have been the subject of requests under paragraphs 1 or 2 of the chapter entitled the 'Human Dimension of the CSCE' of the Vienna Concluding Document or on the results of those procedures, to the participating States through the ODIHR which can equally serve as a venue for bilateral meetings under paragraph 2 or diplomatic channels. Such information may be discussed at Meetings of the CSO, at implementation meetings on Human Dimension issues and review conferences.

(8) Procedures concerning the covering of expenses of expert and rapporteur missions of the Human Dimension Mechanism may be considered by the next review conference in the light of experience gained.

**Implementation**

. . .

(10) The implementation meeting may draw to the attention of the CSO measures to improve implementation which it deems necessary.

(11) The implementation meeting will not produce a negotiated document.

. . .

**CSCE Human Dimension Seminars**

(17) Under the general guidance of the CSO, the ODIHR will organize CSCE Human Dimension seminars which will

address specific questions of particular relevance to the Human Dimension and of current political concern.

. . .

(22) In order to launch the new CSCE Human Dimension Seminars without delay, the participating States decide now at the Helsinki Follow-up Meeting that the ODIHR will organize the following four seminars:

- Migration
- Case Studies on National Minorities Issues: Positive Results
- Tolerance
- Free Media

These seminars will be held before 31 December 1993. The agenda and modalities of the seminars will be decided by the CSO. Seminars on migrant workers and on local democracy will be included in the first annual work programme of seminars. The financial implications of the seminar programme will be kept under consideration by the CSO.

## Enhanced Commitments and Co-operation in the Human Dimension

### *National Minorities*

The participating States

(23) Reaffirm in the strongest terms their determination to implement in a prompt and faithful manner all their CSCE commitments, including those contained in the Vienna Concluding Document, the Copenhagen Document and the Geneva Report, regarding questions relating to national minorities and rights of persons belonging to them;

. . .

(24) Will intensify in this context their efforts to ensure the free exercise by persons belonging to national minorities, individually or in community with others, of their human rights and fundamental freedoms, including the right to participate fully, in accordance with the democratic decision-making procedures of each State, in the political, economic, social and cultural life of their countries including through democratic participation in decision-making and consultative bodies at the national, regional and local level, *inter alia,* through political parties and associations;

. . .

### *Indigenous Populations*

The participating States

(29) Noting that persons belonging to indigenous populations may have special problems in exercising their rights, agree that their CSCE commitments regarding human rights and fundamental freedoms apply fully and without discrimination to such persons.

### Tolerance and Non-discrimination

The participating States

(30) Express their concern over recent and flagrant manifestations of intolerance, discrimination, aggressive nationalism, xenophobia, anti-semitism and racism and stress the vital role of tolerance, understanding and co-operation in the achievement and preservation of stable democratic societies;

. . .

### Migrant Workers

The participating States

(36) Restate that human rights and fundamental freedoms are universal, that they are also enjoyed by migrant workers wherever they live and stress the importance of implementing all CSCE commitments on migrant workers and their families lawfully residing in the participating States;

. . .

### Refugees and Displaced Persons

The participating States

(39) Express their concern over the problem of refugees and displaced persons;

(40) Emphasize the importance of preventing situations that may result in mass flows of refugees and displaced persons and stress the need to identify and address the root causes of displacement and involuntary migration;

(41) Recognize the need for international co-operation in dealing with mass flows of refugees and displaced persons;

. . .

### International Humanitarian Law

The participating States

(47) Recall that international humanitarian law is based upon the inherent dignity of the human person;

(48) Will in all circumstances respect and ensure respect for international humanitarian law including the protection of the civilian population;

(49) Recall that those who violate international humanitarian law are held personally accountable;

(50) Acknowledge the essential role of the International Committee of the Red Cross in promoting the implementation and development of international humanitarian law, including the Geneva Conventions and their relevant Protocols;

. . .

**Democracy at a Local and Regional Level**

The participating States

(53) Will endeavour, in order to strengthen democratic participation and institution building and in developing co-operation among them, to share their respective experience on the functioning of democracy at a local and regional level, and welcome against this background the Council of Europe information and education network in this field;

(54) Will facilitate contacts and encourage various forms of co-operation between bodies at a local and regional level.

. . .

**VII. Economic Co-operation**

. . .

**VIII. Environment**

. . .

**IX. The CSCE and Regional and Transfrontier Co-operation**

. . .

**X. Mediterranean**

. . .

**XI. Programme of Co-ordinated Support for Recently Admitted Participating States**

Further to paragraph 19 of the Summary of Conclusions of the Prague Meeting of the Council, the participating States decide to establish a programme of co-ordinated support for those participating States which have been admitted to the CSCE since 1991.

. . .

**XII. Administrative Decisions**

. . .

*Source:* CSCE summit meeting, Helsinki, 10 July 1992.

## DECISION ON PEACEFUL SETTLEMENT OF DISPUTES

*Stockholm, 14 December 1992*

1. At its Stockholm meeting of 14 and 15 December 1992, the CSCE Council considered the recommendations made by the CSCE Meeting on Peaceful Settlement of Disputes held in Geneva from 12 to 23 October 1992.

2. The Ministers reaffirmed the vital importance of the commitment of all participating States, under Principle V of the Helsinki Final Act, to settle their disputes by peaceful means. In this connection, they recalled other CSCE documents relating to the peaceful settlement of disputes, in particular the Concluding Document of the Vienna Follow-up Meeting, the Charter of Paris for a New Europe, the Report on Peaceful Settlement of Disputes adopted at Valletta and endorsed at the Berlin Meeting of 19 and 20 June 1991, and the Helsinki Document of 1992.

3. The Ministers noted the variety of existing dispute settlement procedures, both within and outside the CSCE. They recalled the important contribution that the potential involvement of an impartial third party can make to the peaceful settlement of disputes and the fact that the Valletta Mechanism enables a participating State, under certain conditions, to seek the mandatory involvement of such a party.

4. The Ministers agreed that in the present circumstances, the principle of the peaceful settlement of disputes assumes particular relevance to problems facing participating States, and that the framework of the CSCE provides a unique opportunity to give impe-

tus to this central aspect of CSCE commitments.

5. In order to further and strengthen their commitment to settle disputes exclusively by peaceful means, and in accordance with paragraphs 57 to 62 of Chapter III of the Helsinki Decisions of 1992 to develop a comprehensive and coherent set of measures available within the CSCE for the peaceful settlement of disputes, the Ministers have:

(a) Adopted measures to enhance the Valletta Provisions through modification of the procedure for selecting Dispute Settlement Mechanisms. This modification is set forth in *Annex 1;*

(b) Adopted the text of a Convention on Conciliation and Arbitration within the CSCE providing for general conciliation and for arbitration on the basis of agreements *ad hoc* or, in advance, on the basis of reciprocal declarations, and declared it open for signature by interested participating States. This text is contained in *Annex 2;*

(c) Adopted a conciliation procedure as an option available to participating States on the basis of agreements *ad hoc* or, in advance, on the basis of reciprocal declarations. This procedure is set forth in *Annex 3;*

(d) Decided that the Council or the Committee of Senior Officials of the CSCE may direct any two participating States to seek conciliation to assist them in resolving a dispute that they have not been able to settle within a reasonable period of time. The provisions relating thereto are set forth in *Annex 4.*

6. The Ministers recalled that nothing stated in the foregoing will in any way affect the unity of the CSCE principles, or the right of participating States to raise within the CSCE process any issue relating to the implementation of any CSCE commitment concerning the principle of the peaceful settlement of disputes, or relating to any other CSCE commitment or provision.

7. Procedures for the peaceful settlement of disputes within the CSCE will be reviewed during the review conference to be held at Budapest in 1994 and periodically thereafter as appropriate.

## Annex 1

### Modification to Section V of the Valletta Provisions for a CSCE Procedure for Peaceful Settlement of Disputes

. . .

## Annex 2

### Convention on Conciliation and Arbitration within the CSCE

The States parties to this Convention, being States participating in the Conference on Security and Co-operation in Europe,

*Conscious* of their obligation, as provided for in Article 2, paragraph 3, and Article 33 of the Charter of the United Nations, to settle their disputes peacefully;

*Emphasizing* that they do not in any way intend to impair other existing institutions or mechanisms, including the International Court of Justice, the European Court of Human Rights, the Court of Justice of the European Communities and the Permanent Court of Arbitration;

*Reaffirming* their solemn commitment to settle their disputes through peaceful means and their decision to develop mechanisms to settle disputes between participating States;

*Recalling* that full implementation of all CSCE principles and commitments constitutes in itself an essential element in preventing disputes between the CSCE participating States;

*Concerned* to further and strengthen the commitments stated, in particular, in the Report of the Meeting of Experts on Peaceful Settlement of Disputes adopted at Valletta and endorsed by the CSCE Council of Ministers of Foreign Affairs at its meeting in Berlin on 19 and 20 June 1991,

*Have agreed* as follows:

### CHAPTER I—GENERAL PROVISIONS

**Article 1**
**Establishment of the Court**

A Court of Conciliation and Arbitration shall be established to settle, by means of conciliation and, where appropriate, arbitration, disputes which are submitted to it in accordance with the provisions of this Convention.

**Article 2**
**Conciliation Commissions and Arbitral Tribunals**

1. Conciliation shall be undertaken by a Conciliation Commission constituted for each dispute. The Commission shall be made up of conciliators drawn from a list established in accordance with the provisions of Article 3.

2. Arbitration shall be undertaken by an Arbitral Tribunal constituted for each dis-

pute. The Tribunal shall be made up of arbitrators drawn from a list established in accordance with the provisions of Article 4.

3. Together, the conciliators and arbitrators shall constitute the Court of Conciliation and Arbitration within the CSCE, hereinafter referred to as 'the Court'.

. . .*

CHAPTER II—COMPETENCE

**Article 18**
**Competence of the Commission and of the Tribunal**

1. Any State party to this Convention may submit to a Conciliation Commission any dispute with another State party which has not been settled within a reasonable period of time through negotiation.

2. Disputes may be submitted to an Arbitral Tribunal under the conditions stipulated in Article 26.

**Article 19**
**Safeguarding the Existing Means of Settlement**

. . .

CHAPTER III—CONCILIATION

**Article 20**
**Request for the Constitution of a Conciliation Commission**

1. Any State party to this Convention may lodge an application with the Registrar requesting the constitution of a Conciliation Commission for a dispute between it and one or more other States parties. Two or more States parties may also jointly lodge an application with the Registrar.

2. The constitution of a Conciliation Commission may also be requested by agreement between two or more States parties or between one or more States parties and one or more other CSCE participating States. The agreement shall be notified to the Registrar.

---

* Articles omitted: 3. Appointment of Conciliators; 4. Appointment of Arbitrators; 5. Independence of the Members of the Court and of the Registrar; 6. Privileges and Immunities; 7. Bureau of the Court; 8. Decision-Making Procedure; 9. Registrar; 10. Seat; 11. Rules of the Court; 12.Working Languages; 13. Financial Protocol; 14. Periodic Report; 15. Notice of Requests for Conciliation or Arbitration; 16. Conduct of Parties—Interim Measures; 17. Procedural Costs.

. . .*

CHAPTER IV—ARBITRATION

. . .**

CHAPTER V—FINAL PROVISIONS

. . .***

**Annex 3**

**Provisions for a CSCE Conciliation Commission**

. . .

**Annex 4**

**Provisions for Directed Conciliation**

. . .

---

* Articles omitted: 21. Constitution of the Conciliation Commission; 22. Procedure for the Constitution of a Conciliation Commission; 23. Conciliation Procedure; 24. Objective of Conciliation; 25. Result of the Conciliation.

** Articles omitted: 26. Request for the Constitution of an Arbitral Tribunal; 27. Cases brought before an Arbitral Tribunal; 28. Constitution of the Arbitral Tribunal; 29. Arbitration Procedure; 30. Function of the Arbitral Tribunal; 31. Arbitral Award; 32. Publication of the Arbitral Award.

*** Articles omitted: 33. Signature and Entry into Force; 34. Reservations; 35. Amendments; 36. Denunciation; 37. Notifications and Communications; 38. Non-Parties. 39. Transitional Provisions.

---

*Source:* CSCE document CSCE/3-C/Dec. 1, Stockholm, 14 Dec. 1992.

## SHAPING A NEW EUROPE—THE ROLE OF THE CSCE

*Summary of Conclusions of the Stockholm Council Meeting, Stockholm, 15 December 1992*

The CSCE Council held its Third Meeting in Stockholm on 14–15 December 1992.

The Ministers consulted on a broad range of issues, in particular the aggression in Bosnia-Herzegovina and Croatia, the crisis in parts of the former Yugoslavia, other regional crises and issues together with the strategy and structure of the CSCE.

In the light of serious threats to peace and security in the CSCE area the Ministers agreed to pursue a strategy of active diplomacy. They will provide the necessary resources.

The Ministers expressed their continuing commitment to use the CSCE to consolidate human rights, democracy, the rule of law and economic freedom as the foundation for peace, security and stability and to prevent, manage and solve conflicts in the CSCE area.

The Ministers condemned the extended use of force in Europe which has bred ever more violence and hatred. They strongly rejected continuing flagrant violations of human rights. They committed themselves to act to counter the growing manifestations of racism, anti-semitism and all forms of intolerance in the CSCE area.

The Ministers agreed to improve co-operation with relevant international organizations. They decided, in particular, to increase co-ordination with the United Nations.

Important aspects of the CSCE strategy include:

– Strengthening the CSCE's operational capabilities through structural reforms and the appointment of a Secretary General;

– Emphasizing the CSCE's ability to provide early warning through the appointment of a High Commissioner on National Minorities who will enjoy the full political support of all participating States;

– Active use of missions and representatives as part of preventive diplomacy to promote dialogue, stability and provide for early warning;

– Enhancing opportunities for the peaceful settlement of disputes through the approval of a comprehensive set of measures to this end. The Ministers stressed their expectations that participating States will avail themselves increasingly of these mechanisms;

– Effective use of missions and representatives in crisis areas as part of a strategy of consultation, negotiation and concerted action to limit conflicts before they become violent;

– Co-operating, as appropriate, with international organizations and with individual participating States to ensure that the broad spectrum of CSCE mechanisms and procedures, including peacekeeping, can be applied;

– Increased efforts at treating the root causes of conflicts by applying all aspects of the human dimension of the CSCE and by involving non-governmental organizations and individual citizens more directly in the work of the CSCE;

– Making all governments accountable to each other for their behaviour towards their citizens and towards neighbouring States and holding individuals personally accountable for war crimes and acts in violation of international humanitarian law;

– Greater use of the Forum for Security Co-operation as a place for negotiation and dialogue which can ensure continued progress in reducing the risks of military conflict and enhancing stability in Europe;

– An active programme to help newly-admitted participating States to participate fully in the structures and work of the CSCE.

---

### Decisions

*Excerpts*

### 1. Regional Issues

. . . *

### 2. The CSCE as a Community of Values

The CSCE's comprehensive concept of security relates peace, security and prosperity directly to the observance of human rights and democratic freedoms. Many of the present problems are linked to the failure to observe CSCE commitments and principles.

---

* Former Yugoslavia; the Baltic States; Moldova; Georgia; Conflict dealt with by the Conference on Nagorno-Karabakh; the Republic of Tajikistan.

The human dimension mechanisms of the CSCE are being used increasingly as a major foundation for the CSCE's efforts at early warning and conflict prevention. Their further elaboration and utilization will strengthen considerably the CSCE's ability to pursue the root causes of tensions and to refine its mechanisms for early warning on potentially dangerous situations.

The Ministers welcomed the strengthened role of the Office for Democratic Institutions and Human Rights and the appointment of the High Commissioner on National Minorities as especially useful steps towards integrating the human dimension more fully into the political consultations and concerted action of the participating States. They also decided to consider ways of using the 1993 Implementation Meeting on Human Dimension Issues to investigate possible new means of utilizing human rights mechanisms for these purposes. They expressed the hope that newly-admitted participating States would make particular use of the opportunities provided by these institutions.

Compliance with CSCE commitments is of fundamental importance. Monitoring of compliance provides governments of participating States with crucial information on which they can formulate policy. The Implementation Meeting on Human Dimension Issues to be held in 1993 offers an opportunity to improve the monitoring of compliance with Human Dimension commitments.

The Ministers expressed their profound concern at the recent manifestations of aggressive nationalism, xenophobia, antisemitism, racism and other violations of human rights. Violations of international humanitarian law and CSCE principles and commitments, such as 'ethnic cleansing', or mass deportation, endangered the maintenance of peace, security and democracy and will not be tolerated. They were convinced that increased attention should be paid by the CSCE, and in particular by the Committee of Senior Officials and the High Commissioner on National Minorities, to these threats to human rights and fundamental freedoms. The CSO will report on this issue to the Council of Ministers at its next session, when the Council will consider developments.

The Ministers also stressed the important role the Human Dimension of the CSCE should play in longer-term conflict prevention. They underlined the need for positive action aimed at fostering understanding, tolerance and national and local preventive action. They emphasized the importance of direct contact between experts, governmental and non-governmental, through the series of Human Dimension seminars successfully begun by the CSCE Seminar on Tolerance and to be followed in 1993 by seminars on national minorities, migration and free media.

The increasing problem of refugees and displaced persons is an issue of major concern to all participating States, particularly in conflicts where the fulfilment of basic human needs is most at risk. The Ministers deplored the plight of civil populations most affected in such conflicts and called on all participating States to contribute to a concerted effort to share the common burden. All Governments are accountable to each other for their behaviour towards their citizens and towards their neighbours. Individuals are to be held personally accountable for war crimes and acts in violation of international humanitarian law.

The Ministers welcomed the rapid convening of the Human Dimension Seminar on Migration as an important contribution to pursuing better understanding of the underlying causes of uncontrolled migration. Another important step towards further implementation of existing human rights standards, including CSCE principles and commitments, will be the United Nations Conference on Human Rights, to be held in Vienna in June 1993. The Ministers expressed their support for the Conference and asked the Chairman-in-Office to represent them there.

The CSCE will continue to give political stimulus to the development of market economies by facilitating, through the March 1993 meeting of the Economic Forum, dialogue and co-operation among participating States and international organizations. The Ministers expressed the view that the Forum's initial meeting would continue the process of co-operation on these issues within the CSCE.

**3. High Commissioner on National Minorities**

The Council appointed Mr. Max van der Stoel as CSCE High Commissioner on National Minorities (HCNM) to strengthen the CSCE's capacity for early warning and preventive diplomacy. The High Commissioner will act within the mandate laid down in the Helsinki Document. The Ministers expressed their support for the High Commissioner and their readiness to co-operate with him in the execution of his complex but crucial task of identifying and containing at the earliest possible stage tensions involving national minority issues which have the potential to develop into a conflict within the CSCE area.

The Ministers encouraged the High Commissioner to analyse carefully potential areas of tension, to visit any participating State and to undertake wide-ranging discussions at all levels with parties directly involved in the issues. In this context, the High Commissioner may discuss the questions with the parties and, where appropriate, promote dialogue, confidence and co-operation between them at all levels, to enhance political solutions in line with CSCE principles and commitments.

The Ministers undertook to provide the High Commissioner with relevant information at their disposal on national minority issues, fully respecting the independence of the High Commissioner in accordance with the mandate.

**4. Peaceful Settlement of Disputes**

The Ministers considered the recommendations made by the CSCE Meeting on Peaceful Settlement of Disputes held in Geneva from 12 to 23 October 1992.

The Ministers reaffirmed the vital importance of the commitment of all participating States, under Principle V of the Helsinki Final Act, to settle their disputes by peaceful means. In this connection, they recalled other CSCE documents relating to the peaceful settlement of disputes, in particular the Concluding Document of the Vienna Follow-up Meeting, the Charter of Paris for a New Europe, the Report on Peaceful Settlement of Disputes adopted at Valletta and endorsed at the Berlin Meeting of 19 and 20 June 1991, and the Helsinki Document 1992.

The Ministers noted the variety of existing dispute settlement procedures, both within and outside the CSCE. They recalled the important contribution that the potential involvement of an impartial third party can make to the peaceful settlement of disputes and the fact that the Valletta Mechanism enables a participating State, under certain conditions, to seek the mandatory involvement of such a party.

The Ministers agreed that in the present circumstances, the principle of the peaceful settlement of disputes assumes particular relevance to problems facing participating States, and that the framework of the CSCE provides a unique opportunity to give impetus to this central aspect of CSCE commitments.

In order to further and strengthen their commitment to settle disputes exclusively by peaceful means, and in accordance with paragraphs (57) to (62) of Chapter III of the Helsinki Decisions of 1992 to develop a comprehensive and coherent set of measures available within the CSCE for the peaceful settlement of disputes, the Ministers have:

(a) Adopted measures to enhance the Valletta Provisions through modification of the procedure for selecting Dispute Settlement Mechanisms;

(b) Adopted the text of a Convention on Conciliation and Arbitration within the CSCE providing for general conciliation and for arbitration on the basis of agreements *ad hoc* or, in advance, on the basis of reciprocal declarations, and declared it open for signature by interested participating States;

(c) Adopted a conciliation procedure as an option available to participating States on the basis of agreements *ad hoc* or, in advance, on the basis of reciprocal declarations;

(d) Decided that the Council or the Committee of Senior Officials of the CSCE may direct any two participating States to seek conciliation to assist them in resolving a dispute that they have not been able to settle within a reasonable period of time and adopted provisions related thereto.

The Ministers recalled that nothing stated in the foregoing will in any way affect the unity of the CSCE principles, or the right of participating States to raise within the CSCE process any issue relating to the implementation of any CSCE commitment concerning the principle of the peaceful settlement of disputes, or relating to any other CSCE commitment or provision.

Procedures for the peaceful settlement of disputes within the CSCE will be reviewed during the review conference to be held at Budapest in 1994 and periodically thereafter as appropriate.

**5. The CSCE Forum for Security Co-operation and Non-Proliferation**

The Ministers welcomed the constructive

work begun in the CSCE Forum for Security Co-operation. They stressed the importance of the contribution to security made by dialogue and negotiations in the Forum and their expectation that further significant progress on the Programme for Immediate Action adopted by the Helsinki Summit should be achieved by the next Meeting of the Council of Ministers. They reaffirmed the importance of full implementation of existing arms control, disarmament and confidence- and security-building provisions agreed within the framework of the CSCE, by all States concerned, including those recently admitted.

Resolved to fully implement the Declaration of the CSCE Council on Non-Proliferation and Arms Transfers adopted at the Prague Council Meeting 30-31 January 1992 and fully committed to CSCE co-operation in respect of non-proliferation, the Ministers agreed as a first step that their States will become original signatories to the Convention on the Prohibition of the Development, Production, Stockpiling and Use of Chemical Weapons and on their Destruction, which will be opened for signature in Paris on 13 January 1993. They also agreed to seek its timely ratification in order for it to enter into force at the earliest date provided for by the Convention. To this end they call upon all other States to sign and ratify the Convention as soon as possible.

They expressed their satisfaction that the Ministers of participating States not yet Parties to the Convention on Bacteriological (Biological) and Toxin Weapons declared that their States intend to become Parties to that Convention as well as to the 1925 Geneva Protocol on the prohibition of the use in war of chemical and biological weapons.

They welcomed that the Ministers of those participating States that are not Parties to the Treaty on the Non-Proliferation of Nuclear Weapons pledged that their States intend to become Parties to that Treaty as non-nuclear weapon States in the shortest time possible. Furthermore, they agree that the Treaty should be extended indefinitely and urged all States that have not yet done so to become Parties to the Treaty.

### 6. Preventive diplomacy and peacekeeping

Consistent with the concept of preventive diplomacy, and while no conflict exists in Estonia, the CSCE is sending a mission to the country, to promote stability and dialogue between the Estonian- and Russian-speaking communities in Estonia.

The Ministers discussed the conflicts that have arisen within the CSCE area, including those in the former Soviet Union, and stressed that they should be resolved by peaceful means.

They reviewed experience with the instruments for early warning, conflict prevention and crisis management, in particular in the field of preventive diplomacy. They noted that, in association with efforts to bring about political solutions, stability can be enhanced by armed contingents for peacekeeping purposes. The deployment and conduct of such operations must be in accordance with the norms of international law and CSCE principles.

The Ministers concluded that the CSCE can play an especially important role in co-operation with mutually reinforcing European and transatlantic organizations by further developing relevant CSCE instruments in the field of preventive diplomacy and peacekeeping.

They requested relevant CSCE institutions, in particular the ODIHR and the CPC, to organize seminars to help share experience and increase knowledge of issues and techniques in the fields of early warning and peacekeeping. Furthermore, they requested the CSO to examine the issues involved in enhancing the capability of all the CSCE instruments.

### 7. Evolution of CSCE structures and institutions

To meet new challenges, the Ministers decided to add further to the improvements in the operational capacity of the CSCE agreed in Paris and Helsinki.

In doing so, they confirmed that the CSCE should retain its flexibility and openness, avoiding the creation of a bureaucracy. Further evolution in CSCE institutions and procedures should be based on the CSCE's democratic rules. It should preserve the strength and diversity afforded by the basic political structure established by the Paris Summit, and should improve the effectiveness of the CSCE's daily work.

Ministers tasked the CSO to conduct a wide ranging review of CSCE structures and operations with a view to establishing organizational arrangements to meet these needs.

As a first step, Ministers have decided to improve further CSCE operations and institutions by establishing the post of Secretary General of the CSCE (Annex 1).

The Ministers also decided to enhance the ability of the CSO to act as their agent and, pending completion of the review mandated above, instructed representatives of the participating States to meet regularly in Vienna in periods between sessions of the CSO. Under the Chairmanship of the Chairman-in-Office, these representatives will conduct consultations on all issues pertinent to the CSCE and undertake preliminary discussion of items suggested for the agenda of the CSO by the Chairman-in-Office. They will decide on matters necessary to ensure prompt and effective implementation of the decisions of the CSO.

To increase the efficiency of the work of the CSCE, the Ministers decided to establish for the Secretariats in Prague and Vienna a single organizational structure under the direction of the Secretary General. The Ministers decided that the CSO should agree the financial and administrative implications of this decision and should adjust staffing, budgets and procedures accordingly.

In implementation of the decision taken by the Heads of Government in Helsinki that the CSO should consider the relevance of an agreement granting an internationally recognized status to the CSCE Secretariat, the Conflict Prevention Centre and the ODIHR, the Ministers instructed the CSO to establish a group of legal and other experts to report through the Committee for decision at the Rome Council Meeting.

The Ministers tasked the Conflict Prevention Centre to take rapid steps to strengthen its ability to provide operational support for CSCE preventive diplomacy missions and peacekeeping activities. The Director of the CPC should present, for approval by the CSO, a proposal setting forth the staffing and budget implications of this decision.

The Ministers emphasized the vital importance of the efficient management of CSCE resources. To this end they instructed the CSO to draw up rules and procedures. They approved the attached Terms of Reference (Annex 2). The Ministers will note progress and take decisions, as necessary, at the Council Meeting.

The Ministers noted that cost-efficiency may also be ensured through seeking new sources of financing CSCE activities. As one innovative possibility they requested the Director of the ODIHR to examine the establishment of a Foundation for Promoting Human Rights in the CSCE.

The Ministers noted that the commitments entered into in Helsinki to expand the role of NGOs have already shown their value. They asked the Chairman-in-Office to examine proposals put forward by NGOs on co-operation between NGOs and the CSCE, and, when appropriate, to submit them to the CSO for consideration.

**8. Improved co-operation and contacts with international organizations, in particular the United Nations**

The new challenges in the CSCE area require improved co-operation and close contacts with relevant international organizations, in accordance with the Helsinki Document. The Ministers expressed their intention to strengthen co-operation in particular with the United Nations.

The CSCE has entered a new phase in its relationship with the United Nations which should be developed further. The Ministers requested the CSO to examine the practical implications of the understanding, expressed in the Helsinki Document, that the CSCE is a regional arrangement in the sense of Chapter VIII of the Charter of the United Nations. In its examination the CSO should also examine the proposal by the United Nations Secretary-General to the CSCE to seek observer status at the United Nations.

The Ministers emphasized that the Chairman-in-Office should keep close contacts with the United Nations in order to promote regular exchanges of information, co-operation and co-ordination and avoid duplication of effort.

They instructed the Chairman-in-Office of the CSO to establish without delay regular contact with the United Nations Secretary-General to ensure that both the United Nations and CSCE participating States are kept informed of relevant activity, especially in the fields of early warning, conflict prevention, management and resolution of conflicts as well as the promotion of democratic values and human rights.

The Ministers decided that a representative of the United Nations Secretary-General will be invited to the meetings of the Council and the Committee of Senior Officials of the CSCE. Furthermore, they decided that the Permanent Mission to the United Nations of the participating State holding the Office of Chairman will serve as a focal point of the CSCE at the United Nations.

**9. Integration of new participating States**

The Ministers decided to intensify their support in conjunction with other institutions, notably the Council of Europe, for the building of democratic institutions to meet the needs identified by newly admitted participating States. They entrusted the Chairman-in-Office, assisted by the CSCE Troika, to consult with newly admitted participating States on useful steps under the Programme of Co-ordinated Support agreed at Helsinki.

The Chairman-in-Office, accompanied by a team of CSCE experts, will conduct a programme of visits to the newly admitted participating States to discuss the CSCE in all its aspects and to explore means of promoting the full involvement of those States in the work and activities of the CSCE. Experts will continue discussions and make an inventory of possible points for further action including ways of promoting information on the CSCE. The Ministers expressed their support for the expansion of CSCE activities in and visits to these States.

**10. Admission of new participating states**

The Ministers agreed that the Czech Republic and the Slovak Republic would be welcomed as participating States from 1 January 1993 following receipt of letters accepting CSCE commitments and responsibilities from each of them according to the draft in Annex 3.

**11. Date and venue of the next Council Meeting**

They agree that the next meeting of the Council will be held in Rome in November/December 1993. They will confirm the specific days for this meeting by silence procedure following the proposal of the host country and recommendation by the CSO not later than March 1993.

**Annex 1**

**The Secretary General of the CSCE**

1. The Ministers decide to establish the post of Secretary General of the Conference on Security and Co-operation in Europe. The Secretary General will derive his/her authority from the collective decisions of the participating States and will act under the guidance of the Chairman-in-Office.

2. The Secretary General will be appointed by the Council by consensus upon recommendation of the CSO and Chairman-in-Office for a period of three years. This period may be extended for one further term of two years.

3. The Chairman-in-Office will be assisted by an open-ended *ad hoc* group in preparing his/her recommendation on the appointment to the CSO and Council.

4. The open-ended *ad hoc* group will assist the Chairman-in-Office in preparing recommendations to the CSO and the Council on the administrative and financial implications of the appointment of a Secretary General, including accommodation, staff requirements and budget.

5. The Ministers agreed on the following mandate for the Secretary General.

**Mandate**

(i) The Secretary General will act as the representative of the Chairman-in-Office and will support him/her in all activities aimed at fulfilling the goals of the CSCE. The Secretary General's tasks will also include the management of CSCE structures and operations; working closely with the Chairman-in-Office in the preparation and guidance of CSCE meetings; and ensuring the implementation of the decisions of the CSCE.

(ii) The Secretary General will oversee the work of the CSCE Secretariat, the CPC Secretariat and the ODIHR. The Secretary General will answer for the effective performance of CSCE staff to the Chairman-in-Office, the Council of Ministers and the CSO.

(iii) The Secretary General will assist the Chairman-in-Office in publicizing CSCE policy and practices internationally, including maintaining contacts with international organizations.

(iv) As the CSCE's Chief Administrative Officer, the Secretary General will advise on the financial implications of proposals and ensure economy in the staff and support services of the institutions.

(v) The Secretary General will prepare an annual report to the CSCE Council.

(vi) The Secretary General will perform such other functions as are entrusted to him/her by the Council or the CSO.

**Annex 2**

**Management of Resources**

. . .

**Annex 3**

**Bratislava/Prague, 1 January 1993**

Your Excellency

The Government of the Slovak/Czech Republic hereby adopts the Helsinki Final Act, the Charter of Paris for a New Europe and all other documents of the Conference on Security and Co-operation in Europe.

The Government of the Slovak/Czech Republic accepts in their entirety all commitments and responsibilities contained in these documents and declares its determination to act in accordance with their provisions. It will assume, in co-operation with the Czech/Slovak Republic as the other successor State to the Czech and Slovak Federal Republic, all CFE obligations of the Czech and Slovak Federal Republic.

The Government of the Slovak/Czech Republic invites and will fully facilitate the visit of a Rapporteur Mission to be arranged by the Chairman of the CSCE Council. This Mission will report to the CSCE participating States on the fulfilment by the Slovak/Czech Republic of CSCE commitments and provide assistance towards their fullest implementation.

The Government of the Slovak/Czech Republic expresses its readiness for signature of the Helsinki Final Act and the Charter of Paris by the Head of State or Government of the Slovak/Czech Republic at the earliest convenience.

I kindly ask Your Excellency to circulate copies of this letter to all CSCE participating States.

Please accept, Your Excellency, the assurances of my highest consideration.

Her Excellency,
Margaretha af Ugglas
Chairman-in-Office of the CSCE Council
Minister of Foreign Affairs
Kingdom of Sweden, Stockholm

---

*Source:* CSCE document CSCE/3-C/Dec. 2, Stockholm, 15 Dec. 1992.

# Part III. Weapons and technology proliferation, 1992

Chapter 6. Nuclear weapon developments and proliferation

Chapter 7. Chemical and biological weapons: developments and proliferation

Chapter 8. Military technology and international security: the case of the USA

# 6. Nuclear weapon developments and proliferation

DUNBAR LOCKWOOD and JON BROOK WOLFSTHAL

## I. Introduction

With the end of the cold war and the growing economic and political pressures to reduce military budgets, the United States and the Russian Federation not only agreed to make deep reductions in their existing strategic nuclear forces but also made unilateral commitments to dramatically scale down their respective strategic nuclear weapon modernization programmes. France and the United Kingdom also made unilateral commitments to curb their nuclear forces but did not abandon the force modernization programmes that could increase the number of their strategic nuclear warheads. All four countries also began to retire tactical (non-strategic) nuclear weapons. China, whose nuclear weapon programmes are shrouded in greater secrecy, continued to modernize its nuclear weapons slowly.

Eight underground nuclear explosions were conducted in 1992—six by the USA and two by China. This was the lowest world total in over 30 years (see appendix 6A for the yields, dates and locations of the tests, and chapter 11 for progress towards a comprehensive test ban).

In 1992 the USA decided to further scale back most of its strategic weapon modernization programmes. These changes included limiting or cancelling production of the B-2 bomber, the advanced cruise missile (ACM), the W-88 warhead for the Trident II (D-5) missile, and MX/Peacekeeper test missiles. In addition, the USA terminated the Midgetman/Small ICBM (SICBM) missile programme.

Russia took several steps in 1992 to reduce its strategic nuclear forces. It stopped producing strategic bombers, long-range cruise missiles, ballistic-missile submarines (SSBNs) and all but one type of intercontinental ballistic missile (ICBM). In addition, all of the former Soviet Union's tactical nuclear weapons, most of which are now scheduled for dismantlement, were consolidated on the territory of Russia during the first half of the year.

The warming of political relations between the USA and Russia and the accompanying cutbacks in their respective nuclear arsenals, along with domestic budgetary constraints, created new pressure on France and the UK to limit the size and scope of their nuclear weapon programmes. France's decision in 1992 to cut back the number of SSBNs it will deploy after the turn of the century reflected these pressures. Similarly, the British Government decided in 1992 to reduce the number of its bomber squadrons and to eliminate its naval tactical nuclear weapons. However, the decisions of the British and French governments to go forward with their strategic modernization programmes indicate that both countries intend to retain their status as members of the 'nuclear club' for many years to come.

Because China continues to be a closed country, it is particularly difficult to gauge trends in Chinese nuclear weapon policy. However, it appears that several new nuclear weapon systems are moving forward, albeit slowly. The Chinese Government has not made any formal public commitments to curb its nuclear weapon programmes.

The tables showing the nuclear forces of the five declared nuclear weapon nations as of January 1993 (tables 6.1–6.5) appear on pages 234–240 of this chapter.

## II. US nuclear weapon programmes

### Strategic nuclear weapons

The United States decided to halt further development of new types of ICBM and to continue retiring older ICBMs in 1992. In his 28 January 1992 State of the Union Address (for an excerpt of the text, see appendix 11A), President George Bush scrapped the only US ICBM under development when he called for the cancellation of the Midgetman/SICBM missile. The programme's estimated acquisition cost of $40–50 billion proved to be prohibitive in a time of declining US defence budgets. In anticipation of the entry into force of the 1991 US–Soviet Treaty on the Reduction and Limitation of Strategic Offensive Arms (START Treaty), the US Air Force continued retiring Minuteman II missiles, a process begun in late 1991. By the end of 1992, it had removed more than 100 of the 450 Minuteman II missiles from their silos[1]—a process scheduled to be completed by 1995.[2] With the retirement of the MX/Peacekeeper missile under the 1993 US–Russian Treaty on Further Reduction and Limitation of Strategic Offensive Arms (START II Treaty; see appendix 11A), the only ICBMs that the USA now plans to deploy beyond the turn of the century are the 500 existing Minuteman III missiles, which will have their loadings reduced from three warheads each to one. To extend the life of the Minuteman III until at least the year 2010, the Air Force plans to replace ageing components in the guidance computer and associated electrical systems and refurbish the second- and third-stage rocket motors. It is also considering providing the Minuteman III missile with advanced guidance technologies that would give it an accuracy similar to that of the MX. The US Defense Department has said that 'there is no technical reason' why the Minuteman III 'cannot be supported to the year 2020'.[3] The projected post-2000 START II Treaty force of 500 Minuteman III missiles downloaded to a

[1] As of 31 Dec. 1992, Malmstrom AFB, Montana, had removed 30 Minuteman II missiles from their silos; Ellsworth AFB, South Dakota, 64; and Whiteman AFB, Missouri, 12.

[2] Dick Cheney, Secretary of Defense, *Annual Report to the President and the Congress* (US Government Printing Office: Washington, DC, Jan. 1993), p. 67.

[3] US Department of Defense Report to Congress, *Minuteman III Life Extension Report,* 29 July 1992, p. 1.

single warhead each—making up less than 15 per cent of the total number of US strategic nuclear warheads planned under the START II Treaty—reflects somewhat of a de-emphasis on ICBMs compared to the existing force structure.

The USA continued to retire older Poseidon submarines, which were built in the 1960s and are now reaching the end of their service lives. The eight remaining Poseidon submarines, which are armed with the Trident I (C-4) missile, are all slated for decommissioning by 1995. In 1997 the US SSBN fleet will consist of 18 Trident submarines—8 in the Pacific and 10 in the Atlantic. The size of the fleet is based not only on budgetary considerations but also on the need to tailor US sea-based nuclear forces to START Treaty limits. Despite these constraints, this fleet is still projected to carry almost 50 per cent of all US strategic warheads under the START II Treaty.

The USA decided in 1992 to cease submarine-launched ballistic missile (SLBM) warhead production, cancelling further production of the 475-kt W-88 warhead largely because the division of the Rocky Flats plant which made plutonium 'pits' for nuclear weapons has been closed since 1989 for environmental and safety reasons. US Navy statements indicate that about 400 W-88 warheads have been produced—enough to arm four Trident submarines based at King's Bay, Georgia, with four warheads per missile. To make up for the shortfall, the Navy plans to transfer 100-kt W-76 warheads from Trident I missiles (taken from retired Poseidon submarines) to Trident II missiles deployed on Trident submarines at King's Bay. Secretary of Defense Dick Cheney said that, as a result of the W-88 cancellation, US 'hard-target kill capability in the SLBM force will be reduced to less than half the planned level'.[4] However, given that the number of Commonwealth of Independent States (CIS)/Russian hardened ICBM silos—which have been among the highest targeting priorities of US planners—would probably decrease to no more than several hundred under the START II Treaty, the loss of such a US capability does not seem particularly significant.

In June 1992 the Air Force released the 'Bomber Road Map', outlining its plans for the future of US strategic aircraft. It has decided, for reasons of cost and obsolescence, to retire all nuclear-role B-52G bombers by the end of 1993[5] and all conventional-role B-52G bombers by 1995,[6] but the US Air Force plans to retain its B-52Hs for both nuclear and conventional missions. B-52Hs are slightly newer than the B-52Gs, have engines that are 30 per cent more efficient, cost less to operate, have a greater range and 'can fight more easily from austere locations on short notice'.[7] There is currently a total of 95 B-52H bombers in the inventory, of which up to 80 are operationally ready for combat at any one time, with the balance in maintenance, training and testing.

[4] Dick Cheney, Secretary of Defense, *Annual Report to the President and the Congress* (US Government Printing Office: Washington, DC, Feb. 1992), p. 63.

[5] Cheney (note 2), p. 67.

[6] Cheney (note 2), p. 150.

[7] US Department of the Air Force, *The Bomber Road Map*, June 1992, p. 8.

In addition to the B-52H bombers, the Air Force plans to retain all the 95 B-1B bombers in the inventory.[8] At any one time, 80 of the 95 bombers are operationally ready. The Air Force has decided to make the B-1B 'the backbone of the conventional bomber force', because of its 'modern capabilities and sheer numbers'.[9] It intends to upgrade the conventional capabilities of the B-1B during the next decade with a host of conventional air-to-surface missiles and bombs.[10]

In his January 1992 State of the Union Address, President Bush called for the cancellation of further B-2 bomber production after 20 aircraft have been built.[11] In response to Bush's proposal, Congress authorized funding to complete the force at 20 bombers, but it attached several conditions that must be fulfilled, including documentation of the bomber's ability to evade radar detection, before the fiscal year (FY) 1992 and FY 1993 procurement funds may be spent for the last five bombers.[12] The first operational B-2s are scheduled to be delivered to Air Combat Command by the end of 1993.[13]

Bush also called for cancellation of further ACM production above the number already funded. Although Congress had authorized funding for 640 ACMs through FY 1992, the ceiling was subsequently reduced to 460 because of cost overruns.[14] The decision to cut back the ACM programme was expected since the programme had been plagued with technical problems, the Air Force already had approximately 1200 relatively new nuclear-armed air-launched cruise missiles (ALCMs), the former Soviet Union's air defence network had deteriorated, and US force structure planning under both START Treaties effectively precludes the deployment of ACMs on the B-1B bomber.

## Non-strategic nuclear weapons

In partial fulfilment of President Bush's 27 September 1991 initiative on tactical nuclear weapons,[15] the USA completed the global withdrawal of all its ground- and sea-launched tactical nuclear weapons from abroad by June 1992.[16] The USA withdrew a total of 1700 ground-launched warheads from

[8] Rockwell built a total of 100 B-1Bs, but 4 crashed; the most recent accident happened on 30 Nov. 1992 in Texas. One of the 96 remaining aircraft is used as a ground trainer at Ellsworth AFB, South Dakota. Although it is START Treaty-accountable, it is not operational.

[9] US Department of the Air Force (note 7), p. 9

[10] US Department of the Air Force (note 7), pp. 10–11; Leopold, G., 'AF expands conventional role of strategic arms', *Defense News*, 27 Apr.–3 May 1992, pp. 12, 20.

[11] In addition to the 20 operational B-2s there will be 1 permanent test plane based at Edwards AFB, California.

[12] For details on B-2 funding restrictions, see *Congressional Record,* 1 Oct. 1992, p. H10221.

[13] US Department of the Air Force (note 7), p. 11.

[14] Cheney (note 2), p. 67.

[15] For the text of the 1991 Bush initiative, see *SIPRI Yearbook 1992: World Armaments and Disarmament* (Oxford University Press: Oxford, 1992), appendix 2A, pp. 85–87.

[16] US Arms Control and Disarmament Agency (ACDA), *Annual Report to the Congress, 1992,* (ACDA: Washington, DC, 1992), pp. 37–38; Williams, P., Assistant Secretary of Defense for Public Affairs, Press Conference, 2 July 1992; Arkin, W. M. and Norris, R. S., *Taking Stock: US Nuclear Deployments at the End of the Cold War* (Greenpeace/Natural Resources Defense Council: Washington, DC, Aug. 1992), p. 4.

abroad: 700 Lance missile warheads and 1000 artillery shells.[17] All of these weapons, plus an additional 150 Lance missile warheads and 300 artillery shells that were already stored in the USA, are scheduled for dismantlement.[18]

The USA also withdrew all 500 warheads routinely deployed at sea on surface ships, attack submarine and aircraft-carriers: 100 W-80 sea-launched cruise missiles (SLCMs) and a combination of 400 B-57 depth strike/bombs and B-61 gravity bombs.[19] In addition, the USA removed from service 350 B-57 depth bombs deployed with land-based naval anti-submarine warfare (ASW) aircraft, including 200 B-57 depth bombs that were withdrawn from abroad.[20] Of these 850 naval tactical nuclear weapons, about half will be dismantled,[21] including all of the B-57 depth bombs.[22]

After NATO fully implements its October 1991 decision to reduce the number of US nuclear gravity bombs in Europe by 50 per cent,[23] only about 700 US B-61s will remain. In February 1992 the first of about 48 F-15E strike aircraft assigned to US squadrons in Europe arrived in the UK at the Lakenheath Air Base. (The F-111E/F aircraft that the F-15Es are replacing will be returned to the USA.)[24] The Bush Administration maintained that the presence of US air-delivered nuclear weapons in Europe was a necessary symbol of the continued US commitment to European security.

Chairman of the Joint Chiefs of Staff (JCS) General Colin Powell announced in January 1992 that the USA planned to retain 1600 tactical nuclear warheads.[25] These forces would apparently consist of 700 B-61 gravity bombs for tactical air forces in Europe and the USA, 550 B-61 gravity bombs stored in the USA for aircraft-carriers, and 350 W-80 Tomahawk SLCM warheads stored in the USA for surface ships and nuclear-powered attack submarines (SSNs). Whether these tactical weapons have any remaining utility at all is a contentious issue in the defence community.[26]

[17] Most of these Lance missile warheads and artillery shells were based in the western part of Germany; see ACDA (note 16) and Williams (note 16).

[18] ACDA (note 16).

[19] ACDA (note 16).

[20] ACDA (note 16), p. 37; Williams (note 16). According to Arkin and Norris (note 16), the 200 B-57 depth bombs withdrawn from abroad came from Italy and the UK and possibly Guam.

[21] ACDA (note 16), p. 38.

[22] Arkin and Norris (note 16).

[23] Smith, R. J., 'NATO approves 50% cut in tactical A-bombs', *Washington Post,* 18 Oct. 1991, pp. 1, 28.

[24] 'Nuclear notebook', *Bulletin of the Atomic Scientists*, Dec. 1992, p. 57.

[25] US Department of Defense, Fact Sheet, 'Total nuclear warheads', 29 Jan. 1992; see also Scarborough, R., 'Cheney proposes cutting B-2, Seawolf', *Washington Times*, 30 Jan. 1992, p. 5, see accompanying chart.

[26] See, for example, Daalder, I. H., 'Nuclear weapons in Europe: why zero is better', *Arms Control Today*, Jan./Feb. 1993, pp. 15–18.

## III. Former Soviet and CIS nuclear weapon programmes

### Strategic nuclear weapons

It appears that the only ICBM still under production in the former Soviet Union is the SS-25, which is assembled at the Votkinsk plant in Russia. In June 1992, Lt. General James Clapper, Director of the Defense Intelligence Agency (DIA), told the Senate that the former Soviet Union continued to deploy SS-25 and SS-18 ICBMs.[27]

The US intelligence community now believes that the follow-on to the SS-25 is the only new Russian ICBM still under development.[28] It expects this missile to be flight-tested and deployed in this decade.[29] Up to 90 of these new missiles may be deployed in converted SS-18 silos under the terms of the START II Treaty. The decision to limit ICBM modernization to the single-warhead SS-25 follow-on missile does not come as a surprise: Russia is confronted with tight budget constraints; the final assembly plant for the existing SS-25 is in Russia (unlike the plants for SS-18s and SS-24s which are in Ukraine); and the START II Treaty bans all MIRVed (equipped with multiple independently targetable re-entry vehicles) ICBMs. Consequently, if the START II Treaty is implemented, the structure of the Russian strategic triad will be more similar to that of the US strategic forces than it has ever been before.

Russia also continued to retire its SS-11 and SS-17 ICBMs.[30] In November 1992, Russian Deputy Defence Minister Andrey Kokoshin said that the remaining 40-odd SS-17s will be removed from service in the next two years.[31]

Russia did not build any new ballistic-missile submarines in 1992, and US intelligence projects that 'none are anticipated before the end of the decade'.[32] The decision to halt SSBN production makes sense for both economic and arms control reasons. With its existing fleet of six Typhoon submarines, seven Delta IVs and 14 Delta IIIs, Russia already has more warheads on submarines

[27] Lt. General James R. Clapper, Jr, Director, US Defense Intelligence Agency (DIA), Statement before the Senate Foreign Relations Committee, 30 June 1992, in *The START Treaty,* Senate Hearing 102-607, Part 2 (US Government Printing Office: Washington, DC, 1992), p. 163.

[28] Clapper (note 27), p. 163; see also Lawrence Gershwin, National Intelligence Officer for Strategic Programs, Central Intelligence Agency, 'Threats to US interests from weapons of mass destruction over the next ten to twenty years', paper presented at the conference on Defense Against Ballistic Missiles: The Emerging Consensus for SDI, sponsored by the Institute for Foreign Political Analysis, Washington, DC, 23 Sep. 1992, p. 5.

[29] Gershwin (note 28), p. 5.

[30] Robert Gates, Director, US Central Intelligence Agency, testimony before the Senate Armed Services Committee, 22 Jan. 1992, *Threat Assessment, Military Strategy, and Defense Planning*, Senate Hearing 102-755 (US Government Printing Office: Washington, DC, 1992), p. 45.

[31] 'Russian Deputy Defense Minister on nuclear missile forces', *Izvestia*, 13 Nov. 1992, in Foreign Broadcast Information Service, *Daily Report–Central Eurasia* (hereafter, *FBIS-SOV),* FBIS-SOV-92-220, 13 Nov. 1992, p. 2; 'Defense official assesses missile forces future', *Krasnaya Zvezda*, 14 Nov. 1992, in FBIS-SOV-92-221, 16 Nov. 1992, p. 2; 'Deputy Defense Minister views future missile forces', *Nezavisimaya Gazeta*, 19 Nov. 1992, pp. 1–2, in FBIS-SOV-92-235, 7 Dec. 1992, pp. 9–11.

[32] Lt. General James R. Clapper, Jr, Director, Defense Intelligence Agency, Statement before the Senate Armed Services Committee, 22 Jan. 1992, in *Threat Assessment, Military Strategy, and Defense Planning* (note 30), p. 33; see also 'Swords and ploughshares', *The Economist*, 16 Jan. 1993, p. 52.

than is permitted under the START II Treaty ceilings. Furthermore, only one to six Russian SSBNs are believed to be on patrol at any given time.[33]

In a November 1992 speech in South Korea, President Boris Yeltsin said that Russia would stop making submarines for 'military purposes' by 1995.[34] Russian officials subsequently clarified Yeltsin's statement, saying that he was referring to further production in the Asian part of Russia, namely, at the Komsomol'sk shipyard, *not* the Severodvinsk shipyard where production of Akula Class SSNs or Delta Class SSBNs may continue.[35] In addition to halting SSBN production, Russia continued retiring Yankee submarines in 1992.[36]

The US intelligence community now also believes that the SS-N-20 follow-on SLBM, which is expected to be more accurate than the SS-N-20, is the only new Russian SLBM likely to become operational during the 1990s.[37] Russia has begun to modify the first Typhoon Class submarine at Severodvinsk to carry the SS-N-20 follow-on SLBM.[38]

President Yeltsin announced in January that Moscow would halt further production of strategic bombers (i.e., Blackjack and Bear-H bombers).[39] Since the 1960s Moscow has placed relatively little emphasis on the bomber leg of its strategic triad. Furthermore, the Bear-H, a propeller-driven aircraft, has 1960s-vintage technology, and the more modern Blackjack has been plagued by numerous technical problems.[40] It was reported in August that the last Blackjack was delivered from the Kazan production line in Russia.[41] Estimates of the number of operational Blackjacks now deployed at the Priluki Air Force Base in Ukraine range from 16 to 20.[42]

In January 1992 Yeltsin announced that Russia would also halt further production of all existing types of long-range ALCM, that is, the AS-15.[43] Ten

[33] Cushman, J. H., Jr, 'US Navy's periscopes still follow Soviet fleet', *New York Times*, 23 Feb. 1992, p. A14; 'Washington outlook: no new subs', *Aviation Week & Space Technology*, 23 Nov. 1992, p. 25; Gordon, M., 'US and Russian submarines collide in the Arctic', *New York Times,* 23 Mar. 1993, p. A12.

[34] Pollack, A., 'Yeltsin plans end to A-sub program', *New York Times,* 20 Nov. 1992, p. A10; see also Shin, P., 'Yeltsin sees end to sub construction', *Washington Post,* 20 Nov. 1992, p. A44.

[35] 'Subs still to be built at Severodvinsk', ITAR-TASS, 20 Nov. 1992, in FBIS-SOV-92-226, 23 Nov. 1992, p. 2; 'Russia sub halt limited to Asia', *Washington Times,* 21 Nov. 1992, p. 2.

[36] Gates (note 30), p. 45.

[37] Clapper (note 27), Part 2, p. 163; see also Gershwin (note 28), p. 5.

[38] Statement of Rear Admiral Edward D. Sheafer, Jr, Director of US Naval Intelligence, before the House Armed Services Committee Subcommittee on Seapower, Strategic, and Critical Materials, 5 Feb. 1992, p. 22; see also Cushman (note 33), p. A14.

[39] 'Yeltsin delivers statement on disarmament', Moscow Teleradiokompaniya Ostankino Television First Program Network, 29 Jan. 1992, in FBIS-SOV-92-019, 29 Jan. 1992, p. 1; *SIPRI Yearbook 1992* (note 15), p. 90.

[40] 'Nuclear notebook: Soviet bomber woes', *Bulletin of the Atomic Scientists*, July 1990, p. 48.

[41] Velovich, A., 'Kazan produces final batch of Blackjacks', *Flight International*, 12–18 Aug. 1992, p. 22.

[42] The Defense Department reported in Sep. 1991 that there were 16 Blackjacks at Priluki (US Department of Defense, *Military Forces in Transition*, Washington, DC, Sep. 1991, p. 34); Velovich (note 41) reported 18; and the IISS estimated that there were 20; International Institute for Strategic Studies (IISS), *Military Balance 1992–1993* (Brassey's: London, 1992).

[43] See 'Yeltsin delivers statement on disarmament' (note 39), p. 1; *SIPRI Yearbook 1992* (note 15), p. 90.

months later he said that Russia will unilaterally halt the production of medium-sized bombers, presumably a reference to Backfire bombers.[44]

### Non-strategic nuclear weapons

In complying with then Soviet President Mikhail Gorbachev's 5 October 1991 commitment, President Yeltsin announced on 29 January 1992 that Russia would destroy all of the nuclear warheads associated with tactical ground-launched weapon systems (artillery shells; land mines; and SS-N-21, FROG and Scud short-range missiles).[45] However, he went further than Gorbachev, saying that Russia would not produce any new warheads to replace them and that Russia would destroy one-third of its tactical sea-launched nuclear warheads, half of its tactical air-launched nuclear warheads, and half of the nuclear warheads for its anti-aircraft missiles.[46]

All tactical nuclear weapons were withdrawn to Russia from Kazakhstan by late January 1992, from Belarus by 28 April and from Ukraine by 6 May.[47] (Tactical nuclear weapons had been withdrawn to Russia from all the other former Soviet republics earlier.) The Russian Defence Ministry announced in February 1993 that all tactical nuclear weapons had been withdrawn from ships and submarines.[48]

## IV. French nuclear weapon programmes

### Strategic nuclear weapons

The future of the land-based leg of France's strategic nuclear forces is in question. In 1991 the French Government decided to cancel the mobile S45 intermediate-range ballistic missile missile (IRBM) that was scheduled to replace the 18 silo-based S3D IRBMs by the end of the 1990s.

The single-warhead S3-Ds, currently deployed on the Plateau d'Albion, are scheduled for retirement in 1996. The French Government, however, is now considering developing an ICBM version of the M-5 SLBM and deploying it

[44] Pollack (note 34), p. A10.

[45] For the texts of the Gorbachev and Yeltsin initiatives, see *SIPRI Yearbook 1992* (note 15), appendix 2A.

[46] SS-N-15s, SS-N-16s, FRAS-1s and torpedoes for anti-submarine warfare (ASW); SS-N-9, SS-N-12, SS-N-19, SS-N-21 and SS-N-22 SLCMs; and AS-4, AS-5 and AS-6 air-to-surface missiles and bombs for naval aircraft; tactical air-launched warheads include AS-4, AS-5 and AS-6 ASMs, AS-16 SRAMs and gravity bombs; and anti-aircraft missile warheads include SA-2s, SA-5s and SA-10s. See 'Yeltsin delivers statement on disarmament' (note 39); *SIPRI Yearbook 1992* (note 15), p. 90.

[47] For Ukraine, see Leonid Kravchuk, National Press Club, Federal News Service Transcript, 7 May 1992, p. 5-1; ACDA (note 16), p. 39; Oberdorfer, D., 'Ukraine agrees to eliminate nuclear arms', *Washington Post,* 7 May 1992, pp. A1, A38. For Belarus, see 'Belarus free of tactical nuclear weapons', *Radio Free Europe/Radio Liberty Research Report,* vol. 1, no. 19 (8 May 1992), 'Military and security notes', p. 49. For Kazakstan, see Yeltsin, 1 Feb. 1992 Camp David Press Conference, Federal News Service Transcript, p. 3; 31 Jan. 1992, Press Conference at the UN, Federal News Service Transcript, p. 3-1.

[48] Shapiro, M., 'Russian Navy rids itself of tactical nuclear arms', *Washington Post,* 5 Feb. 1993, p. A31; 'Tactical nuclear arms removed from vessels', ITAR-TASS, 4 Feb. 1993, in FBIS-SOV-93-022, 4 Feb. 1993, p. 1.

in S3-D silos. If this is done, it would mean that the S3-Ds would have to be kept in service until at least 2005.[49] Reportedly, the alliance of conservative parties which took control of the French Parliament from the Socialists after the national elections in March 1993 strongly supports the deployment of a new land-based missile.[50] When and if the START II Treaty is implemented, the deployment of a land-based version of the M-5 could give France the world's only MIRVed ICBM.

In 1991 France retired its first SSBN, *Le Redoutable*, after nearly 20 years of service.[51] The five remaining submarines are now each equipped with 16 six-warhead M-4 SLBMs. *Le Foudroyant*, which was the last submarine to carry the single-warhead M-20, has been retrofitted with the M-4 and re-entered service in February 1993.[52] France will probably retire all five of these submarines between 2005 and 2010.

Because of budgetary constraints and a reduced threat, France has decided to build four new Triomphant Class SSBNs instead of the six originally planned. The first of these submarines, which will be quieter than the Redoutable Class, will be commissioned in 1995, and work has begun on a second, *Le Téméraire*.[53] The Triomphant Class submarines will initially carry the six-warhead M-45 SLBM—a variant of the M-4—but beginning in 2005 will be armed with the M-5 SLBM, which is expected to carry 6–12 warheads.[54] If 12 warheads are deployed on each M-5 SLBM, France could significantly increase the number of its strategic nuclear warheads. Changing a decade-old policy, the French Navy announced in June 1992 that it had dropped the minimum number of submarines it maintains on patrol from three to two.[55]

France plans to retire some older nuclear-capable bombers, while continuing to develop new ones. The 18 existing Mirage IVPs are slated for retirement in 1996. France now plans to maintain only three of the five Mirage 2000N 15-aircraft squadrons as nuclear-capable aircraft, carrying the 300-km range Air Sol Moyenne Portée (ASMP) air-to-surface missile.[56]

Two versions of the new Rafale dual-capable fighter-bomber are under development: the Rafale-M for the Navy and the Rafale-D for the Air Force. The nuclear-armed versions of the Rafale are scheduled for initial deployment

[49] Barrillot, B., 'French finesse nuclear future', *Bulletin of the Atomic Scientists*, Sep. 1992, p. 25.

[50] Covault, C., 'French $7.7 billion missile to spark nuclear debate', *Aviation Week & Space Technology*, 14/21 Dec. 1992, p. 25.

[51] 'Nuclear notebook: all quiet on the French front', *Bulletin of the Atomic Scientists*, July/Aug. 1992, p. 48.

[52] Lewis, J. A. C., 'M-4 now carried on all French SSBNs', *Jane's Defence Weekly*, 27 Feb. 1993, p. 18.

[53] 'Nuclear notebook: all quiet on the French front' (note 51), p. 48.

[54] Barrillot (note 49), p. 24; Covault (note 50), p. 25. (Barrillot gives the figure of 12 warheads and Covault 6 warheads.) In addition to Barillot and Covault, a BASIC Report projects that the French M-5 SLBM will carry 8–12 warheads each; Butcher, M., Logan, C. and Plesch, D., *French Nuclear Policy Since the Fall of the Wall*, BASIC Report 93-1 (British American Security Information Council (BASIC): London, Feb. 1993), p. 14.

[55] BASIC Report (note 54), p. 11; Norris, R. S. *et al., Nuclear Weapon Databook, Vol. V: British, French and Chinese Nuclear Weapons* (forthcoming); IISS (note 42), p. 33

[56] IISS (note 42), p. 33; BASIC Report (note 54), p. 17.

in about 2005.[57] The French Navy plans to buy a total of 86 Rafale-Ms and the Air Force will buy 235 Rafale-Ds (140 two-seat and 95 single-seat versions).[58] The Rafale is expected to be armed with the new supersonic, 1300-km range Air-Sol-Longue-Portée (ASLP) air-to-surface missile which is currently under development.

### Non-strategic nuclear weapons

Two of the five short-range Pluton missile regiments were removed from operational status in early 1992, and French Defence Minister Pierre Joxe announced in April 1992 that the remaining three regiments would be withdrawn from service in 1993.[59] In September 1991 France announced that it would reduce production of Hadès short-range ballistic missiles, which were intended to replace the Pluton, and store the systems rather than deploy them. On 12 June 1992, the French Government indicated that all of the Hadès missiles and warheads would be destroyed rather than stored.[60] Subsequently, however, it was reported that about 20 Hadès missiles had already been delivered and Aérospatiale was instructed to finish building the remaining 10 missiles, which were delivered by December 1992.[61] These missiles are now being held in 'protective cocoon' storage[62] and could be made operational in two years.[63]

## V. British nuclear weapon programmes

### Strategic nuclear weapons

One of the UK's four Polaris submarines, the *HMS Revenge*, was decommissioned in May 1992.[64] The four Polaris submarines are scheduled to be replaced by four Trident submarines by the turn of the century. The *HMS Vanguard*, the first British SSBN that will carry the Trident II (D-5) missile, rolled out of the Devonshire Dock Hall at Barrow-in-Furness in March 1992. The *Vanguard* is scheduled to become operational in late 1994, but many observers believe that it will not be deployed until 1995.[65] The British Government formally placed an order for the fourth submarine in July.[66] The Ministry of Defence stated in July 1992 that each 16-missile

[57] Norris *et al.* (note 55).
[58] 'Rafale readies for sea trials', *Jane's Defence Weekly,* 17 Oct. 1992, p. 29.
[59] BASIC Report (note 54), p. 16; IISS (note 42), p. 33.
[60] IISS (note 42), p. 33; BASIC Report (note 54), p. 16; 'Hades missile scrapped', *Jane's Defence Weekly*, 20 June 1992, p. 1041.
[61] Norris *et al.* (note 55); Thelari, M., 'Actualites', *Air Fan*, no. 172 (Mar. 1993), p. 4; 'Hades missiles "intact"', *Jane's Defence Weekly*, 20 Feb. 1993, p. 7.
[62] Thelari (note 61).
[63] Isnard, J., 'La France a maintenu une "veille" opérationnelle sur le missile nucléaire Hadès', *Le Monde*, 11 Feb. 1993, p. 10.
[64] Norris *et al.* (note 55).
[65] Norris *et al.* (note 55).
[66] IISS (note 42), p. 33.

Trident submarine 'will carry no more than 128 warheads; that is a self-imposed maximum not a rigid specification. The exact number deployed will reflect our judgement of the minimum required to constitute a credible and effective deterrent. Over time, we may have reason to revise this assessment: for example, if there are significant developments in anti-ballistic missile systems'.[67]

Many observers now believe, however, that with the end of the cold war and US and Russian decisions to dramatically reduce the number of their strategic nuclear warheads, the UK will deploy fewer than eight warheads on each Trident II missile. Furthermore, if US Trident II missiles carry only four warheads each under the START II Treaty—which is considered likely—Britain will be under international political pressure to follow suit. The UK's problems with nuclear materials production may also limit the number of warheads available for the Trident II. Finally, it seems unlikely that the UK will be able to justify deploying a larger number of warheads on its submarines since a decision by Russia to expand its anti-ballistic missile (ABM) system is a remote possibility.

## Non-strategic nuclear weapons

The British Defence Ministry stated in June 1992 that all of the WE-177C nuclear strike/depth bombs (approximately 25), carried by the Royal Navy's Lynx and Sea King ASW helicopters and Sea Harrier carrier-based aircraft, will be removed from service and destroyed.[68] (With the withdrawal of all US tactical naval nuclear weapons, US B-57 depth bombs are no longer available for the UK's Nimrod ASW planes.) As a result, the remaining tactical or 'sub-strategic' British nuclear capability will consist of Royal Air Force Tornado and Buccaneer aircraft armed with the WE-177A/B gravity bomb.

The British Government announced in July that it intends to retire the two remaining Buccaneer squadrons by 1994 and reduce the number of Tornado squadrons from 11 to 8 with four Tornado GR1a/1b squadrons based in the UK and four Tornado GR1 squadrons based in Germany at Bruggen. As a result of the reductions made by the Royal Navy and Air Force, the UK will have reduced the number of its WE-177 gravity nuclear bombs 'by more than half'.[69]

The Ministry of Defence has said that the WE-177 bomb, which was first deployed in 1966, 'will approach the end of its service life around the turn of the century' and needs to be replaced by a tactical air-to-surface missile (TASM).[70] To this end, the UK has signed a $1.6 million contract with the US

[67] *Statement on the Defence Estimates* (Her Majesty's Stationery Office: London, July 1992), CM-1981, p. 28.

[68] 'Reduction in Britain's nuclear weapons', Ministry of Defence Press Release, 15 June 1992; Green, D. and Mauthner, R., 'UK cuts back nuclear forces at sea', *Financial Times*, 16 June 1992; Schmidt, W. E., 'British are planning to remove A-arms from ships and aircraft', *New York Times*, 16 June 1992, p. A-14.

[69] *Statement on the Defence Estimates* (note 67), p. 28.

[70] *Statement on the Defence Estimates* (note 67), p. 28.

defence contractor Martin Marietta Electronics Information & Missiles Group to continue preliminary definition studies for the TASM programme. The UK is also still considering joint development of the French Aérospatiale ASLP missile, among other options.[71]

## VI. Chinese nuclear weapon programmes

### Strategic nuclear weapons

The current Chinese ICBM force may be smaller than previously believed. According to an authoritative article by John Wilson Lewis and Hua Di,[72] only four 13 000-km single-warhead DF-5As have been deployed in silos.[73] Past reports have estimated that there were as many as 10. According to Lewis and Hua, China plans to keep the DF-5 ICBMs in service until the year 2010.[74]

The US intelligence community expects that China will probably field a new mobile ICBM during the 1990s.[75] In what appears to be a reference to the same missile, Lewis and Hua reported that China is developing the DF-31, a mobile land-based version of the JL-2 SLBM, and projected that the new 8000-km range ICBM will become operational in the mid- to late 1990s. The mobile DF-31 will be stored in caves in peacetime and moved on trucks to a pre-selected launching site for rapid response in crises.[76]

Apparently, China plans to field a second type of mobile ICBM sometime after the turn of the century. Lewis and Hua assert that Beijing is developing a three-stage solid propellant missile called the DF-41, which will be deployed between the year 2000 and 2010. It is unclear whether DF-41, which will have an estimated range of 12 000 km, will be road-, rail- and/or river-mobile.[77] This source also states that China plans to develop MIRVs for its mobile ICBMs.[78]

For its part, the Chinese Navy is now expected to deploy the new 8000-km JL-2 SLBM on its second-generation 09-4 Xia Class nuclear-powered submarine in the mid- to late 1990s.[79]

The Chinese Air Force continued work on the H-7, possibly a strategic bomber, in 1992. Reportedly, the H-7, which is expected to eventually replace the Q-5C, is now entering series production at Xian Aircraft Factory.[80]

[71] Leopold, G., 'Britain pursues missile studies', *Defense News,* 14–20 Sep. 1992, p. 10.
[72] Lewis, J. W. and Hua Di, 'China's ballistic missile programs: technologies, strategies, goals', *International Security*, vol. 17, no. 2 (fall 1992), pp. 5–40.
[73] Lewis and Hua (note 72), p. 19.
[74] Lewis and Hua (note 72), p. 29.
[75] Gershwin (note 28), p. 7.
[76] Lewis and Hua (note 72), pp. 28–29.
[77] Lewis and Hua (note 72), p. 29.
[78] Lewis and Hua (note 72), p. 30.
[79] Lewis and Hua (note 72), p. 29.
[80] Sengupta, P., 'China expands air forces', *Military Technology*, Aug. 1992, p. 51.

### Non-strategic nuclear weapons

Lewis and Hua report that in 1985 China initially deployed the 1700- to 1800-km DF-21, a mobile land-based version of the JL-1 SLBM—a system previously not publicly reported. China has six trucks with transporter–erector–launchers (TELs) to transport and launch the two-stage, solid-propellant IRBM.[81]

## VII. Summary of nuclear weapon developments

The United States and Russia are now committed to reducing their existing strategic and tactical nuclear arsenals and to forgo further development and production of a number of new weapons. Many of these force structure decisions have been driven by economic realities and arms control commitments rather than by fundamental changes in military doctrine or basic attitudes towards the utility of nuclear weapons.[82] With the notable exceptions of the US MX/Peacekeeper ICBM and the Russian SS-18 and SS-24 ICBMs, which are scheduled for retirement under the START II Treaty,[83] the USA and Russia have essentially decided to retire their older nuclear systems, retain their more modern ones and finish the production runs of those that have already been mostly or completely paid for. Thus, to some extent it is fair to say that in cutting their nuclear forces the USA and Russia have tried to make a virtue of necessity.

Nevertheless, the recent commitments to denuclearize—specifically to denuclearize the non-Russian former Soviet republics—should have both concrete and symbolic security benefits. The reductions should enhance strategic stability, save resources and help strengthen international consensus for a long-term extension of the Treaty on the Non-Proliferation of Nuclear Weapons (NPT) at the 1995 NPT Extension Conference.

With the two major nuclear weapon powers cutting back their programmes, the size and cost of French and British nuclear forces are likely to be subjected to far more public scrutiny than ever before. While both France and the UK have taken steps to reduce their tactical forces, their planned SLBM modernization programmes could actually increase the number of their deployed strategic warheads, and neither country has indicated interest in giving up its independent nuclear deterrent forces. It appears that the current Chinese Government feels no need to contemplate nuclear force reductions until the other

[81] Lewis and Hua (note 72), pp. 27–28.

[82] See, for example, Finnegan, P. and Leopold, G., 'Budget cuts fuel debate on nuclear strategies', *Defense News*, 20 Jan. 1992, pp. 1, 28. See also Mazarr, M. J., 'Nuclear weapons after the cold war', *Washington Quarterly*, vol. 15, no. 3 (summer 1992), pp. 185–201; Schmitt, E., 'Head of nuclear forces plans for a new world', *New York Times*, 25 Feb. 1993, p. B-7.

[83] The 500 MX warheads that are now slated for retirement make up less than 4% of the warheads the USA had deployed in Sep. 1990 (as reflected in the START Treaty MOU). Congress capped the deployment of the MX missile at 50 in 1989. It should also be noted that both the SS-18 and the SS-24 ICBMs are built in Ukraine, *not* Russia. Thus, Russia may have had little say in their continued production in any case. It is also well documented that the SS-24 missile is not particularly accurate or reliable.

**Table 6.1.** US strategic nuclear forces, January 1993

| Type | Designation | No. deployed | Year first deployed | Range (km) | Warheads x yield | Warheads in stockpile |
|---|---|---|---|---|---|---|
| *Bombers* | | | | | | |
| B-52H[a] | Stratofortress | 95 | 1961 | 16 000 | ALCM 5–150 kt | 1 200 |
| B-1B[b] | Lancer | 95 | 1986 | 19 800 | ACM 5–150 kt | 300 |
| | | | | | Bombs 5 kt–1 Mt | 1 400 |
| *ICBMs* | | | | | | |
| LGM-30F[c] | Minuteman II | 350 | 1966 | 11 300 | 1 x 1.2 Mt | 350 |
| LGM-30G | Minuteman III | | | | | |
| | Mk 12 | 200 | 1970 | 13 000 | 3 x 170 kt | 600 |
| | Mk 12A | 300 | 1979 | 13 000 | 3 x 335 kt | 900 |
| LGM-118 | MX/Peacekeeper | 50 | 1986 | 11 000 | 10 x 300 kt | 500 |
| *SLBMs* | | | | | | |
| UGM-96A[d] | Trident I C-4 | 320 | 1979 | 7 400 | 8 x 100 kt | 2 560 |
| UGM-133A[e] | Trident II D-5 | | | | | |
| | Mk IV/W-76 | 120 | 1992 | 7 400 | 8 x 100 kt | 560 |
| | Mk V/W-88 | | 1990 | | 8 x 475 kt | 400 |

[a] B-52Hs can carry up to 20 ALCMs/ACMs each, but only about 1500 cruise missiles are currently available for deployment.

[b] Rockwell built 100 B1-Bs. Four have crashed and 1 is used as a trainer at Ellsworth AFB, S.D., and is not considered 'operational' (although it is START Treaty-accountable). B1-B weapon loadings vary, depending on the mission. It is assumed here that they would not carry more than 16 gravity bombs each.

[c] Approximately 100 Minuteman IIs had been removed from their silos by the end of 1992. The remaining 350 are slated to be removed from silos by 1995.

[d] By Jan. 1993, 4 of the 12 remaining Poseidon submarines armed with the Trident I (C-4) missile had been deactivated: the *USS James Madison*, the *USS Ben Franklin*, the *USS Henry L. Stimson* and the *USS Francis Scott Key*. The 320 Trident I C-4 missiles are deployed on the remaining 16-missile Poseidon submarines and on the 8 24-missile Ohio Class submarines in the Pacific Fleet.

[e] The 120 Trident II D-5 missiles are deployed on 5 Ohio Class submarines stationed at King's Bay, Ga., the newest of which, the *USS Maryland*, is scheduled to begin patrols in 1993.

*Sources*: Dick Cheney, Secretary of Defense, *Annual Report to the President and the Congress,* Jan. 1993, p. 68; Dick Cheney, *Annual Report to the President and the Congress,* Feb. 1992, p. 60; US Navy Public Affairs, personal communications; US Air Force Public Affairs, personal communications; Natural Resources Defense Council (NRDC); authors' estimates.

nuclear weapon powers have reduced their forces to a level near that of China.[84]

[84] See, for example, Beijing Zhongguo Xinwen She, 30 Jan. 1992, FBIS-CH, 30 Jan. 1992, in Institute for Defense and Disarmament Studies, 'Short-range nuclear forces chronology', *Arms Control Reporter* (IDDS: Brookline, Mass.), sheet 408.B.137, Feb. 1992; see also 'First Supplementary List of

**Table 6.2.** CIS strategic nuclear forces, January 1993

| Type | NATO designation | No. deployed | Year first deployed | Range (km) | Warheads x yield | Warheads in stockpile |
|---|---|---|---|---|---|---|
| *Bombers* | | | | | | |
| Tu-95MS6 | Bear-H6 | 27 | 1984 | 12 800 | 6 x AS-15A ALCMs, bombs | 162 |
| Tu-95MS16[a] | Bear-H16 | 57 | 1984 | 12 800 | 16 x AS-15A ALCMs, bombs | 912 |
| Tu-160[b] | Blackjack | 25 | 1987 | 11 000 | 12 x AS-15B ALCMs or AS-16 SRAMs, bombs | 300 |
| *ICBMs* | | | | | | |
| SS-17[c] | Spanker | 40 | 1979 | 10 000 | 4 x 750 kt | 160 |
| SS-18[d] | Satan | 308 | 1979 | 11 000 | 10 x 550-750 kt | 3 080 |
| SS-19[e] | Stiletto | 300 | 1979 | 10 000 | 6 x 550 kt | 1 800 |
| SS-24 M1/M2[f] | Scalpel | 36/56 | 1987 | 10 000 | 10 x 550 kt | 920 |
| SS-25[g] | Sickle | 378 | 1985 | 10 500 | 1 x 550 kt | 378 |
| *SLBMs*[h] | | | | | | |
| SS-N-8 M2[i] | Sawfly | 280 | 1973 | 9 100 | 1 x 1.5 Mt | 280 |
| SS-N-18 M1 | Stingray | 224 | 1978 | 6 500 | 3 x 500 kt | 672 |
| SS-N-20 | Sturgeon | 120 | 1983 | 8 300 | 10 x 200 kt | 1 200 |
| SS-N-23 | Skiff | 112 | 1986 | 9 000 | 4 x 100 kt | 448 |

[a] According to the START II Treaty MoU, there are 23 Bear-H16 bombers based in Russia, 21 in Ukraine and 13 in Kazakhstan; the 27 Bear-H6s are based in Kazakhstan.

[b] Estimates vary, but the authors estimate that 20 Blackjacks are based in Ukraine and that 5 test planes are stationed in Russia.

[c] The SS-17s are based in Russia and are scheduled for retirement in the next two years.

[d] 104 SS-18s are based in Kazakhstan; 204 are based in Russia.

[e] 130 SS-19s are based in Ukraine; 170 are based in Russia.

[f] Of the 56 silo-based SS-24 M2s, 46 are in Ukraine and 10 are in Russia. All 36 rail-based SS-24 M1s are in Russia.

[g] 81 of the SS-25s are based in Belarus. They are slated to be returned to Russia by the end of 1994.

[h] All CIS SSBNs are based in Russia.

[i] The 280 SS-N-8 M2s are deployed on 18 Delta I and 4 Delta II SSBNs.

*Sources*: START Treaty Memorandum of Understanding, 1 Sep. 1990; US Department of Defense, *Military Forces in Transition*, Washington, DC, Sep. 1991; Natural Resources Defense Council (NRDC); authors' estimates.

Ratifications, Accessions, Withdrawals, Etc. for 1992', presented to the British Parliament by the Secretary of State for Foreign and Commonwealth Affairs by Command of Her Majesty, Oct. 1992 (Her Majesty's Stationery Office: London, Oct. 1992), p. 5.

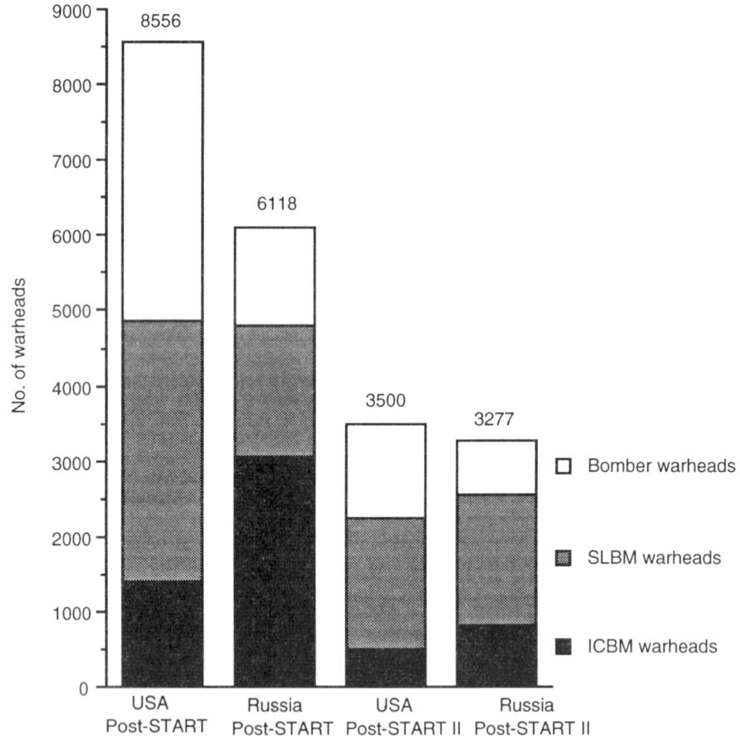


**Figure 6.1.** Projected US and Russian strategic nuclear forces after implementation of the START Treaty and the START II Treaty

*Note:* The estimates of post-START Treaty and post-START II Treaty strategic nuclear force levels rest on assumptions about procurement and modernization decisions and on the assumptions below; actual numbers are likely to vary.

*Bomber warheads*

Projected bomber warhead numbers are based on START Treaty-accountable bombers (that is, the active inventory *minus* test aircraft). The projected figure for post-START II Treaty US bomber warheads assumes the reorientation of the B-1B bomber to a conventional role, the deployment of 20 B-2s with 16 bombs each, and the modification of 41 and 52 B-52H bombers to carry 8 and 12 ALCMs, respectively. The projected figure for post-START II Treaty Russian bomber warheads assumes that Ukraine does not return to Russia any Blackjack or Bear-H aircraft based on its territory, leaving Russia with 27 Bear-Hs (with 6 ALCMs) and 36 Bear-Hs (with 16 ALCMs).

*SLBM warheads*

The projected figure for post-START II Treaty US SLBM warheads assumes that the USA will deploy 18 Trident SSBNs—8 with 24 Trident I (C-4) SLBMs each and 10 with 24 Trident II (D-5) SLBMs each. All of these missiles are assumed to be 'downloaded' to 4 warheads each. The figure for projected post-START II Treaty Russian SLBM warheads assumes that Russia will deploy 12 Delta III SSBNs, each carrying 16 3-warhead SS-N-18 SLBMs; 7 Delta IV SSBNs, each carrying 16 4-warhead SS-N-23 SLBMs; and 6 Typhoon SSBNs, each carrying 20 SS-N-20 follow-on SLBMs 'downloaded' from 10 to 6 warheads.

*ICBM warheads*

The projected figure for post-START II Treaty US ICBM warheads assumes that the USA will deploy 500 Minuteman III ICBMs 'downloaded' to 1 warhead each. The projected figure for post-START II Russian ICBM warheads assumes that Russia will deploy 105 SS-19 ICBMs 'downloaded' to 1 warhead each, 600 single-warhead SS-25 and SS-25 follow-on missiles, and an additional 90 SS-25 follow-on missiles based in modified SS-18 missile silos.

*Sources: For US forces:* START Treaty Memorandum of Understanding, Sep. 1990; US Department of Defense Fact Sheet, 'US Strategic Nuclear Forces', June 1992; Arms Control and Disarmament Agency (ACDA) Fact Sheet, 'The Joint Understanding on the Elimination of MIRVed ICBMs and Further Reduction in Strategic Offensive Arms', 2 July 1992, p. 2; Senate Foreign Relations Committee, *Report on the START Treaty*, Executive Report 102-53, pp. 18–19; *Report of the Secretary of Defense to the President and the Congress*, Feb. 1992, p. 60; *Report of the Secretary of Defense to the President and the Congress*, Jan. 1993, p. 68; Lockwood, D., 'Strategic nuclear forces under START II', *Arms Control Today,* Dec. 1992, p. 13.

*For Russian forces:* Statement of Ted Warner, Senior Defense Analyst, RAND Corporation, before the Senate Foreign Relations Committee, 3 Mar. 1992, in *The START Treaty*, Senate Hearing 102-607, Part 1 (US Government Printing Office: Washington, DC, 1992), pp. 228–29; US Department of Defense, *Military Forces in Transition*, Sep. 1991; START Treaty Memorandum of Understanding, Sep. 1990; Senate Foreign Relations Committee, *Report on the START Treaty*, Executive Report 102-53, p. 22; Lockwood, D., 'Strategic nuclear forces under START II', *Arms Control Today*, Dec. 1992, p. 13.

**Table 6.3.** French nuclear forces, January 1993

| Type | No. deployed | Year first deployed | Range (km)[a] | Warheads x yield | Warheads in stockpile |
|---|---|---|---|---|---|
| *Land-based aircraft* | | | | | |
| Mirage IVP | 18 | 1986 | 1 500 | 1 x 300 kt ASMP | 18 |
| Mirage 2000N | 45[b] | 1988 | 1 570 | 1 x 300 kt ASMP | 42 |
| *Carrier-based aircraft* | | | | | |
| Super Etendard | 24 | 1978 | 650 | 1 x 300 kt ASMP | 20[c] |
| *Land-based missiles* | | | | | |
| S3D | 18 | 1980 | 3 500 | 1 x 1 Mt | 18 |
| Pluton[d] | 21 | 1974 | 120 | 1 x 10-25 kt | 42 |
| Hadès[e] | (15) | – | 480 | 1 x 80 kt | 30 |
| *SLBMs*[f] | | | | | |
| M-4A/B | 64 | 1985 | 6 000 | 6 x 150 kt | 384 |

**Table 6.3** *contd*

[a] Range for aircraft indicates combat radius, without refuelling, and does not include the 90- to 350-km range of the ASMP air-to-surface missile (where applicable).

[b] Only 45 of the 75 Mirage 2000Ns have nuclear missions. On 11 Sep. 1991 President Mitterrand announced that as of 1 Sep. the AN-52 gravity bomb, which had been carried by Jaguar As and Super Etendards, had been withdrawn from service. Forty-two ASMPs are allocated to the 3 squadrons of Mirage 2000Ns.

[c] The Super Etendard used to carry 1 AN 52 bomb. At full strength, the AN 52 equipped 2 squadrons worth (24 of the 36 nuclear-capable aircraft) of Super Etendard: Flottilles 11F, 14F and 17F based at Landivisiau and Hyères, respectively. From Apr. 1989 these squadrons began receiving the ASMP missile. By mid-1990, all 24 aircraft (to be configured to carry the ASMP) were operational. Although originally about 50–55 Super Etendard aircraft were intended to carry the ASMP, because of budgetary constraints the number fell to 24.

[d] The Pluton is in the process of being retired and will be totally withdrawn from service in 1993. The table assumes 3 regiments of the original 5 (with 35 launchers and 70 warheads).

[e] Although the French Government indicated in June 1992 that it would dismantle the Hadès system, it was widely reported in early 1993 that it decided to store 15 Hadès launchers and 30 Hadès missiles.

[f] Upon returning from its 58th and final patrol on 5 Feb. 1991, *Le Redoutable* was retired along with the last M-20 SLBMs. The 5 remaining SSBNs are all deployed with the M-4A/B missile. Although there are 80 launch tubes on the 5 SSBNs, only 4 sets of SLBM were bought and thus the number of TN 70/71 warheads in the stockpile is assumed to be 384, probably with a small number of spares.

*Source*: Norris, R. S., Burrows, A. S. and Fieldhouse, R. W., *Nuclear Weapons Databook Vol. V: British, French and Chinese Nuclear Weapons* (forthcoming).

**Table 6.4.** British nuclear forces, January 1993[a]

| Type | Designation | No. deployed | Date deployed | Range (km)[b] | Warheads x yield | Warheads in stockpile |
|---|---|---|---|---|---|---|
| *Aircraft* | | | | | | |
| GR.1[c] | Tornado | 72 | 1982 | 1 300 | 1–2 x 200–400 kt | 100[e] |
| S2B | Buccaneer | 27 | 1971 | 1 700 | 1 x 200–400 kt[d] | |
| *SLBMs* | | | | | | |
| A3-TK | Polaris | 48 | 1982[f] | 4 700 | 2 x 40 kt | 100[g] |

[a] The US nuclear weapons for certified British systems have been removed from Europe and returned to the USA, specifically for the 11 Nimrod ASW aircraft based at RAF St Magwan, Cornwall, UK, the 1 Army regiment with 12 Lance launchers and the 4 Army artillery regiments with 120 M109 howitzers in Germany. Squadron No. 42, the Nimrod maritime patrol squadron, disbanded in Oct. 1992, but St Magwan will remain a forward base for Nimrods and will have other roles. The 50 Missile Regiment (Lance) and the 56 Special Weapons Battery Royal Artillery will disband by 1 Apr. 1993.

[b] Range for aircraft indicates combat radius, without in-flight refuelling.

[c] The Royal Air Force will eventually operate 8 squadrons of dual-capable strike/attack Tornados. The 3 squadrons at Laarbruch, Germany (Nos 15, 16, 20) were disbanded between Sep. 1991 and May 1992. A fourth squadron there (No. 2) was equipped with the Tornado reconnaissance variant and went to RAF Marham to join a reconnaissance squadron already there (No. 13). The 2 squadrons previously at Marham (Nos 27 and 617) will redeploy to

Lossiemouth, Scotland, in 1993–94, replacing Buccaneer squadrons Nos 12 and 208 in the maritime/strike role. The Tornado squadrons will be redesignated Nos 12 and 617. The 4 squadrons at RAF Bruggen, Germany (Nos 9, 14, 17, 31) will remain. All 8, including the 2 reconnaissance squadrons, will be nuclear-capable, down from 11.

[d] The US Defense Intelligence Agency has confirmed that the RAF Tornados 'use two types of nuclear weapons, however exact types are unknown'. The DIA further concludes that each RAF Tornado is capable of carrying 2 nuclear bombs, 1 on each of the 2 outboard fuselage stations.

[e] The total stockpile of WE-177 tactical nuclear gravity bombs was estimated to have been about 200, of which 175 were versions A and B. The C version of the WE-177 was assigned to selected Royal Navy (RN) Sea Harrier FRS.1 aircraft and ASW helicopters. The WE-177C existed in both a free-fall and depth-bomb modification. There were an estimated 25 WE-177Cs, each with a yield of approximately 10 kt. Following the Bush-Gorbachev initiatives of 27 Sep. and 5 Oct.1991, British Secretary of State for Defence Tom King said that 'we will no longer routinely carry nuclear weapons on our ships'. On 15 June 1992 the Defence Minister announced that all naval tactical nuclear weapons had been removed from surface ships and aircraft, that the nuclear mission would be eliminated and that the 'weapons previously earmarked for this role will be destroyed'. The 1992 White Paper stated that 'As part of the cut in NATO's stockpile we will also reduce the number of British free-fall nuclear bombs by more than half'. A number of British nuclear bombs were returned to the UK. In table 6.4, a total inventory of strike variants of approximately 100 is assumed, including those for training and for spares.

[f] The 2-warhead Polaris A3-TK (Chevaline) was first deployed in 1982 and has now completely replaced the original three-warhead Polaris A-3 missile, first deployed in 1968.

[g] It is now thought that the British produced only enough warheads for 3 full boat-loads of missiles, or 48 missiles, with a total of 96 warheads. In Mar. 1987 French President Mitterrand stated that Britain had '90 to 100 [strategic] warheads'.

Source: Norris, R. S., Burrows, A. S. and Fieldhouse, R. W., Nuclear Weapons Databook Vol. V: British, French and Chinese Nuclear Weapons (forthcoming).

**Table 6.5.** Chinese nuclear forces, January 1993

| Type | NATO designation | No. deployed | Year first deployed | Range (km) | Warheads x yield | Warheads in stockpile |
|---|---|---|---|---|---|---|
| *Bombers*[a] | | | | | | |
| H-5 | B-5 | 30 | 1968 | 1 200 | 1 x bomb | 150 |
| H-6 | B-6 | 120 | 1965 | 3 100 | 1 x bomb | |
| Q-5 | A-5 | 30 | 1970 | 400 | 1 x bomb | |
| H-7 | ? | 0 | 1993? | ? | 1 x bomb | |
| *Land-based missiles*[b] | | | | | | |
| DF-3A | CSS-2 | 50 | 1971 | 2 800 | 1 x 3.3 Mt | 50 |
| DF-4 | CSS-3 | 20 | 1980 | 4 750 | 1 x 3.3 Mt | 20 |
| DF-5A | CSS-4 | 4 | 1981 | 13 000 | 1 x 4–5 Mt | 4 |
| DF-21 | CSS-6 | 36 | 1985-86 | 1 800 | 1 x 200–300 kt | 36 |
| DF-31 | – | 0 | Late 1990s? | 8 000 | 1 x 200–300 kt | 0 |
| DF-41 | – | 0 | 2010? | 12 000 | MIRV | 0 |
| *SLBMs*[c] | | | | | | |
| JL-1 | CSS-N-3 | 24 | 1986 | 1 700 | 1 x 200–300 kt | 24 |
| JL-2 | CSS-N-4 | 0 | Late 1990s? | 8 000 | 1 x 200–300 kt | 0 |

**Table 6.5** *contd*

[a] All figures for bomber aircraft are for nuclear-configured versions only. Assumes 150 bombs for the force. Hundreds of aircraft are deployed in non-nuclear versions. The aircraft bombs are estimated to have yields between 10 kt and 3 Mt.

[b] The Chinese define missile ranges as follows: short-range, < 1000 km; medium-range, 1000–3000 km; long-range, 3000–8000 km; intercontinental-range, > 8000 km.

[c] Two SLBMs are presumed to be available for rapid deployment on the Golf Class submarine (SSB). The nuclear capability of the M-9 is unconfirmed and thus not included.

*Source*: Norris, R. S., Burrows, A. S. and Fieldhouse, R. W., *Nuclear Weapons Databook Vol. V: British, French and Chinese Nuclear Weapons* (forthcoming).

## VIII. Nuclear weapon proliferation

The proliferation of weapons of mass destruction, and particularly nuclear weapons and ballistic missiles, has become one of the major threats, if not the key threat, to global security in the wake of the cold war. Developments in the former Soviet Union, Iraq and other developing nations made it clear in 1992 that the end of the East–West military confrontation increased rather than decreased both the dangers of and incentives for the proliferation of weapons of mass destruction. The crises apparently averted in Iraq and still looming in the former Soviet Union and elsewhere, however, appear to have awakened the international community to the severity of the threat and spurred it to act quickly and co-operatively to strengthen the nuclear non-proliferation regime.

### Developments in the nuclear weapon non-proliferation regime

Although the International Atomic Energy Agency (IAEA) and its system of safeguards is only one of a series of tools to stem the proliferation of nuclear weapons, it is at the forefront of such endeavours. The main purpose of IAEA safeguards is to detect any significant diversion of safeguarded materials from peaceful applications. Confidence in the Agency's system, however, was severely shaken once the vast scope of Iraq's nuclear weapon programme became clear after the Persian Gulf War.[85] As a result of increased awareness of the safeguards system's shortcomings and a renewed political commitment to strengthen the non-proliferation regime, efforts to revitalize and invigorate the IAEA have begun to bear fruit.

In February 1992 the 35-member IAEA Board of Governors 'reaffirmed the Agency's right to undertake special inspections in member states with comprehensive safeguards agreements, when necessary and appropriate'.[86] The Agency's right to conduct these 'special' inspections is contained in the stan-

[85] The IAEA has submitted reports to the UN Security Council on the findings of the 16 nuclear weapon inspection teams which have inspected Iraq from the end of the war until Dec. 1992.

[86] IAEA Press Release 92/12, 26 Feb. 1992.

dard comprehensive safeguards agreement—referred to as INFCIRC/153[87]—agreed in Vienna in 1971 and signed by parties to the 1968 Non-Proliferation Treaty (for the parties as of 1 January 1993, see annexe A), although the Agency has never conducted such an inspection. Special inspections strengthen the IAEA's ability to detect undeclared or clandestine nuclear facilities, thus enlarging its responsibility from one of catching diversion of peaceful nuclear materials to a more challenging one of verifying that states are not engaged in any proscribed nuclear activities.[88] IAEA Director General Hans Blix predicted that the new inspection powers will not only enable the Agency to detect prohibited activities at non-declared facilities but also deter states from engaging in such activities for fear of being discovered.

The key to successfully implementing the stronger inspection system will, according to IAEA officials, be access to intelligence. Responding to criticism about the Agency's failure to uncover the Iraqi nuclear weapon programme, Blix stated that the IAEA had never received intelligence information from member states about suspected violations. To improve IAEA contacts with intelligence organizations, Blix has created a two-person liaison office to receive intelligence information from member states about suspected safeguards violations.

In addition to reaffirming the Inspectorate's inspection powers, the February 1992 Board meeting also strengthened the safeguards system by requiring states to provide the IAEA with design information on future nuclear facilities 'as soon as the decision to construct, to authorize construction or to modify a facility has been taken'.[89] Previously, design information—with which the IAEA develops its site-specific safeguards procedures—was required only 180 days before the introduction of nuclear materials into a facility. The requirement for more advance information will provide the IAEA with more time to prepare and implement safeguards methods and will also prevent states from being able to build nuclear facilities capable of aiding a nuclear weapon programme without declaring its existence to the IAEA, as was the case with Iraq.

The IAEA Board twice reviewed proposals for creating an international register of nuclear-related imports and exports to facilitate monitoring of international commerce in items that could assist states in building a nuclear weapon (see also chapter 11, section IV). Under the proposals, which have been referred back to the IAEA Secretariat for modification, all IAEA member states would be required to report both imports and exports of certain key nuclear and nuclear-related items. The goal of such a register is to prevent a state from clandestinely acquiring a significant amount of nuclear equipment,

[87] The Structure and Content of Agreements Between the Agency and States Required in Connection with the Treaty on the Non-Proliferation of Nuclear Weapons, IAEA document INFCIRC/153 (corrected), (IAEA: Vienna, 1983).

[88] For a review of this major change, see Scheinman, L., 'Nuclear safeguards and non-proliferation in a changing world', *Security Dialogue*, vol. 23, no. 4 (Dec. 1992).

[89] IAEA Press Release 92/12 (note 86).

**Table 6.6.** The Nuclear Suppliers Group list of nuclear-related dual-use equipment and materials and related technology

1. Industrial Equipment
   - 1.1 Spin-forming and flow-forming machines
   - 1.2 'Numerical control' units, machine tools, etc.
   - 1.3 Dimensional inspection systems
   - 1.4 Vacuum induction furnaces
   - 1.5 Isostatic presses
   - 1.6 Robots and end effectors
   - 1.7 Vibration test equipment
   - 1.8 Furnaces – arc remelt, electron beam, and plasma
2. Materials
   - 2.1 Aluminum, high-strength
   - 2.2 Beryllium
   - 2.3 Bismuth (high purity)
   - 2.4 Boron (isotopically enriched in boron-10)
   - 2.5 Calcium (high purity)
   - 2.6 Chlorine trifluoride
   - 2.7 Crucibles made of materials resistant to liquid actinide metals
   - 2.8 Fibrous and filamentary materials
   - 2.9 Hafnium
   - 2.10 Lithium (isotopically enriched lithium-6)
   - 2.11 Magnesium (high purity)
   - 2.12 Maraging steel, high-strength
   - 2.13 Radium
   - 2.14 Titanium alloys
   - 2.15 Tungsten
   - 2.16 Zirconium
3. Uranium Isotope Separation Equipment and Components
   - 3.1 Electrolytic cells for flourine production
   - 3.2 Rotor and bellows equipment
   - 3.3 Centrifugal multiplane balancing machines
   - 3.4 Filament winding machines
   - 3.5 Frequency changers
   - 3.6 Lasers, laser amplifiers, and oscillators
   - 3.7 Mass spectrometers and mass spectrometer ion sources
   - 3.8 Pressure measuring instruments, corrosion-resistant
   - 3.9 Valves, corrosion-resistant
   - 3.10 Superconducting solenoidal electromagnets
   - 3.11 Vacuum pumps
   - 3.12 Direct current high-power supplies (100 V or greater)
   - 3.13 High-voltage direct current power supplies (20,000 V or greater)
   - 3.14 Electromagnetic isotope separators
4. Heavy Water Production Plant Related Equipment (Other Than Trigger List Items)
   - 4.1 Specialized packings for water separation
   - 4.2 Pumps for potassium amide/liquid ammonia
   - 4.3 Water-hydrogen sulfide exchange tray columns
   - 4.4 Hydrogen-cryogenic distillation columns
   - 4.5 Ammonia converters or synthesis reactors

5. Implosion Systems Development Equipment
   - 5.1 Flash x-ray equipment
   - 5.2 Multistage light gas guns/high-velocity guns
   - 5.3 Mechanical rotating mirror cameras
   - 5.4 Electronic streak and framing cameras and tubes
   - 5.5 Specialized instrumentation for hydrodynamic experiments
6. Explosives and Related Equipment
   - 6.1 Detonators and multipoint initiation systems
   - 6.2 Electronic components for firing sets
   - 6.2.1 Switching devices
   - 6.2.2 Capacitors
   - 6.3 Firing sets and equivalent high-current pulsers (for controlled detonators)
   - 6.4 High explosives relevant to nuclear weapons
7. Nuclear Testing Equipment and Components
   - 7.1 Oscilloscopes
   - 7.2 Photomultiplier tubes
   - 7.3 Pulse generators (high speed)
8. Other
   - 8.1 Neutron generator systems
   - 8.2 General nuclear related equipment
   - 8.2.1 Remote manipulators
   - 8.2.2 Radiation shielding windows
   - 8.2.3 Radiation-hardened TV cameras
   - 8.3 Tritium, tritium compounds, and mixtures
   - 8.4 Tritium facilities or plants and components therefor
   - 8.5 Platinized carbon catalysts
   - 8.6 Helium-3
   - 8.7 Alpha-emitting radionuclides

*Source*: IAEA document INFCIRC/254/Rev.1/Part 2, July 1992.

as was done in the case of Iraq. Several countries have formally agreed to provide such information voluntarily.[90]

## The Nuclear Suppliers Group

The members of the Nuclear Suppliers Group (NSG), which first met following India's explosion of a nuclear device in 1974, began to meet again in 1990 for the first time in over 15 years (see also chapter 11, section VI). At the time of its inception, the NSG established a common set of guidelines for exports by the major nuclear suppliers to help prevent the spread of nuclear weapons. Meeting in Warsaw on 3 April 1992, the NSG agreed in the Guidelines for Transfers of Nuclear-Related Dual-Use Equipment, Material and Related Technology (the so-called Warsaw Guidelines)[91] to control 'dual-use' nuclear

[90] IAEA document INFCIRC/415, Dec. 1992.

[91] IAEA document INFCIRC/254/Rev.1/Part 2, July 1992. The members of the NSG are: Australia, Austria, Belgium, Bulgaria, Canada, the Czech Republic, Denmark, Finland, France, Germany, Greece, Hungary, Ireland, Italy, Japan, Luxembourg, Netherlands, Norway, Poland, Portugal, Romania, Russia, Slovakia, Spain, Sweden, Switzerland, the UK and the USA.

items, such as those Iraq used in its nuclear weapon development programme. The members will control materials and equipment that can be used for both legitimate and weapon-oriented nuclear activities. The annexe to the Warsaw Guidelines consists of eight main sections with 67 categories of equipment and materials (see table 6.6).[92]

The members of the NSG also announced that they had unanimously agreed to require recipients of key nuclear facilities and materials to accept comprehensive IAEA safeguards—like those required under the NPT—as a condition of supply. Under the Warsaw Guidelines, none of the members will sell major nuclear components and materials such as reactors or nuclear fuel to any state that does not have a full-scope safeguards agreement with the IAEA, such as India, Israel and Pakistan. This new policy will put additional pressure on several states that have advanced nuclear programmes but are dependent on outside sources of equipment and materials to accept IAEA safeguards.

## IX. Regional developments[93]

### Asia

*North Korea*

North Korea, long thought to be a potential nuclear weapon state, signed a comprehensive safeguards agreement with the IAEA on 30 January 1992. The agreement was ratified by the North Korean Parliament on 9 April and entered into force on 10 April.[94] This step, which came six and a half years after North Korea acceded to the NPT, was as a result of concerted international pressure by the USA, Russia, Japan and China to persuade North Korea to abandon its alleged nuclear intentions and fully comply with its NPT obligations to accept comprehensive IAEA safeguards.[95]

In early 1993, however, Western intelligence agencies provided information to the IAEA about suspected nuclear waste storage sites which North Korea had failed to declare to the Agency. This evidence, as well as other discrepancies in the North Korean declarations to the IAEA, led IAEA Director General Blix on 9 February to request a special inspection of the two sites. Never before had the Agency requested a special inspection. North Korea's refusal to accept the inspection led the IAEA Board of Governors on 25 February to pass a resolution calling on North Korea to accept the inspection and for Blix to report back to the Board at a special session held on 25 March. Citing 'the extraordinary situation prevailing in [North Korea], which jeopardizes its supreme interests', the Minister of Foreign Affairs, Kim Yong Nam,

[92] IAEA document INFCIRC/254/Rev.1/Part 2 (note 91).

[93] For Latin America, see chapter 6, section IX on the Treaty of Tlatelolco.

[94] IAEA Press Release 92/20, 10 Apr. 1992.

[95] Not all efforts to get North Korea to sign the IAEA agreement were coercive. The USA withdrew all of its tactical nuclear weapons in South Korea in late 1991, removing one of North Korea's main security concerns. In addition, states such as the USA and Japan held out the prospect of improved diplomatic and economic ties if North Korea implemented its agreements with the IAEA and South Korea.

announced on 12 March 1993 that North Korea intended to withdraw from the NPT. Under the terms of Article X of the NPT, a country can withdraw from the Treaty three months after notifying the UN Security Council and each member of the Treaty. At the time of writing, not all members of the NPT had been notified.

In addition to its safeguards agreement with the IAEA, which will remain in force until North Korea's withdrawal from the NPT becomes effective, North Korea signed an historic—but as yet unimplemented—agreement with South Korea on 20 January 1992 to denuclearize the Korean Peninsula.[96] The agreement, the terms of which are more stringent than those of the NPT, prohibits the presence, development, manufacture or testing of nuclear weapons anywhere on the Korean Peninsula and also bans both countries from possessing uranium enrichment or plutonium reprocessing facilities.[97] However, disputes over the procedures for verifying the agreement have prevented its full implementation.

In its declaration to the IAEA delivered on 4 May 1992, North Korea acknowledged the possession of a 5-megawatt (MW) gas-cooled, graphite-moderated research reactor at Yongbyon, which began operation in 1986.[98] In addition, after years of denying its existence, North Korea acknowledged that it was building a 'radioisotope laboratory' capable of reprocessing plutonium, a key nuclear weapon ingredient.[99]

The international community's concern about North Korea's nuclear capabilities did not end, however, with the submission of Pyongyang's initial declaration. In fact, concern about the full extent of North Korea's nuclear capabilities continued throughout 1992. IAEA inspectors conducted five missions in North Korea during 1992, visiting all its declared major nuclear facilities. Attention began to focus on the key issue of how much plutonium-bearing fuel North Korea possesses and may have reprocessed.

North Korea declared to the IAEA that it had separated 'gram quantities' of plutonium from some damaged fuel rods removed from its 5-MW reactor and maintained that a facility listed in its declaration was only a 'radiochemical laboratory' and not a full-scale reprocessing plant.[100] The IAEA Director General, after visiting the facility in May 1992, described the construction of the 180-metre long building as 80 per cent complete, with 40 per cent of the equipment installed. He stated that, when completed, he had 'no doubt that it

[96] The Joint Declaration of South and North Korea on the Denuclearization of the Korean Peninsula entered into force on 19 Feb. 1992. For the text, see Conference on Disarmament document CD/1147, 25 Mar. 1992.

[97] See note 96.

[98] The reactor's operating record is of considerable importance and is a matter of debate because of the plutonium-bearing fuel it might have produced. See Washington Council on Nonproliferation and Lawyer's Alliance for World Security, 'North Korea: do they or don't they have the bomb?', conference report (Washington Council on Nonproliferation: Washington, DC, July 1992).

[99] The existence of the building was first publicly disclosed after it was sighted by a French commercial SPOT satellite in 1990; *Nucleonics Week*, 22 Feb. 1990, p. 15.

[100] IAEA Press Release 92/24, 5 May 1992.

would [be] considered a reprocessing *plant* in [IAEA] terminology'[101] not just a laboratory, as North Korea contends.[102]

North Korea also reported to the IAEA that its 5-MW reactor has experienced severe operating problems and still contains its initial loading of natural uranium fuel, minus some removed damaged fuel rods.[103] The IAEA, while stating that it has found no information to contradict North Korea's declaration, will not be able independently to confirm the reactor's operating records without first taking samples from inside the reactor. US Government sources, however, assert that the reactor has been in continuous operation since at least 1987. Such an inspection, which cannot take place until the reactor is taken off line some time between April and October 1993, will allow the IAEA to roughly determine the reactor's operating history and how much plutonium-bearing fuel it may have produced.

During 1992 North Korea not only continued production and development of its reverse-engineered Scud-B and Scud follow-on missiles but also consolidated its role as a major ballistic missile supplier to the developing world, selling missiles to such countries as Iran and Syria. North Korea is known to have sold indigenously produced Scud-B missiles (range, 300 km; payload, 1000 kg) to Iran late in the 1980–88 Iraq–Iran War. In addition, US intelligence officials have publicly stated that North Korea began in March to sell the extended-range Scud-C (range, 500 km; payload, 700 kg),[104] a system which has undoubtedly entered its own arsenal. As many as 300 Scud-C missiles may have been delivered to Iran and Syria in 1992.[105]

According to US officials, North Korea is also developing a longer-range missile known as the No Dong I (range, at least 1000 km) which is expected to enter flight-tests in 1993.[106] US intelligence officials stated in early 1993 that Libya was negotiating with North Korea for the purchase of an unspecified number of these missiles.[107]

[101] Emphasis added in quotation. Director General Blix, 16 May 1992, press conference in Beijing following a visit to North Korea, IAEA transcript.

[102] The distinction between plant and laboratory is important since the Joint Declaration of South and North Korea on the Denuclearization of the Korean Peninsula, signed in Jan. 1992, bans both sides from possessing '*facilities* for nuclear reprocessing and uranium enrichment ' (emphasis added). North Korea, after declaring the 'radiochemical laboratory', said that such 'laboratories' were not covered under the agreement, an interpretation not shared by South Korea.

[103] The plutonium which North Korea declared it had reprocessed came from these damaged fuel rods. See Carnegie Endowment for International Peace Trip Report, Preliminary Report, Carnegie Endowment delegation visit to Pyongyang, 28 Apr.–4 May 1992.

[104] The sales became front-page news when the US press reported thta a North Korean freighter carrying advanced Scud-C missiles was *en route* to Iran. US military forces in the region considered boarding the vessel, but the ship made its delivery to Bandar Abbas on or about 10 Mar. 1992. 'Suspected Scud shipment reaches Iran', *Washington Post,* 11 Mar. 1992, p. 1.

[105] *MedNews,* 25 Jan. 1993.

[106] CIA Director James Woolsey and CIA Nonproliferation Center Director Gordon Oehler, Senate Governmental Affairs Committee, 24 Feb. 1993.

[107] See note 106.

## South Asia

Over the past decade, the rivalry between India and Pakistan has developed into a *de facto* nuclear stand-off, with both states possessing the ability to quickly assemble nuclear weapons. Efforts to induce the two states to join the NPT and accept full-scope safeguards have met with no success, and the focus of international and regional efforts has begun to shift to confidence-building and crisis prevention, with the long-term goal of establishing a South Asian nuclear weapon-free zone (NWFZ).

### *Pakistan*

The ambiguity surrounding Pakistan's nuclear capabilities for much of the 1980s was all but stripped away in 1992 with disclosures by Pakistani and US officials. Pakistan's Foreign Secretary Shahryar Khan stated in February 1992 to Western journalists that Pakistan had the ability to produce at least one nuclear device, saying that 'the capability is there' and that Pakistan possessed 'elements, which if put together, would become a [nuclear] device'.[108] This statement supported US Central Intelligence Agency (CIA) Director Robert Gates's 1992 congressional testimony that both India and Pakistan possessed the ability to assemble nuclear weapons quickly in a crisis. Gates also stated that the CIA had no reason to believe that either country maintained assembled nuclear weapons.[109] Khan stated that Pakistan had stopped the production of highly enriched uranium (HEU) and the construction of nuclear weapon 'cores', assertions that US officials have declined to either confirm or deny.

### *India*

India convincingly displayed its nuclear weapon capability by testing a 12-kt 'device' in 1974. India and the USA, whose relations have warmed somewhat with the end of the cold war (e.g., the two countries conducted joint naval exercises in 1992), held two sets of formal talks on non-proliferation and related issues in 1992, although the talks produced no major results. The USA continued to press India to join the NPT and to accept IAEA comprehensive safeguards, while at the same time it began to suggest confidence-building measures that India could take unilaterally or in co-operation with Pakistan to ease the tension between the two rivals. India has opposed joining the NPT, asserting that it discriminates against the non-nuclear weapon states and legitimizes the right of nuclear weapon states to possess nuclear weapons. It has also refused to pursue regional disarmament measures, arguing that disarma-

[108] 'Pakistan official affirms capacity for nuclear weapon', *Washington Post*, 7 Feb. 1992, p. A18.

[109] Robert Gates, Director of the US Central Intelligence Agency, before the Senate Government Operations Committee, 15 Jan. 1992; Senate Armed Services Committee, 26 Jan. 1992; House Foreign Affairs Committee, 25 Feb. 1992.

ment must take place globally, beginning with the nuclear weapon states[110]—a position reflecting India's security concerns about China.

Pakistan has continued to state that it will sign the NPT, accept IAEA safeguards and take other steps as long as India does the same, measures which India has rejected. Throughout the year, the USA promoted a five-power conference (China, India, Pakistan, Russia and the USA) on non-proliferation in South Asia; India is the only state which has declined to participate.

India and Pakistan did make some progress on confidence-building measures in 1992. The two states implemented the Agreement on the Prohibition of Attack against Nuclear Installations and Facilities by exchanging lists of locations covered by the Agreement on 1 January 1992[111] and bilaterally agreed to become original signatories of the 1993 Chemical Weapons Convention (for the list of signatories, see chapter 14, table 14.1). However, India responded negatively to suggestions made by the USA on steps it could take unilaterally or bilaterally with Pakistan to ease the nuclear stand-off, such as a regional nuclear test ban and a regional cut-off of production of fissile materials.

India's missile development programme received increased attention in 1992. This attention resulted from a series of Indian missile test launches and from the imposition of US sanctions against the Indian Space Research Organization and the Russian space company Glavkosmos following India's purchase from Russia of three liquid-fuelled rocket engines and associated technology. The USA was required under the National Defense Authorization Act of 1990[112] to impose sanctions against the two companies—both of which are state-operated—since the technologies sold are covered by the Equipment and Technology Annex of the Missile Technology Control Regime (MTCR).[113] US law makes no distinction between technology transfers for missile development purposes and those for peaceful space launch purposes. It requires that companies violating the terms of the MTCR be banned for two years from doing business with the US Government and from exporting items to the USA, even if the recipient country is not an adherent to the MTCR.[114]

India and Russia both complained about the imposition of sanctions. India claimed that it had provided adequate end-use assurances that the engines were to be used in peaceful activities and that the technology was not suited to

[110] See Rajiv Ghandi's Address before the UN General Assembly, 1988. UN General Assembly document A/42/PV.4.

[111] The Indian–Pakistani Agreement was signed in Dec. 1988 and entered into force in Jan. 1991. See 'Islamabad reports exchange of nuclear site lists', FBIS-TND-92-002, 31 Jan. 1992, p. 31.

[112] *Congressional Record*, vol. 135, no. 154 (6 Nov. 1989).

[113] The existence of the MTCR was revealed by the 7 founding members in Apr. 1987. The guidelines for the regime were extended at the Oslo Plentary Meeting in July 1992 and entered into force in Jan. 1993, to cover delivery systems intended for use with all weapons of mass destruction (nuclear, chemical and biological). Arms Control Association Factsheet, *The Missile Technology Control Regime*, Jan. 1993; 'Missile Technology Control Regime meets in Canberra', MTCR Departmental Press Release, 11 Mar. 1993.

[114] As of 11 Mar. 1993, the 23 members of the MTCR are: Australia, Austria, Belgium, *Canada*, Denmark, Finland, *France*, *Germany*, Greece, Iceland, Ireland, *Italy*, *Japan*, Luxembourg, the Netherlands, New Zealand, Norway, Portugal, Spain, Sweden, Switzerland, *the UK* and *the USA* (original signatories are shown in italics).

military applications; Russian officials expressed their belief that the sanctions were imposed to keep Russia from competing with US companies in the commercial space launch market. The US State Department maintained, however, that the penalties were imposed in order to 'obtain the broadest possible international cooperation in curbing the dangerous proliferation of missile technology'.[115] India and Russia have announced plans to go through with the sale, valued at $200 million.

India's missile development programme continued despite the imposition of US sanctions. On 29 May 1992, India conducted its second test launch of the Agni IRBM (range, 2500 km; payload, 900 kg), which India refers to as a 'technology demonstrator', from the test range at Balasore. The missile's second stage failed to fire correctly, and the missile landed off target, although it reportedly travelled a distance of over 2400 km.[116] A third test launch is scheduled for March 1993. India also conducted two tests of the shorter-range Prithvi missile (range, 250 km; payload, 500 kg), production and deployment of which is scheduled to begin in 1993. The Prithvi—which was derived from the Soviet SA-2 surface-to-air missile—is suited to tactical military applications because of its range, but the longer-range Agni missile, if deployed, would enable India to target all of Pakistan, much of western China and even Iran and Saudi Arabia.

Pakistan's missile programmes, like its nuclear weapon programme, has focused on acquiring technologies and materials abroad rather than developing them indigenously. Pakistan had contracted with China to purchase an undisclosed number of M-11 missiles (range, 290 km; payload, 800 kg), a deal which was cancelled in 1991 after the USA pressured China to abide by the terms of the MTCR.[117] It appears that Pakistan did receive a number of M-11 mobile missile launchers and some 'dummy' test missiles, but China and Pakistan insist that no missiles had been transferred. It was reported in late 1992, however, that China had delivered 24 M-11 missiles to Pakistan in violation of its oral pledge to visiting Secretary of State James Baker in January and a later written pledge delivered to Washington in June.[118] The US Government was unable to confirm or deny the report by the end of 1992.

## The Middle East

Nowhere are the dangers or the incentives associated with nuclear weapon proliferation as great as they are in the Middle East. No fewer than six countries in the region are either pursuing or already possess nuclear weapons. Several other states in the region are believed to be interested in acquiring a nuclear weapon capability.

[115] Richard Boucher, State Department Briefing, 11 May 1992.

[116] 'India tests controvertial Agni missile; rocket is capable of nuclear payload', *Washington Post,* 30 May 1992; All India Radio Network, in FBIS, TND-92-017, 29 May 1992.

[117] Statement by Secretary of State James Baker, 17 Nov. 1991, Beijing, China, US State Department transcript.

[118] 'China said to sell arms to Pakistan', *Washington Post,* 4 Dec. 1992. p. A1.

In 1992 governments in the Middle East peace process began to address nuclear weapon proliferation in the region, acknowledging that any peace settlement must also consider issues related to weapons of mass destruction. During the year, 13 regional and 11 non-regional parties took part in three multilateral working group meetings. The participants reviewed US–Soviet/Russian, European and South Asian arms control efforts. The most recent session, held in Moscow in September 1992, touched upon potential regional measures, including the creation of a NWFZ in the Middle East. All the states in the region have officially endorsed the idea of a Middle East NWFZ in various forums. At the September meeting the USA obtained an agreement from the Israeli delegation to discuss the NWFZ issue at future meetings, which Israel had been previously unwilling to do. Israel has also expressed its support for a Middle East NWFZ in the UN General Assembly[119] but has stressed that any progress on arms control issues is directly linked to progress in the Middle East peace process.

### *Israel*

Israel has the largest and most advanced arsenal of the three undeclared nuclear weapon states, the other two being India and Pakistan. While reports vary, Israel is believed to possess 50–300 nuclear weapons.[120] A Russian intelligence report stated in early 1993 that Israel may have possessed up to 20 nuclear weapons by 1980, estimated that its current arsenal contained 100–200 weapons, and concluded that 'Tel Aviv's interest in the development of thermonuclear weapons cannot be ruled out'.[121]

Israeli ballistic missile capabilities are also highly advanced. Israel may possess over 100 Jericho II missiles (range, 1500 km; payload, 650 kg). Israel has placed two satellites, Ofek 1–2, into orbit aboard the Shavit space launch vehicle (SLV) and plans to launch Ofek-3 into low earth orbit by the end of 1993.[122] One report estimated that if the Shavit was converted to carry a warhead would have a range of over 7000 km.[123]

### *Iran*

Concerns regarding Iran's nuclear programme continued to grow throughout 1992 as Tehran reportedly embarked on a multi-billion dollar military buildup of both its conventional and non-conventional military capabilities. Then CIA Director Gates testified that Iran—which is an original signatory of the NPT and accepts full-scope IAEA safeguards on its nuclear activities—had a clandestine nuclear weapon programme which could produce a nuclear weapon by

[119] Foreign Minister Shimon Perez, Address to the 47th UN General Assembly, 1 Oct. 1992.

[120] See Spector, L. and Smith, J., *Nuclear Ambitions* (Westview Press: Boulder, Colo., 1990); and Hersch, S., *The Samson Option* (Random House: N.Y., 1991).

[121] Russian Foreign Intelligence Service Report, *Proliferation of Weapons of Mass Destruction*, released 28 Jan. 1993, FBIS draft translation.

[122] FBIS-NES-93-025-HA-ARETZ, in Hebrew, 5 Feb. 1993.

[123] See note 121.

the year 2000.[124] His successor, James Woolsey, testified in February 1993 that 'Iran is pursuing the acquisition of nuclear weapons despite being a signatory of the Non-Proliferation Treaty (NPT)'.[125]

In response to international concern about its nuclear programme, The Iranian Government invited an IAEA mission to visit any nuclear facility in Iran. On 7–12 February 1992 an IAEA official mission—not an inspection team—headed by Deputy Director for Safeguards Jon Jennekens toured several facilities in Iran, including some alleged by an Iranian opposition group to be part of Iran's nuclear weapon development programme; one facility near Mo'allem Kalayeh was of particular concern. The IAEA press release issued after the mission stated that the 'activities reviewed by the team . . . were found to be consistent with the peaceful application of nuclear energy and ionizing radiation',[126] a statement which Iran subsequently characterized as a 'clean bill of health' for its nuclear programme.[127] The IAEA report continued, however, that '[I]t should be clear that the Team's conclusions are limited to facilities and sites visited by it and are of relevance only to the time of the Team's visit',[128] thus avoiding any sweeping conclusions about Iranian nuclear intentions.

US and other Western intelligence agencies, despite Iranian acceptance of the IAEA mission, continued to assert throughout 1992 that Iran was pursuing a nuclear weapon capability. President Bush, in a report to Congress issued shortly before he left office, stated that 'Iran has demonstrated a continuing interest in nuclear weapons and related technology that causes the U.S. to assess that Iran is in the early stages of developing a nuclear weapons program'.[129] On the basis of these concerns, the USA persuaded other countries, including India and Russia, to deny Iran access to sensitive nuclear technologies. In 1992 the USA also convinced the other G-7 countries (Canada, France, Germany, Italy, Japan and the UK) to tighten their nuclear-related exports to Iran, as well as to Iraq and Libya.[130]

### *Iraq*

The IAEA's discovery of Iraq's extensive nuclear weapon development programme has served as a catalyst for improving the nuclear weapon non-proliferation regime's inspection mechanisms over the past two years. IAEA efforts to inspect, destroy and prevent the rebuilding of Iraq's nuclear weapon capa-

[124] In response to a statement by Senator John Glenn during congressional testimony on 15 Jan. 1992 (summing up a German intelligence estimate) that 'Iran will be able to build a nuclear weapon by 2000', CIA Director Gates replied: 'I don't think that we [the CIA] have any reason to disagree with the overall assessment.' See Gates (note 30).

[125] James Woolsey, Director, US Central Intelligence Agency, testimony before the Senate Governmental Affairs Committee, 24 Feb. 1993.

[126] IAEA Press Release 92/11, 14 Feb. 1992.

[127] 'Atomic team reports on Iran probe', *Washington Post*, 15 Feb. 1992, p. 36.

[128] IAEA Press Release 92/11 (note 126).

[129] Arms Control and Disarmament Agency (ACDA), *Adherence to and Compliance with Arms Control Agreements* (ACDA: Washington, DC, 14 Jan. 1993).

[130] Reuter Wire Service Report, 21 Dec. 1992.

bilities, in implementation of UN Security Council Resolution 687,[131] continued, with the assistance and co-operation of the UN Special Commission on Iraq (UNSCOM). The activities of UNSCOM and the IAEA in 1992 included eight nuclear and seven missile site inspections, not all of which went smoothly (see also chapter 13 in this volume). There were major confrontations between the UNSCOM–IAEA inspection teams and Iraq, including one which led to threats of renewed UN Coalition warfare against Iraq. The most publicized event was Iraq's refusal in July 1992 to allow IAEA inspectors and later a missile inspection team to inspect the Ministry of Agriculture building, which Iraq claimed would have constituted a violation of its sovereignty. After a stand-off of over three weeks, a compromise was reached on 26 July 1992 which apparently included a ban on US, British or French inspectors' entering the building. Inspectors of other nationalities were finally allowed into the building, although the inspection revealed no missile-related information.

In another confrontation, Iraq failed to meet a 28 February 1992 UNSCOM deadline for the destruction of key missile production equipment. Iraq wanted to keep the equipment for civilian purposes, but it had been directly linked by UNSCOM to Iraq's missile production programme and included mixers and casting machines for solid rocket fuel.

In April 1992 Iraq again came close to provoking renewed military attacks but then reluctantly agreed to allow the destruction of nuclear weapon design and testing facilities at al-Atheer only hours before the arrival of the IAEA team tasked with overseeing that destruction. The facilities, which included high-technology clean rooms and a high explosive test bunker, were identified by the IAEA as the main centre for Iraq nuclear weapon research and development activities. Valued in the hundreds of millions of dollars, the multi-building facility was destroyed using explosives, and some hardened facilities were filled with reinforced concrete.

By August 1992 Iraq appeared to take a much less confrontational approach to the IAEA–UNSCOM on-site inspection teams. Iraq still refused to fully comply with UN Security Council Resolutions 687 and 715 demanding that it co-operate with the IAEA–UNSCOM teams. By the end of 1992, the main barrier to full implementation of Resolution 687 was Iraq's continued refusal to provide the IAEA and UNSCOM with a full disclosure of the names of foreign companies from which it has purchased equipment and materials.

## Africa

### *South Africa*

One of the long-standing obstacles to achieving an African NWFZ was removed in 1992 as South Africa accepted full-scope safeguards on its extensive nuclear facilities and materials. With Pretoria's accession to the NPT in

[131] For the text of the resolution, see *SIPRI Yearbook 1992* (note 15), appendix 13A, pp. 525–30.

September 1991, IAEA inspectors undertook the task of verifying the declared amount of fissile material in South Africa and placing it under IAEA controls. This was by far the most complex effort ever undertaken by the Safeguards Inspectorate, since South Africa has a highly advanced nuclear capability and has been able to produce weapon-grade uranium for over a decade.[132] By the end of 1992, the IAEA had inspected over 75 sites in South Africa, including a decommissioned uranium enrichment facility at Palindaba and an abandoned nuclear test site in the Kalahari desert.[133]

While the IAEA was unable to make a final determination about South Africa's declaration and the disposition of nuclear materials by the end of 1992, the IAEA Secretariat reported to the 1992 IAEA General Conference that the Agency had 'found no evidence that the inventory of nuclear material . . . was incomplete'.[134] IAEA officials were quick to point out, however, that the Agency would not be able to make any final conclusions for some time.[135] South Africa may have produced as much as 400 kg of HEU at the semi-commercial enrichment plant at Palindaba, enough material for over 20 nuclear weapons. The material was to be blended down to 5 per cent enrichment levels starting by late 1992.[136] At the end of 1992, however, the USA continued to express apprehension about the South African programme, stating that 'the United States has serious questions about South Africa's compliance' with its NPT obligations.[137]

[132] 1992 General Conference document GC(XXXVI)/RES/577.
[133] See note 132.
[134] See note 132.
[135] See note 132.
[136] *Nuclear Fuel,* 28 Sep. 1992.
[137] ACDA (note 129).

# **Appendix 6A.** Nuclear explosions, 1945–92

RAGNHILD FERM

The annual number of nuclear weapon tests has consistently declined since 1988. Eight nuclear explosions were conducted in 1992; by comparison, the average annual number of tests conducted in the preceding 10-year period was 49 (see table 6A.3).

Only two of the declared nuclear weapon states, the United States and China, carried out nuclear tests in 1992. All the six US tests had a yield below 150 kilotons. According to seismic recordings, the Chinese test of 21 May 1992 had a yield of around 660 kilotons, the largest underground test ever conducted by China.

Both the United States and the Russian Federation have a test moratorium in effect until 1 July 1993. Regarding the former Soviet Union and Russia,[1] on 5 October 1991 then Soviet President Mikhail Gorbachev announced a one-year moratorium on nuclear tests. On 19 October 1992, Russian President Boris Yeltsin extended the test moratorium until 1 July 1993, in response to the bill signed by President George Bush on 2 October 1992 to halt US nuclear tests for nine months, to reduce the number of tests for three years and to stop all US testing until 30 September 1996, provided the other states refrain from testing after that date. President Bush later expressed concern over the negative impact on national security of the new restrictive test plans developed by the Department of Defense. President Bill Clinton, who supported a test limitation policy during his election campaign, stated in February 1993 that his Administration is preparing to review questions concerning a comprehensive test ban (CTB) and how to proceed with a limited US test programme after 1 July 1993.[2]

The United Kingdom usually conducts one test per year, in co-operation with the United States at the US test site in the Nevada desert. However, because of the US moratorium, the UK did not conduct a test in 1992. The UK expressed regret regarding the US plans to wind down nuclear testing and considered the case for a moratorium unfounded. It noted that the US plans to limit future tests will allow the UK altogether only three more tests at the US Nevada Test Site, which is not regarded as sufficient to assure the safety of British nuclear weapons.[3]

In April 1992, before its testing programme of the year started, France announced a moratorium on testing until the end of the year. In his speech at the signing ceremony of the Chemical Weapons Convention in Paris in January 1993, President François Mitterrand stated that France would forgo nuclear testing as long as the USA and Russia refrained from testing.[4]

For a further discussion of nuclear testing issues and a comprehensive nuclear test ban, see chapter 11 in this volume.

[1] In tables 6A.2–6A.4 below, only the nuclear explosions conducted by the former Soviet Union are listed, as Russia has not conducted any such tests.

[2] *Arms Control Today*, Mar. 1993, p. 29.

[3] See note 2.

[4] *Le Monde,* 15 Jan. 1993.

**Table 6A.1.** Registered nuclear explosions in 1992

| Date | Origin time (GMT) | Latitude (deg) | | Longitude (deg) | | Region | Body wave magnitude[a] |
|---|---|---|---|---|---|---|---|
| **USA** | | | | | | | |
| 26 Mar. | 163000.0 | 37.272 | N | 116.360 | W | Nevada | 5.6 |
| 30 Apr. | 173000.0 | 37. | N | 116. | W | Nevada | . . |
| 19 June | 164500.0 | 37. | N | 116. | W | Nevada | . . |
| 23 June | 145959.6 | 37.120 | N | 116.041 | W | Nevada | . . |
| 18 Sep. | 170000.0 | 37. | N | 116. | W | Nevada | 4.4 |
| 23 Sep. | 150400.0 | 37. | N | 116. | W | Nevada | . . |
| **China** | | | | | | | |
| 21 May | 050000.0 | 41.6 | N | 88.9 | E | Lop Nor | 7.1 |
| 25 Sep. | 080000.0 | 41.4 | N | 88.9 | E | Lop Nor | 5.4 |

[a] Body wave magnitude ($m_b$) indicates the size of the event. To be able to give a reasonably correct estimate of yield it is necessary to have detailed information, for example on the geological conditions of the area where the test is conducted. Therefore, to give the $m_b$ figure is an unambiguous way of listing the size of an explosion. $m_b$ data for the US and Chinese tests were provided by the Swedish National Defence Research Establishment (FOA).

**Table 6A.2.** Estimated number of nuclear explosions 16 July 1945–5 August 1963 (the signing of the Partial Test Ban Treaty)

a = atmospheric; u = underground

| | USA | | USSR | | UK | | France | | |
|---|---|---|---|---|---|---|---|---|---|
| Year | a | u | a | u | a | u | a | u | Total |
| 1945 | 3 | 0 | | | | | | | **3** |
| 1946 | 2[a] | 0 | | | | | | | **2** |
| 1947 | 0 | 0 | | | | | | | **0** |
| 1948 | 3 | 0 | | | | | | | **3** |
| 1949 | 0 | 0 | 1 | 0 | | | | | **1** |
| 1950 | 0 | 0 | 0 | 0 | | | | | **0** |
| 1951 | 15 | 1 | 2 | 0 | | | | | **18** |
| 1952 | 10 | 0 | 0 | 0 | 1 | 0 | | | **11** |
| 1953 | 11 | 0 | 4 | 0 | 2 | 0 | | | **17** |
| 1954 | 6 | 0 | 7 | 0 | 0 | 0 | | | **13** |
| 1955 | 17[a] | 1 | 5[a] | 0 | 0 | 0 | | | **23** |
| 1956 | 18 | 0 | 9 | 0 | 6 | 0 | | | **33** |
| 1957 | 27 | 5 | 15[a] | 0 | 7 | 0 | | | **54** |
| 1958 | 62[b] | 15 | 29 | 0 | 5 | 0 | | | **111** |
| *1949–58, exact years not available* | | | 18 | | | | | | **18** |
| 1959 | 0 | 0 | 0 | 0 | 0 | 0 | | | **0[d]** |
| 1960 | 0 | 0 | 0 | 0 | 0 | 0 | 3 | 0 | **3[d]** |
| 1961 | 0 | 10 | 50[a] | 1[c] | 0 | 0 | 1 | 1 | **63[d]** |
| 1962 | 39[a] | 57 | 43 | 1[c] | 0 | 2 | 0 | 1 | **143** |
| 1 Jan.– 5 Aug. 1963 | 4 | 25 | 0 | 0 | 0 | 0 | 0 | 2 | **31** |

**Table 6A.2** *contd*

a = atmospheric; u = underground

| | USA | | USSR | | UK | | France | | |
|---|---|---|---|---|---|---|---|---|---|
| Year | a | u | a | u | a | u | a | u | Total |
| **Total** | **217** | **114** | **183**[e] **(214)**[f] | **2**[c] | **21** | **2** | **4** | **4** | **547** **(576)**[f] |

[a] One of these tests was carried out under water.

[b] Two of these tests were carried out under water.

[c] Soviet information released in Sep. 1990 did not confirm whether these were underground or atmospheric tests.

[d] The UK, the USA and the USSR observed a moratorium on testing, Nov. 1958–Sep. 1961.

[e] The total figure for Soviet atmospheric tests includes the 18 additional tests conducted in the period 1949–58, the exact years for which are not available.

[f] The totals in brackets include the (probably atmospheric) explosions revealed by Soviet authorities in Sep. 1990, the exact years for which have still not been announced. See *SIPRI Yearbook 1991*, p. 41. If the two tests in 1961 and 1962 (see note *c*) were atmospheric tests, this figure should read 216, under the column for atmospheric tests.

**Table 6A.3.** Estimated number of nuclear explosions 6 August 1963–31 December 1992

a = atmospheric; u = underground

| | USA[a] | | USSR | | UK[a] | | France | | China | | India | | |
|---|---|---|---|---|---|---|---|---|---|---|---|---|---|
| Year | a | u | a | u | a | u | a | u | a | u | a | u | Total |
| 6 Aug.–31 Dec. 1963 | 0 | 15 | 0 | 0 | 0 | 0 | 0 | 1 | | | | | **16** |
| 1964 | 0 | 38 | 0 | 6 | 0 | 1 | 0 | 3 | 1 | 0 | | | **49** |
| 1965 | 0 | 36 | 0 | 10 | 0 | 1 | 0 | 4 | 1 | 0 | | | **52** |
| 1966 | 0 | 43 | 0 | 15 | 0 | 0 | 6 | 1 | 3 | 0 | | | **68** |
| 1967 | 0 | 34 | 0 | 17 | 0 | 0 | 3 | 0 | 2 | 0 | | | **56** |
| 1968 | 0 | 45[b] | 0 | 15 | 0 | 0 | 5 | 0 | 1 | 0 | | | **66** |
| 1969 | 0 | 38 | 0 | 16 | 0 | 0 | 0 | 0 | 1 | 1 | | | **56** |
| 1970 | 0 | 35 | 0 | 17 | 0 | 0 | 8 | 0 | 1 | 0 | | | **61** |
| 1971 | 0 | 17 | 0 | 19 | 0 | 0 | 6 | 0 | 1 | 0 | | | **43** |
| 1972 | 0 | 18 | 0 | 22 | 0 | 0 | 3 | 0 | 2 | 0 | | | **45** |
| 1973 | 0 | 16[c] | 0 | 14 | 0 | 0 | 5 | 0 | 1 | 0 | | | **36** |
| 1974 | 0 | 14 | 0 | 18 | 0 | 1 | 8 | 0 | 1 | 0 | 0 | 1 | **43** |
| 1975 | 0 | 20 | 0 | 15 | 0 | 0 | 0 | 2 | 0 | 1 | 0 | 0 | **38** |
| 1976 | 0 | 18 | 0 | 17 | 0 | 1 | 0 | 4 | 3 | 1 | 0 | 0 | **44** |
| 1977 | 0 | 19 | 0 | 18 | 0 | 0 | 0 | 8[d] | 1 | 0 | 0 | 0 | **46** |
| 1978 | 0 | 17 | 0 | 27 | 0 | 2 | 0 | 8 | 2 | 1 | 0 | 0 | **57** |
| 1979 | 0 | 15 | 0 | 29 | 0 | 1 | 0 | 9 | 1 | 0 | 0 | 0 | **55** |
| 1980 | 0 | 14 | 0 | 21 | 0 | 3 | 0 | 13 | 1 | 0 | 0 | 0 | **52** |
| 1981 | 0 | 16 | 0 | 22 | 0 | 1 | 0 | 12 | 0 | 0 | 0 | 0 | **51** |
| 1982 | 0 | 18 | 0 | 32 | 0 | 1 | 0 | 6 | 0 | 1 | 0 | 0 | **58** |
| 1983 | 0 | 17 | 0 | 27 | 0 | 1 | 0 | 9 | 0 | 2 | 0 | 0 | **56** |
| 1984 | 0 | 17 | 0 | 29 | 0 | 2 | 0 | 8 | 0 | 2 | 0 | 0 | **58** |
| 1985 | 0 | 17 | 0 | 9[e] | 0 | 1 | 0 | 8 | 0 | 0 | 0 | 0 | **35** |
| 1986 | 0 | 14 | 0 | 0[e] | 0 | 1 | 0 | 8 | 0 | 0 | 0 | 0 | **23** |
| 1987 | 0 | 14 | 0 | 23 | 0 | 1 | 0 | 8 | 0 | 1 | 0 | 0 | **47** |
| 1988 | 0 | 14 | 0 | 17 | 0 | 0 | 0 | 8 | 0 | 1 | 0 | 0 | **40** |
| 1989 | 0 | 11 | 0 | 7 | 0 | 1 | 0 | 8 | 0 | 0 | 0 | 0 | **27** |
| 1990 | 0 | 8 | 0 | 1 | 0 | 1 | 0 | 6 | 0 | 2 | 0 | 0 | **18** |
| 1991 | 0 | 7 | 0 | 0 | 0 | 1 | 0 | 6 | 0 | 0 | 0 | 0 | **14** |
| 1992 | 0 | 6 | 0 | 0 | 0 | 0 | 0 | 0 | 0 | 2 | 0 | 0 | **8** |

| | USA[a] | | USSR | | UK[a] | | France | | China | | India | | |
|---|---|---|---|---|---|---|---|---|---|---|---|---|---|
| Year | a | u | a | u | a | u | a | u | a | u | a | u | Total |
| **Total** | **0** | **611** | **0** | **463 (500)[f]** | **0** | **21** | **44** | **140** | **23** | **15** | **0** | **1** | **1 318 (1 355)[f]** |

[a] See note *a*, table 6A.4.

[b] Five devices used simultaneously in the same test are counted here as one explosion.

[c] Three devices used simultaneously in the same test are counted here as one explosion.

[d] Two of these tests may have been conducted in 1975 or 1976.

[e] The USSR observed a unilateral moratorium on testing, Aug. 1985–Feb. 1987.

[f] The totals in brackets include the explosions revealed by the Soviet authorities in Sep. 1990, the exact years for which have still not been announced. See *SIPRI Yearbook 1991*, p. 41.

**Table 6A.4.** Estimated number of nuclear explosions 16 July 1945–31 December 1992

| USA[a] | USSR[b] | UK[a] | France | China | India | Total |
|---|---|---|---|---|---|---|
| 942 | 648 (715) | 44 | 192 | 38 | 1 | 1 865 (1 931)[b] |

[a] All British tests from 1962 have been conducted jointly with the United States at the Nevada Test Site. Therefore, the number of US tests is actually higher than indicated here.

[b] The figures in brackets for the former Soviet Union include additional tests announced by the Soviet authorities in Sep. 1990 for the period 1949–90. See *SIPRI Yearbook 1991,* p. 41.

## Sources for tables 6A.1–6A.4

Swedish National Defence Research Establishment (FOA), various estimates; Norris, R. S., Cochran, T. B. and Arkin, W. M., 'Known US nuclear tests July 1945 to 31 December 1988', *Nuclear Weapons Databook*, Working Paper no. 86–2 (Rev. 2C) (Natural Resources Defense Council: Washington, DC, Jan. 1989); Reports from the Australian Seismological Centre, Bureau of Mineral Resources, Geology and Geophysics, Canberra; New Zealand Department of Scientific and Industrial Research (DSIR), Geology and Geophysics, Wellington; Cochran, T. B., Arkin, W. M., Norris, R. S. and Sands, J. I., *Nuclear Weapons Databook, Vol. IV, Soviet Nuclear Weapons* (Harper & Row: New York, 1989), chapter 10; Burrows, A. S., *et al.*, 'French nuclear testing, 1960–88', *Nuclear Weapons Databook*, Working Paper no. 89-1 (NRDC: Washington, DC, Feb. 1989); 'Known Chinese nuclear tests, 1964–1988', *Bulletin of the Atomic Scientists*, vol. 45, no. 8 (Oct. 1989), p. 48, see also vol. 45, no. 9 (Nov. 1989), p. 52; and various estimates.

# 7. Chemical and biological weapons: developments and proliferation

THOMAS STOCK*

## I. Introduction

The negotiations on the Chemical Weapons Convention (CWC) culminated in 1992; after approval by the 47th United Nations General Assembly (UNGA), the Convention on the Prohibition of the Development, Production, Stockpiling and Use of Chemical Weapons and on their Destruction was opened for signature in January 1993. The disarmament community is now looking forward to the entry into force of the Convention in 1995. For more than a decade SIPRI has published studies evaluating the negotiations and recommending the CWC (chapter 14 of this volume presents a preliminary analysis of the CWC).

Although the conclusion of the CWC is clearly a positive achievement, it will not solve all of the problems related to chemical warfare. It may be useful for the reader to review the developments of 1992 bearing in mind that the CWC will outlaw the use, development and production of chemical weapons (CW). This chapter deals with matters related to chemical and biological warfare and relevant disarmament undertakings and addresses the following areas:

1. In 1992 several new allegations were made of CW and to a lesser extent biological weapon (BW) use and possession.
2. The future spread of chemical and biological weapons is one of the major concerns of the 1990s, and more effective measures to prevent such proliferation were discussed in 1992. As in the past there was public concern about this issue, and strong arguments were made that international efforts to stop proliferation should be strengthened. This public awareness was partially evoked by the new findings in 1992 of the United Nations Special Commission on Iraq (UNSCOM) concerning foreign support of the former Iraqi chemical and biological warfare (CBW) programme (chapter 13 deals with UNSCOM activities in 1992). New information about the involvement of foreign companies in the buildup of the Iraqi CW and BW capability led to trials and investigations in several countries.

* Anna Hårleman of the SIPRI Chemical and Biological Warfare Programme assisted in preparing references and data for this chapter. The references were gathered from the SIPRI CBW Programme Data Base and were also kindly provided by J. P. Perry Robinson, Science Policy Research Unit, University of Sussex, UK, from the Sussex–Harvard Information Bank.

3. The US CW destruction programme and the experience obtained by UNSCOM's CW disposal efforts in Iraq dramatically increased knowledge about related problems and techniques. Concern about the impact on the environment of such destruction is growing, and there is evidence that destruction costs will be enormous, in some cases 10 times greater than the cost of production. The US demilitarization programme has begun to place greater emphasis on alternative destruction technologies, while Russia is now undertaking the painful process of designing and establishing its destruction programme.

4. New discoveries and information about the former Soviet chemical warfare programme provided evidence that the military overestimated the importance of this aspect of the Soviet weapon programme.

5. The issue of old chemical and conventional ammunition, abandoned or dumped during past decades in the soil or the sea, gained public attention. The withdrawal of troops from bases in Europe brought to light the environmental contamination, particularly of soil and water, caused by former military activities. Redevelopment of areas where troops were formerly stationed will demand immense investment.

6. The experience of the 1991 Persian Gulf War and increasing public awareness of weapon proliferation gave new impetus to research and development (R&D) into nuclear, biological and chemical (NBC) protection.

7. Huge oil spills and oil fires in Kuwait were part of the aftermath of the Persian Gulf War and there was great public concern about their impact on the environment. However, the damage to the environment appears to have been less than initially feared.

8. New information about the former Soviet, now Russian, BW R&D programme confirmed that it had continued until the beginning of 1992, despite earlier official statements to the contrary.

9. The Third Review Conference of the Biological and Toxin Weapons Convention (BWC) was held in 1991, and 1992 was characterized by efforts to discuss future verification measures, especially through the *Ad Hoc* Group of Governmental Experts which was established after the Review Conference. In 1992 the expert group met twice in Geneva and discussed potential verification measures from a scientific and technical standpoint. The 1992 round of information exchange produced much new information, but the number of participants in the exchange did not increase.

## II. Allegations of CW and BW use

In 1992 a number of allegations were made of the use of CW agents or weapons and, in a few cases, of the use of BW agents. These allegations concerned countries or regions of military conflict or high political tension such as the former Yugoslavia, the Middle East, Mozambique and the new independent republics of the former Soviet Union. In some cases it was later clarified that riot control agents had been used.

## The new republics

In 1992 the press reported extensively on the alleged use of chemical weapons in the Nagorno-Karabakh region of Azerbaijan in April–August during the fighting between Armenian factions and Azerbaijani Armed Forces. The following cities or districts were mentioned with respect to the use of chemical warfare agents: (*a*) continuous allegation of CW use, including artillery shells with hydrogen cyanide and cyanogen chloride in the attacks on the city of Shusha,[1] later denied by officials from Nagorno-Karabakh;[2] (*b*) allegation of CW use in the battles of Stepanakert,[3] Agdam, Terter,[4] and the Zangelan, Kubatly and Kelbadjar districts; (*c*) alleged use of cyanide in missile warheads against the village of Mokhratag in the Mardakert district[5] and in the Fizulinskiy district;[6] and (*d*) use of chemical missiles and alleged use of mustard gas in the Nakhichevan Autonomous Republic in the village of Sadarak, close to the borders with Iran and Turkey.[7] In May a representative of the Commonwealth of Independent States (CIS) Joint Armed Forces General Staff categorically denied both that the CIS Armed Forces had chemical weapons and that chemical warfare could be conducted by the use of weapons which are unaccounted for.[8] In July a team of UN experts arrived in Baku to investigate the alleged CW use by the Armenian Armed Forces,[9] but it was unable to confirm such use.[10] The team consisted of three experts appointed by the UN Secretary-General from Belgium, Switzerland and Sweden and two UN staff members. They visited the towns of Fizuly and Kubatly, which had recently been attacked, and interviewed patients in several hospitals in Baku. Later in Yerevan the team reported its conclusions to the Armenians;[11] it interpreted the Azerbaijani discovery of cyanide in soil and other samples from combat areas not as traces of CW agent, but as possible degradation or combustion products from the use of conventional weapons.[12]

In March Azerbaijan alleged that Armenia had used material containing infectious agents of bacteriological origin in the Kelbadzharskiy and Lachinsky

[1] 'Armenians said to stage chemical attack', *Washington Post*, 27 Apr. 1992, p. A22; 'Armenia accused of using chemical weapons', in Foreign Broadcast Information Service, *Daily Report–Soviet Union* (*FBIS-SOV*), FBIS-SOV-92-081, 27 Apr. 1992, p. 66; 'Armenia accused of using chemical weapons', in FBIS-SOV-92-091, 11 May 1992, p. 80; 'Evidence of chemical weapons use noted', in FBIS-SOV-92-095, 15 May 1992, p. 71; 'Baku hosts conference on weapons control', in FBIS-SOV-92-096, 18 May 1992, p. 3 .

[2] 'Use of chemical weapons, aviation denied', in FBIS-SOV-92-091, 11 May 1992, p. 73.

[3] 'Karabakh denies chemical weapons used', in FBIS-SOV-92-086, 4 May 1992, pp. 64–65.

[4] 'Troops warned of chemical weapons', in FBIS-SOV-92-120, 22 June 1992, pp. 84–85.

[5] 'Chemical weapons use noted', in FBIS-SOV-92-111, 9 June 1992, p. 85.

[6] 'Chemical weapons use noted', in FBIS-SOV-92-117, 17 June 1992, pp. 64–65.

[7] 'Chemical weapons use charged', in FBIS-SOV-92-098, 20 May 1992, pp. 65–66; 'Popular front reports chemical warhead tests', in FBIS-SOV-92-099, 21 May 1992, p. 98; 'Nakhichevan reports casualties, mustard gas use', in FBIS-SOV-92-105, 1 June 1992, p. 65.

[8] 'Reports of chemical weapons in Karabakh denied', in FBIS-SOV-92-093, 13 May 1992, pp. 15–16.

[9] 'UN chemical weapons experts to tour provinces', in FBIS-SOV-92-130, 7 July 1992, pp. 70–71

[10] 'Ministry accuses Armenia of using poison gas', in FBIS-SOV-92-181, 17 Sep. 1992, p. 53.

[11] 'UN chemical weapons experts arrive in Yerevan', in FBIS-SOV-92-135, 14 July 1992, p. 35.

[12] 'Azerbaijan accusations Armenian request', *ASA Newsletter*, no. 31 (12 Aug. 1992), p. 8.

districts of Azerbaijan.[13] These allegations were later denied by the Armenian Defence Ministry.[14] In October the reported use of CW shells by the Abkhazian Army against Georgian troops was officially denied.[15] In the conflict between Ossetians and Ingushes the use of chemical shells was reported in November.[16]

## Mozambique

In Mozambique allegations continued to be made that chemical weapons have been and are being used against army forces by Renamo (the Mozambican National Resistance, MNR).[17] Renamo formally denied the allegations.[18] A Mozambican–South African Joint Security Commission was set up to investigate allegations of the use of chemical weapons which affect the nervous system during a military operation at the end of January 1992.[19] Mozambique asked for outside help to determine the nature of the weapons used,[20] and in February Swedish experts conducted an initial investigation. Based upon the results of that investigation an official request was made to the UN Secretary-General, and in March a team of experts from Sweden, Switzerland and the UK conducted investigations in Mozambique. In June the UN Secretary-General presented his report on the mission to the Security Council.[21] Owing to the considerable delay between the previous attack and the investigation, the report pointed out that 'it may not be possible to detect traces of agent if a chemical warfare agent had been used'. In July a somewhat controversial interpretation was presented in another press source, pointing out that 'it can certainly be concluded as possible that an anti-nervous system chemical weapon was used'.[22] This illustrates how difficult it is to provide clear evidence of chemical warfare agent use, especially if much time elapses between the attack and the investigation.

[13] 'Armenia "accused" of bacteriological warfare', in FBIS-SOV-92-053, 18 Mar. 1992, p. 75; 'Bacteriological warfare claimed in insect drop', in FBIS-SOV-92-054, 19 Mar. 1992, pp. 82–83.

[14] 'Defence Ministry denies biological weapons use', in FBIS-SOV-92-054, 19 Mar. 1992, p. 80.

[15] 'Abkhaz defence ministry denies using chemical weapons', in FBIS-SOV-92-207, 26 Oct. 1992, p. 80.

[16] 'Abuse, chemical arms use alleged', in FBIS-SOV-92-213, 3 Nov. 1992, p. 28.

[17] 'Possible Renamo chemical attack investigated', in Foreign Broadcast Information Service, *Daily Report–Africa* (*FBIS-AFR*), FBIS-AFR-92-015, 23 Jan. 1992, pp. 25–26; 'Renamo "deserters" report use of chemical weapons', in FBIS-AFR-92-024, 5 Feb. 1992, pp. 18–19; 'Army chief affirms chemical weapons use by Renamo', in FBIS-AFR-92-034, 20 Feb. 1992, pp. 21–22.

[18] 'Renamo denies reported use of chemical weapons', in FBIS-AFR-92-018, 28 Jan. 1992, p. 19.

[19] 'Chemical attack "kills five"', *The Guardian*, 28 Jan. 1992, p. 5.

[20] 'Letter dated 27 Jan. 1992 from the Permanent Representative of Mozambique to the United Nations addressed to the Secretary General', General Assembly, Security Council document A/47/87, S/23490, 29 Jan. 1992.

[21] 'Report of the mission dispatched by the Secretary-General to investigate an alleged use of chemical weapons in Mozambique', United Nations Security Council document S/24065, 12 June 1992.

[22] 'Experts confirm Renamo use of chemical weapons', in FBIS-AFR-92-140, 21 July 1992, pp. 19–20.

## The former Yugoslavia

Reports of the use of chemical warfare agents or chemical weapons in the military conflict in Bosnia and Herzegovina continued, including allegations that regular Croatian forces might have used such weapons in the bombardment of Sarajevo and on the Trebinje battlefield.[23]

In June a new dimension was added when it was feared that a chemical–industrial complex in Tuzla, north of Sarajevo, might be hit by Serbian artillery shells. Owing to the large quantities of chlorine and mercury stored there, scenarios worse than the Bhopal, India disaster were envisaged.[24] In October there was concern that Bosnian forces might use chlorine deployed in railcars to defend Gradacac,[25] and the use of chemical agents of the irritant type was reported at Gradacac.[26] During the October visit to Iran of President Alija Izetbegovic of Bosnia and Herzegovina, he stated that 'if the arms embargo against Bosnia remains in force, the people of Bosnia—to defend themselves and to stop Serbian crimes—will be forced to use existing poisonous gases'.[27] In late November a gas alarm was sounded after Serbian artillery bombed Tuzla and destroyed some chlorine containers.[28]

It is suspected that the tear-gas CS and the incapacitating chemical warfare agent BZ are being produced by Serbia in Kruselak, near Belgrade, and that CS has already been employed.[29]

## Iraq

In May there were reports that Iraq's President Saddam Hussein might use chemical weapons against the Shiite Arab population in the marshlands district.[30] In light of the very stringent UNSCOM mandate and the plan for further monitoring of all Iraqi CBW activities such allegations seemed highly doubtful.

[23] 'Press reports Croat-Muslim forces using nerve gas', in Foreign Broadcast Information Service, *Daily Report–East Europe* (*FBIS-EEU*), FBIS-EEU-92-116, 16 June 1992, p. 28; 'Sarajevo suburbs under attack; poison gas suspected', in FBIS-EEU-92-122, 24 June 1992, p. 23; 'Sarajevo shelled: chemical agents reportedly used', in FBIS-EEU-92-169, 31 Aug. 1992, p. 28; 'Serbs accuse Croatian army of using poison gas', in FBIS-EEU-92-172, 3 Sep. 1992, p. 18; 'Izetbegovic and Karadzic stellen Friedensgespräche in Frage', *Der Tagesspiegel*, 15 Sep. 1992, p. 1; 'Croat army reportedly using poison gas', in FBIS-EEU-92-178, 14 Sep. 1992, p. 23.

[24] Fitchett, J., '150, 000 are at risk if Serbian gunners hit chemical plant', *International Herald Tribune*, 10 June 1992, pp. 1–2.

[25] AP/Reuters, 'Bosniens UNO-Botschafter: Chlorgas-Einsatz möglich', *Süddeutsche Zeitung*, 14 Oct. 1992, p. 2; 'Serben bieten Abzug ihrer Luftwaffe aus Bosnien an', *Frankfurter Allgemeine Zeitung*, 15 Oct. 1992, p. 1; AP, 'Bosnia chief threatens the use of poison gas', *International Herald Tribune*, 31 Oct.–1 Nov. 1992, p. 2; 'Tuzla forces threaten chemical attacks', in FBIS-EEU-92-198, 13 Oct. 1992, p. 27.

[26] 'Use of chemical agents, napalm alleged', in FBIS-EEU-92-185, 23 Sep. 1992, p. 23.

[27] 'Threatens to use "poisonous gases"', in Foreign Broadcast Information Service, *Daily Report–Near East & South Asia* (*FBIS-NES*), FBIS-NES-92-211, 30 Oct. 1992, p. 34.

[28] 'Serbischer Angriff auf Chemiefabrik Giftgasalarm für Stadt in Nordbosnien', *Der Tagesspiegel*, 22 Nov. 1992, p. 1.

[29] Price, R., 'The Balkan nightmare: an ASA CBW intelligence report', *ASA Newsletter*, no. 32 (15 Oct. 1992), pp. 1, 10.

[30] 'Opposition says Saddam to use chemical weapons', in FBIS-NES-92-088, 6 May 1992, p. 14; 'Iraq threatens Shiites with chemical attacks', in FBIS-NES-92-091, 11 May 1992, p. 50.

## III. Allegations of CW and BW possession

Even as the CWC was being finalized, allegations of CW acquisition programmes and possession, especially in the Middle East, continued to occur. How difficult it will be in the future to ensure that an individual country does not go the way of chemical armament is shown by the example of *Iraq*. Even with the special UN mandate concerning long-term monitoring and with the obligation to destroy all chemical weapons and CW-capable facilities, there are still many doubts about Iraq's total chemical disarmament.[31]

The specific allegations of CW or BW possession made in 1992 are summarized below.

Allegations continued that *Syria* is conducting a CW programme, in particular producing mustard gas and nerve agents and actively developing a CW missile capability.[32] Two location have been mentioned, one near the village of Safiya, in the north-east, close to the Turkish border, and the other to the south of the city of Homs, close to the main road to Damascus.[33] Concern about Syria's CBW programme mounted in August when a German vessel on the way to Syria was stopped in Cyprus with a shipment of 45 tonnes of trimethyl phosphite from Indian United Phosphorus Ltd.[34] Trimethyl phosphite is used in the production of the pesticide dichloro divinyl phosphate but can also be used for nerve gas production. After the USA alerted German authorities, the shipment was stopped in Cyprus and sent back to India. The Indian company had signed an agreement to export a total of 90 tonnes, and the first half of the order reached Damascus in May. An investigation by Indian customs authorities was launched, and after it was found that the company had exported chemicals without government clearance, the company was denied export licences for six months.[35]

As in the past *Iran* was alleged to have an active chemical warfare programme.[36] In February Germany announced that a request by the Iranian Government to participate in the construction of a projected pesticide plant at Qazvin would be refused.[37] In July the Iranian Foreign Minster rejected categorically the allegation that Iran has an active chemical warfare programme and emphasized Iran's rejection of chemical and biological weapons.[38] This

[31] Gaffney, F., 'U.S. foolishly strips capability to deter chemical weapon threat', *Defense News*, vol. 7, no. 9 (2 Mar. 1992); Rowe, T., 'U.N. still "concerned" about Iraq', *Washington Post*, 4 Apr. 1992, p. A19.

[32] Waller, D., 'Sneaking in the Scuds', *Newsweek*, vol. 119, no. 25 (22 June 1992), pp. 20–24; 'Baraq on nuclear, chemical buildup by Syria, Iraq', in FBIS-NES-91-237, 10 Dec. 1991, p. 47; Hoffman, D., 'Israelis say Syrians test-fired new Scud', *Washington Post*, 14 Aug. 1992, p. A25.

[33] 'Syria's secret poison-gas plants', *Foreign Report*, 10 Sep. 1992.

[34] Gordon, M. R., 'India tied to poison gas deal', *International Herald Tribune*, 22 Sep. 1992, p. 5; Rotem, M., 'Indian chemical company won't stop shipment to Syria', *Jerusalem Post*, international edn, 22 Aug. 1992, pp. 1–4;

[35] 'India to prosecute chemical firm', *International Herald Tribune*, 24 Sep. 1992, p. 2.

[36] Timmerman, K. R., *Weapons of Mass Destruction: The Cases of Iran, Syria and Libya*, A Simon Wiesenthal Center Special Report, Aug. 1992.

[37] Hoffmann, W., 'German–Iranian trade: no weapons, says Möllemann', *German Tribune*, no. 1505 (28 Feb. 1992), p. 7.

[38] 'Iran entwickelt keine Nuklearwaffen', *Frankfurter Allgemeine Zeitung*, 31 July 1992, p. 1.

was repeated a few days later by Iran's representative to the UN, who pointed out that Iran does not intend to produce chemical weapons.[39]

*Libya* was again very much in the public eye in 1992 as regards its CBW programme.[40] Libya is alleged to have cleaned up the ruins of its alleged CW production facility at Rabta and to have built a second plant on a site outside Sheba, about 650 km south of Tripoli.[41] However, there is disagreement among experts as to whether a second plant exists. In response to allegations and under pressure from the UN, Colonel Muammar Qadhafi stated that Libya was prepared to consider outside inspection of alleged nuclear and CW sites.[42]

Allegations continued that *North Korea* is conducting a CW programme which may include several facilities for production of nerve gas, blood agents and mustard gas.[43] The annual production capacity of nine plants is said to be approximately 5000 tonnes. The allegations, which have also cited a supposed BW programme, were strongly rejected by North Korean officials,[44] who responded with allegations of South Korean CBW activities and stockpiling.[45]

It was claimed that *Pakistan* is attempting to acquire chemical and biological weapons,[46] but it categorically denied the allegation.[47]

In January the chief of the *Russian* delegation to the Conference on Disarmament (CD) pointed out that all chemical weapons produced in the former Soviet Union are now within the boundaries and under the control of the Russian Federation.[48] However, there may still be some stockpiles of irritants (riot control agents) outside Russia, and certainly the choking gas chloropicrin could still be deployed by chemical defence units of the Russian forces in areas of conflict.[49] In May in Tashkent at the summit meeting of the heads of the CIS states, an agreement on chemical weapons was signed by Armenia, Azerbaijan, Kazakhstan, Kyrgyzstan, Moldova, the Russian Federation, Tajikistan, Turkmenistan and Uzbekistan.[50] The agreement reaffirms that all CW storage and production facilities are on the territory of the Russian Federation. In early July the Russian Parliament adopted a resolution 'On Russia's international obligations on chemical and biological weapons'. Under the resolution Russia assumes responsibility as the legal successor to the Soviet Union

[39] 'UN envoy denies "rumors" on CW production', in FBIS-NES-92-151, 5 Aug. 1992, p. 36.

[40] See Timmerman (note 36).

[41] Sciolino, E. and Schmitt, E., 'U.S. says Tripoli is augmenting and hiding poison weapons', *International Herald Tribune*, 23 Jan. 1992, p. 1.

[42] Drozdiak, W., 'Libya launches bid to boost Western ties', *Washington Post*, 26 Jan. 1992, p. A21; 'Khadhafi: Libyen produziert keine Chemie-Waffen', *Süddeutsche Zeitung*, 6 Feb. 1992, p. 7.

[43] Starr, B., 'DIA warning over North Korean CW', *Jane's Defence Weekly*, vol. 17, no. 2 (11 Jan. 1992), p. 47; 'N. K. building up biochemical arms: NSP', *Korea Newsreview*, 31 Oct. 1992, p. 7.

[44] 'Foreign Ministry rejects chemical weapons charge', in Foreign Broadcast Information Service, *Daily Report–East Asia* (*FBIS-EAS*), FBIS-EAS-92-212, 2 Nov. 1992, p. 12; 'Ministry denies stockpiling chemical weapons', in FBIS-EAS-92-211, 30 Oct. 1992, p. 16.

[45] 'ROK charge of chemical weapons use condemned', in FBIS-EAS-9-221, 16 Nov. 1992, pp. 14–16.

[46] 'U.S. says Pakistan stockpiling chemical weapons', in FBIS-NES-92-054, 19 Mar. 1992, p. 35.

[47] 'Spokesman denies Delhi report on chemical weapons', in FBIS-NES-92-054, 19 Mar. 1992, p. 41.

[48] 'C-Waffen-Einigung in Sicht', *Frankfurter Rundschau*, 10 Jan. 1992, p. 2; *Pacific Research*, vol. 5, no. 1 (Feb. 1992), p. 25.

[49] 'Chemical agents confined to Russian territory', in FBIS-SOV-92-044, 5 Mar. 1992, p. 6.

[50] For the text of the agreement, see *Military News Bulletin*, vol. 1, no. 5 (1992), pp. 2–3; see also entry for 15 May in the chronology in this volume.

with respect to the BWC, the June 1990 bilateral agreement between the former USSR and the USA,[51] and the former Soviet commitment to adhere to the CWC.[52] In August the Ukrainian Foreign Ministry stated that there were no CW stockpiles on Ukrainian territory.[53]

Despite the May 1992 agreement in Tashkent, allegations continued that there are still chemical weapons outside Russia in other CIS states. In May Armenia requested the removal of chemical weapons from a Nagorno-Karabakh CIS troop depot in Azerbaijan.[54] However, earlier in February an official spokesman had stated that the former Soviet troops deployed in Nagorno-Karabakh do not possess a single chemical weapon.[55]

In September a report was published in *Moscow News* by two Russian chemists about the development of a new toxic agent at GSNIIOCT (the State Union Scientific Research Institute for Organic Chemistry and Technology), a chemical technology research institute in Moscow.[56] The *Baltimore Sun* published an expanded version of the article, based on an interview with one of the scientists. According to the two scientists[57] the new agent, Novichok-8 (Russian for 'newcomer'), may considerably surpass the well-known gas VX in toxicity (it may be five to eight times more toxic[58]) and could serve as the basis for a binary weapon—in contrast to the US approach to binary chemical weapons, one component is already a toxic compound. The two components of the binary system are not on the CWC's schedules of controlled chemicals (see chapter 14). The first industrial batch of the agent (5–10 tonnes)[59] was manufactured at the Khimprom plant in Volgograd, and field tests were completed in the first quarter of 1992 at a chemical test site on the Ustyurt plateau near Nukus in Uzbekistan. Both scientists were officially accused of revealing state secrets, and one was arrested and charged with unauthorized disclosure of state secrets. Neither had released the chemical formula of the new agent. The arrested scientist was released after 10 days, but criminal charges against him have not been dropped.[60] The other scientist, who was not arrested, pointed out that they wanted to draw attention to the fact that 'only the production has been stopped, not the research'.[61] The international scientific com-

[51] For the text of the agreement, see SIPRI, *SIPRI Yearbook 1991: World Armaments and Disarmament* (Oxford University Press: Oxford, 1991), pp. 536–39.

[52] 'Resolution adopted on chemical, biological arms', in FBIS-SOV-92-132, 9 July 1992, pp. 55–56.

[53] 'Experts help draft convention', in FBIS-SOV-92-161, 19 Aug. 1992, p. 2.

[54] 'Armenia requests removal of chemical weapons', in FBIS-SOV-992-089, 7 May 1992, p. 4.

[55] 'CIS troops deny chemical weapons possession', in FBIS-SOV-92-040, 28 Feb. 1992, p. 64; 'Spokesman says no chemical weapons in Karabakh', in FBIS-SOV-92-089, 7 May 1992, p. 6.

[56] Mirzayanov, V. and Fyodorov, L., 'A poisoned policy', *Moscow News*, no. 39 (27 Sep.–4 Oct. 1992), p. 9.

[57] 'Russian chemist faces 15 years', *New Scientist*, vol. 136, no. 1847 (14 Nov. 1992), p. 10.

[58] 'Mirzayanov, Federov detail Russian CW production', in FBIS-SOV-92-213, 3 Nov. 1992, pp. 2–7.

[59] See note 58.

[60] 'Officials on disclosure of chemical arms revelations', in FBIS-SOV-92-219, 12 Nov. 1992, pp. 52–54.

[61] Hiatt, F., 'Russia arrests a dissident scientist', *International Herald Tribune*, 27 Sep. 1992, pp. 1–2.

munity expressed great concern about the possibility that one of the scientists might face charges that carry a penalty of up to 15 years in prison.[62]

In 1987 the former Soviet Union made an official disclosure of its CW production. At the same time the alleged development of a new nerve gas was officially denied.[63] On the other hand, it must be noted that, as the head of the newly established Russian Federation Defence Ministry's International Treaty Directorate pointed out, neither Russia nor any other state has pledged to end CW development, and the 1987 decision is only related to CW *production*.[64] The same argument was used by Anatoly Kuntsevich, head of the committee dealing with CW destruction problems, who declared that since 1987 there has been no new CW production in the former USSR or Russia. Until now there has been no international treaty banning offensive CW programmes, and the new CWC will also allow 'science in the sphere of psychologically active, highly toxic chemical compounds'.[65] The debate about using Novichok as a binary weapon continued.[66]

## Non-lethal warfare

Non-lethal warfare is designed to avoid casaulties and long-term damage and to immobilize people rapidly for a short time (see also section VII of chapter 8). From September 1991 to the spring of 1992, allegations continued that in the fighting between Croatian and Serbian forces 'cobwebs' were dropped throughout the countryside by aircraft. The chemical and morphological tests conducted have shown that the fibres employed were a combination of synthetic material with an additional, finer proteinaceous fibre, possibly of natural origin. The fibres are not toxic, infectious or conductive, and are not traditional CW or BW agents.[67] However, the cobwebs were reported to have had a major psychological impact on the population. One explanation was that they may have been used to protect aircraft against anti-aircraft defence.

An August 1992 publication discussed the possibility that non-lethal weapons might be used in Serbia if the UN were to decide to fight there.[68] The possible options include the use of 'carbon-fibre filled warheads' to induce a total breakdown of electricity supply and air defence.

[62] 'Scientists defend Russian whistleblower', *Science*, vol. 258, no. 5085 (13 Nov. 1992), p. 1086; MacKenzie, D., 'Russian chemist faces 15 years', *New Scientist*, vol. 136, no. 1847 (14 Nov. 1992), p. 10.

[63] 'Official denies report on chemical weapons', in FBIS-SOV-92-185, 23 Sep. 1992, p. 2.

[64] 'Chief of international treaty directorate views arms control', in FBIS-SOV-92-218, 10 Nov. 1992, pp. 2–4.

[65] 'CBW aide quizzed on program; secrecy rules questioned', in FBIS-SOV-92-224, 19 Nov. 1992, pp. 2–4.

[66] 'Development of "binary bomb" described', in FBIS-SOV-92-242, 16 Dec. 1992, pp. 23–26.

[67] Garrett, B. C., 'The curious case of the Croatian cobwebs', *ASA Newsletter*, no. 31 (12 Aug. 1992), p. 6; Fuchs, R., Sostaric, B., Plavsic, F., Prodan, I. and Binenfeld, Z., 'Chemical warfare without chemical agents', *Proceedings of the Fourth International Symposium on Protection Against Chemical Warfare Agents*, FOA report A 40067-4.6, 4.7 (National Defence Research Establishment: Umeå, Sweden, June 1992), p. 285.

[68] Fulghum, D. A., 'U.S. weighs use of nonlethal weapons in Serbia if U.N. decides to fight', *Aviation Week & Space Technology*, 17 Aug. 1992, pp. 62–63.

The US Department of Defence (DOD) is co-ordinating a new national security strategy endorsing the use of non-lethal technologies as an alternative to conventional and nuclear weapons and the creation of new options to strengthen the US position in the post-cold war world.[69] One major objective of the use of such new technologies (e.g., blinding lasers, infrasound, non-electromagnetic pulse and neural inhibitors) is to minimize collateral damage and civilian casualties. Also under investigation are techniques which apply chemical compounds to clog machinery and which could be sprayed on to runways to crystallize and destroy aircraft tyres, and microbes that can turn large storage tanks of jet fuel into useless jelly.[70] The US Army's Armament Research, Development and Engineering Center (ARDEC), which among other tasks conducts research on non-conventional, non-lethal munitions, is working on more than a dozen such technologies.[71]

## IV. CBW proliferation and measures to halt it

Public concern about CW and to some extent BW proliferation grew in 1992. In January the Director of the CIA testified to the US Senate on proliferation and stated: 'Today, over 20 countries have, are suspected of having, or are developing nuclear, biological, or chemical weapons and the means to deliver them'.[72] In a White Paper from the British Ministry of Defence, 10 countries were said to have BW programmes and twice that number were alleged to have CW programmes.[73] The number of countries alleged to possess chemical weapons or chemical warfare programmes has remained essentially the same over the past few years;[74] no new evidence became available in 1992.

In the absence of the CWC, counter-proliferation measures such as individual national export control measures, subregional export controls such as those by European Community (EC) countries and co-ordinated export control activities (by the Australia Group) were necessary to contain the spread of chemical and biological weapons. They represented one way of coping with the threat of proliferation of weapons, material and relevant technology.

In the year prior to the conclusion of the CWC there was clear understanding among all of the concerned countries that the export control measures that had already been implemented needed to be tightened. During the December 1991 meeting of the Australia Group two new members, Finland and Sweden,

[69] Opall, B., 'Pentagon forges strategy on non-lethal warfare', *Defense News*, vol. 7, no. 7 (17 Feb. 1992), pp. 1, 50; Opall, B., 'Pentagon units jostle over non-lethal initiative', *Defense News*, vol. 7, no. 9 (2 Mar. 1992), p. 6; Munro, N. and Opall, B., 'Military studies unusual arsenal', *Defense News*, vol. 7, no. 42 (19 Oct. 1992), pp. 3, 44.

[70] See Fulghum (note 68).

[71] Starr, B., 'USA tries to make war less lethal', *Jane's Defence Weekly*, vol. 18, no. 18 (31 Oct. 1992), p. 10.

[72] '15 January', *Chemical Weapons Convention Bulletin*, no. 15 (Mar. 1992), p. 13; AP, Reuters, 'Iraq will quickly rebuild arms program, CIA chief asserts', *International Herald Tribune*, 16 Jan. 1992, p. 3.

[73] Secretary of State for Defence, *Statement on the Defence Estimates 1992* (Her Majesty's Stationery Office: London, July 1992), p. 7.

[74] See SIPRI, *SIPRI Yearbook 1992: World Armaments and Disarmament* (Oxford University Press: Oxford, 1992), pp. 160–61.

were added. At its next meeting in Paris on 2–5 June the Australia Group decided to add four chemicals (sulphur monochloride, sulphur dichloride, triethanolamine hydrochloride and 2-N,N-diisopropylaminoethyl chloride hydrochloride) to its list of 50 chemicals already subject to export control.[75] Additionally a list of 65 biological agents subject to control and a list of 'dual-use' equipment was introduced. However, according to information released from the meeting, not all of the participants were able to provide assurance that their governments would accept an agreement controlling the export of BW equipment. Other issues on the agenda of the June meeting were applications for membership (by Argentina, Czechoslovakia, Hungary and Poland) and the future of the Australia Group *per se*. At the 7–10 December meeting in Paris, the members of the Australia Group agreed to control the export of organisms and the toxins they produce. They also agreed to control equipment usable for BW production.[76] The meeting welcomed the conclusion of the CWC, and the group members reiterated their intention to be included among original signatories. Argentina[77] and Hungary were invited to participate in the next meeting in June 1993 as members. It is perhaps worth mentioning that Hungary held a December seminar in Budapest on CW and BW proliferation for East European countries which are constructing their own export control systems. Turkey also appears likely to join the Australia Group.[78]

During the final stage of the CWC negotiations the Australia Group made a formal statement about the future aim of its activities, noting that its members 'undertake to review, in the light of the implementation of the convention, the measures that they take to prevent the spread of chemical substances and equipment for purposes contrary to the objectives of the convention, with the aim of removing such measures for the benefit of State Parties to the convention acting in full compliance with their obligations under the convention'.[79]

During the first UN Security Council 'summit meeting' at the end of January, 15 heads of states and governments agreed on a communiqué which 'underlines the need for all member states . . . to prevent the proliferation in all its aspects of all weapons of mass destruction. The proliferation of all weapons of mass destruction constitutes a threat to international peace and security'.[80] In Washington officials from the five permanent members of the Security Council met in May to discuss, for the third time after the Persian Gulf War, the control of arms trade especially with the Middle East. They adopted guide-

[75] Odessey, B., 'Chemical, biological weapons export controls agreed', *Wireless File*, no. 113 (United States Information Service, US Embassy: Stockholm, 11 June, 1992), p. 5.

[76] Odessey, B., 'Agreement reached on biological weapon export controls', *Wireless File*, no. 243 (United States Information Service, US Embassy: Stockholm, 16 Dec. 1992), pp. 14–15.

[77] 'Country joins Australian chemical control group', in Foreign Broadcast Information Service, *Daily Report–Latin America* (*FBIS-LAT*), FBIS-LAT-92-240, 14 Dec. 1992, p. 30.

[78] '19 October', *Chemical Weapons Convention Bulletin*, no. 18 (Dec. 1992), p. 18.

[79] Australia, 'Statement made on behalf of the Australia Group', Conference on Disarmament document CD/1164, 7 Aug. 1992.

[80] 'Note by the President of the Security Council', United Nations Security Council document S/23500, 31 Jan. 1992.

lines for control of weapons of mass destruction which also specifically focus on chemical and biological weapons and related technology.[81]

In June, pressed by its Western allies and the US Congress, the Bush Administration changed its position concerning application of the rules of the Coordinating Committee on Multilateral Export Controls (COCOM) to the former Soviet Union. It was agreed that the newly independent republics would be urged to join in the global effort to control the spread of missile technology and NBC weapons.[82]

The EC nations, concerned about the implementation in 1993 of the internal market, intensified their efforts to achieve a co-ordinated policy. A special EC commission worked to harmonize the export control regulations of individual countries which are designed to control chemical substances and sensitive technology with the aim of arriving at a single list of dual-use technologies.[83]

*Germany* took measures to strengthen its export legislation. In January changes were made in the list of countries (country list H) to which German export control measures are applied. Previously the list covered 54 countries, and industry was greatly concerned about the long time which tended to elapse from filing an application to approval. The list now covers only 34 countries.[84] The Federal Assembly (Bundestag) also approved legislation to allow investigators to tap telephones and intercept the mail of individuals suspected of violating export laws.[85] In April the new Federal Export Office (Bundesausfuhramt) was established in Eschborn; it is slated to employ 400 people in 1992. The Federal Export Office is responsible for the control, clarification and approval of all requests for export according to new legislation for foreign trade.[86] Germany's Customs Criminology Institute (ZKI) operates an early-warning data base system called KOBRA which centralizes all documents filed with customs concerning certain categories of technology where there could be suspicion of weapon proliferation. By the end of 1993 a new export list is to be prepared which will be compatible and co-ordinated with new European, Japanese and US lists.[87]

Owing to the involvement of German companies in the buildup of the Iraqi CBW programme and the results of the UNSCOM findings, trials were conducted in Germany to investigate violations of German foreign trade law. In April trials began in Darmstadt against the Karl Kolb Pilot Plant and the WET

[81] Smith, R. J., '5 nations reach arms export accord', *Washington Post*, 30 May 1992, p. A15. For the text of the document, see appendix 10C in this volume.

[82] Auerbach, S., 'Cocom eases rules on equipment sales', *Washington Post*, 3 June 1992, p. A5.

[83] 'Die EG will Schlupflöcher für den Export sensibler Güter stopfen', *Frankfurter Allgemeine Zeitung*, 2 Sep. 1992, p. 1; Bellamy, C., 'EC nations vote for controls on weapon exports', *The Independent*, 19 Sep. 1992, p. 13.

[84] 'Umstrittene Exporte erleichtert', *Frankfurter Rundschau*, 23 Jan. 1992, p. 4.

[85] Vogel, S., 'Bonn to allow wiretaps on arms-related exports', *Washington Post*, 24 Jan. 1992, p. A18; Deupmann, U., 'Ein gutes Gesetz, das Hilfe braucht', *Süddeutsche Zeitung*, 24 Jan. 1992, p. 4.

[86] 'Neues Bundesausfuhramt kontrolliert Exporte', *Der Tagesspiegel*, 2 Apr. 1992, p. 57; 'Bonn erteilt Exporteuren neue Auflagen', *Frankfurter Allgemeine Zeitung*, 22 Apr. 1992, p. 15.

[87] 'Europäer erarbeiten eine gemeinsame Ausfuhrliste', *Frankfurter Allgemeine Zeitung*, 13 Oct. 1992, p. 15.

firms.[88] The court requested the release of UNSCOM documents which might provide additional information about the involvement of German firms, but the German Government denied the request, referring to the political nature of the information involved.[89] Some judges of the Darmstadt court criticized the Ministry of Justice for its lack of co-operation.

In August an appeal was heard against the verdict of the October 1991 trial against three Imhausen company managers who were convicted of involvement in the buildup of the Rabta CW facility in Libya.[90] One of the three managers received a stiffer sentence.[91] In another trial the former head of the Imhausen company confessed that government R&D money had been misused to pay employees.[92]

German companies are estimated to have supplied Iraq with $198 million of so-called dual-use items during 1986–90. Officials from one German company were alleged to have designed four plants in Iraq for CW production, and three other companies made equipment to fill munitions. Six German companies supplied equipment for making botulin toxin and mycotoxins, including laboratory devices and protective equipment. This information is based on German, UN and US sources.[93] As of July only one company had been convicted of exporting to Iraq, and 37 others were under investigation for various violations, not all of which were related to the CBW support of the Iraqi programme.

*Japan* tightened its export controls on 59 chemicals by requiring prior government approval before export.[94] Among them are also chemicals harmful to the environment.

The *Russian* Government established a body, including the heads of its foreign policy, industry, economics, finance and security departments, to control arms exports.[95] In November Russia established rules for control of the export of biological agents that can be used for developing bacteriological (biological) and toxin weapons, thereby making it impossible to export or re-export to states in violation of the 1925 Geneva Protocol or the BWC.[96] Licensing under this legislation[97] is mandatory and a licence can be issued only by the Russian

[88] Müller-Gerbes, H., 'Wie haben deutsche Firmen beim Aufbau des irakischen Giftgas-Arsenals geholfen?', *Frankfurter Allgemeine Zeitung*, 23 Apr. 1992, p. 4; 'Bei Lieferungen nicht an Bomben gedacht', *Frankfurter Allgemeine Zeitung*, 29 Apr. 1992, p. 7.

[89] Müller-Gerbes, H., 'Richter als "Zinnsoldaten der Macht"?', *Frankfurter Allgemeine Zeitung*, 16 June 1992, p. 4; 'Bundesregierung weist Kritik zurück', *Frankfurter Allgemeine Zeitung*, 30 June 1992, p. 4.

[90] 'Imhausen-Prozeß: Zum Teil neu verhandeln', *Frankfurter Allgemeine Zeitung*, 21 Aug. 1992, p. 13.

[91] 'Urteil im Imhausen-Prozeß', *Frankfurter Allgemeine Zeitung*, 7 Oct. 1992, p. 4.

[92] 'Imhausen legt Geständnis ab', *Süddeutsche Zeitung*, 19 Nov. 1992, p. 7; 'Erneut Haft für Hippenstiel-Imhausen', *Süddeutsche Zeitung*, 10 Dec. 1992, p. 7.

[93] Smith, R. J. and Fisher, M., 'Lax Bonn oiled Iraq war machine', *International Herald Tribune*, 24 July 1992, p. 1, 2.

[94] 'Tokyo to tighten controls on chemical exports', in FBIS-EAS-92-117, 17 June 1992, p. 7.

[95] 'Group to control arms export', in FBIS-SOV-92-023, 4 Feb. 1992, p. 36.

[96] 'Statute on control of CBW raw materials', in FBIS-SOV-92-237, 9 Dec. 1992, pp. 8–9.

[97] The law is the Statute on the Procedure for Controlling the Export from the Russian Federation of Pathogens, Their Genetic Variations, and Fragments of Genetic Material Which Could be Used in the Creation of Bacteriological (Biological) and Toxin Weapons. It was approved by the Russian Government as Decree no. 892, dated 20 Nov. 1992; see 'Statute on control of CBW raw materials' (note 96).

Ministry of Foreign Economic Relations.[98] The new legislation lists specific pathogens, viruses, toxins, genetic variations and fragments of genetic material to which the licensing procedure is to be applied.[99]

In the *USA* in testimony before the Joint Economic Committee, Subcommittee on Technology and National Security, Richard Clarke outlined in March the progress in US non-proliferation measures in 1991 and pointed out that more has to be done with respect to chemical and biological weapons.[100] He pointed especially to several regions, including North Korea, Iran and South-East Asia. For South-East Asia, the USA has proposed that China, India, Pakistan, Russia and the United States hold a conference to address regional proliferation problems, but India has not agreed to participate. One of the main achievements of the 1991 Enhanced Proliferation Controls Initiative (EPCI)[101] is its control of CW and BW material. Similar control measures have been adopted or are in the process of being adopted by at least 26 countries.

In July President George Bush outlined a new non-proliferation initiative designed to address the spread of the capability to produce or acquire weapons of mass destruction and the means to deliver them, which were seen to constitute a growing threat to US national security.[102] It suggested four guiding principles for multilateral and regional action, including a demand for harmonization of export controls.[103] In October the US Senate strengthened the 1993 Defence Authorization Bill by adding $56 million for research on non-proliferation and $20 million for international non-proliferation activities.[104] Ultimately, Congress authorized $168 million for fiscal year (FY) 1993 to combat the proliferation of NBC weapons.[105]

[98] 'Government adopts rules for biological weapons export', in FBIS-SOV-92-228, 25 Nov. 1992, p. 2.

[99] The list contains many of the biological agents on the Australia Group's June 1992 list of biological agents; see 'Yeltsin's document on pathogen export control', in FBIS-SOV-92-238, 10 Dec. 1992, pp. 8–10.

[100] Statement of Richard A. Clarke, Assistant Secretary for Politico-Military Affairs, Department of State, before the Joint Economic Committee, Subcommittee on Technology and National Security, 13 Mar. 1992; Mandine, R., 'Iraq's nuclear program said to be put on halt', *Wireless File,* no. 50 (United States Information Service, US Embassy: Stockholm, 13 Mar. 1992), pp. 14–15.

[101] See *SIPRI Yearbook 1992* (note 74) , p. 163.

[102] US Department of State, 'Non-proliferation efforts bolstered', *US Department of State Dispatch,* vol. 3, no. 29 (20 July 1992), pp. 569–71.

[103] The guiding principles are: (*a*) the USA will build on existing global norms against proliferation and where possible strengthen and broaden them; (*b*) the USA will focus special efforts on those areas where the dangers of proliferation remain acute, notably the Middle East, the Persian Gulf, South Asia, and the Korean Peninsula; (*c*) US non-proliferation policy will seek the broadest possible multilateral support while continuing to show leadership on critical issues; and (*d*) the USA will address the proliferation issue through the entire range of political, diplomatic, economic, intelligence, regional security, export controls and other tools available. See US Department of State (note 102).

[104] Towell, P., 'Two major obstacles dissolve as time for talk winds down', *Congressional Quarterly,* vol. 50, no. 38 (26 Sep. 1992), p. 2960.

[105] Towell, P., 'Spending bill trims some now, sets bigger cuts in motion', *Congressional Quarterly,* vol. 50, no. 40 (10 Oct. 1992), pp. 3184–89.

# V. Destruction of chemical weapons

## US–Russian bilateral agreements

During a June summit meeting President George Bush and President Boris Yeltsin stressed their commitment to the global elimination of chemical weapons, as expressed in the Joint Statement on Chemical Weapons.[106] They agreed to instruct their negotiators in Geneva to act so that the CWC could be concluded by the end of August 1992 and pledged to support the 1989 Wyoming Joint Memorandum[107] on confidence-building measures (CBMs) in the area of CW destruction. New provisions for data exchange and inspection under the Joint Memorandum will be implemented as soon as agreed upon. Bush and Yeltsin agreed to update the 1990 bilateral agreement on the destruction of chemical weapons,[108] and to bring it into force.

In the Agreement on the Safe and Secure Transportation, Storage and Destruction of Weapons and the Prevention of Weapons Proliferation,[109] both parties pledged their co-operation to assist Russia to achieve: (a) the destruction of nuclear, chemical and other weapons, (*b*) the safe and secure transportation and storage of such weapons, and (*c*) the establishment of additional verifiable measures against the proliferation of such weapons. Among other things the agreement provides the legal framework for US financial support of Russian CW destruction. It entered into force upon signature in June 1992 and will remain in force for seven years.

In July Russia and the USA began bilateral talks in Geneva about implementing the June 1990 agreement.[110] Under the 1989 Wyoming Joint Memorandum the second phase, data exchange, will start not later than four months prior to the initialling of the text of the CWC.[111]

## The US CW destruction programme

In late November 1991 the US Congress extended the deadline for the destruction of CW stockpiles to July 1999,[112] after earlier having extended the completion date for destruction to 30 April 1997. These deadline changes were further complicated by a six-month shutdown of the Johnston Atoll Chemical Agent Disposal System (JACADS). In April 1992 the destruction deadline

[106] 'Letter dated 3 Aug. 1992 from the Representative of the United States of America addressed to the President of the Conference on Disarmament transmitting documents relating to arms control and disarmament issues agreed on during the summit meeting held by Presidents Bush and Yeltsin in Washington, DC in June 1992', Conference on Disarmament document CD/1162, 12 Aug. 1992.

[107] SIPRI, *SIPRI Yearbook 1990: World Armaments and Disarmament* (Oxford University Press: Oxford, 1990), pp. 531–32.

[108] See note 51.

[109] See note 106.

[110] See note 51.

[111] '20 July', Institute for Defense and Disarmament Studies, *Arms Control Reporter* (IDDS: Brookline, Mass.), sheet 704.B.533, Sep. 1992.

[112] US General Accounting Office, *Chemical Weapons: Stockpile Destruction Cost Growth and Schedule Slippages are Likely to Continue*, Report of the Chairman, Committee on Governmental Affairs, US Senate, GAO/NSIAD-92-18 (General Accounting Office: Washington, DC, Nov. 1991).

was extended to the year 2000, with the cost estimated at nearly $8 billion.[113] Under the leadership and sponsorship of the National Academy of Sciences a study is being conducted to investigate chemical demilitarization (chemdemil) technology alternatives to incineration. These technologies include among others: hydrolysis, aminolysis, thermohydrolysis, bioremediation, supercritical water oxidation, pyrolysis, fluidized-bed combustion, plasma arc and electrochemical techniques.[114] A final report was due by the end of 1992, but in June the Office of Technology Assessment (OTA) was able to present a report on alternatives to on-site incineration for the destruction of CW. The report was prepared partially in response to protests by local community groups and other organizations opposed to the Army's current incineration programme which have suggested that other technologies might be safer.[115] The following destruction techniques were mentioned: chemical neutralization; supercritical water oxidation; steam gasification technology; and plasma arc technology.

In October after intense debate on the FY 1993 defence authorization bill, the US Congress ordered re-examination of alternative destruction technologies and extended the completion date for destruction of all chemical weapons to 31 December 2004.[116] The Army was also requested to establish a Chemical Demilitarization Citizens' Advisory Commission in each state where 5 per cent or less of the US CW stockpile is located. The Secretary of the Army is required to submit to Congress by 31 December 1993 a report assessing possible alternative destruction technologies and to respond to the report by the National Academy of Sciences mentioned above. Additionally, the Secretary of the Army is to submit to Congress not later than 1 May 1993 a report on the condition and integrity of the US CW stockpiles. Table 7.1 shows the current schedule for the US destruction programme.

The US Army established a new agency, the Chemical Material Destruction Agency (USACMDA),[117] based on the former Office of the Program Manager for Chemical Demilitarization, with an expanded mission including the disposal of non-stockpile items such as wastes from earlier disposal efforts, unserviceable munitions, chemical production facilities, sites known to contain significant concentrations of buried chemical weapons and wastes, binary weapons and components.

JACADS continued its operational verification testing (OVT). Phase I of the OVT was completed in February 1991; phase II focused on the disposal of VX

[113] Statement by Susan Livingstone, Assistant Secretary of the Army (Installations, Logistics and Environment), before the Subcommittee on Defense, Committee on Appropriations, US Senate, 102nd Congress, 2nd Session, Chemical Disposal Program, 12 May 1992.

[114] Ember, L., 'Incineration of chemical arms to be studied', *Chemical & Engineering News*, vol. 70, no. 15 (13 Apr. 1992), pp. 29–30.

[115] US Congress, Office of Technology Assessment, *Disposal of Chemical Weapons: Alternative Technologies–Background Paper*, OTA-BP-O-95 (US Government Printing Office: Washington, DC, June 1992).

[116] See Ember (note 114); *National Defense Authorization Act for Fiscal Year 1993: Conference Report to Accompany HR 5006*, 1 Oct. 1992, US House of Representatives, 102nd Congress, 2nd Session, report 102-966 (US Government Printing Office: Washington, DC, 1992), pp. 566–67.

[117] 'PM to convert to agency', *Chemical Demilitarization Update*, vol. 1, no. 6 (May 1992), p. 2; 'US destruction developments', *Pacific Research*, vol. 5, no. 3 (Aug. 1992), pp. 24–25.

**Table 7.1.** US Army plans for destruction of the US chemical stockpile by 2004[a]

| Location | Per cent of total US stockpile | Start of facility construction[b] | Start of system testing[b] | Start of operations[b] | End of operation[b] | New deadline for ceasing operation[c] |
|---|---|---|---|---|---|---|
| Johnston Atoll Chemical Agent Disposal Facility, Pacific Ocean | *6.6* | Nov. 1985 | Aug. 1988 | July 1990 | Oct. 1995 | Dec. 2004 |
| Tooele Army Depot, Tooele, Utah | *42.3* | Sep. 1989 | Aug. 1993 | Feb. 1995 | Apr. 2000 | Dec. 2004 |
| Anniston Army Depot, Anniston, Alabama | *7.1* | June 1993 | Apr. 1996 | Oct. 1997 | Nov. 2000 | Dec. 2004 |
| Umatilla Army Depot Activity, Hermiston, Oregon | *11.6* | Jan. 1994 | Nov. 1996 | May 1998 | Dec. 2000 | Dec. 2004 |
| Pine Bluff Arsenal, Pine Bluff, Arkansas | *12.0* | Jan. 1994 | Sep. 1996 | Mar. 1998 | Nov. 2000 | Dec. 2004 |
| Lexington-Blue Grass Depot Activity, Richmond, Kentucky | *1.6* | May 1994 | Mar. 1997 | Sep. 1998 | Feb. 2000 | Dec. 2004 |
| Pueblo Army Depot Activity, Pueblo, Colorado | *9.9* | May 1994 | Mar. 1997 | Sep. 1998 | May 2000 | Dec. 2004 |
| Newport Army Ammunition Plant, Newport, Indiana | *3.9* | Jan. 1995 | June 1997 | June 1998 | Apr. 1999 | Dec. 2004 |
| Aberdeen Proving Ground, Edgewood, Maryland | *5.0* | Jan. 1995 | June 1997 | June 1998 | June 1999 | Dec. 2004 |

[a] This schedule does not take into account delays from major system failures or litigation and is dependent on funding support.

[b] Planned until mid-1992.

[c] The new deadline is the result of the budgetary constraints of the FY1993 Defense Authorization bill of Oct. 1993.

*Sources*: US Army quoted in *Chemical & Engineering News*, 29 June 1992, p. 20; *National Defense Authorization Act for Fiscal Year 1993*, Conference report to accompany HR 5006, House of Representatives, 102nd Congress, report 102-966 (US Government Printing Office: Washington, DC, 1992), pp. 566–67.

stored in M55 rockets and was concluded in March 1992 with the destruction of 13 876 rockets. On 21 January an explosion occurred in one of the facility's four rotary kiln furnaces and operations were halted.[118] Phase III of the OVT, the disposal of ton-containers filled with mustard gas, was completed in October. During phase III, 67 containers were decontaminated and more than 51 000 kg were destroyed in a two-stage liquid incinerator.[119] The last phase of the OVT, the disposal of mustard gas-filled projectiles, is planned to be concluded in early 1993, more than six months behind schedule.

Construction of the Tooele Chemical Disposal Facility (TOCDF) at Tooele Army Depot in Utah was approximately 61 per cent complete in October.[120] The TOCDF is scheduled to be completed in the summer of 1993 and test runs may start in August 1993. At the TOCDF, disposal of 42.3 per cent of the US stockpile will start in early 1995.

The third destruction facility at Anniston Army Depot, Alabama (where 7.1 per cent of the US CW stockpiles are located) will start destruction according to the previous plan in October 1997. The construction contract is expected to be awarded in the autumn of 1993.[121] However, a new law blocks at least temporarily the Army's use of incineration for its destruction programme, and thus $105 million were removed from the FY 1993 military construction appropriations bill, preventing construction at Anniston. This October 1992 decision is closely related to a review of alternative technologies which the Army is required to conduct and report on by December 1993.[122] At three other sites in Indiana (Newport Army Ammunition Plant), Kentucky (Lexington-Blue Grass Army Depot) and Maryland (Aberdeen Proving Ground), all with less than 5 per cent of the total stockpile, construction was scheduled to start in 1994 and 1995. However, owing to strong opposition to construction of an incineration plant this schedule is unlikely to be kept.

Since 1990 the Army has trained more than 2000 individuals at its Chemical Demilitarization Training Facility at the Aberdeen Proving Ground, another important step in facilitating the US destruction programme.[123]

Some factors have changed since the 1988 decision to use high-temperature incineration at each storage facility, and subsequent to a 1990 congressional decision, the Army is now also conducting a programme to develop and adopt the cryofracture technique. A two-phase cryofracture testing programme began in January 1990. Phase I, non-agent tests, was carried out at the General Atomics facility in San Diego. Phase II, agent-related tests, is being conducted

[118] 'Johnston Atoll blast', *Pacific Research*, vol. 5, no. 1 (Feb. 1992), p. 26; 'Explosion halts chemical arms destruction', *Chemical & Engineering News*, vol. 70, no. 6 (10 Feb. 1992), p. 22.

[119] 'OVT nears completion at JL facility', *Chemical Demilitarization Update*, vol. 1, no. 8 (Oct. 1992), p. 2.

[120] 'Tooele update', *Chemical Demilitarization Update*, vol. 1, no. 8 (Oct. 1992), p. 5.

[121] 'Anniston update', *Chemical Demilitarization Update*, vol. 1, no. 8 (Oct. 1992), p. 5.

[122] Ember, L., 'Chemical arms destruction: Congress puts incineration on hold', *Chemical & Engineering News*, vol. 70, no. 43 (26 Oct. 1992), p. 4.

[123] 'Training is underway at CDTF', *Chemical Demilitarization Update*, vol. 1, no. 8 (Oct. 1992), p. 4.

at the Chemical Agent Munition Disposal System (CAMDS) at Tooele Army Depot in Utah.[124]

## The Russian CW destruction programme

In early 1992 it became apparent that Russia was unable to begin CW destruction.[125] In January it was announced that a state committee for elimination of chemical weapons would be created, and President Yeltsin reported on preparations for a state programme.[126] The then two-year-old draft for the former Soviet state programme will be re-examined by the Supreme Soviet of the Russian Federation.[127]

In February Russian Foreign Minister Andrey Kozyrev addressed the CD in Geneva and stated that the 40 000 tonnes of toxic agents for which Russia has assumed responsibility are difficult to destroy and that, while Russia has the technology for destruction, assistance from other countries would be helpful and welcomed.[128]

By the end of February President Yeltsin had established a Committee on Convention Problems Relating to Chemical and Biological Weapons, with the Russian specialist Professor Anatoly Kuntsevich as its chairman.[129] Among other duties, the committee will deal with implementation of the CWC and organize elimination of the CW stockpiles. Kuntsevich highlighted three major destruction problems: personnel problems, inadequate funding of R&D related to CW elimination, and difficulties in implementing destruction programmes as a result of reactions from local authorities and the public.[130] Under the May 1992 agreement on chemical weapons of other CIS states will co-operate in the CW destruction which will be carried out by Russia.[131] Financial commitments will be regulated by a separate agreement.

In June President Yeltsin issued a directive on Priority Measures for Implementing Russia's Obligations in Destroying Chemical Weapons Stockpiles. Under this order the Committee on Convention Problems Relating to Chemical and Biological Weapons assumed responsibility for organizing the CW destruction programme.[132] In July the Russian Supreme Soviet adopted a resolution which calls for 'draft comprehensive programs for the phased destruction of chemical weapons' to be submitted to the Supreme Soviet by 15 September.[133] A slightly delayed draft plan, the Complex Program of the Stage-

[124] 'Cryo evaluation underway', *Chemical Demilitarization Update*, vol. 1, no. 7 (July 1992), pp. 4–5.

[125] 'Russia not able to destroy chemical arms', *Chemical & Engineering News*, vol. 70, no. 3 (20 Jan. 1992), p. 17; Thränert, O., *Probleme der Abrüstung Chemischer und Biologischer Waffen in der GUS*, no. 53 (Friedrich-Ebert Stiftung: Bonn, Oct. 1992).

[126] 'Chemical weapons elimination group formed', in FBIS-SOV-92-019, 29 Jan. 1992, p. 6.

[127] 'Problems cited in chemical weapons elimination', in FBIS-SOV-92-023, 4 Feb. 1992, p. 7.

[128] 'Further on proposals', in FBIS-SOV-92-030, 13 Feb. 1992, p. 2.

[129] 'Chemical, biological weapons committee set up', in FBIS-SOV-92-040, 28 Feb. 1992, p. 2.

[130] 'Problems of eliminating chemical arms explained', in FBIS-SOV-92-055, 20 Mar. 1992, p. 4.

[131] See note 50.

[132] 'Yeltsin decree on destruction of chemical weapons', in FBIS-SOV-92-117, 17 June 1992, p. 24.

[133] The resolution is entitled Resolution of the Russian Federation Supreme Soviet on Ensuring the Fulfillment of the Russian Federation's International Commitments in the Sphere of Chemical, Bacterio-

By-Stage Elimination of Chemical Weapons in the Russian Federation, was presented to the Russian Parliament in October.[134] According to the plan destruction will be conducted at Novocheboksark in the Chuvashiya region, Kambarka in the Udmurtia region and Gornyy in the Saratov region. At Volks-17 in the Saratov region a pilot industrial facility will be located for recycling the by-products of detoxification. The sarin, soman and VX ammunition (a total of 9800 tonnes), stored at depots in the cities of Shchuchye in the Kurgan region and Kizner in the Udmurtia region, will be transported to the Khimprom facility at Novocheboksark.[135]

The first phase of the programme, which is slated to start in April 1993, will include the following activities: ecological evaluation, feasibility studies of projects, manufacturing and testing of pilot facilities, testing of technology and training of experts. Destruction *per se* will not start until 30 June 1997.[136] According to press reports the first phase of the programme will cost 45 billion roubles (in 1993 prices), of which 4.4 billion roubles will be spent in 1993, and at least $4.5 million will be needed to purchase equipment (e.g., furnaces for thermal treatment) from other countries.[137] The 1993 budget for the various destruction facilities is the following: Kambarka, 320 million roubles; Gornyy, 207 million roubles; Novocheboksark, 100 million roubles; Volks-17, 29 million, 100 million roubles for railroad modernization; and 80 million roubles for a diagnostic and prevention centre. According to the plan, by 2004 some 43 per cent of the CW stocks will be destroyed.[138] In July the overall cost of the programme was estimated at 100 billion roubles;[139] by the end of 1992 the figure had increased to 400–500 billion roubles (approximately $1–$1.25 billion) for destruction of the entire stockpile.[140]

It is currently impossible to evaluate seriously these expenditure figures owing to: (*a*) the enormous inflation in Russia, (*b*) the totally unreliable foreign exchange rates, and (*c*) the disagreement about what kind of expenditures might be included in the calculations (e.g., whether housing costs, approximately 15 per cent of the amount to be spent on infrastructure development, have been included). In phase I of the plan three facilities must be made operational. They are intended to destroy at least 45 per cent of the agents including lewisite, mustard gas and lewisite-mustard gas mixture stored at Kambarka and Gornyy.[141] At Kambarka there are 7000 tonnes of lewisite in

logical (Biological), and Toxin Weapons, no. 3244-1; see 'Resolution on chemical weapons commitments', in FBIS-SOV-92-144, 27 July 1992, pp. 36–37.

[134] 'Deputies view draft bill on eliminating chemical arms', in FBIS-SOV-92-208, 27 Oct. 1992, p. 46; 'Committee examine program for destroying chemical weapons', in FBIS-SOV-92-212, 2 Nov. 1992, p. 56.

[135] 'Panels discuss plan for destruction of chemical weapons', in FBIS-SOV-92-215, 5 Nov. 1992, pp. 49–50.

[136] 'Plans ready for destruction of chemical weapons', in FBIS-SOV-92-188, 28 Sep. 1992, pp. 2–3.

[137] 'Committee examine program for destroying chemical weapons' (note 134).

[138] See note 135.

[139] 'Scrapping of arms to cost 100 billion rubles', in FBIS-SOV-92-132, 9 July 1992, p. 3; 'Delays in chemical weapon disarmament viewed', in FBIS-SOV-92-084, 30 Apr. 1992, pp. 5–6.

[140] Ember, L., 'Russia seeks U.S. expertise, money to destroy its chemical arms', *Chemical & Engineering News*, vol. 70, no. 47 (23 Nov. 1992), pp. 14–15.

[141] 'Official examines destruction of CBW weapons', in FBIS-SOV-92-186, 24 Sep. 1992, pp. 2–4.

containers which have been stored there for more then 40 years.[142] At Gornyy nearly 1200 tonnes of mustard gas and lewisite are stored in barrels (700 tonnes of mustard gas, 230 tonnes of lewisite, 152 tonnes of mustard-lewisite mixture, and 72 tonnes of mustard-lewisite mixture in dichloroethane).[143]

At the Novocheboksark site, production facility no. 3, shop no. 83, for VX was operational from December 1972 to 1987 and is now mothballed.[144] However, there remain artillery shells and rockets filled with sarin, soman and VX of approximately 9800 tonnes under sealed conditions.[145] According to the destruction plan, the facility will be converted into a disposal facility.[146] Destruction operations might also be carried out at the following former production facilities: Berezniki, Chapayevsk and Dzerzhinsk (Gorkiy region) and at Volgograd.[147]

Additional information about the Russian destruction programme and former Soviet CW production activities was presented in October by the two Russian scientists mentioned above.[148] According to their information, 30 000 tonnes of the officially declared 40 000 tonnes of agent are phosphorous organic agents—sarin, soman and VX—and the remaining 10 000 tonnes are composed of 7000 tonnes of lewisite, 1500 tonnes of a mixture of mustard gas and lewisite, and 1500 tonnes of mustard gas. In addition to the previously mentioned former production sites, herbicides were produced in Ufa, psychotropic substances in Volsk and riot control agents in Slavgorod. Mustard gas was produced in the 1940s in Chapayevsk and Dzerzhinsk, as was lewisite. After World War II a confiscated German plant was brought to Dzerzhinsk for mustard gas and lewisite production, which was carried out until 1952. During the 1940s there was also a small production plant for lewisite and mustard gas in Moscow. Production of soman and sarin took place in Volgograd and, in addition, VX was produced at Novocheboksark. In Chapayevsk during the 1940s overall production was of the range of 10 000–15 000 tonnes of mustard gas.

The destruction technologies which might be used by Russia can be readily summarized. Until recently neutralization was the disposal technique of choice. Now other techniques such as chemical destruction, incineration and plasma technologies are under consideration. Lewisite might be reprocessed to extract arsenic, probably in the form of gallium arsenide.[149] In Kambarka the lewisite will likely first be pumped from huge containers into smaller, one cubic metre containers. However, the detoxification of lewisite with molten

142 'Udmurtia discusses chemical weapons disposal', in FBIS-SOV-92-064, 2 Apr. 1992, p. 3.

143 Ember, L., 'Russia seeks U.S. expertise, money to destroy its chemical arms' (note 140); 'No proven technology for chemical disposal', in FBIS-SOV-92-086, 4 May 1992, p. 5.

144 'Chemical weapons destruction sites discussed', in FBIS-SOV-92-190, 30 Sep. 1992, pp. 2–3.

145 See Ember (note 140).

146 'Chemical weapons destruction sites discussed' (note 144); 'Novocheboksark's plant may switch to destroying chemical arms', in FBIS-SOV-92-228, 25 Nov. 1992, pp. 2–3.

147 '"Mixed feelings" over proposed CW Convention', in FBIS-SOV-92-172, 3 Sep. 1992, pp. 2–3.

148 'Fedorov, Mirzayanov article on chemical war against environment', in FBIS-SOV-92-212, 2 Nov. 1992, pp. 2–4; 'Mirzayanov, Fedorov detail Russian CW production' (note 58).

149 'Udmurtia discusses chemical weapons disposal' (note 142); 'Udmurtia plans to destroy chemical weapons', in FBIS-SOV-92-113, 11 June 1992, p. 6; 'Project to convert war gas into metal noted', in FBIS-SOV-063, 1 Apr. 1992, pp. 4–5.

sulphur into a water-insoluble polymer is still being considered,[150] as is a two-step process involving chlorination followed by electrolysis.[151] For mustard gas and the nerve agents, detoxification into reaction products for future commercial use is being studied. As mentioned in the *SIPRI Yearbook 1992*,[152] it has also been proposed that nuclear explosions be used to destroy the Russian CW stockpile.[153] The Moscow-based Chetek Corporation continues its attempt to market technology and assistance, in Russia and elsewhere, for the destruction of toxic wastes using underground explosions. Chetek has proposed using three underground explosions to destroy the entire Russian CW stockpile. However, as international concern and national uncertainty about the method grew, it became increasingly unlikely that such a technique would be utilized.[154] In November it was made known that Russia has abandoned plans to destroy chemical weapons by nuclear explosion.[155]

In December 1991 the US Congress allocated $400 million to support the nuclear and CW disarmament obligations of the former Soviet Union.[156] Of the total amount, $25 million will be allocated to Russia to begin CW destruction as agreed during a visit of Russian officials to Washington on 30 July.[157] The agreement was concluded between the US DOD and the President's Committee on Conventional Problems of Chemical and Biological Weapons of the Russian Federation, headed by Anatoly Kuntsevich. Under it the DOD will provide: (*a*) development of mobile CW destruction systems; (*b*) participation in the establishment of national laboratory complexes, including providing technical equipment; (*c*) control and monitoring systems at destruction site; (*d*) medical facilities at destruction sites; (*e*) joint testing or experimentation related to destruction; and (*f*) other co-operation related to destruction as may be agreed.

The US funds will be used to begin construction of three destruction facilities in Russia. US companies have been invited to participate in the project, but contributions from Germany, Japan and other donors are also sought. In accordance with the above-mentioned July agreement, a Russian delegation

[150] Leonov, G. S. and Sheluchenko, V. V., 'Principal technological and environmental aspects of the destruction of chemical weapons', *Disarmament*, vol. 15, no. 2 (1992), pp. 94–100.

[151] See Ember (note 140).

[152] SIPRI, *SIPRI Yearbook 1992* (note 74), p. 169.

[153] 'Plans to destroy chemical weapons revealed', in FBIS-SOV-92-044, 5 Mar. 1992, pp. 5–6; 'Russia's nuclear business: is the threat real?', *Moscow News*, no. 19 (10–17 May 1992), p. 8.

[154] Hiatt, F., 'Russian nuclear scientists seek business, food', *Washington Post*, 18 Jan. 1992, pp. A1–A20.

[155] DPA, 'Keine atomare Entsorgung von Rußlands Giftmüll', *Süddeutsche Zeitung*, 21 Nov. 1992, p. 2.

[156] *SIPRI Yearbook 1992* (note 74), p. 170.

[157] 'Agreement between the Department of Defense of the United States of America and the President's Committee on Conventional Problems of Chemical and Biological Weapons of the Russian Federation concerning the safe, secure and ecologically sound destruction of chemical weapons', Conference on Disarmament document CD/1161, 5 Aug. 1992; Leopold, G., 'Russia wants early chemical demolition start', *Defense News*, vol. 7, no. 32 (10 Aug. 1992), p. 6; 'USA helfen Rußland bei C-Waffen-Zerstörung', *Süddeutsche Zeitung*, 1–2 Aug. 1992, p. 2.

including local Russian leaders from Cheboksary and Kambarka paid a visit to the Tooele Army Depot in the autumn of 1992.[158]

In September a Russian delegation headed by Kuntsevich visited Germany to talk to officials at German companies about possible involvement in the Russian CW destruction programme. A contract was signed between the German company EST GmbH (set up by Deutsche Aerospace AG and Lurgi-Umwelt-Beteiligungsgesellschaft mbH) and the Russian company Metalchim to construct a destruction facility in Kambarka.[159] During a November visit to Washington, Kuntsevich stated that the Russian legislative committee had approved a destruction plan on 30 October according to which the former CW production facility at Novocheboksark will be converted to a destruction facility and two other destruction facilities will be built.[160] At that time the US Congress was discussing the defence authorization bill for FY1993 and ultimately $800 million was allocated to help former Soviet republics dismantle their arsenals of nuclear and chemical weapons.[161]

In November it was reported that in addition to the $25 million originally earmarked for CW destruction, the USA will provide $30 million more for the Russian destruction programme.[162] Other countries have also considered providing financial assistance to Russia. For 1993 Germany budgeted 10 million DM to support Russian destruction of weapons of mass destruction, particularly nuclear warheads.[163]

## Canadian CW destruction

During Operation 'Swiftsure', Canada spent $14.28 million on the destruction of residual mustard gas stock by incineration. Some 15 tonnes of mustard gas and one-third tonne of assorted nerve agents from World War II, which had been stored at Suffield, were destroyed during the operation. The mustard gas was frozen inside boxes and then fed into the incinerator—technology comparable to the cryofracture technique. The nerve agents were neutralized using an alcohol solution.[164]

[158] 'Russia joins the U.S. in demilitarization effort', *Chemical Demilitarization Update*, vol. 1, no. 8 (Oct. 1992), pp. 1–3.

[159] 'Lurgi will in Rußland Chemiewaffen entsorgen', *Der Tagesspiegel*, 7 Sep. 1992, p. 14; 'Lurgi baut Fabrik zur Waffenvernichtung', *Frankfurter Allgemeine Zeitung*, 8 Sep. 1992, p. 19; 'Contract to destroy chemical arms', *Financial Times*, 9 Sep. 1992, p. 5.

[160] Leopold, G., 'Russia seeks Western, U.S. aid to destroy chemical weapons', *Defense News*, vol. 9, no. 46 (16 Nov. 1992), p. 38

[161] Towell, P., 'Spending bill trims some now, sets bigger cuts in motion', *Congressional Quarterly*, vol. 50, no. 40 (10 Oct. 1992), pp. 3184–89.

[162] See Leopold (note 160).

[163] 'Zehn Millionen Mark als "Abrüstungshilfe" bewilligt', *Süddeutsche Zeitung*, 3 Nov. 1992, p. 2.

[164] Pugliese, D., 'Canada puts new spin on incineration', *Defense News*, vol. 7, no. 30 (27 July 1992), pp. 9–10.

### Iraqi CW destruction

In the autumn of 1992 Iraq began large-scale destruction of its CW stocks under the supervision of UNSCOM.[165] An incineration destruction facility for mustard gas was built by Iraq at Muthanna to UNSCOM specifications. It will also be used for the destruction of chemical precursors. Approximately 400 tonnes of mustard gas will be destroyed by incineration. The incineration facility will be used additionally to destroy chemicals intended for use in missiles and other chemical material found at Muthanna. The nerve agents GB and GB/GF mixtures are being destroyed by controlled hydrolysis in another plant constructed by Iraq at Muthanna. Destruction is closely supervised by UNSCOM. As of December 1992, the following items had been destroyed: 12 000 empty munitions shells, 5000 122-mm rockets filled with sarin, 5500 kg of mustard gas, 40 500 litres mixture of GB/GF, 5000 litres of D4, 1100 litres of dichloroethane and 16.5 tonnes of thiodiglycol (TDG).[166]

## VI. Old CW ammunition and toxic armament wastes

During 1992 there was much public concern about old chemical ammunition dumped in the Baltic Sea. In March a German newspaper reported that as late as 1965 the German Democratic Republic had dumped a large quantity of World War II chemical ammunition (approximately 30 000 tonnes) in the vicinity of the Danish island Bornholm.[167] According to the report the artillery shells contained mustard gas and tabun. It was also reported that a large gas bubble containing warfare agent gas had formed on the bottom of the Baltic Sea near Bornholm. This was soon dismissed by experts as scientifically unfeasible. However, 'rotten' gases can form as a result of the decay of organic material in a marine environment and can be deposited in sediment layers.[168]

During a February visit by a German politician to Königsberg it was discussed that the German Navy and the former Baltic Fleet might together search for dumped chemical ammunition, particularly German munitions, in the Baltic Sea.[169] The pros and cons of raising the dumped CW ammunition were analysed from both a political and scientific point of view. Some estimates place the amount of dumped ammunition in the Baltic Sea at 400 000

[165] 'Beginn der Vernichtung von irakischem Giftgas', *Neue Zürcher Zeitung*, 11 Nov. 1992, p. 3; United Nations press release DH/1227, Geneva, 6 Sep. 1992, p. 3.

[166] Marcaillou, A., 'U.N. Special Commission update: Iraq CBW destruction', *ASA Newsletter*, no. 33 (16 Dec. 1992), pp. 1, 8; United Nations Security Council document S/24984, 17 Dec. 1992, p. 23.

[167] Oberholz, A., 'Eines Tages könnten Giftgasklumpen bis an Bornholms Strände treiben', *Frankfurter Allgemeines Sonntagsblatt*, 1 Mar. 1992, p. 3.

[168] 'Giftgasblase Seifenblase?', *Frankfurter Rundschau*, 3 Mar. 1992; 'Giftgas-Bergung gefährlich', *Frankfurter Rundschau*, 5 Mar. 1992, p. 4; 'Entwarnung für Bornholm', *Frankfurter Rundschau*, 16 Mar. 1992, p. 4; 'Keine Giftgas-Blase südlich von Bornholm', *Süddeutsche Zeitung*, 16 Mar. 1992, p. 5.

[169] 'Hennig: Giftgasgranaten aus der Ostsee bergen', *Der Tagesspiegel*, 2 Mar. 1992, p. 2; 'Dumped chemical weapons, contamination viewed', in FBIS-SOV-92-043, 4 Mar. 1992, pp. 15–16.

tonnes,[170] but a more realistic figure seems to be approximately 300 000 tonnes.[171]

It was reported that the Soviet Union may have dumped chemical bombs in 1946–48 at two sites, 50 nautical miles off Bornholm and 30 nautical miles from the Latvian port Liepaja, which had previously been used in the late 1940s by the Allies for dumping German CW ammunition.[172] Allegations of later dumping by Soviet forces could not be confirmed, and debate continued about the possible effect on the environment of leaking munitions.[173] Not only has former Soviet sea dumping come under public criticism. It was also reported that there is a CW dump left by the former Soviet Army in Nagorno-Karabakh, Azerbaijan, close to the Turkish border.[174]

In March the German Ministry of Traffic established a commission to investigate the status of dumped munitions and to consider future measures,[175] and a conference of Baltic foreign ministers took place in Copenhagen at which the participants agreed to co-operate more fully on environmental questions.[176] In April the Baltic ministers for the environment signed the Convention for the Protection of the Baltic Sea (a revised version of the Baltic Sea Convention) in Helsinki and agreed on a programme for redevelopment of the sea which will cost 36 billion DM.[177] Sweden, which is very much concerned about the ammunition dumped in the Baltic, had in 1991 ordered its National Maritime Administration to conduct investigations on the Swedish continental shelf which resulted in an autumn 1992 report on the current situation of chemical ammunition dumped in the Baltic Sea and the Skagerrak.[178] The report stated that in 1947 the Soviet Union dumped some 5000 tonnes of chemical ammunition in an area off Gotland and approximately 30 000 tonnes east of the island of Christiansö off Bornholm. However, it appears unclear if this dumped ammunition was of Soviet origin or ammunition discovered in Germany after World War II and dumped by the Allies. The former Baltic Fleet participated in the dumping operations carried out by the Allies in the late 1940s.[179]

In February China submitted a document to the CD related to its long-standing dispute with Japan about chemical weapons abandoned by Japan on

[170] 'Dumped chemical weapons, contamination viewed' (note 169); 'St Petersburg official on chemical arms probe', in FBIS-SOV-92-055, 20 Mar. 1992, pp. 35–36.

[171] Laurin, F., 'Massdumpad giftgas i Östersjon', *Svenska Dagbladet* (Stockholm, Sweden), 23 Mar. 1992, p. 11 (in Swedish).

[172] 'St Petersburg official on chemical arms probe', in FBIS-SOV-92-055, 20 Mar. 1992, pp. 35–36.

[173] 'Cleanup of Baltic chemical weapons waste viewed', in FBIS-SOV-92-143, 24 July 1992, pp. 4–7.

[174] 'Chemical weapons dump could pose danger', in FBIS-SOV-92-041, 2 Mar. 1992, p. 65.

[175] Haas, K., 'Seit über vier Jahrzehnten auf Tauchstation', *Süddeutsche Zeitung*, 23 Apr. 1992, p. 9; 'Krause läßt Giftgas in der Ostsee untersuchen', *Süddeutsche Zeitung*, 11 Mar. 1992, p. 6.

[176] 'Außenminster-Konferenz der Ostsee-Anrainer-Staaten', *Frankfurter Allgemeine Zeitung*, 5 Mar. 1992, pp. 1–5; 'Anlieger: Sauberes Meer gemeinsame Aufgabe', *Süddeutsche Zeitung*, 7–8 Mar. 1992, p. 2.

[177] 'EG tritt Ostsee-Konvention bei', *Süddeutsche Zeitung*, 10 Apr. 1992, p. 7.

[178] *Rapport om kartläggning av förekomsten av dumpade kemiska stridsmedel på den svenska delen av kontinentalsockeln* (Sjöfartsverket: Norrköping, Sweden, 1992), (in Swedish); *Summary of a Report on Investigation of the Existence of Dumped Chemical Weapons on the Swedish Part of the Continental Shelf* (National Maritime Administration: Norrköping, Sweden, 1992).

[179] 'Threat of dumped war gases in Baltic denied', in FBIS-SOV-92-059, 26 Mar. 1992, p. 18.

Chinese territory.[180] The document listed 18 dumping sites at six suspected locations. More than 300 000 chemical ammunition pieces and 120 tonnes of bulk CW agents have thus far been recovered including mustard gas, mustard gas-lewisite, diphenylcyanoarsine, hydrogen cyanide, phosgene and chloroacetophenone. In response the Japanese Ambassador to the CD referred to the serious efforts which Japan has made to solve the problem bilaterally.[181]

Austria submitted a paper to the CD describing a 1949–50 campaign where CW munitions, originally German CW munitions from World War II, were collected, sorted and provisionally buried.[182] During 1974–76 more than 28 000 grenades and rocket projectiles were moved to 'long-term' storage (i.e., the individual projectiles were placed in aluminium capsules and then stored in containers). This case influenced the final CWC negotiations because these abandoned CW munitions had to be stored by Austria, which at the time the munitions were dug up was unable to destroy them. Such storage of old CW munitions is not permitted under the CWC.

In Germany reports of recently located old chemical ammunition continued.[183] Chemical warfare agents were discovered on the property of a former German ammunition production facility near Löcknitz in Mecklenburg-Vorpommern that had been used since the 1960s by the National People's Army of the GDR.[184] At Falkenhagen in Brandenburg the former German Wehrmacht had a production facility for chemical warfare agents. After World War II Soviet troops were stationed there. In 1992 experts were still able to find chemical agents in bunkers and dangerous contamination of the soil after the withdrawal of troops.[185]

The problem of conventional ammunition dumping was also in the news in 1992 as allegations were made that the UK has been dumping ammunition off the coast of Cornwall since 1987.[186]

Another important ecological problem which is receiving increasing public attention, particularly in Germany, is related to former troop stationing areas. The withdrawal of CIS troops from the eastern part of Germany has made clear that during the past 40 years no consideration was taken of the possible negative effects of the presence of troops.[187] The German Government has commissioned a private company to work with local authorities to perform an on-site analysis and risk assessment of the troop stationing areas after troop withdrawal. The minimum amount of money which will be needed to

[180] Conference on Disarmament document CD/1127, 18 Feb. 1992.

[181] Conference on Disarmament document CD/PV. 614, 27 Feb. 1992, pp. 18–19.

[182] 'Old chemical weapons: description of a long-term storage facility under safe conditions', Conference on Disarmament document CD/CW/WP. 397, 5 May 1992.

[183] Thomé-Kozmiensky, K. J. *et al.*, *Management zur Sanierung von Rüstungsaltlasten* (EF Verlag für Energie- und Umwelttechik: Berlin, 1992).

[184] 'Kampfstoffreste der Wehrmacht gefunden', *Süddeutsche Zeitung*, 13 Mar. 1992, p. 5; Krispin, S., 'Nichts sehen, nichts riechen, nichts sagen', *Süddeutsche Zeitung*, 11 June 1992, p. 40.

[185] 'Chemische Kampfstoffe in geflutetem Wehrmachtsbunker', *Der Tagesspiegel*, 7 Nov. 1992, p. 7.

[186] 'Greenpeace: Britische Munition im Meer versenkt', *Süddeutsche Zeitung*, 10 Sep. 1992, p. 8.

[187] 'Russisches Roulett', *Der Spiegel*, vol. 46, no. 2 (5 Jan. 1992), pp. 55–58; see also Thomé-Kozmiensky, K. J. *et al.* (note 183); *Fachtagung Rüstungsaltlasten Erkundung und Untersuchung von ehemals und aktuell militärisch genutzten Flächen* (Umweltinstitut Offenbach: Offenbach a.M., Germany, 1992).

redevelop the nearly 44 000 contaminated sites—including 3000 sites contaminated by former Soviet troops and troops from the National People's Army of the GDR—will be of the order of 210 billion DM for a clean-up process which is expected to take 10 years.[188]

The problem of withdrawing troops and cleaning up after them is of general concern for many nations including the USA, which has troops stationed both abroad and within the borders of the USA. In the next 20 years the USA will spend approximately $25 billion for redevelopment and environmental security at its bases within the USA.[189]

## VII. New developments in NBC protection

One of the significant events of 1992 was the Fourth International Symposium on Protection Against Chemical Warfare Agents, held on 8–12 June in Stockholm, Sweden.[190] More than 650 participants from science institutes, government, the defence industry and defence research establishments discussed new trends and developments in chemical defence, influenced partially by early analysis of the Persian Gulf War. There is clear understanding that defence research will continue even when the CWC enters into force and that there will be a link between such research and chemical disarmament.[191] On the first day of the symposium the UNSCOM experts presented their findings and an evaluation of the inspections in Iraq.[192]

It is obvious that defence establishments are drawing their first conclusions about future NBC defence in light of the Persian Gulf War. Until recently military planners focused mainly on fighting under NBC conditions in the cool climate of the North. In the aftermath of the Persian Gulf War, Israel, with its vast experience in desert warfare, is now being asked to share its knowledge about product development. Many Persian Gulf countries are interested in acquiring NBC protective equipment for desert conditions,[193] and the US Army has ordered a new chemical protective undergarment (69 000 suits have already been produced and 100 000 more are to be produced). The new suits reduce heat, stress, weight and bulk, and have been improved by the addition of carbon which acts to trap chemical agents.[194]

As a result of the Persian Gulf War, Israel began distributing more than 800 000 new gas masks to its population.[195] Distribution of the improved

[188] 'IWH: Altlasten kein Investitionshemmnis', *Frankfurter Allgemeine Zeitung*, 17 Nov. 1992, p. 18; 'Umwelt-Altlasten konsten bis zu 400 Milliarden Mark', *Süddeutsche Zeitung*, 19 Nov. 1992, p. 2.

[189] 'Verstöße im Ausland gegen Umweltvorschriften', *Süddeutsche Zeitung*, 23 Nov. 1992, p. 9.

[190] See *Proceedings of the Fourth International Symposium on Protection Against Chemical Warfare Agents* (note 67).

[191] Dunn, P., 'Chemical defence and chemical disarmament: the need for both activities', *Proceedings of the Fourth International Symposium on Protection Against Chemical Warfare Agents* (note 67), pp. 9–23.

[192] 'Chemical warfare agents IV', *ASA Newsletter*, no. 31 (12 Aug. 1992), pp. 1, 10.

[193] Mitchell, B., 'War piques interest in anti-chemical gear', *Jerusalem Post*, 27 June 1992.

[194] 'Army gets new CW clothing', *Jane's Defence Weekly*, vol. 18, no. 22 (28 Nov. 1992), p. 9.

[195] 'Neue Gasmasken für Israel', *Frankfurter Rundschau*, 11 Jan. 1992, p. 2.

masks began in October. Over a period of 10 months, 5 million people will be equipped with the new masks.[196]

In the spring of 1992 the US General Accounting Office (GAO) published a report about the DOD's chemical protection preparedness in the Persian Gulf War. One of the conclusions of that report was that the department was not adequately prepared for chemical warfare. The GAO recommended that the DOD 'develop and implement a long-range action plan with target dates to ensure that required chemical defence equipment is available for all military personnel when needed'.[197] In September the US Army set up a new agency, the Chemical and Biological Defence Agency, to be responsible for R&D and acquisition for chemical and biological defence.[198]

## VIII. New BW developments

The 1991 Third Review Conference of the BWC agreed to establish a new *Ad Hoc* Group of Governmental Experts to discuss future verification measures for the BWC, and its first meeting was held in Geneva on 30 March–10 April 1992.[199] At the second meeting on 23 November–4 December, also in Geneva, the expert group examined 21 potential measures that might be used in the future to enhance compliance under the BWC.[200] The potential measures were grouped in seven categories: (*a*) information monitoring, (*b*) data exchange, (*c*) remote sensing, (*d*) off-site inspections, (*e*) exchange visits, (*f*) on-site inspections, and (*g*) continuous monitoring.

The sixth annual information exchange, and the first since the Third Review Conference where it was decided to improve the information exchange,[201] resulted in declarations from 15 countries by the end of April 1992 (Australia, Austria, Canada, Cyprus, Czechoslovakia, Germany, Japan, Mongolia, New Zealand, Norway, Sweden, Switzerland, the UK, the USA and Yugoslavia).[202] In subsequent months the following countries presented declarations: Argentina (27 August), Belarus (14 May), Belgium (8 July), Bulgaria (26 May),

[196] Reuters, 'Israel starts distributing new gas masks', *International Herald Tribune*, 21 Oct. 1992, p. 5.

[197] 'Chemical warfare protective equipment needs upgrading', *Chemical & Engineering News*, vol. 70, no. 21 (25 May 1992), p. 13; US General Accounting Office, *Operation Desert Storm: DOD Met Need for Chemical Suits and Masks, but Longer Term Actions Needed*, Report to the Chairman, Subcommittee on Readiness, Committee on Armed Services, House of Representatives, GAO/NSIAD-92-116 (General Accounting Office, Washington, DC, Apr. 1992).

[198] Muradian, V., 'U.S. Army creates unit for chemical defense', *Defense News*, vol. 7, no. 41 (12 Oct. 1992), p. 60.

[199] *Arms Control Reporter*, sheet 701.B.91, May 1992.

[200] Newmann, R., 'Nations examine BWC verification compliance measures', *Wireless File*, no. 234 (United States Information Service, US Embassy: Stockholm, 3 Dec. 1992), pp. 9–10.

[201] The improvements were the following: (*a*) to add a declaration on 'nothing to declare' or 'nothing new to declare'; (*b*) to amend and extend the exchange of data on research centres and laboratories and to exchange information on national biological defence research and development programmes; (*c* ) to amend the exchange of information on outbreaks of infectious diseases and similar occurrences caused by toxins; (*d*) to encourage publication of results and promotion of use of knowledge; (*e*) to amend the measure for active promotion of contacts; and (*f*) to add three new confidence-building measures: 'declaration of legislation, regulations and other measures', 'declaration of past activities in offensive and/or defensive research development programmes' and 'declaration of vaccine production facilities'.

[202] United Nations Office for Disarmament Affairs document ref. DDA/4-92/BWIII, 30 Apr. 1992.

China (1 August), Cuba (22 October), Denmark (30 June), Finland (1 June), France (15 June), Hungary (30 April), Japan, additional information (28 April), Jordan (14 August), Malta (30 April), Mexico (2 September), the Netherlands (22 May), Peru (21 September), Russia (3 July), South Korea (1 May), Spain (30 August), Thailand (20 August), Tunisia (7 May) and Ukraine (15 June).[203] By November 36 nations had participated in the information exchange—far fewer than in 1991, the year of the Review Conference.

In 1992 important new information about the former Soviet BW programme became available. During his January visit to Washington President Yeltsin pledged to halt Russian research on biological weapons.[204] He also pledged that no funds would be budgeted for BW research. In April Yeltsin signed a decree committing Russia, as the successor to the USSR, to the BWC.[205]

Official proof of the true nature of the 1979 Sverdlovsk anthrax accident was given in March when draft legislation to provide pensions to families of 64 people who died of anthrax was presented in the Supreme Soviet.[206] In an April interview with the Chief of the Directorate for Protection against Biological Weapons more information was presented about the accident.[207] In 1979 R&D had been conducted at Military Camp 19 in Sverdlovsk, where the outbreak occurred, in an attempt to find a more effective anthrax vaccine. During the June summit meeting in Washington Yeltsin acknowledged that the Sverdlovsk anthrax outbreak was the result of military research to make biological weapons.[208]

Information was also presented about a field test laboratory located on Vozrozhdeniye Island in the Aral Sea,[209] where development and testing of bacteriological weapons had been conducted since World War II.[210] Other sites where bacteriological research had been conducted included: the bacteriological centre in Obolensk near Moscow, the virology centre in Koltsovo near Novosibirsk, the Biological Instrument Building Institute and the Biochemical Machine Project in Moscow, the Institute for Ultrapure Drugs in what was then Leningrad and several Moscow institutions of higher education.[211]

In May, after President Yeltsin had issued a decree forbidding any military biological programme in violation of the BWC, the recently appointed head of the committee on conventional problems of chemical and biological warfare, Anatoly Kuntsevich, confirmed that after the USSR ratified the BWC 'there

[203] United Nations Office for Disarmament Affairs document ref. DDA/4-92/BWIII/Add.1, 12 June 1992; Add. 2, 12 Aug.; Add. 3, 22 Sep. 1992; Add. 4, 3 Nov. 1992,

[204] Devroy, A. and Smith, R. J., 'U.S., Russia pledge new partnership', *Washington Post*, 2 Feb. 1992, pp. A1–A26.

[205] 'Yeltsin signs decree on biological weapons', in FBIS-SOV-92-074, 16 Apr. 1992, p. 3; AP, 'Yeltsin commits to germ warfare ban', *Washington Post*, 17 Apr. 1992, p. A28.

[206] 'Draft laws, congress preparations approved', in FBIS-SOV-92-065, 3 Apr. 1992, pp. 41–42.

[207] 'General quizzed on chemical weapons production', in FBIS-SOV-92-082, 28 Apr. 1992, pp. 4–5; V. Chelikov, 'A weapon against their own people', *Moscow News*, no. 23 (25 May–1 June 1992), p. 8.

[208] Smith, R. J., 'Yeltsin blames '79 anthrax on germ warfare efforts', *Washington Post*, 16 June 1992, pp. A1–A15.

[209] 'Closure of bacteriological test range demanded', in FBIS-SOV-92-010, 15 Jan. 1992, p. 70.

[210] 'Biological weapons program, violations viewed', in FBIS-SOV-92-087, 5 May 1992, pp. 4–6; 'Documents for shutdown of weapons site ready', in FBIS-SOV-92-040, 28 Feb. 1992, pp. 54–55.

[211] 'Biological weapons program, violations viewed' (note 210).

were, legally speaking, violations of it in this country'.[212] In the mid-1980s steps were taken to curtail the offensive programme. He also confirmed that there are now no stockpiles of biological agents in Russia. Additionally, foreign experts have been invited to visit the military research facilities.

Nevertheless in August the UK and the USA expressed concern that Russia might not have fulfilled its earlier promise to discontinue and dismantle its BW programme.[213] After a first official denial[214] and perhaps in consideration of the promised financial aid from the USA, Russia responded to requests from the US State Department to prove that the programme had ceased by admitting that activities banned under the BWC had been conducted from 1946 until March 1992.[215] After talks and consultations in Moscow in September, Russia proved to British and US officials that it has terminated prohibited BW research and closed down the test facilities. President Yeltsin also ordered an examination of the Institute of Specially Pure Biological Preparations in St Petersburg.[216] In a significant step a Joint Statement on Biological Weapons was adopted by Russia, the UK and the USA.

In the spirit of the new climate of openness, Kuntsevich later admitted that the USSR had continued research, testing and production after ratifying the BWC. Methods of preparing biological agents for military purposes and of delivering them via aircraft and missiles were developed in the St Petersburg institute and at the military facilities in Kirov, Yekaterinburg (formerly Sverdlovsk) and Sergiyev Posad, and testing had been carried out on Vozrozhdeniye Island in the Aral Sea.[217] When disclosures about the former Soviet and subsequently Russian BW programme were made, experts also asked about current research in the USA and raised questions about the borderline between defensive and offensive research in light of the growing threat of proliferation.[218]

In October British, Russian and US experts conducted an investigation of the above St Petersburg institute. They found that the institute was 'only indirectly connected in the most general way' with work on bacteriological weapons.[219] Additionally, information was given in September that the Russian Government will reduce by 50 per cent the personnel involved in military biological programmes and will also reduce the financing of such research by 30 per cent.[220]

[212] 'Yeltsin's biological weapons decree assessed', in FBIS-SOV-92-097, 19 May 1992, p. 23.

[213] Smith, R. J., 'U.S. fears Moscow still makes germ weapons', *International Herald Tribune*, 1 Sep. 1992, p. 1.

[214] 'Ministry denies bacteriological weapons charges', in FBIS-SOV-92-172, 3 Sep. 1992, p. 2.

[215] 'Russia broke germ weapons ban', *The Independent*, 15 Sep. 1992, p. 9.

[216] Gordon, M. R., 'Russia and West reach accord on monitoring germ-weapon ban', *New York Times*, 15 Sep. 1992, p. 6; 'Official on biological weapons talks with U.S.', in FBIS-SOV-92-180, 16 Sep. 1992, pp. 15–16.

[217] 'Official examines destruction of CBW weapons', in FBIS-SOV-92-186, 24 Sep. 1992, pp. 2–4.

[218] 'Source cited on U.S. biological weapons', in FBIS-SOV-92-199, 14 Oct. 1992, pp. 3–4; 'Mutual biological warfare concerns pondered', in FBIS-SOV-92-194, 6 Oct. 1992, pp. 2–4.

[219] 'St. Petersburg institute cleared of biological weapons charges', in FBIS-SOV-92-228, 25 Nov. 1992, p. 2.

[220] 'Russian, U.S. statement on biological weapons', in FBIS-SOV-92-178, 14 Sep. 1992, p. 2.

# IX. Environmental implications of the Persian Gulf War

In January the US GAO published two reports dealing with the impact of the Kuwaiti oil fires, which occurred when 611 oil wells were ignited by the Iraqis in late February 1991, on the health of US troops stationed in the Persian Gulf who were exposed to smoke.[221] To investigate the potential long-term health risks, a variety of measures were taken including: (*a*) air and soil sampling to determine whether there were harmful pollutants in areas where US troops were present, (*b*) a biological study of a large number of soldiers, and (*c*) a health risk analysis, which was scheduled to be completed by December 1992. The measures focused on facilitating health risk analysis to project the incidence of illness probably attributable to oil-fire smoke exposure. In a follow-up report,[222] the US Environmental Protection Agency (EPA) reviewed an April 1991 interim report. Its major finding was that no significant quantities of pollutants that would cause severe acute or chronic health effects could be found, except for high levels of certain substances. More important was the assessment that the extent of possible long-term health risks to US troops exposed to pollution remains an unanswered question. At least two years of monitoring a large group which had been exposed is needed.

On the other hand, the sea environment and water quality both seem to have been affected negatively by the oil released during the war. Experts claim that 6–8 billion barrels of oil were released, and the cost of clean-up may amount to as much as 5 billion DM.[223] Experts estimate that 700 km of the Saudi Arabia coast were contaminated by oil.[224] According to investigations conducted by the United Nations Educational, Scientific & Cultural Organization (UNESCO), the coral reefs have survived but with varying degrees of damage.[225] After almost two years the impact of the oil spills is still evident but the ecosystem is starting to recover, owing in part to an enormous redevelopment programme.

# X. Conclusions

The conclusion of the CWC in 1992 was the most significant event of the year. It will provide the basis for CW disarmament via destruction of stockpiled chemical weapons and chemical warfare agents. The destruction must be

[221] US General Accounting Office, *Defense Health Care: Efforts to Address Health Effects of the Kuwait Oil Well Fires*, Report to the Chairman, Legislation and National Security Subcommittee, Committee on Government Operations, House of Representatives, GAO/HRD-92-50 (GAO: Washington, DC, Jan. 1992); US General Accounting Office, *International Environment: Kuwaiti Oil Fires: Chronic Health Risks Unknown but Assessments Are Under Way*, Briefing Report to the Chairman, Legislation and National Security Subcommittee, Committee on Government Operations, House of Representatives, GAO/RCED-92-80BR (GAO: Washington, DC, Jan. 1992).

[222] See US General Accounting Office, *International Environment: Kuwaiti Oil Fires: Chronic Health Risks Unknown but Assessments Are Under Way* (note 221).

[223] 'Ölpest im Golf kostet weitere fünf Milliarden Mark', *Süddeutsche Zeitung*, 27 Apr. 1992, p. 8.

[224] 'Korallenriffe offenbar kaum geschädigt', *Frankfurter Allgemeine* Zeitung, 5 Mar. 1992, p. 12.

[225] 'Korrallenriffe überlebten Ölpest im Golfkrieg', *Süddeutsche Zeitung*, 19 May 1992, p. 48.

carried out within a limited period of time and will be monitored by stringent, intrusive on-site inspection to verify claims of non-compliance.

Destruction presents a challenge to several countries. The US CW destruction programme has been forced to extend its completion date, and costs continue to increase. In the past incineration was considered the most effective destruction method, but a committee of the US National Academy of Science is now studying other technologies for chemical demilitarization with a report expected in summer 1993. The FY 1993 budget appropriations bill contains provisions for revision of the programme, and a new deadline of 31 December 2004 has been set for elimination of US CW stockpiles. The new schedule results in part from the fact that under the CWC more time is allowed for destruction than in the 1990 US–Soviet bilateral destruction agreement. The technologies under study include: chemical neutralization, supercritical water oxidation, steam gasification and plasma arc pyrolysis. Growing awareness in the USA about the possible risks of incineration to health and the environment is another factor which may have contributed to the new situation.

During the June Bush–Yeltsin summit meeting in Washington both leaders reconsidered the validity of the 1990 bilateral destruction agreement and confirmed Russia's role as successor to the former Soviet CW stockpiles. Since all of the chemical weapons of the former Soviet Union are located on Russian territory, Russia now faces a very heavy economic and technical burden and has begun the arduous process of designing and setting up its own destruction programme. In 1992 such a programme was designed by an official State Committee and later approved by the Russian Parliament. In phase I two facilities will be constructed, and a former CW production facility will be converted to a destruction facility.

The cost of the Russian CW destruction programme increased dramatically in 1992. Neutralization technology may be used in the destruction programme, but owing to the need to dispose of a variety of agents and munitions other chemical decomposition techniques are also being considered including incineration, chemical electrolysis, cryofracture and plasma techniques. Foreign financial and technical assistance for the dismantling of Russia's chemical weapons, including the $25 million approved by the US Congress in 1991, is extremely important. Both financial aid and technical expertise are needed, and a number of countries have been asked to help Russia.

New information was revealed in 1992 that the former Soviet Union conducted CW R&D as late as the early 1990s—clear demonstration of the need for the CWC which will outlaw similar activities in the future. The currently existing international legal framework does not provide adequate prohibition. It should be noted that the 1987 Soviet announcement that it had stopped CW production did not encompass its R&D programme.

Official information about the former Soviet biological defence research programme proved that the 1979 anthrax outbreak in Sverdlovsk was related to military research activities. Additionally, there now seems evidence that the programme continued until the beginning of 1992. Only after President Yeltsin intervened were steps taken to dismantle the programme. This information about the continuation of BW R&D highlights the need for improvement of the BWC as regards verification and confidence building.

As in the past various allegations of the use of chemical weapons or chemical warfare agents continued to be made. In 1992 new allegations were made about regions where there is military conflict such as the new republics in the former Soviet Union and in Bosnia and Herzegovina. It remains very difficult to verify such allegations.

CW and to a lesser extent BW proliferation is one of the most threatening developments of the 1990s. The number of countries alleged to possess or to be attempting to acquire chemical weapons has not decreased, and stopping further proliferation appears to be the task which lies ahead. Concerned countries, especially those in Western Europe and North America, have strengthened their national legislation. Export control measures and policies as applied by the Australia Group have been made more specific and elaborated upon. Under the CWC there will be a need for extensive review and revision of both national and international export control policies and legislation.

New findings from the UNSCOM inspections and new information about the involvement of foreign companies in the buildup of the Iraqi CW and BW capability have evoked strong reactions in the concerned countries and led to investigations and trials.

Public interest has increased about the threat posed by old chemical munitions that have been dumped at sea or buried and not yet discovered. There is a need for extensive scientific study and technical evaluation of the many unanswered questions about the state of these munitions, their impact on the environment and the possibility to salvage them.

The problem of soil contamination as a result of chemicals, petrol, oil and other toxic wastes resulting from long-term stationing of troops and military training activities has added a new dimension to study of the ecological impact of military activities. Enormous resources must be allocated to clean up and redevelop the affected areas, as has been shown by the aftermath of the withdrawal of former Soviet troops from eastern Germany.

The enormous oil spills in the Persian Gulf and the environmental impact of the oil fires in Kuwait have been the subject of much investigation. These scientific studies must continue in order to draw conclusions about the possible impact of these occurrences on the environment and on humans.

Defence research will continue under the CWC, and there will be a link between defence research and chemical disarmament. Based on a first analysis of the Persian Gulf War, R&D appears to be needed in the area of protective clothing and gas masks for use under very specific climatic conditions.

The 1991 Third Review Conference of the BWC highlighted the need to strengthen the BWC. Experts are now meeting on a regular basis to prepare verification measures for discussion at the 1996 Review Conference. The 1992 information exchange, which was agreed to at the Third Review Conference, produced much new information, but the number of countries taking part in the mandatory exercise did not increase.

The most important achievement of 1992 was the conclusion of negotiations on the CWC. The main question for the near future will be whether or not the signing and implementation of the CWC will have a significant impact on: (*a*) future proliferation of chemical and biological weapons, (*b*) CW stockpile destruction with respect to ecological, health and safety concerns, (*c*) future research in the area of new toxic agents and (*d*) NBC defence R&D. The 'new environment' will need to be monitored closely to identify any changes which may result from the Chemical Weapons Convention.

# **Appendix 7A.** Benefits and threats of developments in biotechnology and genetic engineering

TAMAS BARTFAI, S. J. LUNDIN and BO RYBECK*

## I. Introduction

Biotechnology, including so-called genetic engineering, has gained tremendous momentum in recent years. While a bright future has been forecast for these techniques for the past 15–20 years, only now do they seem technically applicable on a large scale. While modern biotechniques are revolutionizing medicine and agriculture, the possibility exists of their misuse for political ends, for clandestine production and refinement of biological weapons (BW), and for future development of weapons of mass extermination which could be used for genocide.[1]

Such a weapon system has not yet been developed, but the speed of scientific and technical progress is such that early warnings of potential misuse are justified. There are rumours and allegations of the development of sophisticated biological and toxin weapons and perhaps genetic weapons and of the techniques which could be used to create such weapons.[2] One of the authors of this appendix recently warned that genetic weapons might be developed in the wake of increased knowledge about the human

[1] The term 'weapons of mass extermination' is used instead of the usual 'weapons of mass destruction' (WMD) to emphasize the effect of such weapons on living organisms. WMD encompasses chemical, biological and nuclear weapons. However, as with 'conventional' chemical weapon (CW) and BW use, the aim may be to achieve both mass destruction and a selective, limited effect. The choice of terminology does not discount possible BW use by terrorists.

[2] See, for example, Geissler, E. (ed.), SIPRI, *Biological and Toxin Weapons Today* (Oxford University Press: Oxford, 1986); Wright, S. (ed.), *Preventing a Biological Arms Race* (MIT Press: Cambridge, Mass., 1990); Piller, C. and Yamamoto, K. R., *Gene Wars: Military Control Over the New Genetic Technologies* (Beech Tree Books, William Morrow: New York, 1988); Lundin, S. J., 'Chemical and biological warfare: developments in 1989', SIPRI, *SIPRI Yearbook 1990: World Armaments and Disarmament* (Oxford University Press: Oxford, 1990), p. 132; Douglass, J. D. and Livingstone, N. C., *America the Vulnerable: The Threat of Chemical and Biological Warfare* (D. C. Heath: Lexington, Mass., 1987); Binder, P., 'Biotechnologies et génétique dans le concept de nouvelles formes d'armes biologiques' ['Importance of biotechnologies and genetics in the concept of new biological weapons'], *Médecine et Armées,* 18 July 990, pp. 463–68 (in French). A recent publication dealing *inter alia* with these problems is Roberts, B. (ed.), *Biological Weapons: Weapons of the Future?* (Center for Strategic and International Studies: Washington, DC, 1993). Recent hearings in the US Congress cover the spectrum of the chemical and biological warfare (CBW) threat from the US perspective; see *Countering the Chemical and Biological Weapons Threat in the Post-Soviet World,* Report of the Special Inquiry into the Chemical and Biological Threat, Committee on Armed Services, US House of Representatives, 102nd Congress, 2nd session (US Government Printing Office: Washington, DC, 1993).

* Professor Tamas Bartfai heads the Department of Neurochemistry and Neurotoxicology, University of Stockholm; Dr S. J. Lundin is a retired Senior Director of Research of the Swedish National Defence Research Establishment (FOA) and a former Head of SIPRI's Chemical and Biological Warfare (CBW) Programme; Bo Rybeck, MD, is Director-General of FOA and a former Surgeon General of the Swedish Armed Forces. The opinions expressed are the authors and not necessarily the positions of their institutes or official Swedish policy.

genome and genetic diversity and suggested that efforts should be made to counter such developments.[3]

It appears that such weapon development including the application of genetic engineering techniques took place on a large scale in the former Soviet Union[4] despite the existence of the international agreement prohibiting this—the 1972 Biological Weapons Convention (BWC).[5]

This appendix presents information to help inform the discussion of the possible future misuse of biotechnology techniques, particularly for weapon purposes.[6] It is only a partial overview of the current state of development of genetic engineering. For more extensive reviews, the reader is referred to the articles and books cited in the footnotes.

## II. Advances in scientific knowledge and genome mapping

One of the most important current scientific endeavours is the 'mapping' of the human genome and that of several other organisms (e.g., a nematode, the yeast cell, rats, pigs, insects, etc.).[7]

Mapping the human genome is being done under the auspices of an international organization, the Human Genome Organization (HUGO), which co-ordinates the research and the work on and maintenance of an international data base. HUGO is the largest international project in life sciences sponsored by governments, but this has not hindered national and private enterprises from also setting up their own genome mapping programmes. The aim of the project is to provide insight into the organization and function of genetic material and in the course of this work to base physiology and medicine on solid molecular foundations, to provide the chemical basis for understanding hereditary diseases, and to aid understanding of the mechanisms of immune response and of carcinogenesis (i.e., the appearance of cancer tumours) and the like. It is assumed that insight will be gained not only into the DNA sequences in the chromosomes but also into the mechanisms which govern the expression of certain genes thereby resulting in the synthesis of functional, active protein molecules.

[3] Interview with Bo Rybeck in Westmar, B., 'Varning för genvapenkrig' [Warning for genetic weapon war], *Dagens Nyheter* (Stockholm, Sweden), 16 June 1992, p. A14 (in Swedish).

[4] Barry, J., 'Planning a plague? A secret Soviet network spent decades trying to develop biological weapons', *Newsweek*, 1 Feb. 1993, p. 20; see also chapter 7 in this volume.

[5] See the 1972 Convention on the Prohibition of the Development, Production and Stockpiling of Bacteriological (Biological) and Toxin Weapons and on Their Destruction (the BWC) the text of which is reproduced in Goldblat, J., SIPRI, *Agreements for Arms Control: A Critical Survey* (Taylor & Francis: London, 1982), pp. 193–95.

[6] FOA has periodically reviewed such developments. See, for example, Bovallius, Å. and Nilsson, G., *Genetic Engineering*, FOA rapport D 40011-B5 (Försvarets Forskningsanstalt: Sundbyberg, Sweden, Feb. 1975), (in Swedish); Jaurin, B. *et al.*, *Genteknik och biologiska stridsmedel* [*Gene Technology and Biological Weapons*], FOA rapport A 40058-4.4 (Försvarets Forskningsanstalt: Umeå, Sweden, Nov. 1987), (in Swedish); 'Theme biotechnology', *FOA Tidningen*, vol. 30, no. 3 (Nov. 1992). The annual CBW chapters in the *SIPRI Yearbooks* and other SIPRI publications have also followed the issue. See Geissler (note 2); Geissler, E. (ed.), SIPRI, *Strengthening the Biological Weapons Convention by Confidence-Building Measures*, SIPRI Chemical & Biological Warfare Studies no. 10 (Oxford University Press: Oxford, 1990); Lundin, S. J. (ed.), SIPRI, *Views on Possible Verification Measures for the Biological Weapons Convention*, SIPRI Chemical & Biological Warfare Studies no. 12 (Oxford University Press: Oxford, 1990).

[7] A genome can be defined as the complete set-up and composition of the hereditary substance deoxyribonucleic acid (DNA) in a living organism.

The USA initiated the efforts to fully map the human genome by allocating $3 billion for work on a national basis. The project aims to identify all of the 3 billion base pairs (estimated to encode the 50 000–100 000 human genes) of human deoxyribonucleic acid (DNA) within 15 years. The internationalization of the effort and the impressive development of the necessary technology[8] have resulted in much faster progress than was expected when the plan was first conceived in 1985. Even the scientists involved in the project have been surprised at the unexpected speed of progress—an indication of the rapid rate of advance in this area. The physical and linkage maps of all of the 23 human chromosomes have been determined to 60–98 per cent.[9] These results were reached by the combined efforts of large international research teams. The genome project has required great effort to build up compatible and accessible data bases to store the information which has been gathered from many different sources and acquired in various contexts by a variety of methods. These data bases will be available to all scientists and may thus constitute a valuable basis for openness about the work done in the genome project—thereby, it can be hoped, diminishing the risk for clandestine work for prohibited purposes.

## III. Biotechnology

Some of the most important achievements in biotechnology concern heterologous gene expression, genetic engineering of whole organisms (transgenic animals and plants), protein engineering and human monoclonal antibodies. These are of interest to medicine, industry and defence.

These techniques form the basis of the most rapidly growing industry in the USA and have resulted in products for medical care and agriculture, for the food and petroleum industry, for environmental clean-up, and for chemical and biological warfare (CBW) protection. These developments are of great value to mankind. They also have clear and specific application for defence against biological and chemical weapons (CW) and for the destruction of chemical weapons. However, it may also be possible to use this knowledge for clandestine development of biological weapons, even though this is prohibited. A more detailed description of techniques of possible relevance is given below.[10]

### Heterologous gene expression

Heterologous gene expression is today a widely applied technique to produce proteins in another organism than the one in which they are produced naturally. This is accomplished by inserting the gene encoding the desired protein into an expression vector. The first industrial applications, the production of human growth hormone in the bacterium *Escherichia coli*, were rapidly followed by the production of human

---

[8] See, for example, Anderson, C., 'New French genome centre aims to prove that bigger really is better', *Nature*, vol. 357 (18 June 1992), pp. 526–27.

[9] See, for example, Vollrath, D. *et al.*, 'The human Y chromosome: a 43-interval map based on naturally occurring deletions', *Science*, vol. 258 (2 Oct. 1992), pp. 52–59; Foote, S. *et al.*, 'The human Y chromosome: overlapping DNA clones spanning the euchromatic region', *Science*, vol. 258 (2 Oct. 1992), pp. 60–66; NHI/CEPH Collaborative Mapping Group, 'A comprehensive genetic linkage map of the human genome', *Science*, vol. 258 (2 Oct. 1992), pp. 67–86, Mandel, J. L. *et al.*, 'Genome analysis and the human X chromosome', *Science*, vol. 258 (2 Oct. 1992).

[10] For more comprehensive introductions to the different applications see, for example, Walton, A. G. and Hammer, S. K. (eds), *Genetic Engineering and Biotechnology Yearbook* (Elsevier: Amsterdam, 1988); *The International Biotechnology Handbook* (Euromonitor Publications: London, 1988).

insulin in yeast and the synthesis of human interferon. These three pharmaceutical products showed that from medical, environmental safety and health care points of view it is possible to produce hitherto expensive hormones and proteins by the use of genetic engineering. The list of heterologously expressed or 'recombinant' human proteins has grown, with 10–15 new products/proteins entering the market (e.g., the interferons, different colony-stimulating factors, etc.). The process now has a momentum of its own since generally applicable, effective protein expression systems have been developed which cut the cost of each new product.[11]

Human proteins which improve the healing of wounds, bone resorption, bone marrow formation after radiation damage, and so on are now commercially available as recombinant proteins. This is an important contribution to the healing of injuries and has application in both civilian and military health care.[12]

## Genetically engineered organisms

Genetically engineered organisms (transgenic organisms) are organisms which carry foreign DNA genes owing to artificial introduction of these by man, or organisms whose genetic make-up has been altered by human activity. The introduction of foreign DNA, and thus of foreign genes, occurs naturally every day as viruses (particularly retroviruses) infect humans, other animals, plants and so on and insert their genetic material into the genome of the host. Genetic engineering or modification of bacteria, plants and animals is discussed below.

Genetically engineered bacteria expressing a human gene coding for an important protein may necessitate not only insertion of the human gene which encodes this protein, but also that certain bacterial genes are eliminated, the protein products of which would interfere with the intended production process.[13] In general, the engineered bacteria are less viable and less stable outside the fermenter than are their naturally occurring counterparts. This is one key reason why the preceding decade of large-scale industrial use of engineered bacteria passed without any incidents of the spread of such bacteria outside the containment set-up. The introduction of markers and other safety features helped to prevent the spread of these organisms.

However, from the point of view of possible offensive military use, the situation may be different. In that context, the goal might be to take an already infectious virulent strain and to fortify and make it more virulent rather than weak and anti-viral agents, and by transferring virulence genes. It

By use of a 'combinatorial library approach',[18] it is now possible to select human monoclonal antibodies to deadly toxins without having to expose a human being to these agents to or alteration of genes in these cells would of course imply that the reproduction of new individuals resulting from manipulated sex cells would transmit such changes to future generations with, at least today, unforeseeable consequences.

Another area where medical care gains from new biotechnological and genetic engineering techniques is immunology. Several immune system stimulants (colony-stimulating factors and cytokines) are now available as recombinant proteins. The increased understanding of the mechanisms of immune responses also helps in designing drugs and therapies to fight infection, and enhancing reactions to toxic substances If new weapons of mass extermination emerge, as discussed below, protection against them will have to rely on corresponding medical advances.

## V. Genomic diversity and DNA fingerprinting

Biotechnology influences a number of everyday activities in society apart from industrial and medical applications, which have much debated ethical consequences.[22] A few such activities which may relate indirectly to the development of genetic weapons are discussed below. These activities are devoted to examining, registering and documenting genetic differences between ethnic groups and individuals of the same and of different ethnic groups.

### The Genome Diversity Project

The Genome Diversity Project collects and stores genetic material from 500 populations around the world which anthropologists fear will disappear via urbanization, disease or genocide.[23] It follows the development of human polymorphisms (genetic differences) and studies the effects of large population migrations on the evolution of human population. This is achieved by examining genetic differences in those proteins (gene products) where large differences exist in different races and ethnic groups (i.e., blood group proteins and human leukocyte antigen, HLA-proteins) and in other DNA sequences which do not code for proteins. Only a small portion of genomic DNA codes for proteins, so the probability of finding differences at the total DNA level between races increases dramatically compared to examining only the DNA sequences which code for only a few proteins (HLA, etc.).

---

[22] In 1991–92 the validity of DNA fingerprinting was debated intensively in *inter alia* the two scientific journals *Nature* and *Science*; see, particularly, Abbott, A., 'FBI attaches strings to its DNA database', *Nature*, vol. 357 (25 June 1992), p. 618; Roberts, L., 'Science in court: a culture clash', *Science*, vol. 257 (7 Aug. 1992), pp. 732–36; Anderson, C., 'Courts reject DNA fingerprinting citing controversy after NAS report', *Nature*, vol. 359 (1 Oct. 1992), p. 349. The debate in Germany is presented in, e.g., *Chancen und Risiken der Gentechnologie: Der Bericht der Enquete-Kommission* (Deutscher Bundestag, Referat Öffentlichkeitsarbeit: Bonn, 1987); information about developments in Germany and other countries is covered by *Nature, Science, New Scientist*, etc. The legal and scientific arguments for patenting living organisms, etc. are presented in e.g., Eisenberg, R. S., 'Genes, patents, and product development', *Science*, vol. 257 (14 Aug. 1992), pp. 903–8; Adler, R. G., 'Genome research: fulfilling the public's expectations for knowledge and commercialization', *Science*, vol. 257 (14 Aug. 1992), pp. 908–14; Anderson, C., 'NIH defends gene patents as filing deadline approaches', *Nature*, vol. 357 (28 May 1992), p. 270. See also Müller-Hill, B., 'The shadow of genetic injustice', *Nature*, vol. 362 (8 Apr. 1993), pp. 491–92, which warns of the possible misuse of knowledge from *inter alia* HUGO.

[23] See, for example, Roberts, L., 'Genome Diversity Project; anthropologists climb (gingerly) on board', *Science*, vol. 258 (20 Nov. 1992), p. 1300.

## DNA fingerprinting

So-called DNA fingerprinting may be of great importance in the future. In theory, it is possible to identify the DNA of a single individual from any cell or cell remnant of the person in question. This would make it possible to judge whether a person who is suspected of a crime actually is guilty by comparing cellular traces like blood, hair, sperm and skin, found at the scene of a crime, with those of the suspect.[24] In order for the method to be workable, it is necessary to determine the probability that a certain DNA sequence from one individual is unique and does not occur in another individual. This has triggered large-scale investigation of DNA sequence differences (polymorphisms) between individuals belonging to the same or different races and of the individual variations between individuals not only of the same race, but also of populations living in specific countries or areas. While the statistical validity of the technique is debated, current DNA fingerprint libraries (e.g., that of the US Federal Bureau of Investigation) reveal not only individual but also population differences. Hopefully, data on an individual's genetic make-up will be treated with the same secrecy as that which currently safeguards the medical history of patients, so that such data may only be acquired for diagnostic purposes and used only for medical care.

## VI. The possible use of biotechnology for political and weapon purposes and countermeasures

As the above description of a few recent biotechnological developments has made clear, opportunities exist to misuse this knowledge and these techniques for various purposes on the personal, societal, national, regional or global level. Such a situation is of course not novel. The risk of accidents owing to lack of knowledge, criminal neglect or intentional misuse follows human activities. All positive progress also has negative aspects. It is thus not a new phenomenon; the need to identify and try to eliminate adverse consequences of a new technique is intimately coupled to human risk behaviour. However, judging the probability of serious or even unacceptable damage should misuse take place is another matter. This is particularly difficult as regards the possible misuse of biotechnological and particularly genetic engineering techniques both with respect to unintentional release of genetically modified organisms into the environment with possible deleterious effects,[25] and the active development and subsequent use of new biological weapons. While the techniques described above have triggered a number of ethical considerations, which are not discussed here, the particular ethical problems relating to the possible misuse of biotechnology for weapon purposes is not much discussed outside the scientific circles concerned with arms control and disarmament.[26] Similarly difficult is the evaluation of the risks of the intentional development and use of 'genetic weapons'.[27]

As mentioned above, it has often been suggested that the new biotechnological and genetic engineering techniques could be misused to enhance putative biological and

[24] This has been made possible by the polymerase chain reaction (PCR) method which makes it possible to duplicate a certain DNA sequence a thousandfold into amounts which are possible to analyse chemically to determine its composition.

[25] See, for example, Anderson, J., Brunius, G. and Hermansson, M., 'Ecological risk assessment of transgenic organisms: Sweden', *Ambio*, vol. 21, no. 7 (Nov. 1992), p. 483.

[26] See note 2.

[27] See Geissler (note 2); Wright (note 2); Piller and Yamamoto (note 2); Douglass and Livingstone (note 2); and Barry (note 4).

toxin weapons.[28] It must again be stressed that development, production and stockpiling of such weapons is prohibited by the BWC, which also covers toxin weapons. The BWC was ratified in 1975 and has now been acceded to by 126 countries.[29] Any enhancement activities of putative biological warfare agents would have to be performed clandestinely by civilian agricultural purposes under closely guarded and controlled circumstances, such as the spreading of crop-protecting micro-organisms.[35]

## Heterologous gene expression and protein engineering of toxins

The introduction into

or virus, or to what extent they may do so.[38] Non-pathogenic, genetically engineered organisms used by industry have proven safe in the past decade. This is, however, not sufficient basis to argue that virulent micro-organisms will be equally harmless if released from storage (on purpose or unintentionally) after being subjected to minimal modification in order to withstand certain antibiotics, anti-viral agents or heat.

## Use as weapons

It is clear that the problems associated with a weapon programme and tactical use of possible new biological weapons may be significant, even though there has been previous experience in this area. Comprehensive testing including field testing would be needed if biological weapons were developed for military use, which most probably would not escape notice. Terrorist BW use may involve another scenario.

## Can 'genetic weapons' be developed?

The idea that genetic weapons may be developed and used is not new. In the wake of the ratification of the BWC, in the mid-1970s the Soviet delegation to the Conference of the Committee on Disarmament (CCD) in Geneva suggested that negotiations be conducted on 'new weapons of mass destruction' including 'gene engineering' as biological weapons. The Western delegations rejected the proposal arguing that it would be impossible to agree on weapons which did not exist and further that the suggested biological applications were already covered by the BWC.[39] The current scientific and technical developments described above appear to motivate a new, detailed examination of the possible threat and of the need to strengthen the international measures and prohibitions of the BWC related to the possible development, production and use of such biological weapons as genetic weapons.

It is well known that clear racial differences exist between blood group proteins and histocompatibility proteins. Dramatically different sensitivities to certain infectious agents have also been documented between the races. These genetic differences may in many cases be sufficiently large and stable so as to possibly be exploited by using naturally occurring, selective agents or by genetically engineering organisms and toxins with selectivity for an intended genetic marker. The number of known proteins (mostly from the immune system for which strong polymorphic distinctions are known) is now several dozen. These differences were recorded as the various sensitivities were found or as proteins were typified by antibodies or sequencing, but not as a result of a systematic search on the DNA level, as is now possible and being done.

When HUGO provides data on all protein-coding sequences and moreover on all non-protein-coding sequences the number of well defined genetic differences will increase dramatically compared to those known today. It is, however, scientifically impossible to predict how many regulatory sequences and structure genes are involved in population differences. Only estimates can be made of how large a portion of the human genome may contain sequences which are somehow linked to population

---

[38] But new ways are constantly being developed. For a short discussion see, for example, Hoffman, M., 'New vector puts payload on the outside', *Science*, vol. 256 (24 Apr. 1992), p. 445.

[39] For a short description of the negotiations in the CCD on the question of new weapons of mass destruction, see SIPRI, *World Armaments and Disarmament: SIPRI Yearbook 1978* (Taylor & Francis: London, 1978), p. 382.

characteristics. Such estimates range between 0.1 and 1 per cent but are very arbitrary.[40]

As mentioned above, at the heart of the admissibility of DNA fingerprinting as evidence against an individual in court is the question of whether there are specific variabilities (polymorphisms) in individual DNA to distinguish an individual from all others, whether in the same or another ethnic group, with absolute or very high certainty. Investigations are thus taking place regarding ethnic and individual differences in variable regions of DNA.[41] The results of these investigations may make it possible to identify distinguishing sequences between different ethnic and population groups.

The HUGO project which thus far is mapping genetic material only from Caucasians will not in itself reveal differences between races, but information from the Genome Diversity Project, DNA fingerprinting and HUGO will contribute to the identification of differences between races as variations in the frequency of certain hereditary diseases and in non-coding DNA sequences between ethnic groups are investigated.[42] The existence of racially determined distinctions in human DNA and their stability between ethnic groups sustained by endogamy (even in such 'melting pots' as the USA) is at the focal point of the debate on the forensic use of DNA fingerprinting.

Theoretically, if the aforementioned investigations provide sufficient data on ethnic genetic differences between population groups, it may be possible to use such data to target suitable micro-organisms to attack known rece

## VIII. Summary and conclusions

In summary, a warning is in order to fellow scientists, decision makers and others to be observant and to take both national and international steps to prevent intentional misuse of biotechnology and genetic engineering, particularly attempts to carry out genocide.[45] Both national and international regulatory measures are needed and should be enacted in the context of existing international conventions such as the BWC and the Genocide Convention. It is, however, also vital that the benefits of genetic engineering and sequencing techniques be made internationally available.

Biotechnology can provide tremendous benefits such as vastly improved methods of medical care, increased agricultural yield, the development of new materials, the facilitating of criminal investigations, and the like. The development of protection against chemical and biological weapons, vital for decreasing the threat and usefulness of these weapons, can also benefit from biotechnology.

The science and technology which first lead to positive biotechnology and genetic engineering developments may also increase the chance of misuse. Clandestine development of new and more effective methods of BW and toxin warfare, although prohibited by the BWC, is such an activity. It would be particularly detestable if biological or other weapons were developed which utilize genetic characteristics as the basis for the effect of the weapons.

While the existing safety and regulatory measures concerning biotechnology and genetic engineering have been successful thus far, thought should be given to what would occur if research were conducted that intentionally tried to circumvent them. However, the release of a genetically manipulated organism would probably produce unpredictable results and might be equally disastrous for both its developers and those against whom it would be used.

International efforts should continue in order to reinforce the current understanding that the BWC covers not only existing but also any possible future genetic weapons. This could usefully be reiterated at the next BWC Review Conference in 1996. Today attempts at 'ethnic cleansing' are being carried out with seeming efficiency and impunity by the use of conventional weaponry. However, it is vital to recall that an international Genocide Convention does in fact exist.[46] If it were implemented effectively, it would prohibit not only atrocities caused by conventional weaponry but also those caused by any method, old or new, the aim of which is genocide or damage based on particular genetic characteristics. The BWC and the CWC either directly or implicitly prohibit the development and production of such weaponry. Today it seems valid and urgent to reimplement the Genocide Convention, interpret its coverage and perhaps add interpretations that would explicitly prohibit preparation for genocidal actions, including the development and use of genetic weapons of any type whether conventional, chemical or biological weapons.

---

[45] Current conflicts in the former Yugoslavia and elsewhere constitute a serious threat of such misuse for genocide, although thus far existing measures have prevented such an occurrence.
[46] See Goldblat (note 5), p. 139 and note 43.

# 8. Military technology and international security: the case of the USA

ERIC H. ARNETT and RICHARD KOKOSKI

## I. Introduction

There is a new Western consensus, especially pronounced in the United States, that military research and development (R&D) must be maintained at its cold war level, if not accelerated, even as funding for military personnel and arms procurement programmes are cut. This consensus ensures that the United States will maintain technological supremacy for the foreseeable future but does not rule out deeper cuts in defence expenditure. Even in the two areas of R&D opposed by President Bill Clinton, nuclear weapons and war in space, existing US advantages are likely to remain. Given these advantages, the need for US efforts to deny developing countries access to dual-use technologies may have been overstated, and efforts to stigmatize certain technologies—for example, ballistic missiles—can be seen as self-serving.

This chapter summarizes US R&D policy in the context of Clinton's express aspirations, existing US plans for nuclear and conventional war-fighting, and the major R&D programmes that support them.[1] After highlighting the special nature of changes in US military space programmes, it considers where Clinton's aspirations are likely to be limited by budget constraints and other practicalities of governance and where the new Administration's approach may differ from that of its predecessor. It examines the implications of these policies for developing countries, and concludes with a discussion of the probable nature of war in the future.

## II. R&D and US military supremacy

While most other measures of Western military activity declined in 1992, the resources devoted to R&D on conventional weaponry increased. The renewed commitment reflected the impact of the 1991 Persian Gulf War on military planning and the declining interest in nuclear weapons accompanying the end of the cold war. It also demonstrates that the organizational and economic momentum of technology programmes can keep them on track despite the absence of a specific threat.

[1] This chapter is contributed by SIPRI's Project on Military Technology and International Security. Although the project considers a wide range of issues and is especially concerned with developments and attitudes in the developing world, the unique role of the USA in setting the pace of innovation and specific developments this year, especially the transition from Republican to Democratic control of the executive branch, merits the special case study presented here.

During the year, a consensus emerged that the USA should make a major effort to retain the technological supremacy demonstrated in the war against Iraq. Despite public and academic criticism, the Bush Administration, the US Congress and the Clinton campaign all stood by the R&D, procurement and technology-denial programmes that constitute the effort to remain dominant and deter challengers. This consensus means not only that the United States will continue to devote resources to technological dominance, but also that it will set a pace of military innovation that even the other industrialized states cannot hope to match, except in specialized niches.

This is true despite similar commitments in most industrialized countries to increase R&D spending as procurement budgets decline. Of Europe's top four defence establishments, only Italy's has reduced its R&D budget since the end of the cold war.[2] France is the major spender on military R&D in Europe and the margin is increasing. Between 1985 and 1990, the French military R&D budget rose by 82 per cent.[3] British spending on military R&D has continued to decline in real terms from the mid-1980s, but plans call for a 9.3 per cent real increase in 1992–93.[4] German spending on military R&D increased by 57 per cent in the late 1980s, and by 16 per cent in 1990 alone.[5] Following the same trend, Japan increased funding for the Defence Agency's Technical Research and Development Institute (TRDI) by 10 per cent for 1993, while cutting its military budget.[6]

Few would take issue with the proposition that the USA has amassed unmatched military technology and expertise after decades of tremendous investment to that end—at least seven times as much in any given year as any other state except the USSR. In 1988, a representative example, the USA spent $295 billion on defence, compared to France's $36.1 billion, Germany's $35.1 billion, the UK's $34.6 billion and Japan's $28.5 billion. The same year, Saudi Arabia, Iraq and India, the three leaders in the developing world (excluding China), spent $14.9 billion, $12.9 billion and $9.3 billion, respectively. Estimates of Chinese military expenditure vary between

[2] Italian R&D spending was cut from £390 million ($605 million) in 1989 to £255 million ($395 million) in 1990. Cabinet Office, *Annual Review of Government Funded R&D 1992* (Her Majesty's Stationery Office: London, 1992), table 2.3.2. Figures are in current pounds. In comparison, the Bush Administration requested $39 billion in 1992 for Pentagon R&D and testing in fiscal year (FY) 1993, of which $38.2 was authorized by Congress. Direct comparisons are difficult, because the transition from research to development is rather ambiguous, including as it does applied research. Further, testing can be conducted as part of research, as part of development or in order to assure the reliability of fully developed systems. Finally, secrecy and different rules of accounting and reimbursement for independent R&D further complicate comparisons. Funding for US military R&D has declined slightly since a peak in 1987. Adams, G. and Kosiak, S. M., 'The United States: trends in defence procurement and research and development programmes', ed. H. Wulf, SIPRI, *Arms Industry Limited* (Oxford University Press: Oxford, 1993), p. 31.

[3] The five-year increase was from £1.87 billion to £3.41 billion ($2.90 billion to $5.29 billion). *Annual Review of Government Funded R&D 1992* (note 2), table 2.3.2.

[4] *Annual Review of Government Funded R&D 1992* (note 2), tables 2.3.2 and 2.6.2.

[5] The relevant figures are £630 million in 1985 to £991 million in 1990 ($977 million to $1.54 billion). *Annual Review of Government Funded R&D 1992* (note 2), table 2.3.2.

[6] The 1993 TRDI budget is Y14.0 billion ($110 million). 'Japan will Verteidigungsausgaben erheblich senken', *Frankfurter Allgemeine Zeitung,* 12 Nov. 1992, p. 7. The defence budget was cut by Y580 billion ($4.7 billion). Kensuke E., 'Japanese budget cut despite "destabilizing factors"', *Jane's Defence Weekly*, 9 Jan. 1993, p. 13.

$6 billion and $24 billion. Iraqi defence spending peaked in 1984 at $31.6 billion.[7]

What is more controversial is the notion that this hard-won supremacy could easily be lost without further spending, which is becoming increasingly painful, or without the imposition of control regimes that are often seen as cutting some of the poorer countries off from development. Indeed, some argue that it may be impossible for the USA to retain a technological edge.[8] Furthermore, the tangible benefits of supremacy were thrown into doubt in 1992 by *inter alia* the unwillingness, if not the inability, of the superpower to stop Serbian aggression in the war in the former Yugoslavia. Finally, as shown below, the US conventional preponderance apparently has not convinced military advisers that nuclear war plans emphasizing first use and pre-emption should be retired from US intervention strategy.

## III. President Clinton and US aspirations

Clinton's campaign promises can be seen as his judgement of the nation's aspirations rather than a definite plan, so it is not surprising that he sometimes seemed to contradict himself. While understandable, the contradictions limit the predictive power of these statements for those who sought it in 1992. For example, during the campaign, Clinton vowed he would transfer one-sixth of the nation's military R&D budget (including Defense and Energy Department programmes) to the civilian sector—$7.6 billion per year—by the fourth year of his first term.[9] Yet he also committed himself to 'strengthen . . . the overwhelming superiority of our weapon technology [by] . . . reassigning top priority to research and development, both to keep our edge and to ensure that we have the most advanced technologies available'.[10] As a result, 'We are going to have to spend more money in the future on military technology'.[11]

[7] All figures are in 1988 US dollars and come from Deger, S. and Sen, S., 'World military expenditure', *SIPRI Yearbook 1992: World Armaments and Disarmament* (Oxford University Press: Oxford, 1992), table 7A.2 and pp. 245–50. US Government figures are roughly the same, with Chinese expenditure estimated to be $21.3 billion in 1988. US Arms Control and Disarmament Agency, *World Military Expenditure and Arms Transfers 1989* (US Government Printing Office: Washington, DC, 1989). Total military expenditure captures R&D and procurement investment as well as additional spending on training, operations and maintenance, and infrastructure that affect the ability to field and apply technology.

[8] One of Clinton's closest foreign policy advisers takes this position: 'If current trends continue, the pace of technological diffusion may eventually vitiate the reliance of industrial countries on technological superiority to influence international events. . . . There may be a point of exhaustion in which an increment in technological superiority yields diminishing military returns.' Nolan, J. E., *Trappings of Power: Ballistic Missiles in the Third World* (Brookings Institution: Washington, DC, 1991), p. 7. See also the remarks of M. G. Morgan in American Association for the Advancement of Science (AAAS hereafter), *Arms Transfers, Export Control, and Dual-Use Technology in the Aftermath of the Kuwait War* (AAAS: Washington, DC, 1992), p. 3.

[9] Clinton, B., *Technology: The Engine of Economic Growth—A National Technology Policy for America* (Clinton/Gore '92 Committee: Little Rock, Ark., 1992), p. 15.

[10] Clinton, B., 'Clinton: refocus military role—tailor defense industrial base to future threats', *Defense News*, 26 Oct. 1992, p. 20.

[11] *Jane's Defence Weekly*, 14 Nov. 1992, p. 21. One interpretation of this seeming impossibility is that Clinton means: 'Revitalized science and technology programmes will take a larger share of the [smaller] defence budget.' Yockey, D., 'Responding to new challenges—defense acquisition strategy',

Clinton's first budget would raise US military R&D from $38.2 to $38.6 billion.[12]

Nevertheless, clear themes have emerged from Clinton's circle of security advisers. First, they see the Bush Administration's approach of keeping reductions in military spending small to hedge against uncertainty[13] as less compelling than 'threat-based planning' that tailors the force structure to the most likely contingencies.[14] Les Aspin, Clinton's Defense Secretary and formerly Chairman of the House Committee on the Armed Services, identified the threats on which he would base US military planning: a ground war in the Gulf region, an air war on the Korean Peninsula, and a lesser contingency waged simultaneously. He suggested that China, Cuba, Libya and Syria were 'potential regional aggressors' that must be countered, although only China would present a challenge greater than the one posed by Iraq.[15]

Clinton himself advocated a much stronger reaction—including the use of force—to Serbian aggression in the wake of Yugoslavia's breakup, and is seen as a champion of an idealistic, 'pro-democracy' foreign policy supported by intervention for reasons not directly related to the US national interest. In fact, some observers fear that Clinton's rejection of the previous Administration's cautious *Realpolitik* could aggravate a perceived need to demonstrate his willingness to use force after doubts in that regard were raised during the campaign. These fears were heightened shortly after the election when Aspin opined that the Clinton Administration probably could not sustain popular support for the defence budget unless it occasionally showed 'that [the military] is useful for lesser contingencies'.[16] In the same vein, Aspin expressed the view that the USA may have to demonstrate its resolve even when its vital interests are not at stake and could do so with high-technology air-strikes against politically significant targets.[17]

Although the requirements imposed on military planners by this range of possibilities would seem imposing, Clinton repeatedly vowed during the campaign that US defence spending would be cut by $60 billion during his first term, 4.2 per cent of Bush's planned $1.42 trillion over the remaining period

*NATO's Sixteen Nations*, vol. 37, no. 6 (1992), p. 63. Yockey is the US Under Secretary of Defense for Acquisition. By this measure, 'the science and technology portion of the defence budget (to include the Strategic Defense Initiative) grew from 13% to 21%' in 1992 (p. 64).

[12] Ambush, P., 'Dividing US defense dollars', *Defense News*, 29 Mar.–4 Apr. 1993, p. 1.

[13] Colin Powell, Chairman of the Bush Administration's Joint Chiefs of Staff, wrote the most famous line of the Administration's defence policy: 'The real threat we now face is the threat of the unknown and uncertain.' Powell, C., *US National Military Strategy* (US Government Printing Office: Washington, DC, 1992). Powell clarified the statement in a Senate hearing: 'The primary threat to our security is . . . a crisis or war that no one expected or predicted.' Matthews, W., 'Soviet demise leaves Pentagon wondering who is the foe', *Defense News*, 24 Feb. 1992, p. 34.

[14] Gellman, B., 'Debate over military's future escalates into a war of scenarios', *Washington Post*, 26 Feb. 1992, p. 20.

[15] Gellman (note 14), p. 20.

[16] Graham, G., 'Clinton aides in defence review', *Financial Times*, 13 Nov. 1992.

[17] Congressional critics worry that this approach is reminiscent of what they see as a failed bombing strategy prosecuted against North Viet Nam at the time that Aspin was coming of age as a Pentagon analyst. Towell, P., 'Aspin brings activist views to a changed world', *Congressional Quarterly Weekly*, 9 Jan. 1993, p. 81. Aspin also sees the possibility of using military force 'to physically destroy Third World nuclear facilities' (p. 83).

of the 1992 Future-Year Defense Plan,[18] a promise based on Aspin's earlier positions.[19] The greatest savings would be realized from reductions in staffing levels, the Strategic Defense Initiative (SDI) and nuclear programmes.[20]

On the topic of technology specifically, Clinton said he supports 'the best possible technology in our weaponry, both offensive and defensive'.[21] He suggested that US procurement programmes should focus on mobility and those items 'that improve our technological edge', including the C-17 cargo plane, fast sea-lift ships, the V-22 tilt-rotor utility aircraft, the F-22 Superstar fighter and unspecified new conventional missiles.[22] R&D would focus on *inter alia*, armoured vehicles (especially tank armour and large-calibre gun tubes), submarines (especially propulsion systems), high-performance aircraft, sensors, surveillance, guidance, materials, communications and intelligence.[23] Funding for the Strategic Defense Initiative will continue to decline under Clinton, and interceptors will not be based in space. Further, he implied that much of the savings from reduced defence spending would be directed to 'defense firms and workers' anyway, 'to . . . help [them] diversify into commercial markets'.[24] Some observers see this strategy as a way to prepare the ground for much deeper cuts in defence procurement further in the future.[25]

## IV. Legacies of the cold war

In seeking to meet his and the country's aspirations, Clinton will be limited by a number of factors associated with the cold war and the Bush Administration's first, tentative attempt at coping with the new era. Most obviously, powerful bureaucracies with large constituencies have taken root, and dislodging them would entail political and economic costs that the Bush Administration was largely unwilling to incur. As a result, a first post-cold war consensus on security policy congealed around Bush's positions, and the extent to which Clinton is able or willing to challenge them is not clear.

[18] Bush's 1992 plan, in turn, had been billed as a $50 billion cut (over six years) from his previous position. Bush, G., *State of the Union* (US Government Printing Office: Washington, DC, 1992).

[19] Pine, A., 'Battle shaping up over defense cuts', *Los Angeles Times*, 26 Feb. 1992, p. 4.

[20] These three would account for 30–40% of defence cuts. Capaccio, T., 'Defense cuts vital to Clinton's deficit program', *Defense Week*, 6 July 1992. Some observers expect the consolidation of roles and missions advocated by Sam Nunn, Chairman of the Senate Committee on the Armed Services, could save as much $100 billion. Fulghum, D. A., 'Powell in crossfire over defense cuts', *Aviation Week & Space Technology*, 7 Dec. 1992, p. 22. To realize such savings, the following programmes would be cut: the F-22 Superstar fighter, 'battlefield missile and air defense', space-based missile defences, the Trident submarine force, Army and Air Force close air support, and Navy and Air Force deep interdiction. A Brookings Institution study group had earlier recommended US defence spending be cut to roughly $150 billion per year. Clinton himself had suggested that the defence budget could be cut by one-third before Aspin became active in his campaign. Leopold, G., 'Clinton defense cutbacks would outpace Bush proposals', *Defense News*, 20–26 July 1992, p. 34.

[21] 'Budget put at top of Clinton's list', *Jane's Defence Weekly,* 21 Nov. 1992, p. 9.

[22] Clinton may also maintain production overcapacities for M1 Abrams tanks and F-16 Falcon multirole aircraft. Finnegan, P., 'Analysts: Clinton policies could buoy defense stocks', *Defense News*, 9 Nov. 1992, p. 18.

[23] Clinton (note 10), p. 20.

[24] Clinton (note 10), p. 20.

[25] Finnegan (note 22), p. 18. $1 billion were earmarked for this purpose in Clinton's first budget.

Two 'secret' documents elucidating the Bush Administration's commitment to post-cold war pre-eminence were leaked to the press in early 1992. First, the Strategic Deterrence Study Group organized by the Joint Strategic Targeting Planning Staff, which is responsible for US nuclear war plans, circulated its report of the deliberations of a bipartisan panel convened by Thomas C. Reed, a former Secretary of the Air Force.[26] The 'Reed Study', as it came to be known, not only recommended that nuclear weapons be aimed at 'every reasonable adversary . . . [or] nuclear weapons states [that] are likely to emerge', but also encouraged first use of nuclear weapons or the threat thereof against any country that might have a temporary conventional advantage. To this end, the Reed Study recommended the creation of a 'nuclear expeditionary force' that could be quickly deployed on US warships or air bases, seeming to fly in the face of President Bush's historic initiative removing most nuclear weapons from US warships and bases abroad.[27]

Shortly thereafter, a draft of the US Defense Planning Guidance through the turn of the century was also leaked.[28] This document described a strategy by which the United States not only would seek to maintain its global supremacy, but would also prevent the emergence of rivals, whether friendly or hostile. It envisioned US military responses to a range of developments, from a coup in the Philippines to a 'resurgent/emergent global threat' in the form of an expansionist Russia bent on annexing the Baltics.

Both of these documents were criticized in the US Congress and the international press,[29] but both also found supporters and neither was explicitly disavowed by the Bush Administration.[30] In fact, the positions presented in them continued to be refined. One government-sponsored analyst concluded that the USA should plan for 'rapid though limited redeployment of nuclear weapons on surface ships if they are needed to deter a hostile new nuclear power'. He was also favourably disposed towards 'pre-emption [of nuclear forces in the developing world] using conventional ordnance, if not nuclear weapons', and suggested deploying new nuclear weapons of appropriate yield for nuclear

[26] Reed, T. C. and Wheeler, M. O., *The Role of Nuclear Weapons in the New World Order* (US Department of Defense: Washington, DC, 1991); Smith, R. J., 'US urged to cut 50% of A-arms', *Washington Post*, 6 Jan. 1992, p. 1. Although the Reed Study Group was bipartisan, the fact that the study was leaked indicates a lack of consensus.

[27] Fieldhouse, R., 'Nuclear weapon developments and unilateral reduction initiatives', *SIPRI Yearbook 1992: World Armaments and Disarmament* (Oxford University Press: Oxford, 1992), pp. 66–82.

[28] Tyler, P., 'War in the 1990s: new doubts—Pentagon plans evoke skepticism in Congress', *New York Times*, 18 Feb. 1992, p. 1; and Gellman, B., 'Pentagon war scenario spotlights Russia: study of potential threats presumes US would defend Lithuania', *Washington Post*, 20 Feb. 1992, p. 1.

[29] One of the most prominent critics was Lee Hamilton, a member of the House Committee on Foreign Affairs and Clinton campaign adviser, passed over for the position as Secretary of State. He called the draft *Defense Planning Guidance* 'dead wrong' in a position paper for the journal *Foreign Affairs.* 'A Democrat looks at foreign policy', *Foreign Affairs*, summer 1992, p. 34. The Reed Study also provoked him (p. 40): 'What we do not need are Pentagon experts trying to find new targets and adversaries for existing nuclear weapons.' Some observers believe Hamilton was 'too liberal' for the Clinton Administration (for example, Weisberg, P., 'Nowhere man: a case against Lee Hamilton', *The New Republic*, 14 Dec. 1992, p. 10).

[30] The war scenarios in the draft *Defense Planning Guidance* were disavowed, but not the strategy. Gellman, B., 'Pentagon says war scenario doesn't reflect or predict US policy', *Washington Post*, 21 Feb. 1992, p. 12; and Schmitt, E., 'Some Senators see little risk of war: say the military is inflating its concerns as tactic to inflate its budget, too', *New York Times*, 21 Feb. 1992, p. 11.

retaliation against developing countries.[31] A second government-sponsored analyst advocated nuclear first-use and 'a wider array of graduated nuclear options' to deny 'proliferators the advantage of always getting the best opening nuclear shot'.[32] Analysts at the Energy Department's Los Alamos National Laboratory have recommended a new generation of very-low-yield nuclear weapons for use in wars against developing countries.[33]

## V. Summary of US R&D programmes

Official US thinking on military technology in 1992 had its most tangible manifestation in the form of the Defense Department's budget request for FY 1993. Of Bush's $39 billion R&D and testing budget request, more than $28 billion would be spent by the armed services on programmes closely related to near-term procurement. The two entities that are most interesting from the perspective of longer-term trends are independent of the services and therefore more visionary: the Defense Advanced Research Projects Agency (DARPA) and the Strategic Defense Initiative Organization (SDIO). DARPA's public budget has held steady at about $1.5 billion. The Agency is also responsible for some fraction of the 'black budget', which is not publicly known in detail but is estimated to be $16 billion in the FY 1993 budget request.[34] Of an estimated $6.6 billion spent on black R&D and testing, as much as $1.2 billion may be spent by DARPA and other independent agencies.[35] The Congress granted the SDIO a budget of $3.8 billion of the $5.4 billion requested by the Bush Administration for FY 1993.

### DARPA

The debate over DARPA's role in the post-cold war world came to a head during the year.[36] Critics claimed that DARPA had been 'ghettoized' by the ideol-

[31] Dunn, L., *Containing Nuclear Proliferation*, Adelphi Papers 263 (Brassey's: Oxford, 1991). Dunn—a manager at the Science Applications International Corporation, funded by the Defense and Energy Departments and the Central Intelligence Agency, as well as private foundations—continued to promote these positions in 1992.

[32] Quester, G. H. and Utgoff, V. A., 'U.S. arms reductions and nuclear nonproliferation: the counterproductive possibilities', *Washington Quarterly*, winter 1993, p. 139. Utgoff is a manager at the Institute for Defense Analyses, a federally funded, private corporation.

[33] Dowler, T. and Howard, J. S., 'Countering the threat of the well-armed tyrant: a modest proposal for small nuclear weapons' and Ramos, T. F., 'The future of theater nuclear forces', *Strategic Review*, fall 1991. The Clinton Administration is reportedly considering an exemption of tests below 1kt from the comprehensive test ban they have pledged to negotiate. That exemption would allow the development of such weapons.

[34] The term 'black budget' refers to military expenditures that are not publicly acknowledged, although they are included in the grand total of the Pentagon's budget. Some black budget programmes appear under unexplained code-names or as 'special programmes', while others do not appear at all. The most famous items in the Reagan Administration's black budget were the F-117 and B-2 Stealth bombers. Arnett, E., 'DARPA—many shades of black', *Bulletin of the Atomic Scientists*, Sep. 1992, p. 21.

[35] Sweetman, B., 'Big bucks for black budget', *Jane's Defence Weekly*, 2 May 1992. In comparison, Sweetman estimates that the Air Force share of the black budget is $12 billion; the Navy's, $1.6 billion; and the Army's, $630 million. Sweetman estimates that about half the black budget is spent on reconnaissance satellites.

[36] Ember, L., 'Major change in offing for DARPA', *Chemical and Engineering News*, 14 Sep. 1992.

ogy of the Reagan–Bush Administration that saw federal support for industrial R&D as anathema.[37] In their view—with which Bill Clinton is sympathetic—this phenomenon had acted to the detriment of the US economy. Their proposed solution is to charge DARPA with the task of exploring technologies with both military and civil applications.[38] Critics of this approach countered that the Pentagon was not the place from which the government should support industry, since its primary job imposed incompatible requirements upon it. With Clinton's election, it appears that DARPA will be directed to explore dual-use technologies of industrial interest more energetically and may be joined by a new organization examining technologies with less direct military potential.[39]

DARPA has identified seven areas in which it will make its most strenuous efforts: smarter conventionally armed missiles; air, land and sea warfare; command and control; manufacturing; and training.[40] Little has been publicly acknowledged about DARPA's programmes, but several trends are discernible.

### *Smarter missiles*

DARPA's work on conventionally armed missiles is proceeding on what amount to two separate fronts: developing 'autonomous' missiles and centralized 'reconnaissance-strike complexes'. The former, exemplified by the Thirsty Saber project, envisions weapons that can recognize targets without aid from human operators, a concept that requires software breakthroughs that are not expected by most engineers working in the field.[41] Autonomy is not only technologically very challenging, but also of limited utility in the absence of a major enemy with excellent air defences. The alternative approach puts more emphasis on surveillance and command-and-control networks that would support weapons only slightly smarter than current laser-guided glide bombs. These would be directed to their targets by human operators in airborne and terrestrial nerve-centres collecting and interpreting data collected by sensors dotting the battle area. Warbreaker, the lead programme in this area,

[37] National Academy of Sciences, *The Government Role in Civilian Technology* (National Academy Press: Washington, DC, 1992); and Alic, J. A. *et al.*, *Beyond Spin-off: Military and Commercial Technologies in a Changing World* (Harvard Business School Press: Boston, Mass., 1992).

[38] Previously, DARPA had sought only to identify military technologies with civilian applications and make them available to industry, but had bureaucratic and philosophical problems that limited their success.

[39] The organization would be called either the National Advanced Research Projects Agency (NARPA) or the Advanced Civilian Technology Agency (ACTA). The former is the more popular name, but the latter is the title officially proposed by John Glenn of the Senate Committee on Government Operations in legislation that would fund the enterprise with $500 million over three years. 'Washington outlook: civilian DARPA', *Aviation Week & Space Technology*, 4 Jan. 1993, p. 21. Clinton removed the word 'defense' from DARPA's title on 22 Feb. 1993, but has not created a new civilian agency.

[40] DARPA's terms for these areas are precision strike; air superiority and defence; advanced land combat; sea control and undersea superiority; global surveillance and communications; technologies of affordability; and synthetic environments.

[41] Arnett, E. H., 'The futile quest for autonomy: long-range cruise missiles and the future of strategy', *Security Studies*, autumn 1992.

reportedly seeks the ability to maintain 'situational awareness' over 1 000 000 square kilometres while 'picking out' targets in any given 10 000-square-kilometre sector and directing munitions against them within 10 minutes.[42]

*War in the air*

The air-warfare programmes address the maintenance of US dominance most directly. The two major efforts centre on achieving air superiority over an enemy's territory, as the West has in every conflict of the post-war era, and ensuring that once it has been secured, no counterstrikes can penetrate US defences. The air-superiority mission relies on improvements to the combination of the new F-22 fighter and the older E-3 Sentry AWACS (Airborne Warning and Control System) aircraft, accepting that a successor to the former is unlikely for at least 20 years after its introduction to service early in the next century.[43] Sceptics have expressed doubts that the USA needs or can afford the F-22.[44] Air defences will marry the improved Patriot interceptor and its progeny with a more tightly co-ordinated system of new and existing sensors for detecting and tracking short- and medium-range missiles, as well as any bombers that might survive the battle for air superiority and US attacks on their airfields.

*War on land*

The tank remains the pivotal platform in land warfare. Rival teams of engineers are simultaneously trying to render it obsolete and preserve its central importance. Of the former efforts, the most notable are those to improve the attack helicopter's anti-tank capabilities, and to perfect smart mines and 'indirect-fire' weapons[45] that can deliver munitions accurately against formations of tanks.[46] The latter efforts have sought incremental improvements in tank armour, propulsion, firepower, automation and stealth properties (to avoid detection).[47]

*War at sea*

War at sea is also becoming more centralized and automated, with electronics and sensors absorbing more than half of the investment in advanced vessels

[42] Green, B., 'Technology on five fronts', *Air Force Magazine*, Sep. 1992.

[43] Green (note 42).

[44] Morrison, D., 'Too many birds?', *National Journal*, 4 July 1992. The Air Force plans to build 648 F-22s for a total programme cost of as much as $100 billion. Early in the campaign, Clinton's position was that the F-22 should be 'redesigned'. Reductions in the procurement budget could result in as many as 200 aircraft being cut from the order, according to service personnel. Holzer, R. *et al.*, 'Aspin cuts to redefine the military', *Defense News*, 8–14 Feb. 1993, p. 1; and Leopold (note 20), p. 34.

[45] These include mortars, howitzers and most ballistic missiles, that is, weapons that are not aimed directly at a target by a human operator.

[46] Brower, K. S. and Canby, S. L., 'Weapons for land warfare', ed. E. Arnett, *Future of Smart Weapons* (AAAS: Washington, DC, 1992).

[47] National Research Council, *STAR 21: Strategic Technologies for the Army of the 21st Century* (National Academy Press: Washington, DC, 1992).

like the Seawolf submarine.[48] Ships are more tightly co-ordinated and will rely on extensive sensor arrays, including those deployed on the sea floor. Although the Navy's new strategic concept stresses the maritime contribution to war on land and the littoral,[49] DARPA's efforts are concentrated on undersea warfare, an area that—like strategic bombing—had become almost single-mindedly directed towards missions related to war with the USSR. The challenge of Third World submarines—the significance of which is quite controversial[50]—requires different types of search capability and anti-submarine weapon.

*Command and control*

Global command and control is the heart of the military's operational concept. It is hoped that a network of sensors, computers, platforms and weapons—often referred to as a 'system of systems'—can be developed such that useful information can be extracted from the data obtained by thousands (and later millions) of space-based, airborne, and terrestrial sensors and furnished to the appropriate commanders, personnel and, perhaps, autonomous systems. With some spectacular exceptions—including cueing information provided to Patriot missile batteries from satellites by way of the continental United States and from ground-based back-scatter radars in Australia[51]—such timely, tactical use of the data collected by expensive sensors was disappointingly infrequent during the Persian Gulf War.

*Manufacturing*

DARPA's manufacturing initiatives seek to strengthen communication between procurement programme managers and their contacts in the Defense Department. These steps have much in common with similar efforts in the civilian sector to make producers more responsive to consumers and may facilitate the now-familiar phenomenon of 'spin-on', whereby the military seeks to keep up with developments in the private sector. The primary goals are cost-cutting and waste-reduction, with more rapid delivery of specialized parts produced in small numbers a related benefit.[52] Aspin has been quick to apply advances in manufacturing in hopes that lower unit costs will allow some systems to be bought in the quantities originally planned despite cuts in

[48] Friedman, N., 'Smart weapons, smart platforms: the new economics of defense', ed. E. Arnett, *Science and International Security: New Perspectives for a Changing World Order* (AAAS: Washington, DC, 1991).

[49] A second post-cold war maritime strategy was elucidated in the White Paper *From the Sea: Preparing the Naval Service for the 21st Century, A New Direction for the Naval Service* (Department of the Navy: Washington, DC., Sep. 1992). The Secretary of the Navy, the Chief of Naval Operations and the Commandant of the Marine Corps are the official authors of this paper, which 'defines a combined vision for the Navy and Marine Corps'.

[50] Anthony, I., SIPRI, *The Naval Arms Trade* (Oxford University Press: Oxford, 1990); and Friedman, N., 'Future trends in naval warfare', in Arnett (note 46).

[51] Pike, J., Lang, S. and Stambler, E., 'Military use of outer space', *SIPRI Yearbook 1992* (note 27), pp. 122–27.

[52] Starr, B., 'The Jane's interview: US Deputy Defense Secretary Donald Atwood', *Jane's Defence Weekly*, 28 Nov. 1992.

the procurement budget. He redirected $1.5 billion in defence spending to other agencies for 'improved weapon production' in February 1993.[53]

*Training*

The primary emphasis in the training area is on simulators, much as it is with commercial airlines. Sometimes likened to walk-in video-games or electronic sand-tables, simulators allow personnel to be trained in situations that would be expensive to play out in the field or might otherwise risk personal injury or the destruction of valuable equipment, and to re-examine and recreate those with which they have difficulty. In addition, they may help to familiarize personnel with the more computerized and centralized methods by which future wars may be fought.

## SDIO

Supporters of the SDIO began the year with high hopes as the 'Star Wars' programme's opponents in Washington and Moscow appeared to be readying themselves for compromise and co-operation, respectively. By the end of the year, however, the SDIO's record of test failures had lengthened and been exposed to public scrutiny, Bill Clinton had been elected president having said that the SDIO budget would be cut and no anti-missile systems would be deployed in space,[54] and both the US Congress and the Yeltsin Administration had made it clear that their interest in SDI was very limited. As a final indignity, the Congress moved to transfer the SDIO's original *raison d'être*, research on exotic technologies, back to DARPA.[55] Nevertheless, the near certainty at the end of the year that the 1972 Anti-Ballistic Missile (ABM) Treaty would be honoured was accompanied by broader support for 'low-to-moderate technical risk' ground-based strategic defences that might be deployed by 2002.

*Congress*

In 1991 Congress agreed to fund SDIO activities totalling $4.1 billion and passed the Missile Defense Act which, by suggesting that initial deployments should begin in 1996, seemed to signal a realignment of members in favour of some sort of strategic defence. In hopes of exploiting this opening in 1992, the Bush Administration requested $5.4 billion for the 1993 SDIO budget. Congress rebuffed the request, appropriating $3.8 billion and requiring a 57 per cent cut (from $576 million to $246 million) in the conservatives' favourite

[53] Holzer, R., 'Aspin cuts to redefine the military', *Defense News*, Feb. 1993, p. 1.

[54] In Feb. 1993, Aspin directed the SDIO to cut its proposed first $6.2 billion budget by $2.5 billion. Holzer, 'Aspin cuts to redefine military', p. 1. Clinton's budget contained $73 million in funding for the space-based 'Brilliant Pebbles' interceptor concept. Lawler, A., 'Missile plan downplays space', *Defense News*, 5–11 Apr. 1993, p. 13.

[55] The Congress also placed work on tactical missile defences under a newly established Theater Missile Defense Initiative (TMDI) that may be independent of SDIO. Both decisions depend upon the final determination of the Secretary of Defense.

programme, the 'Brilliant Pebbles' concept for anti-missile satellites.[56] Bill Clinton's promise to base all missile defences on the ground will almost surely mean deep cuts in the programme's funding under the new Administration. In a statement in early 1992 he promised 'to bring a healthy dose of reality to the SDI program . . . [It] should be geared to the real threats we face today and are likely to face in the future, not the fevered rationalizations of a weapons program in search of a mission'.[57] While supporting tactical missile defences and leaving open the option of deploying a single-site ground-based defence, Clinton said that he would consider only modest changes in the ABM Treaty—as opposed to the abrogation or drastic revision advocated during the Reagan–Bush years—and made clear that even minor changes were not needed at present. Clinton's position papers on defence issued before the election suggested that the SDIO budget would remain at about $4 billion.[58]

During the year the initial deployment date for a ground-based defence was repeatedly pushed back. The Missile Defense Act had called for a treaty-compliant ground-based defence ready for deployment 'by the earliest date allowed by the availability of appropriate technology or by Fiscal Year 1996'.[59] Early in 1992, SDIO Director Henry Cooper stated that even with unlimited funding the 1996 date was probably not feasible. By May he had said that 1997 was the earliest technically feasible date.[60] Shortly thereafter, David Chu, the Assistant Secretary of Defense for Program Analysis and Evaluation, warned that a 1997 plan would be 'almost certain to suffer early, significant cost growth and schedule slippage' and that, in order to perform sufficient testing before fielding hardware, deployment should not be planned until early in the next century.[61] None the less in its July report to Congress, the SDIO stated that deployment would most likely begin in 1998.[62] In October Congress repealed the original 1996 date, endorsing instead a new plan for deployment in 2002. It also reiterated that the defence must be compliant with the ABM Treaty, which allows only one site with no more than 100 single-

[56] Lawler, A., 'Pentagon truncates Brilliant Pebbles tests', *Defense News*, 25–31 Jan. 1993, p. 36.

[57] 'The Democrats and arms control: the questions in 1992', *Arms Control Today*, Mar. 1992, p. 6.

[58] Asker, J. R., 'Gore/Quayle face-off foreseen as Clinton offers space plan', *Aviation Week & Space Technology*, 27 July 1992, p. 22. Clinton's first budget includes $3.8 billion for SDIO; Lawler (note 54).

[59] *Aviation Week & Space Technology*, 23 Mar. 1992, p. 20.

[60] Bond, D. F., 'SDIO girds against loss of funds for long-term R&D', *Aviation Week & Space Technology*, 11 May 1992, p. 22.

[61] 'Pentagon critique delays "Star Wars" deployment', *Arms Control Today*, June 1992, p. 24; Asker, J. R., 'SDIO prepares acquisition plan, answers pointed Pentagon questions', *Aviation Week & Space Technology*, 8 June 1992, p. 24; Opall, B., 'Pentagon and SDIO disagree on cost, schedule of GPALS', *Defense News*, 8–14 June 1992, p. 18.

[62] Asker, J. R. and Covault, C., 'SDI will shift funds to ABMs but miss deployment deadline', *Aviation Week & Space Technology*, 23 Mar. 1992, p. 20; Kiernan, V., 'SDI plans to deploy first defense by '98', *Defense News*, 6–12 July 1992, p. 3.

warhead interceptors.[63] The SDIO claims that at least five sites would be needed to provide coverage of the entire USA, including Alaska and Hawaii.[64]

*Russian co-operation*

There was heightened activity throughout 1992 on the possibility of co-operation between Russia and the USA on technologies that might contribute to a strategic defence system. In January President Boris Yeltsin said that Russia was 'ready to develop, then create and jointly operate a global defence system instead of the SDI system'.[65] His statement specified that such a system should comply with the ABM Treaty. Other Russian officials said he was speaking primarily of co-operation on early warning, at least for the foreseeable future. However, despite Yeltsin's original stipulation to the contrary, SDIO Director Cooper and other promoters of strategic defence interpreted Yeltsin's words as indicative of a basic change in Russia's position on the Treaty.

During the June summit meeting it was agreed that the USA and Russia should work with their allies and other interested states to develop a 'concept' for a Global Protection System against ballistic missiles 'as part of an overall strategy regarding the proliferation of ballistic missiles and weapons of mass destruction.' It was agreed that a high-level working group would explore three steps:

> The potential for sharing of early warning information through the establishment of an early warning centre.
>
> The potential for cooperation with participating states in developing ballistic missile defence capabilities and technologies.
>
> The development of a legal basis for cooperation, including new treaties and agreements and possible changes to existing treaties and agreements necessary to implement a Global Protection System.[66]

The US side made it clear that it wanted to move in the direction of altering or amending the ABM Treaty, although Russian negotiators continued their long-standing opposition to changes. Russia also suggested a ban on anti-satellite weapons; such a ban would be incompatible with space-based defences.[67] In July 1992 three working groups were formed. The first will study the structure, features and functions of the system; the second, specific forms of technological co-operation that would be of use; and the third, non-proliferation.[68]

63 *Congressional Quarterly*, 10 Oct. 1992, p. 3186. This site (Grand Forks, North Dakota) has appeared on the Pentagon's list of bases recommended for closure.

64 Three to five sites (depending on whether one is the Grand Forks site stipulated in the ABM Treaty) would be in the contiguous states and one each in Alaska and Hawaii. *Aviation Week & Space Technology*, 3 Feb., p. 28; 13 Apr., p. 25; and 24 Feb. 1992, p. 26; *Defense News*, 6–12 July 1992 p. 3.

65 *Arms Control Today*, Jan./Feb. 1992, p. 38.

66 ACDA text quoted in Institute for Defense & Disarmament Studies, *Arms Control Reporter* (IDDS: Brookline, Mass.), sheet 575.B.418, 1992.

67 Mann, P., 'U.S., Russia slash arms, boost space ventures', *Aviation Week & Space Technology*, 22 June 1992, p. 22.

68 US–Russian Missile Defense Statement, 15 July 1992, reprinted in *Arms Control Reporter*, sheet 575.B.420, 1992.

Another aspect of the SDIO's exploitation of the situation in Russia was a plan to acquire Russian technology in 50 areas, particularly six in which Russian scientists were judged to be ahead: space nuclear power systems, ballistic missile lethality, liquid-fuelled rocket engines, electric thrusters (electric propulsion systems which are potentially much more efficient than conventional rockets), tacitrons (light, high-speed switches that can operate at high temperatures) and neutral particle beams. According to the SDIO, the plan would cost less than $50 million and save $4.5 billion in development costs, while accelerating deployment of US space-based defence systems by several years.[69]

In April the SDIO announced the purchase of four electric thrusters.[70] Shortly thereafter, two Topaz reactors, of a type similar to that used in Soviet radar satellites, were delivered for tests in the USA.[71] The SDIO had planned to demonstrate Topaz's capabilities as early as 1995, when its Electric Propulsion Space Test Satellite was scheduled for launch,[72] but these plans were postponed indefinitely in January 1993. Russian designers from the Topaz programme may contribute to an SDIO-funded programme for thermionic nuclear reactors for space power, in which the Air Force has also expressed an interest.[73] Joint research on tacitrons has been under way in the USA since 1991 and funding was substantially increased in 1992.[74]

### *Tests*

During 1992 the SDIO test programme continued a string of failures that was harshly spotlighted by a congressional audit of its first seven tests. The General Accounting Office found that the SDIO had made 'overstatements' and 'inaccurate' claims.[75] John Conyers, Chairman of the House Government Operations Committee, summarized the report as follows: 'SDI officials have covered up a series of test failures with misleading statements to Congress, chicanery, and outright false claims of success'.[76] An editorial in *Space News*, a staid and usually supportive industry journal, had already taken the SDIO to task for making 'the inexcusable error' of lying about a Brilliant Pebbles test

[69] 'SDIO plans to acquire Russian ABM technology, specialists', *Aviation Week & Space Technology*, 10 Feb. 1992, p. 18.

[70] Asker, J. R., 'Purchase of Russian space hardware signals shift in U.S. trade policy', *Aviation Week & Space Technology*, 6 Apr. 1992, p. 25.

[71] Henderson, B. W., 'SDIO planning mission with Russian Topaz 2 reactor', *Aviation Week & Space Technology*, 29 June 1992, p. 57.

[72] Lenorovitz, J. M., 'SDIO seeks proposals for Topaz 2 launch', *Aviation Week & Space Technology*, 16 Nov. 1992, p. 24.

[73] Henderson, B. W., Russian partners to aid US firms in developing space reactors', *Aviation Week & Space Technology*, 22 June 1992, p. 26.

[74] Henderson, B. W. 'Tacitron research upgrades under way; SDIO officials deny link to Topaz delays', *Aviation Week & Space Technology*, 11 May 1992, p. 26; 'SDIO research contract mix-up delays delivery of Topaz 2 reactors', *Aviation Week & Space Technology*, 11 May 1992, p. 21.

[75] US Congress, General Accounting Office, *Strategic Defense Initiative: Some Claims Overstated for Early Flight Tests of Interceptors* (General Accounting Office: Gaithersburg, Md., Sep. 1992).

[76] Bunn, M., 'GAO reports dispute accuracy of missile defense claims', *Arms Control Today*, Oct. 1992, p. 37.

that it characterized as 'a resounding failure . . . bungled badly'.[77] In the month following the report, the SDIO suffered three launch failures of test missions within a nine-day period.

The last scheduled test of the X-ray laser—the concept that originally helped convince Reagan that nuclear missiles could be made obsolete—was cancelled in 1992.[78] In a further irony, one of the few successful tests was the last for another programme, the High-Altitude Endoatmospheric Defense Interceptor (HEDI), the purpose of which was to develop technology for a ground-based kinetic interceptor of the sort supported by Clinton.[79] Concerns over funding forced a choice between further HEDI testing and the Ground-Based Interceptor—the latter option being the one selected for further development.[80]

## VI. The new era in military space

In 1992 post-cold war US attitudes towards space activities became significantly clearer, but new dilemmas emerged. As noted above, the armed services see a need for large fleets of tactical reconnaissance satellites. Some first steps towards that goal were apparent during the year, including a larger constellation of surveillance satellites and bureaucratic changes that may make the data more useful. In addition, for the first time, the USA may be considering the foreign sale of a surveillance satellite. The tension between non-proliferation and other national goals inherent in that decision was also indicated in the debate over access to satellite navigation data. Finally, satellite programmes relating to the cold war missions of nuclear and space warfare were throttled back, although not as convincingly as was the SDI. Further programme cuts will be necessary to comply with a 15 per cent reduction in funding for military space activities mandated by Congress.[81]

### Surveillance and intelligence

The year began with a record number of intelligence satellites on station, and the USA continued to add to these capabilities during 1992 (see appendix 8A for details of military satellites launched during the year).

A new Central Imagery Office created in the Pentagon is 'tied to attempts to provide greater operational flexibility in determining how pictures taken by advanced KH-11 digital imaging and Lacrosse radar imaging reconnaissance

[77] Arnett, E. H., *Ballistic Missile Defense after the Kuwait War* (AAAS: Washington, DC, 1991), p. 3.

[78] Gordon, M. R., '"Star wars" x-ray laser weapon dies as its final test is cancelled', *New York Times*, 21 July 1992, p. A1.

[79] For more specific details on these programmes see, for example, Pike, J., 'Military use of outer space', *SIPRI Yearbook 1991: World Armaments and Disarmament* (Oxford University Press: Oxford, 1991), pp. 52–53.

[80] 'SDI Interceptor passes test but falls to budget cuts', *Aviation Week & Space Technology*, 7 Sep. 1992, p. 42.

[81] Boyer, W., 'US Space Command chief blasts Titan 4 program', *Space News*, 2–8 Nov. 1992, p. 13.

satellites are targeted and distributed to US forces in the field.'[82] A new Comprehensive Operational Image Architecture is being defined that would allow users to keep up with rapidly developing tactical situations. The global scope of the targeting and analysis operations as set up during the Persian Gulf War hampered acquisition, analysis and distribution of the imagery required, despite unprecedented planning and openness to prevent just that problem. Because of their dissatisfaction with 'tactical employment of national capabilities (TENCAP)' during the war, the services have expressed interest in deploying tactical satellites (TACSATs) under their own control and SDIO has emphasized other applications of its proposed fleet of 50 'Brilliant Eyes' (small, relatively inexpensive reconnaissance satellites).

The Pentagon has officially acknowledged the existence of the National Reconnaissance Office (NRO) which has been responsible for managing satellite programmes and operating the systems for the intelligence community since 1960.[83] Its annual budget is estimated at about $6 billion, but may be as high as $8 billion.[84] The budget for the intelligence community, 'black' but estimated to be about $30 billion, will be cut by about 8 per cent for FY 1993 with the NRO taking the largest reductions. The satellite programme may be reorganized and one of the electronic intelligence (ELINT) satellites eliminated. In addition, the failure of US intelligence to locate Iraqi weapon facilities or predict President Saddam Hussein's actions has led many to conclude that too much emphasis has been put on national technical means (NTM), to the detriment of 'human intelligence' and country analysis, since Stansfield Turner's tenure as Director of Central Intelligence during the Carter Administration, when NTM were given their special stature. If so, other NTM programmes may also be cut.

In 1992 there was also a new round of activity on the 1991 application from the United Arab Emirates to the US Department of State to allow a study of requirements that might culminate in the sale of a reconnaissance satellite, reportedly solicited by a US firm.[85] If the deal goes forward, the USA is expected to limit the capability of the satellite such that the resolution will be no better than 1 metre and the satellite would not be capable of recording images (operators would receive data as the satellite passed within range of their ground station).[86] The satellite would be launched and operated by a US contractor. South Korea and Spain also may have approached the USA with similar interests.[87]

[82] Munro, N., 'Imagery office centralizes oversight of spy data, funds', *Defense News*, 15–21 June 1992, p. 12; *Aviation Week & Space Technology*, 8 June 1992, p. 19.

[83] Richelson, J. T., *The U.S. Intelligence Community*, 2nd edn (Ballinger: Cambridge, Mass., 1989), pp. 26–27.

[84] Sweetman (note 35).

[85] Kiernan, V., 'Itek to limit spy satellite capability', *Space News*, 23–29 Nov. 1992, p. 1; 'U.S. ponders UAE request on photo recon satellite', *Aviation Week & Space Technology*, 23 Nov. 1992, p. 108; Broad, W. J. 'Offers for U.S. spy satellites set off rift', *International Herald Tribune*, 24 Nov. 1992, p. 1.

[86] Kiernan, V., 'Resolution will limit use of UAE satellite', *Defense News*, 23–29 Nov. 1992, p. 8.

[87] Kiernan, V. and Lawler, A., 'Emirates want to buy U.S. spy satellite', *Space News*, 16–22 Nov. 1992, p. 1; Kiernan, V. and Lawler, A., 'UAE satellite plan rattles U.S.', *Defense News*, 16–22 Nov. 1992, p. 3.

## Navigation

The Navstar (navigation satellite with timing and ranging) constellation neared completion in 1992 amid a growing debate over the utility of the Global Positioning System (GPS) to states hostile to US interests.[88] The Pentagon and some private analysts believe that the ability to locate platforms, weapons or targets precisely (with the aid of commercially available satellite imagery) could give the militaries of developing countries a crucial additional capability that could endanger US troops and those of allies[89] or even render the USA newly vulnerable to attack.[90] US forces in the Middle East used GPS extensively during the war against Iraq, and 35 GPS-guided cruise missiles were launched in one attack.[91] The Pentagon advocates a more elaborate system for degrading the GPS signal in times of conflict. Other observers, especially those relying on GPS for civilian applications—including air traffic control, maritime navigation, geological surveying and mapping, and vehicle tracking—question the need for such measures.[92]

## Nuclear and space war-fighting

While most US military satellites would have had a role in the event of a superpower nuclear war in addition to their more mundane functions,[93] the costly and secret Military Strategic and Tactical Relay (Milstar) satellite programme was considered by many to be overly specialized for nuclear war-fighting and as unnecessary in the cold war's aftermath as the B-2 bomber and the Seawolf submarine, given the large fleet of military and civilian communications satellites available already. Costing $850 million apiece by 1990 estimates, Milstar was designed as a nuclear-hardened communications node that could, for example, feed data on mobile targets to bombers operating in Soviet airspace even after a nuclear war had been raging for weeks. Nuclear-hardening and protection from anti-satellite attack reportedly accounted for much of the programme's cost. Three Milstar 1 satellites were built before much public information about the programme was available. The satellite was then modified to the follow-on Milstar 2 design, featuring almost 100 times the original capacity of Milstar 1 without as much protective shielding.[94] As lead service

[88] At the end of 1992, 17 Navstar satellites were in orbit of the 18 necessary for uninterrupted global GPS coverage. Until the 18th is launched, US users will have only 22 hours of 2-dimensional and 16 hours of 3-dimensional coverage a day. The full complement of 24 satellites should be complete in 1994.

[89] Carus, W. C., *Ballistic Missiles in Modern Conflict* (Praeger: New York, 1991); and Lachow, I., *GPS and Cruise Missile Proliferation: Managing the Tensions Between Defense Needs and Civilian Applications* (Carnegie Mellon University: Pittsburgh, Pa., 1992).

[90] Tsipis, K., 'Off-shore threat—cruise missiles', *New York Times*, 1 Apr. 1992.

[91] Arnett (note 41).

[92] Arnett, E., 'The most challenging problem of the 1990s? Cruise missiles in the developing world', eds W. T. Wander and E. H. Arnett, *The Proliferation of Advanced Weaponry: Technology, Motivations and Responses* (AAAS: Washington, DC, 1992); and Lachow (note 89).

[93] Arnett, E., *Antisatellite Weapons* (AAAS: Washington, DC, 1990).

[94] Dornheim, M., 'Milstar 2 brings new program role', *Aviation Week & Space Technology*, 16 Nov. 1992, p. 63.

on the programme, the Air Force suggested cancelling Milstar early in 1993, but was rebuffed by Aspin and the other services.

US anti-satellite programmes remained essentially moribund after the fierce cold war battles between supporters and critics. At the end of 1992, Congress banned testing of the MIRACL (Mid-Infrared Advanced Chemical Laser) against objects in space and cut the ASAT-capable Brilliant Pebbles SDI programme in anticipation of President Clinton's inauguration. The ground-based strategic missile interceptors favoured by Clinton will have some ASAT capability. Beam and rocket ASAT programmes were funded, but no full-scale tests are in the offing.[95]

## VII. Controlling the spread of technology

An important corollary of the US goal of maintaining military supremacy is the necessity to deny technologies to possible enemies. For this reason, many US policy makers see 'technology-denial regimes' as a crucial aspect of the strategy, a consideration more important than others for limiting transfers of military and related technology. For example, Clinton-adviser Janne Nolan, dismissing more altruistic reasons, supports the Missile Technology Control Regime (MTCR) as a 'prudent gesture on the part of the Western countries to stem the deterioration of military environments in which they may have to protect their interests'.[96] Similarly, Seth Carus, an analyst in the Office of the Secretary of Defense, promotes the MTCR as a way to limit 'the risks [to US forces] entailed in military operations in the Third World'.[97] Some observers voice the suspicion that US technology-denial is also self-serving in an economic sense. They see the USA using the combination of its technological clout and international support for control regimes' more idealistic goals selectively to promote its own interests and exports.[98]

Developing countries that seek to emulate the US approach to military planning, in part because of the success of the Coalition against Iraq, have had to take US policies of denial into account. Thus, those who may foresee a clash with the USA, or with a neighbouring state with Western security partners, may have to re-evaluate their chances of developing adequate systems indigenously or through foreign purchases. Of course, creating pressure for such a re-evaluation was one of the goals of the draft US Defense Planning Guidance.

[95] Lockwood, D., 'Congress OKs $274 billion defense budget—2.4 percent off Bush request', *Arms Control Today*, Oct. 1992, p. 33.

[96] Nolan (note 8); on the MTCR see also chapters 6 and 10 in this volume.

[97] Carus (note 89).

[98] Morrison, D. C., 'Washington watching the "have-nots"', *National Journal*, 4 July 1992. According to Morrison, the USA decided to embargo the Indian Space Research Organization for buying cryogenic rocket boosters from the Russian Glavkosmos concern only after the latter had underbid General Dynamics, a US firm. In correspondence later made public by the Indian Government, General Dynamics had assured the ISRO that the technology was not subject to MTCR controls because it was of limited military utility.

Nevertheless, several developing states in the world's most conflict-prone regions continue to strive for force structures incorporating state-of-the-art conventional weaponry, whether through indigenous programmes, imports or a combination of the two. China's effort to acquire contemporary military technology to replace its largely obsolescent forces and improve the quality of future exports is at a relatively early stage and its ability to overcome problems with workmanship and management remains to be proven.[99] India's long-standing programmes to independently produce tanks, missiles, aircraft and warships of a sophistication and quality comparable to those of the West have yet to bear fruit, in the form of operational forces, despite decades of investment.[100] Iraq's defence industrial base was heavily damaged by Coalition bombing, and facilities in certain sectors (nuclear, biological and chemical weapons and missiles) were destroyed after the war. The Iraqi technology base will remain under intense scrutiny and UN and IAEA regulation for the foreseeable future. Of the developing countries,[101] only Israel, which enjoys a unique relationship with the USA, has been able to develop a viable and innovative defence technology base. It appears unlikely that any regional challengers to US interests will have more capable conventional technology than did Iraq in 1991. Nevertheless, some US policy makers fear that 'silver bullet' technologies—for example, land-attack cruise missiles—in the wrong hands may undermine the US advantage or offer a deterrent to US intervention under some circumstances.

## The role of normative arms control

The futility of matching US technological achievements may increase interest in nuclear weapons in some states, a fear voiced by then-Chairman of the House Armed Services Committee Aspin and others.[102] Some states that do not fear the USA as much as they do militarily superior neighbours may draw similar conclusions.[103] Other weapons that have been seen as useful for 'equalizing' or striking back against a superior foe include ballistic and cruise missiles, chemical and biological weapons, submarines, mines and stealth bombers. Efforts to control the transfer of these technologies or to promote norms that characterize them as 'indiscriminate', 'inhumane', 'offensive' or 'destabilizing' can be seen as serving the goal of eliminating the residual

[99] Hua, D., 'China's arms proliferation in perspective: prospects for change due to economic reforms', in Wander and Arnett (note 92), and Bitzinger, R. A., 'Arms to go: Chinese arms sales to the Third World', *International Security*, fall 1992.

[100] Smith, C., SIPRI, *India's Ad Hoc Arsenal* (Oxford University Press: Oxford, forthcoming).

[101] This discussion does not include the Newly Industrialized Countries, such as South Korea and Taiwan, in the category 'developing countries'.

[102] Aspin, L., *From Deterrence to Denuking: Dealing with Proliferation in the 1990s* (US Congress, House Committee on the Armed Services: Washington, DC, 1992). See also Arnett, E. H., 'Power projection issues: can a superpower be "equalized?"', presented at the Science Applications International Corporation, McLean, Va., 1991; and May, M. M. and Speed, R. D., 'Will nuclear weapons be used?', in Wander and Arnett (note 92).

[103] Mazari, S., 'Nuclear weapons and structures of conflict in the developing world', in Wander and Arnett (note 92).

threat posed by such weapons to US intervention forces.[104] One Chinese analyst, for example, claims that US efforts to stigmatize inaccurate ballistic missiles as indiscriminate are hypocritical, given that the USA has only recently implemented techniques for more accurate bombardment and is unwilling to share them.[105]

### *Disabling weaponry*

Promoting such norms without adversely affecting US security policy is an explicit goal of an initiative to develop and deploy 'disabling' or 'non-lethal weaponry'.[106] Some such technologies already exist. Since World War II, for example, electronic warfare systems have disabled the radars and communications equipment on which an enemy's forces and weapons depend. Some proponents of chemical weapons have noted that—though chemical warfare agents are certainly deadly—some types cause fewer deaths for a given number of casualties than do high explosive or anti-personnel munitions, while the tendency of untrained personnel to panic when attacked with chemical weapons yields tactical advantages without taking lives.[107] In 1982, the Royal Navy may have used lasers to blind Argentine sensors. During the 1991 Persian Gulf War, US cruise missiles scattered carbon filaments over Iraqi electrical stations, short-circuiting them and shutting them down, thereby disabling military installations.[108]

The Persian Gulf War and attendant concerns about 'minimizing collateral damage' increased interest sufficiently that many R&D programmes seeking fiscal support were consolidated into 'non-lethal' clusters. Paul Wolfowitz, the Under Secretary of Defense for Policy, formed a Non-Lethal Warfare Study Group in March.[109] Taking the lead, the Army proposed technologies that could 'effectively disable, dazzle or incapacitate aircraft, missiles, armoured vehicles, personnel and other equipment while minimising collateral damage.'[110] The possibilities include a microwave weapon that would produce

[104] Arnett, E., 'Technology and military doctrine: criteria for evaluation', ed. T. Wander, *Advanced Weaponry in the Developing World* (AAAS: Washington, DC, forthcoming).

[105] Hua, D., 'The arms trade and proliferation of ballistic missiles in China', in AAAS, *Arms Sales Versus Nonproliferation: Economic and Political Considerations of Supply, Demand, and Control* (AAAS: Washington, DC, 1992).

[106] Opall, B., 'Pentagon units jostle over non-lethal initiative', *Defense News*, 2 Mar. 1992, p. 6.; see also chapter 9 in this volume.

[107] B. H. Liddell Hart called chemical agents 'the most obstructive yet the least lethal of weapons' making war 'progressively less lethal and more humane', and noted favourably their psychological effects. See Liddell Hart, B. H., 'Is gas a better defence than atomic weapons?', *Marine Corps Gazette*, Jan. 1960, pp. 14–16 and 'Gas in warfare more humane than shells', *Daily Telegraph*, 15 June 1926. The debate is summarized in Adams, V., *Chemical Warfare, Chemical Disarmament* (Indiana University Press: Indianapolis, Ind. 1990), pp. 39–49. The Egyptian chemical warfare programme has also been defended in these terms. Heikal, M., *Illusions of Triumph: an Arab View of the Gulf War* (HarperCollins Publishers: London, 1992), p. 92. Critics of these arguments note that chemical weapons do kill and often permanently maim those who are not killed, and therefore cannot legitimately be termed 'humane'.

[108] Fulghum, D. A., 'Secret carbon-fiber warheads blinded Iraqi air defenses', *Aviation Week & Space Technology*, 27 Apr. 1992. An earlier report that cruise missiles were armed with microwave weapons was informally disavowed by the Pentagon, though such weapons are being developed.

[109] Opall (note 106), p. 6.

[110] Starr, B., 'USA tries to make war less lethal', *Jane's Defence Weekly*, 31 Oct. 1992, p. 10.

an electromagnetic pulse of sufficient intensity to disrupt or damage electrical equipment, 'traction inhibitors, chemical immobilizers, entanglement munitions, . . . neural inhibitors to temporarily incapacitate personnel, infrasound to disorient people',[111] optical weapons to disrupt enemy sensors (including the eyes of personnel), acoustic beams which could direct low-frequency, high-power waves at enemy personnel and pulsed laser systems which would produce a controllable blast wave. Chemical and biological agents might be introduced into petroleum facilities to make their products useless and corrode or soften other items of military equipment.[112]

Although foreseen funding for the initiative amounts to only $148 million over the next five years, some observers hope it could bloom into 'another SDI-like operation where billions of dollars would be available.'[113] The enthusiasm for these projects is such that Sam Nunn, Chairman of the Senate Committee on Armed Services, suggested that 'possibilities in the non-lethal area . . . ought to be looked at carefully and exploited' and might make US military options against Serbia less difficult to accept.[114]

## VIII. Reconciling the aspirations and legacies

As the Clinton Administration struggles to reconcile its aspirations to increase R&D activities with limits on its freedom of action, the greatest constraints are likely to be imposed by the US economy, the focus of both Clinton's election campaign and his security policy.[115] On the one hand, Clinton's desire to cut the federal budget creates an imperative for him to fulfil his promise to reduce military spending, and Aspin has mandated $13.6 billion in cuts for the first year.[116] John Murtha, the conservative Chairman of the Defense Subcommittee of the House Committee on Appropriations, predicted that funding for the C-17, the V-22, the Seawolf and the A/F-X carrier-based bomber would be cut or eliminated by the Congress in Clinton's first year.[117] On the other hand, the slack economy makes Clinton reluctant to cut military spending where it will mean lost jobs.

[111] Opall (note 106), p. 6. Neural inhibitors might pose compliance problems for states parties to the Chemical Weapons Convention.

[112] Fulghum, D. A., 'U.S. weighs use of non-lethal weapons in Serbia if UN decides to fight', *Aviation Week & Space Technology*, 17 Aug. 1992, p. 62; Munro, N. and Opall, B., 'Military studies unusual arsenal', *Defense News*, 19–25 Oct. 1992, p. 3; Ricks, T., 'A kinder, gentler war may be in order', *Globe and Mail* (Toronto), 5 Jan. 1993, p. 1; 'Nonlethal weapons give peacekeepers flexibility', *Aviation Week & Space Technology*, 7 Dec. 1992, p. 50.

[113] Opall (note 106), p. 6.

[114] Fulghum (note 112).

[115] Indeed, Clinton has expanded the National Security Council to include the Treasury Secretary and the Director of the newly formed National Economic Council, as well as the Ambassador to the UN.

[116] He stipulated that R&D be 'fully funded as planned', aside from cuts in the SDIO's budget request. 'Washington outlook: day of reckoning', *Aviation Week & Space Technology*, 8 Feb. 1993, p. 19.

[117] 'Washington outlook: raw truths', *Aviation Week & Space Technology*, 7 Dec. 1992, p. 21. Clinton's first budget eliminates funding for the Seawolf while maintaining or increasing funding for the C-17 and the A/F-X.

### Reconstitution

Since 1990 it has been clear that the USA will not be able to build all of the weapons and platforms planned for during the Carter–Reagan cold war build-up. Nevertheless, US planners have been anxious to continue their emphasis on military security through technological innovation, an approach that requires a commitment to R&D even if the resulting designs are never built. The consequence has been the flourishing of the concept of 'reconstitution', that is, the ability to build forces from existing designs or those developed and tested, but then left on the shelf.[118] Under reconstitution, the USA will continue to improve existing hardware, replace essential items that are indisputably out of date, and make simple modifications of the overall force posture in the direction of light forces that can be deployed abroad quickly, but will only field the advanced designs now in R&D when given incontrovertible warning of a new threat for which the mobilization of the Western defence industrial base is necessary.[119] Aspin advocates building only 'silver bullet' technologies that can make a contribution independent of the broader force and current operational art.[120] For now, in the words of the US Under Secretary for Acquisition, the USA enjoys 'a more-than-adequate battle proven arsenal of arms and equipment'.[121]

## IX. Conclusion

Given the US commitment to military R&D and the difficulty other states will have in sustaining their current levels of support for military R&D, the United States will assuredly succeed in its effort to maintain technological supremacy for the foreseeable future. A number of other issues remain less clear, however. The most basic is whether this investment will yield new generations of weaponry that will be fielded, or simply developed and put on the shelf. If new technologies are fielded, observers are likely to ask the perennial question: Do they better serve the needs of the aggressor or the defender? While promoters see in DARPA's 'system of systems' a world in which defences dominate to such an extent that aggression is rendered futile,[122] others point out that the advantage can never be guaranteed to defence or offence, but will

[118] One of the earliest promoters of reconstitution was the team of Rich Wagner and Ted Gold, both formerly top defence advisors. Wagner, R. and Gold, T., 'Long shadows and virtual swords: managing defense resources in the changing security environment', in Arnett (note 48). See also Joint Chiefs of Staff, *1991 Joint Military Net Assessment* (Department of Defense: Washington, DC, 1991) and a brief but useful critique: Snider, D. M. and Grant, G., 'The future of conventional warfare and U.S. military strategy', *Washington Quarterly*, winter 1992.

[119] Reconstitution is thus the antithesis of the Reagan Administration's controversial strategy of 'concurrent' procurement, in which production began before development and testing had been completed.

[120] The most commonly cited example of a 'silver bullet' technology is the F-117 Stealth bomber. Designed to operate independently of support aircraft, the F-117 was kept in the 'black budget' until 1988. After an embarrassing failure during the invasion of Panama, its successful use in the war against Iraq reinforced planners' enthusiasm for the concept of 'silver bullets' in some applications.

[121] Yockey (note 11), p. 64.

[122] Libicki, M., 'Silicon in the twenty-first century', *Strategic Review*, summer 1992, p. 62.

nearly always fall to the side with more advanced equipment.[123] With the United States boasting the world's most modernized military, observers are anxious to see whether the superpower wields its military might more frequently and under what auspices as foreign policy shifts from Bush's 'realism' to Clinton's 'idealism'.

Regardless of developments in terrestrial warfare, the likelihood of a future war in space must be seen as lower than it has been in a decade. As 1992 drew to a close, US plans for orbiting anti-missile battle-stations and anti-satellite weapons had been put on a slow track. And although some observers fear that growing space-launch capabilities in the developing world might threaten US satellites with destruction,[124] such a prospect is still remote.[125]

Finally, the success of US efforts to stigmatize certain types of weapon cannot be predicted. The signing of the Chemical Weapons Convention in January 1993 marked a hard-won success after years of effort, but also points up a number of difficulties. Observance of the Convention is unlikely to be global, at least at first. Further, the strengths of the Convention will be hard to match if similar norms are to be promoted in other areas. With the exception of chemical and biological weapons and intermediate-range ballistic missiles,[126] the USA has generally been unwilling to eschew the military technologies it hopes to stigmatize when in the hands of others.

Whatever the success of efforts to stigmatize the possession of some technologies, international law limiting the application of technology may be a more important approach to reducing the inhumane use of technologies that the USA and others are unwilling to forswear. Despite its apparent lapses in this regard during the Persian Gulf War,[127] the United States reaffirmed its commitment to such principles in 1992 by justifying its prosecution of the air campaign in those terms.[128]

[123] Biddle, S., 'Weapons for land warfare', in Wander (note 104).

[124] Nolan (note 8).

[125] Arnett (note 93).

[126] Drell, S. D., *Scientists and Security* (AAAS: Washington, DC, 1992); Stone, J., 'Zero ballistic missiles', *FAS Public Interest Report*, summer 1992; and Frye, A., 'Zero ballistic missiles', *Foreign Policy*, winter 1992 and 'The ZBM solution: get rid of all ballistic missiles', *International Herald Tribune*, 3 Jan. 1993, p. 4.

[127] Middle East Watch, *Needless Deaths in the Gulf War: Civilian Casualties During the Air Campaign and Violations of the Laws of War* (Human Rights Watch: Washington, DC, 1991); Arkin, W. M., Durrant, D. and Cherni, M., *On Impact: Modern Warfare and the Environment—A Case Study of the Gulf War* (Greenpeace: Washington, DC, 1991).

[128] US Department of Defense, *Conduct of the Persian Gulf War* (Department of Defense: Washington, DC, 1992).

# **Appendix 8A.** Military satellites launched in 1992

| Type/Country/ Spacecraft name | Alternative name (Host spacecraft) | Designation | Launch date | Booster | Launch site | Mass (kg) | Perigee (km) | Apogee (km) | Inclin. (deg) | Period (min) | Comments |
|---|---|---|---|---|---|---|---|---|---|---|---|
| **Imaging intelligence** | | | | | | | | | | | |
| *CIS* | | | | | | | | | | | |
| THIRD GENERATION | | | | | | | | | | | |
| PHOTO 3M-106 | Cosmos 2207 | 1992-048A | 30 July 92 | SL-4 | PL | 300 | 232 | 317 | 82.33 | 90.1 | |
| FOURTH GENERATION | | | | | | | | | | | |
| PHOTO 4-99 | Cosmos 2175 | 1992-001A | 21 Jan. 92 | SL-4 | PL | 6 500 | 184 | 337 | 67.14 | 89.7 | Focused on Middle East, recovered 20 Mar., 11-day coverage gap |
| PHOTO 4-100 | Cosmos 2182 | 1992-016A | 1 Apr. 92 | SL-4 | PL | 6 500 | 166 | 315 | 67.16 | 89.3 | Ended 11-day coverage gap |
| PHOTO 4-101 | Cosmos 2186 | 1992-029A | 28 May 92 | SL-4 | PL | 6 500 | 182 | 332 | 62.85 | 89.8 | Manœuvred on 6 June |
| PHOTO 4-102 | Cosmos 2203 | 1992-045A | 24 July 92 | SL-4 | PL | 6 500 | 190 | 312 | 62.81 | 89.5 | Replaced by C-2210 |
| PHOTO 4-103 | Cosmos 2210 | 1992-062A | 22 Sep. 92 | SL-4 | PL | 6 500 | 175 | 349 | 67.16 | 89.8 | Replaced C-2203 |
| PHOTO 4-104 | Cosmos 2220 | 1992-077A | 20 Nov. 92 | SL-4 | PL | 6 500 | 167 | 342 | 67.14 | 89.6 | |
| PHOTO 4 ? | Cosmos 2225 | 1992-091A | 22 Dec. 92 | SL-4 | TT | 6 500 | 214 | 306 | 64.90 | 89.7 | Deliberately exploded on 18 Feb. 1993, 4th generation characteristics but possibly 6th generation? |
| FIFTH GENERATION | | | | | | | | | | | |
| PHOTO 5-14 | Cosmos 2183 | 1992-018A | 8 Apr. 92 | SL-4 | TT | 6 800 | 240 | 294 | 64.87 | 89.9 | Recovered 16 Feb. 1993, longest mission to date (314 days ) |
| PHOTO 5-15 | Cosmos 2223 | 1992-087A | 9 Dec. 92 | SL-4 | TT | 6 800 | 240 | 292 | 64.66 | 89.8 | |
| MILITARY MAPPING | | | | | | | | | | | |
| PHOTO 4T-15 | Cosmos 2185 | 1992-025A | 15 May 92 | SL-4 | TT | 6 800 | 211 | 279 | 69.97 | 89.4 | Topographic survey/mapping, recovered 11 June |

| | | | | | | | | | | | |
|---|---|---|---|---|---|---|---|---|---|---|---|
| ***USA*** | | | | | | | | | | | |
| KH-12A /2 | USA-86 | 1992-086A | 28 Nov. 92 | Titan 404 A | WTR | 14 550 | 800 | 800 | 97.00 | 98.0 | Possible KH-12 replacement for KH-11 /7 |
| ***China*** | | | | | | | | | | | |
| FSW-2 1 China 35 | | 1992-051A | 9 Aug. 92 | CZ-2D | SCT | 2 400 | 172 | 335 | 63.07 | 89.6 | Capsule recovered 25 Aug. |
| FSW-2 2 China 38 | | 1992-064B | 6 Oct. 92 | CZ-2D | SCT | 2 400 | 215 | 311 | 63.01 | 89.8 | Capsule recovered 13 Oct. |
| **Electronic intelligence** | | | | | | | | | | | |
| ***CIS*** | | | | | | | | | | | |
| ELINT 3-35 | Cosmos 2221 | 1992-080A | 24 Nov. 92 | SL-14 | PL | 4 375 | 636 | 650 | 82.51 | 97.8 | |
| ELINT 3-36 | Cosmos 2228 | 1992-094A | 25 Dec. 92 | SL-14 | PL | 4 375 | 633 | 669 | 82.53 | 97.8 | |
| ELINT 4-12 A | | FAILURE | 5 Feb. 92 | SL-16 | TT | 12 500 | | | | | Second stage malfunctioned |
| ELINT 4-12 B | Cosmos 2219 | 1992-076A | 17 Nov. 92 | SL-16 | TT | 12 500 | 849 | 855 | 71.01 | 102.0 | First successful SL-16 launch after 3 failures |
| ELINT 4-13 | Cosmos 2227 | 1992-093A | 25 Dec. 92 | SL-16 | TT | 12 500 | 849 | 854 | 71.02 | 102.0 | |
| ***USA*** | | | | | | | | | | | |
| NOSS-F/0 3 | USA-81 | 1992-023A | 25 Apr. 92 | Titan 23G | WTR | 450 | 784 | 805 | 85.14 | 108.2 | |
| **Military communications** | | | | | | | | | | | |
| ***CIS*** | | | | | | | | | | | |
| COM 1-353 | Cosmos 2187 | 1992-030A | 3 June 92 | SL-8 | PL | 45 | 1 402 | 1 480 | 74.01 | 114.7 | |
| COM 1-354 | Cosmos 2188 | 1992-030B | 3 June 92 | SL-8 | PL | 45 | 1 389 | 1 479 | 74.00 | 114.6 | |
| COM 1-355 | Cosmos 2189 | 1992-030C | 3 June 92 | SL-8 | PL | 45 | 1 418 | 1 479 | 74.00 | 114.9 | |
| COM 1-356 | Cosmos 2190 | 1992-030D | 3 June 92 | SL-8 | PL | 45 | 1 431 | 1 480 | 74.00 | 115.1 | |
| COM 1-357 | Cosmos 2191 | 1992-030E | 3 June 92 | SL-8 | PL | 45 | 1 473 | 1 503 | 74.01 | 115.8 | |
| COM 1-358 | Cosmos 2192 | 1992-030F | 3 June 92 | SL-8 | PL | 45 | 1 472 | 1 486 | 74.00 | 115.6 | |
| COM 1-359 | Cosmos 2193 | 1992-030G | 3 June 92 | SL-8 | PL | 45 | 1 447 | 1 480 | 74.01 | 115.2 | |
| COM 1-360 | Cosmos 2194 | 1992-030H | 3 June 92 | SL-8 | PL | 45 | 1 459 | 1 485 | 74.00 | 115.4 | |
| COM 2-48 | Cosmos 2208 | 1992-053A | 12 Aug. 92 | SL-8 | PL | 750 | 788 | 807 | 74.04 | 100.9 | Replaced 1937 |
| COM 3-74 | Cosmos 2197 | 1992-042A | 13 July 92 | SL-14 | PL | 400 | 1 397 | 1 416 | 82.59 | 114.0 | Coplanar with C-2165/70 |
| COM 3-75 | Cosmos 2198 | 1992-042B | 13 July 92 | SL-14 | PL | 400 | 1 410 | 1 416 | 82.59 | 114.1 | |
| COM 3-76 | Cosmos 2200 | 1992-042D | 13 July 92 | SL-14 | PL | 400 | 1 405 | 1 416 | 82.59 | 114.1 | |

| Type/Country/ Spacecraft name | Alternative name (Host spacecraft) | Designation | Launch date | Booster | Launch site | Mass (kg) | Perigee (km) | Apogee (km) | Inclin. (deg) | Period (min) | Comments |
|---|---|---|---|---|---|---|---|---|---|---|---|
| COM 3-77 | Cosmos 2202 | 1992-042F | 13 July 92 | SL-14 | PL | 400 | 1 418 | 1 419 | 82.59 | 114.1 | |
| COM 3-78 | Cosmos 2211 | 1992-068A | 20 Oct. 92 | SL-14 | PL | 400 | 1 400 | 1 415 | 82.59 | 114.0 | |
| COM 3-79 | Cosmos 2212 | 1992-068B | 20 Oct. 92 | SL-14 | PL | 400 | 1 408 | 1 414 | 82.59 | 114.1 | |
| COM 3-80 | Cosmos 2213 | 1992-068C | 20 Oct. 92 | SL-14 | PL | 400 | 1 409 | 1 418 | 82.59 | 114.1 | |
| COM 3-81 | Cosmos 2214 | 1992-068D | 20 Oct. 92 | SL-14 | PL | 400 | 1 414 | 1 422 | 82.59 | 114.2 | |
| COM 3-82 | Cosmos 2215 | 1992-068E | 20 Oct. 92 | SL-14 | PL | 400 | 1 413 | 1 428 | 82.60 | 114.3 | |
| COM 3-83 | Cosmos 2216 | 1992-068F | 20 Oct. 92 | SL-14 | PL | 400 | 1 410 | 1 417 | 82.60 | 114.1 | |
| Molniya 1-83 | | 1992-011A | 4 Mar. 92 | SL-6 | PL | 1 250 | 620 | 39 731 | 62.81 | 717.7 | Replaced Molniya 1-75 |
| Molniya 1-84 | | 1992-050A | 6 Aug. 92 | SL-6 | PL | 1 250 | 632 | 39 721 | 62.85 | 717.7 | |
| ***USA*** | | | | | | | | | | | |
| SDS F/O 2 | USA-87 | 1992-086B | 5 Dec. 92 | STS | ETR | 2 250 | 450 | 40 000 | 63.00 | 720.0 | Similar to USA-40 on STS-28, |
| DSCS III-B 3 | | 1992-006A | 11 Feb. 92 | Atlas 2 | ETR | 825 | 35 780 | 35 780 | 0.00 | 1 436.0 | Qualification model refurbished for flight in 1980 |
| DSCS III-B 6 | | 1992-037A | 7 July 92 | Atlas 2 | ETR | 825 | 35 780 | 35 780 | 0.00 | 1 436.0 | |
| AFSATCOM SCT-3 | (On DSCS III-A3) | 1992-006A | 11 Feb. 92 | Atlas 2 | ETR | 28 | 35 780 | 35 780 | 0.00 | 1 436.0 | |
| AFSATCOM SCT–6 | (On DSCS III-B6) | 1992-037A | 2 July 92 | Atlas 2 | ETR | 28 | 35 780 | 35 780 | 0.00 | 1 436.0 | |
| ***France*** | | | | | | | | | | | |
| Syracuse II-2 | (on Telecom 2B) | 1992-021A | 15 Apr. 92 | Ariane 4 | KO | 545 | 35 777 | 35 797 | 0.07 | 1 436.0 | Uses half of Telecom II capacity |
| **Ballistic missile early warning** | | | | | | | | | | | |
| ***CIS*** | | | | | | | | | | | |
| BMEWS 1-68 | Cosmos 2176 | 1992-003A | 24 Jan. 92 | SL-6 | PL | 1 500 | 626 | 39 730 | 62.80 | 717.8 | Replaced C-2087, 1st BMEWS launch in 14 months |
| BMEWS 1-69 | Cosmos 2196 | 1992-040A | 8 July 92 | SL-6 | PL | 1 500 | 590 | 39 733 | 62.95 | 717.1 | Replaced C-1922, closed 1-year gap in constellation |
| BMEWS 1-70 | Cosmos 2117 | 1992-069A | 21 Oct. 92 | SL-6 | PL | 1 500 | 599 | 39 757 | 62.95 | 717.8 | Replaced C-1903 |

| | | | | | | | | | | | |
|---|---|---|---|---|---|---|---|---|---|---|---|
| BMEWS 1-71 | Cosmos 2222 | 1992-080A | 24 Nov. 92 | SL-6 | PL | 1 500 | 642 | 39 399 | 62.90 | 717.8 | |
| Prognoz-3 | Cosmos 2209 | 1992-059A | 10 Sep. 92 | SL-12 | TT | 2 120 | 35 764 | 35 806 | 1.32 | 1 435.9 | Collocated with C-2133 at 335 East |
| Prognoz-4 | Cosmos 2224 | 1992-088A | 17 Dec. 92 | SL-12 | TT | 2 120 | 35 882 | 35 978 | 2.29 | 1 443.3 | First one announced as Prognoz |
| ***USA*** | | | | | | | | | | | |
| No launches in 1992. | | | | | | | | | | | |
| **Military navigation** | | | | | | | | | | | |
| ***CIS*** | | | | | | | | | | | |
| NAV 3-75 | Cosmos 2180 | 1992-008A | 17 Feb. 92 | SL-8 | PL | 750 | 962 | 1 016 | 82.93 | 104.9 | Replaced C-2004 |
| NAV 3-76 | Cosmos 2184 | 1992-020A | 15 Apr. 92 | SL-8 | PL | 750 | 967 | 1 014 | 82.94 | 105.0 | Replaced C-2182 |
| NAV 3-77 | Cosmos 2195 | 1992-036A | 1 July 92 | SL-8 | PL | 750 | 958 | 1 011 | 82.93 | 104.9 | Replaced C-2026 |
| NAV 3-78 | Cosmos 2218 | 1992-073A | 29 Oct. 92 | SL-8 | PL | 750 | 968 | 1 015 | 82.92 | 105.0 | |
| GLONASS 53 | Cosmos 2177 | 1992-005A | 29 Jan. 92 | SL-12 | TT | 900 | 19 111 | 19 149 | 64.81 | 675.7 | Coplanar with C2109/11 |
| GLONASS 54 | Cosmos 2178 | 1992-005B | 29 Jan. 92 | SL-12 | TT | 900 | 19 088 | 19 172 | 64.79 | 675.7 | Coplanar with C2109/11 |
| GLONASS 55 | Cosmos 2179 | 1992-005C | 29 Jan. 92 | SL-12 | TT | 900 | 19 111 | 19 151 | 64.79 | 675.7 | Coplanar with C2109/11, total of 15 active |
| GLONASS 56 | Cosmos 2204 | 1992-047A | 30 July 92 | SL-12 | TT | 900 | 19 121 | 19 139 | 64.84 | 675.7 | |
| GLONASS 57 | Cosmos 2205 | 1992-047B | 30 July 92 | SL-12 | TT | 900 | 19 117 | 19 142 | 64.87 | 675.7 | |
| GLONASS 58 | Cosmos 2206 | 1992-047C | 30 July 92 | SL-12 | TT | 900 | 19 103 | 19 156 | 64.82 | 675.7 | |
| ***USA*** | | | | | | | | | | | |
| Navstar 2B-23 | USA 79 | 1992-009A | 23 Feb. 92 | Delta 7925 | ETR | 930 | 20 018 | 20 343 | 54.71 | 718.0 | |
| Navstar 2B-24 | USA-80 | 1992-019A | 10 Apr. 92 | Delta 7925 | ETR | 930 | 19 979 | 20 385 | 55.12 | 717.9 | |
| Navstar 2B-25 | USA-83 | 1992-039A | 7 July 92 | Delta 7925 | ETR | 930 | 19 960 | 20 404 | 55.05 | 717.9 | |
| Navstar 2B-26 | USA-84 | 1992-058A | 9 Sep. 92 | Delta 7925 | ETR | 930 | 19 980 | 20 630 | 54.79 | 722.9 | |
| Navstar 2B-27 | USA-85 | 1992-079A | 23 Nov. 92 | Delta 7925 | ETR | 930 | 20 076 | 20 289 | 54.85 | 718.0 | Was on STS-34 5 Oct. 89 |
| Navstar 2B-28 | USA-88 | 1992-089A | 18 Dec. 92 | Delta 7925 | ETR | 930 | 20 039 | 20 325 | 54.90 | 717.9 | Was on STS-34 5 Oct. 89 |
| **Weather** | | | | | | | | | | | |
| ***CIS*** | | | | | | | | | | | |
| No launches in 1992. | | | | | | | | | | | |

| Type/Country/ Spacecraft name | Alternative name (Host spacecraft) | Designation | Launch date | Booster | Launch site | Mass (kg) | Perigee (km) | Apogee (km) | Inclin. (deg) | Period (min) | Comments |
|---|---|---|---|---|---|---|---|---|---|---|---|
| ***USA*** | | | | | | | | | | | |
| No launches in 1992. | | | | | | | | | | | |
| **Nuclear explosion detection** | | | | | | | | | | | |
| ***CIS*** | *Nuclear explosion detection sensors are probably mounted on early warning or navigation satellites.* | | | | | | | | | | |
| ***USA*** | *US nuclear explosion detection sensors are mounted on satellites launched for other primary missions.* | | | | | | | | | | |
| NDS 16 | (Navstar 2B-23) | 1992-009A | 23 Feb. 92 | Delta 7925 | ETR | 135 | 20 018 | 20 343 | 54.71 | 718.0 | Nuclear Detection System (NDS) (EMP, X-ray & optical) |
| NDS 17 | (Navstar 2B-24) | 1992-019A | 10 Apr. 92 | Delta 7925 | ETR | 135 | 19 979 | 20 385 | 55.12 | 717.9 | NDS (EMP, X-ray & optical) |
| NDS 18 | (Navstar 2B-25) | 1992-039A | 7 July 92 | Delta 7925 | ETR | 135 | 19 960 | 20 404 | 55.05 | 717.9 | NDS (EMP, X-ray & optical) |
| NDS 19 | (Navstar 2B-26) | 1992-058A | 9 Sep. 92 | Delta 7925 | ETR | 135 | 19 980 | 20 630 | 54.79 | 722.9 | NDS (EMP, X-ray & optical) |
| NDS 20 | (Navstar 2B-27) | 1992-079A | 23 Nov. 92 | Delta 7925 | ETR | 135 | 20 076 | 20 289 | 54.85 | 718.0 | NDS (EMP, X-ray & optical) |
| NDS 21 | (Navstar 2B-28) | 1992-089A | 18 Dec. 92 | Delta 7925 | ETR | 135 | 20 039 | 20 325 | 54.90 | 717.9 | NDS (EMP, X-ray & optical) |
| **Other military missions** | | | | | | | | | | | |
| ***CIS*** | | | | | | | | | | | |
| GEODETIC | | | | | | | | | | | |
| GEO-IK 12 | Cosmos 2226 | 1992-092A | 22 Dec. 92 | SL-14 | PL | 1 500 | 1 478 | 1 526 | 73.63 | 116.1 | |
| ***USA*** | | | | | | | | | | | |
| BALLISTIC MISSILE DEFENCE | | | | | | | | | | | |
| SDI-E MSTI-1 | Pathfinder | 1992-078A | 21 Nov. 92 | Scout | WTR | | 331 | 444 | 96.75 | 92.4 | Mid-course sensor technology integration |

*Launch facility abbreviations*: EAFB = Edwards Air Force Base, California, USA; ETR = Eastern Test Range, Cape Canaveral, Florida, USA; PL = Plesetsk, Russia; SCT = Shuang Cheng-tsu, China; TT = Tyuratam (Baikonur), Kazakhstan; WTR = Western Test Range, Vandenberg Air Force Base, California, USA

# Part IV. Military expenditure, arms production and trade, 1992

Chapter 9. World military expenditure

Chapter 10. Arms production and arms trade

# 9. World military expenditure

SAADET DEGER

## I. Introduction

World military expenditure continued its downward trend in 1992. After a slow but steady fall which started in 1989, the rate of decline accelerated considerably in 1992. From 1989 to 1992, the fall never exceeded 5 per cent per annum in real terms (i.e., after adjustment for inflation). In 1992 the fall in defence spending in aggregate for the first time accelerated distinctly to about 15 per cent per annum.

Two important features characterize the evolution of world military spending. First, the reductions are unevenly distributed. A very small part of them can be attributed to the North Atlantic Treaty Organization (NATO) and Organization for Economic Co-operation and Development (OECD) countries. In fact, NATO military expenditure actually rose during 1992, partly because of an unexpected spurt in US military outlays. The central reason for the fall in world military expenditure in aggregate is the halving of defence spending in one year by the Commonwealth of Independent States (CIS) countries. This aspect of 'shock therapy' in those countries has made the major contribution to international demilitarization. In the developing world, military spending rose in the Middle East and the Far East but fell in all other regions.

Second, the allocation of expenditure between its constituent parts is changing significantly. In 1992 arms procurement spending fell significantly while personnel costs were relatively protected, and research and development (R&D) exhibited 'resilience' in the sense that its share went up in the aggregate. When hard choices had to be made there was a tendency to postpone procurement but to maintain research and particularly development.

In a period of high uncertainty, military expenditure trends will not be clearly understood unless the purpose and functions of the military forces are clearly defined. In the absence of a proper doctrine—what the military is for in the post-cold war era—allocations to defence will be determined *ad hoc* by economic factors alone, without adequate consideration of the security implications. Traditionally, national defence implied national territorial defence. This function is slowly being eroded, at least in Europe. For the major military spenders, national defence is becoming the defence of national interests—whether for economic or humanitarian or for political reasons. In 1991 and 1992 US forces made significant deployments in the Persian Gulf and Somalia for fundamentally different reasons. Both these commitments were related to the national interests of the only remaining superpower, but neither fitted perfectly with the traditional concepts of security during the cold war. Russian or CIS forces are actively involved in peace-keeping operations which would have been considered imperial interventionism until recently. This complexity and the ambiguity of the role of the military mean that restructuring will take a long time. Thus the changes in military spending will take longer than was thought at the dawn of the post-cold war period.

As mentioned above, the overwhelming proportion of the fall in world military expenditure was accounted for by the CIS, whose military assets, manpower and defence spending are now concentrated in Russia and Ukraine. Both these countries have spent far less than their former respective shares in the Soviet defence budget, while the other countries of the CIS have negligible military spending. More significant have been the changes in the allocation of military expenditure in the CIS. Procurement of new weapon systems has collapsed, by about 80 per cent in Russia and 75 per cent in Ukraine. Military R&D is now down to a bare minimum, and the main items of expenditure are now pensions and personnel costs. The fall in procurement expenditure is so dramatic that it stands out as the most important feature of world military expenditure in 1992. Although large sums are still being expended on the military–industrial complex (MIC), through subsidies from the Central Bank via the newly formed Russian Ministry of Industry, these are not primarily for military production and procurement. The main purpose of these subsidies is the avoidance of mass unemployment (unemployment is rising although slowly) in the MIC and the retention of skilled personnel for the needs of conversion and possibly the future revival of industrial production in general.

Since military expenditure in the former USSR constituted almost 30 per cent of the world total, the changes in this region affect the world aggregate significantly. However, there is considerable difficulty in getting precise information on the defence allocations of the successor states. The reasons for this 'area of darkness', which has paradoxically increased after the collapse of the former system, are reviewed here at some length. They include the high rates of inflation and the impossibility of obtaining adequate data. SIPRI relies on open sources alone to produce its data set. Extensive estimates are made on the basis of official data given by governments through budgets, through media reports and through information supplied to multilateral economic agencies. It has proved impossible to obtain official data produced by the Russian or Ukrainian governments for military expenditure in 1992 and compare them with those available for the last year of the USSR's existence. This creates severe distortion in the analysis and quantification of financial information. Hence, although the qualitative trends are clear enough, the precise expenditure data and changes in them must be treated with extreme caution and considered as preliminary. It is expected that continuing SIPRI research will be able to shed more light in the future on this turbulent transitional period. Although the difficulties of transparency during economic transition are enormous, it is still expected that greater effort will be put by these governments and security organizations such as the Conference on Security and Co-operation in Europe (CSCE) into giving the general public access to data through independent research institutions. Unfortunately, for 1992 such transparency has been conspicuous by its absence.

The USA also allocated less to military expenditure in real terms. The rate of decline, however, is far slower than that of its former superpower rival. It has proved more difficult to cut defence spending dramatically, partly because of the political system which allows for more democratic influence and hence also more bargaining within the domain of public procurement. Since 1992 was an election year, it was particularly difficult to make deep cuts in weapon procurement programmes affecting jobs and regional economic security. In addition, the continuing recession, affecting not only the USA but also all the developed economies of the OECD, made it harder to re-allocate resources from the military to the civilian sectors. Indeed, paradoxically, NATO data showed that in fiscal year (FY) 1992 (October 1991 to September 1992) military expenditure actually increased compared to the previous fiscal year in the USA. Although this is a departure from the general trend, and could have been caused partly by Persian Gulf War spending, the fact should be noted. However, the decline is bound to continue. The last budget of President George Bush sought significant reductions in the next five-year plan, and President Bill Clinton intends to cut defence more than his predecessor.

There is an interesting contrast between the changes in military expenditure in the former USSR and the USA. In all these countries, the end of the cold war has meant that military planners and defence departments are committed to deep quantitative cuts in manpower and armour in the long run as well as to structural adjustments which will improve the qualitative efficiency of smaller forces and will address new requirements, such as increasing reliance on technological advances, command, control, communications and intelligence ($C^3I$) and mobility. Whatever the doctrine and war-fighting capabilities of future armed forces, military expenditure must decline compared to the heights of the 1980s.

The fundamental difference is in the speed at which such reductions are taking place. This speed of adjustment is crucially dependent on the respective economic systems—that of the USA and that of countries of the CIS. Given the recession in advanced capitalist economies, and the fact that in the short run output and employment depend on aggregate demand, a reduction in aggregate demand through cuts in military spending is problematic. Relatively low employment and output in the non-defence sectors also mean that the resources released from the defence sector cannot be easily integrated into the civilian economy. Hence military expenditure needs to be reduced slowly in countries like the USA and in Western Europe which are constrained in the short run by aggregate demand being low. As shown below, the reductions in military budgets in Western Europe, like those in the USA, have been modest in relation to the collapse of the threat from the Warsaw Treaty Organization (WTO) and the plans announced for structural adjustments to the forces. The primary reason has been that re-allocating resources is more difficult during a recessionary period.

In contrast, the transitional economies of the CIS are supply-constrained and military cuts will release resources and inputs for expanding domestic output. This is particularly true of a large economy like Russia's. This does not mean that conversion is costless. However, the costs of resource re-allocation are microeconomic, technological and structural (resulting from the creation of markets, for instance). At the macroeconomic level, the cuts in military expenditure, as part of the government plan to reduce huge budget deficits and concomitant high inflation, can only bring benefits. The economic reform programmes of then Prime Minister Yegor Gaidar, which possibly halved military expenditure in one year for Russia, came out of strict economic necessity. In spite of the short-term problems of the Russian defence and heavy industry which such draconian cuts have caused there is little doubt that in the medium term the impact will be wholly positive.

This leads to general speculation as to world military expenditure in the next few years. It is clear that the USA and Western Europe will have to reduce military expenditure faster than they have done in the immediate aftermath of the end of the cold war. Restructuring will take longer than anticipated since the recession is showing little sign of changing for the better. In addition, the incipient European defence union will require further structural adjustments. The steady decline will mean that by the turn of this decade (or century) NATO military expenditure will be about one-third less than the average for 1990–92 and will have stabilized at that level.

It seems that military expenditure in Ukraine and Russia, which together account for 80–90 per cent of CIS spending on defence, has been halved rapidly and that there is no room for any further reductions. Thus their levels have probably already stabilized and will remain stable under present security scenarios. Developing countries' military expenditure (which comprised about 20 per cent of world military expenditure in 1992) has been falling for the past 10 years and has possibly stabilized with little chance of any further major reductions. These estimates taken together imply that around 1999–2000 world military expenditure should be about $300 billion (at 1990 prices) less per annum than it was in 1990.

The next two sections deal with the USA and Russia, which are still the largest military spenders in the world. The discussion on the USA in section II spells out the arms control and disarmament proposals of the last budget of President Bush and sets them in the historical context of the past 12 years of Republican government. There is also a brief discussion of President Clinton's defence proposal. For the first time in over 10 years a US President is actively interested in military conversion. Section III deals with Russia and briefly with Ukraine and the remainder of the CIS. It explains the nature of the massive military cuts which are taking place currently and which have reduced aggregate defence expenditure in these countries to half the level of that of the former USSR. More importantly, military procurement has been pared to the bone and in 1992 procurement expenditure on major conventional weapons was about 10 per cent of the level seen in 1990.

Sections IV and V discuss European NATO countries and the European Community (EC) countries and mention the possibilities of major restructuring, although these have yet to affect military expenditure significantly. The analyses of countries in Central and Eastern Europe after three years of continuous military reform are important enough to warrant a separate discussion this year and are to be found in appendix 9A. Section VI deals with China and Japan which are at present the two largest military spenders in the Far East—a region of major security implications. Section VII concentrates on the developing world, collectively termed the South, where economic security is as vital as military problems. Two aspects of the interrelationship between economic and military variables became prominent in 1992 for the South. First, the much-vaunted 'peace dividend' failed to materialize, at least so far as the South was concerned. Any benefits from major reductions in the military expenditure of the industrialized countries, such as are occurring in the CIS, would be required first and foremost for domestic development rather than for helping to relieve global economic problems. Second, a more direct link between economic issues, such as foreign aid, and demilitarization is being established as an essential plank of North–South international relations, and in 1992 major changes were instituted in this field. This section analyses the various implications of the proposals made by bilateral donors and multilateral economic agencies on this matter. Section VIII summarizes the conclusions.

## II. The United States

Although US military spending is falling over the long term, the rate of decline is not fast relative to the momentous changes that have taken place in security perceptions. Although by early 1992 it was clear that the USSR no longer existed, that the successor states posed little threat in a conventional sense and that military expenditure was falling in all the former republics, the budgetary response was rather muted. Indeed, according to preliminary data given by NATO for FY 1992, the military expenditure of the USA actually increased by a significant proportion.[1] The massive cuts in actual spending on national defence that were anticipated by many did not take place this year in the USA. As in the Sherlock Holmes story of the dog that did not bark, the mystery is why.

Four points may be made here. It is possible that the economic recession which continued to afflict the country during the year was responsible for the slow build-down, since the civilian sectors could not respond to the release of manpower and capacity from the defence sectors of the economy. Thus, as the economy picks up and restructuring comes into effect, US defence spending could decelerate faster during the next five years or so. Second, in an election year both the President and Congress were wary of making larger cuts which might hurt the defence contractors and the large sector of the economy that is

[1] NATO, *Financial and Economic Data Relating to NATO Defence* (NATO: Brussels, Dec. 1992).

dependent on 'defence jobs'. Third, the international role of the world's largest military power is yet to be defined clearly, and continuing uncertainty produces inertia in cutting down rapidly the inheritance of the early 1980s. Finally, and very importantly, actual military spending partly reflects authorizations given in previous years; thus, even though current budgetary authority is falling, actual expenditure (outlay) does not decline in like manner or even rises, because it is carrying the burden of the past.

It is important to note this distinction in the US budget between authority and outlay. The former is for programme and force expenditure for a specific fiscal year and is the authority provided by federal law to incur financial obligations that will result in actual spending either in the current fiscal year or in the future. It includes funding available for obligations and current spending, the authority required for borrowing, and contract authority (funds only for obligations but not expenditures). Not all of this money will be spent in the year it is authorized, simply because major costs (particularly for procurement) are spread over a number of years. The outlay is the sum expected to be spent within the specific fiscal year and is essentially payments to liquidate obligations, including interest payments on public debt, net of refunds and offsetting collections, but not for the repayment of debts.

On average about 40 per cent of authority is spent in the same year while the rest is spread out over future years. When, as is the case at present, budget authority is falling, then outlays tend to respond slowly, either rising initially to be consistent with previous (large) authorizations or falling more slowly than budget authority.

However, not only is the trend downward, but the fact is inescapable that the *future* anticipated reductions will be large by historical standards. The current restructuring begun for the US armed forces will bring down the size and capabilities of the forces to levels which have not been seen in the post-Viet Nam War era. The speed of decline is slow, but President Clinton is expected to accelerate it.

The FY 1993 budget (running from 1 October 1992 to 30 September 1993) was the second part of a two-year budget request and in a sense needs to be considered in tandem with the authorizations of the previous fiscal year. It was also the last budget presented by President Bush and will be somewhat modified after President Clinton's taking office in early 1993. In addition to the current fiscal appropriations, the budget also lays down the outline of a five-year plan (1992–97) which gives the restructuring options being pursued by the armed forces. This future-oriented programme, which already allows for major reductions, is expected to be changed substantially with further cuts as the Clinton Administration prepares its first budget.

### The budget

In early 1992 when the Bush Administration presented its two-year budget proposals it requested $291 billion in budget authority for FY 1993. In January 1992, presenting its revised proposals for FY 1993, the Adminis-

**Table 9.1.** US Department of Defense and 'national defense' expenditure, budget authority and outlay, FYs 1990–93

Figures are in US $b., current and constant (1993) prices.

| | 1990 | 1991 | 1992 | 1993[a] |
|---|---|---|---|---|
| **Department of Defense** | | | | |
| *Current price* | | | | |
| Budget authority | 293.0 | 290.9 | 276.3 | 267.6 |
| Outlay | 289.8 | 262.4 | 294.7 | 278.3 |
| *Constant price* | | | | |
| Budget authority | 326.4 | 308.8 | 286.6 | 267.6 |
| Outlay | 323.5 | 278.2 | 305.6 | 278.3 |
| **'National defense'** | | | | |
| *Current price* | | | | |
| Budget authority | 303.3 | 303.6 | 289.2 | 281.0 |
| Outlay | 299.3 | 273.3 | 307.3 | 291.4 |
| *Constant price* | | | | |
| Budget authority | 337.9 | 322.3 | 299.6 | 281.0 |
| Outlay | 334.2 | 298.8 | 318.1 | 291.4 |

[a] Initial budget request.

*Source:* US Department of Defense, Office of the Comptroller, *National Defense Budget Estimates for FY 1993* (US Government Printing Office: Washington, DC, Mar. 1992).

tration asked Congress for authorization of $281 billion—a reduction of $10 billion. Congress cut this by a modest $7 billion to produce an authorization of approximately $274 billion for the year. The reduction can thus be calculated either as $17 billion or $7 billion relative to the request made by the President for 'national defense'. Table 9.1 gives data on authority and outlay for the Department of Defense (DOD) budget (which includes all expenditures except those for nuclear weapons financed through the Energy Department) and the wider category of national defence. There is a clearly discernible downward trend.

Analysis of the allocations for procurement, military personnel, operation and maintenance (O&M), research, development, testing and evaluation (RDT&E) and energy defence (for nuclear weapons but not delivery systems) shows how the cuts are being distributed. In the recent past, procurement has fallen fast but O&M has been relatively protected. This year, although procurement was still reduced in real terms following the disappearance of the threat from the former USSR, O&M also bore the brunt of the cuts in the aggregate budget. It is important to note that O&M also includes civilian personnel pay which for over 1 million men is quite substantial. There is also the perennial question of waste and mismanagement, and cutting operational costs is one way of tightening up on wasteful expenditure in the DOD.

It is interesting to note that Congress in 1993 was relatively generous with the President's request in terms of procurement funding of major weapons. Some requests for major weapon systems were accepted; in a few cases more

**Table 9.2.** Comparison of the presidential request for FY 1993 and Congress-approved final appropriations: procurement of major weapons

| | President's request | | Final appropriations | |
|---|---|---|---|---|
| | Number | Value (US $m.) | Number | Value (US $m.) |
| B-2 bomber | 4 | 2 687 | 4 | 2 687 |
| Trident II missiles | 21 | 764 | 21 | 764 |
| Long-range conventional missiles | 340 | 163 | 351 | 167 |
| JSTARS radar plane | 1 | 311 | 2 | 512 |
| C-17 cargo plane | 8 | 2 514 | 6 | 1 810 |
| Aegis destroyer | 4 | 3 347 | 4 | 3 254 |
| Tomahawk cruise missiles | 200 | 404 | 200 | 404 |
| F-16 AF fighter | 24 | 683 | 24 | 615 |
| F/A-18 Navy fighter | 48 | 1 658 | 36 | 1 200 |

*Source: Congressional Quarterly*, vol. 50, no. 41 (17 Oct. 1992), p. 3263.

money was allocated and more weapons authorized than requested by the DOD. Table 9.2 gives data on certain major items of weapon acquisition and shows the numbers requested and the numbers approved by Congress. Only in the case of one item—the advanced F/A-18 stealth Navy fighter—was the request cut substantially.

Production of the much-publicized B-2 bomber, the costliest aeroplane ever built (unit costs now exceed $750 million per plane), will be terminated after the acquisition of a maximum of 20—a reduction from a planned request for 75 two years ago and an initial request for 132. In FY 1993 four such bombers were authorized at a cost of over $2.7 billion. This is now one of the famous relics of the cold war. With the elimination of the Soviet threat it is difficult to foresee an adversary against which a strategic bomber with stealth capabilities could be used in the future. Even now, the cost of this one weapon system exceeds the aggregate military budget of most countries of the world. The eventual termination of production of the B-2 bomber and of the Seawolf Class submarine is one of the largest cost-cutting exercises in the current budget. The trend seems to be to eliminate a few major programmes of weapon purchases and make concentrated savings in specific areas rather than across-the-board cuts affecting all programmes simultaneously.

Another significant feature is the resilience of the military research sector. R&D expenditure has been cut like all others but its share of the total budget has increased. Within a framework of reduced aggregate military spending, research is gaining in relative prominence. Furthermore, as procurement continues to bear the brunt of expenditure reductions, more programmes are being shifted towards the R&D stage rather than actual production. The ability to produce weapons in the future still remains, although production facilities are being limited. The drive towards qualitative arms modernization also remains,

**Table 9.3.** Comparison of the presidential request for military R&D programmes in FY 1993 and Congress-approved final appropriations

Figures are in US $m.

| | President's request | Final appropriations |
|---|---|---|
| SDI R&D | 5 388 | 3 800 |
| of which Brilliant Pebbles | 576 | 300 |
| Comanche Scout helicopter | 443 | 418 |
| Longbow attack helicopter upgrade | 282 | 307 |
| Osprey lilt-rotor | 0 | 755 |
| Sea Wolf Class submarine | 0 | 150 |
| F/A-18 advanced model Navy fighter | 1 134 | 944 |
| F-22 fighter | 2 224 | 2 024 |

*Source: Congressional Quarterly*, vol. 50, no. 41 (17 Oct. 1992), p. 3263.

although in quantitative terms the years of acquisition of increasing numbers of weapons are almost over.

Table 9.3 shows the amounts requested by the President for military R&D and the sums of money allocated by Congress for 1993. All these items are for research alone and do not include any procurement costs. As is clear, except for the Strategic Defense Initiative (SDI), other items remain relatively unscathed.

The SDI was still pursued with vigour through an Administration request for $5.4 billion (up from $4.15 billion the previous year). The R&D that is going on in this area now stresses early deployment of ground-based defences against limited missile attacks (either intercontinental or tactical). Thus the revolutionary (quixotic in the opinion of some commentators) concepts of the earlier Reagan era have been effectively curtailed. Congress allocated around $4 billion for SDI R&D in FY 1993—a major reduction from the original request by the President but a small reduction on 1992.

Overall, the budget this year continues a slow but steady decline, sets in motion larger changes for the future, and shifts the emphasis between procurement and R&D so that acquisitions are reduced (reflecting the end of the Soviet threat) but future capabilities are not eliminated (reflecting growing uncertainties and regional security problems). The Congressional Budget Office puts it succinctly when it states that the DOD:

has chosen to concentrate on developing weapons technology at the expense of producing weapons immediately. The new approach would continue to fund RDT&E, including research into manufacturing and operational risks. It would, however, postpone the actual production of some new weapons until various technical risks are minimized and military threats demand their deployment. The Administration's bud-

**Table 9.4.** Allocation of US 'national defense' budget authority, FYs 1983–92

Figures are in US $b., current prices; figures in italics are percentage shares.

| | 1983 | 1984 | 1985 | 1986 | 1987 | 1988 | 1989 | 1990 | 1991 | 1992 |
|---|---|---|---|---|---|---|---|---|---|---|
| Military personnel | 61.0 | 64.9 | 67.8 | 67.8 | 74.0 | 76.6 | 78.5 | 78.9 | 84.2 | 79.2 |
| *Share of total* | *24.9* | *24.5* | *23.0* | *23.5* | *25.7* | *26.2* | *26.2* | *26.0* | *27.7* | *27.4* |
| O&M[a] | 66.6 | 71.0 | 77.8 | 74.9 | 79.6 | 81.6 | 86.2 | 88.3 | 90.9 | 92.4 |
| *Share of total* | *27.2* | *26.8* | *26.4* | *25.9* | *27.7* | *28.0* | *28.8* | *29.1* | *29.9* | *32.0* |
| Procurement | 80.4 | 86.2 | 96.8 | 92.5 | 80.2 | 80.0 | 79.4 | 81.4 | 71.7 | 60.5 |
| *Share of total* | *32.8* | *32.5* | *32.8* | *32.0* | *27.9* | *27.4* | *26.5* | *26.8* | *23.6* | *20.9* |
| RDT&E[b] | 22.8 | 26.9 | 31.3 | 33.6 | 35.6 | 36.5 | 37.5 | 36.5 | 36.2 | 36.9 |
| *Share of total* | *9.3* | *10.1* | *10.6* | *11.6* | *12.4* | *12.5* | *12.5* | *12.0* | *11.9* | *12.8* |
| Energy, defence | 5.7 | 6.6 | 7.3 | 7.3 | 7.5 | 7.7 | 8.1 | 9.7 | 11.6 | 12.0 |
| *Share of total* | *2.3* | *2.5* | *2.5* | *2.5* | *2.6* | *2.6* | *2.7* | *3.2* | *3.8* | *4.2* |
| Other[c] | 8.5 | 9.6 | 13.7 | 13.0 | 10.5 | 9.6 | 9.9 | 8.5 | 9.0 | 8.2 |
| *Share of total* | *3.5* | *3.6* | *4.6* | *4.5* | *3.7* | *3.3* | *3.3* | *2.8* | *3.0* | *2.8* |
| **Total** | **245.0** | **265.2** | **294.7** | **289.1** | **287.4** | **292.0** | **299.6** | **303.3** | **303.6** | **289.2** |

[a] Operation and maintenance (includes civilian personnel cost).
[b] Research, development, testing and evaluation.
[c] 'Other' includes military and housing construction.

*Sources:* 'Historical Tables' in US Office of Management and Budget, *Budget of the United States Government, Fiscal Year 1993*. Supplement Feb. 1992 (US Government Printing Office: Washington, DC, 1992); author's calculations.

get proposals show evidence of the differential emphasis on RDT&E versus procurement.[2]

## The past

The evolution of military expenditure over the past decade and its allocation between military personnel, O&M, procurement, R&D and nuclear weapons and facilities can be seen either through historical tables of budget authority or through outlays. In table 9.4 the authority data are given, since these reflect the forward-looking perceptions of the Administration and Congress at the time they were approved and show how they have changed.

It is now possible fruitfully to compare the average annual military expenditure in the period 1975–90 (the recent cold war period) and that foreseen in the five-year plan for 1992–97 (the end of restructuring in the first phase of the post-cold war period). The current plans in the FY 1992 and 1993 budgets are essentially the first planning for the post-cold war US military doctrines, posture and force structures. The slow demilitarization set in motion will be completed by 1997. Weapon acquisition costs are similar and the level of

[2] US Congress, Congressional Budget Office, *An Analysis of the President's Budgetary Proposals for Fiscal Year 1993* (US Government Printing Office: Washington, DC, Mar. 1992), p. 62.

**Table 9.5.** Comparison of US defence-related manpower during the cold war (1975–90) and the projected figures for 1997

Figures are in thousands. Figures in italics are percentages.

| | Average for 1975–90 | Projection for 1997 | Change (percentage) |
|---|---|---|---|
| Military | 3 092 | 2 546 | *– 17.7* |
| Active | 2 105 | 1 626 | *– 22.8* |
| Reserve | 987 | 920 | *– 6.8* |
| Civilian | 1 064 | 904 | *– 15.0* |

*Source:* US Congress, Congressional Budget Office, *An Analysis of the President's Budgetary Proposals for Fiscal Year 1993* (US Government Printing Office: Washington, DC, Mar. 1992).

technology roughly comparable. Further, in both periods there exists an all-volunteer force, allowing for direct comparison of real personnel spending (after netting out the effects of inflation), and there was an emphasis on the need for both reserve and active forces to deal with the eventuality of involvement in more than one war at the same time.

Table 9.5 compares the annual average for defence-related manpower between 1975 and 1990 and the projected levels in 1997. Total military manpower is planned to be around 17.7 per cent less in the post-cold war scenario; active-duty servicemen numbers will fall by almost 23 per cent, while the cuts in reserves will be far less. Civilian manpower will also be reduced, but the proportional fall of 15 per cent will be less than the reduction in active-duty forces.

In a similar vein, table 9.6 compares the average of the past and the future projected levels of military spending and its components. Total national defence expenditure is broken down into operating costs and investment costs. These categories are further sub-divided into their various components. It is instructive to see how each category of expenditure fared in the past and their relation to the cold war force structures. Military personnel generally had the highest share of operating costs, since the all-volunteer force after the Viet Nam War tended to cost more. The search for professional soldiers in a competitive labour market implied high pay and better conditions. In like fashion, O&M, which includes civilian pay, was also rather high, taking the second largest share of average budgets during the cold war. Procurement expenditure on major conventional weapons as well as delivery systems for the strategic forces took about 25 per cent of the military budget—a share only exceeded by the other superpower, the USSR, which allocated close to 40 per cent to the same category. R&D was also important, particularly in the 1980s.

Comparison with the second column shows the impact of the end of the cold war. All categories of spending will fall (except for energy defence, i.e., nuclear weapons and other spending of the Department of Energy for military

**Table 9.6.** Comparison of US military expenditure (budget authority) during the cold war (1975–90) and the projected figures for 1997

Figures are in US $ b., constant 1997 prices. Figures in italics are percentages.

| | Average for the cold war (1975–90) | Projected for 1997 | Change (percentage) |
|---|---|---|---|
| Operating costs | 206.7 | 169.5 | *– 18.0* |
| Military personnel | 103.9 | 75.6 | *– 27.2* |
| O&M [a] | 99.0 | 90.2 | *– 8.9* |
| Family housing | 3.7 | 3.7 | – |
| Investment cost | 140.8 | 105.1 | *– 25.4* |
| Procurement | 95.9 | 63.1 | *– 34.2* |
| RDT&E [b] | 36.6 | 36.0 | *– 1.6* |
| Military construction | 6.4 | 5.5 | *– 14.1* |
| Other investment | 1.9 | 0.5 | *– 73.7* |
| Energy defence | 8.4 | 16.0 | *+ 90.5* |
| **Total national defence** | **355.9** | **290.6** | ***– 13.5*** |

[a] Operations and maintenance.

[b] Research, development, testing and evaluation.

*Source:* US Congress, Congressional Budget Office, *An Analysis of the President's Budgetary Proposals for Fiscal Year 1993* (US Government Printing Office: Washington, DC, Mar. 1992).

purposes). Military R&D, however, shows quite remarkable resilience. Its level in constant prices will remain very similar and its share in the defence budget rise from a little over 10 per cent to over 12 per cent. The budget for energy defence looks rather inflated, partly because it does not reflect the provisions of the 1993 Treaty on Further Reduction and Limitation of Strategic Offensive Arms (START II Treaty). Large sums of money have also been allocated for clean-ups of testing sites and other forms of environmental destruction by the military.

Finally, table 9.7 shows the evolution of defence spending by its components—Army, Navy and Air Force—and the total budget authorization during the past five years. It is clear that the Army has suffered the largest reductions. Only in FY 1991 was there positive growth, but that was partly due to Operation Desert Storm. The Navy and Air Force have fared relatively better and this is set to continue.

## The future

At a time of rapid transition and major changes, the past merges into the future and it is not clear where the old epilogue ends and the new prologue begins. The message from this section is that after an unprecedented military buildup during much of the 1980s the USA has begun a steady reduction of military expenditure, capabilities and forces. The greatest impact will be on procure-

**Table 9.7.** US Department of Defense, TOA[a] by service, FYs 1989–93

Figures are in constant FY 1993 US$ m. Figures in italics are percentage changes on the previous year.

| | 1989 | 1990 | 1991 | 1992 | 1993 |
|---|---|---|---|---|---|
| Army | 90 587 | 88 382 | 98 739 | 74 941 | 63 587 |
| | *– 1.2* | *– 2.4* | *11.7* | *–24.1* | *– 15.2* |
| Navy | 112 235 | 112 259 | 100 516 | 92 602 | 84 753 |
| | *– 8.0* | – | *– 5.1* | *–13.1* | *– 8.5* |
| Air Force | 109 253 | 104 090 | 95 712 | 86 806 | 84 162 |
| | *1.3* | *– 4.7* | *– 8.1* | *– 9.3* | *– 3.1* |
| **Total** | **335 276** | **327 316** | **328 202** | **296 628** | **271 347** |
| | ***– 2.6*** | ***– 2.4*** | ***0.3*** | ***– 9.6*** | ***– 8.5*** |

[a] TOA—Total Obligation Authority—is the Department of Defense financial term which expresses the value of the direct defence programme for a fiscal year. It is the sum of: (*a*) all budget authority granted by or requested from Congress; (*b*) amounts authorized to be credited to a specific fund; (*c*) unobligated balances from previous years which remain available for obligation.

*Source:* US Department of Defense, Office of the Comptroller, *National Defense Budget Estimates, FY 1993* (US Government Printing Office: Washington, DC, Mar. 1992).

ment and military personnel, while military research will be spared any cuts for the next five years or so.

The requirements for the next five years will be more internationalized with little need for fixed-formation heavy armour to fight conventional wars in Central Europe. There will be a greater need for mobility and rapid reaction forces either for peace-keeping or for other humanitarian reasons such as protecting aid convoys in Somalia. In addition, the grave uncertainty that characterizes the international security scene requires flexibility and the ability to mobilize quickly if required, but not necessarily maintenance of expensively large standing armies permanently configured for large-scale wars.

How could the active-duty forces be restructured to meet these needs and requirements? This is readily understood by looking at table 9.8, where the 1990 level of forces is compared to the projected levels in 1997, after the current round of restructuring. Army divisions are to be cut the most (from 28 to 20) but the aim is to maintain reserves for contingency planning in an era of uncertainty. The number of aircraft-carriers will be reduced but not substantially, given US global commitments as the only military superpower, and numbers of carrier air wings will be effectively maintained. The maximum reductions will take place in strategic bombers, reflecting the end of the USSR, and tactical fighter wings will also be cut substantially.

Table 9.9 shows the impact of manpower reductions between 1990 and 1997. The plan now is to reduce the US active-duty forces from over 2 million men to around 1.6 million, a drop of over 20 per cent. Civilian personnel are

**Table 9.8.** Comparison of US force capabilities, 1990–97 (projected)

| | 1990 | 1997 | Change (percentage) |
|---|---|---|---|
| Army Divisions | 28 | 20 | *– 28.6* |
| Active | 18 | 12 | *– 33.3* |
| Reserve | 10 | 8 | *– 20.0* |
| Aircraft-carriers | 15 | 12 | *– 20.0* |
| Carrier air wings | 15 | 13 | *– 13.3* |
| Active | 13 | 11 | *– 15.4* |
| Reserve | 2 | 2 | – |
| Tactical fighter wings | 36 | 26 | *– 27.8* |
| Active | 24 | 15 | *– 37.5* |
| Reserve | 12 | 11 | *– 8.3* |
| Strategic Bombers | 268 | 180 | *– 32.8* |

*Source*: US Congress, Congressional Budget Office, *An Analysis of the President's Budgetary Proposals for Fiscal Year 1993* (US Government Printing Office: Washington, DC, Mar. 1992).

not cut as much, and even with around 20 per cent cuts the DOD will have around 3.5 million (total civilian personnel) on its payroll in 1997.

All this information is based on data available at the end of 1992 and represents the current projections of the US Government. There is little hard information available as yet about President Clinton's defence budgetary plans except that the cuts will be somewhat deeper than President Bush's, although by less than was initially thought. The original intention of the President-elect, to reduce the defence budget by about one-third, to about $200 billion, by 1996, would have meant reductions of $80 billion over and above the Bush projections, and was revised in June 1992.

**Table 9.9**. Comparison of US defence-related manpower, 1990–97 (projected)

All figures are in thousands.

| | 1990 | 1997 | Change (percentage) |
|---|---|---|---|
| Military | 3 172 | 2 546 | *– 19.7* |
| Active | 2 044[a] | 1 626 | *– 20.5* |
| Reserve | 1 128 | 920 | *– 18.4* |
| Civilian | 1 073 | 904 | *– 15.8* |
| **Total military and civilian** | **4 245** | **3 450** | ***– 18.7*** |

[a] The active-duty military force of 1990 does not include the 26 000 selected reserves mobilized during Operation Desert Storm.

*Source*: US Congress, Congressional Budget Office, *An Analysis of the President's Budgetary Proposals for Fiscal Year 1993* (US Government Printing Office: Washington, DC, Mar. 1992).

**Table 9.10.** Comparison of the military expenditure plans of Presidents Bush and Clinton, 1993–96

All figures are in current US $ m.

| | 1993 | 1994 | 1995 | 1996 |
|---|---|---|---|---|
| Bush 1993 projections (Authority) | 281.0 | 281.6 | 284.3 | 285.7 |
| Outlay | 292.2 | 283.4 | 282.9 | 286.1 |
| Clinton, military expenditure (proposed) | 271.0 | 270.6 | 271.3 | 266.2 |
| Clinton reduction planned, of which: | 10.0 | 11.0 | 13.0 | 19.5 |
| Defence cuts | 2.0 | 8.8 | 10.5 | 16.5 |
| Reform of procurement management | 5.7 | – | – | – |
| Reform of inventory systems | 2.3 | 2.5 | 2.5 | 2.5 |
| % change from Bush plan | *– 3.6* | *– 3.9* | *– 4.5* | *– 7.0* |

*Source: Congressional Quarterly,* vol. 50, no. 26 (27 June 1992), p. 1901; author's estimates.

In table 9.10 the projections of the last budget of President Bush are shown alongside the Clinton reductions. Over four years the new President proposed about $53.5 billion of additional cuts in aggregate. By FY 1996 this implies that a Clinton budget will be about $266 billion (in current 'then year' prices) compared to the Bush estimate of $286.1 billion—an additional cut of 7 per cent, which is significant but not earth-shaking. The conclusion is that, irrespective of party or Congress, cutting military expenditure very fast is problematic and the transition path is not smooth. Conversion also has initial costs which should not be underestimated.

## III. Russia and the CIS

Before the breakup of the former USSR, the Russian Republic contributed overwhelmingly to the aggregate military expenditure of the Union. In addition it had similar or even larger shares of the MIC enterprises, including establishments involved in military-related R&D. Russia contributed about two-thirds of Soviet military spending between 1989 and 1991, and in 1992 this proportion probably came close to 80 per cent.[3] In 1992 the situation was similar with respect to CIS aggregate military expenditures, and additional shares fell to Russia because of the costs of demobilization as the old military machine was scaled down. Apart from Russia, only Ukraine has relatively

[3] See Deger, S. and Sen, S., 'World military expenditure', *SIPRI Yearbook 1992: World Armaments and Disarmament* (Oxford University Press: Oxford, 1992), pp. 203–25. See also International Institute for Strategic Studies, *The Military Balance 1992–1993* (Brassey's: London, 1992).

high military expenditure (about 25 per cent of Russia's, or roughly 15 per cent of the total for the former USSR). The other countries which were formerly part of the USSR have effectively stopped spending on the military except possibly for maintenance and personnel. In particular, procurement in all the new states, except Russia and Ukraine, is in effect zero. Analysis of military expenditure in the former USSR now essentially means discussing the defence spending of Russia and possibly Ukraine.

In spite of democratization and greater transparency it has become more rather than less difficult to make precise quantitative estimates of Russian military expenditure. The same is true for all countries of the CIS. There are seven reasons for this contradictory state of affairs.

1. Relative price distortions have always characterized the components of military spending in the USSR, and these distortions have actually become more extreme since the start of market reforms in Russia. Some components, such as personnel pay, are essentially valued at market prices. Others, such as purchases from the MIC through procurement orders, still have an element of administered pricing and are yet to be fully liberalized. The same holds for intermediate inputs such as oil consumption by the military and construction material. Formerly, all components of military spending were subject to administrative pricing, which changed little on a year-to-year basis and were heavily subsidized in order to keep within spending limits set artificially low for political reasons. How far these subsidized prices have been liberalized and allowed to reach market levels is not yet known. Not only are the relative prices now distorted, but they are also changing and fluctuating at different rates so that both absolute price levels and relative prices are changing erratically.

2. The absolute level of inflation is high and rising. Price inflation has at times exceeded 50 per cent per month—the formal definition of hyperinflation—and the annual inflation rate over 1992 was at least 1000 per cent although impossible to verify accurately. According to some economists it could even be double this figure.[4] Such inflation rates create havoc for any realistic calculation of budgetary trends and other components of national output. This situation has elements in common with that of Latin America in the 1980s, when the military expenditure estimates of SIPRI needed to be continually updated throughout the year many times over. The additional complication is that high inflation is occurring at the same time as there are transitional economic problems, which means that neither relative prices (the price ratio of one component against the other as stated in the previous paragraph) nor absolute price level (general inflation) can be estimated with any precision.

3. Budgetary forecasts of military expenditure made at the start of the year are vastly different from actual amounts expended by the end of the year. This is partly because the government itself cannot forecast the rate of price

[4] *Russian Economic Trends,* vol. 1, no. 1 (1992), p. 53.

changes and tends to be optimistic about its ability to control the economy. In addition, there are extra-budgetary accounts and indirect subsidies from the Central Bank to enterprises to get rid of inter-enterprise debt, which are not fully accounted for in the military budget. It is not certain whether multiple books of account are being maintained. The publicly announced budget may not be the same as the accounts prepared for ministers and local experts, and the accounts kept for multilateral organizations such as the International Monetary Fund (IMF) may be different. Under these circumstances all categories of the budget are difficult to estimate, and military spending particularly so.

4. The formation of 14 additional states means that comparisons between 1991 and 1992 are not always possible. It is often not clear from published reports whether comparisons made with previous years are made with the then USSR or with the Russian contribution to the USSR. Unless the data reveal comparisons, not only in terms of financial values but also in terms of the regional make-up of the former USSR, it will be difficult to analyse how much military spending has declined in aggregate.

5. Although procurement spending has collapsed, production may not have fallen by the same amount. Military expenditure data in the strict sense should only include procurement expenses rather than the value of weapons produced, but the latter is the crucial variable when demilitarization is the focus of analysis. Stock-building in Russian enterprises producing weapons has important implications for the limits imposed under the Treaty on Conventional Armed Forces in Europe (CFE Treaty), for future budgetary trends, and for the pressure being exerted by the MIC to increase procurement orders in the future or to export these surplus weapons. The security implications of these pressures cannot be ignored. The implications of the divergence between procurement and production are discussed further under *perestroika* below in this section.

6. In the past most estimates made by Western Sovietologists were almost totally independent of Soviet military- and defence-related budgetary data. Independent price systems and purchasing power parities (PPPs) were constructed to produce rouble or dollar estimates of military spending, arms production and exports. SIPRI military expenditure data for the USSR combined (*a*) local data on personnel and O&M; (*b*) weapon production data from Western intelligence sources; (*c*) estimates made on Soviet military R&D based on the science budget; (*d*) the estimated share of the military in all research activities; and (*e*) construction of PPPs to convert rouble values into dollars or vice versa. SIPRI arms trade data for the USSR have hitherto relied exclusively on Western prices since the export prices of the former USSR were virtually unknown. Now, however, these methods, although useful, must be supplemented by information and data coming out from the CIS countries themselves. Just as in all other processes of transition, there exist major difficulties in integrating the old and the new sets of data. The problem is exacerbated when the new sets are themselves inconsistent with each other.

7. It is often not clear whether certain types of expenditure should be classified as military or not. For example, costs of conversion are not easily distinguished from general expenditure on the MIC; the costs of re-settlement and retirement of returning troops are not distinguishable from general payments for army personnel. Furthermore, as nationalist conflicts escalate, particularly among the Central Asian and Caucasus states, it is not clear what is military expenditure as distinct from para-military, internal security or policing expenditures. Although this last problem is generally common in all war economies, it is particularly compounded for the CIS countries by the general lack of data and of institutional structures to promote transparency.

Analysis of Soviet military expenditure in the previous three *SIPRI Yearbooks*[5] has been based on three concepts which constitute a specific framework: (*a*) the question of what precise data can be gleaned from the often confused mass of information that is available; (*b*) the extent of restructuring that has taken place within the military and the progress of demilitarization; and (*c*) the question of what sort of conversion has taken place and what resources have been released to the civilian economy from the defence sectors, including the MIC. All these issues are aspects of the security and economic implications of reductions in military expenditure. They are discussed below under the same headings—*glasnost* (transparency of information), *perestroika* (restructuring) and *konversiya* (transfer of resources to the civilian economy).[6] Even though the USSR no longer exists, many of the economic and security issues are similar to those it faced over the past few years. The same format is therefore retained for clarity of exposition.

### *Glasnost*

To understand the problems of measuring military expenditure in Russia (and the CIS) it is necessary to go back to the last year of the USSR's existence. Its official budgetary defence allocation (OBDA) was announced in early 1991 as 96.6 billion roubles. Although at current prices it was higher than in the previous year (1990) there was little doubt that in real terms there had been a significant reduction of around 10–12 per cent. The OBDA had already made assumptions regarding the impending price reform to be initiated in the first half of 1991. Although no precise figures were available, SIPRI estimates that around 49–50 per cent inflation was built in to the budgetary calculations.

In practice, Soviet Prime Minister Valentin Pavlov's price reform of March–April 1991 raised the consumer price index (CPI) by a factor of 2 in one month. It rose again slightly to about 225 in December 1991.[7] Such inflation

[5] See Deger and Sen (note 3); Deger, S., 'World military expenditure', *SIPRI Yearbook 1991: World Armaments and Disarmament* (Oxford University Press: Oxford, 1991), pp. 115–80; Deger, S., World military expenditure', *SIPRI Yearbook 1990: World Armaments and Disarmament* (Oxford University Press: Oxford, 1990), pp. 165–66.

[6] These terms had a special and specific connotation in the context of the Gorbachev reforms but are used slightly more broadly in this chapter.

[7] *Financial Times*, 22 Apr. 1992, p. 16.

was unprecedented in the recent history of the USSR. Although procurement and R&D budgets have traditionally been shielded from unanticipated price rises through the artificial command price system, it proved difficult to stop personnel and other operating cost categories of the budget increasing over and above the anticipated inflation rate of around 50 per cent.

Taking into account the impact of this on the defence budget announced for operating expenditures, and adding estimates for hidden subsidies and military-related expenditures, SIPRI's own estimate of Soviet defence spending comes to around 140–50 billion roubles in 1991. This represents less than 8 per cent of the official gross national product (GNP), which was about 1.8–1.9 trillion roubles in 1991. However, as indicated above, if price distortions are accounted for, and it is noted that the military sector's overall inflation rate has always been lower than that of the rest of the economy, then even in 1991 the defence burden (the share of military expenditure in gross domestic product, GDP) was at least 10 per cent at the end of successive cuts in the Gorbachev era.

The figures above show that in 1991 Soviet military expenditure was exceptionally high in terms both of its burden on the economy and of its share of the aggregate central government budget, even though it had fallen for three successive years. However, alternative estimates give even higher figures. The International Institute for Strategic Studies (IISS) gives an estimate of almost 172 billion roubles for 1991,[8] and at a conference on conversion organized by the United Nations Industrial Development Organization (UNIDO) the Mayor of St Petersburg, Anatoly Sobchak, claimed that in 1991 the USSR spent 400 billion roubles on the military—about four times the OBDA.[9] Although it is difficult to substantiate such extraordinarily high claims, it is clear that the military expenditure of the former USSR was quite high even after the demilitarization and arms control efforts of the late 1980s.

As regards dollar values, SIPRI calculates its own PPP to convert roubles into dollars, since exchange-rates have been meaningless in the past and currently reflect little. The calculation is complicated by the fact that there is a divergence between the civilian and the military rouble as well as between the domestic and external PPP. Using the 1989 PPP as a benchmark[10] and adjusting for US and Soviet inflation rates for 1989–91, the author estimates that in dollar terms Soviet military expenditure in 1991 was between $210 billion and $220 billion at 1989 prices. As a standard of comparison, in 1989 Soviet military spending was about $270 billion according to SIPRI's estimates.

These figures are important not only for their historical significance but also as an indicator of Russia's military heritage. If in 1991 the Russian Republic contributed around 70 per cent of this military expenditure of the USSR, then its defence spending in 1991 would have been 98–105 billion roubles. Given

[8] International Institute for Strategic Studies (note 3), p. 92.
[9] *Frankfurter Allgemeine Zeitung*, 12 Nov. 1992, p. 11.
[10] See Deger 1990 (note 5).

an estimated GNP of about 1180 billion roubles[11] in 1991, the military burden of Russia (if it had then been a unitary state) would have been between 8.3 and 8.9 per cent. Once again, if all investment costs (procurement, R&D, industrial subsidies, etc.) could be recorded, then the military burden would be far above 10 per cent for Russia—once again an impossibly high cost to the civilian economy.

Military expenditure in Russia for 1992 is difficult to estimate with precision for the reasons discussed above. Even the OBDA has changed rapidly over the year as inflation estimates have had to be revised repeatedly. Moreover, at the beginning of the year Russia was paying the major proportion of the CIS troops on an *ad hoc* basis since it was not clear what the financial allocation among the CIS states would be. Some calculations had been made in the final weeks of 1991 when it was clear that the Union would disintegrate and the former republics would need to pay *pro rata* shares. These calculations were based on the number of troops stationed in each member state, the size of populations, and the value of national income and/or GNP. The respective shares of the total military expenditure of the CIS were estimated to be: Russia, 61–62 per cent; Ukraine, 17 per cent; Kazakhstan, 5.1 per cent; Azerbaijan, 1.9–per cent; Moldova, 1.2 per cent; Turkmenistan, 1.0 per cent; Kyrgyzstan, 0.9 per cent; and Tajikistan, 0.8 per cent.[12] Although these percentage shares do not add up to 100, and there are significant vacant slots such as the contribution of Belarus which has a large stationed force, there was at least a semblance of an orderly division of responsibilities. However, by the first half of the year it was clear that a joint command of general-purpose forces under the CIS was not really possible and that only the strategic nuclear forces would be under some form of central command.

Only in May 1992 was the Yeltsin decree passed which established and set up formally the independent Russian Armed Forces and Ministry of Defence. Russia now has control over the forces in Germany, Poland and the Baltic states, the former Soviet units in Armenia and Azerbaijan, and the 14th Army in Moldova, and pays for the expenses of the CIS forces in the Central Asian republics. The CIS joint command effectively has control only over the Strategic Deterrent Forces, the Strategic Defence Forces and possibly the Border Troops (although independent border troops are being set up). The distribution of costs for the Black Sea Fleet is not clear, nor is it known how the naval assets are to be distributed between Russia, Ukraine and Georgia.

In the first quarter of 1992, when Russia's first budget was presented, military expenditure was reported to be around 384 billion roubles for 1992 or just over 16 per cent of the budget—a substantial reduction in the share going to the military.[13] It was later reported that this figure had risen to about

[11] International Monetary Fund, *Economic Review, Russian Federation* (IMF: Washington, DC, Apr. 1992).

[12] Orlov, A., 'The Army: financial cross section', *Moscow News,* no. 52 (Dec. 1991), p. 9; interview with Lt. General Leonid Ivashov, Chief of the Directorate of the Armed Forces of the CIS, *Izvestia,* 22 Jan. 1992, p. 2.

[13] Interview with Yegor Gaidar on Moscow Russian Television Network in Foreign Broadcast Information Service, *Daily Report–Central Eurasia,* FBIS-SOV-92-059, 26 Mar. 1992, p. 32.

410 billion roubles around the middle of the year.[14] In September the Economy Minister Andrey Nechaev stated that 1992 expenditure would be around 632 billion roubles;[15] in November when the final version of the Russian budget for 1993 was presented to Parliament 1992 expenditure was given as 832.8 billion roubles.[16] At the same time there were press reports that the Russian armed forces, through the Finance Directorate of the Ministry of Defence, were urgently requesting an extra allocation for 1992 so that unprecedented expenses due to high inflation could be met. In September, for example, the government allocated an extra-budgetary amount of 13.6 billion roubles to subsidize 'remuneration of employees of enterprises and organizations, including scientific research institutes and design bureaus, engaged in carrying out the state orders for the development and purchasing of arms and military equipment and meeting the needs of defense and security in 1992'.[17] At the end of October *an additional* 315 billion roubles were requested—200 billion for operating costs, 62.5 billion for the purchase of armaments and 53 billion for construction.[18] Even though such requests are unlikely to be met in full, even partial compensation could raise the aggregate defence spending of Russia substantially.

Taking into account these various reports, and estimating from the force cuts that have taken place, it is possible to estimate Russia's military expenditure at around 800 billion to 1 trillion roubles in current prices as of 1992. This is approximately 8–10 times the nominal value of 1991, but far lower after adjustment for inflation. On the basis of an inflation rate of around 1000 per cent over 1992, the military expenditure of Russia in 1992 is at most two-thirds of the amount it was as part of USSR defence spending in 1991, and could be lower. Alternatively, SIPRI's estimates show that Russia's military spending in 1992 was about 45 per cent of that of the USSR in 1991 in constant prices. According to Western intelligence data, given by the Director of the US Central Intelligence Agency (CIA),[19] Russia's defence spending in 1992 is about one-third that of the whole of the USSR in 1991. The discrepancy between SIPRI's estimates and those of the CIA could arise because the latter tended to overestimate Soviet military spending.[20]

The share of military expenditure in GNP is equally problematic in the absence of reliable data. SIPRI estimates that the military burden of Russia ranged between 6 and 6.5 per cent of GNP in 1992 (estimating GNP at 15 trillion roubles). It would have been even lower if national output itself had not fallen precipitously—by about 20 per cent in 1992.[21] Once again this is a substantial fall, unprecedented in modern history. Other reports quoting

[14] International Institute for Strategic Studies (note 3).
[15] *Radio Free Europe/Radio Liberty Research Report,* vol. 1, no. 40 (9 Oct. 1992), p. 53.
[16] FBIS-SOV-92-212, 2 Nov. 1992, p. 46.
[17] Moscow ITAR-TASS World Service, FBIS-SOV-92-174, 8 Sep. 1992, p. 22.
[18] *Izvestia*, 30 Oct. 1992, p, 2, in FBIS-SOV-92-212, 2 Nov. 1992, p. 46.
[19] *International Herald Tribune*, 26 Feb. 1992, p. 1.
[20] See Deger and Sen 1992 (note 3).
[21] *Radio Free Europe/Radio Liberty Research Report*, vol. 1, no. 48 (4 Dec. 1992), p. 53.

Russian sources claim that the military burden could be as low as 5.7 per cent in 1992 compared to a level of 7.2 per cent in 1991.[22]

Clearly, the structural adjustments required of the military for such traumatic changes have been tremendous and the costs disproportionately high. This is analysed below under *perestroika.* It is now becoming clearer, however, that the reductions have been once and for all. By the end of 1992 there were repeated assurances by the government to both the armed forces and the Ministry of Defence that the military budget will not fall in real terms in 1993. Press reports have even claimed that Prime Minister Gaidar, just prior to his removal, stated in the closed town of Chelyabinsk that arms procurement will rise in real terms by 10 per cent in 1993 as compared to 1992 and that defence spending will remain similar.[23] Public statements by President Yeltsin also claim that the real value of military expenditure will remain the same in 1993 as in 1992.[24] The first version of the budget for 1993 announced a defence spending of 1887.5 billion roubles—a nominal increase of 2.25 times from the OBDA of 832.8 billion in 1992 at an inflation rate of 125 per cent per annum.[25] The aggregate is essentially the same as in 1992. If the inflation forecast is approximately accurate, the value of Russian military expenditure will therefore stabilize from 1993 at around 40 per cent of the defence spending of the former USSR. As a share of a predicted 30 trillion roubles Russian GNP in 1993, the military burden amounts to about 6 per cent—again similar to that of 1992.

Turning now to Ukraine, which accounted for about 15 per cent of former Soviet expenditure, the same sorts of problems are faced in analysing its defence budget. Ukraine took over the assets of the old Soviet armed forces on Ukrainian territory prior to the dissolution of the USSR as well as control over the Black Sea Fleet. It was the first country of the CIS to declare its intention of forming an independent armed force and its own Ministry of Defence. In spite of earlier claims that its armed forces would amount to 450 000, which is also the limit under the CFE Treaty, the actual numbers could be less in the long run and and there has been some modest demobilization. The current manpower level is 230 000, but this excludes the Black Sea Fleet, the strategic nuclear forces' manpower and the para-military (currently very small). Within the ceilings of treaty-limited equipment (TLE) Ukraine has inherited a very large stock of weapons and its procurement expenditure should therefore be minimal for a number of years to come. Apart from Russia, Ukraine now has more tanks and armoured combat vehicles (ACVs) than any country in Europe except Germany, its holdings of attack helicopters are exceeded only by those of Germany and France, and it has more combat aircraft than any other coun-

[22] 'Programme for the deepening of economic reforms', mimeograph, Moscow June 1992, p. 43, reported in Bush, K., 'Russia's latest program for military conversion', *Radio Free Europe/Radio Liberty Research Report,* vol. 1, no. 35 (4 Sep. 1992), pp. 32–35.

[23] *Krasnaya Zvezda*, 15 Sep. 1992, reported in FBIS-SOV-92-182, 18 Sep. 1992 p. 23; *Radio Free Europe/Radio Liberty Research Report* (note 21).

[24] *Süddeutsche Zeitung,* 25 Nov. 1992, p. 12; *Neue Zürcher Zeitung*, 22–23 Nov. 1992, p. 1.

[25] FBIS-SOV-92-212, 2 Nov. 1992, p. 46.

try in Europe.[26] The main expenses now are for personnel, demobilization, the housing and pension costs of returning Ukrainian soldiers and officers from abroad (including Russia and other countries of the CIS) and the O&M costs of the Black Sea Fleet.

Ukraine's first budget[27] in April 1992 assumed expenditure of 458.4 billion roubles, of which 16 per cent would be for defence—a ratio similar to that of Russia. Actual military expenditure as given by the OBDA would thus be around 73 billion roubles. It was also claimed that the deficit would be about 2 per cent of GDP at a level of 17.1 billion roubles, implying a GDP of around 855 billion roubles. This would suggest that the share of defence in GDP would amount to 8.6 per cent, which seems to be high, particularly compared to the defence expenditure of the neighbouring countries with which Ukraine could potentially enter an arms race.

Since Ukraine's share in the GDP of the USSR amounted to about 300 billion roubles in 1991, and inflation was high although less than that of Russia, this budgetary figure was too optimistic. Higher inflation rates in the second half of the year, mostly unanticipated, meant that defence expenditure would be higher than forecast. The IISS gives a figure of 115 billion roubles, while SIPRI's estimates put it around 150 billion roubles or 7–8 per cent of estimated GDP in 1992. This figure, after adjustment for inflation, represents a sizeable fall from the Ukrainian share of what Soviet defence spending would have been in 1992 had the USSR still existed. However, the reduction is less than that of Russia and the concomitant military burden is higher.

### *Perestroika*

Fundamental restructuring of the armed forces of the former USSR took place in 1992 and the most dramatic impact was on military procurement. Traditionally, the largest part of the Soviet budget went to procurement, since weapon acquisition and an emphasis on quantitative superiority in war-fighting capabilities were central planks of defence doctrines: in the latter years of the Gorbachev era when large cuts had begun the share of procurement was about 40 per cent of OBDA. Adding another 20 per cent for military R&D, the overwhelming share of defence expenditure was for the 'investment' category. After the famous Gorbachev speech of May 1989, announcing a 19.5 per cent cut in procurement of new weapons over two years, the situation began to change rapidly. The process accelerated in 1991 when the Soviet procurement budget was reduced by 18–20 per cent compared to the previous year.

In early 1992, Prime Minister Gaidar announced that Russia's procurement spending could be cut by 85 per cent.[28] Essentially this implied that procure-

[26] North Atlantic Assembly, Defence and Security Committee, Sub-Committee on the Future of the Armed Forces, 'The future of the armed forces in the former Soviet Union', in *1992 Reports* (North Atlantic Assembly: Brussels, Nov. 1992).

[27] *Financial Times*, 22 Apr. 1992, p. 3.

[28] *Financial Times*, 21 Feb. 1992; *The Times*, 25 Jan. 1992; *Washington Post*, 23 Jan. 1992; Hummel, C., 'Russian conversion policy encounters opposition', *Radio Free Europe/Radio Liberty Research*

ment expenditure would collapse. Subsequent revisions to the budget, and some further (modest) acquisitions such as the new fighter aircraft, the MiG-31, implied that the reductions initially claimed were far too high. However, there is little doubt that new procurement in real terms in 1992 was about 20–25 per cent of what it was in 1991. There is no evidence of such massive cuts in military procurement ever occurring in peace-time. The only comparisons that can be made are with developments in the immediate post-war era in Europe and the USA.

Even if the other CIS countries were to maintain their procurement (which is impossible) the fact that Russia contributed two-thirds of Soviet defence spending will have a great impact. If it is assumed for the purposes of argument that only Russia cuts procurement expenditure, in 1992 the procurement expenditure of the former Soviet states will still be about 18 per cent of the level of 1990. Because there are large stocks in other countries of the former USSR, however, and because of the economic crisis they are suffering, their procurement budgets have also been cut drastically. Thus in 1990–92 military demand for weapons in these countries has been cut by at least 80 per cent and possibly over 90 per cent. Effectively weapon stocks in 1992 are at the same level as in 1990 and the cuts in procurement spending have been truly deep.

The main reason for the massive cuts has been economic. There was a huge budget deficit unprecedented by Russian historical standards (see table 9.11). Until 1989 Russia had a budget surplus within the overall budget of the USSR. This rapidly degenerated by 1990, and in 1991 and 1992 the deficit turned into a landslide. The 1991 deficit was 10 per cent of GDP; the IMF has given an upper limit of 5 per cent of GDP for 1992 and 1993, which is unlikely to be met in practice. Since additional funding for personnel costs, pensions, housing benefits and looking after the forces returning home had already been promised in order to keep the armed forces reasonably content and to forestall a dangerous explosion, the only cuts that could take place—if any—had to come from the military procurement budget. Hence the huge decline in acquisitions of new arms in 1992—a situation that is to continue in 1993.

Once again, it is difficult to know what exactly was the final level of procurement expenditure in Russia in 1992. Figures ranging from 115 to 150 billion roubles have been quoted by senior officials and ministers.[29] The discrepancies reflect the impact of inflation and the difference between procurement and production. It is, however, clear from SIPRI's estimates that the share of procurement in total military expenditure was less than 15 per cent.

However, one major issue remains, particularly because of its implications for arms control in general—the question of whether production or supply, as distinct from demand emanating from procurement expenditure, has also been reduced by the same amount. Output in the MIC has fallen but by less than

*Report*, vol. 1, no. 32 (14 Aug. 1992), pp. 25–32; *Aviation Week & Space Technology*, 27 Jan. 1992, p. 34.

[29] See the statement by Vice Prime Minister Georgiy Khizha, minister in charge of the military industries, report by Moscow Interfax in FBIS-SOV-92-197, 9 Oct. 1992, p. 20.

**Table 9.11.** Budget deficit of Russia, 1986–92

Figures are given in b. roubles at current values.

| Year | Value of budget deficit |
|---|---|
| 1986 | + 0.2 |
| 1987 | +1.1 |
| 1988 | +3.7 |
| 1989 | + 3.9 |
| 1990 | – 29.8 |
| 1991 | – 109.3 |
| 1992 | approx. – 1000.0 |

*Note*: A positive sign implies a budget surplus and a negative sign implies a budget deficit.

*Source: Sovetskaya Rossia,* 15 Sep. 1992 in FBIS-SOV-92-181, 17 Sep. 1992, p. 27.

would be suggested by the figures on procurement. In other words, output levels, although lower, are still higher than is warranted by demand. It is estimated that military production has fallen by about 65–70 per cent while procurement demand in constant prices has fallen by over 80 per cent. According to Yuri Glybin, a senior official at the Russian Ministry of Industry who is now in overall charge of the MIC, production of armaments in 1992 was only 32 per cent of that of 1991.[30] This implies a fall of 68 per cent, which is close to SIPRI's estimates.

The difference, between a reduction in demand by at least 80 per cent and a fall in production of 70 per cent, is significant. It means that there are unsold goods which are being stocked. Furthermore, the fall in employment in the MIC is far lower than the fall in procurement demand. This means that labour is still being used for military output that is not matched by home demand. The rate of MIC conversion has been fast (see the next two sections for an analysis), but in spite of an increase in civilian output from the defence industrial enterprises it is still possible that the supply of weapons is higher than demand.

There are two questions arising from this excess production of weapons. One is where the excess weapons are going, the second how the excess is being financed if not from military budgets. Simply through CFE Treaty limits, the quantity of weapons to be removed from use (and most of them ultimately destroyed) is large for the Soviet successor states. For example, Russia's current holdings of tanks, ACVs and artillery above CFE Treaty limits are of the order of 4600, 6820 and 2785 respectively.[31] These weapons are held by the Ministry of Defence and are its responsibility. On the other hand, current output, if not sold, is the responsibility of the industrial ministries which produce it. The old Soviet industrial ministries responsible for the MIC are now subsumed under the new Russian Ministry of Industry, and these have

[30] ITAR-TASS report of 10 Sep. 1992 in FBIS-SOV-92-177, 11 Sep. 1992, p. 29.

[31] See note 26.

the authority of disposal. There is now a small department in this ministry in charge of conversion. However, the excess weapons would probably be stored and kept by the enterprises themselves. This could give rise to concern because of the risk of proliferation.

The alternative avenue of disposal is export. In 1991 and 1992 exports were low because of the loss of WTO markets and the insistence on dollar payment, which excluded traditional clients. The emphasis is changing so that weapons will be sold for cash or bartered—understandable since there is a huge shortage of commodities in the economy. There will therefore be strong pressure to export in the transitional period. The connection between cut-backs in procurement which exceed the cut-backs in military production and this incentive to export is important.[32]

The second question is that of who pays for this excess production if not the military budget. Once again the MIC, this time in Russia or the Ukraine, is asking for subsidies from the industry budget in order to continue production. Clearly, there will be implications for the stabilization programme, which seeks to eliminate budget deficits very quickly. The option of privatization of the enterprises in the MIC is discussed below.

The state of the defence procurement budget is essentially similar in Ukraine, which had about 18 per cent of the defence enterprises of the former USSR and contributed an even greater share to arms production. Some of the most sophisticated arms factories were based in Ukraine, including the Arsenal enterprise at Kiev, the Yuzmash enterprise at Dnepropetrovsk and the Nikolaev naval shipyard. It has been reported that government procurement has dropped in 1992 by 75 per cent. Victor Antonov, in charge of the Ukrainian Ministry of Machine Building and overseeing the MIC and conversion in that country, has claimed that the military enterprises have not received 'even a single order' for defence products in 1992.[33]

### *Konversiya*

The dramatic reduction in the procurement budget implies that the defence industrial enterprises have to move much faster towards converting their factories to full-scale civilian production or be phased out and stop a significant part of their production. The need for conversion is now an economic issue as well as an arms control issue. About half of all Russian industry is subordinated to or controlled by the MIC. Industrial restructuring is therefore intimately connected with changing the product mix of the defence industries. Even after massive cuts in 1992, arms procurement in Russia at PPP prices was higher than that of any other country in Europe.

On 20 March 1992 President Yeltsin and the Supreme Council of the Russian Federation approved the Law on Conversion of Defence Industries in the Russian Federation which specified the general principles and a legal

[32] The main discussion about arms exports is in chapter 10.

[33] Hummel, C., 'Ukrainian arms makers are left on their own', *Radio Free Europe/Radio Liberty Research Report,* vol. 1, no. 32 (14 Aug. 1992), p. 34.

framework within which conversion could proceed in Russia. More specific details were given in an unpublished report which was produced at the request of the Sixth Congress of People's Deputies and was drafted by Yevgeni Yasin and Sergey Vasilev from the Experts' Institute of the Russian Union of Manufacturers and Businessmen (an industrial lobbying group) and the Working Centre on Economic Reforms (formed by the government).[34]

Conversion is defined by this law as 'a partial or complete reorientation of production capacities, science and technology potential and labour resources released from military activities of defence and defence-related enterprises, associations and organizations to civilian needs'.[35] There are three basic principles on which the law is based:

1. Conversion should concentrate on high technologies and try to produce non-military goods which are internationally competitive and could be exported in the long run.
2. Production facilities and human resources released from the defence sector after conversion should be used for priority socio-economic programmes so that society can benefit somewhat from the huge investments that have been put into the MIC in the past.
3. Social protection programmes should accompany conversion efforts to protect the personnel currently employed in the defence industrial enterprises.

The fundamental difference from the previous Gorbachev reforms for conversion is that there is currently far greater scope for decentralization and the individual enterprises are allowed far more independence than before. The extensive plans for privatization drawn up by late 1991 are yet to be realized. Current proposals are to begin privatizing 15–20 per cent of assets in 1992, rising to around 75 per cent in 1995.

In the 'World military expenditure' chapters of previous *SIPRI Yearbooks* the predominantly economic problems connected with conversion in the Gorbachev era have been discussed extensively.[36] This historical background is necessary for an understanding of why qualitatively the conversion programme has so far failed to revitalize Soviet industry and reduce the supply constraints which are endemic in the country.

Quantitatively conversion has been successful: the share of civilian production has risen from 40 to 50 per cent of the MIC in 1990 to around 60 per cent in 1991 in the former USSR and is expected to rise to 80 per cent in 1992.[37] The MIC also did better than other sectors of the economy: total industrial output in Russia fell by 20 per cent in 1992 while, according to SIPRI esti-

[34] See 'Programme for the deepening of economic reforms' (note 22); Interfax Report, 10 July 1992, *Ekonomika i Zhizn,* no. 30 (1992) pp. 14–17.

[35] See Obminsky, E., 'Conversion: a Russian case', Paper presented at the Tokyo Conference on Arms Reduction and Economic Development in the Post Cold War Era, 4–6 Nov. 1992, United Nations University, Tokyo.

[36] See Deger and Sen (note 3); Sen, S., 'The economics of conversion: transforming swords into ploughshares', ed. G. Bird, *Economic Reforms in Eastern Europe* (Edward Elgar: Aldershot, 1992); Smart, C., 'Amid the ruins, arms makers raise new threats', *Orbis*, summer 1992, pp. 349–64.

[37] See note 30.

**Table 9.12.** Examples of collaborative ventures by defence industrial enterprises of the CIS countries

| Defence enterprise | Country of co-operation | Type of civilian activity |
|---|---|---|
| Leninets enterprise in St Petersburg | USA | Electronics |
| Nikolaev shipyard | Denmark | Floating hotel |
| ARTA group of Russian defence companies | Germany | Disposal through recycling of munitions |
| Vostokometall in Russian Far East | Japan | Selling scrap aluminium from dismantling old MiG and Sukhoy aircraft |
| Positron of St Petersburg | South Korea | VCRs |

*Source*: *Eye on the East*, vol. 3, no. 18 (11 Sep. 1992); *Jane's Defence Weekly*, 14 Mar. 1992, p. 456; *Financial Times*, 19 Feb. 1992.

mates, the volume of civilian production in the MIC could have risen by as much as 10 per cent, which suggests that in terms of production the sector did relatively well. However, such civilian production has been at high cost, with wastage of resources, inflationary prices, unnecessarily sophisticated quality of products and little understanding of market conditions, sales volumes and profitability. In 1992 the problems continued with an added twist. Because of the disruptions to trade between the former republics and the difficulty of obtaining intermediate goods from suppliers, the MIC was incapable of increasing output at a significant rate. The high expectations built around conversion remain unfulfilled and the fundamental problems—overmanning, low profitability, high inventory-holding and the inability to supply what the markets want at reasonable prices—all remain as before.

There is little doubt that commercialization is proceeding rapidly. Major joint ventures, with private companies across the world, have been set up and these could become the conduit for high technology exports. In table 9.12, a few examples of major joint ventures are shown, signifying the range and diversity of converted enterprises operating profitably in close co-operation with international partners. However, such examples are the exception rather than the rule at present. Unless the structural changes taking place in the overall economy are speeded up the whole conversion programme might be stalled.

The three fields in which Russia could have particular trade advantage internationally are optics, certain types of electronics and the aerospace industry. The Departments of the Defence Industry and the Aviation Industry as part of the overall Ministry of Industry are being substantially maintained to preserve their unique capabilities in terms of facilities and R&D establishments. For example, out of one million people employed in the Defence Industry Department, only 73 000 have lost their jobs over an 18-month period.[38] The job losses announced for the whole MIC are of higher orders of magnitude. These two industrial ministries are also important exporters. The Defence Industry Department is ready to begin serial production of the Nikonov assault rifle as a successor to the Kalashnikov, while the Aviation Industry Department is ready to sell the MiG-29 and other military aircraft.[39]

It is, however, proving increasingly costly to convert industries. Costs include re-tooling, paying social protection to the unemployed, and maintaining personnel when they should be discharged and the infrastructure paid for as in the past by the enterprises themselves. Whole cities such as Perm or Chelyabinsk are at risk, while even in Moscow and St Petersburg a substantial part of the industrial work-force is employed by the branches of the defence complex. Mikhail Malei, a Presidential counsellor and advisor on conversion, claimed that 4 million people work directly for the MIC, 12 million civilian jobs are associated with these industries, and taking families and dependents into account about 37 million Russians or one-fifth of the population are directly or indirectly dependent on the MIC.[40] Even though economic analysis would predict that a significant proportion of the enterprises should simply be shut down and abandoned, in terms of socio-political costs this is not feasible.

There is no simple way of estimating the cost of conversion or how long restructuring expenditure will be required. Initial estimates of the total costs of conversion in Russia of 150 billion roubles made in early 1992 were later changed to $150 billion by Mikhail Malei, who presumably has direct access to the President and may thus be regarded as a reliable source. (The apparent increase in the estimate probably results from the fact that the earlier estimate was made in 1990, before the price reform of 1991 and the price explosion of 1992, when it was normal to quote a one-to-one parity between the rouble and the dollar.[41]) Even over a 10-year period the cost at 1992 prices would be about 150 billion roubles per annum. The Russian Government simply does not have the resources to meet these costs except marginally.

Equally important is the fact that the success or failure of conversion is intimately related to the whole industrial structure of Russia and Ukraine. There are important local political élites who have tied their allegiance to the rejuvenation of Russian industry independent of the fate of the overall economy, and believe that if conversion fails and the broader industrial structure collapses then the future of Russia will be gloomy. Hard economic decisions in the transitional period, such as allowing the bankruptcy of large enterprises, are difficult to carry out in the face of the opposition of such groups. The Russian League of Industrialists and Entrepreneurs headed by Arkadi Volski, Nikolay Travkin, the leader of the Russian Democratic Party, and Vice-

[38] Interview with Gennadiy Yanpolsky in *Krasnaya Zvezda*, 29 Aug. 1992, pp. 1–2, reported in FBIS-SOV-92-173, 4 Sep. 1992, p. 22.

[39] *Rossiyskaya Gazeta*, 28 Feb. 1992, pp. 1–2 in FBIS-SOV-92-043, 4 Mar. 1992, p. 43.

[40] *Rossiyskie Vesti*, 22 May 1992, p. 1; *Rossiyskaya Gazeta,* 28 Feb. 1992, pp. 1-2, in FBIS-SOV-92-043, 4 Mar. 1992, p. 43; Bush (note 22).

[41] 150 billion roubles at 1990 values would in fact at the time of writing (end–1992) be worth $150 billion—1500 billion roubles at then current values, at an exchange-rate of 100 to $1.

**Table 9.13.** Foreign aid provided by the US Government for Central and Eastern Europe and the CIS countries (budget authority, FYs 1992 and 1993)

Figures are in US $m., current prices.

| | FY 1992 | FY 1993 |
|---|---|---|
| International Affairs | 560 | 910 |
| Food for Peace | 10 | 10 |
| Economic Support Fund | 0 | 100 |
| Humanitarian and Technical Assistance | 150 | 350 |
| Support for East European Democracy | 400 | 450 |
| Defense | 500 | 0 |
| Weapons destruction | 400 | 0 |
| Humanitarian aid | 100 | 0 |
| Agriculture | 490 | 490 |
| Cost of subsidy | 490 | 490 |
| **Total** | **1 550** | **1 400** |

*Source*: US Congress, Congressional Budget Office, *Analysis of the President's Budgetary Proposals for Fiscal Year 1993* (US Government Printing Office: Washington, DC, Mar. 1992).

President Alexander Rutskoy have formed the coalition called Civic Union which effectively supports a slow transition towards a market economy and calls for high subsidies to the MIC or for the government to pay the costs of conversion or for the Central Bank to pay inter-enterprise debts accumulated by defence enterprises.

Foreign aid has been discussed as a means to help conversion, including the destruction of weapons. President Bush has offered $400 million in foreign aid, directly from the defence budget, to dismantle warheads and nuclear weapons made redundant by the START Treaty. In addition, there are proposals that US Government officials will closely monitor this aid and its disbursement and uses. In effect this could mean US scientists helping the dismantling process. The industrial ministry of nuclear energy is reluctant to receive US help in the dismantling of weaponry since it claims that the necessary expertise is available within the former USSR. It would like to have the resources instead directly for conversion, elimination, subsidies, and so on. Much of this aid has as yet not been disbursed. Table 9.13 gives data on all foreign aid for Russia, the CIS and Central and Eastern Europe from the defence budget authorized and requested by President Bush in FY 1992 and 1993.

From this exhaustive analysis of the military spending of Russia and the CIS, and bearing in mind the tentative nature of the findings due to the paucity of information, a number of general conclusions can be drawn:

1. Military expenditure in the CIS has been substantially reduced. The defence spending of the CIS countries is about half that of the former USSR.

2. Russia now has the overwhelmingly largest share of the total military spending of these countries—partly because of its obligations towards the up-keep of the CIS forces and troops in foreign countries.

3. Procurement expenditure has been severely cut. In 1992 procurement of new weapons amounted to only 15 per cent of that of the former USSR in 1990. Apart from a few specific acquisitions, in practice 1992 showed an almost total curtailment of new weapon acquisition.

4. Military R&D has suffered less, being now at a level of 40 per cent of the 1991 level. However, even though a considerable amount of R&D personnel and facilities are being maintained, there is little real work in terms of weapon research. Testing and evaluation have all but stopped except in the aerospace industry. The principal loser has been naval research. Almost all the budget is going to personnel and pensions.

5. Republican or national armies will be formed but at lower levels and lower cost. The only effective and real centralized control still existing is that of nuclear weapons, but the situation is chaotic.

6. Conversion has been modestly successful in the sense that output has increased. However, as an economic concept and as something that was expected to increase the supply of consumer and high-tech goods at appropriate prices for domestic and foreign markets, conversion has failed miserably.

7. There is at present a close connection between economic and security issues. Economic factors and crises have forced rapid demilitarization; economics is the predominant arms controller. On the other hand, the rewards of disarmament in terms of additional civilian output are yet to be seen.

## IV. European NATO

### NATO military expenditure

Military expenditure data in current and constant prices for all the countries of NATO are shown in tables 9.14 and 9.15, and their share in GDP in table 9.16. The constant price figures are given in 1991 exchange-rates and prices; because of the depreciation of the dollar the value of European defence spending appears higher than it was in the previous year. In 1992 all NATO countries taken together spent about $493 billion (in 1991 prices), of which the USA had overwhelmingly the largest share (about 60 per cent). This simple measure of burden-sharing was to be used more often in the future by the incoming Administration of the USA to emphasize the importance of the USA in financing collective security. However, more sophisticated measurements of burden-sharing (such as including in calculations indirect measures which show the opportunity costs of conscript armed forces) show that individual European countries also have a high defence burden and contribute 'fair' shares towards collective defence.[42] Overall, as the Soviet threat has disappeared, the economic costs of collective defence and the taking

[42] See Deger 1990 (note 5).

**Table 9.14.** NATO military expenditure, in current price figures, 1983–92

Figures are in local currency, current prices.

| | | 1983 | 1984 | 1985 | 1986 | 1987 | 1988 | 1989 | 1990 | 1991 | 1992 |
|---|---|---|---|---|---|---|---|---|---|---|---|
| *North America* | | | | | | | | | | | |
| Canada | m. dollars | 8 562 | 9 519 | 10 187 | 10 811 | 11 529 | 12 180 | 12 725 | 13 318 | 13 862 | 13 580 |
| USA | m. dollars | 218 084 | 238 136 | 263 900 | 282 868 | 289 391 | 295 841 | 304 607 | 299 701 | 288 791 | 308 489 |
| *Europe* | | | | | | | | | | | |
| Belgium | m. francs | 136 615 | 139 113 | 144 183 | 152 079 | 155 422 | 150 647 | 152 917 | 155 205 | 157 919 | 130 943 |
| Denmark | m. kroner | 12 574 | 13 045 | 13 343 | 13 333 | 14 647 | 15 620 | 15 963 | 16 399 | 17 091 | 16 844 |
| France | m. francs | 165 029 | 176 638 | 186 715 | 197 080 | 209 525 | 215 073 | 224 985 | 232 376 | 239 411 | 241 417 |
| FR Germany | m. D–marks | 56 496 | 57 274 | 58 649 | 60 130 | 61 354 | 61 638 | 63 178 | 68 376 | 65 579 | 66 143 |
| Greece | m. drachmas | 193 340 | 271 922 | 321 981 | 338 465 | 393 026 | 471 820 | 503 032 | 612 344 | 693 846 | 809 387 |
| Italy | b. lire | 13 583 | 15 616 | 17 767 | 19 268 | 22 872 | 25 539 | 27 342 | 28 007 | 30 191 | 30 250 |
| Luxembourg | m. francs | 2 104 | 2 234 | 2 265 | 2 390 | 2 730 | 3 163 | 2 995 | 3 233 | 3 681 | 3 882 |
| Netherlands | m. guilders | 12 149 | 12 762 | 12 901 | 13 110 | 13 254 | 13 300 | 13 571 | 13 513 | 13 548 | 13 822 |
| Norway | m. kroner | 12 395 | 12 688 | 15 446 | 16 033 | 18 551 | 18 865 | 20 248 | 21 252 | 21 316 | 23 763 |
| Portugal | m. escudos | 76 765 | 92 009 | 111 375 | 139 972 | 159 288 | 194 036 | 229 344 | 267 299 | 305 643 | 325 663 |
| Spain | m. pesetas | 540 311 | 594 932 | 674 883 | 715 306 | 852 767 | 835 353 | 920 381 | 922 808 | 947 173 | 984 276 |
| Turkey | b. lire | 557 | 803 | 1 235 | 1 868 | 2 477 | 3 789 | 7 158 | 13 866 | 23 657 | 38 739 |
| UK | m. pounds | 15 605 | 17 104 | 18 156 | 18 581 | 19 125 | 19 439 | 20 474 | 22 067 | 23 988 | 23 942 |

*Notes.* For all NATO countries, in this table and those following, where fiscal year is not the same as calendar year, figures are calculated for calendar year. Figures for recent years are budget estimates.

***Source:*** NATO, *Financial and Economic Data Relating to NATO Defence* (NATO: Brussels, annual); author's calculations. Figures for France are based on national data.

**Table 9.15.** NATO military expenditure, in constant price figures, 1983–92

Figures are in US $m., at 1991 prices and exchange-rates.

| | 1983 | 1984 | 1985 | 1986 | 1987 | 1988 | 1989 | 1990 | 1991 | 1992 |
|---|---|---|---|---|---|---|---|---|---|---|
| *North America* | | | | | | | | | | |
| Canada | 10 651 | 11 349 | 11 683 | 11 899 | 12 164 | 12 351 | 12 295 | 12 278 | 11 576 | 11 658 |
| USA | 298 158 | 312 091 | 334 097 | 351 434 | 346 612 | 340 796 | 334 750 | 312 538 | 288 799 | 299 270 |
| *Europe* | | | | | | | | | | |
| Belgium | 5 111 | 4 894 | 4 839 | 5 038 | 5 069 | 4 857 | 4 783 | 4 693 | 4 625 | 3 746 |
| Denmark | 2 714 | 2 650 | 2 589 | 2 495 | 2 636 | 2 268 | 2 623 | 2 625 | 2 672 | 2 573 |
| France | 39 886 | 39 713 | 39 712 | 40 894 | 42 080 | 42 043 | 42 497 | 42 460 | 42 433 | 41 542 |
| Germany | 39 421 | 39 025 | 39 123 | 40 151 | 40 886 | 40 549 | 40 445 | 42 628 | 39 517 | 38 300 |
| Greece | 3 984 | 4 734 | 4 697 | 4 014 | 4 004 | 4 236 | 3 971 | 4 015 | 3 807 | 3 851 |
| Italy | 18 570 | 19 252 | 20 064 | 20 566 | 23 312 | 24 747 | 24 941 | 24 015 | 24 336 | 23 527 |
| Luxembourg | 76 | 77 | 75 | 78 | 90 | 102 | 94 | 98 | 108 | 110 |
| Netherlands | 7 390 | 7 517 | 7 431 | 7 544 | 7 681 | 7 654 | 7 725 | 7 506 | 7 246 | 7 147 |
| Norway | 3 008 | 2 896 | 3 336 | 3 230 | 3 439 | 3 278 | 3 364 | 3 390 | 3 288 | 3 571 |
| Portugal | 1 560 | 1 446 | 1 467 | 1 651 | 1 717 | 1 909 | 2 003 | 2 061 | 2 116 | 2 044 |
| Spain | 9 122 | 9 027 | 9 411 | 9 168 | 10 386 | 9 707 | 10 044 | 9 407 | 9 115 | 8 989 |
| Turkey | 4 088 | 3 972 | 4 215 | 4 737 | 4 524 | 3 945 | 4 565 | 5 517 | 5 671 | 5 857 |
| UK | 43 413 | 45 313 | 45 358 | 44 893 | 44 363 | 42 954 | 41 994 | 41 326 | 42 442 | 40 690 |
| EC | 171 745 | 174 164 | 175 287 | 177 044 | 182 742 | 181 983 | 181 641 | 181 390 | 179 002 | 173 109 |
| **European NATO total** | **178 262** | **180 516** | **182 316** | **184 459** | **190 187** | **188 669** | **189 047** | **189 738** | **187 375** | **181 947** |

*Note.* This series is based on the data given in the local currency series, deflated to 1991 price levels and converted into dollars at 1991 period-average exchange-rates. Local consumer price indices (CPI) are taken as far as possible from *International Financial Statistics* (*IFS*) (International Monetary Fund: Washington, DC). For the most recent year, the CPI is an estimate based on the first 6–10 months of the year. Period-average exchange-rates are taken as far as possible from *IFS*.

**Table 9.16.** NATO countries' military expenditure as a percentage of GDP, 1983–92

| | 1983 | 1984 | 1985 | 1986 | 1987 | 1988 | 1989 | 1990 | 1991 | 1992 |
|---|---|---|---|---|---|---|---|---|---|---|
| *North America* | | | | | | | | | | |
| Canada | 2.1 | 2.1 | 2.1 | 2.1 | 2.1 | 2.0 | 2.0 | 2.0 | 2.0 | 2.0 |
| USA | 6.4 | 6.3 | 6.5 | 6.6 | 6.4 | 6.0 | 5.8 | 5.4 | 5.1 | 5.4 |
| *Europe* | | | | | | | | | | |
| Belgium | 3.2 | 3.1 | 3.0 | 3.0 | 2.9 | 2.7 | 2.5 | 2.4 | 2.3 | 2.0 |
| Denmark | 2.5 | 2.3 | 2.2 | 2.0 | 2.1 | 2.1 | 2.1 | 2.1 | 2.1 | 1.9 |
| France | 4.1 | 4.1 | 4.0 | 3.9 | 3.9 | 3.8 | 3.6 | 3.6 | 3.5 | 3.4 |
| FR Germany | 3.4 | 3.3 | 3.2 | 3.1 | 3.1 | 2.9 | 2.8 | 2.8 | 2.8 | 2.2 |
| Greece | 6.3 | 7.1 | 7.0 | 6.1 | 6.3 | 6.3 | 5.7 | 5.9 | 5.5 | 5.5 |
| Italy | 2.1 | 2.2 | 2.2 | 2.1 | 2.3 | 2.3 | 2.3 | 2.1 | 2.1 | 2.0 |
| Luxembourg | 1.2 | 1.2 | 1.1 | 1.1 | 1.2 | 1.3 | 1.1 | 1.1 | 1.2 | 1.2 |
| Netherlands | 3.2 | 3.2 | 3.1 | 3.1 | 3.1 | 3.0 | 2.9 | 2.7 | 2.5 | 2.5 |
| Norway | 3.1 | 2.8 | 3.1 | 3.1 | 3.3 | 3.2 | 3.3 | 3.2 | 3.1 | 3.4 |
| Portugal | 3.3 | 3.3 | 3.3 | 3.2 | 3.2 | 3.1 | 3.2 | 3.2 | 3.1 | 2.9 |
| Spain | 2.4 | 2.4 | 2.4 | 2.2 | 2.4 | 2.1 | 2.1 | 1.8 | 1.7 | 1.6 |
| Turkey | 4.8 | 4.4 | 4.5 | 4.8 | 4.2 | 3.8 | 4.3 | 4.9 | 4.0 | 3.9 |
| UK | 5.1 | 5.3 | 5.1 | 4.8 | 4.5 | 4.1 | 4.0 | 4.0 | 4.2 | 4.0 |

*Note*: This series is based on the data given in the local currency series, deflated to 1991 price levels and converted into dollars at 1991 period-average exchange-rates. Local consumer price indices (CPI) are taken as far as possible from *International Financial Statistics* (*IFS*) (International Monetary Fund: Washington, DC). For the most recent year, the CPI is an estimate based on the first 6–10 months of the year. Period-average exchange-rates are taken as far as possible from *IFS*.

*Source:* Author's calculations based on tables 9.14 and 9.15.

over of US army functions as the troops withdraw are bound to become increasingly important with time.

The time series data on European NATO military expenditure over the 10 years 1983–92 show a clear rising trend until 1987 and then an equally clear decline. Military spending rose by 1.6 per cent per annum from 1983 to 1987, the peak year. The fall by approximately 0.9 per cent per annum between 1987 and 1992 is still rather modest, but in 1992 the military expenditure of European NATO in aggregate fell by 2.9 per cent. Currently, it stands at approximately $182 billion per annum (at 1991 prices)—that is, at about the same level as in 1985.

## Procurement

The major impact of limiting military spending after the cold war has been on defence procurement, which has fallen faster in 1992 than any other category of expenditure. Tables 9.17 and 9.18 give data for the 10-year period 1983–92 in current and constant prices. Once again the peak for the 1980s occurred in

1987 when defence procurement spending exceeded $43 billion. In 1992 the value fell to just below $31 billion—a decline of slightly less than 7 per cent per annum during the period 1987–92.

A significant part of this rapid decline has been contributed by Germany, whose procurement spending on major weapons now stands at 64 per cent of the level of 1988. Leaving out Germany, other European NATO countries reduced their defence equipment spending by 5.4 per cent per annum during 1987–92—significant but still slow relative to the limits set by the CFE table Treaty. The rapidity of German reductions is partly the result of surplus material being received from the former East German armed forces—the large number of land-based weapons such as tanks, ACVs and artillery inherited from the war-fighting scenarios of the past in Central Europe but now far higher than the CFE Treaty limits. It is also partly the result of an increase in exports; but much more importantly to faster restructuring of the requirements of the military in the post-cold war era.

The German perception of a threat from conventional war is at its lowest since World War II, and this altered perception is increasingly reflected in the downgrading of Bundeswehr arms purchases. The debate on German participation in the production of the European Fighter Aircraft (EFA) resulted in a final version of the aircraft which will be less expensive and more closely geared to the current threat perception in Europe, which requires less sophisticated technology. It is interesting to note that the value of German purchases of major weapons in 1992 stands at less than 9 per cent of aggregate defence spending—an historic low.

Table 9.19 gives the evidence of structural change in procurement expenditure over the past 10 years, as it rose continuously during 1983–87 and then fell during 1987–92. The three groups of countries of interest are European NATO, the EC member states and European NATO without Germany. The relative importance of the recent German cuts is shown in the differential rates of decline as between the aggregates for European NATO and for NATO without Germany. However, the general trend remains unchanged. The implications for the defence industrial base will be significant.[43]

It became clear in 1992 for the first time that the CFE Treaty limits are having a direct impact on procurement purchases, particularly for land-based systems. Germany, for example, now has about 3000 tanks, 4000 ACVs and 2000 artillery items in its holdings over and above the CFE limits. Other major military purchasers such as France and the UK have brought their holdings closer to the CFE limits.

What is not so clear from the available data for 1992 is whether the qualita tive arms race is still being continued. It is obvious that the adversarial military postures which produced the technological innovations of the past are now ended, so that it is not clear with whom the 'race' might be run, unless it is the requirements of modern warfare and the limits of technology itself. The

[43] See also chapter 10 in this volume.

**Table 9.17.** NATO major weapon procurement expenditure, 1983–92

Figures are in local currency, current prices.

| | | 1983 | 1984 | 1985 | 1986 | 1987 | 1988 | 1989 | 1990 | 1991 | 1992 |
|---|---|---|---|---|---|---|---|---|---|---|---|
| *North America* | | | | | | | | | | | |
| Canada | m. dollars | 1 688 | 1 971 | 1 941 | 2 140 | 2 434 | 2 486 | 2 394 | 2 309 | 2 314 | 2 391 |
| USA | m. dollars | 50 202 | 58 328 | 66 348 | 72 525 | 76 362 | 73 749 | 76 683 | 75 078 | 74 757 | 68 177 |
| *Europe* | | | | | | | | | | | |
| Belgium | m. francs | 18 853 | 18 363 | 18311 | 19 618 | 20 360 | 18 078 | 15 139 | 12 261 | 12 949 | 11 261 |
| Denmark | m. kronor | 2 075 | 2 048 | 1 841 | 1 867 | 2 182 | 2 249 | 2 091 | 2 443 | 2 700 | 2 813 |
| France | m. francs | 39 772 | 42 216 | 46 492 | 49 664 | 55 943 | 56 564 | 60 071 | 58 094 | 56 853 | 57 354 |
| Germany | m. D–mark | 11 299 | 11 455 | 11 730 | 12 267 | 12 332 | 11 896 | 12 004 | 12 103 | 7 214 | 6 151 |
| Greece | m. drachmas | 30 741 | 41 604 | 46 687 | 53 477 | 67 605 | 109 934 | 110 164 | 131 042 | 140 851 | 174 018 |
| Italy | b. lire | 2 664 | 2 843 | 3 494 | 3 693 | 4 900 | 5 235 | 5 605 | 4 901 | 4 921 | 4 175 |
| Luxembourg | m. francs | 36 | 36 | 91 | 74 | 106 | 89 | 114 | 103 | 199 | 175 |
| Netherlands | m. guilders | 2 794 | 3 012 | 3 019 | 2 661 | 2 359 | 2 713 | 2 388 | 2 419 | 2 113 | 2 336 |
| Norway | m. kronor | 2 615 | 2 297 | 3 846 | 3 303 | 3 784 | 3 547 | 5 022 | 4 803 | 4 690 | 5 608 |
| Portugal | m. escudos | 3 761 | 4 416 | 3 675 | 8 818 | 16 088 | 20 374 | 27 292 | 27 532 | 25 980 | 20 842 |
| Spain | m. pesetas | 116 707 | 170 745 | 113 380 | 168 812 | 210 633 | 172 918 | 168 978 | 119 042 | 122 185 | 163 390 |
| Turkey | b. lira | 56 | 105 | 168 | 334 | 553 | 853 | 1 231 | 2 773 | 5 370 | 9 840 |
| UK | m. pounds | 4 122 | 4 629 | 4 907 | 4 762 | 4 744 | 4 904 | 4 668 | 4 164 | 4 588 | 4 330 |

*Sources*: NATO, *Financial and Economic Data Relating to NATO Defence* (NATO: Brussels, annual); author's calculations. Figures for France are based on national data.

**Table 9.18.** NATO and EC major weapon procurement expenditure, 1983–92

Figures are in US $m., at constant (1991) prices.

| | 1983 | 1984 | 1985 | 1986 | 1987 | 1988 | 1989 | 1990 | 1991 | 1992 |
|---|---|---|---|---|---|---|---|---|---|---|
| *North America* | | | | | | | | | | |
| Canada | 2 100 | 2 350 | 2 226 | 2 355 | 2 568 | 2 521 | 2 313 | 2 129 | 2 020 | 2 053 |
| USA | 68 635 | 76 442 | 83 997 | 90 105 | 91 461 | 84 956 | 84 271 | 79 337 | 74 757 | 66 140 |
| *Europe* | | | | | | | | | | |
| Belgium | 705 | 646 | 615 | 650 | 664 | 583 | 474 | 371 | 379 | 322 |
| Denmark | 448 | 416 | 357 | 349 | 393 | 387 | 344 | 391 | 422 | 430 |
| France | 9 613 | 9 491 | 9 888 | 10 305 | 11 235 | 11 057 | 11 347 | 10 615 | 10 077 | 9 869 |
| FR Germany | 7 884 | 7 805 | 7 825 | 8 191 | 8 218 | 7 826 | 7 685 | 7 545 | 4 347 | 3 562 |
| Greece | 633 | 724 | 681 | 634 | 689 | 987 | 870 | 859 | 773 | 828 |
| Italy | 3 642 | 3 505 | 3 946 | 3 942 | 4 994 | 5 075 | 5 112 | 4 202 | 3 967 | 3 247 |
| Luxembourg | 1 | 1 | 3 | 2 | 3 | 3 | 4 | 4 | 6 | 5 |
| Netherlands | 1 700 | 1 774 | 1 739 | 1 531 | 1 367 | 1 561 | 1 359 | 1 344 | 1 130 | 1 208 |
| Norway | 637 | 527 | 834 | 668 | 704 | 619 | 838 | 769 | 727 | 846 |
| Portugal | 76 | 69 | 48 | 104 | 173 | 200 | 238 | 212 | 180 | 131 |
| Spain | 1 970 | 2 591 | 1 581 | 2 164 | 2 565 | 2 009 | 1 838 | 1 214 | 1 176 | 1 492 |
| Turkey | 411 | 519 | 573 | 847 | 1 010 | 888 | 785 | 1 103 | 1 287 | 1 488 |
| UK | 11 467 | 12 263 | 12 259 | 11 505 | 11 004 | 10 842 | 9575 | 7 798 | 8 118 | 7 359 |
| **European NATO total** | **39 189** | **40 333** | **40 349** | **40 894** | **43 021** | **42 038** | **40 468** | **36 427** | **32 587** | **30 787** |
| **NATO total** | **109 923** | **119 125** | **126 572** | **133 354** | **137 050** | **129 514** | **127 052** | **117 892** | **109 364** | **98 979** |
| EC member countries | 38 154 | 39 314 | 38 965 | 39 416 | 41 330 | 40 570 | 38865 | 34 574 | 30 591 | 28 470 |
| **European NATO without Germany** | **30 270** | **31 508** | **31 141** | **31 225** | **33 112** | **32 744** | **31 181** | **27 029** | **26 243** | **24 908** |

*Sources*: NATO, *Financial and Economic Data Relating to NATO Defence* (NATO: Brussels, annual); author's calculations. Figures for France are based on national data.

**Table 9.19.** The rise and fall of European military procurement expenditures, 1983–92

Figures are percentages.

| | % increase per annum 1983–87 | % decrease per annum 1987–92 |
|---|---|---|
| European NATO | 2.0 | 7.0 |
| EC | 1.6 | 7.8 |
| European NATO without Germany | 2.2 | 5.4 |

*Source*: Author's estimates, calculated from tables 9.17 and 9.18 in this chapter.

fact that national territorial defence and the defence of national interests are no longer synonymous also means that wars need to be fought far away and casualties have to be severely limited. This will necessitate more widespread use of the new technologies such as supercomputers, stealth technology, 'smart' weapons, and greater use of electronic and space technologies in $C^3I$, as well as reliance on space-based intelligence and weapons systems. France, for example, even in the face of defence budget austerity, increased its allocation to space defence to 3.6 billion francs in 1992. Although this amount is a very modest proportion of total defence spending, it still represents an increase of 17.6 per cent in nominal terms. In both France and the UK—the two largest R&D spenders in Europe—military research expenditure shares rose in 1992, although the absolute levels are still falling modestly.

Although the data show that the greatest impact on European procurement budgets is being made by Germany, the year 1992 seemed also to be a major turning-point for France. French military expenditure was stable at around $42 billion per annum (at constant 1991 prices) for the period 1987–91. The first sign of change can be seen in 1992 when it fell by around 2 per cent—the first significant reduction of military expenditure in more than a decade. Procurement spending on major weapon systems has now fallen for three successive years although the reductions are far less radical compared to those carried through in Germany and even the UK. The new Loi de Programmation presented in 1992 sets out the framework of defence industrial planning for the next five years. Although the basic principles of defence procurement policy remain similar to those of the past, substantial cuts are envisaged in all areas of military acquisition.

Traditionally, French defence procurement policy has had three features required to satisfy a number of objectives: (*a*) the policy of maintaining an ability to produce domestically a relatively complete spectrum of products from the military industries of 'national champions'; (*b*) arms exports as an important ingredient of overall economic policy and a significant source of funds for military R&D conducted by industry itself; and (*c*) the preservation of close links between various government departments (including defence), the armed forces and the defence industrial base (much of it nationalized). The procurement agency, the Délégation Générale pour l'Armement (DGA), is at the hub of this interconnected and entwined structure. In particular, the government spends about one-third of all R&D funding on the military, a share second only to the UK's within Europe.

The first two aspects of policy are now seriously threatened both by budgetary problems and by the very success of arms control in the new European order. The new plans aim to stretch out production plans and order less of each category of weapons in order to maintain the policy of having a broad-based domestic demand for weapons systems. It is difficult, however, to keep up this façade faced with increasing unit costs because economies of scale are lost if new orders are curtailed. In 1992 it was announced that orders for the Triomphant Class of ballistic missile submarines will be cut from five to four; demand for Rafale fighters will be reduced from a planned 250 to 235; Mirage 2000 orders for 1993 will be cancelled; the number of Franco-German Tiger attack helicopters will be reduced substantially from the planned 215; purchases of Leclerc tanks will be cut from the planned 1000 to 700; and the nuclear-powered aircraft-carrier *Charles de Gaulle* will be procured at a later date than originally planned. Thus, with the fall in exports and an impending substantial reduction in domestic procurement, the French military industries are becoming fully exposed to the general European contraction from which they have been insulated until now.

### The armed forces

European NATO countries have also been reducing the size of the armed forces, although not at a rapid rate, in general reflecting the overall uncertainty about the use of and the requirement for military forces in the future. In 1992 the Belgian Government removed the Gendarmarie from military duties, thus cutting its army by 20 000 in one move. Conscription is to be abolished from 1994 (about 40 per cent of the armed forces are conscripts) and the future manpower requrement will be half that of 1991. The Netherlands also cut its forces by 14 000 in 1992—about half that from the army itself. German forces (including those of the former East German armies) have been steadily cut and today unified Germany has fewer troops than the former Federal Republic of Germany (FRG) had in 1975. France, with a 1992 force of 522 000 men, has cut personnel numbers by 22 000 in 1992 and has fewer people in service than it had in 1970. However, as mentioned above, the cuts are still proceeding cautiously. If the statistical distortions imposed by the special case of the Belgian figures are removed, then European NATO reduced its armed force personnel by only 0.7 per cent in 1992 compared to the previous year. Tables 9.20 and 9.21 give the size of the armies as well as the proportion of military and civilian personnel employed by the military in the total labour force.

The restructuring of NATO continues while new roles are taken on, such as assistance to the CSCE peace-keeping activities. Expenditure is expected to

**Table 9.20.** NATO armed forces, total military personnel, 1983–92

Figures are in thousands.

| | 1983 | 1984 | 1985 | 1986 | 1987 | 1988 | 1989 | 1990 | 1991 | 1992 |
|---|---|---|---|---|---|---|---|---|---|---|
| *North America* | | | | | | | | | | |
| Canada | 81 | 82 | 83 | 85 | 86 | 88 | 88 | 87 | 86 | 83 |
| USA | 2 222 | 2 222 | 2 244 | 2 269 | 2 279 | 2 246 | 2 241 | 2 181 | 2 115 | 1 975 |
| *Europe* | | | | | | | | | | |
| Belgium | 109 | 107 | 107 | 107 | 109 | 110 | 110 | 106 | 101 | 81 |
| Denmark | 30 | 31 | 29 | 28 | 28 | 30 | 31 | 31 | 30 | 29 |
| France | 578 | 571 | 563 | 558 | 559 | 558 | 554 | 550 | 542 | 522 |
| FR Germany | 496 | 487 | 493 | 495 | 495 | 495 | 503 | 545 | 457 | 476 |
| Greece | 177 | 197 | 201 | 202 | 199 | 199 | 201 | 201 | 205 | 202 |
| Italy | 498 | 508 | 504 | 502 | 504 | 506 | 506 | 493 | 473 | 471 |
| Luxembourg | 1 | 1 | 1 | 1 | 1 | 1 | 1 | 1 | 1 | 1 |
| Netherlands | 104 | 103 | 103 | 106 | 106 | 107 | 106 | 104 | 104 | 90 |
| Norway | 41 | 39 | 36 | 38 | 38 | 40 | 43 | 51 | 41 | 41 |
| Portugal | 93 | 100 | 102 | 101 | 105 | 104 | 104 | 87 | 86 | 88 |
| Spain | 355 | 342 | 314 | 314 | 314 | 304 | 277 | 263 | 246 | 228 |
| Turkey | 824 | 815 | 814 | 860 | 879 | 847 | 780 | 769 | 804 | 826 |
| UK | 333 | 336 | 334 | 331 | 328 | 324 | 318 | 308 | 301 | 293 |
| **European NATO** | **3 639** | **3 637** | **3 602** | **3 643** | **3 693** | **3 625** | **3 534** | **3 509** | **3 391** | **3 348** |
| **NATO total** | **5 942** | **5 941** | **5 930** | **5 997** | **6 058** | **5 959** | **5 863** | **5 777** | **5 592** | **5 406** |

*Sources*: NATO, *Financial and Economic Data Relating to NATO Defence* (NATO: Brussels, annual); author's calculations. Figures for France are based on national data.

**Table 9. 21.** NATO military and civilian personnel, as share of total labour force, 1983–92

Figures are percentages.

| | 1983 | 1984 | 1985 | 1986 | 1987 | 1988 | 1989 | 1990 | 1991 | 1992 |
|---|---|---|---|---|---|---|---|---|---|---|
| *North America* | | | | | | | | | | |
| Canada | 1.0 | 1.0 | 1.0 | 1.0 | 1.0 | 1.0 | 0.9 | 0.9 | 0.9 | 0.9 |
| USA | 2.9 | 2.9 | 2.9 | 2.8 | 2.8 | 2.7 | 2.7 | 2.6 | 2.5 | 2.3 |
| *Europe* | | | | | | | | | | |
| Belgium | 2.8 | 2.8 | 2.7 | 2.7 | 2.7 | 2.8 | 2.8 | 2.7 | 2.6 | 2.0 |
| Denmark | 1.5 | 1.5 | 1.4 | 1.4 | 1.4 | 1.4 | 1.4 | 1.4 | 1.4 | 1.4 |
| France | . . | . . | 3.3 | 2.9 | 2.9 | 2.9 | 2.8 | 2.8 | 2.7 | 2.6 |
| FR Germany | 2.4 | 2.4 | 2.4 | 2.4 | 2.4 | 2.3 | 2.3 | 2.6 | 1.9 | 2.0 |
| Greece | 5.2 | 6.0 | 6.1 | 6.0 | 6.0 | 5.6 | 5.8 | 5.8 | 5.8 | 5.7 |
| Italy | 2.3 | 2.3 | 2.4 | 2.4 | 2.3 | 2.3 | 2.3 | 2.2 | 2.2 | 2.1 |
| Luxembourg | 0.9 | 0.9 | 0.9 | 0.9 | 0.8 | 0.8 | 0.8 | 0.7 | 0.7 | 0.7 |
| Netherlands | 2.5 | 2.2 | 2.2 | 2.2 | 2.1 | 2.0 | 2.0 | 1.9 | 1.9 | 1.6 |
| Norway | 2.6 | 2.5 | 2.3 | 2.3 | 2.3 | 2.4 | 2.6 | 2.9 | 2.5 | 2.5 |
| Portugal | 2.3 | 2.5 | 2.6 | 2.6 | 2.6 | 2.5 | 2.5 | 2.1 | 2.0 | 2.0 |
| Spain | 3.0 | 2.9 | 2.6 | 2.5 | 2.4 | 2.3 | 2.1 | 2.0 | 1.9 | 1.8 |
| Turkey | 4.8 | 4.7 | 4.6 | 4.9 | 4.9 | 4.5 | 4.1 | 3.9 | 4.0 | 4.1 |
| UK | 2.0 | 2.0 | 2.0 | 1.9 | 1.8 | 1.8 | 1.7 | 1.7 | 1.7 | 1.6 |
| **European NATO total** | **2.8** | **2.8** | **2.8** | **2.8** | **2.8** | **2.6** | **2.6** | **2.5** | **2.4** | **2.3** |
| **NATO total** | **2.8** | **2.8** | **2.7** | **2.7** | **2.7** | **2.6** | **2.5** | **2.5** | **2.4** | **2.3** |

*Sources*: NATO, *Financial and Economic Data Relating to NATO Defence* (NATO: Brussels, annual); author's calculations. Figures for France are based on national data.

fall, but from a high level (almost $490 billion in 1991 prices with European NATO contributing over $180 billion) and the reduction will take a considerable period of time, possibly until the end of the century. New responsibilities are also cropping up and changes in military doctrines and war-fighting capabilities imply that new expenditure will be sought. The new Allied Command Europe (ACE) Rapid Reaction Force is already being formed to take into account changing circumstances. Although most of the forces are essentially national contributions (with dual functions), such multinational force structures will need new equipment and enhanced capability. Other conversion costs, such as those of pensions or defence industrial subsidies, will also be high. Although the peace dividend is attainable in the long term it will still take a considerable time to achieve large military savings.

## The United Kingdom

Until the end of the cold war, the UK's defence efforts were based on four strategic concepts: the maintenance of an independent nuclear force, both tactical and strategic; the defence of the homeland itself; a major contribution to the defence of Western Europe from the threats posed by the WTO; and the maritime defence of the Eastern Atlantic. It is interesting to note that some of the major operational roles of the British army in the last decade actually fell outside these four roles—the Falklands/Malvinas War campaign, the military presence in Northern Ireland and the Persian Gulf campaign in the war against Iraq.

In 1990 the Ministry of Defence, recognizing early the profound transformation that was taking place in the international security order, produced its 'Options for Change' defence policy statement which would provide the guidelines for restructuring the military forces in accordance with new principles.[44] Although the statements and plans were not billed as a 'defence review' (the last defence review was in 1981) in practice it has become a relatively comprehensive plan for changing the size, composition and role of the British armed forces. A considerable reduction in forces is envisaged since the overall threat of a major inter-state war in Europe is now considered non-existent; however, because uncertainty has increased considerably, there is a new view of the 'insurance function' of the military which stresses flexibility, rapid reaction and new types of capability such as those required in peace-keeping exercises: 'Britain's armed forces are our insurance against the uncertainties of a rapidly changing world. As with any insurance policy, we must ensure

[44] UK House of Commons, Defence Committee, *Options for Change: Royal Navy* (Her Majesty's Stationery Office: London, Feb. 1991), House of Commons paper 1990–91 HC 266; *Options for Change: Royal Air Force* ((Her Majesty's Stationery Office: London, Feb. 1991), House of Commons paper 1990–91 HC 393; *Options for Change: Army—Review of the White Paper, Britain's Army for the 90s, Cm 1595* ((Her Majesty's Stationery Office: London, Feb. 1992), House of Commons paper 1991–92 HC 45; *Options for Change: Reserve Forces* (Her Majesty's Stationery Office: London, Feb. 1992), House of Commons paper 1991–92 HC 163.

that, within the resources available, the cover is right for the nature and the scale of the risks we face.'[45]

The new role of the armed forces has three overlapping aspects: (*a*) the protection of the UK and its dependencies even when there are no major threats; (*b*) insurance against any major threat to the UK and its allies; and (*c*) the promotion of wider security interests by supporting international peace and stability. The fundamental concepts that emerge are that uncertainty requires insuring for new types of capabilities, although there are no major threats in the traditional sense, and that UK forces will assume a broader role within the framework of international peace-keeping organizations such as the CSCE or the UN.

The outlines of the 'Options for Change' were announced in July 1990 and formally stated and amplified in July 1991 in the *Statement on the Defence Estimates* of 1991.[46] The changes it embodies are currently being implemented although with considerable controversy.

The initial reductions proposed implied a cut from 156 000 to 116 000 by the mid-1990s for army personnel;[47] from 55 to 38 infantry battalions; from 19 to 11 armoured regiments; from 14 to 10 field artillery regiments and from 15 to 10 engineer regiments. There were promises that the force would have far better equipment so that the end result would be a 'smaller but better' armed force. However, the cuts are quite severe at least in quantitative terms: a halving of the submarine fleet, a halving of the Tornado bomber force and a halving of the Royal Armoured Corps.[48] The reductions are expected to be completed by 1995 for all three services, with the biggest cuts in the British Army of the Rhine, where 40 garrison stations will be closed and 23 000 of its 56 000 troops withdrawn.[49]

Since the changes are taking place in piecemeal fashion, and the uncertainties of the transition make cost calculations difficult, the British defence budget also shows some erratic fluctuations. After falling for a number of years total real spending in FY 1991/92 (1 April to 31 March) was £21.5 billion—higher in real terms than the previous fiscal year, and with an over-run of over £400 million. Operation Granby, the British contribution to the Allied effort in the Persian Gulf War of 1991, meant additional net costs of around £500 million. (£2 billion of the £2.5 billion total costs were contributed by foreign governments). More important, additional costs are being incurred for restructuring. Over the three-year period 1992–95, £1.3 billion have been allocated to pay for redundancies and £1.1 billion for new construction to rationalize the new force structure and allow troops to return from Germany.

[45] UK House of Commons, *Statement on the Defence Estimates 1992* (Her Majesty's Stationery Office: London, July 1992), Cm 1981, p. 8.

[46] UK House of Commons, *Statement on the Defence Estimates 1991*, vol. 1 (Her Majesty's Stationery Office: London, July 1991), Cm 1559-I.

[47] *Statement on the Defence Estimates 1991*, vol. 1 (note 46), p. 42; *Options for Change: Army*, Cm 1595 (note 42), p. xii.

[48] UK House of Commons, Defence Committee, *Statement on the Defence Estimates 1992* ((Her Majesty's Stationery Office: London, Nov. 1992), House of Commons paper 1992–93 HC 218.

[49] *Statement on the Defence Estimates 1992* (July 1992, note 45), p. 29.

The budget provision for 1992/93 was increased to £24.18 billion (an increase of £831 million over the amount previously announced) and that for 1993–94 to £24.516 billion (an increase of additional £1.12 billion on the sum previously announced). These two increases, totalling almost £2 billion, are to cover the costs of forces being withdrawn from Germany and payments to service personnel being released as part of the 'Options for Change' defence cuts. The budget forecast for 1994/95 is £24.8 billion. The government plans to cut defence spending from 4 per cent of GDP in 1990 to 3.4 per cent by the mid-1990s and to 3 per cent by the end of the decade. The armed forces will be cut by 60 000 by 1995.[50]

The nuclear forces are to be modernized, as planned before the current changes, with the new Vanguard Class Trident missile submarines, which in a couple of years will start taking over from the ageing Polaris fleet. The first Trident rolled out in 1992 and began sea trials while a further three are being constructed. There is considerable questioning of the case for maintaining such a major strategic nuclear role, with its exceptionally high cost, when conventional cuts are so draconian. However, at the sub-strategic and tactical level, even deeper cuts are envisaged—cancellation of Lance missiles, a reduction in the number of nuclear-capable air squadrons, the halving of the number of free-fall nuclear bombs, and elimination of the maritime tactical nuclear weapon capability.

Overall analysis of the British budget shows that under 'Options for Change' few commitments or capabilities have been dropped outright. Rather, there is an element of stretching-out: many capabilities are intended to be maintained but at lower resource cost. On the other hand, reductions are also being phased in over three to five years. There have been none of the major cuts expected, nor the much-desired peace dividend. The 'small but better' armed forces could also be expensive to equip even though a reduction in manpower of around 20 per cent is expected by 1995. The proportion of the budget spent on equipment could then increase from around 37 per cent in 1992–93 to possibly 40 per cent in the mid-1990s.[51]

Among the three new roles mentioned earlier the most important from the point of view of European security is the British contribution to the defence of the European mainland. The most important contribution will be to the multinational ACE Rapid Reaction Corps. This is a newly formed corps in which the UK will make a major contribution of about 55 000 regular soldiers and possibly a further 35 000 reserves which could be mobilized during a crisis. The UK will also provide the permanent command and a substantial part of logistics support.

[50] *Statement on the Defence Estimates 1992* (Nov. 1992, note 48), p. xiii.
[51] 'Country survey: UK', *Jane's Defence Weekly,* vol. 16, no. 18 (2 Nov. 1991), pp. 809–38.

## V. The European Community

Major systemic and structural changes are taking place within the European Community, and although defence issues have been traditionally outside the remit of the EC defence policy is increasingly being discussed and debated within the framework of an eventual European Union. The Maastricht Treaty, signed in February 1992, attempted for the first time to foster a common European approach to the two great symbols of national sovereignty—the authority to control the currency and national security. The Union will 'assert its identity in the international scene, in particular through the implementation of a common foreign and security policy, including the eventual framing of a common defence policy'.[52] The military expenditure patterns of future years in the EC member states, particularly for weapon procurement and policy regarding the defence industrial base, will be affected by progress towards the union. The single market, which began operating in January 1993, will increase integration at the microeconomic level and 'macro' policies—both economic and political—now need to adapt.

In terms of purely defence matters, the proposal to create a European Union was a relatively modest one, and could become more ambitious with time. The main proposal is that the Western European Union (WEU) will be responsible for defence co-ordination between EC member states on the one hand and trans-Atlantic relations embodied in NATO on the other. The operational role of the WEU will be strengthened and a planning cell has been created with the responsibility to (*a*) maintain and update lists of force units which are allocated to the WEU for specific purposes; (*b*) propose recommendations as to the command, control and communications arrangements of such forces; (*c*) prepare contingency plans for the use of such forces. For the time being, the WEU will only be activated for specific purposes: a more ambitious role can only come much later. NATO still has the major responsibility for the defence of Western Europe. Member states of the EC have been asked to accede to the WEU and European NATO members are requested to become associate members. There are now three security-related structures in Western Europe—two explicit (NATO and the WEU) and one implicit (the EC). The position of countries like Denmark, Iceland, Ireland, Norway and Turkey which belong to one or the other but not to all three, creates anomalies which have not been resolved.

The principal role that the WEU can now play is that of co-ordination and channelling the EC response in non-traditional areas such as peace-keeping, in humanitarian assistance to countries at war, and possibly in the longer term in peace-making by military means. The essential requirement is that there should be minimum overlap and no conflict with NATO operations in these fields. When in 1992 the WEU co-ordinated the deployment of naval forces in the Adriatic to help enforce sanctions on Serbia some concern was voiced that

[52] Conference of the Representatives of the Government of the Member States, Treaty on European Union, document no. CONF-UP-UEM-2002/92 (Commission of the European Communities: Brussels, Feb. 1992), p. 2.

there was insufficient co-ordination with NATO. The British Secretary of State in evidence to the House of Commons Defence Select Committee claimed that British forces would be given to WEU operations only if not required by NATO and 'if there was any difference of interest, in our judgement, almost invariably our obligation to NATO should take precedence'.[53]

The Franco-German Corps, announced formally by President François Mitterrand and Chancellor Helmut Kohl in 1992, will be a small embryonic force which could form the basis of a more ambitious European force of the future—particularly if US forces leave in larger numbers. Its headquarters is in Strasbourg and the first commander will be from Germany with command alternating between a German and a French General. Once again the operational relationships with NATO need to be clearly defined, although an agreement governing relations was signed between NATO and the new corps in early 1993.[54] The French attitude is that such attempts to form pan-European forces are a part of the wider question of burden-sharing and of the means by which Europeans can organize themselves better militarily not only for the defence of Europe but also for new security requirements such as peace-keeping. As the conflict in the former Yugoslavia shows, it is not possible for the EC to 'insulate' itself against the fires of conflict raging on its borders. The German view is that the force will bind the French closer to NATO's integrated military command structures.

One important element in the defence and political concerns of the future union is the applications for EC membership of the former neutral and non-aligned countries of Europe, who may join the EC as early as 1995—Austria, Finland, Sweden and Switzerland.[55] Negotiations started in early 1993 for the first three. Although it is difficult to envisage these countries participating in military alliances, a common foreign and security policy appears to be feasible. Clearly there is no possibility of enlarging NATO; nor is it clear whether these countries would wish to join the WEU. However, the goals of security and foreign policy are broad enough to accommodate many variants, as the presence of Ireland in the EC shows. The enlargement of the EC may thus strengthen the movement towards European union, not only in economic matters but also in security policy defined in the broadest possible sense.

Military expenditure in the four applicants for EC membership has been determined in the past by the requirements of independence and the difficulties of 'free riding', given that they were outside NATO. Participation in the EC, and a future Union, would help in the restructuring of the military and could reduce the costs of conversion, while the single European market will make it easier for their defence industries to set up collaboration with EC partners. The trends in defence spending of the four countries and the defence burden are shown in tables 9.22 and 9.23 in local current prices and at constant 1991 prices and exchange-rates.

In the EC today, military expenditure is very much a function of overall

[53] *Statement on the Defence Estimates 1992* (Nov. 1992, note 48).
[54] *Atlantic News*, no. 2495 (3 Feb. 1993).
[55] *Financial Times*, 5 June 1992.

**Table 9.22.** Applicants for EC membership: military expenditure in current prices, 1983–92

Figures are in local currency, current prices. Figures in italics show military expenditure as a percentage of GDP.

| | | 1983 | 1984 | 1985 | 1986 | 1987 | 1988 | 1989 | 1990 | 1991 | 1992[a] |
|---|---|---|---|---|---|---|---|---|---|---|---|
| Austria | m. schillings | 15 362 | 15 554 | 16 786 | 17 940 | 16 972 | 16 597 | 17 850 | 17 537 | 18 208 | 18 274 |
| *Share of GDP* | | *1.3* | *1.2* | *1.2* | *1.3* | *1.2* | *1.1* | *1.1* | *1.0* | *1.0* | *0.93* |
| Finland | m. maarkkaa | 5 656 | 6 082 | 6 555 | 7 245 | 7 636 | 8 419 | 9 226 | 9 672 | 10 235 | 10 771 |
| *Share of GDP* | | *2.1* | *2.0* | *1.9* | *2.0* | *1.9* | *1.9* | *1.9* | *1.8* | *2.0* | *2.2* |
| Sweden [b] | m. kronor | 19 550 | 21 164 | 22 762 | 24 211 | 25 662 | 27 215 | 29 399 | 32 549 | 35 089 | 36 231 |
| *Share of GDP* | | *2.8* | *2.7* | *2.6* | *2.6* | *2.5* | *2.4* | *2.4* | *2.4* | *2.4* | *2.6* |
| Switzerland | m. francs | 3 862 | 4 009 | 4 576 | 4 282 | 4 203 | 4 458 | 4 679 | 5 145 | 5 277 | 5 224 |
| *Share of GDP* | | *1.9* | *1.9* | *2.0* | *1.8* | *1.7* | *1.7* | *1.6* | *1.6* | *1.6* | *1.5* |

**Table 9.23.** Applicants for EC membership: military expenditure in constant prices, 1983–92

Figures are in US $ at constant (1991) prices.

| | 1983 | 1984 | 1985 | 1986 | 1987 | 1988 | 1989 | 1990 | 1991 | 1992[a] |
|---|---|---|---|---|---|---|---|---|---|---|
| Austria | 1 650 | 1 581 | 1 653 | 1 737 | 1 621 | 1 555 | 1 631 | 1 552 | 1 559 | 1 507 |
| Finland | 2 124 | 2 131 | 2 171 | 2 332 | 2 361 | 2 476 | 2 546 | 2 514 | 2 556 | 2 629 |
| Sweden [b] | 5 543 | 5 556 | 5 563 | 5 679 | 5 775 | 5 789 | 5 875 | 5 888 | 5 802 | 5 888 |
| Switzerland | 3 436 | 3 436 | 3 823 | 3 549 | 3 436 | 3 578 | 3 640 | 3 797 | 3 680 | 3 514 |

[a] Latest year figure is an estimate.

[b] The Swedish budgets for the years 1991 and 1992 contain a special item called 'price rise adjustment' which is not present in the previous years' budgets. The figures given here are therefore not strictly comparable with those for previous years.

*Note.* For all countries, where fiscal year is not the same as calendar year, figures are calculated for calendar year.

*Sources (tables 9.22 and 9.23)*: National budgets. The percentages of GDP are based on the data given in the local currency series, deflated to 1991 price levels and converted into dollars at 1991 period-average exchange-rates. Local consumer price indices (CPI) are taken as far as possible from *International Financial Statistics (IFS)* (International Monetary Fund: Washington, DC). For the most recent year, the CPI is an estimate based on the first 6–10 months of the year. Period-average exchange-rates are taken as far as possible from *IFS*.

**Table 9.24.** Comparative economic and military indicators of the European Community countries, applicants for EC membership, the USA and Japan, 1991

US $ values are in constant (1991) prices.

| Country | GNP (US $b.) | Population (thousands) | Per capita GNP (US $) | ODA[a]/GNP (%) | Military expenditure (US $m.) | Armed forces (thou.) | Military expenditure per capita (US $) | Unem- ployment (%) | Infla- tion rate (%) | Economic growth (%) |
|---|---|---|---|---|---|---|---|---|---|---|
| Germany | 1 690.1 | 79 819 | 21 200 | 0.41 | 39 517 | 457 | 495 | 6.7 | 3.8 | *+3.7* |
| France[b] | 1 199.1 | 57 050 | 21 100 | 0.62 | 42 433 | 542 | 744 | 9.6 | 3.1 | *+1.2* |
| Italy | 1 134.4 | 56 411 | 19 700 | 0.30 | 24 336 | 473 | 431 | 10.9 | 6.7 | *+1.4* |
| UK[b] | 1 008.8 | 57 411 | 17 500 | 0.32 | 42 442 | 301 | 739 | 8.3 | 7.2 | –2.2 |
| Spain | 522.0 | 38 900 | 13 400 | 0.23 | 9 115 | 246 | 234 | 16.3 | 6.3 | *+2.4* |
| Netherlands[b] | 287.0 | 15 065 | 19 900 | 0.88 | 7 246 | 104 | 482 | 5.9 | 3.3 | *+2.1* |
| Belgium | 200.0 | 9 979 | 20 000 | 0.42 | 4 625 | 101 | 463 | 8.8 | 2.8 | *+2.1* |
| Denmark | 125.5 | 5 146 | 24 400 | 0.96 | 2 672 | 30 | 519 | 10.4 | 2.4 | *+1.2* |
| Greece[b, c] | 67.7 | 10 210 | 6 631 | 0.07 | 3 807 | 205 | 373 | 8.2 | 18.5 | *+1.8* |
| Portugal | 68.9 | 10 600 | 6 500 | 0.31 | 2 116 | 86 | 200 | 4.1 | 12.0 | *+2.1* |
| Ireland | 38.4 | 3 520 | 10 900 | 0.19 | 585 | 13 | 166 | 15.8 | 3.2 | *+2.5* |
| Luxembourg | 9.0 | 365 | 24 658 | 0.28 | 108 | 1 | 296 | 1.3 | 3.2 | *+3.1* |
| **EC total** | **6 350.9** | **344 476** | **17 157**[d] | **0.42**[d] | **179 002** | **2 559** | **520** | **9.2** | **5.0** | ***+1.5*** |
| Austria | 279.7 | 17 335 | 20 800 | 0.34 | 1 559 | 52 | 90 | 3.5 | 3.4 | *+3.1* |
| Finland | 122.0 | 5 014 | 24 300 | 0.76 | 2 556 | 33 | 510 | 7.6 | 5.3 | *–6.5* |
| Norway | 103.2 | 4 274 | 24 300 | 1.14 | 3 288 | 41 | 769 | 5.5 | 3.6 | *+1.9* |
| Sweden | 230.3 | 8 621 | 25 800 | 0.92 | 5 543 | 61 | 643 | 2.7 | 10.2 | *–-1.4* |
| Switzerland | 238.2 | 6 832 | 34 900 | 0.36 | 3 680 | 35 | 539 | 1.3 | 5.7 | *–0.1* |
| USA[b] | 5 685.8 | 253 000 | 22 500 | 0.20 | 288 799 | 2 115 | 1 141 | 6.7 | 4.3 | *–1.2* |
| Japan[b, e] | 3 391.5 | 124 949 | 24 000 | 0.32 | 32 559 | 240 | 261 | 2.1 | 2.6 | *+4.4* |

[a] Official development aid.

[b] Includes forgiveness of non-ODA debt as follows:

i. Export credits claims: Japan $7 million, UK $17.million

ii. Military debt; USA $1 855 m.

Exclusion of thse amounts would change the ratio for the USA to 0.17.

[c] Greece has also received US $36m. in aid from the member countries of the Development Assistance Committee (DAC) of the OECD, multilateral organizations and Arab countries. Greece contributes to EC aid programmes and to multilateral organizations but is not a member of the DAC.

[d] Average figure.

[e] The figure for Japanese military expenditure is not strictly comparable with figures for NATO countries, as it is defined by somewhat different criteria.

*Note:* For all countries where fiscal year is not the same as calendar year, figures are calculated for calendar year.

*Source:* Author's calculations based on OECD, IMF and NATO publications and national statistical information and White Papers.

economic well-being. It is also partly determined by the costs of restructuring. As the discussion on France and the UK in the previous section shows, the costs of structural change could be high initially, and could place some limits on the reductions in forces proposed. At the same time, Europe is suffering from a general recession which means that conversion is more difficult. SIPRI in recent *Yearbooks* has provided background economic and security-related data to assist informed debate. Table 9.24 gives new and updated information on these issues. It has been expanded from previous years to include the EC member states, the applicants for EC membership, the USA and Japan.

## VI. China and Japan

The fastest-growing countries of the world today are those of the Asia–Pacific region. In recent years, as global threats have diminished but internal security problems have increased across the world, military expenditure has been increasingly susceptible to economic trends. In particular, in the absence of any particular country-specific threats, defence budgets have tended to follow economic prosperity and failure. It is not surprising therefore that the countries of the Asia–Pacific region, taken as a whole, have shown sustained increases in real military expenditures in aggregate, even though there are major exceptions. The share of the military in GDP may even have fallen as a result of high growth rates, even while defence expenditure in absolute terms has risen. Japan for example has consistently spent around 1 per cent of GNP on defence while in real terms its expenditure has consistently grown throughout the 1980s.

The region has high procurement expenditure, since force modernization is continuing and domestic production is expanding. Total procurement spending of the countries in the Asia–Pacific region (including Australia) could be as high as $15 billion per year.

### China

Until 1989, China was a major exception to this trend of rising defence spending in the region. After 1979, when military expenditure peaked as a result of the short war with Viet Nam, Chinese defence expenditure fell with minor fluctuations for 10 years. When the pattern was broken in 1990 and the military was given a substantial increase in the defence budget over the sum allocated in 1989 it was initially believed that this was a reward for the loyalty of the People's Liberation Army (PLA) during the Tiananmen Square uprising of 1989 and the subsequent imposition of martial law. It was also thought that the most significant part of the extra allocation would go for personnel expenditure, since the condition of ordinary soldiers had deteriorated significantly in the period of defence cuts and morale was particularly low. However, in 1991 the Chinese Government increased defence spending by another generous 15 per cent in nominal terms. Since economic stabilization policies during

**Table 9.25.** Official figures for China's military expenditure, 1983–92

Figures in italics are percentage shares.

| | b. yuan, current values | US $b., constant (1988 prices) | Share of national income (%) |
|---|---|---|---|
| 1983 | 17.7 | 7.7 | *3.7* |
| 1984 | 18.1 | 7.6 | *3.2* |
| 1985 | 19.1 | 7.2 | *2.7* |
| 1986 | 20.1 | 7.1 | *2.6* |
| 1987 | 21 | 6.8 | *2.5* |
| 1988 | 21.8 | 5.9 | *1.9* |
| 1989 | 25.2 | 5.8 | *1.9* |
| 1990 | 29.0 | 6.6 | *2.0* |
| 1991 | 32.5 | 7.2 | *2.0* |
| 1992 | 37.0 | 7.8 | *2.0* |

*Source*: People's Republic of China, State Statistical Bureau, *China Statistical Yearbook*, various years (China Statistical Information and Consultancy Service Centre, Beijing); *Jane's Defence Weekly,* 2 May 1992, p. 745, author's estimates.

1990 and 1991 were reducing inflation rates significantly, such an increase in defence expenditures could only be construed as a substantial real rise of the order of almost 10 per cent.

In 1992, for the third year running, defence spending has been raised by 13.8 per cent in nominal terms and now stands at around 37 billion yuan.[56] SIPRI estimates that in real terms this amounts to a rise of over 8 per cent. Thus between 1989 and 1992 real defence spending has risen by over one-third. The level of defence spending after adjusting for inflation is in 1992 about the same as it was in 1982. Table 9.25 gives the data for official Chinese defence spending for the 10 years 1983–92 in current yuan and constant dollars to facilitate comparison, as well as the share in national income.

It is now well known that the official budget is a substantial underestimate, for four reasons. A number of defence items are not included; off-budgetary items are not adequately costed; the industrial subsidies for defence enterprises are not accounted for; and the revenue earned by semi-autonomous arms sellers from the MIC who are liable to pay a part of their income for military spending, specifically weapon acquisition, is not accounted for. Indeed, one reason for the increase in military expenditure in recent years could be that arms sales are declining and hard currency earnings, particularly from the Gulf states, are drying up.

A major problem with any analysis of defence economics for China is the veil of secrecy shrouding military allocations. There is no official information on details of military budgets even in the crudest aggregated form. There is only a single-line entry for defence in what is called the state budget but little

[56] Japan, Defense Agency, *Defense of Japan, 1992* [Japan Defense Agency: Tokyo, 1992], p. 49; *Far Eastern Economic Review,* 20 Aug. 1992, p. 9; *Jane's Defence Weekly,* 2 May 1992, p. 745.

is known officially about what it includes. It clearly covers personnel costs and O&M, as in the case of the former USSR. According to most analysts it leaves out military R&D, which was standard practice in the budgeting of centrally planned economies. It is not clear how much of procurement expenditure (25–40 per cent of all estimated spending) is actually defence procurement. Presenting some conservative estimates at hearings of the US Congress, a noted analyst states: 'While Beijing claims it spends less than 1.8 per cent of its gross national product on defense, the actual number may be closer to 3.5 per cent'.[57] For the sake of greater transparency and confidence-building measures, China should at least publish defence White Papers or submit its detailed budgets to the United Nations Reduction of Military Expenditure programmes. Otherwise speculation will continue to undermine the fragile security environment in Asia.

Chinese defence cuts of the mid-1980s were accompanied by substantial conversion.[58] Over 70 per cent of the military enterprises' output is for civilian purposes and significant revenue is earned from sales of non-military products. It is estimated that in 1991 the volume of net output was 7 billion yuan and sales revenue 6.6 billion yuan. These sums represent a rise of 21–25 per cent on the corresponding figures for 1990. The defence industry produces and sells a wide range of products, over 800 in number, divided into 18 categories, and including industrial goods such as vehicles, mining machinery and electronics.[59] Very recently, Chinese defence enterprises—particularly the research establishments—have moved into areas of high-technology goods and substantial benefits from conversion are discernible. For example, the '502 Institute' of the Ministry of Aeronautics and Astronautics (one of the defence industrial ministries)[60] used to specialize in military space technology. It has now formed a semi-autonomous enterprise called Kangtuo Science and Technology Corporation, based in the industrial development zone of Beijing, and is now a major producer of industrial control computers. The conversion process has slowly moved from consumer goods to emphasizing machinery production and finally reached high technology products.

Conversion has been costly here too, although the costs were not carried by the military budget. One of the most expensive 'experiments' of the Maoist era was to prove to be the move of significant numbers of the arms-producing factories to the remote regions of the south-west in preparation for the 'people's war'.[61] In addition to defence industrial enterprises, other capital-

[57] Kaufman, R.F., 'Overview', in *Report on the Chinese Economy. Hearings before the Joint Economic Committee*, 103rd Congress (US Government Printing Office: Washington, DC, 1991), pp. 645–47. See also in same volume Harris, J., 'Interpreting trends in Chinese defence spending', pp. 676–84.

[58] The analytical framework of conversion is discussed in Sen (note 36). For details of Chinese conversion in the 1980s see Deger, S. and Sen, S., SIPRI, *Military Expenditure: The Political Economy of International Security* (Oxford University Press: Oxford, 1990), pp. 86–104.

[59] Xinhua News Agency, 7 Feb. 1992, reported in *News Review on East Asia*, vol. 6, no. 3 (Mar. 1992), p. 196.

[60] For a description of the structure of such ministries see Deger, 1991 (note 5).

[61] For a discussion of the various doctrines influencing military spending in China, see Deger, 1991 (note 5).

intensive industrial production (called machine-building) was shifted to the hinterland from the coastal industrial regions, to prevent its destruction if all-out war actually did break out. This was called the Third Capital Construction Front and enormous resources were poured into such projects during the Maoist era, particularly the late 1960s. It is believed that there were around 2000 companies at the peak of production on the Front. Over the past five years, the reverse process has started, and these industrial enterprises are being shifted back to the economically more profitable industrial regions of Manchuria and the south-east province of Guangdong, adjacent to Hong Kong. These factories will all be converted and be utilized for the production of civilian goods. However, the cost has been high. According to reports in 1992, about $566 million have already been spent to convert 121 armaments factories. The next plan is to move and convert another 115 at a total cost of over $1 billion.[62]

In 1992 it also became clear that after 30 years China's co-operation with Russia (and Ukraine) in the field of defence procurement would begin to expand rapidly. China's arms imports from Russia have reportedly surged with a $1 billion contract for buying arms, including Su-27 fighters. Given the relatively small procurement budget of the PLA it is possible that only part of the deal will be paid in hard currency while the rest would be a barter arrangement of consumer products. According to Major General Sergey Karaoglanov, chairman of the Oboron-Export organization formed by the Russian Government to handle arms trade, about one-third of the fighter contract is payable in hard currency and the rest in consumer goods.[63] If such transactions are acceptable to Russia and Ukraine then in future there could be much more joint and collaborative effort, recalling the 1950s when Soviet technology effectively armed the PLA. Force modernization, which has suffered dramatically in the 1980s, could then be achieved at lower cost.

There is clearly great scope for collaborative ventures between China and Russia (and Ukraine), to produce a new generation of military equipment products domestically, with Chinese enterprises producing with Russian blueprints. The most likely areas for joint production are aircraft engines and even highly sophisticated radar-evading stealth technology. China is developing its next-generation fighter (the F-10) and such technology transfers at relatively concessional prices would be of considerable help.

It seems that in the future Chinese defence spending and procurement will continue to rise. The era of conversion could be drawing to a close after a successful beginning. However, budget constraints will always be present and it is impossible to sacrifice economic reforms to pay the military. One way of cutting costs would be to limit the size of the standing army, which at 3.2 million men is still too large for the country's requirements. The Central Military Commission, which is the paramount decision making body on all military matters and is still chaired by paramount leader Deng Xiao Ping, met

[62] For instance, *Hongkong Standard,* 6 Dec. 1991, reported in IDSA, *News Review on East Asia,* vol. 6, no. 2 (Feb. 1992), p. 119.

[63] *Far Eastern Economic Review,* 3 Sep. 1992, p. 21.

**Table 9.26.** Allocations of the Japanese national defence budget, FYs 1990–92

Figures are in b. yen, current prices; figures in italics are percentage shares.

| | 1990 | 1991 | 1992 |
|---|---|---|---|
| Personnel provisions | 1 668.0 | 1 756.8 | 1 880.8 |
| *Share of total* | *40.1* | *40.1* | *41.3* |
| Equipment acquisition | 1 140.3 | 1 216.2 | 1 141.9 |
| *Share of total* | *27.4* | *27.7* | *25.1* |
| R&D[a] | 92.9 | 102.9 | 114.8 |
| *Share of total* | *2.2* | *2.3* | *2.5* |
| Facility improvement | 132.9 | 136.0 | 161.7 |
| *Share of total* | *3.2* | *3.1* | *3.6* |
| Maintenance | 669.7 | 696.9 | 745.7 |
| *Share of total* | *16.1* | *15.9* | *16.4* |
| Base countermeasures | 406.1 | 424.7 | 451.5 |
| *Share of total* | *9.8* | *9.7* | *9.9* |
| Others | 49.4 | 52.6 | 55.4 |
| *Share of total* | *1.2* | *1.2* | *1.2* |
| **Total** | **4 159.3** | **4 386.0** | **4 551.8** |
| | *100* | *100* | *100* |

[a] Research and development.

*Source:* Japan, Defense Agency, *Defense of Japan 1992* [Tokyo, 1992].

in special session in April to consider the role of the armed forces in the altered international security scenario. It is thought that a decision was taken to reduce the size of the PLA in order to release funds for procurement. The level of troop cuts has not been declared, although some sources claim that the reduction could amount to 1 million men.[64] If this is so, it would parallel the 1985 reforms when military forces were cut from 4 to 3.2 million in about five years. However, such large reductions may not be feasible again since the army leadership carries great political weight and would not allow drastic pruning. With the end of Communism across the globe, the Chinese political leadership cannot afford to ignore the wishes of the military élites.

## Japan

Developments in military expenditure in Japan are the reverse of those in China. After a sustained rise in defence spending throughout the 1980s, the FY 1992 budget (running from 1 April 1992 to 31 March 1993) asks for 4551.8 billion yen, a modest rise of 3.8 per cent in current prices, which with inflation could mean no real increase. The level of military expenditure in 1992 will still be modestly higher than in 1991, but a plateau has been reached. It is possible that after the sustained rises of the 1980s the time has

[64] *Jane's Defence Weekly*, 2 May 1992, p. 745.

**Table 9.27.** Japanese defence expenditure, FYs 1975–92

| | Defence budget in b. yen, current values | Defence budget in US $b., constant 1991 prices | Increase over 12 months (%) | Defence budget's share of GNP (%) |
|---|---|---|---|---|
| 1975 | 1 327.3 | 17 184 | *21.4* | *0.84* |
| 1976 | 1 512.4 | 17 886 | *13.9* | *0.90* |
| 1977 | 1 690.6 | 18 498 | *11.8* | *0.88* |
| 1978 | 1 901.0 | 19 974 | *12.4* | *0.90* |
| 1979 | 2 094.5 | 21 192 | *10.2* | *0.90* |
| 1980 | 2 230.2 | 20 936 | *6.5* | *0.90* |
| 1981 | 2 400.0 | 21 473 | *7.6* | *0.91* |
| 1982 | 2 586.1 | 22 523 | *7.8* | *0.93* |
| 1983 | 2 754.2 | 23 561 | *6.5* | *0.98* |
| 1984 | 2 934.6 | 24 541 | *6.6* | *0.99* |
| 1985 | 3 137.1 | 25 710 | *6.9* | *0.997* |
| 1986 | 3 343.5 | 27 238 | *6.6* | *0.993* |
| 1987 | 3 517.4 | 28 626 | *5.2* | *1.004* |
| 1988 | 3 700.3 | 29 907 | *5.2* | *1.013* |
| 1989 | 3 919.8 | 30 978 | *5.9* | *1.006* |
| 1990 | 4 159.3 | 31 887 | *6.1* | *0.997* |
| 1991 | 4 386.0 | 32 559 | *5.4* | *0.954* |
| 1992 | 4 551.8 | 33 292 | *3.8* | *0.941* |

*Source:* Japan, Defense Agency, *Defense of Japan* 1992, [Tokyo, 1992].

come to rethink the options for the country, in a region where the old security threat could have vanished and the new security threats are yet to be identified.

In the 1992 budget, personnel expenditure rises by 7 per cent, O&M by 7 per cent, and military R&D, although a small fraction of total spending, by over 10 per cent. The only significant cut is in procurement spending, which contributes to the overall restraint in the budget increases. Military procurement of major weapons has until now accounted for about 25 per cent of the defence budget: in FY 1992, procurement of major weapons was 22 per cent of total defence expenditure. Its fall signifies a major policy reversal. It should however be remembered that Japan's procurement expenditure rose sharply in the 1980s, so that, as in the USA, the decline is occurring from high levels. Furthermore, with the cuts in European (including Russian) procurement expenditure, Japan's ranking in the world tables of military expenditure could be rising. If cuts of the scale discussed above for Russia are indeed taking place, Japanese procurement spending on major weapons could now be the third highest in the world after that of USA and France, exceeding that of the UK, Germany and Russia.

In the Japanese defence White Paper for 1992 it is explicitly stated for the first time that the threat from Moscow—which has guided defence policy since the formation of the Self Defense Force—is now considered reduced and possibly irrelevant.[65] However, the existence of stocks of modern weapons including strategic and nuclear forces in the far eastern parts of Russia creates a danger of instability and uncertainty. Nearer home, force modernization continues in China, North Korea, South Korea and the Association of South-East Asian Nations (ASEAN) countries. Although all of them have cordial relations with Japan (except of course North Korea), the underlying security concerns remain in a region which is increasingly becoming heavily armed.

The Mid-Term Defence Programme for 1991–95 allocated funding for five years and laid down guiding principles on arms purchases and force levels.[66] The Prime Minister and the Japanese Defense Agency have agreed to a new review of this mid-term plan which was set up and approved in 1991. The plan proposes to spend 22.75 trillion yen over five years (on average about $36 billion per annum). Some adjustments will need to be made and modest cuts have been announced. It is possible that the aggregate allocated by the plan will remain the upper limit, but the structure of allocations will change. Personnel and other operating costs will increase their share of the budget while the share of procurement will go down.

Table 9.26 gives data for the past three years for operating costs and investment costs, while table 9.27 gives Japanese military expenditure from 1975 to 1992 in current billion yen and constant 1991 dollar values. The share of military expenditure in GNP has been on average around 1 per cent since the early 1980s, while its rate of growth has been falling for the past two decades.

## VII. The developing world

Military expenditure in the developing world has been falling in aggregate from the early 1980s, although there are significant regional variations. Between 1985 and 1991 military spending in developing countries fell by over 10 per cent.[67] Developing countries spent around 5 per cent of GDP on defence in 1985; this ratio fell to 3.8 per cent in 1991. In 1985, the share of defence in their central government expenditure was 17 per cent; by 1991 the ratio had declined slightly to a little over 16 per cent.[68]

Military spending patterns in developing countries and across regions are significantly affected by external security relations, internal threats and economic constraints. At the end of the cold war, with the absence of superpower rivalry and the increasingly assertive role of the UN, external security is improving and inter-state relations are becoming better. Although there are exceptional cases, such as the Iraq–Kuwait conflict in 1990–91, in general the

[65] See note 56.

[66] For discussions of the Mid-Term Defence Plans see the chapters on military expenditure in previous *SIPRI Yearbooks*.

[67] These estimates are based on military expenditure data given in *SIPRI Yearbook 1992* (note 3) and the author's estimates.

[68] SIPRI does not collect central government expenditure ratios. However, estimates made by the IMF using basic SIPRI data on military expenditure give the above shares. See Hewitt, D. P., 'Military expenditure in developing countries', Paper presented at the OECD Expert Workshop on Military Expenditure in Developing Countries, Paris, 1–2 Feb. 1993.

**Table 9.28**. Long-term debt and financial flows, growth rates per capita and military expenditure as a share of GDP/GNP of low- and middle-income economies, 1990–91

Figures in italics are percentages.

| | 1990 | 1991 |
|---|---|---|
| Debt outstanding (US $b.) | 1 047.0 | 1 064.5 |
| Debt as percentage of GNP (%) | *32.1* | *29.5* |
| Debt service (US $b.)[a] | 123.1 | 134.9 |
| Debt service ratio (%)[b] | *19.8* | *21.2* |
| Official development finance (US $b.) | 49.3 | 50.7 |
| Net transfers on long-term lending (US $b.)[c] | – 21.6 | – 18.4 |
| Growth rate of GDP per capita (%) | *0.3* | *– 0.1* |
| Military expenditure as share of per capita GDP (%) | *3.9* | *3.8* |

[a] Debt service includes interest and principal payments.

[b] Debt service ratio is debt service as percentage of exports.

[c] Net transfers are disbursements minus principal repayments and interest payments.

*Source:* World Bank, *Annual Report 1992* (World Bank: Washington, DC, 1992); author's estimates.

security climate seems to be improving. However, internal security threats are still present in spite of major successes in conflict resolution, as in Ethiopia and Central America. Often internal security problems such as civil war or the loss of legitimacy of the government or even the state are directly attributable to economic crises, as in much of Africa.

The central problem in the developing world is that of economic security. The main reason for the reductions in defence spending in Africa or South America for example has been the continuing crisis that followed on from the lost decade of the 1980s. One of the major indicators of this crisis is the debt problem which still leaves a debt overhang in those countries which became heavily indebted in the 1970s. In spite of major rehabilitation efforts made since the beginning of the debt crisis, the total stock of debt and the concomitant obligations to service that debt have had a crippling effect on such poor economies. The nominal value of outstanding debt in 1991 was around $1.28 trillion, or almost 40 per cent of the combined GDP of all low- and middle-income countries (which includes the countries of Central and Eastern Europe). This was very similar to the level of a year earlier. Of this, outstanding long-term debt amounted to $1.06 trillion. Table 9.28 shows the level of long-term debt and other indicators of resource flows for the developing countries for the years 1990–91.[69] It also shows that the net transfer on long-term lending is negative—new disbursements to the South are less than repayments of principal and interest payments made—and that the growth rates

[69] Earlier figures are to be found in chapters on 'World military expenditure' of previous *SIPRI Yearbooks*. See also Sen, S., 'Debt, financial flows and international security', SIPRI, *SIPRI Yearbooks 1990, 1991* (Oxford University Press: Oxford, 1990, 1991), pp. 203–17 and 181–95, respectively.

of per capita GDP for the low- and middle-income countries are almost zero or even negative. Income levels are not rising at all.

The economic crisis has reduced military spending, and this must be beneficial for international security. On the other hand, in terms of the broader concepts of security, which include economic factors also, the position of many people in the developing world has possibly worsened. Without economic security, it will not be possible for conflict resolution to be successful and the vision of the new world order will not be realized.

One of the major sources of development finance that the developing world can receive is foreign aid, which for the poorest countries, particularly in sub-Saharan Africa, has become a life-blood essential to finance vital imports. Official development assistance (ODA), sometimes called foreign aid, has actually increased during the 1980s from about $43 billion (at 1990 prices and exchange-rates) in 1980 to about $53 billion in 1990.[70] However, with the dramatic fall in other forms of financial flows to the developing countries (such as export credit or bank lending) the dependency on aid has increased. According to OECD data, in 1980 foreign aid amounted to 23.3 per cent of total net resource flows to developing countries. In 1991 this ratio had increased to 42.4 per cent.[71]

The relationship between foreign aid and military expenditure was of central concern among donors and recipients during 1991–92. Bilateral donors such as Japan and Germany began questioning 'excessive military expenditures' in selected recipient countries. Other important aid donors, such as Sweden, focused on the role of good governance and the optimum use of aid which also have indirect implications for military spending. Multilateral aid agencies such as the World Bank and the Development Assistance Committee (DAC) of the OECD introduced the issue in their policy dialogue with developing countries. Extensive discussions were held regarding what can be termed 'aid conditionality'.[72] At the same time wider issues of the impact of defence spending on development and policy coherence were also discussed in policy forums.[73] At the major multilateral forum of the high-level meeting of the DAC it was asserted:

> Reduction in excessive military expenditure is a key element in sound economic policy and good governance and frees scarce resources for sustainable economic and social development. DAC members welcome the growing attention paid by multilateral institutions, such as the World Bank, IMF and UNDP, in monitoring public

[70] OECD, *Development Co-operation,* Report on the efforts and policies of the members of the Development Assistance Committee, 1992 (OECD: Paris, 1992).

[71] *Development co-operation* (note 68), p. 78.

[72] For a theoretical analysis see Deger, S. and Sen, S., 'Military expenditure, aid and economic development', *Proceedings of the the World Bank Annual Conference on Development Economics 1991* (IBRD and World Bank: [New York], 1992), pp. 159–86.

[73] Deger, S. 'The impact of military expenditure on economic development', Paper presented at the OECD Expert Workshop on Military Expenditure in Developing Countries, Paris, 1–2 Feb. 1993. See also Sen, S., 'Policy coherence and cooperation in policies towards "excessive" military expenditure', presented at the same workshop.

expenditure programmes for opportunities to reduce unproductive expenditures especially excessive military expenditures. They note that several DAC members take the size and trends in military expenditures increasingly into account in their aid allocation decisions.[74]

Similar and stronger sentiments have been expressed by some individual donor countries such as Germany and Japan. Japan introduced a new Official Development Assistance Charter in June 1992. It strongly emphasizes the links between ODA and military expenditure. The two basic principles it states are (*a*) 'any use of ODA for military purpose or for aggravation of international conflicts should be avoided' and (*b*) 'full attention should be paid to trends in recipient countries' military expenditure, their development and production of mass destruction weapons and missiles, their exports and imports of arms, etc., so as to maintain and strengthen international peace and stability, and from the viewpoint that developing countries should place appropriate priorities in the allocation of their resources for their own economic and social development'.[75] Japan's intellectual role and leadership in this area of North–South relations is extremely important since it is one of the two largest donors in the world and has consistently emphasized the importance of economic security in bringing about global peace.

In terms of actual implementation, Germany has probably done most. In 1991 Chancellor Kohl told the Bundestag that 'our aid can only be successful if favourable framework conditions exist in the recipient countries: respect for human rights and human dignity; forms of government based on democracy and the rule of law; economic development instead of the stockpiling of military potential'.[76] Economic aid has been used both as an incentive and as a sanction to reduce militarization and cut defence-related expenditure. The volume of aid available to Mozambique and Uganda has been raised to support disarmament efforts after long periods of conflict and to help rehabilitation efforts and integrating former soldiers into civilian life. Aid to Cameroon, China, India, Pakistan and Zaire was reduced in 1992 as a sign of disapproval of their military spending policies.

It is clear that such efforts—unprecedented in the history of North–South relations—will ultimately have an overall beneficial impact on both security and development. However, the increasing use of conditionality in foreign aid also raises the cost of such assistance, and some developing countries might also treat this as infringement of national sovereignty. One way of balancing the costs and benefits of conditionality is to increase the level of foreign aid

[74] Conclusions of the Development Assistance Committee (DAC) 1992 High-Level Meeting (SG/PRESS (91)72), para. 16, reproduced in OECD, Public Policy Statements on Participatory Development/Good Governance, OECD/GD (92)67 (OECD: Paris, 1992), p. 8.

[75] Kawakami, T., 'Japan's ODA policies for a peace initiative', address given at the Tokyo Conference on Arms Reduction and Economic Development in the Post Cold War Era, Tokyo, 4–6 Nov. 1992.

[76] Sand, K. van der, 'Approaches towards reducing military expenditure in developing countries in the framework of German development co-operation', Statement for the paper presented at the OECD Expert Workshop on Military Expenditure in Developing Countries, Paris, 1–2 Feb. 1993.

somewhat so that there are greater incentives for poor countries. This combination of sanctions and incentives—the carrot and stick policy—could be the most effective. However, at present the possibility of increased foreign aid is remote. There were hopes that the peace dividend would be a unique source of additional resources from which some amounts could be transferred to the less fortunate. Just as aid conditionality to reduce defence spending is a special way of forging links between political and economic aspects of North–South relations, the peace dividend could be a complementary way of forging another link. However, as has been mentioned, the costs of conversion have outweighed the peace dividend. Only in the long term is there any hope of cuts in military spending in the North large enough to have a reasonable impact on the total volume of foreign aid.

## VIII. Conclusion

There is no doubt that broader international security considerations are gaining ground against the narrower older doctrines which emphasized territorial national defence. Concern for global security is now far greater than it was in the recent past and there is greater awareness that security has to be defined in a broader sense to include economic, political and environmental factors. Even though the decline in military expenditure has accelerated during 1992, specifically because of a rapid fall in the spending of the CIS countries, more fundamental problems of peace and conflict remain. Consolidation of the gains made in the past few years requires a rapid solution to external security threats, internal conflicts and the resolution of the deeper economic threats that come from poverty and environmental degradation.

The realization of the peace dividend in the long run among the richer countries of the world could be a major force in that direction. Restructuring among the West European and US militaries has not led to immediate savings yet, nor helped to narrow the poverty gap between North and South through resource transfers. The speed of reduction is expected to accelerate in future. In the meantime the process of conversion in the West, whereby resources are released from the military to the civilian sectors, has proved costly, particularly because of continuing recession. Conversion is similar to an investment process where net benefits are negative in the short run but turn positive in the long run. Developing countries in aggregate have seen declining military expenditure for the past decade. The process will continue, albeit slowly, with the linkages now established between economic aid and reductions in military spending.

The falls in defence spending seen in 1992 partly reflect the complex interaction of the three facets of international security—external threats, internal conflicts and economic security. The first requires security guarantees under the auspices of the UN and the major powers. In turn the UN needs to be financially independent and able to command extra resources for peace-keeping and peace-making. The second requires good governance and adherence to the rule of law in internal affairs. The third requires a more equitable and sustainable level of international economic development. Military expenditure reductions, and the gains from the peace dividend in the long run, could make a substantial contribution in all these three spheres.

# **Appendix 9A.** Military expenditure in the Central and East European countries

EVAMARIA LOOSE-WEINTRAUB

## I. Introduction

This appendix examines the development of military expenditure in the new security environment of Central and Eastern Europe. It covers Bulgaria, the former Czechoslovakia, Hungary, Poland and Romania. The order in which they are examined reflects the adequacy of the budgetary information available, those on which there is most information being examined first. Albania and Yugoslavia are not included in this appendix as the process of disintegration in the former and the specific development of the latter make them separate cases necessitating special analysis.

The pursuit of economic change from a centrally planned economy to a free market economy is still very much more a struggle than a matter of following a steady path. Huge losses of output have resulted from the inevitable disruptions of a period of transition consequent on the dismantling of the command system, the breakdown of trade among the former members of the Council for Mutual Economic Assistance (CMEA) (intra-regional trade is now carried out on the basis of world prices and convertible currency settlements), the obsolescence of much of the capital stock, rigidities in capital and labour markets, and financing constraints.

The unevenness of the path to economic reforms is shown in table 9A.1. While Hungary and the former Czechoslovakia made some progress in macroeconomic policies and achieved some improvement in their current account and their balance of payments, it remains to be seen how the division of the Czech and Slovak Republic will effect the economic, political and military transformation in the two republics from 1993 onwards. The late starters, Bulgaria and Romania, are still far from achieving macroeconomic stabilization. In Poland, reforms started with strong 'shock therapy' and achieved some positive results. The country is now facing macroeconomic instability, although the political environment is more stable than before.

The collapse of trade among the countries of the former CMEA is the major problem underlying their economic crisis. These countries cannot produce goods at competitive prices and have few markets to sell in, unless the West liberalizes trade for them. Taking care of external indebtedness is another problem for the countries of Central and Eastern Europe, is as shown by table 9A.2. With the exception of the former Czechoslovakia and Romania, they have had problems in managing external debt. Hungary has succeeded in servicing its debts in a timely way, but Bulgaria and Poland have not, and the problems appear likely to get worse. It may become necessary to demand debt forgiveness and the rescheduling of interest and principal repayments.

The pull-out of Soviet troops from Central and Eastern Europe has created a substantial strategic and military vacuum for the new democracies, for example in terms of the Treaty on Conventional Armed Forces in Europe limiting military equipment and providing for a reduction in arms production in general. In this period of great uncertainty, military expenditure trends will not be clear unless the purpose and

**Table 9A.1.** Macroeconomic indicators in Central and East European countries, 1990 and 1991

| | Real growth of GDP (%)[a] | | Consumer prices (% change on 12 months) | | Unemployment[b] (% of labour force, Dec.) | | Current account (% of GDP) | |
|---|---|---|---|---|---|---|---|---|
| | 1990 | 1991 | 1990 | 1991 | 1990 | 1991 | 1990 | 1991 |
| Bulgaria | *– 10.6* | *– 23.0* | 64.0 | 400.0 | *1.4* | *11.1* | *– 5.3* | *– 12.1* |
| Czechoslovakia | *– 0.4* | *– 15.9* | 10.0 | 55.0 | *1.0* | *7.5* | *– 2.9* | *2.1* |
| Hungary | *– 4.1* | *– 11.0* | 28.9 | 38.0 | *1.7* | *8.0* | *1.2* | *1.4* |
| Poland | *– 11.6* | *– 7.0* | 553.6 | 65.0 | *6.1* | *12.0* | *4.0* | *– 2.1* |
| Romania | *– 7.4* | *– 13.0* | 4.7 | 164.3 | *– 1.0* | *6.0* | *– 8.7* | *– 7.7* |

[a] GDP definition from the International Monetary Fund (IMF).
[b] Based on registered unemployment figures.

*Source:* International Monetary Fund, *World Economic Outlook* (IMF: Washington, DC, Oct. 1992); Organization for Economic Co-operation and Development, *Reforming the Economies of Central and Eastern Europe* (OECD: Paris, 1992).

functions of the military forces are clearly defined, and any replacement of military hardware will place a substantial burden on the already stretched economies of these countries.

## Availability of information on military expenditure

Calculating military expenditure in the former Warsaw Treaty Organization (WTO) countries is fraught with difficulty. Until recently military expenditure was disclosed but mistrusted, even by national parliaments, and no information was published or shared among the WTO allies. Until 1989, for some of them, official figures covered only expenses for personnel, construction and operation and maintenance (O&M), not weapon procurement or research and development (R&D). There was always a tendency to underestimate; defence procurement expenditure was often by convention not reported in defence budgets, and when it was in some countries it was given at distorted prices. The allocation of funds in excess of existing budgets seems to have been permitted, if necessary through subsidies or deficit financing. Deceptive accounting practices, off-budget financing and bartering of military imports also meant that official figures often veiled significant parts of these expenditures.

Even with greater transparency, the artificial prices set administratively by the former WTO countries render any calculations of estimates difficult. The North Atlantic Treaty Organization (NATO) has a common methodology used for purposes of comparison and publishes annual statistics on defence expenditure,[1] and the United Nations has set ground rules for obtaining defence budget information from member countries.[2] NATO and the former WTO countries have now agreed to develop a

[1] *Financial and Economic Data Relating to NATO Defence* (NATO: Brussels), annual.
[2] United Nations, General Assembly, *Reduction of Military Budgets, Military Expenditure in Standardized Form,* Report of the Secretary-General (United Nations: New York), annual.

**Table 9A.2.** Indicator of external indebtedness in Central and East European countries, 1990[a]

Figures in italics are percentage shares.

| | Net external debt[b] end year | | | Net interest payments | | | |
|---|---|---|---|---|---|---|---|
| | (a) in US $b. | (b) as % of GDP | (c) as % of exports | (a) in US $m. | (b) as % of exports | Debt service ratio[c] | Official debt as % of total |
| Bulgaria | 10.4 | *42* | *497* | 896 | *43* | 77 | *18* |
| Czechoslovakia | 6.6 | *18* | *116* | 598 | *10* | 25 | *14* |
| Hungary | 19.4 | *59* | *328* | 2 068 | *35* | 65 | *12* |
| Poland | 40.8 | *72* | *408* | 4 080 | *41* | 71 | *75* |
| Romania | – | – | – | 26 | *1* | 10 | – |

[a] Figures are for external debt and exports of goods in convertible currency.

[b] Gross debt less deposits in Bank for International Settlements (BIS) area banks.

[c] All annual interest and amortization on medium- and long-term terms as a percentage of annual exports.

*Source:* Organization for Economic Co-operation and Development, *Reforming the Economies of Central and Eastern Europe* (OECD: Paris, 1992), p. 65.

single method of calculating defence budgets to ensure that accurate comparisons can be made in the future.[3]

Since 1989, two fundamental new elements have been gradually incorporated into the budgetary processes—*(a)* public debate and therefore openness of the process, and *(b)* the involvement of the civilian element in decision making on defence policies. The methodology of defence appropriations has, however, remained basically unchanged: parliaments do not select and allocate resources on a project basis, as is done in many countries, but cut or increase expenditure on broader items such as personnel, construction and O&M.

The introduction of democratic control over the defence ministries and the institution of general staffs as the sole organs of command over the armed forces have already taken place in the former Czechoslovakia, Hungary and Poland. However, in any of these countries it will probably take some years before management and command functions become clearly separated in reality and co-operation between defence ministries and general staffs becomes smoothly functional.

Problems in comparing East and Central with West European budgets now arise not only in establishing how budget items are categorized and reported, but also in the basic analysis of defence needs. In the absence of military doctrines, as now, this is not easy, and it seems as if military expenditure allocations will be determined by economic factors alone instead of true long-term security needs. In order to avoid raising tensions among the fractious new democracies it is important that they understand each other's reasoning before determining their own military needs.

There are two further problems: *(a)* the surplus stocks of old armaments in the era after implementation of the CFE Treaty and *(b)* the fact that production may not have dropped as sharply as procurement expenditure has. The figures given in the military

[3] Hitchens, T., 'NATO, ex-Warsaw Pact nations agree on budget format', *Defense News*, vol. 7, no. 41 (12–18 Oct. 1992), p. 62.

budgets do not distinguish between procurement expenditure or demand and production expenditure or supply. It is unclear whether actual production expenditure has been reduced as much as procurement expenditure. There are no controls on production of treaty-limited armaments inside the territory covered by the Treaty. (Inadequate efforts were made to establish a mechanism for data exchange and reporting requirements for US and former Soviet territory outside the area of application of the treaty.[4]) As matters now stand, treaty-limited equipment (TLE) can be stockpiled indefinitely at production plants so long as the equipment is formally still in production. It is thus possible that stocks are being built up which will have implications for budgetary trends and could lead to pressure to trade surplus armaments abroad in future.

Although more official information is now available, the level of disaggregation is not detailed enough to evaluate hidden categories of expenditure. SIPRI bases its estimates on open sources. Obviously, the quality of the data depends on the original source, which if distorted involves underestimates. The chaotic nature of statistical information-gathering in many of these countries in general and especially during the present economic, political and military transition period makes it even more difficult to construct historical series.

## II. The Czech and Slovak republics

### The dissolution of the Federation

On 26 August 1992, Prime Ministers Vaclav Klaus of the Czech Republic and Vladimir Meciar of the Slovak Republic agreed on a timetable for the dissolution of the Czech and Slovak Federal Republic and the establishment of two independent states from 1 January 1993. The Czech and Slovak republics are the successors to the federal state. All federal institutions and the Czechoslovak armed forces were to be divided and have now ceased to exist in their old form.

According to the agreement on the division of the Czechoslovak Army,[5] military assets will be divided on a 2:1 ratio, two-thirds of the assets going to the Czech Republic. Since only about one-fifth of the former Czechoslovakia's troops and virtually none of its air force are currently stationed on the territory of Slovakia, large numbers of troops, vehicles, aircraft and weapons and more than 100 000 tons of *matériel* are being transferred to Slovakia. Both republics have also begun forming their own ministries of defence.[6]

The two new states were to form a common economic sphere through a customs union and a monetary union, which at the time of writing has collapsed. In some basic commodities, there is mutual strategic interdependence between the two republics: this is the case with oil and gas transport across Slovak territory to the Czech Republic and for deliveries of coal and electricity from the Czech to the Slovak Republic. A new tax system was introduced in January 1993, which makes it

[4] Dean, J. and Forsberg, R.W. 'CFE and beyond: the future of conventional arms control', *International Security,* vol. 17, no. 1 (summer 1992), pp. 96–97.

[5] Schulte, B., General Rapporteur, Germany, 'Social and Political Impact of Force Reductions in Central and Eastern Europe' (North Atlantic Assembly: Brussels, Nov. 1992), document AJ 234 CC(92), pp. 16–17.

[6] 'Lidove noviny', 21 Nov. 1992; *Radio Free Europe/Radio Liberty Research Report*, vol. 1, no. 48 (4 Dec. 1992), pp. 1–5.

difficult to forecast budget revenues. The hard currency assets and debt of the former Czechoslovakia will also be divided on the basis of a 2:1 ratio. The proceeds from the sale of vouchers, which constitutes the backbone of the large privatization scheme, as it is called—more than 8.5 million Czechoslovak citizens are taking part in the mass distribution programme which aims to turn over $9.9 billion worth of state-owned equity to private shareholders[7]—will be divided on a ratio of 2.29:1, reflecting the fact that 2.29 times more vouchers were purchased in the Czech Republic than in Slovakia.[8]

## Economic development

Czechoslovakia is one of the countries most affected by the reorganization of the former Soviet market, formerly the most important outlet for sales of Czechoslovak machines and equipment, which has deprived it of considerable foreign currency earnings. This market was closed when the Soviet debt to Czechoslovakia was estimated at $4000 million, and the former almost total dependence must therefore be offset by turning to other markets. Growth in trade with the West, however, has not yet been strong enough to compensate fully for the collapse of the CMEA.

Total exports of 321 billion koruna ($11 billion) in 1991 were 94 per cent of their 1990 level and nearly 30 per cent down on the $14.3 billion value for 1989, while imports have shrunk even more sharply to $10.1 billion from $13.2 billion in 1990.[9]

The collapse of exports to the former USSR in particular deepened the domestic recession, especially in Slovakia with its heavy concentration of arms factories and other plants geared specifically to the old Soviet market. The value of Czech arms production, for example, was 11 557 million koruna in 1987, falling to 4561 million koruna in 1991. Slovak arms production was 17 741 million koruna in 1987 and 6509 million koruna in 1991. The shares of exports going to former socialist countries also fell from 58.2 per cent in 1987 to 20.6 per cent in 1991.[10] A weakening of the mechanism that until recently ensured a large transfer of resources could possibly impose additional painful adjustments on Slovakia.[11] The true size of this transfer is difficult to judge as most taxes up to the end of 1992 were collected at the federal level and distributed to the two republics according to a negotiated ratio.

Unemployment figures show that in January 1992 the unemployment rate in the Czech republic was 4.31 per cent while in the Slovak republic it was roughly three times higher at 12.74 per cent. These figures confirm the growing disparity in the impact of economic reforms as between the two republics.[12]

With high unemployment, high inflation and the hesitation of foreign investors who fear political instability, even if conversion of the defence industries to civilian production were seriously contemplated, the new Slovak Government cannot afford the investment that is necessary to Slovakia's economic revitalization and is likely to

[7] *Financial Times*, 5 June 1992.

[8] *Radio Free Europe/Radio Liberty Research Report*, (note 6).

[9] *Financial Times,* 15 May 1992.

[10] Czechoslovak Federal Ministry of Economy, 'Defence conversion and armament production in the Czech and Slovak Federal Republic', paper presented at the NATO Central and East European Countries' Defence Industry Conversion Seminar, Brussels, 20–22 May 1992.

[11] Winiarski, M., 'Hopp om västkapital lindrar oro för delning' [Hope of western capital eases fears about division ], *Dagens Nyheter* (Stockholm, Sweden), 29 Dec. 1992, pp. 39–40.

[12] Economist Intelligence Unit, *Czechoslovakia,* Report no. 2 (Business International: London, 1992), p. 19.

revive the defence industry and rely even more heavily on defence production and export.

Thus, while the independent Czech state may be expected to reap benefits from exposure to the west, geography and economic weakness could force Slovakia to look to the east. A mutually beneficial solution for both the Czech and Slovak sides would therefore be the maintenance of a common market between the two new independent states in order to prevent a collapse of inter-republican trade. For the two republics and for the region as a whole the European Community (EC) with open markets and financial help could be a driving force for integration.

### Military postures and new military doctrine

Military posture in general was of an offensive character in the former WTO. The Czech Army was the WTO's first strategic echelon, and this was reflected both in its deployment and in its organization.

The new doctrine adopted by the Federal Assembly on 20 March 1990[13] is based on the principle of territorial defence and the exclusively defensive nature of the armed forces. Czechoslovakia's new doctrine endeavours to re-balance the country's defence by moving from West to East, and more particularly to Slovakia's frontier with the Commonwealth of Independent States (CIS), 80 per cent of the forces then stationed in the Czech Republic.[14]

However, restructuring and relocation of military forces will have to take place under severe budgetary and in the shorter run administrative constraints. It is therefore safe to assume on the one hand that defence postures in general will not correspond fully to the true political/military requirements much before the turn of the century, and on the other hand that substantive modernization and re-equipment will be delayed for some time to come.

### Trends in military expenditure

Military expenditure has been on a downward trend for the past four years. Part of the saving was produced by a reduction in the period of military service in 1991 from 24 to 18 months and from 19 to 12 months for conscripts with higher education certificates. However, if unidentified spending (such as earnings from arms exports) were added, the total could be higher. For example, in 1989 exports of conventional weapons amounted to $543 million, while in 1991 arms exports fell sharply. On the one hand, the collapse in demand was due to the international situation—the disarmament process that started in the late 1980s and led to the CFE Treaty, the emergence of new competitors and the loss of the Soviet market.[15] On the other hand, Czecho-

[13] Statement by Major-General Karel Pezl, Chief of the General Staff of the Czech Army and Deputy Defence Minister at the Second Seminar on Military Doctrine, Vienna, 9 Oct. 1991.

[14] Obrman, J., 'From idealism to realism', *Report on Eastern Europe*, 20 Dec. 1991, pp. 9–13.

[15] Western European Union Assembly, *Defence Industry in Czechoslovakia, Hungary, Poland.* Report submitted on behalf of the Technological and Aerospace Committee by Mr Atkinson, Rapporteur, 37th ordinary session, Document no. 1289, Paris, 8 Nov. 1991, p 7.

**Table 9A.3.** Czechoslavakia's military expenditure allocation, official figures 1989–92

Figures are in current m. korunas. Figures in italics are percentage shares.

| | 1989 | 1990 | 1991 | 1992 |
|---|---|---|---|---|
| Personnel | 9 611 | 7 674 | 8 647 | 10 690 |
| *Share of total* | *27.4* | *23.8* | *31.0* | *37.5* |
| O&M[a] | 10 105 | 12 214 | 14 163 | 14 288 |
| *Share of total* | *28.8* | *37.8* | *50.8* | *50.1* |
| Procurement | 12 205 | 9 989 | 3 146 | 1 245 |
| *Share of total* | *34.8* | *30.9* | *11.2* | *4.4* |
| Construction | 1 812 | 1 346 | 1 235 | 1 680 |
| *Share of total* | *5.1* | *4.7* | *4.4* | *5.9* |
| R&D | 1 329 | 1 065 | 677 | 657 |
| *Share of total* | *3.8* | *3.3* | *2.4* | *2.3* |
| **Total** | **35 062** | **32 288** | **27 868** | **28 500** |
| | *100* | *100* | *100* | *100* |

[a] Operation and maintenance (includes civilian personnel costs).

*Sources:* Figures supplied by the Federal Ministry of Economy, the Federal Ministry of Defence 1989–92 and the Federal Statistics Bureau, Prague. Military expenditure data for 1992 are estimates from the Federal Ministry of Economy.

slovakia was reported to have plans to export large amounts of TLE, which otherwise would have to be destroyed under the CFE Treaty,[16] outside the European region.[17]

Table 9A.3 shows the substantial shrinking of actual disbursements for weapon procurement in Czechoslovakia over the past four years. While in 1989 procurement consumed 34.8 per cent of the military budget, in 1992 the share was only 4.3 per cent. The decline in the Czechoslovak defence expenditure over 1990–92 was so rapid that lack of financial resources stopped weapon development projects and led to substantial restraints in the procurement of equipment and troop training.[18] On the other hand, costs for personnel and maintenance have been increasing. In real terms, the purchasing power of these allocations diminished proportionately faster because of high inflation (54 per cent, for instance, in 1991)[19] and the introduction of hard currency settlement in military trade with the former USSR. Although military R&D fell from 3.8 per cent in 1989 to 2.3 per cent in 1992, this share is still the highest in the whole Central and East European region. This is due to the fact that, now that the German Democratic Republic (GDR) has ceased to exist, the two successor states to Czechoslovakia have technologically the most advanced arms industries in the region.

[16] TLE includes tanks, armoured combat vehicles (ACV), heavy artillery, combat aircraft and attack helicopters.

[17] Anthony, I. *et al.*, 'The trade in major conventional weapons', SIPRI, *SIPRI Yearbook 1992: World Armaments and Disarmament* (Oxford University Press: Oxford, 1992), pp. 290–91.

[18] Cechak, O., Selesovsky, J. and Stembera, M, 'Czechoslovakia: reductions in arms production at a time of economic and political transformation', ed. H. Wulf, SIPRI, *Arms Industry Limited* (Oxford University Press: Oxford, 1993), pp. 242–43.

[19] Fucik, J., 'The Czechoslovak armament industry', *Military Technology*, no. 7 (1991), pp. 98–101.

Although Czechoslovakia designed and developed a number of weapons and types of military equipment indigenously, and its arms industry is relatively well-developed, the over-reliance on the former Soviet sources of weapons supply, accounting for about 50 per cent of procurement, will not vanish quickly, partly because most Russian weaponry is incompatible with Western equivalents, partly because of continuing economic constraints. Limited budgetary resources for new weapons and technical acquisition will hardly permit reorientation to Western hardware on a large scale. On the other hand, both the republics are increasing their efforts to co-operate in other European arms programmes, and Slovakia, where the largest armament production is concentrated, is trying to sell weapons to the West.

## III. Hungary

### Economic development

There are signs that the Hungarian economy is performing in some respects better and in some respects worse than previously expected. In particular, it is now clear that the recession in domestic industry is much more serious than predicted. According to revised statistics for 1991, industrial production dropped by 21.5 per cent and the corresponding drop in gross domestic product (GDP) was 10.2 per cent—significantly above the 7–9 per cent given in preliminary estimates.[20] The National Bank of Hungary is now braced for a fall of 5 per cent in GDP for 1992, a drastic revision of its original forecast of zero growth.[21]

The expectation that output will continue to fall is increasing pressure to stimulate the economy while at the same time depriving the government of the budgetary means to do so. The severity of the recession is undermining the government's efforts to contain the budget deficit within the limits demanded by the International Monetary Fund (IMF). Shortfalls in tax revenues have resulted from a combination of recession, tax and customs evasion, and the debts of enterprises which are currently being liquidated; that and extra spending on unemployment (expected to rise from 11 per cent of total budget in 1992 to 17–20 per cent in 1993) threaten to put the 1992 budget deficit up to 220 billion forints ($2.8 billion), 8 per cent of GDP.[22]

However, the effects of the domestic recession on foreign trade performance have been overshadowed by the much greater disturbances caused by the loss of the former Soviet and East European markets. Preliminary data for 1992 confirm that reduced demand at home is forcing exporters to boost sales abroad, for example by stimulating Russian bilateral trade, which grew by 28.8 per cent to $2.28 billion in the first quarter of 1992 after falling by more than 50 per cent in 1991,[23] and is at the same time dramatically curtailing imports as well. According to joint preliminary data from the Ministry of Foreign Economic Relations and the Central Statistical Office, exports in the first quarter of 1992 were $2433 million, 13.4 per cent higher than hard currency exports in the same period in 1991, while imports were $2039 million for the same period, down by 10.5 per cent on the same period. The effects of recession

[20] Economist Intelligence Unit, *Hungary, Economic Intelligence Country Report,* no. 3 (Business International: London, Aug. 1992), p. 41.

[21] Denton, N., 'Hungary expects recession to worsen', *Financial Times,* 9 Sep. 1992, p. 4.

[22] *Financial Times,* 1 Sep. 1992.

[23] Denton, N., 'Hungary takes arms in debt deal', *Financial Times,* 12 Nov. 1992, p. 3.

on import demand had been underestimated by the Hungarian Government in much the same way as were its effects on state budget revenues.

Hungary has experienced more favourable external conditions than the other countries of the region, with a hard currency reserve of more than $5 billion, while gross foreign debt has fallen slightly to $21.6 billion since the end of 1991. These positive developments have strengthened the case for introducing limited convertibility for the forint in 1993.[24]

A word of caution is perhaps warranted when surveying the apparently good news in foreign trade and finance. Low import demand resulting from falling industrial output was also experienced in 1991, but was obscured by the drastic increases in energy prices caused by the Persian Gulf War and the switch to hard currency for oil and gas imports from the former USSR. This trade surplus will also be very sensitive to any revival in the domestic economy and will fall as soon as revival occurs.[25]

## Military postures and new military doctrine

The former Hungarian People's Army was also entirely integrated in the WTO military structure and was under strict Communist Party control. After withdrawal of the Soviet troops, numbering about 65 000, by mid-1991, the bases and orientation of Hungary's defence changed. It is now national and aims to ward off any attack from any direction ('defence of all frontiers against potential aggressors').[26] The new defensive military doctrine aims to preserve the integrity of national territory without conducting military operations in foreign territory.

As in the case of the former Czechoslovakia, restructuring and relocation of the military forces will have to be conducted under budgetary constraints. To overcome this lack of resources, the Hungarian Government is obliged to retain conscription in the armed forces (although the period of military service was reduced from 18 to 12 months in 1992) in parallel with some degree of professionalization of the armed forces.

## Trends in military expenditure

If one bears in mind that total Hungarian military expenditure for 1990 was about 52.4 billion forints ($829 million) and that, according to the Hungarian military, an adequate start to modernization would alone need some 70 billion forints ($1107 million), new procurement seems improbable, since about 80 per cent of the whole military budget was already earmarked for operations and current expenditure, as shown in table 9A.4, leaving only about 20 per cent for development and procurement. This trend continues: while shares in aggregated operating costs have increased by 12.6 per cent between 1989 and 1992, devoted mostly to the army's day-to-day needs, investment costs have fallen by 13 per cent during the same period.

[24] Lorinc, H. I., 'Economic and social consequences of restructuring in Hungary. Foreign debt, debt management policy and implications for Hungary's development', *Soviet Studies*, vol. 44, no. 6 (1992), pp. 1002–1004.

[25] Richter, S., 'Economic and social consequences of restructuring in Hungary: Hungary's changing patterns of trade and their effects', *Soviet Studies,* vol. 44, no. 6 (1992), pp. 965–82.

[26] Western European Union, 'Defence: Central Europe in evolution', Report submitted on behalf of the Defence Committee by Mr. Cox, Rapporteur, Assembly of the Western European Union, 38th ordinary session (second part), (WEU: Paris, 5 Nov. 1992), Document no. 1336, pp. 17–18.

**Table 9A.4.** Hungary's military expenditure allocation, official figures 1989–92
Figures are in current b. forints. Figures in italics are percentage shares.

| | 1989 | 1990 | 1991 | 1992 |
|---|---|---|---|---|
| Operating cost[a] | 36.2 | 41.5 | 47.6 | 53.9 |
| *Share of total* | *75.8* | *79.3* | *88.1* | *88.4* |
| Investment cost[b] | 11.5 | 10.8 | 6.4 | 6.8 |
| *Share of total* | *24.1* | *20.7* | *11.9* | *11.1* |
| **Total** | **47.8** | **52.4** | **54.0** | **61.0** |
| | *100* | *100* | *100* | *100* |

[a] Operating cost includes maintenance, personnel (military and civilian), pensions and other social expenditure.
[b] Investment cost includes procurement, construction and R&D.

*Source:* Federal Ministry of Defence, *Törvények és Rendeletek Itivaltalos Gyöjteménye 1989–92* [Official Collection of Laws and Regulations, 1989–92], (Budapest, [1992]). Submitted by the Hungarian Library of Parliament, 6 Aug. 1992.

According to Csaba Kiss, special adviser to the Hungarian Defence Minister Lajos Fur, Hungary's armament stock is obsolete even by East European standards. The Hungarian Government is engaged in developing a long-term plan to upgrade the military. Most of the fighter fleet will have to be retired by 1996–2002. The government is seeking to diversify its procurement by considering Western sources for several high-priority requirements, including modern air defence radars and other electronic equipment from the USA worth $12.9 million. Under this deal, US suppliers will deliver four ground-based radar stations and electronics for 118 military aircraft.[27]

Given the rather low allocation for military procurement and the much higher cost of Western equipment, Hungary will primarily have to update its Soviet-made equipment. This was demonstrated when in November 1992 an agreement was reached between Russia and Hungary to take $800 million worth of military equipment and technology in part settlement of Russia's $1.7 billion debt to this former WTO country.[28]

# IV. Poland

## Economic development

Alone among the former WTO countries, Poland is showing signs of sustained recovery from the recession that has plagued the region since 1989. The 22.3 per cent per annum consumer price inflation rate for the first half of 1992 was the lowest recorded since 1988, when many prices were still controlled. Labour productivity registered an 11.5 per cent increase during the first five months of 1992, and the decline in employment ceased during the first half of the year, so that the unemployment rate seemed to have stabilized in the 12–13 per cent range.[29] At the end of

[27] Denton, N., 'Hungary buys US military hardware', *Financial Times*, 9 Dec. 1992, p. 8.
[28] See Denton (note 23).
[29] *Post Soviet/East European Report,* vol. 11, no. 34 (15 Sep 1992), p. 2.

June 1992, the balance of trade surplus was $946 million, and the current account surplus stood at $389 million. These figures contrast with a balance of trade deficit of $332 million and a current account deficit of $1.67 billion in mid-1991.[30]

The rapid growth of private economic activity and of exports seems to be the key factor behind the Polish recovery. Poland now has the largest private economy among the former Soviet bloc countries, estimated at 44–55 per cent of GDP.[31] This growth reflects the rapid expansion of new private firms rather than the privatization of existing state enterprises; programmes for privatizing large state enterprises or attracting foreign investment have not hitherto been very successful. The private sector's growth is not only reversing declines in production but also contributing to industrial restructuring and the development of capital markets.

While there are several factors that suggest that the recovery of 1992 may be more durable than previous upturns, there are also major problems in the way of a sustained recovery.

On the one hand one can say that the recovery this time has been accompanied by falling inflation rates and healthy surpluses on the current account and trade balance; that there seem to be no more external shocks similar to the collapse of CMEA trade or the Persian Gulf War on the horizon; and that the private sector and the extent of Poland's integration into the international economy are much larger now than 1990.

On the other hand, strikes are currently threatening the finance plans of the government. Striking workers have already forced wage increases in, for example, the coal industry, and it remains to be seen whether the government can keep its tight economic policy in order to restrain the growth in the budget deficit.[32] The second major problem is the budget deficit itself. The limit of 65.5 trillion zloty ($4.4 billion or 5 per cent of GDP) agreed upon with the IMF in May 1992 seems to have been too low. Limiting the deficit to 7–8 per cent of GDP is a more realistic assumption.[33] Third, the fiscal problems of state enterprises and the banking system have continued to worsen in 1992. This is perhaps the most serious indication of the Polish economy's basic weakness; the recovery has so far been unable to redress many of the long-term structural problems. The wave of strikes seems likely to complicate the task of dealing with bankrupt state enterprises, while the government's ability to finance the restructuring of these firms is reduced in proportion to the growth in the budget deficit.

## Military postures and new military doctrine

In accordance with a declared sufficiency principle,[34] Poland decided to phase out heavily equipped units, air force units and logistic support units, while retaining an anti-tank and anti-aircraft capability. It therefore gave priority to rapid reaction forces and, above all, the air defence system. The army will be composed of operational forces and regional defence units equipped with light weapons. It is expected that Polish troops will be redeployed in the eastern parts of the country. However, the

[30] 'Statistics of Poland', *Rzeczpospolita*, 4 Aug. 1992, p. 2.

[31] Slay, B., 'Privatization in Poland: an overview', *Radio Free Europe/Radio Liberty Research Report*, no. 17 (24 Apr. 1992).

[32] 'Wage increases from negative profits?', *Rzeczpospolita*, 24 Aug. 1992.

[33] Slay, B., 'The Polish economy: between recession and recovery', *Radio Free Europe/Radio Liberty Research Report*, vol. 1, no. 36 (11 Sep. 1992), p. 57.

[34] Statement by Dr Janusz Onyszkiewicz, Deputy Minister for Defence of the Republic of Poland, at the Second Seminar on Military Doctrine, Vienna, 8 Oct. 1991.

**Table 9A.5.** Poland's military expenditure allocation, official figures 1989–92
Figures are in current b. zlotys. Figures in italics are percentage shares.

| | 1989 | 1990 | 1991 | 1992 |
|---|---|---|---|---|
| Personnel | 874 | 4 913 | 8 046 | 9 666 |
| *Share of total* | *39.4* | *32.9* | *44.0* | *39.7* |
| O&M[a] | 676 | 5 034 | 5 441 | 10 060 |
| *Share of total* | *30.5* | *33.7* | *29.7* | *41.3* |
| Procurement | 502 | 3 312 | 2 813 | 2 926 |
| *Share of total* | *22.7* | *22.2* | *15.4* | *12.0* |
| Construction | 110 | 1 320 | 1 532 | 1 271 |
| *Share of total* | *5.2* | *8.8* | *8.4* | *5.2* |
| R&D | 52 | 366 | 468 | 451 |
| *Share of total* | *2.3* | *2.4* | *2.6* | *1.9* |
| **Total** | **2 214** | **14 945** | **18 300** | **24 374** |
| | *100* | *100* | *100* | *100* |

[a] Operation and maintenance (includes civilian personnel cost).

*Source:* United Nations General Assembly, *Reduction of Military Budgets, Military Expenditure in Standardized Form*, Report of the Secretary-General, annual. Submitted by the Polish Ministry of National Defence, Warsaw, 7 Oct. 1992.

costs of relocation, including those of transporting several thousand tons of equipment and *matériel*, the erection of new barracks and technical infrastructure, new communication lines, roads, training grounds, schools and health care centres will inevitably slow down the redeployment process.

## Trends in military expenditure

The new democracies in general, and Poland in particular, are plagued by the fact that they have little money to spend on defence. Plummeting defence budgets are generally being spent on military salaries and day-to-day operations. Poland, as shown in table 9A.5, has assigned 39.7 per cent of its 1992 budget for personnel costs and only 12 per cent for equipment procurement, while R&D is reduced to 1.9 per cent of the budget in 1992. The share of procurement fell dramatically by 10.7 per cent between 1989 and 1992, while of O&M has been increasing by about 11 per cent over the same four years, and personnel outlays remain high. These are general trends in all the countries of Central and Eastern Europe because, under the general pressure for economies and rationalization in military outlays, cuts in *matériel* are much easier to achieve than reductions in salaries, pensions and other payments to personnel. The need to keep pay in line with rising inflation outweighed certain other factors influencing the structure of and the overall decline in military spending, including a substantial reduction in the size of the army itself.

Even equipping the army with spare parts and technical *matériel* is proving difficult. The former USSR was reluctant to provide military and technical spares to its former allies.[35] Unable to secure full supply guarantees, Polish defence planners examined alternative possibilities, including establishing co-operative ties with Czechoslovakia and Hungary. Czechoslovakia, Hungary and Poland agreed in February 1991 at Visegrad to create mechanisms for economic and political co-operation. In a broader context this agreement offered these three (now four) states a different venue for security co-ordination and integration into European security organizations.

The Polish armed forces have systematically reduced procurement from abroad as well as from domestic suppliers because of defence expenditure cuts. This reduction in purchases and the decline of co-operative exports and sales to the former East bloc nations have caused severe problems for the Polish arms industry. For example, the Polish aviation industry exported to the former USSR military equipment worth a total of $100 million, for which it still has not received payment. This coincided with the loss of the traditional Middle Eastern market because of the 1991 Persian Gulf War and the embargo on arms supplies to Iraq, which resulted in the loss of contracts worth a total value of $70 million. Iraq is also overdue with payments for earlier deliveries.[36] Arms exports were regarded as the most profitable branch of export activity. The income from exports covered the cost of imports of both armaments and raw materials and licences for the arms industry.

## V. Bulgaria

### Economic development

In contrast to Hungary and Poland, Bulgaria did little prior to 1990 to relax its tight central planning system or to orient its economic activities towards world market opportunities. Its trade policy was geared to specialization between the members of the CMEA and foreign trade was managed as a state monopoly. Bulgaria has thus been more susceptible than other East and Central European countries to the demand shock from the collapse of the Soviet market over the past two years and to the price shock accompanying the shift of all trade prices to world prices. The drop in Bulgaria's exports to former CMEA countries in 1991 was equivalent to 12 per cent of GDP and was thus two to three times greater than the drop experienced by any other former CMEA member.[37]

Since Bulgaria embarked on market reforms in February 1991 it has made progress towards a market-oriented economy. However, the country has been unable to overcome the macro-economic instability that threatens the progress made thus far and hinders foreign investment. The slow progress in structural reforms and the $11.5 billion foreign debt owed to more than 300 commercial banks and $2 billion to the official creditors in the Paris Club[38] remained a heavy burden on the economy. Bulgaria has started the second phase of its economic transformation, including restoration of land to its original ownership, banking regulation and foreign debt repayment. The decline in industrial output has continued at an annual rate of over

[35] Podbielski, P.J., 'Whence security? Polish defense and security after the Warsaw Pact', *Journal of Soviet Military Studies*, vol. 5, no. 1 (Mar. 1992), pp. 97–114.

[36] *Warsaw Voice-News,* 26 Apr. 1992, p. 4.

[37] European Bank for Reconstruction and Development, *Quarterly Economic Review*, 30 Sep. 1992, pp. 46–47.

[38] *Financial Times*, 25 Mar. 1992, p 2.

**Table 9A.6.** Bulgaria's military expenditure allocation, official figures 1989–92

Figures are in current m. levas. Figures in italics are percentage shares.

| | 1989 | 1990 | 1991 | 1992 |
|---|---|---|---|---|
| Operating cost[a] | 481.0 | 980.0 | 3 298.0 | 5 103.4 |
| *Share of total* | *28.6* | *59.1* | *83.5* | *88.4* |
| Investment cost[b] | 1 201.2 | 678.3 | 650.1 | 667.5 |
| *Share of total* | *71.4* | *41.0* | *16.3* | *12.0* |
| **Total** | **1 682.2** | **1 658.4** | **3 949.1** | **5 771.0** |
| | *100* | *100* | *100* | *100* |

[b] Operating cost includes maintenance, personnel (military and civilian), pensions and other social expenditure.

[a] Investment cost includes procurement, construction and R&D.

*Source*: Bulgarian Ministry of Defence, Financial Department; United Nations General Assembly, *Reduction of Military Budgets, Military Expenditure in Standardized Form*, Report of the Secretary-General, annual.

20 per cent, however, inflation in 1992 was over 80 per cent, the fall of real income continues, there is no final agreement on rescheduling the country's external debt and the amount of external assistance has been disappointing. All these factors have had seriously adverse effects on the budget deficit, which was expected to be more than 6 per cent of GDP for 1992.[39]

## Security, defence and the armed forces

Bulgaria's geographical position at the heart of the Balkans gave it strategic importance in the former WTO. It is now without the protection of a major ally in the region just when it is having to cope with an unstable regional situation. Above all, Bulgaria is seeking integration in wider European structures. NATO is considered the only credible alternative that could guarantee the country's security.[40]

The restructuring and depoliticization of the army have begun, with the creation of smaller and more mobile forces. The Ministry of Defence is being restructured to form several directorates (e.g., administration, economics and political questions) and two new departments to handle social questions and ecology. Finally, two study groups comprising both civilians and military personnel have been instructed to define the various stages in the development of the armed forces from 1994 to 2000. Mobile brigades should replace traditional structures and military headquarters should be given increased power at the expense of the Ministry of Defence, but there are no plans to reduce troop levels (at present 100 000 men). These new concepts will be the subject of a basic law on security, defence and the armed forces.[41]

[39] *Frankfurter Allgemeine Zeitung*, 19 Oct. 1992, p. 18.

[40] Interviews with President Julio Jelev by B. Bollaert in *Le Figaro*, 19 Nov. 1991, and by S. Kauffman in *Le Monde*, 6 Dec. 1991.

[41] Engelbrekt, K., 'Reforms reach the Bulgarian armed forces', *Radio Free Europe/Radio Liberty Research Report*, vol. 1, no. 4 (24 Jan. 1992), pp. 54–56.

### Trends in military expenditure

Because of the present restructuring of the Bulgarian military sector and the inadequacies of the available official aggregated budget data, these data probably do not accurately represent total military outlay. From table 9A.6 a general observation can be made: shares of investment costs (including procurement, construction and R&D) have dropped dramatically by 59.4 per cent between 1989 and 1992. The increase in the share of personnel costs (including military and civilian, O&M, pensions and other social expenditure) has been approximately 60 per cent during the same period. It is clear that the authorities are aware of the dominance of procurement and R&D in the military budget and the debilitating loss they represent to the civilian economy. This structural problem has to be solved because of the disproportionate share of the military sector in the economy. While 20 000–30 000 employees, 25 per cent of the arms industry, found themselves out of work between 1989 and 1991, this sector also owes the state between 1500 and 2000 million leva ($1875–2500 million in 1989 and $93.7–125 million at 1991 official exchange-rates).[42] The fall in investment cost and increase in personnel costs can be observed in all the Central and East European countries.

## VI. Romania

### Economic development

Because of its geographical position between Yugoslavia and the CIS with all the difficulties caused by the collapse of existing structures there, Romania is seeking economic (access to raw materials and new markets) and security guarantees both among its former CMEA partners and among Western countries. The future of Romania's economic reform programme has become uncertain because the incoming government under Prime Minister Theodor Stolojan has pledged to slow down and revise it. The immediate goal is to stabilize the economy, which is performing below 65 per cent of 1989 levels. More specifically, the new strategy aims to halt the decline in production, control inflation which has averaged more than 14 per cent per month during 1992, and build up the country's gold and hard currency reserves. The government hopes that one possible source of hard currency is the $2.7 billion owed to Romania by developing nations. Iraq, Romania's main debtor, accounts for $1.7 billion, Sudan for $172 million, Syria for $119 million and Libya for $46 million.[43] The prospect of recovering debts from those countries seems rather bleak.

### Security, defence and the armed forces

Romanian defence policy is based on the general principle of 'adequate sufficiency of defence and optimum gradual response'[44] and is affected by its leaders' nervousness about the country's isolation and divisions of opinion concerning the

[42] Western European Union, Institute for Security Studies, 'A new security order in Europe', Document prepared for the WEU Assembly Symposium (WEU: Paris, 3 Mar. 1992), p. 95.

[43] *Radio Free Europe/Radio Liberty Research Report* (note 6).

[44] Statement by Major-General Dumitru Cioflina, State Secretary and Chief of the General Staff, at the Second Seminar on Military Doctrine, Vienna, 10 Oct. 1991.

**Table 9A.7.** Romanian military expenditure allocation, official figures 1990–92

Figures are in current m. lei. Figures in italics are percentage shares.

| | 1990 | 1991 | 1992 |
|---|---|---|---|
| Personnel | 5 917 | 10 764 | 42 000 |
| *Share of total* | *17.5* | *33.2* | *26.5* |
| O&M[a] | 5 749 | 7 704 | 61 068 |
| *Share of total* | *17.0* | *23.8* | *38.5* |
| Procurement | 21 151 | 12 807 | 52 901 |
| *Share of total* | *62.6* | *39.5* | *33.4* |
| Construction | 527 | 653 | 959 |
| *Share of total* | *1.6* | *2.0* | *0.6* |
| R&D | 448 | 450 | 1 590 |
| *Share of total* | *1.3* | *1.4* | *1.0* |
| **Total** | **33 792** | **32 378** | **158 518**[b] |
| | *100* | *100* | *100* |

[a] Operation and maintenance (includes civilian personnel cost).

[b] The 1992 submission to the United Nations gives the total figure of 138.558 m. lei, but this does not include an additional 20 m. lei that was approved by the Parliament in July 1992. 5 m. lei of this was for O&M and 15 m. lei for capital expenditure.

*Source:* Romanian Ministry of National Defence, *Laws of Military Budgets 1982–92*, submitted through the Romanian Embassy, Stockholm, 30 Nov. 1992; discussions at the North Atlantic Cooperation Council (NACC) Economic Committee Meeting with Co-operating Partners, Brussels, 30 Sep.–2 Oct. 1992.

external dangers. Romania wishes to develop closer relations with NATO at the highest level acceptable to NATO.[45]

To a greater extent than in the other countries of the region, the armed forces had in the past a role of political and social control. Limited resources and technology held up essential modernization.

The restructuring policy is only just starting to be applied. The number of troops has been reduced to 200 000 and the restructuring programme provides for mobile, lightly equipped forces, adequate air defence and rapid intervention units.

### Trends in military expenditure

As shown in table 9A.7, personnel receives 26.5 per cent of military outlay for 1992, while arms procurement with 33 per cent has a surprisingly high share of the budget. It is not known whether it is spent on acquisitions from the domestic defence industry. Production in Romania under Western licence has proved to be extremely costly. Apart from the cost of obtaining the licences, the weapons had to be modernized, for which more money was needed. Political uncertainty has also led to some foreign companies putting off discussions on collaboration with the aviation industry. On the other hand, Romaero, the state-owned aircraft manufacturer, has received its

[45] Carp, M., 'New foreign policy initiatives', *Report on Eastern Europe*, vol. 2, no. 36 (6 Sep. 1991), pp. 26–29.

**Table 9A.8.** Official estimates of Central and East European countries' CFE-limited weapons in Europe before and after the CFE Treaty implementation

| | Tanks | ACVs | Artillery | Aircraft | Helicopters | **Total** |
|---|---|---|---|---|---|---|
| Bulgaria | | | | | | |
| Before CFE | 2 145 | 2 204 | 2 116 | 243 | 44 | **6 752** |
| Change | – 670 | – 204 | – 366 | – 8 | 23 | **– 1 225** |
| After CFE | 1 475 | 2 000 | 1 750 | 235 | 67 | **5 527** |
| Czechoslovakia | | | | | | |
| Before CFE | 1 797 | 2 538 | 1 566 | 348 | 56 | **6 305** |
| Change | – 362 | – 488 | – 416 | – 3 | 19 | **– 1 250** |
| After CFE | 1 435 | 2 050 | 1 150 | 345 | 75 | **5 055** |
| Hungary | | | | | | |
| Before CFE | 1 345 | 1 720 | 1 047 | 110 | 39 | **4 261** |
| Change | – 510 | – 20 | – 207 | 70 | 69 | **– 598** |
| After CFE | 835 | 1 700 | 840 | 180 | 108 | **3 663** |
| Poland | | | | | | |
| Before CFE | 2 850 | 2 377 | 2 300 | 551 | 29 | **8 107** |
| Change | – 1 120 | – 227 | – 690 | – 91 | 101 | **– 2 027** |
| After CFE | 1 730 | 2 150 | 1 610 | 460 | 130 | **6 080** |
| Romania | | | | | | |
| Before CFE | 2 851 | 3 102 | 3 789 | 505 | 13 | **10 260** |
| Change | – 1 476 | – 1 002 | – 2 314 | – 75 | 107 | **– 4 760** |
| After CFE | 1 375 | 2 100 | 1 475 | 430 | 120 | **5 500** |

ACV = Armoured combat vehicle.

*Source:* Institute for Defense and Disarmament Studies, *Vienna Fax,* vol. 3, nos. 10, 11 (20 Dec. 1991), p. 4.

first order from a Western company to produce 11 BAC-1-11 aircraft, with an option for five more, since it began making the medium- to-short-haul jets under licence from British Aerospace in 1982. The companies have not disclosed the value of the order.

As in all the countries in the region, the share of procurement in the budget is falling dramatically and there is clearly almost no acquisition of new weapons. One reason, as explained above, is the budgetary and economic constraints that these countries are facing. Another could be the surplus of stocks of older armaments in the period since the CFE Treaty. All these countries will have to dispose of some of the TLE as shown in table 9A.8, and, in the absence of modernization, procurement expenditure is clearly unnecessary so long as the TLE limits remain valid. There is however still no external control, such as treaties, on the production of TLE inside the Central and East European countries. Even though arms procurement spending has been falling dramatically in the region, expenditure on production may not have fallen by the same amount.

# 10. Arms production and arms trade

IAN ANTHONY, PAUL CLAESSON, ELISABETH SKÖNS and SIEMON T. WEZEMAN

## I. Introduction

Governments which can no longer justify the level of military expenditure sustained during the cold war nevertheless wish to retain industrial capacities to the degree possible. Companies which had no need to seek foreign sales in order to be profitable in the 1980s are doing so in the early 1990s as the level of national arms procurement expenditure falls in real terms. From the perspective of governments and the arms industry, international sales are becoming increasingly important.

In October 1992 the Secretary-General of the United Nations underlined the need for the international community to address the issue of arms transfer control seriously in the new political environment. In his review of recommendations and decisions relating to disarmament, Boutros Boutros-Ghali stressed that the successful conclusion of arms control in Europe and the end of the cold war should not lead to increased efforts to export arms.[1]

This chapter primarily reflects the traditional focus of SIPRI research in the areas of arms production and arms trade.[2] The discussion of production is dominated by the Organization for Economic Co-operation and Development (OECD) countries and, to a lesser extent, those in the developing world. However, the arms industry most affected by reduced domestic arms procurement expenditure is that of the former Soviet Union—especially in Russia and Ukraine, where most of the former Soviet arms industry was concentrated. Russia and Ukraine have adopted different strategies to manage the crisis in their arms industries following the failure of the ambitious conversion plans drawn up between 1989 and 1991.[3] The division of Czechoslovakia into two sovereign states is the latest manifestation of the dramatic political transformation of this region. This transformation will require further changes in an important centre of arms production.[4]

[1] United Nations, *New Dimensions of Arms Regulation and Disarmament in the post-Cold War Era*, Report of the Secretary-General of the United Nations, General Assembly document A/C.1/47/7 (United Nations: New York, 23 Oct. 1992), p. 9.

[2] This year SIPRI has consolidated the discussion of arms production and the international arms trade in one chapter.

[3] Izyumov, A., 'The Soviet Union: arms control and conversion—plan and reality', ed. H. Wulf, SIPRI, *Arms Industry Limited* (Oxford University Press: Oxford, 1993), chapter 6, pp. 109–20; *M.I.C. Newsletter*, Jan. 1993, pp. 1–2.

[4] The future of the defence industries in Central and Eastern Europe (including Russia and Ukraine) is the subject of a SIPRI research project initiated in 1993, the first results of which will be presented in the *SIPRI Yearbook 1994*.

A reduction in the volume of global arms production is reflected in the SIPRI list of the arms sales of the 100 largest arms-producing companies in the OECD and in developing countries for 1991. The combined value of arms sales by the 100 companies with the highest arms sales in 1991 decreased from $183.7 billion in 1990 to $178.8 billion in 1991, representing a fall of 3 per cent in current prices. Moreover, the decline in the volume of arms sales by the arms industry in these countries is sharper than these statistics suggest. These data include the effect of inflation, and the average rate of inflation in the OECD area in 1991 was 3.9 per cent. If military goods experienced the same rate of inflation, this would suggest a fall of 7 per cent in real terms in the arms sales of the 100 companies on the SIPRI list for 1991.

The consequences of this reduction in sales for employment in the arms industry are also suggested by data from the SIPRI arms industry data base. Almost 80 per cent of those companies among the 100 largest arms-producing companies and their largest subsidiaries that depended on arms sales for more than half of total sales in 1988 had reduced their employment levels by 1991.

The declining market for military equipment reinforces incentives for industrial concentration and collaboration. International take-overs and mergers have normally been preceded by national concentration, and the pace of restructuring in the arms industry accelerated in 1992 with many of the largest arms-producing companies participating in collaborative arrangements of some kind. International mergers and acquisitions in the arms industry also continued in 1992. While most of this activity took place in Western Europe, US companies were involved in three major international take-overs and two mergers/joint ventures. Two of these were still pending at the end of 1992 and one foreign take-over was blocked by US legislators during the year. This was the bid by the French electronics company Thomson-CSF for the missile division of the US corporation LTV which was ultimately acquired by a US company. These findings relating to company activities are discussed in more detail in section III of this chapter.

The global value of foreign deliveries of major conventional weapons in 1992 is estimated by SIPRI to have been $18 405 million in 1990 US dollars.[5] This figure—roughly 25 per cent less than the value recorded for 1991—continues the trend reported in the *SIPRI Yearbook 1992*.[6]

The United States retained the position it achieved in 1991 as the dominant arms-exporting country, accounting for 46 per cent of all deliveries. This reflected the continued reduction in the size of Russian arms exports rather than an increase in US deliveries of major conventional weapons. Russia

[5] SIPRI arms trade statistics do not reflect purchase prices. They are not comparable with national account or foreign trade statistics, nor with arms sales reported in the parts of this chapter dealing with arms production. The methods used in compiling SIPRI arms trade statistics are described in appendix 10D.

[6] Anthony, I., Courades Allebeck, A., Miggiano, P., Sköns, E. and Wulf, H., 'The trade in major conventional weapons', *SIPRI Yearbook 1992: World Armaments and Disarmament* (Oxford University Press: Oxford, 1992), chapter 8, pp. 271–301.

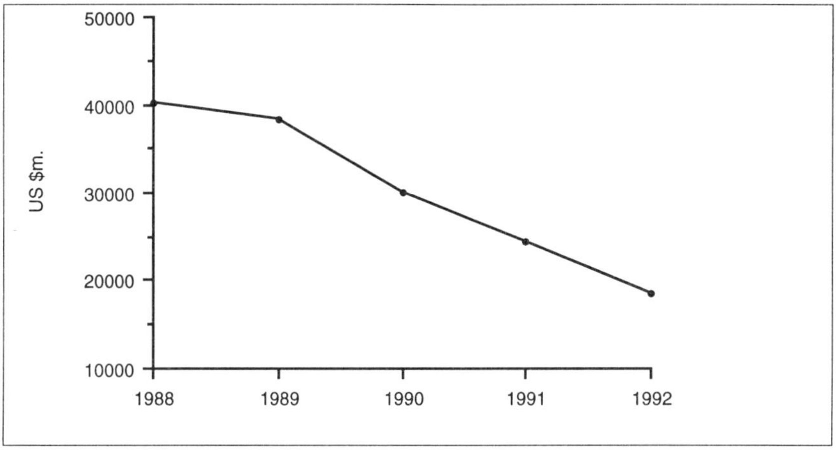


**Figure 10.1.** The downward trend in the aggregate value of deliveries of major conventional weapon systems, 1988–92

Data are SIPRI trend-indicator values.

*Source*: SIPRI arms trade data base.

accounted for 11 per cent of deliveries recorded for 1992, although this figure probably includes some equipment delivered from Ukraine. The level of Russian arms exports is now comparable to that of other European countries and can no longer be considered alongside that of the United States. This fact reflects the inability of Russia to subsidize arms exports in the manner of the former Soviet Union in support of a global foreign and security policy. The countries of the European Community (EC) collectively account for 26 per cent of all deliveries. Three countries—France, Germany and the UK—account for 85 per cent of this figure, with Germany emerging as the predominant West European arms-exporting country for the first time. German arms exports are at a level comparable to that of Russia and considerably larger than that of both France and the UK. In 1992 China, which accounted for 8 per cent of all deliveries recorded for the year, has come to occupy a more important position in the global market. This reflects the fact that Chinese sales are not declining in a market which is shrinking. These findings are discussed in more detail in section IV.

The Middle East was the least important of the three primary markets for international deliveries of major conventional weapons—Asia, Europe and the Middle East. The Middle East was replaced by Europe and Asia as the primary market for major conventional weapons after the Iraq–Iran War ended in 1988. According to SIPRI estimates the Middle East accounted for 22 per cent of deliveries of such weapons in 1992 against 33 per cent in 1983. Asia accounted for 30 per cent of deliveries in 1992 and 19 per cent in 1983 while Europe accounted for 36 per cent in 1992 and 24 per cent in 1983. A discussion of arms procurement by Asian countries is included in section IV. This regional pattern may change since, during 1990–92, a number of large orders for arms were placed by Egypt, Israel and several states located around the Persian Gulf. Other such agreements are under consideration and, as these are implemented, the Middle East may once again emerge as the most important single international market for conventional weapons.[7] However, two Asian countries—India and Japan—received a larger volume of major conventional weapons in 1992 than any Middle Eastern country.

There has been a progressive reduction in imports by African and South American countries. Neither these regions nor Central America and Oceania have been of more than minor importance as arms importers throughout the period.

North America is not a major arms-importing region as the United States meets almost all of its equipment needs from local production and the armed forces of Canada and Mexico are of limited size.

The largest volume of equipment delivered in 1992 went to Greece and Turkey—contiguous with the Middle East but located in the south-eastern corner of Europe. This resulted from the transfer of major weapons within NATO—including the implementation of the 'cascade', associated with the 1990 Treaty on Conventional Armed Forces in Europe (CFE), and the continuation of bilateral military assistance programmes established when Greece and Turkey were part of NATO's southern flank. There is little evidence that attitudes towards arms transfers among NATO countries have been modified in the new political circumstances in spite of the resignation of German Defence Minister Gerhard Stoltenberg in March 1992 when German weapons were used by the Turkish Government against Kurdish separatists.[8]

Elsewhere in Europe, France and the UK emerged as major arms importers in 1991–92 largely because of the delivery of E-3 Airborne Early Warning aircraft from the USA, underlining the fact that national producers cannot meet all of the requirements that their governments establish for the armed forces.

## II. Important developments of 1992

### The United Nations embargo on Yugoslavia

On 25 September 1991 the UN Security Council established a mandatory arms embargo on Yugoslavia (including the territories of all six republics).[9] Finding a means of terminating the series of wars under way on the territory of the former Yugoslavia has come to absorb much of the time and attention of senior decision makers in governments of the participating states of the Con-

[7] Platt, A. (ed.), *Report of the Study Group on Multilateral Arms Transfer Guidelines for the Middle East* (Stimson Center: Washington, DC, May 1992); Williams, J., *The Middle East Peace Process and the Arms Trade: A Fatal Contradiction?* (Saferworld Foundation: Bristol, Aug. 1992).

[8] Hollis, R., 'Sitting on a Middle East fence', *Military Technology*, vol. 17, no. 2 (Feb. 1993), pp. 18–27.

[9] The text of Resolution 713 of 25 Sep. 1991 is reproduced in the *US Department of State Dispatch*, 30 Sep. 1991, pp. 724–25. See also the discussion of the conflict in the former Yugoslavia in chapter 3 of this volume.

ference on Security and Co-operation in Europe (CSCE).[10] At the end of 1992 the incoming Clinton Administration actively considered supplying arms to Muslim forces fighting in Bosnia and Herzegovina as an alternative to US military intervention in the conflict.[11] No clear rationale for such an action was offered and at the time of writing no action—which would require either lifting or circumventing the UN arms embargo—had been taken.

Responsibility for implementing the embargo initially rested with the individual member states. While Security Council Resolution 724 of 15 December 1991 established a Special Committee of the Security Council linked to the arms embargo, the terms of reference of the Committee permit it only an information-collection and distribution function. As the UN has no independent means of collecting information, it depends on submissions by government and non-government sources. The Committee may consider this information and bring to the attention of member states allegations that the arms embargo has been violated. However, if members reject these allegations, the Committee has no power to question them further or to impose sanctions on the country concerned.

In February 1992 Security Council Resolution 740 expressed concern that the embargo was not being observed.[12] These concerns were repeated in Resolution 787 of 16 November 1992.[13] However, no evidence that the arms embargo was being breached has been published. Media allegations about embargo violations have named private arms dealers operating from Austria, Bulgaria, Czechoslovakia, Germany, Greece, Iran, Lebanon and Ukraine as sources of supply to one or more of Bosnia and Herzegovina, Croatia and Serbia. None of these allegations has been substantiated. The first clear and documented case of a violation of the arms embargo occurred in January 1993, when the *Dolphin One*, a ship carrying the flag of St Vincent and the Grenadines, was stopped in the Adriatic Sea. The ship was carrying artillery rockets and small-calibre ammunition manufactured in the former Soviet Union and China.[14] Even if allegations were proved to be correct, the volume of such 'black market' sales would represent a small fraction of the volume of arms and military equipment available to warring parties from the stockpiles already in Yugoslavia either from local production or from foreign suppliers prior to the imposition of the United Nations embargo.

At the end of May 1992 the arms embargo was supplemented by a mandatory trade embargo on Serbia and Montenegro. The embargo applies to all goods and transactions except those concerning medical supplies and foodstuffs. While the arms embargo remains in force on all of the territories of the former Yugoslavia—including Bosnia and Herzegovina, Croatia, the Former Yugoslav Republic of Macedonia and Slovenia—the trade embargo applies

[10] International efforts are discussed more fully in chapter 2 of this volume.

[11] *International Herald Tribune*, 23–24 Jan. 1993, p. 1.

[12] UN Security Council Resolution 740, UN document S/RES/740 (1992) (United Nations: New York, 7 Feb. 1992).

[13] UN Security Council Resolution 787, UN document S/RES/787 (1992) (United Nations: New York, 16 Nov. 1992).

[14] *Frankfurter Allgemeine Zeitung*, 21 Jan. 1993, p. 2.

only to the Federal Republic of Yugoslavia—that is, Serbia and Montenegro.[15] The same Committee of the Security Council established under Resolution 724 also has responsibility for collecting and distributing information relating to the implementation of the general trade embargo but still has no independent means of doing this.

The first multilateral mechanism for sanctions implementation was created through decisions taken at the Conference on the Former Yugoslavia held in London on 26–27 August 1992. Among the decisions taken by the Conference was an agreement to 'implement an agreed action plan to ensure the rigorous application of sanctions'.[16] Under this agreement the UN Security Council was asked to take steps to enforce sanctions in areas away from the immediate geographical vicinity of the former Yugoslavia while the EC and the CSCE were asked to co-ordinate all necessary practical assistance to the countries bordering Serbia and Montenegro.[17] This was to include experts drawn from CSCE member countries to participate in monitoring missions established in neighbouring countries.

The UK, which at the time chaired the EC, co-sponsor of the Conference with the United Nations, drew up a specific plan to implement Conference decisions. On the invitation of the governments of Bulgaria, Hungary and Romania, Sanctions Assistance Missions were established in these three countries to implement the EC action plan.[18] The missions became operational in October 1992. While there is a CSCE Liaison Group established in Geneva the Sanctions Assistance Missions report directly to an office in Brussels where the Commission of the European Communities has established an office (SAMCOMM) within the Directorate General for Customs and Indirect Taxation to manage the work of the missions.[19] A review of early reports by SAMCOMM in December underlined several points. First, the review recommended establishing new missions in Albania, Croatia and Ukraine. This would be dependent on invitations being extended by those countries, which had not occurred at the time of writing. Second, the group recommended that those countries situated on the banks of the River Danube—Bulgaria, Romania, Hungary and Austria—should stop and inspect foreign vessels.[20]

[15] UN Security Council Resolution 757, UN document S/RES/757 (1992) (United Nations: New York, 30 May 1992).

[16] 'Texts of statements approved 26–27 Aug. 1992, at the London Conference on Yugoslavia, London, United Kingdom', *US Department of State Dispatch*, vol. 3, supplement no. 7 (Sep. 1992), pp. 3–6.

[17] The involvement of the EC was a practical matter. The European Community Monitoring Mission (ECMM) had been active in the former Yugoslavia since July 1991, and had already established a logistics system and an understanding of local conditions.

[18] 'Decision on sanctions monitoring', Report on the Sixteenth Meeting of the Committee of Senior Officials (CSO) of the CSCE, Prague, 18 Sep. 1992.

[19] *Second SAMCOMM Situation Report* (Commission of the European Communities: Brussels, 30 Oct. 1992).

[20] *Chairman's Report of the Sanctions Assistance Missions Liaison Group Meeting in Vienna 4 Dec. 1992*, CSCE Communication No. 410, Prague, 11 Dec. 1992. Sanctions Assistance Mission reports indicate that many vessels on the Danube, when challenged, refused to stop. The London Conference expressed the view that riparian states had the right to enforce sanctions by halting ships in international waters to inspect their cargoes. This right was incorporated in UN Security Council Resolution 787 of 16 Nov. 1992 (note 13).

From late 1992 the Romanian Government began taking steps to stop ships carrying goods to and from the former Yugoslavia and detaining ships deemed to be in breach of United Nations sanctions and noted that several were flying the Ukrainian flag.[21] Earlier the Ukrainian Government had issued an official statement to the effect that the embargo was being strictly enforced.[22]

Multilateral enforcement of the economic sanctions has improved the chance that the arms embargo will be enforced on Serbia and Montenegro. Trade sanction enforcement measures do not apply to Bosnia and Herzegovina, Croatia or Slovenia, although the mandatory arms embargo imposed under Security Council Resolution 713 remains in force. At the third meeting of the CSCE Council of Foreign Ministers in Stockholm on 14–15 December 1992 Bosnia recommended that the Security Council lift the arms embargo on Bosnia and Herzegovina. While the Bosnian Foreign Minister initially made this a condition for agreeing to the final communiqué at the meeting, the communiqué did not ask the UN to lift the arms embargo.[23]

## Fighter aircraft decisions in 1992

Decisions taken in 1992 relating to fighter aircraft programmes illustrate many of the factors currently reshaping arms production and arms trade. These include heightened economic incentives to sell arms stemming from the decline in domestic demand; reviews of strategy; the influence of international co-operation at the government-to-government level and at the company level; the success or failure of products in foreign markets; and the multilateral discussion of arms transfer control for the Middle East.

In the USA decisions about mergers and acquisitions taken in 1992 by companies which are consistently among the 10 largest arms-producing companies in the world—General Dynamics, Lockheed and Northrop—will lead to a dramatic restructuring of the US aircraft industry.

The announcement by President Bush that the US Government supported the sales of 72 F-15XP and 150 F-16 fighter aircraft to Saudi Arabia and Taiwan, respectively, briefly placed the issue of arms exports at the centre of the Presidential election campaign. However, the manner of the announcement concentrated attention on the employment and balance-of-payments aspects of arms exports. McDonnell Douglas (manufacturer of the F-15) and General Dynamics (manufacturer of the F-16) both stressed that these sales were important to the future viability of the company.

The commercial imperative to export is even more significant for producers in countries where domestic markets are smaller than that of the USA. In 1992 this was especially true for the Russian aircraft industry in the face of reduced

[21] 'Aide memoire on the actions and concrete steps taken by the Government of Romania for observing the embargo against Yugoslavia; the needs for assistance to Romania in this respect', mimeo, Delegation of Romania to the CSO Meeting, Prague, 2–4 Feb. 1993.

[22] 'Statement by the Ministry for Foreign Affairs of Ukraine', mimeo, 23 Jan. 1993.

[23] *Atlantic News*, 18 Dec. 1992, p. 3.

orders by the Russian Ministry of Defence.[24] In France the next-generation Rafale fighter aircraft is expected to be built from 1996.[25] The French Air Force reduced its projected procurement of the Mirage-2000DA from 225 aircraft to 192 and then to 168 before the programme was terminated in 1992 at 153 aircraft.[26] Dassault Aviation will have to rely on exports and upgrading current French Air Force aircraft until 1996 and it was in these circumstances that the French Government decided to permit the export of 60 Mirage-2000-5 fighters to Taiwan in 1992. In the UK the decision not to buy further Tornado aircraft for the Royal Air Force meant that the British Aerospace assembly line for the Tornado was due to close in January 1993.[27] Under these circumstances Prime Minister John Major travelled to Riyadh to confirm that Saudi Arabia would move ahead with contracts agreed in principle under a 1988 memorandum of understanding.[28] In Sweden aircraft manufacturer Saab, which rejected the need for sales of the JAS-39 Gripen beyond Europe for much of 1992, has investigated the global market and identified Argentina, Brazil, Chile, Ecuador, Malaysia, Singapore and Thailand as potential customers.[29]

Also in Europe the question of the future of the European Fighter Aircraft (EFA) illustrated the changed strategic and budget environment after the cold war. Critics of EFA maintained that the rationale for deploying the aircraft—designed when two alliances faced one another across the inter-German border—was weak once Germany faced no direct military threat from a neighbour.[30] As the government cut military spending in the face of its difficulties in containing public expenditure, EFA was projected to eat up more and more of the shrinking German defence budget.

The sale of US F-16 fighter aircraft to Taiwan provided the immediate cause for China to suspend its participation in multilateral discussions of arms transfers to the Middle East. China argued that the United States had violated the guidelines agreed in the meeting of the five permanent members of the UN Security Council in London in October 1991.

These and other aspects of current fighter aircraft programmes are discussed further below.

[24] Kislov, A., 'Conversion: Russian experience and perspectives', *Peace and the Sciences*, Sep. 1992, pp. 52–53.

[25] *Atlantic News*, 13 Jan. 1993, p. 4; *Air & Cosmos*, 11–17 Jan. 1993, pp. 24–29; *Interavia Aerospace World*, Feb. 1993, p. 46.

[26] *Avis présenté au nom de la commission de la défense nationale et des forces armées sur le projet de loi de finances pour 1993 par M. J. Briane, Tome X, Défense: Air*, Assemblée nationale, no. 2948 (14 Oct. 1992), p. 19.

[27] *Interavia Aerospace World*, Dec. 1992, p. 9. An assembly line for the Tornado would have remained in Italy. The German line closed in Feb. 1992; *Interavia Air Letter*, 3 Feb. 1992, p. 5.

[28] *Defense News*, 1–7 Feb. 1993, p. 3; *World Weapons Review*, 10 Feb. 1993, p. 14.

[29] *Dagens Nyheter* (Stockholm), 19 Mar. 1993, p. A5. Saab had hoped to secure sales to Austria, Denmark, Finland, Germany and Hungary, but none of these discussions had led to contracts by the end of 1992.

[30] Albrecht, U., 'The European Fighter Aircraft', *Bulletin of Arms Control*, no. 8 (Nov. 1992), pp. 2–6.

### *The F-22*

Tactical aircraft have consistently accounted for a high percentage of both annual arms procurement expenditure and the international arms trade. Seven of the 10 largest arms-producing companies in the world are aircraft manufacturers and in the USA procurement and research and development (R&D) for new tactical aircraft for the Air Force, Army, Navy and Marines were worth $6 billion in 1992.[31]

In the United States the Fiscal Year (FY) 1993 Defense Authorization Bill restricts spending on the development of the A/FX and the F/A-18E/F fighter aircraft to 65 per cent of the authorized amount until the Department of Defense (DOD) submits a 'comprehensive affordability assessment' of its long-range modernization plans for tactical aircraft. The Congressional Budget Office anticipates a funding gap since DOD plans would cost an average of $9.6 billion per year over the next 20 years on the basis of current assumptions about programme costs. Annual funding for the programmes under currently anticipated budget plans would amount to $8.3 billion.[32]

In these circumstances General Colin Powell, Chairman of the Joint Chiefs of Staff, underlined that no future procurement decisions should be taken for granted. He stated:

> If you are not expecting a war to begin in two weeks, and if you're not being chased by someone else's technology, or you're not chasing theirs, you can take an entirely new look at the research-and-development and acquisition and procurement system.
>
> If our F-15s, for example, are at a certain level of capability and the Russians have capped out below us because their industrial base has gone to pot, then that F-15 fleet of ours is good for a long time to come. We only go up to the next family, the F-22, if the threat to justify it really exists at the time it is ready or if the F-15s are falling out of the sky.[33]

US Air Force thinking on tactical aircraft force planning was presented by General Merrill McPeak, Air Force Chief of Staff, in testimony before the House Armed Services Committee. McPeak suggested funding several parallel prototype/technology demonstrator programmes but postponing a procurement decision until the year 2000. Meanwhile, the Air Force would replace retiring aircraft with upgraded versions of existing fighters.[34] This position was subsequently reaffirmed by Air Force Secretary Donald Rice in a paper for the Undersecretary of Defense for Acquisition, Donald Yockey.[35]

[31] According to the US Congress House Armed Services Committee, procurement of aircraft of all kinds in 1993 will be worth $17.5 billion of a total procurement budget of $53 billion. The pattern of US arms procurement is described in chapter 9 of this volume.

[32] US Congressional Budget Office (CBO), *Balance and Affordability of the Fighter and Attack Aircraft Fleets of the Department of Defense*, CBO Papers (US Government Printing Office: Washington, DC, Apr. 1992), p. 28.

[33] *Proceedings of the US Naval Institute*, vol. 118, no. 1073 (July 1992), p. 15.

[34] *National Defense Authorization Act for Fiscal Year 1993—Conference Report to Accompany H. R. 5006*, US House of Representatives, 102nd Congress, 2nd Session (US Government Printing Office: Washington, DC, 1 Oct. 1992), Title IX, Sec. 902, pp. 161–63.

[35] Morrocco, J. D., 'USAF offers options to stretch F-22 program', *Aviation Week & Space Technology*, 23 Nov. 1992, p. 30.

The programme for the F-22, selected in July 1991 as the next-generation US Air Force air superiority fighter, is that most directly affected by the review of tactical aircraft. In July 1991 it was planned to begin production in 1997.[36] However, it now appears that the decision on whether to begin production will be postponed until the year 2000.

Tactical aircraft procurement decisions will have a significant influence over the size and structure of the US arms industry. In February 1992 US Defense Secretary Les Aspin, at that time Chairman of the Armed Services Committee of the House of Representatives, tabled a five-year defence plan under which the number of aircraft types being produced in the United States should shrink from 25 in 1992 to 6 by 1998.[37] In anticipation of changes of this magnitude the restructuring of the US arms industry accelerated in 1992.

The F-22 is being developed by a consortium of Boeing, General Dynamics and Lockheed under a $9.5 billion development contract in which Lockheed is the prime contractor.[38] In 1992 General Dynamics agreed to merge its military aircraft operations with Lockheed. The primary motive was to control costs in the F-22 programme and to ensure that the development phase of the programme is profitable for the companies regardless of the production decision. The strategy of rationalization through consolidation was devised by General Dynamics Chief Executive Officer William Anders who stated that it is 'critical that we address the massive and growing over-capacity which plagues most sectors of the Defense Industrial Base—especially at the prime supplier level. . . . [T]his overcapacity translates directly into expensive, unproductive overhead and production inefficiencies.'[39] The company created will be smaller than the combined size of Lockheed and General Dynamics but will have sufficient business to survive until a production contract for the F-22 is awarded.

### *The European Fighter Aircraft*

One element of the debate over the F-22 has been the discussion of the size and structure of the US Air Force after the year 2000. In Europe strategic reviews have also been conducted on a national basis even though procurement choices increasingly involve collaborative equipment programmes.

[36] The 1997 date was announced in Apr. 1990 and was itself a postponement of the original date for the planned start of production, which was 1995. For a review of the early stages of the programme, see US General Accounting Office (GAO), *Reasons for Recent Cost Growth in the Advanced Tactical Fighter Aircraft*, report no. GAO/NSIAD-91-138 (US Government Printing Office: Washington, DC, 1 Feb. 1991).

[37] Aspin, L., *Tomorrow's Defense From Today's Industrial Base: Finding the Right Resource Strategy for a New Era*, mimeo, House Armed Services Committee, 12 Feb. 1992, p. 4.

[38] The full R&D costs of the F-22 are expected to be around $11 billion, of which $7.7 billion had been spent by the end of 1992. In 1992 the total programme was expected to cost $72 billion, although total programme costs (albeit highly speculative) of up to $100 billion for as few as 100 aircraft have been mentioned. The true cost cannot be known until final decisions have been made about the number of aircraft to be produced. Dudney, R. S., 'The F-22 enters the fray', *Air Force*, vol. 74, no. 7 (July 1991), pp. 32–38.

[39] Anders, W. A., *Revisiting the Rationalization of America's Defense Industrial Base*, speech to the Aerospace Industries Association Human Resources Council, 27 Oct. 1992.

The Maastricht Treaty, signed on 7 February 1992, decided that a common foreign and security policy for the EC should include the eventual framing of a common defence policy which might in time lead to a common defence. In the Treaty the Western European Union (WEU) was asked to implement EC decisions which have defence implications. The WEU is examining the feasibility of a European armaments agency but this will inevitably be a long process.[40] Meanwhile, as the Assistant Secretary (Air) from the British Ministry of Defence pointed out to the House of Commons Defence Committee, a programme such as EFA 'means four nations taking decisions at the same time and co-ordinating all their activities'.[41] In the budget circumstances prevailing in Germany, balancing domestic needs and international commitments is difficult.

At the end of 1992 the following decisions had been taken:

1. None of the four countries participating in the programme would withdraw from the production phase of the project. On 10 December 1992 the Defence Ministers of Germany, Italy, Spain and the UK, all in Brussels to attend a meeting of NATO Defence Ministers, announced that the production phase of the programme would be deferred but the project would go ahead under the name Eurofighter 2000.[42]

2. The versions of EFA produced for each country will differ significantly but a common airframe and two-engine configuration will be kept. There will not be a new design—such as the 'EFA-Light' proposed by Germany earlier in 1992. Cost reductions will be achieved by omitting certain sub-systems from aircraft bought by Germany.

3. Planned procurement of the aircraft for producer countries has fallen from 765 to 667 aircraft. Germany, Italy and Spain have reduced their orders and the UK may do the same. In Germany reductions below the revised figures are possible. Late in 1991 it was suggested that 175 aircraft may be a more realistic total for German procurement than the then official figure of 200. In July 1992, after meeting with British Defence Minister Malcolm Rifkind in London, German Defence Minister Volker Rühe stated that his future budget plan included funding for only 140 aircraft.[43]

[40] Western European Union, Assembly, *European Armaments Co-operation after Maastricht*, Report submitted on behalf of the Technological and Aerospace Committee, Assembly of the WEU, document no. 1332 (WEU: Paris, 23 Oct. 1992).

[41] Testimony of Derek Dreher reproduced in *European Fighter Aircraft*, Sixth Report of the House of Commons Defence Committee, Session 1991–92, HC Paper 1991–92 299 (Her Majesty's Stationery Office: London, 11 Mar. 1992).

[42] 'Germans back new fighter aircraft', *The Independent*, 11 Dec. 1992, p. 11.

[43] *Interavia Aerospace Review*, Nov. 1991, p. 9; 'Germans call for Light EFA, partners say plan will not fly', *Aviation Week & Space Technology*, 6 July 1992, pp. 20–22; *The Guardian*, 7 July 1992, p. 1. The House of Commons Defence Committee concluded that since it remained 'open to question' how far the Royal Air Force still requires an aircraft with the full level of performance offered by EFA, 'It may also be possible for the MoD to reduce the cost of the programme by purchasing fewer than the currently planned 250 EFAs'; *European Fighter Aircraft* (note 41), para. 98(e). It has been suggested that Italy may reduce its procurement to 'around 100'; *Jane's Defence Weekly*, 21 Mar. 1992, p. 478.

**Table 10.1.** EFA-related decisions taken in 1992

| Date | Location | Event |
|---|---|---|
| Mar. | Bonn | German Defence Minister Gerhard Stoltenberg says EFA is the most cost-effective option for the FRG. The Luftwaffe endorses EFA as agreed in the 1987 European Staff Requirement. |
| 31 Mar. | Bonn | Volker Rühe becomes new German Defence Minister. |
| 30 June | Bonn | The German Government announces its intention to withdraw from the production phase of EFA. |
| 6 July | London | Rühe tells British Defence Minister Malcolm Rifkind of his desire to redirect unspent EFA money to develop a cheaper aircraft. |
| 4 Aug. | Madrid | Rühe declares that the EFA is 'dead'. |
| 9 Sep. | Farnborough | Rifkind announces that the UK will build EFA alone if necessary. |
| 16 Sep. | Bonn | The German Government officially informs its EFA partners of the decision to withdraw from the production phase of EFA. |
| 10 Oct. | Genoa | Italian Defence Minister Salvo Andò suspends EFA work in Italy. |
| | Madrid | The Spanish Ministry of Defence announces a 'slowdown in the investment programme' for EFA. |
| 20 Oct. | London | Rühe says of EFA that 'the aircraft will not be built'. |
| 11 Nov. | Ditchley Park | British Prime Minister John Major and German Chancellor Helmut Kohl say that both are committed to produce EFA. |
| 20 Nov. | Gleneagles | Rühe re-states Germany's decision to withdraw from the EFA programme. |
| 10 Dec. | Brussels | A joint statement by the Defence Ministers of Germany, Italy, Spain and the UK states that EFA will go ahead. |

*Sources*: *Frankfurter Allgemeine Zeitung*, 12 Dec. 1992, p. 2; *Süddeutsche Zeitung*, 12–13 Dec. 1992, p. 4; *Atlantic News*, 16 Dec. 1992, pp. 1, 3.

Within the EFA programme work is shared between participating nations according to procurement.[44] How to make this allocation was the subject of much debate. Spain argued that companies from each country should be represented in each aspect of production. In fact, contracts are awarded on a competitive basis with the understanding that the work-share should be balanced over the total programme. If any country is 'too successful' in the early stages of the programme it is likely to find its companies excluded from bidding on later contracts. By the end of 1992 British companies won almost 40 per cent of contracts awarded (by value), similar to the percentage which would be awarded to British companies after recent EFA decisions assuming that current plans for Britain to buy 250 aircraft are implemented.

While not a new phenomenon, the discussion of EFA in 1992 underlined the growing difficulty governments face in aligning national procurement choices to permit projects carried out by multinational consortia of 'national champion' companies.

[44] The details of collaborative arrangements were the subject of UK House of Commons, Public Accounts Committee, *The European Fighter Aircraft*, Fourteenth Report of the Public Accounts Committee of the House of Commons, Session 1990–91, HC Paper 1990–91 380 (Her Majesty's Stationery Office: London, 22 Apr. 1991).

**Table 10.2.** Revised work-share for the EFA based on announced plans

| Country | Original planned procurement (no.) | Original work-share (%) | 1992 planned procurement (no.) | Work-share 1992 (%) |
|---|---|---|---|---|
| Germany | 250 | *33* | 200 | *30.0* |
| UK | 250 | *33* | 250 | *37.5* |
| Italy | 165 | *21* | 130 | *19.5* |
| Spain | 100 | *13* | 87 | *13.0* |
| **Total** | **765** | ***100*** | **667** | ***100*** |

*Sources*: Adapted from UK House of Commons, *European Fighter Aircraft*, Sixth Report of the House of Commons Defence Committee Session 1991–2 (Her Majesty's Stationary Office: London, 11 Mar. 1992); 'Germans claim EFA cost will soar', *Defense News*, 30 Mar.–5 Apr. 1992, pp. 2, 21; *Military Technology*, vol. 16, no. 3 (Mar. 1992), p. 88; 'Britain launches a final EFA sortie to keep Bonn on board', *Financial Times*, 11 June 1992.

# III. Arms production in the OECD and developing countries

The global arms industry is currently characterized by major structural adjustment. Production cuts and plant closures, rationalization, mergers and acquisitions, diversification and privatization are major themes in this process of industrial change. As noted above, the arms industries located in Central and Eastern Europe (including Russia and Ukraine) are experiencing the most serious crisis.

Within the OECD and developing countries, arms sales and arms industry employment are falling. The process of national restructuring in Western Europe continued in 1992. The most fundamental changes in arms industry ownership in 1992 took place in the USA, where this development is expected to continue at least during 1993.

The process of internationalization in Europe documented in previous *SIPRI Yearbooks* continued in 1992. The competitive advantage of companies in the USA that are even larger than those which exist today may increase the pace of developments in Western Europe, where national markets are relatively small.

## The 'SIPRI 100': developments in the SIPRI list of companies

### *Arms sales*

Arms sales data for the 100 largest arms-producing companies in the OECD and developing countries reflect in 1991 for the first time the expected decline of the arms industry (see appendix 10A). The value of the combined arms sales of the 100 companies with the highest arms sales in 1991 fell from \$183.7 billion in 1990 to \$178.8 billion in 1991, representing a fall of 3 per cent in current prices.[45]

[45] Financial data on companies in the SIPRI arms industry data base are given in current prices, as reported by companies in their annual reports and by governments, industrial associations and other

**Table 10.3.** Regional/national shares of arms sales for the 100 largest arms-producing companies, 1991 compared to 1990

| Number of companies 1991 | Region/ country | Share of total arms sales (%) 1990 | 1991 | Arms sales 1991 ($b.) |
|---|---|---|---|---|
| 47 | USA | *60.2* | *60.9* | 108.9 |
| 40 | West European OECD | *33.0* | *32.8* | 58.6 |
| 13 | UK | *10.4* | *10.3* | 18.4 |
| 11 | France | *12.0* | *11.9* | 21.4 |
| 8 | FRG | *5.0* | *4.9* | 8.7 |
| 3 | Italy | *3.4* | *3.1* | 5.6 |
| 2 | Switzerland | *1.1* | *1.0* | 1.9 |
| 2 | Sweden | *0.3* | *0.8* | 1.4 |
| 1 | Spain | *0.9* | *0.8* | 1.3 |
| 7 | Other OECD | *3.8* | *3.8* | 6.7 |
| 5 | Japan | *3.2* | *3.1* | 5.5 |
| 2 | Canada | *0.6* | *0.7* | 1.2 |
| 6 | Developing countries | *3.0* | *2.5* | 4.6 |
| 3 | Israel | *1.2* | *1.3* | 2.4 |
| 2 | India | *1.1* | *0.8* | 1.5 |
| 1 | South Africa | *0.7* | *0.4* | 0.7 |
| 100 | **Total** | ***100.0*** | ***100.0*** | **178.8** |

*Source:* Appendix 10A.

The decline in the volume of arms sales of the arms industry as a whole in these countries is sharper than these statistics suggest. If arms industry output experienced the same rate of inflation as the average inflation in the OECD area, this would suggest a fall of 7 per cent in real terms in the arms sales of the 100 companies on the SIPRI list.[46] In addition, corporate concentration tends to exert an upward pressure on the trend. When companies which have expanded enter the list and other companies leave, the net effect is that the 100 companies on the list represent a higher share of the total market. This effect is particularly strong in periods with a high level of merger and acquisition activities.

A total of 13 countries are represented on the list in 1991 (table 10.3). US companies represent the largest group, with 47 firms taking a 61 per cent share of the combined arms sales of all 100. Forty West European companies in seven countries account for 33 per cent of overall arms sales. French and British firms dominate, with about one-third each of West European arms sales. The remaining 13 companies on the list are located in two other OECD countries (Japan and Canada) and three developing countries (India, Israel and South Africa). The national distribution of combined arms sales was remarkably stable during 1990 and 1991.

sources. For the applied methodology and sources, see appendix 10D. These data cannot be compared with the SIPRI data on arms trade, since the latter do not reflect actual financial flows.

[46] The average rate of inflation in the OECD area as measured by the GDP deflator was 3.9 per cent in 1991. *OECD Economic Outlook,* Dec. 1991, p. 43.

**Table 10.4.** Companies whose arms sales changed the most in 1991

Figures are in US$ m.

| Company | Country | Arms sales 1991 | Change 1990–91 |
|---|---|---|---|
| *Companies with decreased arms sales* | | | |
| General Dynamics | USA | 7 620 | – 680 |
| Armscor | South Africa | 710 | – 620 |
| Lockheed | USA | 6 900 | – 600 |
| Thomson S.A. | France | 4 800 | – 450 |
| EFIM | Italy | 1 270 | – 440 |
| ITT | USA | 1 200 | – 410 |
| Mitsubishi Heavy Industries | Japan | 2 630 | – 410 |
| Raytheon | USA | 5 100 | – 400 |
| *Companies with increased arms sales* | | | |
| Celsius | Sweden | 870 | + 690 |
| Loral | USA | 2 600 | + 680 |
| Aérospatiale | France | 3 450 | + 590 |

*Source:* Appendix 10A.

Companies whose arms sales experienced the biggest decreases or increases from 1990 to 1991 are listed in table 10.4. There were more cases of decreased than increased sales in 1991, reversing the trend from 1990.[47] Of the eight companies whose sales decreased by more than $400 million in 1991, four were involved in major restructuring in 1992: General Dynamics, Armscor, Lockheed and EFIM. Two of the three companies that increased their arms sales in 1991 did so through acquisitions: Celsius and Loral. The increase in arms sales by Aérospatiale is probably related to the creation of Eurocopter in January 1991, when Aérospatiale merged its helicopter division with that of the German company MBB.

Three companies on the SIPRI 100 list for 1990 have left the arms industry entirely. In the United States the Ford Motor Company sold its defence division, Ford Aerospace, to Loral in October 1990. Emerson Electric spun off its six defence divisions into Esco Electronics, a new independent company, in December 1990. In Sweden, the last arms-producing division of FFV became a direct subsidiary of Celsius Industries in June 1991, while Nobel Industries merged its ordnance, ammunition and missile activities with those of FFV into a new company, Swedish Ordnance.

### *Employment*

Arms industry employment world-wide fluctuated at around 15–16 million employees in the 1980s, with a downward trend in the second half of the 1980s. The seven largest arms-production centres—China, France, Germany, Russia, Ukraine, the UK and the USA—account for nearly 90 per cent of all employment in arms manufacturing. Around 40 additional countries have arms-producing facilities which account for the remaining employment in the

[47] Miggiano, P., Sköns, E. and Wulf, H., 'Arms production', *SIPRI Yearbook 1992* (note 6), table 9.2, p. 364.

arms industry (around 1.5 million jobs). Global arms industry employment dropped by about 6 per cent between the mid-1980s and the early 1990s. This decline was mainly concentrated in the USA and the EC.[48]

Table 10.5 shows the downward trend in arms industry employment during 1988–91. The table lists the 36 arms-producing companies on the SIPRI 100 list whose arms sales accounted for more than half of their total sales in 1988, and for which 1991 data are available.[49] In 28 of the 36 listed companies, employment was reduced between 1988 and 1991. For US manufacturers, 10 out of 11 companies listed reported employment cuts, reflecting the severity of the budget cuts that had been implemented in the USA by 1991. In companies predominantly active in the aerospace sector, 17 out of 21 companies reported employment reductions, indicative of the particular difficulties of the aircraft and missile industries facing a shrinking market.

Increased employment was registered for 8 of the 36 companies listed. However, for all but a few this is the result of company mergers and acquisitions, rather than of the creation of new jobs in the arms industry. Instead of generating jobs, restructuring allows for significant reductions of staff over the long term as divisions are consolidated and redundant units and capacities eliminated. This applies in particular to major take-overs and mergers, as in the cases of GIAT, Loral and Agusta.

Most of the employment cuts registered in the table occurred in 1990–91, indicative of the inherent time lag between procurement cuts and reductions in production. The full impact of the increasingly empty order books on employment will be felt as the buffer of backlog orders is whittled away. The implementation of the employment cuts announced in 1991 and 1992 will give a more accurate measure of the extent of the crisis. If plans announced by industry in 1991–92 are implemented, the rate of decline in arms industry employment will accelerate in the 1990s.[50]

In the USA the FY 1993 Defence Budget called for the termination of or a reduction in orders for a number of employment-sensitive weapon programmes involving most prime contractors and a host of dependent subcontractors.[51] Available estimates of future reductions in US arms industry

[48] Wulf, H., 'Arms industry limited: the turning point in the 1990s', ed. H. Wulf (note 3), pp. 13–18.

[49] Among the 125 companies and company subsidiaries that ranked highest in 1988, there were 43 whose share of arms sales was more than half of their total sales. Of these, 7 companies have been liquidated, bought or restructured, so that comparisons are not possible.

[50] See Miggiano, P., Sköns, E. and Wulf, H., 'Arms production', *SIPRI Yearbook 1992* (note 6), table 9.3, pp. 366–67, for a listing of employment cuts in arms production announced in 1991.

[51] US Secretary of Defence Dick Cheney, *Report of the Secretary of Defense to the President and Congress* (US Government Printing Office: Washington, DC, Feb. 1992), p. 21; Boatman, J., Lopez, R. and Starr, B., 'Cuts deeper than feared leave firms reeling', *Jane's Defence Weekly*, 15 Feb. 1992, pp. 222–23.

**Table 10.5.** Employment changes among the largest arms-producing companies in OECD and developing countries, 1988–91

Figures in italics are percentages.

| Name | Country | Product sector[a] | Arms sales as % of total | Employment 1988 | Employment 1991 | Change |
|---|---|---|---|---|---|---|
| Armscor | South Africa | L | *90* | 26 000 | 16 000 | *– 38* |
| Israel Military Industries | Israel | L | *98* | 12 150 | 8 500 | *– 30* |
| Grumman | USA | A | *82* | 32 000 | 23 600 | *– 26* |
| General Dynamics | USA | A | *84* | 102 800 | 80 600 | *– 22* |
| Hollandse Signaalapparaten | Netherlands | E | *90* | 5 300 | 4 265 | *– 20* |
| Lockheed | USA | A | *79* | 86 800 | 71 300 | *– 18* |
| Dassault Electronique | France | E | *76* | 4 100 | 3 416 | *– 17* |
| EN Bazan | Spain | S | *81* | 10 908 | 9 149 | *– 16* |
| VSEL Consortium | UK | S | *100* | 15 520 | 13 028 | *– 16* |
| Dassault | France | A | *70* | 13 818 | 11 914 | *– 14* |
| Northrop | USA | A | *90* | 42 000 | 36 200 | *– 14* |
| CASA | Spain | A | *72* | 10 372 | 9 338 | *– 10* |
| Martin Marietta | USA | A | *75* | 67 500 | 60 500 | *– 10* |
| McDonnell Douglas | USA | A | *60* | 121 000 | 109 123 | *– 10* |
| Thiokol | USA | A | *54* | 12 600 | 11 500 | *– 9* |
| Eidgenössische Rüstungsbetriebe | Switzerland | L | *92* | 4 900 | 4 495 | *– 8* |
| Hindustan Aeronautics | India | A | *97* | 43 663 | 40 336 | *– 8* |
| Oto Melara | Italy | L | *98* | 2 329 | 2 149 | *– 8* |
| Hughes Electronics | USA | A | *61* | 100 000 | 93 000 | *– 7* |
| British Aerospace | UK | A | *55* | 131 300 | 123 200 | *– 6* |
| Litton Industries | USA | E | *60* | 55 000 | 52 300 | *– 5* |
| Raytheon | USA | A | *67* | 75 000 | 71 600 | *– 5* |
| Dornier | FRG | A | *52* | 9 800 | 9 527 | *– 3* |
| Ordnance Factories | India | L | *99* | 177 863 | 173 000 | *– 3* |
| Krauss-Maffei | FRG | L | *52* | 5 100 | 5 004 | *– 2* |
| FIAT Aviazione | Italy | A | *82* | 4 749 | 4 719 | *– 1* |
| MTU | FRG | A | *52* | 17 200 | 17 052 | *– 1* |
| Westland Group | UK | A | *70* | 9 163 | 9 060 | *– 1* |
| Israel Aircraft Industries | Israel | A | *75* | 16 500 | 17 100 | *+ 4* |
| DCN | France | S | *100* | 28 000 | 30 000 | *+ 7* |
| Hunting | UK | L | *62* | 6 834 | 7 302 | *+ 7* |
| Saab Aircraft | Sweden | A | *57* | 6 490 | 6 909 | *+ 7* |
| Thomson-CSF | France | A | *77* | 41 400 | 44 500 | *+ 8* |
| GIAT Industries | France | L | *100* | 14 740 | 17 000 | *+ 15* |
| Loral | USA | E | *88* | 14 000 | 22 000 | *+ 57* |
| Agusta | Italy | A | *72* | 4 285 | 8 426 | *+ 97* |

[a] A = Aerospace; E = Electronics; L = Land systems/infantry weapons; S = Shipbuilding.
*Source*: SIPRI arms industry data base.

employment suggest cuts by about one-third over a four- to six-year period from 1990 or 1991.[52]

[52] The US Office of Technology Assessment (OTA) predicts a reduction by 530 000–920 000 positions by 1995 from a fiscal year 1990 level of 2.9 million as a direct consequence of announced and anticipated cutbacks in defence spending; OTA, *After the Cold War: Living with Lower Defense Spending* (US Government Printing Office: Washington, DC, Feb. 1992), p. 3 and table A-2, p. 230. The US General Accounting Office (GAO) forecasts cuts in employment levels from 3.1 million in 1991 to

SIPRI estimates place European arms industry employment in 1990–92 at around 1.2 million.[53] According to François le Ministrel, President of the French Defence Industries Council (CIDEF), Western Europe stands to lose 500 000 jobs by 1995 as a result of reductions in national procurement and exports, unless governments and producers succeed in greater collaboration in manufacturing and procurement.[54] According to a 1992 study, the EC arms industry, employing some 1.04 million in 1991, may lose 614 000–740 000 jobs by 1998.[55]

In France, government efforts to reduce defence spending combined with reduced arms exports, in particular in the dominant aerospace sector, point towards drastic reductions in future employment levels. Arms industry employment in France totalled 283 000 in 1991.[56] According to forecasts issued by the trade associations for the French aeronautics and space, electronics, and land systems industries, 60 000–105 000 of these jobs will be lost by 1994.[57]

## National restructuring

The long-term process of national concentration of arms industries through take-overs and mergers continues. In 1992 fundamental industrial reorganization of ownership took place in Italy, Sweden and the United States. Among the 20 leading arms-producing companies in 1991, seven were involved in major acquisition deals in 1992: General Dynamics, General Motors (through its subsidiary Hughes Aircraft, part of Hughes Electronics), Lockheed, General Electric, Northrop, Martin Marietta and IRI. In France some potentially important national mergers and acquisitions were under discussion.[58]

3 million in 1992 and to 2.7 million in 1993, on the basis of approved and requested defence authorizations; GAO, *Defence Procurement: Trends for 1985–93 in DOD's Spending, Employment and Contractors*, Report no. GAO/NSIAD-92-274BR (US Government Printing Office: Washington, DC, 1992), p. 7. According to a study by the Defense Budget Project, an independent research organization, the US arms industry stands to lose between 0.9 and 1.2 million jobs between 1991 and 1997, roughly one-third of these in 1993 alone; Schmidt, C. P. and Kosiak, S., Defense Budget Project (DBP), *Potential Impact of Defense Spending Reductions on the Defense Industrial Labor Force by State* (DBP: Washington, DC, Mar. 1992), p. 5.

[53] Wulf (note 48), table 1.1, pp. 14–15.

[54] Cited in Reed, C., 'Case for Fortress Europe', *Jane's Defence Weekly*, 26 Sep. 1992, p. 27; a forecast made by SIPRI in 1990 was that arms industry employment in the European NATO countries would decline by 355 000–505 000 jobs; Anthony, I. and Wulf, H., 'The future of the industry: a prognosis', eds I. Anthony, A. Courades Allebeck and H. Wulf, SIPRI, *West European Arms Production: Structural Changes in the New Political Environment*, SIPRI Research Report, Stockholm, Oct. 1990, pp. 60–61.

[55] Adam, B., 'Les perspectives du marché de l'armement en Europe: ébauche de scénario jusqu'en 1988', Institut Européen de Recherche sur la Paix et la Sécurité (GRIP), 'Memento défense–désarmement 1992: L'Europe et la sécurité internationale', *Les Dossiers du GRIP*, no. 168–171 (Apr.–July 1992), pp. 211–14.

[56] Assemblée Nationale, *Avis présenté au nom de la commission de la défense nationale et des forces armées sur le projet de loi de finances pour 1993 par M. Jean-Guy Branger, Tome VI, Défense, recherche et industrie d'armement,* Assemblée nationale, no. 2948 (14 Oct. 1992), p. 23.

[57] Sparaco, P., 'French aerospace industry to slash 32 000 jobs', *Aviation Week & Space Technology*, 24 Aug. 1992, p. 20; Reed (note 54); Casamayou, J.-P., 'La production d'armements en chute libre', *Air & Cosmos*, 23 Nov. 1992, p. 10–11.

[58] Notably between two of the largest arms-producing companies in France, Aérospatiale and Dassault Aviation.

Arms-producing companies in several developing countries have also been forced to restructure and reduce their production capacities. In 1992 this was true of Brazil, Israel and South Africa.

*The United States*

Seven major acquisitions, each valued at more than $200 million, contributed to a dramatic transformation of the US military aerospace and electronics industry in 1992 (table 10.6). Some of the companies created will become giant arms producers. Martin Marietta's purchase of the aerospace and three other units from General Electric will create a company with 94 000 employees and with annual revenues expected to reach $11 billion.[59] Lockheed's acquisition of the military aircraft division of General Dynamics will create a company with predicted annual military aircraft sales of $6.5 billion.[60] A new entrant in the arms industry was the Carlyle Group, a US private merchant banking firm, which in 1991 and 1992 acquired several US arms-producing companies and company divisions, including BDM International, the electronics divisions of General Dynamics and a majority share of the aircraft division of LTV.[61]

Hughes' acquisition of the General Dynamics missile division will significantly reduce US missile production capacity. Two of the six production plants acquired will be closed, resulting in cuts of 1900 jobs. Part of the production line at two other plants will also be closed.[62]

Military aerospace and electronics are the sectors with the sharpest fall in contracts received from the US DOD. While prime contracts for ships, land systems, weapons and ammunition have remained constant or shown only a slight decline since 1985, prime contracts for electronics and communications equipment, missiles and space and in particular aircraft—the sectors of the US arms industry that grew fastest during the first half of the 1980s—showed a sharp decline for the same period.[63] Other arms-producing sectors are not immune to the forces spurring national concentration. A major example is the 1992 agreement in principle between the Defense Systems Group of FMC and the BMY Combat Systems Division of Harsco, both leading producers of land systems, to merge their activities into a new, jointly owned company.[64]

Large take-overs are expected to continue in the US arms industry as part of the effort to adjust to a rapidly declining home market for military

[59] 'GE Aerospace to merge into Martin Marietta', *Aviation Week & Space Technology,* 30 Nov. 1992, pp. 20–24.

[60] *Interavia Air Letter*, 11 Dec. 1992, p. 1.

[61] The annual arms sales of its subsidiaries total about $1750 million. *World Weapons Review,* 28 Oct. 1992, p. 12.

[62] 'Hughes to close ex-GD sites', *Jane's Defence Weekly,* 19 Sep. 1992, p. 27.

[63] The adjustment difficulties facing the US aerospace industry are described in OTA (note 52), figure 7.2, p. 194.

[64] The 1993 sales of the new company have been estimated at $1.2 billion. *Jane's Defence Weekly,* 12 Dec. 1992, p. 10.

**Table 10.6.** Major acquisitions in the US arms industry announced 1992

Figures are in US $m.

| Buyer | Seller | Purchased unit | Contract value |
|---|---|---|---|
| Martin Marietta | General Electric | Aerospace division | 3 050 |
| Lockheed | General Dynamics | Tactical military aircraft | 1 525 |
| Textron Inc. | General Dynamics | Cessna Aircraft | 600 |
| Hughes Electronics | General Dynamics | Missiles divisions | 450 |
| Carlyle Group | General Dynamics | Electronics division (GDE Systems) | . . |
| Loral | LTV | Missiles division (Loral Vought Systems) | 260 |
| Carlyle Group (51%)/ Northrop (49%) | LTV | Aircraft division (Vought Aircraft) | c. 230 |

*Sources*: *Military Technology*, vol. 16, no. 5 (May 1992); *Defence Industry Digest*, 1 Oct. 1992; *Jane's Defence Weekly*, 29 Aug. and 7 Nov. 1992.

equipment.[65] The sharp fall in prices paid for companies being acquired reflects the general deterioration of the market. Prices paid by acquiring companies are reported to have declined from $1.27 per dollar of arms sales in 1987 to $0.45 per dollar of arms sales in 1991.[66] The announcement by incoming US Secretary of Defense Les Aspin that the defence budget should be reduced by 4 per cent increases the probability of greater turbulence in the US arms industry in 1993.[67]

One factor which may slow down the speed of restructuring is the body of US anti-trust legislation. The rules applied under the Clayton Antitrust Act, as updated by Congress in 1990, to measure the impact of impending mergers on competition are criticized for being ill-suited for application on the US arms market.[68] Rather than being a free market, this is a monopsony with many suppliers but only one buyer, the DOD. In 1992, attempts to block the acquisition by Alliant Techsystems of the tank ammunition division of Olin resulted in the withdrawal of Alliant Techsystems from this purchase.[69]

A Defense Conversion Commission was appointed by Congress in April 1992 to study the economic impact of falling arms procurement budgets and to find measures to soften these impacts. One of its main recommendations was to accelerate the integration of military and commercial manufacturing and development with the dual goal of preserving military technology capabilities and at the same time promoting economic growth. The first step

[65] The different adjustment strategies adopted by US companies are analysed in Reppy, J., 'The United States: unmanaged change in the defence industry', ed. Wulf (note 3), chapter 3, pp. 50–65.

[66] According to Pierre Chao, Director of Mergers and Acquisitions for a US consulting group, cited in *Defense News*, 30 Nov.–6 Dec. 1992, p. 20.

[67] 'Aspin seeks $10.8 billion cut in Pentagon's spending', *Congressional Quarterly Weekly Report*, vol. 51, no. 6 (6 Feb. 1993), p. 275.

[68] President of the American Defense Preparedness Association General Skibbie (Ret.), in *Aviation Week & Space Technology*, 16 Nov. 1992, p. 61.

[69] *Jane's Defence Weekly*, 21 Nov. 1992, p. 15; *Interavia Air Letter*, 11 Dec. 1992, p. 6.

towards integration would be to harmonize arms procurement laws and regulations with commercial practices.[70]

*Italy*

The Italian arms industry is composed of many relatively small producers largely controlled by three entities: the state-owned IRI and EFIM groups and the private group FIAT.[71]

In July 1992 the Italian Government disclosed plans to privatize many state-owned industries, including some of the IRI group of companies, and to liquidate the most unprofitable state holding, EFIM. In October, after a failed attempt on the part of EFIM's state-appointed board to default on group debts worth $7 billion (provoking criticism from foreign creditors and undermining both the Lira and the credit-worthiness of other state-owned interests) the government presented plans for the sale of the group's holdings. In December, following bidding that focused on EFIM's arms, aerospace and rail industries, the Government declared that the group's arms and aerospace companies—Agusta, Agusta Omi, Agusta Sistemi, Oto Melara, Breda Meccanica Bresciana, Officine Galileo and Sma—should remain under state control.

Despite objections from the state anti-trust board (which criticized the further concentration of arms and aerospace assets) and the EC Commission (concerned that the plan violates EC subsidy rules) the government decided to transfer these companies to IRI. IRI already controls several other arms producers, including Alenia, Italy's largest arms contractor. Finmeccanica (owned by IRI) is to lease these seven companies from EFIM from 1 January to 30 June 1993, at which time the transfer is to be converted into a sale if a price can be established. For the duration of the lease—designed to exempt Finmeccanica from assuming the $3.6 billion debt owed by these companies—debt, investment and financing costs are covered by the Italian Government.[72]

*Sweden*

Although small by international standards, the Swedish arms industry produces modern equipment in a wide range of product categories. A series of ownership changes in 1990–92 have produced one dominant 'national cham-

[70] 'Defense Conversion Panel urges dramatic changes', *Aviation Week & Space Technology*, 25 Jan. 1993, pp. 64–65.

[71] For a profile of the Italian arms industry, see Perani, G. and Pianta, M., 'The slow restructuring of the Italian arms industry', eds. M. Brzoska and P. Lock, SIPRI, *Restructuring of Arms Production in Western Europe* (Oxford University Press: Oxford, 1992), chapter 12, pp. 140–53.

[72] Heuzé, R., 'L'Efim, mouton noir des holdings d'État', *La Tribune de l'Expansion*, 15 July 1992, p. 7; 'Efim, ecco le regole per l'affitto all'Iri', *Il Sole–24 Ore*, 24 Oct. 1992; 'Antitrust all'attaco sul caso Efim: il decreto va contro la concorrenza', *Il Sole–24 Ore*, 28 Oct. 1992; Simonian, H., 'Efim receives first offer for assets', *Financial Times*, 2 Nov. 1992; 'EFIM shake-up deadline due', *Jane's Defence Weekly*, 21 Nov. 1992, p. 13; Webb, S., 'Efim upsets the bankers', *Financial Times*, 15 Dec. 1992, p. 27; Borriello, E., 'Agusta e Oto Melara sei mesi in affitto all'Iri—Il canone? Una lira sola', *La Repubblica*, 17 Dec. 1992; Baccaro, A., 'Finmeccanica con le stellette', *Corriere della Sera*, 17 Dec. 1992; Borriello, E., 'La ritirata Efim comincia dai boiardi', *La Repubblica*, 19 Dec. 1992; de Briganti, G. and Politi, A., 'Italy's Finmeccanica to take over failed EFIM', *Defense News*, 21–27 Dec. 1992, p. 8.

pion' in nearly all product sectors.[73] The restructuring has reduced both production capacity and employment.

Ordnance activities have been concentrated in Bofors AB, owned by the state-owned Celsius Group.[74] Celsius has become by far the largest arms-producing company in Sweden, incorporating the Swedish military shipbuilding industry in the shape of the Kockums Group. The future strategy of Celsius includes continued restructuring, privatization and active participation in projects with foreign arms-producing companies.[75]

Military electronics are largely concentrated in NobelTech, a wholly owned subsidiary of Nobel Industries. The arms-producing subsidiaries of NobelTech, NobelTech Electronics (previously Bofors Electronics) and NobelTech Systems (previously Bofors Aerotronics) were acquired by Celsius in February 1993 and renamed CelsiusTech. The other significant producer of military electronics in Sweden is Ericsson. Its subsidiary Ericsson Radar Electronics is a relatively large producer of military radars, and Ericsson Radio Systems produces military communications systems.

Military aircraft production has always been concentrated in one company—Saab-Scania. Military and civil aircraft production were separated in April 1992, when Saab Aircraft was split into two units. A major reason for this reorganization was the different marketing conditions for the two types of aircraft. Saab Military Aircraft is concentrated on the development and production of the JAS-39 Gripen fighter aircraft. Swedish aeroengine construction and maintenance was concentrated in January 1991 in Volvo Aerosupport, a joint venture between FFV Aerotech and Volvo Flygmotor, in which the latter owns a 90 per cent share.

### *Brazil*

In Brazil the arms industry, which was built up over a 30-year period between the mid-1960s and the mid-1980s, suffered from the decline in overseas markets for armoured vehicles, artillery rockets and aircraft. The loss of Iraq as a major customer was particularly important for Brazil.[76]

The state-owned company Embraer has seen a sharp decline in sales of both commercial and military aircraft.[77] In 1990 and 1991 the company experienced heavy losses. Military aircraft produced by Embraer are the EMB-312 Tucano military trainer aircraft and the AMX light fighter. The AMX was developed

[73] The exceptions are missiles and military vehicles. Missile production takes place in two companies, Bofors and Saab Combitech. Military vehicles are produced primarily by Bofors and Hägglunds and, to some extent, by Volvo and Saab-Scania.

[74] The gun system, missile and ammunition operations of Nobel Industries and the state-owned company FFV Ordnance were merged in Jan. 1991 into a new company, Swedish Ordnance, owned jointly by FFV and Nobel. Celsius acquired FFV and its share in Swedish Ordnance in June 1991 and the remaining half of Swedish Ordnance from Nobel Industries in Feb. 1992. The name of Swedish Ordnance was changed to the better known name Bofors AB in Mar. 1992. The military electronics part of the former Bofors Industries remained within Nobel Industries in its subsidiary NobelTech.

[75] Celsius Annual Report 1991, pp. 2–3.

[76] Gupta, A., 'Third World militaries: new suppliers, deadlier weapons', *Orbis*, vol. 37, no. 1 (winter 1993), pp. 57–68.

[77] The total sales of Embraer declined from US$700 million in 1989 to $400 million in 1991.

jointly with the Italian firms Alenia and Aermacchi. Production plans for 1992 included 14 of each of these aircraft, leaving idle capacity estimated at 30 per cent.[78] Employment has been cut from a peak of 12 600 in 1989 to 8300 in 1991, and a further cut of 2500 people was announced in May 1992.[79] The company has tried to compensate for the decline in final sales by increased refit and maintenance work in its military unit and by more sub-contract work in its commercial unit. Foreign investors have been sought since at least January 1991, when privatization of 80 per cent of the company was approved, including a 40 per cent quota for foreign investors.[80]

### *Israel*

The 30-per cent employment loss registered for Israel Military Industries (IMI) in table 10.5 illustrates the crisis in the Israeli arms industry in the face of falling sales both at home and abroad. Producing almost exclusively for military markets with a range of mainly low-technology weapon systems, IMI has suffered acutely from the world-wide decline in arms sales.[81]

The state-run arms producers—IMI, Israel Aircraft Industries and the Rafael Arms Development Authority—were the focus of critical government scrutiny in 1992. While not reflected in the figures for 1991, low productivity, cost overruns in programmes such as the Arrow missile and cancelled orders (many in civil markets into which these companies had tried to diversify) contributed to poor results for 1992. With limited prospects for exports, the Labour Government of Prime Minister and Defence Minister Yitzhak Rabin has chosen a policy of retrenchment in an effort to cut costs and raise productivity. The policy of heavy subsidization has ended and what aid remains has been redirected (from IAI and Rafael to the floundering IMI). Employment levels have been reduced in all three companies, at considerable political cost in a period of record unemployment.[82]

### *South Africa*

The South African arms industry is in the midst of a thorough restructuring process. Armscor, the state-owned military–industrial complex, was broken up

[78] *Aviation Week & Space Technology,* 1 June 1992, p. 17; 'Silva soothes bruised Embraer', *Interavia Aerospace Review,* Mar. 1992, p. 44.

[79] 'Embraer cuts workforce as aircraft demand falls', *Financial Times,* 27 May 1992, p. 20.

[80] The change of government in Brazil in Oct. 1992 and the subsequent change in privatization policy have raised a question about the Embraer privatization programme, originally planned for Mar. 1993. *Financial Times,* 21 Oct. 1992, p. 6; *Air & Cosmos*, 9–15 Nov. 1992.

[81] 'Firms forced to diversify', *Jane's Defence Weekly*, 15 Feb. 1992, p. 15; 'IMI seeks cuts as sales fall', *Jane's Defence Weekly*, 20 Apr. 1992, p. 669; 'IMI to cut workforce by 2000', *Interavia Air Letter*, 15 July 1992, p. 5; Meyer, C., 'Tempête sur l'industrie de defense Israélienne', *Air & Cosmos*, 5–11 Oct. 1992, p. 13.

[82] 'Schwerer Nackenschlag für den Luftfahrtkonzern IAI', *Handelsblatt*, 17 Sep. 1992, p. 18; Meyer (note 81), p. 13; Parnes, S., 'Losses spur study of Israel's IMI, state-owned firms', *Defense News*, 12–18 Oct. 1992, p. 82; Odenheimer, A., 'Rabin, Shohat reject IAI recovery plan', *Jerusalem Post* (international edition), 12 Dec. 1992, p. 20; Odenheimer, A., 'High wages blamed for IMI crisis', *Jerusalem Post* (international edition), 26 Dec. 1992, p. 21; 'Israel reallocated defence spending', *Interavia Air Letter*, 5 Jan. 1993, p. 6; 'State-run sector fights for a future', *Jane's Defence Weekly*, 6 Feb. 1993, pp. 29–30.

in April 1992. All arms-producing companies under Armscor were transferred to a new organization, Denel, owned by the state but expected to operate as a profit-based manufacturing group in which each of 23 business units have independent responsibility for sales.[83]

In spite of the contraction of the world arms market and the remaining mandatory UN sanctions against South Africa, company officials hope to expand sales in Africa and Asia to compensate for falling domestic sales.[84]

## Internationalization

The declining market for military equipment reinforces incentives for concentration and collaboration also across borders. International take-overs and joint ventures have normally been preceded by national concentration, since governments tend to protect their arms industries from foreign participation for as long as possible. Where national markets are small, structural change is more likely to cross borders. This is illustrated by the Belgian arms industry, which is dominated by foreign owners, and by the entire European arms industry, characterized by a high degree of cross-border take-overs and joint ventures.[85]

### *Changing policies on foreign ownership*

In 1992 Sweden decided to facilitate foreign ownership of arms-producing companies, while US policy on foreign acquisitions in the arms industry became more restrictive.

Since 1 January 1993 the former restrictions on foreign corporate ownership of Swedish arms-producing companies no longer exist. Thus, when NobelTech was put up for sale in early 1993 it was the first time that a Swedish arms-producing company could have come under foreign ownership. This is both part of a liberalization of policy on foreign acquisitions in general and a response to changing conditions for arms-production activities. The new act on military equipment adopted in 1992[86] preserves government authority to prevent foreign influence considered incompatible with Swedish security interests.[87]

The issue of foreign ownership of the US arms industry was highlighted by the bid for the missile division of the LTV Corporation by French electronics

[83] Cock, J., 'Rocks, snakes and South Africa's arms industry', unpublished paper distributed by the Military Research Group, Pretoria, South Africa, Nov. 1992.

[84] 'Rüstungsindustrie Südafrikas mit neuer Struktur', *Wehrtechnik,* Jan. 1993, p. 60; 'South African business: back in the arms bazaar', *Jane's Defence Weekly,* 14 Nov. 1992, pp. 33–45.

[85] Sköns, E., 'Western Europe: internationalization of the arms industry', ed. Wulf (note 3), pp. 160–90.

[86] *Lag om krigsmateriel* (Swedish Military Equipment Act), SFS 1992:1300, 28 Dec. 1992 (in force since 1 Jan. 1993).

[87] The compulsory licence needed to manufacture military equipment must include specific provisions on the degree of foreign influence which is acceptable; Military Equipment Act, Press release from Ministry for Foreign Affairs, Trade Department, 10 June 1992. See also interview with M. Sahlin, Sweden's Undersecretary of State for Defence, in *Military Technology*, vol. 16, no. 8 (Aug. 1992), pp. 22–26.

company Thomson-CSF. Opposition from the Congress and the government Committee on Foreign Investment in the USA forced Thomson to withdraw from the competition. Particularly sensitive was the fact that the French Government owns a 60 per cent share of Thomson.

Several new pieces of legislation were adopted by Congress in 1992 to increase control over foreign investment in US arms-producing companies. New DOD guidelines are also being prepared for the same purpose.[88] Congress adopted a prohibition on merger with or purchase of certain US defence contractors by entities controlled by foreign governments. This relates to contractors which have been awarded Department of Defense prime contracts (or Department of Energy national security contracts) in excess of $500 million in the previous year and with access to classified information. Under an amendment to the US Defense Production Act foreign acquisitions now require investigation in all cases where a foreign government-controlled entity seeks to engage in any merger, acquisition or take-over which could affect US national security. Legislation was also adopted requiring the Departments of Defense and Energy to establish data bases containing all of their contractors controlled by foreign persons who were awarded contracts exceeding $100 000 in any single year since 1988.[89]

*International take-overs and joint ventures*

The more significant international acquisitions and joint ventures in the arms industry decided in 1992 are presented in tables 10.7 and 10.8. Most of this activity took place within Western Europe, although US companies were involved in three major international take-overs and two joint ventures. Two of these take-over deals were still awaiting US approval at the end of 1992: these were the acquisition of the Allison Transmission division from General Motors by Zahnradfabrik Friedrichshafen, ZF (Germany)[90] and the Electronics Manufacturing Center from General Dynamics by Elbit Computers (Israel).[91] The ZF purchase was, however, blocked by the German Federal Cartel Office in early 1993.[92]

The military aerospace and electronics industries have been most affected by international acquisitions and joint ventures. One reason is vulnerability to budget cuts created by their high and rising R&D costs, both in absolute terms and per unit. In 1992 both McDonnell Douglas and British Aerospace courted the newly created Taiwan Aerospace Corporation (TAC) in an attempt to obtain capital for R&D in return for technology transfers to Taiwan.[93]

[88] 'US Air Force officials rap DoD's foreign investment rule', *Defense News*, 19–25 Oct. 1992, p. 42.

[89] *National Defense Authorization Act for Fiscal Year 1993—Conference Report to Accompany H. R. 5006* (note 34), Title VIII, Sec. 835, pp. 153–60.

[90] 'Industry shake-out and reshuffles', *Defence Industry Digest*, Oct. 1992, p. 13.

[91] *Jane's Defence Weekly*, 14 Nov. 1992, p. 5.

[92] 'Kartellamt gegen Übernahme', *Handelsblatt*, 27 Jan. 1993, p. 21.

[93] TAC justified its decision against a partial acquisition of MDC with the alleged undercapitalization of MDC; *Handelsblatt*, 16–17 Oct. 1992.

**Table 10.7.** International take-overs in the arms industry decided in 1992

| Buyer | | Seller | | Purchased company | | |
|---|---|---|---|---|---|---|
| Name | Country | Name | Country | Name | Country | Comments |
| DASA | FRG | Netherlands Government | Netherlands | Fokker | Netherlands | Share raised to 51% |
| Bodenseewerke Gerätetechnik | FRG | | | VDO Luft | FRG | EC approved in Dec. 1992 |
| Sextant Avionique | France | | | | | |
| SNPE | France | BPD | Italy | Sipe Nobel | Italy | |
| Thomson Sintra | France | Inisel/Bazan | Spain | SAES | Spain | 49% share |
| Simrad | Norway | Osprey | UK | Osprey | UK | Price £3.5 m. |
| Meggitt Aerospace | UK | Allied Signal | USA | Endevco | USA | Price $53 m. |

*Source*: SIPRI arms industry data base.

An additional factor for military aerospace is the simultaneous downturn of the commercial aircraft market with which it is closely inter-linked. The number of major aircraft manufacturers has decreased dramatically during the post-war period and is currently under 15, all of which produce military aircraft.[94] Manufacture of military aircraft takes place almost exclusively within companies that also produce commercial aircraft, although within separate divisions.[95] While the entire aerospace industry is affected by national and international concentration, the military part has traditionally been more protected than the commercial part. International acquisition activities are also high below the prime contractor level.

The most significant new foreign acquisition in 1992 was the agreement in principle signed in October 1992 for the German company DASA to take control of the Netherlands Fokker aircraft company in a $520 million purchase.[96] This acquisition was motivated by developments in the civil aircraft sector but will have implications also for military aircraft production. The future of the Fokker 50 aircraft (including military versions, development of which was supported by Netherlands Government loans of about US $300 million) was

[94] See Golich, V. L., 'From competition to collaboration; the challenge of commercial-class aircraft manufacturing', *International Organization,* vol. 46, no. 4 (autumn 1992), pp. 899–934.

[95] This created a problem in the aborted purchase by Taiwan Aerospace Corporation of part of the commercial aircraft activities of McDonnell Douglas (MDC). MDC would have had to avoid sharing military technologies by separating the military businesses from the commercial. See US General Accounting Office, *Issues Raised by Taiwan's Proposed Investment in McDonnell Douglas,* report no. GAO/NSIAD-92-120 (US Government Printing Office: Washington, DC, 1992).

[96] Subject to approval by the EC Commission and the Netherlands Parliament, a new holding company will acquire 51% of the Fokker shares. The holding company will be owned 78% by DASA and 22% by the Netherlands Government. After a 3-year period, the Netherlands Government will sell its shares. *Military Technology*, vol. 16, no. 9 (Sep. 1992), p. 111.

**Table 10.8.** International joint ventures in the arms industry decided in 1992

| Companies | Countries | Merged/new company | Purpose |
|---|---|---|---|
| Fairchild<br>Galram Technology Industries (Rafael) | USA<br>Israel | Oramir | Produce and sell new machine for manufacture of semi-conductors |
| Short Brothers<br>Thomson-CSF | UK<br>France | Star | Produce and market short-range SAMs |
| Hispano Suiza (Snecma)<br>Alenia (IRI) | France<br>Italy | Euronacelle (France) | Produce and market aircraft nacelles |
| Hurel-Dubois<br>Short Borthers | France<br>UK | INS | Broader manufacture of air-craft and engine nacelles |
| Elettronica (EIS)<br>Syseca | Italy<br>France | Eisys | Development of software for missiles |
| MaK Systems (Rheinmetall)<br>Panhard (Peugeot) | FRG<br>France | Euro-LAV | Joint development of a light armoured vehicle |
| Chantiers de l'Atlantique<br>Bremer Vulkan | France<br>FRG | Eurocorvette | Marketing and sales of the jointly designed BRECA family of ships |
| SNPE<br>Kaman | France<br>USA | Advanced Energetic Materials Corp. of Europe | Market advanced insensi-tive armour systems |
| IMBEL<br>Royal Ordnance | Brazil<br>UK | South American Ordnance (Brazil) | Munitions |
| Racal Radio<br>Sapura | UK<br>Malysia | Sapura Radio | Manufacture of military communications equipment |

*Source*: SIPRI arms industry data base.

one of the stumbling-blocks in the negotiations, which continued through March 1993. In the final settlement, the price paid by DASA was set at $368 million.[97]

In the military electronics sector, also closely tied to its civil counterpart, international restructuring in Western Europe has been intense in the period 1988–92. Most major producers of military electronics—including Alcatel-Alsthom, Bofors Electronics, DASA, Ferranti International, GEC-Marconi, Hollandse Signaalapparaten, Philips, Plessey, Sextant Avionique, Siemens, Thomson-CSF and Thorn EMI—have been affected by this process.[98]

In May 1991 the two largest European defence electronics companies—GEC and Thomson-CSF—formed GTAR (GEC/Thomson Airborne Radar) for

[97] 'Fokker finds a way to flex its wings', *Financial Times*, 2 Nov. 1992, p. 17; 'Fokker-Vertrag in Amsterdam paraphiert', *Die Tagesspiegel*, 18 Mar. 1993, p. 25.

[98] Sandström, M., FOA (Swedish National Defence Research Establishment), *Strukturförändringar inom den europeiska försvarselektronikindustrin* ['Structural changes in the European defence electronics industry'], FOA-Rapport A 10036-1.3 (FOA 1, Huvudavd. för försvarsanalys: Sundbyberg, Sweden, Sep. 1992).

**Table 10.9.** Teaming arrangements to bid for the US JPATS contract

| Teaming companies | Country | Contending aircraft |
|---|---|---|
| FMA<br>Vought Aircraft | Argentina<br>USA | IA-63 modified to Pampa 2000 |
| Embraer<br>Northrop | Brazil<br>USA | EMB-312H Super Tucano |
| DASA<br>Rockwell | FRG<br>USA | Fan Ranger |
| Aermacchi<br>Lockheed | Italy<br>USA | MB-339D |
| Agusta<br>Grumman | Italy<br>USA | S-211 |
| Pilatus<br>Beech | Switzerland<br>USA | PC-9 Mk II |
| Cessna Aircraft<br>Williams International<br>Flight Safety International | USA<br>USA<br>USA | Unnamed design, based on components from commercial Citation Jet |

*Sources:* 'The JPATS contenders', *US Naval Institute Proceedings,* Sep. 1992, pp. 9-91; 'Cessna team joins JPATS trainer fray', *Defense News,* 14–20 Dec. 1992, p. 22; *Wehrtechnik,* Oct. 1992, p. 38.

joint development of radar systems for fighter aircraft. In 1992 DASA opened negotiations to join GTAR.[99] The Anglo-German company Siemens-Plessey will work with the US company Hughes Electronics to develop a ground-based radar to meet British and US requirements.[100]

International industrial co-operation takes many different forms. Two common forms in the arms industry are consortia and teaming arrangements. International consortia are usually formed for the management of a large weapon programme in collaboration with two or more national procurement agencies. The basis for such organizations is government-to-government agreements on collaborative projects, which are a traditional form of collaboration within NATO. Several consortia of this type were formed in 1991–92: Euroflag (for the development of a military cargo aircraft) with the participation of Belgium, France, Germany, Italy, Portugal, Spain, Turkey and the UK; Europatrol (for the development of maritime patrol aircraft) with the participation of France, Germany, Italy, the Netherlands, Spain and the UK; and Apache-MAW (for the modification of the French Apache air-to-surface missile to German requirements).

Teaming arrangements are joint bids between two or more companies for contracts to develop and produce major weapons as a means of sharing the risks involved. The companies within a team retain their individual autonomy and a team usually consists of companies with complementary skills—one

[99] 'Dasa wants to join radar consortium' *Jane's Defence Weekly,* 21 Nov. 1992, p. 7; *Air & Cosmos,* 23–29 Nov. 1992, p. 6.

[100] 'UK/USA clear radar teaming', *Jane's Defence Weekly,* 19 Sep. 1992, p. 17.

company taking the lead as prime contractor. With a smaller number of new large arms contracts available this can also represent an attempt to share the market, ensuring the survival of companies. Teaming may thus delay the rationalization of the arms industry which most observers believe is necessary. Examples of large competitions with international teaming arrangements formed in 1992 include those for the JPATS—a jet trainer to meet the requirement for a Joint Primary Aircraft Training System for the US Navy and Air Force (table 10.9)—for a British Army attack helicopter and for a British short-range air-to-air missile. By early 1993 the teaming model was being reconsidered as too expensive an acquisition form for JPATS.[101]

# IV. The trade in major conventional weapons

The value of foreign deliveries of major conventional weapons in 1992 is estimated by SIPRI to have been $18 405 million in 1990 US dollars.[102] This figure—roughly 25 per cent less than the value recorded for 1991—continues the downward trend in the aggregate value of the arms trade since 1987, as reported in the *SIPRI Yearbook 1992*.[103]

In this section the major arms-exporting countries as identified by SIPRI are discussed individually below. Table 10.12 includes arms export data as reported by governments. Arms imports are dealt with on a regional basis and the region discussed in this *Yearbook* is Asia.

In 1991 SIPRI published a list of official arms export data to illustrate the fact that very little such information is available and that which does exist is compiled in a manner that prevents comparative analysis. In 1992 this remains the case.

After 40 years of providing no official data on arms exports, in 1992 Russia provided three figures, all of them different. It is possible that all three are correct but measure different things and the fact that official discussions of the arms trade are becoming available is a significant step forward in the formulation of an export policy in Russia. The official figures available for Russia are discussed below.

## The major arms exporters

### *The United States*

There were no significant changes in US arms transfer policy during the final months of the Bush Administration and at the time of its inauguration no new policy had been formulated by the Clinton Administration.

[101] 'Yockey may clear obstacles to JPATS', *Aviation Week & Space Technology*, 11 Jan. 1993, p. 24.

[102] Since the SIPRI arms trade statistics do not reflect purchase prices, they are not comparable with economic statistics such as national accounts or foreign trade statistics, nor with the arms sales data reported in the sections of this chapter dealing with arms production. The methods used for the valuation of SIPRI arms trade statistics are described in appendix 10D.

[103] Anthony *et al.* (note 6).

**Table 10.10.** The leading exporters of major conventional weapons, 1988–92

The countries are ranked according to 1988–92 aggregate exports. Figures are in US $m., at constant (1990) prices. Figures may not add up to totals due to rounding.

| | Exporters | 1988 | 1989 | 1990 | 1991 | 1992 | 1988–92 |
|---|---|---|---|---|---|---|---|
| *To the industrialized world* | | | | | | | |
| 1 | USA | 7 710 | 8 186 | 6 200 | 7 519 | 5 355 | 34 968 |
| 2 | USSR/Russia | 4378 | 3 962 | 3 109 | 461 | 139 | 12 048 |
| 3 | Germany, FR | 957 | 606 | 820 | 2 106 | 1 632 | 6 121 |
| 4 | France | 734 | 795 | 335 | 96 | 800 | 2 761 |
| 5 | Czechoslovakia | 644 | 494 | 583 | 0 | 0 | 1 722 |
| 6 | UK | 199 | 717 | 292 | 104 | 295 | 1 607 |
| 7 | German DR | 367 | 367 | 245 | 0 | 0 | 980 |
| 8 | Sweden | 326 | 142 | 104 | 59 | 111 | 741 |
| 9 | Switzerland | 41 | 130 | 157 | 341 | 56 | 725 |
| 10 | Netherlands | 183 | 66 | 73 | 176 | 210 | 708 |
| 11 | Poland | 359 | 116 | 152 | 55 | 0 | 681 |
| 12 | Italy | 143 | 98 | 23 | 114 | 288 | 666 |
| 13 | Spain | 7 | 312 | 6 | 27 | 19 | 371 |
| 14 | Israel | 22 | 100 | 66 | 74 | 25 | 287 |
| 15 | Norway | 20 | 92 | 6 | 37 | 17 | 172 |
| | Others | 256 | 329 | 119 | 60 | 140 | 904 |
| | **Total** | **16 346** | **16 510** | **12 290** | **11 230** | **9 086** | **65 461** |
| *To the developing world* | | | | | | | |
| 1 | USSR/Russia | 10 280 | 10 348 | 6 615 | 3 987 | 1 904 | 33 135 |
| 2 | USA | 4 494 | 3 662 | 4 622 | 4 147 | 3 075 | 20 000 |
| 3 | China | 2 097 | 945 | 1 249 | 1 705 | 1 535 | 7 531 |
| 4 | France | 1 668 | 2 051 | 1 794 | 724 | 351 | 6 588 |
| 5 | UK | 1 505 | 1 993 | 1 163 | 697 | 658 | 6 016 |
| 6 | Germany, FR | 284 | 208 | 857 | 425 | 296 | 2 069 |
| 7 | Czechoslovakia | 282 | 221 | 85 | 74 | 779 | 1 442 |
| 8 | Netherlands | 443 | 458 | 154 | 189 | 95 | 1 340 |
| 9 | Brazil | 505 | 291 | 167 | 21 | 34 | 1 019 |
| 10 | Italy | 550 | 139 | 162 | 49 | 47 | 947 |
| 11 | Yugoslavia | 4 | 0 | 60 | 661 | 21 | 746 |
| 12 | Sweden | 281 | 233 | 117 | 42 | 2 | 675 |
| 13 | Spain | 228 | 297 | 77 | 23 | 18 | 643 |
| 14 | Korea, North | 155 | 0 | 0 | 86 | 313 | 554 |
| 15 | Israel | 146 | 221 | 37 | 45 | 41 | 489 |
| | Others | 766 | 555 | 522 | 365 | 152 | 2 360 |
| | **Total** | **23 688** | **21 623** | **17 682** | **13 240** | **9 320** | **85 552** |
| *To all countries* | | | | | | | |
| 1 | USA | 12 204 | 11 848 | 10 822 | 11 666 | 8 429 | 54 968 |
| 2 | USSR/Russia | 14 658 | 14 310 | 9 724 | 4 448 | 2 043 | 45 182 |
| 3 | France | 2 403 | 2 846 | 2 129 | 820 | 1 151 | 9 349 |
| 4 | Germany, FR | 1 241 | 814 | 1 677 | 2 530 | 1 928 | 8 190 |
| 5 | China | 2 161 | 1 009 | 1 249 | 1 705 | 1 535 | 7 658 |
| 6 | UK | 1 704 | 2 710 | 1 456 | 801 | 952 | 7 623 |
| 7 | Czechoslovakia | 927 | 715 | 669 | 74 | 779 | 3 163 |
| 8 | Netherlands | 626 | 525 | 226 | 365 | 305 | 2 048 |
| 9 | Italy | 693 | 237 | 185 | 163 | 335 | 1 613 |
| 10 | Sweden | 606 | 375 | 221 | 101 | 113 | 1 416 |
| 11 | German DR | 367 | 510 | 245 | 0 | 0 | 1 123 |
| 12 | Brazil | 507 | 293 | 169 | 23 | 36 | 1 028 |
| 13 | Spain | 235 | 608 | 83 | 50 | 37 | 1 014 |
| 14 | Switzerland | 76 | 154 | 192 | 369 | 83 | 874 |
| 15 | Israel | 168 | 321 | 103 | 119 | 66 | 777 |
| | Others | 1 459 | 859 | 822 | 1 234 | 614 | 4 987 |
| | **Total** | **40 034** | **38 133** | **29 972** | **24 470** | **18 405** | **151 013** |

*Source*: SIPRI arms trade data base.

**Table 10.11.** The leading importers of major conventional weapons, 1988–92

The countries are ranked according to 1988–92 aggregate imports. Figures are in US $m., at constant (1990) prices. Figures may not add up to totals due to rounding.

| Importers | | 1988 | 1989 | 1990 | 1991 | 1992 | 1988–92 |
|---|---|---|---|---|---|---|---|
| *By the industrialized world* | | | | | | | |
| 1 | Japan | 2 544 | 2 673 | 1 915 | 998 | 1 095 | 9 224 |
| 2 | Greece | 814 | 1 470 | 960 | 1 035 | 1 918 | 6 197 |
| 3 | Turkey | 1 447 | 1 177 | 808 | 1 224 | 1 511 | 6 167 |
| 4 | Germany, FR | 514 | 1 186 | 1 351 | 1 278 | 144 | 4 473 |
| 5 | Spain | 1 653 | 912 | 725 | 88 | 370 | 3 747 |
| 6 | Czechoslovakia | 1 122 | 1 492 | 835 | 47 | 4 | 3 501 |
| 7 | USSR/Russia | 1 421 | 1 016 | 891 | 55 | 0 | 3 383 |
| 8 | Poland | 1 299 | 1 225 | 334 | 143 | 0 | 3 001 |
| 9 | Australia | 692 | 827 | 437 | 250 | 398 | 2 604 |
| 10 | UK | 165 | 116 | 101 | 892 | 1 051 | 2 326 |
| 11 | Canada | 634 | 159 | 186 | 865 | 234 | 2 079 |
| 12 | German DR | 614 | 636 | 649 | 0 | 0 | 1 899 |
| 13 | USA | 154 | 560 | 109 | 294 | 726 | 1 843 |
| 14 | Netherlands | 258 | 787 | 266 | 274 | 181 | 1 765 |
| 15 | France | 121 | 169 | 45 | 1 207 | 86 | 1 626 |
| | Others | 2 893 | 2 106 | 2 679 | 2 581 | 1 370 | 11 629 |
| | **Total** | **16 346** | **16 510** | **12 290** | **11 230** | **9 086** | **65 461** |
| *By the developing world* | | | | | | | |
| 1 | India | 3 709 | 4 437 | 1 410 | 1 483 | 1 197 | 12 235 |
| 2 | Saudi Arabia | 2 441 | 1 931 | 2 537 | 898 | 883 | 8 690 |
| 3 | Afghanistan | 1 264 | 2 622 | 2 414 | 1 215 | 0 | 7 515 |
| 4 | Iraq | 2 845 | 1 526 | 596 | 0 | 0 | 4 967 |
| 5 | Iran | 648 | 372 | 833 | 902 | 877 | 3 632 |
| 6 | Korea, South | 1 125 | 1 114 | 524 | 347 | 414 | 3 524 |
| 7 | Pakistan | 334 | 773 | 947 | 1 000 | 432 | 3 486 |
| 8 | Thailand | 518 | 536 | 419 | 929 | 869 | 3 269 |
| 9 | Egypt | 540 | 213 | 1 175 | 745 | 621 | 3 295 |
| 10 | Korea, North | 1 382 | 1 066 | 636 | 15 | 24 | 3 123 |
| 11 | Israel | 561 | 209 | 43 | 1 246 | 709 | 2 768 |
| 12 | Syria | 1 393 | 395 | 28 | 86 | 716 | 2 618 |
| 13 | Taiwan | 363 | 384 | 641 | 561 | 285 | 2 234 |
| 14 | United Arab Emirates | 69 | 774 | 936 | 155 | 131 | 2 065 |
| 15 | Angola | 1 171 | 92 | 748 | 0 | 0 | 2 011 |
| | Others | 5 326 | 5 178 | 3 797 | 3 657 | 2 162 | 20 118 |
| | **Total** | **23 688** | **21 623** | **17 682** | **13 240** | **9 320** | **85 552** |
| *By all countries* | | | | | | | |
| 1 | India | 3 709 | 4 437 | 1 410 | 1 483 | 1 197 | 12 235 |
| 2 | Japan | 2 544 | 2 673 | 1 915 | 998 | 1 095 | 9 224 |
| 3 | Saudi Arabia | 2 441 | 1 931 | 2 537 | 898 | 883 | 8 690 |
| 4 | Afghanistan | 1 264 | 2 622 | 2 414 | 1 215 | 0 | 7 515 |
| 5 | Greece | 814 | 1 470 | 960 | 1 035 | 1 918 | 6 197 |
| 6 | Turkey | 1 447 | 1 177 | 808 | 1 224 | 1 511 | 6 167 |
| 7 | Iraq | 2 845 | 1 526 | 596 | 0 | 0 | 4 967 |
| 8 | Germany, FR | 514 | 1 186 | 1 351 | 1 278 | 144 | 4 473 |
| 9 | Spain | 1 653 | 912 | 725 | 88 | 370 | 3 747 |
| 10 | Iran | 648 | 372 | 833 | 902 | 877 | 3 632 |
| 11 | Korea, South | 1 125 | 1 114 | 524 | 347 | 414 | 3 524 |
| 12 | Czechoslovakia | 1 122 | 1 492 | 835 | 47 | 4 | 3 501 |
| 13 | Pakistan | 334 | 773 | 947 | 1 000 | 432 | 3 486 |
| 14 | USSR/Russia | 1 421 | 1 016 | 891 | 55 | 0 | 3 383 |
| 15 | Thailand | 518 | 536 | 419 | 929 | 869 | 3 271 |
| | Others | 17 636 | 14 895 | 12 810 | 12 971 | 8 693 | 67 003 |
| | **Total** | **40 034** | **38 133** | **29 972** | **24 470** | **18 405** | **151 013** |

*Source*: SIPRI arms trade data base.

**Table 10.12.** Official arms export data

Comments are worded as closely as possible to the details given in the source documents.

| Country | Year | Value | Comments |
|---|---|---|---|
| Canada | 1991 | C$189.2 m. | Value of export permits for military goods |
| Czechoslovakia | 1991 | CSK 5.2 b. | Value of arms production carried out for foreign customers |
| France | 1991 | FFr 34.2 b. | Value of orders of defence *matériel* |
| | 1991 | FFr 29.1 b. | Value of deliveries of defence *matériel* |
| Poland | 1991 | $396.2 m. | Value of exports of arms equipment, spare parts and ammunition |
| Russia | 1991 | $7.8 b. | Minister for Foreign Economic Relations |
| | 1992 | $3 b. | |
| | 1991 | $1.55 b. | Ministry for Foreign Economic Relations |
| | 1992 | $4 b. | Office of the Chief of Staff, Supreme Military Command |
| Sweden | 1991 | SEK 2 559 m. | Value of export licences for war *matériel* |
| | 1991 | SEK 2 705 m. | Value of exports of war *matériel* |
| Switzerland | 1992 | SFR 258.8 m. | Value of exports of war *matériel* |
| UK | 1991 | £1 862 m. | Value of defence equipment which passed the British Customs barrier |
| USA | 1991 | $22 981 m. | Value of Foreign Military Sales (FMS) accepted in FY 1991 |
| | 1991 | $8 845 m. | Value of FMS deliveries in FY 1991 |
| | 1991 | $39 109 m. | Value of licences approved for commercially sold defence articles and services in FY 1991 |
| | 1991 | $3 829 m. | Value of commercial arms deliveries in 1991 |

*Source*: SIPRI arms industry data base.

According to SIPRI estimates the value of deliveries of major conventional weapons by the USA declined from $11.6 billion in 1991 to $8.4 billion in 1992. This figure is expected to rise over the next few years as agreements involving transfers of major items of equipment are fulfilled. In particular, much of the equipment agreed after the war against Iraq in 1991 has not yet been delivered—over $13 billion of the value of Foreign Military Sales (FMS) agreements in 1991 was for Saudi Arabia alone.

The discussion of arms transfers by the United States in 1992 was dominated by a small number of high-profile government-to-government agreements. Three such agreements (with Greece, Saudi Arabia and Taiwan) were announced in the final stages of the presidential election campaign, provoking criticism that export policy was motivated more by short-term political factors than anything else. There is no doubt that campaign considerations influenced both the timing of the announcements and their form—President Bush announced the sale of F-16 fighter aircraft to Taiwan during a visit to the plant where the aircraft are manufactured. While President Bush certainly

**Table 10.13.** Value of US Foreign Military Sales agreements and deliveries, 1982–91

Figures are in US $m., at current prices.

| | 1982 | 1983 | 1984 | 1985 | 1986 | 1987 | 1988 | 1989 | 1990 | 1991 |
|---|---|---|---|---|---|---|---|---|---|---|
| Agreements | 16 422 | 14 339 | 12 784 | 10 482 | 6 540 | 6 449 | 11 739 | 10 747 | 13 948 | 22 982 |
| Deliveries | 8 765 | 10 790 | 8 198 | 7 520 | 7 268 | 11 103 | 8 844 | 6 991 | 7 389 | 8 845 |

*Source*: US Defense Security Assistance Agency, *Foreign Military Sales, Foreign Military Construction Sales and Military Assistance Facts* (DSAA: Washington, DC, 30 Sep. 1991).

calculated the announcement for maximum political effect, none of these transactions departs from the pattern of US arms transfer policy. This is true even for the sale of fighter aircraft to Taiwan—which has long been an importer of US military technology. The possibility that a military reconnaissance satellite might be sold to the United Arab Emirates (UAE) and rumours that similar discussions are under way with South Korea and Spain represent potential new departures in US policy.[104]

Discussion in the USA in 1992 focused on the massive increase in the value of agreements recorded in official data for both US arms transfers under the FMS programme and US commercial arms sales. The official data are difficult to interpret.

The value of FMS agreements has grown considerably since 1987 while the value of FMS deliveries has not. Therefore, while the value of FMS deliveries could be expected to grow significantly in the future, there may be a time-lag of several years before the large increase in 1991 is translated into deliveries. In the interim many things may change and the scale of future US transfers is still a matter of speculation. Moreover, the term Foreign Military Sales is not a synonym for arms sales as it includes all support services provided along with weapon systems including training, instruction in how to perform maintenance and even English-language training for military personnel. Moreover, the value of the agreements represents an estimate. Items transferred under FMS are managed by the DOD, and contracts are between private sector arms producers and the DOD. The DOD negotiates FMS agreements with foreign governments on the basis of its own estimate of what equipment will cost and not on the basis of the real price, which may not be known for several years. When it comes to negotiating a price with a manufacturer, the DOD combines foreign orders with orders for the US armed forces in order to achieve economies of scale.

In order to avoid having to re-negotiate agreements with foreign governments because it has been unable to secure equipment within the agreed ceiling, the DOD makes its cost estimates deliberately high and then refunds money to the purchasing government at the end of the transaction. The value of FMS agreements declared by the Defense Security Assistance Agency (DSAA) is, therefore, usually higher than the actual monies which will be received.

[104] This issue is discussed further in chapter 8 in this volume.

The value of commercial arms deliveries fell from $8.4 billion in 1989 to $5.6 billion in 1990 and to $3.8 billion in 1991.[105] While there was a massive increase in the value reported for commercial sales in 1992 (to over $28.7 billion), this figure represents the value of licences approved and not contracts signed or items delivered. Normally, only a small percentage of licences lead to firm orders after discussions with foreign governments.

*Russia and Ukraine*

According to SIPRI data the estimated value of arms exports from the territory of the former Soviet Union continued to decline in 1992 although less steeply than in the period 1989–91. The great majority of these deliveries came from Russia—where the bulk of existing weapon inventories and arms production capacity are located.[106] While it is impossible to quantify the amount precisely, some deliveries were probably made from Ukraine—where most of the non-Russian arms production capacity of the former Soviet Union was located. Ukrainian industrial and government representatives were active in 1992 in important Russian arms markets such as India. After several rounds of discussions India and Ukraine concluded a trade deal including military equipment on 17 October 1992.[107]

Russian data concerning arms exports are beginning to become available but precisely how these data were compiled remains unclear.

In September 1992 Gennadiy Yanpolsky, General Director of the Defense Industry Department of the Ministry of Industry, stated that while exports accounted for 30 per cent of sales by the Russian defence industry in 1991 they accounted for 7.2 per cent of sales in the first half of 1992.[108] In November 1992 Peter Aven, then Russian Minister of Foreign Economic Relations, told the Russian Supreme Soviet that the value of Russian arms sales for 1991 was $7.8 billion—a reduction from a high point of $23 billion in 1989. The estimated value of sales for 1992 was $3 billion.[109] These figures were different from earlier data supplied from the same Ministry which had said that arms deliveries in 1991 were worth $1.55 billion of which $20 million was in the form of grants.[110] It is possible that Aven was referring to the value of new agreements rather than deliveries of equipment. A spokes-

[105] US Department of State and Department of Defense data, reproduced in 'US commercial arms sales', *Defense & Economy World Report*, 24 June 1991.

[106] The percentage of arms production capacity from the former Soviet Union located in Russia is in the region of 65–70%. The lower estimate is by the US Defense Intelligence Agency, the higher by Professor J. Cooper of the University of Birmingham and B. Horrigan, in *Radio Free Europe/Radio Liberty (RFE/RL) Research Report*, 21 Aug. 1992.

[107] *Asia–Pacific Defence Reporter*, June–July 1992, p. 25; *RFE/RL Research Report*, 30 Oct. 1992, p. 61. The deal is to be financed in part through the barter of Indian consumer goods and in part in hard currency.

[108] 'Defense industry's status, future eyed', *Krasnaya Zvezda*, 29 Aug. 1992, in Foreign Broadcast Information Service, *Daily Report–Central Eurasia (FBIS-SOV)*, FBIS-SOV-92-173, 4 Sep. 1992, p. 23.

[109] Sneider, D., 'Russian armsmakers take off on their own', *Christian Science Monitor*, 25 Nov. 1992, p. 6; *Defense News*, 7–13 Dec. 1992, p. 44.

[110] *Nezavisimaya Gazeta*, 29 Sep. 1992; *RFE/RL Research Report*, 9 Oct. 1992, p. 52.

**Table 10.14.** Regional distribution of deliveries of arms and military equipment by the former Soviet Union in 1991

| Region | Number of deliveries |
|---|---|
| Near East | 8 |
| Middle East | 61 |
| Europe | 12 |
| Africa | 1 |
| Latin America | 1 |
| Asia | 17 |

*Source*: *Nezavisimaya Gazeta*, 29 Set. 1992.

man for the Russian Ministry of Foreign Affairs subsequently stated that the estimates given by Aven for 1991 and 1992 had no official status.[111]

Chairman of the Russian Committee for Defence Industries Viktor Glukhikh stated that Russian arms exports in 1992 were worth $4 billion.[112] This figure has also been used by Lieutenant General Andrey Nikolayev, First Deputy Chief of Staff of the Supreme Military Command.[113] Finally, on 2 December 1992 in a speech to the Russian Supreme Soviet then Prime Minister Yegor Gaidar stated that Russia had concluded agreements worth a total of $2.2 billion in 1992 with three countries—China, India and Iran.[114]

Whereas in the past official economic data were provided in convertible roubles all of the estimates for the value of the arms trade have been given in US dollars and it is likely that at least some part of the payment for arms will be made in hard currency. However, the arms transfers to China and India agreed in 1992 were part of broad economic packages that included payment in commodities and, in the case of India, a soft currency financing arrangement. Therefore whether the values declared will correspond to money received cannot yet be known even by the parties to the agreement.

Of the official data released the most detailed were provided by the Ministry for Foreign Economic Relations which released to the public data that had previously been given to the other permanent members of the UN Security Council in the context of the discussion of arms control in the Middle East (discussed below). According to these data the value of arms deliveries by the former Soviet Union in 1991 was $1.55 billion. The following information was also provided regarding the regional distribution and the distribution across weapon categories of deliveries of arms and military equipment by the former Soviet Union in 1991.

The depression in arms sales reflects the wider trade pattern of Russia and Ukraine, both of which have experienced a significant decline in foreign trade

[111] Peter Litavrin, Department for Export Control and Conversion, Ministry of Foreign Affairs of Russia, private communication with the authors, 12 Jan. 1993.

[112] *East Defence & Aerospace Update*, 16–31 Jan. 1993, p. 1; *Frankfurter Allgemeine Zeitung*, 12 Feb. 1993, p. 16.

[113] Moosa, E., 'Russia proposes demilitarized zones in Far East', Reuters, Tokyo, 24 Feb. 1993.

[114] Of the $2.2 billion, sales to China account for $1 billion, to India $650 million, and to Iran $600 million. *Defense News*, 7–13 Dec. 1992, p. 3.

**Table 10.15.** Distribution by weapon category of deliveries of arms and military equipment by the former Soviet Union in 1991

| Weapon category | Number of deliveries |
|---|---|
| Tanks | 553 |
| Armoured combat vehicles | 658 |
| Large-calibre artillery | 381 |
| Combat aircraft | 40 |
| Combat helicopters | 1 |
| Surface ships | 3 |
| Missiles | 1 783 |
| Air defence complexes | 1 |

*Source*: *Nezavisimaya Gazeta*, 29 Sep. 1992.

since the dissolution of the socialist trading bloc. The European Bank for Reconstruction and Development (EBRD) has reported that 'the slump in the Russian oil sector, the concomitant balance of payments crisis, the disintegration of the Union and the dismantling of the old system of economic management have triggered a dramatic decline in Russia's foreign trade'.[115]

Three countries—China, India and Iran—now dominate the discussion of Russian arms exports. Although Russian officials have held discussions with several countries regarding arms sales few new agreements have been concluded.[116] Russia has found new customers after 1989—most notably China, Iran and Turkey. Two traditionally important relationships with Cuba and North Korea were resumed in 1992, although only through the provision of spare parts.[117]

Ukraine depends on machine-building and metal-working industries that make sub-assemblies for shipment to Russia rather than having an independent capacity for system integration.[118] Not only has Ukraine lost much of its traditional market, but the nature of its industrial activity further complicates the formation of new relationships. Bilateral agreements within the Commonwealth of Independent States (CIS) should in theory have allowed continuity in inter-republic trade. However, the breakdown of the administrative system of the former USSR has meant that few agreements have been implemented.

In October 1992 Leonid Kuchma was confirmed as the new Prime Minister of Ukraine. He was formerly the director of the Southern Machine Construction Plant in Dnepropetrovsk—the largest rocket and missile production plant in the former Soviet Union. While Kuchma apparently favours privatization in

[115] European Bank for Reconstruction and Development, *Quarterly Economic Review* (EBRD: London, 30 Sep. 1992), p. 64.

[116] Countries where Russia is marketing arms include Brazil, Greece, Indonesia, Malaysia, Oman, the Philippines, South Korea, Turkey, the United Arab Emirates and the United Kingdom. Of these countries only Turkey and the UAE have placed orders for equipment. Pakistan and Taiwan both denied reports that arms sales are under discussion, though Taiwan and Russia have discussed technical and scientific co-operation in aerospace.

[117] 'Cuba signs oil, sugar cane and arms "accessories" agreements', *Interfax*, 3 Nov. 1992, in FBIS-SOV-92-214, 4 Nov. 1992, p. 14; *Far Eastern Economic Review*, 20 Aug. 1992, p. 7.

[118] *Quarterly Economic Review of the EBRD* (note 115), p. 70.

the service sector, light industry and agriculture, he has stated that nuclear, energy-related and military industries must remain in state ownership.[119]

India and Iran took their own initiatives to establish contact with arms producers both in Russia and in Ukraine to reassure themselves that arms agreements would be fulfilled. The Indian armed forces in particular depend heavily on the industry of the former Soviet Union for equipment and spare parts.

### *Germany*

While the alleged role of German companies in the chemical and nuclear programmes of several developing countries has attracted much international attention, the growing importance of Germany as an exporter of major conventional weapons has not. According to SIPRI estimates Germany is now the third largest exporter of major conventional weapons. German arms exports are at a level comparable to those of Russia and considerably larger than those of either France or the UK. In 1992 Germany accounted for 41 per cent of deliveries by EC countries against 24 per cent by France and 20 per cent by the UK.

German arms exports are dominated by four elements of which the most important in financial terms is the sale of naval systems—including conventional submarines, frigates, fast attack craft and naval auxiliaries. In the past two years there have been significant transfers to NATO allies through the 'cascade' of equipment that must be eliminated in order to comply with the CFE Treaty as well as under the 'Materialhilfe' assistance programmes. These programmes make a very limited contribution to the overall balance of payments but represent a significant indirect subsidy paid to German companies. Germany has also been disposing of assets owned by the armed forces of the former German Democratic Republic (GDR), but this equipment has often been sold at scrap value to save Germany the cost and trouble of destroying it. Finally, German companies make a significant contribution to many weapons produced in joint programmes together with allies, other European countries and developing countries.

The SIPRI methodology does not include any of the final category of German equipment transfers but prices second-hand equipment at 40 per cent of its new value. This may be an over-estimate. Nevertheless, SIPRI data indicate that Germany has become a very significant source of major conventional weapons for many countries. One reason for the lack of attention to these exports is that 75 per cent of deliveries in the period 1988–92 were made to industrialized countries—either members of NATO or countries such as Switzerland—which are not considered sensitive destinations. However, transfers to a NATO ally forced the resignation of Defence Minister Gerhard Stoltenberg in March 1992 when armoured vehicles supplied by Germany were used by the Turkish Government against Kurdish separatists.

[119] Solchanyk, R., 'Ukraine: the politics of reform', *RFE/RL Research Report*, 20 Nov. 1992, p. 4.

Two of the most important customers for German major conventional weapons are Greece and Turkey. Much of this equipment is transferred under programmes formulated during the cold war and intended to strengthen the southern flank of NATO in the context of possible conflict with the Warsaw Treaty Organization (WTO). In 1992, when the strategic and political environment had changed entirely, the arms transfer policy remained largely unchanged. Major armed conflicts are under way in Bosnia and Herzegovina and Croatia and the south-eastern part of Europe is characterized by political instability. The southern and eastern borders of Turkey are politically unsettled while across the Caspian Sea the nature of future political developments within and between Afghanistan, Iran, Pakistan and the newly independent countries of Central Asia is also uncertain.[120]

### *China*

China is a small arms exporter compared with the United States or Russia or the countries of Western Europe. According to SIPRI estimates China accounted for 8 per cent of deliveries of major conventional weapons in 1992. However, China (along with North Korea) has become the central focus for governments interested in preventing the transfer of medium- and long-range ballistic missiles. Recent reports have alleged Chinese sales of both ballistic missiles and the production technologies needed to develop and produce them to Iran, Pakistan and Syria. The Chinese Government has denied selling missiles to any of these countries except Pakistan, to which it acknowledged the sale of 'a very small number of short-range tactical missiles' in 1991.[121] In early 1993 reports from Beijing suggested that Chinese officials were prepared to acknowledge transfers of missiles of the type widely known as the M-11, while in Pakistan the former Army Chief of Staff, General Mirza Aslam Beg, also apparently confirmed that these deliveries had taken place.[122] At the official level none of the three countries named as recipients has acknowledged the transfers.

One reason that missile transfers attract so much international attention is that they would, if confirmed, require US sanctions against China unless the President ruled that waiving sanctions was essential to US national security.[123] However, such transfers represent a small proportion of Chinese exports. The most important elements of Chinese arms exports are fighter aircraft, armoured vehicles and naval vessels. South Asia is the most important market for the A-5 ground attack aircraft and the F-6 and F-7 fighter aircraft, with Bangladesh, Myanmar and Pakistan being the most important customers.

[120] The strategic environment of Turkey and its impact on arms procurement are discussed in appendix 10E.

[121] Recent summaries of the issue of Chinese missile sales are contained in Kan, S. A., Congressional Research Service, *Chinese Missile and Nuclear Proliferation*, CRS Issue Brief (US Library of Congress: Washington, DC, 16 Nov. 1992); McCarthy, T. V., *A Chronology of PRC Missile Trade and Developments* (Monterey Institute of International Studies: Monterey, Calif., 12 Feb. 1992).

[122] *Interavia Air Letter*, 9 Dec. 1992, p. 5; *Far Eastern Economic Review*, 7 Jan. 1993, p. 6.

[123] In Mar. 1992, President Bush lifted sanctions imposed on China in June 1991 in the wake of the use of armed force against students demonstrating in Tiananmen Square.

Sales of armoured vehicles have been concentrated on Pakistan and Thailand. Bangladesh and Thailand have been the most important customers for Chinese naval vessels.

Efforts to market Chinese weapons in Africa and Latin America in the 1980s did not lead to any significant market penetration outside Asia. Chinese sales to Middle Eastern countries appear to have reflected the inability of regional countries to buy weapons from other sources rather than a preference for Chinese systems.[124]

In spite of explanations which cast Chinese arms exports in terms of efforts to secure hard currency, in many cases Chinese weapons are transferred on concessional terms. Transfers to Myanmar and Thailand have largely been paid for through commodity exchange. Moreover, arms transfers have helped build better political relations with countries that were suspicious of or hostile towards China either because of their anti-communist sentiments (such as Iran, Thailand and Saudi Arabia) or because of past Chinese support for domestic insurgencies (such as Myanmar and Thailand). This is not to say that China is indifferent to monies received for weapons.

## Arms imports by Asian countries

Given the size of the Asian region—defined to include all of the countries located on mainland Asia from Afghanistan to China along with the island states of Brunei, Indonesia, Japan, the Philippines, Singapore, Sri Lanka and Taiwan—the fact that the region overtook the Middle East as a market for international arms transfers in the late 1980s is not surprising. The changing regional distribution is indicated in appendix 10C.

Asia is more commonly divided into the three sub-regions: North-East Asia, South-East Asia and South Asia for purposes of analysis. However, pan-Asian relations (between China, India and the former USSR) have influenced arms procurement choices in the region. Moreover, the sub-regions contain bilateral relationships which influence procurement choices (such as relations between India and Pakistan, China and Taiwan or North and South Korea). Given that arms procurement policies in Asia have been shaped by global, regional, sub-regional and domestic factors it is not possible to do more than sketch the broad outlines of these policies and note some specific developments that occurred in 1992.[125]

Asian countries are making arms procurement choices in an environment of great uncertainty about the path of future political development within the region and between regional and extra-regional countries (especially the United States). A number of territorial disputes between states in the region

[124] The pattern of Chinese arms transfers is described in Bates Gill, R., 'Curbing Beijing's arms sales', *Orbis*, vol. 36, no. 3 (summer 1992), pp. 379–96; Bitzinger, R. A., 'Arms to go: Chinese arms sales to the Third World', *International Security*, vol. 17, no. 2 (fall 1992), pp. 84–111.

[125] An introduction to the discussion of whether and how arms control might enhance Asian regional security is contained in UN Department for Disarmament Affairs, *Confidence-building Measures in the Asia–Pacific Region* and *Confidence and Security-building Measures: From Europe to other Regions*, Disarmament Topical Papers 6 and 7 (United Nations: New York, 1991).

remain unresolved and there is no multilateral framework for conflict resolution in Asia. Nevertheless, the probability of a major regional conflict is low in comparison with Africa, the Middle East or Europe.

At the regional level, it is possible that major powers—China, India, Indonesia, Japan, Russia and the USA—will compete with one another for political and economic influence. In addition there are unresolved territorial disputes between China and India, China and Japan, and China and Russia. The unpredictability of domestic political development in the large countries is another primary source of insecurity in Asia. Japan and the USA are the only two large countries in which an upheaval in domestic politics is very unlikely in the next decade. Problems in relations between large states and small neighbouring states are another possible source of insecurity. For the two Korean states, the countries of South-East Asia and Taiwan, relations with China are a particular source of concern. These countries have the problem of preparing collective approaches to China without bringing about the outcome that they seek to avoid—a deterioration of relations with their giant neighbour.[126] Finally, some bilateral relationships have led to major conflicts in the past. In addition to the persistent bilateral conflicts noted above there are periodic tensions such as those between Bangladesh and India, Laos and Thailand, Myanmar and Thailand, and Thailand and Viet Nam.

Under these circumstances it is natural that many regional governments regard the preparedness of their armed forces as the most important element of their security policy.

Procurement choices are being shaped by relative levels of economic development across the region. Many Asian countries began to develop successful economies in the 1970s and are replacing equipment produced in the 1950s (in some cases the 1940s) which was transferred as military assistance by the USA and the former USSR. However, there are examples of countries introducing new capabilities into their armed forces for the first time. South Korea is developing a significant conventional submarine fleet while both Malaysia and Thailand are seeking to establish submarine squadrons. Many countries in the region have recently established a capability for maritime reconnaissance and several are upgrading the fighter aircraft deployed in their air forces.[127] These are expensive capabilities involving the purchase of advanced systems from abroad. Nevertheless, modernization is affordable within the existing framework of military expenditure/gross domestic product (GDP) ratios because of the economic growth enjoyed by many of the countries in Asia in the 1980s. The share of GDP allocated to defence in the period 1981–90 was stable or declining in Indonesia, Japan, South Korea,

[126] On the other hand, Indonesia—which is large and geographically removed from China—has consistently expressed concern about Chinese policies. Recent statements include those by Indonesian Defence Minister Benny Murdani, in Richardson, M., 'Energy plus military menace', *International Herald Tribune*, 2 Dec. 1992, p. 1, and by Foreign Minister Ali Alatas, in Richardson, M., 'A call to control arms race in Asia', *International Herald Tribune*, 29 Oct. 1992, p. 6.

[127] Details of these programmes are contained in appendix 10C.

Malaysia, Singapore, Taiwan and Thailand while in India growth in this share has been less than 1 per cent.[128]

Procurement is also being shaped by industrial policies chosen in the region. In contrast with the Middle East—where imports of finished systems seem certain to remain the dominant form of arms procurement—several Asian countries will sustain and may increase their indigenous arms-production capacities.[129] China, India and Japan have arms industries that are significant even in global terms and all three continue to invest in further research and development of advanced major conventional systems, notably in the naval and aerospace sectors.[130] Of the smaller Asian countries both South Korea and Taiwan have made significant investments in their arms industries and the same is true to a lesser extent of Indonesia and Singapore. While the quality of empirical data is poor, the evidence available suggests that India, Japan and South Korea increased their military research and development expenditure at a rapid rate in the 1980s (albeit from a relatively low baseline level).

This preference for licensed production and technology transfer can be seen in several of the largest recent transactions involving countries in the region. In 1991 and 1992 China has taken delivery of at least 24 Su-27 Flanker fighter aircraft bought from Russia and the eventual Chinese requirement is believed to be around 72 aircraft.[131] However, China has also been conducting wide-ranging discussions with the Russian arms industry concerning transfers of advanced technologies into the Chinese aerospace industry in particular.[132] The most ambitious modernization programmes in South Korea all involve local production—the US F-16 fighter and the German Type-209/3 submarine are being produced under licence while the local main battle tank, the K-1 Rokit, incorporates many features of the US M-1 Abrams tank.

Even looking ahead 10–15 years it is very unlikely that the place occupied by Asian countries in the global arms market will be comparable to the place they now occupy in the manufacture of civilian goods. However, if the countries of Asia are unlikely to rival the United States or European arms producers, they are likely to play a greater role than they do today.

Indonesia, Singapore, South Korea and Taiwan all have a limited domestic demand for arms and military equipment and the existence of production capacities in excess of domestic demand will add to commercial pressures to

[128] Deger, S., Loose-Weintraub, E. and Sen, S., 'Tables of world military expenditure', *SIPRI Yearbook 1992* (note 6), appendix 7A.3, pp. 265–66.

[129] The crisis in the Israeli arms industry is discussed in this chapter. Elsewhere in the Middle East, Egypt has a limited arms industry and Iran is investing in developing such an industry.

[130] For recent discussions of the Chinese arms industry, see Schichor, Y., 'The conversion of military technology to civilian use in China: from the 1980s to the 1990s', Paper presented to the Conference on China: Science and Technology towards the Year 2000, Beijing, 25–31 Oct. 1992; and Frankenstein, J., 'The People's Republic of China: arms production, industrial strategy and problems of history', ed. Wulf (note 3), chapter 14, pp. 271–319. For recent developments in the Japanese arms industry, see Ikegami-Andersson, M., 'Japan: a latent but large supplier of dual-use technology', ed. Wulf (note 3), chapter 15, pp. 320–44. For a recent discussion of the Indian arms industry, see Anthony I., *The Arms Trade and Medium Powers: Case studies of India and Pakistan, 1947–90* (Harvester Wheatsheaf: Hemel Hempstead, 1992).

[131] *Asia Pacific Defense Reporter*, Oct.–Nov. 1992, p. 21; *Asian Recorder*, 23–31 Dec. 1992, p. 22800.

[132] *Defense News*, 20–26 July 1992, p. 3; *Armed Forces Journal International*, Jan. 1993, p. 16.

export. In the case of Japan, where domestic demand is also limited, there are political constraints on arms exports which would be difficult for any government to override. However, these pressures may not be as strong with regard to dual-use goods and technologies.

## Land-system retrofits

As noted in previous *SIPRI Yearbooks*, one form of arms trade which remains largely undocumented is the retrofit market.[133] In this section the term retrofit means modifying an already existent platform through the addition of new engines, electronics and/or weapons. While most major conventional weapon platforms are modified several times during their life these modifications tend to be small in size and in value. Sometimes, however, a weapon platform undergoes a major modification or an almost total rebuild—reflecting the fact that the aircraft, ship or armoured vehicle has a frame that lasts much longer than the engines, electronics or weapons on it. In the major arms-producing countries this activity is routine and provision for through-life improvements in weapon systems is an important part of the weapon design and planning process. However, retrofitting is also an international activity and this international process is the focus of the discussion in this section.

Retrofitting involves trade in sub-systems, not complete systems, and it is not fully reflected in the SIPRI arms trade registers.[134] While impossible to evaluate precisely, this market is likely to grow in importance as large numbers of aircraft, armoured vehicles and artillery are taken out of service from the inventories of countries in NATO and the former WTO. This form of trade is important not only in terms of its value but as a form of technology transfer—the upgrade is often carried out in the country where the platform is located. It also has important repercussions for evaluating military capabilities and formulating arms control policy.

Retrofitting is increasingly being seen as a cost-effective procurement option. Some upgrades offer almost the same combat power for an estimated 20–35 per cent of the cost of a new system. While major new platforms were probably designed with the specific needs of the armed forces of the producer country in mind, retrofit packages can be tailored to the conditions and purchasing power of the buyer. Since there is a range of companies in several countries able to supply retrofit kits, the degree of buyer dependency on a single supplier is reduced.

In addition to cost considerations, this may also be the only way for some countries to acquire modern weapons. Israel and South Africa, for example, have relied heavily on retrofits. In general this is a much less visible trade than the trade in major weapon systems and less likely to provoke political debate in the exporting country. A case in point is the contrast between the upgrading

[133] The issue of fighter aircraft retrofits was discussed in Anthony, I., Courades Allebeck, A., Hagmeyer-Gaverus, G., Miggiano, P. and Wulf, H., 'The trade in major conventional weapons', *SIPRI Yearbook 1991: World Armaments and Disarmament* (Oxford University Press: Oxford, 1991), pp. 226–27. This section is confined to a discussion of land system retrofits.

[134] The registers include some naval upgrades such as radars and missile launchers.

of Turkish tanks by German companies—a process which has no political repercussions in Turkey and the FRG—and the sale of weapons from the inventory of the former GDR, which became a major problem in 1992 when Turkey used them against the Kurds.

Conversely, the relatively low visibility of the process of retrofitting equipment means that this form of trade lacks the political symbolism which may be part of the motivation for some arms transfer activities.

From an industry point of view, the retrofit market is often attractive to small- and medium-sized companies which lack the technical and financial resources to develop major new systems. Some countries may see this as the best way to preserve their arms industries in a difficult financial environment. Other countries—such as Greece and Spain whose plans to develop an arms industry were disrupted by the political changes stemming from the end of the cold war—may find this to be the only way of developing a defence industry. The complexity of this market is likely to grow as more companies (including those in countries such as Israel) learn how to modernize the tens of thousands of armoured vehicles, artillery pieces and land-based air defence systems produced in the former Soviet Union in service around the world.

As table 10.16 indicates, retrofitting land-based systems is a significant industrial activity. The growing importance of retrofits is a function of technological advances—especially reductions in the size and weight of electronically and mechanically engineered products. The effectiveness of a tank is no longer linked to size and weight alone. As important are its fire control system, ammunition, mobility and armour—all of which can be upgraded in almost all existing types of tank. Bolt-on composite, reactive or spaced armour; more compact but more powerful engines; new transmissions and suspensions; new guns; digital computers connected to new sensors and target acquisition systems can transform a 20-year-old tank into a potent adversary for even the newest tanks.

Artillery systems are also retrofitted including self-propelled artillery—such as the US M-109 and M-44 systems—and towed artillery pieces such as the US M-114. Most prominent in these upgrades is the fitting of a longer barrel, increasing the range by more than 30 per cent which also makes it possible to fire new types of ammunition, increasing the range even further.[135] The effectiveness of self-propelled artillery can be improved by changing engines and adding new fire-control systems. Towed guns can be fitted with a small engine to offer limited autonomous movement and to power automated loading systems.

Air defence systems—both anti-aircraft guns and missiles—have been modified to include improved radars and command systems. The widely used US MIM-23 Hawk surface-to-air missile system is one of the main systems to have been upgraded. For example, Egypt has ordered an upgrade of its air

[135] For example, exchanging the barrel of an M-114 increases its range from 15 to 24 km. With special shells this range can be increased to 32 km.

**Table 10.16.** Selected international retrofits of armoured vehicles in progress in 1992

Values are in current US $m.

| Recipient | System | Major upgrades | Supplier | Quantity | Value |
|---|---|---|---|---|---|
| Austria | M-48 | Rebuilding to ARV | FRG | 25–35 | . . |
| Brunei | Ferret | Power pack | UK | . . | . . |
| Denmark | Leopard 1 | Fire-control system | FRG | 100 | . . |
| Egypt | T-55 | Fire-control system; transmission; additional armour | FRG | 200 | . . |
| | T-55 | L-7 105-mm gun | UK | . . | . . |
| Finland | T-55 | Fire-control optics | | 185 | 85 |
| Greece | M-48-A5 | Fire-control system | FRG | 200 | . . |
| India | Vijayanta | Engine | FRG | . . | . . |
| | | Fire-control system | UK | (400) | . . |
| | T-55 | L-7 105-mm gun | UK | 500 | . . |
| | | Fire-control system | Yugoslavia | . . | . . |
| | T-72M | Fire-control system | Yugoslavia | 1 400 | . . |
| Israel | M-60 | Engine | USA | (1 600) | . . |
| Malaysia | Ferret | Power pack | UK | . . | . . |
| Norway | Leopard 1 | Fire-control system; additional armour; electric turret drive | FRG | 104 | . . |
| | M-109 | 39-calibre barrel | FRG | 126 | . . |
| Saudi Arabia | Shahine | Digital electronics; radio data links | France | . . | 438 |
| Singapore | M-113 | Engine | USA | 720 | . . |
| | AMX-13 | Engine | USA | 350 | . . |
| | | Transmission | FRG | . . | . . |
| | | Fire-control system | Israel | . . | . . |
| Spain | AMX-30 | Power pack | FRG | 210 | . . |
| | | Fire-control system | USA/Spain | . . | . . |
| | M-109 | Barrel; electric systems | USA | 96 | 175 |
| Switzerland | Pz-68 | Fire-control system | FRG | 195 | . . |
| | M-113 | Additional armour; engines | USA | 382 | 86 |
| Taiwan | M-48-A5 | Fire-control computer | Canada | . . | . . |
| Thailand | M-113A1 | To M-113A2 standard | USA | 100 | . . |
| Turkey | M-48 | Fire-control system | USA | 760 | 760 |
| | M-48 | To M-48-A5 standard | USA | 402 | |
| | M-44 | 39-calibre barrel; diesel engine | FRG | 168 | . . |
| | M-48-A5 | Fire-control computer | Canada | . . | . . |
| Venezuela | AMX-30 | Fire-control system | Belgium | 81 | . . |
| UAE | AMX-F-3 | Power pack | Netherlands | 20 | . . |
| | AMX-VCI | Power pack | Netherlands | . . | . . |
| Uruguay | M-41 | Diesel engines; fire-control system | Brazil | 20 | . . |
| USA | AAAV-7A1 | Applique armour | Israel | 1 137 | . . |

*Source:* SIPRI arms trade data base.

defence system worth a total of $646 million, including $146 million for 12 Hawk missile batteries.

# V. Arms transfer control initiatives

The *SIPRI Yearbook 1992* contained a description of national and multilateral arms transfer control initiatives. This section is confined to describing how these processes developed in 1992. For an earlier description of the processes the reader is referred to chapter 8 of the *SIPRI Yearbook 1992*. In a related development, in August 1992 the UN Secretary-General submitted a Report on the Register of Conventional Arms detailing the technical procedures necessary for the effective operation of the Register.[136] The Register is not a control mechanism and is discussed in full in appendix 10F of this *Yearbook*.

The discussion of arms transfer control is an item on the agenda of almost every major inter-governmental organization. Many governments also have significant political capital invested in the process. In these circumstances it is likely that the issues controlling the trade in arms and related technologies will become an area of security concern to which governments will increasingly turn their attention. Interaction among suppliers—to co-ordinate export policies and improve their enforcement—is likely to become more extensive. The possibility of a dialogue between suppliers and recipients—which has yet to be attempted—is also likely to be explored.

There is no consensus on what kinds of practical arms transfer control measure are desirable and it is therefore predictable that any progress in multilateral arms control in these areas will occur over the long term (if any can be achieved at all). The problems which need to be solved before a meaningful arms control agreement could be contemplated involve basic issues—not least the ultimate objective of the exercise—and it remains to be seen whether the various interests of concerned governments can be reconciled.

## Multilateral export control processes in 1992

In 1992 multilateral discussions of arms export regulation occurred at the United Nations, and within the European Community, the CSCE and the Group of Seven (G7) country groupings. In addition, the members of the Missile Technology Control Regime (MTCR) agreed certain changes in the document defining their area of activity.

### *The United Nations*

As agreed at their meeting in London in October 1991 the five permanent members of the UN Security Council (the 'P5') met twice in 1992 in Washington to discuss arms transfers and the proliferation of 'weapons of mass destruction' in the Middle East.

[136] UN General Assembly, *Report on the Register of Conventional Arms*, UN document A/47/342 (United Nations: New York, 14 Aug. 1992).

A meeting took place between a group of government experts on 20–21 February 1992 and a plenary meeting of the P5 on 28–29 May was attended by high-level officials.[137] The final communiqué from the May meeting focused exclusively on issues related to weapons of mass destruction.[138] The absence of any reference to arms transfers reflected the failure of the P5 to move beyond the positions they adopted on this issue in 1991.

The P5 meetings stumbled over irreconcilable disagreements about both their purpose and scope. In terms of the purpose of the discussions there is no agreement over whether the intention is to aim at increasing transparency in arms transfers or to try and limit weapon flows. In terms of the scope of the discussions, China has argued from the beginning that the P5 process should try and establish guidelines for global application while the United States has presented this as a process confined to the Middle East.

The failure to establish a coherent agenda has made it impossible to resolve more specific arguments. There is consensus among the five countries on the need to prohibit transfers of biological, chemical and nuclear weapons. However, the question whether or not to limit transfers of potential weapon delivery systems led to arguments about how to classify certain systems in the discussions. China has pointed out the inconsistency of the US approach of isolating one sub-category of equipment (ballistic missiles) for limitation while presenting the main focus of the talks as being about transparency.[139] While arguing that all ballistic missile transfers should be stopped, none of the other four parties is prepared to discuss limitations on transfers of combat aircraft or naval systems other than on a case-by-case basis.

The five governments agree that transparency measures are useful but disagree about how these should be implemented. The USA has argued for advance notification of arms sales into the region. China on the other hand has not gone beyond a commitment to provide retrospective information about items delivered. In 1992 the P5 did exchange information with one another on a confidential basis about deliveries of equipment to the region in the year 1991.[140]

In the context of the dispute over geographical scope there is disagreement not only about global versus regional approaches to arms transfer control but also about the specific boundaries of the Middle East. At the Washington meeting the USA apparently argued for the inclusion of Libya in the region while China wanted to include both Greece and Turkey.

As a result of these fundamental disagreements the future of the P5 talks was already in doubt after the May meeting. However, it was tentatively agreed that there would be a follow-on meeting in Moscow—although no date

[137] *Arms Sales Monitor*, no. 13–14 Mar.–Apr. 1992, p. 5; *Arms Control Today*, June 1992, p. 21.

[138] The communiqué is reproduced in appendix 10G.

[139] While it isnot an official position, the rationale for China's approach is explained in Hua Di, 'The arms trade and proliferation of ballistic missiles in China', ed. E. H. Arnett, *Arms Sales versus Non-proliferation: Economic and Political Considerations of Supply, Demand and Control*, PSIS Proceedings (American Association for the Advancement of Science: Washington, DC: 1992), pp. 3–6.

[140] *Aviation Week & Space Technology*, 8 June 1992; *Aviation Week & Space Technology*, 15 June 1992; BASIC Report 22 (British American Security Information Council (BASIC): London, 3 June 1992).

was set. In October 1992 the Chinese Government suspended its participation in the process (it did not withdraw) when the United States announced the sale of 150 F-16 fighter aircraft to Taiwan.[141]

At the end of 1992 no further meetings of the P5 had been held and none were scheduled. Whether the discussions will resume as a four-country initiative was unknown. There is no evidence that the USA, France or the UK will modify its position to attract China back to the talks or that China can be persuaded to change its position. Coercing a change in Chinese policy seems impossible. China's important position on the Security Council and its central position in Asian affairs make political isolation impossible. The Chinese economy is growing quickly and foreign financial and industrial engagement in China is expanding rather than contracting.[142] The P5 process seems to have an uncertain future at best.

### *The European Community*

Three of the organs of the EC—the Council of Ministers, the European Parliament and the Commission of the EC—have played an active role in the debate on aspects of arms export policy.

The European Parliament expressed support for the development of an EC arms export policy in 1989, 1990, 1991 and, on 17 September 1992, passed yet another resolution urging action in this area.[143] More important than initiatives originating with the European Parliament have been the deliberations of the inter-governmental process of European Political Co-operation (EPC) and the efforts of the EC Commission to devise an export control process compatible with the completion of the internal market.

At the June 1991 Luxembourg European Council meeting the Ministers of Foreign Affairs of EC countries agreed on seven common criteria for arms exports in the framework of EPC.[144] At the meeting on 26–27 June 1992 in Lisbon the European Council added an eighth criterion—in making decisions on whether or not to permit arms exports members should consider the compatibility of arms exports with the technical and economic capacity of the recipient country, taking into account the desirability that states should achieve their legitimate needs of security and defence with the least diversion for armaments of human and economic resources.

As with the seven criteria agreed in 1991, the interpretation of this guideline will be decided by the competent authority in the country from which any export originates. The inclusion of this criterion reflected the interest of some EC member states—Germany and the Netherlands—in discussing the linkage between security and economic development.

[141] *Peace* (Beijing), no. 28 (Dec. 1992), pp. 16–17.

[142] It was not coincidental that the Chinese Foreign Minister announced the decision during a visit to Israel. This underlined that in the post-cold war period, Chinese political and diplomatic contacts are growing, not contracting.

[143] European Parliament, *Resolution on the Community's Role in the Supervision of Arms Exports and the Armaments Industry*, document no. PE 161.873, mimeo, 17 Sep. 1992.

[144] Reproduced in Anthony *et. al.* (note 6), pp. 295–96.

In 1992 the question whether the issue of arms exports should be brought into the competence of the Commission of the European Communities was effectively decided.[145] At the Maastricht summit meeting of the EC the suggestion that Article 223—which gives national governments exclusive jurisdiction over questions of arms and military equipment—should be deleted from the Treaty of Rome was discussed and rejected. In the political circumstances of 1992 (when ensuring ratification of the Maastricht Treaty became the main priority for all Community institutions) efforts to dilute national control over all aspects of foreign and security policy were overridden.

By contrast, the Commission made progress in devising a regulation on the control of exports of dual-use goods and technologies. In January 1992, following the presentation of a Commission report on the issue, an *ad hoc* working group was established by the European Council to help prepare a final proposal for the regulation. The proposal was completed on 31 August 1992.[146]

Implementation of export controls will be by the competent authorities in each member state. The Commission will collect and distribute information. The harmonized EC regulation as proposed will contain five essential elements:

1. A common list of dual-use goods and technologies subject to control by all EC member states. This list is a modified version of the industrial list employed in enforcing the Co-ordinating Committee for Multilateral Export Controls (COCOM) embargo;
2. A common list of destinations to which exports should be controlled. Whether there are destinations to which all exports should be proscribed had not been decided;
3. Common criteria for issuing licences;
4. The establishment of a permanent forum or mechanism for co-ordinating licensing and enforcement policies and procedures;
5. The establishment of procedures for administrative co-operation between licensing and enforcement agencies including a system for information exchange.

At the time of writing the member governments had not considered the regulation.

### *The Conference on Security and Co-operation in Europe*

In 1992 the issue of arms transfer control also became a more important aspect of CSCE activity. In 1991 the arms trade-related activity of the CSCE was linked to the question of conflict resolution in the former Yugoslavia. In 1992

[145] See Courades Allebeck, A., 'The European Community: from the EC to the European Union', ed. Wulf (note 3).

[146] European Communities, Commission, 'Proposal for a Council Regulation (EEC) on the control of exports of certain dual-use goods and technologies and of certain nuclear products and technologies', document no. COM(92) 317 final (Office for Official Publications of the European Communities: Luxembourg, 31 Aug. 1992).

this remained an important issue on the CSCE agenda. However, the CSCE has also been linked to the issue of developing a multilateral arms transfer control process.

In January 1992 at their Prague Council meeting, CSCE Foreign Ministers agreed a Declaration on Non-Proliferation and Arms Transfers.[147] The declaration commits all CSCE members 'to provide full information to the United Nations Register of Conventional Arms'. In addition, members declared that arms transfers should be included as a matter of priority in the work programme for the arms control process.

Defining how this issue will be dealt with as an arms control problem will be a task for the Forum for Security Co-operation established within the CSCE in the framework of the Helsinki Document 1992.[148] One of the objectives of the Forum will be to support and enhance regimes on non-proliferation and arms transfers. Initially this might be expected to focus on sharing expertise and information which will permit all CSCE members to establish effective import and export regulations on a national basis. However, over the longer term one task for the CSCE may be to define acceptable criteria. This is hinted at in the Programme for Immediate Action, which notes the need to establish 'a responsible approach to international armaments transfers'.

### *The Group of Seven*

At the meeting of the heads of government and heads of state of the G7 in London in July 1991 the countries represented made a Declaration on Conventional Arms Transfers and NBC [nuclear, biological and chemical] Non-Proliferation.[149]

The issue also found its way into the Political Declaration issued after the G7 meeting in Munich on 7 July 1992. The Declaration included the following commitments:

> We will continue to encourage all countries to adopt the guidelines of the Missile Control Technology Regime (MTCR) and welcome the recent decision by the plenary session of the MTCR to extend the scope of guidelines to cover missiles capable of delivering all kinds of weapons of mass destruction. Each of us will continue our efforts to improve transparency and consultation in the transfer of conventional weapons and to encourage restraint in such transfers. Provision of full and timely information on the UN Arms Register is an important element in these efforts.
>
> We will continue to intensify our co-operation in the area of export controls of sensitive items in the appropriate fora to reduce threats to international security. A major element of this effort is the informal exchange of information to improve and harmonize these export controls.[150]

[147] The Declaration is reproduced in appendix 10G.

[148] A description of the Forum is contained in chapter 5 of Helsinki Document 1992, released at the Helsinki CSCE summit meeting in July 1992. See appendix 5A.

[149] Reproduced in Anthony *et al.* (note 6), appendix 8A, pp. 303–304.

[150] *Political Declaration issued at the Group of Seven Economic Summit*, Munich, 7 July 1992, articles II.5 and II.6; reproduced in *US Department of State Dispatch*, vol. 3, no. 5 (Aug. 1992), p. 7.

*The Missile Technology Control Regime*

At the Oslo Plenary Meeting of the MTCR members Greece, Ireland, Portugal and Switzerland attended for the first time. The meeting agreed a joint appeal to all countries to adopt the Guidelines for Sensitive Missile-Relevant Transfers together with its Equipment and Technology Annex—which are the basic documents underpinning the MTCR. However, because the MTCR has no such thing as observer status, it was not possible for China, Israel and Russia—all of which have done this—to attend plenary meetings.[151]

The MTCR participants agreed to extend the scope of the Equipment and Technology Annex to reflect concern about possible proliferation of delivery systems for chemical and biological weapons. The MTCR originally focused on delivery systems for nuclear weapons only and applied to missiles with a range in excess of 300 km and a payload of 500 kg or greater.[152] In 1992 missiles capable of a maximum range equal or superior to 300 km were included in a new item 19 under category II of the Equipment and Technology Annex. This extension in the coverage of the MTCR guidelines had been implemented in national legislation by all members by 7 January 1993.

At the MTCR Plenary Meeting in Canberra, Australia on 8–11 March 1993 Iceland joined, expanding the membership of the Regime to 23 states. In addition, the MTCR welcomed applications from Argentina and Hungary to participate in the MTCR and agreed to invite these countries to become partners.[153]

*The Co-ordinating Committee for Multilateral Export Controls*

Most of the items whose export control was co-ordinated through the COCOM have now been taken into the terms of reference of other multilateral bodies—the International Atomic Energy Agency (IAEA), the Australia Group, the international organization responsible for implementing the Chemical Weapons Convention and the MTCR. The future of COCOM appears to be as a mechanism which will remain in being for an interim period before the Committee is disbanded. During this interim period the function of COCOM will be to introduce new members—especially those of Central and Eastern Europe—into these other multilateral export control agencies. In 1992 this reorientation of COCOM away from being an embargo towards becoming a multilateral export regulation continued.

These changes have had direct consequences for several Central and East European countries. Hungary is not a member of COCOM. However, the export regulations introduced by Hungary in 1990 have been progressively modified to prevent unauthorized re-transfers of technology sold to Hungary by COCOM members.

[151] *Missile Technology Control Regime*, Oslo Plenary Meeting, 29 June–2 July 1992, Norwegian Royal Ministry of Foreign Affairs Press Release 119/92, 2 July 1992.

[152] The MTCR is described in Anthony, I., 'The Missile Technology Control Regime', ed. I. Anthony, SIPRI, *Arms Export Regulations* (Oxford University Press: Oxford, 1991), pp. 219–27.

[153] *Missile Technology Control Regime Meets in Canberra*, News Release by the Australian Department of Foreign Affairs and Trade, Canberra, 11 Mar. 1993.

This was a pre-condition established by COCOM members before granting a generalized 'presumption of approval' for many categories of goods currently contained in COCOM control lists. This practice will apply to Czechoslovakia and Poland once they have implemented export control measures which satisfy the COCOM group.[154] In Czechoslovakia the implementation of export control measures has been complicated by the judicial separation into two sovereign states.[155] Hungary became the first former WTO member to benefit from the relaxation of the COCOM embargo in February 1992.[156] In April 1992 the Hungarian Government introduced the COCOM International Munitions List and the International Atomic Energy List into its national regulations. This creates the conditions under which Hungary may purchase arms and some dual-use technologies from members of COCOM or countries, such as Sweden, which apply certain export control measures derived from COCOM.

On 2 June 1992 COCOM members decided to propose widening the geographical coverage of the relaxed export controls to include not only Czechoslovakia, Hungary and Poland but also members of the CIS and other Central and East European countries. A new body, to be called the COCOM Co-operation Forum, was proposed.[157] Technology transfers and sales of controlled items would be conditional on the establishment of effective export regulations and enforcement procedures. Members of the Forum would offer advice and assistance to Co-operation Forum partners on how to establish such procedures. The Co-operation Forum met for the first time in Paris on 23–24 November 1992 where 42 countries were represented.[158]

## National arms transfer control initiatives in 1992

Most of the changes made to national export regulations in 1992 were taken in order to implement mandatory decisions reached in United Nations arms embargoes or consensus decisions taken by members in the framework of COCOM and the MTCR. On 27 April 1992 Argentinian President Carlos Menem established a Sensitive Exports Regime by Decree. Through this decree Argentina established a control mechanism which applies to nuclear, biological and chemical materials and technologies. However, the regime also represents the administrative implementation of the decision taken in May 1991 to apply the Missile Technology Control Regime in Argentinian law.[159] However, there have been several other national initiatives in 1992.

[154] Recent changes in COCOM are described in US General Accounting Office, *Export Controls: Multilateral Efforts to Improve Enforcement*, Report no. GAO/NSIAD-92-167 (US Government Printing Office: Washington, DC: May 1992).

[155] *RFE/RL Research Report*, 27 Nov. 1992, pp. 58–59.

[156] *RFE/RL Research Report*, 18 Sep. 1992, p. 59.

[157] 'State Department statement on COCOM', *US Department of State Dispatch*, 8 June 1992, p. 457.

[158] *Defense News*, 30 Nov.–6 Dec. 1992, p. 4.

[159] The Decree establishing the Sensitive Exports Regime was reproduced by the United Nations as UN document A/47/371/Add.2 (United Nations: New York, 10 Nov. 1992).

*Russia*

Russia moved quickly to create a new administrative export control apparatus in 1992. The dissolution of the USSR left the armed forces of the CIS with an ambiguous legal status and many both in Russia and outside were concerned that this ambiguity would permit an unconstrained proliferation of the vast inventory of arms and equipment located in Russia. The power to control exports was established by a Presidential Decree of 22 February 1992, banning trade in a range of 'strategic goods and commodities' (including precious stones and metals as well as defence materials and equipment) without authorization.[160] Yeltsin had earlier ordered the preparation of a draft law on arms exports, intending that it be ready by 1 October 1992.[161] By the end of 1992, however, no such legislation had come before the Russian Parliament.

In January 1992 a parliamentary body was established to draft legislation to oversee Russian arms exports. Members of this body are drawn from four permanent Committees—on Industry and Energy; International Affairs; Defence and Security; and Budget, Taxation and Pricing—although the Committee on Industry and Energy played a dominant role. This body will apparently review all export decisions where transfers are made on credit or free of charge and all deals with a value in excess of $50 million.[162] It is unclear whether the oversight function of this parliamentary body is already established under the February Decree or whether it is to be established by the awaited export control law.

By the end of 1992 this law—first promised by then Soviet Foreign Minister Shevardnadze in a letter to the UN in August 1990—had not been presented. However, there is no reason to question the intention to introduce such legislation. The Russian Conversion Law, enacted on 20 March 1992, included a reference to a forthcoming law on arms exports and established a directory of manufacturers, enterprises, associations and organizations that may conduct foreign trade negotiations and conclude a contract.[163] This directory of legitimate exporters is seen as a necessary complement to export legislation.

On 12 May President Yeltsin approved regulations governing the procedures for state control of exports and imports of arms, military equipment, work and services in this field.[164] Under these regulations controlled goods require an export licence issued by the Ministry of Foreign Economic Relations. The list of controlled items to which the regulations apply was

[160] 'Decree bans free trade in strategic goods', *Interfax*, 27 Feb. 1992, in FBIS-SOV-92-040, 28 Feb. 1992, p. 27.

[161] Interview with Vladimir Shibayev, deputy chairman of the Committee for Foreign Economic Relations, in 'Official describes arms export controls', *Nezavisimaya Gazeta*, 19 Feb. 1992, p. 2, in FBIS-SOV-92-039, 27 Feb. 1992, pp. 31–34.

[162] Malyshkin, A., 'Russia: Parliament to control arms sales', *Military News Bulletin*, no. 1 (Jan. 1992), pp. 1–2; Interview with General S. A. Karaoglanov, Chairman of VO Oboronexport (General Defence Export Corporation), *Military Technology*, vol. 16, no. 10 (Oct. 1992), pp. 53–56.

[163] 'Russian law on conversion: defence industry joins in economic reform', *Military News Bulletin*, no. 3 (Mar. 1992), pp. 1–2.

[164] Interview with Vitaly Vitebsky, vice-chairman of the Committee for Industry and Energy of the Russian Supreme Soviet, *Military News Bulletin*, no. 5 (May 1992), pp. 6–7; *Aviation Week & Space Technology*, 15 June 1992, p. 34.

inherited from the Soviet Union—it was apparently established on 20 March 1989.[165] While the Ministry of Foreign Economic Relations issues licences, the decision to approve or deny a request for a licence are made by the KVTS (Commission for Military Technical and Economic Co-operation) headed by the Prime Minister. The KVTS also includes as members the Ministers of Foreign Economic Relations, Foreign Affairs, Defence, Economics, Finance and Security and the Head of the External Intelligence Service.[166]

Late in 1992 the export control apparatus of the Russian Federation was expanded to include items which could be used for military production. A new committee, with representatives of the Ministry of Economics, the State Customs Committee and the Ministry of Foreign Economic Relations, was established to co-ordinate the export of raw materials, equipment, technology and information which may be used to produce weapons of mass destruction.[167]

### *The United States*

The 1993 Defense Authorization Act contained two provisions which will directly affect US arms export control.

On the initiative of the US Senate Title XVI of the Authorization Act was included as the Iran–Iraq Arms Non-Proliferation Act of 1992. The provision imposes sanctions on countries and individuals which supply nuclear, chemical, biological or advanced conventional weapons and related technology to Iran or Iraq. Where such supplies occur, sanctions are mandatory. However, it is for the President to determine when such transfers have taken place and he can extend a waiver on sanctions where it is deemed to be essential to the national interest of the United States.

Section 1365 of the Authorization Act imposed a one-year moratorium on transfers of anti-personnel land mines abroad. As the United States is not a major exporter of such systems, the moratorium was intended to 'set an example' and to establish a policy of seeking a verifiable international agreement prohibiting the sale, transfer or export of such mines and limiting their use and deployment.

### *Sweden*

A new law on war *matériel* was adopted by the Swedish Parliament on 9 December 1992 to take effect on 1 January 1993. The principle underpinning the law remains unchanged—a ban on arms exports to which the government is entitled to make exceptions within the framework of certain guidelines. Important changes have been made in the classification of equipment subject to control and to the guidelines.[168]

165 *RFE/RL Research Report*, 5 June 1992, p. 49.

166 Interview with General S. A. Karaoglanov (note 162).

167 'Moscow tightens control on export of military materials', *Interfax*, 8 Nov. 1992, in FBIS-SOV-92-218, 10 Nov. 1992.

168 *Regeringens proposition med förslag till lag om krigsmateriel* [Swedish Government Bill proposal for the Military Equipment Act], Bill 1991/92:174 (Government Printer: Stockholm, 1992); *Utrikesutskottets betänkande om Krigsmaterielexport* [Swedish Parliamentary Foreign Affairs

The list of items classified as war-fighting *matériel* has been broadened and divided into two categories: war-fighting *matériel* and other war *matériel.* Data and intellectual property have been included in the list of controlled items for the first time.

The 'two-tier' classification will mean a considerable liberalization of export rules for items previously contained in the undifferentiated list and now classified as other war *matériel.* All goods remain subject to three unconditional criteria which disallow an export. These are: where an export would be in conflict with international agreements by Sweden, with decisions of the UN Security Council or with international law pertaining to exports by a neutral state in time of war.[169] Whereas four additional restrictions to export previously applied to all regulated goods, they now apply only to war-fighting *matériel.* Other *matériel* is subject to only one additional restriction: extensive and brutal violations of human rights by the recipient country. The three additional restrictions which apply to war-fighting *matériel* are that they may not be exported to recipients engaged in armed interstate conflict; states involved in international conflict which may lead to armed conflict; or states with internal armed disturbances.

The law also established rules for export of goods produced by Swedish companies in collaboration with other countries. Exports to the partner country are permitted subject only to the three unconditional criteria; exports of the goods produced to a third country should be subject to the national regulations of the country where the product received its predominant identity. If a collaborative project is deemed to be a vital Swedish interest, exports may be permitted according to the regulations of the partner country even if the item produced has a predominantly Swedish identity.[170]

### *Canada*

In September 1992 the Sub-Committee on Arms Export of the Committee on External Affairs and International Trade of the Canadian Parliament presented a report recommending changes to the Export and Import Permits Act as amended which is the current framework for Canadian export regulation. The report recommended 20 amendments to Canadian export practices.[171] The government has 150 days to consider the report and make what it considers to be an appropriate response.

Committee Report on Exports of War Materiel], Report no. 1992/93:UU1 (Government Printer: Stockholm, 26 Nov. 1992).

[169] The arms export restrictions of the 1907 Hague Conventions on Neutrality were interpreted by Parliament to apply only to exports by state agencies, not to exports by wholly state-owned companies.

[170] *Regeringens proposition med förslag till lag om krigsmateriel* (note 168), p. 42.

[171] Bosley, J. and McCreath, P., *The Future of Canadian Military Goods Production and Export*, Report of the Sub-Committee on Arms Export of the Standing Committee on External Affairs and International Trade, House of Commons Canada, Sep. 1992.

# **Appendix 10A.** The 100 largest arms-producing companies, 1991

IAN ANTHONY, PAUL CLAESSON, GERD HAGMEYER-GAVERUS, ELISABETH SKÖNS and SIEMON T. WEZEMAN

Table 10A contains information on the 100 largest arms-producing companies in the OECD and the developing countries ranked by their arms sales in 1991.[1] Companies with the designation *S* in the column for rank in 1991 are subsidiaries; their arms sales are included in the figure in column 6 for the holding company. Subsidiaries are listed in the position where they would appear if they were independent companies. In order to facilitate comparison with data for the previous year, the rank order and arms sales figures for 1990 are also given. Where new data for 1990 have become available, this information is included in the table; thus the 1990 rank order and the arms sales figures for some companies which appeared in table 9A in the *SIPRI Yearbook 1992* have been revised.

## Sources and methods

*Sources of data.* The data in the table are based on the following sources: company reports, a questionnaire sent to over 400 companies, and corporation news published in the business sections of newspapers and military journals. Company archives, marketing reports, government publication of prime contracts and country surveys were also consulted. In many cases exact figures on arms sales were not available, mainly because companies often do not report their arms sales or lump them together with other activities. Estimates were therefore made.

*Definitions.* Data on total sales, profits and employment are for the entire company, not for the arms-producing sector alone. Profit data are after taxes in all cases when the company provides such data. Employment data are either a year-end or a yearly average figure as reported by the company. Data are reported on the fiscal year basis reported by the company in its annual report.

*Exchange-rates.* To convert local currency figures into US dollars, the period-average of market exchange-rates of the International Monetary Fund, *International Financial Statistics,* was used.

*Key to abbreviations in column 5.* A = artillery, Ac = aircraft, El = electronics, Eng = engines, Mi = missiles, MV = military vehicles, SA/O = small arms/ordnance, Sh = ships, and Oth = other.

[1] The 24 member countries of the Organization for Economic Co-operation and Development are: Australia, Austria, Belgium, Canada, Denmark, Germany, Finland, France, Greece, Iceland, Ireland, Italy, Japan, Luxembourg, the Netherlands, New Zealand, Norway, Portugal, Spain, Sweden, Switzerland, Turkey, the UK and the USA (Yugoslavia participates with special status). For the countries in the developing world, see appendix 10B.

**Table 10A.** The 100 largest arms-producing companies in the OECD and developing countries, 1991[a]

Figures in columns 6, 7, 8 and 10 are in US $ million.

| 1 | 2 | 3 | 4 | 5 | 6 | 7 | 8 | 9 | 10 | 11 |
|---|---|---|---|---|---|---|---|---|---|---|
| Rank | | | | | Arms sales | | | | | |
| 1991 | 1990[b] | Company[c] | Country | Industry | 1991 | 1990[d] | Total sales 1991 | Col. 6 as % of col. 8 | Profit 1991 | Employment 1991 |
| 1 | 1 | McDonnell Douglas | USA | AC EL MI | 10 200 | 9 890 | 18 448 | *55* | 423 | 109 123 |
| 2 | 2 | General Dynamics | USA | AC EL MI MV SH | 7 620 | 8 300 | 9 548 | *80* | 505 | 80 600 |
| 3 | 3 | British Aerospace | UK | AC A EL MI SA/O | 7 550 | 7 520 | 18 687 | *40* | – 269 | 123 200 |
| 4 | 5 | General Motors | USA | AC ENG EL MI | 7 500 | 7 380 | 123 056 | *6* | – 4 452 | 756 000 |
| 5 | 4 | Lockheed | USA | AC | 6 900 | 7 500 | 9 809 | *70* | 308 | 71 300 |
| | | Hughes Electronics (General Motors) | USA | AC EL | 6 600 | 6 700 | 11 541 | *57* | 559 | 93 000 |
| 6 | 6 | General Electric | USA | ENG | 6 120 | 6 450 | 60 236 | *10* | 2 636 | 284 000 |
| 7 | 10 | Northrop | USA | AC | 5 100 | 4 930 | 5 694 | *90* | 201 | 36 200 |
| 8 | 7 | Raytheon | USA | EL MI | 5 100 | 5 500 | 9 274 | *55* | 592 | 71 600 |
| 9 | 9 | Boeing | USA | AC EL MI | 5 100 | 5 100 | 29 314 | *17* | 1 567 | 159 100 |
| 10 | 8 | Thomson S.A. | France | EL MI | 4 800 | 5 250 | 12 634 | *38* | 479 | 105 000 |
| | | Thomson-CSF (Thomson S.A.) | France | EL MI | 4 800 | 5 250 | 6 235 | *77* | 416 | 44 500 |
| 11 | 11 | Martin Marietta | USA | MI | 4 560 | 4 600 | 6 080 | *75* | 313 | 60 500 |
| 12 | 14 | Rockwell International | USA | AC EL MI | 4 000 | 4 100 | 11 927 | *34* | 601 | 87 000 |
| 13 | 13 | United Technologies | USA | AC EL MI | 4 000 | 4 100 | 20 840 | *19* | – 1 021 | 185 100 |
| 14 | 12 | GEC | UK | EL | 3 960 | 4 280 | 16 693 | *24* | 975 | 104 995 |
| 15 | 15 | Daimler Benz | FRG | AC ENG EL MI | 3 920 | 4 020 | 57 252 | *7* | 1 170 | 379 252 |
| 16 | 16 | DCN | France | SH | 3 710 | 3 830 | 3 715 | *100* | . . | 30 000 |
| | | DASA (Daimler Benz) | FRG | AC ENG EL MI | 3 620 | 3 720 | 7 441 | *49* | 30 | 56 465 |
| 17 | 22 | Aérospatiale | France | AC MI | 3 450 | 2 860 | 8 614 | *40* | 38 | 43 287 |
| 18 | 19 | Litton Industries | USA | EL SH | 3 150 | 3 000 | 5 219 | *60* | 64 | 52 300 |

| | | | | | | | | | | |
|---|---|---|---|---|---|---|---|---|---|---|
| 19 | 17 | IRI | Italy | AC ENG EL SH | . . | 3 270 | 54 794 | *5* | – 541 | 368 267 |
| 20 | 21 | Grumman | USA | AC EL | 2 900 | 2 900 | 4 038 | *72* | 99 | 23 600 |
| 21 | 20 | TRW | USA | MV OTH | 2 800 | 3 000 | 7 913 | *35* | – 140 | 71 300 |
| 22 | 18 | Mitsubishi Heavy Industries | Japan | AC MI MV SH | 2 630 | 3 040 | 18 441 | *14* | . . | . . |
| 23 | 28 | Loral | USA | EL | 2 600 | 1 920 | 2 882 | *90* | 122 | 22 000 |
| 24 | 23 | Westinghouse Electric | USA | EL | 2 300 | 2 330 | 12 794 | *18* | – 1 086 | 113 664 |
| 25 | 26 | Tenneco | USA | SH | 2 220 | 2 110 | 13 662 | *16* | – 732 | 89 000 |
| | | Newport News (Tenneco) | USA | SH | 2 220 | 2 110 | 2 216 | *100* | 225 | 28 100 |
| | | Alenia (IRI) | Italy | AC EL MI | 2 140 | 1 640 | 3 879 | *55* | 45 | 30 099 |
| | | Pratt & Whitney (United Technologies) | USA | ENG | 2 100 | 2 000 | 7 171 | *29* | . . | 41 000 |
| 26 | 25 | Texas Instruments | USA | EL MI OTH | 1 950 | 2 120 | 6 784 | *29* | – 249 | 62 939 |
| 27 | 24 | Dassault Aviation | France | AC | 1 870 | 2 260 | 2 544 | *74* | 18 | 11 914 |
| 28 | 36 | LTV | USA | AC MV EL | 1 800 | 1 490 | 5 986 | *30* | 74 | 34 600 |
| 29 | 29 | Textron | USA | AC ENG MV | 1 800 | 1 900 | 7 840 | *23* | 300 | 52 000 |
| 30 | 31 | CEA Industrie | France | OTH | 1 750 | 1 810 | 6 895 | *25* | 233 | 37 300 |
| 31 | 27 | Unisys | USA | EL | 1 750 | 2 000 | 8 696 | *20* | – 1 393 | 60 300 |
| 32 | 30 | Rolls Royce | UK | ENG | 1 680 | 1 830 | 6 219 | *27* | 90 | 57 100 |
| 33 | 40 | E-Systems | USA | EL | 1 550 | 1 400 | 1 991 | *78* | 110 | 18 622 |
| | | MBB (DASA) | FRG | AC EL MI | 1 540 | 1 790 | 2 994 | *51* | . . | 20 730 |
| 34 | 48 | Israel Aircraft Industries | Israel | AC EL MI | 1 410 | 1 120 | 1 606 | *88* | 22 | 17 100 |
| 35 | 35 | INI | Spain | AC A EL MV SH | 1 330 | 1 560 | 17 971 | *7* | – 1 | 142 295 |
| 36 | 37 | SNECMA Groupe | France | ENG OTH | 1 320 | 1 490 | 4 241 | *31* | – 12 | 27 236 |
| 37 | 34 | IBM | USA | EL OTH | 1 300 | 1 600 | 64 792 | *2* | – 2 827 | 373 815 |
| 38 | 32 | EFIM | Italy | AC EL MV | 1 270 | 1 710 | 4 433 | *29* | . . | 35 489 |
| 39 | 38 | GIAT Industries | France | A MV SA/O | 1 220 | 1 430 | 2 003 | *61* | – 71 | 17 000 |
| 40 | 44 | GTE | USA | EL | 1 200 | 1 250 | 19 621 | *6* | 1 580 | 175 000 |
| 41 | 33 | ITT | USA | EL | 1 200 | 1 610 | 20 421 | *6* | 817 | 110 000 |
| 42 | 41 | Oerlikon-Bührle | Switzerl. | AC A EL SA/O | 1 170 | 1 340 | 2 527 | *46* | – 130 | 19 138 |

| 1 | 2 | 3 | 4 | 5 | 6 | 7 | 8 | 9 | 10 | 11 |
|---|---|---|---|---|---|---|---|---|---|---|
| Rank | | | | | Arms sales | | | | | |
| 1991 | 1990[b] | Company[c] | Country | Industry | 1991 | 1990[d] | Total sales 1991 | Col. 6 as % of col. 8 | Profit 1991 | Employment 1991 |
| 43 | 49 | FMC | USA | MV SH OTH | 1 170 | 1 070 | 3 924 | *30* | 164 | 23 150 |
| 44 | 45 | FIAT | Italy | ENG | 1 140[e] | 1 180 | 38 933 | *3* | 898 | 287 957 |
| 45 | 39 | Ordnance Factories | India | A SA/O OTH | 1 120 | 1 430 | 1 166 | *96* | . . | 173 000 |
| 46 | 55 | Gencorp | USA | ENG EL SA/O OTH | 1 110 | 870 | 1 993 | *56* | 32 | 14 500 |
| 47 | 60 | Alcatel-Alsthom | France | EL | 1 100 | 760 | 28 373 | *4* | 1 546 | 213 100 |
| 48 | 47 | Alliant Tech Systems | USA | SA/O | 1 100 | 1 150 | 1 186 | *93* | 39 | 6 700 |
| 49 | 43 | Allied Signal | USA | AC EL OTH | 1 100 | 1 300 | 11 831 | *9* | – 273 | 98 300 |
| | | Aerojet (Gencorp) | USA | ENG EL SA/O OTH | 1 090 | 850 | 1 142 | *95* | 75 | . . |
| 50 | 46 | Matra Groupe | France | EL MI OTH | 1 050 | 1 180 | 4 024 | *26* | 45 | 21 334 |
| 51 | 51 | Kawasaki Heavy Industries | Japan | AC ENG SH | 1 050 | 1 010 | 6 914 | *15* | . . | . . |
| 52 | 54 | VSEL Consortium | UK | MV SH | 920 | 930 | 920 | *100* | 85 | 13 028 |
| 53 | 52 | Siemens | FRG | EL | 900 | 990 | 43 994 | *2* | 1 080 | 402 000 |
| | | Matra Défense (Matra Groupe) | France | MI | 890 | 920 | 886 | *100* | . . | 2 500 |
| 54 | 138 | Celsius | Sweden | A MI MV SA/O SH | 870 | 180 | 1 832 | *47* | 70 | 14 508 |
| | | SNECMA (SNECMA Groupe) | France | ENG | 850 | 650 | 2 566 | *33* | 15 | 13 816 |
| | | Telefunken Systemtechnik (DASA) | FRG | EL | 810 | 920 | 985 | *82* | – 4 | 8 846 |
| 55 | 56 | Diehl | FRG | EL SA/O | 800 | 860 | 1 817 | *44* | . . | 15 529 |
| 56 | 50 | Bremer Vulkan | FRG | EL SH | 780 | 1 050 | 2 006 | *39* | 45 | 15 021 |
| 57 | 61 | Rheinmetall | FRG | A SA/O | 770 | 750 | 2 092 | *37* | 20 | 13 661 |
| 58 | 62 | Thyssen | FRG | MV SH | 770 | 710 | 22 032 | *3* | 313 | 148 557 |
| 59 | 59 | Harris | USA | EL | 760 | 790 | 3 040 | *25* | 20 | 30 700 |
| | | Swedish Ordnance-Bofors | | | | | | | | |

| | | | | | | | | | | |
|---|---|---|---|---|---|---|---|---|---|---|
| | | (Celsius/Nobel) | Sweden | A MI MV SA/O | 740 | 0 | 744 | *99* | 10 | 6 274 |
| 60 | 58 | Ishikawajima–Harima | Japan | ENG SH | 740 | 800 | 7 822 | *9* | 183 | . . |
| 61 | 69 | CAE Industries | Canada | EL | 730 | 640 | 908 | *80* | 29 | 10 000 |
| | | Eurocopter France (Aérospatiale) | France | AC | 720 | 0 | 1 723 | *42* | 38 | 7 525 |
| 62 | 78 | Bath Iron Works | USA | SH | 720 | 550 | 750 | *96* | . . | 10 000 |
| 63 | 67 | Mitsubishi Electric | Japan | EL MI | 710 | 690 | 19 383 | *4* | . . | . . |
| 64 | 42 | Armscor | S. Africa | AC A EL MV SA/O | 710 | 1 330 | 1 016 | *70* | . . | 20 000 |
| | | MTU (DASA) | FRG | ENG | 690 | 760 | 2 148 | *32* | 533 | 17 052 |
| 65 | 66 | Eidgenössische Rüstungsbetriebe | Switzerl. | AC A ENG SA/O | 690 | 700 | 730 | *95* | 10 | 4 495 |
| 66 | 72 | Thiokol | USA | ENG MI SA/O OTH | 690 | 620 | 1 270 | *54* | 53 | 11 500 |
| 67 | 75 | Science Applications Intl. | USA | AC ENG EL | 680 | 570 | 680 | *100* | 37 | 13 100 |
| | | CASA (INI) | Spain | AC | 670 | 780 | 893 | *75* | – 93 | 9 338 |
| 68 | 63 | AT&T | USA | EL | 650 | 700 | 63 089 | *1* | 522 | 317 100 |
| 69 | 76 | Computer Sciences | USA | EL | 620 | 560 | 2 113 | *29* | 68 | 26 500 |
| 70 | 64 | Sequa | USA | ENG EL OTH | 610 | 700 | 1 879 | *32* | – 6 | 15 700 |
| 71 | 85 | Smiths Industries | UK | EL | 600 | 490 | 1 160 | *52* | 130 | 12 100 |
| 72 | 57 | Hercules | USA | ENG EL SA/O OTH | 600 | 800 | 2 929 | *20* | 95 | 17 324 |
| 73 | 68 | Motorola | USA | EL | 600 | 650 | 11 341 | *5* | 454 | 102 000 |
| 74 | 74 | SAGEM Groupe | France | EL | 590 | 570 | 2 081 | *28* | 75 | 15 076 |
| 75 | 71 | Lucas Industries | UK | AC | 570 | 630 | 4 184 | *14* | 91 | 54 900 |
| 76 | 81 | Westland Group | UK | AC | 530 | 510 | 827 | *64* | 34 | 9 060 |
| 77 | 77 | Avondale Industries | USA | SH | 530 | 560 | 776 | *68* | – 14 | 8 200 |
| 78 | 90 | Saab-Scania | Sweden | AC ENG EL MI | 520 | 450 | 4 845 | *11* | 231 | 29 329 |
| 79 | 83 | Teledyne | USA | ENG EL MI | 500 | 500 | 3 207 | *16* | – 25 | 29 400 |
| 80 | 79 | Dassault Electronique | France | EL | 490 | 530 | 685 | *72* | 11 | 3 416 |
| 81 | 70 | Israel Military Industries | Israel | A MV SA/O | 490 | 640 | 520 | *94* | – 239 | 8 500 |
| | | Oto Melara (EFIM) | Italy | A MV MI | 480 | 480 | 484 | *99* | – 35 | 2 149 |
| 82 | 91 | Thorn EMI | UK | EL | 470 | 450 | 6 996 | *7* | 311 | 53 757 |
| 83 | 94 | Rafael | Israel | SA/O OTH | 450 | 420 | 460 | *98* | . . | 5 100 |

| 1 | 2 | 3 | 4 | 5 | 6 | 7 | 8 | 9 | 10 | 11 |
|---|---|---|---|---|---|---|---|---|---|---|
| Rank | | | | | Arms sales | | | | | |
| 1991 | 1990[b] | Company[c] | Country | Industry | 1991 | 1990[d] | Total sales 1991 | Col. 6 as % of col. 8 | Profit 1991 | Employment 1991 |
| | | Hollandse Signaalapparaten (Thomson-CSF, France) | Netherl. | EL | 450 | 490 | 481 | *94* | . . | 4 265 |
| | | AVCO (Textron) | USA | AC | 450 | 550 | . . | . . | . . | . . |
| 84 | 86 | Bombardier | Canada | AC | 440 | 490 | 2 670 | *16* | 94 | 26 692 |
| | | Fincantieri (IRI) | Italy | SH | 440 | 300 | 2 238 | *20* | – 140 | 19 750 |
| 85 | 87 | Racal Electronics | UK | EL | 440 | 480 | 2 843 | *15* | 98 | 35 384 |
| 86 | | Esco Electronics | USA | EL | 440 | 520 | 481 | *91* | – 65 | . . |
| | | Systemtechnik Nord (Bremer Vulkan) | FRG | EL | 430 | 470 | 570 | *75* | . . | 2 441 |
| 87 | 89 | Devonport Management | UK | SH | 430 | 470 | 430 | *100* | . . | 11 460 |
| 88 | 96 | Toshiba | Japan | EL MI | 420 | 410 | 35 056 | *1* | 293 | 168 000 |
| 89 | 88 | Hawker Siddeley | UK | EL | 420 | 480 | 3 843 | *11* | . . | 40 500 |
| | | Agusta (EFIM) | Italy | AC | 410 | 660 | 574 | *71* | . . | 6 998 |
| | | Blohm & Voss (Thyssen) | FRG | MV SH | 400 | 250 | 816 | *49* | . . | 5 758 |
| 90 | 95 | Mannesmann | FRG | MV | 400 | 410 | 14 652 | *3* | 158 | 125 188 |
| | | Krauss-Maffei (Mannesmann) | FRG | MV | 400 | 410 | 853 | *47* | 11 | 5 004 |
| | | SAGEM (SAGEM Groupe) | France | EL | 400 | 390 | 940 | *43* | 30 | 6 006 |
| 91 | 93 | Hunting | UK | SA/O | 400 | 420 | 1 326 | *30* | 24 | 7 302 |
| 92 | 97 | Honeywell | USA | EL MI | 400 | 400 | 6 193 | *6* | 331 | 58 182 |
| 93 | 103 | Penn Central | USA | OTH | 400 | 360 | 1 699 | *24* | 3 | 12 100 |
| 94 | 98 | Lürssen | FRG | SH | 390[e] | 400 | . . | . . | . . | 1 000 |
| 95 | 80 | Dowty Group | UK | AC EL | 390 | 520 | 1 229 | *32* | 58 | 13 000 |
| 96 | 104 | Dyncorp | USA | AC EL | 390 | 360 | . . | . . | . . | . . |
| 97 | 102 | Mitre | USA | EL | 390 | 370 | . . | . . | . . | . . |
| 98 | 108 | Olin | USA | AC EL SA/O OTH | 380 | 360 | 2 275 | *17* | – 13 | 14 400 |

| | | | | | | | | | | |
|---|---|---|---|---|---|---|---|---|---|---|
| 99 | 84 | Hindustan Aeronautics | India | ÁC MI | 370 | 500 | 379 | *98* | 20 | 35 000 |
| 100 | 99 | Sundstrand | USA | AC | 370 | 390 | 1 670 | *22* | 109 | 12 800 |

. . Data not available.

[a] Both the rank designation and the arms sales figures for 1990 are also given, in columns 2 and 7, respectively, for comparison with the data for 1991 in columns 1 and 6.

[b] The rank designation in this column may not correspond to that given in table 9A in the *SIPRI Yearbook 1992*. A dash (–) in this column indicates either that the company did not produce arms in 1990, in which case there is a zero (0) in column 7, or that it did not rank among the 100 largest companies in table 9A in the *SIPRI Yearbook 1992*, in which case figures for arms sales in 1990 do appear in column 7. A figure above 100 in this column shows the actual rank order in 1990, although the company was not included in the SIPRI 100 table in the *SIPRI Yearbook 1992*.

[c] Company names in parentheses after the name of the ranked company are the names of the holding companies. The parent companies, with data pertaining to them, appear in their rank order for 1991.

[d] A zero (0) in this column indicates that the company did not produce arms in 1990, but began arms production in 1991, or that in 1990 the company did not exist as it was structured in 1991.

[e] Data are for 1990

*Note:* The authors acknowledge financial assistance to operate the SIPRI arms production data bank from The John D. and Catherine T. MacArthur Foundation and assistance in the data collection provided by Anthony Bartzokas (Athens), Centre d'Estudis sobre la Pau i el Desarmament (Barcelona), Defence Research & Analysis (London), Ken Epps (Ontario), Ernst Gülcher (Antwerp), Peter Hug (Bern), Keidanren (Tokyo), Rudi Leo (Vienna), Rita Manchanda (New Delhi), Reuven Padhatzur (Tel Aviv), Giulio Perani (Rome), Paul Rusman (Haarlem), Gülay Günlük-Senesen (Istanbul) and Werner Voß (Bremen).

# **Appendix 10B.** Tables of the value of the trade in major conventional weapons

IAN ANTHONY, PAUL CLAESSON, GERD HAGMEYER-GAVERUS, ELISABETH SKÖNS and SIEMON T. WEZEMAN

**Table 10B.1.** Values of imports of major conventional weapons, 1983–92

Figures are SIPRI trend-indicator values, as expressed in US $m., at constant (1990) prices.

| | 1983 | 1984 | 1985 | 1986 | 1987 | 1988 | 1989 | 1990 | 1991 | 1992 |
|---|---|---|---|---|---|---|---|---|---|---|
| World total | 45 006 | 43 098 | 40 106 | 42 964 | 46 555 | 40 034 | 38 133 | 29 972 | 24 470 | 18 405 |
| Developing world | 30 584 | 29 345 | 26 356 | 28 295 | 31 775 | 23 688 | 21 623 | 17 682 | 13 240 | 9 320 |
| LDCs | 1 008 | 1 171 | 1 017 | 1 700 | 1 346 | 2 232 | 3 328 | 2 983 | 1 666 | 350 |
| Industrialized world | 14 422 | 13 752 | 13 750 | 14 668 | 14 780 | 16 346 | 16 510 | 12 290 | 11 230 | 9 086 |
| Europe | 10 808 | 10 409 | 10 136 | 11 050 | 11 358 | 12 411 | 12 366 | 9 697 | 8 901 | 6 583 |
| EC | 3 756 | 3 778 | 2 447 | 3 325 | 3 095 | 4 249 | 5 395 | 3 913 | 6 225 | 4 015 |
| Other Europe | 7 052 | 6 631 | 7 689 | 7 726 | 8 262 | 8 163 | 6 971 | 5 784 | 2 676 | 2 569 |
| Americas | 5 847 | 5 544 | 3 777 | 2 939 | 3 385 | 2 068 | 2 248 | 1 533 | 1 927 | 1 582 |
| North | 1 096 | 1 131 | 1 420 | 1 077 | 1 221 | 929 | 782 | 317 | 1 162 | 960 |
| Central | 1 203 | 759 | 824 | 694 | 338 | 243 | 278 | 314 | 152 | . . |
| South | 3 547 | 3 654 | 1 533 | 1 167 | 1 826 | 896 | 1 188 | 902 | 613 | 622 |
| Africa | 4 091 | 4 718 | 3 727 | 3 605 | 3 195 | 2 367 | 2 017 | 1 346 | 338 | 168 |
| Sub-Saharan | 1 926 | 2 596 | 2 425 | 2 282 | 2 540 | 1 884 | 495 | 1 202 | 234 | 140 |
| Asia | 8 557 | 7 410 | 9 709 | 11 986 | 12 040 | 12 533 | 14 733 | 10 011 | 8 252 | 5 468 |
| Middle East | 14 865 | 14 350 | 12 350 | 12 489 | 16 003 | 9 901 | 5 912 | 6 918 | 4 714 | 4 138 |
| Oceania | 837 | 667 | 407 | 895 | 574 | 754 | 856 | 467 | 338 | 467 |
| OECD | 8 869 | 8 715 | 7 915 | 8 331 | 8 674 | 10 526 | 11 563 | 8 158 | 10 477 | 8 937 |
| CSCE | 11 760 | 11 539 | 11 551 | 12 073 | 12 481 | 13 200 | 13 085 | 9 993 | 10 060 | 7 543 |
| NATO | 5 488 | 5 937 | 4 663 | 5 037 | 5 832 | 6 730 | 7 638 | 5 330 | 8 821 | 6 656 |
| OPEC | 10 091 | 10 983 | 9 830 | 9 199 | 10 278 | 7 376 | 6 341 | 5 675 | 3 403 | 2 414 |
| ASEAN | 1 272 | 1 352 | 1 160 | 1 137 | 1 398 | 1 417 | 955 | 1 048 | 1 560 | 1 060 |

*The following countries are included in each region:*[a]

*Developing world*: Afghanistan, Algeria, Angola, Argentina, Bahrain, Bangladesh, Barbados, Bahamas, Belize, Benin, Bolivia, Botswana, Brazil, Brunei, Burkina Faso, Burundi, Cameroon, Cape Verde, Central African Republic, Cambodia, Chad, Chile, China, Colombia, Comoros, Congo, Costa Rica, Côte d'Ivoire, Cuba, Cyprus, Dominica, Djibouti, Dominican Republic, Ecuador, Egypt, Equatorial Guinea, Ethiopia, Fiji, Gabon, Gambia, Ghana, Guatemala, Guinea, Guinea-Bissau, Guyana, Haiti, Honduras, India, Indonesia, Iran, Iraq, Israel, Jamaica, Jordan, Kenya, Kiribati, North Korea, South Korea, Kuwait, Laos, Lebanon, Lesotho, Liberia, Libya, Madagascar, Malawi, Malaysia, Mauritania, Mali, Marshall Islands, Mauritius, Mexico, Fed. States of Micronesia, Mongolia, Morocco, Mozambique, Myanmar, Namibia, Nepal, Nicaragua, Niger, Nigeria, Oman, Pakistan, Panama, Papua New Guinea, Paraguay, Peru, Philippines, Qatar, Rwanda, St Vincent & the Grenadines, El Salvador, Samoa, Saudi Arabia, Senegal, Seychelles, Sierra Leone, Singapore, Solomon Islands, Somalia, South Africa, Sri Lanka, Sudan, Suriname, Swaziland, Syria, Taiwan, Tanzania, Thailand, Togo, Tonga, Trinidad & Tobago, Tunisia, Tuvalu, Uganda, United Arab Emirates, Uruguay, Vanuatu, Venezuela, Viet Nam, Yemen, North Yemen, South Yemen, Zaire, Zambia, Zimbabwe.

*Less developed countries (LDCs)*[b] Afghanistan, Bangladesh, Benin, Botswana, Burkina Faso, Burundi, Cape Verde, Central African Republic, Chad, Comoros, Djibouti, Equatorial Guinea, Ethiopia, Gambia, Guinea, Guinea-Bissau, Haiti, Kiribati, Laos, Lesotho, Liberia, Malawi, Mali, Mauritania, Mozambique, Myanmar, Nepal, Niger, Rwanda, Samoa, Sierra Leone, Somalia, Sudan, Tanzania, Togo, Uganda, Vanuatu, Yemen, North Yemen, South Yemen.

**Table 10B.2.** Values of exports of major conventional weapons, 1983–92

Figures are SIPRI trend-indicator values, as expressed in US $m., at constant (1990) prices.

| | 1983 | 1984 | 1985 | 1986 | 1987 | 1988 | 1989 | 1990 | 1991 | 1992 |
|---|---|---|---|---|---|---|---|---|---|---|
| World total | 45 006 | 43 098 | 40 106 | 42 964 | 46 555 | 40 034 | 38 133 | 29 972 | 24 470 | 18 405 |
| Developing world | 3 913 | 3 462 | 2 565 | 2 657 | 4 846 | 3 688 | 1 921 | 1 719 | 2 163 | 2 031 |
| LDCs | . . | 27 | . . | 34 | 99 | 3 | . . | . . | 2 | . . |
| Industrialized world | 41 092 | 39 636 | 37 542 | 40 307 | 41 709 | 36 345 | 36 212 | 28 252 | 22 307 | 16 375 |
| Europe | 26 297 | 27 241 | 27 259 | 28 272 | 27 957 | 24 040 | 24 287 | 17 251 | 10 554 | 7 839 |
| EC | 9 999 | 11 426 | 8 655 | 8 138 | 7 604 | 6 939 | 7 827 | 5 889 | 4 739 | 4 717 |
| Other Europe | 16 298 | 15 815 | 18 604 | 20 134 | 20 353 | 17 101 | 16 460 | 11 361 | 5 816 | 3 122 |
| Americas | 15 210 | 12 677 | 10 475 | 12 279 | 14 431 | 12 840 | 12 215 | 11 074 | 11 710 | 8 481 |
| North | 14 744 | 12 296 | 10 229 | 12 020 | 13 733 | 12 296 | 11 919 | 10 890 | 11 681 | 8 446 |
| Central | . . | . . | . . | . . | 1 | . . | 1 | 3 | 2 | . . |
| South | 466 | 381 | 246 | 259 | 697 | 544 | 295 | 181 | 28 | 36 |
| Africa | 158 | 99 | 109 | 85 | 273 | 125 | . . | 35 | 36 | 53 |
| Sub-Saharan Africa | 20 | 52 | 78 | 48 | 180 | 69 | . . | 7 | 36 | 53 |
| Asia | 2 055 | 2 161 | 1 786 | 1 700 | 3 207 | 2 388 | 1 138 | 1 358 | 1 971 | 1 967 |
| Middle East | 1 258 | 840 | 442 | 623 | 668 | 621 | 487 | 143 | 137 | 66 |
| Oceania | 28 | 79 | 35 | 5 | 18 | 10 | 6 | 112 | 62 | . . |
| OECD | 25 235 | 24 129 | 19 525 | 20 629 | 21 927 | 19 972 | 20 393 | 17 334 | 17 041 | 13 489 |
| CSCE | 41 040 | 39 538 | 37 488 | 40 292 | 41 691 | 36 336 | 36 206 | 28 141 | 22 235 | 16 284 |
| NATO | 24 825 | 23 737 | 18 926 | 20 169 | 21 388 | 19 255 | 19 838 | 16 789 | 16 456 | 13 179 |
| OPEC | 245 | 98 | 66 | 98 | 244 | 246 | 35 | 40 | 18 | . . |
| ASEAN | 7 | 58 | 65 | 31 | 52 | 33 | 14 | 9 | . . | 12 |

*Industrialized world*: Albania, Austria, Australia, Belgium, Bulgaria, Canada, Czechoslovakia, Denmark, Finland, France, FR Germany, German DR, Greece, Hungary, Iceland, Ireland, Italy, Lithuania, Japan, Luxembourg, Malta, Netherlands, Norway, New Zealand, Poland, Portugal, Romania, Russia, Slovenia, Spain, Sweden, Switzerland, Turkey, UK, USA, USSR, Yugoslavia.

*Europe*: Albania, Austria, Belgium, Bulgaria, Cyprus, Czechoslovakia, Denmark, Finland, France, FR Germany, German DR, Greece, Hungary, Iceland, Ireland, Italy, Lithuania, Luxembourg, Malta, Netherlands, Norway, Poland, Portugal, Romania, Russia, Slovenia, Spain, Sweden, Switzerland, Turkey, UK, USSR, Yugoslavia.

*European Community (EC)*: Belgium, Denmark, France, FR Germany, Greece, Ireland, Italy, Luxembourg, Netherlands, Portugal, Spain, UK.

*Other Europe*: Albania, Austria, Bulgaria, Cyprus, Czechoslovakia, Finland, German DR, Hungary, Iceland, Lithuania, Malta, Norway, Poland, Romania, Russia, Slovenia, Sweden, Switzerland, Turkey, USSR, Yugoslavia.

*Americas*: Argentina, Barbados, Bahamas, Belize, Bolivia, Brazil, Canada, Chile, Colombia, Costa Rica, Cuba, Dominican Republic, Ecuador, Guatemala, Guyana, Haiti, Honduras, Jamaica, Mexico, Nicaragua, Panama, Paraguay, Peru, El Salvador, Suriname, St Vincent & the Grenadines, Trinidad & Tobago, Uruguay, USA, Venezuela.

*North America*: Canada, Mexico, USA.

*Central America*: Barbados, Bahamas, Belize, Costa Rica, Cuba, Dominican Republic, Guatemala, Haiti, Honduras, Jamaica, Nicaragua, Panama, El Salvador, St Vincent & the Grenadines, Trinidad & Tobago.

*South America*: Argentina, Bolivia, Brazil, Chile, Colombia, Ecuador, Guyana, Paraguay, Peru, Suriname, Uruguay, Venezuela.

*Africa*: Algeria, Angola, Burundi, Benin, Burkina Faso, Botswana, Cameroon, Cape Verde, Central African Republic, Chad, Comoros, Congo, Djibouti, Equatorial Guinea, Ethiopia, Gabon, Gambia, Guinea-Bissau, Ghana, Guinea, Côte d'Ivoire, Kenya, Lesotho, Liberia, Libya, Madagascar, Malawi, Mali, Mauritania, Mauritius, Morocco, Mozambique, Namibia, Niger, Nigeria, Rwanda, Senegal, Seychelles, Sierra Leone, Somalia, South Africa, Sudan, Swaziland, Tanzania, Togo, Tunisia, Uganda, Zaire, Zambia, Zimbabwe.

*Sub-Saharan Africa*: Angola, Burundi, Benin, Burkina Faso, Botswana, Cameroon, Cape Verde, Central African Republic, Chad, Comoros, Congo, Djibouti, Equatorial Guinea, Ethiopia, Gabon, Gambia, Guinea-Bissau, Ghana, Guinea, Côte d'Ivoire, Kenya, Lesotho, Liberia, Madagascar, Malawi, Mali, Mauritania, Mauritius, Mozambique, Namibia, Nigeria, Niger, Rwanda, Senegal, Seychelles, Sierra Leone, Somalia, South Africa, Sudan, Swaziland, Tanzania, Togo, Uganda, Zaire, Zambia, Zimbabwe.

*Asia*: Afghanistan, Bangladesh, Brunei, Cambodia, China, India, Indonesia, Japan, North Korea, South Korea, Laos, Malaysia, Mongolia, Myanmar, Nepal, Pakistan, Philippines, Singapore, Sri Lanka, Taiwan, Thailand, Viet Nam.

*Middle East*: Bahrain, Egypt, Iran, Iraq, Israel, Jordan, Kuwait, Lebanon, Oman, Qatar, Saudi Arabia, Syria, United Arab Emirates, Yemen, North Yemen, South Yemen.

*Oceania*: Australia, Fiji, Kiribati, Marshall Islands, Fed. States of Micronesia, New Zealand, Papua New Guinea, Samoa, Solomon Islands, Tonga, Tuvalu, Vanuatu.

*Organization for Economic Co-operation and Development (OECD)*: Australia, Austria, Belgium, Canada, Denmark, Finland, France, FR Germany, Greece, Iceland, Ireland, Italy, Japan, Luxembourg, Netherlands, New Zealand, Norway, Portugal, Spain, Sweden, Switzerland, Turkey, UK, USA.

*Conference on Security and Co-operation in Europe (CSCE)*:[c] Albania, Austria, Belgium, Bulgaria, Canada, Cyprus, Czechoslovakia, Denmark, Finland, France, FR Germany, German DR, Greece, Hungary, Iceland, Ireland, Italy, Lithuania, Luxembourg, Malta, Netherlands, Norway, Poland, Portugal, Romania, Spain, Slovenia, Sweden, Switzerland, Turkey, UK, USA, USSR, Yugoslavia.

*NATO*: Belgium, Canada, Denmark, France, FR Germany, Greece, Iceland, Italy, Luxembourg, Netherlands, Norway, Portugal, Spain, Turkey, UK, USA.

*Organization of Petroleum Exporting Countries (OPEC)*: Algeria, Ecuador, Gabon, Indonesia, Iran, Iraq, Kuwait, Libya, Nigeria, Qatar, Saudi Arabia, United Arab Emirates, Venezuela.

*Association of South East Asian Nations (ASEAN)*: Brunei, Indonesia, Malaysia, Philippines, Singapore, Thailand.

[a] Only countries for which there is an entry in the SIPRI arms trade data base are included.
[b] As defined by the International Monetary Fund.
[c] For a complete listing of CSCE participating states see the glossary.

*Source:* SIPRI data base.

**Table 10B.3.** World trade in major conventional weapon systems, 1988–92

Figures are SIPRI trend-indicator values for the total five-year period, as expressed in US $m., at constant (1990) prices. Figures may not add up to totals due to rounding. The table lists recipient countries only if the total value of their arms imports amounted to US $25 m. or more for the five-year period.

| | Seller | | | | | | | | | | | |
|---|---|---|---|---|---|---|---|---|---|---|---|---|
| Recipient | USA | USSR/Russia | France | FRG | China | UK | Czech. | Netherlands | Italy | Sweden | Others | Total |
| Afghanistan | 54 | 7 369 | 2 | – | 47 | – | 9 | – | – | – | 35 | 7 515 |
| Algeria | – | 756 | – | – | 112 | 51 | 71 | – | – | – | 7 | 997 |
| Angola | 1 | 1 958 | 20 | – | – | – | – | – | – | – | 32 | 2 011 |
| Argentina | 123 | – | 45 | 148 | – | – | – | – | 52 | – | 95 | 462 |
| Australia | 2 380 | – | 54 | – | – | 68 | – | – | 12 | – | 90 | 2 604 |
| Austria | 29 | – | – | – | – | – | – | – | – | 184 | – | 213 |
| Bahamas | 44 | – | – | – | – | – | – | – | – | – | – | 44 |
| Bahrain | 527 | – | 60 | 146 | – | – | – | – | – | 5 | – | 738 |
| Bangladesh | 12 | – | – | – | 900 | – | – | – | – | – | 191 | 1 104 |
| Belgium | 709 | – | 54 | – | – | – | – | – | 69 | – | 102 | 933 |
| Bolivia | 64 | – | – | – | – | – | – | – | – | – | 11 | 75 |
| Botswana | 1 | – | – | – | – | 21 | – | – | – | – | 55 | 77 |
| Brazil | 505 | – | 394 | 107 | – | 1 | – | – | 70 | 9 | 18 | 1 103 |
| Bulgaria | – | 1 225 | – | – | – | – | 89 | – | – | – | – | 1 314 |
| Cambodia | – | 170 | – | – | 20 | – | – | – | – | – | 81 | 271 |
| Canada | 1 428 | 3 | – | 15 | – | 70 | – | 62 | 48 | 14 | 440 | 2 079 |
| Chad | 54 | – | – | – | – | – | – | – | – | – | – | 54 |
| Chile | 142 | – | 303 | 31 | 14 | 141 | – | – | – | – | 203 | 832 |
| China | 84 | 1 286 | 130 | – | – | – | – | – | – | – | 19 | 1 519 |
| Colombia | 186 | – | – | – | – | – | – | – | – | – | 179 | 365 |
| Cuba | – | 616 | – | – | – | – | – | – | – | – | – | 616 |
| Cyprus | – | – | 110 | – | – | – | – | – | 143 | – | 190 | 443 |
| Czechoslovakia | – | 3 501 | – | – | – | – | – | – | – | – | – | 3 501 |
| Denmark | 204 | – | 12 | 49 | – | 286 | – | – | – | 22 | 21 | 596 |

| | Seller | | | | | | | | | | | |
|---|---|---|---|---|---|---|---|---|---|---|---|---|
| Recipient | USA | USSR/Russia | France | FRG | China | UK | Czech. | Netherlands | Italy | Sweden | Others | Total |
| Ecuador | 47 | – | – | – | – | 193 | – | – | – | – | 25 | 265 |
| Egypt | 2 774 | – | 222 | 47 | – | 3 | – | – | 86 | – | 163 | 3 295 |
| Ethiopia | 46 | 162 | – | – | – | – | 227 | – | – | – | 59 | 494 |
| Finland | 1 | 173 | 248 | 118 | – | 35 | – | – | – | 283 | – | 857 |
| France | 1 577 | – | – | – | – | 13 | – | – | – | – | 36 | 1 626 |
| Gabon | – | – | 139 | – | – | – | – | – | – | – | 4 | 143 |
| German DR | – | 1 879 | – | – | – | – | – | – | – | – | 20 | 1 899 |
| Germany, FR | 4 279 | – | 67 | – | – | 80 | – | 32 | – | – | 15 | 4 473 |
| Greece | 3 309 | – | 1 365 | 987 | – | 24 | – | 254 | 15 | – | 244 | 6 197 |
| Guinea | – | 83 | 1 | – | – | – | – | – | – | – | – | 84 |
| Guinea-Bissau | – | 62 | – | – | – | – | – | – | – | – | – | 62 |
| Honduras | 69 | – | – | – | – | – | – | – | – | – | – | 69 |
| Hungary | – | 71 | – | 71 | – | – | – | – | – | – | – | 143 |
| India | 10 | 9 364 | 387 | 261 | – | 1 044 | – | 484 | – | 569 | 117 | 12 235 |
| Indonesia | 390 | – | 43 | 156 | – | 201 | – | 341 | – | – | 43 | 1174 |
| Iran | – | 1 820 | – | – | 976 | – | 177 | – | – | – | 659 | 3 632 |
| Iraq | 228 | 3 164 | 397 | 41 | 234 | – | 75 | – | 28 | – | 799 | 4 967 |
| Ireland | 23 | – | – | – | – | 30 | – | – | 3 | – | 16 | 71 |
| Israel | 2 676 | – | – | 19 | – | – | – | 63 | – | – | 9 | 2 768 |
| Italy | 494 | – | 17 | 58 | – | – | – | – | – | 19 | 100 | 688 |
| Japan | 9 017 | – | 50 | – | – | 157 | – | – | – | – | – | 9 224 |
| Jordan | 28 | 22 | 79 | – | – | 81 | – | – | – | – | 156 | 365 |
| Kenya | – | – | 27 | 1 | – | 20 | – | – | – | – | 21 | 70 |
| Korea, North | – | 2 816 | – | – | 307 | – | – | – | – | – | – | 3 123 |
| Korea, South | 3 238 | – | 53 | 96 | – | 106 | – | – | 3 | – | 28 | 3 524 |
| Kuwait | 220 | 211 | – | – | – | 28 | – | – | 48 | – | 736 | 1 243 |
| Lebanon | – | – | – | – | – | – | – | – | – | – | 67 | 67 |
| Libya | – | 604 | – | – | – | – | 18 | – | – | – | 78 | 700 |

| | | | | | | | | | | | | |
|---|---|---|---|---|---|---|---|---|---|---|---|---|
| Lithuania | – | 100 | – | – | – | – | – | – | – | – | – | 100 |
| Malaysia | – | – | 11 | – | – | 78 | – | 5 | 28 | – | 9 | 131 |
| Mali | – | 31 | – | – | – | – | – | – | – | – | – | 31 |
| Mexico | 214 | – | 13 | – | – | – | – | – | – | – | – | 227 |
| Morocco | 84 | – | 48 | – | – | – | – | – | – | – | 381 | 513 |
| Myanmar | – | – | – | – | 407 | – | – | – | – | – | 102 | 509 |
| Netherlands | 1 734 | – | – | 14 | – | 3 | – | – | – | – | 13 | 1 765 |
| New Zealand | 23 | – | – | – | – | 2 | – | 19 | 96 | – | 54 | 195 |
| Nicaragua | – | 188 | – | – | – | – | – | – | – | – | 18 | 207 |
| Nigeria | 3 | – | 74 | 4 | – | 75 | 134 | – | 143 | – | – | 433 |
| Norway | 467 | – | – | 658 | – | 19 | – | – | – | 217 | – | 1 361 |
| Oman | 39 | – | 59 | – | – | 367 | – | – | – | – | 5 | 471 |
| Pakistan | 1 119 | – | 55 | – | 1 935 | 158 | – | – | 17 | 41 | 161 | 3 486 |
| Panama | 19 | – | – | – | – | – | – | – | – | – | 17 | 36 |
| Papua New Guinea | – | 19 | – | – | – | – | – | – | – | – | 14 | 33 |
| Peru | 86 | 172 | – | 4 | 7 | – | 81 | – | 2 | – | 27 | 379 |
| Philippines | 75 | – | – | 5 | – | 3 | – | – | 36 | – | 7 | 126 |
| Poland | 6 | 2 820 | – | 172 | – | – | – | – | – | – | 2 | 3 001 |
| Portugal | 449 | – | 36 | 836 | – | 10 | – | 43 | – | – | – | 1 374 |
| Qatar | – | – | 272 | – | – | – | – | – | – | – | 13 | 285 |
| Romania | – | 1 014 | 121 | – | – | 27 | – | – | – | – | 32 | 1 193 |
| Saudi Arabia | 2 783 | – | 1 572 | 44 | 858 | 3 116 | – | – | 154 | – | 163 | 8 690 |
| Singapore | 856 | – | 55 | 376 | – | – | – | – | – | 29 | – | 1 316 |
| South Africa | – | – | – | – | – | – | – | – | – | – | 37 | 37 |
| Spain | 3 040 | – | 372 | 30 | – | 19 | – | – | 126 | – | 159 | 3 747 |
| Sri Lanka | – | – | – | – | 96 | 5 | 8 | – | 1 | – | 54 | 164 |
| Sudan | 4 | – | – | – | 16 | – | – | – | 12 | – | 68 | 101 |
| Sweden | 42 | – | 92 | 6 | – | 26 | – | – | – | – | 2 | 168 |
| Switzerland | 49 | – | 109 | 812 | – | 185 | – | – | – | – | – | 1 154 |
| Syria | – | 1 816 | – | – | – | – | 630 | – | – | – | 172 | 2 618 |
| Taiwan | 1 473 | – | – | 217 | – | – | – | 333 | – | – | 211 | 2 234 |

| | Seller | | | | | | | | | | | |
|---|---|---|---|---|---|---|---|---|---|---|---|---|
| Recipient | USA | USSR/Russia | France | FRG | China | UK | Czech. | Netherlands | Italy | Sweden | Others | Total |
| Thailand | 1 409 | – | 80 | 36 | 1 482 | 167 | – | 19 | 33 | – | 45 | 3 271 |
| Tunisia | 55 | – | 5 | 12 | – | – | – | – | – | – | – | 72 |
| Turkey | 3 635 | 21 | 23 | 1 832 | – | 10 | – | 265 | 100 | – | 281 | 6 167 |
| Uganda | – | 12 | – | – | – | – | – | – | 19 | – | – | 31 |
| UK | 2 074 | – | 121 | 32 | – | – | – | 33 | – | – | 65 | 2 326 |
| United Arab Emirates | 66 | 71 | 1 482 | 293 | – | 23 | – | 8 | 48 | 14 | 60 | 2 065 |
| Uruguay | 22 | – | 69 | 12 | – | 2 | – | – | – | – | 10 | 115 |
| USA | – | – | 3 | 429 | 128 | 543 | – | – | 199 | 1 | 540 | 1 843 |
| USSR/Russia | – | – | – | – | – | – | 1 632 | – | – | – | 1 750 | 3 383 |
| Venezuela | 123 | – | 285 | – | – | 76 | – | 63 | 21 | 8 | 41 | 617 |
| Yemen, North[a] | – | 27 | – | – | 42 | – | – | – | – | – | 6 | 75 |
| Yemen, South[a] | – | 304 | – | – | – | – | – | – | – | – | – | 304 |
| Yugoslavia | – | 1 240 | 20 | – | – | – | – | – | – | – | 7 | 1 267 |
| Zimbabwe | – | – | 5 | – | 70 | 45 | – | – | – | – | 17 | 136 |
| Others | 47 | 30 | 94 | 11 | 11 | 11 | 12 | 24 | 4 | – | 92 | 336 |
| **Total** | **54 968** | **45 182** | **9 349** | **8 190** | **7 660** | **7 623** | **3 163** | **2 048** | **1 613** | **1 416** | **9803** | **151 013** |

[a] North Yemen and South Yemen were joined on 22 May 1990; no trade in major conventional weapons is recorded for Yemen for the period 1990–92.

*Source*: SIPRI data base.

# **Appendix 10C.** Register of the trade in and licensed production of major conventional weapons in industrialized and developing countries, 1992

IAN ANTHONY, PAUL CLAESSON, GERD HAGMEYER-GAVERUS, ELISABETH SKÖNS and SIEMON T. WEZEMAN

This register lists major weapons on order or under delivery, or for which the licence was bought and production was under way or completed during 1992. 'Year(s) of deliveries' includes aggregates of all deliveries and licensed production since the beginning of the contract. Entries are alphabetical, by recipient, supplier and licenser. Abbreviations, acronyms and conventions are explained at the end of the register. Sources and methods are explained in Appendix 10D.

| Recipient/ supplier (S) or licenser (L) | No. ordered | Weapon designation | Weapon description | Year of order/ licence | Year(s) of deliveries | No. delivered/ produced | Comments |
|---|---|---|---|---|---|---|---|
| **I. Industrialized countries** | | | | | | | |
| **Australia** | | | | | | | |
| **S:** Canada | 97 | LAV | APC | 1992 | | | Deal worth $200 m incl spares and training |
| Italy | (10) | HSS-1 | Surveillance radar | 1986 | 1988–92 | (10) | Deal worth $20 m |
| Sweden | 8 | 9LV | Fire control radar | (1991) | | | For 8 Meko-200 Type frigates |
| | 8 | Sea Giraffe | Surveillance radar | 1991 | | | For 8 Meko-200 Type frigates |
| UK | (128) | Sea Skua | Anti-ship missile | 1992 | | | Arming 16 Seahawk helicopters |
| USA | 4 | Boeing-707-320C | Transport aircraft | (1990) | 1991–92 | (4) | Modified as tankers by Israel Aircraft Industries |
| | 4 | CH-47D Chinook | Helicopter | 1992 | | | Exchanged for 11 CH-47C in Australian service |
| | 18 | F-111 | Fighter/bomber | 1992 | | | For storage |
| | 8 | Phalanx | CIWS | 1991 | | | Arming 8 Meko-200 Type frigates |
| | 2 | RGM-84A launcher | ShShM launcher | 1983 | 1992 | 1 | For 2 FFG-7 Class frigates produced under licence |
| | 2 | RIM-66A launcher | ShAM launcher | 1985 | 1992 | 1 | For 2 FFG-7 Class frigates produced under licence |
| | 8 | Seasparrow VLS | ShAM launcher | (1991) | | | For 8 Meko-200 Type frigates |
| | (128) | NATO Seasparrow | ShAM | (1991) | | | Arming 8 Meko-200 Type frigates |
| | (48) | RGM-84A Harpoon | ShShM | 1987 | 1992 | (24) | Arming 2 FFG-7 Class frigates |
| | (64) | RIM-67C/SM-2 | ShAM | (1987) | 1992 | (32) | Arming 2 FFG-7 Class frigates; deal worth $50 m |

| Recipient/ supplier (S) or licenser (L) | No. ordered | Weapon designation | Weapon description | Year of order/ licence | Year(s) of deliveries | No. delivered/ produced | Comments |
|---|---|---|---|---|---|---|---|
| L: Germany, FR | 10 | Meko-200 Type | Frigate | 1989 | | | 8 for Australia; 2 for New Zealand; option for 2 more |
| Sweden | 6 | Type-471 | Submarine | 1987 | | | Deal worth $2.8 b |
| UK | 129 | Hamel 105mm | Towed gun | (1982) | 1987–92 | (129) | Incl 24 for New Zealand; deal worth $112 m |
| USA | 2 | FFG-7 Class | Frigate | 1983 | 1992 | (1) | |
| **Austria** | | | | | | | |
| S: Sweden | 500 | RBS-56 Bill | Anti-tank missile | 1989 | 1989–92 | (450) | Deal worth $80 m |
| UK | 2 | BAe-146 | Transport aircraft | 1991 | | | For Austrian UN relief activities |
| USA | 24 | M-109-A2 155mm | Self-propelled gun | 1988 | 1989–91 | (18) | Deal worth $36 m |
| **Belgium** | | | | | | | |
| S: France | 714 | Mistral | Portable SAM | 1988 | 1991–92 | (200) | Deal worth $93 m (offsets 75%) incl 118 launchers |
| | 290 | Mistral | Portable SAM | 1991 | | | Second order; including 24 launchers |
| USA | 545 | AIM-9M Sidewinder | Air-to-air missile | 1988 | | | Arming F-16 fighters; deal worth $49 m |
| | 940 | AIM-9M Sidewinder | Air-to-air missile | 1989 | | | Deal worth $80 m |
| | (224) | BGM-71A TOW | Anti-tank missile | (1989) | 1992 | (224) | Arming 28 A-109A Mk-2 helicopters |
| L: Italy | 46 | A-109A Mk-2 | Helicopter | 1988 | 1992 | (30) | Deal worth $317 m (offsets 40%) |
| **Bulgaria** | | | | | | | |
| L: USSR | . . | MT-LB | APC | (1970) | 1972–92 | (1 200) | |
| **Canada** | | | | | | | |
| S: Italy | 35 | EH-101 ASW | Helicopter | 1992 | | | Part of deal incl 15 SAR versions from the UK |
| Netherlands | 4 | DA-08 | Surveillance radar | 1986 | 1991 | 2 | For 4 retrofitted Tribal Class destroyers |
| | 4 | LW-08 | Surveillance radar | 1986 | 1991 | 2 | |
| | 8 | STIR | Fire control radar | 1986 | 1991 | 4 | |
| | 24 | STIR | Fire control radar | (1985) | 1990–92 | (6) | For 12 City Class frigates |

| | | | | | | | |
|---|---|---|---|---|---|---|---|
| Sweden | 12 | Sea Giraffe | Surveillance radar | (1985) | 1988–92 | (3) | For 12 City Class frigates |
| Switzerland | 36 | ADATS | SAM system | 1986 | 1988–92 | (21) | Deal worth $1 b incl SAMs, AA guns and fire control radars |
| UK | 15 | EH-101 SAR | Helicopter | 1992 | | | Part of deal incl 35 ASW versions from Italy |
| USA | 12 | AN/SPS-49 | Surveillance radar | 1985 | 1990–92 | (3) | For 12 City Class frigates |
| | 2 | AN/TPS-70 | Surveillance radar | 1990 | 1992 | (1) | Deal worth $23 m |
| | 4 | Phalanx | CIWS | 1987 | 1991–92 | (2) | Arming 4 Tribal Class frigates |
| | 6 | Phalanx | CIWS | 1986 | 1988–92 | (3) | Arming 6 City Class frigates |
| | 6 | Phalanx | CIWS | 1990 | | | Deal worth $32 m; arming 2nd batch of 6 City Class frigates |
| | 12 | RGM-84A launcher | ShShM launcher | 1983 | 1988–92 | (3) | For 12 City Class frigates |
| | 12 | Seasparrow VLS | ShAM launcher | 1983 | 1988–92 | (3) | For 12 City Class frigates; deal worth $75 m incl missiles |
| | 4 | Standard VLS | Fire control radar | 1986 | 1991–92 | 2 | For 4 Tribal Class destroyers |
| | 3 | AGM-84A Harpoon | Anti-ship missile | 1991 | 1992 | 3 | |
| | . . | RGM-84A Harpoon | ShShM | 1988 | 1988–92 | (72) | Arming 4 City Class frigates |
| | 116 | RIM-67C/SM-2 | ShAM | 1986 | 1991–92 | (58) | Arming 4 Tribal Class destroyers |
| | 336 | Seasparrow | ShAM | 1984 | 1988–92 | (84) | Arming 12 City Class frigates; deal worth $75 m |
| **L:** France | 5 000 | Eryx | Anti-tank missile | 1992 | | | Deal worth $151 m incl 400 launchers |
| UK | 40 | L-119 105mm | Towed gun | 1990 | | | |
| USA | 100 | Model 412 | Helicopter | 1992 | | | Deal worth $844 m |
| **Czechoslovakia** | | | | | | | |
| **L:** USSR | . . | T-72 | Main battle tank | 1978 | 1981–92 | (762) | |
| **Denmark** | | | | | | | |
| **S:** France | (9) | TRS-2106 3D | Surveillance radar | 1991 | | | |
| | (9) | TRS-2620 | Surveillance radar | 1991 | | | |
| Germany, FR | 140 | Leopard-1 | Main battle tank | (1991) | 1992 | (70) | CFE cascade |
| | 6 | TRS-3D | Surveillance radar | 1990 | 1992 | 1 | Arming 2nd batch of 6 Stanflex-300 Type patrol craft |
| Sweden | 13 | 9LV | Fire control radar | (1988) | 1989–92 | (7) | For 13 Stanflex-300 Type patrol craft |

| Recipient/ supplier (S) or licenser (L) | No. ordered | Weapon designation | Weapon description | Year of order/ licence | Year(s) of deliveries | No. delivered/ produced | Comments |
|---|---|---|---|---|---|---|---|
| USA | 12 | M-110 203mm | Self-propelled gun | (1991) | 1992 | (12) | CFE cascade |
| | 1 | RGM-84A CDS | Coast defence system | 1991 | | | |
| | 162 | AGM-65D Maverick | Air-to-surface missile | 1989 | | | Arming F-16 fighters; deal worth $24 m |
| | 840 | FIM-92A Stinger | Portable SAM | 1991 | | | |
| | (24) | RGM-84A Harpoon | ShShM | 1991 | | | Arming coastal defence battery |
| **Finland** | | | | | | | |
| **S:** France | 20 | Crotale NG | SAM system | 1990 | 1992 | (10) | Deal worth $230 m |
| | 10 | TRS-2230/15 | Surveillance radar | 1990 | | | Deal worth $200 m |
| | (360) | Mistral | Portable SAM | 1989 | 1990–91 | (180) | |
| | (480) | VT-1 | SAM | 1990 | 1992 | (240) | |
| Germany, FR | (290) | D-30 122mm | Towed gun | 1992 | 1992 | (290) | Former GDR equipment |
| | 90 | T-72 | Main battle tank | 1992 | 1992 | (90) | Former GDR equipment |
| Sweden | 4 | 9LV | Fire control radar | (1988) | 1990–92 | 4 | For 4 Rauma Class fast attack craft |
| | 4 | Giraffe 100 | Surveillance radar | 1991 | | | |
| | 4 | RBS-15 launcher | ShShM launcher | 1987 | 1990–92 | (4) | For 4 Rauma Class fast attack craft |
| | 64 | RBS-15 | ShShM | (1987) | 1990–92 | 64 | Arming 4 Rauma Class fast attack craft |
| UK | 7 | Hawk | Jet trainer aircraft | 1990 | | | |
| | . . | Marksman | AAV(G) | 1992 | | | 2nd order |
| USA | 64 | F/A-18 Hornet | Fighter aircraft | 1992 | | | 57 C and 7 D version; limited assembly in Finland |
| | (128) | AIM-120A AMRAAM | Air-to-air missile | 1992 | | | Arming 64 F/A-18 fighters |
| | (384) | AIM-9M Sidewinder | Air-to-air missile | 1992 | | | Arming 64 F/A-18 fighters |
| **France** | | | | | | | |
| **S:** Brazil | 50 | EMB-312 Tucano | Trainer aircraft | 1991 | | | Deal worth $170 m option for 30 more |
| Germany, FR | (30) | Alpha Jet | Jet trainer aircraft | 1992 | | | |
| Spain | 6 | CN-235 | Transport aircraft | 1991 | | | In addition to 2 delivered; option on 7 more |
| Switzerland | 5 | PC-6 | Utility aircraft | 1990 | 1992 | 5 | |
| USA | 1 000 | VT-1 | SAM | 1988 | 1990–92 | (625) | 700 for re-export |

| | | | | | | | |
|---|---|---|---|---|---|---|---|
| **L:** USA | 55 | MLRS 227mm | MRL | 1985 | 1985–92 | (46) | |
| | . . | VT-1 | SAM | 1991 | | | |
| **Germany, FR** | | | | | | | |
| **S:** France | 23 | TRS-3050 | Surveillance radar | 1987 | 1987–92 | (16) | Retrofit for 20 Type-148 fast attack craft |
| | 200 | Apache | Air-to-surface missile | 1992 | | | Arming Tornado aircraft as MAW |
| Netherlands | 4 | LW-08 | Surveillance radar | (1989) | | | Equipping 4 Type-123 frigates |
| | 5 | Smart | Surveillance radar | 1989 | | | For 4 Type-123 frigates |
| | 8 | STIR | Fire control radar | 1989 | | | For 4 Type-123 frigates |
| USA | 10 | D-500 Egrett | AEW&C aircraft | 1992 | | | Deal worth $795 m incl ground station, spares and support; status uncertain |
| | 3 | AN/FPS-117 | Surveillance radar | 1988 | 1991–92 | (3) | |
| | 5 | AN/FPS-117 | Surveillance radar | 1992 | | | Deal worth $94 m incl 2 simulators and spares |
| | 4 | Seasparrow VLS | ShAM launcher | 1989 | | | For 4 Type-123 frigates |
| | 1 182 | AGM-88 Harm | Anti-radar missile | 1987 | 1988–92 | (900) | Arming Tornado fighters |
| | 175 | AIM-120A AMRAAM | Air-to-air missile | 1991 | | | Arming F-4F fighters |
| | 804 | MIM-104 Patriot | SAM | 1984 | 1989–91 | 804 | |
| | (64) | Seasparrow | ShAM | 1989 | | | Arming 4 Type-123 frigates |
| **L:** USA | 150 | MLRS 227mm | MRL | 1985 | 1989–92 | 120 | |
| | . . | AIM-120A AMRAAM | Air-to-air missile | 1989 | | | Deal worth $81 m |
| | 4 500 | FIM-92 Stinger | Portable SAM | 1987 | 1992 | 250 | |
| | (1 500) | RIM-116 A RAM | ShAM | 1985 | 1989–92 | 350 | |
| **Greece** | | | | | | | |
| **S:** France | 40 | Mirage-2000 | Fighter aircraft | 1985 | 1988–92 | 40 | |
| | (240) | Magic-2 | Air-to-air missile | 1986 | 1988–92 | (240) | Arming Mirage-2000 fighters |
| Germany, FR | . . | RF-4E Phantom | Recce aircraft | (1991) | | | Status uncertain |
| | 150 | RM-70 122mm | MRL | (1991) | | | Former GDR equipment |
| | 75 | Leopard-1 | Main battle tank | (1991) | 1992 | (25) | CFE cascade |
| | 200 | M-113 | APC | (1991) | | | CFE cascade |
| | 312 | M-60-A3 Patton | Main battle tank | (1990) | 1990–92 | (253) | CFE cascade |

| Recipient/ supplier (S) or licenser (L) | No. ordered | Weapon designation | Weapon description | Year of order/ licence | Year(s) of deliveries | No. delivered/ produced | Comments |
|---|---|---|---|---|---|---|---|
| | (64) | NATO Seasparrow | ShAM | (1988) | 1992 | (16) | Arming 4 Meko-200 Type frigates |
| | 1 | Meko-200 Type | Frigate | 1988 | 1992 | 1 | Deal worth $1.2 b (offsets $250 m) incl 3 to be built under licence |
| | 5 | Thetis Class | Corvette | 1989 | 1991–92 | 5 | Ex-French Navy |
| | 8 | Type 520 | Landing craft | 1989 | 1989–92 | (8) | Ex-French Navy |
| Netherlands | 171 | M-30 107mm | Mortar | (1991) | | | CFE cascade |
| | 170 | Leopard-1-A4 | Main battle tank | 1991 | 1992 | 170 | CFE cascade |
| | 177 | M-113 | APC | 1991 | | | |
| | 4 | DA-08 | Surveillance radar | 1988 | 1992 | 1 | For 4 Meko-200 Type frigates |
| | 4 | MW-08 | Surveillance radar | (1989) | 1992 | 1 | For 4 Meko-200 frigates |
| | 3 | RGM-84A launcher | ShShM launcher | 1992 | | | For 3 Kortenaer Class frigates |
| | 3 | Seasparrow launcher | ShAM launcher | 1992 | | | For 3 Kortenaer Class frigates |
| | 8 | STIR | Fire control radar | 1989 | 1992 | 2 | For 4 Meko-200 Type frigates |
| | 3 | Kortenaer Class | Frigate | 1992 | | | Ex-Royal Netherlands Navy; deal worth $211 m |
| UK | 32 | F-4 Phantom | Fighter aircraft | 1992 | | | Ex-Royal Air Force |
| | 2 | Martello 743-D | Surveillance radar | 1990 | 1992 | (2) | |
| USA | (36) | A-7E Corsair-2 | Fighter/ground attack | 1991 | | | Deal worth $120 m incl 14 spare engines and spares |
| | 8 | AH-64 Apache | Helicopter | (1991) | | | Deal worth $505 m incl 3 spare engines, electronic warfare systems, support and spares |
| | 5 | C-130B Hercules | Transport aircraft | (1991) | | | |
| | (40) | F-16C | Fighter aircraft | 1992 | | | Deal worth $1.8 b incl 10 spare engines and 40 Lantirn navigation pods |
| | 12 | P-3A Orion | Maritime patrol | 1990 | 1992 | 12 | 4 in reserve |
| | 5 | SH-60B Seahawk | Helicopter | 1991 | | | For 4 Meko-200 Type frigates; deal worth $161 m; option on 7 more |
| | 72 | M-110 203mm | Self-propelled gun | (1991) | 1992 | (72) | CFE cascade |
| | 100 | M-30 107mm | Mortar | 1991 | 1992 | (35) | CFE cascade |
| | 150 | M-113 | APC | (1991) | 1992 | (50) | CFE cascade |
| | 359 | M-60-A1 Patton | Main battle tank | (1990) | 1991–92 | (359) | CFE cascade |
| | 8 | Phalanx | CIWS | (1987) | 1992 | 2 | Arming 4 Meko-200 Type frigates |

| | | | | | | | |
|---|---|---|---|---|---|---|---|
| | 4 | RGM-84A launcher | ShShM launcher | 1991 | 1991–92 | 4 | For 4 Adams Class destroyers |
| | 4 | RGM-84A launcher | ShShM launcher | 1989 | 1992 | 1 | For 4 Meko-200 Type frigates |
| | 3 | RGM-84A launcher | ShShM launcher | 1992 | 1992 | 2 | Arming 3 Knox Class frigates |
| | 4 | RIM-67A launcher | ShAM launcher | 1991 | 1991–92 | 4 | For 4 Adams Class destroyers |
| | 3 | Seasparrow launcher | ShAM launcher | 1992 | 1992 | 2 | Arming 3 Knox Class frigates |
| | 4 | Seasparrow VLS | ShAM launcher | 1988 | 1992 | 1 | For 4 Meko-200 Type frigates |
| | 446 | AGM-114A Hellfire | Anti-tank missile | 1991 | | | Arming AH-64 Apache helicopters |
| | 1 500 | FIM-92A Stinger | Portable SAM | 1988 | 1989–92 | (1 000) | Deal worth $124 m incl 500 launchers |
| | 16 | RGM-84A Harpoon | ShShM | 1989 | 1992 | 16 | Arming 1st of 4 Meko-200 Type frigates; deal worth $19 m |
| | 24 | RGM-84A Harpoon | ShShM | 1991 | 1991–92 | (24) | Arming 4 Adams Class destroyers; part of deal worth $100 m incl 64 Standard SAMs, 56 Mk 48 torpedoes, 10 000 rounds ammunition and support |
| | (24) | RGM-84A Harpoon | ShShM | 1992 | 1992 | (16) | Arming 3 Knox Class frigates |
| | (64) | RIM-67A/SM-1 | ShAM | 1991 | 1991–92 | (64) | Arming 4 Adams Class destroyers |
| | 4 | Adams Class | Destroyer | 1990 | 1991–92 | 4 | Ex-US Navy |
| | 3 | Knox Class | Frigate | 1992 | 1992 | 2 | Ex-US Navy |
| **L:** Austria | 324 | Steyr-4K 7FA | APC | (1987) | 1991–92 | 120 | 3rd order |
| Germany, FR | 3 | Meko-200 Type | Frigate | 1988 | | | In addition to 1 delivered direct; deal worth $1.2 b; financial aid from FRG and USA |
| **Ireland** | | | | | | | |
| **S:** Spain | 2 | CN-235MPA | Maritime patrol | 1991 | 1992 | 1 | Deal worth $37 m incl 1 transport version |
| **Italy** | | | | | | | |
| **S:** Germany, FR | 8 | Do-228-200 | Transport aircraft | 1990 | 1991 | 2 | |
| | . . | Kormoran-2 | Anti-ship missile | (1986) | 1990–91 | (30) | Arming Tornado fighters |
| USA | 13 | AV-8B Harrier 2-Plus | Fighter/ground attack | 1990 | | | Deal worth $392 m; option on 8 more |
| | 4 | AN/FPS-117 | Surveillance radar | 1990 | | | |
| | 2 | RIM-67A launcher | ShAM launcher | (1987) | 1992 | 1 | For 2 Animoso Class destroyers |
| | 74 | AGM-88 Harm | Anti-radar missile | 1991 | 1992 | (74) | Arming Tornado fighter/bombers |

| Recipient/ supplier (S) or licenser (L) | No. ordered | Weapon designation | Weapon description | Year of order/ licence | Year(s) of deliveries | No. delivered/ produced | Comments |
|---|---|---|---|---|---|---|---|
| | 446 | AGM-88 Harm | Anti-radar missile | 1992 | | | Deal worth $145 m; 2nd order |
| | (3 900) | BGM-71D TOW-2 | Anti-tank missile | 1987 | 1990–92 | (600) | Arming A-129 Mangusta helicopters |
| | (80) | RIM-67C/SM-2 | ShAM | 1987 | 1992 | (40) | Arming 2 Animoso Class destroyers |
| **L:** France | . . | Aster | SAM | 1988 | | | Co-development |
| USA | . . | AB-206B | Helicopter | 1972 | 1978–92 | (675) | |
| | . . | AB-212 | Helicopter | 1970 | 1971–92 | (183) | |
| | . . | AB-212ASW | Helicopter | 1975 | 1975–91 | (105) | |
| | . . | AB-412 Griffon | Helicopter | 1980 | 1982–92 | (67) | Italy holds marketing rights |
| | 50 | Model 500E | Helicopter | 1987 | 1987–92 | (30) | |
| | . . | S-61R | Helicopter | 1990 | 1991–92 | (6) | Production restarted 1990 to produce 15 more aircraft |
| | 22 | MLRS 227mm | MRL | 1985 | 1990–92 | (18) | |
| | 20 | Patriot SAMS | SAM system | 1988 | | | Part of $2.9 b deal incl 1 280 missiles |
| | 1 280 | MIM-104 Patriot | SAM | 1988 | | | Part of deal worth $2.9 b |
| **Japan** | | | | | | | |
| **S:** UK | 3 | BAe-125-800 | Transport aircraft | 1989 | 1992 | 3 | |
| | 3 | BAe-125-800 | Transport aircraft | 1991 | | | Follow-on order for up to 24 expected |
| USA | 3 | Beechjet 400T | Transport aircraft | 1992 | 1992 | 3 | |
| | 3 | E-2C Hawkeye | AEW aircraft | 1989 | 1992 | 3 | Deal worth $214 m incl spares |
| | 2 | E-2C Hawkeye | AEW aircraft | 1990 | | | Deal worth $170 m |
| | 2 | EP-3C Orion | Elint aircraft | 1992 | | | 2nd order |
| | 36 | MLRS 227mm | MRL | (1991) | 1992 | 9 | Deal worth $362 m; status of Japanese production uncertain |
| | 1 | AN/SPY-1D | Surveillance radar | 1988 | 1992 | 1 | Part of Aegis phased-array radar system for 1st Kongo Class destroyer; deal worth $17.7 m |
| | 2 | AN/SPY-1D | Surveillance radar | 1992 | | | Part of Aegis phased-array radar system for 2nd and 3rd Kongo Class destroyers |
| | 6 | Phalanx | CIWS | 1988 | | | Arming Kongo Class destroyers |

| | | | | | | | |
|---|---|---|---|---|---|---|---|
| | 3 | RGM-84A launcher | ShShM launcher | 1988 | | | For 3 Kongo Class destroyers |
| | 3 | Standard VLS | Fire control radar | 1988 | | | For 3 Kongo Class destroyers |
| | 75 | AGM-84A Harpoon | Anti-ship missile | 1990 | 1991–92 | (39) | Deal worth $125 m |
| | 32 | RGM-84A Harpoon | ShShM | 1988 | | | Arming Kongo Class destroyers |
| | 14 | RGM-84A Harpoon | ShShM | 1992 | | | Deal worth $35 m incl spare parts |
| | . . | RIM-66C/SM-2 | ShAM | 1988 | 1991 | 24 | For Kongo Class destroyers |
| **L:** France | . . | TB-120mm | Mortar | 1992 | | | |
| Italy | 3 | Sparviero Class | Fast attack craft | 1990 | | | Deal worth $170 m; option on 3 more |
| UK | 176 | FH-70 155mm | Towed gun | 1984 | 1989–92 | 125 | Incl direct delivery of 20 |
| USA | . . | CH-47D Chinook | Helicopter | (1984) | 1986–92 | 35 | |
| | 2 | EP-3C Orion | Elint aircraft | 1987 | 1991–92 | 2 | Deal worth $91 m |
| | 55 | F-15J Eagle | Fighter aircraft | 1985 | 1988–92 | 48 | MoU signed Dec. 1984; in addition to 100 ordered previously |
| | (130) | FS-X | Fighter aircraft | 1988 | | | Based on F-16C; US firms guaranteed 42% of work |
| | . . | Model-205 Kai | Helicopter | 1991 | 1992 | 13 | |
| | 78 | Model 209 AH-1S | Helicopter | 1982 | 1984–92 | 65 | |
| | 135 | OH-6D | Helicopter | 1977 | 1978–92 | 144 | |
| | 70 | P-3C Orion | Maritime patrol | 1985 | 1987–92 | 39 | In addition to 45 ordered previously |
| | 49 | SH-60J Seahawk | Helicopter | 1988 | 1990–92 | 22 | Order likely to reach 100 |
| | 46 | UH-60J Blackhawk | Helicopter | 1988 | 1990–92 | 13 | |
| | 1 330 | AIM-7M Sparrow | Air-to-air missile | 1990 | 1990–92 | 475 | Arming F-15 fighters; deal worth $477 m |
| | . . | BGM-71C I-TOW | Anti-tank missile | (1983) | 1985–92 | 4 974 | Total requirement: up to 10 000 |
| | 980 | MIM-104 Patriot | SAM | 1984 | 1989–92 | 644 | |
| | . . | MIM-23B Hawk | SAM | 1978 | 1978–92 | 3 104 | |
| **Lithuania** | | | | | | | |
| **S:** Russia | 2 | Grisha-3 Class | Frigate | 1992 | 1992 | 2 | Lithuania will build houses in Russia in exchange |
| | 2 | Stenka Class | Fast attack craft | 1992 | 1992 | 2 | |
| | 2 | Turya Class | Fast attack craft | 1992 | 1992 | 2 | |
| **Netherlands** | | | | | | | |
| **S:** Germany, FR | 25 | Buffel | ARV | 1990 | 1992 | (6) | Option on 10–15 more |

| Recipient/ supplier (S) or licenser (L) | No. ordered | Weapon designation | Weapon description | Year of order/ licence | Year(s) of deliveries | No. delivered/ produced | Comments |
|---|---|---|---|---|---|---|---|
| Italy | 3 | AB-412 Griffon | Helicopter | 1992 | | | Deal worth $22.8 m; for search and rescue |
| USA | 4 | Patriot SAMS | SAM system | 1985 | | | In addition to 4 delivered earlier |
| | 8 | RGM-84A launcher | ShShM launcher | 1988 | 1991–92 | 3 | Arming 8 Karel Doorman Class frigates |
| | 8 | Seasparrow VLS | ShAM launcher | 1985 | 1991–92 | 3 | For 8 Karel Doorman Class frigates |
| | (40) | AGM-84A Harpoon | Anti-ship missile | 1988 | | | Status uncertain |
| | 290 | AIM-9M Sidewinder | Air-to-air missile | 1988 | | | Arming F-16 fighters; deal worth $27 m |
| | 256 | MIM-104 Patriot | SAM | 1985 | | | |
| | (192) | RGM-84A Harpoon | ShShM | 1988 | 1991–92 | (72) | Arming 8 Karel Doorman Class frigates |
| | (128) | Seasparrow | ShAM | 1985 | 1991–92 | (48) | Arming 8 Karel Doorman Class frigates |
| **L:** USA | 57 | F-16A | Fighter aircraft | 1983 | 1987–92 | (57) | 4th order |
| **New Zealand** | | | | | | | |
| **S:** Australia | 2 | Meko-200 Type | Frigate | 1989 | | | Deal worth $554.7 m; option on 2 more |
| Italy | 18 | MB-339C | Jet trainer aircraft | 1990 | 1991–92 | 12 | Deal worth $125 m |
| Netherlands | 2 | LW-08 | Surveillance radar | (1991) | 1991–92 | (2) | For 2 Leander Class frigates |
| Sweden | 2 | 9LV | Fire control radar | 1991 | | | For 2 Meko-200 Type frigates |
| | 2 | Sea Giraffe | Surveillance radar | 1991 | | | For 2 Meko-200 Type frigates |
| USA | 2 | Phalanx | CIWS | 1991 | | | Arming 2 Meko-200 Type frigates |
| | 2 | Seasparrow VLS | ShAM launcher | 1992 | | | For 2 Meko-200 Type frigates |
| | . . | NATO Seasparrow | ShAM | (1991) | | | Arming 2 Meko-200 Type frigates |
| **Norway** | | | | | | | |
| **S:** France | 400 | Mistral | Portable SAM | 1990 | | | Deal worth $60 m (offsets 75%) |
| Germany, FR | 92 | Leopard-1 | Main battle tank | 1991 | 1992 | (46) | CFE cascade |
| | 6 | Type-210 | Submarine | 1982 | 1989–92 | 6 | Norwegian designation Ula Class |
| Sweden | (9) | Giraffe | Surveillance radar | 1989 | 1992 | (3) | Deal worth $90 m |
| | (360) | RBS-70 | Portable SAM | 1989 | 1991–92 | (180) | Deal worth $80 m (offsets 45%); 6th order |
| UK | 1 | SH-3D Sea King | Helicopter | 1989 | 1992 | 1 | Deal worth $18 m incl upgrade of 8 delivered earlier |

| | | | | | | | |
|---|---|---|---|---|---|---|---|
| USA | 136 | M-113 | APC | 1991 | | | CFE cascade |
| | 100 | AIM-120A AMRAAM | Air-to-air missile | 1989 | | | Arming F-16 fighters; deal worth $75 m |
| | 7 612 | BGM-71D TOW-2 | Anti-tank missile | 1985 | 1987–92 | (5 000) | Deal worth $126 m incl 300 launchers and spares |
| **Poland** | | | | | | | |
| **S:** USA | . . | PA-34-200T | Transport aircraft | 1977 | | | First military version entered production in 1992 |
| **L:** USSR | . . | 2S1 122mm | Self-propelled gun | (1980) | 1982–91 | 490 | Some built for export |
| **Portugal** | | | | | | | |
| **S:** Germany, FR | . . | LARS 110mm | MRL | (1991) | | | CFE cascade |
| Netherlands | 104 | M-113 | APC | 1991 | | | CFE cascade |
| | 24 | YP-408 | APC | 1991 | | | CFE cascade |
| UK | 5 | Super Lynx | Helicopter | 1990 | | | For 3 Meko-200 Type frigates; deal worth $81 m (offsets 25%) |
| USA | 20 | F-16A/B | Fighter aircraft | 1990 | | | 17 A versions and 3 B versions |
| | . . | Model 205 UH-1H | Helicopter | 1989 | | } | In partial payment of US base rights in the Azores; |
| | . . | Model 209 AH-1G | Helicopter | 1989 | | | ex-US Air Force; part of a total of 52 helicopters |
| **Romania** | | | | | | | |
| **S:** Bulgaria | (42) | 2S1-122mm | Turret | 1988 | 1989–92 | (42) | To be fitted on Romanian chassis |
| USSR | . . | SA-7 Grail | Portable SAM | (1978) | 1978–92 | (375) | |
| **L:** UK | . . | BN-2A Islander | Transport aircraft | 1968 | 1969–92 | (450) | |
| USSR | . . | Yak-52 | Trainer aircraft | 1976 | 1979–92 | (1 620) | |
| **Spain** | | | | | | | |
| **S:** France | 840 | Mistral | Portable SAM | 1991 | 1992 | (150) | Deal worth $154 m (offsets 50%) incl 200 firing posts |
| USA | 8 | AV-8B Harrier 2-Plus | Fighter/ground attack | 1992 | | | |
| | 1 | F/A-18 Hornet | Fighter aircraft | 1990 | | | Attrition replacement |
| | 8 | S-76C | Helicopter | 1991 | 1991–92 | 6 | |

| Recipient/ supplier (S) or licenser (L) | No. ordered | Weapon designation | Weapon description | Year of order/ licence | Year(s) of deliveries | No. delivered/ produced | Comments |
|---|---|---|---|---|---|---|---|
| | 6 | SH-60B Seahawk | Helicopter | 1991 | 1992 | 2 | For 6 FFG-7 Class frigates; deal worth $251 m |
| | 1 | TAV-8B Harrier | Fighter/ground attack | 1992 | | | Deal worth $25 m |
| | 83 | M-110-A2 203mm | Self-propelled gun | 1991 | 1992 | (28) | CFE cascade |
| | 100 | M-113 | APC | 1991 | | | CFE cascade |
| | 160 | M-60-A1 Patton | Main battle tank | 1991 | 1992 | (133) | CFE cascade |
| | 260 | M-60-A3 Patton | Main battle tank | 1991 | 1992 | (166) | CFE cascade |
| | 4 | RGM-84A launcher | ShShM launcher | 1988 | | | Coastal defence version |
| | 2 | RGM-84A launcher | ShShM launcher | 1989 | | | For 2 FFG-7 Class frigates |
| | 2 | RIM-67A launcher | ShAM launcher | 1989 | | | For 2 FFG-7 Class frigates |
| | 250 | AGM-65F/G Maverick | Air-to-surface missile | 1989 | 1990–92 | (250) | Arming F/A-18 fighters; mix of F and G versions |
| | 200 | AIM-120A AMRAAM | Air-to-air missile | 1990 | | | Deal worth $132 m |
| | 16 | RGM-84A Harpoon | ShShM | 1989 | | | Arming coastal defence battery |
| | (16) | RGM-84A Harpoon | ShShM | 1989 | | | Arming 2 FFG-7 Class frigates |
| | 150 | RIM-67A/SM-1 | ShAM | (1989) | | | Arming 6 FFG-7 Class frigates; deal worth $88 m |
| L: UK | 4 | Sandown Class | MCM | (1988) | | | |
| USA | 2 | FFG-7 Class | Frigate | 1990 | | | In addition to 4 ordered previously |
| **Sweden** | | | | | | | |
| S: France | . . | TRS-2620 | Surveillance radar | 1990 | 1991 | (1) | |
| USA | 1 | Gulfstream-4 | Transport aircraft | 1992 | | | |
| | 2 | Gulfstream-4 | Transport aircraft | 1992 | | | |
| | 700 | AGM-114A Hellfire | Anti-tank missile | 1987 | 1990–92 | (450) | Deal worth $65 m; Hellfire coastal defence version |
| **Switzerland** | | | | | | | |
| S: France | 12 | AS-332 Super Puma | Helicopter | 1989 | 1991–92 | (12) | Deal worth $190 m (offsets 100%) |
| UK | 3 | Watchman | Surveillance radar | 1990 | 1992 | (1) | |
| USA | 34 | F/A-18 Hornet | Fighter aircraft | 1988 | | | Deal worth $2.5 b (offsets 100%) incl 26 C versions, 8 D versions and spares; status uncertain |

| | | | | | | | |
|---|---|---|---|---|---|---|---|
| | (500) | AGM-65B Maverick | Air-to-surface missile | 1991 | | | Arming F-5 fighters |
| | 204 | AIM-7M Sparrow | Air-to-air missile | 1988 | | | Arming F/A-18 Hornet fighters; status uncertain |
| | (204) | AIM-9L Sidewinder | Air-to-air missile | (1988) | | | Arming F/A-18 Hornet fighters; status uncertain |
| | 12 000 | BGM-71D TOW-2 | Anti-tank missile | (1985) | 1988–92 | (3 450) | Deal worth $209 m incl 400 launchers and night-vision sights |
| | 3 500 | FIM-92A Stinger | Portable SAM | 1988 | | | Licensed production under discussion |
| **L:** Germany, FR | 345 | Leopard-2 | Main battle tank | 1983 | 1987–92 | 336 | Deal worth $1400 m incl 35 delivered direct |
| **Turkey** | | | | | | | |
| **S:** France | 5 | Stentor | Surveillance radar | 1987 | 1988–92 | (5) | |
| | 14 | TRS-22XX | Surveillance radar | 1987 | | | Deal worth $150 m |
| Germany, FR | (46) | F-4F Phantom | Fighter aircraft | (1991) | 1992 | 46 | Part of *Materialhilfe* aid programme; 16 for spares |
| | 46 | RF-4E Phantom | Recce aircraft | (1991) | | | Part of *Materialhilfe* aid programme |
| | 131 | LARS 110mm | MRL | (1991) | 1992 | (50) | CFE cascade |
| | 131 | M-110-A2 203mm | Self-propelled gun | (1991) | | | CFE cascade |
| | 300 | BTR-60P | APC | (1990) | 1990–92 | (300) | CFE cascade; former GDR equipment |
| | 100 | Leopard-1-A1 | Main battle tank | (1991) | 1992 | 15 | CFE cascade |
| | 20 | M-48 ARV | ARV | (1991) | | | Part of *Materialhilfe* aid programme |
| | 10 | M-48 AVLB | Bridge layer | (1991) | 1992 | (10) | Part of *Materialhilfe* aid programme |
| | 100 | Ratac-S | Battlefield radar | 1992 | | | Most for local assembly |
| | 1 | Meko-200 Type | Frigate | 1990 | | | Part of deal worth $465 m incl 1 to be built in Turkey |
| | 1 | Meko-200 Type | Frigate | 1992 | | | Part of deal worth $330 m incl 1 built in Turkey |
| Italy | 14 | SF-260D | Trainer aircraft | 1990 | 1990–92 | 14 | Assembled from knock-down kits |
| | 100 | M-113 | APC | (1991) | | | CFE cascade |
| | 4 | Seaguard | Fire control radar | 1990 | | | For 2 Meko-200 Type frigates |
| | (48) | Aspide | ShAM | 1990 | | | Arming 2 Meko-200 Type frigates |
| Russia | 17 | Mi-17 Hip-H | Helicopter | 1992 | 1992 | (3) | Deal worth $75 m incl BTR-60 APCs and other equipment |
| | 10 | BTR-60P | APC | 1992 | 1992 | 10 | Part of deal worth $75 m; for the Gendarmerie |
| USA | 23 | Model 209 AH-1S | Helicopter | 1990 | 1992 | (9) | In addition to 5 supplied in 1990 |
| | 10 | R-22 | Helicopter | 1991 | 1992 | 10 | For training |
| | 45 | UH-60 Blackhawk | Helicopter | 1992 | 1992 | 5 | Part of deal worth $1.1 b incl 50 produced in Turkey |

| Recipient/ supplier (S) or licenser (L) | No. ordered | Weapon designation | Weapon description | Year of order/ licence | Year(s) of deliveries | No. delivered/ produced | Comments |
|---|---|---|---|---|---|---|---|
| | 72 | M-110-A2 203mm | Self-propelled gun | (1991) | | | CFE cascade |
| | 300 | M-113 | APC | 1990 | | | |
| | (250) | M-113 | APC | (1991) | | | CFE cascade |
| | (164) | M-60-A1 Patton | Main battle tank | (1991) | | | CFE cascade |
| | 600 | M-60-A3 Patton | Main battle tank | (1990) | 1992 | (300) | Southern Region Amendment aid programme |
| | 658 | M-60-A3 Patton | Main battle tank | (1991) | | | CFE cascade |
| | (40) | V-150 Commando | APC | 1992 | | | Order may be up to 120; for Police and Gendarmerie |
| | 1 | AN/FPS-117 | Surveillance radar | 1991 | | | Deal worth $15 m; options on 2 more |
| | 5 | AN/TPQ-36 | Tracking radar | 1992 | | | Deal worth $28 m |
| | 2 | RGM-84A launcher | ShShM launcher | 1990 | | | For 2 Meko-200 Type frigates |
| | 2 | Seasparrow launcher | ShAM launcher | 1990 | | | For 2 Meko-200 Type frigates |
| | 350 | AGM-65D Maverick | Air-to-surface missile | 1991 | | | |
| | 100 | AGM-88 Harm | Anti-radar missile | 1991 | | | Deal worth $29 m incl support |
| | 20 | AIM-120A AMRAAM | Air-to-air missile | (1992) | | | Arming F-16 fighters; deal worth $17 m |
| | 310 | AIM-9E Sidewinder | Air-to-air missile | 1990 | | | Deal worth $30 m incl training missiles |
| | 469 | FIM-92A Stinger | Portable SAM | 1991 | 1992 | (469) | Deal worth $33 m incl 150 launchers |
| | (48) | RGM-84A Harpoon | ShShM | (1990) | | | Arming 2 Meko-200 Type frigates |
| L: Germany, FR | 2 | FPB-57 | Fast attack craft | 1991 | | | Improved Dogan Class |
| | 1 | Meko-200 Type | Frigate | 1990 | | | Part of deal worth $465 m |
| | 1 | Meko-200 Type | Frigate | 1992 | | | Part of deal worth $330 m incl 1 delivered direct |
| | 2 | Type-209/3 | Submarine | 1987 | | | Option on 4 more |
| Italy | 26 | SF-260D | Trainer aircraft | 1990 | 1992 | 4 | In addition to 14 delivered direct |
| Spain | 50 | CN-235M | Transport aircraft | 1990 | | | Part of deal worth $500 m; 2 delivered direct |
| USA | 152 | F-16C | Fighter aircraft | 1984 | 1987–92 | 114 | Part of deal worth $4 b incl 8 C and D versions delivered direct |
| | 40 | F-16C | Fighter aircraft | 1992 | | | Deal worth $2.8 b incl 12 spare engines; in addition to component manufacture for another 40 |
| | 50 | UH-60 Blackhawk | Helicopter | 1992 | | | In addition to 45 supplied direct; option on 125 more |
| | 120 | MLRS 227mm | MRL | 1988 | 1991–92 | 16 | Including 36 000 rockets |

| | | | | | | | |
|---|---|---|---|---|---|---|---|
| | 1 698 | AIFV | AIFV | 1988 | 1990–92 | 311 | Deal worth $1 b (offsets $700 m) |
| | (4 800) | FIM-92A Stinger | Portable SAM | 1989 | | | Part of NATO Stinger programme |
| **UK** | | | | | | | |
| **S:** USA | 6 | E-3D Sentry | AEW&C aircraft | 1986 | 1991–92 | 6 | Offsets worth 130% |
| | 1 | E-3D Sentry | AEW&C aircraft | 1987 | 1992 | 1 | Deal worth $120 m (offsets 130%); option on 8th declined |
| | 2 | S-70C | Helicopter | 1992 | | | Deal worth $23.8 m incl spares, support and training; for Hong Kong |
| | 210 | AIM-120A AMRAAM | Air-to-air missile | 1992 | | | |
| | 220 | AIM-9S Sidewinder | Air-to-air missile | 1990 | 1992 | (50) | Deal worth $23 m incl spare parts and support equipment |
| **L:** Brazil | 128 | EMB-312 Tucano | Trainer aircraft | 1985 | 1987–92 | (113) | Deal worth $145–50 m; option on 15 more |
| Switzerland | (1 000) | Piranha | APC | 1991 | 1992 | (50) | Produced for export to unnamed customer |
| USA | . . | WS-70 | Helicopter | 1987 | 1987 | (1) | |
| | 57 | MLRS 227mm | MRL | 1985 | 1989–92 | (38) | |
| | . . | BGM-71A TOW | Anti-tank missile | 1980 | 1982–92 | (26 201) | |
| **USA** | | | | | | | |
| **S:** Angola | 250 | FIM-92A Stinger | Portable SAM | 1992 | 1992 | 250 | Returned from UNITA |
| Australia | 12 | CH-47C Chinook | Helicopter | 1991 | | | In exchange for 4 CH-47D |
| Germany, FR | 12 | MiG-23BN Flogger | Fighter/ground attack | 1991 | 1991–92 | (12) | Former GDR equipment |
| | 48 | Tpz-1 Fuchs | APC | (1991) | 1991–92 | 28 | US designation M-93 Fox |
| | 181 | SSN-2 Styx | ShShM | 1992 | 1992 | 181 | Arming ex-GDR Tarantul Class corvette |
| Italy | 10 | G-222 | Transport aircraft | 1990 | 1991–92 | 8 | Deal worth $157 m; option on 10 more |
| | 4 | Spada battery | SAM system | 1988 | | | For defence of US air bases in Italy |
| | 16 | Skyguard | Fire control radar | 1990 | | | For defence of US air bases in Italy |
| | (144) | Aspide | SAM | 1988 | | | For Spada SAM system |
| Japan | (183) | Beechjet 400T | Transport aircraft | 1990 | 1992 | 17 | US designation T-1A Jayhawk |
| Norway | 64 | Penguin-2-7 | Anti-ship missile | 1990 | 1992 | (14) | Option for 200 more |
| | 82 | Penguin-2-7 | Anti-ship missile | 1992 | | | 3rd order |

| Recipient/ supplier (S) or licenser (L) | No. ordered | Weapon designation | Weapon description | Year of order/ licence | Year(s) of deliveries | No. delivered/ produced | Comments |
|---|---|---|---|---|---|---|---|
| Spain | (6) | C-212-300 Aviocar | Transport aircraft | 1989 | 1990–92 | (5) | Test bed for tactical reconnaissance radar |
| UK | 38 | Firefly-160 | Trainer aircraft | 1992 | | | Deal worth $12 m; option on additional 75 |
| | 10 | Sherpa | Transport aircraft | 1988 | 1990–92 | (10) | In addition to 18 ordered previously |
| **L:** Israel | 86 | AGM-142A Have Nap | Air-to-surface missile | 1988 | 1989–92 | (84) | For co-production with Martin Marietta; US designation AGM-142A Popeye |
| Italy | 17 | Osprey Class | MCM | 1986 | 1992 | (1) | Improved Lerici Class MCM |
| Switzerland | . . | ADATS LOS-FH | SAM | 1987 | 1991–92 | (198) | |
| UK | 302 | T-45 Hawk | Jet trainer aircraft | 1986 | 1988–92 | (15) | Deal worth $512 m incl 32 simulators |
| | 436 | M-119 105mm | Towed gun | 1987 | 1990–92 | (106) | Arming US light divisions |
| | 13 | Cyclone Class | Patrol craft | 1990 | 1992 | (4) | Based on Ramadan Class fast attack craft |
| **II. Developing countries** | | | | | | | |
| **Algeria** | | | | | | | |
| **S:** China | 7 | Chui-E Class | Patrol craft | 1989 | 1990–92 | 7 | |
| **L:** UK | 3 | Kebir Class | Patrol craft | (1990) | | | |
| **Angola** | | | | | | | |
| **S:** Spain | 2 | C-212-300 Aviocar | Maritime patrol | (1990) | | | |
| | 4 | C-212-300 Aviocar | Maritime patrol | 1991 | | | |
| | (3) | Cormoran Class | Fast attack craft | 1989 | | | Status uncertain |
| Switzerland | 8 | PC-7 | Trainer aircraft | (1989) | 1990 | 6 | |
| **Argentina** | | | | | | | |
| **S:** Canada | 150 | Model 212 | Helicopter | 1990 | 1992 | (4) | Limited local assembly |
| USA | 36 | A-4M Skyhawk-2 | Fighter/bomber | 1992 | 1992 | 36 | Sale of additional 16 possible |

| | | | | | | | |
|---|---|---|---|---|---|---|---|
| **L:** Canada | . . | Model-412 | Helicopter | 1991 | | | Licence authorizes sales to Latin American countries |
| Germany, FR | 6 | Meko-140 Type | Frigate | 1980 | 1985–90 | 4 | Last 2 will be available for export |
| | 2 | Type TR-1700 | Submarine | 1977 | | | In addition to 2 delivered direct; will be available for export; plans for 2 more abandoned |
| Italy | . . | A-109 Hirundo | Helicopter | 1988 | | | Deal worth $120 m |
| **Bahrain** | | | | | | | |
| **S:** USA | 8 | AH-64 Apache | Helicopter | 1991 | | | |
| | 9 | MLRS 227mm | MRL | 1990 | 1992 | 9 | Deal worth $50 m |
| | 450 | AGM-114A Hellfire | Anti-tank missile | 1990 | | | Arming AH-64 Apache helicopters |
| **Bangladesh** | | | | | | | |
| **S:** China | (40) | F-6 | Fighter aircraft | 1992 | 1992 | (40) | Replacing aircraft lost in 1991 cyclone |
| | (21) | F-7M Airguard | Fighter aircraft | 1992 | | | Replacing aircraft lost in 1991 cyclone |
| | 2 | Hai Ying-2 L | ShShM launcher | 1988 | 1989 | 1 | For 2 Jianghu Class frigates |
| | 2 | Hai Ying-2 L | ShShM launcher | 1992 | | | For 2 Huangfen Class fast attack craft |
| | (24) | Hai Ying-2 | ShShM | 1988 | 1989 | (12) | Arming 2 Jianghu Class frigates |
| | (8) | Hai Ying-2 | ShShM | 1992 | | | Arming 2 Huangfen Class fast attack craft |
| | 2 | Huangfen Class | Fast attack craft | 1992 | | | Designation uncertain |
| | 2 | Jianghu Class | Frigate | (1988) | 1989 | 1 | Status of 2nd uncertain |
| **Brazil** | | | | | | | |
| **S:** France | 20 | AS-550 Fennec | Helicopter | 1992 | 1992 | (5) | Deal worth $25 m |
| Italy | . . | FILA | Fire control radar | (1987) | 1989–92 | (12) | Fire control for RBS-70 SAM system |
| UK | (7) | Super Lynx | Helicopter | 1991 | | | For Inhauma Class corvettes |
| | 4 | L119 105mm gun | Towed gun | 1991 | 1992 | 4 | |
| **L:** Austria | . . | GHN-45 155mm | Towed gun | (1985) | | | Status uncertain |
| France | 10 | AB-565 Panther | Helicopter | 1988 | 1990–92 | 10 | Part of deal worth $249 m |
| Germany, FR | . . | SNAC-1 | Nuclear submarine | 1989 | | | Nuclear reactor designed and built in Brazil |
| | 3 | Type-209/3 | Submarine | 1982 | | | |

| Recipient/ supplier (S) or licenser (L) | No. ordered | Weapon designation | Weapon description | Year of order/ licence | Year(s) of deliveries | No. delivered/ produced | Comments |
|---|---|---|---|---|---|---|---|
| Italy | . . | MSS-1 | Anti-tank missile | 1986 | 1988–91 | 130 | Italian designation MAF |
| Singapore | (4) | Grauna Class | Offshore patrol ship | 1987 | 1992 | 2 | Option on 4 more |
| UK | . . | L119 105mm gun | Towed gun | 1991 | | | |
| **Brunei** | | | | | | | |
| **S:** Germany, FR | (96) | AIM-9L Sidewinder | Air-to-air missile | 1989 | | | Arming Hawk-100 fighters |
| Indonesia | 3 | CN-235 | Maritime patrol | 1989 | | | |
| UK | 16 | Hawk-100 | Jet trainer aircraft | 1989 | | | Deal worth $260 m |
| USA | 1 | UH-60 Blackhawk | Helicopter | (1989) | | | VIP version |
| **Chile** | | | | | | | |
| **S:** France | 4 | AS-332 Super Puma | Helicopter | 1988 | 1988 | 2 | Part of deal worth $77 m |
| | 2 | AS-532 Cougar-2 | Helicopter | 1992 | | | In addition to 4 AS-332 ordered in 1988; replacing 1988 order for 4 AS-565 |
| | 12 | Mygale | SAM system | (1990) | 1991–92 | (18) | |
| | . . | AM-39 Exocet | Anti-ship missile | 1992 | | | Arming 2 Navy AS-532 Cougar-2 helicopters |
| | (1 400) | Mistral | Portable SAM | (1990) | 1990–92 | (600) | |
| Germany, FR | (30) | Bo-105CB | Helicopter | 1985 | 1986–92 | (18) | |
| Israel | (6) | Barak launcher | ShAM launcher | 1989 | | | For 2 County Class destroyers and 4 Leander Class frigates; replacing Seacat |
| | 2 | Phalcon | AEW&C radar | (1989) | | | Part of deal worth $500 m |
| | (256) | Barak | ShAM | 1989 | | | |
| UK | 1 | Leander Class | Frigate | 1992 | 1992 | 1 | Ex-Royal Navy |
| USA | 10 | A-37B Dragonfly | Close support aircraft | 1992 | 1992 | 10 | Ex-US Air Force |
| | 2 | Boeing-707 | Transport aircraft | 1991 | 1992 | 2 | |
| | 2 | C-130B Hercules | Transport aircraft | 1992 | 1992 | (2) | Deal worth $3 m; 1st sale after arms embargo lifted |
| **L:** South Africa | (400) | G-5 155 mm | Towed gun | 1989 | 1990 | (6) | |
| | . . | G-6 155mm | Self-propelled gun | 1989 | 1991 | (1) | Prototype completed but no full-scale production |

| | | | | | | | |
|---|---|---|---|---|---|---|---|
| Switzerland | . . | Piranha | APC | 1980 | 1981–92 | (249) | |
| UK | . . | Rayo | MRL | 1986 | | | |
| **China** | | | | | | | |
| **S:** Russia | 2 | Su-27 Flanker | Fighter aircraft | 1992 | 1992 | 2 | Trainer version; original order incl 10 fighter version |
| USA | 6 | CH-47D Chinook | Helicopter | 1989 | | | Deliveries suspended June 1989 |
| | 4 | AN/TPQ-37 | Tracking radar | (1987) | 1988 | 2 | Deal incl avionics, 4 Mk 46 torpedoes and 155mm ammunition; deliveries suspended June 1989 |
| USSR | 40 | MiG-29 Fulcrum | Fighter aircraft | 1991 | | | |
| | 12 | Su-24 Fencer | Fighter/bomber | (1990) | | | |
| | 24 | Su-27 Flanker | Fighter aircraft | 1991 | 1991–92 | 24 | Deal worth $700 m (offsets 40%) |
| | . . | AA-10 Alamo | Air-to-air missile | 1991 | 1991–92 | (144) | Arming 24 Su-27 Flankers |
| | (96) | AA-8 Aphid | Air-to-air missile | 1991 | 1991–92 | (96) | Arming 24 Su-27 Flankers |
| **L:** France | (30) | AS-365N Dauphin | Helicopter | 1992 | | | |
| Israel | . . | PL-8H | SAM | (1989) | 1990–92 | (1 385) | Based on Python III air-to-air missile |
| | . . | PL-9 | Air-to-air missile | (1989) | 1990–92 | (3 731) | Based on Python III air-to-air missile |
| **Colombia** | | | | | | | |
| **S:** Brazil | 2 | EMB-110 Bandeirante | Transport aircraft | 1992 | 1992 | 2 | |
| | 14 | EMB-312 Tucano | Trainer aircraft | 1992 | 1992 | 14 | |
| USA | . . | Citation-2 | Transport aircraft | (1990) | 1990 | 1 | |
| **Cyprus** | | | | | | | |
| **S:** France | 36 | AMX-30-B2 | Main battle tank | 1989 | | | Deal worth $115 m |
| | . . | MM-40 CDS | Coast defence system | 1989 | | | |
| | . . | MM-40 Exocet | ShShM | 1989 | | | Arming coastal defence batteries |
| Greece | 75 | Steyr-4K 7FA | APC | (1990) | 1990–92 | (48) | Options for 65 more |
| **Ecuador** | | | | | | | |
| **S:** UK | 3 | Jaguar | Fighter/ground attack | 1991 | 1992 | 3 | Ex- Royal Air Force; order may be for 6 |

| Recipient/ supplier (S) or licenser (L) | No. ordered | Weapon designation | Weapon description | Year of order/ licence | Year(s) of deliveries | No. delivered/ produced | Comments |
|---|---|---|---|---|---|---|---|
| **Egypt** | | | | | | | |
| S: Czechoslovakia | 48 | L-59 Albatross | Jet trainer aircraft | 1991 | 1992 | (20) | Deal worth $204 m; improved L-39 |
| USA | 24 | AH-64 Apache | Helicopter | 1990 | | | Deal worth $488 m incl Hellfire missiles |
| | 2 | Commuter-1900 | Transport aircraft | (1989) | 1991–92 | 2 | In addition to 6 delivered earlier |
| | 2 | E-2C Hawkeye | AEW aircraft | 1989 | 1990 | 1 | Deal worth $84 m |
| | 42 | F-16C | Fighter aircraft | 1987 | 1991–92 | (22) | 3rd order |
| | 46 | F-16C/D | Fighter aircraft | 1991 | | | Deal worth $1.6 b incl spare engines and armament; mix of C and D versions; assembled in Turkey |
| | 40 | M-88-A1 | ARV | 1990 | 1991–92 | (40) | Deal worth $70 m |
| | 492 | AGM-114A Hellfire | Anti-tank missile | 1990 | | | Arming AH-64 Apache helicopters |
| | 144 | AGM-65D Maverick | Air-to-surface missile | 1988 | 1991 | 80 | Arming F-16 fighters; deal worth $27 m incl training missiles, parts and electronic countermeasure pods |
| | 40 | AGM-65D Maverick | Air-to-surface missile | 1991 | 1992 | 40 | Arming F-16 fighters |
| | 40 | AGM-65G Maverick | Anti-ship missile | 1991 | 1992 | 40 | Arming F-16 fighters |
| | 282 | AIM-7M Sparrow | Air-to-air missile | (1987) | 1992 | (150) | Arming F-16 fighters; deal worth $42 m |
| | 695 | BGM-71D TOW-2 | Anti-tank missile | 1992 | 1992 | 695 | Deal worth $28 m incl 152 launchers and support |
| | 7 511 | BGM-71D TOW-2 | Anti-tank missile | 1988 | 1989–91 | (600) | Incl 180 launchers, 504 night-vision sights and spares |
| | 29 | UGM-84A Harpoon | SuShM | 1990 | | | Arming 4 Romeo Class submarines; deal worth $69 m |
| | 3 | Swiftships MCM | MCM | 1991 | | | Order number may be up to 6 |
| L: Germany, FR | . . | Fahd | APC | 1978 | 1988–92 | 550 | |
| UK | . . | Swingfire | Anti-tank missile | 1977 | 1979–92 | 8 168 | |
| USA | (530) | M-1-A1 Abrams | Main battle tank | 1988 | | | Deal worth $2 b; in addition to 25 delivered direct |
| | 26 | AN/TPS-63 | Surveillance radar | 1986 | 1988–92 | 26 | Deal worth $190 m incl 8 delivered direct |
| | . . | AIM-9P Sidewinder | Air-to-air missile | (1988) | 1990–92 | 1 121 | In addition to 37 assembled from kits |
| **Ethiopia** | | | | | | | |
| S: USSR | 1 | Natya Class | MCM | (1990) | | | Status uncertain |

| | | | | | | | |
|---|---|---|---|---|---|---|---|
| | 2 | Sonya Class | MCM | (1990) | | | Status uncertain |
| **Fiji** | | | | | | | |
| **S:** Australia | 3 | ASI-315 | Patrol craft | 1992 | | | Pacific Forum aid |
| France | 1 | AS-365N Dauphin | Helicopter | 1990 | 1992 | 1 | |
| **Gabon** | | | | | | | |
| **S:** France | (5) | Mygale | SAM system | (1990) | 1992 | (2) | |
| **India** | | | | | | | |
| **S:** France | . . | PSM-33 | Surveillance radar | 1988 | 1990–92 | (3) | |
| Germany, FR | 1 | Rajaba Class | Support ship | 1987 | | | Option on 1 more |
| Russia | 20 | MiG-29 Fulcrum | Fighter aircraft | 1992 | | | Deal worth $500 m; order may be up to 30 |
| | . . | 2S6 | AAV(M) | 1992 | | | Part of larger deal incl aircraft and tanks |
| | . . | SA-19 | SAM | 1992 | | | Part of 2S6 system |
| USA | 2 | AN/TPQ-37 | Tracking radar | (1990) | 1992 | (2) | Deal worth $22 m |
| USSR | 10 | Mi-26 Halo | Helicopter | 1988 | 1992 | 1 | 2nd order |
| | 8 | Bass Tilt | Fire control radar | 1983 | 1989–91 | 4 | For 8 Khukri Class corvettes |
| | 8 | SSN-2 Styx L | ShShM launcher | 1983 | 1989–91 | 4 | For 8 Khukri Class corvettes |
| | 6 | SSN-2 Styx L | ShShM launcher | 1987 | 1991–92 | 2 | Arming 6 Vibhuti Class corvettes |
| | . . | AT-4 Spigot | Anti-tank missile | 1983 | 1991–92 | (600) | Arming BMP-2 AIFVs |
| | (400) | SA-16 Gimlet | Portable SAM | (1990) | 1991–92 | (400) | |
| | . . | SA-N-5 Grail | ShAM | (1983) | 1989–91 | (160) | Arming 8 Khukri Class corvettes |
| | . . | SA-N-5 Grail | ShAM | 1987 | 1991–92 | (80) | Arming 6 Vibhuti Class corvettes |
| | (200) | SA-N-5 Grail | ShAM | 1983 | 1989–91 | (160) | Arming 5 Pauk Class patrol craft |
| | . . | SSN-2 Styx | ShShM | 1983 | 1989–91 | (48) | Arming 8 Khukri Class corvettes |
| | . . | SSN-2 Styx | ShShM | 1987 | 1991–92 | (24) | Arming 6 Vibhuti Class corvettes |
| | 5 | Pauk Class | Patrol craft | 1983 | 1989–91 | 4 | |
| **L:** France | . . | SA-316B Chetak | Helicopter | (1962) | 1964–92 | 209 | Also produced for civilian use |
| | 27 112 | Milan | Anti-tank missile | 1982 | 1985–92 | 27 112 | Production switched to Milan-2 |
| | (15 000) | Milan-2 | Anti-tank missile | 1992 | | | |

| Recipient/ supplier (S) or licenser (L) | No. ordered | Weapon designation | Weapon description | Year of order/ licence | Year(s) of deliveries | No. delivered/ produced | Comments |
|---|---|---|---|---|---|---|---|
| Germany, FR | 103 | Do-228 | Transport aircraft | 1983 | 1987–92 | 46 | |
| | 2 | Type-1500 | Submarine | 1981 | 1992 | 1 | In addition to 2 delivered direct |
| Korea, South | 7 | Sukanya Class | Offshore patrol ship | 1987 | 1990–92 | 4 | In addition to 3 delivered direct |
| Netherlands | 212 | Flycatcher | Fire control radar | (1987) | 1988–92 | 82 | In addition to direct deliveries |
| UK | 2 | Magar Class | Landing ship | (1979) | 1987–92 | 2 | |
| USSR | (200) | MiG-27 Flogger | Fighter/ground attack | 1983 | 1984–92 | 117 | |
| | . . | BMP-2 | AIFV | 1983 | 1987–92 | 184 | Indian designation Sarath |
| | 500 | T-72 | Main battle tank | (1980) | 1987–92 | 346 | Including 175 knock-down kits with very low Indian content; in addition to 500 delivered direct |
| | . . | AA-8 Aphid | Air-to-air missile | (1986) | | | Indian designation Astra |
| | 6 | Vibhuti Class | Corvette | 1987 | 1991–92 | 2 | Order may reach 15 |
| **Indonesia** | | | | | | | |
| **S:** Germany, FR | (128) | SA-N-5 Grail | ShAM | 1992 | | | Arming 16 Parchim Class corvettes, former GDR equipment |
| | 12 | Frosch Class | Landing ship | 1992 | | 2 | Part of deal for 39 former GDR ships; transferred without armament |
| | 2 | Frosch II Class | Supply ship | 1992 | | 2 | |
| | 9 | Kondor Class | MCM | 1992 | | 2 | |
| | 16 | Parchim Class | Corvette | 1992 | | | Part of deal for 39 former GDR ships |
| Netherlands | . . | F-27 Mk-100 Friendship | Transport aircraft | 1990 | | | |
| UK | (14) | AR-325 | Surveillance radar | 1989 | 1991–92 | (6) | |
| | 1 | Rover Class | Supply ship | 1991 | 1992 | 1 | Ex-Royal Navy; delivered after refit |
| USA | 1 | B-737 Surveiller | Maritime patrol | 1991 | | | |
| **L:** France | . . | AS-332 Super Puma | Helicopter | 1983 | 1985–91 | 10 | |
| Germany, FR | (100) | NBo-105 | Helicopter | 1987 | 1988–92 | 60 | |
| | 6 | PB-57 Type | Patrol craft | 1982 | 1988–92 | 6 | |
| Spain | (80) | CN-212 Aviocar | Transport aircraft | 1976 | 1978–92 | 40 | |
| UK | (14) | Hawk-100 | Jet trainer aircraft | 1992 | | | |
| | (10) | Hawk-200 | Fighter/ground attack | 1992 | | | |

| | | | | | | | |
|---|---|---|---|---|---|---|---|
| USA | . . | Model 412 | Helicopter | 1982 | 1986–91 | 16 | Production suspended 1992 |
| **Iran** | | | | | | | |
| **S:** China | (72) | F-7M Airguard | Fighter aircraft | (1991) | 1992 | (18) | |
| | (8) | HQ-2B SAMS | SAM system | (1989) | 1990–92 | (6) | For coastal air defence batteries |
| | (96) | HQ-2B | SAM | 1989 | 1990–92 | (72) | |
| | (10) | Hegu Class | Fast attack craft | (1991) | | | Order may be for 12 |
| Korea, North | . . | Scud-C launcher | Mobile SSM system | (1991) | 1992 | (5) | |
| | (170) | Scud-C | SSM | (1991) | 1992 | (100) | Agreement apparently includes production equipment; number may be up to 200 |
| Russia | 2 | A-50 Mainstay | AEW&C aircraft | 1992 | | | Part of deal worth $2.5 b; signed in July 1992; status uncertain |
| USSR | (500) | T-72 | Main battle tank | 1989 | 1990–92 | (200) | |
| | 2 | Kilo Class | Submarine | 1991 | 1992 | 1 | Deal reportedly worth $750 m; option on 1 more |
| **Israel** | | | | | | | |
| **S:** Germany, FR | 1 | SA-6 SAMS | SAM system | 1991 | | | Former GDR equipment |
| | . . | SA-6 Gainful | SAM | 1991 | | | Former GDR equipment |
| | 2 | Dolphin Class | Submarine | 1991 | | | Deal worth $570 m |
| USA | 7 | AH-64 Apache | Helicopter | (1992) | | | Deal worth $140 m |
| | (24) | AH-64 Apache | Helicopter | 1992 | | | From US forces in Europe; funded by US grant |
| | 15 | F-15A Eagle | Fighter aircraft | 1990 | 1991–92 | (15) | Ex-US Air Force |
| | 10 | F-15A Eagle | Fighter aircraft | 1991 | 1992 | 10 | In addition to 15 leased in 1990 |
| | 60 | F-16C/D | Fighter aircraft | 1988 | 1991–92 | (30) | 30 C versions and 30 D versions; follow-on order for 60 more under negotiation |
| | . . | S-65A Stallion | Helicopter | (1992) | | | |
| | (10) | UH-60 Blackhawk | Helicopter | (1992) | | | |
| | 3 | RGM-84A launcher | ShShM launcher | (1988) | | | Arming Saar-5 Class corvettes |
| | 539 | AGM-114A Hellfire | Anti-tank missile | 1990 | 1990–92 | (300) | Arming 18 AH-64 Apache helicopters |
| | 300 | AIM-9S Sidewinder | Air-to-air missile | 1990 | | | Deal worth $32 m incl support |
| | . . | FIM-92A Stinger | Portable SAM | 1990 | | | |
| | (48) | RGM-84A Harpoon | ShShM | (1988) | | | Arming Saar-5 Class corvettes |

| Recipient/ supplier (S) or licenser (L) | No. ordered | Weapon designation | Weapon description | Year of order/ licence | Year(s) of deliveries | No. delivered/ produced | Comments |
|---|---|---|---|---|---|---|---|
| | 3 | Saar-5 Class | Corvette | 1988 | | | Israeli design; fully financed with FMS credits worth $300 m; some sub-systems to be fitted in Israel |
| **Kenya** | | | | | | | |
| S: France | 100 | Mistral | Portable SAM | 1990 | 1990–92 | (60) | |
| **Kiribati** | | | | | | | |
| S: Australia | 1 | ASI-315 | Patrol craft | 1992 | | | Pacific Forum aid |
| **Korea, North** | | | | | | | |
| S: China | . . | Romeo Class | Submarine | 1973 | 1975–92 | 14 | In addition to 7 directly from China |
| **Korea, South** | | | | | | | |
| S: France | 1 000 | Mistral | Portable SAM | 1992 | | | Deal worth $180 m (offsets 25%) |
| Germany, FR | 1 | Type-209/3 | Submarine | 1987 | | | Deal worth $600 m |
| Netherlands | . . | Goalkeeper | CIWS | 1991 | | | Arming KDX-2000 Type frigates |
| | 1 | STIR | Fire control radar | (1992) | | | For 1st KDX-2000 Type frigate |
| Spain | 12 | CN-235 | Transport aircraft | 1992 | | | Deal worth $164 m |
| UK | 20 | Hawk | Jet trainer aircraft | 1990 | | | Deal worth $140 m |
| USA | 37 | AH-64 Apache | Helicopter | 1992 | | | Deal worth $997 m incl 775 Hellfire ATMs and 8 spare engines |
| | 30 | F-16C | Fighter aircraft | 1981 | 1987–92 | (30) | Deal worth $931 m incl 6 F-16D |
| | 48 | F-16C | Fighter aircraft | 1991 | | | Deal worth $2.52 b incl 12 delivered direct, 36 assembled locally and 72 produced in Korea, 12 spare engines and 20 Lantirn navigation pods |
| | 8 | P-3C Update-3 Orion | Maritime patrol | 1990 | | | Deal worth $840 m incl spare engines, training and spares |
| | 90 | UH-60 Blackhawk | Helicopter | 1990 | 1991–92 | (7) | Deal worth $500 m; for local assembly |

| | | | | | | | |
|---|---|---|---|---|---|---|---|
| | 3 | AN/FPS-117 | Surveillance radar | 1990 | 1992 | (1) | In addition to 5 delivered earlier |
| | 1 | RGM-84A launcher | ShShM launcher | (1992) | | | For 1st KDX-2000 Type frigate |
| | 4 | Seasparrow VLS | ShAM launcher | 1990 | | | For KDX-2000 Type frigates |
| | 775 | AGM-114A Hellfire | Anti-tank missile | 1992 | | | Arming 37 AH-64A Apache helicopters |
| | 28 | AGM-84A Harpoon | Anti-ship missile | 1992 | | | Arming P-3C Update-3 Orion aircraft; deal worth $58 m incl technical assistance |
| | 40 | AGM-88 Harm | Anti-radar missile | 1992 | | | |
| | 179 | AIM-7M Sparrow | Air-to-air missile | 1991 | 1992 | (90) | Deal worth $31 m |
| | 704 | BGM-71D TOW-2 | Anti-tank missile | 1987 | 1990–92 | (704) | |
| | (24) | RGM-84A Harpoon | ShShM | (1992) | | | Arming 1st KDX-2000 Type frigate |
| | 21 | Seasparrow | ShAM | 1990 | | | Arming KDX-2000 Type frigates; deal worth $33 m incl training rounds and support |
| **L:** Germany, FR | 2 | Type-209/3 | Submarine | 1987 | 1992 | 1 | In addition to 1 delivered direct |
| | 3 | Type-209/3 | Submarine | 1989 | | | 2nd order for 3 |
| | 3 | Type-209/3 | Submarine | 1992 | | | 3rd order for 3; total order may reach 18 |
| Italy | 6 | Lerici Class | MCM | (1986) | 1986 | 1 | |
| Japan | 30 | BK-117 | Helicopter | 1990 | 1991–92 | 20 | For local assembly |
| USA | 72 | F-16C | Fighter aircraft | 1991 | | | Part of deal worth $2.52 b |
| | (150) | H-76 Eagle | Helicopter | 1986 | 1991 | 12 | |
| | 250 | M-109-A2 155mm | Self-propelled gun | 1990 | 1991–92 | 100 | Deal worth $260 m |
| | (620) | K-1 | Main battle tank | 1980 | 1985–92 | 560 | |
| | . . | M-167 Vulcan | CIWS | (1986) | 1986–91 | 66 | |
| **Kuwait** | | | | | | | |
| **S:** Australia | 2 | ASI-315 | Patrol craft | 1992 | | | 1st order for 2 |
| | 2 | ASI-315 | Patrol craft | 1992 | | | 2nd order for 2 |
| Egypt | 10 | Skyguard SAMS | SAM system | 1988 | 1990–92 | 10 | Part of Amoun air defence system |
| | (320) | Aspide | SAM | 1988 | 1990–92 | (320) | For Skyguard SAM system |
| France | 4 | MM-40 launcher | ShShM launcher | 1992 | | | For 4 Combattante-4 Type fast attack craft |
| | . . | Mistral | Portable SAM | 1992 | | | Sadral system; arming 4 Combattante-4 Type fast attack craft |
| | (96) | MM-40 Exocet | ShShM | 1992 | | | Arming 4 Combattante-4 Type fast attack craft |

| Recipient/ supplier (S) or licenser (L) | No. ordered | Weapon designation | Weapon description | Year of order/ licence | Year(s) of deliveries | No. delivered/ produced | Comments |
|---|---|---|---|---|---|---|---|
| | 4 | Combattante-4 | Fast attack craft | 1992 | | | Part of deal worth $500 m; option on 4 more |
| Netherlands | 4 | Goalkeeper | CIWS | 1992 | | | Arming 4 Combattante-4 Type fast attack craft |
| USA | 40 | F/A-18C/D Hornet | Fighter aircraft | 1988 | 1991–92 | 6 | Deal worth $1.9 b incl 32 C and 8 D versions and armament |
| | 256 | M-1-A2 Abrams | Main battle tank | 1992 | | | Deal worth $4 b incl spares |
| | 125 | M-113-A2 | APC | 1992 | | | Part of deal worth $4 b |
| | 52 | M-577-A2 | APC command post | 1992 | | | Part of deal worth $4 b |
| | 46 | M-88-A1 | ARV | 1992 | | | Part of deal worth $4 b |
| | 1 | AN/FPS-117 | Surveillance radar | 1992 | | | Deal worth $92 m |
| | 6 | I-Hawk SAMS | SAM system | 1992 | | | Deal worth $2.5 b incl Patriot missile systems, training and support |
| | 1 | Patriot SAMS | SAM system | 1992 | | | Part of deal worth $2.5 b |
| | 300 | AGM-65G Maverick | Anti-ship missile | 1988 | | | Arming F/A-18 Hornet fighters |
| | 40 | AGM-84A Harpoon | Anti-ship missile | 1988 | | | Arming F/A-18 Hornet fighters |
| | 200 | AIM-7F Sparrow | Air-to-air missile | 1988 | 1992 | (50) | Arming F/A-18 Hornet fighters |
| | 120 | AIM-9L Sidewinder | Air-to-air missile | 1988 | 1992 | (30) | Arming F/A-18 Hornet fighters |
| | 210 | MIM-104 Patriot PAC-2 | SAM | 1992 | | | Part of deal worth $2.5 b |
| | 342 | MIM-23B Hawk | SAM | 1992 | | | Part of deal worth $2.5 b |
| **Malaysia** | | | | | | | |
| **S:** France | 2 | MM-40 launcher | ShShM launcher | 1992 | | | For 2 Yarrow Type frigates; deal worth $181 m |
| | . . | Mistral | Portable SAM | (1991) | | | |
| | (48) | MM-40 Exocet | ShShM | 1992 | | | Arming 2 Yarrow Type frigates |
| Italy | 4 | Skyguard | Fire control radar | 1988 | 1989–92 | (4) | |
| Netherlands | 2 | DA-08 | Surveillance radar | 1992 | | | For 2 Yarrow Type frigates |
| Sweden | 2 | Sea Giraffe | Surveillance radar | 1992 | | | For 2 Yarrow Type frigates |
| UK | 10 | Hawk-100 | Jet trainer aircraft | 1990 | | | Part of deal worth $740 m incl 18 Hawk-200 aircraft, weapons, training and services |
| | 18 | Hawk-200 | Fighter/ground attack | 1990 | | | |

| | | | | | | | |
|---|---|---|---|---|---|---|---|
| | 2 | Martello 743-D | Surveillance radar | 1990 | 1992 | (2) | Deal worth $190 m |
| | 12 | DN-181 Rapier | SAM system | 1988 | | | |
| | 2 | Seawolf VLS | ShAM launcher | 1992 | | | For 2 Yarrow Type frigates |
| | 576 | Improved Rapier | SAM | 1988 | | | |
| | (96) | Seawolf-2 | ShAM | 1992 | | | Arming 2 Yarrow Type frigates |
| | 2 | Yarrow Type | Frigate | 1992 | | | Deal worth $600 m incl spares, training and support |
| USA | 4 | B-200T Maritime | Maritime patrol | 1990 | | | |
| **Mexico** | | | | | | | |
| **S:** USA | 10 | Model 530MG Defender | Helicopter | 1992 | | | |
| **Morocco** | | | | | | | |
| **S:** Spain | 2 | F-30 Class | Frigate | 1991 | | | Provisional order |
| **Myanmar** | | | | | | | |
| **S:** China | 12 | F-7M Airguard | Fighter aircraft | 1990 | 1990–92 | (12) | Incl 2 trainer versions |
| | (2) | Y-8 | Transport aircraft | (1991) | | | |
| | . . | PL-2A | Air-to-air missile | 1990 | 1990–92 | (48) | Arming F-6 and F-7 fighters |
| Poland | 10 | Mi-2 Hoplite | Helicopter | (1992) | 1992 | 10 | |
| | 12 | W-3 Sokol | Helicopter | 1990 | 1991–92 | (12) | |
| **Namibia** | | | | | | | |
| **S:** France | 1 | Falcon-900 | Transport aircraft | (1991) | 1992 | 1 | VIP version |
| **Nigeria** | | | | | | | |
| **S:** Czechoslovakia | 27 | L-39Z Albatross | Jet trainer aircraft | 1991 | 1991–92 | (27) | Deal worth $100 m incl support |
| France | 12 | AS-332 Super Puma | Helicopter | 1985 | 1989–90 | 6 | |
| UK | 80 | MBT Mk-3 | Main battle tank | 1990 | 1991–92 | 50 | Deal worth $282 m |
| **L:** USA | . . | Air Beetle | Trainer aircraft | 1988 | 1988–92 | 3 | |

| Recipient/ supplier (S) or licenser (L) | No. ordered | Weapon designation | Weapon description | Year of order/ licence | Year(s) of deliveries | No. delivered/ produced | Comments |
|---|---|---|---|---|---|---|---|
| **Oman** | | | | | | | |
| S: France | 2 | Crotale NG Navale | ShAM system | 1992 | | | For 2 Muheet Class corvettes |
| | 2 | MM-40 launcher | ShShM launcher | 1992 | | | For 2 Muheet Class corvettes |
| | (48) | MM-40 Exocet | ShShM | 1992 | | | Arming 2 Muheet Class corvettes |
| | (48) | VT-1 | SAM | 1992 | | | For 2 Muheet Class corvettes |
| Netherlands | 2 | MW-08 | Surveillance radar | 1992 | | | For 2 Muheet Class corvettes |
| UK | 4 | Hawk-100 | Jet trainer aircraft | 1989 | | | Deal worth $225 m incl 12 Hawk-200 versions |
| | 12 | Hawk-200 | Fighter/ground attack | 1990 | | | |
| | . . | Improved Rapier | SAM | 1992 | | | Part of deal worth $63 m |
| | 2 | Muheet Class | Corvette | 1992 | | | Deal worth $237 m |
| USA | (96) | AIM-9L Sidewinder | Air-to-air missile | 1990 | | | Arming 16 Hawk-100 and Hawk-200 aircraft; possibly from European production line |
| **Pakistan** | | | | | | | |
| S: China | 98 | A-5 Fantan-A | Fighter/ground attack | 1984 | | | 2nd order |
| | 40 | F-7M Airguard | Fighter aircraft | 1988 | 1992 | (20) | Includes 20 FT-7 trainer versions |
| | 40 | F-7P Skybolt | Fighter aircraft | 1992 | | | |
| | 25 | K-8 Karakorum 8 | Jet trainer aircraft | 1987 | | | |
| France | 12 | SA-315B Lama | Helicopter | 1992 | | | |
| | 1 | Eridan Class | MCM | 1992 | 1992 | 1 | Ex-French Navy |
| | 1 | Eridan Class | MCM | 1992 | | | In addition to 1 ex-French Navy and 1 licence-produced |
| USA | 6 | SH-2F/G Seasprite | Helicopter | 1989 | 1989 | 3 | 3 F and 3 G versions |
| | (20) | M-109-A2 155mm | Self-propelled gun | 1988 | | | Deal worth $40 m incl M-198 howitzers and support equipment |
| | . . | AN/TPQ-36 | Tracking radar | (1990) | | | Deal worth $65 m |
| | 4 | AN/TPQ-37 | Tracking radar | (1985) | 1987–89 | (3) | |
| | 2 386 | BGM-71D TOW-2 | Anti-tank missile | 1987 | | | 1st Pakistani order for TOW-2; incl 144 launchers |

| | | | | | | | |
|---|---|---|---|---|---|---|---|
| **L:** China | . . | T-69II | Main battle tank | (1989) | 1991–92 | 160 | Deal worth $1.2 b |
| | . . | Anza | Portable SAM | (1988) | 1989–92 | 350 | |
| | . . | Red Arrow-8 | Anti-tank missile | 1989 | 1990–92 | 150 | |
| France | 1 | Eridan Class | MCM | 1992 | | | In addition to 2 delivered direct |
| Sweden | . . | Supporter | Trainer aircraft | 1974 | 1975–92 | 212 | |
| USA | . . | LAADS | Surveillance radar | (1989) | | | Lead items delivered from 1989 |
| **Panama** | | | | | | | |
| **S:** USA | 4 | Cape Class | Patrol craft | 1991 | 1991 | 1 | Ex-US Coast Guard |
| **Papua New Guinea** | | | | | | | |
| **S:** Spain | 4 | CN-235 | Transport aircraft | 1991 | 1992 | 2 | May be from Indonesian production line |
| **Peru** | | | | | | | |
| **S:** Czechoslovakia | 100 | T-55 | Main battle tank | 1992 | 1992 | (100) | |
| USA | 6 | B-200T Maritime | Maritime patrol | (1990) | 1991–92 | (2) | |
| USSR | 18 | Mi-17 Hip-H | Helicopter | 1989 | 1990 | 14 | In addition to 15 delivered earlier |
| **Philippines** | | | | | | | |
| **S:** Australia | 3 | PC-57M | Patrol craft | 1990 | | | Option on 3 more |
| France | 3 | MM-40 launcher | ShShM launcher | 1991 | | | For 3 Cormoran Class corvettes |
| | . . | MM-40 Exocet | ShShM | 1991 | | | Arming 3 Cormoran Class Corvettes |
| Italy | 36 | S-211 | Jet trainer aircraft | 1988 | 1989–92 | (24) | |
| | 16 | SF-260TP | Trainer aircraft | 1992 | | | Order may be for up to 30 |
| Spain | 3 | Cormoran Class | Fast attack craft | 1991 | | | Deal worth $100 m |
| USA | 22 | Model 500D Defender | Helicopter | 1988 | 1990–92 | 22 | Deal worth $25 m; military aid |
| | 8 | Model 530MG Defender | Helicopter | 1992 | 1992 | 6 | Deal worth $11 m |
| | 24 | OV-10F Bronco | Close support aircraft | 1991 | 1991 | 5 | |
| | 2 | Besson Class | Landing ship | 1992 | | | Deal worth $32.2 m; 4 more to be ordered |
| **L:** UK | 150 | FS-100 Simba | APC | 1992 | | | Deal worth $57 m |

| Recipient/ supplier (S) or licenser (L) | No. ordered | Weapon designation | Weapon description | Year of order/ licence | Year(s) of deliveries | No. delivered/ produced | Comments |
|---|---|---|---|---|---|---|---|
| **Qatar** | | | | | | | |
| S: France | 4 | Crotale NG Navale | ShAM system | 1992 | | | For 4 Vita Class fast attack craft |
| | 4 | MM-40 launcher | ShShM launcher | 1992 | | | For 4 Vita Class fast attack craft |
| | 500 | Mistral | Portable SAM | 1990 | | | |
| | (24) | Mistral | Portable SAM | 1992 | | | Arming 4 Vita Class fast attack craft |
| | (96) | MM-40 Exocet | ShShM | 1992 | | | Arming 4 Vita Class fast attack craft |
| | . . | VT-1 | SAM | 1992 | | | For 4 Vita Class fast attack craft |
| Netherlands | 4 | Goalkeeper | CIWS | 1992 | | | Arming 4 Vita Class fast attack craft |
| UK | 4 | Vita Class | Fast attack craft | 1992 | | | Deal worth $200 m |
| **Saudi Arabia** | | | | | | | |
| S: Canada | 1 117 | LAV | APC | 1990 | | | Deal worth $700 m incl 384 LAV-25 and 733 other versions |
| France | 3 | MM-40 launcher | ShShM launcher | 1990 | | | For 3 La Fayette Class frigates |
| | 1 200 | Mistral | Portable SAM | 1989 | 1991–92 | (800) | |
| | (72) | MM-40 Exocet | ShShM | 1990 | | | Arming 3 La Fayette Class frigates |
| | 3 | La Fayette Class | Frigate | 1992 | | | Part of deal worth $4 b (offsets 30%) |
| UK | 12 | BAe-125-800 | Transport aircraft | 1988 | 1988–92 | (12) | Part of 1988 Tornado deal; for VIP use |
| | 20 | Hawk-100 | Jet trainer aircraft | 1988 | | | Part of 1988 Tornado deal |
| | 40 | Hawk-200 | Fighter/ground attack | 1988 | | | Part of 1988 Tornado deal |
| | 48 | Tornado IDS | Fighter/ground attack | 1988 | | | |
| | (50) | WS-70 | Helicopter | 1988 | | | |
| | 461 | Piranha | APC | 1990 | 1992 | (50) | Deal worth $400 m |
| | 200 | ALARM | Anti-radar missile | 1986 | 1991–92 | (120) | Arming Tornado IDS fighters |
| | (480) | Sea Eagle | Anti-ship missile | 1985 | | | Arming Tornado IDS fighters |
| | 3 | Sandown Class | MCM | 1988 | 1991–92 | 2 | Option on 3 more |
| USA | 12 | AH-64 Apache | Helicopter | 1990 | 1992 | (12) | Deal worth $300 m incl 155 Hellfire missiles; follow-on order for 36 probable |
| | 24 | F-15C Eagle | Fighter aircraft | 1990 | 1991–92 | 22 | |

| | | | | | | | |
|---|---|---|---|---|---|---|---|
| | 72 | F-15XP Eagle | Fighter/ground attack | 1992 | | | Deal worth $9 b incl 24 spare engines, 48 Lantirn navigation pods and armament |
| | 7 | KC-130H Hercules | Tanker/transport | 1990 | | | Part of deal worth $750 m incl C-130H Hercules transport aircraft |
| | 8 | UH-60 Blackhawk | Helicopter | 1990 | 1991–92 | 8 | Medevac version, deal worth $121 m |
| | 8 | UH-60 Blackhawk | Helicopter | 1992 | | | Medevac version; deal worth $225 m |
| | 27 | M-198 155mm | Towed gun | 1990 | 1991–92 | (27) | |
| | 150 | M-1-A1 Abrams | Main battle tank | 1990 | | | |
| | 315 | M-1-A2 Abrams | Main battle tank | 1990 | | | Deal worth $1.5 b; status uncertain |
| | 207 | M-113-A2 | APC | 1990 | 1991–92 | (207) | Part of deal worth $3.1 b |
| | 400 | M-2 Bradley | AIFV | 1990 | 1992 | (140) | In addition to 220 ordered previously |
| | 50 | M-548 | APC | 1991 | | | Part of deal worth $3.1 b |
| | 9 | M-577-A2 | APC command post | 1990 | | | Part of deal worth $3.1 b |
| | 43 | M-578 | ARV | 1991 | | | Part of deal worth $3.1 b |
| | 12 | M-88-A1 | ARV | 1990 | 1992 | (12) | Deal worth $26 m |
| | (6) | AN/TPS-43 | Surveillance radar | 1985 | 1987–92 | (6) | |
| | 8 | Patriot SAMS | SAM system | 1990 | | | Deal worth $984 m incl 384 missiles, 6 radars and support |
| | 14 | Patriot SAMS | SAM system | 1991 | | | Deal worth $3.1 b incl 758 missiles |
| | 362 | AGM-114A Hellfire | Anti-tank missile | 1992 | | | Arming 12 Apache helicopters; deal worth $606 m incl 3 500 rockets, 40 trucks and a simulator |
| | 900 | AGM-65D/G Maverick | Air-to-surface missile | 1992 | | | Arming 48 F-15XP fighters; mix of D and G versions |
| | 770 | AIM-7M Sparrow | Air-to-air missile | 1991 | | | Part of deal worth $365 m incl laser-guided bombs |
| | 300 | AIM-7M Sparrow | Air-to-air missile | 1992 | | | Arming 72 F-15XP fighters |
| | 300 | AIM-9S Sidewinder | Air-to-air missile | 1992 | | | Arming 72 F-15XP fighters |
| | 4 460 | BGM-71D TOW-2 | Anti-tank missile | 1988 | 1989–92 | (2 000) | |
| | 1 750 | BGM-71D TOW-2 | Anti-tank missile | 1990 | | | Deal worth $55 m incl 116 launchers |
| | 384 | MIM-104 Patriot PAC-2 | SAM | 1990 | | | |
| | 758 | MIM-104 Patriot PAC-2 | SAM | 1991 | | | |
| **Singapore** | | | | | | | |
| **S:** France | 20 | AS-550 Fennec | Helicopter | 1989 | 1991–92 | (20) | |
| | 36 | LG-1 105mm | Towed gun | 1990 | 1991–92 | (24) | |

| Recipient/ supplier (S) or licenser (L) | No. ordered | Weapon designation | Weapon description | Year of order/ licence | Year(s) of deliveries | No. delivered/ produced | Comments |
|---|---|---|---|---|---|---|---|
| | 22 | AMX-10 PAC-90 | AIFV | 1990 | 1992 | (22) | In addition to 22 AMX-10P versions |
| | 22 | AMX-10P | AIFV | (1990) | 1991–92 | (22) | |
| | (200) | Milan-2 | Anti-tank missile | 1989 | 1990–92 | (200) | Deal incl 30 launchers; order may be for 400 |
| | 150 | Mistral | Portable SAM | 1992 | | | Supplied with 30 launchers; follow-on order for Navy probable |
| Netherlands | 4 | F-50 Enforcer | Maritime patrol | 1991 | | | Deal worth $52 m; option on 2 more |
| Sweden | 4 | Landsort Class | MCM | 1991 | | | |
| USA | 11 | F-16A | Fighter aircraft | 1992 | | | Deal worth $330 m |
| | 20 | AGM-84A Harpoon | Anti-ship missile | 1991 | | | |
| | (240) | BGM-71C I-TOW | Anti-tank missile | 1989 | 1991–92 | (240) | Arming AS-550 Fennec helicopters |
| **Sri Lanka** | | | | | | | |
| S: Argentina | 4 | IA-58A Pucara | Close support aircraft | 1992 | 1992 | 4 | Ex-Argentine Air Force |
| Czechoslovakia | (25) | T-55 | Main battle tank | 1991 | 1992 | (25) | |
| UK | 2 | HS-748-2 | Transport aircraft | (1991) | 1992 | 2 | In addition to 2 delivered earlier |
| **Syria** | | | | | | | |
| S: Czechoslovakia | (252) | T-72 | Main battle tank | 1991 | 1992 | (252) | Order may be for up to 300 and may include 90 T-55s; status of deliveries uncertain |
| Korea, North | (8) | Scud-C launcher | Mobile SSM system | 1991 | | | May be up to 20 |
| | (150) | Scud-C | SSM | 1989 | 1991–92 | (80) | At least 20 delivered via Iran |
| USSR | 3 | Kilo Class | Submarine | (1987) | | | Status uncertain |
| **Taiwan** | | | | | | | |
| S: France | 60 | Mirage-2000-5 | Fighter aircraft | 1992 | | | Deal worth $2.6 b |
| | (1 000) | Magic-2 | Air-to-air missile | 1992 | | | Arming 60 Mirage-2000-5 fighters |
| | (500) | Mica | Air-to-air missile | (1992) | | | Arming 60 Mirage-2000-5 fighters |
| | 6 | La Fayette Class | Frigate | 1991 | | | Status uncertain |

| | | | | | | | |
|---|---|---|---|---|---|---|---|
| Netherlands | 8 | DA-08 | Surveillance radar | (1989) | 1989–92 | 8 | For 8 Gearing Class destroyers |
| | 8 | STIR | Fire control radar | (1989) | 1989–92 | 8 | For 8 Gearing Class destroyers |
| USA | (4) | E-2C Hawkeye | AEW aircraft | 1990 | | | Upgraded ex-US Navy E-2B |
| | 150 | F-16A | Fighter aircraft | 1992 | | | Deal worth $5.8 b incl spare engines and missiles |
| | 18 | Model-209 AH-1W | Helicopter | 1992 | | | Option on 24 more |
| | 26 | OH-58D Kiowa | Helicopter | 1991 | | | |
| | 12 | SH-2F Seasprite | Helicopter | 1992 | | | Deal worth $161 m incl spare engines |
| | 110 | M-60-A3 Patton | Main battle tank | 1991 | | | Deal worth $119 m incl overhaul, spares and logistics |
| | 8 | Phalanx | CIWS | (1989) | 1989–92 | 8 | Arming 8 Gearing Class destroyers; deal worth $15 m |
| | . . | AN/MPQ-53 | Fire control radar | (1992) | | | Fire control radar for Sky Bow missile |
| | 3 | Phalanx | CIWS | 1992 | 1992 | 3 | Arming 3 Knox Class frigates |
| | 6 | Phalanx | CIWS | 1991 | | | Arming 6 FFG-7 Class frigates |
| | 3 | RGM-84A launcher | ShShM launcher | 1992 | 1992 | 3 | For 3 Knox Class frigates |
| | 6 | RIM-67A launcher | ShAM launcher | 1989 | | | For 6 FFG-7 Class frigates |
| | 8 | W-160 | Fire control radar | (1989) | 1989–92 | 8 | For 8 Gearing Class destroyers |
| | (144) | AGM-114A Hellfire | Anti-tank missile | (1991) | | | Arming OH-58D Kiowa helicopters |
| | 600 | AIM-7M Sparrow | Air-to-air missile | 1992 | | | Arming 150 F-16 fighters |
| | 900 | AIM-9S Sidewinder | Air-to-air missile | 1992 | | | Arming 150 F-16 fighters |
| | (24) | RGM-84A Harpoon | ShShM | 1992 | 1992 | (24) | Arming 3 Knox Class frigates |
| | 80 | RIM-67A/SM-1 | ShAM | (1989) | 1989–92 | (80) | Arming 8 Gearing Class destroyers |
| | 97 | RIM-67A/SM-1 | ShAM | 1991 | | | Arming 6 FFG-7 Class frigates; deal worth $55 m incl spares and support |
| | 3 | Knox Class | Frigate | 1992 | 1992 | 2 | 5-year lease |
| **L:** Israel | . . | Gabriel-2 | ShShM | (1978) | 1980–92 | 583 | Taiwanese designation Hsiung Feng |
| USA | 6 | FFG-7 | Frigate | 1989 | | | |
| **Thailand** | | | | | | | |
| **S:** China | 9 | AS-365N Dauphin | Helicopter | 1992 | | | |
| | (450) | T-69 | Main battle tank | 1987 | 1989–92 | (450) | Part of deal involving over 1 500 armoured vehicles |
| | 4 | C-801 launcher | ShShM launcher | 1988 | 1991–92 | 4 | For 4 Jianghu Class frigates |

| Recipient/ supplier (S) or licenser (L) | No. ordered | Weapon designation | Weapon description | Year of order/ licence | Year(s) of deliveries | No. delivered/ produced | Comments |
|---|---|---|---|---|---|---|---|
| | (25) | Type-311B | Fire control radar | 1991 | 1991–92 | (25) | |
| | 96 | C-801 | ShShM | 1988 | 1991–92 | (96) | Arming 4 Jianghu Class frigates; deal worth $40 m |
| | (900) | HN-5A | Portable SAM | 1991 | | | Deal worth $46 m incl 90 launchers |
| | 4 | Jianghu Class | Frigate | 1988 | 1991–92 | 4 | Part of deal worth $272 m; Thai designation Chao Phraya Class |
| | 2 | Naresuan Class | Frigate | 1989 | | | Jianghu E Class; in addition to 4 Jianghu Class frigates |
| Czechoslovakia | 36 | L-39Z Albatross | Jet trainer aircraft | 1992 | | | Upgraded with Israeli avionics |
| France | 20 | Crotale NG | SAM system | 1991 | | | |
| | (480) | VT-1 | SAM | 1991 | | | |
| Netherlands | 2 | STIR | Fire control radar | 1992 | | | For 2 Naresuan Class frigates |
| Spain | 1 | ASS Bazan | Landing ship | 1992 | | | Deal worth $228 m for unarmed vessel |
| Switzerland | 20 | PC-9 Turbo Trainer | Trainer aircraft | 1990 | 1991–92 | (20) | Deal worth $90 m incl spares and training; follow-on order for 10 expected |
| UK | 2 | Martello 743-D | Surveillance radar | 1991 | 1992 | (2) | |
| USA | 38 | A-7E Corsair-2 | Fighter/ground attack | 1991 | | | Deal worth $30 m; incl 6 TA-7 trainers and 8 to be broken up for spares |
| | 4 | C-130H-30 | Transport aircraft | 1991 | | | |
| | 18 | F-16A/B | Fighter aircraft | 1991 | | | 12 A and 6 B versions; deal worth $547 m incl 4 spare engines, 6 Lantirn navigation pods, spares, logistics and support |
| | 25 | Model 212 | Helicopter | 1990 | 1991–92 | (23) | |
| | 3 | P-3B Orion | Maritime patrol | 1989 | | | Ex-US Navy, deal worth $140 m incl Harpoon ASMs |
| | 2 | SH-2F Seasprite | Helicopter | 1989 | | | For 2 Naresuan Class frigates |
| | 20 | M-109 155mm | Self-propelled gun | (1991) | 1992 | (5) | Deal worth $63 m |
| | 350 | M-48-A5 Patton | Main battle tank | 1990 | | | |
| | 300 | M-60-A1 Patton | Main battle tank | 1990 | 1991–92 | (300) | |
| | 2 | Seasparrow VLS | ShAM launcher | (1991) | | | For 2 Naresuan Class frigates |
| | 16 | AGM-84A Harpoon | Anti-ship missile | 1990 | | | Arming 3 P-3 Orion aircraft |
| | 48 | Seasparrow | ShAM | (1991) | | | Arming 2 Naresuan Class frigates |

| | | | | | | | |
|---|---|---|---|---|---|---|---|
| **L:** UK | 3 | Khamronsin Class | Fast attack craft | 1987 | 1992 | 3 | |
| | 1 | Province Class | Patrol craft | 1989 | 1992 | 1 | Patrol version of Khamronsin Class fast attack craft; for Royal Thai Marine Police |
| **Tunisia** | | | | | | | |
| **S:** Germany, FR | 4 | Kondor Class | MCM | 1992 | 1992 | 4 | Former GDR ships; transfer incl 5 small patrol craft |
| **Tuvalu** | | | | | | | |
| **S:** Australia | 1 | ASI-315 | Patrol craft | 1992 | | | Pacific Forum aid |
| **United Arab Emirates** | | | | | | | |
| **S:** France | 500 | Mistral | Portable SAM | 1988 | 1991–92 | (240) | |
| Indonesia | 7 | CN-235 | Transport aircraft | 1992 | | | |
| Russia | 500 | BMP-3 | AIFV | 1992 | 1992 | (50) | 300 for Abu Dhabi, 200 for Dubai |
| | (4 000) | AT-10 Bastion | Anti-tank missile | 1992 | 1992 | (400) | Arming 500 BMP-3 AIFVs |
| South Africa | 78 | G-6 155mm | Self-propelled gun | 1990 | 1991–92 | (49) | |
| UK | 18 | Hawk-100 | Jet trainer aircraft | 1989 | 1992 | 2 | For Abu Dhabi; part of deal worth $340 m |
| USA | 20 | AH-64 Apache | Helicopter | 1991 | | | Deal worth $680 m incl Hellfire missiles |
| | 2 | C-130H Hercules | Transport aircraft | 1991 | | | Deal worth $54.9 m |
| | 620 | AGM-114A Hellfire | Anti-tank missile | 1991 | | | Arming AH-64 Apache helicopter |
| **Uruguay** | | | | | | | |
| **S:** Switzerland | 6 | PC-7 Turbo Trainer | Trainer aircraft | 1992 | 1992 | 6 | Deal worth $10.5 m |
| UK | 2 | Wessex | Helicopter | 1992 | 1992 | 2 | Ex-Royal Navy |
| **Venezuela** | | | | | | | |
| **S:** Brazil | 100 | EE-11 Urutu | APC | 1988 | 1989–92 | (40) | |
| France | 18 | Mirage-5OEV | Fighter/ground attack | 1988 | 1991–92 | (16) | |
| | (50) | AM-39 Exocet | Anti-ship missile | (1988) | | | Arming Mirage-5OEV fighters |
| | (100) | Magic-2 | Air-to-air missile | 1988 | 1991–92 | (40) | Arming Mirage-5OEV fighters; deal worth $30 m |

| Recipient/ supplier (S) or licenser (L) | No. ordered | Weapon designation | Weapon description | Year of order/ licence | Year(s) of deliveries | No. delivered/ produced | Comments |
|---|---|---|---|---|---|---|---|
| Spain | 2 | C-212-300 Aviocar | Transport aircraft | (1991) | | | Attrition replacements |
| USA | 18 | OV-10E Bronco | Close support aircraft | 1991 | | | |
| | (6) | RGM-84A launcher | ShShM launcher | 1989 | | | |
| | 18 | RGM-84A Harpoon | ShShM | 1989 | | | Deal worth $50 m |
| **Zaire** | | | | | | | |
| **S:** France | 13 | AMX-13 | Light tank | 1989 | | | |
| **Zimbabwe** | | | | | | | |
| **S:** UK | 5 | Hawk | Jet trainer aircraft | 1990 | 1991–92 | (5) | |

**Abbreviations and acronyms:**

| | |
|---|---|
| AAV(G) | Anti-aircraft vehicle (gun-armed) |
| AAV(M) | Anti-aircraft vehicle (missile-armed) |
| ADATS | Air defence and anti-tank system |
| AEW | Airborne early-warning (system) |
| AEW&C | Airborne early-warning and control |
| AIFV | Armoured infantry fighting vehicle |
| ALARM | Air-launched anti-radar missile |
| AMRAAM | Advanced medium-range air-to-air missile |
| APC | Armoured personnel carrier |
| ARV | Armoured recovery vehicle |
| ASW | Anti-submarine warfare |
| CDS | Coastal defence system |
| CIWS | Close-in support system |
| Elint | Electronic intelligence |
| FMS | Foreign Military Sales (USA) |
| incl | Including/includes |
| LAV | Light armoured vehicle |
| MCM | Mine countermeasures (ship) |
| MLRS | Multiple-launch rocket system |
| MRL | Multiple rocket launcher |
| RAM | Rolling airframe missile |
| Recce | Reconnaissance |
| SAM | Surface-to-air missile |
| SAMS | Surface-to-air missile system |
| ShAM | Ship-to-air missile |
| ShShM | Ship-to-ship missile |
| SuShM | Submarine-to-ship missile |
| UNITA | National Union for the Total Independence of Angola |
| VIP | Very important person |
| VLS | Vertical launch system |

**Conventions:**

| | |
|---|---|
| . . | Data not available or not applicable |
| – | Negligible figure (< 0.5) or none |
| ( ) | Uncertain data or SIPRI estimate |

# **Appendix 10D.** Sources and methods

## I. The SIPRI sources

The sources of the data presented in the arms trade registers are of five general types: newspapers; periodicals and journals; books, monographs and annual reference works; official national documents; and documents issued by international and intergovernmental organizations. The registers are largely compiled from information contained in around 200 publications searched regularly.

Published information cannot provide a comprehensive picture because the arms trade is not fully reported in the open literature. Published reports provide partial information, and substantial disagreement among reports is common. Therefore, the exercise of judgement and the making of estimates are important elements in compiling the SIPRI arms trade data base. Order dates and the delivery dates for arms transactions are continuously revised in the light of new information, but where they are not disclosed the dates are estimated. Exact numbers of weapons ordered and delivered may not always be known and are sometimes estimated—particularly with respect to missiles. It is common for reports of arms deals involving large platforms—ships, aircraft and armoured vehicles—to ignore missile armaments classified as major weapons by SIPRI. Unless there is explicit evidence that platforms were disarmed or altered before delivery, it is assumed that a weapons fit specified in one of the major reference works such as the *Jane's* or *Interavia* series is carried.

## II. Selection criteria

SIPRI arms trade data cover five categories of major weapons or systems: aircraft, armour and artillery, guidance and radar systems, missiles, and warships. Statistics presented refer to the value of the trade in these five categories only. The registers and statistics do not include trade in small arms, artillery under 100-mm calibre, ammunition, support items, services and components or component technology, except for specific items. Publicly available information is inadequate to track these items satisfactorily.

There are two criteria for the selection of major weapon transfers for the registers. The first is that of military application. The aircraft category excludes aerobatic aeroplanes and gliders. Transport aircraft and VIP transports are included only if they bear military insignia or are otherwise confirmed as military registered. Micro-light aircraft, remotely piloted vehicles and drones are not included although these systems are increasingly finding military applications.

The armour and artillery category includes all types of tanks, tank destroyers, armoured cars, armoured personnel carriers, armoured support vehicles, infantry combat vehicles as well as multiple rocket launchers, self-propelled and towed guns and howitzers with a calibre equal to or above 100 mm. Military lorries, jeeps and other unarmoured support vehicles are not included.

The category of guidance and radar systems is a residual category for electronic-tracking, target-acquisition, fire-control, launch and guidance systems that are either (*a*) deployed independently of a weapon system listed under another weapon category (e.g., certain ground-based SAM launch systems) or (*b*) shipborne missile-launch or point-defence (CIWS) systems. The values of acquisition, fire-control, launch and guidance systems on aircraft and armoured vehicles are included in the value of the respective aircraft or armoured vehicle. The reason for treating shipborne systems separately is that a given type of ship is often equipped with numerous combinations of different surveillance, acquisition, launch and guidance systems.

The missile category includes only guided missiles. Unguided artillery rockets, man-portable anti-armour rockets and free-fall aerial munitions (e.g., 'iron bombs') are excluded. In the naval sphere, anti-submarine rockets and torpedoes are excluded.

The ship category excludes small patrol craft (with a displacement of less than 100 t), unless they carry cannon with a calibre equal to or above 100 mm; missiles or torpedoes; research vessels; tugs and ice-breakers. Combat support vessels such as fleet replenishment ships are included.

The second criterion for selection of items is the identity of the buyer. Items must be destined for the armed forces, paramilitary forces, intelligence agencies or police of another country. Arms supplied to guerrilla forces pose a problem. For example, if weapons are delivered to the Contra rebels they are listed as imports to Nicaragua with a comment in the arms trade register indicating the local recipient. The entry of any arms transfer is made corresponding to the five weapon categories listed above. This means that missiles and their guidance/launch vehicles are often entered separately under their respective category in the arms trade register.

## III. The value of the arms trade

The SIPRI system for arms trade evaluation is designed as a *trend-measuring device*, to permit measurement of changes in the total flow of major weapons and its geographic pattern. Expressing the evaluation in monetary terms reflects both the quantity and quality of the weapons transferred. Aggregate values and shares are based only on *actual deliveries* during the year/years covered in the relevant tables and figures.

The SIPRI valuation system is not comparable to official economic statistics such as gross domestic product, public expenditure and export/import figures. The monetary values chosen do not correspond to the actual prices paid, which vary considerably depending on different pricing methods, the length of production runs and the terms involved in individual transactions. For instance, a deal may or may not cover spare parts, training, support equipment, compensation, offset arrangements for the local industries in the buying country, and so on. Furthermore, to use only actual sales prices—even assuming that the information were available for all deals, which it is not—military aid and grants would be excluded, and the total flow of arms would therefore not be measured.

Production under licence is included in the arms trade statistics in such a way as to reflect the import share embodied in the weapon. In reality, this share is normally high in the beginning, gradually decreasing over time. However, as SIPRI makes a single estimate of the import share for each weapon produced under licence, the value of arms produced under licence agreements may be slightly overstated.

## IV. Conventions

The following conventions are used in appendices 10B and 10C:

| | |
|---|---|
| . . | Data not available or not applicable |
| – | Negligible figure (<0.5) or none |
| ( ) | Uncertain data or SIPRI estimate |

# **Appendix 10E.** An overview of the arms industry modernization programme in Turkey

GÜLAY GÜNLÜK-SENESEN

## I. Introduction

In a period of falling investment in the military sector among NATO countries, Turkey emerges as a significant exception. This appendix describes Turkey's arms industry modernization programme, including its political and financial set-up, as well as developments in the aerospace industry—so far the most developed sector.

The average annual growth rate of military expenditures among European NATO states in the decade 1982–91 was about 0.6 per cent. In the same period Turkey's military expenditures grew at an average rate of 4 per cent per annum. Arms procurement expenditures for European NATO countries shrank by 0.1 per cent per year in this period, while in Turkey such expenditure increased by 10 per cent per year. Turkey spent a total of $2.7 billion on arms procurement in 1992—$1.59 billion from the defence budget and $760 million from the Defence Industry Support Fund (DISF).[1] Although high, these financial indicators seem to understate the flow of major conventional weapons into Turkey in quantitative terms.

According to SIPRI estimates, deliveries of major conventional weapons to Turkey increased by about 14 per cent per year, while the average annual growth rate among European NATO countries for the same period was 0.4 per cent. SIPRI arms import data indicate that Turkey ranks highest among the NATO arms importers, while in the Middle East only Saudi Arabia and Iraq rank higher for the period 1987–91.[2] By contrast, comparative general trade statistics for Turkey do not suggest the volume of arms imported by Turkey to be particularly great.

There could be several explanations for this discrepancy. Foreign military assistance, combined with subsidized transfers under the 1990 Treaty on Conventional Armed Forces in Europe (CFE), account for some of the difference. Financial arrangements involving long-term credit which defer payment might also contribute, as might export subsidies by suppliers seeking to benefit their domestic industries. Finally, national income accounting data may not include military transactions. These possible explanations have not been documented, however, and the question of how Turkey and its suppliers distribute the financial costs of arms procurement remains unanswered.

Under the 10-year modernization programme started in 1985, Turkey is beginning to play a part as a producer of a variety of military equipment. In the period 1975–84, the Turkish arms industry was classified as being of limited sophistication and size when compared with 32 developing nations with some arms production capacity. With regard to its industrial base, Turkey ranked fifth among these countries—sug-

[1] *Defense News*, 31 Aug.–6 Sep. 1992, pp. 6 and 12.

[2] *SIPRI Yearbook 1992: World Armaments and Disarmament* (Oxford University Press: Oxford, 1992), table 8.2, p. 273.

gesting that local industry provided an infrastructure which assisted military modernization.[3]

## II. The strategic context

The dominant factor influencing Turkey's military endeavours is its geopolitical position.[4] Turkey does not expect a diminishing role for NATO after the end of the cold war and expects to play a crucial role in Europe.[5] Turkey is the only NATO ally adjoining the Middle East, and Turkish bases played a key role in the operations against Iraq after August 1990. This alone explains much of the weapons flow during the past two years. However, Turkey's militarization is also linked to the unrest and level of armament in bordering countries.[6]

The yet unresolved problem of the PKK (People's Army for the Liberation of Kurdistan) insurgency, mainly in the south-east and east of Turkey, is one of the problems. Turkish forces carried out operations against the Kurdish guerrillas in south-eastern Turkey and northern Iraq in 1992.[7] Tensions with Iran, Iraq, Syria and Cyprus over their alleged tolerance of PKK camps are exacerbated by the conflict over the water supply from the Tigris and Euphrates rivers—the headwaters of which are controlled by Turkey. The Turkish South East Anatolia Development Project—including the Atatürk Dam and others over the Tigris and Euphrates—has given rise to Syrian and Iraqi anxieties over the availability of water in spite of Turkish assurances.[8]

Turkish intervention in clashes between Azerbaijan and Armenia in Transcaucasia has been exclusively non-military. Ties have been established with the five Islamic republics of Turkistan (formerly part of the Soviet Union) mainly for economic reasons—in part a reaction to frustration over the delays in the approval of its full membership in the Economic Community (EC).[9] The Black Sea Economic Co-operation Pact, initiated by Turkey in 1990 and signed by Albania, Armenia, Azerbaijan, Bulgaria, Georgia, Greece, Moldova, Romania, Russia, Turkey and Ukraine in June 1992, is an indication that Turkey is seeking a new role in the region, which, with this diversity of countries, cannot be attributed to primarily military goals.

[3] Brauer, J., 'Arms production in developing nations: the relation to industrial structure, industrial diversification, and human capital formation', *Defence Economics*, vol. 2, no. 2 (1991), pp. 165–75.

[4] Turkish policy is discussed in a special edition of *Nato's Sixteen Nations,* Special Edition, 1989/1990, p. 10 and in Wyllie, J., 'Turkey—adapting to new strategic realities', *Jane's Intelligence Review*, Oct. 1992, p. 450.

[5] Interview with H. Dogan (then National Defence Minister ), *Nato's Sixteen Nations*, Special Edition, vol. 36, no. 2 (1991), pp. 18–20.

[6] Interview with Turkish National Defence Minister Nevzat Ayaz, *Nato's Sixteen Nations*, vol. 37, no. 4 (1992), pp. 57–62.

[7] *Financial Times*, 11 Sep. 1992; *The Economist,* 10 Oct. 1992, p. 35; *International Herald Tribune*, 28 Oct. 1992, p. 2.

[8] Starr, J. R., 'Water wars', *Foreign Policy*, no. 82 (1991), pp. 17–36; Anderson, E., 'Water conflict in the Middle East—a new initiative', *Jane's Intelligence Review*, May 1992, p. 227; Wyllie, J., 'Turkey—adapting to new strategic realities' (note 4), p. 451.

[9] Krause, A., 'Turkey looks beyond the EC', *International Herald Tribune*, 14 Nov. 1990; Rugman, J., 'Turkey hopes its ship is coming in', *The Guardian*, 3 Feb. 1992; 'The Turkish question', *The Independent*, 1 Apr. 1992, p. 26; Erginsoy, U., 'Turkey renews friendships with Turkic neighbors', *Defense News*, 31 Aug.–6 Sep. 1992, p. 10; *The Commonwealth of Independent States and the Middle East,* vol. 17, no. 6 (1992), pp. 10–12.

Turkey has occasional disputes with Greece over territorial water zones in the Aegean Sea and the rights of the Turkish minority in Greece. However, the most important dispute between Greece and Turkey is over the partition of Cyprus, the northern part of which has been controlled by Turkish troops since 1974.

Given this strategic context Turkish Defence Minister Nevzat Ayaz has stated that the 'development towards disarmament throughout the world does not require a modification of the Turkish defence policy'.[10]

## III. The modernization programme

### Background

The US embargo on arms deliveries to Turkey between the Cyprus Operation of 1974 and 1978 raised the issue of secure access to arms within the Turkish armed forces for the first time.[11] The main goal of arms production in Turkey was import substitution—with the intention of escaping dependence on foreign military equipment. The achievement of this goal faces serious obstacles, and Turkish policy has increasingly stressed the need for foreign industrial linkages.

Aeronautics, artillery and ammunition production in Turkey along with maintenance and overhaul facilities, although with a weak heavy industry infrastructure, were established during the early years of the Republic (i.e., over half a century ago). However, NATO membership resulted in the provision of relatively modern military equipment, mainly from the USA, which slowed down industrialization efforts and raised Turkish dependence on arms imports.[12]

The Cyprus Operation was the first active engagement by the Turkish armed forces after the Korean War and revealed their capabilities and shortcomings. It created an awareness of the need for self-sufficiency in arms production as outmoded equipment had to be replaced in the face of restrictions on external military assistance and a severe scarcity of foreign exchange. Turkey aimed to develop a modern arms industry, the export revenues of which would help ease the burden on the defence budget.[13] After 1974 initial steps in capital accumulation were taken including donations and grants—indicators of public support—to various foundations established to develop the military in general, or the Air Forces, Land Forces and the Navy in particular. This provided a core of funds for investment, and these funds were merged in 1987 under the name of the Turkish Armed Forces Foundation.

The political and economic instability of the late 1970s slowed the development of a national arms industry until the military take-over on 12 September 1980. Between 1980 and 1983 the armed forces prepared a comprehensive modernization plan with the support of bureaucrats. Negotiations on the production of F-16 fighter aircraft in Turkey in late 1983 mark the first step in the implementation of this plan. The F-16

[10] *Nato's Sixteen Nations* (note 6).

[11] US credit to finance co-production of helicopters in Turkey is still provided on the condition that the aircraft are not used in Cyprus. *Defense News*, 7–13 Sep. 1992, p. 32.

[12] Erdem, V., 'Defence industry policy of Turkey', *Nato's Sixteen Nations,* Special Edition, 1989/1990, p. 10; Erdem, V., 'History of the Turkish defence industry', *Nato's Sixteen Nations,* Special Edition, vol. 36, no. 2 (1991), p. 28. Erdem has been Head of the Defence Industries Development and Support Administration (DIDA) and the Undersecretariat for Defence Industries (UDI) since their inception.

[13] de Briganti, G., *Defense News*, 14 May 1990, p. 17; *Jane's Defence Weekly*, 28 July 1990, p. 126; Erdem, V., 'Defence industry policy of Turkey', *Nato's Sixteen Nations,* Special Edition, 1989/1990, p. 10.

indirect offset financing programme included investment in the communications and tourism sectors, mining of copper and cobalt as well as marketing of Turkish traditional export goods by the subsidiary companies of General Dynamics of the USA.[14] This was the first time that the economic and thus welfare benefits which the installation of military complexes would bring to the country were presented to the public.[15]

## The administrative basis

The ambitious modernization programme required a reorganization of defence decision making to plan, finance, implement and supervise its fulfilment. The resulting structure is a unique example of its kind—although it may share some characteristics with administrative bodies in Egypt and Saudi Arabia.[16]

The Defence Industries Development and Support Administration (DIDA) was formed at the end of 1985 to administer the 10-year, $10 billion programme. DIDA was restructured as the Undersecretariat for Defence Industries (UDI) within the Ministry of National Defence in 1989. UDI monitors the implementation of decisions taken by the Defence Industry Supreme Board of Co-ordination (DISBC) and the Defence Industry Executive Committee (DIEC). The administrative structure is summarized in table 10E.1.

The DISBC is charged with planning and funding general policy for the acquisition of military equipment. Chaired by the Prime Minister, the DISBC has 13 additional members: the Chief of General Staff, three Armed Forces Commanders, the Commander of the Gendarmes, five ministers (including the Minister of Defence) and three senior civil servants.

The DIEC makes specific acquisition decisions in line with the general policy set by the DISBC. The DIEC—also chaired by the Prime Minister—includes the Chief of General Staff and the Minister of Defence with the Undersecretary of UDI acting as the secretary of the committee.

The UDI implements decisions under the following terms of reference:

- To ensure the execution of the large-scale projects of the Turkish Armed Forces involving supply of defence equipment, within the framework of co-operation between national industry and foreign technology and capital in Turkey;
- To organize and integrate the existing national industry in line with the needs and requirements of the defence industry;
- To encourage and guide any new investments in keeping with the current requirements;
- To finance any research and development activities, prototype manufacturing and similar works as well as any related investments and operational activities.[17]

[14] *Cumhuriyet*, 21 Sep. 1983.

[15] When conflict over the partial implementation of the indirect offset agreement arose in 1989, the stress on military priorities prevented this from becoming a critical issue. *Armed Forces Journal International,* June 1989, p. 58. By the end of 1991, however, the UDI announced Industry Offset Guidelines which were seen by foreign contractors as too demanding. *Jane's Defence Weekly*, 9 Nov. 1991, pp. 883–90; Silverberg, D., 'U.S. contractors say Turkey misfires with offset obligations', *Defense News*, 24 Feb. 1992, p. 16.

[16] Sayigh, Y., *Arab Military Industry, Capability, Performance and Impact* (Brassey's: London 1992), pp. 167 and 210.

[17] Erdem (note 13), p. 17; *Savunma Sanayii* [Defence Industry] Handbook prepared by the Office of the Turkish Prime Minister (Basbakanlik: Türkiye Cumhuriyeti, 1990).

**Table 10E.1.** Administration of the Turkish military modernization programme, in order of decision and implementation, since 1989

| Body | Task | Members |
|---|---|---|
| DISBC: Defence Industry Supreme Board of Co-ordination | General defence policy; overall co-ordination, planning and funding | Prime Minister, Chief of General Staff, Commanders of Land, Air, Naval and Gendarmerie forces, Minister of National Defence, 4 ministers, 3 under-secretaries |
| DIEC: Defence Industry Executive Committee | Decision-making body on the basis of the general policy of the DISBC; procurement and funding decisions | Prime Minister, Chief of General Staff, Minister of National Defence, Under-secretary of the UDI |
| UDI: Undersecretariat for Defence Industry | Monitoring and implementation of decisions of DIEC; management of the DISF | A partially independent body, with its own budget, under the Ministry of National Defence |
| DISF: Defence Industries Support Fund | Provides funding for investments in the arms industry | Under the auspices of the Central Bank; managed by the UDI; audited by a special committee |

*Source*: Office of the Turkish Prime Minister (T. C. Basbakanlik), *Savunma Sanayii* [Defence Industry] (T. C. Basbakanlik: Ankara, 1990).

## Financing

The UDI also controls the Defence Industries Support Fund (DISF). The DISF was established under Law 3238 to provide continuous and stable financial support for arms production. Along with the initial assets, the main sources of income of the Fund are transfers from the foundations mentioned above and taxes levied on income (5 per cent goes to the Fund), on fuel consumption (10 per cent goes to the Fund) and on alcoholic beverages and tobacco consumption. DISF receives a proportion of revenue from the national lottery, gambling, payments for exemption from military service, bank interest revenue, as well as support from the National Defence Ministry budget and the General Budget. The Fund is unique in Turkey as it is almost entirely exempt from Turkish accounting and bidding laws in order to ensure secrecy and rapid procedures. It is audited and supervised by a three-member board drawn from the Prime Minister's Office, the Ministry of Defence and the Ministry of Finance and Customs.

UDI also initiated the formation of the Defence Industry Producers Association (SASAD in Turkish) by leading local firms in 1990. Membership is restricted to arms producers: the number of members increased from 21 in 1991 to 33 in 1992, reflecting a growing interest in arms production by local firms. Table 10E.2 lists these leading arms producers.

**Table 10E.2.** Leading arms producers in Turkey, on the basis of SASAD membership

| Company | Year founded | Major activity |
|---|---|---|
| Aremsan | 1945 | Diesel generator sets for military purposes |
| Aselsan | 1976 | Military communication, electronics for F-16s |
| Asil Celik | 1974 | Steel |
| Baris | 1986 | Tubes for rocket launchers, helicopter blades |
| CAM IS MAKINA | 1970 | Glass moulds, mechanical systems |
| Coskunoz | 1973 | Hydraulic and mechanical presses, automotive spare parts |
| ERMEKS-ER | 1988 | Parts for CASA aircraft |
| FMC-Nurol (FNSS) | 1988 | Armoured (combat) infantry fighting vehicles |
| Hema Elektronik | 1986 | Electronic equipment |
| Hema Hidrolik | 1973 | Hydraulic systems |
| Kale Kalip | 1969 | Various moulds and spare parts |
| KOSGEB | 1990 | Small- and medium-scale industry support administration |
| Marconi (MKAS) | 1988 | HF-SSB Radio Communications |
| Mercedes-Benz | 1967 | Tactical vehicles |
| MES | 1965 | Precision parts, rocket motors, tubes, cartridge cases |
| Mikes | 1987 | Electronic warfare sets for F-16s |
| MKEK | 1950 | Artillery, small arms ammunition, anti-tank rocket launchers, machine-guns |
| MKEK-AV Fisek | 1930 | Ammunition |
| MKEK-Barutsan | 1989 | Explosives, propellants |
| NETAS | 1967 | Communication electronics |
| NUROL | 1982 | Electronics |
| Otokar | 1963 | Diesel engines, Land Rover chassis |
| Roketsan | 1989 | Propellants and rocket motors |
| SGS-PROFILO | 1988 | Mobile telephones |
| SIMKO | 1955 | Electronics |
| STFA-Savronik | 1986 | Fire control and secure communications |
| TUSAS-TAI | 1984 | F-16 aircraft |
| TUSAS-TEI | 1985 | F-110 engine components for F-16s |
| TDI | 1991 | Association of a group of exporting firms in defence industry |
| Teletas | 1984 | Communications electronics |
| TESTAS | 1976 | Electronic components |
| TRANSVARO | 1988 | Night vision systems |
| TSKGV | 1987 | Funds defence projects |

*Sources*: Milli Savunma Bakanligi (MSB, Ministry of National Defence), *Savunma Sanayii Envanteri* [Defence Industry Inventory], SAGEB/1 (MSB: Ankara, 1988); information supplied by SASAD and the companies listed in the table.

These developments represent a radical deviation from the public sector monopoly which shaped procurement policy for 50 years. This policy change has two main implications: (*a*) the public sector is taking over the planning, co-ordination and financing of defence production and is becoming indirectly involved in production through partnerships; and (*b*) the USA is no longer automatically assumed to be the major arms supplier for Turkey.

Law 3238 lays out the need for co-operation between the public and private sector in Turkey and with foreign producers. In the words of the head of DIDA: 'It is believed that the co-operation of the Turkish private sector with foreign partners will contribute to the establishment and implementation of the basic industrial infrastructure that will allow Turkey to develop its own defence industry technology over the long term and a considerable emphasis is accordingly being placed on the Turkish private sector's participation in this field.'[18]

Establishing a defence industry which 'has a certain export potential, which can easily adapt to new technologies and is capable of updating itself in line with technological developments and improvements and thus ensures balanced co-operation with other, particularly, NATO countries'[19] required foreign participation. Moreover,

> Difficulties created by technology transfers, under license arrangements oriented to new developments and improvements, have led to the adoption of a 'joint venture' model instead of license production. With the new approach, joint venture partners on both the foreign and national sides, are held jointly responsible at each level of production, and advantages accruing from foreign investment, in terms of quality control, cost and offset commitments, flow into sales to third-party countries.[20]

## Domestic spill-over expectations

The officially expected spill-over effects of a modern arms industry include more diverse industrial production, efficiency in production, product quality improvement, foreign exchange savings, acceleration of economic growth, increased value added, decreasing unemployment, increases in the overall technology level, and improvement in the quality of the labour force and university education, especially engineering.[21] The true causality of these effects, where they occur, remains subject to dispute. The Turkish Chief of General Staff made an interesting statement on the reverse link in 1989, noting that in order to achieve military modernization the army demanded that the inflation rate be curbed, that the population growth rate and unemployment be decreased and that the economy should grow at an optimal rate with a balanced budget, increased exports and reduced foreign debt.[22]

This discussion has been on the agenda of economists for a long time. While econometric findings from time series data sometimes contradict those from cross-sectional research, it is clear that the military burden—while perhaps a public good through the satisfaction of a security need—is in the long run paid at the expense of social welfare. With the reservation that military expenditures cannot be identified with arms production activities but can be one indicator of the magnitude of the new potential demand to be injected into the economy, some findings can be given for the Turkish case.

A macro-econometric simultaneous equation model for the period 1964–85 in Turkey indicates that defence expenditure has no significant effect on the growth of

[18] Erdem (note 13), p. 23.
[19] Erdem (note 13), p. 10.
[20] Erdem (note 13), p. 10.
[21] Interview with V. Erdem, *Ankara Sanayi Odasi Dergisi*, no. 110 (1991), p. 19; Gencler, R., *Savunma Sanayi Sektör Raporu* [Sectoral Report for the Defence Industry], 1991 Sanayi Kongresi [1991 Congress on Industry], no. 149-3 (TMMOB–MMO), p. 209; *Armed Forces Journal International,* June 1989, p. 58.
[22] *Cumhuriyet*, 17 Sep. 1989.

investment, and that the overall impact on the economic growth rate is negative.[23] Similar models estimating the impact of military expenditure on the growth rates of defence-related industries for the period 1972–86 show that the impact will be adverse, except for the basic metals industry.[24] Noting that these are the key industries in the Turkish manufacturing sector and also the most import-dependent ones, these findings do not support the optimistic expectations for the modernization programme. Specific inputs of the defence industry such as metal alloys and rubber are all imported. Turkey lags far behind in terms of aviation and electronics technologies.[25] The present level of research and development (R&D) activities does not provide a promising environment for innovation.[26] The oldest state enterprise in ordnance and ammunition production, MKEK, withdrew from the Middle East arms market during the Persian Gulf crisis when other suppliers included products of the type developed by MKEK as free bonuses in broader arms packages.

Nevertheless, the Turkish authorities remain committed to the modernization programme and there is consensus on this in Parliament as well as between the government and the armed forces.

## The modernization programme in practice

It is too early to make an overall evaluation of the implementation of the modernization programme as investment started only in late 1989 after a period of evaluation and selection between competing foreign bids. However, an overview of the initial projects undertaken and the stage which they have reached illustrates the emergence of Turkey as a significant arms importer and as an aircraft producer.

Tables 10E.3 and 10E.4 summarize the stages of the implementation of the modernization programme by the UDI. Turkey's first international arms fair, the International Defence Equipment and Avionics Exhibition (IDEA), hosted 306 participating companies in 1987; peak attendance was reached in 1989, with the participation of 700 exhibitors. At IDEA 1991 the number of companies fell to 300 and by the end of 1991 doubts had arisen about Turkey as 'one of the world's most lucrative defense markets'.[27]

According to Ministry of Defence information there were 983 arms-producing establishments located in 45 different cities in Turkey in 1988.[28] More than one-third of these are in the Marmara region, the most industrialized part of Turkey. The Trade Union in this sector, HARB-IS, has 41 500 members, but, with the exception of a few leading, mostly new, establishments, the overall technological capability is limited. High-technology companies have foreign partners to provide technology.

[23] Günlük-Senesen, G., 'Yerli Silah Sanayiinin Kurulmasinin Ekonomiye Olasi Etkileri' [Probable Economic Impacts of the Installation of the (Turkish) Domestic Arms Industry], *1989 Sanayi Kongresi Bildirileri*, I, MMO/134, 1989, pp. 267–74.

[24] Günlük-Senesen, G., 'An econometric model for the arms industry of Turkey', paper presented at the 11th European Congress on Operational Research, 16–19 July 1991, Aachen, Germany; Günlük-Senesen, G., 'An evaluation of the arms industry in Turkey', poster paper presented at the Economics of International Security Conference, 21–23 May 1992, The Hague, Netherlands.

[25] Çakmakçi, A., *Savunma Sanayii* [Defence industry], (Seminar/Lecture Notes), (I.T.Ü. Isletme Fakültesi [Istanbul Technical University, Faculty of Management]: Istanbul, 1989).

[26] Public and private R&D expenditure accounts for 0.24 per cent of GNP.

[27] de Briganti, G. 'Turkey begins to lose shine as arms market', *Defense News*, 11 Nov. 1991, p. 14.

[28] Milli Savunma Bakanligi (MSB, Ministry of National Defence), *Savunma Sanayii Envanteri* [Defence Industry Inventory], SAGEB/1 (MSB: Ankara, 1988).

**Table 10E.3.** Military projects initiated in Turkey as of mid-1992

| Project | Local producer (licenser) | Date of contract |
|---|---|---|
| Armoured Infantry Fighting Vehicle | FMC (USA)<br>Nurol | 15 Aug. 1989 |
| Propellants and rocket motors (for Stinger (under licence), MLRS, Maverick) | Roketsan<br>ARC | 14 June 1989 |
| F-16 electronic warfare | Loral (USA)<br>Mikes | 20 Sep. 1989 |
| HF/SSB radio communications system | Marconi (UK)<br>HAS<br>CIHAN<br>ELIT | 9 Jan. 1990 |
| Basic trainer aircraft (SF-260D) | Agusta (Italy)<br>TAI<br>KIBM | 21 Mar. 1990 |
| Mobile Radar Complex Project | | |
| for $C^3$ | AYDIN (USA)<br>Hema Elektronik | 8 Oct. 1990 |
| for radar (TRS-22XX) | Thomson-CSF (France)<br>Tekfen | |
| Light Transport Aircraft (CN-235M) | CASA (Spain)<br>TAI<br>KIBM | Feb. 1990 |
| Multiple Launch Rocket System | LTV (USA)<br>Roketsan | Feb. 1990 |
| General-purpose helicopter (UH-60) | Sikorsky (USA) | 21 Sep. 1992 |
| Unmanned air vehicles (UAV) | AAI (USA)<br>General Atomics (USA)<br>TAI | Oct. 1992 |

*Sources*: Office of the Turkish Prime Minister (T. C. Basbakanlik), *Savunma Sanayii* [Defence Industry] (T. C. Basbakanlik: Ankara, 1990); information supplied by UDI, 1992.

**Table 10E.4.** Military projects under negotiation as of mid-1992

| |
|---|
| Low Level Air Defense System |
| 35-mm Anti-Aircraft Fire Control System |
| MCM Vessels |
| Coast Guard Vessels |
| Advanced Technology Industrial Park |
| Aviation Center and Airport Construction |

*Sources*: Office of the Turkish Prime Minister (T. C. Basbakanlik), *Savunma Sanayii* [Defence Industry] (T. C. Basbakanlik: Ankara, 1990); *Nato's Sixteen Nations*, Special Edition, vol. 36, no. 2 (1991); *Jane's Defence Weekly*, 26 Sep. 1992, p. 5.

Conforming to equipment specifications laid down in NATO Allied Quality Assurance Publications (AQAP)—started for the first time in Turkey in 1988—can be cited as problematic for local industry. Local producers have requested increased government funding for R&D as well as protective measures for local industry, state-guaranteed long-term financial funding and accelerated administrative procedures. Since the present market is limited to a single buyer—the Turkish Armed Forces—and the export prospects of most firms are dim, large-scale projects might not turn out to be as profitable as was expected. It has been proposed that companies might be established on the basis of specialization in components rather than on a project basis.[29]

Present doubts concerning the continuity of the modernization aspirations and the effectiveness of the UDI notwithstanding, the production of armoured infantry fighting vehicles, Scimitar tactical radio sets and radar are the major ongoing programmes for the modernization of land forces. The Turkish Armed Forces Integrated Communications Systems (TAFICS) is an ambitious attempt to modernize the military communications network to NATO standards.

The modernization of naval equipment seems to be following a more settled and better defined route—perhaps because of the long collaboration with German shipyards after the late 1970s. German shipyards (mainly Blohm & Voss, HDW and Lürssen Werft) have sold frigates, submarines and fast-attack craft (FAC) worth approximately $4 billion to Turkey, orders which form a significant part of these companies' overall business. Sales are supported by both the German Government and banks. Although second-hand warships were previously received from the US Navy at a nominal cost, Turkey's warships are largely German-designed and built locally with technical assistance from the parent yards in Germany.[30] The ships are built at the Gölcük and Taskizak yards, both owned by the Turkish Navy. Gölcük deals with frigates and submarines while Taskizak accommodates FACs and landing craft.[31] The Turkish Navy plans to order up to eight minehunters from France, some of them to be built in Turkey.[32]

Some major equipment decisions were postponed until after the results of the elections in October 1991. Moreover, the 1991 Persian Gulf War strengthened the resolve of the Turkish armed forces to restructure themselves into a mobile, technologically sophisticated, professional force with increased firepower, air defence, communications, electronics and electronic warfare capabilities.[33] A review of the overall procurement plan contributed to further delays in finalizing decisions. Nevertheless, a wide range of major procurement projects are under way.

[29] Aris, H., 'Creating a defence industry', *Nato's Sixteen Nations,* vol. 36, no. 3 (1991), p. 78; Minutes of the discussions held between industrialists and UDI representatives, *Ankara Sanayi Odasi Dergisi*, no. 110 (1991), pp. 21–35; Tosun, A.,'Savunma Sanayii ve Düsündürdükleri' [Thoughts on the defence industry], *Ankara Sanayi Odasi Dergisi*, no. 110 (1991), pp. 50–51.

[30] *Defense and Foreign Affairs Weekly*, 5–11 Feb. 1990, p. 5; *Jane's Defence Weekly*, 28 July 1990, p. 131; *Military Technology*, No. 4 1991, pp. 15–16; *Naval Intelligence,* 3 July 1992, p. 4; *Wehrtechnik*, no. (1992), p. 54.

[31] Todd, D., 'Mediterranean naval shipbuilding, challenges and prospects', *International Defense Review*, no. 10 (1992).

[32] *Jane's Fighting Ships, 1992–93* (Jane's Information Group: Coulsdon, 1992), p. 873.

[33] Interview with Turkish National Defence Minister Nevzat Ayaz, *Defense News,* 31 Aug.–6 Sep. 1992, p. 46; *Nato's Sixteen Nations*, vol. 37, no. 4 (1992), pp. 57–62.

## The aerospace industry

The formation of Turkish Aerospace Industries (TAI) in May 1984 underlined Turkish aspirations in a product sector where the technological requirements are most challenging. It is centred on the programme to produce F-16 fighter aircraft for the Turkish Air Force and for export, and is being advanced by a recent decision to produce light transport aircraft and trainers in collaboration with CASA of Spain and Agusta of Italy, respectively.

TAI was the organizational model for subsequent arms production in Turkey. It was formed as a joint venture with a 51 per cent Turkish share (49 per cent TUSAS-TAI, 1.9 per cent Turkish Armed Forces Foundation and 0.1 per cent Turkish Air League) and a 49 per cent US share (42 per cent General Dynamics and 7 per cent General Electric). At the end of 1991 TAI employed 2266 people, with a US General Manager.[34]

TAI delivered over 80 F-16s between 1987 and April 1992 and expects to complete the initial production order for 152 aircraft on schedule in 1994. TAI has started local production of fuselage sections that were previously supplied by European suppliers Sonaca and Fokker. The forward fuselage, cockpit and fins are still supplied by foreign suppliers. In 1992 70 per cent of each airframe is produced locally and this proportion is planned to rise to 90–95 per cent in the second production order. Parts of the F110-GE-100 jet engine which powers the F-16 are assembled and tested by TUSAS Engine Industries (TEI) which was formed in 1985 by TUSAS (51 per cent) and General Electric (49 per cent)—another joint venture.

The second production order depends on financing worth $3.5 billion from Saudi Arabia, Kuwait, the United Arab Emirates and the USA in compensation for Turkish economic losses incurred through UN economic sanctions on Iraq.[35] When the export of 46 Turkish-made F-16s to Egypt by 1995 was announced in 1991, Turkey's hopes for entering the international market were revived. In addition, this order might be considered as confirmation of the sought-after significant role in the market, with the expectation that the Turkish F-16 plant might become a major international supplier of F-16 fighter aircraft plus parts and components after the USA stops production in the mid-1990s.[36]

The dispute between Turkey and the USA over an electronic countermeasures (ECM) system for the F-16 underlines the difficulty of achieving self-sufficiency in defence production. The USA initially rejected Turkish requests for access to system software, noting that this would be an unlawful technology transfer. In August 1992 the State Department stated that Turkey would have the right to modify the device's software, although the extent of the modification was not revealed.[37]

[34] Company records; 'Türk Havacilik ve Uzay Sanayi (TAI)' [The Turkish aerospace industry], *Ankara Sanayi Odasi Dergisi*, no. 110 (1991), pp. 37–39.

[35] Aris, H., 'A new player in the making, Turkish defence industry', *Military Technology*, no. 4 (1991), pp. 11–15; *Jane's Defence Weekly*, 12 Oct. 1991; *Defense News*, 11 Nov. 1991; *Military Technology*, no. 4 (1992), p. 95.

[36] 'Where East meets West', *Armada International* , no. 2 (1989), editorial, p. 2; Boyle, D. and Salvy, R., 'Turkish defense modernisation', *International Defense Review*, no. 6 (1989), p. 847; *Defence Industry Digest*, Oct. 1991, p. 16; Enginsoy, Ü., 'Turkey seeks more F-16s with Arab, U.S. funds', *Defense News*, 11 Nov. 1991, p. 12; Sariibrahimoglu, L., 'Building an industry', *Jane's Defence Weekly*, 9 Nov. 1991, p. 881; *Arms Sales Monitor*, no. 13–14 (Mar.–Apr. 1992); *Military Technology*, no. 4 (1992), p. 95.

[37] Consequently, the installation of Loral AN/ALQ-178 Rapport III integrated electronic countermeasures systems on Turkish Air Force F-16s started in July 1992 at the TAI-Mürted factory in Ankara.

## IV. Conclusions

While Turkish arms production seems bound to suffer from the re-evaluation of defence and security around the world, the determination of Turkish authorities—both the armed forces and politicians—combined with the need for foreign producers to find new markets should be expected to carry the modernization plan further. The trend towards arms reduction and drawing down of arms production in the industrialized world is not mirrored in Turkey. Stockpiling conventional weapons, improving their quality and encouraging local production might all serve to upgrade the Turkish inventory. Steered by new public policy, Turkish industry is reorganizing towards a militarization of civilian industries.

The flow of armaments through the NATO 'cascade',[38] along with supplies during and after the 1991 Persian Gulf War and long-term modernization programmes, have already served to upgrade Turkey's equipment. The USA and Germany are the two major suppliers under the CFE Treaty-related cascade and these supplies might reduce the incentives for local production. However, both countries have attached conditions to the use of their equipment.[39] The CFE zone of application does not include that part of Turkey which was a battlefield at the end of 1992. In that region operations of the Turkish land, air and gendarmerie forces against the PKK have accelerated the acquisition of helicopters and armoured vehicles which are to be bought from Russia—with no restrictions on their use.[40]

The Turkish case illustrates the contradiction between the incentive to develop a local defence industry—escaping dependence—and the reality that production also depends on foreign technology and resources. Moreover, the armament programme of Turkey continues to put pressure on the economy, and the competition for resources between defence and welfare items such as health and education seems to have been resolved in favour of the former.

The potential negative political, economic and social consequences of this trend are not matters of public discussion in Turkey but neither, it seems, are they considered in the international arena. The Turkish case is one example among several where the disarmament process advocated by the Western world is undermined by actions taken in close collaboration with the Western world. The security-concerned circles, then, must include this in their agenda before it is too late.

*Jane's Defence Weekly*, 9 May 1992, p. 796; Erginsoy, Ü., 'Pratt, GE vie for Turkish F-16 engine contract', *Defense News*, 31 Aug.– 6 Sep. 1992, p. 8.

[38] *Financial Times*, 27 June 1991, 'Arms windfall dilemma for Turkey'; Hitchens, T., 'Nato arms transfer benefits Turkish military', *Defense News*, 31 Aug.–6 Sep. 1992, p. 7.

[39] In Mar. 1992 Gerhard Stoltenberg, German Minister of Defence, was forced to resign after it became clear that German armoured vehicles were being used in combat against the Kurds.

[40] *Financial Times*, 11 Sep. 1992, p. 5; *The Independent*, 9 Nov. 1992, p. 8; Foreign Broadcast Information Service, *Daily Report–Central Eurasia*, FBIS-SOV-92-221, 16 Nov. 1992, p. 14.

# **Appendix 10F.** The United Nations Register of Conventional Arms

HERBERT WULF

## I. Introduction

Important steps have been taken within the United Nations to increase transparency in armaments and to maintain international peace and security as set out in Article I of the UN Charter. On 9 December 1991, in Resolution 46/36 L, the UN General Assembly voted to establish a 'universal and non-discriminatory Register of Conventional Arms'.[1] UN member countries are requested to report, voluntarily and on an annual basis, their arms transfers—both imports and exports—in the following categories: battle tanks, armoured combat vehicles, large-calibre artillery systems, combat aircraft, attack helicopters, warships, and missiles and missile launchers.[2]

If the Register is adequately implemented, this will be a significant step forward towards transparency and confidence building. Although the Register is not designed to control the flow of arms, the mere fact that it will increase publicly available information on which weapons are transferred by which countries could restrain 'excessive and destabilizing accumulation of arms', as stated in the UN resolution (paragraph 12). For the first time since the reports of the League of Nations,[3] official government information on global arms transfers will be made public. Currently available data on arms transfers are either not comprehensive[4] or are based on unofficial sources, such as the registers of the trade in major conventional weapons which have been published annually in the *SIPRI Yearbook* since 1969.[5]

Two events in 1991 had a profound impact on the arms transfer control debate in the United Nations: the war fought against Iraq and the dissolution of the Soviet Union.

Iraq had been accumulating destabilizing amounts of military equipment and technology, especially during the war with Iran in the 1980s. When suppliers delivered

[1] United Nations General Assembly document A/RES/46/36 L, 9 Dec. 1991; excerpts of the resolution are reproduced in *SIPRI Yearbook 1992: World Armaments and Disarmament* (Oxford University Press: Oxford, 1992), pp. 305–307.

[2] For the definitions of these weapon categories, see section IV of this appendix.

[3] A UN study mentions the publication of arms trade data in the *Statistical Yearbook* of the League of Nations from 1925 to 1938 and its limited success. United Nations, General Assembly, *Report by the Secretary-General, Study on Ways and Means of Promoting Transparency in International Transfers of Conventional Arms,* UN document A/46/301, 9 Sep. 1991, pp. 16–17. See also SIPRI, 'Proposals concerning the arms trade', *The Arms Trade With the Third World* (Almqvist & Wiksell: Stockholm, 1971), chapter 2, pp. 86–132, especially pp. 87, 93–94.

[4] A few governments report their arms exports to their respective parliaments. The US Arms Control and Disarmament Agency (ACDA) publishes global data, based on US intelligence sources; the published information is restricted to dollar value estimates and highly aggregated numbers of transferred equipment.

[5] SIPRI's long-term efforts in reporting trends in the trade in major conventional weapons were instrumental in the establishment of the UN Register. The experience of the SIPRI arms transfers research has been taken into consideration, and the present author served as a consultant to the UN Panel of Governmental Technical Experts in 1992 in his capacity as Leader of the SIPRI arms transfers research project.

arms to Iraq, they presented the sales as a stabilizing factor in view of the perceived threat of Iranian fundamentalism. In the wake of Iraq's invasion of Kuwait, however, significant attention has been paid to the negative consequences of the trade in arms which allowed such a scenario to develop. Several multilateral groups began to act or to call for arms transfer regulations—among them the five permanent members of the UN Security Council and the Group of Seven (G7) leading industrial nations.[6]

The USSR—the largest single supplier of major conventional weapons for most of the 1980s—ceased to exist at the end of 1991. With the dissolution of the Soviet Union, the superpower competition—often articulated in rivalry for spheres of influence and arms exports to client states—disappeared.

In this new security environment the UN General Assembly passed Resolution 46/36 L by a vote of 150 to 0, with two abstentions (Cuba and Iraq). China—one of the major arms-exporting countries—was among the countries that did not participate in the vote.

The General Assembly drew the consequences of the realization that arms buildups can pose a threat to security. The Register's stated goals include: increasing openness and transparency; reducing mistrust and tension; enhancing confidence; promoting national, regional and international peace and stability; assisting in resolving conflicts; and supporting member states in restraining their arms imports and exports.[7]

The Register is thus seen as part of a larger family of international arms control and conflict prevention measures and it aims at the same time at increasing transparency and openness concerning arms transfers. Effective as of 1 January 1992, the first reports of UN member governments on exports and imports of conventional arms for calendar year 1992 are due by 30 April 1993.

## II. The history of the Register and transparency in the armaments process

The establishment of the Register in 1991 and its implementation in 1992 were preceded by several unsuccessful attempts. Recurrent initiatives concerning conventional arms transfers were made, all of which failed to reach fruition. Debates on the control of arms transfers within the United Nations were usually initiated by the group of 'Western industrialized countries' and a few 'non-aligned and neutral countries'. The point of departure in most cases was the experience of the League of Nations.[8]

At the 20th session of the General Assembly in 1965, Malta submitted a draft resolution to invite the Eighteen-Nation Disarmament Committee (ENDC) to consider the transfers of arms between states.[9]

In 1968 Denmark, together with Iceland, Malta and Norway as co-sponsors, submitted a draft resolution requesting the Secretary-General to ascertain the positions of governments on contributing arms transfer data to a UN register. The proposal was not put for a vote because of opposition mainly from the non-aligned countries.

[6] The G7 includes Canada, France, Germany, Italy, Japan, the UK and the USA. The UN and G7 declarations are reprinted in *SIPRI Yearbook 1992* (note 1), pp. 302–305.

[7] Preamble, Resolution 46/36 L; see *SIPRI Yearbook 1992* (note 1), p. 305.

[8] See note 3.

[9] The proposal was rejected by a vote of 18 in favour with 19 against and 39 abstentions. For the history and a review of previous proposals for an arms transfer register, see United Nations (note 3); and Corradini, A., *Considerations of the Question of International Arms Transfers by the United Nations*, Disarmament Topical Papers 3 (United Nations: New York, 1990), pp. 44–59.

In 1976, at the 31st session of the General Assembly, 13 countries sponsored a draft resolution in which a factual study of the international transfer of conventional arms was requested. A majority voted to adjourn the debate.

In 1978 it was agreed that the United Nations should undertake a study on the subject of a register of arms transfers and control of the illicit trade in arms.[10] At the request of the General Assembly in various resolutions, the Secretary-General, with the assistance of groups of governmental experts, carried out several studies on arms transfers and related issues.[11]

While a number of these studies requested the establishment of a United Nations conventional arms transfer register, there was no agreement on joint action until 1991. The main reason why a number of developing countries rejected the arms transfer register was its claimed partiality. Critics rightly pointed out that in a transfer register recipient countries without domestic arms production facilities would have to report all their arms imports, while the major arms producers would not have to report much at all. Such a one-sided register would over a long period of time have given a fairly detailed picture of the weapon inventory of the importing countries, while the major arms-producing countries could continue to conceal their weapon inventories. Consequently, governments of the developing countries asked to establish a more comprehensive register which would include both transfers and production of arms, but no compromise was reached.

One reason for the reservations on the part of some Western countries, particularly the US Government, was their belief that some of the proposed controls of arms transfers were perceived as not desirable, while desirable controls (that is, on Soviet arms exports) were non-negotiable. In addition, the Register was seen as a rhetorical exercise with little or no relevance to real arms control. Only when it became apparent that the Register would be approved by the UN in 1991 did the US Government join the group of co-sponsors and support it.

## III. The 1991 UN General Assembly decision

Although the overwhelming majority which voted for the establishment of the Register seems to signal a consensus that had failed so often before, the debate among the UN member states indicates that many governments were not entirely satisfied with the resolution.[12]

To reach agreement, in contrast to the result of previous efforts, one important alteration had to be made in the resolution sponsored by Japan and co-sponsored by Western countries.[13] In addition to information on the export and import of conventional arms, countries are also invited to provide information 'on military holdings,

[10] United Nations, General Assembly Resolution S-10/2.

[11] United Nations, *Relationship between Disarmament and Development* (UN document A/36/356); *Comprehensive Study on Confidence-Building Measures* (A/36/474); *Relationship between Disarmament and International Security* (A/36/597); *Economic and Social Consequences of the Arms Race and Military Expenditures* (A/8469/Rev.1, A/32/88/Rev.1, A/37/386, A/43/368); *Study on Conventional Disarmament* (A/39/348); *Study on All Aspects of Regional Disarmament* (A/35/416); *Reduction of Military Budgets* (A/35/479, A/S-12/7, A/40/421).

[12] Moodie, M., 'Transparency in armaments: a new item for the new security agenda', *Washington Quarterly*, summer 1992, pp. 75–82, accurately depicts the opposition to the UN First Committee version of the original resolution by such countries as Argentina, Brazil, Egypt and Pakistan.

[13] The minutes of the debate in the First Committee of the 46th session of the General Assembly (UN document A/C.1/46/PV.37), pp. 18–22, show that last-minute revisions were introduced into the draft resolution to comply with some of the requests of countries in favour of an expanded register.

procurement through national production and relevant policies'.[14] The Secretary-General will then include in the Register an index of countries which supplied information on production and other import, export or production policies, under certain headings.

It was furthermore agreed, 'with a view to future expansion', to review the operation of the Register in 1994. Member states are invited to provide the Secretary-General with their views on 'the addition of further categories of equipment and the elaboration of the Register to include military holdings and procurement through national production'.[15] The agreement reached—to discuss expansion of the Register to include data on arms production—is the result of the consistent criticism of some member states and is meant as a signal that the goal must be an impartial and universal register. The Register is not a 'transfer' register but, as its name clearly indicates, a 'Register of Conventional Arms'.

As a result of this compromise, the Register is planned to be implemented in two stages. During the first stage, governments will report their arms exports and imports for 1992 and 1993.[16] In the second stage, beginning in 1994, a possible expansion of the Register is on the agenda, to include data on the stock of military equipment, production, technology transfer and weapons of mass destruction.

Despite this compromise to accommodate some of the complaints of member states, many government representatives voiced reservations. Before the vote was taken, the Cuban delegate raised the often mentioned criticism of partiality and pointed out that his government was not satisfied with just the possibility for expansion: 'What we are against is the establishment of a partial and selective registry which relates only to international transfers of conventional weapons and not to production and stockpiling, to the transfer of advanced military technology, to research and development activities or to weapons of mass destruction.'[17] After the vote several delegates stated for the record their continued reservations.[18] Their criticism concentrated on the following issues:

1. Indigenous production capabilities and advanced production technology need to be taken into account simultaneously with arms transfers (Algeria, North Korea, Pakistan and Uganda).
2. The national security interests of member states have to be taken into consideration (Pakistan and Singapore).
3. All types of weapons, particularly those with more devastating effects (weapons of mass destruction), need to be included (Algeria, Iran and Syria).

The background to these statements for the UN record are, of course, often regional conflicts. Pakistan, for example, would want Indian arms production to be registered; North Korea would want South Korean production registered; and Algeria, Iran and

[14] Paragraph 7 of General Assembly Resolution 46/36 L (note 1).

[15] Paragraph 11 of General Assembly Resolution 46/36 L (note 1).

[16] In addition to data on arms transfers, governments are asked during the first stage to provide other relevant information.

[17] Statements by Member States in the 46th session of the General Assembly (UN document A/46/PV.66, Agenda Item 60), p. 43.

[18] Among them were (in chronological order) Pakistan, North Korea, Algeria, Iran, Singapore, Syria and Uganda. See Statement by Member States in the 46th session of the General Assembly (note 17), pp. 47–55. In addition, the representative of Lithuania stated that his country might not be able to fulfil the requirements of the Register since a considerable number of Soviet forces were still stationed in Lithuania without the agreement or consent of his government.

Syria want regional chemical and nuclear arsenals registered, particularly Israel's nuclear weapons. This suggests that the agenda of disagreement is as much regional and South–South as it is North–South disagreement. What effect this will have on the 1994 review is uncertain.

## IV. Technical procedures

The General Assembly resolution requested the Secretary-General, with the assistance of a Panel of Governmental Technical Experts, 'to elaborate the technical procedures and to make any adjustments to the annex of the present resolution [describing the categories of weapons to be reported] necessary for the effective operation of the Register, and to prepare a report on the modalities for early expansion of the scope of the Register by addition of further categories of equipment and inclusion of data on military holdings and procurement through national production'.[19]

The result of the Panel's work was published on 14 August 1992 in a UN report, after three sessions in 1992.[20] The differences of opinion on the function and content of the Register that had emerged during the General Assembly debate continued to dominate the deliberations of the Panel—although the report was approved by consensus. The divergencies are illustrative, as they indicate how governments view the purpose and the future of the Register. They are summarized in the sections below.

### The mandate of the Panel

Differences first emerged in interpreting the mandate of the Panel. While a number of Panel members (most outspokenly those from China and Egypt but also those from most of the other developing countries) suggested making adjustments to the seven categories of weapon system in the General Assembly resolution by defining the weapons and adding to them, a narrower interpretation was favoured by others (most decisively by France, the UK and the USA). They suggested sticking as closely as possible to the text of the resolution and insisted that it was not the mandate of the Panel to add other categories of weapon system (neither conventional nor weapons of mass destruction nor production technology). As can be seen by comparing the original seven categories described in the 1991 General Assembly resolution with the definitions as adjusted by the Panel,[21] the Panel made some changes but did not add new categories of weapon or production technology. In the compromise reached, the Panel decided to leave this issue for the 1994 Register review.

The definitions of categories of weapon system as revised by the Panel are as follows:

[19] UN General Assembly Resolution A/46/36 L of 9 Dec. 1991 (note 1), paragraph 8.

[20] UN Secretary-General, *Report on the Register of Conventional Arms,* UN document A/47/342 (United Nations: New York, 14 Aug. 1992). The following countries were represented on the Panel: the Netherlands (chairman), Argentina, Brazil, Canada, China, the Czech and Slovak Republic, Egypt, France, Ghana, India, Italy, Japan, Malaysia, Mexico, Russia, the UK and the USA. In addition, a UN political affairs officer served as secretary and two independent researchers as consultants to the Panel.

[21] The Annex of Resolution 46/36 L (note 1) has been revised. The revised version (Resolution 47/342, note 20) appears in the quotation below.

I. *Battle tanks*
Tracked or wheeled self-propelled armoured fighting vehicles with high cross-country mobility and a high level of self-protection, weighing at least 16.5 metric tonnes unladen weight, with a high muzzle velocity direct fire main gun of at least 75 millimetres calibre.

II. *Armoured combat vehicles*
Tracked, semi-tracked or wheeled self-propelled vehicles, with armoured protection and cross-country capability, either: (a) designed and equipped to transport a squad of four or more infantrymen, or (b) armed with an integral or organic weapons of at least 12.5 millimetres calibre or a missile launcher.

III. *Large calibre artillery systems*
Guns, howitzers, artillery pieces, combining the characteristics of a gun or a howitzer, mortars or multiple-launch rocket systems, capable of engaging surface targets by delivering primarily indirect fire, with a calibre of 100 millimetres and above.

IV. *Combat aircraft*
Fixed-wing or variable-geometry wing aircraft designed, equipped or modified to engage targets by employing guided missiles, unguided rockets, bombs, guns, cannons, or other weapons of destruction, including versions of these aircraft which perform specialized electronic warfare, suppression of air defence or reconnaissance missions. The term "combat aircraft" does not include primary trainer aircraft, unless designed, equipped or modified as described above.

V. *Attack helicopters*
Rotary-wing aircraft designed, equipped or modified to engage targets by employing guided or unguided anti-armour, air-to-surface, air-to-subsurface, or air-to-air weapons and equipped with an integrated fire control and aiming system for these weapons, including versions of these aircraft which perform specialized reconnaissance or electronic warfare missions.

VI. *Warships*
Vessels or submarines armed and equipped for military use with a standard displacement of 750 metric tonnes or above, and those with a standard displacement of less than 750 metric tonnes, equipped for launching missiles with a range of at least 25 kilometres or torpedoes with similar range.

VII. *Missiles and missile launchers*
Guided or unguided rockets, ballistic or cruise missiles capable of delivering a warhead or weapon of destruction to a range of at least 25 kilometres, and means designed or modified specifically for launching such missiles or rockets, if not covered by categories I through VI. For the purpose of the Register, this category:

(a) Also includes remotely-piloted vehicles with the characteristics for missiles as defined above;

(b) Does not include ground-to-air missiles.[22]

## Adjustment and definitions of weapon categories

Agreement about the exact definitions was problematic for some of the weapon categories. A consensus report could only be achieved by many compromises until the very last day of the five weeks of deliberations. The most critical disagreements were the following:[23]

[22] UN Secretary-General (note 20), paragraph 14, pp. 11–12.

[23] For other, less important changes, compare the original definitions of the 7 weapon categories and the revised versions. Both are printed in the Secretary-General's report of 14 Aug. 1992 (note 20), pp. 11–12 (revised) and pp. 25–26 (original).

*Combat aircraft* (category IV) and *attack helicopters* (category V). The Chinese member of the Panel (with varying degrees of support from several others) suggested adding to the combat aircraft category reconnaissance and electronic warfare aircraft, air-refuelling aircraft, command and early-warning aircraft. The reasoning was that such aircraft could add considerable offensive capabilities to the armed forces and should thus be reported. These proposals were rejected until a last-minute compromise was found to include versions of combat aircraft 'which perform specialized electronic warfare, suppression of air defence or reconnaissance missions'. In other words, not all but *versions of combat aircraft* for reconnaissance and electronic warfare are included and air-refuelling aircraft are excluded. A similar compromise was found for attack helicopters.

*Warships* (category VI). The original definition gives a displacement of 850 metric tonnes as a minimum level for reporting. At the suggestion of the Malaysian member of the Panel, the displacement was lowered to 750 metric tonnes to include certain types of corvette. An important category of highly effective ships, fast attack missile craft, which usually have a displacement of about 250 tonnes or less would thus not be included in the Register. These warships are transferred in larger numbers than the heavier ships of 750 tonnes and more. It was much more difficult to reach agreement to include fast missile attack craft. The British member of the Panel (with support from the USA) argued that the important point was to report the missiles (which had to be done under category VII). The ships could be considered as a missile platform. A majority of the Panel favoured the inclusion of this type of ship. It was agreed to include ships 'with a standard displacement of less than 750 metric tonnes, equipped for launching missiles with a range of at least 25 kilometres or torpedoes with similar range'. This adjustment to a weapon category was the most far-reaching adjustment made by the Panel.

*Missiles and missile launchers* (category VII). Two issues were of major concern in this category. First, China suggested excluding all missiles of a defensive nature, particularly surface-to-air missiles. The categorization of weapon systems as offensive or defensive caused difficulties and differing opinions within the Panel, as in many other forums. After long debates and as part of a general compromise, the Panel accepted the exclusion of all 'ground-to-air missiles',[24] on the basis that these missiles were used for defensive purposes and that not many ground-to-air missiles had a range of 25 kilometres or more.

Second, and more importantly, a long debate emerged on the issue of whether missiles and missile launchers were to be reported in one single figure. It was clear to the panelists that this was like 'adding apples and pears'. Nevertheless, the countries who insisted on this, particularly the United Kingdom and the United States, were not willing to accept the disaggregation of missiles and missile launchers. In contrast to the rest of the report, an example is given (in paragraph 16a) to explain this unusual way of reporting: 'For example, if a country imports six missile launchers and 100 associated missiles as well as 500 missiles associated with other launchers covered under categories I to VI, the number 606 will be entered.' One hundred missiles of one type plus 500 missiles of another type plus 6 missile launchers make a total of 606 'items', but it is unclear how each is defined. The observer is left with a confusing picture, and it is clear that this counting method will in fact conceal more than it will reveal. It will not increase transparency, and since the panelists were

[24] Note that not all surface-to-air missiles (SAMs) are excluded; SAMs mounted on ships are to be reported.

aware of this fact, it must be assumed that the intention was not to reveal detailed information in this sensitive category of weapons.

## Designations of weapon systems

The most divergent views were expressed on whether or not weapon designations would have to be mandatory in the report. The consequences of this decision have important implications for the value of the Register. The alternative was the following. If all transfers in one category, for example combat aircraft, are lumped into a single figure when a country reports its exports to another country, it would be left to the reader of the Register to speculate which aircraft might have been transferred. If designations or descriptions of the combat aircraft are reported, the observer will have information about the particular sale and can distinguish between modern, technically advanced and often expensive aircraft on the one hand and cheap, simple, often second-hand equipment on the other. The Brazilian and Italian members of the Panel and others argued that it is the opposite of transparency to treat a light attack aircraft, armed with a machine-gun and with a limited radius of operation, in the same way as a modern, missile-carrying, multi-purpose fighter.

A similar argument was made about whether a remarks column should be provided to allow governments to explain a particular import or export.

While many developing countries had reservations against the establishment of the Register in the first place, once it was established this group of delegates pleaded for detailed reporting to make the Register a valuable instrument of transparency. In contrast, several members from the industrialized countries argued that—in addition to reasons of national security—it was necessary to have a simple system of reporting to make the Register a success. Reporting should be as simple as possible to allow all countries to participate.

The compromise found at the suggestion of the Brazilian, Canadian and Italian Panel members is clearly reflected in the reporting forms (see figure 10F.1). Two remarks columns—separated by a small margin from the other columns—are placed at the right side of both the import and the export forms. This graphic separation of the data is intended to make clear that information in this column is not only voluntary but, as one member phrased it during the deliberations, 'very voluntary'. Paragraph 19 of the report explains the intentions:

> The right hand column on the form, divided into two parts, 'description of item' and 'comments on the transfer', is designed to accommodate additional information on the transfers. Since the provision of such information might be affected by security and other relevant concerns of Member States, this column should be filled in at Member States' discretion; no specific patterns are prescribed. To aid the understanding of the international transfers reported, Member States may wish to enter designation, type or model of categories I to VII, which also serve as guides to describe equipment transferred. Member States may also use this column to clarify, for example, that a transfer is of obsolete equipment, the result of co-production, or for other such explanatory remarks as Member States see fit.[25]

[25] UN Secretary-General (note 20), p. 13.

*EXPORTS [IMPORTS]*

Report of international conventional arms transfers (according to United Nations General Assembly resolution 46/36 L)

Reporting country: ____________________

Calendar year: 1992

| A | B | C | D* | E* | REMARKS** | |
|---|---|---|---|---|---|---|
| Category (I–VII) | Final importer State(s) | Number of items | State of origin (if not exporter) | Intermediate location (if any) | Description of item | Comments on the transfer |
| I. Battle tanks | 1)<br>2)<br>3) | | | | | |
| II. Armoured combat vehicles | 1)<br>2)<br>3) | | | | | |
| III. Large calibre artillery systems | 1)<br>2)<br>3) | | | | | |
| IV. Combat aircraft | 1)<br>2)<br>3) | | | | | |
| V. Attack helicopters | 1)<br>2)<br>3) | | | | | |
| VI. Warships | 1)<br>2)<br>3) | | | | | |
| VII. Missiles and missile launchers | 1)<br>2)<br>3) | | | | | |

Background information provided: yes/no

* See para. 18 of the present report.

** See para. 19 of the present report.

**Figure 10F.1.** Standardized forms for reporting international transfers of conventional arms

The remarks columns on each of the two forms are likely to be the most interesting part of the Register. Since there was no agreement on a compulsory weapon designation or remarks column, it is up to governments whether or not they provide data for the columns. Some might report only numbers in the relevant categories and thus comply with the Register. Others might provide details on arms imports that the arms suppliers are hesitant to reveal.

### Modalities for expansion of the Register

The issue that had for such a long time prevented the establishment of an arms transfer register—the inclusion of arms production and weapons of mass destruction—was raised again in the discussion of the expansion of the scope of the Register in 1994.[26] Many of the issues where no agreement could be found in the 1992 Panel of Governmental Technical Experts are listed in section II of the report as non-binding suggestions for consideration of the group of governmental experts to be formed in 1994. This relates particularly to including new categories of weapon systems, to taking account of significant technological developments, to including weapons not covered by existing categories and a list of seven weapon systems or categories, with several sub-categories.[27]

Once the Panel had produced a consensus report, the prospects for adoption were almost certain. On 15 December 1992 the General Assembly adopted Resolution 47/69 without a vote, endorsing the Panel of Governmental Technical Experts' report, requesting all member states to provide data on exports and imports of weapons (as well as other relevant information).

## V. The objectives of the Register and the prospects for implementation

Compared to the wider goals of enhancing peace and stability, increasing openness, transparency and confidence, and supporting the restraint in arms imports and exports, the first step of the Register is a very modest one. In judging the relevance of the Register, however, it should be observed that international arms transfers were a taboo subject for a long time, and the concept of the registration of conventional armaments was considered to be a 'non-starter' in the United Nations.[28] The Register is none the less a modest step for several reasons.

1. In its present form the Register will not significantly facilitate a judgement about the military capabilities of countries which contribute to it. In addition to the treatment of arms production capabilities, there is no obligation to report on an entire range of weapons: small arms, bombs, munition, mortars, guns below a certain calibre, missiles below a range of 25 kilometres, support ships, non-combat planes and helicopters, and others. Moreover, many other forms of arms-related trade are outside the scope of the Register, including major sub-systems (especially engine and electronic upgrades), dual-use items and production technologies.

[26] The result is contained in part II of the report; UN Secretary-General (note 20), pp. 15–18.

[27] See UN Secretary-General (note 20), pp. 16–17.

[28] This conclusion was drawn by the former UN Under-Secretary General for Disarmament Affairs, Yasushi Akashi, 'An overview of the situation', *Transparency in International Arms Transfers*, Disarmament Topical Papers 3 (United Nations: New York, 1990), p. 3.

The Register will be of no assistance in understanding the economics of the arms trade since the value of given transactions and the details of their financing are outside its scope.

2. The Register and its aim of transparency are not synonymous with a restriction on arms and were not intended to be. Whether the Register will be an 'action-oriented tool' and an 'instrument of preventive diplomacy', in the words of the Secretary-General in his foreword to the report,[29] has to be proven in practice. In the aftermath of Iraq's invasion of Kuwait, the calls by many governments for arms trade control and tightening of export control systems went far beyond the goals of the Register. However, the Register was deliberately never intended to establish a new control mechanism, and reduction of the arms trade is not the primary purpose of the Register.

3. There is no verification mechanism. It is a voluntary exercise of member states which will not be controlled or verified. The Panel of Governmental Technical Experts made it clear that the task of the UN Secretariat is to file and distribute the incoming reports from member states but not to check or verify them. This shortcoming should, however, not be overestimated since, if importers and exporters report separately on the same deal, this will provide a signal of a discrepancy and a reference-point for cross-checking. Furthermore, the research community will certainly thoroughly scrutinize the reports of the first two years.

In an analysis of the Register process, Edward J. Laurance has presented the rationales for compliance or non-compliance with the Register.[30] The reasons for *compliance* are several. There is a certain inertia; the step-by-step approach of establishing the Register on a consensus basis is hopefully a guarantee for universal or nearly universal compliance. Nearly 60 countries—all the Conference on Security and Co-operation in Europe (CSCE) states, Japan and several of the less-developed countries—have in published statements committed themselves to making a report. Furthermore, in a number of countries domestic legislation requires more information on arms transfers than is required for the UN Register.[31] In addition, much information already exists in the public domain, not least that published annually in the SIPRI register of the trade in major conventional weapons. Also, reporting data for the first two years presents a minimal security risk for states. Reporting to the Register is favoured—even by arms-producing countries—over the more extreme control proposals that might be introduced instead. Finally, compliance with the Register might even serve to increase the legitimacy of arms transfers: governments can point to their reports and explain to their domestic publics that their arms transfers are not illegal.

Rationales for *non-compliance*, according to Laurance, might result from the reluctance of states to be transparent for security reasons. Some states may wait to see how others respond. Others, for example Russia and other East and Central European states, have just established or plan to introduce new arms export control systems which might at present not be sufficiently equipped to control their arms transfers.

[29] UN Secretary-General (note 20), pp. 2–3.

[30] Laurance, E., 'The UN Register of Conventional Arms: rationales and prospects for compliance and effectiveness', *Washington Quarterly*, vol. 16, no. 2 (spring 1993), pp. 163–72.

[31] See Anthony, I., SIPRI, *Arms Export Regulations* (Oxford University Press: Oxford, 1991).

The Register could well be a success due to its novel approach. It is assumed that transparency will have an early-warning effect, even though the data concern arms transfers that have already taken place, because certain developments will become apparent. This is a modest aim and not an ideal situation, but it might work better than previous attempts to control or reduce the arms trade, all of which more or less failed or worked for only a limited period of time.

The future of the UN Register depends on governments' sincerity and willingness to participate. As Laurance has pointed out:

> Assuming that these states [those which complained that the original resolution was discriminatory] were sincere in their desire to have universal and nondiscriminatory transparency, they would have a strong incentive to submit the requested data during the first two years of the Register, to see if the world is serious about taking their concerns into account. Lack of participation in the Register by these states would provide powerful evidence for the naysayers who felt that the whole exercise was just more of the same rhetorical approach to international security problems.[32]

Even if not all of the 180 UN member states report to the Register, it might still be a worthwhile and an informative exercise. All of the 10 leading arms exporters—except China—have publicly anounced their willingness to participate, and China has indicated that, as a result of its participation in the consensus report of the group of governmental technical experts, it would be difficult to imagine that it would not report.[33] If the 10 leading arms exporters do report, over 95 per cent of the weapon systems in the seven categories will be recorded.[34]

The Register, if implemented in the envisaged two stages and complied with universally, could even develop into a far-reaching international control mechanism which could create unprecedented transparency both in the international trade in arms and in the national production of arms. It is a framework for dialogue in a concrete area of military activity and a basis for future verifiable limitations and reductions.

[32] Laurance (note 30), pp. 168–69.

[33] The announcement of US sales of F-16 aircraft to Taiwan in the autumn of 1992, however, has been used by China as an argument for not participating 'for the time being' in the five permanent Security Council members' talks on arms transfer restrictions.

[34] This estimate is based on the SIPRI arms trade statistics; SIPRI reports slightly different categories than those required in the UN Register (see chapter 10 in this volume).

# Appendix 10G. Documents on arms export control in 1992

## DECLARATION OF THE CSCE COUNCIL ON NON-PROLIFERATION AND ARMS TRANSFERS

The Ministers reiterated the commitment of their Governments to the prevention of the proliferation of weapons of mass destruction and the control of missile technology. They underlined their willingness to contribute to the ongoing efforts and international co-operation to this end. In this context, they expressed their support for the Treaty on the Non-Proliferation of Nuclear Weapons and for universal adherence to it. They welcomed the intention of all those CSCE-States not yet party to the NPT to accede to it and urged other States, who are not yet party to it, to do so as well. They also renewed their support for a global, comprehensive and effectively verifiable chemical weapons convention to be concluded in 1992. They also reaffirmed their support for the biological weapons convention, welcomed the results of the September 1991 review conference and called for universal adherence to it.

They expressed their view that excessive build-ups of conventional weapons beyond legitimate defensive needs pose a threat to international peace and security in particular in regions of tension. Based on the principles of transparency, consultation and restraint, they declared their commitment to address the threat of excessive accumulations of conventional weapons and committed themselves to exercise responsibility, in particular with regard to arms transfers to States engaging in such excessive accumulations and to regions of tension.

They confirmed their support for and firmly committed themselves to provide full information to the United Nations Register of Conventional Arms. They called upon all other States to take the same action.

They agreed that effective national control over weapons and equipment transfer is acquiring the greatest importance. They declared their readiness to exchange views and to provide mutual assistance in the establishment of efficient national control mechanisms.

They agreed that in this connection the conversion of arms production to civilian production is also acquiring special importance.

The Ministers decided that the question of non-proliferation, including the transfer of sensitive expertise, and the establishment of a responsible approach to international armaments transfers should be included as a matter of priority in the work programme for the post Helsinki arms control process.

---

*Source*: Second Meeting of the CSCE Council, Prague, CSCS/2–C/Dec. 1, 31 Jan. 1992.

---

## INTERIM GUIDELINES RELATED TO WEAPONS OF MASS DESTRUCTION

The People's Republic of China, the French Republic, the Russian Federation, the United Kingdom of Great Britain and Northern Ireland, and the United States of America

Reaffirming their objectives and commitments as expressed in the communiques following their meetings in Paris and London,

Determined to work towards maintaining world peace and freeing mankind from the threat of weapons of mass destruction,

Affirming that international non-proliferation efforts should not prejudice the legitimate rights and interests of states in the exclusively peaceful use of science and technology for development,

Recalling the announcement made by each of the parties of its commitment to or support for the Missile Technology Control Regime (MTCR),

Recalling their respective positions on the application of International Atomic Energy Agency (IAEA) safeguards to nuclear co-operation with non-nuclear weapon states,

Calling upon states that have not yet done so to accede to the Treaty on Non-Proliferation of Nuclear Weapons,

Declare that they will observe and consult upon the following guidelines:

1. Not assist, directly or indirectly, in the development, acquisition, manufacture, testing, stockpiling, or deployment of nuclear weapons by any non-nuclear-weapons state;

2. Promptly notify the International

Atomic Energy Agency of the export to a non-nuclear weapons state of any nuclear materials, equipment, or facilities and place them under IAEA safeguards;

3. Exercise restraint in the transfer of sensitive nuclear facilities, technology and weapons-usable material, services or technology which could be used in the manufacture of nuclear-weapons-usable material except when satisfied that such exports would not contribute to the development or acquisition of nuclear weapons or to any nuclear activity not subject to safeguards;

4. Not assist, directly or indirectly, in the development, acquisition, manufacture, testing, stockpiling, or deployment of chemical weapons by any recipient whatever;

5. Not export equipment, material, services or technology which could be used in the manufacture of chemical weapons except when satisfied, for example, by recipient country guarantees or confirmation by the recipient, that such exports would not contribute to the development or acquisition of chemical weapons;

6. Strictly abide by the provision of the Convention on the Prohibition of the Development, Production and Stockpiling of Bacteriological (Biological) and Toxin Weapons and on their Destruction, undertake to maintain and support efforts for enhancing the effectiveness of the convention and implement in earnest the confidence-building measures adopted by the Third Review Conference of the Parties to the Convention;

7. Not export equipment, material, services or technology which could be used in the manufacture of biological weapons except when satisfied, for example, by recipient country guarantees or confirmation by the recipient, that such exports would not contribute to the development or acquisition of biological weapons;

8. In considering whether to authorize the export for permitted purposes of the relevant items which might be of use in the manufacture of weapons of mass destruction, take into account:

(a) the capabilities, objectives, policies and practices of the recipients, and any related proliferation concerns;

(b) the significance and appropriateness of the items to be transferred;

(c) an assessment of the proposed end-use, including relevant assurances by the government of the recipient state and controls on retransfer;

9. Maintain export control systems in accordance with their national laws or regulations to enable these guidelines to be effectively implemented;

10. Work together to increase the effectiveness of export controls pursuant to these guidelines.

*Source*: Washington Communiqué of the five permanent members of the Security Council, 29 May 1992.

# Part V. Arms control and disarmament, 1992

Chapter 11. Nuclear arms control

Chapter 12. Conventional arms control in Europe

Chapter 13. The United Nations Special Commission on Iraq: activities in 1992

Chapter 14. The Chemical Weapons Convention: the success of chemical disarmament negotiations

# 11. Nuclear arms control

DUNBAR LOCKWOOD

## I. Introduction

The political and economic climate in 1992 was clearly conducive to progress in nuclear arms control. The year was highlighted by the completion of negotiations on the landmark US–Russian Treaty on Further Reduction and Limitation of Strategic Offensive Arms (START II Treaty), which was signed on 3 January 1993, and by progress towards a comprehensive nuclear test ban—agreements that have important ramifications for the future of nuclear arms control. If the START II Treaty is ratified and implemented and a comprehensive test ban (CTB) treaty agreed, these accords should help build a strong consensus for long-term extension of the 1968 Treaty on the Non-Proliferation of Nuclear Weapons (NPT) at the 1995 NPT Extension Conference. However, there were also setbacks in nuclear arms control in 1992, particularly the failure to bring the 1991 US–Soviet Treaty on the Reduction and Limitation of Strategic Offensive Arms (START Treaty) into force and the failure to make a nuclear warhead dismantlement regime with strict verification provisions a top priority.

In addition to the successes and failures, there was a growing recognition that traditional approaches to arms control were becoming anachronistic in both style and substance. The adversarial days of the superpowers competing to wring greater concessions out of each other and using weapon programmes as bargaining chips appear to be over. The future US–Russian arms control agenda seems rather to be moving towards discussions on how to facilitate greater levels of co-operation in order to attain common goals. Moreover, the new agenda will focus less on limiting launchers and the nuclear warheads attributed to them and more on how to dismantle those warheads and dispose of the fissile materials they contain.

Furthermore, assuming that the two START treaties enter into force, it is clear that the international community will in the future put less emphasis on cutting the existing nuclear weapon arsenals and more emphasis on efforts to halt the proliferation of nuclear weapons to additional countries.

## II. The 1991 START Treaty

Although the START Treaty was signed in July 1991, it had still not entered into force by the end of 1992. The dissolution of the Soviet Union in 1991 severely complicated the fortunes of the START Treaty by creating four states with strategic nuclear weapons based on their territories—Russia, Belarus,

Kazakhstan and Ukraine. As a result, the Bush Administration decided not to push for Senate approval of the START Treaty until key issues had been worked out with and among those four states.[1]

Until April 1992, the Bush Administration favoured keeping the START Treaty a bilateral treaty, with Russia as the sole 'agent' for working out Treaty implementation arrangements with Belarus, Kazakhstan and Ukraine. However, after the failure of two high-level meetings of the Commonwealth of Independent States (CIS)—a 20 March summit meeting of heads of state in Kiev and an 11 April ministerial meeting in Moscow—to reach agreement on these arrangements, it became clear that Ukraine and Kazakhstan insisted on being formally made equal parties to the Treaty. The challenge then was to accommodate this desire without seeming to grant these new nations the status of nuclear weapon states.

In April 1992, then US Secretary of State James Baker actively undertook to arrange a five-nation settlement, intervening personally with numerous phone calls to Moscow, Kiev, Alma-Ata and Minsk. Baker's intensive diplomatic efforts eventually bore fruit. Ukrainian President Leonid Kravchuk and Kazakh President Nursultan Nazarbayev met with President George Bush in Washington on 6 and 19 May 1992, respectively, when the text of letters from the Ukrainian and Kazakh leaders to President Bush were agreed on, committing Ukraine and Kazakhstan to eliminate all of the nuclear weapons on their soil within seven years of the entry into force of the START Treaty. Belarussian parliamentary chairman Stanislav Shushkevich—who did not pay an official visit to Washington but who had consistently indicated that Belarus would follow Ukraine's lead—made the same commitment in a letter of 20 May 1992 to President Bush.[2]

These developments paved the way for a 23 May 1992 ministerial meeting in Lisbon, Portugal, where the USA, Russia, Belarus, Kazakhstan and Ukraine signed a protocol to the START Treaty (the Lisbon Protocol), making all five states parties to the Treaty and committing the three non-Russian former Soviet republics to accede to the NPT 'in the shortest possible time' as non-nuclear weapon states. (For the text of the Lisbon Protocol, see appendix 11A.)

As of February 1993, the legislative branches of Kazakhstan, the USA, Russia and Belarus had approved the START Treaty. Kazakhstan ratified it on 2 July 1992 but had not acceded to the NPT by the end of 1992. The US Senate approved the START Treaty on 1 October 1992 with the understanding that the Lisbon Protocol and the accompanying three letters to President Bush carry the same legal obligation as the Treaty.[3]

[1] For the text of the START Treaty, see *SIPRI Yearbook 1992: World Armaments and Disarmament* (Oxford University Press: Oxford, 1992), appendix 1A.

[2] For the texts of the letters, see Institute for Defense and Disarmament Studies, *Arms Control Reporter* (IDDS: Brookline, Mass.), sheet 611.D.92–94, June 1992; and *Arms Control Today,* June 1992, pp. 35–36.

[3] Senate Foreign Relations Committee (SFRC) hearings: *The START Treaty,* Senate Hearing 102-607, Feb., Mar. and June 1992, Parts 1 and 2 (US Government Printing Office: Washington, DC, 1992); Senate Armed Services Committee (SASC hearings): *Military Implications of START I and II,* Senate

The US Senate's resolution of ratification also included a somewhat controversial condition sponsored by Senator Joe Biden. The condition, originally adopted by the Senate Foreign Relations Committee in July 1992, requires the US President, 'in connection with any further agreement reducing strategic offensive arms', to seek an appropriate arrangement including the use of reciprocal inspections, data exchanges and other co-operative measures to monitor stockpiled nuclear warheads and facilities that produce fissile material for weapons.

Initially, the Bush Administration strongly opposed the Biden condition, arguing that efforts to monitor non-deployed warheads and fissile material on a reciprocal basis would be 'exceedingly intrusive, complex, and expensive' and that any formal agreement would probably prove unverifiable. In addition, the Administration asserted that reciprocal arrangements for monitoring were unnecessary, since proliferation risks in Russia were being dealt with in the Safety, Security and Dismantlement (SSD) Talks (see section VII) and the USA already has adequate safeguards in place.[4]

Responding to the US Administration's concerns, the Senate Foreign Relations Committee report explained that the condition sought only 'a good faith effort' from the Administration and noted that 'the committee's language is intended to leave it up to the discretion of the President to determine the scope and the terms of this arrangement'.[5] Once the report made it clear that the condition did not include any deadline or require the implementation of any specific measures, the Administration dropped its opposition and the condition was retained as part of the resolution of ratification. Despite the fact that the language in the Committee's report watered down the condition to the point where it had the effect of a non-binding resolution of ratification, it is a harbinger of future US–Russian nuclear arms control endeavours.

The Russian Supreme Soviet ratified the START Treaty on 4 November 1992 but stipulated that the actual exchange of instruments of ratification would not occur until after the other former Soviet republics with nuclear weapons on their soil accede to the NPT as non-nuclear weapon states and agree to START Treaty implementation measures.[6]

Hearing 102-953, July–Aug. 1992 (US Government Printing Office: Washington, DC, 1992); Senate Report: SFRC together with additional views, *The START Treaty*, Executive Report 102-53, Sep. 1992 (US Government Printing Office: Washington, DC, 1992); Senate Floor Debate: *Congressional Record*, 29 Sep. 1992, pp. S15497–567; 30 Sep. 1992, pp. S15706–867; and 1 Oct. 1992, p. S15939. For more information about the US Senate START hearings on ratification of the 1991 START Treaty, see Lockwood, D., 'Senate Foreign Relations Committee approves START', *Arms Control Today*, July/Aug. 1992, p. 24, and Lockwood, D., 'Senate ratifies START agreement; sets groundwork for deeper cuts', *Arms Control Today*, Oct. 1992, p. 30.

[4] *Military Implications of START I and START II*, Testimony of Robert Galluci, Assistant Secretary of State for Politico-Military Affairs, to the Senate Armed Services Committee, 4 Aug. 1992, Senate Hearing 102-953 (note 3), pp. 249, 266; also in *US State Department Dispatch*, vol. 3, no. 32 (10 Aug. 1992), p. 631. See also Gordon, M. R., 'Senate calls for monitoring of the warheads in Russia', *New York Times*, 3 July 1992, p. A2.

[5] SFRC together with additional views, *The START Treaty*, Executive Report 102-53 (note 3), p. 92.

[6] For more information about the Russian Supreme Soviet's ratification of the 1991 START Treaty, see Lockwood, D., 'Russia ratifies START; Ukraine reaffirms conditions for approval', *Arms Control Today*, Nov. 1992, pp. 26, 31–32. See also Text of the Supreme Soviet Resolution on the Ratification of

On the question of implementation measures, Russian Foreign Minister Andrey Kozyrev had indicated earlier that Russia's understanding of how the START Treaty should be implemented may be significantly different from that of Belarus, Kazakhstan and Ukraine. Several months before the Russian Parliament ratified the Treaty, Kozyrev stated that Russia 'proceeds on the assumption' that seven years after the START Treaty enters into force there will be neither nuclear warheads nor strategic nuclear delivery vehicles on the territories of Belarus, Kazakhstan or Ukraine.[7] Kazakhstan and Ukraine, however, have indicated that they would like to retain some of the land-based missiles on their territories—without their warheads—to use as boosters to launch commercial satellites.[8]

Kozyrev also cited Ukraine's commitment, under the CIS Minsk Agreement of 30 December 1991,[9] to eliminate all of the strategic forces based on its territory by the end of 1994, implying that Russia still expects Ukraine to honour that pledge.[10] Somewhat paradoxically, however, under the terms of the START Treaty, the Lisbon Protocol and accompanying 7 May 1992 letter, Ukraine's legal obligation is to eliminate strategic forces on its territory within seven years of the entry into force of the START Treaty. Thus, assuming that the START Treaty enters into force in 1993, Ukraine could retain nuclear weapons until the year 2000—six years longer than the commitment made in the 1991 Minsk Agreement.

In addition to disagreements over questions of START Treaty implementation, Russia and Ukraine continued to dispute a number of issues related to strategic weapons on Ukrainian territory. Ukraine insists that it is the legal owner of the nuclear weapons on its territory and should have 'administrative control' over those systems.[11] Administrative control entails having troops at the intercontinental ballistic missile (ICBM) and bomber bases take an oath of allegiance to Ukraine and providing those troops with housing, food and salaries. Russia has argued that it should have jurisdiction over the weapons and has complained that Ukraine has blocked its efforts to service the liquid-fuelled SS-19 ICBMs in Ukraine—a development that could have severe

the Treaty Between the Union of Soviet Socialist Republics and the United States of America on the Reduction and Limitation of Strategic Offensive Arms, 4 Nov. 1992.

[7] 'Written statement by the Russian side at the signing of the Protocol to the START Treaty on 23 May 1992 in Lisbon', *Arms Control Today*, June 1992, p. 36; see also Kozyrev's remarks in a Moscow TV interview, 6 Jan. 1993, in 'Kozyrev on START II, Yugoslavia', Foreign Broadcast Information Service, *Daily Report–Central Eurasia* (hereafter, *FBIS-SOV)*, FBIS-SOV-93-004, 7 Jan. 1993, pp. 1–4.

[8] In order to remove an ICBM system from accountability under the 1991 START Treaty, the parties are required to destroy the silos or mobile launchers but not the missiles themselves, except for non-deployed mobile ICBMs that exceed the negotiated numerical ceiling.

[9] *SIPRI Yearbook 1992* (note 1), appendix 14A.

[10] See note 7.

[11] Boris Krivoshey and Leonid Timofeyev, ITAR-TASS (Moscow), in 'Ministers hold separate, joint meetings', FBIS-SOV-92-129, 6 July 1992, p. 8; Radio Ukraine World Service broadcast, in 'Ukraine's position on strategic forces noted', FBIS-SOV-92-129, 6 July 1992, p. 9. See also Smith, R. J., 'Ukraine likelier to retain nuclear arms, US believes', *International Herald Tribune*, 21 Dec. 1992, p. 3.

environmental consequences.[12] Furthermore, Ukrainian President Kravchuk has claimed that Ukraine has the technical capability to block the launch of missiles from Ukrainian territory, while Marshal Yevgeny Shaposhnikov, Commander-in-Chief of the CIS Joint Forces, has said that this is only a political veto, not a technical one.[13]

The Parliament of Belarus ratified the START Treaty on 4 February 1993. The Parliament also voted to accede to the NPT as a non-nuclear weapon state.[14] Belarus has made a commitment to withdraw the 81 SS-25 ICBMs now on its territory to Russia by the end of 1994.[15]

After promising on several occasions that Ukraine would promptly ratify the START Treaty, Ukrainian Foreign Minister Anatoly Zlenko announced in December 1992 that Ukraine would not be able to ratify the START Treaty in 1992.[16] President Kravchuk said that 'serious people understand that before agreeing to anything, all matters must be studied thoroughly'.[17] He pointed out that the US Senate 'needed more than a year to study the START Treaty and all the consequences of its implementation, for the country's security and economy before ratification'.[18]

In its efforts to persuade Ukraine to comply with the Lisbon Protocol, ratify the 1991 START Treaty and accede to the NPT, the Bush Administration put together a two-part package. First, the USA offered Ukraine $175 million from the $800 million of re-programmed US defence funds authorized by Congress (see section VII) to assist primarily in the destruction of SS-19 and SS-24 missiles and silos on Ukrainian soil.[19] Ukrainian President Kravchuk has called that only a 'drop' of what Ukraine needs but has not specified a precise amount.[20] Second, the USA has informed Ukraine that it (in parallel

[12] 'Discord reigns on eve of top-level CIS meeting', *Washington Post*, 22 Jan. 1993, p. A24; Freeland, C. and Lloyd, J., 'Russia "trying to paralyse Ukraine"', *Financial Times*, 19 Feb. 1993; Kiselyov, S., 'Ukraine: stuck with the goods', *Bulletin of the Atomic Scientists*, Mar. 1993, p. 32.

[13] At the Dec. 1991 CIS summit meetings at Alma-Ata and Minsk, the CIS heads of state agreed that the Russian President's decision to use former Soviet nuclear weapons would be taken 'in agreement with the heads of' Ukraine, Belarus and Kazakhstan. Although Kravchuk has often hinted that there are technical safeguards that would block the launch of former Soviet strategic missiles from Ukrainian territory without his consent, Ukrainian Defence Minister Konstantin Mozorov has acknowledged that there are in fact no such safeguards. Tsikora, S., 'Ukrainian Defense Minister: we will tackle the problems of the army our own way', *Izvestia*, 25 Mar. 1992, in FBIS-SOV-92-158, 25 Mar. 1992, p. 59. For the Shaposhnikov statement on CIS control over nuclear weapons in Ukraine, see 'Discord reigns on eve of top-level CIS meeting', *Washington Post*, 22 Jan. 1993, p. A24.

[14] Reuter, 'Belarus approves first arms limitation pact', *New York Times*, 5 Feb. 1993, p. 8; Belinform, 4 Feb. 1993, in 'Legislature ratifies START I, military pacts with Russia', FBIS-SOV-93-023, 5 Feb. 1993, p. 50.

[15] Interfax, 26 Oct. 1992, in FBIS-SOV-92-208, 27 Oct. 1992, p. 3.

[16] Radio Ukraine World Service (Kiev), 14 Dec. 1992, in 'Foreign Minister addresses diplomats on nuclear policy', in FBIS-SOV-92-240, 14 Dec. 1992, p. 12.

[17] Khotin, R., Reuter, Kiev, 'Ukraine's Kravchuk pleads for time on START Treaty', 15 Dec. 1992.

[18] Khotin (note 17).

[19] Lockwood, D., 'Ukraine stalls on START I vote; presses U.S. on aid, security issues', *Arms Control Today*, Jan./Feb. 1993, p. 22.

[20] When Ukrainian Foreign Minister Anatoly Zlenko met with President Clinton on 25 Mar. 1993, he reportedly requested $2.8 billion to help Ukraine dismantle the nuclear weapons on its territory. Erlanger, S., 'Ukraine and arms accords: Kiev reluctant to say "I do"', *New York Times*, 31 Mar. 1993, p. A8; 'A persistently nuclear nightmare', *The Economist*, 3 Apr. 1993, p. 34; Giacomo, C., 'Clinton presses Ukraine to ratify arms treaty', Reuter, 25 Mar. 1993.

with Russia and the UK) is prepared to offer Ukraine security assurances along the lines of three commitments it has made in the past.[21]

In addition to these security assurances and the funds authorized for destruction assistance, the USA has informed Ukraine that, if final agreements can be worked out with Russia, Ukraine will receive a portion—reportedly about 10–20 per cent—of the proceeds from the sale to the USA of highly enriched uranium (HEU) extracted from former Soviet nuclear warheads dismantled in Russia. (For details of the HEU agreement, see section VII.)[22]

## III. The 1993 START II Treaty

Presidents Bush and Yeltsin signed the START II Treaty in Moscow on 3 January 1993, concluding the most sweeping nuclear arms reduction treaty in history. The Treaty, which will not enter into force until the 1991 START Treaty does, will require the USA and Russia to eliminate their MIRVed (equipped with multiple independently targetable re-entry vehicles) ICBMs and reduce the number of their deployed strategic nuclear warheads to 3000–3500 each. These reductions are to be carried out by 1 January 2003 or by the end of the year 2000 if the USA can help finance the elimination of strategic offensive arms in Russia. (For the text of the START II Treaty, see appendix 11A.)

The Treaty also limits the number of submarine-launched ballistic missile (SLBM) warheads to 1700–1750 each. The rules for 'downloading' warheads from existing types of ballistic missiles under the START II Treaty will be less restrictive than those called for by the 1991 START Treaty.[23] The START II Treaty, unlike the 1991 START Treaty which heavily under-counts bomber weapons, will count strategic bombers as having the number of nuclear weapons for which they are 'actually equipped'. Each side, however, will be permitted to exempt up to 100 strategic bombers from the Treaty limits—provided that they have never been equipped with long-range nuclear

[21] In 1968 the USA made a commitment, in connection with the NPT, to seek immediate UN Security Council action to assist any non-nuclear weapon party to the NPT that is subjected to aggression or the threat of aggression involving nuclear weapons. UN document S/RES/255 (1968). In 1975 the USA and other members of the CSCE signed the Helsinki Final Act, which recognizes existing borders and permits changes in those borders 'only by peaceful and consensual means'. In 1978 the USA declared in a so-called 'negative security assurance' that it would not use nuclear weapons against non-nuclear weapon parties to the NPT unless they were engaged in an attack, in association with a nuclear weapon power, against the USA or its allies. UN document A/S-10/AC.1/30, 13 June 1978. See also Oberdorfer, D., 'Bush details assurances for security of Ukraine', *Washington Post*, 9 Jan. 1993, p. A18; Lockwood (note 19), p. 22.

[22] The criterion for determining the 'appropriate and equitable' share of the proceeds for Ukraine, Belarus and Kazakhstan has not yet been established, but it could be based on either the percentage of the former USSR's nuclear warheads located in each country or on the percentage of the former USSR's foreign debt that each state is willing to pay. For a discussion of this and other estimates, see Lepingwell, J. W. R., 'How much is a warhead worth?', *Radio Free Europe/Radio Liberty Research Report*, vol. 2, no. 8 (19 Feb. 1993), pp. 62–64.

[23] In the START II Treaty, unlike the 1991 START Treaty, there will be no aggregate limit on the number of warheads that may be downloaded; a missile that is downloaded by more than 2 warheads does not have to have its re-entry vehicle platform or 'bus' destroyed and replaced with a new one that conforms to the reduced loading; up to 105 ICBMs may be downloaded by as many as 5 warheads each—a provision that in practice applies only to the Russian SS-19 missile.

air-launched cruise missiles (ALCMs), by 'reorienting them to conventional roles'.[24]

The Treaty will require the USA to reduce its deployed strategic nuclear warheads by more than 70 per cent from its September 1990 level and by almost 60 per cent from the number it had planned to deploy under the START Treaty. Under the START II Treaty, Russia is expected to reduce its strategic forces by approximately 70 per cent from the number the USSR deployed in September 1990 and by about 50 per cent from the number which the Defense Intelligence Agency (DIA) projected Russia would deploy under the START Treaty.[25]

The bidding between the USA and Russia on the START II Treaty began in January 1992—almost exactly six months after the 1991 START Treaty had been signed and about one month after the Soviet Union had collapsed. In his 28 January 1992 State of the Union Address, President Bush proposed a new agreement requiring far deeper cuts than those required by START. In his new proposal, Bush offered to reduce US SLBM warheads by 'about a third' below the number (3456) of warheads which the USA planned to deploy under the START Treaty if the CIS states ('the former Soviet Union') agreed to a ban on MIRVed ICBMs. The next day, General Colin Powell, chairman of the Joint Chiefs of Staff (JCS), estimated that the Bush proposals, if accepted, would leave the USA with approximately 4700 deployed strategic warheads: 500 on ICBMs; 2300 on SLBMs; and 1900 on bombers.[26]

Russian President Boris Yeltsin, who had been told of Bush's proposals in advance, responded the day after Bush's Address by proposing that the two sides cut their strategic nuclear warheads to 2000–2500 each.

Baker and Kozyrev held ministerial meetings in February, March, May and June 1992, paving the way for a Washington summit meeting between Bush and Yeltsin in June. On 17 June, Bush and Yeltsin signed the Joint Understanding on Further Reductions in Strategic Offensive Arms (the De-MIRVing Agreement) to form the basis for a follow-on to the START Treaty.[27] The Agreement included the numerical ceilings and timeframe for reductions. Less than a week after the De-MIRVing Agreement was signed, Baker predicted that the START II Treaty would be completed at the latest by 1 September 1992.[28] However, the combination of a US presidential election, Russia's preoccupation with its domestic problems, bureaucratic inertia on both sides and differences over several implementation issues slowed the negotiations.

[24] Reoriented bombers do not have to be physically re-configured, but they must be based separately from bombers with nuclear roles; they must have external observable differences from nuclear bombers of the same type; bomber nuclear weapons may not be stored at their bases; and their crews may not train or conduct exercises for nuclear missions.

[25] Lt. General James R. Clapper, Jr, Director of the Defense Intelligence Agency (DIA), Statement before the Senate Foreign Relations Committee, 30 June 1992, in *The START Treaty,* Senate Hearing 102-607, Part 2 (note 3), p. 164. (Clapper projected that Russia would deploy 6700 strategic warheads after the START Treaty was implemented.)

[26] Statement of General Colin L. Powell, Chairman, Joint Chiefs of Staff, on the FY 1993 Defense Budget, before the Senate Armed Services Committee, 31 Jan. 1992.

[27] For the text of the De-MIRVing Agreement, see appendix 11A in this volume.

[28] James Baker, Secretary of State, Statement before the Senate Foreign Relations Committee, 23 June 1992, *The START Treaty* (note 3), Part 2, p. 9.

In late July 1992, six weeks after the Washington summit meeting, the USA submitted a draft treaty to Moscow. In late November, Russia responded with a formal draft treaty of its own, reflecting some of the issues that Russia had raised at a 24 September meeting at the UN between then acting Secretary of State Lawrence Eagleburger and Kozyrev.[29]

While endorsing the provisions of the De-MIRVing Agreement, Russia asked for an easing of some of the terms of implementation for economic reasons. Russia asked if it could convert—rather than destroy—its SS-18 silos to hold single-warhead ICBMs, such as the SS-25. As a general principle, the START Treaty permits both sides to change the type of missile based in a silo so long as the silo is no longer capable of launching the previous type of missile.[30] However, it explicitly requires the destruction of half of the former Soviet Union's SS-18 silos (i.e., 154),[31] but whether this provision would apply to the second set of 154 silos in the START II Treaty was not specifically discussed at the June summit meeting.

Russia also asked if the downloading rules that were apparently agreed in June could be changed to allow it to retain a single-warhead version of the SS-19. Russia now deploys 170 SS-19s—older, relatively inaccurate missiles that were first deployed in the 1970s and are no longer in production. According to Bush Administration officials, it was agreed at the Washington summit meeting to carry over the 1991 START Treaty rule that an individual missile may not be downloaded by more than four warheads.[32] Since only single-warhead missiles would be permitted under the START II Treaty and the SS-19 has six warheads, Russia could not retain this missile unless the USA agreed that a missile could be downloaded by five warheads rather than four.

In addition, Russia put forward several proposals intended to help it verify limits on US bombers with more confidence. Among other things, Russia proposed a rule prohibiting any strategic bomber designated as conventional (and exempted from the aggregate START II Treaty ceiling of 3500 nuclear warheads) from being re-designated as a nuclear bomber. Russia also proposed that conventional and nuclear bombers have some external observable differences so that they can be distinguished from each other. Finally, Russia insisted on inspecting the US B-2 'stealth' bomber to determine that it is not equipped with more nuclear weapons than the number (16) which the USA had attributed to it.

[29] Goshko, J. M. and Diehl, J. 'New arms cuts readied', *Washington Post*, 25 Sep. 1992, p. A27; Friedman, T. L., 'US–Russia accord on arms hits snag', *New York Times*, 15 Oct. 1992, p. 13.

[30] The USA, for example, has already begun deploying Minuteman IIIs in empty Minuteman II silos at Malmstrom AFB, Montana.

[31] This commitment is contained not in the Treaty text but in an exchange of letters on 30 July 1991 between the chief US and Soviet negotiators, Linton Brooks and Yuri Nazarkin.

[32] Whether this was agreed to at the June summit meeting is unclear. The START Treaty prohibits downloading individual missiles by more than 4 warheads each and START provisions apply to the START II Treaty except where the latter treaty explicitly states otherwise, but the two pertinent documents that were produced at the June summit meeting—the Joint Understanding and a letter from Baker to Kozyrev—do not specifically address whether missiles may be downloaded by more than 4 warheads.

In late December 1992 Presidents Bush and Yeltsin as well as President-elect Bill Clinton made it clear that they supported the completion of the START II Treaty before Bush left office on 20 January 1993. If they waited until after inauguration day, the negotiating process would be delayed by a number of months as the new US Administration scrambled to get its personnel in place and its policy positions in order. Furthermore, it might have been far more difficult politically for a Democrat to show flexibility on the remaining issues. If Clinton had made compromises on the SS-18 silo conversion and SS-19 downloading issues, for example, he might have been accused of 'giving away the store'. In addition, Bush had his presidential foreign policy legacy to think of, and Clinton did not want to be distracted from his domestic policy agenda.

President Yeltsin, for his part, had a host of political and economic reasons for completing the Treaty quickly. First, given the political challenges coming from the Russian Parliament, it was in his interest to appear to the Russian people as a respected statesman and the leader in Moscow with whom the United States negotiated. Second, the Russian economy needs some breathing space, and Yeltsin and his advisers have estimated that the savings on operations and maintenance costs for Russian strategic forces would exceed the dismantlement costs. Finally, the completion of the Treaty would engender goodwill in the West, promoting a climate more conducive to granting economic assistance to Russia.

With these incentives to finish the START II Treaty quickly, Bush and Yeltsin exchanged telephone calls on 20 and 21 December 1992.[33] After their conversations produced some progress, a team of US and Russian technical specialists met in Geneva on 22–24 December to try to complete the final details. The last issues were finally resolved during high-level meetings in Geneva on 28 and 29 December between Secretary of State Eagleburger, on the US side, and Foreign Minister Kozyrev and Defence Minister Pavel Grachev, on the Russian side.

### SS-18 missile silos

The SS-18 missile issue posed more of a political problem than a technical one. The USA ultimately agreed to let Russia convert 90 of the 154 SS-18 missile silos that will remain after the 1991 START Treaty is implemented; the other 64 silos must be destroyed. In order to ensure that the silos cannot be quickly reconverted to launch banned SS-18 missiles, Russia agreed to pour 5 metres of concrete on the floor of the silos.[34] In a second measure designed to make the silos incapable of launching SS-18s, a 'restrictive ring' with a

[33] Following a meeting between Eagleburger and Kozyrev on the START II Treaty held in Stockholm on 13 Dec. 1992, Yeltsin made a surprise announcement in Beijing on 18 Dec. 1992 that START II would be ready for signature in early Jan. 1993. Mineyev, A. and Spirin, P., ITAR-TASS, 18 Dec. 1992, in 'Comments on SALT II [sic] Treaty signing', in FBIS-SOV-92-244, 18 Dec. 1992, p. 6; Wines, M., 'A-treaty is ready, Yeltsin declares', *New York Times*, 19 Dec. 1992, p. A1.

[34] The total length of an SS-18 missile with its launch canister and front section is over 35 m and the total length of the SS-25 with its launch canister and front section is 22.3 m.

diameter smaller than that of the SS-18 must be installed into the upper portion of the silo.[35] The USA will be permitted to observe the entire process of pouring concrete into the silos and to measure the diameter of the restrictive ring. The USA will also be allowed to conduct four re-entry vehicle (RV) inspections each year of converted SS-18 silos in addition to the 10 RV inspections it may conduct each year under the 1991 START Treaty.

Importantly, Russia will also be required to destroy all of its SS-18 missiles and their canisters, both deployed and non-deployed. The missile's stages must be cut up into pieces or the missiles may be destroyed by using them as space launch vehicles.[36]

## SS-19 downloading

The USA and Russia agreed that Russia may retain 105 of its 170 SS-19 missiles by removing five of the six warheads from each missile. Russia is not required to change the missile's RV platform or 'bus': the cost would have defeated the purpose of retaining the missiles. In practice, the SS-19 will be the only missile that may be downloaded by more than four warheads. As of 1991, Russia had 110 SS-19s deployed at Tatishchevo and 60 at Kozelsk.[37] (Presumably, the latter base will be closed to save money.)

## Bomber issues

The USA and Russia agreed that heavy bombers 'reoriented' to conventional roles (and exempted from the START II Treaty nuclear warhead limits) may be returned to a nuclear role but may not subsequently be reoriented to a conventional role. Consequently, the US Air Force will have the option of declaring all of its B-1B bombers as conventional weapon carriers and then later returning them to the strategic nuclear force as older B-52Hs are retired. (In practice, this provision does not apply to either of Russia's most modern bombers—the Blackjack and the Bear-H—because both of them have already been equipped with long-range nuclear ALCMs.) The two sides also agreed that heavy bombers reoriented to a conventional role will have observable external differences from nuclear bombers of the same type.[38]

Finally, the two sides agreed to exhibit one heavy bomber of each type specified in the START II Treaty Memorandum of Understanding (MOU)—including the US B-2 bomber—to demonstrate to the other party the number of nuclear weapons which each bomber is actually equipped to carry. During these exhibitions, which must be held only once, the inspection team may look at the exhibited bomber's weapon bays and those portions of the exterior

[35] The restrictive ring may have a diameter of no more than 2.9 m; the SS-18 has a diameter of 3 m.

[36] The START I Treaty rules for removing an ICBM from accountability apply to the START Treaty; see note 8. However, unlike the START Treaty, START II also requires the elimination of all SS-18 missiles.

[37] START Memorandum of Understanding (MOU), Sep. 1990; and US Department of Defense (DOD), *Military Forces in Transition* (DOD: Washington, DC, Sep. 1991).

[38] See note 24.

equipped for carrying weapons. At the discretion of the inspected party, however, all the other portions of the bomber may be shrouded to conceal technological secrets. The US decision to permit Russia to inspect the B-2 bomber was a concession that surprised many observers, especially considering that the USA had essentially refused to agree to such an inspection in the START Treaty.[39]

## The START II Treaty benefits both countries

Russia's willingness to ban MIRVed ICBMs, which make up the backbone of Russia's strategic forces, constituted a significant concession to the USA, which deploys a far smaller fraction of its strategic nuclear warheads on ICBMs. The suggestion that the START II Treaty represents a US 'negotiating victory' over Russia, however, has been grossly overstated. First, after START II Treaty reductions are implemented, Russia will still retain 3000–3500 strategic warheads, which is more than sufficient for a robust deterrent. Second, Russia received several important concessions in the negotiations from the USA, including: a ceiling on US SLBM warheads 50 per cent below the number the USA had planned to deploy under the 1991 START Treaty; new bomber counting rules that abandon the steep discounting of bomber weapons and count bombers as having the number of weapons for which they are 'actually equipped';[40] the right to inspect the B-2 'stealth' bomber; limits on the number of times bombers may be 'reoriented' between nuclear and conventional roles; and the right to convert 90 SS-18 silos and download 105 SS-19 missiles.[41] More important, however, is the central fact that the START II Treaty is in Russia's security interest. On a mutual basis, the Treaty will enhance strategic stability, increase predictability and transparency, improve prospects for a long-term extension of the NPT at the 1995 Extension Conference and potentially save a significant amount of money—all of which will serve Russian interests.

## Savings from the START II Treaty

It is not entirely clear how much money the USA will actually save as a result of the START II Treaty. While the direct savings from the Treaty will be small—perhaps less than $100 million per year—the improvement in the political climate and the predictability in the strategic relationship that the Treaty will foster could lead to substantial cost reductions. The US Congressional Budget Office (CBO) estimated in June 1992 that the USA could save more than $50 billion over the next 15 years by taking steps not neces-

[39] Under the 1991 START Treaty, Russia would be permitted to inspect B-2s only if they were tested with long-range nuclear ALCMs, which the USA does not plan to do.

[40] As a result of the change in bomber counting rules, the projected gap between the number of US and Russian warheads will drop from about 2500 under the START Treaty to fewer than 500 warheads under the START II Treaty; see chapter 6, figure 6.1.

[41] See, for example, Erlanger, S., 'Concessions on arms pact made by US', *New York Times*, 3 Jan. 1993, p. 8.

sarily mandated by but consistent with the spirit of the Treaty.[42] Such steps would include reducing the size of the Department of Energy's warhead production complex, cutting Trident II missile production, and reducing command, control and communications and intelligence activities focused on strategic weapons.

Presumably, Russia could take similar steps for commensurate savings. Russian officials have estimated that, while dismantlement costs may be substantial, they will be exceeded in the long term by savings from reduced operations and support costs.[43]

### The future of US–Russian strategic arms control

It now appears that further strategic arms control negotiations will be put on hold indefinitely, with most US and Russian efforts focused on bringing the START Treaty and the START II Treaty into force. Despite this, there seems to be a consensus in the US arms control community that 3000–3500 strategic warheads each is far more than the USA and Russia need to maintain minimum levels of deterrence.[44] In September 1991, several months before the dissolution of the Soviet Union, the US National Academy of Sciences released a study which concluded that if positive trends continue and 'other nuclear powers have accepted appropriate strategic arms limitations', then the USA and Russia could reduce their strategic arsenals to 1000–2000 warheads.[45] Former Secretary of Defense Robert McNamara argued in February 1993 that 100–200 warheads would be sufficient for deterrence.[46] Such proposals have also raised the question of including France, the UK and China in a legally binding nuclear arms reduction regime.

## IV. The Non-Proliferation Treaty

By 1 January 1993, over 150 states were members of the NPT (see annexe A). China deposited instruments of accession to the NPT on 9 March 1992. In the past, China has been accused of helping a number of countries, including Iran and Pakistan, to develop the capability to produce nuclear weapons.

France acceded to the NPT on 3 August 1992. Now all five declared nuclear weapon states (which are also the permanent members of the UN Security Council) have made legally binding commitments to adhere to the terms of the

[42] Hall, R., *Memorandum for the Record: Budgetary Impact of Bush/Yeltsin Accord* (US Congressional Budget Office: Washington, DC, 29 June 1992), pp. 1–4.

[43] O'Brien, C., Associated Press, 'Russian hardliners set high hurdle for START II', *Washington Times*, 12 Jan. 1993, p. A8; Yeltsin, B., Moscow Press Conference with George Bush, Federal News Service transcript, 3 Jan. 1993, p. 3.

[44] See Dean, J. and Gottfried, K., *Nuclear Security in a Transformed World* (Union of Concerned Scientists: Washington, DC, 1992); McNamara, R., Rathjens, G. and Kaysen, C., 'Nuclear weapons after the cold war', *Foreign Affairs*, vol. 70, no. 4 (fall 1991), pp. 95–110; Bundy, M., Crowe, W. and Drell, S., 'Reducing nuclear danger', *Foreign Affairs*, vol. 72, no. 2 (spring 1993), pp. 140–55.

[45] National Academy of Sciences, *The Future of the US–Soviet Nuclear Relationship* (National Academy Press: Washington, DC, 1991), pp. 3, 30.

[46] McNamara, R. S., 'Nobody needs nukes', *New York Times*, 23 Feb. 1993, p. A21.

NPT—a development that will probably improve chances for a long-term extension of the NPT at the 1995 Extension Conference. In addition, new nuclear arms control agreements, such as the START II Treaty and the US and Russian commitments to negotiate a CTB, may be perceived as a good-faith effort by the nuclear weapon states to fulfil their obligations under NPT Article VI to reduce their nuclear arsenals 'at an early date'.

In addition to China and France, Azerbaijan, Estonia, Latvia, Myanmar (formerly Burma), Namibia, Niger, Slovenia and Uzbekistan all acceded to the NPT in 1992.[47] In January 1993, the Bush Administration issued a report raising questions about the NPT compliance of China, Iraq, Iran, North Korea, Libya and South Africa.[48] In March 1993, North Korea became the first country to announce its intention to withdraw from the NPT (see chapter 6, section IX).

### The IAEA

The 35-member Board of Governors of the International Atomic Energy Agency (IAEA) met in February 1992 and 'reaffirmed the Agency's right to undertake special inspections in member states with comprehensive safeguard agreements' and agreed to strengthen the Agency's safeguards system. The right to conduct special inspections is already contained in the IAEA safeguards agreement that all non-nuclear weapon states are required to sign after joining the NPT. The IAEA, however, has never conducted a suspect site inspection. It has said that it will conduct such inspections based on intelligence information received from member states.

The Board of Governors met again in June 1992 and approved the first IAEA budget increase since 1984. The Board members deferred action for a second time on proposals to create a registry of nuclear-related transfers (see also chapter 6, section VIII).

## V. A comprehensive test ban

### The United States

In 1992 the Bush Administration continued to oppose a CTB and no formal negotiations were held.[49] Under international and congressional pressure, however, the Administration officially changed its testing policy. In July 1992 the Administration announced that, for the next five years, the USA would conduct no more than six tests per year and no more than three tests per year above 35 kt, and that all tests would be conducted for 'safety and reliability'

[47] US Arms Control and Disarmament Agency (ACDA), *Annual Report to the Congress, 1992*, p. 109.

[48] US Arms Control and Disarmament Agency (ACDA), *Adherence to and Compliance with Arms Control Agreements and the President's Report to Congress on Soviet Non-compliance with Arms Control Agreements,* 14 Jan. 1993, pp. 16–18.

[49] CTB talks held in Geneva among the USA, the UK and the USSR were adjourned in Nov. 1980. The Reagan Administration formally withdrew from the CTB talks in 1982.

purposes.[50] Many critics in the US Congress argued that these were only cosmetic changes and successfully pushed through sweeping legislation limiting US nuclear tests.

On 2 October 1992, President Bush signed into law the Fiscal Year (FY) 1993 Energy and Water Development Appropriations Act,[51] which contained a provision mandating a permanent ban on all US nuclear tests after 1996, unless another country tests after that date. Bush called the provision limiting tests 'highly objectionable' but decided not to veto the bill because it included $517 million for the Superconducting Super Collider, an $8 billion project located in Texas, a key state in Bush's re-election effort. During and after the presidential campaign, however, Clinton indicated that he would support the nuclear test ban legislation.[52]

The Act states that 'no underground test of nuclear weapons may be conducted by the USA after 30 September 1996, unless a foreign state conducts a nuclear test after this date'.[53] The legislation requires the suspension of all US tests from 1 October 1992 until at least 1 July 1993. For tests to be conducted between the end of the nine-month moratorium and the cut-off date, the Administration must submit reports to Congress. The three reports—which will cover the last quarter of FY 1993 and all of FYs 1994, 1995 and 1996—require a description of all proposed tests and a plan for installing modern safety features (insensitive high explosives, enhanced detonation safety systems or fire-resistant fissile material 'pits') in the warheads slated for testing. Only those warheads that have been re-designed to include a modern safety feature that they previously did not have, in accordance with the reports submitted by the Administration, may be tested.

In the period covered by these three reports, between the end of the moratorium and the 1996 cut-off date, a total of no more than 15 tests may be conducted, with no more than five in any one report period. All of these tests must be conducted for safety purposes except for one 'reliability test' per report period, which must be approved by Congress.[54] The UK, which also conducts its tests at the US Nevada Test Site, is permitted to conduct one test per report period, but each test would count towards the report period and overall test limits.

[50] Gordon, M., 'U.S. tightens limits on nuclear tests', *New York Times*, 15 July 1992, p. A1; see also Porth, J., 'Nuclear testing ban won't aid arms control', *Wireless File*, no. 136 (15 July 1992), pp. 2–3.

[51] For the text of the Act, see 'US Congress nuclear testing limits', Institute for Defense and Disarmament Studies (IDDS), *Arms Control Reporter,* sheet 608.D.1–2, Oct. 1992; *Congressional Record,* 24 Sep. 1992, p. H9424.

[52] 'Remarks by Governor Bill Clinton, A roundtable discussion with employees of Sandia National Laboratories, Albuquerque, N.M., 18 September 1992', Transcript from Clinton–Gore Campaign, p. 9; Letter dated 12 Feb. 1993, from President Bill Clinton to Senate Majority Leader George Mitchell, *Congressional Record*, 16 Feb. 1993, p. S1513; Smith, R. J., 'Environmental cleanup role considered for A-weapons lab', *Washington Post,* 9 Mar. 1993, p. A10.

[53] Due to a drafting error, the legislation also cites 1 Jan. 1997 as a cut-off date, but the language in the Act appears to prohibit testing after 30 Sep. 1996. In any case, the legislation may be modified in the future to clear up the drafting error.

[54] While it was clearly the intent of the legislation's sponsors to have the reliability tests count towards the limit of 15 tests, some Bush Administration officials have interpreted the legislation differently; that is, a total of 18 tests could be conducted, with 15 tests for safety and 3 more for reliability.

The law also directs the Administration to submit to Congress a schedule for the resumption of test talks with Russia and a 'plan for achieving a multilateral comprehensive ban on the testing of nuclear weapons by September 30, 1996'.[55]

## Russia

In 1992 Russia adhered to the commitment made by former Soviet President Mikhail Gorbachev in October 1991 not to conduct nuclear tests for one year, and on 19 October 1992 President Yeltsin announced that Russia would extend its moratorium at least until 1 July 1993.[56] In November 1992 Yeltsin reiterated his support for the negotiation of a CTB treaty[57] and announced that he would urge China to join the nuclear test moratorium.[58]

Yeltsin, however, under pressure from Russian nuclear weapon laboratories, had called on 27 February 1992 for preparations to be made at the Russian test site at Novaya Zemlya for a resumption of testing at a rate of two to four tests per year if the moratorium expires.[59] CIS military commander Marshal Shaposhnikov said in September 1992 that 'if our partners in the West don't stop these nuclear explosions, I think we would have to part with the moratorium and resume nuclear testing, maybe in a less intensive manner'.[60] Defence Minister Grachev reiterated this in October, noting that after 1 July 1993 'everything will . . . depend on the American side'. Grachev added that the Russian Parliament was under pressure from Russia's nuclear laboratories not to ban all nuclear testing.[61]

## France

In April 1992 France announced a nuclear test moratorium through the end of 1992 and said that it would decide whether to resume testing in 1993 based on other countries' testing practices. This announcement marked a significant departure from past French policy. For the previous three years, France had conducted more tests per year than any country except the USA. Moreover, until 1992 France had opposed or abstained from resolutions in the UN General Assembly to ban nuclear testing. Noting that the announcement came just a

[55] See note 51.

[56] ITAR-TASS, 'Yeltsin extends nuclear test moratorium', p. 1, in FBIS-SOV-92-202, 19 Oct. 1992, p. 2; 'Text of decree extending nuclear test moratorium', *Rossiyskaya Gazeta*, 21 Oct. 1992, FBIS-SOV-92-205, 22 Oct. 1992, p. 12.

[57] Robinson, E., 'Yeltsin vows to remain in control', *Washington Post*, 11 Nov. 1992, p. A31.

[58] Pollack, A., 'Yeltsin plans end to A-sub program', *New York Times*, 20 Nov. 1992, p. A10.

[59] 'Secret decree may reopen nuclear test site', *Nezavisimaya Gazeta*, 24 Mar. 1992, p. 6, in FBIS-SOV-92-060, 27 Mar. 1992, p. 1; 'Preparations at nuclear test site "going ahead"', *Rossiyskaya Gazeta*, 18 June 1992, p. 2, in FBIS-SOV-92-121, 23 June 1992, p. 4; Higgins, A., 'Yeltsin orders resumption of nuclear testing', *The Independent*, 15 Apr. 1992.

[60] Shargorodsky, S., AP Moscow, 'Russian missiles aimed at US, marshal says', *Boston Globe*, 26 Sep. 1992.

[61] Hiatt, F., 'Russia extends test ban', *Washington Post*, 14 Oct. 1992, pp. 1 and 25. See also Burbyga, N., 'Inspection in Novaya Zemlya', *Izvestia*, 25 Sep. 1992, p. 2, in FBIS-SOV-92-190, 30 Sep. 1992, p. 3.

month after the Socialists had suffered a serious defeat in regional and local French elections and that two environmentalist parties—the Greens and the Generation Ecologie—had gained in the polls, some observers suggested that the moratorium was motivated primarily by domestic politics.[62] On the other hand, France took several other important initiatives in 1992, including accession to the NPT, suggesting that the moratorium reflected a real change in the French leadership's thinking.

In November 1992 then Foreign Minister Roland Dumas called for five-power talks on nuclear testing, proposing that China, France, Russia, the United Kingdom and the USA 'engage next in a common reflection on the question of nuclear tests'.[63] In January 1993 President François Mitterrand said that France would extend its moratorium for as long as the USA and Russia refrained from testing.[64] According to one account, a one- or two-year suspension of French tests would not delay the pace of French nuclear modernization programmes since prospective tests are planned to develop warheads for the M-5 SLBM missile, which is not scheduled to be deployed until the year 2005.[65]

## China

China has not yet demonstrated that it is seriously interested in negotiating a CTB in the near term. In response to the November 1992 French statement, however, a Chinese Foreign Ministry spokesman stated that the Chinese Government was willing to discuss nuclear test issues with all the members of the Conference on Disarmament (CD), 'within the existing framework of the conference'.[66] China conducted as many tests in 1992 (two) as it had in the three previous years combined. On 21 May 1992, China conducted the largest underground test in the history of its underground nuclear testing programme.[67]

In March 1992, when it acceded to the NPT, China appeared to lay out conditions for CTB participation, saying that states 'with the largest nuclear arsenals', such as the United States and Russia, should take the lead in 'halting . . . testing, production, deployment . . . and drastically reducing those

[62] 'Testing, testing', *The Economist*, 11 Apr. 1992, p. 30; Barrillot, B., 'French finesse nuclear future', *Bulletin of the Atomic Scientists*, Sep. 1992, p. 23; Butcher, M., Logan, C. and Plesch, D., *French Nuclear Policy Since the Fall of the Wall*, BASIC Report 93-1 (British American Security Information Council (BASIC): London, Feb. 1993), p. 28.

[63] 'France proposes five-power test ban', *Washington Times*, 4 Nov. 1992, p. 2; 'Paris seeks five-power N-weapons test talks', *Financial Times*, 4 Nov. 1992.

[64] Drozdiak, W., 'Historic pact bans chemical weapons', *Washington Post*, 14 Jan. 1993, p. A24; Reuters, Paris, 'France to maintain nuclear test ban—Mitterrand', 13 Jan. 1993.

[65] BASIC Report (note 62), p. 27.

[66] Xinhua News Agency, 12 Nov. 1992, in Foreign Broadcast Information Service, *Daily Report–China* (FBIS-CH), 12 Nov. 1992, in IDDS, *Arms Control Reporter*, sheet 608.B.246, Nov. 1992.

[67] Crossette, B., 'Chinese set off their biggest explosion', *New York Times*, 22 May 1992, p. A1; Vidale, J. E. and Benz, H. M., 'Seismological mapping of fine structure near the base of the earth's mantle', *Nature*, 11 Feb. 1993, p. 529. The Chinese explosion of 21 May 1992 consisted of one test of 660 kt, according to Vidale and Benz. See also IDDS, *Arms Control Reporter*, sheet 608.B.236, July 1992.

weapons'. After 'tangible progress' by those states, Beijing would be prepared to participate in a nuclear disarmament conference.[68]

### The United Kingdom

The UK, which has conducted its nuclear tests jointly with the USA in Nevada since 1962, did not support the new US law limiting nuclear tests. In fact, in August 1992, before the US Congress passed the law, the British ambassador to the United States, Sir Robin Renwick, wrote to the legislation's Senate sponsors that the total of three tests permitted the UK under the legislation would be 'insufficient to assure the safety of U.K. warheads for the indefinite future'. He also said that the UK 'cannot exclude the need to modernise' its warheads and that 'neither a moratorium nor the complete phasing out of testing, as currently contemplated' would allow Britain to maintain an 'effective deterrent'.[69] In November 1992, Viscount Cranborne, the British Under-Secretary for Defence, stated that the new US testing law was 'unfortunate and misguided', arguing that Britain would need to continue testing not for the 'safety of the Trident system' but for the safety of 'future systems'.[70]

Despite the criticism of the new US position on testing, Tristan Garel-Jones, the press spokesman for Prime Minister John Major, said in October that the UK had 'always accepted the long term goal of a comprehensive test ban to be achieved on a step-by-step basis' and that the UK was not considering any alternative underground nuclear test site to the US site in Nevada.[71]

## VI. The Nuclear Suppliers Group

The Nuclear Suppliers Group (NSG), also known as the 'London Club'—which includes most of the world's major suppliers of nuclear equipment—announced in April 1992 that it had established guidelines putting new limits on exports of nuclear-related 'dual-use' items. Under this agreement, dual-use items were defined as those which have 'legitimate non-nuclear uses, but if diverted, could make a major contribution to nuclear explosive and unsafeguarded nuclear fuel activities'. The NSG members also established a list of dual-use items to be subject to controls. In addition, they agreed that they will not supply other countries with significant nuclear equipment and materials,

[68] 'First Supplementary List of Ratifications, Accessions, Withdrawals, Etc. for 1992', presented to the British Parliament by the Secretary of State for Foreign and Commonwealth Affairs by Command of Her Majesty, Oct. 1992 (Her Majesty's Stationery Office: London, Oct. 1992), p. 5.

[69] 'Text of letter sent from British Embassy Washington to Senators Hatfield, Mitchell, and Exon in August 1992', British American Security Information Council (BASIC) Report no. 28 (18 Feb. 1993), p. 5; *Arms Control Today,* Mar. 1993, p. 29.

[70] British American Security Information Council (BASIC), *A Comprehensive Test Ban Treaty: Britain's Public Position, 1962–1992*, Jan. 1993, p. 4; see also 22 Feb. 1993 letter from Congressman Mike Kopetski to Prime Minister John Major.

[71] Written Answers, House of Commons, Parliamentary Debates, *Official Report,* 23 Oct. 1992, cols 407 and 408.

such as reactors, unless the recipient countries accept full-scope safeguards (see also chapter 6, section VIII).[72]

## VII. The Safety, Security and Dismantlement Talks

The failed Soviet coup in August 1991 underlined the potential nuclear weapon-related dangers attending the breakup of the Soviet Union and the need to accelerate the arms control process. As a result, US and Russian arms control efforts began to focus increasingly on rapid implementation measures to consolidate former Soviet nuclear weapons in Russia, to strengthen central control over those weapons, and to improve their physical security and safety.[73]

In his September 1991 initiative on tactical nuclear weapons, President Bush proposed that the USA and the USSR explore 'joint technical cooperation on the safe and environmentally responsible storage, transportation, dismantling, and destruction of nuclear warheads'.[74] He also called for the two states to discuss ways in which 'existing arrangements for the physical security and safety of nuclear weapons' could be enhanced. On 5 October 1991, then Soviet President Mikhail Gorbachev acknowledged that Moscow was amenable to such discussions. In November the US Congress passed the Soviet Nuclear Threat Reduction Act of 1991. This legislation, sponsored by Senators Sam Nunn and Richard Lugar, authorized $400 million in re-programmed funds from the US Defense Department budget to facilitate 'the transportation, storage, safeguarding, and destruction of nuclear and other weapons in the Soviet Union'. (In 1992 Congress passed the Former Soviet Union Demilitarization Act, adding another $400 million to the Nunn–Lugar funding and broadening the scope of programmes for which the money may be used to include defence industry conversion and military-to-military contacts.) A new nuclear arms control forum, the Safety, Security and Dismantlement (SSD) talks, grew out of these initiatives. The forum helps to institutionalize continuous co-operation between US and former Soviet authorities on nuclear weapon issues.

In 1992 the USA held SSD talks on a bilateral basis with Russia, Belarus, Kazakhstan and Ukraine. By the end of February 1993, the USA had earmarked $303.45 million of the $800 million set aside for helping the former Soviet Union to dismantle its weapons of mass destruction. (This amount did not include the $175 million pledged to Ukraine if it ratifies the START Treaty and accedes to the NPT as a non-nuclear weapon state.) Of the $303.54 million, only $133.3 million had been committed through formal agreements.

[72] Wolfsthal, J. B., 'Nuclear Suppliers Group agrees on "dual-use" export controls', *Arms Control Today,* Apr. 1992, p. 19; see also UN General Assembly document A/47/467, 24 Sep. 1992; Conference on Disarmament document CD/1175, 10 Sep. 1992.

[73] For a comprehensive discussion of policy measures for attenuating the potential dangers, see Allison, G., Carter, A. B., Miller, S. E. and Zelikow, P. (eds), *Cooperative Denuclearization: From Pledges to Deeds,* CSIA Studies in International Security No. 2 (Center for Science and International Affairs (CSIA), John F. Kennedy School of Government, Harvard University: Cambridge, Mass., 1993).

[74] *SIPRI Yearbook 1992* (note 1), appendix 2A, pp. 86.

Furthermore, as of the end of 1992, once administrative costs had been subtracted, less than $20 million had actually been obligated—all of which is earmarked for Russia.

## Russia

The USA and Russia signed an SSD 'umbrella' agreement on 17 June 1992, establishing the legal framework for the USA to transfer the Nunn–Lugar legislation funds from the US Defense Department to Russia. By early 1993, the USA had earmarked $242.2 million for Russia, primarily to: (*a*) facilitate safe and secure transportation of former Soviet nuclear weapons and materials, (*b*) assist Russia in the design and construction of a fissile material storage facility, and (*c*) establish an export control system and a fissile material control and accounting system.

The USA and Russia agreed that the USA would produce 10 000 storage containers for the transport and storage of fissile material from dismantled Russian warheads. Production of these containers is scheduled to begin in early 1994 and delivery is to be completed by 31 December 1995.[75] The Defense Department has estimated that each container would cost in the range of $3000–5000, bringing the total cost of 10 000 containers to up to $50 million. Russia has stated that it will eventually need an additional 35 000 containers, but the USA has said that the decision regarding the production of additional containers will depend on decisions made about the ultimate disposition of fissile material from dismantled warheads.[76]

In July 1992 the USA completed the transfer of 250 special nylon blankets to Russia. The USA also made a commitment to produce an additional 250 sets of 10 Kevlar blankets each, with deliveries to Russia starting in the spring of 1993.[77] The blankets, designed to protect nuclear containers from small arms fire, will cost about $5 million.

The USA has pledged to transfer to Russia over 1000 pieces of 'emergency response' protective clothing and equipment, worth an estimated $15 million, in FY 1993. The USA intends to provide training to Russian experts in the use of such equipment. The package includes communications equipment, high-energy radiography equipment to examine the inside of a damaged warhead after an accident and liquid foam to ensure that warhead components do not move after a weapon has been damaged. The USA began delivering the first set of equipment in January 1993.[78]

The USA also agreed to modify Russian railway cars to provide safer and more secure transport of nuclear warheads and their components. Washington plans to deliver up to 100 cargo railcar conversion kits and 15 guard railcar conversion kits by April 1994, at a cost of up to $20 million.

[75] General William Burns, head of the US SSD delegation, testimony before the Senate Foreign Relations Committee, 9 Feb. 1993, p. 5.

[76] General Burns, testimony before the Senate Foreign Relations Committee, 27 July 1992, p. 4.

[77] Burns (note 75), p. 4.

[78] Burns (note 75), p. 4.

In October 1992 the USA agreed to provide Russia with up to $15 million to help scientists there design a facility for long-term storage of fissile material from dismantled Russian warheads. The facility, which the USA may ultimately help build, will probably be located in Tomsk. If an agreement is reached to build as well as design the facility together, the USA has indicated that it would contribute an additional $75 million for construction costs for a total of $90 million. (Russia had estimated that the facility would cost $150 million but the USA has questioned this figure.) According to the General Accounting Office (GAO), Russia has agreed in principle to permit the USA to monitor the storage facility under the terms of the SSD umbrella agreement.[79] The conclusion of an agreement to help build the storage facility appears likely since Russia has argued that, without a new storage facility for fissile materials from dismantled warheads, there will be a 'bottleneck' that will slow down the process dramatically. Specifically, Russian officials have told the USA, in the words of the GAO, that 'they are overloading their existing storage space and that they need a new storage facility by 1997 to meet their dismantlement schedule'.[80]

There are four Russian warhead dismantlement facilities: a large plant at Nizhnyaya Tura in the Urals, 200 km north of Yekaterinburg (formerly Sverdlovsk); a much smaller facility at Yuryuzan, 85 km south-west of Zlatoust; a small component fabrication and assembly plant at Penza, 350 km south-east of Nizhni Novgorod (formerly Gorkiy); and a small facility at Arzamas, one of Russia's nuclear warhead design laboratories.[81]

In January 1992 then director of US Central Intelligence Robert Gates projected that it would take ten years for Russia to dismantle the 15 000 warheads Russian officials 'claim they are going to take down'. He added that 'based on the variety of problems that they are having internally right now . . . [dismantling] 1,500 warheads a year is probably an optimistic assessment'.[82] In February 1993, however, Lawrence Gershwin, the CIA's National Intelligence Officer for Strategic Programs, said that although the US intelligence community's 'best judgement' is that Russia is dismantling 'something less than 2,000 [warheads] per year, they themselves have said, and we think that

[79] Kelley, J. K., Director-in-Charge, International Affairs Issues, National Security and International Affairs Division, *Soviet Nuclear Weapons: US Efforts to Help Former Soviet Republics Secure and Destroy Weapons*, GAO/T-NSIAD-93-5, 9 Mar. 1993, p. 8.

[80] See note 79, p. 7.

[81] Lt. General James R. Clapper, Jr, Director, Defense Intelligence Agency, Statement before the Senate Armed Services Committee, 22 Jan. 1992, in *Threat Assessment, Military Strategy, and Defense Planning*, Senate Hearing 102-755 (US Government Printing Office: Washington, DC, 1992), pp. 55–56; 'Nuclear notebook' *Bulletin of the Atomic Scientists*, Jan./Feb. 1993, p. 56; Cochran, T. B. and Norris, R. S., 'Russian/Soviet nuclear warhead production', Natural Resources Defense Council (NRDC) Working Paper 93-1 (forthcoming).

[82] Robert Gates, Director, US Central Intelligence Agency, Statement before the Senate Governmental Affairs Committee, 15 Jan. 1992, Senate Hearing 102-720 (US Government Printing Office: Washington, DC, 1992), p. 17. Eleven months after his statement to the Committee, Gates said: 'even under the best circumstances, it will take [Russia] more than 10 years' to dismantle the warheads that it is committed to dismantle; see Proposed Remarks before the Comstock Club, Sacramento, Calif., 15 Dec. 1992, p. 6.

is entirely possible, that they could dismantle 4,000 or 5,000 per year with the available complexes that they have'.[83]

The USA and Russia discussed but had as of February 1993 not yet reached agreement on US assistance for establishing a control and accountability system for fissile material. While a specific programme had not yet been worked out, the Pentagon did earmark $10 million for this purpose.

The USA allocated $25 million to help develop an international science and technology centre in Moscow, intended to alleviate a potential 'brain drain' by putting former Soviet weapon scientists and engineers to work on peaceful projects.[84] The USA has also earmarked $25 million to assist Russia in dismantling its chemical weapon stockpile. Finally, $2.26 million has been set aside to help Russia develop export control laws.

In the future, the USA plans to give top priority to assistance in the dismantlement of ICBMs and other strategic nuclear delivery vehicles (SNDVs). Washington has an incentive to conclude such an agreement since in the START II Treaty the USA and Russia agreed that the Treaty could be implemented by the end of the year 2000 rather than by 2003 if the USA could help finance the dismantlement of Russia's SNDVs.

### Belarus

The USA and Belarus signed an umbrella agreement on 2 October 1992, establishing the legal basis for the transfer of Nunn–Lugar legislation money.[85] The USA has committed up to $5 million to Belarus for 'emergency response' equipment, such as dosimeters, and clothing. It has earmarked another $2.26 million to assist Belarus in developing export control laws plus another $2.3 million to establish a continuous communications link between Minsk and Washington.[86] None of this money was obligated as of early 1993.[87]

### Ukraine

As of February 1993, Ukraine was not legally eligible to receive Nunn–Lugar legislation money since the USA and Ukraine had not signed an SSD umbrella agreement. The Bush Administration, however, promised Ukraine a total of $175 million, mainly for assistance in dismantling the 176 ICBM missiles and silos located in Ukraine, if the Rada (Parliament) ratifies the START Treaty

[83] Lawrence Gershwin, National Intelligence Officer for Strategic Programs, Central Intelligence Agency, Statement before the Senate Governmental Affairs Committee, 24 Feb. 1993, Federal News Service Transcript, p. 51; see also Lockwood, D., 'CIA sheds new light on nuclear control in CIS', *Arms Control Today,* Mar. 1993, p. 25.

[84] For details on the centre, see interview with Ambassador Robert Gallucci, 'Redirecting the Soviet weapons establishment', *Arms Control Today*, June 1992, pp. 3–6.

[85] ACDA (note 47), p. 36.

[86] Burns (note 75), Annex A; Kelley (note 79), attachment I.

[87] Kelley (note 79), attachment I.

and votes to accede to the NPT. In this context, the US and Ukrainian delegations have focused their discussions on developing methods for dismantling the 130 liquid-fuelled SS-19s based on Ukrainian soil in ways that are not harmful to the environment.

In addition to assistance for ICBM dismantlement, the US and Ukrainian SSD delegations have tentatively agreed to the following US assistance: $5 million for emergency response equipment, $2.4 million for a continuous communications link betwen Kiev and Washington, $7.5 million for material controls, and $10 million for a science centre in Kiev.

### Kazakhstan

The USA and Kazakhstan had not signed an umbrella agreement by early 1993. However, the two countries discussed various forms of assistance, including emergency response equipment, export control, a government-to-government communications link, and material control and accounting. In addition, they discussed US aid for dismantling the 104 SS-18 missiles in Kazakhstan.[88]

### The 1993 US–Russian HEU agreement

As part of the SSD Talks, on 18 February 1993 the USA and Russia signed the Agreement Concerning the Disposition of Highly Enriched Uranium Extracted from Nuclear Weapons, committing the USA to purchase, over the next 20 years, 500 metric tons of highly enriched uranium (HEU) extracted from nuclear warheads of the former Soviet arsenal.[89] The process will not actually begin, however, until a detailed US–Russian contract is negotiated and until arrangements are agreed between Russia and Belarus, Kazakhstan and Ukraine on the division of the proceeds.

The agreement is intended to address a gap in existing nuclear arms control agreements which do not require the destruction of nuclear warheads. The HEU will be recovered from warheads Russia is committed to retire under the October 1991 Soviet unilateral initiative on tactical nuclear weapons and planned reductions under the START Treaty, the START II Treaty and the 1987 Intermediate-range Nuclear Forces (INF) Treaty. Then Soviet President Gorbachev announced in 1989 that the USSR had stopped producing HEU for weapons (see section VIII below).

The HEU will be 'blended down' to low enriched uranium (LEU) in Russia, rendering it unuseable for nuclear weapons but making it suitable for use in commercial nuclear reactors to produce electricity. LEU is defined in the agreement as uranium which contains less than 20 per cent of the fissile isotope uranium-235.

[88] Burns (note 75), p. 10.

[89] For information about the HEU agreement, see Burns Statement before the Senate Armed Services Committee, 24 Feb. 1993; Burns (note 75); see also Lockwood, D., 'US, Russia reach agreement on sale of nuclear weapons material', *Arms Control Today,* Mar. 1993, p. 22.

Under the agreement, the USA has agreed to purchase no less than 10 tons of HEU per year in the first five years and at least 30 tons each year thereafter until it has bought a total of 500 tons. The US Department of Energy (DOE) will act as the 'executive agent' to implement the agreement; for Russia, it will be the Ministry of Atomic Energy (Minatom).

The price of the transaction has not yet been established, but the deal is estimated to be worth $7–10 billion over as many as 20 years. Although the agreement is intended to provide hard currency to the four former Soviet republics with nuclear weapons on their territory (as well as to give them an incentive to dismantle their strategic nuclear warheads, which is not required under the START treaties), it is also meant to be 'revenue neutral' for the USA—namely, the 'DOE would finance the purchase with receipts of sales to utilities and with the savings derived because the availability of the Russian LEU reduces the need to enrich U.S. uranium in DOE facilities'.[90]

## VIII. Fissile material production cut-off

As part of a 'non-proliferation initiative', President Bush announced on 13 July 1992 that the USA would not 'produce plutonium or HEU for nuclear explosive purposes'.[91] However, this initiative simply had the effect of changing a *de facto* situation into a *de jure* one, turning current US practice into officially stated policy. The USA has not produced any plutonium since 1988—because of environmental and safety problems at the Savannah River production site—or any HEU for nuclear weapons since 1964.

In January 1992 President Yeltsin reiterated a proposal made earlier by then Soviet President Gorbachev to negotiate an agreement with the USA to cease the production of fissile materials for weapons. He added that, regardless of such an agreement, Russia would stop producing weapon-grade plutonium by the year 2000. (Only three of the original 13 plutonium production reactors are still operating in Russia, and these are apparently used for civilian electric power generation.)

## IX. The Treaty of Tlatelolco

Argentina and Brazil, which for many years appeared to be heading for a nuclear arms competition, agreed on 13 December 1991, as part of a four-party arrangement, to accept comprehensive IAEA safeguards on their nuclear activities. The agreement established the joint Brazilian–Argentine Agency for Accounting and Control of Nuclear Material (ABACC).

Eighteen members of the 1967 Treaty of Tlatelolco (the Latin American Nuclear-Weapon Free Zone Treaty; for parties, see annexe A) met at the special General Conference of the Agency for the Prohibition of Nuclear

[90] Burns (note 75), p. 5.

[91] *US Department of State Dispatch,* vol. 3, no. 29 (20 July 1992), p. 569; Wolfsthal, J. B., 'White House formalizes end to fissionable materials production', *Arms Control Today*, July/Aug. 1992, p. 25.

Weapons in Latin America (OPANAL) in Mexico City on 26 August 1992 and adopted amendments to the Treaty proposed by Argentina, Brazil and Chile that will pave the way for them to become parties. The Treaty of Tlatelolco bans the possession, acquisition, testing and storage of nuclear weapons in Latin America and the Caribbean. The amendments, which will affect the Tlatelolco Treaty's inspection procedures and reporting requirements, were designed to protect nuclear industrial secrets of member states.[92] All Treaty member states must approve the changes before they can enter into force.

The Cuban Government has said that it would adhere to the Tlatelolco Treaty once Argentina, Brazil and Chile do.[93] In addition, France deposited its instrument of ratification to the Treaty's Protocol I on 24 August 1992, committing it to apply the regulations of the Treaty to its territorial possessions within the Treaty's zone of application—which includes French Guyana, Martinique and Guadeloupe.[94]

## X. Conclusion

The year 1992 marked a change from the pace and style of past nuclear arms control efforts. The 1991 START Treaty was negotiated over a nine-year period in which technical specialists haggled seemingly *ad infinitum* over details in Geneva until high-level political officials intervened to resolve the final issues. The 1993 START II Treaty, by contrast, was negotiated over a period of less than one year, in which high-level officials worked out agreements in principle and then let technical specialists work out the final details.

In 1992 some traditional approaches to arms control were de-emphasized. With fears of cheating on the wane but concerns about political instability in the former Soviet Union waxing, the USA and Russia continued a trend begun at the end of 1991 and placed less emphasis on agreements with extensive verification provisions and more emphasis on parallel unilateral initiatives that could be implemented quickly. Moreover, with the START II Treaty, Washington and Moscow are no longer focusing on the question of 'how low can we go?' but on issues related to the physical security of and central control over nuclear weapons.

Some new arms control paradoxes emerged in 1992. Plans to dismantle nuclear delivery vehicles and warheads to comply with agreements and unilateral initiatives seem to pose a much more formidable challenge, economically and environmentally, than most observers ever imagined they would. Russia argues that it does not have sufficient funds to dismantle its ICBM silos under the START II Treaty without US aid, and thus far no one has identified a safe

[92] UN General Assembly document A/47/467, 24 Sep. 1992.

[93] IDDS, *Arms Control Reporter,* sheet 452.B.137, June 1992; ACDA (note 47), pp. 113–14.

[94] Wolfsthal, J. B., 'Argentina, Brazil, and Chile to implement Tlatelolco Treaty', *Arms Control Today,* Sep. 1992, p. 27; ACDA (note 47), p. 114.

and cost-effective method for disposing of plutonium extracted from dismantled warheads.[95]

The year was a watershed for nuclear arms control, with the signing of the Lisbon Protocol, the completion of the START II Treaty for signature on 3 January 1993, simultaneous nuclear test moratoria in the USA, Russia and France in conjunction with movement towards a CTB, and the accession of 10 states, notably including China and France, to the NPT.

Despite these accomplishments, the arms control community should not rest on its laurels in the year ahead. Neither the 1991 START Treaty nor the START II Treaty has entered into force, and the implementation of both is now in serious question: the Ukrainian Rada has indicated that it might not ratify the START Treaty, a decision that would preclude both the START and START II Treaty from entering into force, and a number of members of the Russian Supreme Soviet have expressed strong opposition to the START II Treaty, arguing that its terms are too favourable to the USA. Formal CTB negotiations, which last took place in 1980, have not yet resumed. There are still no existing agreements that establish a legally binding obligation to dismantle nuclear warheads, nor is there a formal, legally binding agreement to ban the production of fissile material for weapons. In addition, the proposal put forward in February 1992 by Russian Foreign Minister Kozyrev to reduce the alert levels of strategic forces has not yet been given the serious consideration that it deserves. The political instability in the new states of the former Soviet Union continues to raise the spectre of nuclear weapons, materials or expertise being sold to the highest bidder on the black market. Developing countries such as Iran and North Korea continue to pursue the capability to build a nuclear weapon. In sum, while 1992 was a banner year in nuclear arms control, it is too soon to become complacent—the nuclear arms control agenda is still full.

[95] See, for example, Kelley (note 79), p. 3. The Russian Ministry of Atomic Energy (MINATOM) wants to use plutonium in commercial breeder and conventional reactors to produce energy.

# **Appendix 11A.** Documents on nuclear arms control

## STATE OF THE UNION ADDRESS BY PRESIDENT GEORGE BUSH

*Delivered on 28 January 1992*

*Excerpt*

. . .

Two years ago I began planning cuts in military spending that reflected the changes of the new era. But now, this year, with imperial communism gone, that process can be accelerated.

Tonight I can tell you of dramatic changes in our strategic nuclear force. These actions we are taking on our own—because they are the right thing to do.

After completing 20 planes for which we have begun procurement, we will shut down further production of the B-2 bomber. We will cancel the small ICBM program. We will cease production of new warheads for our sea-based ballistic missiles. We will stop all new production of the Peacekeeper missile. And we will not purchase any more advanced cruise missiles.

This weekend I will meet at Camp David with Boris Yeltsin of the Russian Federation. I have informed President Yeltsin that if the Commonwealth—the former Soviet Union—will eliminate all land-based multiple warhead ballistic missiles, I will do the following:

We will eliminate all Peacekeeper missiles. We will reduce the number of warheads on Minuteman missiles to one, and reduce the number of warheads on our sea-based missiles by about one-third. And we will convert a substantial portion of our strategic bombers to primarily conventional use.

President Yeltsin's early response has been very positive, and I expect our talks at Camp David to be fruitful.

. . .

The reductions I have approved will save us an additional $50,000 million over the next five years. By 1997 we will have cut defense by 30 percent since I took office. These cuts are deep, and you must know my resolve: This deep, and no deeper.

. . .

*Source:* US Information Service, USIS Documentation Center, American Embassy, Stockholm, Sweden.

## PROTOCOL TO FACILITATE THE IMPLEMENTATION OF THE START TREATY (LISBON PROTOCOL)

*Signed on 23 May 1992*

*Excerpt*

The Republic of Byelarus, the Republic of Kazakhstan, the Russian Federation, Ukraine, and the United States of America, hereinafter referred to as the Parties,

Reaffirming their support for the Treaty Between the United States of America and the Union of Soviet Socialist Republics on the Reduction and Limitation of Strategic Offensive Arms of July 31, 1991, hereinafter referred to as the Treaty,

Recognizing the altered political situation resulting from the replacement of the former Union of Soviet Socialist Republics with a number of independent states,

Recalling the commitment of the member states of the Commonwealth of Independent States that the nuclear weapons of the former Union of Soviet Socialist Republics will be maintained under the safe, secure, and reliable control of a single unified authority,

Desiring to facilitate implementation of the Treaty in this altered situation,

Have agreed as follows:

### Article I

The Republic of Byelarus, the Republic of Kazakhstan, the Russian Federation, and Ukraine, as successor states of the former Union of Soviet Socialist Republics in connection with the Treaty, shall assume the obligations of the former Union of Soviet Socialist Republics under the Treaty.

**Article II**

The Republic of Byelarus, the Republic of Kazakhstan, the Russian Federation, and Ukraine shall make such arrangements among themselves as are required to implement the Treaty's limits and restrictions; to allow functioning of the verification provisions of the Treaty equally and consistently throughout the territory of the Republic of Byelarus, the Republic of Kazakhstan, the Russian Federation, and Ukraine; and to allocate costs.

**Article III**

1. For purposes of Treaty implementation, the phrase "Union of Soviet Socialist Republics" shall be interpreted to mean the Republic of Byelarus, the Republic of Kazakhstan, the Russian Federation, and Ukraine.

2. For purposes of Treaty implementation, the phrase "national territory", when used in the Treaty to refer to the Union of Soviet Socialist Republics, shall be interpreted to mean the combined national territories of the Republic of Byelarus, the Republic of Kazakhstan, the Russian Federation, and Ukraine.

3. For inspections and continuous monitoring activities on the territory of the Republic of Byelarus, the Republic of Kazakhstan, the Russian Federation or Ukraine, that state shall provide communications from the inspection site or continuous monitoring site to the Embassy of the United States in the respective capital.

4. For purposes of Treaty implementation, the embassy of the Inspecting Party referred to in Section XVI of the Protocol on Inspections and Continuous Monitoring Activities Relating to the Treaty Between the United States of America and the Union of Soviet Socialist Republics on the Reduction and Limitation of Strategic Offensive Arms shall be construed to be the embassy of the respective state in Washington or the embassy of the United States of America in the respective capital.

5. The working languages for Treaty activities shall be English and Russian.

**Article IV**

Representatives of the Republic of Byelarus, the Republic of Kazakhstan, the Russian Federation, and Ukraine will participate in the Joint Compliance and Inspection Commission on a basis to be worked out consistent with Article I of this Protocol.

**Article V**

The Republic of Byelarus, the Republic of Kazakhstan, and Ukraine shall adhere to the Treaty on the Non-Proliferation of Nuclear Weapons of July 1, 1968 as non-nuclear weapons states in the shortest possible time, and shall begin immediately to take all necessary actions to this end in accordance with their constitutional practices.

**Article VI**

1. Each Party shall ratify the Treaty together with this Protocol in accordance with its own constitutional procedures. The Republic of Byelarus, the Republic of Kazakhstan, the Russian Federation, and Ukraine shall exchange instruments of ratification with the United State of America. The Treaty shall enter into force on the date of the final exchange of instruments of ratification.

2. This Protocol shall be an integral part of the Treaty and shall remain in force throughout the duration of the Treaty.

. . .

---

*Source*: US Arms Control and Disarmament Agency (ACDA), ACDA document (mimeo).

## JOINT US–RUSSIAN UNDERSTANDING ON FURTHER REDUCTIONS IN STRATEGIC OFFENSIVE ARMS (DE-MIRVING AGREEMENT)

*Signed on 17 June 1992*

The President of the United States of America and the President of the Russian Federation have agreed to substantial further reductions in strategic offensive arms. Specifically, the two sides have agreed upon and will promptly conclude a Treaty with the following provisions:

Within the seven-year period following entry into force of the START Treaty, they will reduce their strategic forces to no more than:

– An overall total number of warheads for each between 3,800 and 4,250 (as each nation shall determine) or such lower number as each nation shall decide.

– 1,200 MIRVed ICBM warheads.

– 650 heavy ICBM warheads.

– 2,160 SLBM warheads.

By the year 2003 (or by the end of the year 2000 if the United States can contribute to the financing of the destruction or elimination of strategic offensive arms in Russia), they will:

– Reduce the overall total to no more than a number of warheads for each between 3,000 and 3,500 (as each nation shall determine) or such lower number as each nation shall decide.

– Eliminate all MIRVed ICBMs.

– Reduce SLBM warheads to no more than 1,750.

For the purpose of calculating the overall totals described above:

The number of warheads counted for heavy bombers with nuclear roles will be the number of nuclear weapons they are actually equipped to carry.

Under agreed procedures, heavy bombers not to exceed 100 that were never equipped for long-range nuclear ALCMs and that are reoriented to conventional roles will not count against the overall total established by this agreement.

– Such heavy bombers will be based separately from heavy bombers with nuclear roles.

– No nuclear weapons will be located at bases for heavy bombers with conventional roles.

– Such aircraft and crews will not train or exercise for nuclear missions.

– Current inspection procedures already agreed in the START Treaty will help affirm that these bombers have conventional roles. No new verification procedures are required.

– Except as otherwise agreed, these bombers will remain subject to the provisions of the START Treaty, including the inspection provisions.

The reductions required by this agreement will be carried out by eliminating missile launchers and heavy bombers using START procedures, and, in accordance with the plans of the two sides, by reducing the number of warheads on existing ballistic missiles other than the SS-18. Except as otherwise agreed, ballistic missile warheads will be calculated according to START counting rules.

The two Presidents directed that this agreement be promptly recorded in a brief Treaty document which they will sign and submit for ratification in their respective countries. Because this new agreement is separate from but builds upon the START Treaty, they continue to urge that the START Treaty be ratified and implemented as soon as possible.

---

*Source*: Conference on Disarmament document CD/1162, 12 August 1992

---

## TREATY BETWEEN THE UNITED STATES OF AMERICA AND THE RUSSIAN FEDERATION ON FURTHER REDUCTION AND LIMITATION OF STRATEGIC OFFENSIVE ARMS (START II TREATY)

*Signed on 3 January 1993*

The United States of America and the Russian Federation, hereinafter referred to as the Parties,

Reaffirming their obligations under the Treaty Between the United States of America and the Union of Soviet Socialist Republics on the Reduction and Limitation of Strategic Offensive Arms of July 31, 1991, hereinafter referred to as the START Treaty,

Stressing their firm commitment to the Treaty on the Non-Proliferation of Nuclear Weapons of July 1, 1968, and their desire to contribute to its strengthening,

Taking into account the commitment by the Republic of Belarus, the Republic of Kazakhstan, and Ukraine to accede to the Treaty on the Non-Proliferation of Nuclear Weapons of July 1, 1968, as non-nuclear-weapon States Parties,

Mindful of their undertakings with respect to strategic offensive arms under Article VI of the Treaty on the Non-Proliferation of Nuclear Weapons of July 1, 1968, and under the Treaty Between the United States of America and the Union of Soviet Socialist Republics on the Limitation of Anti-Ballistic Missile Systems of May 26, 1972, as well as the provisions of the Joint Understanding signed by the Presidents of the United States of America and the Russian Federation on June 17, 1992, and of the Joint Statement on a Global Protection System signed by the Presidents of the United States of America and the Russian Federation on June 17, 1992,

Desiring to enhance strategic stability and predictability, and, in doing so, to reduce further strategic offensive arms, in addition to the reductions and limitations provided for in the START Treaty,

Considering that further progress toward that end will help lay a solid foundation for a world order built on democratic values that would preclude the risk of outbreak of war,

Recognizing their special responsibility as permanent members of the United Nations Security Council for maintaining international peace and security,

Taking note of United Nations General Assembly Resolution 47/52K of December 9, 1992,

Conscious of the new realities that have transformed the political and strategic relations between the Parties, and the relations of partnership that have been established between them,

Have agreed as follows:

**Article I**

1. Each Party shall reduce and limit its intercontinental ballistic missiles (ICBMs) and ICBM launchers, submarine-launched ballistic missiles (SLBMs) and SLBM launchers, heavy bombers, ICBM warheads, SLBM warheads, and heavy bomber armaments, so that seven years after entry into force of the START Treaty and thereafter, the aggregate number for each Party, as counted in accordance with Articles III and IV of this Treaty, does not exceed, for warheads attributed to deployed ICBMs, deployed SLBMs, and deployed heavy bombers, a number between 3800 and 4250 or such lower number as each Party shall decide for itself, but in no case shall such number exceed 4250.

2. Within the limitations provided for in paragraph 1 of this Article, the aggregate numbers for each Party shall not exceed:

(a) 2160, for warheads attributed to deployed SLBMs;

(b) 1200, for warheads attributed to deployed ICBMs of types to which more than one warhead is attributed; and

(c) 650, for warheads attributed to deployed heavy ICBMs.

3. Upon fulfillment of the obligations provided for in paragraph 1 of this Article, each Party shall further reduce and limit its ICBMs and ICBM launchers, SLBMs and SLBM launchers, heavy bombers, ICBM warheads, SLBM warheads, and heavy bomber armaments, so that no later than January 1, 2003, and thereafter, the aggregate number for each Party, as counted in accordance with Articles III and IV of this Treaty, does not exceed, for warheads attributed to deployed ICBMs, deployed SLBMs, and deployed heavy bombers, a number between 3000 and 3500 or such lower number as each Party shall decide for itself, but in no case shall such number exceed 3500.

4. Within the limitations provided for in paragraph 3 of this Article, the aggregate numbers for each Party shall not exceed:

(a) a number between 1700 and 1750, for warheads attributed to deployed SLBMs or such lower number as each Party shall decide for itself, but in no case shall such number exceed 1750;

(b) zero, for warheads attributed to deployed ICBMs of types to which more than one warhead is attributed; and

(c) zero, for warheads attributed to deployed heavy ICBMs.

5. The process of reductions provided for in paragraphs 1 and 2 of this Article shall begin upon entry into force of this Treaty, shall be sustained throughout the reductions period provided for in paragraph 1 of this Article, and shall be completed no later than seven years after entry into force of the START Treaty. Upon completion of these reductions, the Parties shall begin further reductions provided for in paragraphs 3 and 4 of this Article, which shall also be sustained throughout the reductions period defined in accordance with paragraphs 3 and 6 of this Article.

6. Provided that the Parties conclude, within one year after entry into force of this Treaty, an agreement on a program of assistance to promote the fulfillment of the provisions of this Article, the obligations provided for in paragraphs 3 and 4 of this Article and in Article II of this Treaty shall be fulfilled by each Party no later than December 31, 2000.

**Article II**

1. No later than January 1, 2003, each Party undertakes to have eliminated or to have converted to launchers of ICBMs to which one warhead is attributed all its deployed and non-deployed launchers of ICBMs to which more than one warhead is attributed under Article III of this Treaty (including test launchers and training launchers), with the exception of those launchers of ICBMs other than heavy ICBMs at space launch facilities allowed under the START Treaty, and not to have thereafter launchers of ICBMs to which more than one warhead is attributed. ICBM launchers that have been

converted to launch an ICBM of a different type shall not be capable of launching an ICBM of the former type. Each Party shall carry out such elimination or conversion using the procedures provided for in the START Treaty, except as otherwise provided for in paragraph 3 of this Article.

2. The obligations provided for in paragraph 1 of this Article shall not apply to silo launchers of ICBMs on which the number of warheads has been reduced to one pursuant to paragraph 2 of Article III of this Treaty.

3. Elimination of silo launchers of heavy ICBMs, including test launchers and training launchers, shall be implemented by means of either:

(a) elimination in accordance with the procedures provided for in Section II of the Protocol on Procedures Governing the Conversion or Elimination of the Items Subject to the START Treaty; or

(b) conversion to silo launchers of ICBMs other than heavy ICBMs in accordance with the procedures provided for in the Protocol on Procedures Governing Elimination of Heavy ICBMs and on Procedures Governing Conversion of Silo Launchers of Heavy ICBMs Relating to the Treaty Between the United States of America and the Russian Federation on Further Reduction and Limitation of Strategic Offensive Arms, hereinafter referred to as the Elimination and Conversion Protocol. No more than 90 silo launchers of heavy ICBMs may be so converted.

4. Each Party undertakes not to emplace an ICBM, the launch canister of which has a diameter greater than 2.5 meters, in any silo launcher of heavy ICBMs converted in accordance with subparagraph 3(b) of this Article.

5. Elimination of launchers of heavy ICBMs at space launch facilities shall only be carried out in accordance with subparagraph 3(a) of this Article.

6. No later than January 1, 2003, each Party undertakes to have eliminated all of its deployed and non-deployed heavy ICBMs and their launch canisters in accordance with the procedures provided for in the Elimination and Conversion Protocol or by using such missiles for delivering objects into the upper atmosphere or space, and not to have such missiles or launch canisters thereafter.

7. Each Party shall have the right to conduct inspections in connection with the elimination of heavy ICBMs and their launch canisters, as well as inspections in connection with the conversion of silo launchers of heavy ICBMs. Except as otherwise provided for in the Elimination and Conversion Protocol, such inspections shall be conducted subject to the applicable provisions of the START Treaty.

8. Each Party undertakes not to transfer heavy ICBMs to any recipient whatsoever, including any other Party to the START Treaty.

9. Beginning on January 1, 2003, and thereafter, each Party undertakes not to produce, acquire, flight-test (except for flight tests from space launch facilities conducted in accordance with the provisions of the START Treaty), or deploy ICBMs to which more than one warhead is attributed under Article III of this Treaty.

**Article III**

1. For the purposes of attributing warheads to deployed ICBMs and deployed SLBMs under this Treaty, the Parties shall use the provisions provided for in Article III of the START Treaty, except as otherwise provided for in paragraph 2 of this Article.

2. Each Party shall have the right to reduce the number of warheads attributed to deployed ICBMs or deployed SLBMs only of existing types, except for heavy ICBMs. Reduction in the number of warheads attributed to deployed ICBMs and deployed SLBMs of existing types that are not heavy ICBMs shall be carried out in accordance with the provisions of paragraph 5 of Article III of the START Treaty, except that:

(a) the aggregate number by which warheads are reduced may exceed the 1250 limit provided for in paragraph 5 of Article III of the START Treaty;

(b) the number by which warheads are reduced on ICBMs and SLBMs, other than the Minuteman III ICBM for the United States of America and the SS-N-18 SLBM for the Russian Federation, may at any one time exceed the limit of 500 warheads for each Party provided for in subparagraph 5(c)(i) of Article III of the START Treaty;

(c) each Party shall have the right to reduce by more than four warheads, but not by more than five warheads, the number of warheads attributed to each ICBM out of no more than 105 ICBMs of one existing type of ICBM. An ICBM to which the number of warheads attributed has been reduced in accordance with this paragraph shall only be deployed in an ICBM launcher in which an ICBM of that type was deployed as of the date of signature of the START Treaty; and

(d) the reentry vehicle platform for an ICBM or SLBM to which a reduced number of warheads is attributed is not required to be destroyed and replaced with a new reentry vehicle platform.

3. Notwithstanding the number of warheads attributed to a type of ICBM or SLBM in accordance with the START Treaty, each Party undertakes not to:

(a) produce, flight-test, or deploy an ICBM or SLBM with a number of reentry vehicles greater than the number of warheads attributed to it under this Treaty; and

(b) increase the number of warheads attributed to an ICBM or SLBM that has had the number of warheads attributed to it reduced in accordance with the provisions of this Article.

**Article IV**

1. For the purposes of this Treaty, the number of warheads attributed to each deployed heavy bomber shall be equal to the number of nuclear weapons for which any heavy bomber of the same type or variant of a type is actually equipped, with the exception of heavy bombers reoriented to a conventional role as provided for in paragraph 7 of this Article. Each nuclear weapon for which a heavy bomber is actually equipped shall count as one warhead toward the limitations provided for in Article I of this Treaty. For the purpose of such counting, nuclear weapons include long-range nuclear air-launched cruise missiles (ALCMs), nuclear air-to-surface missiles with a range of less than 600 kilometers, and nuclear bombs.

2. For the purposes of this Treaty, the number of nuclear weapons for which a heavy bomber is actually equipped shall be the number specified for heavy bombers of that type and variant of a type in the Memorandum of Understanding on Warhead Attribution and Heavy Bomber Data Relating to the Treaty Between the United States of America and the Russian Federation on Further Reduction and Limitation of Strategic Offensive Arms, hereinafter referred to as the Memorandum on Attribution.

3. Each Party undertakes not to equip any heavy bomber with a greater number of nuclear weapons than the number specified for heavy bombers of that type or variant of a type in the Memorandum on Attribution.

4. No later than 180 days after entry into force of this Treaty, each Party shall exhibit one heavy bomber of each type and variant of a type specified in the Memorandum on Attribution.

The purpose of the exhibition shall be to demonstrate to the other Party the number of nuclear weapons for which a heavy bomber of a given type or variant of a type is actually equipped.

5. If either Party intends to change the number of nuclear weapons specified in the Memorandum on Attribution, for which a heavy bomber of a type or variant of a type is actually equipped, it shall provide a 90-day advance notification of such intention to the other Party. Ninety days after providing such a notification, or at a later date agreed by the Parties, the Party changing the number of nuclear weapons for which a heavy bomber is actually equipped shall exhibit one heavy bomber of each such type or variant of a type. The purpose of the exhibition shall be to demonstrate to the other Party the revised number of nuclear weapons for which heavy bombers of the specified type or variant of a type are actually equipped. The number of nuclear weapons attributed to the specified type and variant of a type of heavy bomber shall change on the ninetieth day after the notification of such intent. On that day, the Party changing the number of nuclear weapons for which a heavy bomber is actually equipped shall provide to the other Party a notification of each change in data according to categories of data contained in the Memorandum on Attribution.

6. The exhibitions and inspections conducted pursuant to paragraphs 4 and 5 of this Article shall be carried out in accordance with the procedures provided for in the Protocol on Exhibitions and Inspections of Heavy Bombers Relating to the Treaty Between the United States of America and the Russian Federation on Further Reduction and Limitation of Strategic Offensive Arms, hereinafter referred to as the Protocol on Exhibitions and Inspections.

7. Each Party shall have the right to reorient to a conventional role heavy bombers equipped for nuclear armaments other than long-range nuclear ALCMs. For the purposes of this Treaty, heavy bombers reoriented to a conventional role are those heavy bombers specified by a Party from among its heavy bombers equipped for nuclear armaments other than long-range nuclear ALCMs that have never been accountable under the START Treaty as heavy bombers equipped for long-range nuclear ALCMs. The reorienting Party shall provide to the other Party a

notification of its intent to reorient a heavy bomber to a conventional role no less than 90 days in advance of such reorientation. No conversion procedures shall be required for such a heavy bomber to be specified as a heavy bomber reoriented to a conventional role.

8. Heavy bombers reoriented to a conventional role shall be subject to the following requirements:

(a) the number of such heavy bombers shall not exceed 100 at any one time;

(b) such heavy bombers shall be based separately from heavy bombers with nuclear roles;

(c) such heavy bombers shall be used only for non-nuclear missions. Such heavy bombers shall not be used in exercises for nuclear missions, and their aircrews shall not train or exercise for such missions; and

(d) heavy bombers reoriented to a conventional role shall have differences from other heavy bombers of that type or variant of a type that are observable by national technical means of verification and visible during inspection.

9. Each Party shall have the right to return to a nuclear role heavy bombers that have been reoriented in accordance with paragraph 7 of this Article to a conventional role. The Party carrying out such action shall provide to the other Party through diplomatic channels notification of its intent to return a heavy bomber to a nuclear role no less than 90 days in advance of taking such action. Such a heavy bomber returned to a nuclear role shall not subsequently be reoriented to a conventional role.

Heavy bombers reoriented to a conventional role that are subsequently returned to a nuclear role shall have differences observable by national technical means of verification and visible during inspection from other heavy bombers of that type and variant of a type that have not been reoriented to a conventional role, as well as from heavy bombers of that type and variant of a type that are still reoriented to a conventional role.

10. Each Party shall locate storage areas for heavy bomber nuclear armaments no less than 100 kilometers from any air base where heavy bombers reoriented to a conventional role are based.

11. Except as otherwise provided for in this Treaty, heavy bombers reoriented to a conventional role shall remain subject to the provisions of the START Treaty, including the inspection provisions.

12. If not all heavy bombers of a given type or variant of a type are reoriented to a conventional role, one heavy bomber of each type or variant of a type of heavy bomber reoriented to a conventional role shall be exhibited in the open for the purpose of demonstrating to the other Party the differences referred to in subparagraph 8(d) of this Article. Such differences shall be subject to inspection by the other Party.

13. If not all heavy bombers of a given type or variant of a type reoriented to a conventional role are returned to a nuclear role, one heavy bomber of each type and variant of a type of heavy bomber returned to a nuclear role shall be exhibited in the open for the purpose of demonstrating to the other Party the differences referred to in paragraph 9 of this Article. Such differences shall be subject to inspection by the other Party.

14. The exhibitions and inspections provided for in paragraphs 12 and 13 of this Article shall be carried out in accordance with the procedures provided for in the Protocol on Exhibitions and Inspections.

**Article V**

1. Except as provided for in this Treaty, the provisions of the START Treaty, including the verification provisions, shall be used for implementation of this Treaty.

2. To promote the objectives and implementation of the provisions of this Treaty, the Parties hereby establish the Bilateral Implementation Commission. The Parties agree that, if either Party so requests, they shall meet within the framework of the Bilateral Implementation Commission to:

(a) resolve questions relating to compliance with the obligations assumed; and

(b) agree upon such additional measures as may be necessary to improve the viability and effectiveness of this Treaty.

**Article VI**

1. This Treaty, including its Memorandum on Attribution, Elimination and Conversion Protocol, and Protocol on Exhibitions and Inspections, all of which are integral parts thereof, shall be subject to ratification in accordance with the constitutional procedures of each Party. This Treaty shall enter into force on the date of the exchange of instruments of ratification, but not prior to the entry into force of the START Treaty.

2. The provisions of paragraph 8 of Article II of this Treaty shall be applied provisionally by the Parties from the date of its signature.

3. This Treaty shall remain in force so long as the START Treaty remains in force.

4. Each Party shall, in exercising its national sovereignty, have the right to withdraw from this Treaty if it decides that extraordinary events related to the subject matter of this Treaty have jeopardized its supreme interests. It shall give notice of its decision to the other Party six months prior to withdrawal from this Treaty. Such notice shall include a statement of the extraordinary events the notifying Party regards as having jeopardized its supreme interests.

**Article VII**

Each Party may propose amendments to this Treaty. Agreed amendments shall enter into force in accordance with the procedures governing entry into force of this Treaty.

**Article VIII**

This Treaty shall be registered pursuant to Article 102 of the Charter of the United Nations. Done at Moscow on January 3, 1993, in two copies, each in the English and Russian languages, both texts being equally authentic.

## Protocol on Procedures Governing Elimination of Heavy ICBMs and on Procedures Governing Conversion of Silo Launchers of Heavy ICBMs Relating to the Treaty Between the United States of America and the Russian Federation on Further Reduction and Limitation of Strategic Offensive Arms

Pursuant to and in implementation of the Treaty Between the United States of America and the Russian Federation on Further Reduction and Limitation of Strategic Offensive Arms, hereinafter referred to as the Treaty, the Parties hereby agree upon procedures governing the elimination of heavy ICBMs and upon procedures governing the conversion of silo launchers of such ICBMs.

### I. Procedures for Elimination of Heavy ICBMs and Their Launch Canisters

1. Elimination of heavy ICBMs shall be carried out in accordance with the procedures provided for in this Section at elimination facilities for ICBMs specified in the START Treaty or shall be carried out by using such missiles for delivering objects into the upper atmosphere or space. Notification thereof shall be provided through the Nuclear Risk Reduction Centers (NRRCs) 30 days in advance of the initiation of elimination at conversion or elimination facilities, or, in the event of launch, in accordance with the provisions of the Agreement Between the United States of America and the Union of Soviet Socialist Republics on Notifications of Launches of Intercontinental Ballistic Missiles and Submarine-Launched Ballistic Missiles of May 31, 1988.

2. Prior to the confirmatory inspection pursuant to paragraph 3 of this Section, the inspected Party:

(a) shall remove the missile's reentry vehicles;

(b) may remove the electronic and electromechanical devices of the missile's guidance and control system from the missile and its launch canister, and other elements that shall not be subject to elimination pursuant to paragraph 4 of this Section;

(c) shall remove the missile from its launch canister and disassemble the missile into stages;

(d) shall remove liquid propellant from the missile;

(e) may remove or actuate auxiliary pyrotechnic devices installed on the missile and its launch canister;

(f) may remove penetration aids, including devices for their attachment and release; and

(g) may remove propulsion units from the self-contained dispensing mechanism.

These actions may be carried out in any order.

3. After arrival of the inspection team and prior to the initiation of the elimination process, inspectors shall confirm the type and number of the missiles to be eliminated by making the observations and measurements necessary for such confirmation. After the procedures provided for in this paragraph have been carried out, the process of the elimination of the missiles and their launch canisters may begin. Inspectors shall observe the elimination process.

4. Elimination process for heavy ICBMs:

(a) missile stages, nozzles, and missile interstage skirts shall each be cut into two pieces of approximately equal size; and

(b) the self-contained dispensing mechanism as well as the front section, including the reentry vehicle platform and the front section shroud, shall be cut into two pieces of approximately equal size and crushed.

5. During the elimination process for launch canisters of heavy ICBMs, the launch canister shall be cut into two pieces of approximately equal size or into three pieces in such a manner that pieces no less than 1.5 meters long are cut from the ends of the body of such a launch canister.

6. Upon completion of the above requirements, the inspection team leader and a member of the in-country escort shall confirm in a factual, written report containing the results of the inspection team's observation of the elimination process that the inspection team has completed its inspection.

7. Heavy ICBMs shall cease to be subject to the limitations provided for in the Treaty after completion of the procedures provided for in this Section. Notification thereof shall be provided in accordance with paragraph 3 of Section I of the Notification Protocol Relating to the START Treaty.

**II. Procedures for Conversion of Silo Launchers of Heavy ICBMs, Silo Training Launchers for Heavy ICBMs, and Silo Test Launchers for Heavy ICBMs**

1. Conversion of silo launchers of heavy ICBMs, silo training launchers for heavy ICBMs, and silo test launchers for heavy ICBMs shall be carried out *in situ* and shall be subject to inspection.

2. Prior to the initiation of the conversion process for such launchers, the missile and launch canister shall be removed from the silo launcher.

3. A Party shall be considered to have initiated the conversion process for silo launchers of heavy ICBMs, silo training launchers for heavy ICBMs, and silo test launchers for heavy ICBMs as soon as the silo launcher door has been opened and a missile and its launch canister have been removed from the silo launcher. Notification thereof shall be provided in accordance with paragraphs 1 and 2 of Section IV of the Notification Protocol Relating to the START Treaty.

4. Conversion process for silo launchers of heavy ICBMs, silo training launchers for heavy ICBMs, and silo test launchers for heavy ICBMs shall include the following steps:

(a) the silo launcher door shall be opened, the missile and the launch canister shall be removed from the silo launcher;

(b) concrete shall be poured into the base of the silo launcher up to the height of five meters from the bottom of the silo launcher; and

(c) a restrictive ring with a diameter of no more than 2.9 meters shall be installed into the upper portion of the silo launcher. The method of installation of the restrictive ring shall rule out its removal without destruction of the ring and its attachment to the silo launcher.

5. Each Party shall have the right to confirm that the procedures provided for in paragraph 4 of this Section have been carried out. For the purpose of confirming that these procedures have been carried out:

(a) the converting Party shall notify the other Party through the NRRCs:

(i) no less than 30 days in advance of the date when the process of pouring concrete will commence; and

(ii) upon completion of all of the procedures provided for in paragraph 4 of this Section; and

(b) the inspecting Party shall have the right to implement the procedures provided for in either paragraph 6 or paragraph 7, but not both, of this Section for each silo launcher of heavy ICBMs, silo training launcher for heavy ICBMs, and silo test launcher for heavy ICBMs that is to be converted.

6. Subject to the provisions of paragraph 5 of this Section, each Party shall have the right to observe the entire process of pouring concrete into each silo launcher of heavy ICBMs, silo training launcher for heavy ICBMs, and silo test launcher for heavy ICBMs that is to be converted, and to measure the diameter of the restrictive ring. For this purpose:

(a) the inspecting Party shall inform the Party converting the silo launcher no less than seven days in advance of the commencement of the pouring that it will observe the filling of the silo in question;

(b) immediately prior to the commencement of the process of pouring concrete, the converting Party shall take such steps as are necessary to ensure that the base of the silo launcher is visible, and that the depth of the silo can be measured;

(c) the inspecting Party shall have the right to observe the entire process of pouring concrete from a location providing an unobstructed view of the base of the silo launcher, and to confirm by measurement that concrete has been poured into the base of the silo launcher up to the height of five meters from the bottom of the silo launcher. The measurements shall be taken from the level of the lower edge of the closed silo launcher door to the base of the silo launcher, prior to the pouring of the concrete, and from the level of the lower edge of the closed silo launcher door to the top of the concrete fill, after the concrete has hardened;

(d) following notification of completion of the procedures provided for in paragraph 4 of this Section, the inspecting Party shall be permitted to measure the diameter of the restrictive ring. The restrictive ring shall not be shrouded during such inspections. The Parties shall agree on the date for such inspections;

(e) the results of measurements conducted pursuant to subparagraphs (c) and (d) of this paragraph shall be recorded in written, factual inspection reports and signed by the inspection team leader and a member of the in-country escort;

(f) inspection teams shall each consist of no more than 10 inspectors, all of whom shall be drawn from the list of inspectors under the START Treaty; and

(g) such inspections shall not count against any inspection quota established by the START Treaty.

7. Subject to the provisions of paragraph 5 of this Section, each Party shall have the right to measure the depth of each silo launcher of heavy ICBMs, silo training launcher for heavy ICBMs, and silo test launcher for heavy ICBMs that is to be converted both before the commencement and after the completion of the process of pouring concrete, and to measure the diameter of the restrictive ring. For this purpose:

(a) the inspecting Party shall inform the Party converting the silo launcher no less than seven days in advance of the commencement of the pouring that it will measure the depth of the silo launcher in question both before the commencement and after the completion of the process of pouring concrete;

(b) immediately prior to the commencement of the process of pouring concrete, the converting Party shall take such steps as are necessary to ensure that the base of the silo launcher is visible, and that the depth of the silo launcher can be measured;

(c) the inspecting Party shall measure the depth of the silo launcher prior to the commencement of the process of pouring concrete;

(d) following notification of completion of the procedures provided for in paragraph 4 of this Section, the inspecting Party shall be permitted to measure the diameter of the restrictive ring, and to remeasure the depth of the silo launcher. The restrictive ring shall not be shrouded during such inspections. The Parties shall agree on the date for such inspections;

(e) for the purpose of measuring the depth of the concrete in the silo launcher, measurements shall be taken from the level of the lower edge of the closed silo launcher door to the base of the silo launcher, prior to the pouring of the concrete, and from the level of the lower edge of the closed silo launcher door to the top of the concrete fill, after the concrete has hardened;

(f) the results of measurements conducted pursuant to subparagraphs (c), (d), and (e) of this paragraph shall be recorded in written, factual inspection reports and signed by the inspection team leader and a member of the in-country escort;

(g) inspection teams shall each consist of no more than 10 inspectors, all of whom shall be drawn from the list of inspectors under the START Treaty; and

(h) such inspections shall not count against any inspection quota established by the START Treaty.

8. The converting Party shall have the right to carry out further conversion measures after the completion of the procedures provided for in paragraph 6 or paragraph 7 of this Section or, if such procedures are not conducted, upon expiration of 30 days after notification of completion of the procedures provided for in paragraph 4 of this Section.

9. In addition to the reentry vehicle inspections conducted under the START Treaty, each Party shall have the right to conduct, using the procedures provided for in Annex 3 to the Inspection Protocol Relating to the START Treaty, four additional reentry vehicle inspections each year of ICBMs that are deployed in silo launchers of heavy ICBMs that have been converted in accordance with the provisions of this Section. During such inspections, the inspectors also shall have the right to confirm by visual observation the presence of the restrictive ring and that the observable portions of the launch canister do not differ externally from the observable portions of the launch canister that was exhibited pursuant to paragraph 11 of Article XI of the START Treaty. Any shrouding of the upper portion of the silo launcher shall not obstruct visual observation of the upper portion of the launch canister and shall not obstruct visual observation of

the edge of the restricted ring. If requested by the inspecting Party, the converting Party shall partially remove any shrouding, except for shrouding of instruments installed on the restrictive ring, to permit confirmation of the presence of the restrictive ring.

10. Upon completion of the procedures provided for in paragraph 6 or paragraph 7 of this Section or, if such procedures are not conducted, upon expiration of 30 days after notification of completion of the procedures provided for in paragraph 4 of this Section, the silo launcher of heavy ICBMs being converted shall, for the purposes of the Treaty, be considered to contain a deployed ICBM to which one warhead is attributed.

### III. Equipment; Costs

1. To carry out inspections provided for in this Protocol, the inspecting Party shall have the right to use agreed equipment, including equipment that will confirm that the silo launcher has been completely filled up to the height of five meters from the bottom of the silo launcher with concrete. The Parties shall agree in the Bilateral Implementation Commission on such equipment.

2. For inspections conducted pursuant to this Protocol, costs shall be handled pursuant to paragraph 19 of Section V of the Inspection Protocol Relating to the START Treaty.

This Protocol is an integral part of the Treaty and shall enter into force on the date of entry into force of the Treaty and shall remain in force as long as the Treaty remains in force. As provided for in subparagraph 2(b) of Article V of the Treaty, the Parties may agree upon such additional measures as may be necessary to improve the viability and effectiveness of the Treaty. The Parties agree that, if it becomes necessary to make changes in this Protocol that do not affect substantive rights or obligations under the Treaty, they shall use the Bilateral Implementation Commission to reach agreement on such changes, without resorting to the procedure for making amendments set forth in Article VII of the Treaty.

Done at Moscow on January 3, 1993, in two copies, each in the English and Russian languages, both texts being equally authentic.

## Protocol on Exhibitions and Inspections of Heavy Bombers Relating to the Treaty Between the United States of America and the Russian Federation on Further Reduction and Limitation of Strategic Offensive Arms

Pursuant to and in implementation of the Treaty Between the United States of America and the Russian Federation on Further Reduction and Limitation of Strategic Offensive Arms, hereinafter referred to as the Treaty, the Parties hereby agree to conduct exhibitions and inspections of heavy bombers pursuant to paragraphs 4, 5, 12, and 13 of Article IV of the Treaty.

### I. Exhibitions of Heavy Bombers

1. For the purpose of helping to ensure verification of compliance with the provisions of the Treaty, and as required by paragraphs 4, 5, 12, and 13 of Article IV of the Treaty, each Party shall conduct exhibitions of heavy bombers equipped for nuclear armaments, heavy bombers reoriented to a conventional role, and heavy bombers that were reoriented to a conventional role and subsequently returned to a nuclear role.

2. The exhibitions of heavy bombers shall be conducted subject to the following provisions:

(a) the location for such an exhibition shall be at the discretion of the exhibiting Party;

(b) the date for such an exhibition shall be agreed upon between the Parties through diplomatic channels, and the exhibiting Party shall communicate the location of the exhibition;

(c) during such an exhibition, each heavy bomber exhibited shall be subject to inspection for a period not to exceed two hours;

(d) the inspection team conducting an inspection during an exhibition shall consist of no more than 10 inspectors, all of whom shall be drawn from the list of inspectors under the START Treaty;

(e) prior to the beginning of the exhibition, the inspected Party shall provide a photograph or photographs of one of the heavy bombers of a type or variant of a type reoriented to a conventional role and of one of the heavy bombers of the same type and variant of a type that were reoriented to a conventional role and subsequently returned to a nuclear role, so as to show all of their differences that are observable by national technical means of verification and visible during inspection; and

(f) such inspections during exhibitions shall not count against any inspection quota established by the START Treaty.

**II. Inspections of Heavy Bombers**

1. During exhibitions of heavy bombers, each Party shall have the right to perform the following procedures on the exhibited heavy bombers; and each Party, beginning 180 days after entry into force of the Treaty and thereafter, shall have the right, in addition to its rights under the START Treaty, to perform, during data update and new facility inspections conducted under the START Treaty at air bases of the other Party, the following procedures on all heavy bombers based at such air bases and present there at the time of the inspection:

(a) to conduct inspections of heavy bombers equipped for long-range nuclear ALCMs and heavy bombers equipped for nuclear armaments other than long-range nuclear ALCMs, in order to confirm that the number of nuclear weapons for which a heavy bomber is actually equipped does not exceed the number specified in the Memorandum on Attribution. The inspection team shall have the right to visually inspect those portions of the exterior of the inspected heavy bomber where the inspected heavy bomber is equipped for weapons, as well as to visually inspect the weapons bay of such a heavy bomber, but not to inspect other portions of the exterior or interior;

(b) to conduct inspections of heavy bombers reoriented to a conventional role, in order to confirm the differences of such heavy bombers from other heavy bombers of that type or variant of a type that are observable by national technical means of verification and visible during inspection. The inspection team shall have the right to visually inspect those portions of the exterior of the inspected heavy bomber having the differences observable by national technical means of verification and visible during inspection, but not to inspect other portions of the exterior or interior; and

(c) to conduct inspections of heavy bombers that were reoriented to a conventional role and subsequently returned to a nuclear role, in order to confirm the differences of such heavy bombers from other heavy bombers of that type or variant of a type that are observable by national technical means of verification and visible during inspection, and to confirm that the number of nuclear weapons for which a heavy bomber is actually equipped does not exceed the number specified in the Memorandum on Attribution. The inspection team shall have the right to visually inspect those portions of the exterior of the inspected heavy bomber where the inspected heavy bomber is equipped for weapons, as well as to visually inspect the weapons bay of such a heavy bomber, and to visually inspect those portions of the exterior of the inspected heavy bomber having the differences observable by national technical means of verification and visible to inspection, but not to inspect other portions of the exterior or interior.

2. At the discretion of the inspected Party, those portions of the heavy bomber that are not subject to inspection may be shrouded. The period of time required to carry out the shrouding process shall not count against the period allocated for inspection.

3. In the course of an inspection conducted during an exhibition, a member of the in-country escort shall provide, during inspections conducted pursuant to subparagraph 1(a) or subparagraph 1(c) of this Section, explanations to the inspection team concerning the number of nuclear weapons for which the heavy bomber is actually equipped, and shall provide, during inspections conducted pursuant to subparagraph 1(b) or subparagraph 1(c) of this Section, explanations to the inspection team concerning the differences that are observable by national technical means of verification and visible during inspection.

This Protocol is an integral part of the Treaty and shall enter into force on the date of entry into force of the Treaty and shall remain in force so long as the Treaty remains in force. As provided for in subparagraph 2(b) of Article V of the Treaty, the Parties may agree upon such additional measures as may be necessary to improve the viability and effectiveness of the Treaty. The Parties agree that, if it becomes necessary to make changes in this Protocol that do not affect substantive rights or obligations under the Treaty, they shall use the Bilateral Implementation Commission to reach agreement on such changes, without resorting to the procedure for making amendments set forth in Article VII of the Treaty.

Done at Moscow on January 3, 1993, in two copies, each in the English and Russian languages, both texts being equally authentic.

## Memorandum of Understanding on Warhead Attribution and Heavy Bomber Data Relating to the Treaty Between the United States of America and the Russian Federation on Further Reduction and Limitation of Strategic Offensive Arms

Pursuant to and in implementation of the Treaty Between the United States of America and the Russian Federation on Further Reduction and Limitation of Strategic Offensive Arms, hereinafter referred to as the Treaty, the Parties have exchanged data current as of January 3, 1993, on the number of nuclear weapons for which each heavy bomber of a type and a variant of a type equipped for nuclear weapons is actually equipped. No later than 30 days after the date of entry into force of the Treaty, the Parties shall additionally exchange data, current as of the date of entry into force of the Treaty, according to the categories of data contained in this Memorandum, on heavy bombers equipped for nuclear weapons; on heavy bombers specified as reoriented to a conventional role, and on heavy bombers reoriented to a conventional role that are subsequently returned to a nuclear role; on ICBMs and SLBMs to which a reduced number of warheads is attributed; and on data on the elimination of heavy ICBMs and on conversion of silo launchers of heavy ICBMs.

Only those data used for purposes of implementing the Treaty that differ from the data in the Memorandum of Understanding on the Establishment of the Data Base Relating to the START Treaty are included in this Memorandum.

### I. Number of Warheads Attributed to Deployed Heavy Bombers Other than Heavy Bombers Reoriented to a Conventional Role

1. Pursuant to paragraph 3 of Article IV of the Treaty each Party undertakes not to have more nuclear weapons deployed on heavy bombers of any type or variant of a type than the number specified in this paragraph. Additionally, pursuant to paragraph 2 of Article IV of the Treaty, for each Party the numbers of warheads attributed to deployed heavy bombers not reoriented to a conventional role as of the date of signature of the Treaty or to heavy bombers subsequently deployed are listed below. Such numbers shall only be changed in accordance with paragraph 5 of Article IV of the Treaty. The Party making a change shall provide a notification to the other Party 90 days prior to making such a change. An exhibition shall be conducted to demonstrate the changed number of nuclear weapons for which heavy bombers of the listed type or variant of a type are actually equipped:

(a) United States of America

| Heavy Bomber Type and Variant of a Type* | Number of Warheads |
|---|---|
| B-52G | 12 |
| B-52H | 20 |
| B-1B | 16 |
| B-2 | 16 |

Aggregate Number of Warheads Attributed to Deployed Heavy Bombers, Except for Heavy Bombers Reoriented to a Conventional Role —

(b) Russian Federation

| Heavy Bomber Type and Variant of a Type | Number of Warheads |
|---|---|
| Bear B | 1 |
| Bear G | 2 |
| Bear H6 | 6 |
| Bear H16 | 16 |
| Blackjack | 12 |

Aggregate Number of Warheads Attributed to Deployed Heavy Bombers, Except for Heavy Bombers Reoriented to a Conventional Role —

### II. Data on Heavy Bombers Reoriented to a Conventional Role and Heavy Bombers Reoriented to a Conventional Role that Have Subsequently Been Returned to a Nuclear Role

1. For each Party, the numbers of heavy bombers reoriented to a conventional role are as follows:

(a) United States of America

| Heavy Bomber Type and Variant of a Type | Number |
|---|---|
| — | — |
| — | — |

---

* Heavy bombers of the type and variant of a type designated B-52C, B-52D, B-52E, and B-52F, located at the Davis-Monthan conversion or elimination facility as of September 1, 1990, as specified in the Memorandum of Understanding to the START Treaty, will be eliminated, under the provisions of the START Treaty, before the expiration of the seven-year reductions period.

(b) Russian Federation

| Heavy Bomber Type and Variant of a Type | Number |
|---|---|
| — | — |
| — | — |

2. For each Party, the numbers of heavy bombers reoriented to a conventional role as well as data on related air bases are as follows:

(a) United States of America
Air Bases:

| Name/Location | Bomber Type and Variant of a Type |
|---|---|
| — | — |

| Heavy Bombers Reoriented to a Conventional Role | Number |
|---|---|
| | — |

(b) Russian Federation
Air Bases:

| Name/Location | Bomber Type and Variant of a Type |
|---|---|
| — | — |

3. For each Party, the differences observable by technical means of verification for heavy bombers reoriented to a conventional role are as follows:

(a) United States of America

| Heavy Bomber Type and Variant of a Type | Difference |
|---|---|
| — | — |

(b) Russian Federation

| Heavy Bomber Type and Variant of a Type | Difference |
|---|---|
| — | — |

4. For each Party, the differences observable by national technical means of verification for heavy bombers reoriented to a conventional role that have subsequently been returned to a nuclear role are as follows:

(a) United States of America

| Heavy Bomber Type and Variant of a Type | Difference |
|---|---|
| — | — |

(b) Russian Federation

| Heavy Bomber Type and Variant of a Type | Difference |
|---|---|
| — | — |

**III. Data on Deployed ICBMs and Deployed SLBMs to Which a Reduced Number of Warheads is Attributed**

For each Party, the data on ICBM bases or submarine bases, and on ICBMs or SLBMs of existing types deployed at those bases, on which the number of warheads attributed to them is reduced pursuant to Article III of the Treaty are as follows:

(a) United States of America

| | Type of ICBM or SLBM |
|---|---|
| | — |
| Deployed ICBMs or Deployed SLBMs, on Which the Number of Warheads Is Reduced | — |
| Warheads Attributed to Each Deployed ICBM or Deployed SLBM After Reduction in the Number of Warheads on It | — |
| Number of Warheads by Which the Original Attribution of Warheads for Each ICBM or SLBM Was Reduced | — |
| Aggregate Reduction in the Number of Warheads Attributed to Deployed ICBMs or Deployed SLBMs of that Type | — |

ICBM Bases at Which the Number of Warheads on Deployed ICBMs Is Reduced:

| Name/Location | ICBM Type on Which the Number of Warheads Is Reduced |
|---|---|
| — | — |
| Deployed ICBMs on Which the Number of Warheads Is Reduced | — |
| Warheads Attributed to Each Deployed ICBM After Reduction in the Number of Warheads on It | — |
| Number of Warheads by Which the Original Attribution of Warheads for Each ICBM Was Reduced | — |
| Aggregate Reduction in the Number of Warheads Attributed to Deployed ICBMs of that Type | — |

SLBM Bases at Which the Number of Warheads on Deployed SLBMs Is Reduced:

| Name/Location | SLBM Type on Which the Number of Warheads Is Reduced |
|---|---|
| — | — |

| | |
|---|---|
| Deployed SLBMs on Which the Number of Warheads Is Reduced | — |
| Warheads Attributed to Each Deployed SLBM After Reduction in the Number of Warheads on It | — |
| Number of Warheads by Which the Original Attribution of Warheads for Each SLBM Was Reduced | — |
| Aggregate Reduction in the Number of Warheads Attributed to Deployed SLBMs of that Type | — |

(b) Russian Federation

| | Type of ICBM or SLBM |
|---|---|
| | — |
| Deployed ICBMs or Deployed SLBMs, on Which the Number of Warheads Is Reduced | — |
| Warheads Attributed to Each Deployed ICBM or Deployed SLBM After Reduction in the Number of Warheads on It | — |
| Number of Warheads by Which the Original Attribution of Warheads for Each ICBM or SLBM Was Reduced | — |
| Aggregate Reduction in the Number of Warheads Attributed to Deployed ICBMs or Deployed SLBMs of that Type | — |

ICBM Bases at Which the Number of Warheads on Deployed ICBMs Is Reduced:

| Name/Location | ICBM Type on Which the Number of Warheads Is Reduced |
|---|---|
| — | — |

| | |
|---|---|
| Deployed ICBMs on Which the Number of Warheads Is Reduced | — |
| Warheads Attributed to Each Deployed ICBM After Reduction in the Number of Warheads on It | — |
| Number of Warheads by Which the Original Attribution of Warheads for Each ICBM Was Reduced | — |
| Aggregate Reduction in the Number of Warheads Attributed to Deployed ICBMs of that Type | — |

SLBM Bases at Which the Number of Warheads on Deployed SLBMs Is Reduced:

| Name/Location | SLBM Type on Which the Number of Warheads Is Reduced |
|---|---|
| — | — |

| | |
|---|---|
| Deployed SLBMs on Which the Number of Warheads Is Reduced | — |
| Warheads Attributed to Each Deployed SLBM After Reduction in the Number of Warheads on It | — |
| Number of Warheads by Which the Original Attribution of Warheads for Each SLBM Was Reduced | — |
| Aggregate Reduction in the Number of Warheads Attributed to Deployed SLBMs of that Type | — |

**IV. Data on Eliminated Heavy ICBMs and Converted Silo Launchers of Heavy ICBMs**

1. For each Party, the numbers of silo launchers of heavy ICBMs converted to silo launchers of ICBMs other than heavy ICBMs are as follows:

(a) United States of America

| | |
|---|---|
| Aggregate Number of Converted Silo Launchers | — |

| ICBM Base for Silo Launchers of ICBMs: Name/Location | ICBM Type Installed in a Converted Silo Launcher |
|---|---|
| — | — |

Silo Launcher Group: (designation)

—

Silo Launchers:

| | |
|---|---|
| — | — |

(b) Russian Federation

| | |
|---|---|
| Aggregate Number of Converted Silo Launchers | — |

| ICBM Base for Silo Launchers of ICBMs: Name/Location | ICBM Type Installed in a Converted Silo Launcher |
|---|---|
| — | |

Silo Launcher Group: (designation)

—

Silo Launchers:

| | |
|---|---|
| — | — |

2. For each Party, the aggregate numbers of heavy ICBMs and eliminated heavy ICBMs are as follows:

| (a) United States of America | Number |
|---|---|
| Deployed Heavy ICBMs | — |
| Non-Deployed Heavy ICBMs | — |
| Eliminated Heavy ICBMs | — |
| **(b) Russian Federation** | **Number** |
| Deployed Heavy ICBMs | — |
| Non-Deployed Heavy ICBMs | — |
| Eliminated Heavy ICBMs | — |

**V. Changes**

Each Party shall notify the other Party of changes in the attribution and data contained in this Memorandum.

The Parties, in signing this Memorandum, acknowledge the acceptance of the categories of data contained in this Memorandum and the responsibility of each Party for the accuracy only of its own data.

This Memorandum is an integral part of the Treaty and shall enter into force on the date of entry into force of the Treaty and shall remain in force so long as the Treaty remains in force. As provided for in subparagraph 2(b) of Article V of the Treaty, the Parties may agree on such additional measures as may be necessary to improve the viability and effectiveness of the Treaty. The Parties agree that, if it becomes necessary to change the categories of data contained in this Memorandum or to make other changes to this Memorandum that do not affect substantive rights or obligations under the Treaty, they shall use the Bilateral Implementation Commission to reach agreement on such changes, without resorting to the procedure for making amendments set forth in Article VII of the Treaty.

Done at Moscow on January 3, 1993, in two copies, each in the English and Russian languages, both texts being equally authentic.

---

*Source:* US Arms Control and Disarmament Agency (ACDA), 'Official Text: Treaty between the United States of America and the Russian Federation on Further Reduction and Limitation of Strategic Offensive Arms' (ACDA: Washington, DC, 3 Jan. 1993), mimeo.

# 12. Conventional arms control in Europe

JANE M. O. SHARP

## I. Introduction

Three of the arms control agreements on which negotiations began at the end of the cold war were signed in 1992. The Vienna Document 1992 was signed on 4 March, establishing a new set of confidence- and security-building measures, and the multilateral Treaty on Open Skies was signed on 24 March.[1] On 10 July the 29 signatories to the 1990 Treaty on Conventional Armed Forces in Europe (CFE) signed the Concluding Act of the Negotiation on Personnel Strength of Conventional Armed Forces in Europe (the CFE-1A Agreement), limiting military personnel.[2]

Ratification of the CFE Treaty, signed on 20 November 1990, was finally completed on 30 October. It entered into force *de facto* on 17 July 1992 after the original 22 CFE signatories, plus seven former Soviet republics with territory in the Atlantic-to-Urals (ATTU) zone of application, signed the Provisional Application of the CFE Treaty on 10 July.[3] The Treaty entered into force *de jure* on 9 November, 10 days after the last signatory states deposited their instruments of ratification in The Hague. In September, the Conference on Security and Co-operation in Europe (CSCE) participating states opened the Forum for Security Co-operation in Vienna to co-ordinate future negotiations on regional security and harmonize the rights and obligations of CSCE countries with respect to arms control.[4]

This chapter deals primarily with the ratification and implementation of the CFE Treaty.[5] It begins with an account of the ratification debate, focusing on the problems associated with the dissolution of the USSR. These include the opting out of Treaty obligations by the three Baltic states and the difficulties of re-allocating equipment ceilings initially allowed the USSR to the former Soviet republics with territory in the ATTU zone. The different attitudes to the Treaty among the new states parties are compared and the prospects for compliance are examined. Russia's loss of status compared with that of the former USSR not only made ratification problematic but also casts doubt on future

[1] See appendix 12C for the texts of the Vienna Document 1992 of the Negotiations on Confidence- and Security-Building Measures convened in accordance with the relevant provisions of the Concluding Document of the Vienna Meeting of the Conference on Security and Co-operation in Europe (Vienna Document 1992) and the 1992 Treaty on Open Skies.

[2] See appendix 12C for the text of the Concluding Act of the Negotiation on Personnel Strength of Conventional Armed Forces in Europe.

[3] See appendix 12C for the text of the Provisional Application of the CFE Treaty.

[4] See chapter 5 in this volume.

[5] See appendix 12A on the implementation of the Vienna Document 1992 and appendix 12B on the status of implementation of the Treaty on Open Skies.

**Table 12.1.** Ratification of the CFE Treaty

| State party | Date of ratification | State party | Date of ratification |
|---|---|---|---|
| Czechoslovakia | 5 Aug. 1991 | Romania | 21 Apr. 1992 |
| Hungary | 4 Nov. 1991 | Italy | 22 Apr. 1992 |
| Netherlands | 8 Nov. 1991 | Spain | 1 June 1992 |
| Bulgaria | 12 Nov. 1991 | Georgia | 6 July 1992 |
| United Kingdom | 19 Nov. 1991 | Moldova | 6 July 1992 |
| Canada | 22 Nov. 1991 | Greece | 8 July 1992 |
| Poland | 26 Nov. 1991 | Turkey | 8 July 1992 |
| Norway | 29 Nov. 1991 | Azerbaijan | 9 July 1992 |
| Belgium | 17 Dec. 1991 | Ukraine | 9 July 1992 |
| Germany | 23 Dec. 1991 | Portugal | 14 Aug. 1992 |
| Iceland | 24 Dec. 1991 | Russia | 3 Sep. 1992 |
| Denmark | 30 Dec. 1991 | Armenia | 12 Oct. 1992 |
| Luxembourg | 22 Jan. 1992 | Belarus | 30 Oct. 1992 |
| United States | 29 Jan. 1992 | Kazakhstan | 30 Oct. 1992 |
| France | 24 Mar. 1992 | | |

*Source:* Netherlands Embassy, Stockholm.

implementation, with problems already surfacing in early 1993 about information exchange, access for inspection teams, and requests for simpler and less expensive destruction procedures. The chapter ends with a brief account of the CFE-1A Agreement limiting military personnel.

## II. Ratification of the CFE Treaty

The CFE Treaty was remarkable for the speed with which it was negotiated and the conciliatory negotiating style of the Soviet delegates in Vienna during 1989–90 under the leadership of Mikhail Gorbachev and Eduard Shevardnadze.[6] The rapport between Washington and Moscow that made speedy signature possible was bitterly resented in the upper echelons of the Soviet (and later Russian) military command, and not always welcome to the other CFE states parties, who tended to worry about a superpower condominium. These tensions surfaced during the ratification process. Although signed in November 1990 the Treaty could not enter into force until 10 days after ratification by all signatories, on 9 November 1992. As table 12.1 shows, at the beginning of 1992 only 12 of the 22 signatories had deposited their instruments of ratification. The parliaments of three signatory states had ratified but not deposited their instruments of ratification and the remaining seven had for a variety of reasons not yet submitted the Treaty to their legislatures. At the end of 1991 the USSR dissolved into its constituent republics of which three opted out of the Treaty, four were outside the application zone and eight would eventually become signatories. The ratification process was difficult not least because the Treaty codified a balance of forces (between NATO and former

[6] For the history of the negotiation of the CFE Treaty, 1989–90, see *SIPRI Yearbooks 1989*, *1990* and *1991*, chapters 11, 13 and 13, respectively.

WTO states) that was meaningless once the WTO and the USSR collapsed. For Russia, in particular, bereft of both allies and colonies, the Treaty codified a humiliating loss of status and military power.

Several sets of disputes delayed ratification. Initially these were primarily differences between the USSR and the other 21 original signatories. Later, the main problems were among and between the former Soviet republics.

Three problems with respect to ambiguous Soviet behaviour were resolved in 1991. The first was the attempt in 1990 to evade Treaty destruction requirements by transferring equipment east of the Urals, outside the zone of application. The second involved data, presented by the Soviet delegates immediately prior to CFE Treaty signature in November 1990, that were disputed by Western intelligence agencies. The third was the attempt to avoid limitations on a set of Soviet land-based equipment by redefining it as equipment for naval forces. Revised Soviet data were submitted in May 1991, and in a statement to the Joint Consultative Group (JCG) in Vienna on 14 June 1991 the Head of the Soviet CFE delegation pledged to destroy some of the equipment transferred east of the Urals, to use some to replace old equipment and to store the rest in such a way as not to create a strategic reserve.[7]

## The Baltic states opt out of the CFE Treaty

Estonia, Latvia and Lithuania became independent in September 1991, before the formal dissolution of the USSR. On 18 October the JCG, which monitors CFE Treaty compliance, agreed that these states should no longer be considered part of the Soviet Baltic Military District and thus were not in the ATTU zone of application of the CFE Treaty. In effect this freed the Baltic states from CFE Treaty obligations, except to open their territory to CFE inspectors as long as former Soviet troops remained there.

Although this was formally agreed by all signatories it was a solution imposed by the US and Soviet delegates and not welcomed by any of the others. Poland and Hungary would have preferred all three Baltic states to be included in the same arms control regime as all the other states of Central and Eastern Europe.[8] Resentment by the smaller states of the bilateral resolution of problems was a recurrent problem throughout the CFE Negotiaion. Hungary and Poland, for example, were also concerned about the James Baker–Eduard Shevardnadze deals that had resolved Soviet data discrepancies in June 1991.[9]

## Former Soviet republics become parties to the CFE Treaty

On 21 December 1991, at a meeting in Alma-Ata (Kazakhstan), the leaders of Armenia, Azerbaijan, Belarus, Kazakhstan, Kyrgyzstan, Moldova, Russia,

[7] These problems are described by Sharp, J. M. O., 'Conventional arms control in Europe', *SIPRI Yearbook 1991: World Armaments and Disarmament* (Oxford University Press: Oxford, 1991), pp. 428–33 and Sharp, J. M. O., 'Conventional arms control in Europe', *SIPRI Yearbook 1992: World Armaments and Disarmament* (Oxford University Press: Oxford, 1992), pp. 461–68.

[8] See Sharp, 1992 (note 7), pp. 465–66.

[9] See Sharp, 1992 (note 7), p. 463.

Tajikistan, Turkmenistan, Uzbekistan and Ukraine signed a Protocol to the 8 December Agreement on the Commonwealth of Independent States (CIS). The 11 republics agreed that Russia should inherit the permanent seat on the UN Security Council held by the USSR since 1945.[10]

At the meeting of CSCE Foreign Ministers in Prague on 30–31 January 1992 the CSCE admitted the new CIS states. In their letters of accession, the CIS governments underlined the need for those states with territory in the area of application to move forward promptly with ratification of the CFE Treaty.

The CFE Treaty could not be ratified or implemented, however, until all the former Soviet republics had resolved their differences over force allocations, and until these allocations had been approved by the other signatories.

*Re-allocation of Soviet treaty-limited equipment: the role of the High Level Working Group*

NATO states welcomed the formation of the CIS, especially to the extent that it provided a vehicle for the unified command of strategic nuclear weapons and an interlocutor for both nuclear and conventional arms control. At a meeting of the North Atlantic Council in Rome in November 1991, NATO ministers invited all the former WTO states, including the former Soviet republics, to participate in a North Atlantic Cooperation Council (NACC). Their main motivation was to pre-empt applications from the former WTO states for full membership in NATO and to maintain a *droit de regard* over the restructuring of the former Soviet armed forces. In particular they wanted to encourage the co-operative denuclearization of the newly sovereign republics that had inherited Soviet nuclear weapons and to increase East–West military contacts in the hope of encouraging more democratic control over the military in the former communist states. At the first NACC meeting in Brussels in December 1991, German Foreign Minister Hans-Dietrich Genscher proposed the formation of a High Level Working Group (HLWG) to facilitate the entry into force of the CFE Treaty in the wake of the dissolution of the USSR. The HLWG was open to all NATO and former WTO countries, including the 11 former Soviet republics with territory in the CFE zone of application (Georgia, the three Baltic states and the seven CIS states with territory west of the Urals: Armenia, Azerbaijan, Belarus, Kazakhstan, Moldova, Russia and Ukraine).

States in the flank zone argued that some provisions would need clarification as a result of the proliferation of new states. There was a potential loophole in Article XII of the Treaty, which permitted each state party 1000 armoured infantry fighting vehicles (AIFVs) for use with forces engaged in 'peacetime internal security functions'. This was clearly intended to apply to the entire USSR, not to each of its constituent republics. This problem was

[10] On the formation of the CIS, see text of the Minsk Agreement Establishing a Commonwealth of Independent States, 8 Dec. 1991; text of the Ashkhabad Declaration, 13 Dec. 1991; text of the Alma-Ata Declaration, 21 Dec. 1991; and Agreement on Joint Measures on Nuclear Weapons, Alma-Ata, 21 Dec. 1991, all reproduced in *SIPRI Yearbook 1992* (note 7), appendix 14A, pp. 558–62.

temporarily resolved in May 1992 when each of the former Soviet republics agreed to deploy no more than 100 AIFVs with internal security forces.

A nine-point statement was released after the first HLWG meeting on 10 January confirming *inter alia* that Treaty obligations would be apportioned between former Soviet republics in a manner acceptable to all CFE Treaty signatories; that all newly independent states in the ATTU zone should ratify the Treaty; that the deadline for ratification should be as soon as possible in view of the Helsinki follow-up meeting of the CSCE scheduled for March to July 1992; and that CFE should form the basis for further progress in fostering a common security forum in which all CSCE states should participate.[11]

Intra-CIS problems emerged at CIS meetings in January and February in Minsk. On 16 January, Russia demanded two-thirds of all ground-force equipment and three-quarters of the aircraft and helicopters. Ukraine objected and tried to claim equality with Russia.[12] On 14 February Russian Minister of Defence Konstantin Kobets outlined the Russian rationale for dividing up the treaty-limited equipment (TLE). He said that the status quo was not a good starting-point because the USSR had deployed so much of its equipment in the border republics during the cold war. Kobets suggested that a better way to arrive at an equitable distribution would be to calculate and compare the lengths of borders and numbers of people that each state had to defend. On this basis he would have distributed tanks as follows: Russia, 54 per cent (7114); Ukraine, 21.8 per cent (2867); Armenia, Azerbaijan, Georgia and Moldova, 17.5 per cent (2301); and Belarus, 6.6 per cent (868).[13] Little headway was made at the February CIS meeting on this or any other military issue, however, because the smaller republics were uncertain whether and how they would subordinate their military forces to a joint CIS command.

At the next HLWG meeting on 21 February in Brussels delegates agreed on a 'road map' for bringing the CFE Treaty into force by the Helsinki summit meeting of the CSCE in early July. In the first phase the CIS states would agree among themselves about TLE allocation and all CFE Treaty signatories would convene an extraordinary conference to provide a basis for the entry into force of the Treaty. It would enter into force 10 days after all instruments of ratification were deposited, after which states parties would meet to agree on any technical adjustments and make any necessary formal amendments.[14]

In February CIS Commander-in-Chief Marshal Yevgeniy Shaposhnikov suggested that all problems relating to TLE allocation would be solved by the next CIS summit meeting in Kiev on 20 March.[15] At the NACC meeting on 10 March NATO delegates urged their CIS partners to expedite their CFE decisions so as to meet the timetable set out in the February 'road map', but the CIS states were still undecided about whether and how to restructure their

[11] See Sharp, 1992 (note 7), pp. 467–68.

[12] Institute for Defense and Disarmament Studies, *Arms Control Reporter* (IDDS: Brookline, Mass.), sheet 407.B.465, 1992.

[13] Konstantin Kobets, interview in *Izvestia*, 13 Feb. 1992; reprinted in Foreign Broadcast Information Service, *Daily Report–Central Eurasia (FBIS-SOV)*, FBIS-SOV-92-031, 14 Feb. 1992, pp. 20–22.

[14] Text of the HLWG road map reprinted in *Arms Control Reporter*, sheets 407.B.466–67, 1992.

[15] Cited in *Krasnaya Zvezda*, 22 Feb. 1992, FBIS-SOV-92-036, 24 Feb. 1992, p. 10.

armed forces, specifically whether or not to have national armies instead of, or as well as, a CIS army.

In early 1992 senior Russian military officers called for a single CIS security space and a unitary military command. President Boris Yeltsin supported them, saying repeatedly that Russia would not create its own army unless forced into it by the other CIS states breaking away to form national armed forces. His decree of 16 March establishing a Russian Ministry of Defence recognized the failure of the CIS to forge a common defence policy.

Two approaches were hotly debated at the Kiev summit meeting on 20 March. Russia wanted a quota system based on its view of an optimum set of future regional balances, whereas the non-Russian states wanted to appropriate most of the former Soviet equipment on their territories.[16] The meeting ended without agreement and at the third HLWG meeting on 3 April Western delegates urged the CIS states to keep to the schedule outlined in the February 'road map', that is, to bring the Treaty into force by early July.

A few days later Russia announced that it would assume responsibility for all CIS forces stationed abroad, meaning not only Germany, Poland and the three Baltic states but also Moldova and the three states of the former Transcaucasus Military District (Armenia, Azerbaijan and Georgia).[17] These latter four countries, however, insisted on allocations of former Soviet equipment. Russia was reluctant to hand over Soviet assets to these states but eventually agreed to do so after some energetic 'shuttle diplomacy' by Lynn Hansen between Moscow and the capitals of the four smaller former republics.[18]

### *The Tashkent Document on the CFE Treaty*

The CIS summit meeting in Tashkent on 15 May 1992 saw agreement on allocations of TLE between the eight former Soviet republics when Armenia, Azerbaijan, Belarus, Georgia, Kazakhstan, Moldova, Russia and Ukraine signed the Agreement on the Principles and Procedures of Implementation of the Treaty on Conventional Armed Forces in Europe (the Tashkent Document).[19] The TLE allocations were presented in the relevant CIS capitals immediately after the Tashkent meeting and to the HLWG on 25 May in Brussels (see table 12.2).

### *The Oslo Document*

On 2 June the JCG in Vienna negotiated and agreed the language changes required in the 1990 Treaty text to accommodate the newly independent republics. The Extraordinary Meeting called for in the February road map

[16] Rogov, S. (ed.), Russian Defence Policy: Challenges and Developments, an occasional paper by the Institute of USA and Canada (Moscow) and the Center for Naval Analyses (Alexandria, Va.), Feb. 1993, pp. 4–6.

[17] Volkov, D., *Financial Times*, 7 Apr. 1992.

[18] Falkenrath, R., 'Ratification of the CFE Treaty in the former USSR', chapter 8 of unpublished PhD dissertation draft, King's College, University of London.

[19] See appendix 12C for the text of the Tashkent Document.

**Table 12.2.** Allocation of treaty-limited equipment entitlements among the former Soviet republics

| | Main battle tanks | | | Armoured combat vehicles | | | | | Artillery | | | | |
|---|---|---|---|---|---|---|---|---|---|---|---|---|---|
| | Total | Active | Stored | Total | Active | Stored | AIFVs | HACVs | Total | Active | Stored | Aircraft | Helicopters |
| Total for Russia | 6 400 | 4 975 | 1 425 | 11 480 | 10 525 | 955 | 7 030 | 574 | 6 415 | 5 105 | 1 310 | 3 450 | 890 |
| Total for Ukraine | 4 080 | 3 130 | 950 | 5 050 | 4 350 | 700 | 3 095 | 253 | 4 040 | 3 240 | 800 | 1 090 | 330 |
| *Total for former republics in zone IV.2* | 10 300 | 8 650 | 1 650 | 17 400 | 16 120 | 1 280 | n.a. | n.a. | 9 500 | 8 050 | 1 450 | n.a. | n.a. |
| Russia | 5 100 | 4 275 | 825 | 10 100 | 9 945 | 155 | n.a. | n.a. | 4 735 | 3 825 | 910 | n.a. | n.a. |
| Ukraine | 3 400 | 2 850 | 550 | 4 700 | 4 000 | 700 | n.a. | n.a. | 3 150 | 2 850 | 300 | n.a. | n.a. |
| Belarus | 1 800 | 1 525 | 275 | 2 600 | 2 175 | 425 | 1 590 | 130 | 1 615 | 1 375 | 240 | 260 | 80 |
| *Total for former republics in the flank zone* | 2 850 | 1 850 | 1 000 | 2 600 | 1 800 | 800 | n.a. | n.a. | 3 675 | 2 775 | 900 | n.a. | n.a. |
| Russia | 1 300 | 700 | 600 | 1 380 | 580 | 800 | n.a. | n.a. | 1 680 | 1 280 | 400 | n.a. | n.a. |
| Ukraine | 680 | 280 | 400 | 350 | 350 | 0 | n.a. | n.a. | 890 | 390 | 500 | n.a. | n.a. |
| Moldova | 210 | 210 | 0 | 210 | 210 | 0 | 130 | 10 | 250 | 250 | 0 | 50 | 50 |
| Georgia | 220 | 220 | 0 | 220 | 220 | 0 | 135 | 11 | 285 | 285 | 0 | 100 | 50 |
| Armenia | 220 | 220 | 0 | 220 | 220 | 0 | 135 | 11 | 285 | 285 | 0 | 100 | 50 |
| Azerbaijan | 220 | 220 | 0 | 220 | 220 | 0 | 135 | 11 | 285 | 285 | 0 | 100 | 50 |
| **Total** | **13 150** | **10 500** | **2 650** | **20 000** | **17 920** | **2 080** | **12 250** | **1 000** | **13 175** | **10 825** | **2 350** | **5 150** | **1 500** |

*Note:* AIFVs = armoured infantry fighting vehicles; HACVs = heavy armoured combat vehicles.

*Source:* Chairman's Summary of HLWG Meeting on 25 May 1992, *NATO Press Release,* vol. 92, no. 50 (25 May 1992).

occurred on 5 June at the NACC meeting in Oslo at which the 29 CFE signatories formally approved these changes and the allocation of TLE among the CIS states.[20] Annex A of the Oslo Document contains the understandings and changes to the Treaty necessitated by the dissolution of the USSR into its constituent republics, and Annex B contains notifications, confirmations and commitments, in most cases clarifying earlier unilateral statements. These include commitments to the exchanges of information required by the Treaty, clarification of the commitments made by the USSR in June 1991 relating to land-based naval forces, and the temporary resolution of differences over Article XII of the CFE Treaty limiting the number of AIFVs.

*The Provisional Application of the CFE Treaty*

On 10 July 1992, at the close of the CSCE summit meeting, the 29 states parties signed the Provisional Application of the CFE Treaty, which brought the Treaty provisionally into force. This met the schedule set out in the HLWG road map.

While the group of NATO states parties to the CFE Treaty breathed a sigh of relief that its provisions could now be applied from 17 July, there were some misgivings among the former WTO allies, articulated most forcefully by Poland and Hungary, that they had not been consulted about reallocation of TLE among the CIS states. The 3 November 1990 Budapest Agreement by WTO foreign ministers had allocated TLE ceilings among the WTO group of states parties and stipulated that all members of the group must be consulted before the ceilings were re-adjusted. (On 5 February 1993, for example, the new Czech and Slovak Republics formally asked permission of all CFE states parties to split their TLE allotments at an Extraordinary Meeting of the JCG.) Article VII of the CFE Treaty stipulates how states can adjust their holdings within the limits set for each equipment category permitted each of the two main groups of states parties. In effect maximum levels of TLE can be increased in a group only if another member of the group is willing to decrease its TLE by the same amount. Paragraph 7 of Article VII stipulates that 'States belonging to the same group of States Parties shall consult in order to ensure that the maximum levels for holdings notified . . . do not exceed the limitations set forth in Articles IV, V and VI'.

Poland and Hungary complained not only that they were not consulted prior to the May 1992 Tashkent meeting but also that the Tashkent Document stated that any further reallocations will be agreed among the CIS states only, implying no need to consult other members of the Budapest group of states parties. For the moment the European members of the group have been given informal reassurances that they will be consulted prior to any future ceiling adjustments.[21]

[20] The Final Document of the Extraordinary Conference of the States Parties to the Treaty on Conventional Armed Forces in Europe (Oslo Document), Oslo, 5 June 1992, is reproduced in appendix 12C.

[21] Falkenrath (note 18).

## Attitudes to the CFE Treaty among the former Soviet republics

Throughout the ratification debates, Western government spokesmen, as well as those from Central Europe, emphasized the relevance of the CFE Treaty for the future stability and security of Europe. The old NATO–WTO balances codified by the Treaty were rendered meaningless by the collapse of the WTO and breakup of the USSR. Nevertheless, the European states parties repeatedly told their CIS partners in the HLWG that the implementation regime attached to the Treaty will be the basis of trust and confidence between the states parties in the post-cold war era. For the former Soviet republics who want good relations with the Western democracies, especially for those that aspire to membership in Western institutions such as NATO, the Western European Union (WEU) and the European Community (EC), adherence to the CFE Treaty thus offered an important opportunity to emerge from the shadow of Russian domination and to participate in a serious enterprise with their Western partners. Moreover, the Treaty implementation regime gives a *droit de regard* over Russian military planning to ensure compliance with CFE limits.

Ukraine was perhaps the most enthusiastic supporter of the CFE Treaty and the most anxious to distance itself from Moscow.[22] The Treaty was seen as a useful vehicle to engage the West in support of Ukrainian claims to what it considered to be its fair share of the former Soviet military assets in establishing its own national army. As table 12.2 shows, Ukraine achieved entitlement to more tanks and artillery, and Russia far fewer, than Defence Minister Kobets and other Russian military spokesmen had anticipated in February 1992. Moreover, Ukraine's entitlements are higher than any other former WTO state except Russia and more than the sum of entitlements for Poland, Hungary, Slovakia and the Czech Republic, a preponderance of military capability that was disconcerting to defence planners in Budapest and Warsaw.[23]

Belarus was unique among the former Soviet republics parties to the CFE Treaty in judging that it had inherited too much military equipment from the USSR. Its government was slower than that of Ukraine to ratify the Treaty and slower to create its own independent army. Initially Belarussian leaders had hoped that Russia would subsidize its defence via a joint CIS army, but Russia wanted a larger contribution from Belarus for the CIS than the Minsk Parliament would support. The Parliament authorized an independent army on 20 March 1992 but did not ratify the CFE Treaty until October. On several occasions during the summer of 1992 President Stanislav Shushkevich reassured his Western partners that Belarus could be counted on to ratify the Treaty.[24] The delay in formal ratification was not so much opposition to the terms of the Treaty *per se*, as unwillingness on the part of the President to reassemble a Parliament he feared might derail his reform programme.

[22] Ukrainian statements in support of the CFE Treaty in *Holos Ukrayiny* (Kiev), no. 232 (30 Nov. 1991), in FBIS-SOV-91-240, 13 Dec. 1991, p. 76.

[23] Author's interviews in Warsaw in Jan. 1993.

[24] On Belarus, see Mihalisko, K., 'Belarus', *RFE/RL Research Report*, vol. 1, no. 7 (14 Feb. 1992), pp. 6–10; 'Belarus moves to assert its own military policy', *RFE/RL Research Report*, vol. 1, no. 11 (13 Mar. 1992), pp. 47–50.

It was difficult for the smaller countries in the ATTU zone to assess the costs and benefits of the CFE Treaty, because none of them had a technical community able to analyse the zonal sub-limits, complex destruction procedures, exchange of information requirements, and schedules for active and passive inspections. Like Ukraine, however, most of them hoped to exploit the CFE Treaty to legalize their control over the Soviet military assets on their territory.

Apart from Russia, Armenia was the former Soviet republic that was least enthusiastic about the CFE Treaty. It felt highly vulnerable, sandwiched between two countries (Turkey and Azerbaijan) that were potential allies with each other and hostile to Armenia. An agreement that accorded all the smaller states in the region equal allotments of TLE thus seemed unfair, but Armenia was less interested in independent control over former Soviet assets than in some kind of collective arrangement with Russia. Armenia was one of the most enthusiastic supporters of a unified CIS force at the February 1992 CIS summit meeting in Minsk, and signed the CIS Treaty on Collective Security on 15 May 1992. The government in Yerevan was disappointed later, however, by the lack of Russian support for its bid to control Nagorno-Karabakh. On 9 August, President Levon Ter-Petrosyan tried to invoke the CIS security pact against Azerbaijan, but was told by the Chief of CIS Joint Forces that the pact had not been ratified so it could not be considered in force.

Azerbaijan, however, had refused to sign the CIS Treaty on Collective Security, and on 7 October the parliament in Baku even voted to withdraw from the CIS altogether.[25] Azerbaijan was the most anxious to establish its own army and tried to do so by taking over the assets of the Russian 4th Army. The Russians were not enthusiastic about handing over equipment to the Azeris, however, 'until a peaceful settlement had been reached with Armenia with respect to the future control over Nagorno-Karabakh'.[26] The CFE Treaty was obviously seen in Baku as one way to try to force the Russians to relinquish control over equipment.[27] In the event, after the Tashkent Document in May, General Pavel Grachev claimed the three states of the Transcaucasus simply stole most of the equipment belonging to the former Soviet units on their territory.[28]

During 1992, Georgia seemed one of the least stable of the smaller newly independent states as it was embroiled in both internal and border crises. Georgia sat on the HLWG and was sometimes an observer to (but not a member of) the CIS. When Eduard Shevardnadze returned to Georgia to head its State Council, he advocated the formation of a national army but appeared anxious to play down differences with Russia. Russian intervention with

[25] Agence France Presse, 'Azerbaijan Parliament bars commonwealth membership', *International Herald Tribune*, 8 Oct. 1992.

[26] Lt General Yuri Grekov, former First Deputy Commander of the Transcaucasian Military District, cited in *Krasnaya Zvezda*, 22 Feb. 1992.

[27] Fuller, E., 'Nagorno-Karabakh: internal conflict becomes international', *RFE/RL Research Report*, vol. 1, no. 11 (13 Mar. 1992), pp. 1–5.

[28] Rogov (note 16), pp. 25–26.

peace-keeping forces that appeared partial to the Abkhazian separatists exacerbated relations after the 29 July declaration of sovereignty by Abkhazia.

Throughout the CFE Treaty ratification debate Moldova was also embroiled in a civil war and anxious to gain control over the former Soviet military equipment on its territory.[29] The Treaty was valuable to Moldova mainly as a way to focus Western attention on Russian 14th Army support for the separatist movement in Trans-Dniester. In the Tashkent Agreement the Moldova ceilings correspond to those owned by the 14th Army on Moldovan soil. During summer 1992 Moldova appropriated most of the equipment of the 14th Soviet Army on the right bank of the Dniester River, while the equipment of the 14th Army on the left bank remained formally under Russian jurisdiction.[30]

## Russian problems with the CFE Treaty

Of all the former Soviet republics, Russia was the least enthusiastic about the CFE Treaty since it codified its lost status as a military superpower. In addition to its absolute loss of military power compared to that of the former Soviet armed forces (see table 12.3), Russian territory will also be subject to inspections by all the other 28 states parties to the Treaty.

After the dissolution of the USSR into its constituent republics, the Russian Government was reluctant to withdraw forces from the newly independent states in the ATTU zone, to recognize the other former republics as independent parties to the Treaty or to reallocate the assets of the Soviet armed forces among the other states. The Russian General Staff in particular appeared to regard the entire territory of the former USSR as its strategic space and all Russian-speaking people as in need of military protection.[31]

Beyond the need to divide up military assets in compliance with international arms control obligations, the CIS states also had to restructure, and reallocate responsibility for, the former Soviet forces now spread over 15 different republics. Decisions about who should acquire and control former Soviet military units varied according to the degree of independence the different republics wanted from Moscow as well as their desire and ability to raise national armies.

These were difficult negotiations because during the cold war years the Soviet leadership had deployed its most up-to-date offensive equipment on its western border facing NATO. This policy left larger amounts of modern equipment in Belarus and Ukraine than Russia felt to be acceptable after the breakup of the Union. In addition, Russia, Belarus, Kazakhstan and Ukraine

[29] For details of the dispute in Moldova, see chapter 3 in this volume; see also Socor, V., 'Moldova's "Dniester" ulcer', *RFE/RL Research Report*, vol. 2, no.1 (1 Jan. 1993).

[30] Rogov (note 16), p. 9.

[31] The new draft Russian doctrine is outlined in *Foundations of Russia's Defence Doctrine*, a special issue of *Voennaya mysl'* [Military Thought] summarized by Gross, N., 'Reflections on Russia's new military doctrine', *Jane's Intelligence Review*, Aug. 1992, pp. 339–41.

**Table 12.3.** Treaty-limited equipment in the former Soviet republics compared with Soviet holdings in 1988

Allocations are those adopted at Tashkent, 15 May 1992.

| State | Tanks | Artillery | ACVs | Helicopters | Aircraft |
|---|---|---|---|---|---|
| Russia | 6 400 | 6 415 | 11 480 | 890 | 3 450 |
| Belarus | 1 800 | 1 615 | 2 600 | 80 | 260 |
| Ukraine | 4 080 | 4 040 | 5 050 | 330 | 1 090 |
| Moldova | 210 | 250 | 210 | 50 | 50 |
| Kazakhstan | 0 | 0 | 0 | 0 | 0 |
| Georgia | 220 | 285 | 220 | 50 | 100 |
| Armenia | 220 | 285 | 220 | 50 | 100 |
| Azerbaijan | 220 | 285 | 220 | 50 | 100 |
| **Total** | **13 150** | **13 175** | **20 000** | **1 500** | **5 150** |
| *Soviet holdings in December 1988* | | | | | |
| **Total** | **41 580** | **52 400** | **57 800** | **–** [a] | **8 395** |

[a] Helicopter assets not reported

*Source*: Crawford, D., *Conventional Armed Forces in Europe (CFE): A Reprise of Key Treaty Elements* (ACDA: Washington, DC, 1993).

were trying to resolve control over former Soviet nuclear weapons as well as ground force equipment.[32] Russia and Ukraine also disagreed over who should control the Crimea and the Black Sea Fleet, and several of the former republics were at war and in no mood to accept limits on their armed forces.

In early 1993 the future of the CIS looked very bleak. Georgia remained aloof, Armenia and Moldova were decidedly lukewarm, and Azerbaijan voted to withdraw from the CIS in October 1992. Of the original Slavic trio, Ukraine and Belarus had agreed to join primarily to ensure the end of the USSR and both states clearly hope for closer ties with Western democracies than with Russia.

On the other hand, the Central Asian states were growing increasingly apprehensive about their neighbours and began to yearn for the old reassurances of economic union with and military protection from Russia.[33] The five Central Asian states thus established what they called a 'common security space' with Russia, not least because in addition to their local and internal problems all saw potential threats from outside the borders of the old Soviet Union and felt the need for continued protection from Moscow. Where conflicts had erupted over border disputes or ethnic rivalries, the governments of some republics wanted Russian troops to serve as peace-keeping forces to separate the warring factions.

[32] On the problems of dismantling the former Soviet nuclear arsenal, see Miller, S. E. *et al.*, *Co-operative Denuclearization: From Pledges to Deeds* (Center for Science and International Affairs (CSIA), Harvard University: Harvard, Jan. 1993); and Potter, W. C., 'Nuclear exports from the former Soviet Union: what's new, what's true?', *Arms Control Today*, vol. 23, no. 1 (Jan. 1993), pp. 3–10.

[33] McElvoy, A., 'Republics seek reunion with Moscow', *The Times*, 8 Oct. 1992.

**Peace-keeping or military intervention?**

On 15 May 1992 in Tashkent, at the same meeting which re-allocated CFE ceilings among the former Soviet republics, six CIS states also signed a new Treaty on Collective Security that pledged each of the signatories to come to the aid of any other that was attacked.[34] This pledge was soon put to the test as breakaway factions in several republics took up arms against their governments. In early July, CIS leaders meeting in Moscow agreed to establish joint CIS peace-keeping forces to disengage warring factions in the Commonwealth. On 16 July in Tashkent the foreign and defence ministers of seven CIS states (Armenia, Kazakhstan, Kyrgyzstan, Moldova, Russia, Tajikistan and Uzbekistan) signed a Protocol committing them to contribute men and equipment to CIS peace-keeping operations.

Patterned on the UN 'blue berets', the CIS peace-keepers were to wear white helmets with a blue stripe, would be mustered on a case-by-case basis and should preferably be volunteers not conscripts. In the latter part of 1992, varying combinations of these national contingents were deployed as CIS 'peace-keepers' in a number of conflicts within and between the former Soviet republics. In most cases, however, these actions were less peace-keeping operations than post-imperialist military interventions.[35]

During 1992 two schools of thought emerged in Moscow about how to handle conflicts on the periphery of Russia.[36] The 'Atlanticists' in the Foreign Ministry appeared to prefer peace-keeping actions to be conducted under the auspices of the UN, the CSCE or NACC.[37] Conservatives in Parliament and the military cited the inability of international institutions to end the conflict in Bosnia and Herzegovina and, together with the military, apparently want to handle peripheral conflicts as any post-imperialist power would.[38]

NATO governments watched this process cautiously. In early 1993 the Clinton Administration sought Russian help in resolving the Bosnian crisis, but few, if any, Western leaders wanted to share the burden of restoring order in the former Soviet republics even under the auspices of the UN, the CSCE or NACC.[39] On the other hand Western governments have mixed feelings about acquiescing to a new Russian imperialism, and urged Russia to negotiate firm timetables for the withdrawal of its forces from the newly independent former Soviet republics as they did from the former non-Soviet WTO allies Czecho-

[34] Galeotti, M., 'Decline and fall: the Tashkent summit', *Jane's Intelligence Review*, vol. 4, no. 9 (Sep. 1992), p. 386.

[35] Rogov (note 16), pp. 53–55.

[36] Crow, S., 'Competing blueprints for Russian foreign policy', *RFE/RL Research Reports*, vol. 1, no. 50 (18 Dec. 1992), pp. 45–50.

[37] Kozyrev, A., 'The new Russia and the Atlantic Alliance', *NATO Review*, vol. 41, no. 1 (Feb. 1993), pp. 3–6.

[38] Karaganov, S., 'A strategy for Russia', *Nezavisimaya Gazeta*, 19 Aug. 1992.

[39] At its July summit meeting the CSCE was redefined as a regional security organization under Chapter VIII of the UN Charter with a view to co-ordinating peace-keeping (but not peace-enforcing) efforts in the CSCE area, an area that now stretches from Vancouver to Vladivostock. The *Helsinki Document 1992* (see appendix 5A) states that the CSCE may call upon other organizations such as the EC, NATO, WEU and the peace-keeping mechanisms of the CIS to support peace-keeping in the CSCE region.

slovakia, the German Democratic Republic (GDR), Hungary and Poland.[40] Russia found these negotiations difficult to conduct, especially in the Baltic states, Moldova and the Transcaucasus because some Russian parliamentarians questioned the pace of Russian withdrawals, suggesting that the former Soviet republics should not be considered fully sovereign until they had guaranteed the rights of Russian and Russian-speaking minorities.

### Russian troop withdrawals from the Baltic States

The Baltic Military District was formally disbanded in November 1991 by Soviet President Mikhail Gorbachev and the Soviet forces there redesignated the Northwestern Group of Forces to serve, until withdrawal home to Russia, on the same basis as Soviet troops had in the former allied territories of Czechoslovakia, the GDR, Hungary and Poland. In 1992 Russia took over responsibility for these forces and for the negotiation of withdrawals begun under Gorbachev. Withdrawal schedules proved difficult to negotiate, however, because the Baltic states wanted the Russian troops to leave immediately, whereas Russia was short of housing for returning army families.

The Baltic governments appealed to the international community for support on the withdrawal issue, asking both the CSCE and the UN to pressure the Russians to withdraw more quickly and to provide observers to monitor the withdrawals. This succeeded in some measure. In October President Bush signed a US Foreign Aid bill that made aid to Russia conditional on Russian commitments to the removal of troops from the Baltic states.[41]

In September a withdrawal timetable was signed by the Lithuanian and Russian defence ministers specifying that all Russian troops should be withdrawn by 31 August 1993, but agreements with Estonia and Latvia proved more difficult. Russian withdrawals continued throughout 1992 but Russia would not sign timetables with these governments because, as President Yeltsin said on 7 October on Russian television, they had not brought their human and civil rights into line with international standards.[42] On 29 October Yeltsin issued a decree suspending all troop withdrawals from the Baltic states, citing lack of housing in Russia and denial of basic rights to Russian nationals in the Baltic states.[43] This announcement followed a decision on 20 October by the Russian Defence Ministry to suspend withdrawals, suggesting that Yeltsin's announcement was simply rubber-stamping the decisions of the military leadership. Deputy Defence Minister Boris Gromov said immediately after Yeltsin's 29 October announcement that, although the

[40] For details of Soviet withdrawals from Germany, Poland, Hungary and the Czech and Slovak Republics, see Sharp, 1991 (note 7), pp. 433–39; and Brandenburg, U., *The Friends are Leaving: Soviet and Post-Soviet Troops in Germany after Unification* (Bundesinstitut für ostwissenschaftliche und internationale Studien: Cologne, 1992).

[41] Doherty, C., *US Congressional Quarterly Weekly Report*, 10 Oct. 1992, cited in *Arms Control Reporter*, sheet 407.E-1.90, 1992.

[42] Cited in Bungs, D., 'Latvia: toward full independence', *RFE/RL Research Reports*, vol. 2, no. 1 (1 Jan. 1993), pp. 96–98.

[43] Barber, T., 'Fury in Baltics over Yeltsin troop decree', *The Independent*, 31 Oct. 1992.

President said the suspension was not intended as a form of pressure on the Baltic states, Gromov himself thought pressure should be applied and that the optimum time for withdrawal would be the end of 1995.[44]

In September 1992 Gromov had announced that 74 000 Russian troops were in the Baltic states: 35 000 in Lithuania, 15 000 in Latvia and 24 000 in Estonia.[45] At the end of the year estimates varied as to how many Russian troops remained. In December, in a ceremony to mark the withdrawal of Russian troops from the Vilnius region, Lithuanian Defence Minister Audrius Butkevicius said that he anticipated that the remaining 15 000 Russian troops in Lithuania would leave by the end of August 1993 as agreed earlier.[46] In March 1993, however, Russian Foreign Minister Andrey Kozyrev, perhaps in an effort to conciliate the conservative right wing in the Russian Parliament, warned the Baltic states that gross violations of human rights against ethnic Russians would require the dispatch of vast numbers of Russian 'peace-keeping forces' in a 'new Yugoslavia'.[47] This brought further condemnations from NATO ministers at a meeting of NACC defence ministers in Brussels in March 1993.[48] General Grachev, at the NACC meeting, retorted that the Russians were under no legal obligation to withdraw from the Baltic states.[49]

### Russian troop withdrawals from the Transcaucasus and Moldova

The former Transcaucasus Military District (covering the territory of Armenia, Azerbaijan and Georgia) was disbanded in September 1992.[50] Withdrawal of Russian troops was complicated by the many conflicts that erupted in the region. General Grachev suggested that while Russia will not exceed its national ceilings for manpower and equipment as laid down in the CFE Treaty and CFE-1A Agreement, he cannot guarantee not to exceed some of the zonal sub-limits in the Transcaucasus and Moldova region as long as Russian peace-keeping forces are required there. This raises the question, certain to be debated in the CSCE Forum for Security Co-operation, of the extent to which paramilitary forces, and forces engaged in national or multinational missions that might be defined as 'peace-keeping', might be exempt from national CFE limits.

Armenia differed from the other three states in not wanting Russian troops to withdraw, or at least wanting to stretch out the withdrawal period as long as possible. When the 366th Russian motor-rifle regiment began to withdraw from Stepanakert in March, Armenians in Nagorno-Karabakh tried to block

[44] Cited by Davis, M. T., 'The suspension of the Baltic troop withdrawal', SHAPE, ref CND/1078, Supreme Headquarters Allied Powers Europe, Belgium, 19 Nov. 1992.

[45] *Arms Control Reporter*, sheets 407.E-1.89–90, 1992.

[46] Radio Vilnius cited in FBIS-SOV-92-251, 30 Dec. 1992, *Arms Control Reporter*, sheet 407.E-1.104, 1992.

[47] Bridge, A., 'Kozyrev warns Balts of a "new Yugoslavia"', *The Independent*, 17 Mar. 1993.

[48] Reuters, 'NATO assails Moscow failure to pull troops from Baltics', *Financial Times*, 30 Mar. 1993.

[49] Lucas, E., 'Kremlin halts pullout—again', *Baltic Independent*, vol. 3, no. 155 (2–8 Apr. 1993), p. 1.

[50] *Moscow Interfax*, in English, 22 May 1992 in FBIS-SOV-92-100, 22 May 1992, pp. 54–55.

the convoys, claiming that the unit offered the last shred of security in the region.[51] In April the Armenian Defence Ministry information service reported that the 15th division of the Russian 7th Army had begun to withdraw, and in May General Grachev said that Russian troops were only temporarily in the Transcaucasus and would all be withdrawn over the next two or three years.[52] In June, however, the Russian Defence Ministry announced suspension of troop withdrawals from Armenia, Azerbaijan and Georgia at the request of Armenia and Georgia.

The Azeris were quite aggressive during 1992 in their demands to take over equipment from departing Russian units, but the Russians were reluctant to make the transfers or to take sides in the dispute between Armenia and Azerbaijan over the enclave of Nagorno-Karabakh.

Georgia was anxious for the complete withdrawal of Russian troops, who were supporting the separatists in Abkhazia. In December the Georgian National Security and Defence Council wanted to suspend the Russian withdrawals, claiming that the Russians were taking equipment that now rightly belonged to Georgia under the terms of the CFE Treaty. In response Russia argued that negotiating allocations for permitted equipment under the Treaty in no way obliges Russia to provide that equipment. In March 1993 Grachev said that Russia must maintain a presence in Georgia indefinitely, 'otherwise we lose the Black Sea'.[53]

In November the Supreme Headquarters, Allied Powers Europe (SHAPE) reported that when Yeltsin approved delay of withdrawals from the Baltic states he also approved the formation (by 1 January 1993) of a new Russian Group of Forces-Transcaucasus, comprising three Russian divisions stationed in Armenia and Georgia. At the same time Yeltsin also approved the indefinite stationing of the Russian 14th Army in Moldova.[54] In Moldova, as in Georgia, Russian withdrawals were complicated by the fact that Russia has taken sides in the civil war there, in this case supporting the Trans-Dniester separatists.[55]

## III. Implementation of the CFE Treaty

Although Russian military spokesmen were unabashed about their dissatisfaction with the terms of the CFE Treaty, in July 1992 analysts at the Centre for National Security and International Relations, a subcommittee of the Committee of International Affairs and Foreign Economic Ties of the Supreme Soviet, produced a report that emphasizes the benefits to Russian security of strict compliance with the Treaty.[56] The report acknowledges the loss of capability from 50–60 per cent of all European military power once held by the USSR to 15 per cent now held by Russia. The report cited as advantages of the Treaty,

[51] Schmemann, S., *New York Times*, 4 Mar. 1992.
[52] *Arms Control Reporter*, sheet 407.E-1.82, 1992.
[53] Ablodia, T., 'Shevardnadze says Moscow backs rebels', *The Independent*, 17 Mar. 1993.
[54] Davis (note 44).
[55] For details of the conflict in Moldova, see chapter 3 in this volume.
[56] *Conventional Arms Treaty: Consequences for Russia*, summarized in *Krasnaya Zvezda*, 17 July 1992, FBIS-SOV-92-142, 23 July 1992, p. 38.

however, that it would force Russia to make reductions that would help the country to meet its economic goals and would enhance Russian security by cutting the military capability of neighbouring states.

In November 1992 the Russian Defence Ministry announced that it would eliminate its excess TLE in three stages: 25 per cent in the first six months, 60 per cent in the second stage of 12 months and 15 per cent in the final 12 months of the 40-month reduction period.

### Exchange of data

As required by the CFE Treaty, information on national equipment holdings was exchanged on 14 August 1992 (30 days after provisional entry into force on 17 July) and again on 15 December 1992. Tables 12.4 and 12.5 compare these data with data exchanged at signature in November 1990 and ceilings to be complied with at the end of the three-year reduction period in August 1995.

### Inspections

For inspection purposes the Treaty is divided into four phases (as illustrated in figure 12.1): phase I was the base-line validation period which lasted 120 days from the provisional entry into force on 17 July 1992 until 14 November 1992; phase II is the three-year reduction period that began on 15 November 1992; phase III is the 120-day residual validation period from 15 November 1995 until 15 March 1996, which will entail intensive inspection of the new reduced levels; and phase IV lasts for the (unlimited) duration of the Treaty. The Inspection Protocol lays down how many inspections each state must accept in each phase. This schedule provides for the number of inspections to be 20 per cent of previously defined objects of verification (OOVs) in Phase I, 10 per cent per annum in Phase II, 20 per cent again in Phase III, and 15 per cent per annum in Phase IV.[57] Table 12.6 estimates a schedule of inspections based on a re-allocation of Soviet assets according to the Tashkent Document.

The actual numbers of inspections in Phase I (mid-July to mid-November 1992—the initial baseline validation period) were 238 NATO inspections of non-NATO sites, 128 inspections by non-NATO parties and 17 intra-NATO inspections.[58] Between mid-November 1992 and mid-March 1993 over 100 Phase II inspections were conducted.[59] As the number of obligatory inspections relates to the number of declared OOVs, the schedule will be adjusted after each exchange of information to the extent the states parties change their declarations of OOVs, either because of equipment reductions or redeployment. The information exchanged on 15 December 1992, for example, shows substantial differences in OOV declarations from the previous information exchange (compare tables 12.6 and 12.7). This schedule was adjusted once in 1991 after the USSR revised its data, and again in June 1992 after the Tashkent Document.

[57] See *SIPRI Yearbook 1991* (note 7), table 13.2, p. 411.
[58] *Arms Control Reporter*, sheet 407.B.483, 1992.
[59] *Atlantic News*, no. 2509 (19 Mar. 1993), p. 2.

**Table 12.4.** NATO TLE holdings in 1990–92 and 1995 CFE Treaty ceilings

| CFE party | Date | Tanks | ACVs | Artillery | Aircraft | Helicopters | Total |
|---|---|---|---|---|---|---|---|
| Belgium | Nov. 90 | 359 | 1 381 | 376 | 191 | 0 | 2 307 |
| | Aug. 92 | 362 | 1 383 | 378 | 202 | 8 | 2 333 |
| | Dec. 92 | 362 | 1 267 | 378 | 202 | 10 | 2 219 |
| | Aug. 95 | 334 | 1 099 | 320 | 232 | 46 | 2 031 |
| Canada | Nov. 90 | 77 | 277 | 38 | 45 | 12 | 449 |
| | Aug. 92 | 76 | 136 | 32 | 28 | 0 | 272 |
| | Dec. 92 | 60 | 72 | 32 | 24 | 0 | 188 |
| | Aug. 95 | 77 | 277 | 38 | 90 | 13 | 495 |
| Denmark | Nov. 90 | 419 | 316 | 553 | 106 | 3 | 1 397 |
| | Aug. 92 | 499 | 316 | 553 | 106 | 12 | 1 486 |
| | Dec. 92 | 499 | 293 | 553 | 106 | 12 | 1 463 |
| | Aug. 95 | 353 | 316 | 553 | 106 | 12 | 1 340 |
| France | Nov. 90 | 1 343 | 4 177 | 1 360 | 699 | 418 | 7 997 |
| | Aug. 92 | 1 335 | 4 387 | 1 436 | 695 | 366 | 8 219 |
| | Dec. 92 | 1 335 | 4 154 | 1 392 | 688 | 376 | 7 945 |
| | Aug. 95 | 1 306 | 3 820 | 1 292 | 800 | 352 | 7 570 |
| Germany | Nov. 90 | 7 000 | 8 920 | 4 602 | 1 018 | 258 | 21 798 |
| | Aug. 92 | 7 170 | 9 099 | 4 735 | 1 040 | 256 | 22 300 |
| | Dec. 92 | 6 733 | 8 626 | 4 369 | 946 | 250 | 20 924 |
| | Aug. 95 | 4 166 | 3 446 | 2 705 | 900 | 306 | 11 523 |
| Greece | Nov. 90 | 1 879 | 1 641 | 1 908 | 469 | 0 | 5 897 |
| | Aug. 92 | 1 971 | 1 432 | 1 975 | 455 | 0 | 5 833 |
| | Dec. 92 | 2 276 | 1 430 | 2 149 | 458 | 1 | 6 314 |
| | Aug. 95 | 1 735 | 2 534 | 1 878 | 650 | 18 | 6 815 |
| Italy | Nov. 90 | 1 246 | 3 958 | 2 144 | 577 | 168 | 8 093 |
| | Aug. 92 | 1 232 | 3 774 | 2 013 | 542 | 176 | 7 737 |
| | Dec. 92 | 1 276 | 3 746 | 2 041 | 542 | 177 | 7 782 |
| | Aug. 95 | 1 348 | 3 339 | 1 955 | 650 | 142 | 7 434 |
| Netherlands | Nov. 90 | 913 | 1 467 | 837 | 196 | 91 | 3 504 |
| | Aug. 92 | 913 | 1 445 | 837 | 176 | 90 | 3 461 |
| | Dec. 92 | 813 | 1 445 | 837 | 175 | 90 | 3 360 |
| | Aug. 95 | 743 | 1 080 | 607 | 230 | 69 | 2 729 |
| Norway | Nov. 90 | 205 | 146 | 531 | 90 | 0 | 972 |
| | Aug. 92 | 205 | 124 | 544 | 89 | 0 | 962 |
| | Dec. 92 | 205 | 124 | 544 | 88 | 0 | 961 |
| | Aug. 95 | 170 | 225 | 527 | 100 | 0 | 1 022 |
| Portugal | Nov. 90 | 146 | 244 | 343 | 96 | 0 | 829 |
| | Aug. 92 | 146 | 280 | 354 | 92 | 0 | 872 |
| | Dec. 92 | 146 | 280 | 354 | 91 | 0 | 871 |
| | Aug. 95 | 300 | 430 | 450 | 160 | 26 | 1 366 |
| Spain | Nov. 90 | 854 | 1 256 | 1 373 | 242 | 28 | 3 753 |
| | Aug. 92 | 858 | 1 223 | 1 368 | 178 | 28 | 3 655 |
| | Dec. 92 | 896 | 1 057 | 1 219 | 175 | 28 | 3 375 |
| | Aug. 95 | 794 | 1 588 | 1 310 | 310 | 71 | 4 073 |
| Turkey | Nov. 90 | 2 823 | 1 502 | 3 442 | 511 | 5 | 8 283 |
| | Aug. 92 | 3 008 | 2 059 | 3 107 | 360 | 11 | 8 545 |
| | Dec. 92 | 3 234 | 1 862 | 3 210 | 355 | 11 | 8 672 |
| | Aug. 95 | 2 795 | 3 120 | 3 523 | 750 | 43 | 10 231 |
| UK | Nov. 90 | 1 198 | 3 193 | 636 | 842 | 368 | 6 237 |
| | Aug. 92 | 1 159 | 3 206 | 534 | 757 | 369 | 6 025 |
| | Dec. 92 | 1 078 | 3 003 | 502 | 717 | 340 | 5 640 |
| | Aug. 95 | 1 015 | 3 176 | 636 | 900 | 384 | 6 111 |
| USA | Nov. 90 | 5 904 | 5 747 | 2 601 | 626 | 243 | 15 121 |
| | Aug. 92 | 5 163 | 4 963 | 1 973 | 398 | 349 | 12 846 |
| | Dec. 92 | 4 511 | 4 800 | 1 773 | 334 | 341 | 11 759 |
| | Aug. 95 | 4 006 | 5 372 | 2 492 | 784 | 518 | 13 172 |
| **Totals** | **Dec. 92** | **23 424** | **32 159** | **19 353** | **4 901** | **1 636** | **81 473** |
| **for NATO** | **Aug. 95** | **19 142** | **29 822** | **18 286** | **6 662** | **2 000** | **75 912** |

**Table 12.5.** Former WTO TLE holdings in 1990–92 and 1995 CFE Treaty ceilings

| CFE party | Date | Tanks | ACVs | Artillery | Aircraft | Helicopters | Total |
|---|---|---|---|---|---|---|---|
| Bulgaria | Nov. 90 | 2 145 | 2 204 | 2 116 | 243 | 44 | 6 752 |
| | Aug. 92 | 2 269 | 2 232 | 2 154 | 335 | 44 | 7 034 |
| | Dec. 92 | 2 209 | 2 232 | 2 085 | 335 | 44 | 6 905 |
| | Aug. 95 | 1 475 | 2 000 | 1 750 | 234 | 67 | 5 526 |
| Czech. | Nov. 90 | 1 198 | 1 692 | 1 044 | 232 | 37 | 4 203 |
| | Aug. 92 | 1 803 | 2 515 | 1 723 | 228 | 37 | 6 306 |
| | Dec. 92 | 1 703 | 2 462 | 1 612 | 231 | 37 | 6 045 |
| | Aug. 95 | 957 | 1 367 | 767 | 230 | 50 | 3 371 |
| Hungary | Nov. 90 | 1 345 | 1 720 | 1 047 | 110 | 39 | 4 261 |
| | Aug. 92 | 1 345 | 1 731 | 1 047 | 143 | 39 | 4 305 |
| | Dec. 92 | 1 331 | 1 731 | 1 037 | 143 | 39 | 4 281 |
| | Aug. 95 | 835 | 1 700 | 840 | 180 | 108 | 3 663 |
| Poland | Nov. 90 | 2 850 | 3 377 | 2 300 | 551 | 29 | 9 107 |
| | Aug. 92 | 2 850 | 2 396 | 2 315 | 509 | 30 | 8 100 |
| | Dec. 92 | 2 807 | 2 416 | 2 309 | 508 | 30 | 8 070 |
| | Aug. 95 | 1 730 | 2 150 | 1 610 | 460 | 130 | 6 080 |
| Romania | Nov. 90 | 2 851 | 3 102 | 3 789 | 505 | 13 | 10 260 |
| | Aug. 92 | 2 967 | 3 171 | 3 942 | 508 | 15 | 10 603 |
| | Dec. 92 | 2 960 | 3 143 | 3 928 | 505 | 15 | 10 551 |
| | Aug. 95 | 1 375 | 2 100 | 1 475 | 430 | 120 | 5 500 |
| Slovakia | Nov. 90 | 559 | 846 | 522 | 116 | 19 | 2 062 |
| | Aug. 92 | 901 | 1 258 | 861 | 114 | 19 | 3 153 |
| | Dec. 92 | 851 | 1 231 | 806 | 116 | 18 | 3 022 |
| | Aug. 95 | 478 | 683 | 383 | 115 | 25 | 1 684 |
| **Totals for** | **Dec. 92** | **11 861** | **13 215** | **11 777** | **1 838** | **183** | **38 874** |
| **former NSWTO** | **Aug. 95** | **6 850** | **10 000** | **6 825** | **1 649** | **500** | **25 824** |
| Armenia | Nov. 90 | 258 | 641 | 357 | 0 | 7 | 1 263 |
| | Aug. 92 | n.a. | n.a. | n.a. | n.a. | n.a. | n.a. |
| | Dec. 92 | 77 | 189 | 160 | 3 | 13 | 442 |
| | Aug. 95 | 220 | 220 | 285 | 100 | 50 | 875 |
| Azerbaijan | Nov. 90 | 391 | 1 285 | 463 | 124 | 24 | 2 287 |
| | Aug. 92 | 134 | 113 | 126 | 15 | 9 | 397 |
| | Dec. 92 | 278 | 338 | 294 | 50 | 6 | 966 |
| | Aug. 95 | 220 | 220 | 285 | 100 | 50 | 875 |
| Belarus | Nov. 90 | 2 263 | 2 776 | 1 396 | 243 | 82 | 6 760 |
| | Aug. 92 | 3 457 | 3 824 | 1 562 | 335 | 76 | 9 254 |
| | Dec. 92 | 3 457 | 3 947 | 1 610 | 389 | 79 | 9 482 |
| | Aug. 95 | 1 800 | 2 600 | 1 615 | 260 | 80 | 6 355 |
| Georgia | Nov. 90 | 850 | 1 054 | 363 | 245 | 48 | 2 560 |
| | Aug. 92 | 77 | 28 | 0 | 0 | 0 | 105 |
| | Dec. 92 | 75 | 49 | 24 | 4 | 3 | 155 |
| | Aug. 95 | 220 | 220 | 285 | 100 | 50 | 875 |
| Moldova | Nov. 90 | 155 | 392 | 248 | 0 | 0 | 795 |
| | Aug. 92 | 0 | 98 | 108 | 30 | 0 | 236 |
| | Dec. 92 | 0 | 118 | 108 | 29 | 0 | 255 |
| | Aug. 95 | 210 | 210 | 250 | 50 | 50 | 770 |
| Russia | Nov. 90 | 10 333 | 16 589 | 7 719 | 4 161 | 1 035 | 39 837 |
| | Aug. 92 | 9 338 | 19 399 | 8 326 | 4 624 | 1 005 | 42 692 |
| | Dec. 92 | 7 993 | 16 469 | 7 003 | 4 387 | 989 | 36 841 |
| | Aug. 95 | 6 400 | 11 480 | 6 415 | 3 450 | 890 | 28 635 |
| Ukraine | Nov. 90 | 6 475 | 7 153 | 3 392 | 1 431 | 285 | 18 736 |
| | Aug. 92 | 6 128 | 6 703 | 3 591 | 1 648 | 271 | 18 341 |
| | Dec. 92 | 6 052 | 6 627 | 3 602 | 1 650 | 274 | 18 205 |
| | Aug. 95 | 4 080 | 5 050 | 4 040 | 1 090 | 330 | 14 590 |
| **Totals for** | **Dec. 92** | **17 932** | **27 737** | **12 801** | **6 512** | **1 364** | **66 346** |
| **former USSR** | **Aug. 95** | **13 150** | **20 000** | **13 175** | **5 150** | **1 500** | **52 975** |

Sources for tables 12.4 and 12.5 are given at the foot of the next page.

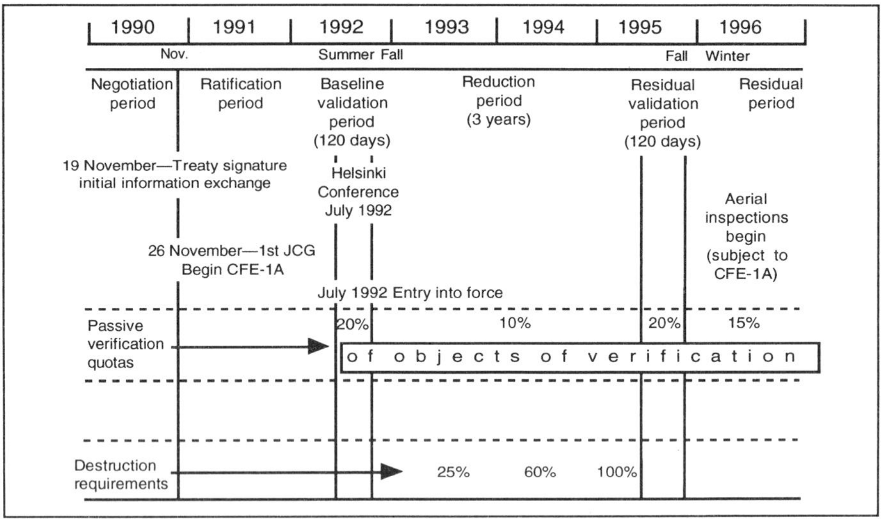


**Figure 12.1.** Implementation and verification of the CFE Treaty and the CFE-1A Agreement

*Source*: Adapted from a schedule supplied by the Office of the US Secretary of Defense, June 1992.

The Treaty inspection provisions are both a challenge and an opportunity for the states parties. The challenge is that compliance with the provisions on inspection and destruction, as well as on information exchanges, is a litmus test of the new co-operative relationship between former adversaries. The opportunities come not only from the potential to increase confidence and trust through transparency and co-operative compliance, but also for the less technologically endowed parties to share the more sophisticated technologies of the Western powers via East–West co-operative inspection teams.

Some parties of course see dangers and insecurities in the intrusive nature of the inspections. Some senior Russian military officers seem to resent the fact that not only former adversaries but also former subordinate allies and colonies now have the right to inspect hitherto secret military installations in Russia. Offers by some Western states to lead joint East–West inspection teams, although made largely in a spirit of co-operation and confidence-building, nevertheless risk exacerbating the paranoia of some Russian conservatives. These joint inspection teams were reportedly first suggested by Poland and taken up with varying degrees of enthusiasm by different NATO delegations.

*Sources for tables 12.4 and 12.5*: Feinstein, L., 'Factfile: weapons in Europe before and after CFE', *Arms Control Today*, Mar. 1993, p. 28; Crawford, D., *Conventional Armed Forces in Europe (CFE): A Reprise of Key Treaty Elements* (US Arms Control and Disarmament Agency: Washington, DC, 1993).

**Table12.6.** Estimated passive and challenge inspections, June 1992
Figures apply after re-allocation of TLE ceilings by former Soviet states.

| | | Inspections[a] each state must accept in: | | | |
|---|---|---|---|---|---|
| State | OOVs | Phase I | Phase II | Phase III | Phase IV |
| *NATO group* | | | | | |
| Belgium | 50 | 10 (2) | 5 (1) | 10 (2) | 8 (2) |
| Canada | 13 | 3 (1) | 1 (1) | 3 (1) | 2 (1) |
| Denmark | 64 | 13 (2) | 6 (1) | 13 (2) | 10 (2) |
| France | 257 | 51 (8) | 26 (4) | 51 (8) | 39 (9) |
| Germany | 470 | 94 (14) | 47 (7) | 94 (14) | 70 (16) |
| Greece | 60 | 12 (2) | 6 (1) | 12 (2) | 9 (2) |
| Iceland | – | – (1) | – (1) | – (1) | – (1) |
| Italy | 190 | 38 (6) | 19 (3) | 38 (6) | 28 (6) |
| Luxembourg | 2 | – (1) | – (1) | – (1) | – (1) |
| Netherlands | 88 | 18 (3) | 9 (1) | 18 (3) | 13 (3) |
| Norway | 59 | 12 (2) | 6 (1) | 12 (2) | 9 (2) |
| Portugal | 28 | 6 (1) | 3 (1) | 6 (1) | 4 (1) |
| Spain | 93 | 19 (3) | 9 (1) | 19 (3) | 14 (3) |
| Turkey | 150 | 30 (4) | 15 (2) | 30 (4) | 22 (5) |
| UK | 226 | 45 (7) | 23 (3) | 45 (7) | 34 (8) |
| USA | 169 | 34 (5) | 17 (3) | 34 (5) | 25 (6) |
| **Total (16)** | **1 919** | **385 (62)** | **192 (32)** | **385 (62)** | **287 (68)** |
| *Budapest/Tashkent group* | | | | | |
| Bulgaria | 93 | 19 (3) | 9 (1) | 19 (3) | 14 (3) |
| Czech/Slovak | 179 | 36 (5) | 18 (3) | 36 (5) | 27 (6) |
| Hungary | 59 | 12 (2) | 6 (1) | 12 (2) | 9 (2) |
| Poland | 134 | 27 (4) | 13 (2) | 27 (4) | 20 (5) |
| Romania | 127 | 25 (4) | 13 (2) | 25 (4) | 19 (4) |
| USSR | 910 | 182 (27) | 91 (14) | 182 (27) | 136 (31) |
| of which: | | | | | |
| Russia | 491 (54%) | 98 (14) | | | |
| Ukraine | 253 (28%) | 50 (6) | | | |
| Belarus | 102 (12%) | 22 (3) | | | |
| Armenia | 16(1.5%) | 3 (1) | | | |
| Azerbaijan | 16(1.5%) | 3 (1) | | | |
| Georgia | 16(1.5%) | 3 (1) | | | |
| Moldova | 16(1.5%) | 3 (1) | | | |
| **Total (12)** | **1 502** | **301 (45)** | **150 (23)** | **301 (45)** | **225 (51)** |

[a] Challenge inspections are in parentheses.

*Sources*: 19 Feb. 1991 corrections to original CFE documents; Dunay, P., 'Verifying conventional arms limitations: the case of the November 19, 1990 Treaty on Conventional Armed Forces in Europe', *Bochumer Schriften*, no. 6, Bochum, FRG, 1991, p. 139; Agreement on the Principles and Procedures for implementing the Treaty on Conventional Armed Forces in Europe (see appendix 12C); Factfile, *Arms Control Today*, vol. 22, no. 5 (June 1992), p. 32.

During 1991 and 1992, parallel to the ratification process, a number of signatories to the CFE Treaty conducted trial inspections partly to test the procedures and partly as confidence-building measures. Some were multinational, both West–West and East–West. The multinational teams began in late 1990 with informal Anglo-French co-operation, followed in 1991 by East–West

**Table 12.7.** CFE Treaty declared sites and objects of verification

| NATO group | | | Budapest/Tashkent group | | |
|---|---|---|---|---|---|
| State | OOVs | Declared sites | State | OOVs | Declared sites |
| Belgium | 59 | 35 | Armenia | 10 | 8 |
| Canada | 4 | 3 | Azerbaijan | (not yet reported) | |
| Denmark | 63 | 32 | Belarus | 74 | 51 |
| France | 211 | 168 | Bulgaria | 114 | 94 |
| Germany | 255 | 215 | Czech Republic | 79 | 62 |
| Greece | 82 | 71 | Georgia | 6 | 6 |
| Italy | 186 | 180 | Hungary | 46 | 35 |
| Netherlands | 62 | 41 | Moldova | 8 | 5 |
| Norway | 47 | 32 | Poland | 149 | 124 |
| Portugal | 39 | 37 | Romania | 164 | 130 |
| Spain | 95 | 92 | Russia | 431 | 299 |
| Turkey | 120 | 102 | Slovakia | 40 | 31 |
| UK | 180 | 152 | Ukraine | 207 | 135 |
| USA | 105 | 70 | | | |

*Note:* Iceland, Kazakhstan and Luxembourg have no declared sites nor objects of verification in the area of application.
*Source*: Reported at the annual information exchange, 15 Dec. 1992.

practice teams including a Netherlands–Polish team and several bilateral teams involving Hungary with a NATO partner.[60]

In July 1992 baseline inspections included inspections by Russia of British bases, by France in Ukraine, by the UK in Bulgaria and by the Netherlands in Belarus. In August, Norway inspected Russian units in the Kola Peninsula. Inspections of equipment destruction included British and French inspections of the destruction of Ukrainian tanks in August. In November US and Netherlands officials inspected a motorized rifle division in the Volga region of Russia.

In most cases the inspection teams reported positive results and expressed confidence that states parties were doing their best to comply with Treaty provisions. Russia and Ukraine complained, however, that the approved destruction procedures were unnecessarily expensive. The JCG responded by promising to explore other methods, but some Western delegates cautioned that procedures must not be so degraded that they became reversible; weapons must be rendered unambiguously inoperable. There were also some instances of Western inspectors complaining about violations in the form of denial of inspection rights at certain sites in Russia, Ukraine and Belarus.[61] Normally, when teams arrive for inspection at a declared site, they will choose one or two OOVs to inspect and also have access to other common areas at the sites (e.g., ammunition stores and helicopter landing pads). Problems arose for Western inspectors when Russia started to redefine common areas as OOVs, thereby cutting the area between OOVs that should have been available for

[60] Colonel Roy Giles, Joint Arms Control Implementation Group, UK, personal communication.
[61] US Arms Control and Disarmament Agency, *Report to Congress on Soviet Non-Compliance with Arms Control Agreements*, cited in *Arms Control Reporter*, sheet 407.B.485, 1993.

inspection. In its last report the Bush Administration's Arms Control and Disarmament Agency also cited as violations by some of the former Soviet republics: under-declaring reduction obligations, inaccurate data declarations and exporting excess TLE.[62] In any event these problems were all satisfactorily resolved in the JCG.

NATO's Committeee on Coordination and Verification held a two-day seminar on CFE verification in Brussels in January 1993 to which each state party (then 30 after the division of Czechoslovakia) sent two delegates. NATO officials judged that verification had so far proved a very positive experience in terms of transparency and co-operation. Difficulties were more often caused by communication problems than by deliberate evasion, and to accommodate such difficulties the notification period of an inspection was extended from 15 to 30 days. During the 120-day baseline phase Russia conducted all the inspections it was allowed, but some other former Soviet republics did not carry out any at all.

At the seminar, based on the very positive experiences of the trial inspections with multilateral teams, NATO offered its co-operation partners from the former WTO countries the opportunity to join NATO inspection teams in order to save money and increase the inspection efficiency of the less well-endowed CFE parties. NATO also offered training in verification as well as peace-keeping to its Eastern partners at the NATO school in Oberammergau. All the CFE states parties responded positively by mail to the NATO offer, although Russia was noticeably cooler to the idea of multinational inspections than the other parties. Not all the Western parties are equally enthusiastic either, some fearing that forms of industrial espionage might result. Nevertheless NATO has apparently agreed to offer to form multinational teams for 20 per cent of its inspections. The first of these was conducted in mid-March 1993 by a team of Azeri, Hungarian, Italian and Polish inspectors at a site in Romania.[63] British officials report that on 20 per cent of their inspections they will offer three of the nine inspection team places to co-operation partners.

Some effort has been made to make Western data available to former WTO states with the usual intra-NATO disputes about who does what. France, predictably, objects to NATO qua NATO facilitating this data distribution, claiming that the CSCE is thereby undermined. Germany has an expensive but not very efficient data distribution system that it wants to use. The UK and the USA would prefer to use a newly developed NATO data base.[64]

## IV. The CFE-1A Agreement

The CFE-1A Agreement signed at the 10 July CSCE summit meeting in Helsinki sets ceilings on various categories of military personnel in the terri-

[62] *Arms Control Reporter*, sheet 407.B.485, 1992.
[63] NATO Press Office, 'First joint multinational inspection under the CFE Treaty', *Press Release* (93), no. 26 (16 Mar. 1993).
[64] *Atlantic News*, no. 2494 (29 Jan. 1993), p. 2.

**Table 12.8.** CFE-1A manpower limitations

| NATO group | | | Budapest/Tashkent group | | |
|---|---|---|---|---|---|
| State | Ceilings | Holdings | State | Ceilings | Holdings |
| Belgium | 70 000 | 76 088 | Armenia | (not reported) | 7 101 |
| Canada | 10 660 | 4 077 | Azerbaijan | (not yet reported) | |
| Denmark | 39 000 | 29 256 | Belarus | 100 000 | 143 865 |
| France | 325 000 | 341 988 | Bulgaria | 104 000 | 99 404 |
| Germany | 345 000 | 401 102 | Czech Republic | 93 333 | 110 010 |
| Greece | 158 621 | 165 400 | Georgia | 40 000 | (not reported) |
| Italy | 315 000 | 294 900 | Hungary | 100 000 | 76 226 |
| Netherlands | 80 000 | 69 324 | Moldova | (not yet reported) | |
| Norway | 32 000 | 29 500 | Poland | 234 000 | 273 050 |
| Portugal | 75 000 | 39 700 | Romania | 230 000 | 244 807 |
| Spain | 300 000 | 177 078 | Russia | 1 450 000 | 1 298 299 |
| Turkey | 530 000 | 575 045 | Slovakia | 46 667 | 55 005 |
| UK | 260 000 | 288 626 | Ukraine | 450 000 | 509 531 |
| USA | 250 000 | 175 070 | | | |

*Note*: Iceland and Kazakhstan have no military manpower in the area of application; Luxembourg reported a strength of 618 against its ceiling of 900.
*Source*: Concluding Act of the Negotiation on Personnel Strength of Conventional Armed Forces in Europe (CFE-1A), 10 July 1992, Helsinki.

tories of the 29 states parties within the zone of application defined by the CFE Treaty, that is, the ATTU zone.[65] The primary purpose of the CFE-1A Agreement was to avoid a singular limitation on the Germans who agreed to limit the armed forces of a unified Germany to 370 000 in the context of making unification acceptable to the USSR in the summer of 1990.

Negotiations were conducted by the CFE Treaty signatories in Vienna from 26 November 1990 until July 1992. They got off to a slow start until the data disputes surrounding the Treaty were resolved in the summer of 1991. Thereafter the talks went relatively smoothly. Once the USSR broke up the Western states were anxious to conclude an agreement quickly so as to preclude proliferation of armed forces in the former Soviet republics. None of the parties was anxious to do more than register force levels. There was no pressure to make deep cuts; indeed for some parties the CFE-1A limits are higher than current force levels, allowing for growth (see table 12.8).

The Agreement comprises eight articles. Article I lists seven categories of full-time and one category of reserve manpower to be limited, and three categories not subject to limitation: peacetime security forces, personnel in transit in one place for less than seven days and personnel serving under UN command. Article II lists the national personnel limits. Article III deals with the required notification to make revisions in national limits: 42 days in most cases. Article IV deals with information exchange requirements. Article V provides for stabilizing measures. Article VI deals with verification and

[65] For the text of the Concluding Act of the Negotiation on Personnel Strength of Conventional Armed Forces in Europe (CFE-1A), 10 July 1992, Helsinki, see appendix 12C.

evaluation, Article VII deals with review mechanisms and Article VIII notes that the limits are politically rather than legally binding, will have the same duration as the CFE Treaty and may be supplemented, modified or superseded.

### Issues in the negotiations

There was some debate about verification. NATO countries are sceptical about whether manpower numbers can be adequately verified, a position they held throughout the fruitless years of Mutual and Balanced Force Reduction (MBFR) talks aimed at an inter-alliance agreement on conventional manpower from 1973 until 1989.

There was also some debate about which kinds of manpower should be limited. Russia, for example, did not want to limit paramilitary forces nor count manpower in units that did not field CFE Treaty-limited equipment. In April, for example, the Russian delegate in Vienna said that Russia was willing to limit land forces, most air forces, air defence aviation forces, naval infantry and coastal defence but not logistics, command and long-range transport forces. Russia also wanted an exemption for what it called strategic forces, presumably strategic rocket forces assigned to operate strategic missile installations.

Like the CFE Treaty, the CFE-1A Agreement requires an annual exchange of information (Article XIII).

Two states are assigned zero forces: Iceland, which has no armed forces, and Kazakhstan, whose territory lies almost wholly east of the Urals. In the text published on 10 July four states currently at war had not yet agreed to any limits on personnel: Armenia, Azerbaijan, Georgia and Moldova.

## V. Conclusion

The CFE process began as a *de facto* inter-alliance negotiating forum in the late 1980s and progressed to a genuinely pan-European arms control regime as the 1990 CFE Treaty imposed the rule of law over military force planning in Europe between the Atlantic Ocean and the Ural Mountains. This was no small achievement given the history of violent warfare on the continent in past centuries and the extraordinary military buildup on both sides of the iron curtain during more than four decades of cold war. Some of the manpower and equipment would certainly have been reduced unilaterally once the cold war ended, but the compliance mechanism, with its regular exchanges of information and schedule of inspections, is designed to build confidence and trust between the states parties and reduce insecurities that stem from unpredictability in force planning. This should strengthen the trend towards co-operation and conciliation between former adversaries in Europe.

The CFE process has been especially beneficial to the former allies of the USSR in Europe by providing an international legal framework in which they extricated themselves from Russian hegemony. These newly sovereign countries can now begin to develop their own foreign and defence policies as independent powers, able if they so wish to pool their sovereignty in new alliances or to remain non-aligned. The CFE Treaty also strengthened the hands of the non-Russian former Soviet republics. Without the international scrutiny inherent in the Treaty compliance mechanism Russia might not have been as willing to give up as much control as it has over former Soviet military assets.

As always in multilateral negotiations, the smaller powers (in this case especially, but not exclusively, Hungary and Poland) complained throughout the negotiations of the tendency of the two big powers, the USA and the USSR/Russia, to settle bilaterally issues that affect all the parties. Without strong US leadership, however, some CFE problems might never have been solved. Indeed on other issues during 1992, most notably the wars in the former Yugoslavia, the smaller European powers were crying out for more responsible and more intrusive leadership from the USA.

If the general, and overwhelming, view of the CFE Treaty in Europe and North America is positive there are also some shortcomings to consider. The Treaty solves many of the old problems that plagued cold war relationships on the continent, but does nothing to prepare (indeed may hinder) European governments to deal with the new post-cold war problems such as genocidal aggression in Bosnia and Herzegovina and unrest within and between former Soviet republics in the zone of application.

By codifying a balance between NATO and the WTO countries just as the WTO and the USSR collapsed, the Treaty also codified a humiliating loss of status for Russia, leading many conservative Russians to regard the CFE Treaty as the Versailles Treaty of the cold war. One way for Russia to overcome this sense of humiliation would be to engage in joint peace-keeping activity with Western powers, although the idea of joint action with the West seemed controversial in Moscow in early 1993.

The former unitary state of Yugoslavia was not a party to the CFE Treaty and the new Federal Republic of Yugoslavia (Serbia and Montenegro) was suspended from the CSCE (and from the United Nations) indefinitely because of its genocidal aggression in Bosnia and Herzegovina. Thus none of the elaborate procedures worked out in the arms control treaties and documents for transparency of military force postures (through regular exchanges of information and inspections) apply to the belligerent states in the Balkan war. Even if the former Yugoslavia had been a party to the CFE Treaty it would not have limited the kind of small-calibre artillery which took most lives in Bosnia and Herzegovina.[66]

With respect to controlling conflicts among and between former Soviet republics, the armed forces of eight of these are limited by the CFE Treaty and the CFE-1A Agreement, but most of the manpower and equipment in Russia

[66] The CFE Treaty limits artillery defined as large (over 100-mm) calibre systems (CFE Treaty Article IIF.).

and Kazakhstan beyond the Urals and in the four other Central Asian republics of Kyrgyzstan, Tajikistan, Turkmenistan and Uzbekistan are outside CFE jurisdiction.

Much unfinished arms control business thus remains for the new CSCE Forum for Security Co-operation. NATO has already offered joint peace-keeping training to its NACC partners, but given the wars raging in many of the former communist countries there is an urgent need to recruit and train more effective forces than most of those which served in the UN operations in Croatia and Bosnia and Herzegovina. In those conflicts, traditional 'blue beret' forces, with their open lines of communication and passive rules of engagement, not only failed to halt aggression but proved counter-productive, because they were seen by local populations in Bosnia and Herzegovina primarily as agents of Serbia's ethnic cleansing operation. European governments will need to recruit and train forces for effective military operations against unruly armed gangs. This is especially necessary as the USA is cutting its manpower in Europe to 100 000 men, well below the level permitted by the CFE-1A Agreement.

The need for more rather than less trained military manpower on the continent suggests that the CFE-1A Agreement has also been overtaken by events. The manpower limits in the CFE-1A Agreement were designed primarily to satisfy a German desire not to be the only European state to have accepted numerical limits on military personnel. In order to raise substantial European peace-keeping or peace-making forces, CFE-1A limits may have to be raised substantially or special exemptions made for national contributions to multilateral organizations.

# Appendix 12A. The Vienna confidence- and security-building measures in 1992

ZDZISLAW LACHOWSKI

## I. Introduction

The year 1992 bore further witness to the fact that confidence- and security-building measures (CSBMs) of the type agreed at the 1984–86 Stockholm Conference and the Vienna CSBM Negotiations are losing ground. They are increasingly overshadowed by the debate on other co-operative security arrangements. This is partly the result of the inadequacy of measures designed for an old political and military configuration, the former bloc division, in the face of new challenges and requirements and partly the result of the inherent limitations of the CSBMs themselves. Other security arrangements, such as conflict prevention and crisis management, peace-keeping and peace-making, as well as the search for new CSBMs, have become the focus of international attention.[1]

Events in the former Yugoslavia, in particular, have shown that 'classic' CSBMs are of little use in a new, non-bloc type of conflict that characteristically starts out as an intra-state conflict. Thus, in the run-up to and in the course of the Helsinki follow-up meeting (24 March–8 July 1992) of the Conference on Security and Co-operation in Europe (CSCE), the participating states adopted a two-track approach and worked on (*a*) the further improvement and supplementing of CSBMs and (*b*) new solutions and arrangements to be further elaborated within the framework of the new CSCE Forum for Security Co-operation (FSC). As a result of the former group of activities a new accord was reached.

The Vienna Document 1992 of the Negotiations on Confidence- and Security-Building Measures, agreed on 4 March[2] just prior to the Helsinki CSCE follow-up meeting, entered into force on 1 May. Like its forerunner, the Vienna Document 1992 is politically binding and not a treaty. It develops and builds upon the CSBMs established by the Vienna Document 1990 and supplements them with more detailed parameters and some additional measures. The major changes and additions are as follows:

1. Under the heading *annual exchange of military information,* the states undertake to provide additional information on military forces, planned increases in personnel, and temporary activation of non-active combat units and non-active formations. It is

[1] See the United Nations Security Council, *An Agenda for Peace: Preventive Diplomacy, Peace-making and Peace-keeping*, Report of the Secretary-General pursuant to the statement adopted by the Summit Meeting of the Security Council on 31 Jan. 1992, UN document A/47/277 (S/241111) (United Nations: New York, 17 June 1992), (reproduced as appendix 2A in this volume).

[2] *Vienna Document 1992 of the Negotiations on Confidence- and Security-Building Measures Convened in Accordance with the Relevant Provisions of the Concluding Document of the Vienna Meeting of the Conference on Security and Co-operation in Europe.* Vienna, 4 Mar. 1992, reproduced as appendix 12C in this volume. For major CSBM proposals in the period of negotiating the Vienna Document see Brauch, H. G. and Neuwirth, G. (eds), *Confidence and Security Building Measures in Europe II. From Vienna 1990 to Vienna 1992—Documents*, Arbeitsgruppe Friedensforschung und Europäische Sicherheitspolitik (AFES-PRESS) Report no. 28 (Mosbach, 1992).

required that detailed data (plus appropriate photographs) relating to major weapon and equipment systems be provided once to all other CSCE participating states.

2. Under *risk reduction,* provisions are strengthened by encouraging states to host visits to allay other participating states' concerns about military activities.

3. Under *military contacts,* new types of major weapon and equipment system are to be demonstrated to representatives of all other participating states.

4. Parameters for *prior notification of certain military activities* are further changed and supplemented in the pursuit of filling the 'transparency gap'.

5. Under *observation of certain military activities,* the numbers of troops subject to observation are also reduced and an additional category (battle tanks) is introduced.

6. Under *constraining measures,* there are further limitations on the size, number and notification requirements for major manœuvres.

7. Under *verification,* there is a possibility of an inspecting state inviting other participating states to take part in an inspection (multinational teams). The main novelty is that non-active formations and temporarily activated combat units are now subject to evaluation.[3]

In the run up to the 1992 Helsinki follow-up meeting, in parallel to the Vienna negotiations, consultations on new CSBMs were held in Vienna (17 September 1991–19 March 1992). The consultations did not result in a final document; the FSC set up in the wake of the Helsinki summit meeting is carrying on this work.[4]

During the year, the number of CSCE participating states and adherents to the Vienna Document 1992 rose to 53.[5] At the same time, the area of application was extended to cover the territories of several former Soviet republics,[6] thus embracing the areas defined in the Concluding Document of the 1983 CSCE follow-up meeting in Madrid ('the whole of Europe as well as the adjoining sea area and air space') plus the territories of Kazakhstan, Kyrgyzstan, Tajikistan, Turkmenistan and Uzbekistan.[7]

[3] For a detailed enumeration of the new measures in the Vienna Document 1992 see Palmisano, S., 'KSZE/VVSBM: vom Wiener Dokument 1990 zum Wiener Dokument 1992 Chronik', *Studien und Berichte* (Landesverteidigungsakademie: Vienna, Apr. 1992), pp. 59–68.

[4] At the consultations, the 'host' state of each session summed up the current state of debate from its own point of view. The last 'host's perception' summary of conclusions of the Norwegian delegation was circulated on 18 Mar. 1992.

[5] On 30 Apr. Bosnia and Herzegovina was accepted as the 52nd CSCE participating state. As of 8 July the CSCE suspended Yugoslavia (Serbia and Montenegro) from the CSCE for three months and then prolonged its membership suspension indefinitely. On 15 Dec. the Czech Republic and the Slovak Republic were admitted as separate participants to the CSCE and as of 1 Jan. 1993, after Czechoslovakia formally split, the CSCE membership rose to 53, including the suspended Yugoslavia.

[6] Russia took over CSCE membership and CSBM commitments from the former USSR. The Baltic states joined the Helsinki process on 10 Sep. 1991, and Georgia did so on 24 Mar. 1992. The CSCE Council of Foreign Ministers in Prague on 30 Jan. 1992 admitted the other 10 former Soviet republics on the basis of their 'Letters of Accession', submitted the day before, which stated that each government 'agrees to apply all the provisions of the Vienna Document on CSBMs, and to an understanding that the geographic scope of its application should be revised as soon as possible in order to ensure full effect of the rules of transparency, predictability and conflict prevention on its territory.' See *Journal No. 1, CSCE Second Meeting of the Council,* Prague, 30 Jan. 1992.

[7] See Annex I to the Vienna Document 1992 (note 2). It states that the commitments undertaken by the remaining 10 former Soviet republics have the effect of extending the application of CSBMs to the territories of those states 'insofar as their territories were not covered already by them.' Japan was also invited to co-operate and develop relations with the CSCE by attending CSCE meetings and contributing (bar participating in and adopting decisions) on subjects of direct interest. See *Helsinki Document 1992: The Challenges of Change*, Helsinki summit meeting, Helsinki, 1992, Helsinki Decisions, chapter IV, paras. 9–11 (for excerpts see appendix 5A in this volume).

The Norwegian paper of 18 Mar. 1992 proposed the extension of the area of application of CSBMs to longitude 90°E (the Yenisey River) to cover the Siberian area of military significance. Since the USA

On 21 March 1992, the Open Skies talks concluded in Vienna with the initialling of a treaty by 24 CSCE states. The Open Skies Treaty, providing for the aerial observation of an area from Vancouver to Vladivostok, has been welcomed as another measure of potential use for confidence-building and arms control (e.g., to aid verification of the 1990 Treaty on Conventional Armed Forces in Europe—CFE).

## II. Implementation

The effectiveness of the Vienna Document 1990 and the Vienna Document 1992 has been critically assessed by the participating states. CSBMs have been criticized for not adequately responding to developments in Europe, especially as regards preventing conflicts. In the former Yugoslavia massive flagrant infringements of principles of the 1975 Helsinki Final Act, CSBMs and other international norms continued, with violence spreading in spring 1992 to Bosnia and Herzegovina. As a result, the participation of Serbia and Montenegro in the CSCE was suspended. In the face of failure to apply conventional instruments effectively, a number of states began to lose interest in continuing work on the improvement of traditional CSBMs,[8] instead showing more interest in elaborating the structure and agenda of the new Forum for Security Co-operation which was to be called into being by the 1992 Helsinki summit meeting.

However, other CSCE participating states called for better implementation of new and existing measures, for example by making more effective use of the mechanism for consultation and co-operation as regards unusual military activities, multilateral inspections and the hosting of voluntary visits; co-ordinating the CSBM regime with instruments for monitoring, fact-finding or peace-keeping, and early warning, and widening its application; considering the applicability of existing CSBMs for internal and regional situations (e.g., lowering the thresholds for notification); generalizing some measures already carried out between neighbouring countries as regional measures; and harmonizing the CSBM and CFE Treaty regimes.[9]

### Military activities

The downward trends in the level of military activities that have been discernible since 1989, continued in 1992, and are certain to be sustained in 1993, too (see table 12A.1).

In 1992 only six military exercises subject to notification were conducted.[10] The activities subject to notification and observation continue to be reduced in numbers thanks to the change in the international political climate, profound transformations

rejected the Russian demand in Helsinki to make a 'compensation' concerning North American territory, a package deal was agreed whereby east-of-Urals data would be given by Russia on a voluntary basis.

[8] 'There's not much left to negotiate', was the prevailing conviction among CSCE officials at the time of the adoption of the Vienna Document 1992; see *Arms Control Reporter,* sheet 402.B.300, 1992.

[9] Following the second annual assessment meeting of the CSCE states, the CPC compiled an informal list of suggestions made by participants, entitled *Survey of Suggestions made at the Annual Implementation Assessment Meeting*, Vienna, 9–11 Nov. 1992.

[10] France conducted a notifiable multilateral exercise, 'FARFADET 92', on 9–18 June 1992 with Italy and Spain (amphibious landing phase only, 15–18 June) involving 12 500 troops (France—10 400; Italy—2100). It had not been reported earlier to SIPRI by any of the states concerned and so is not recorded in the *SIPRI Yearbook 1992*.

**Table 12A.1.** Annual numbers of military exercises conducted by NATO, the WTO/former WTO, and the neutral and non-aligned countries in 1989–92 and forecast for 1993

| | 1989 | 1990 | 1991 | 1992 | 1993 |
|---|---|---|---|---|---|
| NATO | 10 | 4 | 4 | 6 | 4 |
| WTO/former WTO | 13 | 5 | 0 | 0 | 0 |
| NNA | 3 | 3 | 1 | 0 | 1 |
| **Total** | **26** | **12** | **5** | **6** | **5** |

**Table 12A.2.** Notifiable military activities which were scaled down in 1992

| State(s)/Location | No. of troops reduced from—to | Exercise no. in *SIPRI Yearbook 1992*[a] |
|---|---|---|
| FRG, Netherlands Norway, UK, USA in Norway (FTX 'Teamwork 92') | 21 400—21 000 | 1 |
| Denmark, Netherlands, UK, USA, Norway in Norway | 8 000—7 200 | 2 |
| USA in Germany ('Reforger') | c. 8 000—6 500 | 4 |

[a] See Lachowski, Z., 'Implementation of the Vienna Document 1990 in 1991', SIPRI, *SIPRI Yearbook 1992: World Armaments and Disarmament* (Oxford University Press: Oxford, 1992), table 12.A.2, p. 495.

in military strategies, structures and doctrines, and reductions in military manpower and expenditures. Environmental factors and the burden placed by military movements and exercises on the populations are also increasingly taken into consideration. Many activities are at present command post or staff exercises that make use of computer simulations or rapid reaction force manœuvres requiring smaller amounts of manpower or equipment, and thus fall below the thresholds subject to notification under the CSBM provisions.

According to information received by SIPRI, all the exercises in 1992 were held as planned, with some changes in the numbers of troops involved. Details of activities that were scaled down are given in table 12A.2.

Five major military exercises are planned for 1993.[11] No NATO manœuvres will employ more than 15 000 troops. Sweden, which holds its military exercise every two years, is going to carry out a manœuvre involving 13 500 troops.

Neither Russia, other former Soviet republics nor the Central European states will carry out any notifiable military activities in 1993. As in previous years, this is because of budgetary constraints and new military policies which do not require large-scale military exercises.

[11] The 'Dragon Hammer-93' exercise involves a notifiable amphibious landing part. Originally, seven manœuvres were envisaged, but two of them have been either cancelled (command field exercise (CFX)/command post exercise (CPX) 'EM-FOR' in Germany) or scaled down (field training exercise (FTX) 'Display Determination' in Turkey) below the notifiable level.

## Annual exchange of military information

All CSCE states with military forces, except Yugoslavia, complied with the annual exchange of military data among CSCE participating states in 1991. Some of the newly admitted states failed to take part in the subsequent exchange to be completed by 15 December 1992 because the development of their military structures and implementation procedures was still under way. The annual exchange of military information combined with the relevant data exchange and procedures under the 1990 CFE Treaty, which entered into force in November 1992, represent a significant contribution to confidence- and security-building.

The Vienna Document 1992 introduced a number of changes concerning information exchange. In connection with the annual evaluation quota,[12] participating states should furnish a statement indicating their total number of units.[13] Armoured personnel carrier (APC) look-alikes and armoured infantry fighting vehicle (AIFV) look-alikes should be included in the categories of weapons within the formations or combat units to be reported.

The participating states undertook to provide additional information on planned personnel increases in excess of 1500 (active combat units) or 5000 (active formations) troops for more than 21 days and to exchange information on temporary activation of non-active combat units and non-active formations which exceed 2000 troops for more than three weeks. The problem of activation was a bone of contention between NATO and some neutral states, such as Switzerland, which have their force structures based on non-active formations. Originally NATO wanted this measure to cover any activation of over 7 days. Eventually, the 21-day period was found satisfactory by the neutral and non-aligned (NNA) states whose activations last generally up to 2 weeks.

A new section is included in the chapter on annual exchange of military data, entitled 'Data relating to major weapon and equipment systems'. It calls upon the participating states to furnish these data (including photographs of the weapons and equipment[14]) once to all other participants by 15 December 1992. By and large, there was no major violation or circumvention of the Vienna Document provisions in this regard. Any delays were the result of technical rather than political considerations. There are some recommendations to improve efficiency and transparency, including the streamlining and updating of information concerning changes in forces and weapons, expanding the time-frame for planned weapon deployments, making the exchange of information smoother and more open to analysis, and so on. Accordingly, it is recommended that participants stick to the rule of centralizing the information exchange in the CPC data bank in Vienna.

A number of suggestions have also been made concerning budget information. It is proposed that budget information cover more than one year, and that ways be found to make individual costs more transparent (e.g., defence expenditure-to-total budget ratio) and more quickly available when significant budgetary changes are made after the regular information has been submitted.

[12] Each state is obliged to accept a quota of one evaluation visit per calendar year for every 60 units; see Vienna Document 1992 (note 2), para. 114.

[13] To a large degree, the innovations were made on the basis of the Polish proposal (CSCE document CSCE/WV.20, 3 July 1991).

[14] A number of states have failed to attach photographs to the military information provided for 1992.

**Table 12A.3.** Calendar of planned notifiable military activities in 1993, as required by the Vienna Document 1992[a]

| States/ Location | Dates/ Start window | Type/Name of activity | Area | Level of command | No. of troops | Type of forces or equipment | No. and type of divisions | Comments |
|---|---|---|---|---|---|---|---|---|
| 1. Canada, Germany, Netherlands, Norway, UK and USA in Norway | 6 days, 4-13 Mar. | FTX 'Battle Griffin' | NW Elgsnes–Kjeoeya–Vinje–E. Brandmoen–Lunde–Raudvatnet–Rombaksbotn–Narvik–Baroey–S. Rotvaer–Botntind (965)–Roeykenes | Division | 11 700[b] | Ground, air and amphibious forces | 1 light inf. div. (reduced) and 1 brig. equivalent | Exercise forces in deployment operations and practice co-operation and interoperability between Norway and allied formations |
| 2. Germany, Italy, France, Greece, Turkey, Spain, Portugal, Netherlands, UK and USA in Italy | 62 days 15 Apr.- 15 June | FTX 'Dragon Hammer 93' | Northern Italy and Capo Teulada, Sardinia | Army group | n.a., 1 500 marine troops in amphibious landing | Marine, ground and aviation and air forces | 1 US arm.div., 1 inf. div., 1 German mech. inf. div. | Reinforce European theatre; receive CONUS-based forces in theatre; use pre-positioned equipment, conduct computer-assisted CFX, and redeploy forces to home station |
| 3. Belgium, Denmark, Germany, Italy, Luxembourg, Netherlands, UK, USA and France in Denmark | 27 days, 5 Sep. – 1 Oct. | Livex/FTX 'Exercise Action Express–93' | Zealand group of islands and surrounding waters, excluding Swedish territory | Corps level and performed at brigade level | 14 860[c] | Naval, marine, land and air forces | Both external and Danish forces are under corps command, no division level formations | Provide training for the AMF LandZealand and other external forces; amphibious element scheduled, not notifiable |
| 4. Sweden in Sweden | 7 days, 23–30 Sep. | FTX 'Orkan' | Norrtälje–Enköping–Eskilstuna–Katrineholm–Oxelösund | Corps | 13 500 | Ground, naval and air force units | 1 div. HQ and elements of subordinated units | Joint arms operation for units from all armed services. Limited number of battle tanks involved |

[a] Because of lack of consensus among CSCE participating states on the public availability of military information the data compiled by SIPRI from CSCE government responses are not comprehensive. Italy failed to provide any information, which meant that the 'Ardente 93' exercise to be conducted in Italy could not be covered.

[b] Canada–100, Germany–300, Netherlands–100, Norway–4 850, UK–2 350, USA–4 000.

[c] Belgium–1 250, Denmark–5 500, France–300, Germany–2 000, Italy–1 050, Luxembourg–180, Netherlands–1 500, UK–2 600, USA–480.

## Risk reduction

There has been much criticism of failures to make use of risk reduction capabilities and it has been claimed that the unusual military activity mechanism should be more extensively employed. In drafting the Vienna Document 1992, CSCE participating states decided to supplement the risk reduction chapter with a section on voluntary hosting of visits to dispel concerns about military activities. Participating states are encouraged to invite other participating states to designate personnel 'to take part in visits to areas on the territory of the host State in which there may be cause for such concerns'(para. 19).[15] It is felt that hosting such visits should be made obligatory and become more operational, but it should not duplicate the inspection regime.

The other risk reduction mechanism concerning hazardous military incidents was invoked for the first time on 14 January 1992 by Portugal, on behalf of the European Community (EC), in the wake of the downing by Yugoslav National Army (JNA) fighter planes of a helicopter carrying the EC monitor team over Croatian territory. The JNA apologized, while also partially blaming the EC monitors themselves. The explanations given by the former Yugoslavia have not satisfied the EC states.

In general, there is agreement that risk reduction mechanisms should be used more frequently and efficiently, having been adapted to new circumstances and types of conflict. They could also play some role in arrangements to 'restore confidence'.

## Military contacts

In 1992, the NATO countries continued to develop and intensify military contacts with the former WTO states as well as with the new states on former Soviet territory. Numerous officers from those states visited and studied at military schools and academies in Western Europe and the USA. In particular, Western countries stressed contacts with the Russian Federation, and the extension of contacts with other new republics (Ukraine, Belarus and Kazakhstan) is planned for 1993.[16]

Some new arrangements were made in this area. Regarding visits to air bases, the Vienna Document 1992 stipulates that a state invited to visit an air base may decide whether to send military and/or civilian visitors, including personnel accredited to the host state. It was agreed that outlines of schedules given by states concerning visits to air bases for the coming year(s) may be discussed in advance at annual implementation assessment meetings. Seven air bases were visited in 1992: Istrana (Italy), Murted (Turkey), Lechfeld (Germany), Rygge (Norway), Rovaniemi (Finland), Powidz (Poland) and Dijon (France), and all were found to be satisfactory.

There is also a new category: 'Demonstration of new types of major weapon and equipment system'. The first participating state to deploy a new type of major weapon and equipment system with its forces must arrange a demonstration for all other CSCE participants at the earliest opportunity. If other states deploy the same type of weapon or equipment system later on, no demonstration is required. The first demonstration of a new weapon system in 1992 took place in France, which dis-

[15] This provision was inspired by a more detailed British/Bulgarian proposal (CSCE document CSCE/WV.27, 11 Dec. 1991). Accordingly, Bulgaria invited an overflight of its territory from Yugoslavia when the latter suspected that the former was massing troops on the common frontier; *Arms Control Reporter*, sheet 402B.299, 1992.

[16] A second special course was organized by the NATO Defence College in Rome from 25 Apr. to 1 May 1992 for some 70 military and civil officials from CSCE states, half of them from Central and Eastern Europe; *Atlantic News*, no. 2417 (23 Apr. 1992). The first course was held in October 1991.

played a new model of the 'Mirage F-1-CT' fighter plane on the occasion of the visit to the Dijon air base on 7–9 October. Italy, in turn, demonstrated the 'Centoro' battle tank. It is felt that improvements in the organization of such demonstrations, such as cost-saving solutions, timely notification and responses, and other technical measures, could contribute to better implementation.

In early 1992, the CPC sponsored two specialized seminars in Prague and Vienna on Conversion of military industry to civilian production (Prague, 19–21 February) and Armed forces in democratic societies (Vienna, 4–6 March). NATO was invited to participate in both seminars. On 7–8 November 1992, a seminar was organized in Vienna by the CPC for new CSCE participating states on the implementation of agreed CSBMs.

## Notification and observation

The problem of parameters on notification thresholds has haunted the participating states since the onset of the CSCE process. In the cold war era, the Western states insisted on lowering the thresholds in order to make the WTO military activities more transparent. In the 1990s East European and NNA participants, expressing much concern about a security vacuum east of the NATO area, are in turn anxious to lower the thresholds still further, while NATO states argue that were the parameters to go any lower they would affect and actually impede daily training activities. Furthermore they point to the fact that the original aim of notification and observation was to dispel concerns about a possible threat, and the lowering of thresholds would contradict this goal.

A compromise was eventually reached. In the Vienna Document 1992 the parameters for prior notification of certain military activities were changed and supplemented. Military activities involving at least 9000 troops (including support troops), or 3000 troops in an amphibious landing/parachute drop or at least 250 battle tanks if organized into a divisional structure or at least two brigades/regiments, are subject to notification (in the Vienna Document 1990, the numbers were 13 000, 3000 and 300, respectively). The numbers of troops subject to observation were also lowered—to 13 000, or 3500 in an amphibious landing/parachute assault by airborne forces (compared with Vienna Document 1990 provisions of 17 000 and 5000, respectively). A novelty is that the engagement of 300 battle tanks also entitles other states to observe the activity.

To meet concerns about a widening 'transparency gap' in the number of observable exercises, it is proposed that the following alternative steps be taken in future:

1. Lowering the threshold even further but with due regard to its military significance (this would require additional parameters);
2. Making observation dependent on the percentage of the total forces of a state (e.g., 10–15 per cent) engaged in the exercise;
3. Adopting a solution that in any case one exercise is to be observed within a certain period of time.

It is also proposed that measures of transparency should pertain not only to military exercises with a certain number of troops but also to staff exercises, command post exercises and manœuvres involving high mobility (e.g., rapid reaction forces) or a greater level of preparedness. These measures can be also supplemented with such

**Table 12A.4.** CBM/CSBM notification and observation thresholds, 1975–92

| Document | Notification | Observation |
|---|---|---|
| Helsinki Final Act 1975 | 25 000 troops (voluntary, 21 days in advance; area: European states, USSR and Turkey—250-km strip east of the western borders) | No parameters, voluntary, on a bilateral basis |
| Stockholm Document 1986 | 13 000 troops or 300 battle tanks, or 3 000 troops in amphibious landing or parachute drop (obligatory, 42 days in advance; area: 'from the Atlantic to the Urals'). Air force included in notification if at least 200 sorties by aircraft, excluding helicopters, are flown | 17 000 troops or 5 000 troops in amphibious landing or parachute drop |
| Vienna Document 1990 | Ditto | Ditto |
| Vienna Document 1992 | 9 000 troops or 250 battle tanks, or 3 000 in amphibious landing or parachute drop (obligatory, 42 days in advance; area: Europe plus new Central Asian republics). Air force included in notification if at least 200 sorties by aircraft, excluding helicopters, are flown | 13 000 troops or 3 500 in airborne landing or parachute drop, or 300 battle tanks |

sub-regional or bilateral solutions as the military confidence-building measures contained in the Bulgarian–Turkish 'Edirne Document' and the Bulgarian-Greek accord on CSBMs of 12 November 1992 and 3 December 1992, respectively.[17] Such measures might secure approval from all CSCE participants, but not apply to all of them.

## Constraining provisions

The Vienna Document 1992 introduced the first real constraints on military activities: manœuvres involving more than 40 000 troops or 900 battle tanks can be held only once every two years; only six exercises involving more than 13 000 troops or 300 battle tanks but less than 40 000 troops or 900 battle tanks can be carried out yearly; of these six manœuvres, only three can involve more than 25 000 troops or 400 battle tanks; and only three military activities can take place simultaneously, each involving

[17] In the Edirne Document the two states agreed to notify each other in advance of activities near the border whenever they involved at least 7000 troops, or at least 150 battle tanks, or at least 150 artillery pieces of 100-mm calibre. They would invite observers when these numbers reached 10 000 troops, 200 tanks or 200 artillery pieces. The two countries set up a fax connection to provide the exchange of information; *Arms Control Reporter,* sheet 850.362, 1992. Bulgaria and Greece agreed on: expanding the border area where CBMs would apply; inviting observers to military exercises when more than 9000 troops or more than 250 battle tanks are involved, or more than 200 100-mm artillery pieces are involved; inviting, once per year, observers to attend an exercise conducted at battalion level or higher; and seeking further improvements with the aim of signing a new agreement in 1993; *Arms Control Reporter,* sheet 402.B.314, 1993.

more than 13 000 troops or 300 battle tanks. Each country is bound to communicate information to all other participating states on exercises involving more than 40 000 troops or 900 battle tanks which it plans to carry out or host in the second subsequent year; otherwise it will not be allowed to carry out such an activity.[18]

None of the replies to SIPRI's requests for CSBM information on military activities in 1993 and planned for 1994 envisage manœuvres exceeding the thresholds of 40 000 troops or 900 battle tanks. Since 1991, exercises by NATO and Central and East European states have been falling well below the ceilings set by the constraining measures, which, though less felt in military terms, constitute a further commitment to enhancing trust and confidence.

## Compliance and verification

As asserted by many observers at the successive annual implementation assessment meetings, evaluation visits and inspections have become a useful instrument for verification and for gaining insight into military activities and checking other states' compliance with the agreed CSBMs. With the number of notifiable military activities steadily decreasing, there is a widespread feeling that increasing the quota of inspections would help maintain the standard of openness and transparency achieved so far; also, by forming larger multinational inspection teams more countries are able to take part in compliance and verification activities.

The main innovations introduced in this field by the Vienna Document 1992 are: (*a*) a possibility of forming multinational inspection teams headed by the inspecting states (in addition the host state is encouraged to provide a map depicting the area specified, and the matter of expenses incurred for evaluation visits was worked out in more detail); (*b*) making non-active formations and combat units temporarily activated available for evaluation during the period of such an activation; in such cases the provisions for the evaluation of active formations and units are applicable *mutatis mutandis*. This kind of visit counts against the established quotas for evaluation.

Four inspections were carried out in 1992—by Russia in Norway ('Teamwork 92' exercise) and Germany, by the UK in Russia and by Germany in Russia.

During the year 47 evaluation visits were paid by 16 states. Two-thirds of the visits were conducted by NATO states; Germany alone carried out eight visits, mostly to former WTO states, being the most active evaluator. Russia paid the same number of visits. Among the states receiving evaluation teams, Russia received nine visits, Germany five (including three to allied bases—US, Canadian and Belgian), Albania three, and Romania and Ukraine two each. The results of those visits were generally satisfactory, even if some minor discrepancies occurred between figures given in the annual exchanges and actual ones; in most cases this was the result of reorganization in the units inspected or evaluated.

Given the growing significance of this form of checking compliance such improvements are recommended as: increasing the size of the two categories of teams; allowing for more openness; making more use of multinational inspection teams; making it possible to form multinational evaluation teams; and increasing the number of evaluation visits either (*a*) by lowering the quota threshold, or (*b*) by increasing the minimum to two visits per year.

[18] Vienna Document 1992 (note 2), paras 71 to 74. This section was the subject of a lively debate and a number of states submitted proposals. See Brauch and Neuwirth (note 2), pp. 25–26, 28.

## Communications

The establishment of a 24-hour communications network with terminals to receive and exchange information and notification is one of the challenges facing the CSCE. By the end of 1992, 27 participating states were connected to the network, including seven states that did not yet have operational terminals. However, it will be some time before the CSCE participating states will be able to rely fully on the network. Some countries prefer to rely on bilateral communications systems (chiefly via embassies), others are bound by financial or technical obstacles; some smaller states let it be known that they do not intend (in spite of earlier pledges) to connect their capitals to the network. Nevertheless, all militarily significant states (i.e., those with the most important information to send) are integrated in the system. Various solutions are being sought to make the network comprehensive and embrace all participating states as soon as possible (e.g., using private firms establishing communications links, and satellite communications). The network should not only support CSBM information exchange but also all other communications; CFE Treaty-related information is also being sent via the system.

Another problem is a need for rapid transmission of messages and their immediate interpretation. The participating states reiterated their agreement on formats with headings in all six CSCE languages and on two working languages in which they would like to receive the translation, if need be.[19]

## Annual assessment of implementation

The second annual implementation assessment meeting (AIAM) of the CSCE participating states, attended by several new participants, was held on 9–11 November 1992 at the CPC in Vienna.

The meeting reviewed and assessed the implementation of CSBMs and, unlike the previous year, no essential criticism was voiced. The participants drew up a few dozen recommendations geared to supplement and streamline the implementation of existing measures. Generally, the suggestions aimed to deepen, strengthen and broaden the regime by harmonizing CFE and CSBM information exchanges; shaping new CSBMs to prevent conflicts among and within states; making adherence to the rules more stringent; and covering a wide spectrum of military activities as soon as they are of military relevance (e.g., new stabilizing measures). The meeting recommended that the next AIAM should be held in spring 1993, and concentrate on information exchange with a specialized meeting of the Consultative Committee held prior to it. No public report of the meeting was issued.

## New participants

When admitting former Soviet republics to the CSCE, in early 1992, the participating states were aware that there would be a host of problems for the new members. Many of them are not closely familiar with CSCE practice, including its security dimension, and consequently deserve a period of education and special treatment. It was agreed that, exceptionally, practical problems that may arise at the initial stages in implementing confidence- and security-building measures on their territories should be

[19] See Annex II and the Chairman's statement in Annex III to the Vienna Document 1992 (note 2).

'taken into consideration' by other participants.[20] Accordingly, a seminar on the implementation of agreed CSBMs was held for the new CSCE participating states on 7–8 November 1992 at the Vienna Conflict Prevention Centre. Another question is the case of contiguous areas of the former Soviet East European and Central Asian republics which share frontiers with non-European non-participating states; this matter has been referred to future annual implementation assessment meetings.[21]

It is proposed that to ensure stability the CSBM principles should be extended to areas beyond the original zone of application, and that the provisions concerning problems confronting the new participants, as contained in Annexes IV and V of the Vienna Document, be made compatible with the provisions of the Helsinki Final Act on the area of application. It is also suggested that future 'regional tables' should be harmonized with the CSBM regime.

## The data bank and public access to the data

The Helsinki Decisions of the CSCE, presented at the 1992 summit meeting, outline measures intended to 'increase openness of the CSCE institutions and structures and ensure wide dissemination of information on the CSCE'.[22] However there have been disquieting signs that this pledge of transparency is not being honoured. There have even been some backward steps—the press centre at Vienna has been closed, accreditation of media representatives abandoned and access to information on the work of the Forum for Security Co-operation has been curtailed.[23]

With the collapse of the cold war division, progressive removal of the shroud of secrecy and significant broadening of the flow of military data among states, two general problems have arisen concerning (*a*) elaborating, systematizing and disseminating the unprecedented amount of information among participating states, and (*b*) keeping track of all the information and making some of it available to the broader public. The Conflict Prevention Centre was to be responsible for the organization of a data bank and making information available to the public. However, for a number of reasons, the CPC failed to produce its yearbook in 1991 and 1992. States are not in agreement on the form and contents of such a yearbook: some participants view the data they provide to the Centre as classified or strictly confidential and there are fears that sensitive information could be used for terrorist or other hostile purposes. In effect, the publication reached deadlock.

SIPRI has published its own reports on the implementation of CSBMs since the Stockholm Conference 1986 and has tried to fill the gap in public knowledge of the vicissitudes of CSBM implementation. Now, like other independent institutes, it is in a paradoxical situation. Countries that in the past demanded maximum transparency and openness are taking a very restrictive or even negative position on public access to CSBM information. It is fortunate that, at the same time, other participants are fairly forthcoming and co-operative in their responses to requests for CSBM data. Because there is no agreed or consistent policy in this regard, information received by the public is becoming scarce, and even provision of 'neutral', non-sensitive data is

[20] It was acknowledged that the new participants' different geographical premises should be further discussed after the Helsinki follow-up meeting, *CSCE/CSBM Journal*, no. 314 (28 Feb. 1992).

[21] See the Chairman's statements in Annexes IV and V to the Vienna Document 1992 (note 2).

[22] Helsinki Document 1992 (note 7), Helsinki Decisions, Chapter IV, paras 12–18.

[23] *Focus on Vienna*, no. 28, Nov. 1992, p. 11.

not uncommonly denied by CSCE governments. This situation requires urgent review and redress, to avoid the risk of depreciating the very notion of confidence building.

## III. Assessment and outlook

Negotiation on CSBMs is a continuous process which has recently been accelerating to keep up with the new political situation facing the international community. After the end of the cold war, two CSBM agreements were hurriedly reached by CSCE participating states within one and a half years, thus completing the work begun in the early 1970s. The new international situation requires a thoroughly new approach. The Helsinki Document 1992 took the CSCE process one step further in its adjustment to the new era. A major task of the FSC, the only conventional arms control forum in Europe, which opened on 22 September 1992, is to improve and develop the CSBMs contained in the Vienna Document 1992, including measures with regional or border area-related application.[24]

The 'classic' CSBMs should survive because one cannot rule out aggression against one state by another, although this likelihood has decreased. Furthermore, in this time of disorder and low stability, the co-operative regime embracing CSBMs ensures a measure of continuity and a pattern of conduct. Another problem is the applicability and adequate use of the Vienna Document in times of crisis. Measures so far applied to Europe as a whole might also be developed for temporary, sub-regional or internal use (e.g., multilateral inspections of force concentrations, the hosting of voluntary visits, the unusual military activity mechanisms, the lowering of thresholds). The CSBM regime can and should be harmonized with early-warning, fact-finding, monitoring or peace-keeping instruments. The CSBM and CFE Treaty regimes should also be harmonized and streamlined, without damage to either. On the other hand, it is argued that the burgeoning number of measures, the cost of their implementation and the growing complexity and intrusiveness of the CSBM regime may in effect decrease rather than increase confidence among states, and participants may begin to see the negotiation as an aim in itself rather than as a means to enhance security.[25]

It is clear that new measures are required, adequate to meet the new challenges stemming from the changes on the European scene. They do not necessarily have to have state-to-state application or a status quo-preserving function but should serve to strengthen confidence and stability in an environment in which violations of commitments stem mostly from intra-state, national and ethnic roots. In such an environment the established parameters of notification, transparency, verification or constraining measures are of relatively minor value.

The FSC has the task of working out new confidence-strengthening and stabilizing measures for regional and sub-regional application and international monitoring tools to prevent and defuse crises and conflicts while simultaneously managing change in various parts of the CSCE area.

[24] Helsinki Document 1992 (note 7), Helsinki Decisions, chapter V, para 14. For more on the FSC see Kuglitsch, F.J., 'Das KSZE-Forum für Sicherheitskooperation. Mandat und Verhandlungsverlauf', *Österreichische Militärische Zeitschrift*, vol. 6 (1992), pp. 485–90.

[25] See more on this in Macintosh, J., 'Future CSBM options: post-Helsinki arms control negotiations', eds H. Chestnutt and S. Mataija, *Towards Helsinki 1992: Arms Control and the Verification Process* (Centre for International and Strategic Studies, York University: Toronto, 1991).

The 'Programme for Immediate Action' included in the Helsinki Document 1992 for the FSC makes harmonization its prime item—the aim of the negotiations would be to harmonize the commitments and rights of all participating states derived from the three agreements on the military dimension reached in the CSCE framework, that is, the 1990 CFE Treaty, the 1992 CFE-1A Agreement and the Vienna Document 1992. Harmonization, which is likely to take the shape of a politically binding act rather than a treaty, will place all CSCE states on a common footing for setting goals and working out agreements on future military matters.

In autumn 1992 a number of proposals were tabled in the FSC. A position paper by Czechoslovakia, Hungary and Poland was submitted on 7 October, proposing harmonization in five areas (national levels, verification, information exchange and CFE and CSBM notifications, review mechanisms for the 'harmonized regime', and areas of application). On 14 October the NATO states introduced their own harmonization proposition, which extends the provisions of the Vienna Document 1992 to require more information on a number of things (e.g., types and numbers of treaty-limited items, changes in organizational structures or force levels, locations of troops and equipment). The North Atlantic allies tabled another proposal on 21 October on sharing NATO-type defence planning information. Countries would be asked to provide information about their national defence policies, any major changes in their forces or deployments, major investment projects (especially procurement of major equipment) and possible shifts in plans and priorities. They would also be asked to provide detailed budgetary data for the next three years, as well as more general information on budgets for the two years after that.[26]

Another goal of the FSC, that of developing and improving upon the CSBMs contained in the Vienna Document 1992, has not yet been addressed thoroughly in the Forum. A number of CSBM-related matters such as the proposals concerning military contacts (UK), global information exchange (Russia), the code of conduct (Poland, the UK and Turkey) as well as non-proliferation (Iceland) and arms transfers have been submitted.[27] At the last FSC session of 1992, on 16 December, the EC states submitted a proposal regarding a military code of conduct. It seeks to establish norms for the internal conduct of military institutions regarding their own nations, such as their treatment of ethnic minorities.[28]

[26] For the text see *BASIC Reports*, 9 Nov. 1992 (no. 26), pp. 4–5.
[27] For other items on the agenda of the FSC see chapter 5 in this volume.
[28] *Arms Control Reporter,* sheet 402.B.315, 1993.

# **Appendix 12B.** The Treaty on Open Skies

RICHARD KOKOSKI

## I. Introduction

The concept of Open Skies was first put forward by President Dwight D. Eisenhower on 21 July 1955 at the Geneva Conference of Heads of Government. On 12 May 1989 US President George Bush relaunched the Open Skies proposal for an agreement that could allow flights by unarmed reconnaissance aircraft over the territories of the USA, the USSR and their allies.

Four rounds of negotiations were held: the first in Ottawa in early 1990 and two subsequent rounds in Budapest and Vienna in 1991. In the fourth and final round negotiators met continuously from 13 January 1992 until well into March, in order to complete the final 100-page treaty text.[1]

Noting its potential 'to improve openness and transparency, to facilitate the monitoring of compliance with existing or future arms control agreements and to strengthen the capacity for conflict prevention and crisis management in the framework of the Conference on Security and Co-operation in Europe and in other relevant institutions',[2] the Treaty on Open Skies was signed on 24 March 1992. The Treaty had been initialled on 21 March by the 16 members of NATO, the five former members of the WTO (Bulgaria, Czechoslovakia, Hungary, Poland and Romania), as well as Russia, Ukraine and Belarus. The admission of Georgia to the CSCE during the opening of the Helsinki follow-up meeting brought the total number of signatories to 25.[3]

Covering the area from Vancouver to Vladivostok, the Treaty on Open Skies represents 'the most wide ranging international effort to date to promote openness and transparency in military forces and activities'.[4]

## II. The Consultative Commission

An Open Skies Consultative Commission (OSCC) was established by Article X of the Treaty to consider compliance questions, resolve ambiguities which may arise, consider applications for, and agree to technical and administrative measures upon accession by other states. It may also propose amendments to and agree on improvements to the Treaty. At least four regular sessions per year are to be held.

The OSCC has in fact already shown itself capable of operating effectively by eliminating some of the issues involving costs and sensors which were left for it to resolve at the time the Treaty was signed.

[1] For an outline of the key provisions of the Treaty on Open Skies see *SIPRI Yearbook 1992*, chapter 12, p. 477–79.

[2] Treaty on Open Skies preamble. The text of the Treaty is reproduced in appendix 12C of this volume.

[3] White, D., 'Old enemies agree to surveillance flights', *Financial Times,* 21 Mar. 1992, p. 2; Agence France Presse, 24 Mar. 1992.

[4] Statement by US State Department spokesman Richard A. Boucher quoted in 'Agreement will open skies to reconnaissance flights', *New York Times*, 21 Mar. 1992.

During 1992 two meetings of the OSCC were held in Vienna (2 April–17 July and 21 September–17 December). The first session, chaired by Canada, sought to resolve issues related to sensor calibration and various cost considerations. In order to deal with the issues in a sufficiently satisfactory fashion so that the Treaty could be opened for ratification the commission remained in session after its 30 June deadline by simply stopping the clock on that date and continuing its work for the necessary additional time.[5] The resulting decisions covered the types of camera and film to be used, film processing methods and minimum requirements for camera operations.[6]

The second meeting, chaired by Denmark, discussed sensor calibration mechanisms given the results of the test overflights made in November 1992. Cost issues were also discussed and agreement reached on a number of points including which countries would pay for such items as fuel, oil and servicing under various circumstances. Concerning information exchange the intent was to have a format resembling that used in the Conventional Armed Forces in Europe (CFE) Treaty and for confidence- and security-building measures (CSBMs), and by the close of the session formats had been agreed to for some 35—all but two—of the required notifications and reports. Those still undecided were to be agreed upon during the first session in 1993. In addition, a method of calibration for synthetic-aperture radar resolution was agreed. An environmental seminar was also held to explore ways in which the Open Skies Treaty could contribute to monitoring such aspects as measurement of ozone levels, global warming, water and air pollution and deforestation.[7]

The third round of the OSCC was opened on 21 January 1993. Items on the agenda included the methodologies for operating infra-red and video cameras (these issues for optical cameras and SAR having already been resolved in the second meeting), the distribution of overflight quotas between the Czech Republic and Slovakia and further discussion on environmental monitoring.[8]

## III. Trial overflights in 1992

Among the many trial overflights conducted during 1992 after the signing of the Treaty the first was made over Poland on 2 April by a Belgian Air Force C-130 Hercules aircraft. Primary sensors for the flight were a panoramic camera as well as a FLIR (forward-looking infra-red) sensor. The latter, while not strictly allowed by the Treaty at present, was none the less employed to perfect procedures for the future. Operating at an altitude of between about 600 and 2000 metres, the aircraft was used to observe two Polish air bases, an army training ground, a chemical plant and three Russian facilities.[9] A reciprocal flight was made over the Benelux countries on behalf of Poland on 8 April.

In June flight tests took place from Boscombe Down in the UK in order to settle issues relating to sensor resolution.[10] The flights aided in reaching the above-

[5] *Arms Control Reporter*, sheet 409B.33, 1992; sheet 409.A.2, 1993.

[6] *Disarmament Bulletin*, no. 19 (winter 1992/93) p. 15.

[7] Jones, P., 'Open Skies: events in 1992', VERTIC, *Verification Report 1993: Yearbook on Arms Control and Environmental Agreements* (Verification Technology Information Centre: London, 1993, forthcoming); *Arms Control Reporter*, sheets 409.B.32-35, 37, 1992.

[8] *Arms Control Reporter*, sheet 409.B.37, 1992.

[9] Participating observers (in addition to the three-person crew) were from Belgium, Canada, Czechoslovakia, France, Germany, Hungary, Italy, Luxembourg, the Netherlands, Poland, Portugal, Spain, the UK and the USA. *Arms Control Reporter*, sheet 409.B.32-33, 1992.

[10] *Trust and Verify*, no. 28 (May 1992), p. 4; *Arms Control Reporter*, sheet 409.B.33, 1992.

mentioned agreement on film and camera types and processing methods during the first OSCC.

In September overflights of Russia and Belarus were conducted by an Andover aircraft from the UK and a Russian An-30. Three flights were made involving approximately 10 hours of total flying time. The exercise tested procedures which included ensuring that sensors were sealed while overflying countries other than those slated for inspection. In addition to the British and Russian crews, observers from all of the Western European Union (WEU) countries as well as from Sweden and the USA were present. The series of flights was considered a success.[11]

In October three aircraft, provided by Denmark, Russia and Canada, tested synthetic-aperture radars (SAR) over specially designed targets in Hungary in order to demonstrate calibration of three very different SAR sensors. This successful experiment was hailed as 'a milestone in technical co-operation among parties to the Open Skies Treaty' and it was noted that the 'monumental task of negotiating such complicated issues as SAR parameters was a vivid example of the confidence-building intent of the Treaty at work'.[12]

Further calibration flights also occurred in November involving Canadian, Russian and Danish aircraft flying over a testing ground.[13]

## IV. Status of the Treaty and conclusion

At the end of 1992, 27 countries had signed the Treaty[14] but only three had ratified it—Canada, Denmark and, just before it split up at the end of the year, Czechoslovakia.[15] However, the ratification process has begun for most signatories. Although all former Soviet republics which had not done so from the beginning are entitled to accede to the Treaty only Kyrgyzstan had done so. Future problems may arise should the substantial number of states eligible to do so decide to join the regime, most probably necessitating increases in quotas and certainly increasing costs.[16] Entry into force of the Treaty is expected sometime in 1993.[17]

In summary no major or insurmountable problems were encountered during 1992. The trial overflights conducted proved useful for a variety of testing and demonstration purposes. Further, and perhaps most importantly, the OSCC was able to show that it could function in an efficient and constructive manner by resolving key issues on which it focused attention—this bodes well for its future usefulness as the Treaty proceeds through full ratification and on to successful implementation.

[11] *Atlantic News*, no. 2543 (9 Sep. 1992), p. 4; *Trust and Verify*, no. 32 (Oct. 1992), p. 3.

[12] *Disarmament Bulletin*, no. 19 (winter 1992/93), p. 15.

[13] *Arms Control Reporter*, sheet 409.B.34, 1992.

[14] Signatories are Belarus, Belgium, Bulgaria, Canada, Czech Republic, Denmark, France, Georgia, Germany, Greece, Hungary, Iceland, Italy, Kyrgyzstan, Luxembourg, Netherlands, Norway, Poland, Portugal, Romania, Russia, Slovakia, Spain, Turkey, Ukraine, the UK and the USA .

[15] *Arms Control Reporter*, sheets 402.A.5 and 490.A.2. 1993.

[16] Jones (note 7).

[17] Lenorovitz, J., 'Flight tests, training begin for Open Skies Treaty', *Aviation Week & Space Technology*, 22 Feb. 1993, p. 57.

# **Appendix 12C.** Documents on conventional arms control in Europe, 1992

## VIENNA DOCUMENT 1992 OF THE NEGOTIATIONS ON CONFIDENCE- AND SECURITY-BUILDING MEASURES CONVENED IN ACCORDANCE WITH THE RELEVANT PROVISIONS OF THE CONCLUDING DOCUMENT OF THE VIENNA MEETING OF THE CONFERENCE ON SECURITY AND CO-OPERATION IN EUROPE

*Signed 4 March 1992, Vienna*

(1) Representatives of the participating States of the Conference on Security and Co-operation in Europe (CSCE), Albania, Armenia, Austria, Azerbaijan, Belarus, Belgium, Bulgaria, Canada, Cyprus, the Czech and Slovak Federal Republic, Denmark, Estonia, Finland, France, Germany, Greece, the Holy See, Hungary, Iceland, Ireland, Italy, Kazakhstan, Kyrgyzstan, Latvia, Liechtenstein, Lithuania, Luxembourg, Malta, Moldova, Monaco, the Netherlands, Norway, Poland, Portugal, Romania, the Russian Federation, San Marino, Spain, Sweden, Switzerland, Tajikistan, Turkey, Turkmenistan, Ukraine, the United Kingdom, the United States of America, Uzbekistan and Yugoslavia, met in Vienna in accordance with the provisions relating to the Conference on Confidence- and Security-Building Measures and Disarmament in Europe contained in the Concluding Documents of the Madrid and Vienna Follow-up Meetings of the CSCE.

(2) The Negotiations were conducted from 9 March 1989 to 4 March 1992.

(3) The participating States recalled that the aim of the Conference on Confidence- and Security- Building Measures and Disarmament in Europe is, as a substantial and integral part of the multilateral process initiated by the Conference on Security and Co-operation in Europe, to undertake, in stages, new, effective and concrete actions designed to make progress in strengthening confidence and security and in achieving disarmament, so as to give effect and expression to the duty of States to refrain from the threat or use of force in their mutual relations as well as in their international relations in general.

(4) The participating States recognized that the mutually complementary confidence- and security-building measures which are adopted in the present document and which are in accordance with the mandates of the Madrid and Vienna Follow-up Meetings of the CSCE serve by their scope and nature and by their implementation to strengthen confidence and security among the participating States.

(5) The participating States recalled the declaration on Refraining from the Threat or Use of Force contained in paragraphs (9) to (27) of the Document of the Stockholm Conference and stressed its continuing validity as seen in the light of the Charter of Paris for a New Europe.

(6) From 8 to 18 October 1991, the participating States held discussions in a seminar setting on military doctrine in relation to the posture, structure and activities of conventional forces in the zone of application for confidence- and security-building measures*. The discussions built on the results of the first such seminar, which had been held in Vienna from 16 January to 5 February 1990.

(7) On 17 November 1990, the participating States adopted the Vienna Document 1990, which built upon and added to the confidence- and security-building measures contained in the Document of the Stockholm Conference 1986.

(8) In fulfilment of the Charter of Paris for a New Europe of November 1990, they continued the CSBM negotiations under the same mandate, and have adopted the present document which integrates a set of new confidence- and security-building measures with measures previously adopted.

(9) The participating States have adopted the following:

### 1. ANNUAL EXCHANGE OF MILITARY INFORMATION

#### Information on Military Forces

(10) The participating States will exchange annually information on their military forces concerning the military organization, manpower and major weapon and equipment sys-

*Annex 1

tems, as specified below, in the zone of application for confidence- and security-building measures (CSBMs).

(11) The information will be provided in an agreed format to all other participating States not later than 15 December of each year. It will be valid as of 1 January of the following year and will include:

(11.1) 1. Information on the command organization of those military forces referred to under points 2 and 3 specifying the designation and subordination of all formations* and units** at each level of command down to and including brigade/regiment or equivalent level.

(11.1.1) Each participating State providing information on military forces will include a statement indicating the total number of units contained therein and the resultant annual evaluation quota as provided for in paragraph (114).

(11.2) 2. For each formation and combat unit*** of land forces down to and including brigade/regiment or equivalent level the information will indicate:

(11.2.1) – the designation and subordination;

(11.2.2) – whether it is active or non-active****;

(11.2.3) – the normal peacetime location of its headquarters indicated by exact geographic terms and/or co-ordinates;

(11.2.4) – the peacetime authorized personnel strength;

(11.2.5) – the major organic weapon and equipment systems, specifying the numbers of each type of:

(11.2.5.1) – battle tanks;

(11.2.5.2) – helicopters;

(11.2.5.3) – armoured combat vehicles (armoured personnel carriers, armoured infantry fighting vehicles, heavy armament combat vehicles);

(11.2.5.4) – armoured personnel carrier look-alikes and armoured infantry fighting vehicle look-alikes;

(11.2.5.5) – anti-tank guided missile launchers permanently/integrally mounted on armoured vehicles;

(11.2.5.6) – self-propelled and towed artillery pieces, mortars and multiple rocket launchers (100 mm calibre and above);

(11.2.5.7) – armoured vehicle launched bridges.

(11.3.1) For planned increases in personnel strength above that reported under paragraph (11.2.4) for more than 21 days by more than 1,500 troops for each active combat unit and by more than 5,000 troops for each active formation, excluding personnel increases in the formation's subordinate formations and/or combat units subject to separate reporting under paragraph (11.2); as well as

(11.3.2) for each non-active formation and non-active combat unit which is planned to be temporarily activated for routine military activities or for any other purpose with more than 2,000 troops for more than 21 days

(11.3.3) the following additional information will be provided in the annual exchange of military information:

(11.3.3.1) – designation and subordination of the formation or combat unit;

(11.3.3.2) – purpose of the increase or activation;

(11.3.3.3) – for active formations and combat units the planned number of troops exceeding the personnel strength indicated under paragraph (11.2.4) or for non-active formations and combat units the number of troops involved during the period of activation;

(11.3.3.4) – start and end dates of the envisaged increase in personnel strength or activation;

(11.3.3.5) – planned location/area of activation;

(11.3.3.6) – the numbers of each type of the major weapon and equipment systems as listed in paragraphs (11.2.5.1) to (11.2.5.7) which are planned to be used during the period of the personnel increase or activation.

(11.3.4) In cases where the information required under paragraphs (11.3.1) to (11.3.3.6) cannot be provided in the annual exchange of military information, or in cases of changes in the information already provided, the required information will be communicated at least 42 days prior to such a

* In this context, formations are armies, corps and divisions and their equivalents.

** In this context, units are brigades, regiments and their equivalents.

*** In this context, combat units are infantry, armoured, mechanized, motorized rifle, artillery, combat engineer and army aviation units. Those combat units which are airmobile or airborne will also be included.

**** In this context, non-active formations or combat units are those manned from zero to fifteen per cent of their authorized combat strength. This term includes low strength formations and units.

personnel increase or temporary activation taking effect or, in cases when the personnel increase or temporary activation is carried out without advance notice to the troops involved, at the latest at the time the increase or the activation has taken effect.

(11.4) For each amphibious formation and amphibious combat unit* permanently located in the zone of application down to and including brigade/regiment or equivalent level, the information will include the items as set out above.

(11.5) 3. For each air formation and air combat unit** of the air forces, air defence aviation and of naval aviation permanently based on land down to and including wing/ air regiment or equivalent level the information will include:

(11.5.1) – the designation and subordination;

(11.5.2) – the normal peacetime location of the headquarters indicated by exact geographic terms and/or co-ordinates;

(11.5.3) – the normal peacetime location of the unit indicated by the air base or military airfield on which the unit is based, specifying:

(11.5.3.1) – the designation or, if applicable, name of the air base or military airfield and

(11.5.3.2) – its location indicated by exact geographic terms and/or co-ordinates;

(11.5.4) – the peacetime authorized personnel strength***;

(11.5.5) – the numbers of each type of:

(11.5.5.1) – combat aircraft;

(11.5.5.2) – helicopters

organic to the formation or unit.

**Data Relating to Major Weapon and Equipment Systems**

(12) The participating States will exchange data relating to their major weapon and equipment systems as specified in the provisions on Information on Military Forces within the zone of application for CSBMs.

(12.1) Data on existing weapon and equipment systems will be provided once to all other participating States not later than 15 December 1992.

* Combat units as defined above.

** In this context, air combat units are units, the majority of whose organic aircraft are combat aircraft.

***As an exception, this information need not be provided on air defence aviation units.

(12.2) Data on new types or versions of major weapon and equipment systems will be provided by each State when its deployment plans for the systems concerned are provided for the first time in accordance with paragraphs (14) and (15) below or, at the latest, when it deploys the systems concerned for the first time in the zone of application for CSBMs. If a participating State has already provided data on the same new type or version, other participating States may, if appropriate, certify the validity of those data as far as their system is concerned.

(13) The following data will be provided for each type or version of major weapon and equipment systems:

(13.1) BATTLE TANKS

(13.1.1) Type

(13.1.2) National Nomenclature/Name

(13.1.3) Main Gun Calibre

(13.1.4) Unladen Weight

(13.1.5) Data on new types or versions will, in addition, include:

(13.1.5.1) Night Vision Capability yes/no

(13.1.5.2) Additional Armour yes/no

(13.1.5.3) Track Width cm

(13.1.5.4) Floating Capability yes/no

(13.1.5.5) Snorkelling Equipment yes/no

(13.2) ARMOURED COMBAT VEHICLES

(13.2.1) Armoured Personnel Carriers

(13.2.1.1) Type

(13.2.1.2) National Nomenclature/Name

(13.2.1.3) Type and Calibre of Armaments, if any

(13.2.1.4) Data on new types or versions will, in addition, include:

(13.2.1.4.1) Night Vision Capability yes/no

(13.2.1.4.2) Seating Capacity

(13.2.1.4.3) Floating Capability yes/no

(13.2.1.4.4) Snorkelling equipment yes/no

(13.2.2) Armoured Infantry Fighting Vehicles

(13.2.2.1) Type

(13.2.2.2) National Nomenclature/Name

(13.2.2.3) Type and Calibre of Armaments

(13.2.2.4) Data on new types or versions will, in addition, include:

(13.2.2.4.1) Night Vision Capability yes/no

(13.2.2.4.2) Additional Armour yes/no

(13.2.2.4.3) Floating Capability yes/no

(13.2.2.4.4) Snorkelling Equipment yes/no

(13.2.3) Heavy Armament Combat Vehicles

(13.2.3.1) Type
(13.2.3.2) National Nomenclature/Name
(13.2.3.3) Main Gun Calibre
(13.2.3.4) Unladen Weight
(13.2.3.5) Data on new types or versions will, in addition, include:
(13.2.3.5.1) Night Vision Capability yes/no
(13.2.3.5.2) Additional Armour yes/no
(13.2.3.5.3) Floating Capability yes/no
(13.2.3.5.4) Snorkelling Equipment yes/no

(13.3) ARMOURED PERSONNEL CARRIER LOOK-ALIKES AND ARMOURED INFANTRY FIGHTING VEHICLE LOOK-ALIKES
(13.3.1) Armoured Personnel Carrier Look-Alikes
(13.3.1.1) Type
(13.3.1.2) National Nomenclature/Name
(13.3.1.3) Type and Calibre of Armaments, if any
(13.3.2) Armoured Infantry Fighting Vehicle Look-Alikes
(13.3.2.1) Type
(13.3.2.2) National Nomenclature/Name
(13.3.2.3) Type and Calibre of Armaments, if any

(13.4) ANTI-TANK GUIDED MISSILE LAUNCHERS PERMANENTLY/INTEGRALLY MOUNTED ON ARMOURED VEHICLES
(13.4.1) Type
(13.4.2) National Nomenclature/Name

(13.5) SELF-PROPELLED AND TOWED ARTILLERY PIECES, MORTARS AND MULTIPLE ROCKET LAUNCHERS (100 mm CALIBRE AND ABOVE)
(13.5.1) Artillery pieces
(13.5.1.1) Type
(13.5.1.2) National Nomenclature/Name
(13.5.1.3) Calibre
(13.5.2) Mortars
(13.5.2.1) Type
(13.5.2.2) National Nomenclature/Name
(13.5.2.3) Calibre
(13.5.3) Multiple Launch Rocket Systems
(13.5.3.1) Type
(13.5.3.2) National Nomenclature/Name
(13.5.3.3) Calibre
(13.5.3.4) Data on new types or versions will, in addition, include:
(13.5.3.4.1) Number of Tubes

(13.6) ARMOURED VEHICLE LAUNCHED BRIDGES
(13.6.1) Type
(13.6.2) National Nomenclature/Name
(13.6.3) Data on new types or versions will, in addition, include:
(13.6.3.1) Span of the Bridge — m
(13.6.3.2) Carrying Capacity/Load Classification — metric tons

(13.7) COMBAT AIRCRAFT
(13.7.1) Type
(13.7.2) National Nomenclature/Name
(13.7.3) Data on new types or versions will, addition, include:
(13.7.3.1) Type of Integrally Mounted Armaments, if any

(13.8) HELICOPTERS
(13.8.1) Type
(13.8.2) National Nomenclature/Name
(13.8.3) Data on new types or versions will, in addition, include:
(13.8.3.1) Primary Role (e.g. specialized attack, multi-purpose attack, combat support, transport)
(13.8.3.2) Type of Integrally Mounted Armaments, if any

(13.9) Each participating State will, at the time the data are presented, ensure that other participating States are provided with photographs presenting the right or left side, top and front views for each of the types of major weapon and equipment systems concerned.

(13.10) Photographs of armoured personnel carrier look-alikes and armoured infantry fighting vehicle look-alikes will include a view of such vehicles so as to show clearly their internal configuration illustrating the specific characteristic which distinguishes each particular vehicle as a look-alike.

(13.11) The photographs of each type will be accompanied by a note giving the type designation and national nomenclature for all models and versions of the type which the photographs represent. The photographs of a type will contain an annotation of the data for that type.

**Information on Plans for the Deployment of Major Weapon and Equipment Systems**

(14) The participating States will exchange annually information on their plans for the deployment of major weapon and equipment systems as specified in the provisions on Information on Military Forces within the zone of application for CSBMs.

(15) The information will be provided in an agreed format to all other participating States not later than 15 December of each year. It will cover plans for the following year and will include:

(15.1) – the type and name of the weapon/equipment systems to be deployed;

(15.2) – the total number of each weapon/equipment system;

(15.3) – whenever possible, the number of each weapon/equipment system planned to be allocated to each formation or unit;

(15.4) – the extent to which the deployment will add to or replace existing weapon/equipment systems.

**Information on Military Budgets**

(16) The participating State will exchange annually information on their military budgets for the forthcoming fiscal year, itemising defence expenditures on the basis of the categories set out in the United Nations 'Instrument for Standardised International Reporting of Military Expenditures' adopted on 12 December 1980.

(16.1) The information will be provided to all other participating States not later than two months after the military budget has been approved by the competent national authorities.

(16.2) Each participating State may ask for clarification from any other participating State of the budgetary information provided. Questions should be submitted within a period of two months following the receipt of a participating State's budgetary information. Participating States will make every effort to answer such questions fully and promptly. The questions and replies may be transmitted to all other participating States.

## II. RISK REDUCTION

### Mechanism for Consultation and Co-operation as Regards Unusual Military Activities

(17) Participating States will, in accordance with the following provisions, consult and co-operate with each other about any unusual and unscheduled activities of their military forces outside their normal peacetime locations which are militarily significant, within the zone of application for CSBMs and about which a participating State expresses its security concern.

(17.1) The participating State which has concerns about such an activity may transmit a request for an explanation to other participating State where the activity is taking place.

(17.1.1) The request will state the cause, or causes, of the concern and, to the extent possible, the type and location, or area, of the activity.

(17.1.2) The reply will be transmitted within not more than 48 hours.

(17.1.3) The reply will give answers to questions raised, as well as any other relevant information which might help to clarify the activity giving rise to concern.

(17.1.4) The request and the reply will be transmitted to all other participating States without delay.

(17.2) The requesting State, after considering the reply provided, may then request a meeting to discuss the matter.

(17.2.1) The requesting State may ask for a meeting with the responding State.

(17.2.1.1) Such a meeting will be convened within not more than 48 hours.

(17.2.1.2) The request for such a meeting will be transmitted to all participating States without delay.

(17.2.1.3) The responding State is entitled to ask other interested participating States, in particular those which might be involved in the activity, to participate in the meeting.

(17.2.1.4) Such a meeting will be held at a venue to be mutually agreed upon by the requesting and the responding States. If there is no agreement, the meeting will be held at the Conflict Prevention Centre.

(17.2.1.5) The requesting and responding States will, jointly or separately, transmit a report of the meeting to all other participating States without delay.

(17.2.2) The requesting State may ask for a meeting of all participating States.

(17.2.2.1) Such a meeting will be convened within not more than 48 hours.

(17.2.2.2) The Conflict Prevention Centre will serve as the forum for such a meeting.

(17.2.2.3) Participating States involved in the matter to be discussed undertake to be represented at such a meeting.

(17.3) The communications between participating States provided for above will be transmitted preferably through the CSBM communications network.

### Co-operation as Regards Hazardous Incidents of a Military Nature

(18) Participating States will co-operate by reporting and clarifying hazardous incidents of a military nature within the zone of application for CSBMs in order to prevent possible misunderstandings and mitigate the effects on another participating State.

(18.1) Each participating State will designate a point to contact in case of such hazardous incidents and will so inform all other participating States. A list of such points will be kept available at the Conflict Prevention Centre.

(18.2) In the event of such a hazardous incident the participating State whose military forces are involved in the incident should provide the information available to other participating States in an expeditious manner. Any participating State affected by such an incident may also request clarification as appropriate. Such requests will receive a prompt response.

(18.3) Communications between participating States will be transmitted preferably through the CSBM communications network.

(18.4) Matters relating to information about such hazardous incidents may be discussed by participating States at the Conflict Prevention Centre, either at the annual implementation assessment meeting at the Centre, or at additional meetings convened there.

(18.5) These provisions will not affect the rights and obligations of participating States under any international agreement concerning hazardous incidents, nor will they preclude additional methods of reporting and clarifying hazardous incidents.

### Voluntary Hosting of Visits to Dispel Concerns about Military Activities

(19) In order to help to dispel concerns about military activities in the zone of application for CSBMs, participating States are encouraged to invite, at their discretion, other participating States to designate personnel accredited to the host State or other representatives to take part in visits to areas on the territory of the host State in which there may be cause for such concerns. Such invitations will be without prejudice to any action taken under paragraphs (17) to (17.3).

(19.1) States invited to participate in such visits will include those which are understood to have concerns. At the time invitations are issued, the host State will communicate to all other participating States its intention to conduct the visit, indicating the reasons for the visit, the area to be visited, the States invited and the general arrangements to be adopted.

(19.2) Arrangements for such visits, including the number of the representatives from other participating States to be invited, will be at the discretion of the host State, which will bear the in-country costs. However, the host State should take appropriate account of the need to ensure the effectiveness of the visit, the maximum amount of openness and transparency and the safety and security of the invited representatives. It should also take account, as far as practicable, of the wishes of visiting representatives as regards the itinerary of the visit. The host State and the States which provide visiting personnel may circulate joint or individual comments on the visit to all other participating States.

## III. CONTACTS

### Visits to Air Bases

(20) Each participating State with air combat units reported under paragraph (11) will arrange visits for representatives of all other participating States to one of its normal peacetime air bases* on which such units are located in order to provide the visitors with the opportunity to view activity at the air base, including preparations to carry out the functions of the air base and to gain an impression of the appropriate number of air sorties and type of missions being flown.

(21) No participating State will be obliged to arrange more than one such visit in any five-year period. Prior indications given by participating States of forthcoming schedules for such visits for the subsequent year(s) may be discussed at the annual implementation assessment meetings.

(22) As a rule, up to two visits from each participating State will be invited.

(23) Invitations will be extended to all participating States 42 days or more in advance of the visit. The invitation will indicate a preliminary programme, including: place, date and time of assembly; planned duration; languages to be used; arrangements for board, lodging and transportation; equipment permitted to be used during the visit; and any other information that may be considered useful.

(24) When the air base to be visited is located on the territory of another participating State, the invitations will be issued by the participating State on whose territory the air base is located. In such cases, the responsibilities as host delegated by this State to the participating States arranging the visit will be specified in the invitation.

---

* In this context, the term normal peacetime air base is understood to mean the normal peacetime location of the air combat unit indicated by the air base or military airfield on which the unit is based.

(25) The invited State may decide whether to send military and/or civilian visitors, including personnel accredited to the host State. Military visitors will normally wear their uniforms and insignia during the visit..

(26) Replies, indicating whether or not the invitation is accepted, will be given not later than 21 days after the issue of the invitation. Participating States accepting an invitation will provide the names and ranks of the visitors in their replies. If the invitation is not accepted in time, it will be assumed that no visitors will be sent

(27) The visit to the air base will last for a minimum of 24 hours. In the course of the visit, the visitors will be given a briefing on the purpose and functions of the air base and on current activity at the air base. They will have the opporunity to communicate with commanders and troops, including those of support/logistic units located at the air base.

(28) The visitors will be provided with the opportunity to view all types of aircraft located at the air base.

(29) At the close of the visit, the host State will provide an opporunity for the visitors to meet together and also with host State officials and senior air base personnel to discuss the course of the visit.

(30) The host State will determine the programme for the visit and access granted to visitors at the air base. The visitors will follow the instructions issued by the host State in accordance with the provisions set out in this document.

(31) The visitors will be provided with appropriate accommodation in a location suitable for carrying out the visit.

(32) The invited State will cover the travel expenses of its representatives to and from the place of assembly specified in the invitation.

(33) Participating States should, in due cooperation with the visitors, ensure that no action is taken which could be harmful to the safety of visitors.

## Military Contacts

(34) To improve further their mutual relations in the interest of strengthening the process of confidence- and security-building, the participating States will, as appropriate, promote and facilitate:

(34.1) – exchanges and visits between senior military/defence representatives;

(34.2) – contacts between relevant military institutions;

(34.3) – attendance by military representatives of other participating States at courses of instruction;

(34.4) – exchanges between military commanders and officers of commands down to brigade/regiment or equivalent level;

(34.5) – exchanges and contacts between academics and experts in military studies and related areas;

(34.6) – sporting and cultural events between members of their armed forces.

## Demonstration of New Types of Major Weapon and Equipment Systems

(35) The first participating State which deploys with its military forces in the zone of application a new type of major weapon and equipment system as specified in the provisions on Information on Military Forces will arrange at the earliest opportunity (e.g. during an observation) a demonstration for representatives of all other participating States.*

(35.1) The host State will determine the duration, the programme and other modalities of the demonstration.

(35.2) Invitations will be extended to all participating States 42 days or more in advance of visits. The invitation will indicate a preliminary programme, including: the number of visitors invited from each participating State; the type(s) of major weapon and equipment system(s) to be viewed; place, date and time of assembly; planned duration; languages to be used; arrangements for board, lodging and transportation, where necessary; equipment permitted to be used during the visit; and any other information that may be considered useful.

(35.3) Replies, indicating whether or not the invitation is accepted, will be given not later than 21 days after the issue of the invitation. Participating States accepting an invitation will provide the names and ranks of the visitors in their replies. If the invitation is not accepted in time, it will be assumed that no visitors will be sent.

(35.4) The invited State will cover the travel expenses of its representatives to and from the place of assembly and, if applicable, costs for accommodation during the visit.

---

*This provision will not apply if another participating State has already arranged a demonstration of the same type of major weapon and equipment system.

## IV. PRIOR NOTIFICATION OF CERTAIN MILITARY ACTIVITIES

(36) The participating States will give notification in writing through diplomatic channels in an agreed form of content, to all other participating States 42 days or more in advance of the start of notifiable* military activities in the zone of application for CSBMs.

(37) Notification will be given by the participating State on whose territory the activity in question is planned to take place even if the forces of that State are not engaged in the activity or their strength is below the notifiable level. This will not relieve other participating States of their obligation to give notification, if their involvement in the planned military activity reaches the notifiable level.

(38) Each of the following military activities in the field conducted as a single activity in the zone of application for CSBMs at or above the levels defined below, will be notified:

(38.1) The engagement of formations of land forces** of the participating States in the same exercise activity conducted under a single operational command independently or in combination with any possible air or naval components.

(38.1.1) This military activity will be subject to notification whenever it involves at any time during the activity:

– at least 9,000 troops, including support troops, or

– at least 250 battle tanks if organized into a divisional structure or at least two brigades/ regiments, not necessarily subordinate to the same division.

(38.1.2) The participation of air forces of the participating States will be included in the notification if it is foreseen that in the course of the activity 200 or more sorties by aircraft, excluding helicopters, will be flown.

(38.2) The engagement of military forces either in an amphibious landing or in a parachute assault by airborne forces in the zone of application for CSBMs.

(38.2.1) These military activities will be subject to notification whenever the amphibious landing involves at least 3,000 troops or whenever the parachute drop involves at least 3,000 troops.

(38.3) The engagement of formations of land forces of the participating States in a transfer from outside the zone of application for CSBMs to arrival points in the zone, or from inside the zone of application for CSBMs to points of concentration in the zone, to participate in a notifiable exercise activity or to be concentrated.

(38.3.1) The arrival or concentration of these forces will be subject to notification whenever it involves, at any time during the activity:

– at least 9,000 troops, including support troops, or

– at least 250 battle tanks

if organized into a divisional structure or at least two brigades/regiments, not necessarily subordinate to the same division.

(38.3.2) Forces which have been transferred into the zone will be subject to all provisions of agreed CSBMs when they depart their arrival points to participate in a notifiable exercise activity or to be concentrated within the zone of application for CSBMs.

(39) Notifiable military activities carried out without advance notice to the troops involved, are exceptions to the requirement for prior notification to be made 42 days in advance.

(39.1) Notification of such activities, above the agreed thresholds, will be given at the time the troops involved commence such activities.

(40) Notification will be given in writing of each notifiable military activity in the following agreed form:

**(41) A—General Information**

(41.1) The designation of the military activity;

(41.2) The general purpose of the military activity;

(41.3) The names of the States involved in the military activity,

(41.4) The level of command, organizing and commanding the military activity;

(41.5) The start and end dates of the military activity.

**(42) B—Information on different types of notifiable military activities**

(42.1) The engagement of formations of land forces of the participating States in the same exercise activity conducted under a single operational command independently or in combination with any possible air or naval components:

(42.1.1) The total number of troops taking part in the military activity (i.e. ground troops, amphibious troops, airmobile and air-

---

* In this document, the term notifiable means subject to notification.

** In this context, the term land forces includes amphibious, airmobile and airborne forces.

borne troops) and the number of troops participating for each State involved, if applicable;

(42.1.2) The designation, subordination, number and type of formations and units participating for each State down to and including brigade/regiment or equivalent level;

(42.1.3) The total number of battle tanks for each State and the total number of anti-tank guided missile launchers mounted on armoured vehicles;

(42.1.4) The total number of artillery pieces and multiple rocket launchers (100 mm calibre or above);

(42.1.5) The total number of helicopters, by category;

(42.1.6) Envisaged number of sorties by aircraft, excluding helicopters;

(42.1.7) Purpose of air missions;

(42.1.8) Categories of aircraft involved;

(42.1.9) The level of command, organizing and commanding the air force participation;

(42.1.10) Naval ship-to-shore gunfire;

(42.1.11) Indication of other naval ship-to-shore support;

(42.1.12) The level of command, organizing and commanding the naval force participation.

(42.2) The engagement of military forces either in an amphibious landing or in a parachute assault by airborne forces in the zone of application for CSBMs:

(42.2.1) The total number of amphibious troops involved in notifiable amphibious landings, and/or the total number of airborne troops involved in notifiable parachute assaults;

(42.2.2) In the case of a notifiable amphibious landing, the point or points of embarkation, if in the zone of application for CSBMs.

(42.3) The engagement of formations of land forces of the participating States in a transfer from outside the zone of application for CSBMs to arrival points in the zone, or from inside the zone of application for CSBMs to points of concentration in the zone, to participate in a notifiable exercise activity or to be concentrated:

(42.3.1) The total number of troops transferred;

(42.3.2) Number and type of divisions participating in the transfer;

(42.3.3) The total number of battle tanks participating in a notifiable arrival or concentration;

(42.3.4) Geographical co-ordinates for the points of arrival and for the points of concentration.

**(43) C—The envisaged area and time-frame of the activity**

(43.1) The area of the military activity delimited by geographic features together with geographic co-ordinates, as appropriate;

(43.2) The start and end dates of each phase (transfers, deployment, concentration of forces, active exercise phase, recovery phase) of activities in the zone of application for CSBMs of participating formations, the tactical purpose and corresponding geographical areas (delimited by geographical co-ordinates) for each phase;

(43.3) Brief description of each phase.

**(44) D—Other information**

(44.1) Changes, if any, in relation to information provided in the annual calendar regarding the activity;

(44.2) Relationship of the activity to other notifiable activities.

## V. OBSERVATION OF CERTAIN MILITARY ACTIVITIES

(45) The participating States will invite observers from all other participating States to the following notifiable military activities:

(45.1) – The engagement of formations of land forces* of the participating States in the same exercise activity conducted under a single operational command independently or in combination with any possible air or naval components.

(45.2) – The engagement of military forces either in an amphibious landing or in a parachute assault by airborne forces in the zone of application for CSBMs.

(45.3) – In the case of the engagement of formations of land forces of the participating States in a transfer from outside the zone of application for CSBMs to arrival points in the zone, or from inside the zone of application for CSBMs to points of concentration in the zone, to participate in a notifiable exercise activity or to be concentrated, the concentration of these forces. Forces which have been transferred into the zone will be subject to all provisions of agreed confidence- and security-building measures when they depart their arrival points to participate in a notifiable exercise activity or to be concentrated within the zone of application for CSBMs.

(45.4) The above-mentioned activities will be subject to observation whenever the

*In this context, the term land forces includes amphibious, airmobile and airborne forces.

number of troops engaged meets or exceeds 13,000 or where the number of battle tanks engaged meets or exceeds 300, except in the case of either an amphibious landing or a parachute assault by airborne forces, which will be subject to observation whenever the number of troops engaged meets or exceeds 3,500.

(46) The host State will extend the invitations in writing through diplomatic channels to all other participating States at the time of notification. The host State will be the participating State on whose territory the notified activity will take place.

(47) The host State may delegate some of its responsibilities as host to another participating State engaged in the military activity on the territory of the host State. In such cases, the host State will specify the allocation of responsibilities in its invitation to observe the activity.

(48) Each participating State may send up to two observers to the military activity to be observed.

(49) The invited State may decide whether to send military and/or civilian observers, including personnel accredited to the host State. Military observers will normally wear their uniforms and insignia while performing their tasks.

(50) Replies, indicating whether or not the invitation is accepted, will be given in writing not later than 21 days after the issue of the invitation.

(51) The participating States accepting an invitation will provide the names and ranks of their observers in their reply to the invitation. If the invitation is not accepted in time, it will be assumed that no observers will be sent.

(52) Together with the invitation the host State will provide a general observation programme, including the following information:

(52.1) – the date, time and place of assembly of observers;

(52.2) – planned duration of the observation programme;

(52.3) – languages to be used in interpretation and/or translation;

(52.4) – arrangements for board, lodging and transportation of the observers;

(52.5) – arrangements for observation equipment which will be issued to the observers by the host State;

(52.6) – possible authorization by the host State of the use of special equipment that the observers may bring with them;

(52.7) – arrangements for special clothing to be issued to the observers because of weather or environmental factors.

(53) The observers may make requests with regard to the observation programme. The host State will, if possible, accede to them.

(54) The host State will determine a duration of observation which permits the observers to observe a notifiable military activity from the time that agreed thresholds for observation are met or exceeded until, for the last time during the activity, the thresholds for observation are no longer met.

(55) The host State will provide the observers with transportation to the area of the notified activity and back. This transportation will be provided from either the capital or another suitable location to be announced in the invitation, so that the observers are in position before the start of the observation programme.

(56) The invited State will cover the travel expenses for its observers to the capital, or another suitable location specified in the invitation, of the host State, and back.

(57) The observers will be provided equal treatment and offered equal opportunities to carry out their functions.

(58) The observers will be granted, during their mission, the privileges and immunities accorded to diplomatic agents in the Vienna Convention on Diplomatic Relations.

(59) The participating States will ensure that official personnel and troops taking part in an observed military activity, as well as other armed personnel located in the area of the military activity, are adequately informed regarding the presence, status and functions of observers. Participating States should, in due co-operation with the observers, ensure that no action is taken which could be harmful to the safety of observers.

(60) The host State will not be required to permit observation of restricted locations, installations or defence sites.

(61) In order to allow the observers to confirm that the notified activity is non-threatening in character and that it is carried out in conformity with the appropriate provisions of the notification, the host State will:

(61.1) – at the commencement of the observation programme give a briefing on the purpose, the basic situation, the phases of the activity and possible changes as compared with the notification and provide the observers with an observation programme with a daily schedule;

(61.2) – provide the observers with a map with a scale of 1 to not more than 250,000 depicting the area of the notified military activity and the initial tactical situation in this area. To depict the entire area of the notified military activity, smaller-scale maps may be additionally provided;

(61.3) – provide the observers with appropriate observation equipment; in addition, the observers will be permitted to use their own binoculars, maps, photo and video cameras, dictaphones and hand-held passive night-vision devices. The above-mentioned equipment will be subject to examination and approval by the host State. It is understood that the host State may limit the use of certain equipment in restricted locations, installations or defence sites;

(61.4) – be encouraged, whenever feasible and with due consideration for the security of the observers, to provide an aerial survey, preferably by helicopter, of the area of the military activity. If carried out, such a survey should provide the observers with the opportunity to observe from the air the disposition of forces engaged in the activity in order to help them gain a general impression of its scope and scale. At least one observer from each participating State represented at the observation should be given the opportunity to participate in the survey. Helicopters and/or aircraft may be provided by the host State or by another participating State at the request of and in agreement with the host State;

(61.5) – in the course of the observation programme give the observers daily briefings with the help of maps on the various phases of the military activity and their development and inform the observers about their positions geographically; in the case of a land force activity conducted in combination with air or naval components, briefings will be given by representatives of these forces;

(61.6) – provide opportunities to observe directly forces of the State(s) engaged in the military activity so that the observers get an impression of the flow of the entire activity; to this end, the observers will be given the opportunity to observe combat and support units of all participating formations of a divisional or equivalent level and, whenever possible, to visit units below divisional or equivalent level and communicate with commanders and troops. Commanders and other senior personnel of the participating formations as well as of the visited units will inform the observers of the mission and disposition of their respective units;

(61.7) – guide the observers in the area of the military activity; the observers will follow the instructions issued by the host State in accordance with the provisions set out in this document;

(61.8) – provide the observers with appropriate means of transportation in the area of the military activity;

(61.9) – provide the observers with opportunities for timely communication with their embassies or other official missions and consular posts; the host State is not obligated to cover the communication expense of the observers;

(61.10) – provide the observers with appropriate board and lodging in a location suitable for carrying out the observation programme and, when necessary, medical care;

(61.11) – at the close of each observation, provide an opportunity for the observers to meet together and also with host State officials to discuss the course of the observed activity. Where States other than the host State have been engaged in the activity, military representatives of those States will also be invited to take part in this discussion.

(62) The participating States need not invite observers to notifiable military activities which are carried out without advance notice to the troops involved unless these notifiable activities have a duration of more than 72 hours. The continuation of these activities beyond this time will be subject to observation while the agreed thresholds for observation are met or exceeded. The observation programme will follow as closely as practically possible all the provisions for observation set out in this document.

(63) The participating States are encouraged to permit media representatives from all participating States to attend observed military activities in accordance with accreditation procedures set down by the host State. In such instances, media representatives from all participating States will be treated without discrimination and given equal access to those facets of the activity open to media representatives.

(64) The presence of media representatives will not interfere with the observers carrying out their functions nor with the flow of the military activity.

## VI. ANNUAL CALENDARS

(65) Each participating State will exchange, with all other participating States, an annual calendar of its military activities

subject to prior notification,* within the zone of application for CSBMs, forecast for the subsequent calendar year. A participating State which is to host military activities subject to prior notification conducted by any other participating State(s) will include these activities in its annual calendar. It will be transmitted every year, in writing, through diplomatic channels, not later than 15 November for the following year.

(66) If a participating State does not forecast any military activity subject to prior notification it will so inform all other participating States in the same manner as prescribed for the exchange of annual calendars.

(67) Each participating State will list the above-mentioned activities chronologically and will provide information on each activity in accordance with the following model:

(67.1) – type of military activity and its designation;

(67.2) – general characteristics and purpose of the military activity;

(67.3) – States involved in the military activity;

(67.4) – area of the military activity, indicated by geographic features where appropriate and defined by geographic co-ordinates;

(67.5) – planned duration of the military activity, indicated by envisaged start and end dates;

(67.6) – the envisaged total number of troops* engaged in the military activity. For activities involving more than one State, the host State will provide such information for each State involved;

(67.7) – the types of armed forces involved in the military activity;

(67.8) – the envisaged level of the military activity and designation of direct operational command, under which this military activity will take place;

(67.9) – the number and type of divisions whose participation in the military activity is envisaged;

(67.10) – any additional information concerning, *inter alia,* components of armed forces, which the participating State planning the military activity considers relevant.

(68) – Should changes regarding the military activities in the annual calendar prove necessary, they will be communicated to all other participating States no later than in the appropriate notification.

(69) – Should a participating State cancel a military activity included in its annual calendar or reduce it to a level below notification thresholds, that State will inform the other participating States immediately.

(70) – Information on military activities subject to prior notification not included in an annual calendar will be communicated to all participating States as soon as possible, in accordance with the model provided in the annual calendar.

## VII. CONSTRAINING PROVISIONS

(71.1) No participating State will carry out within two calendar years more than one military activity subject to prior notification* involving more than 40,000 troops or 900 battle tanks.

(71.2) No participating State will carry out within a calendar year more than six military activities subject to prior notification* each one involving more than 13,000 troops or 300 battle tanks but not more than 40,000 troops or 900 battle tanks.

(71.2.1) Of these six military activities no participating State will carry out within a calendar year more than three military activities subject to prior notification* each one involving more than 25,000 troops or 400 battle tanks.

(71.3) No participating State will carry out simultaneously more than three military activities subject to prior notification* each one involving more than 13,000 troops or 300 battle tanks.

(72) Each participating State will communicate, in writing, to all other participating States, by 15 November each year, information concerning military activities subject to prior notification* involving more than 40,000 troops or 900 battle tanks, which it plans to carry out or host in the second subsequent calendar year. Such a communication will include preliminary information on the activity, as to its general purpose, timeframe and duration, area, size and States involved.

(73) No participating State will carry out a military activity subject to prior notification* involving more than 40,000 troops or 900 battle tanks, unless it has been the object of a communication as defined above and unless it has been included in the annual calendar, not later than 15 November each year.

(74) If military activities subject to prior notification* are carried out in addition to those contained in the annual calendar, they should be as few as possible.

---

*as defined in the provisions on Prior Notification of Certain Military Activities.

## VIII. COMPLIANCE AND VERIFICATION

(75) According to the Madrid Mandate, the confidence- and security-building measures to be agreed upon 'will be provided with adequate forms of verification which correspond to their content' .

(76) The participating States recognize that national technical means can play a role in monitoring compliance with agreed confidence- and security-building measures.

### Inspection

(77) In accordance with the provisions contained in this document each participating State has the right to conduct inspections on the territory of any other participating State within the zone of application for CSBMs. The inspecting State may invite other participating States to participate in an inspection.

(78) Any participating State will be allowed to address a request for inspection to another participating State on whose territory, within the zone of application for CSBMs, compliance with the agreed confidence- and security-building measures is in doubt.

(79) No participating State will be obliged to accept on its territory within the zone of application for CSBMs, more than three inspections per calendar year.

(80) No participating State will be obliged to accept more than one inspection per calendar year from the same participating State.

(81) An inspection will not be counted if, due to force majeure, it cannot be carried out.

(82) The participating State which requests an inspection will state the reasons for such a request.

(83) The participating State which has received such a request will reply in the affirmative to the request within the agreed period of time, subject to the provisions contained in paragraphs (79) and (80).

(84) Any possible dispute as to the validity of the reasons for a request will not prevent or delay the conduct of an inspection.

(85) The participating State which requests an inspection will be permitted to designate for inspection on the territory of another State within the zone of application for CSBMs, a specific area. Such an area will be referred to as the 'specified area'. The specified area will comprise terrain where notifiable military activities are conducted or where another participating State believes a notifiable military activity is taking place. The specified area will be defined and limited by the scope and scale of notifiable military activities but will not exceed that required for an army level military activity.

(86) In the specified area the inspection team accompanied by the representatives of the receiving State will be permitted access, entry and unobstructed survey, except for areas or sensitive points to which access is normally denied or restricted, military and other defence installations, as well as naval vessels, military vehicles and aircraft. The number and extent of the restricted areas should be as limited as possible. Areas where notifiable military activities can take place will not be declared restricted areas, except for certain permanent or temporary military installations which, in territorial terms, should be as small as possible, and consequently those areas will not be used to prevent inspection of notifiable military activities. Restricted areas will not be employed in a way inconsistent with the agreed provisions on inspection.

(87) Within the specified area, the forces of participating States other than the receiving State will also be subject to the inspection.

(88) Inspection will be permitted on the ground, from the air, or both.

(89) The representatives of the receiving State will accompany the inspection team, including when it is in land vehicles and an aircraft from the time of their first employment until the time they are no longer in use for the purposes of inspection.

(90) In its request, the inspecting State will notify the receiving State of:

(90.1) – the reasons for the request;

(90.2) – the location of the specified area defined by geographical co-ordinates;

(90.3) – the preferred point(s) of entry for the inspection team;

(90.4) – mode of transport to and from the point(s) of entry and, if applicable, to and from the specified area;

(90.5) – where in the specified area the inspection will begin;

(90.6) – whether the inspection will be conducted from the ground, from the air, or both simultaneously;

(90.7) – whether aerial inspection will be conducted using an airplane, a helicopter, or both;

(90.8) – whether the inspection team will use land vehicles provided by the receiving State or, if mutually agreed, its own vehicles;

(90.9) – other participating States participating in the inspection, if applicable;

(90.10) – information for the issuance of diplomatic visas to inspectors entering the receiving State.

(91) The reply to the request will be given in the shortest possible period of time, but within not more than twenty-four hours. Within thirty-six hours after the issuance of the request, the inspection team will be permitted to enter the territory of the receiving State.

(92) Any request for inspection as well as the reply thereto will be communicated to all participating States without delay.

(93) The receiving State should designate the point(s) of entry as close as possible to the specified area. The receiving State will ensure that the inspection team will be able to reach the specified area without delay from the point(s) of entry.

(94) All participating States will facilitate the passage of the inspection teams through their territory.

(95) Within 48 hours after the arrival of the inspection team at the specified area, the inspection will be terminated.

(96) There will be no more than four inspectors in an inspection team. The inspecting State may invite other participating States to participate in an inspection. The inspection team will be headed by a national of the inspecting State, which will have at least as many inspectors in the team as any invited State. The inspection team will be under the responsibility of the inspecting State, against whose quota the inspection is counted. While conducting the inspection, the inspection team may divide into two sub-teams.

(97) The inspectors and, if applicable, auxiliary personnel, will be granted during their mission the privileges and immunities in accordance with the Vienna Convention on Diplomatic Relations.

(98) The participating States will ensure that troops, other armed personnel and officials in the specified area are adequately informed regarding the presence, status and functions of inspectors and, if applicable, auxiliary personnel. The receiving State will ensure that no action is taken by its representatives which could endanger inspectors and, if applicable, auxiliary personnel. In carrying out their duties, inspectors and, if applicable, auxiliary personnel will take into account safety concerns expressed by representatives of the receiving State.

(99) The receiving State will provide the inspection team with appropriate board and lodging in a location suitable for carrying out the inspection, and, when necessary, medical care; however this does not exclude the use by the inspection team of its own tents and rations.

(100) The inspection team will have use of its own maps and charts, photo and video cameras, binoculars, hand-held passive night vision devices and dictaphones. Upon arrival in the specified area the inspection team will show the equipment to the representatives of the receiving State. In addition, the receiving State may provide the inspection team with a map depicting the area specified for the inspection.

(101) The inspection team will have access to appropriate telecommunications equipment of the receiving State for the purpose of communicating with the embassy or other official missions and consular posts of the inspecting State accredited to the receiving State.

(102) The receiving State will provide the inspection team with access to appropriate telecommunications equipment for the purpose of continuous communication between the sub-teams.

(103) Inspectors will be entitled to request and to receive briefings at agreed times by military representatives of the receiving State. At the inspectors' request, such briefings will be given by commanders of formations or units in the specified area. Suggestions of the receiving State as to the briefings will be taken into consideration.

(104) The inspecting State will specify whether aerial inspection will be conducted using an airplane, a helicopter or both. Aircraft for inspection will be chosen by mutual agreement between the inspecting and receiving States. Aircraft will be chosen which provide the inspection team with a continuous view of the ground during the inspection.

(105) After the flight plan, specifying, *inter alia,* the inspection team's choice of flight path, speed and altitude in the specified area, has been filed with the competent air traffic control authority the inspection aircraft will be permitted to enter the specified area without delay. Within the specified area, the inspection team will, at its request, be permitted to deviate from the approved flight plan to make specific observations provided such deviation is consistent with paragraph (86) as well as flight safety and air traffic requirements. Directions to the crew will be given through a representative of the receiv-

ing State on board the aircraft involved in the inspection.

(106) One member of the inspection team will be permitted, if such a request is made, at any time to observe data on navigational equipment of the aircraft and to have access to maps and charts used by the flight crew for the purpose of determining the exact location of the aircraft during the inspection flight.

(107) Aerial and ground inspectors may return to the specified area as often as desired within the 48-hour inspection period.

(108) The receiving State will provide for inspection purposes land vehicles with cross country capability. Whenever mutually agreed taking into account the specific geography relating to the area to be inspected, the inspecting State will be permitted to use its own vehicles.

(109) If land vehicles or aircraft are provided by the inspecting State, there will be one accompanying driver for each land vehicle, or accompanying aircraft crew.

(110) The inspecting State will prepare a report of its inspection and will provide a copy of that report to all participating States without delay.

(111) The inspection expenses will be incurred by the receiving State except when the inspecting State uses its own aircraft and/or land vehicles. The inspecting State will be responsible for travel expenses to and from the point(s) of entry.

**Evaluation**

(112) Information provided under the provisions on Information on Military Forces and on Information on Plans for the Deployment of Major Weapon and Equipment Systems will be subject to evaluation.

(113) Subject to the provisions below each participating State will provide the opportunity to visit active formations and units in their normal peacetime locations as specified in point 2 and 3 of the provisions on Information on Military Forces to allow the other participating States to evaluate the information provided.

(113.1) Non-active formations and combat units temporarily activated will be made available for evaluation during the period of temporary activation and in the area/location of activation indicated under paragraph (11.3.3). In such cases the provisions for the evaluation of active formations and units will be applicable, *mutatis mutandis*. Evaluation visits conducted under this provision will count against the quotas established under paragraph (114).

(114) Each participating State will be obliged to accept a quota of one evaluation visit per calendar year for every sixty units, or portion thereof, reported under paragraph (11). However, no participating State will be obliged to accept more than fifteen visits per calendar year. No participating State will be obliged to accept more than one fifth of its quota of visits from the same participating State; a participating State with a quota of less than five visits will not be obliged to accept more than one visit from the same participating State during a calendar year. No formation or unit may be visited more than twice during a calendar year and more than once by the same participating State during a calendar year.

(115) No participating State will be obliged to accept more than one visit at any given time on its territory.

(116) If a participating State has formations or units stationed on the territory of other participating States (host States) in the zone of application for CSBMs, the maximum number of evaluation visits permitted to its forces in each of the States concerned will be proportional to the number of its units in each State. The application of this provision will not alter the number of visits this participating State (stationing State) will have to accept under paragraph (114).

(117) Requests for such visits will be submitted giving 5 days notice.

(118) The request will specify:

(118.1) – the formation or unit to be visited;

(118.2) – the proposed date of the visit;

(118.3) – the preferred point(s) of entry as well as the date and estimated time of arrival for the evaluation team;

(118.4) – the mode of transport to and from the point(s) of entry and, if applicable, to and from the formation or unit to be visited;

(118.5) – the names and ranks of the members of the team and, if applicable, information for the issue of diplomatic visas;

(119) If a formation or unit of a participating State is stationed on the territory of another participating State, the request will be addressed to the host State and sent simultaneously to the stationing State.

(120) The reply to the request will be given within 48 hours after the receipt of the request.

(121) In the case of formations or units of

a participating State stationed on the territory of another participating State, the reply will be given by the host State in consultation with the stationing State. After consultation between the host State and the stationing State the host State will specify in its reply any of its responsibilities which it agrees to delegate to the stationing State.

(122) The reply will indicate whether the formation or unit will be available for evaluation at the proposed date at its normal peacetime location.

(123) Formations or units may be in their normal peacetime location but be unavailable for evaluation. Each participating State will be entitled in such cases not to accept a visit; the reasons for the non-acceptance and the number of days that the formation or unit will be unavailable for evaluation will be stated in the reply. Each participating State will be entitled to invoke this provision up to a total of five times for an aggregate of no more than 30 days per calendar year.

(124) If the formation or unit is absent from its normal peacetime location, the reply will indicate the reasons for and the duration of its absence. The requested State may offer the possibility of a visit to the formation or unit outside its normal peacetime location. If the requested State does not offer this possibility, the requesting State will be able to visit the normal peacetime location of the formation or unit. The requesting State may however refrain in either case from the visit.

(125) Visits will not be counted against the quotas of receiving States, if they are not carried out. Likewise, if visits are not carried out, due to force majeure, they will not be counted.

(126) The reply will designate the point(s) of entry and indicate, if applicable, the time and place of assembly of the team. The point(s) of entry and, if applicable, the place of assembly will be designated as close as possible to the formation or unit to be visited. The receiving State will ensure that the team will be able to reach the formation or unit without delay.

(127) The request and the reply will be communicated to all participating States without delay.

(128) Participating States will facilitate the passage of teams through their territory.

(129) The team will have no more than two members. It may be accompanied by an interpreter as auxiliary personnel.

(130) The members of the team and, if applicable, auxiliary personnel, will be granted during their mission the privileges and immunities in accordance with the Vienna Convention on Diplomatic Relations.

(131) The visit will take place in the course of a single working day and last up to 12 hours.

(132) The visit will begin with a briefing by the officer commanding the formation or unit, or his deputy, in the headquarters of the formation or unit, concerning the personnel as well as the major weapon and equipment systems reported under paragraph (11).

(132.1) In the case of a visit to a formation, the receiving State may provide the possibility to see personnel and major weapon and equipment systems reported under paragraph (11) for that formation, but not for any of its formations or units, in their normal locations.

(132.2) In the case of a visit to a unit, the receiving State will provide the possibility to see the personnel and the major weapon and equipment systems of the unit reported under paragraph (11) in their normal locations.

(133) Access will not have to be granted to sensitive points, facilities and equipment.

(134) The team will be accompanied at all times by representatives of the receiving State.

(135) The receiving State will provide the team with appropriate transportation during the visit to the formation or unit.

(136) Personal binoculars and dictaphones may be used by the team.

(137) The visit will not interfere with activities of the formation or unit.

(138) The participating States will ensure that troops, other armed personnel and officials in the formation or unit are adequately informed regarding the presence, status and functions of members of teams and, if applicable, auxiliary personnel. Participating States will also ensure that no action is taken by their representatives which could endanger the members of teams and, if applicable, auxiliary personnel. In carrying out their duties, members of teams and, if applicable, auxiliary personnel will take into account safety concerns expressed by representatives of the receiving State.

(139) Travel expenses to and from the point(s) of entry, including expenses for refuelling, maintenance and parking of aircraft and/or land vehicles of the visiting State, will be borne by the visiting State according to existing practices established under the CSBM inspection provisions.

(139.1) Expenses for evaluation visits

incurred beyond the point(s) of entry will be borne by the receiving State, except when the visiting State uses its own aircraft and/or land vehicles in accordance with paragraph (118.4).

(139.2) The receiving State will provide appropriate board and, when necessary, lodging in a location suitable for carrying out the evaluation as well as any urgent medical care which may be required.

(139.3) In the case of visits to formations or units of a participating State stationed on the territory of another participating State, the stationing State will bear the costs for the discharge of those responsibilities which have been delegated to it by the host State under the terms of paragraph (121).

(140) The visiting State will prepare a report of its visit which will be communicated to all participating States expeditiously.

(141) Each participating State will be entitled to obtain timely clarification from any other participating State concerning the application of agreed confidence- and security-building measures. Communications in this context will, if appropriate, be transmitted to all other participating States.

(142) The communications concerning compliance and verification will be transmitted preferably through the CSBM communications network.

## IX. COMMUNICATIONS

(143) The participating States have established a network of direct communications between their capitals for the transmission of messages relating to agreed measures. The network will complement the existing use of diplomatic channels. Participating States undertake to use the network flexibly, efficiently and in a cost-effective way.

(144) Each participating State will designate a point of contact capable of transmitting and receiving such messages from other participating States on a 24-hour-a-day basis and will notify in advance any change in this designation.

(145) Cost-sharing arrangements are set out in documents CSCE/WV/Dec. 2 and CSCE/WV/Dec. 4.

(146) Communications may be in any one of the six working languages of the CSCE.

(147) Details on the use of these six languages are set out in Annex II. The provisions of this annex have been elaborated for the practical purposes of the communication system only. They are not intended to change the existing use of all six working languages of the CSCE according to established rules and practice as set out in the Final Recommendations of the Helsinki Consultations.

(148) Messages will be considered official communications of the sending State. If the content of a message is not related to an agreed measure, the receiving State has the right to reject it by so informing the other participating States.

(149) Participating States may agree among themselves to use the network for other purposes.

(150) All aspects of the implementation of the network may be discussed at the annual implementation assessment meeting.

## X. ANNUAL IMPLEMENTATION ASSESSMENT MEETING

(151) The participating States will hold each year a meeting to discuss the present and future implementation of agreed CSBMs. Discussion may extend to:

(151.1) – clarification of questions arising from such implementation;

(151.2) – operation of agreed measures;

(151.3) – implications of all information originating from the implementation of any agreed measures for the processs of confidence- and security-building in the framework of the CSCE.

(152) Before the conclusion of each year's meeting the participating States will normally agree upon the agenda and dates for the subsequent year's meeting. Lack of agreement will not constitute sufficient reason to extend a meeting, unless otherwise agreed. Agenda and dates may, if necessary, be agreed between meetings.

(153) The Conflict Prevention Centre will serve as the forum for such meetings.

* * *

(154) The participating States will implement this new set of mutually complementary confidence- and security-building measures in order to promote security co-operation and to reduce the risk of military conflict.

(155) Reaffirming the relevant objectives of the Final Act and the Charter of Paris, the participating States are determined to continue building confidence and to enhance security for all.

(156) The measures adopted in this document are politically binding and will come into force on 1 May 1992.

(157) The Government of Austria is requested to transmit the present document to

the Helsinki Follow-up Meeting of the CSCE. The Government of Austria is also requested to transmit the present document to the Secretary-General of the United Nations and to the Governments of the non-participating Mediterranean States.

(158) The text of this document will be published in each participating State, which will disseminate it and make it known as widely as possible.

(159) The representatives of the participating States express their profound gratitude to the Government and people of Austria for the excellent arrangements they have made for the Vienna CSBM Negotiations and the warm hospitality they have extended to the delegations which participated in the Negotiations.

*Vienna, 4 March 1992*

## ANNEX I

Under the terms of the Madrid mandate, the zone of application for CSBMs is defined as follows:

'On the basis of equality of rights, balance and reciprocity, equal respect for the security interests of all CSCE participating States, and of their respective obligations concerning confidence- and security-building measures and disarmament in Europe, these confidence- and security-building measures will cover the whole of Europe as well as the adjoining sea area* and air space. They will be of military significance and politically binding and will be provided with adequate forms of verification which correspond to their content.

As far as the adjoining sea area* and air space is concerned, the measures will be applicable to the military activities of all the participating States taking place there whenever these activities affect security in Europe as well as constitute a part of activities taking place within the whole of Europe as referred to above, which they will agree to notify. Necessary specifications will be made through the negotiations on the confidence- and security-building measures at the Conference.

Nothing in the definition of the zone given above will diminish obligations already undertaken under the Final Act. The confidence- and security-building measures to be agreed upon at the Conference will also be applicable in all areas covered by any of the provisions in the Final Act relating to confidence-building measures and certain aspects of security and disarmament.

Wherever the term 'the zone of application for CSBMs' is used in this document, the above definition will apply. The following understanding will apply as well:

The commitments undertaken in letters to the Chairman-in-Office of the CSCE Council by Armenia, Azerbaijan, Belarus, Kazakhstan, Kyrgyzstan, Moldova, Tajikistan, Turkmenistan, Ukraine and Uzbekistan on 29 January 1992 have the effect of extending the application of CSBMs in the Vienna Document 1992 to the territories of the above-mentioned States insofar as their territories were not covered already by the above.

*In this context, the notion of adjoining sea area is understood to refer also to ocean areas adjoining Europe.

## ANNEX II

### Use of the six CSCE working languages

Messages will, wherever possible, be transmitted in formats with headings in all six CSCE working languages.

Such formats, agreed among the participating States with a view to making transmitted messages immediately understandable by reducing the language element to a minimum, are annexed to document CSCE/WV/Dec. 4. The formats may be subject to agreed modifications as required. Partcipating States will co-operate in this respect.

Any narrative text, to the extent it is required in such formats, and messages that do not lend themselves to formatting will be transmitted in the CSCE working language chosen by the transmitting State.

Each participating State has the right to ask for clarification of messages in cases of doubt.

## ANNEX III

### Chairman's Statement

The participating States, in order to facilitate an efficient use of the communications network, will give due consideration to practical needs of rapid transmission of their messages and of immediate understandability. A translation into another CSCE working language will be added where needed to meet that principle. The participating States have indicated at least two CSCE working languages in which they would prefer to receive the translation.

These provisions do not prejudice in any way the future continued use of all six working languages of the CSCE according to established rules and practice as set out in the Final Recommendations of the Helsinki Consultations.

This statement will be an annex to the Vienna Document 1992 and will be published with it.

*Vienna, 4 March 1992*

## ANNEX IV

### Chairman's Statement

It is understood that the implementation aspects of CSBMs in the case of contiguous areas of participating States specified in the understanding of Annex I which share frontiers with non-European non-participating States may be discussed at future Annual Implementation Assessment Meetings.

This statement will be an annex to the Vienna Document 1992 and will be published with it.

*Vienna, 4 March 1992*

## ANNEX V

### Chairman's Statement

It is understood that the participating States will take into consideration practical problems which may arise at an initial stage in implementing CSBMs on the territories of new participating States.

This statement will not constitute a precedent.

This statement will be an annex to the Vienna Document 1992 and will be published with it.

---

*Source:* Vienna, 4 Mar. 1992

# TREATY ON OPEN SKIES

*Signed on 24 March 1992, Helsinki*

The States concluding this Treaty, hereinafter referred to collectively as the States Parties or individually as a State Party,

Recalling the commitments they have made in the Conference on Security and Co-operation in Europe to promoting greater openness and transparency in their military activities and to enhancing security by means of confidence- and security-building measures,

Welcoming the historic events in Europe which have transformed the security situation from Vancouver to Vladivostok,

Wishing to contribute to the further development and strengthening of peace, stability and co-operative security in that area by the creation of an Open Skies regime for aerial observation,

Recognizing the potential contribution which an aerial observation regime of this type could make to security and stability in other regions as well,

Noting the possibility of employing such a regime to improve openness and transparency, to facilitate the monitoring of compliance with existing or future arms control agreements and to strengthen the capacity for conflict prevention and crisis management in the framework of the Conference on Security and Co-operation in Europe and in other relevant international institutions,

Envisaging the possible extension of the Open Skies regime into additional fields, such as the protection of the environment,

Seeking to establish agreed procedures to provide for aerial observation of all the territories of States Parties, with the intent of observing a single State Party or groups of States Parties, on the basis of equity and effectiveness while maintaining flight safety,

Noting that the operation of such an Open Skies regime will be without prejudice to States not participating in it,

Have agreed as follows:

### Article I. General Provisions

1. This Treaty establishes the regime, to be known as the Open Skies regime, for the conduct of observation flights by States Parties over the territories of other States Parties, and sets forth the rights and obligations of the States Parties relating thereto.

2. Each of the Annexes and their related Appendices constitutes an integral part of this Treaty.

**Article II. Definitions**

For the purposes of this Treaty:

1. The term 'observed Party' means the State Party or group of States Parties over whose territory an observation flight is conducted or is intended to be conducted, from the time it has received notification thereof from an observing Party until completion of the procedures relating to that flight, or personnel acting on behalf of that State Party or group of States Parties.

2. The term 'observing Party' means the State Party or group of States Parties that intends to conduct or conducts an observation flight over the territory of another State Party or group of States Parties, from the time that it has provided notification of its intention to conduct an observation flight until completion of the procedures relating to that flight, or personnel acting on behalf of that State Party or group of States Parties.

3. The term 'group of States Parties' means two or more States Parties that have agreed to form a group for the purposes of this Treaty.

4. The term 'observation aircraft' means an unarmed, fixed wing aircraft designated to make observation flights, registered by the relevant authorities of a State Party and equipped with agreed sensors. The term 'unarmed' means that the observation aircraft used for the purposes of this Treaty is not equipped to carry and employ weapons.

5. The term 'observation flight' means a flight of the observation aircraft conducted by an observing Party over the territory of an observed Party, as provided in the flight plan, from the point of entry or Open Skies airfield to the point of exit or Open Skies airfield.

6. The term 'transit flight' means a flight of an observation aircraft conducted by or on behalf of an observing Party over the territory of a third State Party en route to or from the territory of the observed Party.

7. The term 'transport aircraft' means an aircraft other than an observation aircraft that, on behalf of the observing Party, conducts flights to or from the territory of the observed Party exclusively for the purposes of this Treaty.

8. The term 'territory' means the land, including islands, and internal and territorial waters, over which a State Party exercises sovereignty.

9. The term 'passive quota' means the number of observation flights that each State Party is obliged to accept as an observed Party.

10. The term 'active quota' means the number of observation flights that each State Party has the right to conduct as an observing Party.

11. The term 'maximum flight distance' means the maximum distance over the territory of the observed Party from the point at which the observation flight may commence to the point at which that flight may terminate, as specified in Annex A to this Treaty.

12. The term 'sensor' means equipment of a category specified in Article IV, paragraph 1 that is installed on an observation aircraft for use during the conduct of observation flights.

13. The term 'ground resolution' means the minimum distance on the ground between two closely located objects distinguishable as separate objects.

14. The term 'infra-red line-scanning device' means a sensor capable of receiving and visualizing thermal electro-magnetic radiation emitted in the invisible infra-red part of the optical spectrum by objects due to their temperature and in the absence of artificial illumination.

15. The term 'observation period' means a specified period of time during an observation flight when a particular sensor installed on the observation aircraft is operating.

16. The term 'flight crew' means individuals from any State Party who may include, if the State Party so decides, interpreters and who perform duties associated with the operation or servicing of an aircraft or transport aircraft.

17. The term 'pilot-in-command' means the pilot on board the observation aircraft who is responsible for the operation of the observation aircraft, the execution of the flight plan, and the safety of the observation aircraft.

18. The term 'flight monitor' means an individual who, on behalf of the observed Party, is on board an observation aircraft provided by the observing Party during the observation flight and who performs duties in accordance with Annex G to this Treaty.

19. The term 'flight representative' means an individual who, on behalf of the observing Party, is on board an observation aircraft provided by the observed Party during an observation flight and who performs duties in accordance with Annex G to this Treaty.

20. The term 'representative' means an individual who has been designated by the observing Party and who performs activities on behalf of the observing Party in accordance with Annex G during an observation flight on an observation aircraft designated by a State Party other than the observing Party or the observed Party.

21. The term 'sensor operator' means an individual from any State Party who performs duties associated with the functioning, operation and maintenance of the sensors of an observation aircraft.

22. The term 'inspector' means an individual from any State Party who conducts an inspection of sensors or observation aircraft of another State Party.

23. The term 'escort' means an individual from any State Party who accompanies the inspectors of another State Party.

24. The term 'mission plan' means a document which is in a format established by the Open Skies Consultative Commission, presented by the observing Party that contains the route, profile, order of execution and support required to conduct the observation flight, which is to be agreed upon with the observed Party and which will form the basis for the elaboration of the flight plan.

25. The term 'flight plan' means a document elaborated on the basis of the agreed mission plan in the format and with the content specified by the International Civil Aviation Organization, hereinafter referred to as the ICAO, which is presented to the air traffic control authorities and on the basis of which the observation flight will be conducted.

26. The term 'mission report' means a document describing an observation flight completed after its termination by the observing Party and signed by both the observing and observed Parties, which is in a format established by the Open Skies Consultative Commission.

27. The term 'Open Skies airfield' means an airfield designated by the observed Party as a point where an observation flight may commence or terminate.

28. The term 'point of entry' means a point designated by the observed Party for the arrival of personnel of the observing Party on the territory of the observed Party.

29. The term 'point of exit' means a point designated by the observed Party for the departure of personnel of the observing Party from the territory of the observed Party.

30. The term 'refuelling airfield' means an airfield designated by the observed Party used for fuelling and servicing of observation aircraft and transport aircraft.

31. The term 'alternate airfield' means an airfield specified in the flight plan to which an observation aircraft or transport aircraft may proceed when it becomes inadvisable to land at the airfield of intended landing.

32. The term 'hazardous airspace' means the prohibited areas, restricted areas and danger areas, defined on the basis of Annex 2 to the Convention on International Civil Aviation, that are established in accordance with Annex 15 to the Convention on International Civil Aviation in the interests of flight safety, public safety and environmental protection and about which information is provided in accordance with ICAO provisions.

33. The term 'prohibited area' means an airspace of defined dimensions, above the territory of a State Party, within which the flight of aircraft is prohibited.

34. The term 'restricted area' means an airspace of defined dimensions, above the territory of a State Party, within which the flight of aircraft is restricted in accordance with specified conditions.

35. The term 'danger area means an airspace of defined dimensions within which activities dangerous to the flight of aircraft may exist at specified times.

### Article III. Quotas

SECTION I. GENERAL PROVISIONS

1. Each State Party shall have the right to conduct observation flights in accordance with the provisions of this Treaty.

2. Each State Party shall be obliged to accept observation flights over its territory in accordance with the provisions of this Treaty.

3. Each State Party shall have the right to conduct a number of observation flights over the territory of any other State Party equal to the number of observation flights which that other State Party has the right to conduct over it.

4. The total number of observation flights that each State Party is obliged to accept over its territory is the total passive quota for that State Party. The allocation of the total passive quota to the States Parties is set forth in Annex A, Section I to this Treaty.

5. The number of observation flights that a State Party shall have the right to conduct each year over the territory of each of the other States Parties is the individual active

quota of that State Party with respect to that other State Party. The sum of the individual active quotas is the total active quota of that State Party. The total active quota of a State Party shall not exceed its total passive quota.

6. The first distribution of active quotas is set forth in Annex A, Section II to this Treaty.

7. After entry into force of this Treaty, the distribution of active quotas shall be subject to an annual review for the following calendar year within the framework of the Open Skies Consultative Commission. In the event that it is not possible during the annual review to arrive within three weeks at agreement on the distribution of active quotas with respect to a particular State Party, the previous year's distribution of active quotas with respect to that State Party shall remain unchanged.

8. Except as provided for by the provisions of Article VIII, each observation flight conducted by a State Party shall be counted against the individual and total active quotas of that State Party.

9. Notwithstanding the provisions of paragraphs 3 and 5 of this Section, a State Party to which an active quota has been distributed may, by agreement with the State Party to be overflown, transfer a part or all of its total active quota to other States Parties and shall promptly notify all other States Parties and the Open Skies Consultative Commission thereof. Paragraph 10 of this Section shall apply.

10. No State Party shall conduct more observation flights over the territory of another State Party than a number equal to 50 per cent, rounded up to the nearest whole number, of its own total active quota, or of the total passive quota of that other State Party, whichever is less.

11. The maximum flight distances of observation flights over the territories of the States Parties are set forth in Annex A, Section III to this Treaty.

SECTION II. PROVISIONS FOR A GROUP OF STATES PARTIES

1. (A) Without prejudice to their rights and obligations under this Treaty, two or more States Parties which hold quotas may form a group of States Parties at signature of this Treaty and thereafter. For a group of States Parties formed after signature of this Treaty, the provisions of this Section shall apply no earlier than six months after giving notice to all other States Parties, and subject to the provisions of paragraph 6 of this Section.

(B) A group of States Parties shall cooperate with regard to active and passive quotas in accordance with the provisions of either paragraph 2 or 3 of this Section.

2. (A) The members of a group of States Parties shall have the right to redistribute amongst themselves their active quotas for the current year, while retaining their individual passive quotas. Notification of the redistribution shall be made immediately to all third States Parties concerned.

(B) An observation flight shall count as many observation flights against the individual and total active quotas of the observing Party as observed Parties belonging to the group are overflown. It shall count one observation flight against the total passive quota of each observed Party.

(C) Each State Party in respect of which one or more members of a group of States Parties hold active quotas shall have the right to conduct over the territory of any member of the group 50 per cent more observation flights, rounded up to the nearest whole number, than its individual active quota in respect of that member of the group or to conduct two overflights if it holds no active quota in respect of that member of the group.

(D) In the event that it exercises this right the State Party concerned shall reduce its active quotas in respect of other members of the group in such a way that the total sum of observation flights it conducts over their territories shall not exceed the sum of the individual active quotas that the State Party holds in respect of all the members of the group in the current year.

(E) The maximum flight distances of observation flights over the territories of each member of the group shall apply. In case of an observation flight conducted over several members, after completion of the maximum flight distance for one member all sensors shall be switched off until the observation aircraft reaches the point over the territory of the next member of the group of States Parties where the observation flight is planned to begin. For such follow-on observation flight the maximum flight distance related to the Open Skies airfield nearest to this point shall apply.

3. (A) A group of States Parties shall, at its request, be entitled to a common total passive quota which shall be allocated to it and common individual and total active quotas shall be distributed in respect of it.

(B) In this case, the total passive quota is

the total number of observation flights that the group of States Parties is obliged to accept each year. The total active quota is the sum of the number of observation flights that the group of States Parties has the right to conduct each year. Its total active quota shall not exceed the total passive quota.

(C) An observation flight resulting from the total active quota of the group of States Parties shall be carried out on behalf of the group.

(D) Observation flights that a group of States Parties is obliged to accept may be conducted over the territory of one or more of its members.

(E) The maximum flight distances of each group of States Parties shall be specified pursuant to Annex A, Section III and Open Skies airfields shall be designated pursuant to Annex E to this Treaty.

4. In accordance with the general principles set out in Article X, paragraph 3, any third State Party that considers its rights under the Provisions of Section 1, paragraph 3 of this Article to be unduly restricted by the operation of a group of States Parties may raise this problem before the Open Skies Consultative Commission.

5. The group of States Parties shall ensure that procedures are established allowing for the conduct of observation flights over the territories of its members during one single mission, including refuelling if necessary. In the case of a group of States Parties established pursuant to paragraph 3 of this Section, such observation flights shall not exceed the maximum flight distance applicable to the Open Skies airfields at which the observation flights commence.

6. No earlier than six months after notification of the decision has been provided to all other States Parties:

(A) a group of States Parties established pursuant to the provisions of paragraph 2 of this Section may be transformed into a group of States Parties pursuant to the provisions of paragraph 3 of this Section;

(B) a group of States Parties established pursuant to the provisions of paragraph 3 of this Section may be transformed into a group of States Parties pursuant to the provisions of paragraph 2 of this Section;

(C) a State Party may withdraw from a group of States Parties; or

(D) a group of States Parties may admit further States Parties which hold quotas.

7. Following entry into force of this Treaty, changes in the allocation or distribution of quotas resulting from the establishment of or an admission to or a withdrawal from a group of States Parties according to paragraph 3 of this Section shall become effective on 1 January following the first annual review within the Open Skies Consultative Commission occurring after the six-month notification period. When necessary, new Open Skies airfields shall be designated and maximum flight distances established accordingly.

**Article IV. Sensors**

1. Except as otherwise provided for in paragraph 3 of this Article, observation aircraft shall be equipped with sensors only from amongst the following categories:

(A) optical panoramic and framing cameras;

(B) video cameras with real-time display;

(C) infra-red line-scanning devices; and

(D) sideways-looking synthetic aperture radar.

2. A State Party may use, for the purposes of conducting observation flights, any of the sensors specified in paragraph 1 above, provided that such sensors are commercially available to all States Parties, subject to the following performance limits:

(A) in the case of optical panoramic and framing cameras a resolution of no better than 30 centimetres at the minimum height above ground level determined in accordance with the provisions of Annex D, Appendix 1, obtained from no more than one panoramic camera, one vertically-mounted framing camera and two obliquely-mounted framing cameras, one on each side of the aircraft, providing coverage, which need not be continuous, of the ground up to 50 kilometres of each side of the flight path of the aircraft;

(B) in the case of video cameras, a ground resolution of no better than 30 centimetres determined in accordance with the provisions of Annex D, Appendix 1;

(C) in the case of infra-red line-scanning devices, a ground resolution of no better than 50 centimetres at the minimum height level determined in accordance with the provisions of Annex D, Appendix 1, obtained from a single device; and

(D) in the case of sideways-looking synthetic aperture radar, a ground resolution of no better than three metres calculated by the impulse response method, which, using the object separation method, corresponds to the ability to distinguish on a radar image two corner reflectors, the distance between the centres of which is no less than five metres, over a swath width of no more than 25 kilometres, obtained from a single radar unit capable of looking from either side of the aircraft, but not both simultaneously.

3. The introduction of additional categories and improvements to the capabilities of existing categories of sensors provided for in this Article shall be addressed by the Open Skies Consultative Commission pursuant to Article X of this Treaty.

4. All sensors shall be provided with aperture covers or other devices which inhibit the operation of sensors so as to prevent collection of data during transit flights or flights to points of entry or from points of exit over the territory of the observed Party. Such covers or other devices shall be removable or operable only from outside the observation aircraft.

5. Equipment that is capable of annotating data collected by sensors in accordance with Annex B, Section II shall be allowed on observation aircraft. The State Party providing the observation aircraft for an observation flight shall annotate the data collected by sensors with the information provided for in Annex B, Section II to this Treaty.

6. Equipment that is capable of displaying data collected by sensors in real-time shall be allowed on observation aircraft for the purposes of monitoring the function and operation of the sensors during the conduct of an observation flight.

7. Except as required for the operation of the agreed sensors, or as required for the operation of the observation aircraft, or as provided for in paragraphs 5 and 6 of this Article, the collection, processing, retransmission or recording of electronic signals from electro-magnetic waves are prohibited on board the observation aircraft and equipment for such operations shall not be on that observation aircraft.

8. In the event that the observation aircraft is provided by the observing Party, the observing Party shall have the right to use an observation aircraft equipped with sensors in each sensor category that do not exceed the capability specified in paragraph 2 of this Article.

9. In the event that the observation aircraft used for an observation flight is provided by the observed Party, the observed Party shall be obliged to provide an observation aircraft equipped with sensors from each sensor category specified in paragraph 1 of this Article, at the maximum capability and in the numbers specified in paragraph 2 of this Article, subject to the provisions of Article XVIII, Section II, unless otherwise agreed by the observing and observed Parties. The package and configuration of such sensors shall be installed in such a way so as to provide coverage of the ground provided for in paragraph 2 of this Article. In the event that the observation aircraft is provided by the observed Party, the latter shall provide a sideways-looking synthetic aperture radar with a resolution of no worse than six metres, determined by the object separation method.

10. When designating an aircraft as an observation aircraft pursuant to Article V of this Treaty, each State Party shall inform all other States Parties of the technical information on each sensor installed on such aircraft as provided for in Annex B to this Treaty.

11. Each State Party shall have the right to take part in the certification of sensors installed on observation aircraft in accordance with the provisions of Annex D. No observation aircraft of a given type shall be used for observation flights until such type of observation aircraft and its sensors has been certified in accordance with the provisions of Annex D to this Treaty.

12. A State Party designating an aircraft as an observation aircraft shall, upon 90-day prior notice to all other States Parties and subject to the provisions of Annex D to this Treaty, have the right to remove, replace or add sensors, or amend the technical information it has provided in accordance with the provisions of paragraph 10 of this Article and Annex B to this Treaty. Replacement and additional sensors shall be subject to certification in accordance with the provisions of Annex D to this Treaty prior to their use during an observation flight.

13. In the event that a State Party or group of States Parties, based on experience with using a particular observation aircraft, considers that any sensor or its associated equipment installed on an aircraft does not correspond to those certified in accordance with the provisions of Annex D, the interested States Parties shall notify all other States Parties of their concern. The State Party that designated the aircraft shall:

(A) take the steps necessary to ensure that the sensor and its associated equipment installed on the observation aircraft correspond to those certified in accordance with the provisions of Annex D, including, as necessary, repair, adjustment or replacement of the particular sensor or its associated equipment; and

(B) at the request of an interested State Party, by means of a demonstration flight set up in connection with the next time that the aforementioned observation aircraft is used, in accordance with the provisions of Annex F, demonstrate that the sensor and its associated equipment installed on the observation aircraft correspond to those certified in accordance with the provisions of Annex D. Other States Parties that express concern regarding a sensor and its associated equipment installed on an observation aircraft shall have the right to send personnel to participate in such a demonstration flight.

14. In the event that, after the steps referred to in paragraph 13 of this Article have been taken, the States Parties remain concerned as to whether a sensor or its associated equipment installed on an observation aircraft correspond to those certified in accordance with the provisions of Annex D, the issue may be referred to the Open Skies Consultative Commission.

### Article V. Aircraft Designation

1. Each State Party shall have the right to designate as observation aircraft one or rare types or models of aircraft registered by the relevant authorities of a State Party.

2. Each State Party shall have the right to designate types or models of aircraft as observation aircraft or add new types or models of aircraft to those designated earlier by it, provided that it notifies all other States Parties 30 days in advance thereof. The notification of the designation of aircraft of a type or model shall contain the information specified in Annex C to this Treaty.

3. Each State Party shall have the right to delete types or models of aircraft designated earlier by it, provided that it notifies all other States Parties 90 days in advance thereof.

4. Only one exemplar of a particular type and model of aircraft with an identical set of associated sensors shall be required to be offered for certification in accordance with the provisions of Annex D to this Treaty.

5. Each observation aircraft shall be capable of carrying the flight crew and the personnel specified in Article VI, Section III.

### Article VI. Choice of Observation Aircraft, General Provisions for the Conduct of Observation Flights, and Requirements for Mission Plannning

#### SECTION I. CHOICE OF OBSERVATION AIRCRAFT AND GENERAL PROVISIONS FOR THE CONDUCT OF OBSERVATION FLIGHTS

1. Observation flights shall be conducted using observation aircraft that have been designated by a State Party pursuant to Article V. Unless the observed Party exercises its right to provide an observation aircraft that it has itself designated, the observing Party shall have the right to provide the observation aircraft. In the event that the observing Party provides the observation aircraft, it shall have the right to provide an aircraft that it has itself designated or an aircraft designated by another State Party. In the event that the observed Party provides the observation aircraft, the Party shall have the right to be provided with an aircraft capable of achieving a minimum unrefuelled range, including the necessary fuel reserves, equivalent to one-half of the flight distance, as notified in accordance with paragraph 5, subparagraph (G) of this Section.

2. Each State Party shall have the right, pursuant to paragraph 1 of Section, to use an observation aircraft designated by another State Party for observation flights. Arrangements for the use of such aircraft shall be worked out by the States Parties involved to allow for active participation in the Open Skies regime.

3. States Parties having the right to conduct observation flights may co-ordinate their plans for conducting observation flights in accordance with Annex H to this Treaty. No State Party shall be obliged to accept more than one observation flight at any one time during the 96-hour period specified in paragraph 9 of this Section, unless that State Party has requested a demonstration flight pursuant to Annex F to this Treaty. In that case, the observed Party shall be obliged to accept an overlap for the observation flights of up to 24 hours. After having been notified of the results of the co-ordination of plans to conduct observation flights, each State Party over whose territory observation flights are to be conducted shall inform other States Parties, in accordance with the provisions of Annex H, whether it will exercise, with regard to each specific observation flight, its

right to provide its own observation aircraft.

4. No later than 90 days after signature of this Treaty, each State Party shall provide notification to all other States Parties:

(A) of the standing diplomatic clearance number for Open Skies observation flights, flights of transport aircraft and transit flights; and

(B) of which languages of the Open Skies Consultative Commission specified in Annex L, Section I, paragraph 7 to this Treaty shall be used by personnel for all activities associated with the conduct of observation flights over its territory, and for completing the mission plan and mission report, unless the language to be used is the one recommended in Annex 10 to the Convention on International Civil Aviation, Volume II, paragraph 5.2.1.1.2.

5. The observing Party shall notify the observed Party of its intention to conduct an observation flight, no less than 72 hours prior to the estimated time of arrival of the observing Party at the point of entry of the observed Party. States Parties providing such notifications shall make every effort to avoid using the minimum pre-notification period over weekends. Such notification shall include:

(A) the desired point of entry and, if applicable, Open Skies airfield where the observation flight shall commence;

(B) the date and estimated time of arrival of the observing Party at the point of entry and the date and estimated time of departure for the flight from the point of entry to the Open Skies airfield, if applicable, indicating specific accommodation needs;

(C) the location, specified in Annex E, Appendix 1, where the conduct of the pre-flight inspection is desired and the date and start time of such pre-flight inspection in accordance with the provisions of Annex F;

(D) the mode of transport and, if applicable, type and model of the transport aircraft used to travel to the point of entry in the event that the observation aircraft used for the observation flight is provided by the observed Party;

(E) The diplomatic clearance number for the observation flight or for the flight of the transport aircraft used to bring the personnel in and out of the territory of the observed Party to conduct an observation flight;

(F) the identification of the observation aircraft, as specified in Annex C;

(G) the approximate observation flight distance; and

(H) the names of the personnel, their gender, date and place of birth, passport number and issuing State Party, and their function.

6. The observed Party that is notified in accordance with paragraph 5 of this Section shall acknowledge receipt of the notification within 24 hours. In the event that the observed Party exercises its right to provide the observation aircraft, the acknowledgement shall include the information about observation aircraft specified in paragraph 5, subparagraph (F) of this Section. The observing Party shall be permitted to arrive at the point of entry at the estimated time of arrival as notified in accordance with paragraph 5 of this Section. The estimated time of departure for the flight from the point of entry to the Open Skies airfield where the observation flight shall commence and the location, the date and the start time of the pre-flight inspection shall be subject to confirmation by the observed Party.

7. Personnel of the observing Party may include personnel designated pursuant to Article XIII by other States Parties.

8. The observing Party, when notifying the observed Party in accordance with paragraph 5 of this Section, shall simultaneously notify all other States Parties of its intention to conduct the observation flight.

9. The period from the estimated time of arrival at the point of entry until completion of the observation flight shall not exceed 96 hours, unless otherwise agreed. In the event that the observed Party requests a demonstration flight pursuant to Annex F to the Treaty, it shall extend the 96-hour period pursuant to Annex F, Section III, paragraph 4, if additional time is required by the observing Party for the unrestricted execution of the mission plan.

10. Upon arrival of the observation aircraft at the point of entry, the observed Party shall inspect the covers for sensor apertures or other devices that inhibit the operation of sensors to confirm that they are In their proper position pursuant to Annex E, unless otherwise agreed by all States Parties involved.

11. In the event that the observation aircraft is provided by the observing Party, upon the arrival of the observation aircraft at the point of entry or at the Open Skies airfield where the observation flight commences, the observed Party shall have the right to carry out the pre-flight inspection pursuant to Annex F, Section I. In the event that, in accordance with paragraph 1 of this Section, an observation aircraft is provided

by the observed Party, the observing Party shall have the right to carry out the pre-flight inspection pursuant to Annex F, Section II. Unless otherwise agreed, such inspections shall terminate no less than four hours prior to the scheduled commencement of the observation flight set forth in the flight plan.

12. The observing Party shall ensure that its flight crew includes at least one individual who has the necessary linguistic ability to communicate freely with the personnel of the observed Party and its air traffic control authorities in the language or languages notified by the observed Party in accordance with paragraph 4 of this Section.

13. The observed Party shall provide the flight crew, upon its arrival at the point of entry or at the Open Skies airfield where the observation flight commences, with the most recent weather forecast and air navigation information and information on flight safety, including Notices to Airmen. Updates of such information shall be provided as requested. Instrument procedures, and information about alternate airfields along the flight route shall be provided upon approval of the mission plan in accordance with the requirements of Section II of this Article.

14. While conducting observation flights pursuant to this Treaty, all observation aircraft shall be operated in accordance with the provisions of this Treaty and in accordance with the flight plan. Without prejudice to the provisions of Section II, paragraph 2 of this Article, observation flights shall also be conducted in compliance with:

(A) published ICAO standards and recommended procedures; and

(B) published national air traffic control rules, procedures and guidelines on flight safety of the State Party whose territory is being overflown.

15. Observation flights shall take priority over any regular air traffic. The observed Party shall ensure that its air traffic control authorities facilitate the conduct of observation flights in accordance with this Treaty.

16. On board the aircraft the pilot-in-command shall be the sole authority for the safe conduct of the flight and shall be responsible for the execution of the flight plan.

17. The observed Party shall provide:

(A) a calibration target suitable for confirming the capability of sensors in accordance with the procedures set forth in Annex D, Section III to this Treaty, to be overflown during the demonstration flight or the observation flight upon the request of either Party, for each sensor that is to be used during the observation flight. The calibration target shall be located in the vicinity of the airfield at which the pre-flight inspection is conducted pursuant to Annex F to this Treaty;

(B) customary commercial aircraft fuelling and servicing for the observation aircraft or transport aircraft at the point of entry, at the Open Skies airfield, at any refuelling airfield, and at the point of exit specified in the flight plan, according to the specifications that are published about the designated airfield;

(C) meals and the use of accommodation for the personnel of the observing Party; and

(D) upon the request of the observing Party, further services, as may be agreed upon between the observing and observed Parties, to facilitate the conduct of the observation flight.

18. All costs involved in the of the observation flight, including the costs of the recording media and the processing of the data collected by sensors, shall be reimbursed in accordance with Annex L, Section I, paragraph 9 to this Treaty.

19. Prior to the departure of the observation aircraft from the point of exit, the observed Party shall confirm that the covers for sensor apertures or other devices that inhibit the operation of sensors are in their proper position pursuant to Annex E to this Treaty.

20. Unless otherwise agreed, the observing Party shall depart from the point of exit no later than 24 hours following completion of the observation flight, unless weather conditions or the air worthiness of the observation aircraft or transport aircraft do not permit, in which case the flight shall commence as soon as practicable.

21. The observing Party shall compile a mission report of the observation flight using the appropriate format developed by the Open Skies Consultative Commission. The mission report shall contain pertinent data on the date and time of the observation flight, its route and profile, weather conditions, time and location of each observation period for each sensor, the approximate amount of data collected by sensors, and the result of inspection of covers for sensor apertures or other devices that inhibit the operation of sensors in accordance with Article VII and Annex E. The mission report shall be signed by the observing and observed Parties at the point of exit and shall be provided by the observing Party to all other States Parties within seven days after departure of the observing Party from the point of exit.

## SECTION II. REQUIREMENTS FOR MISSION PLANNING

1. Unless otherwise agreed, the observing Party shall, after arrival at the Open Skies airfield, submit to the observed Party a mission plan for the proposed observation flight that meets the requirements of paragraphs 2 and 4 of this Section.

2. The mission plan may provide for an observation flight that allows for the observation of any point on the entire territory of the observed Party, including areas designated by the observed Party as hazardous airspace in the source specified in Annex I. The flight path of an observation aircraft shall not be closer than, but shall be allowed up to, ten kilometres from the border with an adjacent State that is not a State Party.

3. The mission plan may provide that the Open Skies airfield where the observation flight terminates, as well as the point of exit, may be different from the Open Skies airfield where the observation flight commences or the point of entry. The mission plan shall specify, if applicable, the commencement time of the observation flight, the desired time and place of planned refuelling stops or rest periods, and the time of continuation of the observation flight after a refuelling stop or rest period within the 96-hour period specified in Section I, paragraph 9 of this Article.

4. The mission plan shall include all information necessary to file the flight plan and shall provide that:

(A) the observation flight does not exceed the relevant maximum flight distance as set forth in Annex A, Section I;

(B) the route and profile of the observation flight satisfies observation flight safety conditions in conformity with ICAO standards and recommended practices, taking into account existing differences in national flight rules, without prejudice to the provisions of paragraph 2 of this Section;

(C) the mission plan takes into account information on hazardous airspace, as provided in accordance with Annex I;

(D) the height above ground level of the observation aircraft does not permit the observing Party to exceed the limitation on ground resolution for each sensor, as set forth in Article IV, paragraph 2;

(E) the estimated time of commencement of the observation flight shall be no less than 24 hours after the submission of the mission plan, unless otherwise agreed;

(F) the observation aircraft flies a direct route between the co-ordinates or navigation fixes designated in the mission plan in the declared sequence; and

(G) the flight path does not intersect at the same point more than once, unless otherwise agreed, and the observation aircraft does not circle around a single point, unless otherwise agreed. The provisions of this subparagraph do not apply for the purposes of taking off, flying over calibration targets, or landing by the observation aircraft.

5. In the event that the mission plan filed by the observing Party provides for flights through hazardous airspace, the observed Party shall:

(A) specify the hazard to the observation aircraft;

(B) facilitate the conduct of the observation flight by co-ordination or suppression of the activity specified pursuant to subparagraph (A) of this paragraph; or

(C) propose an alternative flight altitude, route, or time.

6. No later than four hours after submission of the mission plan, the observed Party shall accept the mission plan or propose changes to it in accordance with Article VIII, Section I, paragraph 4 and paragraph 5 of this Section. Such changes shall not preclude observation of any point on the entire territory of the observed Party, including areas designated by the observed Party as hazardous airspace in the source specified in Annex I to this Treaty. Upon agreement, the mission plan shall be signed by the observing and observed Parties. In the event that the Parties do not reach agreement on the mission plan within eight hours of the original mission plan, the observing Party shall have the right to decline to conduct the observation flight in accordance with the provisions of Article VIII of this Treaty.

7. If the planned route of the observation flight approaches the border of other States Parties or other States, the observed Party may notify that State or those States of the estimated route, date and time of the observation flight.

8. On the basis of the agreed mission plan the State Party providing the observation aircraft shall, in co-ordination with the other State Party, file the flight plan immediately, which shall have the content specified in Annex 2 to the Convention on International Civil Aviation and shall be in the format specified by ICAO Document No. 4444-RAC/501/12, 'Rules of the Air and Air

Traffic Services', as revised or amended.

SECTION III. SPECIAL PROVISIONS

1. In the event that observation aircraft is provided by the observing Party, the observed Party shall have the right to have on board the observation aircraft two flight monitors and one interpreter, in addition to one flight monitor for each sensor control station on board the observation aircraft, unless otherwise agreed. Flight monitors and interpreters shall have the rights and obligations specified in Annex G to this Treaty.

2. Notwithstanding paragraph 1 of this Section, in the event that an observing Party uses an observation aircraft which has a maximum take-off gross weight of no more than 35,000 kilograms for an observation flight distance of no more than 1,500 kilometres as notified in accordance with Section I, paragraph 5, subparagraph (G) of this Article, it shall be obliged to accept only two flight monitors and one interpreter on board the observation aircraft, unless otherwise agreed.

3. In the event that the observation aircraft is provided by the observed Party, the observed Party shall permit the personnel of the observing Party to travel to the point of entry of the observing Party in the most expeditious manner. The personnel of the observing Party may elect to travel to the point of entry using ground, sea, or air transportation, including transportation by an aircraft owned by any State Party. Procedures regarding such travel are set forth in Annex E to this Treaty.

4. In the event that the observation aircraft is provided by the observed Party, the observing Party shall have the right to have on board the observation aircraft two flight representatives and one interpreter, in addition to one flight representative for each control station on the aircraft, unless otherwise agreed. Flight representatives and interpreters shall have the rights and obligations set forth in Annex G to this Treaty.

5. In the event that the observing State Party provides an observation aircraft designated by a State Party other than the observing or observed Party, the observing Party shall have the right to have on board the observation aircraft two representatives and one interpreter, in addition to one representative for each sensor control station on the aircraft, unless otherwise agreed. In this case, the provisions on flight monitors set forth in paragraph 1 of this Section shall also apply. Representatives and interpreters shall have the rights and obligations set forth in Annex G to this Treaty.

**Article VII. Transit Flights**

1. Transit flights conducted by an observing Party to and from the territory of an observed Party for the purposes of this Treaty shall originate on the territory of the observing Party or of another State Party.

2. Each State Party shall accept transit flights. Such transit flights shall be conducted along internationally recognized Air Traffic Services routes, unless otherwise agreed by the States Parties involved, and in accordance with the instructions of the national air traffic control authorities of each State Party whose airspace is transited. The observing Party shall notify each State Party whose airspace is to be transited at the same time that it notifies the observed Party in accordance with Article VI.

3. The operation of sensors on an observation aircraft during transit flights is prohibited. In the event that, during the transit flight, the observation aircraft lands on the territory of a State Party, that State Party shall, upon landing and prior to departure, inspect the covers of sensor apertures or other devices that inhibit the operation of sensors to confirm that they are in their proper position.

**Article VIII. Prohibitions, Deviations from Flight Plans and Emergency Situations**

SECTION I. PROHIBITION OF OBSERVATION FLIGHTS AND CHANGES TO MISSION PLANS

1. The observed Party shall have the right to prohibit an observation flight that is not in compliance with the provisions of this Treaty.

2. The observed Party shall have the right to prohibit an observation flight prior to its commencement in the event that the observing Party fails to arrive at the point of entry within 24 hours after the estimated time of arrival specified in the notification provided in accordance with Article VI, Section I, paragraph 5, unless otherwise agreed between the States Parties involved.

3. In the event that an observed State Party prohibits an observation flight pursuant to this Article or Annex F, it shall immediately state the facts for the prohibition in the mission plan. Within seven days the observed Party shall provide to all States Parties,

through diplomatic channels, a written explanation for this prohibition in the mission report provided pursuant to Article VI, Section 1, paragraph 21. An observation flight that has been prohibited shall not be counted against the quota of either State Party.

4. The observed Party shall have the right to propose changes to the mission plan as a result of any of the following circumstances:

(A) the weather conditions affect flight safety;

(B) the status of the Open Skies airfield to be used, alternate airfields, or refuelling airfields prevents their use; or

(C) the mission plan is inconsistent with Article VI, Section II, paragraphs 2 and 4.

5. In the event that the observing Party disagrees with the proposed changes to the mission plan, it shall have the right to submit alternatives to the proposed changes. In the event that agreement on a mission plan is not reached within eight hours of the submission of the original mission plan, and if the observing Party considers the changes to the mission plan to be prejudicial to its rights under this Treaty with respect to the conduct of the observation flight, the observing Party shall have the right to decline to conduct the observation flight, which shall not be recorded against the quota of either State Party.

6. In the event that an observing Party declines to conduct an observation flight pursuant to this Article or Annex F, it shall immediately provide an explanation of its decision in the mission plan prior to the departure of the observing Party. Within seven days after departure of the observing Party, the observing Party shall provide to all other States Parties, through diplomatic channels, a written explanation for this decision in the mission report provided pursuant to Article VI, Section I, paragraph 21.

SECTION II. DEVIATIONS FROM THE FLIGHT PLAN

1. Deviations from the flight plan shall be permitted during the observation flight if necessitated by:

(A) weather conditions affecting flight safety;

(B) technical difficulties relating to the observation aircraft;

(C) a medical emergency of any person on board; or

(D) air traffic control instructions related to circumstances brought about by *force majeure.*

2. In addition, if weather conditions prevent effective use of optical sensors and infra-red line-scanning devices, deviations shall be permitted, provided that:

(A) flight safety requirements are met;

(B) in cases where national rules so require, permission is granted by air traffic control authorities; and

(C) the performance of the sensors does not exceed the capabilities specified in Article IV, paragraph 2, unless otherwise agreed.

3. The observed Party shall have the right to prohibit the use of a particular sensor during a deviation that brings the observation aircraft below the minimum height above ground level for operating that particular sensor, in accordance with the limitation on ground resolution specified in Article IV, paragraph 2. In the event that a deviation requires the observation aircraft to alter its flight path by more than 50 kilometres from the flight path specified in the flight plan, the observed Party shall have the right to prohibit the use of all the sensors installed on the observation aircraft beyond that 50-kilometre limit.

4. The observing Party shall have the right to curtail an observation flight during its execution in the event of sensor malfunction. The pilot-in-command shall have the right to curtail an observation flight in the event of technical difficulties affecting the safety of the observation aircraft.

5. In the event that a deviation from the flight plan permitted by paragraph 1 of this Section results in curtailment of the observation flight, or a curtailment occurs in accordance with paragraph 4 of this Section, an observation flight shall be counted against the quotas of both States Parties, unless the curtailment is due to:

(A) sensor malfunction on an observation aircraft provided by the observed Party;

(B) technical difficulties relating to the observation aircraft provided by the observed Party;

(C) a medical emergency of a member of the flight crew of the observed Party or of flight monitors; or

(D) air traffic control instructions related to circumstances brought about by *force majeure.*

In such cases the observing Party shall have the right to decide whether to count it against the quotas of both States Parties.

6. The data collected by the sensors shall be retained by the observing Party only if the

observation flight is counted against the quotas of both States Parties.

7. In the event that a deviation is made from the flight plan, the pilot-in-command shall take action in accordance with the published national flight regulations of the observed Party. Once the factors leading to the deviation have ceased to exist, the observation aircraft may, with the permission of the air traffic control authorities, continue the observation flight in accordance with the flight plan. The additional flight distance of the observation aircraft due to the deviation shall not count against the maximum flight distance.

8. Personnel of both States Parties on board the observation aircraft shall be immediately informed of all deviations from the flight plan.

9. Additional expenses resulting from provisions of this Article shall be in accordance with Annex L, Section I, paragraph 9 to this Treaty.

SECTION III. EMERGENCY SITUATIONS

1. In the event that an emergency situation arises, the pilot-in-command shall be guided by 'Procedures for Air Navigation Services—Rules of the Air and Air Traffic Services', ICAO Document No. 4444-RAC/501/12, as revised or amended, the national flight regulations of the observed Party, and the flight operation manual of the observation aircraft.

2. Each observation aircraft declaring an emergency shall be accorded the full range of distress and navigational facilities of the observed Party in order to ensure the most expeditious recovery of the aircraft to the nearest suitable airfield.

3. In the event of an aviation accident involving the observation aircraft on the territory of the observed Party, search and rescue operations shall be conducted by the observed Party in accordance with its own regulations and procedures for such operations.

4. Investigation of an aviation accident or incident involving an observation aircraft shall be conducted by the observed Party, with the participation of the observing party, in accordance with the ICAO recommendations set forth in Annex 13 to the Convention on International Civil Aviation ('Investigation of Aviation Accidents') as revised or amended and in accordance with the national regulations of the observed Party.

5. In the event that the observation aircraft is not registered with the observed Party, at the conclusion of the investigation all wreckage and debris of the observation aircraft and sensors, if found and recovered, shall be returned to the observing Party or to the Party to which the aircraft belongs, if so requested.

**Article IX. Sensor Output from Observation Flights**

SECTION I. GENERAL PROVISIONS

1. For the purposes of recording data collected by observation flights, the following recording media shall be used:

(A) in the case of optical panoramic and framing cameras, black and white photographic film;

(B) in the case of video cameras, magnetic tape;

(C) in the case of infra-red line-scanning devices, black and white photographic film or magnetic tape; and

(D) in the case of sideways-looking synthetic aperture radar, magnetic tape.

The agreed format in which such data is to be recorded on other recording media shall be decided within the Open Skies Consultative Commission during the period of provisional application of this Treaty.

2. Data collected by sensors during observation flights shall remain on board the observation aircraft until completion of the observation flight. The transmission of data collected by sensors from the observation aircraft during the observation flight is prohibited.

3. Each roll of photographic film and cassette or reel of magnetic tape used to collect data by a sensor during an observation flight shall be placed in a container and sealed in the presence of the States Parties as soon as is practicable after it has been removed from the sensor.

4. Data collected by sensors during observation flights shall be made available to States Parties in accordance with the provisions of this Article and shall be used exclusively for the attainment of the purposes of this Treaty.

5. In the event that, on the basis of data provided pursuant to Annex B, Section I to this Treaty, a data recording medium to be used by a State Party during an observation flight is incompatible with the equipment of another State Party for handling that data recording medium, the States Parties

involved shall establish procedures to ensure that all data collected during observation flights can be handled, in terms of processing, duplication and storage, by them.

## SECTION II. OUTPUT FROM SENSORS THAT USE PHOTOGRAPHIC FILM

1. In the event that output from duplicate optical cameras is to be exchanged, the film and film processing shall be of an identical type.

2. Provided that the data collected by a single optical camera is subject to exchange, the States Parties shall consider, within the Open Skies Consultative Commission during the period of provisional application of this Treaty, the issue of whether the responsibility for the development of the original film negative shall be borne by the observing Party or by the State Party providing the observation aircraft. The State Party developing the original film negative shall be responsible for the quality of processing the original negative film and producing the duplicate positive or negative. In the event that States Parties agree that the film used during the observation flight conducted on an observation aircraft provided by the observed Party shall be processed by the observing Party, the observed Party shall bear no responsibility for the quality of the processing of the original negative film.

3. All the film used during the observation flight shall be developed:

(A) in the event that the original film negative is developed at a film processing facility arranged for by the observed Party, no later than three days, unless otherwise agreed, after the arrival of the observation aircraft at the point of exit; or

(B) in the event that the original film negative is developed at a film processing facility arranged for by the observing Party, no later than ten days after the departure of the observation aircraft from the territory of the observed Party.

4. The State Party that is developing the original film negative shall be obliged to accept at the film processing facility up to two officials from the other State Party to monitor the unsealing of the film cassette or container and each step in the storage, processing, duplication and handling of the original film negative, in accordance with the provisions of Annex K, Section II to this Treaty. The State Party monitoring the film processing and duplication shall have the right to designate such officials from among its nationals present on the territory on which the film processing facility arranged for by the other State Party is located, provided that such individuals are on the list of designated personnel in accordance with Article XIII, Section I of this Treaty. The State Party developing the film shall assist the officials of the other State Party in their function provided for in this paragraph to the maximum extent possible.

5. Upon completion of an observation flight, the State Party that is to develop the original film negative shall attach a 21-step sensitometric test strip of the same film type used during the observation flight or shall expose a 21-step optical wedge onto the leader or trailer of each roll of original film negative used during the observation flight. After the original film negative has been processed and duplicate film negative or positive has been produced, the States Parties shall assess the image quality of the 21-step sensitometric test strips or images of the 21-step optical wedge against the characteristics provided for that type of original film negative or duplicate film negative or positive in accordance with the provisions of Annex K, Section I to this Treaty.

6. In the event that only one original film negative is developed:

(A) the observing Party shall have the right to retain or receive the original film negative; and

(B) the observed Party shall have the right to select and receive a complete first generation duplicate or part thereof, either positive or negative, of the original film negative. Unless otherwise agreed, such duplicate shall be:

(1) of the same format and film size as the original film negative;

(2) produced immediately after development of the original film negative; and

(3) provided to the officials of the observed Party immediately after the duplicate has been produced.

7. In the event that two original film negatives are developed:

(A) if the observation aircraft is provided by the observing Party, the observed Party shall have the right, at the completion of the observation flight, to select either of the two original film negatives, and the original film negative not selected shall be retained by the observing Party; or

(B) if the observation aircraft is provided by the observed Party, the observing Party shall have the right to select either of the original film negatives, and the original film negative not selected shall be retained by the observed Party.

SECTION III. OUTPUT FROM SENSORS THAT USE OTHER RECORDING MEDIA

3. The State Party that provides the observation aircraft shall record at least one original set of data collected by sensors using other media.

2. In the event that only one original set is made:

(A) if the observation aircraft is provided by the observing Party, the observing Party shall have the right to retain the original set and the observed Party shall have the right to receive a first generation duplicate copy; or

(B) if the observation aircraft is provided by the observed Party, the observing Party shall have the right to receive the original set and the observed Party shall have the right to receive a first generation duplicate copy.

3. In the event that two original sets are made:

(A) if the observation aircraft is provided by the observing Party, the observed Party shall have the right, at the completion of the observation flight, to select either of the two sets of recording media, and the set not selected shall be retained by the observing Party; or

(B) if the observation aircraft is provided by the observed Party, the observing Party shall have the right to select either of the two sets of recording media, and the set not selected shall be retained by the observed Party.

4. In the event that the observation aircraft is provided by the observing Party, the observed Party shall have the right to receive the data collected by a sideways-looking synthetic aperture radar in the form of either initial phase information or a radar image, at its choice.

5. In the event that the observation aircraft is provided by the observed Party, the observing Party shall have the right to receive the data collected by a sideways-looking synthetic aperture radar in the form of either initial phase information or a radar image, at its choice.

SECTION IV. ACCESS TO SENSOR OUTPUT

Each State Party shall have the right to request and receive from the observing Party copies of data collected by sensors during an observation flight. Such copies shall be in the form of first generation duplicates produced from the original data collected by sensors during an observation flight. The State Party requesting copies shall also notify the observed Party. A request for duplicates of data shall include the following information:

(A) the observing Party ;

(B) the observed Party ;

(C) the date of the observation flight;

(D) the sensor by which the data was collected;

(E) the portion or portions of the observation period during which the data was collected; and

(F) the type and format of duplicate recording medium, either negative or positive film, or magnetic tape.

**Article X. Open Skies Consultative Commission**

1. In order to promote the objectives and facilitate the implementation of the provisions of this Treaty, the States Parties hereby establish an Open Skies Consultative Commission.

2. The Open Skies Consultative Commission shall take decisions or make recommendations by consensus. Consensus shall be understood to mean the absence of any objection by any State Party to the taking of a decision or the making of a recommendation.

3. Each State Party shall have the right to raise before the Open Skies Consultative Commission, and have placed on its agenda, any issue relating to this Treaty, including any issue related to the case when the observed Party provides an observation aircraft.

4. Within the framework of the Open Skies Consultative Commission the States Parties to this Treaty shall:

(A) consider questions relating to compliance with the provisions of this Treaty;

(B) seek to resolve ambiguities and differences of interpretation that may become apparent in the way this Treaty is implemented;

(C) consider and take decisions on applications for accession to this Treaty; and

(D) agree as to those technical and administrative measures, pursuant to the provisions of this Treaty, deemed necessary following the accession to this Treaty by other States.

5. The Open Skies Consultative Commission may propose amendments to this Treaty

for consideration and approval in accordance with Article XVI. The Open Skies Consultative Commission may also agree on improvements to the viability and effectiveness of this Treaty, consistent with its provisions. Improvements relating only to modification of the annual distribrution of active quotas pursuant to Article III and Annex A, to updates and additions to the categories or capabilities of sensors pursuant to Article IV, to revision of the share of costs pursuant to Annex L, Section I, paragraph 9, to arrangements for the sharing and availability of data pursuant to Article IX, Sections III and IV and to the handling of mission reports pursuant to Article VI, Section I, paragraph 21, as well as to minor matters of an administrative or technical nature, shall be agreed upon within the Open Skies Consultative Commission and shall not be deemed to be amendments to this Treaty.

6. The Open Skies Consultative Commission shall request the use of the facilities and administrative support of the Conflict Prevention Centre of the Conference on Security and Co-operation in Europe, or other existing facilities in Vienna, unless it decides otherwise.

7. Provisions for the operation of the Open Skies Consultative Commission are set forth in Annex L to this Treaty.

### Article XI. Notifications and Reports

The States Parties shall transmit notifications and reports required by this Treaty in written form. The States Parties shall transmit such notifications and reports through diplomatic channels or, at their choice, through other official channels, such as the communications network of the Conference on Security and Co-operation in Europe.

### Article XII. Liability

A State Party shall, in accordance with international law and practice, be liable to pay compensation for damage to other States Parties, or to their natural or juridical persons or their property, caused by it in the course of the implementation of this Treaty.

### Article XIII. Designation of Personnel and Privileges and Immunities

#### SECTION I. DESIGNATION OF PERSONNEL

1. Each State Party shall, at the same time that it deposits its instrument of ratification to either of the Depositaries, provide to all other States Parties, for their review, a list of designated personnel who will carry out all duties relating to the conduct of observation flights for that State Party, including monitoring the processing of the sensor output. No such list of designated personnel shall include more than 400 individuals at any time. It shall contain the name, gender, date of birth, place of birth, passport number, and function for each individual included. Each State Party shall have the right to amend its list of designated personnel until 30 days after entry into force of this Treaty and once every six months thereafter.

2. In the event that any individual included on the original or any amended list is unacceptable to a State Party reviewing the list, that State Party shall, no later than 30 days after receipt of each list, notify the State Party providing that list that such individual shall not be accepted with respect to the objecting State Party. Individuals not declared unacceptable within that 30-day period shall be deemed accepted. In the event that a State Party subsequently determines that an individual is unacceptable, that State Party shall so notify the State Party that designated such individual. Individuals who are declared unacceptable shall be removed from the list previously submitted to the objecting State Party.

3. The observed Party shall provide visas and any other documents to ensure that each accepted individual may enter and remain on the territory of that State Party for the purpose of carrying out duties relating to the conduct of observation flights, including monitoring the processing of the sensor output. Such visas and any other necessary documents shall be provided either:

(A) no later than 30 days after the individual is deemed to be accepted, in which case the visa shall be valid for a period of no less than 24 months; or

(B) no later than one hour after the arrival of the individual at the point of entry, in which case the visa shall be valid for the duration of that individual's duties; or

(C) at any other time, by mutual agreement of the States Parties involved.

#### SECTION II. PRIVILEGES AND IMMUNITIES

1. In order to exercise their functions effectively, for the purpose of implementing this Treaty and not for their personal benefit, personnel designated in accordance with the provisions of Section I, paragraph 1 of this

Article shall be accorded the privileges and immunities enjoyed by diplomatic agents pursuant to Article 29; Article 30, paragraph 2; Article 31, paragraphs 1, 2 and 3; and Articles 34 and 35 of the Vienna Convention on Diplomatic Relations of 18 April 1961, hereinafter referred to as the Vienna Convention. In addition, designated personnel shall be accorded the privileges enjoyed by diplomatic agents pursuant to Article 36, paragraph 1, subparagraph (b) of the Vienna Convention, except in relation to articles, the import or export of which is prohibited by law or controlled by quarantine regulations.

2. Such privileges and immunities shall be accorded to designated personnel for the entire period between arrival on and departure from the territory of the observed Party, and thereafter with respect to acts previously performed in the exercise of their official functions. Such personnel shall also, when transiting the territory of other States Parties, be accorded the privileges and immunities enjoyed by diplomatic agents pursuant to Article 40, paragraph 1 of the Vienna Convention.

3. The immunity from jurisdiction may be waived by the observing Party in those cases when it would impede the course of justice and can be waived without prejudice to this Treaty. The immunity of personnel who are not nationals of the observing Party may be waived only by the States Parties of which such personnel are nationals. Waiver must always be express.

4. Without prejudice to their privileges and immunities or the rights of the observing Party set forth in this Treaty, it is the duty of designated personnel to respect the laws and regulations of the observed Party.

5. The transportation means of the personnel shall be accorded the same immunities from search, requisition, attachment or execution as those of a diplomatic mission pursuant to Article 22, paragraph 3 of the Vienna Convention, except as otherwise provided for in this Treaty.

### Article XIV. Benelux

1. Solely for the purposes of Articles II to IX and Article XI, and of Annexes A to I and Annex K to this Treaty, the Kingdom of Belgium, the Grand Duchy of Luxembourg, and the Kingdom of the Netherlands shall be deemed a single State Party, hereinafter referred to as the Benelux.

2. Without prejudice to the provisions of Article XV, the above-mentioned States Parties may terminate this arrangement by notifying all other States Parties thereof. This arrangement shall be deemed to be terminated on the next 31 December following the 60-day period after such notification.

### Article XV. Duration and Withdrawal

1. This Treaty shall be of unlimited duration.

2. A State Party shall have the right to withdraw from this Treaty. A State Party intending to withdraw shall provide notice of its decision to withdraw to either Depositary at least six months in advance of the date of its intended withdrawal and to all other States Parties. The Depositaries shall promptly inform all other States Parties of such notice.

3. In the event that a State Party provides notice of its decision to withdraw from this Treaty in accordance with paragraph 2 of this Article, the Depositaries shall convene a conference of the States Parties no less than 30 days and no more than 60 days after they have received such notice, in order to consider the effect of the withdrawal on this Treaty.

### Article XVI. Amendments and Periodic Review

1. Each State Party shall have the right to propose amendments to this Treaty. The text of each proposed amendment shall be submitted to either Depositary, which shall circulate it to all States Parties for consideration. If so requested by no less than three States Parties within a period of 90 days after circulation of the proposed amendment, the Depositaries shall convene a conference of the States Parties to consider the proposed amendment. Such a conference shall open no earlier than 30 days and no later than 60 days after receipt of the third of such requests.

2. An amendment to this Treaty shall be subject to the approval of all States Parties, either by providing notification, in writing, of their approval to a Depositary within a period of 90 days after circulation of the proposed amendment, or by expressing their approval at a conference convened in accordance with paragraph 1 of this Article. An amendment so approved shall be subject to ratification in accordance with the provisions of Article XVII, paragraph 1, and shall enter into force 60 days after the deposit of

instruments of ratification by the States Parties.

3. Unless requested to do so earlier by no less than three States Parties, the Depositaries shall convene a conference of the States Parties to review the implementation of this Treaty three years after entry into force of this Treaty and at five-year intervals thereafter.

**Article XVII. Depositaries, Entry into Force and Accession**

1. This Treaty shall be subject to ratification by each State Party in accordance with its constitutional procedures. Instruments of ratification and instruments of accession shall be deposited with the Government of Canada or the Government of the Republic of Hungary or both, hereby designated the Depositaries. This Treaty shall be registered by the Depositaries pursuant to Article 102 of the Charter of the United Nations.

2. This Treaty shall enter into force 60 days after the deposit of 20 instruments of ratification, including those of the Depositaries, and of States Parties whose individual allocation of passive quotas as set forth in Annex A is eight or more.

3. This Treaty shall be open for signature by Armenia, Azerbaijan, Georgia, Kazakhstan, Kirgistan, Moldova, Tajikistan, Turkmenistan and Uzbekistan and shall be subject to ratification by them. Any of these States which do not sign this Treaty before it enters into force in accordance with the provisions of paragraph 2 of this Article may accede to it at any time by depositing an instrument of accession with one of the Depositaries.

4. For six months after entry into force of this Treaty, any other State participating in the Conference on Security and Co-operation in Europe may apply for accession by submitting a written request to one of the Depositaries. The Depositary receiving such a request shall circulate it promptly to all States Parties. The States applying for accession to this Treaty may also, if they so wish, request an allocation of a passive quota and the level of this quota.

The matter shall be considered at the next regular meeting of the Open Skies Consultative Commission and decided in due course.

5. Following six months after entry into force of this Treaty, the Open Skies Consultative Commission may consider the accession to this Treaty of any State which, in the judgement of the Commission, is able and willing to contribute to the objectives of this Treaty.

6. For any State which has not deposited an instrument of ratification by the time of entry into force, but which subsequently ratifies or accedes to this Treaty, this Treaty shall enter into force 60 days after the date of deposit of its instrument of ratification or accession.

7. The Depositaries shall promptly inform all States Parties of:

(A) the date of deposit of each instrument of ratification and the date of entry into force of this Treaty;

(B) the date of an application for accession, the name of the requesting State and the result of the procedure;

(C) the date of deposit of each instrument of accession and the date of entry into force of this Treaty for each State that subsequently accedes to it;

(D) the convening of a conference pursuant to Articles XV and XVI;

(E) any withdrawal in accordance with Article XV and its effective date;

(F) the date of entry into force of any amendments to this Treaty; and

(G) any other matters of which the Depositaries are required by this Treaty to inform the States Parties.

**Article XVIII. Provisional Application and Phasing of Implementation of the Treaty**

In order to facilitate the implementation of this Treaty, certain of its provisions shall be provisionally applied and others shall be implemented in phases.

SECTION I. PROVISIONAL APPLICATION

1. Without detriment to Article XVII, the signatory States shall provisionally apply the following provisions of this Treaty:

(A) Article VI, Section I, paragraph 4;

(B) Article X, paragraphs 1, 2, 3, 6 and 7;

(C) Article XI;

(D) Article XIII, Section I, paragraphs 1 and 2;

(E) Article XIV; and

(F) Annex L, Section I.

2. This provisional application shall be effective for a period of 12 months from the date when this Treaty is opened for signature. In the event that this Treaty does not enter into force before the period of provisional application expires, that period may be extended if all the signatory States so decide. The period of provisional application shall in any event terminate when this Treaty enters

into force. However, the States Parties may then decide to extend the period of provisional application in respect of signatory States that have not ratified this Treaty.

SECTION II. PHASING OF IMPLEMENTATION

1. After entry into force, this Treaty shall be implemented in phases in accordance with the provisions set forth in this Section. The provisions of paragraphs 2 to 6 of this Section shall apply during the period from entry into force of this Treaty until 31 December of the third year following the year during which entry into force takes place.

2. Notwithstanding the provisions of Article IV, paragraph 1, no State Party shall during the period specified in paragraph 1 above use an infra-red line-scanning device if one is installed on an observation aircraft, uness otherwise agreed between the observing and observed Parties. Such sensors shall not be subject to certification in accordance with Annex D. If it is difficult to remove such sensor from the observation aircraft, then it shall have covers or other devices that inhibit its operation in accordance with the provisions of Article IV, paragraph 4 during the conduct of observation flights.

3. Notwithstanding the provisions of Article IV, paragraph 9, no State Party shall, during the period specified in paragraph 1 of this Section, be obliged to provide an observation aircraft equipped with sensors from each sensor category, at the maximum capability and in the numbers specified in Article IV, paragraph 2, provided that the observation aircraft is equipped with:

(A) a single optical panoramic camera; or

(B) not less than a pair of optical framing cameras.

4. Notwithstanding the provisions of Annex B, Section II, paragraph 2, subparagraph (A) to this Treaty, data recording media shall be annotated with data in accordance with existing practice of States Parties during the period specified in paragraph 1 of this Section.

5. Notwithstanding the provisions of Article VI, Section I, paragraph 1, no State Party during the period specified in paragraph 1 of this Section shall have the right to be provided with an aircraft capable of achieving any specified unrefuelled range.

6. During the period specified in paragraph 1 of this Section, the distribution of active quotas shall be established in accordance with the provisions of Annex A, Section II, paragraph 2 to this Treaty.

7. Further phasing in respect of the introduction of additional categories of sensors or improvements to the capabilities of existing categories of sensors shall be addressed by the Open Skies Consultative Commission in accordance with the provisions of Article IV, paragraph 3 concerning such introduction or improvement.

**Article XIX. Authentic Texts**

The originals of this Treaty, of which the English, French, German, Italian, Russian and Spanish texts are equally authentic, shall be deposited in the archives of the Depositaries. Duly certified copies of this Treaty shall be transmitted by the Depositaries to all the States Parties.

---

*Source:* Helsinki, 24 Mar. 1992.

---

**TASHKENT DOCUMENT**
(unofficial translation)

*Signed 15 May 1992, Tashkent, Uzbekistan*

**JOINT DECLARATION**

of the Azerbaijan Republic, the Republics of Armenia, Belarus, Kazakhstan and Moldova, the Russian Federation, Ukraine and the Georgian Republic, in connection with the Treaty on Conventional Armed Forces in Europe.

With the aim of furthering the implementation of the CFE Treaty and documents related to it, the Azerbaijan Republic, the Republics of Armenia, Belarus, Kazakhstan and Moldova, the Russian Federation, Ukraine and the Georgian Republic declare the following:

1. The Azerbaijan Republic, the Republics of Armenia, Belarus, Kazakhstan and Moldova, the Russian Federation, Ukraine and the Georgian Republic confirm their adherence to the provisions of the Declaration of the States Parties to the CFE Treaty with regard to personnel strength.

2. The Azerbaijan Republic, the Republics of Armenia, Belarus, Kazakhstan and Moldova, the Russian Federation, Ukraine and the Georgian Republic adhere to the pro-

visions of the Declaration of the States Parties to the CFE Treaty with regard to land-based naval aircraft and recognize that the number of permanently land-based naval combat aircraft, specified in paragraph 1 of the Declaration, refers to the Russian Federation and Ukraine, with the Russian Federation having not more than 300 combat aircraft in the area of application of the Treaty, and Ukraine not more than 100 combat aircraft.

3. The Russian Federation shall fulfill the provisions of the Statement of the Representative of the USSR in the Joint Consultative Group of the 14 June 1991 with regard to armaments and equipment withdrawn beyond the area of application in the period before the signing of the Treaty.

4. Taking account of the politically-binding nature of the documents referred to in paragraphs 1, 2, and 3 of this Statement, the obligations that arise from these paragraphs have a politically-binding nature.

5. The Azerbaijan Republic, the Republics of Armenia, Belarus, Kazakhstan and Moldova, the Russian Federation, Ukraine and the Georgian Republic confirm that all decisions taken previously within the framework of the Joint Consultative Group are binding on all these states.

Done in Tashkent this 15th day of May 1992 in one original copy in the Russian language. The original copy shall be held in the archives of the Government of the Republic of Belarus, which will send a certified copy to all States that have signed this Statement, and to the Depositary and the States Parties of the Treaty on Conventional Armed Forces in Europe.

## AGREEMENT ON THE PRINCIPLES AND PROCEDURES FOR IMPLEMENTING THE TREATY ON CONVENTIONAL ARMED FORCES IN EUROPE

The Azerbaijan Republic, the Republic of Armenia, the Republic of Belarus, the Republic of Kazakhstan, the Republic of Moldova, the Russian Federation, Ukraine and the Republic of Georgia as the successors of the USSR with regard to the Treaty on Conventional Armed Forces in Europe and its associated documents, henceforth called the Contracting Parties,

confirming their adherence to the aims and tasks of the Conference on Security and Co-operation in Europe,

regarding the Treaty on Conventional Armed Forces in Europe of 19 November 1990, henceforth called the Treaty, as one of the fundamental elements of the new security system in Europe,

striving to achieve the consistent fulfillment of all obligations arising from the Treaty and its associated documents,

taking account of the security interests of all Contracting Parties,

have agreed the following:

### Article 1

1. Each Contracting Party fully exercises the rights and fulfills the obligations provided for in the Treaty and its associated documents unless otherwise provided for in paragraph 2 of this Article.

2. The Russian Federation exercises the rights and fulfills the obligations of the Treaty and its associated documents with regard to forces, and also conventional armaments and equipment, stationed on the territories of the Republic of Latvia and the Republic of Lithuania, the Republic of Poland, the Federal Republic of Germany and the Republic of Estonia, and subject to withdrawal to the territory of the Russian Federation.

In the case of their withdrawal to the territory of another Contracting Party, the exercising of the rights and the fulfilling of the Treaty commitments is placed on this Contracting Party.

3. The Contracting Parties cooperate in exercising the rights and fulfilling the obligations arising from the Treaty and its associated documents.

### Article 2

1. Within the obligations arising from the provisions of the Treaty, the following are laid down in the corresponding Protocols for each of the Contracting Parties:

(a) maximum ceilings for holdings of conventional armaments and equipment;

(b) the number of armoured vehicle launched bridges in active units;

(c) the number of Mi-24R and Mi-24K helicopters, equipped for conducting reconnaissance, spotting or chemical/biological/radiological sampling, which are not subject to the limits on attack helicopters.

2. The maximum ceilings for holdings of conventional armaments and equipment, defined for each Contracting Party, does not exceed in total the maximum ceilings laid down for the Union of Soviet Socialist

Republics in the Treaty and the Agreement on maximum ceilings for holdings of conventional armaments and equipment of the Peoples' Republic of Bulgaria, the Hungarian Republic, the Republic of Poland, Romania, the Union of Soviet Socialist Republics and the Czech and Slovak Federal Republic in connection with the Treaty on Conventional Armed Forces in Europe of 3 November 1990.

3. The numbers of armaments and equipment, enumerated in subparagraphs (b) and (c) of paragraph 1 of this Article, do not exceed in total the ceilings and numbers, laid down in the Treaty and its associated documents, for the USSR.

4. After the Treaty takes effect the Contracting Parties will coordinate their efforts in matters of implementation of the Treaty and its associated documents, including questions of distributing maximum ceilings for holdings of armaments and equipment of each Party by the adherence to the provisions of paragraphs 2 and 3 of this Article, with the aim of ensuring regional, national and collective security in Europe.

**Article 3**

1. The Contracting Parties adhere to the provisions of the Declaration of the Government of the USSR of 14 June 1991, in the case of conventional armaments and equipment, relating to categories limited by the Treaty and located in the Coastal Defence, Naval Infantry and Strategic Rocket forces.

2. Within the obligations arising out of the above-mentioned Declaration, the corresponding Protocol lays down for the Contracting Parties maximum ceilings for conventional armaments and equipment, relating to categories limited by the Treaty and located in the Coastal Defence, Naval Infantry and Strategic Rocket forces.

**Article 4**

1. The Contracting Parties shall transmit to one another, by mutual agreement and in compliance with the reduction liability and other demands of the Treaty and its associated documents, lists of conventional armaments and equipment subject to reduction.

2. In order to achieve the optimum organisation of the process of reducing conventional armaments and equipment limited by the Treaty and a decrease in reduction costs, the Contracting Parties shall jointly utilise reduction sites. The sequence of utilisation of the reduction sites and reduced armaments and equipment shall be determined by the Contracting Parties on the basis of appropriate agreements.

**Article 5**

1. The Contracting Parties shall cooperate as necessary on matters of the preparation and transmission of information and notification, stipulated in the Treaty and its associated documents, during the period of its provisional application and after its entry into force.

2. The Contracting Parties affirm that the information on conventional armed forces declared upon the signing of the Treaty by the Union of Soviet Socialist Republics, including technical information and photographs of conventional armaments and equipment, shall remain in force.

3. The Contracting Parties, simultaneously with the transfer to the Depositary of the instruments of ratification of the Treaty, shall transmit to all other States Parties notifications, the provision of which is stipulated before the Treaty comes into force, and acknowledge the information provided earlier required for the implementation of the Treaty.

**Article 6**

1. The Contracting Parties acknowledge that during the first 120 days after entry into force of the Treaty their total passive quota of inspections shall not be less than the passive quota of the former Union of Soviet Socialist Republics, determined according to the data on the numbers of objects of verification provided at the time of signature of the Treaty in accordance with the Protocol on Notification and Exchange of Information.

2. The Contracting Parties shall cooperate in the implementation of inspection activities. They shall also cooperate in the formation of multinational inspection teams for carrying out inspections on the territories of Participating States other than Contracting Parties.

**Article 7**

1. At the proposal of any Contracting Party the Depositary of this Agreement shall convene consultations of all Contracting Parties for discussions connected with the implementation of this Agreement. Such consultations shall take place no later than 15 days after despatch of their notification to all Contracting Parties.

**Article 8**

1. In the event of a withdrawal of a Contracting Party from the Treaty it shall cease to be a party to this Agreement.

2. Each Contracting Party has the right to withdraw from this Agreement. A Contracting Party intending to withdraw from the Agreement shall give notice of its decision to all other Contracting Parties at least 90 days prior to the intended withdrawal. The Depository of this Agreement shall convene, not later than 21 days after receipt of such a notification, consultations of Contracting Parties for the discussion of matters connected with such a withdrawal.

**Article 9**

Nothing in this Agreement may be interpreted as infringing the sovereign rights of the Contracting Parties, including anything arising from Bills passed by them concerning their governmental sovereignty and independence.

**Article 10**

The following are an integral part of this Agreement: the Protocol on maximum levels for holdings of conventional armaments and equipment limited by the Treaty; the Protocol on armoured vehicle launched bridges in active units; the Protocol on the combat helicopters Mi-24R and Mi-24K which are not subject to the limitations on attack helicopters; the Protocol on conventional armaments and equipment relating to the categories limited by the Treaty which are serving as part of the Coastal Defence forces, Naval Infantry or Strategic Rocket forces.

**Article 11**

Each Contracting Party, as a rightful successor to the Union of Soviet Socialist Republics, regarding the Treaty signed by the USSR on 19 November 1990, undertakes as quickly as possible to ratify the Treaty and deposit the instruments of ratification in the custody of the government of the Netherlands.

**Article 12**

This Agreement is subject to ratification in accordance with the constitutional procedures of each Contracting Party simultaneous with ratification of the Treaty.

Documents on ratification of this Agreement are to be deposited in the custody of the Depositary.

This Agreement shall come into force 10 days after depositing with the Depositary the documents on ratification of this Agreement by all Contracting Parties and shall remain in force while the Treaty is in force.

This Agreement shall be registered in accordance with Article 102 of the United Nations Charter.

Completed in the city of Tashkent on 15 May 1992 in one example each of the Azerbaijani, Armenian, Belarussian, Kazakh, Moldovan, Russian, Ukrainian and Georgian languages, all texts, however, carry equal authority. The original copy of the Agreement shall be held in the archives of the Government of the Republic of Belarus (which is appointed as the Depositary). Three copies of the Agreement shall be sent by the Depositary to Contracting Parties and to other Participating States Parties of the Treaty.

**PROTOCOL**

On conventional armaments and equipment relating to categories limited by the Treaty in service with Coastal Defence forces, Naval Infantry and Strategic Rocket forces.

1. Confirming all the commitments of the USSR as set forth in USSR government's declaration of 14 June 1991, regarding conventional armaments and equipment relating to categories limited by the Treaty in service with the Coastal Defence forces, Naval Infantry and Strategic Rocket forces, the Contracting Parties agree to take on the responsibility for the implementation of the aforementioned declaration:

– Regarding conventional armaments and equipment in the Coastal Defence forces and Naval Infantry—in the Russian Federation and Ukraine;

– Regarding conventional armaments and equipment (armoured personnel carriers) in the Strategic Rocket forces—in Belarus, the Russian Federation and Ukraine.

2. On the territory of the Russian Federation and Ukraine within the area of application of the Treaty, stationed conventional armaments and equipment relating to categories limited by the Treaty shall not exceed the following totals:

(a) Coastal Defence forces: on Russian Federation territory—542 battle tanks, 407 armoured combat vehicles (ACV), 686 artillery pieces; in Ukraine—271 battle tanks, 470 armoured combat vehicles and 160 artillery pieces.

(b) Naval Infantry: on Russian Federation territory—120 tanks, 583 ACVs, 186 artillery pieces; in Ukraine—265 ACVs, 48 artillery pieces.

3. The destruction or conversion of conventional armaments and equipment set out in paragraph 3 of Section III of the Declaration shall be carried out in the following way:

(a) by the Russian Federation—as foreseen in paragraph 3 of Section III of the Declaration;

(b) by Ukraine—in full within the area of application of the Treaty.

4. The Republic of Belarus, the Russian Federation and Ukraine within the boundaries of the area of application of the Treaty shall have conventional armaments and equipment (armoured combat vehicles) in the Strategic Rocket forces in quantities not exceeding:

| | |
|---|---|
| Belarus | 585 ACVs |
| Russian Federation | 700 |
| Ukraine | 416 |

For Belarus and Ukraine the figures set out here shall remain in force until the complete removal of the Strategic Rocket forces from their territory. The quota of permitted ACVs for these States shall be transferred to the Russian Federation in proportion to the removal of Strategic Rocket forces from their territory.

## PROTOCOL

On the maximum levels for holdings of conventional arms and technical equipment of the Azerbaijan Republic and the Republics of Armenia, Belarus, Kazakhstan, Moldova, the Russian Federation, Ukraine and the Republic of Georgia in relation to the Treaty on Conventional Armed Forces in Europe.

The Contracting Parties, as rightful heirs to the Union of Soviet Socialist Republics, in connection with the Treaty on Conventional Armed Forces in Europe, referred to hereafter as the Treaty, and the Agreement about the maximum holding levels of conventional weapons of the People's Republic of Bulgaria, the Hungarian Republic, the Polish Republic, Romania, the Union of Soviet Socialist Republics and the Czech and Slovak Federal Republic, in connection with the Treaty on Conventional Forces in Europe, hereafter referred to as the Budapest Agreement, hereby confirm that their maximum level for holdings of conventional equipments in total will not exceed the maximum level set for the former USSR by the Treaty and the Budapest Agreement.

1. In accordance with the provisions of the Treaty and Article 1 of the Budapest Agreement each of the Contracting Parties has a maximum set level for holdings of conventional armaments limited by the Treaty.

**Azerbaijan Republic**

| | |
|---|---|
| Battle tanks | not more than 220 units |
| in active units | not more than 220 units |
| Armoured Combat Vehicles (ACV) | not more than 220 units |
| in active units | not more than 220 units |
| of which Armoured Infantry Fighting Vehicles (AIFV) and Heavy Armoured Combat Vehicles (HACV) | not more than 135 units |
| of which HACV | not more than 11 units |
| Artillery | not more than 285 units |
| in active units | not more than 285 units |
| Combat aircraft | not more than 100 units |
| Attack helicopters | not more than 50 units |

**Republic of Armenia**

| | |
|---|---|
| Battle tanks | not more than 220 units |
| in active units | not more than 220 units |
| ACVs | not more than 220 units |
| in active units | not more than 220 units |
| of which AIFV and HACV | not more than 135 units |
| of which HACV | not more than 11 units |
| Artillery | not more than 285 units |
| in active units | not more than 285 units |
| Combat aircraft | not more than 100 units |
| Attack helicopters | not more than 50 units |

**Republic of Belarus**

| | |
|---|---|
| Battle tanks | not more than 1800 units |
| in active units | not more than 1525 units |
| ACVs | not more than 2600 units |
| in active units | not more than 2175 units |
| of which AIFV and HACV | not more than 1590 units |
| of which HACV | not more than 130 units |

| | |
|---|---|
| Artillery | not more than 1615 units |
| in active units | not more than 1375 units |
| Combat aircraft | not more than 260 units |
| Attack helicopters | not more than 80 units |

**Republic of Kazakhstan (in the area of application)**

Not more than 0 in all categories

**Republic of Moldova**

| | |
|---|---|
| Battle tanks | not more than 210 units |
| in active units | not more than 210 units |
| ACVs | not more than 210 units |
| in active units | not more than 210 units |
| of which AIFV and HACV | not more than 130 units |
| of which HACV | not more than 10 units |
| Artillery | not more than 250 units |
| in active units | not more than 250 units |
| Combat aircraft | not more than 50 units |
| Attack helicopters | not more than 50 units |

**Russian Federation (in the area of application)**

| | |
|---|---|
| Battle tanks | not more than 6400 units |
| in active units | not more than 4975 units |
| ACVs | not more than 11 480 units |
| in active units | not more than 10 525 units |
| of which AIFV and HACV | not more than 7030 units |
| of which HACV | not more than 574 units |
| Artillery | not more than 6415 units |
| in active units | not more than 5105 units |
| Combat aircraft | not more than 3450 units |
| Attack helicopters | not more than 890 units |

**on the territory of the Russian Federation within the limits of the area defined in paragraph 1 of Article V of the Treaty**

| | |
|---|---|
| Battle tanks | not more than 1300 units |
| in storage | not more than 600 units |
| ACVs | not more than 1380 units |
| in storage | not more than 800 units |
| Artillery | not more than 1680 units |
| in storage | not more than 400 units |

**Ukraine**

| | |
|---|---|
| Battle tanks | not more than 4080 units |
| in active units | not more than 3130 units |
| ACVs | not more than 5050 units |
| in active units | not more than 4350 units |
| of which AIFV and HACV | not more than 3095 units |
| of which HACV | not more than 253 units |
| Artillery | not more than 4040 units |
| in active units | not more than 3240 units |
| Combat aircraft | not more than 1090 units |
| Attack helicopters | not more than 330 units |

**on the territory of Ukraine within the limits of the area defined in paragraph 1 of Article V of the Treaty**

| | |
|---|---|
| Battle tanks | not more than 680 units |
| in storage | not more than 400 units |
| ACVs | not more than 350 units |
| Artillery | not more than 890 units |
| in storage | not more than 500 units |

**Republic of Georgia**

| | |
|---|---|
| Battle tanks | not more than 220 units |
| in active units | not more than 220 units |
| ACVs | not more than 220 units |
| in active units | not more than 220 units |
| of which AIFV and HACV | not more than 135 units |
| of which HACV | not more than 11 units |
| Artillery | not more than 285 units |
| in active units | not more than 285 units |
| Combat aircraft | not more than 100 units |
| Attack helicopters | not more than 50 units |

2. In accordance with the provisions of the Treaty and of this Protocol, the Contracting Parties have the right to change their maximum holdings of conventional equipment limited by the Treaty on Conventional Forces in Europe.

A Contracting Party, intending to increase its maximum level of conventional equipment limited by the Treaty, has the right to effect such an increase only with the agreement of all interested Contracting Parties, in such a way as to avoid any infringement of the relevant provisions of the Treaty.

In the case where a Contracting Party gives notification of its intention to increase its maximum level of actual holdings of conventional armaments and equipment limited by the Treaty, all interested Contracting Parties, not later than 14 days after the receipt of such a notification, must notify all other Parties as to their position on the proposed increase contained in the notification. In the absence of a consensus, the Contracting Party wishing to effect the increase of its maximum levels in respect of conventional armaments and equipment limited by the Treaty, must, not later than 21 days after the receipt of the lack of consensus, call a consultation of all interested Contracting Parties in order to examine the questions raised in the original notification.

A decrease in the quantity of conventional armaments and equipment limited by the Treaty by Contracting Parties owning such equipment, does not in itself give the right to another Contracting Party to increase its holdings of conventional armaments and equipment limited by the Treaty: the use of its maximum levels for holdings of conventional armaments and equipment limited by the Treaty is the exclusive prerogative of each Contracting Party.

**ANNEX**

On armoured vehicle launched bridges (AVLB) in regular units.

1. Proceeding from the assumption that, in accordance with the Budapest Agreement of 3 November 1990, the USSR could hold no more than 462 AVLB in regular units, the Contracting Parties agreed to limit their numbers as follows:

| | |
|---|---|
| Azerbaijan | 8 units |
| Armenia | 8 |
| Belarus | 64 |
| Kazakhstan | 0 |
| Moldova | 7 |
| Russian Federation | 223 |
| Ukraine | 144 |
| Georgia | 8 |

2. An increase in any of the Contracting Parties' totals mentioned above of AVLB must, by agreement with the Contracting Parties, be announced beforehand and be accompanied by a corresponding reduction of one or more of the Contracting Parties' totals of their AVLBs.

**PROTOCOL**

On combat helicopters Mi-24R and Mi-24K not subject to the limitations on attack helicopters.

1. Proceeding from the assumption that, in accordance with the Treaty, the USSR could possess an aggregate total of Mi-24R and Mi-24K not exceeding 100, equipped for reconnaissance, spotting or chemical/biological/radiological sampling, which are not subject to the limitations on attack helicopters, the Contracting Parties agreed to limit their number in the following manner:

| | |
|---|---|
| Azerbaijan | 4 units |
| Armenia | 4 |
| Belarus | 16 |
| Kazakhstan | 0 |
| Moldova | 4 |
| Russian Federation | 50 |
| Ukraine | 18 |
| Georgia | 4 |

2. An increase in the above-mentioned totals by any one of the Contracting Parties concerning Mi-24R or Mi-24K helicopters not subject to the limitations on attack helicopters, must, by agreement with the Contracting Parties, be announced beforehand or be accompanied by corresponding reductions by one or more of the Contracting Parties in their totals of specified helicopters.

*Source:* CIS summit meeting, Tashkent, 15 May 1992.

---

## FINAL DOCUMENT OF THE EXTRAORDINARY CONFERENCE OF THE STATES PARTIES TO THE CFE TREATY (OSLO DOCUMENT)

*Signed 5 June 1992, Oslo*

The Republic of Armenia, the Republic of Azerbaijan, the Republic of Belarus, the Kingdom of Belgium, the Republic of Bulgaria, Canada, the Czech and Slovak Federal Republic, the Kingdom of Denmark, the French Republic, the Republic of Georgia, the Federal Republic of Germany, the Hellenic Republic, the Republic of Hungary, the Republic of Iceland, the Italian Republic, the Republic of Kazakhstan, the Grand Duchy of Luxembourg , the Republic of Moldova, the Kingdom of the Netherlands, the Kingdom of Norway, the Republic of Poland, the Portuguese Republic, Romania, the Russian Federation, the Kingdom of Spain, the Republic of Turkey, Ukraine, the United Kingdom of Great Britain and Northern Ireland and the United States of America, which are the States Parties to the Treaty on Conventional Armed Forces in Europe of November 19, 1990, hereinafter referred to as the States Parties,

Reaffirming their determination to bring into force the Treaty on Conventional Armed Forces in Europe of November 19, 1990, hereinafter referred to as the Treaty, by the time of the Helsinki Summit Meeting of the Conference on Security and Cooperation in Europe on July 9–10, 1992,

Desiring to meet the objectives and requirements of the Treaty while responding to the historic changes which have occurred in Europe since the Treaty was signed,

Recalling in this context the undertaking in paragraph 4 of the Joint Declaration of Twenty-Two States signed in Paris on November 19, 1990, to maintain only such military capabilities as are necessary to prevent war and provide for effective defence and to bear in mind the relationship between military capabilities and doctrines, and confirming their commitment to that undertaking,

Having met together at an Extraordinary Conference chaired by the Kingdom of Spain in Oslo on June 5, 1992, pursuant to Article XXI, paragraph 2, of the Treaty, as provisionally applied,

Have agreed as follows:

1. The understandings, notifications, confirmations and commitments contained or referred to in this Final Document and its Annexes A and B, together with the deposit of instruments of ratification by all the States Parties, shall be deemed as fulfilling the requirements for the entry into force of the Treaty in accordance with its provisions. Accordingly, the Treaty shall enter into force 10 days after the last such instrument has been deposited.

2. In this context, the States Parties note the Agreement of May 15, 1992, on the Principles and Procedures of Implementation of the Treaty on Conventional Armed Forces in Europe, the four Protocols to that Agreement and the Joint Declaration of May 15, 1992, in relation to the Treaty on Conventional Armed Forces in Europe, as transmitted on June 1, 1992, by that Agreement's Depositary to all States Parties to the Treaty. In this regard, Articles 1, 2, 3, 4, 5, 6, 10, 11 and 12 of that Agreement, the four Protocols to that Agreement, and the Joint Declaration of May 15, 1992, in relation to the Treaty on Conventional Armed Forces in Europe contain necessary confirmations and information.

3. The States Parties confirm the understandings as elaborated in the Joint Consultative Group, and specified in Annex A of this Final Document.

4. The States Parties confirm all decisions and recommendations adopted in the Joint Consultative Group.

5. This Final Document in no way alters the rights and obligations of the States Parties as set forth in the Treaty and its associated documents.

6. This Final Document shall enter into force upon signature by all of the States Parties.

7. This Final Document, together with its Annexes A and B, which are integral to it, in all the official languages of the Conference on Security and Co-operation in Europe, shall be deposited with the Government of the Kingdom of the Netherlands, as the designated Depositary for the Treaty, which shall circulate copies of this Final Document to all the States Parties.

**Annex A: Understandings**

1. The first paragraph of the Preamble of the Treaty shall be understood to read:

'The Republic of Armenia, the Republic of Azerbaijan, the Republic of Belarus, the Kingdom of Belgium, the Republic of Bulgaria, Canada, the Czech and Slovak Federal Republic, the Kingdom of Denmark, the French Republic, the Republic of Georgia, the Federal Republic of Germany, the Hellenic Republic, the Republic of Hungary, the Republic of Iceland, the Italian Republic, the Republic of Kazakhstan, the Grand Duchy of Luxembourg, the Republic of Moldova, the Kingdom of the Netherlands, the Kingdom of Norway, the Republic of Poland, the Portuguese Republic, Romania, the Russian Federation, the Kingdom of Spain, the Republic of Turkey, Ukraine, the United Kingdom of Great Britain and Northern Ireland and the United States of America, hereinafter referred to as the States Parties,'.

2. The second paragraph of the Preamble of the Treaty shall be understood to read:

'Guided by the Mandate for Negotiation on Conventional Armed Forces in Europe of January 10, 1989,'.

The third paragraph of the Preamble of the Treaty shall be understood to read:

'Guided by the objectives and the purposes of the Conference on Security and Cooperation in Europe, within the framework of which the negotiation of this Treaty was conducted in Vienna beginning on March 9, 1989,'.

3. With regard to the ninth paragraph of the Preamble of the Treaty, it is noted that the Treaty of Warsaw of 1955 is no longer in force, and that some of the States Parties in the first group specified in paragraph 4 of this Annex did not sign or accede to that Treaty.

4. The 'groups of States Parties' referred to in paragraph 1(A) of Article II of the Treaty shall be understood to consist of:

'the Republic of Armenia, the Republic of Azerbaijan, the Republic of Belarus, the Republic of Bulgaria, the Czech and Slovak Federal Republic, the Republic of Georgia, the Republic of Hungary, the Republic of Kazakhstan, the Republic of Moldova, the Republic of Poland, Romania, the Russian Federation and Ukraine,'

and

'the Kingdom of Belgium, Canada, the Kingdom of Denmark, the French Republic, the Federal Republic of Germany, the Hellenic Republic, the Republic of Iceland, the Italian Republic, the Grand Duchy of Luxembourg, the Kingdom of the Netherlands, the Kingdom of Norway, the Portuguese Republic, the Kingdom of Spain, the Republic of Turkey, the United Kingdom of Great Britain and Northern Ireland and the United States of America.'.

5. The first two sentences of paragraph 1(B) of Article II of the Treaty shall be understood to read:

'The term 'area of application' means the entire land territory of the States Parties in Europe from the Atlantic Ocean to the Ural Mountains, which includes all the European island territories of the States Parties, including the Faroe Islands of the Kingdom of Denmark, Svalbard including Bear Island of the Kingdom of Norway, the islands of Azores and Madeira of the Portuguese Republic, the Canary Islands of the Kingdom of Spain and Franz Josef Land and Novaya Zemlya of the Russian Federation. In the case of the Russian Federation and the Republic of Kazakhstan, the area of application includes all territory lying west of the Ural River and the Caspian Sea.'.

6. In Article IV of the Treaty, in accordance with the map provided by the former Union of Soviet Socialist Republics at signature of the Treaty:

– the second sentence of the second part of paragraph 1 shall be understood to read:

'Such designated permanent storage sites may also be located in the Republic of Moldova, that part of Ukraine comprising the portion of the former Odessa Military District on its territory, and that part of the territory of the Russian Federation comprising the southern part of the Leningrad Military District.'

– the first sentence of paragraph 2 shall be understood to read:

'Within the area consisting of the entire land territory in Europe, which includes all the European island territories, of the Republic of Belarus, the Kingdom of Belgium, the Czech and Slovak Federal Republic, the Kingdom of Denmark, including the Faroe Islands, the French Republic, the Federal Republic of Germany, the Republic of Hungary, the Italian Republic, that part of the Republic of Kazakhstan within the area of application, the Grand Duchy of Luxembourg, the Kingdom of the Netherlands, the Republic of Poland, the Portuguese Republic including the islands of Azores and Madeira, that part of the Russian Federation comprising the portion of the former Baltic Military District on its territory, the Moscow Military District and the portion of the Volga–Ural Military District on its territory west of the Ural Mountains, the Kingdom of Spain including the Canary Islands, that part of the territory of Ukraine comprising the former Carpathian and former Kiev Military Districts and the United Kingdom of Great Britain and Northern Ireland, each State Party shall limit and, as necessary, reduce its battle tanks, armoured combat vehicles and artillery so that, 40 months after entry into force of this Treaty and thereafter, for the group of States Parties to which it belongs the aggregate numbers do not exceed:'

– the first sentence of paragraph 3 shall be understood to read:

'Within the area consisting of the entire land territory in Europe, which includes all the European island territories, of the Republic of Belarus, the Kingdom of Belgium, the Czech and Slovak Federal Republic, the

Kingdom of Denmark, including the Faroe Islands, the French Republic, the Federal Republic of Germany, the Republic of Hungary, the Italian Republic, the Grand Duchy of Luxembourg, the Kingdom of the Netherlands, the Republic of Poland, that part of the Russian Federation comprising the portion of the former Baltic Military District on its territory, that part of the territory of Ukraine comprising the former Carpathian and former Kiev Military Districts and the United Kingdom of Great Britain and Northern Ireland, each State Party shall limit and, as necessary, reduce its battle tanks, armoured combat vehicles and artillery so that, 40 months after entry into force of this Treaty and thereafter, for the group of States Parties to which it belongs the aggregate numbers in active units do not exceed:'

– the first sentence in paragraph 3(D) shall be understood to read:

'in that part of Ukraine comprising the former Kiev Military District, the aggregate numbers in active units and designated permanent storage sites together shall not exceed:'.

7. The first sentence of paragraph 1(A) of Article V of the Treaty shall be understood, in accordance with the map provided by the former Union of Soviet Socialist Republics at signature of the Treaty, to read:

'Within the area consisting of the entire land territory in Europe, which includes all the European island territories, of the Republic of Armenia, the Republic of Azerbaijan, the Republic of Bulgaria, the Republic of Georgia, the Hellenic Republic, the Republic of Iceland, the Republic of Moldova, the Kingdom of Norway, Romania, that part of the Russian Federation comprising the Leningrad and North Caucasus Military Districts, the part of the Republic of Turkey within the area of application and that part of Ukraine comprising the portion of the former Odessa Military District on its territory, each State Party shall limit and, as necessary, reduce its battle tanks, armoured combat vehicles and artillery so that, 40 months after entry into force of this Treaty and thereafter, for the group of States Parties to which it belongs the aggregate numbers in active units do not exceed the difference between the overall numerical limitations set forth in Article IV, paragraph 1 and those in Article IV, paragraph 2, that is:'.

8. Paragraph 3 of Section I of the Protocol on Procedures Governing the Categorisation of Combat Helicopters and the Recategorisation of Multi-Purpose Attack Helicopters shall be understood to read:

'Notwithstanding the provisions in paragraph 2 of this Section and as a unique exception to that paragraph, the Republic of Armenia, the Republic of Azerbaijan, the Republic of Belarus, the Republic of Georgia, the Republic of Kazakhstan, the Republic of Moldava, the Russian Federation and Ukraine may hold an aggregate total not to exceed 100 Mi-24R and Mi-24K helicopters equipped for reconnaissance, spotting or chemical/biological/radiological sampling which shall not be subject to the limitations on attack helicopters in Articles IV and VI of the Treaty. Such helicopters shall be subject to exchange of information in accordance with the Protocol on Information Exchange and to internal inspection in accordance with Section VI, paragraph 30 of the Protocol on Inspection. Mi-24R and Mi-24K helicopters in excess of this limit shall be categorised as specialised attack helicopters regardless of how they are equipped and shall count against the limitations on attack helicopters in Articles IV and VI of the Treaty.'

9. With reference to paragraph 11 of the Protocol on the Joint Consultative Group, the proportion of the expenses of the Joint Consultative Group allocated to the Union of Soviet Socialist Republics shall become the collective responsibility of the Republic of Armenia, the Republic of Azerbaijan, the Republic of Belarus, the Republic of Georgia, the Republic of Kazakhstan, the Republic of Moldova, the Russian Federation and Ukraine.

## Annex B: Notifications, Confirmations and Commitments

### I: Notifications

1. The States Parties note that each State Party has provided to all other States Parties notifications of maximum levels for its holdings of conventional armaments and equipment limited by the Treaty (Article VII, paragraph 2) in advance of the Extraordinary Conference.

2. Each State Party shall provide the following notifications and information, where applicable, to all other States Parties no later than July 1, 1992:

(A) in view of the inspection requirements in the Treaty, information on its objects of verification and declared sites effective as of November 19, 1990 (Protocol on Notification and Exchange of Information, Section V and Annex on the Format for the Exchange of Information, Section V);

(B) list of its points of entry/exit (Annex on Format for the Exchange of Information, Section V, paragraph 3);

(C) notification of changes to its points of entry/exit (Protocol on Inspection, Section III, paragraph 11);

(D) lists of its proposed inspectors and transport crew members (Protocol on Inspection, Section III, paragraph 3);

(E) notification of deletions from the lists of inspectors and transport crew members (Protocol on Inspection, Section III, paragraphs 4 and 7);

(F) notification of its standing diplomatic clearance numbers for transportation means (Protocol on Inspection, Section III, paragraph 9);

(G) notification of the official language or languages to be used by inspection teams (Protocol on Inspection, Section III, paragraph 12);

(H) notification of its active inspection quota for the baseline validation period (Protocol on Inspection, Section II, paragraph 24);

(I) notification of entry into service of new types, models or versions of conventional armaments and equipment subject to the Treaty (Protocol on Existing Types, Section IV, paragraph 3);

(J) notification in the event of destruction by accident, and documentary evidence supporting destruction by accident, of conventional armaments and equipment limited by the Treaty (Protocol on Reduction, Section IX, paragraphs 2 and 3).

**II: Confirmations**

1. With regard to Article VIII, paragraph 7, of the Treaty, the States Parties confirm that, except as otherwise provided for in the Treaty, their respective reduction liabilities in each category shall be no less than the difference between their respective holdings notified, in accordance with the Protocol on Information Exchange, as of the signature of the Treaty, and their respective maximum levels for holdings notified pursuant to Article VII. In this regard, for those States Parties that have jointly confirmed the validity for them of holdings as of the signature of the Treaty, the sum of their reduction liabilities in each category shall, except as otherwise provided for in the Treaty, be no less than the difference between the jointly confirmed holdings and the sum of their maximum levels for holdings notified pursuant to Article VII.

2. The States Parties confirm their commitment, in the Declaration of the States Parties to the Treaty on Conventional Armed Forces in Europe with Respect to Personnel Strength of November 19, 1990, not to increase during the period of the negotiations referred to in Article XVIII of the Treaty the total peacetime authorised personnel strength of their conventional armed forces pursuant to the Mandate in the area of application.

3. The States Parties confirm their commitment to the Declaration of the States Parties to the Treaty on Conventional Armed Forces in Europe with Respect to Land-Based Naval Aircraft of November 19, 1990.

4. The States Parties confirm their adherence to the Agreement set out in the Statement by the Chairman of the Joint Consultative Group on October 18, 1991.

**III: Commitments**

**A: Costs**

1. In accordance with Article XVI, paragraph 2(F), of the Treaty, and with reference to paragraph 11 of the Protocol on the Joint Consultative Group, the Joint Consultative Group shall review its scale of distribution of expenses after entry into force of the Treaty in the light of decisions taken on the scale of distribution of expenses of the Conference on Security and Cooperation in Europe.

**B: Article XII**

1. In order to meet the security interests of all States Parties in light of new circumstances in Europe, the States Parties shall as a first priority seek to reach agreement, immediately after entry into force of the Treaty, on Article XII, paragraph 1, of the Treaty.

2. In this context, the States Parties will cooperate to respect the security objectives of Article XII within the area of application of the Treaty. In particular, no State Party will increase, within the area of application, its holdings of armoured infantry fighting vehicles held by organisations designed and structured to perform in peacetime internal security functions above that aggregate number held by such organisations at the time of signature of the Treaty, as notified pursuant to the information exchange effective as of November 19, 1990.

3. Notwithstanding the political commitment set forth in paragraph 2 above, any State Party that had an aggregate number of

armoured infantry fighting vehicles held by organisations designed and structured to perform in peacetime internal security functions on its territory, as notified effective as of November 19, 1990, that was less than five percent of its maximum levels for holdings for armoured combat vehicles, as notified pursuant to Article VII, paragraph 2, of the Treaty, or less than 100 such armoured infantry fighting vehicles, whichever is greater, will have the right to increase its holdings of such armoured infantry fighting vehicles to an aggregate number not to exceed five percent of its maximum levels for holdings for armoured combat vehicles, as notified pursuant to Article VII, paragraph 2, of the Treaty, or to an aggregate number not to exceed 100, whichever is greater.

---

*Source:* Extraordinary Conference of the States Parties to the CFE Treaty, Oslo, 15 June 1992.

---

## PROVISIONAL APPLICATION OF THE TREATY ON CONVENTIONAL ARMED FORCES IN EUROPE OF 19 NOVEMBER 1990

*Signed in Helsinki, 10 July 1992*

The Republic of Armenia, the Republic of Azerbaijan, the Republic of Belarus, the Kingdom of Belgium, the Republic of Bulgaria, Canada, the Czech and Slovak Federal Republic, the Kingdom of Denmark, the French Republic, the Republic of Georgia, the Federal Republic of Germany, the Hellenic Republic, the Republic of Hungary, the Republic of Iceland, the Italian Republic, the Republic of Kazakhstan, the Grand Duchy of Luxembourg, the Republic of Moldova, the Kingdom of the Netherlands, the Kingdom of Norway, the Republic of Poland, the Portuguese Republic, Romania, the Russian Federation, the Kingdom of Spain, the Republic of Turkey, Ukraine, the United Kingdom of Great Britain and Northern Ireland, and the United States of America, which are the States Parties to the Treaty on Conventional Armed Forces in Europe of November 19, 1990, hereinafter referred to as the States Parties,

Recalling the Final Document of the Extraordinary Conference of the States Parties of June 5, 1992, wherein they reaffirmed their determination to bring into force the Treaty on Conventional Armed Forces in Europe of November 19, 1990, hereinafter referred to as the Treaty, by the time of the Helsinki Summit meeting of the Conference on Security and Co-operation in Europe on July 9–10, 1992,

Recognising that the Treaty is an important achievement on which to build the new Europe proclaimed by the Charter of Paris,

Having due regard to the ratification procedures of their parliaments and governments,

Taking note of the signing of the Concluding Act of the Negotiation on Personnel Strength of Conventional Armed Forces in Europe,

Having met together at an Extraordinary Conference chaired by the French Republic in Helsinki on July 10, 1992, pursuant to Article XXI, paragraph 2 of the Treaty, as provisionally applied,

Have agreed as follows:

1. Without prejudice to the provisions of Article XXII of the Treaty and notwithstanding the Protocol on Provisional Application of the Treaty, the states parties shall apply provisionally all of the provisions of the Treaty, beginning on July 17, 1992, on the basis of the agreement reached by all States Parties expressed hereby. The States Parties deem that such provisional application constitutes an improvement to the Treaty.

2. Such provisional application of the Treaty shall be for a period of 120 days but shall terminate upon entry into force of the Treaty if the Treaty enters into force before such 120-day period expires.

3. In order to enhance the operation of the Treaty, during such period of provisional application as well as following entry into force of the Treaty, the date set forth in paragraph 1 above shall be used as the basis for determining the timing of all rights and obligations of the States Parties that are specifically tied to the date of entry into force of the Treaty.

4. An extraordinary conference shall be convened, in accordance with Article XXI, paragraph 2 of the Treaty, in connection with entry into force of the Treaty in order to assess the implementation of the Treaty in light of its provisional application pursuant hereto.

5. This document, in all the official languages of the Conference on Security and Co-operation in Europe, shall be deposited

with the Government of the Kingdom of the Netherlands, as the designated Depositary for the Treaty, which shall circulate copies of this document to all the States Parties.

---

*Source:* CSCE summit meeting, Helsinki, 10 July 1992.

---

## CONCLUDING ACT OF THE NEGOTIATION ON PERSONNEL STRENGTH OF CONVENTIONAL ARMED FORCES IN EUROPE (CFE-1A AGREEMENT)

*Signed in Helsinki, 10 July 1992*

The Republic of Armenia, the Republic of Azerbaijan, the Republic of Belarus, the Kingdom of Belgium, the Republic of Bulgaria, Canada, the Czech and Slovak Federal Republic, the Kingdom of Denmark, the French Republic, the Republic of Georgia, the Federal Republic of Germany, the Hellenic Republic, the Republic of Hungary, the Republic of Iceland, the Italian Republic, the Republic of Kazakhstan, the Grand Duchy of Luxembourg, the Republic of Moldova, the Kingdom of the Netherlands, the Kingdom of Norway, the Republic of Poland, the Portuguese Republic, Romania, the Russian Federation, the Kingdom of Spain, the Republic of Turkey, Ukraine, the United Kingdom of Great Britain and Northern Ireland, and the United States of America, hereinafter referred to as the participating States,

Recalling the obligations undertaken in the Treaty on Conventional Armed Forces in Europe of November 19, 1990, hereinafter referred to as the CFE Treaty, and the important achievements attained in that Treaty,

In accordance with the obligation in Article XVIII of the CFE Treaty to continue the negotiations on conventional armed forces with the same Mandate and with the goal of building on the CFE treaty and with the objective of concluding an agreement, no later than the Helsinki 1992 Follow-up Meeting of the Conference on Security and Co-operation in Europe (CSCE), on additional measures aimed at further strengthening security and stability in Europe,

Guided by the Mandate for Negotiation on Conventional Armed Forces in Europe of January 10, 1989, and having conducted negotiations in Vienna,

Having decided to limit and, if applicable, reduce, on a national basis, the personnel strength of their conventional armed forces within the area of application,[*]

Guided by the objectives and the purposes of the CSCE, within the framework of which these negotiations were conducted,

Looking forward to a more structured co-operation among all CSCE participating States on security matters and to new negotiations on disarmament and confidence and security building in accordance with their commitment in the Charter of Paris for a New Europe, and, accordingly, to the possibility, within the context of those new negotiations, for all CSCE participating States to subscribe to a common regime based upon the measures adopted in this Concluding Act, hereinafter referred to as the Act,

Taking into account the principle of sufficiency, and recalling the undertaking of the military capabilities as are necessary to pre-participating States to maintain only such military capabilities as are necessary to prevent war and provide for effective defence, bearing in mind the relationship between military capabilities and doctrines,

Recognising the freedom of each participating State to choose its own security arrangements,

Have adopted the following:

### SECTION I. Scope of Limitation

1. Each participating State will limit, as specified in Section II of this Act, its personnel based on land within the area of application in the following categories of conventional armed forces:

(A) all full-time military personnel serving with land forces, including air defence formations and units subordinated at or below the military district or equivalent level, as specified in Section I of the Protocol on Information Exchange of the CFE Treaty;

---

* The area of application of the measures adopted in this Act is the area of application of the CFE Treaty as defined in Article II, paragraph 1, subparagraph (B) of the CFE Treaty, taking into account the understanding specified in Annex A, paragraph 5 of the Final Document of the Extraordinary Conference of the States Parties to the Treaty on Conventional Armed Forces in Europe of June 5, 1992.

(B) all full-time military personnel serving with air and air defence aviation forces, including long-range aviation forces reported pursuant to Section I of the Protocol on Information Exchange of the CFE Treaty, as well as military transport aviation forces;

(C) all full-time military personnel serving with air defence forces other than those specified in subparagraphs (A) and (B) of this paragraph;

(D) all full-time military personnel, excluding naval personnel, serving with all central headquarters, command and staff elements;

(E) all full-time military personnel, excluding naval personnel, serving with all centrally-controlled formations, units and other organisations, including those of rear services;

(F) all full-time military personnel serving with all land-based naval formations and units which hold battle tanks, armoured combat vehicles, artillery, armoured vehicle launched bridges, armoured infantry fighting vehicle look-alikes or armoured personnel carrier look-alikes as defined in Article II of the CFE Treaty or which hold land-based naval combat aircraft referred to in the Declaration of the States Parties to the Treaty on Conventional Armed Forces in Europe with Respect to Land-Based Naval Aircraft of November 19, 1990;

(G) all full-time military personnel serving with all other formations, units and other organisations which hold battle tanks, armoured combat vehicles, artillery, combat aircraft or attack helicopters in service with its conventional armed forces, as defined in Article II of the CFE Treaty; and

(H) all reserve personnel who have completed their initial military service or training and who are called up or report voluntarily for full-time military service or training in conventional armed forces for a continuous period of more than 90 days.

2. Notwithstanding the provisions of paragraph 1 of this Section, the following categories of personnel are not included within the scope of limitation specified in this Act:

(A) personnel serving with organisations designed and structured to perform in peacetime internal security functions;

(B) personnel in transit from a location outside the area of application to a final destination outside the area of application who are in the area of application for no longer than seven days; and

(C) personnel serving under the command of the United Nations.

3. If, after the date on which this Act comes into effect, any land-based formations or units are formed within the area of application which, according to their structure and armaments, have a capability for ground combat outside national borders against an external enemy, a participating State may raise in the Joint Consultative Group any issue regarding personnel serving with such formations and units. The Joint Consultative Group will consider any such issue on the basis of all available information, including information provided by the participating States concerned, with a view to determining whether the above-mentioned criteria are applicable to such formations and units; if such criteria are deemed to apply, the personnel serving with such formations and units will be included within the scope of limitation specified in this Act.

**SECTION II. National Personnel Limits**

1. Each participating State will limit its military personnel based on land within the area of application in the categories of conventional armed forces specified in Section I, paragraph 1 of this Act so that, 40 months after entry into force of the CFE Treaty and thereafter, the aggregate number of such personnel will not exceed the number representing its national personnel limit as specified in this paragraph:

| | |
|---|---|
| The Republic of Armenia | |
| The Republic of Azerbaijan | |
| The Republic of Belarus | 100,000 |
| The Kingdom of Belgium | 70,000 |
| The Republic of Bulgaria | 104,000 |
| Canada | 10,660 |
| The Czech and Slovak Federal Republic | 140,000 |
| The Kingdom of Denmark | 39,000 |
| The French Republic | 325,000 |
| The Republic of Georgia | |
| The Federal Republic of Germany | 345,000 |
| The Hellenic Republic | 158,621 |
| The Republic of Hungary | 100,000 |
| The Republic of Iceland | 0 |
| The Italian Republic | 315,000 |
| The Republic of Kazakhstan | 0 |
| The Grand Duchy of Luxembourg | 900 |
| The Republic of Moldova | |
| The Kingdom of the Netherlands | 80,000 |

| | |
|---|---|
| The Kingdom of Norway | 32,000 |
| The Republic of Poland | 234,000 |
| The Portuguese Republic | 75,000 |
| Romania | 230,000 |
| The Russian Federation | 1,450,000 |
| The Kingdom of Spain | 300,000 |
| The Republic of Turkey | 530,000 |
| Ukraine | 450,000 |
| The United Kingdom of Great Britain and Northern Ireland | 260,000 |
| The United States of America | 250,000 |

2. For the purpose of recording changes to the information specified in paragraph 1 of this Section, the Government of the Kingdom of the Netherlands will distribute to all the participating States a revised version of the information in that paragraph.

3. Each participating State may revise its national personnel limit in accordance with the provisions of Section III of this Act.

## SECTION III. Revision Procedures

1. A participating State may revise downward its national personnel limit by providing a notification of its revised limit to all other participating States. Such notification will specify the date on which the revised limit will become effective.

2. A participating State intending to revise upward its national personnel limit will provide notification of such intended revision to all other participating States. Such notification will include an explanation of the reasons for such a revision. Any participating State may raise any question concerning the intended revision. A revised national personnel limit will become effective 42 days after notification has been provided, unless a participating State raises an objection to such revision by providing notification of its objection to all other participating States.

3. If an objection is raised, any participating State may request the convening of an extraordinary conference which will examine the intended revision in the light of the explanations provided and seek to decide on a future national personnel limit. The extraordinary conference will open no later than 15 days after receipt of the request and, unless it decides otherwise, will last no longer than three weeks.

## SECTION IV. Information Exchange

1. Each participating State will provide to all other participating States, in accordance with the provisions of this Section, the following information in respect of its personnel based on land within the area of application:

(A) in respect of all personnel specified in Section I, paragraph 1 of this Act, the aggregate number;

(B) in respect of all full-time military personnel serving with land forces, including air defence formations and units subordinated at or below the military district or equivalent level, as specified in Section I of the Protocol on Information Exchange of the CFE Treaty, the aggregate number and the number in each formation, unit and other organisation down to the brigade/regiment or equivalent level, specifying the command organisation, designation, subordination and peacetime location, including the geographic name and coordinates, for each such formation, unit and organisation;

(C) in respect of all full-time military personnel serving with air and air defence aviation forces, including long-range aviation forces reported pursuant to Section I of the Protocol on Information Exchange of the CFE Treaty, as well as military transport aviation forces, the aggregate number and the number in each formation, unit and other organisation of conventional armed forces down to the wing/air regiment or equivalent level, specifying the command organisation, designation, subordination and peacetime location, including the geographic name and coordinates, for each such formation, unit and organisation;

(D) in respect of all full-time military personnel serving with air defence forces other than those specified in subparagraphs (B) and (C) of this paragraph, the aggregate number and the number in each formation and other organisation down to the next level of command above division or equivalent level (i.e., air defence army or equivalent), specifying the organisation, designation, subordination and peacetime location, including the geographic name and coordinates, for each such formation and organisation;

(E) in respect of all full-time military personnel of conventional armed forces, excluding naval personnel, serving with all central headquarters, command and staff elements, the aggregate number;

(F) in respect of all full-time military personnel of conventional armed forces, excluding naval personnel, serving with all centrally-controlled formations, units and other organisations, including those of rear services, the aggregate number and the number in each formation, unit and other organisation down to the brigade/regiment, wing/air regiment or equivalent level, specifying the command organisation, designation, subordination and peacetime location, including the geographic name and coordinates, for each such formation, unit and organisation;

(G) in respect of all full-time military personnel serving with all land-based naval formations and units which hold conventional armaments and equipment in the categories specified in Section III of the Protocol on Information Exchange of the CFE Treaty or which hold land-based naval combat aircraft referred to in the Declaration of the States Parties to the Treaty on Conventional Armed Forces in Europe with Respect to Land-Based Naval Aircraft of November 19, 1990, the aggregate number and the number in each formation and unit down to the brigade/regiment, wing/air regiment or equivalent level, as well as units at the next level of command below the brigade/regiment, wing/air regiment level which are separately located or independent (i.e., battalions/squadrons or equivalent), specifying the designation and peacetime location, including the geographic name and coordinates, for each such formation and unit;

(H) in respect of all full-time military personnel serving with all formations, units and other organisations of conventional armed forces specified in Section III of the Protocol on Information Exchange of the CFE Treaty, the number in each such formation, unit and organisation down to the brigade/regiment, wing/air regiment or equivalent level, as well as units at the next level of command below the brigade/regiment, wing/air regiment level which are separately located or independent (i.e., battalions/squadrons or equivalent), specifying the designation and peacetime location, including the geographic name and coordinates, for each such formation, unit and organisation;

(I) in respect of all personnel serving with all formations and units down to the independent or separately located battalion or equivalent level which hold battle tanks, artillery, combat aircraft or specialised attack helicopters as well as armoured infantry fighting vehicles as specified in Article XII of the CFE Treaty, in organisations designed and structured to perform in peacetime internal security functions, the number in each such formation and unit at each site at which such armaments and equipment are held, specifying the national level designation of each such organisation and the location, including the geographic name and coordinates, of each site at which such armaments and equipment are held;

(J) in respect of all personnel serving with all formations and units in organisations designed and structured to perform in peacetime internal security functions, excluding unarmed or lightly armed civil police forces and protective services, the aggregate number and the aggregate number in each administrative region or equivalent;

(K) in respect of all reserve personnel who have completed their initial military service or training and who have been called up or have reported voluntarily for military service or training in conventional armed forces since the most recent exchange of information provided in accordance with this Section, the aggregate number, specifying the number, if any, of those who have been called up or have reported voluntarily for full-time military service or training in conventional armed forces for a continuous period of more than 90 days;

(L) in respect of all military personnel serving under the command of the United Nations, the aggregate number; and

(M) in respect of all military personnel, excluding naval personnel, serving with all other formations, units and other organisations of conventional armed forces, the aggregate number, specifying the designation of such formations, units and organisations.

2. In providing information on personnel strengths in accordance with this Section, each participating State will provide the peacetime authorized personnel strength, which will approximate the number of personnel serving within the area of application with each of the formations, units and other organisations specified in paragraph 1 of this Section.

3. The provisions of this Section will not apply to personnel who are in transit through the area of application from a location outside the area of application to a final destination outside the area of application. Personnel in the categories specified in paragraph 1 of this Section who entered the area of application in transit will be subject to the provisions of this Section if they remain within the area of application for a period longer than seven days.

4. Each participating State will be responsible for its own information; receipt of such information will not imply validation or acceptance of the information provided.

5. The participating States will provide the

information specified in this Section in accordance with the formats and procedures to be agreed in the Joint Consultative Group.

6. Prior to the date on which national personnel limits become effective in accordance with Section II of this Act, each participating State will provide to all other participating States the information specified in paragraph 1, subparagraphs (A), (D), (E) and (G) to (M) of this Section, as well as the information on aggregate numbers of personnel in the categories specified in subparagraphs (B), (C) and (F) of that paragraph, in written form, in one of the official CSCE languages, using diplomatic channels or other official channels designated by them, in accordance with the following timetable:

(A) no later than 30 days following entry into force of the CFE Treaty, with the information effective as of the date of entry into force of that Treaty; and

(B) on the 15th day of December of the year in which the CFE Treaty comes into force (unless entry into force of that Treaty occurs within 60 days of the 15th day of December), and on the 15th day of December of every year thereafter, with the information effective as of the first day of January of the following year.

7. Beginning with the date on which national personnel limits become effective in accordance with Section II of this Act, each participating State will provide to all other participating States all the information specified in paragraph 1 of this Section in written form, in one of the official CSCE languages, using diplomatic channels or other official channels designated by them, in accordance with the following timetable:

(A) on the date on which national personnel limits become effective in accordance with Section II of this Act, with the information effective as of that date; and

(B) on the 15th day of December of the year in which the national personnel limits become effective in accordance with Section II of this Act, and on the 15th day of December of every year thereafter, with the information effective as of the first day of January of the following year.

8. The participating States will, at the first review of the operation of this Act in accordance with Section VII, paragraph 3 of this Act, consider issues relating to the adequacy and effectiveness of the disaggregation of the information specified in paragraph 1, subparagraphs (B), (C) and (F) of this Section.

## SECTION V. Stabilising Measures

### NOTIFICATION OF INCREASES IN UNIT STRENGTHS

1. Each participating State will notify all other participating States at least 42 days in advance of any permanent increase in the personnel strength of any formation, unit or other organisation which was reported in the most recent exchange of information at the brigade/regiment, wing/air regiment or equivalent level in accordance with Section IV of this Act when such increase equals 1,000 or more at the brigade/regiment level, or 500 or more at the wing/air regiment level, or equivalent levels.

### NOTIFICATION OF CALL-UP OF RESERVE PERSONNEL

2. Any participating State intending to call up reserve personnel of its conventional armed forces based on land within the area of application will notify all other participating States whenever the cumulative total of the personnel called up and retained on full-time military service will exceed a threshold of 35,000.

3. Such notification will be provided at least 42 days in advance of such threshold being exceeded. As an exception, in the case of emergency situations where advance notification is not practical, notification will be provided as soon as possible and, in any event, no later than the date such threshold is exceeded.

4. Such notification will include the following information:

(A) the total number of reserve personnel to be called up, specifying the number to be called up for more than 90 days;

(B) a general description of the purpose of the call-up;

(C) the planned start and end dates of the period during which such threshold will be exceeded; and

(D) the designation and location of any formation in which more than 7,000 at the division or equivalent level or more than 9,000 at the army/army corps or equivalent level of the personnel so called up will serve.

### RESUBORDINATION OF UNITS

5. After the first exchange of information in accordance with Section IV of this Act, a participating State intending to resubordinate formations, units or other organisations

whose personnel are subject to limitation in accordance with Section I of this Act to a formation, unit or other organisation whose personnel would not otherwise be subject to limitation will notify all other participating States of the planned resubordination no later than the date on which such resubordination will become effective.

6. Such notification will include the following information:

(A) the date on which such resubordination will become effective;

(B) the subordination, designation and peacetime location of each formation, unit and organisation to be resubordinated, both before and after such resubordination will become effective;

(C) the peacetime authorized personnel strength for each formation, unit and organisation to be resubordinated, both before and after such resubordination will become effective; and

(D) the number, if any, of battle tanks, armoured infantry fighting vehicles, artillery, combat aircraft, attack helicopters and armoured vehicle launched bridges as defined in Article II of the CFE Treaty held by each formation, unit and organisation to be resubordinated, both before and after such resubordination will become effective.

7. Personnel serving with formations, units or other organisations resubordinated after the date on which national personnel limits become effective in accordance with Section II of this Act will remain subject to limitation in accordance with Section I of this Act until the date of the exchange of information in accordance with Section IV of this Act one year subsequent to the year in which such resubordination becomes effective, after which time the procedure specified in paragraph 8 of this Section will apply.

8. Forty-two days prior to the end of the one-year period specified in paragraph 7 of this Section, the participating State resubordinating such formations, units or other organisations will provide to all other participating States notification of the planned exclusion. Upon the request of any other participating State, the participating State resubordinating such formations, units or other organisations will provide all relevant information supporting such exclusion.

## SECTION VI. Verification/Evaluation

1. For the purpose of evaluating observance of national personnel limits and the other provisions of this Act, participating States will apply Section VII and Section VIII of the Protocol on Inspection of the CFE Treaty and other relevant provisions of that Treaty, together with the provisions set out in this Section.

2. In the case of an inspection pursuant to Section VII of the Protocol on Inspection of the CFE Treaty, the pre-inspection briefing will include information on the number of personnel serving with any formation, unit or other organisation which was notified in the most recent exchange of information in accordance with Section IV of this Act and which is located at that inspection site. If the number of such personnel differs from the number of personnel notified in that most recent exchange of information, the inspection team will be provided with an explanation of such difference. The pre-inspection briefing will also include information on the number of personnel serving with any other formation or unit down to the brigade/regiment, wing/air regiment or equivalent level, as well as independent units at the battalion/squadron or equivalent level, in the categories specified in Section IV, paragraph 1, subparagraphs (B), (C) and (F) of this Act, which is located at that inspection site.

3. In the case of an inspection pursuant to Section VIII of the Protocol on Inspection of the CFE Treaty, the escort team will provide, if requested by the inspection team, information on the number of personnel serving with any formation, unit or other organisation which was notified in the most recent exchange of information in accordance with Section IV of this Act, which is located at that inspection site and whose facilities are being inspected. If the number of such personnel differs from the number of personnel notified in that most recent exchange of information, the inspection team will be provided with an explanation of such difference.

4. During an inspection pursuant to Section VII or Section VIII of the Protocol on Inspection of the CFE Treaty, inspectors may have access, consistent with the provisions of that Protocol, to all facilities subject to inspection at the inspection site, including those used by all formations, units and other organisations located at that inspection site. During such an inspection, the escort team will specify, if requested by the inspection team, whether a particular building on the inspection site is a personnel barracks or messing facility.

5. Inspectors will include in the inspection report prepared pursuant to Section XII of

the Protocol on Inspection of the CFE Treaty information provided to the inspection team in accordance with paragraphs 2 and 3 of this Section in a format to be agreed in the Joint Consultative Group. Inspectors may also include in that report written comments pertaining to the evaluation of personnel strengths.

6. Evaluation of observance of the provisions of this Act will be further facilitated through confidence- and security-building measures that have been developed and that may be developed in the context of the new negotiations on disarmament and confidence and security building following the Helsinki Follow-up Meeting. In this context, participating States are prepared to join in considering ways and means to refine the evaluation provisions specified in the Vienna Document 1992.

**SECTION VII. Review Mechanisms**

1. The participating States will review the implementation of this Act in accordance with the procedures set out in this Section, using the relevant bodies and channels within the framework of the CSCE process.

2. In particular, any participating State may at any time raise and clarify questions relating to the implementation of this Act within the framework, as appropriate, of the Joint Consultative Group. The participating States will consider in the context of the new negotiations on disarmament and confidence and security building which will be conducted following the Helsinki Follow-up Meeting, the role of the Conflict Prevention Centre in this regard, as appropriate.

3. Six months after the date on which national personnel limits become effective in accordance with Section II of this Act and at five-year intervals thereafter, the participating States will conduct a review of the operation of this Act.

4. The participating States will meet in an extraordinary conference if requested to do so by any participating State which considers that exceptional circumstances relating to this Act have arisen. Such a request will be transmitted to all other participating States and will include an explanation of exceptional circumstances relating to this Act, e.g., an increase in the number of military personnel in categories listed in Section I of this Act in a manner or proportion which the participating State requesting such an extraordinary conference deems to be prejudicial to security and stability within the area of application. The conference will open no later than 15 days after receipt of the request and, unless it decides otherwise, will last no longer than three weeks.

**SECTION VIII. Closing Provisions**

1. The measures adopted in this Act are politically binding. Accordingly, this Act is not eligible for registration under Article 102 of the Charter of the United Nations. This Act will come into effect simultaneously with the entry into force of the CFE Treaty.

2. This Act will have the same duration as the CFE Treaty and may be supplemented, modified or superseded.

3. The Government of the Kingdom of the Netherlands will transmit true copies of this Act, the original of which is in English, French, German, Italian, Russian and Spanish, to all participating States, and bring this Act to the attention of the Secretariat of the CSCE and the Secretary General of the United Nations.

---

*Source:* CSCE summit meeting, Helsinki, 10 July 1992.

# 13. The United Nations Special Commission on Iraq: activities in 1992

ROLF EKÉUS

## I. Introduction

The implementation of United Nations Security Council Resolution 687[1] which established the cease-fire after the 1991 Persian Gulf War continued in 1992 with the same intensity as in 1991.[2] Under the cease-fire resolution all weapons of mass destruction in Iraq should be declared, identified, located and disposed of and a monitoring system to ensure that no new weapons be reintroduced to Iraq was to be established. The prohibited weapons are nuclear weapons, biological weapons (BW), chemical weapons (CW) and ballistic missiles with a range greater than 150 km. Furthermore, in October 1991 UN Security Council Resolution 715 approved two plans, one for nuclear items and one for non-nuclear items, for monitoring Iraqi compliance with the obligations under the cease-fire regime not to use, develop, construct or acquire any of the prohibited weapons.[3]

It should be recalled that in provisions of the cease-fire resolution, the United Nations Special Commission on Iraq (UNSCOM) was established as a subsidiary organ of the Security Council to carry out the tasks of supervising and executing the elimination of Iraq's BW, CW and ballistic missile capabilities and of monitoring Iraq's compliance. The Director General of the International Atomic Energy Agency (IAEA) was requested to carry out the corresponding tasks regarding Iraq's nuclear capability with the assistance and co-operation of the Special Commission.

For chemical weapons, in 1992 there was a shift of emphasis and resources towards destruction activities. While inspection of both declared and undeclared sites continued, UNSCOM teams supervised the destruction of chemical ammunition and the completion of construction of two chemical destruction facilities. The facilities—one a hydrolysis plant for the destruction of nerve agent and the other an incinerator for mustard gas—were commissioned in late 1992 and early 1993, respectively, and are operating at full capacity.

In 1992 missile inspections intensified and diversified. Traditional weapon searches, document and computer investigation, and highly specialized inspections focusing on components, fuel and production elements were carried out with the aim of establishing whether all of Iraq's missiles and related capabil-

[1] United Nations Security Council document S/RES/687 (1991), 8 Apr. 1991; for the text of the resolution, see SIPRI, *SIPRI Yearbook 1992: World Armaments and Disarmament* (Oxford University Press: Oxford, 1992), appendix 13A, pp. 525–30.

[2] See 'The United Nations Special Commission on Iraq', *SIPRI Yearbook 1992* (note 1), pp. 509–24.

[3] United Nations Security Council document S/RES/715 (1991), 11 Oct. 1991.

ities have been accounted for as Iraq states.[4] A number of production facilities and equipment were destroyed in the early months of 1992, and tangible progress was made in obtaining information from Iraq about its operational use of missiles.

While doubt continues to be expressed about the completeness of Iraq's declarations concerning its BW programme, there was little development in this area in 1992. Inspections continued with only limited concrete results.

As of February 1993 the IAEA has carried out a total of 17 nuclear inspections with the assistance and co-operation of UNSCOM. These have entailed inspections at more than 70 sites. The results of these activities during 1992 and early 1993 have given good insight into and understanding of Iraq's nuclear programme. The conclusion arrived at is that the nuclear programme was intended to produce enriched uranium and to develop a nuclear weapon capability.

## II. Iraqi non-compliance

Owing to consistent refusal by Iraq to accept the initiation and practical implementation of Resolution 715 for compliance monitoring, on 19 February 1992 the Security Council declared that Iraq was in material breach of the cease-fire resolution and dispatched to Iraq a high-level mission led by the Chairman of UNSCOM, armed with a statement demanding that Iraq give the necessary assurances of compliance with the Security Council resolutions or face serious consequences. The mission visited Iraq from 21 to 24 February. In its report to the Security Council,[5] it concluded that unconditional compliance by Iraq had not been provided and that therefore the implementation of Resolution 715 could not be initiated. Partly coinciding with the mission, Iraq refused to permit the start of the destruction of the facilities and equipment associated with its missile production programme. The Security Council, having received the report, condemned Iraq's failure to comply with its obligations to accept the destruction as required and to make the necessary declarations under Resolution 715. In response the Government of Iraq requested that it be allowed to present its views directly to the Security Council.

During the debate on 11 and 12 March in the Security Council at which Iraq's Deputy Prime Minister presented the position of the Iraqi Government, the President of the Security Council, speaking on behalf of all members of the Security Council, stated that the Government of Iraq must immediately take steps to comply fully and unconditionally with its obligations under the relevant resolutions.[6] Following the debate, teams of ballistic missile experts from the Special Commission visited Iraq in March and April and supervised the destruction of 10 large buildings for the production of ballistic missiles and of a large amount of equipment for missile production. These events

[4] United Nations Security Council document S/PV.3139 (Resumption 1).

[5] United Nations Security Council document S/23643, 26 Feb. 1992.

[6] United Nations Security Council documents S/PV.3059, S/PV.3059 (resumption 1), and S/PV.3059 (Resumption 2).

concluded a period of confrontation between the Government of Iraq and the United Nations on the question of whether items other than proscribed weapons were to be destroyed under the cease-fire regime. After this development, uncontested destruction took place of key technical installations comprising buildings and equipment at the nuclear weapon development complex at Al Atheer–Al Haytham, including 8 large buildings.

A major political problem developed on 5 July when Iraq refused an inspection team access to the Ministry of Agriculture. UNSCOM had reliable information from two sources that the building contained archives related to proscribed activities. These archives were of relevance to the work of the Special Commission and their retention by Iraq was also prohibited. The Government of Iraq claimed that UNSCOM had no right to enter the building as it contained nothing of relevance to the weapon systems proscribed under Resolution 687 and that to allow access would be to undermine Iraq's sovereignty and national security. The Special Commission established round-the-clock surveillance of the building by members of the UNSCOM inspection team. The Security Council issued a statement on 6 July declaring that Iraq's refusal to permit the inspection team entry to the Ministry of Agriculture constituted a material breach of the provisions of the cease-fire resolution.[7] Despite this, Iraq continued in its refusal. At the request of the Security Council, the Chairman of the Special Commission visited Baghdad on 17–19 July. Talks with the Iraqi authorities—the Deputy Prime Minister and the Minister for Foreign Affairs—did not resolve the situation, as reported to the Security Council on 20 July by the Chairman of the Special Commission. UNSCOM surveillance of the Ministry of Agriculture had to be terminated on 22 July when an escalation of violence against the UNSCOM inspectorate in Baghdad culminated in an attempt to kill UNSCOM inspectors guarding the building.

Against the background of continued demonstration, violence and threats directed at the UNSCOM inspection team in Baghdad and growing impatience among Security Council members, intensive talks took place on 24–26 July in New York between the Chairman of the Special Commission and the Permanent Representative of Iraq to the United Nations. Agreement was reached on 26 July and the inspection of the building was finally carried out on 28 July, parallel with new talks in Baghdad between the Chairman of the Special Commission and the Iraqi Deputy Prime Minister. No items relevant to Security Council resolutions were found in the Ministry of Agriculture building which had stood unguarded by UNSCOM inspectors for six days. The solution to this crisis, which at times nearly led to military action, established to the satisfaction of the Special Commission its unequivocal right to enter any site or building in Iraq, without exception.

In the month that followed, tension remained high as some Iraqi officials made statements that ministry buildings and other installations of similar significance were off-limits to UNSCOM. However, the strain gradually subsided as the matter was not raised by the Iraqi Cabinet members directly responsible.

[7] United Nations Security Council document S/24240, 6 July 1992.

The Chairman of the Special Commission reiterated the policy that UNSCOM would continue its inspection activities in Iraq with due regard to Iraq's legitimate concern for its dignity and sovereignty. The principle of unrestricted access was upheld through a subsequent inspection carried out on 22 October without major difficulty.

After the Ministry of Agriculture stand-off, a campaign of harassment, threat and physical attack against UNSCOM personnel in Baghdad continued, albeit at a lower level of intensity. It was clear from the nature of the campaign that it was centrally co-ordinated and not, as Iraqi authorities insisted, a spontaneous outburst of public indignation. Security for UNSCOM personnel has improved or worsened in a pattern conveniently corresponding to the political needs of the authorities.

## III. UNSCOM surveillance activities

Following a 9 April letter in which Iraq called for a halt to all of the Special Commission's high-altitude U-2 aerial surveillance flights and warned that their continuation would endanger the aircraft and its pilot, the Security Council held consultations, after which the President issued a statement on behalf of its members[8] in which it was pointed out that the 'surveillance flights are carried out under the authority of Security Council Resolutions 687 (1991), 707 (1991) and 715 (1991)' and that the right of UNSCOM to conduct such flights was reaffirmed. The Security Council called upon the Government of Iraq to give assurances on the security and safety of the flights and warned of various consequences if Iraq did not comply with its obligations in this regard. In a 12 April letter to the Security Council, the Foreign Minister of Iraq affirmed that the Government of Iraq 'did not intend and does not intend to carry out any military operation aimed at the Commission's aerial surveillance flights'.[9]

During 1992 aerial surveillance activities intensified. The regular flights of the high-altitude U-2 aircraft—flying approximately three times per week—were supplemented by aerial inspections conducted from UNSCOM helicopters based at Rasheed Airbase in the Baghdad area. The helicopter inspections commenced on 21 June and were carried out to supplement the high-altitude photography surveillance in the planning of inspections, monitoring of sites, preparation of inspection teams and identification of inspection targets. These operations were added to the original task of the helicopters—the rapid transport of inspection teams to sites supposed to contain time-sensitive data. The combination of high-altitude surveillance operations and helicopter aerial inspection has proven effective as the former offers the advantages of longer flight time, wider surveillance coverage and maintaining uncertainty of the exact sites which are being photographed, while the latter offers better oblique photography, higher resolution, 360° video coverage and faster response time.

[8] United Nations Security Council document S/23803, 10 Apr. 1992.

[9] United Nations Security Council document S/23806, 13 Apr. 1992.

From initiation of this new surveillance procedure, Iraq embarked upon a campaign to achieve an early end to the helicopter operations by placing limits on the way the helicopters might be used. It thus claimed limits with regard to so-called sensitive sites, tried to designate narrow flight patterns and routes, and accepted the use of helicopter surveillance only for well-defined sites, rather than for area or route surveys.

In September, upon imposition by the Persian Gulf War Coalition of the no-fly zone south of the 32nd parallel, Iraq sought to prevent UNSCOM from flying its C-160 fixed-wing aircraft from Bahrain across the no-fly zone and suggested that the aircraft cross into Iraqi airspace above the 32nd parallel. UNSCOM rejected this idea because it constituted an infringement of its rights and would in the long run be operationally impracticable given that the C-160 aircraft's base was in Bahrain. UNSCOM continued using the short route through the no-fly zone.

At the end of 1992 the Iraqi Government refused to let UNSCOM utilize helicopter surveillance over Baghdad. Although the Special Commission has itself decided not to route flights over central Baghdad, important operational needs require that flights be made over the large military and industrial areas on the outskirts of the city. The Special Commission has pointed out that Security Council Resolution 707 explicitly sanctions the use by UNSCOM of fixed- or rotary-wing aerial surveillance over all Iraqi territory.[10] This matter is still unresolved.

A serious development occurred on 7 January 1993 when the Government of Iraq informed the Special Commission that it would no longer be permitted to land its aircraft in Iraq.[11] Instead Iraq offered the use of either Iraqi aircraft or the overland route from Amman, Jordan, for the transport of UNSCOM personnel and equipment. This was reported by the Special Commission to the Security Council which dealt with the matter in conjunction with a number of other problems in relation to Iraq. In a 8 January 1993 statement,[12] the Security Council demanded that Iraq permit UNSCOM to use its own aircraft in Iraq. The statement noted that the restrictions placed on UNSCOM flights constituted an unacceptable and material breach of the relevant provisions of Security Council Resolution 687 and contained a warning to Iraq about the serious consequences which would ensue from failure to comply with its obligations. Upon the statement by the Security Council, the Special Commission on 9 January 1993 notified Iraq of the intended flights of the UNSCOM C-160 for the following days. In a letter to the President of the Security Council on the same day, the Iraqi Minister for Foreign Affairs reiterated Iraq's position.[13]

On 11 January in a Presidential Statement the Security Council demanded that Iraq co-operate fully with UNSCOM and warned Iraq of the serious con-

[10] United Nations Security Council document S/RES/707 (1991), 15 Aug. 1991.
[11] United Nations Security Council document S/25172, 29 Jan. 1993, p. 3.
[12] United Nations Security Council document S/25081, 8 Jan. 1993.
[13] United Nations Security Council document S/25086, 10 Jan. 1993.

sequences that would attend continued defiance.[14] On 13 January Iraq's Foreign Minister again refused to allow flights under the normal procedures, stating that flights would be accepted on a case-by-case basis but that Iraq could bear no responsibility for the safety of UNSCOM aircraft.[15] On 14 January the Special Commission delivered a new note to the Iraqi Foreign Ministry containing notification of flight plans for the UNSCOM C-160 aircraft for the coming days.[16] In a 15 January note from the Foreign Ministry Iraq reiterated that it would not be responsible for the safety of UNSCOM flights, and extended this condition to cover any confusion or error on the part of Iraq.[17]

The Chairman of the Special Commission informed the President of the Security Council on 16 January that the response by the Government of Iraq constituted a refusal on the part of Iraq of UNSCOM's notification because it abdicated Iraq's responsibility for ensuring the security and safety of UNSCOM personnel.[18] However, the Special Commission made further efforts to achieve a peaceful settlement by providing Iraq with a new set of flight notifications to be acknowledged in accordance with Iraq's obligations. Later the same day, Iraq rejected the request for ensuring the safety and security of the notified flights. Guarantees to this effect would only be given if the UNSCOM aircraft entered Iraqi airspace from Jordanian airspace. The Special Commission responded the same evening by stating that it could not carry out its operations using the route indicated by Iraq and that it intended to fly the direct route between its base in Bahrain and Baghdad (Habbaniyah Airfield).[19] Iraq was also notified accordingly. The following day (17 January) Iraq expressed its readiness to guarantee the safety of the flights provided the Special Commission in its turn would guarantee that Coalition aircraft (i.e., those of France, the UK and the USA) did not fly in Iraqi airspace while UNSCOM's aircraft were in the air. In a response the same day, the Special Commission stated that it was not in a position to provide the guarantees requested by Iraq.[20]

This development took place against the background of a tense situation between Iraq and the members of the Coalition. Thus on 13 January some member states of the Coalition carried out air attacks against military targets in Iraq. On 15–17 January, the military pressure on Iraq escalated until the United States on 17 January carried out an attack with cruise missiles on a major industrial installation on the outskirts of Baghdad. This installation is well-known to both the IAEA and UNSCOM as it has been subject to detailed inspection both by nuclear and missile expert teams.

[14] United Nations Security Council document S/25091, 11 Jan. 1993.
[15] See note 11, p. 4.
[16] See note 11, pp. 4, 9.
[17] See note 11, p. 4.
[18] See note 11, pp. 4, 10–11.
[19] See note 11, pp. 4, 13.
[20] See note 11, p. 14.

On 19 January, Iraq informed the Special Commission that it would allow the resumption of UNSCOM flights in accordance with established procedure and with the necessary guarantees to ensure the safety of the aircraft.[21]

It can be seen from the above that the crisis concerning the flights of the UNSCOM aircraft was brought about by Iraq's initial refusal on 7 January 1993 not to permit the use of C-160s to transport UNSCOM personnel and equipment into Iraq and the continuing and insistent obstruction despite repeated opportunities to modify its position. This was a breach of Iraq's obligations under Security Council Resolutions 687, 707 and 715—all adopted under Chapter VII of the UN Charter.[22]

## IV. UNSCOM inspection activities

### Nuclear weapons

Considering that Iraq's initial declaration in May 1991[23] under Resolution 687 stated that it had no nuclear weapons, no nuclear weapon programme, no weapon grade materials, no knowledge of or activity related to nuclear weapon subsystems, components or manufacture, and no research and development (R&D) facilities related to the production of nuclear weapons, it is quite remarkable that one and a half years after this declaration it is now proven that—with the exception of nuclear weapons—Iraq had all of the above.

All of the fresh fuel for Iraq's Russian research reactor has been transferred to Russia and transformed into enriched uranium to slightly less than 20 per cent in uranium-235. The material is stored under IAEA safeguards in a facility in Russia pending its resale. The IAEA is in the process of finding the ways and means for the removal and transportation from Iraq, and final disposal outside Iraq, of irradiated fuel under seal and verification by the IAEA. The IAEA has reported that all fuel assemblies now are accessible and can be removed without major technical difficulties. The IAEA continues to pursue its inquiry into inconsistencies in the nuclear material flow declarations. As mentioned above, the IAEA has carried out large-scale destruction of R&D facilities at Al Atheer. Furthermore, at Tarmiya the IAEA has destroyed all of the electromagnetic isotope separation (EMIS) production buildings and associated electrical-power distribution capability. Also at other EMIS-related facilities, relevant capabilities have been eliminated. Table 13.1 below lists the nuclear and other UNSCOM inspections which have been carried out as of 31 December 1992.

The IAEA considers that it is now able to draw a reasonably coherent and consistent picture of Iraq's nuclear programme, even if doubts remain as to

[21] See note 11, p. 5.

[22] See 'Chapter VII of the United Nations Charter', SIPRI, *SIPRI Yearbook 1991: World Armaments and Disarmament* (Oxford University Press: Oxford, 1991), appendix 18B, pp. 636–37.

[23] United Nations Security Council document S/22614, 17 May 1991.

**Table 13.1.** The UNSCOM inspection schedule, as of 31 December 1992

| Date and type of inspection | Team |
|---|---|
| *Nuclear* | |
| 15–21 May 1991 | IAEA1/UNSCOM1 |
| 22 June–3 July 1991 | IAEA2/UNSCOM4 |
| 7–18 July 1991 | IAEA3/UNSCOM5 |
| 27 July–10 Aug. 1991 | IAEA4/UNSCOM6 |
| 14–20 Sep. 1991 | IAEA5/UNSCOM14 |
| 21–30 Sep. 1991 | IAEA6/UNSCOM16 |
| 11–22 Oct. 1991 | IAEA7/UNSCOM19 |
| 11–18 Nov. 1991 | IAEA8/UNSCOM22 |
| 11–14 Jan. 1992 | IAEA9/UNSCOM25 |
| 5–13 Feb. 1992 | IAEA10/UNSCOM27+30 |
| 7 –15 Apr. 1992 | IAEA11/UNSCOM33 |
| 26 May–4 June 1992 | IAEA12/UNSCOM37 |
| 14–21 July 1992 | IAEA13/UNSCOM41 |
| 31 Aug.–7 Sep. 1992 | IAEA14/UNSCOM43 |
| 8–19 Nov. 1992 | IAEA15/UNSCOM46 |
| 6–14 Dec. 1992 | IAEA16/UNSCOM47 |
| *Chemical* | |
| 9–15 June 1991 | CW1/UNSCOM2 |
| 15–22 Aug. 1991 | CW2/UNSCOM9 |
| 31 Aug.–8 Sep. 1991 | CW3/UNSCOM11 |
| 31 Aug.–5 Sep. 1991 | CW4/UNSCOM12 |
| 6 Oct.–9 Nov. 1991 | CW5/UNSCOM17 |
| 22 Oct.–2 Nov. 1991 | CW6/UNSCOM20 |
| 18 Nov.–1 Dec. 1991 | CBW1/UNSCOM21 |
| 27 Jan.–5 Feb. 1992 | CW7/UNSCOM26 |
| 15–29 Apr. 1992 | CW8/UNSCOM35 |
| 21–29 Sep. 1992 | CW9/UNSCOM44 |
| 26 June–10 July 1992 | CBW2/UNSCOM39 |
| 6–14 Dec. 1992 | CBW3/UNSCOM47 |
| 21 Feb.–24 Mar. 1992 | CD1/UNSCOM29 |
| 5–13 Apr. 1992 | CD2/UNSCOM32 |
| 18 June 1992–ongoing | CDG/UNSCOM38 |
| *Biological* | |
| 2–8 Aug. 1991 | BW1/UNSCOM7 |
| 20 Sep.–3 Oct. 1991 | BW2/UNSCOM15 |
| *Ballistic missile* | |
| 30 June–7 July 1991 | BM1/UNSCOM3 |
| 18–20 July 1991 | BM2/UNSCOM10 |
| 8–15 Aug. 1991 | BM3/UNSCOM8 |
| 6–13 Sep. 1991 | BM4/UNSCOM13 |
| 1–9 Oct. 1991 | BM5/UNSCOM18 |
| 1–9 Dec. 1991 | BM6/UNSCOM23 |
| 9–17 Dec. 1991 | BM7/UNSCOM24 |
| 21–29 Feb. 1992 | BM8/UNSCOM28 |
| 21–29 Mar. 1992 | BM9/UNSCOM31 |
| 13–21 Apr. 1992 | BM10/UNSCOM34 |

| Date and type of inspection | Team |
|---|---|
| 14–22 May 1992 | BM11/UNSCOM36 |
| 11–29 July 1992 | BM12/UNSCOM40A+B |
| 7–18 Aug. 1992 | BM13/UNSCOM42 |
| 16–30 Oct. 1992 | BM14/UNSCOM45 |
| *Special missions* | |
| 30 June–3 July 1991 | . . |
| 11–14 Aug. 1991 | . . |
| 4–6 Oct. 1991 | . . |
| 11–15 Nov. 1991 | . . |
| 27–30 Jan. 1992 | . . |
| 21–24 Feb. 1992 | . . |
| 17–19 July 1992 | . . |
| 28–29 July 1992 | . . |
| 6–12 Sep. 1992 | . . |
| 4–9 Nov. 1992 | . . |

*Source*: United Nations Security Council document S/24984, 17 Dec. 1992.

whether the picture is complete. Furthermore, during the period under review efforts to implement the provisions of the cease-fire resolution with regard to destruction, removal and the rendering harmless of all nuclear-related prohibited items have continued, largely successfully. Thus, key buildings and equipment have been demolished under the supervision of the IAEA inspection teams at the Al Atheer, Tarmiya and Al Sharqat sites. All currently known nuclear-weapon-usable material has been verified and is being kept under seal awaiting removal from Iraq. In addition, numerous other materials, equipment and components have been either destroyed, removed from Iraq or placed under seal.

## Ballistic missiles

Iraq's initial declarations of its ballistic missile programme included incorrect numbers for its missile holdings, launchers and some support equipment. The original declarations did not contain information about such important elements of the programme as major missile parts, test, decoy and training missiles and launchers, and production, testing and repair equipment and facilities.

In late March Iraq declared a large number of ballistic missiles, not earlier accounted for, along with certain associated equipment. This information, which meant that the assessed number of destroyed missiles had to be radically revised upwards, signified something of a change in the previously uncooperative and confrontational posture of Iraq.

Upon learning in March 1992 that UNSCOM had photographic proof that its declarations were false, Iraq increased from 52 to 144 its declaration of the Scud missiles and Scud variants held at the end of the war. All the missiles thus accounted for have been destroyed by Iraq either under the supervision of UNSCOM or secretly without international control. In subsequent inspections UNSCOM has been able to verify that the destruction of all identified missiles has indeed been carried out. The number of declared fixed launchers has increased over time from 30 to 53, of mobile launchers from 10 to 19, of conventional missile warheads from 23 to 113, and of chemical missile warheads from 30 to 75.

In 1992 UNSCOM inspections and analytical activities increasingly concentrated on production capabilities such as reverse engineering and modification of Scud missiles by extending the range and rocket fuel capabilities. This has radically improved UNSCOM's understanding of the ballistic missile programme. Several inspections included seminar-type meetings with Iraq's weapon experts in order to resolve complicated questions arising from inspection activities. During this process the Special Commission obtained detailed information relating to the scope and extent of programmes to acquire or produce prohibited ballistic missiles and components, including data on previously undisclosed projects for computer support and missile fuel production. No evidence was found that Iraq had the capability indigenously to produce fuel for prohibited ballistic missiles. The interrelationship between the various projects in the ballistic missile programme and the involvement of different organizations in the programme has been mapped out. Despite Iraq's resistance, important information on foreign involvement in certain aspects of the programme was acquired.

### Chemical weapons

Gradually during 1992 Iraq admitted possession of additional chemical weapons. Thus Iraq's possession of chemicals at the end of the Persian Gulf War amounted to 150 000 filled and unfilled munitions, 300 tonnes of bulk agent and 3000 tonnes of precursor chemicals. The agents which Iraq possessed were mustard gas agent, the nerve agents GB and GF (and about 70 tonnes of GA) as well as small research quantities of three other nerve agents. Iraq's facilities include the substantial CW production complex Al Muthanna and three CW-related production plants in the vicinity of Al Fallujah. In addition to the central storage of filled chemical munitions, warfare agents and precursor chemicals in bulk at the Al Muthanna facility, filled chemical munitions, often damaged and leaking, were stored at various sites throughout the country. Those which were judged safe to move were transported to Al Muthanna, which has been designated as the central location for CW destruction. Those which could not be moved, a limited number of 122-mm rockets, were destroyed through explosive demolition incineration.

As a result of its chemical inspection programme, UNSCOM now has considerable information on Iraq's chemical agents and munitions. The munitions include various kinds of aerial bombs, mortar bombs, artillery shells and rockets, rocket-propelled grenades and 75 Scud (Al Hussein) missile chemical warheads, of which 45 were destroyed unilaterally by Iraq, destruction that subsequently was verified by UNSCOM. However, Iraq's continued refusal in its declarations to admit the use during the 1980s of chemical weapons against Iran and internally against the Kurdish population makes it impossible for UNSCOM to establish a material balance of chemical weapons in Iraq and thus to identify fully Iraq's CW programme.

The destruction of mustard gas agent is carried out in an incinerator specially built by Iraq to meet UNSCOM requirements and commissioned in January 1993. Precursor chemicals and missile-related chemicals, which have been moved to Al Muthanna, and various other chemicals found at Al Muthanna will also be incinerated. The nerve agents GB and GB/GF mixtures are being destroyed by controlled hydrolysis in a plant constructed by Iraq to UNSCOM specifications, commissioned in September 1992. After the aqueous waste has partially evaporated, cement will be added. The concrete blocks produced will then be buried on site.

Most 122-mm rockets are assessed as being too dangerous to drill and drain; they are therefore destroyed by a combination of high-temperature incineration and simultaneous explosive opening.

In order to minimize the danger of exposure to CW agents a health and safety regime has been set up by establishing remote agent detector arrays at the hydrolysis plant and at the rocket destruction site. As yet no downwind hazard has been recorded.[24]

## Biological weapons

The inspection activities related to Iraq's BW capability, which initially focused on the major R&D site at Salman Pak, have been dispersed to a number of additional sites. Conclusive evidence that Iraq was engaged in a military BW research programme has been collected. The facility at which this programme was carried out was unilaterally destroyed by the Iraqi authorities. UNSCOM was therefore prevented from obtaining detailed information on the programme. No evidence of an actual weapon programme has been found, but the inspections have provided a sound data base for future monitoring of BW capabilities in Iraq. Undeclared sites known to have been related to BW research have been inspected without any new information having been obtained by the UNSCOM inspectors.

## Inspection developments

The Special Commission and the IAEA have developed innovative inspection procedures. The document inspection in September 1991, during which the parking lot incident took place,[25] proved very successful by generating a wealth of information on the Iraqi nuclear weapon programme and to a lesser

[24] United Nations Security Council document S/24984, 17 Dec. 1992.
[25] United Nations Security Council document S/23122, 8 Oct. 1991.

degree on Iraqi missile activities. The translation and analysis of approximately 60 000 pages have not yet been concluded. A number of inspections carried out by UNSCOM and the IAEA to find additional documents were not successful. Apparently as a result of earlier experiences, on occasion Iraq has taken forceful countermeasures, the most serious of which led to the incident at the Ministry of Agriculture. On another occasion in December 1992 Iraqi personnel removed documents from a building in Baghdad during an inspection. An innovation which proved rather successful was the seminar-type discussions held between the inspection team and its Iraqi counterparts in the context of an inspection activity. Important clarifications of Iraq's weapon programmes have frequently been made during these exchanges. The use of helicopters for aerial inspection has given important corroborating information.

In January 1993 the Special Commission introduced continuous inspection by deploying an interim monitoring team in Baghdad. A small group of inspectors, specialized in missile technology, has been given the task of closely following the work at a facility for R&D related to missiles in order to ascertain that no missiles of the type forbidden by the cease-fire resolution are being developed. Early indications are that such an approach can facilitate the work of the Special Commission.

Through the strengthening of the information assessment unit, the Special Commission has considerably increased its ability to make good use of the inspection reports and other information made available to it in the UNSCOM operations.

## V. Conclusion

It goes without saying that the complex and sometimes intrusive activities carried out in Iraq by UNSCOM and the IAEA would be unnecessary if Iraq were to change its policy towards the Special Commission and adopt a genuinely co-operative and forthcoming attitude. Iraq has failed to substantiate the information provided about its prohibited programmes—information which has not infrequently proven to be misleading in character. The Special Commission has repeatedly urged Iraq to provide access to authentic documents that would substantiate the data provided by Iraq. Iraqi authorities have claimed that they destroyed all documents related to prohibited activities after the adoption of Resolution 687 and that no records have been kept of the destroyed documents. The Special Commission has difficulty in accepting this claim. On rare occasions Iraq has produced documents to support data it has provided. It is necessary that Iraq meet the long-standing requirement for credible and verifiable data on all of its prohibited programmes.

The Government of Iraq has issued an order that certain types of documents must be protected from inspection by UNSCOM, including by removing them from the sites under inspection. Inspection teams have visited a number of sites which have been sanitized. Within the context of the declarations it has submitted, the Government of Iraq has stated that it declines to divulge information indicating the names of foreign companies from which it has purchased equipment and materials.[26] It is alleged that this decision has been taken on moral grounds. Although the Commission is in possession of some evidence of procurement through elaborate third-party arrangements, the picture of Iraq's supplier network is far from complete. Accurate and full information about Iraq's foreign procurement networks and suppliers is essential if the Special Commission and the IAEA are to establish a complete, coherent and credible picture of Iraq's programmes for weapons of mass destruction as they existed in January 1991 and to decide in a realistic manner whether all of the proscribed items have been accounted for.

The plans for future monitoring and verification of Iraq's non-acquisition of proscribed weapons were approved by the Security Council in Resolution 715. Although the resolution was adopted under Chapter VII of the UN Charter, and is thus enforceable and obviously binding on all UN members, Iraq has challenged the Security Council by refusing to accept it. The position of the Government of Iraq, as outlined by its Deputy Prime Minister, is that the plans approved by the Security Council are unlawful. Iraq's position is that it may possibly accept the technical elements of the plans, but not the general provisions of the plans and the resolution. The general provisions grant to the Special Commission far-reaching rights as regards authority to carry out on-site inspections of Iraq and aerial surveillance without any limitations as to site or area. The Chairman of the Special Commission declared on repeated occasions in Security Council meetings in March and November 1992 and in meetings with representatives of the Government of Iraq that, provided Iraq demonstrates a forthcoming attitude with regard to monitoring and verification, UNSCOM will exercise its rights under Resolution 715 with due regard to the dignity and national sovereignty of Iraq.[27] The Special Commission has expressed concern that it is impossible for it to commence the full monitoring and verification of Iraq's dual capabilities if it is not allowed to exercise its rights under Resolution 715. Iraq can use any pretext for failing to co-operate if it is allowed to maintain that the provisions of Resolution 715 do not apply to it. The non-recognition by Iraq of Resolution 715 is a major obstacle to quick development of the implementation of the cease-fire arrangements, including the plans for monitoring Iraq's weapon capability.

Together with the continuing refusal by Iraq to provide UNSCOM and the IAEA with full information on its foreign procurement, progress in the implementation of the relevant resolutions will be very slow. In spite of the open Iraqi challenge of the Security Council and its resolutions there is no sign of a weakening of the resolve of the Security Council members to demand full and complete implementation of all of the relevant resolutions.

[26] See note 24; United Nations Security Council document S/24002, 26 May 1992.
[27] See note 24.

# 14. The Chemical Weapons Convention: the success of chemical disarmament negotiations

J. P. PERRY ROBINSON, THOMAS STOCK and RONALD G. SUTHERLAND

## I. Introduction

The Convention on the Prohibition of the Development, Production, Stockpiling and Use of Chemical Weapons and on their Destruction was opened for signature in Paris on 13 January 1993.[1] After the completion of the negotiations at the Conference on Disarmament (CD) in Geneva on 3 September 1992, the final text of the Chemical Weapons Convention (CWC) was sent to the United Nations General Assembly (UNGA), where a draft resolution commending the CWC was adopted by the First Committee on 12 November. This was followed by UNGA's adoption of the draft resolution on 30 November, without vote.[2] The work of more than two decades of negotiations and deliberations had come to an end. It was far from obvious in 1968, when the idea of a convention banning chemical weapons (CW) was conceived, that there would be any real chance of achieving success. Even as late as the beginning of 1992 the draft CWC text contained many footnotes and brackets, marking areas of disagreement.

Several developments led to finalization of the CWC. First, the end of the cold war increased mutual trust and confidence among states. Other positive factors included changes in the US negotiating position, the collapse of the Soviet Union, the outcome of the 1991 Persian Gulf War (which clearly demonstrated that chemical weapons are no longer politically desirable), and not least the clear political will of the majority of states to totally prohibit chemical weapons.

The CWC is an historic agreement, banning all chemical weapons worldwide, imposing a wide spectrum of inspections to verify the ban, outlawing any use of these weapons (a goal previously established by the 1925 Geneva Protocol) and imposing a strict ban on all activities to develop new chemical weapons.

[1] In August 1992 France confirmed its earlier invitation to host a signatory conference in Paris at the beginning of 1993. The United Nations General Assembly under Resolution A/47/39 on 30 Nov. 1992 at the 74th plenary meeting adopted the report of the First Committee on Chemical and Bacteriological (Biological) Weapons including the draft resolution on the Convention (see United Nations General Assembly, A/47/690, 25 Nov. 1992, pp. 6–8). Under Paragraph 2 the Secretary-General, as Depositary of the Convention was requested to open it for signature in Paris on 13 Jan. 1993.

[2] The resolution was adopted by consensus on 30 Nov. 1992, co-sponsored by 144 nations.

The Convention is a major disarmament and arms control achievement constructed as a balance between the rights and obligations of States Parties and between benefits and costs. It is also a multilateral agreement where States Parties will benefit from a better security environment and from sharing the cost for facilitating compliance monitoring and verification activities. According to the last chairman of the *Ad Hoc* Committee on Chemical Weapons, Ambassador Adolf Ritter von Wagner of Germany,[3] the achievements of the CWC are the following:[4]

1. Article I, General Obligations, and the Preamble outline the total ban on chemical weapons and all of the activities which are prohibited by the CWC. A non-discriminatory approach has been achieved by applying the general obligations to each State Party on an equal basis.
2. The Convention balances substantial verification, by the use of provisions providing sufficient deterrence against any potential violator, with protection of national security interests, by application of a mechanism which allows suspicions to be transformed from a bilateral concern to a multilateral verification undertaking.
3. The rights of individual States Parties are also balanced against multilateral Convention obligations by applying verification procedures in such a way that they do not interfere with national security concerns unrelated to the CWC.
4. There is a balance between the interests of industrial and less developed states, which have expressed interest in promoting increased co-operation under the Convention, by the obligation that States Parties review any restrictive measures, including export controls, in the field of chemical industry with the aim of removing such restrictions for States Parties which are in full compliance with their Convention obligations.
5. The difficult problem of membership on the Executive Council was solved by allocating seats to different regional groups. There was awareness that the majority of States Parties will be less developed countries. The interests of the industrialized countries were served by introducing the approach of so-called industrial seats (i.e., in each group countries with the most significant national chemical industries will be given special consideration).
6. The Convention also addresses the difference between CW-possessor states and non-possessor states by limiting the destruction period to 10 years (in exceptional cases an extension may be granted). Possessor states are obliged to share destruction and verification costs. Any extension of the destruction period must be compensated by greater openness and an increased number of inspections.

The Convention takes an approach to balancing national and multinational costs and benefits which is unique in the history of disarmament. On the one

[3] Ambassador von Wagner was appointed by the CD as Chairman of the *Ad Hoc* Committee on Chemical Weapons at its plenary meeting on 21 Jan. 1992 and served as its last chairman. The committee held 32 meetings from 24 Jan. to 26 Aug. 1992.

[4] Conference on Disarmament document CD/1173, 3 Sep. 1992, pp. 37–39.

hand, individual States Parties must provide declarations, adopt general measures for disclosure, open their chemical industry, accept the rules for challenge inspections and pay costs related to the Convention. On the other hand, they benefit by increased security, confidence and international behaviour. There are other benefits including better prospects for trade in chemical products and technology, specific protection against chemical weapons and a provision for international assistance.

The CWC does not solve all of the problems related to verification, improvement of future international co-operation and exchange of chemical products, technology and know-how. The consensus which was ultimately achieved involved compromise on the part of many states.

Verification under the Convention relies on several types of inspection methods and monitoring activities. The final agreed procedure for challenge inspections to resolve concerns about possible non-compliance and to re-establish confidence among States Parties is mandatory on-site inspection any time, anywhere. Procedures and rules to govern the conduct, timing and carrying out of such inspections by the use of various techniques, the clearly defined role of the International Inspectorate and the participation of the challenged state in the inspection process constitute a framework in which misuse of inspection will be minimized.

The two years following the Paris Conference will be crucial to implementation of the CWC in two ways. First, although 65 states may have ratified the Convention by 1995 (a prerequisite for its entry into force),[5] the goal is the broadest possible adherence. States Parties which have already stated that they are not prepared to sign should perhaps review this position. Second, the work of the Preparatory Commission will start immediately in The Hague.[6] The Preparatory Commission is responsible for the organization of the Convention and also assumes major tasks and responsibilities for developing all of the procedures for the Organization for the Prohibition of Chemical Weapons (OPCW). States Parties which are still uncertain about joining the Convention may be influenced by the desirability of participation in the early organization of this process.

If the Convention enters into force in 1995, States Parties must destroy their chemical weapons by the year 2005, or by 2010 at the latest. Destruction will be a costly undertaking. States Parties which possess chemical weapons must be prepared to pay destruction costs that will be over 10 times greater than the cost of producing these weapons.[7]

This chapter is a first attempt to analyse the CWC, to provide information about the destruction obligations including the verification provisions, and to illustrate the interplay between national implementation obligations and international compliance measures. A short historical overview indicates how it

[5] See Article XXI, Entry into Force, in Conference on Disarmament document CD/1173, 3 Sep. 1992, Appendix I, p. 50.

[6] In June the *Ad Hoc* Committee on Chemical Weapons agreed that The Hague will be the seat of the Organization for the Prohibition of Chemical Weapons. Belgrade, Geneva and Vienna had also been considered.

[7] See chapter 7 in this volume.

was possible to achieve the Convention. In addition, the next steps after signing are discussed. Appendix 14A reproduces the text of the CWC, and table 14.1 lists the signatory status as of 8 February 1993 (the opening of the first meeting of the Preparatory Commission in the Hague).

## II. Historical overview of the CWC negotiation

### Introduction

The CWC is the product of two primary influences. On the one hand, it stands squarely within a normative tradition: it is the latest expression of what is evidently an ancient sentiment widespread throughout many cultures, that fighting with poison is somehow reprehensible, immoral, wrong—that to resort to chemical warfare is to violate a taboo of a peculiarly deep kind, a taboo that can nevertheless condone other categories of weapon, even ones capable of inflicting the most hideous injuries. This social norm had previously found its fullest expression in the 1925 Geneva Protocol, the treaty to which some 150 states are now parties which prohibits 'the use in war of asphyxiating, poisonous or other gases, and of all analogous liquids, materials or devices', and which also prohibits 'bacteriological methods of warfare'. Earlier prohibitions of toxic warfare occur in Article 23(a) of the Hague Regulations of 1907 and 1899, in the Hague Gas Projectile Declaration of 1899, and in article 13(a) of the Brussels Declaration of 1874.

On the other hand, the CWC is a security agreement, a form of collective protection against a particular type of threat. The Convention clearly reflects the way states have assessed that threat—how they have judged the military and political usefulness of toxic weapons both to themselves and to potential adversaries. The course of the negotiation is likewise explicable in terms of how those assessments varied from state to state over the quarter-century of intergovernmental talks.

### Security aspects of toxic weapons

The military usefulness seen for toxic weapons during the decades after World War II may be summarized as follows. In contrast to their predecessors, the new 'nerve gases' were so powerful, it seemed, that a little could go a long way, meaning that to acquire nerve gas weapons could be to gain a force-multiplier. The 'human wave' assaults which UN forces faced in Korea were a pressing reminder of the value of weapons which would allow a small force to prevail over a much larger one. It was true that antichemical protection in the form of gas masks, special clothing, detectors, decontaminants and medical antidotes might negate such value, but not all potential adversaries would be protected. Besides this there were the morale effects offered by poison gas: the huge psychological impact of the unseen killer, and the terror it could instil

**Table 14.1.** Signatory status of the Chemical Weapons Convention as of 8 February 1993

*Signed in Paris*

| | | |
|---|---|---|
| Afghanistan | Gabon | Netherlands |
| Albania | Gambia | New Zealand |
| Algeria | Georgia | Niger |
| Argentina | Germany | Nigeria |
| Australia | Ghana | Norway |
| Austria | Greece | Pakistan |
| Azerbaijan | Guatemala | Papua New Guinea |
| Bangladesh | Guinea | Paraguay |
| Belarus | Guinea-Bissau | Peru |
| Belgium | Haiti | Philippines |
| Benin | Holy See | Poland |
| Bolivia | Honduras | Portugal |
| Brazil | Hungary | Romania |
| Brunei Darussalam | Iceland | Russian Federation |
| Bulgaria | India | Samoa (Western) |
| Burkina Faso | Indonesia | San Marino |
| Burundi | Iran | Senegal |
| Cambodia | Ireland | Seychelles |
| Cameroon | Israel | Sierra Leone |
| Canada | Italy | Singapore |
| Cape Verde | Japan | Slovak Republic |
| Central African Republic | Kazakhstan | Slovenia |
| Chile | Kenya | South Africa |
| China | Korea, South | Spain |
| Colombia | Liberia | Sri Lanka |
| Comoros | Lithuania | Sweden |
| Congo | Luxembourg | Switzerland |
| Cook Islands | Madagascar | Tajikistan |
| Costa Rica | Malawi | Thailand |
| Côte d'Ivoire | Malaysia | Togo |
| Croatia | Mali | Tunisia |
| Cuba | Malta | Turkey |
| Cyprus | Marshall Islands | Uganda |
| Czech Republic | Mauritania | Ukraine |
| Denmark | Mauritius | United Kingdom |
| Dominican Republic | Mexico | United States |
| Ecuador | Micronesia | Uruguay |
| El Salvador | Moldova | Venezuela |
| Equatorial Guinea | Monaco | Viet Nam |
| Estonia | Mongolia | Zaire |
| Ethiopia | Morocco | Zambia |
| Fiji | Myanmar (formerly Burma) | Zimbabwe |
| Finland | Namibia | |
| France | Nauru | |

**Table 14.1** *contd*

| | | |
|---|---|---|
| *Signed in New York* | | |
| Kuwait | Qatar | Yemen |
| Nepal | Saudi Arabia | |
| Oman | United Arab Emirates | |
| *Non-signatories* | | |
| Angola | Jamaica | St Vincent and the Grenadines |
| Antigua and Barbuda | Jordan | Sao Tome and Principe |
| Armenia | Korea, North | Solomon Islands |
| Bahamas | Kyrgyzstan | Somalia |
| Bahrain | Laos | Sudan |
| Barbados | Latvia | Suriname |
| Belize | Lebanon | Swaziland |
| Bhutan | Lesotho | Syria |
| Bosnia and Herzegovina | Libya | Taiwan |
| Botswana | Liechtenstein | Tanzania |
| Chad | Maldives | Trinidad and Tobago |
| Djibouti | Mozambique | Turkmenistan |
| Dominica | Nicaragua | Uzbekistan |
| Egypt | Panama | Vanuatu |
| Grenada | Rwanda | Yugoslavia |
| Guyana | St Kitts and Nevis | |
| Iraq | St Lucia | |

*Source*: 'Update of chemical weapons treaty signatories/parties', *Wireless File*, no. 63 (United States Information Service, US Embassy: Stockholm, 5 Apr. 1993), pp. 6–7.

into military as well as civilian populations untrained and unequipped to protect themselves against it. Toxic weapons were seen to have value, then, for economy of force against an unprotected enemy or for terrorization; weapons that after all might not be so specialized in their usefulness that occasion for exploiting them in preference to more conventional means would rarely arise.

Such utilities were in principle precluded by the 1925 Geneva Protocol and associated international law, the practical effect of which was to prohibit at least the first use of chemical and biological warfare (CBW) weapons. However, the emerging dogma of deterrence gave them cover, for the spread of deterrence ideas had made people more inclined to believe that it was not the military shortcomings of toxic warfare which had kept it out of World War II, or the rather low level of institutional preparedness for it, but the threat of retaliation in kind. People could be persuaded, in other words, that the development and stockpiling of CBW weapons was an act of common prudence.

Such mixtures of motives undoubtedly impelled the rich-country chemical armament efforts of the post-1945 decades: the British nerve gas programme that ran until the mid-1950s, and the French one until the 1960s; the Soviet one, to which President Gorbachev called a halt in 1987; and the US one, finally collapsing in 1990. What brought them to an end was, most probably, much the same in all four cases: (*a*) a slowly building realization that the promise held out by those wartime discoveries was actually a false promise;

(*b*) a realization that scientific developments might not, after all, be capable of overcoming the inherent technical limitations of toxic warfare to the point where its weapons had more than marginal utility; and (*c*) an awareness that there was more which they could gain by giving the weapons up than by continuing to retain and develop them. Possessing a huge range of other armaments, their need for additional weapons of terror or force-multiplication was hardly overwhelming. Nerve gas might add something to their overall deterrent postures, but only at cost of implying diminished resolve to use nuclear weapons *in extremis.*

Although CBW capability might today be thought to have rather low value to rich countries of the North, the same was not necessarily the case for countries of the South, particularly where it was believed that deterrence relationships could keep the peace. In a world of sharpening North–South polarization, it was well into the 1980s before it was realized that the countries whose armed forces stood to benefit most from the new nerve gas and related technologies might not be at all the same as the countries which were doing the pathfinding research and development. One or another of those CBW-pathfinder countries might find themselves in military confrontation with a less developed country, in which case widespread availability of powerful new force-multipliers would hardly be in their best interests. If it was really true, as some commentators were claiming, that CBW armament had no important part to play in the high-technology, heavily militarized, East–West confrontation, then it would be better to get out of the technologies altogether and if possible to seek their suppression. As a policy objective, such counter-proliferation made excellent security sense for rich countries of East and West alike. The problem would be to persuade key less developed countries that it could be in their best interests also.

## Entry of CBW into the agenda of the Geneva disarmament conference

The idea that renouncing CBW weapons might bring more benefit to the national security than keeping the weapons could began to surface slowly in countries of the North during the latter 1960s. The East–West arms talks that had been proceeding in various fora since the Korean War, chiefly it seemed for purposes of cold war rhetoric, provided occasion for the debate, especially after they had passed into a phase of genuine negotiation. Before that phase began—before, that is to say, the United States and the Soviet Union had agreed to join forces in pursuing a partial nuclear-weapon test ban and then a nuclear Non-Proliferation Treaty (NPT)—CBW weapons had occasionally been mentioned, but generally only for the purpose of vilifying the cold war adversary. This changed somewhat in 1968 when the two superpowers accepted the proposal of Sweden that CBW weapons should be placed on the agenda of the Geneva multilateral disarmament conference. The background to the Swedish proposal was the resort to poison gas warfare by Egypt in the Yemen and, more conspicuously, the upsurge in the chemical warfare which the United States, not yet a party to the Geneva Protocol, was conducting with herbicides and irritants in Viet Nam.

Perhaps with a view to taking the anti-American sting out of the talks, the United Kingdom almost immediately proposed that biological weapons be considered ahead of chemical weapons, later in July 1969,[8] tabling a draft biological disarmament treaty. The US Government, meanwhile, had been subjecting the country's CBW armament programmes to their first full-scale interdepartmental review for a decade. In November 1969 President Richard M. Nixon announced that his Government had decided to join the 1925 Geneva Protocol, to close down the US biological weapon (BW) programme, and to associate itself with the aims and objectives of the British draft treaty. In this President Nixon, advised by Henry Kissinger, was doing at least two things at once, so the public record now shows. He was signalling to the Soviet Union that the United States was serious about arms control, which then meant SALT, the Strategic Arms Limitation Talks. He was also acting upon the aforementioned evaluation of biological warfare: that there was no point in putting the huge resources of US technical genius into a technology that could provide poor weak countries with cheap powerful force-multipliers. The fact that biological warfare had this potential was camouflaged—remarkably successfully—behind a depiction of biological weapons as unreliable and militarily useless, a depiction sedulously propagated not least by the negotiators in Geneva.

On biological weapons the Soviet leadership no doubt had an agenda as multifarious as that of the USA. How seriously in fact it contemplated closure of its biological armament programmes, and whether it had a single or several competing views on the subject, are still unclear. One may observe only that in 1992 the Russian leadership was saying publicly that the programmes had still been continuing, albeit in diminished state, when the Soviet Union itself came to an end.[9] Be that as it may, the Soviet Union and the United States had by August 1971 agreed the text of a Biological Weapons Convention (BWC) that modified the British draft in certain key respects—in fact drastically weakening it—and the Soviet Union had, to the dismay of the non-aligned countries at the conference, withdrawn its opposition to biological weapons being treated separately from chemical ones. The UNGA duly endorsed the Convention, which was thereupon opened for signature on 10 April 1972. The BWC entered into force three years later, on 26 March 1975, a fortnight before the United States formally became party to the Geneva Protocol.

A concession made to those who opposed the separation of biological and chemical weapons during the 1969–71 negotiation is to be found in Article IX of the BWC, which reads as follows:

> Each State Party to this Convention affirms the recognized objective of effective prohibition of chemical weapons and, to this end, undertakes to continue negotiations in good faith with a view to reaching early agreement on effective measures for the

[8] United Kingdom, Eighteen-Nation Disarmament Committee document ENDC/255, 10 July 1969.
[9] See chapter 7 in this volume.

prohibition of their development, production and stockpiling and for their destruction, and on appropriate measures concerning equipment and means of delivery specifically designed for the production or use of chemical agents for weapons purposes.

The Geneva disarmament conference was accordingly obliged to retain chemical weapons on its agenda; which it did. A succession of drafts for the projected CWC began to emerge, led in March 1972 by one from the Soviet Union and its allies modelled on the BWC.[10]

## The exploratory talks on chemical weapons

The first important step towards agreement, although not widely recognized as such at the time, came in April 1973 with the outline draft convention put forward jointly by the neutral and non-aligned countries represented at the conference save India and Pakistan.[11] The draft proposed a declarations-based international control system for a ban on chemical weapons comprehensively defined, thereby opening up for what proved to be constructive and sustained international discussion the delicate issue of verification, on which the USA and the USSR had entrenched themselves behind apparently irreconcilable positions.

The next important milestone came in July 1974: the unexpected Nixon–Brezhnev communiqué from the US–Soviet summit meeting in Moscow envisaging a 'joint initiative' focused on 'the most dangerous, lethal means of chemical warfare' to be submitted in due course to the Geneva conference. The United States and the Soviet Union were thus promising Geneva a leadership on the issue that could hardly be gainsaid. The effect was to freeze multilateral activity, which by now included consideration of a draft convention proposed by Japan[12] and much attention to the possibilities for verifying non-production of chemical weapons in the civil chemical industry. However, nothing was heard of the joint initiative for two years and more, until after the USA's closest ally at the conference had broken ranks and put forward a draft convention of its own. This was the British draft of August 1976,[13] apparently timed to stimulate whatever new administration might shortly be taking command in Washington.

If that was indeed the main purpose of the British draft, it succeeded. The bilateral US–Soviet working group promoted by the Carter Administration in March 1977 to work on the joint initiative subsequently had a lasting and profound effect on the way the talks evolved. By the time its work had stopped in 1980—cut short by the demise of *détente* and the onset of a new cold war—

[10] Bulgaria, Czechoslovakia, Hungary, Mongolia, Poland, Romania, USSR, Conference of the Committee on Disarmament document CCD/361, 28 Mar. 1972.

[11] Argentina, Brazil, Burma, Egypt, Ethiopia, Mexico, Morocco, Nigeria, Sweden and Yugoslavia, Conference of the Committee on Disarmament document CCD/400*, 26 Apr. 1973.

[12] Japan, Conference of the Committee on Disarmament document CCD/420, 30 Apr. 1974.

[13] United Kingdom, Conference of the Committee on Disarmament document CCD/512, 6 Aug. 1976.

it had entrenched such key ideas as precursor control, convention oversight through a consultative committee of states parties with a permanent secretariat, verification by challenge (though not yet with mandatory inspection), the single small-scale facility, and the use of systematic international on-site inspection for verification in a routine mode.[14]

There had in the meanwhile been moves within the Geneva conference, by then restructured as the Committee on Disarmament under rotating chairmanship of its members (increased in 1980 from 35 to 40 states), to intensify multilateral attention to chemical weapons. A key document in this process was a questionnaire distributed by The Netherlands in July 1979 which sought, and in many cases subsequently obtained, the written views of other participants about the projected CWC.[15] Poland submitted a draft outline for the convention.[16] However, without constructive US and Soviet involvement, rather little could be achieved. The principal development was the decision in March 1980 to establish an *Ad Hoc* Working Group on Chemical Weapons to 'define, through substantive examination, issues to be dealt with in the negotiation' on the convention.[17]

## The start of negotiations on chemical weapons

At the beginning of 1984, as an early sign of improving superpower relations, the Conference on Disarmament (as it came to be called from February 1984 onwards) agreed that it should now move away from exploratory discussion and start its 'final elaboration' of a CW ban, mandating its *Ad Hoc* Committee on Chemical Weapons accordingly.[18] The negotiation gradually got under way, impelled partly by the submission from the United States of a new draft convention,[19] but more especially by reaction to the verification by the UN Secretary-General that Iraq had been using chemical weapons against Iran.[20] As consensus on more and more of the content of the projected CWC developed, it was registered every six months in the so-called 'rolling text'—a progressively expanding draft convention whose outline was first agreed in 1984.[21] Bilateral US–Soviet talks on the CW prohibition also resumed that year, but this time in the margin of the multilateral talks.

[14] The Geneva conference received three progress reports from the bilateral talks culminating in the joint USA–Soviet paper Conference on Disarmament document CD/112, 7 July 1980.

[15] The Netherlands, Conference on Disarmament document CD/41, 25 July 1979. Several of the responses took the form of non-papers.

[16] Poland, Conference on Disarmament document CD/44, 26 July 1979.

[17] 'Decision adopted at the 69th plenary meeting held on 17 March 1980', Conference on Disarmament document CD/80, 17 Mar. 1980.

[18] 'Decision on the re-establishment of an *ad hoc* subsidiary body on chemical weapons', Conference on Disarmament document CD/440, 28 Feb. 1984.

[19] USA, Conference on Disarmament document CD/500, 18 Apr. 1984.

[20] United Nations General Assembly document S/16433, 26 Mar. 1984.

[21] The rolling texts were appended to the formal end-of-session reports to the CD from its *Ad Hoc* Committee (formerly Working Group) on Chemical Weapons, the chairmanship of which rotated each year through each of the three main political groups of delegations. The rolling texts submitted by the successive chairmen are contained in the following Conference documents: CD/539, 28 Aug. 1984 (Ekéus of Sweden); CD/636*, 23 Aug. 1985 (Turbanski of Poland); CD/734, 29 Jan. 1987 (Cromartie of the UK); CD/795*, 2 Feb. 1988 (Ekéus of Sweden); CD/881, 3 Feb. 1989 (Sujka of Poland); CD/961,

In 1986 came the first solid token of Western commitment to the goal of chemical disarmament. There began then a succession of international meetings between representatives of the civil chemical industry, on the one hand, and the negotiating diplomats on the other, climaxing in September 1989 at the Government–Industry Conference Against Chemical Weapons which the Australian Government convened in Canberra.[22] Non-governmental organizations in the forms of SIPRI and Pugwash had worked to promote these contacts during the exploratory talks phase in the 1970s and early 1980s.

Then, in August 1987, speeches by Soviet Foreign Minister Eduard Shevardnadze and his ambassador in Geneva[23] opened up the Soviet negotiating position, announced *glasnost* in the Soviet military chemical programmes, and accepted the proposal for mandatory challenge inspection and other such intrusive verification measures which the United States had put forward in its 1984 draft convention. There were those who saw this Soviet action simply as the calling of a US bluff, for it had seemed, even to the less cynical observers, as though the chief function of the US draft had been to justify to the US Congress the quest for authority and funding to modernize the US CW arsenal (by acquiring the newly developed 'binary munitions') on which the Reagan Administration had embarked at its commencement. Such modernization, it had been argued, would provide the country with powerful bargaining chips. There were many people in Washington, undoubtedly, who saw the talks in this way, just as there were many people who argued that no CWC could ever have sufficient verifiability to be acceptable.

It took Washington until July 1989 to decide an appropriately responsive policy, by which time East–West relations were thawing very rapidly. The essence of the policy was stated by President George Bush at the UN General Assembly two months later.[24] No longer would the United States judge the acceptability of chemical arms control in terms of whether it was or was not verifiable. Instead, it would seek a 'level of verification that gives us confidence to go forward'.

President Bush was thus signifying a new suppleness to the US negotiating position. The old criterion—that no agreement on chemicals could be worthwhile if compliance with it could not be assured—was to be superseded, replaced by an altogether different means of assessment. The judgement would now be a relative one: would the USA feel more confident in a world regulated by the particular chemical warfare arms control package on offer or

1 Feb. 1990 (Morel of France); CD/1046, 18 Jan. 1991 (Hyltenius of Sweden); CD/1116, 20 Jan. 1992 (Batsanov of Russia); and CD/1170, 26 Aug. 1992 (von Wagner of Germany). They can also be found in the annual reports of the Conference on Disarmament to the UN General Assembly. See also the proto-rolling texts: CD/131/Rev.1, 4 Aug. 1980 (Okawa of Japan); CD/220, 17 Aug. 1981 (Lidgard of Sweden); CD/334, 15 Sep. 1982 (Sujka of Poland); and CD/416, 22 Aug. 1983 plus CD/429, 7 Feb. 1984 (McPhail of Canada).

[22] SIPRI, *SIPRI Yearbook 1990: World Armaments and Disarmament* (Oxford University Press: Oxford, 1990), pp. 535–38 and p. 544.

[23] Conference on Disarmament document CD/PV.428, 6 Aug. 1987, pp. 10–11; Conference on Disarmament document CD/PV.429, 11 Aug. 1987, pp. 2–7.

[24] *SIPRI Yearbook 1990* (note 22), p. 532.

in the world of the status quo? Would the USA, in other words, be better off inside than outside the proposed convention regime?

The importance of this change lay in its express recognition of what had always been the case, that no international ban on chemical weapons could ever be fully verifiable. This was so because a great many chemicals and chemical manufacturing technologies can be used as well for chemical warfare as for peaceful purposes—'dual use' attributes first demonstrated so vividly during World War I, and being demonstrated once again by the disclosures of exactly how Iraq had acquired its chemical weapons. States could hardly deny themselves such technologies altogether; and, though they might renounce any intention of ever exploiting their chemical industries for CW purposes, no amount of international inspection could generate complete confidence that their renunciations were in fact being adhered to. By continuing to negotiate for a CWC, governments were therefore declaring that a worthwhile convention might nevertheless be attainable: that somewhere along the continuum ranging from 0 to 100 per cent confidence in compliance, an acceptable balance of benefit over risk could perhaps be struck. President Bush was now saying, in effect, that the US Government was no exception.

Somewhat overshadowing this major development was another feature of the President's new policy. He also told the UN General Assembly in that September 1989 address that the United States would want the multilateral chemical convention to allow it to retain 2 per cent of its stockpile until such time as 'all nations capable of building chemical weapons' had become parties. A throw-back to an earlier (and abandoned) French proposal[25] to allow states to retain small 'security stocks' to tide them over the supposed dangers of the transitional period, this bizarre idea at least replaced an even more problematic negotiating position, namely that the United States would not join the convention at all until all other 'chemical capable' states had done so. Since a consequence of the proposal would surely be to stimulate proliferation, it was rather widely seen, not as a real negotiating position, but as a domestic political accommodation that would be reconsidered once the negotiations had advanced further. This is indeed what happened.

There was a third element in the package of US policy changes. President Bush announced that the United States was willing to conclude, ahead of the multilateral treaty, an interim bilateral agreement with the Soviet Union under which each side would reduce its stocks to about 20 per cent of the existing US holdings. Such an agreement was signed by the two sides in June 1990.[26] Under it the two countries were to begin destroying their stockpiles by the end of 1992, and neither country was to be permitted more than 5000 agent tonnes of chemical weapons after the year 2002. Russia has since succeeded to this agreement, which, however, at the time of writing has not yet entered into force. It is officially stated that all the chemical weapons produced by the

[25] France, Conference on Disarmament document CD/757, 11 June 1987.

[26] For the text of the agreement, see SIPRI, *SIPRI Yearbook 1991: World Armaments and Disarmament* (Oxford University Press: Oxford, 1991), pp. 536–39.

former Soviet Union are now held within Russia,[27] but economic and other difficulties have been standing in the way of destruction commencing.

## The end of the negotiations

All this fading away of East–West disagreement naturally had the effect of exposing as-yet-unreconciled differences in the hinterland of the negotiation, especially along its North–South dimension. Of these, the chief one was the lesser military utility and political value of chemical weapons to the rich industrialized countries as compared with the rest of the world. Countries asked to give up much might reasonably expect much in return, a consideration which had already assumed prominence when, in January 1989, the League of Arab States had collectively linked progress on banning chemical weapons to progress on nuclear disarmament. This particular linkage continues to be asserted, some League countries making their participation in the CWC conditional at least upon Israel acceding to the NPT.

However, with this one exception, the supposedly significant security value of chemical weapons in the less developed countries was rarely referred to directly. It found expression, instead, in a variety of proxy issues, these typically being portrayed in Western commentary as sticks or carrots for increasing adherence to the convention. They included the provision of assistance to victims of convention violation, the exemption of parts of civil chemical industry from liability to routine international inspection, a weakening of the mandatory challenge inspection provisions, the apportionment of responsibility for chemical weapons abandoned by one state on the territory of another and, especially, the role of export controls and other trade barriers in the eventual international regime. While mutual accommodations on these issues contributed to the final reconciliation, the key bargain seems to have been struck from trade-offs across a more fundamental linkage: on the one side, the future of the CW counterproliferation device which countries of the North had already fashioned outside the Geneva talks, namely the Australia Group,[28] and, on the other side, the degree of real power to be devolved upon the new international organization which the convention would be creating, an organization which countries of the South might have some expectation of controlling through sheer weight of numbers.

These developments took place during the final months of the negotiation in 1992. The start of the end-game had been signalled in May 1991 when President Bush announced another package of US policy changes.[29] Of these the central one was his declaration of US willingness to renounce the option of retaliation in kind against chemical warfare attack once the convention was in

[27] See chapter 7.

[28] The only detailed account currently available in published form of this widely misunderstood body is: Robinson, J. P. P., 'The Australia Group: a description and assessment', eds H. G. Brauch, H. J. van der Graaf, J. Grin and W. A. Smit, *Controlling the Development and Spread of Military Technology*, (VU University Press: Amsterdam, 1992), pp. 157–76.

[29] SIPRI, *SIPRI Yearbook 1992: World Armaments and Disarmament* (Oxford University Press: Oxford, 1992), p. 155.

force (a policy change thought to have originated in the US decision during the 1991 Persian Gulf War not to equip Desert Storm forces with stocks of retaliatory chemical weapons). Now that the rolling text could incorporate a ban on use of chemical weapons under any circumstances, there was no reason why all stockpiles, even down to the last 2 per cent, should not be placed under international control (pending destruction) as soon as the convention entered into force. The fact that destruction of stockpiles might take 10 years to complete need no longer be a source of confidence-sapping doubts about compliance.

President Bush also called for a one-year deadline now to be imposed on the negotiations. This proved acceptable to most of the negotiators. It placed an enormous burden on the 1992 chairman of the *ad hoc* negotiating body, Ambassador von Wagner of Germany. Having apparently been goaded into hastier action than he had initially intended by Australia tabling a draft convention in March 1992,[30] he distributed a draft of his own two months later.[31] This incorporated into the latest rolling text his and his bureau's 'visions' of possible compromise solutions for the many remaining points of disagreement. The majority of these had to do with matters on which North–South consensus had yet to be achieved, so when, on 4 June, a set of proposed amendments[32] was submitted jointly by the principal less developed country participants, it became clear that a breakthrough had been achieved. Controversial though the proposed amendments certainly were, their submission carried the clear implication that the rest of the draft was acceptable. In two hectic final rounds of negotiation, the draft went through two revisions, tabled on 22 June[33] (with an explanatory note[34] distributed later) and 7 August,[35] before being proposed for formal adoption by the *ad hoc* negotiating body and then by the conference plenary. On 3 September the draft was accepted into the report of the conference to the UN General Assembly[36] where it was accompanied by various national statements commenting on its content, as well as a negotiated text setting out the chairman's understanding of certain contentious provisions.[37]

As is to be expected of so long and complex a document, finalized to an artificial deadline and accommodating so many competing interests, the text of the CWC is not flawless, but it is, at the same time, a magnificent achievement.

[30] Australia, Conference on Disarmament document CD/1143, 12 Mar. 1992.

[31] Conference on Disarmament document CD/CW/WP.400, 18 May 1992.

[32] Conference on Disarmament documents CD/CW/WP.402, 403, 404, 405, 406, 407, 408 and 409, 4 June 1992.

[33] Conference on Disarmament document CD/CW/WP.400/Rev.1, 22 June 1992

[34] Conference on Disarmament document CD/CW/WP.414, 26 June 1992.

[35] Conference on Disarmament document CD/CW/WP.400/Rev.2, 10 Aug. 1992

[36] Conference on Disarmament document CD/1173, 3 Sep. 1992.

[37] Conference on Disarmament document CD/1173, 3 Sep. 1992, Appendix I, pp. 61–63.

## III. The National Implementation provision

Under Article VII, National Implementation Measures, individual States Parties will be required to take measures to adhere to the Convention.[38] Verification in all its aspects is the international side of the compliance undertaking, however, it depends heavily on the national side—information provided by the States Parties. On-site verification activities are an indispensable element of verifying compliance, and national implementation and the providing of all necessary information via declarations are prerequisite to successful international verification.

Since 1984 the article related to the International Organization, Article VIII, has been extensively elaborated to develop the machinery required to verify compliance. In contrast, Article VII remained unchanged until 1989. The situation was very much a reflection of the past East–West debate on verification, in which the former Soviet Union and its allies opposed any on-site verification activities.

Under Article VII, a State Party is required to adopt all necessary measures to implement its obligations in accordance with its constitutional processes. A State Party is obliged in particular to adhere to the Convention by adopting measures to prohibit any natural or legal person on its territory or in any place under its jurisdiction from undertakings which are prohibited by the CWC. The State Party is called upon to adjust or apply its penal legislation to any national who might be involved in prohibited activities. In order to fulfil its obligations under the Convention, each State Party is obliged to designate or establish a National Authority to serve as the national focal point for liaison with the OPCW and other States Parties. The State Party must notify the OPCW of its National Authority at the time the Convention enters into force for that particular State Party.

Taking into account the fact that the chemical industry is at different stages of development in the various States Parties and the main objective of a National Authority,[39] the major implementation requirements are the following: (*a*) to provide general declarations under Article III concerning chemical weapons, chemical weapons production facilities (CWPFs), other facilities, old and abandoned chemical weapons and riot control agents; (*b*) to provide annual declarations concerning destruction of chemical weapons under Article IV, and of CWPFs under Article V; (*c*) to carry out destruction of chemical weapons and CWPFs within a specified timeframe; (*d*) to provide initial and annual declarations of activities not prohibited under the Convention under Article VI; (*e*) to provide access for different on-site verification activities including the support of on-challenge verification investigations; (*f*) to support the OPCW and its suborgans as outlined in Article VIII; (*g*) to provide assistance and support the voluntary fund as

[38] See note 37, pp. 22–23.

[39] Stock, T. and Sutherland, R. G. (eds.), SIPRI, *National Implementation of the Future Chemical Weapons Convention*, SIPRI Chemical & Biological Warfare Studies, no. 11 (Oxford University Press: Oxford, 1990).

outlined in Article X; (*h*) to co-operate and have the right to participate in exchange of chemicals, equipment and scientific and technical information relating to the development and application of chemistry for purposes not prohibited under the Convention; and (*i*) to develop national legislative measures in the context of the need to collect the national information as required under the Convention.

Additionally, and not of minor importance, are the requirements under Article IX, Consultations, Co-operation and Fact-finding, which are of significance for the proper functioning of the Convention. Article IX outlines the general provisions, requirements and obligations for the handling of on-challenge inspections, calls for every State Party to 'consult and cooperate, directly among themselves, or through the Organization or other appropriate international procedures . . . on any matter which may be raised relating to the object and purpose, or the implementation of the provisions, of this Convention' and presents the obligation to 'whenever possible, first make every effort to clarify and resolve . . . any matter which may cause doubt about compliance with this Convention, or which gives rise to concerns about a related matter which may be considered ambiguous'. This, together with the general obligation to co-operate, serves as one of the major confidence-building measures (CBMs) of the Convention.

Implementation under Article VII is also embedded in the clear understanding that the CWC verification regime will be meaningless without the active participation of individual States Parties. States Parties which are among the original signatories will have to give highest priority to the preparation of their national implementation obligations. In particular, they must prepare the requested declarations with respect to chemical weapons, old and abandoned chemical weapons, CWPFs and other facilities. In addition, Article VI requires initial declaration of relevant chemicals and facilities, which will put a heavy burden on the National Authority. Several countries have already accumulated experience in preparing for the tasks and duties of the National Authority.[40]

In general, national legal implementation of the Convention will be complicated and will not only affect existing national legislation but may also require the enacting of new laws. A wide range of issues may need review, as indicated by the following examples: (*a*) if destruction is required environmental standards will have to be reviewed; (*b*) for the required declaration of data concerning chemicals and facilities the rules, procedures and laws for data collecting and data handling will have to re-evaluated; and (*c*) the probability of on-challenge inspection will require a general survey of existing domestic legislation if applicable to a wide range of enterprises. Last but not least, every concerned enterprise must soon begin to prepare for the eventuality of a visit by an inspection team.

[40] 'Australian National Secretariat: survey of chemical industry', Conference on Disarmament document CD/1129, 20 Feb. 1992; Australia, 'Strategy for preparing for the implementation of the Chemical Weapons Convention', Conference on Disarmament document CD/1055, Feb. 1991; see also *SIPRI Yearbook 1991* (note 26), p. 527.

## IV. The international organization

The international organization now has a specific title that clearly outlines its function—the Organization for the Prohibition of Chemical Weapons, whose headquarters will be located in The Hague. Its mandate among others is to conduct verification activities as provided for under the CWC in an unobtrusive manner. It must do this while protecting the confidentiality of the data collected (Confidentiality Annex[41]) making use of advances in science and technology. Its costs will be apportioned among States Parties in accordance with a modified UN scale of assessment.

The OPCW will be established by the States Parties; its major organs will be the Conference of the States Parties, the Executive Council and the Technical Secretariat.

### The Conference of the States Parties

The Conference of the States Parties will be composed of all States Parties and each member will have one representative who may be accompanied by alternates and advisers. The first session will be convened not later than 30 days after entry into force and will thereafter meet annually. Under certain circumstances special sessions can be convened; under Article XV the Conference may be convened as an Amendment Conference. All sessions will be held in The Hague unless the Conference decides otherwise. The Conference will adopt its own rules of procedure and elect a chairman and other officers at the beginning of each regular session. A simple majority of the States Parties will constitute a majority for the purposes of a quorum with each member having one vote. Matters of procedure require only a simple majority; matters of substance should be dealt with by consensus where possible but if not decided in 24 hours, the matter will be decided by a two-thirds majority. The Conference is the principal organ of the CWC and can consider any matter within the scope of the Convention including the powers and functions of the Executive Council and the Technical Secretariat. It may make recommendations and decisions on any issues related to the Convention brought before it by a State Party of the Executive Council. Its major roles relate to implementation and compliance with the CWC.

Among its duties at the regular sessions are to: (*a*) adopt reports, programmes and budget; (*b*) decide on the scale of financial contributions; (*c*) elect the Executive Council; (*d*) appoint the Director-General of the Technical Secretariat; (*e*) constitute relevant subsidiary organs; and (*f*) direct the Director-General to establish a Scientific Advisory Board.

[41] See note 37, pp. 169–74.

## The Executive Council

The Executive Council will consist of 41 members. Each State Party has the right to serve and election will be for a two-year period. After much delicate negotiation, the distribution of seats was as follows:

(*a*) 9 States Parties from Africa, 3 of whom have the most significant chemical industry;
(*b*) 9 States Parties from Asia, 4 of whom have the most significant chemical industry;
(*c*) 5 States Parties from Eastern Europe, 1 of whom has the most significant chemical industry;
(*d*) 7 States Parties from Latin America and the Caribbean, 3 of whom have the most significant chemical industry;
(*e*) 10 States Parties from among Western European and other states, 5 of whom have the most significant chemical industry; and
(*f*) 1 State Party to be designated consecutively by States Parties from Asia and from Latin America and the Caribbean.

At the first election 20 members will be elected to the Executive Council for one year. After full implementation of Articles IV and V, the composition of the Executive Council may be reviewed. The Council's rules of procedure must be approved by the Conference of the States Parties, but it will elect its own chairman. It will meet as often as required with each member having one vote and decisions on matters of substance being taken on a two-thirds majority. The Council is the executive organ of the Convention. Among its duties will be: (*a*) supervision of the Technical Secretariat; (*b*) co-operation with National Authorities; (*c*) concluding agreements with states and international organizations with the approval of the Conference; and (*d*) approving arrangements for implementation of verification activities. It will have special responsibilities in the consideration of concerns relating to compliance and non-compliance. In cases of particular gravity it can bring an issue to the UN General Assembly and the Security Council.

## The Technical Secretariat

The Technical Secretariat will be the operational arm of the OPCW. It will carry out the verification measures provided for, all other functions entrusted to it and those functions delegated to it. Among its functions will be to: (*a*) negotiate agreements relating to implementation with the approval of the Executive Council; (*b*) establish the stockpiles of supplies for emergency and humanitarian assistance required under Article X; (*c*) inform the Executive Council of any problems especially those relating to uncertainty about compliance; and (*d*) provide technical assistance to States Parties concerning implementation of the CWC.

The Technical Secretariat will be comprised of a Director-General, inspectors and such technical and scientific personnel as required. The Inspectorate will be a specific unit which acts under the supervision of the Director-General. The OPCW will enjoy the legal capacity and such privileges and immunities as are necessary to carry out its function in the territory of or any place under the jurisdiction and control of a State Party.

## The Inspectorate

The Inspectorate is that part of the Technical Secretariat that is engaged in verifying that States Parties are implementing the CWC and, by means of inspection, are in compliance with their obligations under it. An inspection team will be made up of inspectors and inspection assistants as defined in the Verification Annex to the Convention.[42] Not later than 30 days after entry into force the Technical Secretariat will communicate the names, ranks and nationalities of all inspectors and assistants to the States Parties who then can have 30 days to state which of these are unacceptable to them and therefore cannot be designated for that State Party. There is an inherent continuing right not to accept a particular individual unless an inspection team has already been named for that State Party. Designated inspectors and assistants have the same inviolability and protection as diplomatic agents as required to exercise their functions effectively. Each State Party will be expected to provide the appropriate visas within 30 days of acknowledging the list of inspectors and assistants, and the documents will be valid for at least two years. An inspection team will have the same privileges while transiting the territory of a non-inspected State Party. The Director-General may waive the immunity from jurisdiction if it would impede the course of justice. Each State Party will have to develop appropriate standing arrangements to allow access and egress.

The size and cost of an Inspectorate are still matters of a debate that is not yet based on full disclosure of what has to be verified and where. It is also complicated by provisions in Articles IV and V concerning (*a*) 'unnecessary duplication of bilateral and multilateral agreements on verification of chemical weapons storage and their destruction among States Parties', and (*b*) the provision that the State Party must also meet the verification costs of complying with the provisions of Articles IV and V. The delegation of the Russian Federation indeed has proposed alternate language for paragraph 16 of Article IV and paragraph 19 of Article V.[43] The latest public estimates were described in the *SIPRI Yearbook 1991* where the Canadian estimate was $120 million annually for the first 10 years, and the estimated cost for on-site inspection and compliance with the bilateral CW agreement was estimated at $15–70 million annually.[44]

[42] See note 37, pp. 61–168.

[43] Russian Federation, 'Proposed amendments to CD/CW/WP.400/Rev. 1', Conference on Disarmament document CD/CW/WP.479, 27 July 1992.

[44] *SIPRI Yearbook 1991* (note 26), pp. 525–26.

## V. Destruction requirements under the CWC

In essence, the destruction requirements of the CWC form the disarmament component of this convention, and the obligations under Articles IV and V together with the corresponding annexes apply to those who declare possession of either chemical weapons and/or CWPFs under Article III.

The obligations for CW possessors are simple: they must declare possession and whether they have transferred or received any chemical weapons since 1 January 1946. This has to be done within 30 days after entry into force. The declaration must specify location, aggregate quantity and detailed inventory, and provide general plans for destruction. Declarations are also required for 'old chemical weapons and abandoned chemical weapons'. These declarations include an obligation to state whether there are abandoned chemical weapons on the territory of a State Party and whether the State Party has abandoned chemical weapons on the territory of other states.

Declarations concerning CWPFs are retrospective to 1 January 1946 even if such facilities are no longer extant. This declaration includes CWPFs on the territory of a State Party, under the control of another State Party or under its control on the territory of another State Party. This is required 30 days after entry into force and must include information on closure, temporary conversion and the general plan for destruction of the facility(ies). The declaration also requires information on equipment transferred or received since 1 January 1946. There is a requirement to declare any facility primarily used for the development of chemical weapons since 1 January 1946. The provisions of the CWC do not apply to land burial before 1 January 1977 or sea dumping before 1 January 1985.

As far as riot control agents are concerned the declaration merely specifies the nature of the chemicals used, and this should be updated within 30 days of any change in the chemicals used.

Immediately after a State Party has made its declaration regarding chemical weapons there is a requirement to permit access to the locations specified. The actual destruction is to begin not later than 2 years after entry into force following an agreed rate and order of destruction. Detailed plans for any destruction campaign must be available 60 days before it begins, and declarations regarding the state of implementation are required within 60 days of completion of the campaign. The State Party is to certify that all declared chemical weapons are destroyed not later than 30 days after completion of destruction. Any chemical weapons on the territory of a State Party but under the control of another state should be removed not later than one year after entry into force. In all cases destruction should be completed not later that 10 years after entry into force. An extension up to 5 years is possible, but has to be approved by the Conference of the States Parties based on a request of the individual state. A complication is the draft provision that any State Party which has chemical weapons to destroy is obligated to cover the cost of verification.

Any State Party declaring possession of a CWPF must cease operation immediately and close the facility within 90 days, and access must be provided after closure for verification purposes. The facilities should be destroyed in accordance with an agreed rate and sequence that begins 1 year after entry into force, and be complete 9 years later. Detailed destruction plans are due 180 days before facility destruction commences, with declarations relating to plan implementation not later than 90 days after the campaign is completed. Certification is required within 30 days of completion of the destruction of the facilities declared. Converted facilities must be destroyed as soon as their projects are finished but, in any case, within the 10-year period. It is possible to present a case for the use of a former production facility for purposes not prohibited by the CWC; such a facility must not be capable of reconversion. Again there is a provision for the Organization to recover the verification costs associated with the destruction processes.

Each State Party with a sizeable destruction task involving chemical weapons or CWPFs will require a comprehensive National Authority under Article VII to oversee the destruction process and also to validate any declarations before they are sent to the OPCW. The State Party has an obligation to co-operate with the OPCW in all its functions and to provide appropriate assistance to the Technical Secretariat. The National Authority must be in a position to provide the OPCW with all necessary declarations within 30 days of entry into force including general plans for destruction.

Co-operation among States Parties is discussed briefly in regard to destruction activities; States Parties agree to share information leading to the safe and efficient destruction of chemical weapons either on a bilateral basis or via the Technical Secretariat. There is also a brief mention of possible bilateral or multilateral agreements on the destruction of CWPFs especially as they may affect verification. There is currently, in principle, a bilateral agreement on the destruction of chemical weapons between the USA and the former USSR,[45] but the successor Russian Federation still has not developed its demilitarization plans. Iraq has agreed to carry out the destruction of its CW stocks under the supervision of UNSCOM.[46]

One of the final sticking points in the draft Convention was the problem of 'old chemical weapons' and 'abandoned chemical weapons'. The first problem was a useful definition and the second was how to deal with these contentious categories. The definition adopted for old chemical weapons related to time—those produced before 1925 or weapons that were produced between 1925 and 1946 but no longer capable of use. Abandoned chemical weapons were defined as weapons left on the territory of another state without its consent at any time after 1925. The declarations under Article III have to specify whether a State Party had either old or abandoned weapons on its territory with all appropriate information and also whether it had abandoned any chemical weapons on the territory of another state. As noted above there was a

[45] See note 26.
[46] *SIPRI Yearbook 1992* (note 29), pp. 509–30.

caveat relating to weapons previously disposed of either by irreversible burying or sea dumping.

## VI. The verification regime under the CWC

The discussions and the following complex negotiations that led to the successful conclusion of the CWC are history; the question now is will the verification provisions work? The negotiators have left behind them a labyrinthine series of detailed verification provisions that have to be converted into reality first by the Preparatory Commission and finally the Technical Inspectorate of the OPCW. There are four dominant verification tasks: (*a*) the control of all CW stockpiles; (*b*) the destruction of chemical weapons; (*c*) the control and destruction of all CWPFs except single small-scale production facilities (SSSFs); and (*d*) the control of chemicals deemed to pose a significant risk to the Convention. All of this presents a formidable technical challenge; to ensure compliance will be costly and funds must be available for at least a decade to oversee the destruction phase while there is no limit to the inspection phase for the chemical industry.

The definition of a chemical weapon in the CWC has been crafted to cover both chemicals and potential delivery systems.[47] The formulation is broad, and because of the many uses of toxic chemicals it depends on intent to use such chemicals as weapons (general purpose criteria). In this regard not all possible chemicals for weaponization are included on Schedule 1, only those that are currently thought to pose a significant risk to the CWC. The definitions of old or abandoned chemical weapons are time-based but also should be taken as only referring to munitions that are no longer serviceable. The definition of a riot control agent is extremely wide and only limited in this by the declaration requirement in Article III. The definition of a CWPF[48] is essentially functional and time-based from the perspective of the CWC although the phrase 'any other chemical that has no use, above 1 tonne per year' is ambiguous in that it is unrelated to any schedule. The overall definition of a CWPF does include filling facilities, but SSSFs are excluded.

The key to Article VI verification is the Schedules of Chemicals;[49] in principle the CWC embraces all toxic chemicals and their precursors while in practice 14 families and 29 chemicals are listed. There are 12 entries on Schedule 1, 14 on Schedule 2, and 17 on Schedule 3. It has been estimated that approximately 108 000 chemicals could be covered by entries 1–3, inclusive, of Schedule 1, and this will have implications for the chemical analyses involved in routine inspections. Schedule 1 effectively removes a substance from commercial use; a State Party may possess an aggregate of 1 tonne of such chemicals and any production above 10 kg must be carried out in the designated SSSF. Chemicals on Schedule 2 must be reported when production

[47] See note 37, p. 9.
[48] See note 37, pp. 10–11.
[49] See note 37, pp. 22–23 and pp. 53–59.

exceeds a threshold, but there are no limits on production. Schedule 2 production facilities are subject to initial and routine inspections; reporting thresholds are in the 1 kg–1 tonne range depending on the chemical. Schedule 3, which is used for chemicals in large-scale production or for specific precursors to Schedule 2 chemicals, has a threshold reporting range of 30–200 tonnes.

A major achievement was the elaboration of the Annex on Implementation and Verification (Verification Annex).[50] This is a complete codification of the tasks of the Technical Secretariat defining the rights and responsibilities of the Inspectorate in 11 chapters and 108 pages. The parts are arranged as follows: (*a*) definitions; (*b*) general rules of verification; (*c*) general provisions (Articles III, IV and V); (*d*) destruction of chemical weapons and verification (Article IV); (*e*) destruction of CWPFs and verification (Article V); (*f*) activities not prohibited under the CWC (Article VI) including Schedules 1, 2, 3 and other chemical production facilities; (*g*) challenge inspection (Article IX); and (*h*) investigation in cases of allegation of use.

The first two parts define the terminology to be applied to the verification activities of the Inspectorate together with the formal rules for the designation of inspectors, their relationship with the inspected State Party and the general conduct of inspections. Parts III, IV and V are concerned with the destruction of chemical weapons and CWPFs. Parts VI, VII, VIII and IX separate inspection activities as they relate to Schedules 1, 2 and 3 together with activities associated with 'other chemical production facilities'. Part X is concerned with the modalities of challenge inspection while allegations of use are dealt with in Part XI which of necessity overlaps with some aspects of challenge.

There are many obligations that States Parties will assume under the CWC, but the important verification provisions are that: (*a*) all CW stockpiles and CWPFs must be destroyed within a specific timeframe; (*b*) the chemical industry must not be used for the production of new chemical weapons; and (*c*) the declarations must reveal the full extent of stockpiles and production facilities. The verification of the first of these aims is by far the most straightforward. Verification will be accomplished by on-site inspection and monitoring. Access by the Inspectorate to stockpiles and their destruction is essentially unlimited. This is also true of facility destruction but does become more complex when a CWPF is temporarily converted to a destruction facility. The main problem currently is estimating the real size of the effort needed to monitor destruction and the ability of States Parties which possess both chemical weapons and CWPFs to maintain the integrity of the environment throughout the process. At this point it appears doubtful that the major possessors will be able to meet the 10-year destruction requirement.

## Verification of activities not prohibited by the Convention

Verification of activities not prohibited by the Convention will be a very complex mix of science, technology, commerce and politics; it will also depend

[50] See note 42.

upon whether the information collected can lead to an unambiguous finding confirming compliance with the obligations of the CWC. The verification strategy is based upon the schedules of Article VI and the procedural annexes. Schedule 1 chemicals, being those of greatest hazard to the CWC, are subject to the greatest scrutiny. The permitted uses of Schedule 1 chemicals are few and so this verification task will be the most straightforward; these are applications in research, medicine and pharmacy and for protective purposes. A nation's aggregate holding cannot exceed one tonne and can only be produced in one designated facility except for: (*a*) 10 kg production for protective purposes, and (*b*) 100 g per year at facilities involved in research, pharmacy or medicine with an aggregate of 10 kg. Similar uses requiring aggregate quantities of less than 100 g are not subject to reporting. Verification of the activities at a SSSF should be relatively unambiguous but there could well be difficulties as the quantities involved become smaller at other facilities. Difficulties can be anticipated with Schedule 2 verification with the varying thresholds of 1 kg, 100 kg and 1 tonne, and considerable resources will have to be allocated to this activity. The inspection aims will be difficult to achieve, especially: (*a*) absence of *any* Schedule 1 chemical and, more important, (*b*) non-diversion of Schedule 2 chemicals for prohibited activities. The latter aim would appear to require more information than it is planned for the Inspectorate to collect. All plants producing Schedule 3 chemicals have to be declared in ranges from 30 tonnes to above 100 000 tonnes. On-site inspection will be restricted to operations above 200 tonnes with a maximum of two inspections per year at any one plant site. Inspections relate to the declared chemicals, the absence of Schedule 1 chemicals, and there will be no requirement for a facility attachment.

In addition to plants producing scheduled chemicals there is a requirement for declaration of facilities that produce more than 30 tonnes of organic compounds containing phosphorus, sulphur and fluorine (PSF)—chlorine is not included! Verification will be carried out at locations where PSF production exceeds 200 tonnes. Implementation of PSF inspections is to commence in the fourth year after entry into force.

It will be some time before an unqualified yes can be given to the question of whether the verification procedures will produce confidence in compliance. The routine inspection required is limited to: (*a*) two per facility per year, and (*b*) a maximum of 20 routine annual inspections per State Party. Given that some countries could have more than 1000 chemically capable production sites the latter number is inadequate, although budget restrictions may well reinforce these low limits of industrial inspection in the early years after entry into force.

### Challenge inspection

Challenge inspection will have to fill the gaps in the routine inspection infrastructure. States Parties are required to accept challenge but have a chance to delay the process (i.e., each State Party has the right to modify a challenge request, but it has an obligation to demonstrate compliance and enable the inspection team to fulfil its mandate). Essentially the State Party has the right to protect sensitive installations for security and commercial reasons and may do this via restrictions on perimeters of sites to be inspected and by managed access within sites. The built-in time delays are also problematic for verification purposes since these delays could add up to 120 hours. The Executive Council may block any inspection which it feels to be frivolous or abusive. It will be some time before the true value of the process within the CWC will become apparent. Nevertheless the overall principle of having an arms control agreement which encompasses the right of an international body to request a challenge inspection without right of refusal is important and no doubt will be a precedent to be followed in other such agreements.

## Investigation of alleged use

The CWC has also developed procedures for the investigation of allegations of use; these procedures build on those previously developed by the United Nations in its investigations in South-East Asia and in the 1980–88 Iraq–Iran War. An allegation of use is the worst possible case under the Convention since it will prove that other procedures have failed and that a State Party has evaded all its obligations under the Convention. If the allegation is founded in fact, then a State Party has retained either a CW stockpile or the capacity to produce chemical weapons and has incorporated chemical weapons into its military procedures. Unless a non-State Party is involved, all other verification activities will have failed. Investigations of allegations of use flow from Article X and are essentially a variant on Challenge Inspections. The actual use of chemical weapons would either involve a traditional agent and munitions or a novel (unknown) agent. It is likely that the inspection team will require the assistance of experts who are not regular members of the Inspectorate in the latter case. The novel agent situation is very complex in that it will involve unknown substances of unknown toxicity which are not on Schedule 1. The timeframe that the team will operate under would have to be extended when the initial investigation shows that novel agents are likely to be involved. The investigation would, of course, also be hazardous if hostilities are occurring and the safety of the team cannot be guaranteed.

There is an obvious tension in all of the verification modalities described in the CWC; there is to be a balance between the rights and obligations of the States Parties and the OPCW. There will be difficulties in implementation at the national level of the data requirements of the Convention with consequent difficulties for the OPCW in assessing compliance initially. Even a state in complete compliance with its obligations may resent the necessarily intrusive verification process. This will be at its worst in the highly competitive international chemical industry. Political tension will be at its greatest in challenge inspections carried out at these industrial sites. There is no way around this since chemicals by their nature are often dual-purpose and can be misused. The Convention gives the Executive Council paramountcy in concerns over non-compliance but does not specify how decisions on non-compliance should be made. Both the Executive Council and the Conference of the States Parties can call on a participating state to take steps to address situations of non-compliance and to inform the UN Security Council in extremely dangerous situations. There are obvious weaknesses in the verification provisions and procedures within the Convention as it currently stands. However, it is only after ratification and entry into force that the participants will be made to face up to these weaknesses and consider the appropriate amendment procedures.

## VII. The Preparatory Commission

After the signatory ceremony in Paris the Preparatory Commission will begin its work in The Hague not later than 30 days after 50 States have signed the Convention.[51] A Chairman and an Executive Secretary will be selected and other tasks include: decisions on budget and personnel questions, elaborating procedures for conducting inspections and investigations, and establishing rules of procedure for the OPCW. To ensure the timely and correct implementation of the Convention many preliminary tasks which are on the agenda of the Preparatory Commission must be completed between signature and entry into force.

For the first official session of the Preparatory Commission for the OPCW at The Hague at least eight meetings are planned for 8–12 February 1993.[52] The United Nations has agreed to make available $500 000 for the first official session of the Preparatory Commission; these funds are to be reimbursed by the Preparatory Commission within 90 days. The First Committee of the 47th General Assembly noted that this 'Convention would be a convention of States Parties and its associated costs would therefore be met in accordance with the financial arrangements to be made by the signatories'.[53] No future payments are intended to be made by the UN. In December the UN Secretary-General transmitted a 'Draft resolution establishing the Preparatory Commission for the Organization for the Prohibition of Chemical Weapons'[54] to which was annexed the 'Text on the establishment of a Preparatory Commission' together with three annexes, as adopted by the CD at its 1992 session and contained in the conference report.[55]

The Commission is *inter alia* requested to establish a provisional Technical Secretariat (PTS), to which only nationals of signatory states are to be ap-

[51] See note 37, p. 177.

[52] General Assembly, 'Programme budget for the biennium 1992–1993, chemical and bacteriological (biological) weapons: Convention on the Prohibition of the Development, Production, Stockpiling and Use of Chemical Weapons and on their Destruction, programme budget implications of draft resolution A/C.1/47/L.1/Rev.2', document A/C.5/47/49, 13 Nov. 1992.

[53] See note 52.

[54] United Nations, C.N.406.1992, Treaties-1 (Depositary Notification); and C.N.406.1992, Treaties-1 (Annex), 17 Dec. 1992.

[55] See note 37, pp. 175–91.

pointed, in order to make the necessary arrangements for the first session of the Conference of the States Parties, including preparation of a draft agenda and draft rules of procedure. The Commission will also: (*a*) elaborate the staffing pattern for the Technical Secretariat; (*b*) make assessments of personnel requirements; (*c*) elaborate staff rules for recruitment and service conditions; (*d*) recruit and train the technical personnel and support staff; (*e*) organize the office and administration services; (*f*) prepare administrative and financial regulations; (*g*) purchase and standardize equipment; (*h*) prepare a programme of work and a budget for the first year of OPCW activities; (*i*) prepare detailed budgetary provisions for the OPCW, especially for the categories 'administrative and other costs' and 'verification costs'; (*j*) prepare the scale of financial contributions to the OPCW; (*k*) prepare OPCW administrative and financial regulations; and (*l*) develop arrangements to facilitate the election of 20 members for a term of one year to the first Executive Council.

According to Article VIII, the Preparatory Commission is also recommended to develop draft agreements, provisions and guidelines for consideration and approval by the Conference of the States Parties. The Convention text on the establishment of the Preparatory Commission contains at least 23 specific items. Major areas where guidelines or draft agreements are recommended are *inter alia*: (*a*) procedures for verification and conduct of inspections; (*b*) models for facility agreements; (*c*) guidelines to determine the status of chemical weapons produced between 1925 and 1946; (*d*) guidelines for the release of classified information by the OPCW; (*e*) agreements between the OPCW and the States Parties; (*f*) a list of items to be stockpiled for emergency and humanitarian assistance; (*g*) guidelines for determining the frequency of systematic on-site inspections; and (*h*) procedures for handling and securing samples. Further, the Preparatory Commission is committed to follow closely the signature and ratification process and to establish a framework for facilitating the exchange between signatory states concerning legal and administrative measures for the implementation of the Convention and the establishment of the National Authority.

In the beginning there will be 20 to 40 Preparatory Commission staff members, whose number will increase to 40 to 70 staff members in the last phase of the Commission's work.[56] During the organizational period it may be useful to consider not only the text on establishment of the Preparatory Commission but also practical experience. Financial matters are likely to present difficulties since funding for the Commission will be supplied by States Parties to a future Convention to which they are not yet parties. Taking this into account, it may be advisable to keep staff and expenses to as modest a scale as possible yet sufficient to perform the duties and functions of the Preparatory Commission.

Important documents from earlier drafts of the CWC which will be of essential importance for the Commission in elaborating the recommended guide-

[56] Meerburg, A. J., 'Structure and timing factors of the Preparatory Commission', *OPCW: The First Five Years: Symposium Report*, Symposium on the Establishment of the Organisation for the Prohibition of Chemical Weapons (OPCW) under the Future Chemical Weapons Convention, The Hague, 8–9 May 1992, pp. 34–38.

lines, draft agreements and provisions were forwarded to UNGA in September 1992 in an annex to the draft Convention; the annex is a compilation of material to be transmitted to the Preparatory Commission.[57]

The work of the Preparatory Commission will take at least two years and will be essential for implementation of the Convention. For many countries it will also be important to join in the complex work of the Commission. By so doing, they may be in a position to influence or at least take part in the decision process for establishing the necessary framework before the OPCW starts its normal work. The Preparatory Commission's work and performance will be watched closely by States Parties and will be essential for ensuring their confidence in the forthcoming Convention.[58]

## VIII. Entry into force

The CWC was opened for signature on 13 January 1993 by its Depositary, the Secretary-General of the United Nations, at a ceremony in Paris hosted by the French Government. The ceremony ended on 15 January, by which time 130 states had signed the Convention. It is to remain open for signature at UN headquarters in New York until its entry into force.

The Convention provides that entry into force can be no earlier than two years after the opening for signature, and then only if 180 days have elapsed after 65 states have ratified their signatures and deposited their instruments of ratification.

The sheer number of signatures during the Paris ceremony, and the assurances of speedy ratification made in so many of the accompanying national statements, suggest that the Convention will enter into force on or soon after 13 January 1995. Signatory and non-signatory states as of 8 February 1993 are identified in table 14.1.

## IX. The CWC: some conclusions

After more than a generation the CWC has been finalized and will enter into force by 1995 at the earliest. The long negotiations at the CD were very much affected by the cold war and North–South distrust; successful finalization of the Convention is owing to changed attitudes, shock at the use of chemical weapons in the 1980–88 Iraq–Iran War, the Iraqi threat of CW use in the 1991 Persian Gulf War and increasing awareness that chemical weapons have little military value. Conclusion of the Convention was also accelerated by the growing threat of CW proliferation, one of the major security concerns of the 1990s, and increased understanding that CW stockpiles are both costly to maintain and pose many risks to mankind and the environment. The enormous destruction costs which the USA and Russia face may be another contributing factor.

[57] See note 37, p. 191.
[58] Gizowski, S., 'Tasks of the Prepcom', *OPCW: The First Five Years* (note 56), pp. 32–33.

The major achievement of the Convention—delegitimization of chemical weapons in all their aspects linked to an obligation of total destruction—is a disarmament undertaking which will create a much higher level of confidence among States Parties. States adhering to the Convention will renounce the acquisition of a CW capability.

The verification of non-production in the chemical industry is a novel achievement in the history of disarmament. The technological and technical basis for CW capability is linked to the huge chemical industry which has developed over the past 100 years, and which the verification concept addresses. Of course, no arms control agreement can provide total assurance against small-scale cheating. The Convention must cope with the entire chemical industry, and verification under it is based on monitoring and inspection activities which rely on the readiness of States Parties to participate actively in verification by providing the necessary information in the form of various declarations.

The concept of challenge inspections is a new approach for a multilateral disarmament treaty, and it represents an element of deterrence against possible Convention violations. States Parties are required to grant access to OPCW inspectors to any site at which another party has requested a challenge inspection to resolve questions of possible non-compliance and to re-establish compliance. In an attempt to handle the conflict between the objectives of effective verification and the legitimate needs and rights of States Parties to protect sensitive installations and to guard commercial secrets of industries, the concept of mandatory on-site inspection, any time, anywhere was developed. Under the leadership of the OPCW, using specific procedures and a complex set of rules, the on-challenge concept clearly takes into consideration the special situation of the chemical industry and the interests of individual States Parties. It also takes a co-operative approach which calls for active participation in the process by both the challenging and the challenged party.

The conclusion of the negotiations on the CWC proved that it is possible to remove political obstacles if there is sufficient motivation. Two examples are the problems related to herbicides and riot control agents. It was clear that only a compromise could solve these controversial issues. For herbicides this was done by inserting a special paragraph in the Preamble. The Second Review Conference of the Convention on the Prohibition of Military or any Other Hostile Use of Environmental Modification Techniques (the Enmod Convention) in September addressed the herbicide issue by stating 'that the military or any other hostile use of herbicides as an environmental modification technique . . . is a method of warfare [and] prohibited by Article I [of the Enmod Convention]. . . '.[59]

Riot control agents were a problem throughout the history of the negotiations. While largely used for law enforcement and crowd control by police and other organizations responsible for maintaining law and order, they could con-

[59] Second Review Conference of the Parties to the Convention on the Prohibition of Military or any Other Hostile Use of Environmental Modification Techniques, *Final Document*, Part II, *Final Declaration*, Article II, p. 11.

stitute a risk to the CWC if developed into a new generation of non-lethal, but effective, warfare agents. The compromise finally agreed was to include a paragraph in Article I to prohibit the use of 'riot control agents as a method of warfare' together with a definition of riot control agents in Article II. In addition, Article III incorporates the obligation that each State Party must declare the specifications of each chemical it possesses for riot control purposes. Taking into account that some States Parties voiced strong objections to certain verification provisions and more detailed declaration obligations, the compromise can be seen as a measure to avoid disruptive verification activities. It may also contribute to greater transparency. Once the Convention has entered into force, this is an area which will need close monitoring.

Achieving the broadest possible adherence to the CWC is a major political challenge. States not adhering to it should seriously review its benefits and watch carefully the possible changes which may occur by way of trade restrictions and even political isolation if they do not become States Parties to the Convention.

Another important disarmament and arms control issue is related to the definition of research for protective purposes under the Convention. In contrast to the BWC, a clear distinction has been made between research for protective (defensive) purposes and non-permitted (offensive) research. This careful attention to definition under the Convention will minimize possible mistrust and provide for effective verification.

Now that the CWC exists, the intense discussion of the need to strengthen the BWC will receive new stimulus. Many lessons can be learned from the 'chemical exercise', but simply copying its concepts for the BWC could be meaningless. However, future improvement of the BWC might be stimulated by the organizational set-up of the OPCW and national implementation experiences.

There are weaknesses in the Convention, but the negotiation of a multilateral disarmament treaty of such complexity must be viewed as a process of achieving the best possible compromise. Co-operation is one of the most important aspects of the CWC, and the actual way in which co-operative efforts meet the obligations and requirements of the Convention will need to be tested in practical terms. World security will be strengthened when the Convention enters into force. Its functioning will provide many opportunities for confidence-building in areas of tension.

# **Appendix 14A.** The Convention on the Prohibition of the Development, Production, Stockpiling and Use of Chemical Weapons and on their Destruction

## Preamble

The States Parties to this Convention,

*Determined* to act with a view to achieving effective progress towards general and complete disarmament under strict and effective international control, including the prohibition and elimination of all types of weapons of mass destruction,

*Desiring* to contribute to the realization of the purposes and principles of the Charter of the United Nations,

*Recalling* that the General Assembly of the United Nations has repeatedly condemned all actions contrary to the principles and objectives of the Protocol for the Prohibition of the Use in War of Asphyxiating, Poisonous or Other Gases, and of Bacteriological Methods of Warfare, signed at Geneva on 17 June 1925 (the Geneva Protocol of 1925),

*Recognizing* that this Convention reaffirms principles and objectives of and obligations assumed under the Geneva Protocol of 1925, and the Convention on the Prohibition of the Development, Production and Stockpiling of Bacteriological (Biological) and Toxin Weapons and on their Destruction signed at London, Moscow and Washington on 10 April 1972,

*Bearing in mind* the objective contained in Article IX of the Convention on the Prohibition of the Development, Production and Stockpiling of Bacteriological (Biological) and Toxin Weapons and on their Destruction,

*Determined* for the sake of all mankind, to exclude completely the possibility of the use of chemical weapons, through the implementation of the provisions of this Convention, thereby complementing the obligations assumed under the Geneva Protocol of 1925,

*Recognizing* the prohibition, embodied in the pertinent agreements and relevant principles of international law, of the use of herbicides as a method of warfare,

*Considering* that achievements in the field of chemistry should be used exclusively for the benefit of mankind,

*Desiring* to promote free trade in chemicals as well as international cooperation and exchange of scientific and technical information in the field of chemical activities for purposes not prohibited under this Convention in order to enhance the economic and technological development of all States Parties,

*Convinced* that the complete and effective prohibition of the development, production, acquisition, stockpiling, retention, transfer and use of chemical weapons, and their destruction, represent a necessary step towards the achievement of these common objectives,

*Have agreed* as follows:

## Article I. General Obligations

1. Each State Party to this Convention undertakes never under any circumstances:

(a) To develop, produce, otherwise acquire, stockpile or retain chemical weapons, or transfer, directly or indirectly, chemical weapons to anyone;

(b) To use chemical weapons;

(c) To engage in any military preparations to use chemical weapons;

(d) To assist, encourage or induce, in any way, anyone to engage in any activity prohibited to a State Party under this Convention.

2. Each State Party undertakes to destroy chemical weapons it owns or possesses, or that are located in any place under its jurisdiction or control, in accordance with the provisions of this Convention.

3. Each State Party undertakes to destroy all chemical weapons it abandoned on the territory of another State Party, in accordance with the provisions of this Convention.

4. Each State Party undertakes to destroy any chemical weapons production facilities it owns or possesses, or that are located in any place under its jurisdiction or control, in accordance with the provisions of this Convention.

5. Each State Party undertakes not to use riot control agents as a method of warfare.

## Article II. Definitions and Criteria

For the purposes of this Convention:

1. 'Chemical Weapons' means the following, together or separately:

(a) Toxic chemicals and their precursors, except where intended for purposes not prohibited under this Convention, as long as the types and quantities are consistent with such purposes;

(b) Munitions and devices, specifically designed to cause death or other harm through the toxic properties of those toxic chemicals specified in subparagraph (a), which would be released as a result of the employment of such munitions and devices;

(c) Any equipment specifically designed for use directly in connection with the employment of munitions and devices specified in subparagraph (b).

2. 'Toxic Chemical' means:

Any chemical which through its chemical action on life processes can cause death, temporary incapacitation or permanent harm to humans or animals. This includes all such chemicals, regardless of their origin or of their method of production, and regardless of whether they are produced in facilities, in munitions or elsewhere.

(For the purpose of implementing this Convention, toxic chemicals which have been identified for the application of verification measures are listed in Schedules contained in the Annex on Chemicals.)

3. 'Precursor' means:

Any chemical reactant which takes part at any stage in the production by whatever method of a toxic chemical. This includes any key component of a binary or multicomponent chemical system.

(For the purpose of implementing this Convention, precursors which have been identified for the application of verification measures are listed in Schedules contained in the Annex on Chemicals.)

4. 'Key Component of Binary or Multicomponent Chemical Systems' (hereinafter referred to as 'key component') means:

The precursor which plays the most important role in determining the toxic properties of the final product and reacts rapidly with other chemicals in the binary or multicomponent system.

5. 'Old Chemical Weapons' means:

(a) Chemical weapons which were produced before 1925; or

(b) Chemical weapons produced in the period between 1925 and 1946 that have deteriorated to such extent that they can no longer be used as chemical weapons.

6. 'Abandoned Chemical Weapons' means:

Chemical weapons, including old chemical weapons, abandoned by a State after 1 January 1925 on the territory of another State without the consent of the latter.

7. 'Riot Control Agent' means:

Any chemical not listed in a Schedule, which can produce rapidly in humans sensory irritation or disabling physical effects which disappear within a short time following termination of exposure.

8. 'Chemical Weapons Production Facility':

(a) Means any equipment, as well as any building housing such equipment, that was designed, constructed or used at any time since 1 January 1946:

(i) As part of the stage in the production of chemicals ('final technological stage') where the material flows would contain, when the equipment is in operation:

(1) Any chemical listed in Schedule 1 in the Annex on Chemicals; or

(2) Any other chemical that has no use, above 1 tonne per year on the territory of a State Party or in any other place under the jurisdiction or control of a State Party, for purposes not prohibited under this Convention, but can be used for chemical weapons purposes; or

(ii) For filling chemical weapons, including, *inter alia*, the filling of chemicals listed in Schedule 1 into munitions, devices or bulk storage containers; the filling of chemicals into containers that form part of assembled binary munitions and devices or into chemical submunitions that form part of assembled unitary munitions and devices, and the loading of the containers and chemical submunitions into the respective munitions and devices;

(b) Does not mean:

(i) Any facility having a production capacity for synthesis of chemicals specified in subparagraph (a) (i) that is less than 1 tonne;

(ii) Any facility in which a chemical specified in subparagraph (a) (i) is or was produced as an unavoidable by-product of activities for purposes not prohibited under this Convention, provided that the chemical does not exceed 3 per cent of the total product and that the facility is subject to declaration and inspection under the Annex on Implementation and Verification (hereinafter referred to as 'Verification Annex'); or

(iii) The single small-scale facility for production of chemicals listed in Schedule 1 for purposes not prohibited under this Convention as referred to in Part VI of the Verification Annex.

9. 'Purposes Not Prohibited Under this Convention' means:

(a) Industrial, agricultural, research, medical, pharmaceutical or other peaceful purposes;

(b) Protective purposes, namely those purposes directly related to protection against toxic chemicals and to protection against chemical weapons;

(c) Military purposes not connected with the use of chemical weapons and not dependent on the use of the toxic properties of chemicals as a method of warfare;

(d) Law enforcement including domestic riot control purposes.

10. 'Production Capacity' means:

The annual quantitative potential for manufacturing a specific chemical based on the technological process actually used or, if the process is not yet operational, planned to be used at the relevant facility. It shall be deemed to be equal to the nameplate capacity or, if the nameplate capacity is not available, to the design capacity. The nameplate capacity is the product output under conditions optimized for maximum quantity for the production facility, as demonstrated by one or more test-runs. The design capacity is the corresponding theoretically calculated product output.

11. 'Organization' means the Organization for the Prohibition of Chemical Weapons established pursuant to Article VIII of this Convention.

12. For the purposes of Article VI:

(a) 'Production' of a chemical means its formation through chemical reaction;

(b) 'Processing' of a chemical means a physical process, such as formulation, extraction and purification, in which a chemical is not converted into another chemical;

(c) 'Consumption' of a chemical means its conversion into another chemical via a chemical reaction.

### Article III. Declarations

1. Each State Party shall submit to the Organization, not later than 30 days after this Convention enters into force for it, the following declarations, in which it shall:

(a) With respect to chemical weapons:

(i) Declare whether it owns or possesses any chemical weapons, or whether there are any chemical weapons located in any place under its jurisdiction or control;

(ii) Specify the precise location, aggregate quantity and detailed inventory of chemical weapons it owns or possesses, or that are located in any place under its jurisdiction or control, in accordance with Part IV (A), paragraphs 1 to 3, of the Verification Annex, except for those chemical weapons referred to in sub-subparagraph (iii);

(iii) Report any chemical weapons on its territory that are owned and possessed by another State and located in any place under the jurisdiction or control of another State, in accordance with Part IV (A), paragraph 4, of the Verification Annex;

(iv) Declare whether it has transferred or received, directly or indirectly, any chemical weapons since 1 January 1946 and specify the transfer or receipt of such weapons, in accordance with Part IV (A), paragraph 5, of the Verification Annex;

(v) Provide its general plan for destruction of chemical weapons that it owns or possesses, or that are located in any place under its jurisdiction or control, in accordance with Part IV (A), paragraph 6, of the Verification Annex;

(b) With respect to old chemical weapons and abandoned chemical weapons:

(i) Declare whether it has on its territory old chemical weapons and provide all available information in accordance with Part IV (B), paragraph 3, of the Verification Annex;

(ii) Declare whether there are abandoned chemical weapons on its territory and provide all available information in accordance with Part IV (B), paragraph 8, of the Verification Annex;

(iii) Declare whether it has abandoned chemical weapons on the territory of other States and provide all available information in accordance with Part IV (B), paragraph 10, of the Verification Annex;

(c) With respect to chemical weapons production facilities:

(i) Declare whether it has or has had any chemical weapons production facility under its ownership or possession, or that is or has been located in any place under its jurisdiction or control at any time since 1 January 1946;

(ii) Specify any chemical weapons production facility it has or has had under its ownership or possession or that is or has been located in any place under its jurisdiction or control at any time since 1 January 1946, in accordance with Part V, paragraph 1, of the Verification Annex, except for those facilities referred to in sub-subparagraph (iii);

(iii) Report any chemical weapons production facility on its territory that another State has or has had under its ownership and possession and that is or has been located in any place under the jurisdiction or control of another State at any time since 1 January 1946, in accordance with Part V, paragraph 2, of the Verification Annex;

(iv) Declare whether it has transferred or received, directly or indirectly, any equipment for the production of chemical weapons since 1 January 1946 and specify the transfer or receipt of such equipment, in accordance with Part V, paragraphs 3 to 5, of the Verification Annex;

(v) Provide its general plan for destruction of any chemical weapons production facility it owns or possesses, or that is located in any place under its jurisdiction or control, in accordance with Part V, paragraph 6, of the Verification Annex;

(vi) Specify actions to be taken for closure of any chemical weapons production facility it owns or possesses, or that is located in any place under its jurisdiction or control, in accordance with Part V, paragraph 1 (i), of the Verification Annex;

(vii) Provide its general plan for any temporary conversion of any chemical weapons production facility it owns or possesses, or that is located in any place under its jurisdiction or control, into a chemical weapons destruction facility, in accordance with Part V, paragraph 7, of the Verification Annex;

(d) With respect to other facilities:

Specify the precise location, nature and general scope of activities of any facility or establishment under its ownership or possession, or located in any place under its jurisdiction or control, and that has been designed, constructed or used since 1 January 1946 primarily for development of chemical weapons. Such declaration shall include, *inter alia*, laboratories and test and evaluation sites;

(e) With respect to riot control agents: Specify the chemical name, structural formula and Chemical Abstracts Service (CAS) registry number, if assigned, of each chemical it holds for riot control purposes. This declaration shall be updated not later than 30 days after any change becomes effective.

2. The provisions of this Article and the relevant provisions of Part IV of the Verification Annex shall not, at the discretion of a State Party, apply to chemical weapons buried on its territory before 1 January 1977 and which remain buried, or which had been dumped at sea before 1 January 1985.

**Article IV. Chemical Weapons**

1. The provisions of this Article and the detailed procedures for its implementation shall apply to all chemical weapons owned or possessed by a State Party, or that are located in any place under its jurisdiction or control, except old chemical weapons and abandoned chemical weapons to which Part IV (B) of the Verification Annex applies.

2. Detailed procedures for the implementation of this Article are set forth in the Verification Annex.

3. All locations at which chemical weapons specified in paragraph 1 are stored or destroyed shall be subject to systematic verification through on-site inspection and monitoring with on-site instruments, in accordance with Part IV (A) of the Verification Annex.

4. Each State Party shall, immediately after the declaration under Article III, paragraph 1 (a), has been submitted, provide access to chemical weapons specified in paragraph 1 for the purpose of systematic verification of the declaration through on-site inspection. Thereafter, each State Party shall not remove any of these chemical weapons, except to a chemical weapons destruction facility. It shall provide access to such chemical weapons, for the purpose of systematic on-site verification.

5. Each State Party shall provide access to any chemical weapons destruction facilities and their storage areas, that it owns or possesses, or that are located in any place under its jurisdiction or control, for the purpose of systematic verification through on-site inspection and monitoring with on-site instruments.

6. Each State Party shall destroy all chemical weapons specified in paragraph 1 pursuant to the Verification Annex and in accordance with the agreed rate and sequence of destruction (hereinafter referred to as 'order of destruction'). Such destruction shall begin not later than two years after this Convention enters into force for it and shall finish not later than 10 years after entry into force of this Convention. A State Party is not precluded from destroying such chemical weapons at a faster rate.

7. Each State Party shall:

(a) Submit detailed plans for the destruction of chemical weapons specified in paragraph 1 not later than 60 days before each annual destruction period begins, in accordance with Part IV (A), paragraph 29, of the Verification Annex; the detailed plans shall encompass all stocks to be destroyed during the next annual destruction period;

(b) Submit declarations annually regarding the implementation of its plans for destruction of chemical weapons specified in paragraph 1, not later than 60 days after the end of each annual destruction period; and

(c) Certify, not later than 30 days after the destruction process has been completed, that all chemical weapons specified in paragraph 1 have been destroyed.

8. If a State ratifies or accedes to this Convention after the 10 year period for destruction set forth in paragraph 6, it shall destroy chemical weapons specified in paragraph 1 as soon as possible. The order of destruction and procedures for stringent verification for such a State Party shall be determined by the Executive Council.

9. Any chemical weapons discovered by a State Party after the initial declaration of chemical weapons shall be reported, secured and destroyed in accordance with Part IV (A) of the Verification Annex.

10. Each State Party, during transportation, sampling, storage and destruction of chemical weapons, shall assign the highest priority to ensuring the safety of people and to protecting the environment. Each State Party shall transport, sample, store and destroy chemical weapons in accordance with its national standards for safety and emissions.

11. Any State Party which has on its territory chemical weapons that are owned or possessed by another State, or that are located in any place under the jurisdiction or control of another State, shall make the fullest efforts to ensure that these chemical weapons are removed from its territory not later than one year after this Convention enters into force for it. If they are not removed within one year, the State Party may request the Organization and other States Parties to provide assistance in the destruction of these chemical weapons.

12. Each State Party undertakes to cooperate with other States Parties that request information or assistance on a bilateral basis or through the Technical Secretariat regarding methods and technologies for the safe and efficient destruction of chemical weapons.

13. In carrying out verification activities pursuant to this Article and Part IV (A) of the Verification Annex, the Organization shall consider measures to avoid unnecessary duplication of bilateral or multilateral agreements on verification of chemical weapons storage and their destruction among States Parties.

To this end, the Executive Council shall decide to limit verification to measures complementary to those undertaken pursuant to such a bilateral or multilateral agreement, if it considers that:

(a) Verification provisions of such an agreement are consistent with the verification provisions of this Article and Part IV (A) of the Verification Annex;

(b) Implementation of such an agreement provides for sufficient assurance of compliance with the relevant provisions of this Convention; and

(c) Parties to the bilateral or multilateral agreement keep the Organization fully informed about their verification activities.

14. If the Executive Council takes a decision pursuant to paragraph 13, the Organization shall have the right to monitor the implementation of the bilateral or multilateral agreement.

15. Nothing in paragraphs 13 and 14 shall affect the obligation of a State Party to provide declarations pursuant to Article III, this Article and Part IV (A) of the Verification Annex.

16. Each State Party shall meet the costs of destruction of chemical weapons it is obliged to destroy. It shall also meet the costs of verification of storage and destruction of these chemical weapons unless the Executive Council decides otherwise. If the Executive Council decides to limit verification measures of the Organization pursuant to paragraph 13, the costs of complementary verification and monitoring by the Organization shall be paid in accordance with the United Nations scale of assessment, as specified in Article VIII, paragraph 7.

17. The provisions of this Article and the relevant provisions of Part IV of the Verification Annex shall not, at the discretion of a State Party, apply to chemical weapons buried on its territory before 1 January 1977 and which remain buried, or which had been dumped at sea before 1 January 1985.

### Article V. Chemical Weapons Production Facilities

1. The provisions of this Article and the detailed procedures for its implementation shall apply to any and all chemical weapons production facilities owned or possessed by a

State Party, or that are located in any place under its jurisdiction or control.

2. Detailed procedures for the implementation of this Article are set forth in the Verification Annex.

3. All chemical weapons production facilities specified in paragraph 1 shall be subject to systematic verification through on-site inspection and monitoring with on-site instruments in accordance with Part V of the Verification Annex.

4. Each State Party shall cease immediately all activity at chemical weapons production facilities specified in paragraph 1, except activity required for closure.

5. No State Party shall construct any new chemical weapons production facilities or modify any existing facilities for the purpose of chemical weapons production or for any other activity prohibited under this Convention.

6. Each State Party shall, immediately after the declaration under Article III, paragraph 1 (c), has been submitted, provide access to chemical weapons production facilities specified in paragraph 1, for the purpose of systematic verification of the declaration through on-site inspection.

7. Each State Party shall:

(a) Close, not later than 90 days after this Convention enters into force for it, all chemical weapons production facilities specified in paragraph 1, in accordance with Part V of the Verification Annex, and give notice thereof; and

(b) Provide access to chemical weapons production facilities specified in paragraph 1, subsequent to closure, for the purpose of systematic verification through on-site inspection and monitoring with on-site instruments in order to ensure that the facility remains closed and is subsequently destroyed.

8. Each State Party shall destroy all chemical weapons production facilities specified in paragraph 1 and related facilities and equipment, pursuant to the Verification Annex and in accordance with an agreed rate and sequence of destruction (hereinafter referred to as 'order of destruction'). Such destruction shall begin not later than one year after this Convention enters into force for it, and shall finish not later than 10 years after entry into force of this Convention. A State Party is not precluded from destroying such facilities at a faster rate.

9. Each State Party shall:

(a) Submit detailed plans for destruction of chemical weapons production facilities specified in paragraph 1, not later than 180 days before the destruction of each facility begins;

(b) Submit declarations annually regarding the implementation of its plans for the destruction of all chemical weapons production facilities specified in paragraph 1, not later than 90 days after the end of each annual destruction period; and

(c) Certify, not later than 30 days after the destruction process has been completed, that all chemical weapons production facilities specified in paragraph 1 have been destroyed.

10. If a State ratifies or accedes to this Convention after the 10-year period for destruction set forth in paragraph 8, it shall destroy chemical weapons production facilities specified in paragraph 1 as soon as possible. The order of destruction and procedures for stringent verification for such a State Party shall be determined by the Executive Council.

11. Each State Party, during the destruction of chemical weapons production facilities, shall assign the highest priority to ensuring the safety of people and to protecting the environment. Each State Party shall destroy chemical weapons production facilities in accordance with its national standards for safety and emissions.

12. Chemical weapons production facilities specified in paragraph 1 may be temporarily converted for destruction of chemical weapons in accordance with Part V, paragraphs 18 to 25, of the Verification Annex. Such a converted facility must be destroyed as soon as it is no longer in use for destruction of chemical weapons but, in any case, not later than 10 years after entry into force of this Convention.

13. A State Party may request, in exceptional cases of compelling need, permission to use a chemical weapons production facility specified in paragraph 1 for purposes not prohibited under this Convention. Upon the recommendation of the Executive Council, the Conference of the States Parties shall decide whether or not to approve the request and shall establish the conditions upon which approval is contingent in accordance with Part V, Section D, of the Verification Annex.

14. The chemical weapons production facility shall be converted in such a manner that the converted facility is not more capable of being reconverted into a chemical weapons production facility than any other facility used for industrial, agricultural, research, medical, pharmaceutical or other

peaceful purposes not involving chemicals listed in Schedule 1.

15. All converted facilities shall be subject to systematic verification through on-site inspection and monitoring with on-site instruments in accordance with Part V, Section D, of the Verification Annex.

16. In carrying out verification activities pursuant to this Article and Part V of the Verification Annex, the Organization shall consider measures to avoid unnecessary duplication of bilateral or multilateral agreements on verification of chemical weapons production facilities and their destruction among States Parties.

To this end, the Executive Council shall decide to limit the verification to measures complementary to those undertaken pursuant to such a bilateral or multilateral agreement, if it considers that:

(a) Verification provisions of such an agreement are consistent with the verification provisions of this Article and Part V of the Verification Annex;

(b) Implementation of the agreement provides for sufficient assurance of compliance with the relevant provisions of this Convention; and

(c) Parties to the bilateral or multilateral agreement keep the Organization fully informed about their verification activities.

17. If the Executive Council takes a decision pursuant to paragraph 16, the Organization shall have the right to monitor the implementation of the bilateral or multilateral agreement.

18. Nothing in paragraphs 16 and 17 shall affect the obligation of a State Party to make declarations pursuant to Article III, this Article and Part V of the Verification Annex.

19. Each State Party shall meet the costs of destruction of chemical weapons production facilities it is obliged to destroy. It shall also meet the costs of verification under this Article unless the Executive Council decides otherwise. If the Executive Council decides to limit verification measures of the Organization pursuant to paragraph 16, the costs of complementary verification and monitoring by the Organization shall be paid in accordance with the United Nations scale of assessment, as specified in Article VIII, paragraph 7.

**Article VI. Activities not Prohibited under this Convention**

1. Each State Party has the right, subject to the provisions of this Convention, to develop, produce, otherwise acquire, retain, transfer and use toxic chemicals and their precursors for purposes not prohibited under this Convention.

2. Each State Party shall adopt the necessary measures to ensure that toxic chemicals and their precursors are only developed, produced, otherwise acquired, retained, transferred, or used within its territory or in any other place under its jurisdiction or control for purposes not prohibited under this Convention. To this end, and in order to verify that activities are in accordance with obligations under this Convention, each State Party shall subject toxic chemicals and their precursors listed in Schedules 1, 2 and 3 of the Annex on Chemicals, facilities related to such chemicals, and other facilities as specified in the Verification Annex, that are located on its territory or in any other place under its jurisdiction or control, to verification measures as provided in the Verification Annex.

3. Each State Party shall subject chemicals listed in Schedule 1 (hereinafter referred to as 'Schedule 1 chemicals') to the prohibitions on production, acquisition, retention, transfer and use as specified in Part VI of the Verification Annex. It shall subject Schedule 1 chemicals and facilities specified in Part VI of the Verification Annex to systematic verification through on-site inspection and monitoring with on-site instruments in accordance with that Part of the Verification Annex.

4. Each State Party shall subject chemicals listed in Schedule 2 (hereinafter referred to as 'Schedule 2 chemicals') and facilities specified in Part VII of the Verification Annex to data monitoring and on-site verification in accordance with that Part of the Verification Annex.

5. Each State Party shall subject chemicals listed in Schedule 3 (hereinafter referred to as 'Schedule 3 chemicals') and facilities specified in Part VIII of the Verification Annex to data monitoring and on-site verification in accordance with that Part of the Verification Annex.

6. Each State Party shall subject facilities specified in Part IX of the Verification Annex to data monitoring and eventual on-site verification in accordance with that Part of the Verification Annex unless decided otherwise by the Conference of the States Parties pursuant to Part IX, paragraph 22, of the Verification Annex.

7. Not later than 30 days after this Convention enters into force for it, each State

Party shall make an initial declaration on relevant chemicals and facilities in accordance with the Verification Annex.

8. Each State Party shall make annual declarations regarding the relevant chemicals and facilities in accordance with the Verification Annex.

9. For the purpose of on-site verification, each State Party shall grant to the inspectors access to facilities as required in the Verification Annex.

10. In conducting verification activities, the Technical Secretariat shall avoid undue intrusion into the State Party's chemical activities for purposes not prohibited under this Convention and, in particular, abide by the provisions set forth in the Annex on the Protection of Confidential Information (hereinafter referred to as 'Confidentiality Annex').

11. The provisions of this Article shall be implemented in a manner which avoids hampering the economic or technological development of States Parties, and international co-operation in the field of chemical activities for purposes not prohibited under this Convention including the international exchange of scientific and technical information and chemicals and equipment for the production, processing or use of chemicals for purposes not prohibited under this Convention.

### Article VII. National Implementation Measures

General undertakings

1. Each State Party shall, in accordance with its constitutional processes, adopt the necessary measures to implement its obligations under this Convention. In particular, it shall:

(a) Prohibit natural and legal persons anywhere on its territory or in any other place under its jurisdiction as recognized by international law from undertaking any activity prohibited to a State Party under this Convention, including enacting penal legislation with respect to such activity;

(b) Not permit in any place under its control any activity prohibited to a State Party under this Convention; and

(c) Extend its penal legislation enacted under subparagraph (a) to any activity prohibited to a State Party under this Convention undertaken anywhere by natural persons, possessing its nationality, in conformity with international law.

2. Each State Party shall cooperate with other States Parties and afford the appropriate form of legal assistance to facilitate the implementation of the obligations under paragraph 1.

3. Each State Party, during the implementation of its obligations under this Convention, shall assign the highest priority to ensuring the safety of people and to protecting the environment, and shall cooperate as appropriate with other States Parties in this regard.

Relations between the State Party and the Organization

4. In order to fulfil its obligations under this Convention, each State Party shall designate or establish a National Authority to serve as the national focal point for effective liaison with the Organization and other States Parties. Each State Party shall notify the Organization of its National Authority at the time that this Convention enters into force for it.

5. Each State Party shall inform the Organization of the legislative and administrative measures taken to implement this Convention.

6. Each State Party shall treat as confidential and afford special handling to information and data that it receives in confidence from the Organization in connection with the implementation of this Convention. It shall treat such information and data exclusively in connection with its rights and obligations under this Convention and in accordance with the provisions set forth in the Confidentiality Annex.

7. Each State Party undertakes to cooperate with the Organization in the exercise of all its functions and in particular to provide assistance to the Technical Secretariat.

### Article VIII. The Organization

A. General Provisions

1. The States Parties to this Convention hereby establish the Organization for the Prohibition of Chemical Weapons to achieve the object and purpose of this Convention, to ensure the implementation of its provisions, including those for international verification of compliance with it, and to provide a forum for consultation and cooperation among States Parties.

2. All States Parties to this Convention shall be members of the Organization. A State Party shall not be deprived of its membership in the Organization.

3. The seat of the Headquarters of the Organization shall be The Hague, Kingdom of the Netherlands.

4. There are hereby established as the organs of the Organization: the Conference of the States Parties, the Executive Council, and the Technical Secretariat.

5. The Organization shall conduct its verification activities provided for under this Convention in the least intrusive manner possible consistent with the timely and efficient accomplishment of their objectives. It shall request only the information and data necessary to fulfil its responsibilities under this Convention. It shall take every precaution to protect the confidentiality of information on civil and military activities and facilities coming to its knowledge in the implementation of this Convention and, in particular, shall abide by the provisions set forth in the Confidentiality Annex.

6. In undertaking its verification activities the Organization shall consider measures to make use of advances in science and technology.

7. The costs of the Organization's activities shall be paid by States Parties in accordance with the United Nations scale of assessment adjusted to take into account differences in membership between the United Nations and this Organization, and subject to the provisions of Articles IV and V. Financial contributions of States Parties to the Preparatory Commission shall be deducted in an appropriate way from their contributions to the regular budget. The budget of the Organization shall comprise two separate chapters, one relating to administrative and other costs, and one relating to verification costs.

8. A member of the Organization which is in arrears in the payment of its financial contribution to the Organization shall have no vote in the Organization if the amount of its arrears equals or exceeds the amount of the contribution due from it for the preceding two full years. The Conference of the States Parties may, nevertheless, permit such a member to vote if it is satisfied that the failure to pay is due to conditions beyond the control of the member.

B. The Conference of the States Parties

Composition, procedures and decision-making

9. The Conference of the States Parties (hereinafter referred to as 'the Conference') shall be composed of all members of this Organization. Each member shall have one representative in the Conference, who may be accompanied by alternates and advisers.

10. The first session of the Conference shall be convened by the depositary not later than 30 days after the entry into force of this Convention.

11. The Conference shall meet in regular sessions which shall be held annually unless it decides otherwise.

12. Special sessions of the Conference shall be convened:

(a) When decided by the Conference;

(b) When requested by the Executive Council;

(c) When requested by any member and supported by one third of the members; or

(d) In accordance with paragraph 22 to undertake reviews of the operation of this Convention.

Except in the case of subparagraph (d), the special session shall be convened not later than 30 days after receipt of the request by the Director-General of the Technical Secretariat, unless specified otherwise in the request.

13. The Conference shall also be convened in the form of an Amendment Conference in accordance with Article XV, paragraph 2.

14. Sessions of the Conference shall take place at the seat of the Organization unless the Conference decides otherwise.

15. The Conference shall adopt its rules of procedure. At the beginning of each regular session, it shall elect its Chairman and such other officers as may be required. They shall hold office until a new Chairman and other officers are elected at the next regular session.

16. A majority of the members of the Organization shall constitute a quorum for the Conference.

17. Each member of the Organization shall have one vote in the Conference.

18. The Conference shall take decisions on questions of procedure by a simple majority of the members present and voting. Decisions on matters of substance should be taken as far as possible by consensus. If consensus is not attainable when an issue comes up for decision, the Chairman shall defer any vote for 24 hours and during this period of deferment shall make every effort to facilitate achievement of consensus, and shall report to the Conference before the end of this period. If consensus is not possible at the end of 24 hours, the Conference shall take the decision by a two-thirds majority of members present

and voting unless specified otherwise in this Convention. When the issue arises as to whether the question is one of substance or not, that question shall be treated as a matter of substance unless otherwise decided by the Conference by the majority required for decisions on matters of substance.

Powers and functions

19. The Conference shall be the principal organ of the Organization. It shall consider any questions, matters or issues within the scope of this Convention, including those relating to the powers and functions of the Executive Council and the Technical Secretariat. It may make recommendations and take decisions on any questions, matters or issues related to this Convention raised by a State Party or brought to its attention by the Executive Council.

20. The Conference shall oversee the implementation of this Convention, and act in order to promote its object and purpose. The Conference shall review compliance with this Convention. It shall also oversee the activities of the Executive Council and the Technical Secretariat and may issue guidelines in accordance with this Convention to either of them in the exercise of their functions.

21. The Conference shall:

(a) Consider and adopt at its regular sessions the report, programme and budget of the Organization, submitted by the Executive Council, as well as consider other reports;

(b) Decide on the scale of financial contributions to be paid by States Parties in accordance with paragraph 7;

(c) Elect the members of the Executive Council;

(d) Appoint the Director-General of the Technical Secretariat (hereinafter referred to as 'the Director-General');

(e) Approve the rules of procedure of the Executive Council submitted by the latter;

(f) Establish such subsidiary organs as it finds necessary for the exercise of its functions in accordance with this Convention;

(g) Foster international cooperation for peaceful purposes in the field of chemical activities;

(h) Review scientific and technological developments that could affect the operation of this Convention and, in this context, direct the Director-General to establish a Scientific Advisory Board to enable him, in the performance of his functions, to render specialized advice in areas of science and technology relevant to this Convention, to the Conference, the Executive Council or States Parties. The Scientific Advisory Board shall be composed of independent experts appointed in accordance with terms of reference adopted by the Conference;

(i) Consider and approve at its first session any draft agreements, provisions and guidelines developed by the Preparatory Commission;

(j) Establish at its first session the voluntary fund for assistance in accordance with Article X;

(k) Take the necessary measures to ensure compliance with this Convention and to redress and remedy any situation which contravenes the provisions of this Convention, in accordance with Article XII.

22. The Conference shall not later than one year after the expiry of the fifth and the tenth year after the entry into force of this Convention, and at such other times within that time period as may be decided upon, convene in special sessions to undertake reviews of the operation of this Convention. Such reviews shall take into account any relevant scientific and technological developments. At intervals of five years thereafter, unless otherwise decided upon, further sessions of the Conference shall be convened with the same objective.

C. The Executive Council

Composition, procedure and decision-making

23. The Executive Council shall consist of 41 members. Each State Party shall have the right, in accordance with the principle of rotation, to serve on the Executive Council. The members of the Executive Council shall be elected by the Conference for a term of two years. In order to ensure the effective functioning of this Convention, due regard being specially paid to equitable geographical distribution, to the importance of chemical industry, as well as to political and security interests, the Executive Council shall be composed as follows:

(a) Nine States Parties from Africa to be designated by States Parties located in this region. As a basis for this designation it is understood that, out of these nine States Parties, three members shall, as a rule, be the States Parties with the most significant national chemical industry in the region as determined by internationally reported and published data; in addition, the regional group shall agree also to take into account other regional factors in designating these three members;

(b) Nine States Parties from Asia to be designated by States Parties located in this region. As a basis for this designation it is understood that, out of these nine States Parties, four members shall, as a rule, be the States Parties with the most significant national chemical industry in the region as determined by internationally reported and published data; in addition, the regional group shall agree also to take into account other regional factors in designating these four members;

(c) Five States Parties from Eastern Europe to be designated by States Parties located in this region. As a basis for this designation it is understood that, out of these five States Parties, one member shall, as a rule, be the State Party with the most significant national chemical industry in the region as determined by internationally reported and published data; in addition, the regional group shall agree also to take into account other regional factors in designating this one member;

(d) Seven States Parties from Latin America and the Caribbean to be designated by States Parties located in this region. As a basis for this designation it is understood that, out of these seven States Parties, three members shall, as a rule, be the States Parties with the most significant national chemical industry in the region as determined by internationally reported and published data; in addition, the regional group shall agree also to take into account other regional factors in designating these three members;

(e) Ten States Parties from among Western European and Other States to be designated by States Parties located in this region. As a basis for this designation it is understood that, out of these ten States Parties, five members shall, as a rule, be the States Parties with the most significant national chemical industry in the region as determined by internationally reported and published data; in addition, the regional group shall agree also to take into account other regional factors in designating these five members;

(f) One further State Party to be designated consecutively by States Parties located in the regions of Asia and Latin America and the Caribbean. As a basis for this designation it is understood that this State Party shall be a rotating member from these regions.

24. For the first election of the Executive Council 20 members shall be elected for a term of one year, due regard being paid to the established numerical proportions as described in paragraph 23.

25. After the full implementation of Articles IV and V the Conference may, upon the request of a majority of the members of the Executive Council, review the composition of the Executive Council taking into account developments related to the principles specified in paragraph 23 that are governing its composition.

26. The Executive Council shall elaborate its rules of procedure and submit them to the Conference for approval.

27. The Executive Council shall elect its Chairman from among its members.

28. The Executive Council shall meet for regular sessions. Between regular sessions it shall meet as often as may be required for the fulfillment of its powers and functions.

29. Each member of the Executive Council shall have one vote. Unless otherwise specified in this Convention, the Executive Council shall take decisions on matters of substance by a two-thirds majority of all its members. The Executive Council shall take decisions on questions of procedure by a simple majority of all its members. When the issue arises as to whether the question is one of substance or not, that question shall be treated as a matter of substance unless otherwise decided by the Executive Council by the majority required for decisions on matters of substance.

Powers and functions

30. The Executive Council shall be the executive organ of the Organization. It shall be responsible to the Conference. The Executive Council shall carry out the powers and functions entrusted to it under this Convention, as well as those functions delegated to it by the Conference. In so doing, it shall act in conformity with the recommendations, decisions and guidelines of the Conference and assure their proper and continuous implementation.

31. The Executive Council shall promote the effective implementation of, and compliance with, this Convention. It shall supervise the activities of the Technical Secretariat, cooperate with the National Authority of each State Party and facilitate consultations and cooperation among States Parties at their request.

32. The Executive Council shall:

(a) Consider and submit to the Conference the draft programme and budget of the Organization;

(b) Consider and submit to the Conference the draft report of the Organization on the implementation of this Convention, the report on the performance of its own activities and such special reports as it deems necessary or which the Conference may request;

(c) Make arrangements for the sessions of the Conference including the preparation of the draft agenda.

33. The Executive Council may request the convening of a special session of the Conference.

34. The Executive Council shall:

(a) Conclude agreements or arrangements with States and international organizations on behalf of the Organization, subject to prior approval by the Conference;

(b) Conclude agreements with States Parties on behalf of the Organization in connection with Article X and supervise the voluntary fund referred to in Article X;

(c) Approve agreements or arrangements relating to the implementation of verification activities, negotiated by the Technical Secretariat with States Parties.

35. The Executive Council shall consider any issue or matter within its competence affecting this Convention and its implementation, including concerns regarding compliance, and cases of non-compliance, and, as appropriate, inform States Parties and bring the issue or matter to the attention of the Conference.

36. In its consideration of doubts or concerns regarding compliance and cases of non-compliance, including, *inter alia*, abuse of the rights provided for under this Convention, the Executive Council shall consult with the States Parties involved and, as appropriate, request the State Party to take measures to redress the situation within a specified time. To the extent that the Executive Council considers further action to be necessary, it shall take, *inter alia*, one or more of the following measures:

(a) Inform all States Parties of the issue or matter;

(b) Bring the issue or matter to the attention of the Conference;

(c) Make recommendations to the Conference regarding measures to redress the situation and to ensure compliance.

The Executive Council shall, in cases of particular gravity and urgency, bring the issue or matter, including relevant information and conclusions, directly to the attention of the United Nations General Assembly and the United Nations Security Council. It shall at the same time inform all States Parties of this step.

D. The Technical Secretariat

37. The Technical Secretariat shall assist the Conference and the Executive Council in the performance of their functions. The Technical Secretariat shall carry out the verification measures provided for in this Convention. It shall carry out the other functions entrusted to it under this Convention as well as those functions delegated to it by the Conference and the Executive Council.

38. The Technical Secretariat shall:

(a) Prepare and submit to the Executive Council the draft programme and budget of the Organization;

(b) Prepare and submit to the Executive Council the draft report of the Organization on the implementation of this Convention and such other reports as the Conference or the Executive Council may request;

(c) Provide administrative and technical support to the Conference, the Executive Council and subsidiary organs;

(d) Address and receive communications on behalf of the Organization to and from States Parties on matters pertaining to the implementation of this Convention;

(e) Provide technical assistance and technical evaluation to States Parties in the implementation of the provisions of this Convention, including evaluation of scheduled and unscheduled chemicals.

39. The Technical Secretariat shall:

(a) Negotiate agreements or arrangements relating to the implementation of verification activities with States Parties, subject to approval by the Executive Council;

(b) Not later than 180 days after entry into force of this Convention, coordinate the establishment and maintenance of permanent stockpiles of emergency and humanitarian assistance by States Parties in accordance with Article X, paragraphs 7 (b) and (c). The Technical Secretariat may inspect the items maintained for serviceability. Lists of items to be stockpiled shall be considered and approved by the Conference pursuant to paragraph 21 (i) above;

(c) Administer the voluntary fund referred to in Article X, compile declarations made by the States Parties and register, when requested, bilateral agreements concluded between States Parties or between a State Party and the Organization for the purposes of Article X.

40. The Technical Secretariat shall inform the Executive Council of any problem that has arisen with regard to the discharge of its functions, including doubts, ambiguities or uncertainties about compliance with this Convention that have come to its notice in the performance of its verification activities and that it has been unable to resolve or clarify through its consultations with the State Party concerned.

41. The Technical Secretariat shall comprise a Director-General, who shall be its head and chief administrative officer, inspectors and such scientific, technical and other personnel as may be required.

42. The Inspectorate shall be a unit of the Technical Secretariat and shall act under the supervision of the Director-General.

43. The Director-General shall be appointed by the Conference upon the recommendation of the Executive Council for a term of four years, renewable for one further term, but not thereafter.

44. The Director-General shall be responsible to the Conference and the Executive Council for the appointment of the staff and the organization and functioning of the Technical Secretariat. The paramount consideration in the employment of the staff and in the determination of the conditions of service shall be the necessity of securing the highest standards of efficiency, competence and integrity. Only citizens of States Parties shall serve as the Director-General, as inspectors or as other members of the professional and clerical staff. Due regard shall be paid to the importance of recruiting the staff on as wide a geographical basis as possible. Recruitment shall be guided by the principle that the staff shall be kept to a minimum necessary for the proper discharge of the responsibilities of the Technical Secretariat.

45. The Director-General shall be responsible for the organization and functioning of the Scientific Advisory Board referred to in paragraph 21 (h). The Director-General shall, in consultation with States Parties, appoint members of the Scientific Advisory Board, who shall serve in their individual capacity. The members of the Board shall be appointed on the basis of their expertise in the particular scientific fields relevant to the implementation of this Convention. The Director-General may also, as appropriate, in consultation with members of the Board, establish temporary working groups of scientific experts to provide recommendations on specific issues. In regard to the above, States Parties may submit lists of experts to the Director-General.

46. In the performance of their duties, the Director-General, the inspectors and the other members of the staff shall not seek or receive instructions from any Government or from any other source external to the Organization. They shall refrain from any action that might reflect on their positions as international officers responsible only to the Conference and the Executive Council.

47. Each State Party shall respect the exclusively international character of the responsibilities of the Director-General, the inspectors and the other members of the staff and not seek to influence them in the discharge of their responsibilities.

E. Privileges and Immunities

48. The Organization shall enjoy on the territory and in any other place under the jurisdiction or control of a State Party such legal capacity and such privileges and immunities as are necessary for the exercise of its functions.

49. Delegates of States Parties, together with their alternates and advisers, representatives appointed to the Executive Council together with their alternates and advisers, the Director-General and the staff of the Organization shall enjoy such privileges and immunities as are necessary in the independent exercise of their functions in connection with the Organization.

50. The legal capacity, privileges, and immunities referred to in this Article shall be defined in agreements between the Organization and the States Parties as well as in an agreement between the Organization and the State in which the headquarters of the Organization is seated. These agreements shall be considered and approved by the Conference pursuant to paragraph 21 (i).

51. Notwithstanding paragraphs 48 and 49, the privileges and immunities enjoyed by the Director-General and the staff of the Technical Secretariat during the conduct of verification activities shall be those set forth in Part II, Section B, of the Verification Annex.

## Article IX. Consultations, Cooperation and Fact-Finding

1. States Parties shall consult and cooperate, directly among themselves, or through the Organization or other appropriate international procedures, including procedures within the framework of the United

Nations and in accordance with its Charter, on any matter which may be raised relating to the object and purpose, or the implementation of the provisions, of this Convention.

2. Without prejudice to the right of any State Party to request a challenge inspection, States Parties should, whenever possible, first make every effort to clarify and resolve, through exchange of information and consultations among themselves, any matter which may cause doubt about compliance with this Convention, or which gives rise to concerns about a related matter which may be considered ambiguous. A State Party which receives a request from another State Party for clarification of any matter which the requesting State Party believes causes such a doubt or concern shall provide the requesting State Party as soon as possible, but in any case not later than 10 days after the request, with information sufficient to answer the doubt or concern raised along with an explanation of how the information provided resolves the matter. Nothing in this Convention shall affect the right of any two or more States Parties to arrange by mutual consent for inspections or any other procedures among themselves to clarify and resolve any matter which may cause doubt about compliance or gives rise to a concern about a related matter which may be considered ambiguous. Such arrangements shall not affect the rights and obligations of any State Party under other provisions of this Convention.

Procedure for requesting clarification

3. A State Party shall have the right to request the Executive Council to assist in clarifying any situation which may be considered ambiguous or which gives rise to a concern about the possible non-compliance of another State Party with this Convention. The Executive Council shall provide appropriate information in its possession relevant to such a concern.

4. A State Party shall have the right to request the Executive Council to obtain clarification from another State Party on any situation which may be considered ambiguous or which gives rise to a concern about its possible non-compliance with this Convention. In such a case, the following shall apply:

(a) The Executive Council shall forward the request for clarification to the State Party concerned through the Director-General not later than 24 hours after its receipt;

(b) The requested State Party shall provide the clarification to the Executive Council as soon as possible, but in any case not later than 10 days after the receipt of the request;

(c) The Executive Council shall take note of the clarification and forward it to the requesting State Party not later than 24 hours after its receipt;

(d) If the requesting State Party deems the clarification to be inadequate, it shall have the right to request the Executive Council to obtain from the requested State Party further clarification;

(e) For the purpose of obtaining further clarification requested under subparagraph (d), the Executive Council may call on the Director-General to establish a group of experts from the Technical Secretariat, or if appropriate staff are not available in the Technical Secretariat, from elsewhere, to examine all available information and data relevant to the situation causing the concern. The group of experts shall submit a factual report to the Executive Council on its findings;

(f) If the requesting State Party considers the clarification obtained under subparagraphs (d) and (e) to be unsatisfactory, it shall have the right to request a special session of the Executive Council in which States Parties involved that are not members of the Executive Council shall be entitled to take part. In such a special session, the Executive Council shall consider the matter and may recommend any measure it deems appropriate to resolve the situation.

5. A State Party shall also have the right to request the Executive Council to clarify any situation which has been considered ambiguous or has given rise to a concern about its possible non-compliance with this Convention. The Executive Council shall respond by providing such assistance as appropriate.

6. The Executive Council shall inform the States Parties about any request for clarification provided in this Article.

7. If the doubt or concern of a State Party about a possible non-compliance has not been resolved within 60 days after the submission of the request for clarification to the Executive Council, or it believes its doubts warrant urgent consideration, notwithstanding its right to request a challenge inspection, it may request a special session of the Conference in accordance with Article VIII, paragraph 12 (c). At such a special session, the Conference shall consider the matter and may recommend any measure it deems appropriate to resolve the situation.

Procedures for Challenge Inspections

8. Each State Party has the right to request an on-site challenge inspection of any facility or location in the territory or in any other place under the jurisdiction or control of any other State Party for the sole purpose of clarifying and resolving any questions concerning possible non-compliance with the provisions of this Convention, and to have this inspection conducted anywhere without delay by an inspection team designated by the Director-General and in accordance with the Verification Annex.

9. Each State Party is under the obligation to keep the inspection request within the scope of this Convention and to provide in the inspection request all appropriate information on the basis of which a concern has arisen regarding possible non-compliance with this Convention as specified in the Verification Annex. Each State Party shall refrain from unfounded inspection requests, care being taken to avoid abuse. The challenge inspection shall be carried out for the sole purpose of determining facts relating to the possible non-compliance.

10. For the purpose of verifying compliance with the provisions of this Convention, each State Party shall permit the Technical Secretariat to conduct the on-site challenge inspection pursuant to paragraph 8.

11. Pursuant to a request for a challenge inspection of a facility or location, and in accordance with the procedures provided for in the Verification Annex, the inspected State Party shall have:

(a) The right and the obligation to make every reasonable effort to demonstrate its compliance with this Convention and, to this end, to enable the inspection team to fulfil its mandate;

(b) The obligation to provide access within the requested site for the sole purpose of establishing facts relevant to the concern regarding possible non-compliance; and

(c) The right to take measures to protect sensitive installations, and to prevent disclosure of confidential information and data, not related to this Convention.

12. With regard to an observer, the following shall apply:

(a) The requesting State Party may, subject to the agreement of the inspected State Party, send a representative who may be a national either of the requesting State Party or of a third State Party, to observe the conduct of the challenge inspection.

(b) The inspected State Party shall then grant access to the observer in accordance with the Verification Annex.

(c) The inspected State Party shall, as a rule, accept the proposed observer, but if the inspected State Party exercises a refusal, that fact shall be recorded in the final report.

13. The requesting State Party shall present an inspection request for an on-site challenge inspection to the Executive Council and at the same time to the Director-General for immediate processing.

14. The Director-General shall immediately ascertain that the inspection request meets the requirements specified in Part X, paragraph 4, of the Verification Annex, and, if necessary, assist the requesting State Party in filing the inspection request accordingly. When the inspection request fulfils the requirements, preparations for the challenge inspection shall begin.

15. The Director-General shall transmit the inspection request to the inspected State Party not less than 12 hours before the planned arrival of the inspection team at the point of entry.

16. After having received the inspection request, the Executive Council shall take cognizance of the Director-General's actions on the request and shall keep the case under its consideration throughout the inspection procedure. However, its deliberations shall not delay the inspection process.

17. The Executive Council may, not later than 12 hours after having received the inspection request, decide by a three-quarter majority of all its members against carrying out the challenge inspection, if it considers the inspection request to be frivolous, abusive or clearly beyond the scope of this Convention as described in paragraph 8. Neither the requesting nor the inspected State Party shall participate in such a decision. If the Executive Council decides against the challenge inspection, preparations shall be stopped, no further action on the inspection request shall be taken, and the States Parties concerned shall be informed accordingly.

18. The Director-General shall issue an inspection mandate for the conduct of the challenge inspection. The inspection mandate shall be the inspection request referred to in paragraphs 8 and 9 put into operational terms, and shall conform with the inspection request.

19. The challenge inspection shall be conducted in accordance with Part X or, in the case of alleged use, in accordance with

Part XI of the Verification Annex. The inspection team shall be guided by the principle of conducting the challenge inspection in the least intrusive manner possible, consistent with the effective and timely accomplishment of its mission.

20. The inspected State Party shall assist the inspection team throughout the challenge inspection and facilitate its task. If the inspected State Party proposes, pursuant to Part X, Section C, of the Verification Annex, arrangements to demonstrate compliance with this Convention, alternative to full and comprehensive access, it shall make every reasonable effort, through consultations with the inspection team, to reach agreement on the modalities for establishing the facts with the aim of demonstrating its compliance.

21. The final report shall contain the factual findings as well as an assessment by the inspection team of the degree and nature of access and cooperation granted for the satisfactory implementation of the challenge inspection. The Director-General shall promptly transmit the final report of the inspection team to the requesting State Party, to the inspected State Party, to the Executive Council and to all other States Parties. The Director-General shall further transmit promptly to the Executive Council the assessments of the requesting and of the inspected States Parties, as well as the views of other States Parties which may be conveyed to the Director-General for that purpose, and then provide them to all States Parties.

22. The Executive Council shall, in accordance with its powers and functions, review the final report of the inspection team as soon as it is presented, and address any concerns as to:

(a) Whether any non-compliance has occurred;

(b) Whether the request had been within the scope of this Convention; and

(c) Whether the right to request a challenge inspection had been abused.

23. If the Executive Council reaches the conclusion, in keeping with its powers and functions, that further action may be necessary with regard to paragraph 22, it shall take the appropriate measures to redress the situation and to ensure compliance with this Convention, including specific recommendations to the Conference. In the case of abuse, the Executive Council shall examine whether the requesting State Party should bear any of the financial implications of the challenge inspection.

24. The requesting State Party and the inspected State Party shall have the right to participate in the review process. The Executive Council shall inform the States Parties and the next session of the Conference of the outcome of the process.

25. If the Executive Council has made specific recommendations to the Conference, the Conference shall consider action in accordance with Article XII.

### Article X. Assistance and Protection against Chemical Weapons

1. For the purposes of this Article, 'Assistance' means the coordination and delivery to States Parties of protection against chemical weapons, including, *inter alia*, the following: detection equipment and alarm systems; protective equipment; decontamination equipment and decontaminants; medical antidotes and treatments; and advice on any of these protective measures.

2. Nothing in this Convention shall be interpreted as impeding the right of any State Party to conduct research into, develop, produce, acquire, transfer or use means of protection against chemical weapons, for purposes not prohibited under this Convention.

3. Each State Party undertakes to facilitate, and shall have the right to participate in, the fullest possible exchange of equipment, material and scientific and technological information concerning means of protection against chemical weapons.

4. For the purposes of increasing the transparency of national programmes related to protective purposes, each State Party shall provide annually to the Technical Secretariat information on its programme, in accordance with procedures to be considered and approved by the Conference pursuant to Article VIII, paragraph 21 (i).

5. The Technical Secretariat shall establish, not later than 180 days after entry into force of this Convention and maintain, for the use of any requesting State Party, a data bank containing freely available information concerning various means of protection against chemical weapons as well as such information as may be provided by States Parties.

The Technical Secretariat shall also, within the resources available to it, and at the request of a State Party, provide expert advice and assist the State Party in identifying how its programmes for the development and improvement of a protective capacity against chemical weapons could be implemented.

6. Nothing in this Convention shall be interpreted as impeding the right of States Parties to request and provide assistance bilaterally and to conclude individual agreements with other States Parties concerning the emergency procurement of assistance.

7. Each State Party undertakes to provide assistance through the Organization and to this end to elect to take one or more of the following measures:

(a) To contribute to the voluntary fund for assistance to be established by the Conference at its first session;

(b) To conclude, if possible not later than 180 days after this Convention enters into force for it, agreements with the Organization concerning the procurement, upon demand, of assistance;

(c) To declare, not later than 180 days after this Convention enters into force for it, the kind of assistance it might provide in response to an appeal by the Organization. If, however, a State Party subsequently is unable to provide the assistance envisaged in its declaration, it is still under the obligation to provide assistance in accordance with this paragraph.

8. Each State Party has the right to request and, subject to the procedures set forth in paragraphs 9, 10 and 11, to receive assistance and protection against the use or threat of use of chemical weapons if it considers that:

(a) Chemical weapons have been used against it;

(b) Riot control agents have been used against it as a method of warfare; or

(c) It is threatened by actions or activities of any State that are prohibited for States Parties by Article I.

9. The request, substantiated by relevant information, shall be submitted to the Director-General, who shall transmit it immediately to the Executive Council and to all States Parties. The Director-General shall immediately forward the request to States Parties which have volunteered, in accordance with paragraphs 7 (b) and (c), to dispatch emergency assistance in case of use of chemical weapons or use of riot control agents as a method of warfare, or humanitarian assistance in case of serious threat of use of chemical weapons or serious threat of use of riot control agents as a method of warfare to the State Party concerned not later than 12 hours after receipt of the request. The Director-General shall initiate, not later than 24 hours after receipt of the request, an investigation in order to provide foundation for further action. He shall complete the investigation within 72 hours and forward a report to the Executive Council. If additional time is required for completion of the investigation, an interim report shall be submitted within the same time-frame. The additional time required for investigation shall not exceed 72 hours. It may, however, be further extended by similar periods. Reports at the end of each additional period shall be submitted to the Executive Council. The investigation shall, as appropriate and in conformity with the request and the information accompanying the request, establish relevant facts related to the request as well as the type and scope of supplementary assistance and protection needed.

10. The Executive Council shall meet not later than 24 hours after receiving an investigation report to consider the situation and shall take a decision by simple majority within the following 24 hours on whether to instruct the Technical Secretariat to provide supplementary assistance. The Technical Secretariat shall immediately transmit to all States Parties and relevant international organizations the investigation report and the decision taken by the Executive Council. When so decided by the Executive Council, the Director-General shall provide assistance immediately. For this purpose, the Director-General may cooperate with the requesting State Party, other States Parties and relevant international organizations. The States Parties shall make the fullest possible efforts to provide assistance.

11. If the information available from the ongoing investigation or other reliable sources would give sufficient proof that there are victims of use of chemical weapons and immediate action is indispensable, the Director-General shall notify all States Parties and shall take emergency measures of assistance, using the resources the Conference has placed at his disposal for such contingencies. The Director-General shall keep the Executive Council informed of actions undertaken pursuant to this paragraph.

### Article XI. Economic and Technological Development

1. The provisions of this Convention shall be implemented in a manner which avoids hampering the economic or technological development of States Parties, and international cooperation in the field of chemical activities for purposes not prohibited under this Convention including the international

exchange of scientific and technical information and chemicals and equipment for the production, processing or use of chemicals for purposes not prohibited under this Convention.

2. Subject to the provisions of this Convention and without prejudice to the principles and applicable rules of international law, the States Parties shall:

(a) Have the right, individually or collectively, to conduct research with, to develop, produce, acquire, retain, transfer, and use chemicals;

(b) Undertake to facilitate, and have the right to participate in, the fullest possible exchange of chemicals, equipment and scientific and technical information relating to the development and application of chemistry for purposes not prohibited under this Convention;

(c) Not maintain among themselves any restrictions, including those in any international agreements, incompatible with the obligations undertaken under this Convention, which would restrict or impede trade and the development and promotion of scientific and technological knowledge in the field of chemistry for industrial, agricultural, research, medical, pharmaceutical or other peaceful purposes;

(d) Not use this Convention as grounds for applying any measures other than those provided for, or permitted, under this Convention nor use any other international agreement for pursuing an objective inconsistent with this Convention;

(e) Undertake to review their existing national regulations in the field of trade in chemicals in order to render them consistent with the object and purpose of this Convention.

### Article XII. Measures to Redress a Situation and to Ensure Compliance, Including Sanctions

1. The Conference shall take the necessary measures, as set forth in paragraphs 2, 3 and 4, to ensure compliance with this Convention and to redress and remedy any situation which contravenes the provisions of this Convention. In considering action pursuant to this paragraph, the Conference shall take into account all information and recommedations on the issues submitted by the Executive Council.

2. In cases where a State Party has been requested by the Executive Council to take measures to redress a situation raising problems with regard to its compliance, and where the State Party fails to fulfil the request within the specified time, the Conference may, *inter alia*, upon the recommendation of the Executive Council, restrict or suspend the State Party's rights and privileges under this Convention until it undertakes the necessary action to conform with its obligations under this Convention.

3. In cases where serious damage to the object and purpose of this Convention may result from activities prohibited under this Convention, in particular by Article I, the Conference may recommend collective measures to States Parties in conformity with international law.

4. The Conference shall in cases of particular gravity, bring the issue, including relevant information and conclusions, to the attention of the United Nations General Assembly and the United Nations Security Council.

### Article XIII. Relation to Other International Agreements

Nothing in this Convention shall be interpreted as in any way limiting or detracting from the obligations assumed by any State under the Protocol for the Prohibition of the Use in War of Asphyxiating, Poisonous or Other Gases, and of Bacteriological Methods of Warfare, signed at Geneva on 17 June 1925, and under the Convention on the Prohibition of the Development, Production and Stockpiling of Bacteriological (Biological) and Toxin Weapons and on Their Destruction, signed at London, Moscow and Washington on 10 April 1972.

### Article XIV. Settlement of Disputes

1. Disputes that may arise concerning the application or the interpretation of this Convention shall be settled in accordance with the relevant provisions of this Convention and in conformity with the provisions of the Charter of the United Nations.

2. When a dispute arises between two or more States Parties, or between one or more States Parties and the Organization, relating to the interpretation or application of this Convention, the parties concerned shall consult together with a view to the expeditious settlement of the dispute by negotiation or by other peaceful means of the parties' choice, including recourse to appropriate organs of this Convention and, by mutual consent, referral to the International Court of Justice

in conformity with the Statute of the Court. The States Parties involved shall keep the Executive Council informed of actions being taken.

3. The Executive Council may contribute to the settlement of a dispute by whatever means it deems appropriate, including offering its good offices, calling upon the States Parties to a dispute to start the settlement process of their choice and recommending a time-limit for any agreed procedure.

4. The Conference shall consider questions related to disputes raised by States Parties or brought to its attention by the Executive Council. The Conference shall, as it finds necessary, establish or entrust organs with tasks related to the settlement of these disputes in conformity with Article VIII, paragraph 21 (f).

5. The Conference and the Executive Council are separately empowered, subject to authorization from the General Assembly of the United Nations, to request the International Court of Justice to give an advisory opinion on any legal question arising within the scope of the activities of the Organization. An agreement between the Organization and the United Nations shall be concluded for this purpose in accordance with Article VIII, paragraph 34 (a).

6. This Article is without prejudice to Article IX or to the provisions on measures to redress a situation and to ensure compliance, including sanctions.

### Article XV. Amendments

1. Any State Party may propose amendments to this Convention. Any State Party may also propose changes, as specified in paragraph 4, to the Annexes of this Convention. Proposals for amendments shall be subject to the procedures in paragraphs 2 and 3. Proposals for changes, as specified in paragraph 4, shall be subject to the procedures in paragraph 5.

2. The text of a proposed amendment shall be submitted to the Director-General for circulation to all States Parties and to the Depositary. The proposed amendment shall be considered only by an Amendment Conference. Such an Amendment Conference shall be convened if one third or more of the States Parties notify the Director-General not later than 30 days after its circulation that they support further consideration of the proposal. The Amendment Conference shall be held immediately following a regular session of the Conference unless the requesting States Parties ask for an earlier meeting. In no case shall an Amendment Conference be held less than 60 days after the circulation of the proposed amendment.

3. Amendments shall enter into force for all States Parties 30 days after deposit of the instruments of ratification or acceptance by all the States Parties referred to under subparagraph (b) below:

(a) When adopted by the Amendment Conference by a positive vote of a majority of all States Parties with no State Party casting a negative vote; and

(b) Ratified or accepted by all those States Parties casting a positive vote at the Amendment Conference.

4. In order to ensure the viability and the effectiveness of this Convention, provisions in the Annexes shall be subject to changes in accordance with paragraph 5, if proposed changes are related only to matters of an administrative or technical nature. All changes to the Annex on Chemicals shall be made in accordance with paragraph 5. Sections A and C of the Confidentiality Annex, Part X of the Verification Annex, and those definitions in Part I of the Verification Annex which relate exclusively to challenge inspections, shall not be subject to changes in accordance with paragraph 5.

5. Proposed changes referred to in paragraph 4 shall be made in accordance with the following procedures:

(a) The text of the proposed changes shall be transmitted together with the necessary information to the Director-General. Additional information for the evaluation of the proposal may be provided by any State Party and the Director-General. The Director-General shall promptly communicate any such proposals and information to all States Parties, the Executive Council and the Depositary;

(b) Not later than 60 days after its receipt, the Director-General shall evaluate the proposal to determine all its possible consequences for the provisions of this Convention and its implementation and shall communicate any such information to all States Parties and the Executive Council;

(c) The Executive Council shall examine the proposal in the light of all information available to it, including whether the proposal fulfils the requirements of paragraph 4. Not later than 90 days after its receipt, the Executive Council shall notify its recommendation, with appropriate explanations, to all States Parties for consideration. States

Parties shall acknowledge receipt within 10 days;

(d) If the Executive Council recommends to all States Parties that the proposal be adopted, it shall be considered approved if no State Party objects to it within 90 days after receipt of the recommendation. If the Executive Council recommends that the proposal be rejected, it shall be considered rejected if no State Party objects to the rejection within 90 days after receipt of the recommendation;

(e) If a recommendation of the Executive Council does not meet with the acceptance required under subparagraph (d), a decision on the proposal, including whether it fulfils the requirements of paragraph 4, shall be taken as a matter of substance by the Conference at its next session;

(f) The Director-General shall notify all States Parties and the Depositary of any decision under this paragraph;

(g) Changes approved under this procedure shall enter into force for all States Parties 180 days after the date of notification by the Director-General of their approval unless another time period is recommended by the Executive Council or decided by the Conference.

**Article XVI. Duration and Withdrawal**

1. This Convention shall be of unlimited duration.

2. Each State Party shall, in exercising its national sovereignty, have the right to withdraw from this Convention if it decides that extraordinary events, related to the subject matter of this Convention, have jeopardized the supreme interests of its country. It shall give notice of such withdrawal 90 days in advance to all other States Parties, the Executive Council, the Depositary and the United Nations Security Council. Such notice shall include a statement of the extraordinary events it regards as having jeopardized its supreme interests.

3. The withdrawal of a State Party from this Convention shall not in any way affect the duty of States to continue fulfilling the obligations assumed under any relevant rules of international law, particularly the Geneva Protocol of 1925.

**Article XVII. Status of the Annexes**

The Annexes form an integral part of this Convention. Any reference to this Convention includes the Annexes.

**Article XVIII. Signature**

This Convention shall be open for signature for all States before its entry into force.

**Article XIX. Ratification**

This Convention shall be subject to ratification by States Signatories according to their respective constitutional processes.

**Article XX. Accession**

Any State which does not sign this Convention before its entry into force may accede to it at any time thereafter.

**Article XXI. Entry into Force**

1. This Convention shall enter into force 180 days after the date of the deposit of the 65th instrument of ratification, but in no case earlier than two years after its opening for signature.

2. For States whose instruments of ratification or accession are deposited subsequent to the entry into force of this Convention, it shall enter into force on the 30th day following the date of deposit of their instrument of ratification or accession.

**Article XXII. Reservations**

The Articles of this Convention shall not be subject to reservations. The Annexes of this Convention shall not be subject to reservations incompatible with its object and purpose.

**Article XXIII. Depositary**

The Secretary-General of the United Nations is hereby designated as the Depositary of this Convention and shall, *inter alia*:

(a) Promptly inform all signatory and acceding States of the date of each signature, the date of deposit of each instrument of ratification or accession and the date of the entry into force of this Convention, and of the receipt of other notices;

(b) Transmit duly certified copies of this Convention to the Governments of all signatory and acceding States; and

(c) Register this Convention pursuant to Article 102 of the Charter of the United Nations.

**Article XXIV. Authentic Texts**

This Convention, of which the Arabic, Chinese, English, French, Russian and Spanish texts are equally authentic, shall be deposited with the Secretary-General of the United Nations.

IN WITNESS WHEREOF the undersigned, being duly authorized to that effect, have signed this Convention.

Done at . . . on . . .

## ANNEX ON IMPLEMENTATION AND VERIFICATION (VERIFICATION ANNEX)

Contents


**Part I. Definitions**

**Part II. General Rules of Verification**

A. Designation of inspectors and inspection assistants
B. Privileges and immunities
C. Standing arrangements ( Points of entry, Arrangements for use of non-scheduled aircraft, Administrative arrangements, Approved equipment)
D. Pre-inspection activities (Notification, Entry into the territory of the inspected State Party or Host State and transfer to the inspection site, Pre-inspection briefing)
E. Conduct of inspections (General rules, Safety, Communications, Inspection team and inspected State Party rights, Collection, handling and analysis of samples, Extension of inspection duration, Debriefing)
F. Departure
G. Reports
H. Application of general provisions

**Part III. General provisions for verification measures pursuant to Articles IV, V and VI, paragraph 3**

A. Initial inspections and facility agreements
B. Standing arrangements
C. Pre-inspection activities

**Part IV (A). Destruction of Chemical Weapons and its verification pursuant to Article IV**

A. Declarations (Chemical weapons, Declaration of chemical weapons pursuant to Article III, paragraph 1 (a) (iii), Declaration of past transfers and receipts, Submission of the general plan for destruction of chemical weapons)
B. Measures to secure the storage facility and storage facility preparation
C. Destruction (Principles and methods for destruction of chemical weapons, Order of destruction, Modification of intermediate destruction deadlines, Extension of the deadline for completion of destruction, Detailed annual plans for destruction, Annual reports on destruction)
D. Verification (Verification of declarations of chemical weapons through on-site inspection, Systematic verification of storage facilities, Inspections and visits, Systematic verification of the destruction of chemical weapons, Chemical weapons storage facilities at chemical weapons destruction facilities, Systematic on-site verification measures at chemical weapons destruction facilities)

**Part IV (B). Old chemical weapons and abandoned chemical weapons**

A. General
B. Regime for old chemical weapons
C. Regime for abandoned chemical weapons

**Part V. Destruction of chemical weapons production facilities and its verification pursuant to Article V**

A. Declarations (Declarations of chemical weapons production facilities, Declarations of chemical weapons production facilities pursuant to Article III, paragraph 1 (c) (iii), Declarations of past transfers and receipts, Submission of general plans for destruction, Submission of annual plans for destruction and annual reports on destruction)
B. Destruction (General principles for destruction of chemical weapons production facilities, Principles and methods for closure of a chemical weapons production facility, Technical maintenance of chemical weapons production facilities prior to their destruction, Principles and methods for temporary conversion of chemical weapons production facilities into chemical weapons destruction facilities, Principles and methods related to destruction of a chemical weapons production facility, Order of destruction, Detailed plans for destruction, Review of detailed plans)
C. Verification (Verification of declarations of chemical weapons production facilities through on-site inspection, Systematic verification of chemical weapons production facilities and cessation of their activities, Verification of destruction of chemical weapons production facilities, Verification of temporary conversion of a chemical weapons production facility into a chemical weapons destruction facility)
D. Conversion of chemical weapons production facilities to purposes not prohibited under this Convention (Procedures for requesting conversion, Actions pending a decision, Conditions for conversion, Decisions

by the Executive Council and the Conference, Detailed plans for conversion, Review of detailed plans)

**Part VI. Activities not prohibited under this Convention in accordance with Article VI: Regime for Schedule 1 chemicals and facilities related to such chemicals**

A. General provisions
B. Transfers
C. Production (General principles for production, Single small-scale facility, Other facilities)
D. Declarations (Single small-scale facility, Other facilities referred to in paragraphs 10 and 11)
E. Verification (Single small-scale facility, Other facilities referred to in paragraphs 10 and 11)

**Part VII. Activities not prohibited under this Convention in accordance with Article VI: Regime for Schedule 2 chemicals and facilities related to such chemicals**

A. Declarations (Declarations of aggregate national data, Declarations of plant sites producing, processing or consuming Schedule 2 chemicals, Declarations on past production of Schedule 2 chemicals for chemical weapons purposes, Information to States Parties )
B. Verification (General, Inspection aims, Initial inspections, Inspections, Inspection procedures, Notification of Inspection)
C. Transfers to States not Party to this Convention

**Part VIII. Activities not prohibited under this Convention in accordance with Article VI: Regime for Schedule 3 chemicals and facilities related to such chemicals**

A. Declarations (Declarations of aggregate national data, Declarations of plant sites producing Schedule 3 chemicals, Declarations on past production of Schedule 3 chemicals for chemical weapons purposes, Information to States Parties)
B. Verification (General, Inspection aims, Inspection procedures, Notification of inspection)
C. Transfers to States not Party to this Convention

**Part IX. Activities not prohibited under this Convention in accordance with Article VI: Regime for other chemical production facilities**

A. Declarations (List of other chemical production facilities, Assistance by the Technical Secretariat, Information to States Parties)
B. Verification (General, Inspection aims, Inspection procedures, Notification of inspection)
C. Implementation and review of Section B (Implementation, Review)

**Part X. Challenge inspections pursuant to Article IX**

A. Designation and selection of inspectors and inspection assistants
B. Pre-inspection activities (Notification, Entry into the territory of the inspected State Party or the Host State, Alternative determination of final perimeter, Verification of location, Securing the site, exit monitoring, Pre-inspection briefing and inspection plan, Perimeter activities)
C. Conduct of inspections (General rules, Managed access, Observer, Duration of inspection)
D. Post-inspection activities (Departure, Reports)

**Part XI. Investigations in cases of alleged use of chemical weapons**

A. General
B. Pre-inspection activities (Request for an investigation, Notification, Assignment of inspection team, Dispatch of inspection team, Briefings)
C. Conduct of inspections (Access, Sampling, Extension of inspection site, Extension of inspection duration, Interviews)
D. Reports (Procedures, Contents)
E. States not Party to this Convention

# Annexes

Annexe A. Major multilateral arms control agreements

Annexe B. Chronology 1992

# **Annexe A.** Major multilateral arms control agreements

RAGNHILD FERM

For the texts of the arms control agreements, see Goldblat, J., SIPRI, *Agreements for Arms Control: A Critical Survey* (Taylor & Francis: London, 1982); for the Treaty of Rarotonga, see SIPRI, *World Armaments and Disarmament: SIPRI Yearbook 1986* (Oxford University Press: Oxford, 1986), pp. 509–19; for the CFE Treaty, see SIPRI, *SIPRI Yearbook 1991: World Armaments and Disarmament* (Oxford University Press: Oxford, 1991), pp. 461–74.

## I. Summaries of the agreements

### Protocol for the prohibition of the use in war of asphyxiating, poisonous or other gases, and of bacteriological methods of warfare (Geneva Protocol)

*Signed at Geneva on 17 June 1925; entered into force on 8 February 1928.*

Declares that the parties agree to be bound by the above prohibition, which should be universally accepted as part of international law, binding alike the conscience and the practice of nations.

### Antarctic Treaty

*Signed at Washington on 1 December 1959; entered into force on 23 June 1961.*

Declares the Antarctic an area to be used exclusively for peaceful purposes. Prohibits any measure of a military nature in the Antarctic, such as the establishment of military bases and fortifications, and the carrying out of military manœuvres or the testing of any type of weapon. Bans any nuclear explosion as well as the disposal of radioactive waste material in Antarctica, subject to possible future international agreements on these subjects.

At regular intervals consultative meetings are convened to exchange information and hold consultations on matters pertaining to Antarctica, as well as to recommend to the governments measures in furtherance of the principles and objectives of the Treaty. A Protocol on the protection of the Antarctic environment was signed on 4 October 1991.

### Treaty banning nuclear weapon tests in the atmosphere, in outer space and under water (Partial Test Ban Treaty—PTBT)

*Signed at Moscow on 5 August 1963; entered into force on 10 October 1963.*

Prohibits the carrying out of any nuclear weapon test explosion or any other nuclear explosion: (*a*) in the atmosphere, beyond its limits, including outer space, or under water, including territorial waters or high seas; (*b*) in any other environment if such explosion causes radioactive debris to be present outside the territorial limits of the state under whose jurisdiction or control the explosion is conducted.

**Treaty on principles governing the activities of states in the exploration and use of outer space, including the moon and other celestial bodies (Outer Space Treaty)**

*Signed at London, Moscow and Washington on 27 January 1967; entered into force on 10 October 1967.*

Prohibits the placing into orbit around the earth of any objects carrying nuclear weapons or any other kinds of weapons of mass destruction, the installation of such weapons on celestial bodies, or the stationing of them in outer space in any other manner. The establishment of military bases, installations and fortifications, the testing of any type of weapons and the conduct of military manœuvres on celestial bodies are also forbidden.

**Treaty for the prohibition of nuclear weapons in Latin America (Treaty of Tlatelolco)**

*Signed at Mexico City on 14 February 1967; entered into force on 22 April 1968.*

Prohibits the testing, use, manufacture, production or acquisition by any means, as well as the receipt, storage, installation, deployment and any form of possession of any nuclear weapons by Latin American countries.

The parties should conclude agreements with the IAEA for the application of safeguards to their nuclear activities.

Under *Additional Protocol I* the extra-continental or continental states which, *de jure* or *de facto*, are internationally responsible for territories lying within the limits of the geographical zone established by the Treaty (France, the Netherlands, the UK and the USA) undertake to apply the statute of military denuclearization, as defined in the Treaty, to such territories.

Under *Additional Protocol II* the nuclear weapon states undertake to respect the statute of military denuclearization of Latin America, as defined and delimited in the Treaty, and not to contribute to acts involving a violation of the Treaty, nor to use or threaten to use nuclear weapons against the parties to the Treaty.

In 1990 the General Conference of the Agency for the Prohibition of Nuclear Weapons in Latin America decided that the official name of the Treaty should be changed by adding the words 'and the Caribbean'; in 1991, it decided to modify the wording of Article 25, paragraph 2, which determines which states may become parties to the Treaty; and, in 1992, it decided that Articles 14, 15, 16, 19 and 20, dealing with verification of compliance (in particular, with special inspections) should be replaced by a new text. On 1 January 1993, none of these amendments was in force.

**Treaty on the non-proliferation of nuclear weapons (NPT)**

*Signed at London, Moscow and Washington on 1 July 1968; entered into force on 5 March 1970.*

Prohibits the transfer by nuclear weapon states, to any recipient whatsoever, of nuclear weapons or other nuclear explosive devices or of control over them, as well as the assistance, encouragement or inducement of any non-nuclear weapon state to manufacture or otherwise acquire such weapons or devices. Prohibits the receipt by non-nuclear weapon states from any transferor whatsoever, as well as the manufacture or other acquisition by those states of nuclear weapons or other nuclear explosive devices.

Non-nuclear weapon states undertake to conclude safeguard agreements with the International Atomic Energy Agency (IAEA) with a view to preventing diversion of nuclear energy from peaceful uses to nuclear weapons or other nuclear explosive devices.

The parties undertake to facilitate the exchange of equipment, materials and scientific and technological information for the peaceful uses of nuclear energy and to ensure that potential benefits from peaceful applications of nuclear explosions will be made available to non-nuclear weapon parties to the Treaty. They also undertake to pursue negotiations in good faith on effective measures relating to cessation of the nuclear arms race at an early date and to nuclear disarmament, and on a treaty on general and complete disarmament.

Twenty-five years after the entry into force of the Treaty (1995), a conference shall be convened to decide whether the Treaty shall continue in force indefinitely or shall be extended for an additional fixed period or periods.

**Treaty on the prohibition of the emplacement of nuclear weapons and other weapons of mass destruction on the seabed and the ocean floor and in the subsoil thereof (Seabed Treaty)**

*Signed at London, Moscow and Washington on 11 February 1971; entered into force on 18 May 1972.*

Prohibits emplanting or emplacing on the seabed and the ocean floor and in the subsoil thereof beyond the outer limit of a 12-mile seabed zone any nuclear weapons or any other types of weapons of mass destruction as well as structures, launching installations or any other facilities specifically designed for storing, testing or using such weapons.

**Convention on the prohibition of the development, production and stockpiling of bacteriological (biological) and toxin weapons and on their destruction (BW Convention)**

*Signed at London, Moscow and Washington on 10 April 1972; entered into force on 26 March 1975.*

Prohibits the development, production, stockpiling or acquisition by other means or retention of microbial or other biological agents, or toxins whatever their origin or method of production, of types and in quantities that have no justification of prophylactic, protective or other peaceful purposes, as well as weapons, equipment or means of delivery designed to use such agents or toxins for hostile purposes or in armed conflict. The destruction of the agents, toxins, weapons, equipment and means of delivery in the possession of the parties, or their diversion to peaceful purposes, should be effected not later than nine months after the entry into force of the Convention.

**Convention on the prohibition of military or any other hostile use of environmental modification techniques (Enmod Convention)**

*Signed at Geneva on 18 May 1977; entered into force on 5 October 1978.*

Prohibits military or any other hostile use of environmental modification techniques having widespread, long-lasting or severe effects as the means of destruction, damage or injury to states party to the Convention. The term 'environmental modification techniques' refers to any technique for changing—through the deliberate manipulation of natural processes—the dynamics, composition or structure of the Earth, including its biota, lithosphere, hydrosphere and atmosphere, or of outer space. The understandings reached during the negotiations, but not written into the Convention, define the terms 'widespread', 'long-lasting' and 'severe'.

**Convention on the prohibitions or restrictions on the use of certain conventional weapons which may be deemed to be excessively injurious or to have indiscriminate effects ('Inhumane Weapons' Convention)**

*Signed at New York on 10 April 1981; entered into force on 2 December 1983.*

The Convention is an 'umbrella treaty', under which specific agreements can be concluded in the form of protocols.

Protocol I prohibits the use of weapons intended to injure by fragments which are not detectable in the human body by X-rays.

Protocol II prohibits or restricts the use of mines, booby-traps and similar devices.

Protocol III restricts the use of incendiary weapons.

**South Pacific Nuclear Free Zone Treaty (Treaty of Rarotonga)**

*Signed at Rarotonga, Cook Islands, on 6 August 1985; entered into force on 11 December 1986.*

Prohibits the manufacture or acquisition by other means of any nuclear explosive device, as well as possession or control over such device by the parties anywhere inside or outside the zone area described in an annex. The parties also undertake not to supply nuclear material or equipment, unless subject to IAEA safeguards, and to prevent in their territories the stationing as well as the testing of any nuclear explosive device. Each party remains free to allow visits, as well as transit, by foreign ships and aircraft.

Under Protocol 1, France, the UK and the USA would undertake to apply the treaty prohibitions relating to the manufacture, stationing and testing of nuclear explosive devices in the territories situated within the zone, for which they are internationally responsible.

Under Protocol 2, China, France, the UK, the USA and the USSR would undertake not to use or threaten to use a nuclear explosive device against the parties to the Treaty or against any territory within the zone for which a party to Protocol 1 is internationally responsible.

Under Protocol 3, China, France, the UK, the USA and the USSR would undertake not to test any nuclear explosive device anywhere within the zone.

**Treaty on Conventional Armed Forces in Europe (CFE Treaty)**

*Signed at Paris on 19 November 1990; entered into force on 9 November 1992.*

Sets ceilings on five categories of military equipment (battle tanks, armoured combat vehicles, artillery pieces, combat aircraft and attack helicopters) in an area stretching from the Atlantic Ocean to the Ural Mountains (the ATTU zone). The CFE–1A Agreement, limiting personnel strength of conventional armed forces in the same area, was signed at Helsinki on 10 July and entered into force on 17 July 1992.

# II. Status of the implementation of the major multilateral arms control agreements, as of 31 December 1992

## Number of parties

| | | | |
|---|---|---|---|
| 1925 Geneva Protocol | 132 | Seabed Treaty | 88 |
| Antarctic Treaty | 40 | BW Convention | 126 |
| Partial Test Ban Treaty | 120 | Enmod Convention | 57 |
| Outer Space Treaty | 93 | 'Inhumane Weapons' Convention | 35 |
| Treaty of Tlatelolco | 24 | Treaty of Rarotonga | 11 |
| Additional Protocol I | 4 | Protocol 1 | 0 |
| Additional Protocol II | 5 | Protocol 2 | 2 |
| Non-Proliferation Treaty | 156 | Protocol 3 | 2 |
| NPT safeguards agreements (non-nuclear weapon states) | 94 | CFE Treaty | 29 |

## Notes

1. The Russian Federation, constituted in 1991 as an independent sovereign state, has confirmed the continuity of international obligations assumed by the Union of Soviet Socialist Republics (USSR).

2. The Federal Republic of Germany and the German Democratic Republic merged into one state in 1990. The dates of entry into force of the treaties listed in the table for the united Germany are the dates previously given for FR Germany.

3. The Yemen Arab Republic and the People's Democratic Republic of Yemen merged into one state in 1990. According to a statement by the united Yemen state, all agreements which either state has entered into are in force for Yemen. The dates of entry into force of the treaties listed in the table for Yemen are the earliest dates previously given for either of the former Yemen states.

4. The table records year of ratification, accession or succession.

5. The Partial Test Ban Treaty, the Outer Space Treaty, the Non-Proliferation Treaty, the Seabed Treaty and the BW Convention provide for three depositaries—the governments of the UK, the USA and the USSR. For these agreements, the dates indicated are the earliest dates on which countries deposited their instruments of ratification, accession or succession—whether in London, Washington or Moscow. The dates given for other agreements (for which there is only one depositary) are the dates of the deposit of the instruments of ratification, accession or succession with the relevant depositary, except in the case of the 1925 Geneva Protocol, where the dates refer to the date of notification by the depositary.

6. The 1925 Geneva Protocol, the Partial Test Ban Treaty, the Outer Space Treaty, the Non-Proliferation Treaty, the Seabed Treaty, the BW Convention, the Enmod Convention and the 'Inhumane Weapons' Convention are open to all states for signature.

The Antarctic Treaty is subject to ratification by the signatories and is open for accession by UN members or by other states invited to accede with the consent of all the contracting parties whose representatives are entitled to participate in the consultative meetings provided for in Article IX.

The Treaty of Tlatelolco is open for signature by all the Latin American republics; all other sovereign states situated in their entirety south of latitude 35° north in the western hemisphere; and (except for a political entity the territory of which is the subject of an international dispute) all such states which become sovereign, when they have been admitted by the General Conference; Additional Protocol I—by 'all extra-continental or continental states having *de jure* or *de facto* international responsibility for territories situated in the zone of application of the Treaty', that is, France, the Netherlands, the UK and the USA; Additional Protocol II—by 'all powers possessing nuclear weapons', that is, the USA, the USSR, the UK, France and China.

The Treaty of Rarotonga is open for signature by members of the South Pacific Forum; Protocol 1—by France, the UK and the USA; Protocol 2—by France, China, the USSR, the UK and the USA; Protocol 3—by France, China, the USSR, the UK and the USA.

The CFE Treaty was negotiated and signed by the members of NATO and the then WTO. In the Tashkent Document of 15 May 1992 all former Soviet republics with territory in the ATTU zone (Armenia, Azerbaijan, Belarus, Georgia, Kazakhstan, Moldova, Russia and Ukraine), as successor states of the USSR regarding the Treaty, agreed to become parties to the Treaty.

7. Key to abbreviations used in the table:

| | |
|---|---|
| S | Signature without further action |
| PI, PII | Additional Protocols to the Treaty of Tlatelolco |
| P1, P2, P3 | Protocols to the Treaty of Rarotonga |
| CP | Party entitled to participate in the consultative meetings provided for in Article IX of the Antarctic Treaty |
| SA | Nuclear safeguards agreement in force with the International Atomic Energy Agency as required by the Non-Proliferation Treaty or the Treaty of Tlatelolco, or concluded by a nuclear weapon state on a voluntary basis. |

8. Footnotes with summaries of the most important reservations/declarations are listed at the end of the table and are grouped separately under the heading for the respective agreements. Not all reservations for all treaties are given. The texts of the statements contained in the footnotes have been abridged, but the wording is close to the original version.

9. A complete list of UN member states and year of membership appears in section III.

| State | Geneva Protocol | Antarctic Treaty | Partial Test Ban Treaty | Outer Space Treaty | Treaty of Tlatelolco | Non-Proliferation Treaty | Seabed Treaty | BW Convention | Enmod Convention | 'Inhumane Weapons' Convention | Treaty of Rarotonga | CFE Treaty |
|---|---|---|---|---|---|---|---|---|---|---|---|---|
| **Afghanistan** | 1986 | | 1964 | 1988 | | 1970<br>SA | 1971 | 1975 | 1985 | S | | |
| **Albania** | 1989 | | | | | 1990 | | 1992 | | | | |
| **Algeria** | 1992 | | S | 1992 | | | 1992 | | 1991 | | | |
| **Angola** | 1990[1] | | | | | | | | | | | |
| **Antigua and Barbuda** | 1988 | | 1988 | 1988 | 1983[1] | 1985 | 1988 | | 1988 | | | |
| **Argentina** | 1969 | 1961<br>CP | 1986 | 1969 | S[2] | | 1983[1] | 1979 | 1987 | S | | |
| **Armenia** | | | | | | | | | | | | 1992 |
| **Australia** | 1930[1] | 1961<br>CP | 1963 | 1967 | | 1973<br>SA | 1973 | 1977 | 1984 | 1983 | 1986 | |
| **Austria** | 1928 | 1987 | 1964 | 1968 | | 1969<br>SA | 1972 | 1973 | 1990 | 1983 | | |
| **Azerbaijan** | | | | | | 1992 | | | | | | 1992 |
| **Bahamas** | | | 1976 | 1976 | 1977[1] | 1976 | 1989 | 1986 | | | | |

| State | Geneva Protocol | Antarctic Treaty | Partial Test Ban Treaty | Outer Space Treaty | Treaty of Tlatelolco | Non-Proliferation Treaty | Seabed Treaty | BW Convention | Enmod Convention | 'Inhumane Weapons' Convention | Treaty of Rarotonga | CFE Treaty |
|---|---|---|---|---|---|---|---|---|---|---|---|---|
| **Bahrain** | 1988[1] | | | | | 1988 | | 1988 | | | | |
| **Bangladesh** | 1989[1] | | 1985 | 1986 | | 1979<br>SA | | 1985 | 1979 | | | |
| **Barbados** | 1976[2] | | | 1968 | 1969[1] | 1980 | | 1973 | | | | |
| **Belarus** | 1970[3] | | 1963 | 1967 | | | 1971 | 1975 | 1978 | 1982 | | 1992 |
| **Belgium** | 1928[1] | 1960<br>CP | 1966 | 1973 | | 1975<br>SA | 1972 | 1979 | 1982 | S | | 1991 |
| **Belize** | | | | | S | 1985 | | 1986 | | | | |
| **Benin** | 1986 | | 1964 | 1986 | | 1972 | 1986 | 1975 | 1986 | 1989[1] | | |
| **Bhutan** | 1979 | | 1978 | | | 1985<br>SA | | 1978 | | | | |
| **Bolivia** | 1985 | | 1965 | S | 1969[1] | 1970 | S | 1975 | S | | | |
| **Botswana** | | | 1968 | S | | 1969 | 1972 | 1991 | | | | |
| **Brazil** | 1970 | 1975<br>CP | 1964 | 1969[1] | 1968[3] | | 1988[2] | 1973 | 1984 | | | |
| **Brunei Darussalam** | | | | | | 1985<br>SA | | 1991 | | | | |

| | | | | | | | | | | | | |
|---|---|---|---|---|---|---|---|---|---|---|---|---|
| **Bulgaria** | 1934[1] | 1978 | 1963 | 1967 | | 1969 SA | 1971 | 1972 | 1978 | 1982 | | 1991 |
| **Burkina Faso** | 1971 | | S | 1968 | | 1970 | | 1991 | | | | |
| **Burma see: Myanmar** | | | | | | | | | | | | |
| **Burundi** | | | S | S | | 1971 | S | S | | | | |
| **Cambodia** | 1983[4] | | | | | 1972 | S | 1983 | | | | |
| **Cameroon** | 1989 | | S | S | | 1969 | S | | | | | |
| **Canada** | 1930[1] | 1988 | 1964 | 1967 | | 1969 SA | 1972[3] | 1972 | 1981 | S | | 1991 |
| **Cape Verde** | 1992 | | 1979 | | | 1979 | 1979 | 1977 | 1979 | | | |
| **Central African Rep.** | 1970 | | 1964 | S | | 1970 | 1981 | S | | | | |
| **Chad** | | | 1965 | | | 1971 | | | | | | |
| **Chile** | 1935[1] | 1961 CP | 1965 | 1981 | 1974[4] | | | 1980 | | | | |

| State | Geneva Protocol | Antarctic Treaty | Partial Test Ban Treaty | Outer Space Treaty | Treaty of Tlatelolco | Non-Proliferation Treaty | Seabed Treaty | BW Convention | Enmod Convention | 'Inhumane Weapons' Convention | Treaty of Rarotonga | CFE Treaty |
|---|---|---|---|---|---|---|---|---|---|---|---|---|
| **China** | 1952[5] | 1983<br>CP | | 1983 | PII: 1974[5] | 1992[1] | 1991[4] | 1984 | | 1982[2] | P2: 1989<br>P3: 1989 | |
| **Colombia** | | 1989 | 1985 | S | 1972[1]<br>SA | 1986 | S | 1983 | | | | |
| **Congo** | | | | | | 1978 | 1978 | 1978 | | | | |
| **Cook Islands** | | | | | | | | | | | 1985 | |
| **Costa Rica** | | | 1967 | | 1969[1]<br>SA[14] | 1970<br>SA | S | 1973 | | | | |
| **Côte d'Ivoire** | 1970 | | 1965 | | | 1973<br>SA | 1972 | S | | | | |
| **Cuba** | 1966 | 1984 | | 1977 | | | 1977 | 1976 | 1978 | 1987 | | |
| **Cyprus** | 1966 | | 1965 | 1972 | | 1970<br>SA | 1971 | 1973 | 1978 | 1988[3] | | |
| **Czecho-slovakia** | 1938[6] | 1962 | 1963 | 1967 | | 1969<br>SA | 1972 | 1973 | 1978 | 1982 | | 1991 |
| **Denmark** | 1930 | 1965 | 1964 | 1967 | | 1969<br>SA | 1971 | 1973 | 1978 | 1982 | | 1991 |
| **Dominica** | | | | | S | 1984 | | | 1992 | | | |

| | | | | | | | | | | | | |
|---|---|---|---|---|---|---|---|---|---|---|---|---|
| **Dominican Rep.** | 1970 | | 1964 | 1968 | 1968[1] SA[14] | 1971 SA | 1972 | 1973 | | | | |
| **Ecuador** | 1970 | 1987 CP | 1964 | 1969 | 1969[1] SA[14] | 1969 SA | | 1975 | | 1982 | | |
| **Egypt** | 1928 | | 1964 | 1967 | | 1981[2] SA | | S | 1982 | S | | |
| **El Salvador** | S | | 1964 | 1969 | 1968[1] SA[14] | 1972 SA | | S | | | | |
| **Equatorial Guinea** | 1989 | | 1989 | 1989 | | 1984 | 1992 | 1989 | | | | |
| **Estonia** | 1931 | | | | | 1992 | | | | | | |
| **Ethiopia** | 1935 | | S | S | | 1970 SA | 1977 | 1975 | S | | | |
| **Fiji** | 1973[1] | | 1972 | 1972 | | 1972 SA | | 1973 | | | 1985 | |
| **Finland** | 1929 | 1984 CP | 1964 | 1967 | | 1969 SA | 1971 | 1974 | 1978 | 1982 | | |
| **France** | 1926[1] | 1960 CP | | 1970 | PI: 1992[6] PII: 1974[7] | 1992 SA[3] | | 1984 | | 1988[4] | | 1992 |
| **Gabon** | | | 1964 | | | 1974 | | S | | | | |

| State | Geneva Protocol | Antarctic Treaty | Partial Test Ban Treaty | Outer Space Treaty | Treaty of Tlatelolco | Non-Proliferation Treaty | Seabed Treaty | BW Convention | Enmod Convention | 'Inhumane Weapons' Convention | Treaty of Rarotonga | CFE Treaty |
|---|---|---|---|---|---|---|---|---|---|---|---|---|
| Gambia | 1966 | | 1965 | S | | 1975<br>SA | S | S | | | | |
| Georgia | | | | | | | | | | | | 1992 |
| Germany | 1929 | 1979<br>CP | 1964 | 1971 | | 1975[4]<br>SA | 1975 | 1983[1] | 1983 | 1992 | | 1991 |
| Ghana | 1967 | | 1963 | S | | 1970<br>SA | 1972 | 1975 | 1978 | | | |
| Greece | 1931 | 1987 | 1963 | 1971 | | 1970<br>SA | 1985 | 1975 | 1983 | 1992 | | 1992 |
| Grenada | 1989 | | | | 1975[1] | 1975 | | 1986 | | | | |
| Guatemala | 1983 | 1991 | 1964 | | 1970[1]<br>SA[14] | 1970<br>SA | S | 1973 | 1988 | 1983 | | |
| Guinea | | | | | | 1985 | S | | | | | |
| Guinea-Bissau | 1989 | | 1976 | 1976 | | 1976 | 1976 | 1976 | | | | |
| Guyana | | | | S | | | | S | | | | |
| Haiti | | | S | S | 1969[1] | 1970 | | S | | | | |

| | | | | | | | | | | | | |
|---|---|---|---|---|---|---|---|---|---|---|---|---|
| **Holy See** | 1966 | | | S | | 1971[5] SA | | | S | | | |
| **Honduras** | | | 1964 | S | 1968[1] SA[14] | 1973 SA | S | 1979 | | | | |
| **Hungary** | 1952 | 1984 | 1963 | 1967 | | 1969 SA | 1971 | 1972 | 1978 | 1982 | | 1991 |
| **Iceland** | 1967 | | 1964 | 1968 | | 1969 SA | 1972 | 1973 | S | S | | 1991 |
| **India** | 1930[1] | 1983 CP | 1963 | 1982 | | | 1973[5] | 1974[2] | 1978 | 1984 | | |
| **Indonesia** | 1971 | | 1964 | S | | 1979[6] SA | | 1992 | | | | |
| **Iran** | 1929 | | 1964 | S | | 1970 SA | 1971 | 1973 | S | | | |
| **Iraq** | 1931[1] | | 1964 | 1968 | | 1969 SA | 1972 | 1991 | S | | | |
| **Ireland** | 1930[7] | | 1963 | 1968 | | 1968 SA | 1971 | 1972[3] | 1982 | S | | |
| **Israel** | 1969[8] | | 1964 | 1977 | | | | | | | | |

| State | Geneva Protocol | Antarctic Treaty | Partial Test Ban Treaty | Outer Space Treaty | Treaty of Tlatelolco | Non-Proliferation Treaty | Seabed Treaty | BW Convention | Enmod Convention | 'Inhumane Weapons' Convention | Treaty of Rarotonga | CFE Treaty |
|---|---|---|---|---|---|---|---|---|---|---|---|---|
| **Italy** | 1928 | 1981 CP | 1964 | 1972 | | 1975[7] SA | 1974[6] | 1975 | 1981 | S[5] | | 1992 |
| **Jamaica** | 1970 | | 1991 | 1970 | 1969[1] SA[14] | 1970 SA | 1986 | 1975 | | | | |
| **Japan** | 1970 | 1960 CP | 1964 | 1967 | | 1976[8] SA | 1971 | 1982 | 1982 | 1982 | | |
| **Jordan** | 1977[9] | | 1964 | S | | 1970 SA | 1971 | 1975 | | | | |
| **Kazakhstan** | | | | | | | | | | | | 1992 |
| **Kenya** | 1970 | | 1965 | 1984 | | 1970 | | 1976 | | | | |
| **Kiribati** | | | | | | 1985 SA | | | | | 1986 | |
| **Korea (North)** | 1989[1, 10] | 1987 | | | | 1985 SA | | 1987 | 1984 | | | |
| **Korea (South)** | 1989[1] | 1986 CP | 1964 | 1967 | | 1975[9] SA | 1987 | 1987 | 1986[1] | | | |
| **Kuwait** | 1971[11] | | 1965 | 1972 | | 1989 | | 1972 | 1980 | | | |
| **Laos** | 1989 | | 1965 | 1972 | | 1970 | 1971 | 1973 | 1978 | 1983 | | |

| | | | | | | | | | | | | |
|---|---|---|---|---|---|---|---|---|---|---|---|---|
| **Latvia** | 1931 | | | | | 1992 | 1992 | | | | | |
| **Lebanon** | 1969 | | 1965 | 1969 | | 1970<br>SA | S | 1975 | S | | | |
| **Lesotho** | 1972 | | | S | | 1970<br>SA | 1973 | 1977 | | | | |
| **Liberia** | 1927 | | 1964 | | | 1970 | S | S | S | | | |
| **Libya** | 1971[12] | | 1968 | 1968 | | 1975<br>SA | 1990 | 1982 | | | | |
| **Liechtenstein** | 1991 | | | | | 1978[10]<br>SA | 1991 | 1991 | | 1989 | | |
| **Lithuania** | 1932 | | | | | 1991<br>SA | | | | | | |
| **Luxembourg** | 1936 | | 1965 | S | | 1975<br>SA | 1982 | 1976 | S | S | | 1992 |
| **Madagascar** | 1967 | | 1965 | 1968[2] | | 1970<br>SA | S | S | | | | |
| **Malawi** | 1970 | | 1964 | | | 1986<br>SA | | S | 1978 | | | |
| **Malaysia** | 1970 | | 1964 | S | | 1970<br>SA | 1972 | 1991 | | | | |

| State | Geneva Protocol | Antarctic Treaty | Partial Test Ban Treaty | Outer Space Treaty | Treaty of Tlatelolco | Non-Proliferation Treaty | Seabed Treaty | BW Convention | Enmod Convention | 'Inhumane Weapons' Convention | Treaty of Rarotonga | CFE Treaty |
|---|---|---|---|---|---|---|---|---|---|---|---|---|
| **Maldives** | 1966 | | | | | 1970<br>SA | | | | | | |
| **Mali** | | | S | 1968 | | 1970 | S | S | | | | |
| **Malta** | 1964 | | 1964 | | | 1970<br>SA | 1971 | 1975 | | | | |
| **Mauritania** | | | 1964 | | | | | | | | | |
| **Mauritius** | 1970 | | 1969 | 1969 | | 1969<br>SA | 1971 | 1972 | 1992 | | | |
| **Mexico** | 1932 | | 1963 | 1968 | 1967[1, 8]<br>SA | 1969[11]<br>SA | 1984[7] | 1974[4] | | 1982 | | |
| **Moldova** | | | | | | | | | | | | 1992 |
| **Monaco** | 1967 | | | | | | | | | | | |
| **Mongolia** | 1968[13] | | 1963 | 1967 | | 1969<br>SA | 1971 | 1972 | 1978 | 1982 | | |
| **Morocco** | 1970 | | 1966 | 1967 | | 1970<br>SA | 1971 | S | S | S | | |
| **Mozambique** | | | | | | 1990 | | | | | | |

| | | | | | | | | | | | | |
|---|---|---|---|---|---|---|---|---|---|---|---|---|
| **Myanmar (Burma)** | | | 1963 | 1970 | | 1992 | S | S | | | | |
| **Namibia** | | | | | | 1992 | | | | | | |
| **Nauru** | | | | | | 1982 SA | | | | | 1987 | |
| **Nepal** | 1969 | | 1964 | 1967 | | 1970 SA | 1971 | S | | | | |
| **Netherlands** | 1930[14] | 1967 CP | 1964 | 1969 | PI: 1971 SA[15] | 1975 SA | 1976 | 1981 | 1983[2] | 1987[6] | | 1991 |
| **New Zealand** | 1930[1] | 1960 CP | 1963 | 1968 | | 1969 SA | 1972 | 1972 | 1984 | S | 1986 | |
| **Nicaragua** | 1990 | | 1965 | S | 1968[1, 9] SA[14] | 1973 SA | 1973 | 1975 | S | S | | |
| **Niger** | 1967 | | 1964 | 1967 | | 1992 | 1971 | 1972 | | 1992 | | |
| **Nigeria** | 1968[1] | | 1967 | 1967 | | 1968 SA | | 1973 | | S | | |
| **Niue** | | | | | | | | | | | 1986 | |
| **Norway** | 1932 | 1960 CP | 1963 | 1969 | | 1969 SA | 1971 | 1973 | 1979 | 1983 | | 1991 |

| State | Geneva Protocol | Antarctic Treaty | Partial Test Ban Treaty | Outer Space Treaty | Treaty of Tlatelolco | Non-Proliferation Treaty | Seabed Treaty | BW Convention | Enmod Convention | 'Inhumane Weapons' Convention | Treaty of Rarotonga | CFE Treaty |
|---|---|---|---|---|---|---|---|---|---|---|---|---|
| **Oman** | | | | | | | | 1992 | | | | |
| **Pakistan** | 1960 | | 1988 | 1968 | | | | 1974 | 1986 | 1985 | | |
| **Panama** | 1970 | | 1966 | S | 1971[1] SA | 1977 | 1974 | 1974 | | | | |
| **Papua New Guinea** | 1981[1] | 1981 | 1980 | 1980 | | 1982 SA | | 1980 | 1980 | | 1989 | |
| **Paraguay** | 1933[15] | | S | | 1969[1] SA[14] | 1970 SA | S | 1976 | | | | |
| **Peru** | 1985 | 1981 CP | 1964 | 1979 | 1969[1] SA[14] | 1970 SA | | 1985 | | | | |
| **Philippines** | 1973 | | 1965 | S | | 1972 SA | | 1973 | | S | | |
| **Poland** | 1929 | 1961 CP | 1963 | 1968 | | 1969 SA | 1971 | 1973 | 1978 | 1983 | | 1991 |
| **Portugal** | 1930[1] | | S | | | 1977 SA | 1975 | 1975 | S | S | | 1992 |
| **Qatar** | 1976 | | | | | 1989 | 1974 | 1975 | | | | |

| | | | | | | | | | | | | |
|---|---|---|---|---|---|---|---|---|---|---|---|---|
| **Romania** | 1929[1] | 1971[1] | 1963 | 1968 | | 1970<br>SA | 1972 | 1979 | 1983 | S[7] | | 1992 |
| **Russia** | 1928[16] | 1960<br>CP | 1963 | 1967 | PII: 1979[10] | 1970<br>SA[12] | 1972 | 1975 | 1978 | 1982 | P2: 1988<br>P3: 1988 | 1992 |
| **Rwanda** | 1964 | | 1963 | S | | 1975 | 1975 | 1975 | | | | |
| **Saint Kitts and Nevis** | 1989 | | | | | | | 1991 | | | | |
| **Saint Lucia** | 1988 | | | | S | 1979<br>SA | | 1986 | | | | |
| **Saint Vincent and the Grenadines** | | | | | 1992 | 1984<br>SA | | | | | | |
| **Samoa, Western** | | | 1965 | | | 1975<br>SA | | | | | 1986 | |
| **San Marino** | | | 1964 | 1968 | | 1970 | | 1975 | | | | |
| **Sao Tome and Principe** | | | | | | 1983 | 1979 | 1979 | 1979 | | | |
| **Saudi Arabia** | 1971 | | | 1976 | | 1988 | 1972 | 1972 | | | | |
| **Senegal** | 1977 | | 1964 | | | 1970<br>SA | S | 1975 | | | | |

| State | Geneva Protocol | Antarctic Treaty | Partial Test Ban Treaty | Outer Space Treaty | Treaty of Tlatelolco | Non-Proliferation Treaty | Seabed Treaty | BW Convention | Enmod Convention | 'Inhumane Weapons' Convention | Treaty of Rarotonga | CFE Treaty |
|---|---|---|---|---|---|---|---|---|---|---|---|---|
| Seychelles | | | 1985 | 1978 | | 1985 | 1985 | 1979 | | | | |
| Sierra Leone | 1967 | | 1964 | 1967 | | 1975 | S | 1976 | S | S | | |
| Singapore | | | 1968 | 1976 | | 1976<br>SA | 1976 | 1975 | | | | |
| Slovenia | | | 1992 | | | 1992 | 1992 | 1992 | | 1992 | | |
| Solomon Islands | 1981 | | | | | 1981 | 1981 | 1981 | 1981 | | 1989 | |
| Somalia | | | S | S | | 1970 | | S | | | | |
| South Africa | 1930[1] | 1960<br>CP | 1963 | 1968 | | 1991<br>SA | 1973 | 1975 | | | | |
| Spain | 1929[17] | 1982<br>CP | 1964 | 1968 | | 1987<br>SA | 1987 | 1979 | 1978 | S | | 1992 |
| Sri Lanka | 1954 | | 1964 | 1986 | | 1979<br>SA | | 1986 | 1978 | | | |
| Sudan | 1980 | | 1966 | | | 1973<br>SA | S | | | S | | |
| Suriname | | | | | 1977[1]<br>SA[14] | 1976<br>SA | | | | | | |

| | | | | | | | | | | | |
|---|---|---|---|---|---|---|---|---|---|---|---|
| **Swaziland** | 1991 | | 1969 | | | 1969 SA | 1971 | 1991 | | | |
| **Sweden** | 1930 | 1984 CP | 1963 | 1967 | | 1970 SA | 1972 | 1976 | 1984 | 1982 | |
| **Switzerland** | 1932 | 1990 | 1964 | 1969 | | 1977[10] SA | 1976 | 1976[5] | 1988 | 1982 | |
| **Syria** | 1968 | | 1964 | 1968 | | 1969 SA | | S | S | | |
| **Taiwan** | | | 1964 | 1970 | | 1970 | 1972 | 1973 | | | |
| **Tanzania** | 1963 | | 1964 | | | 1991 | S | S | | | |
| **Thailand** | 1931 | | 1963 | 1968 | | 1972 SA | | 1975 | | | |
| **Togo** | 1971 | | 1964 | 1989 | | 1970 | 1971 | 1976 | | S | |
| **Tonga** | 1971 | | 1971 | 1971 | | 1971 | | 1976 | | | |
| **Trinidad and Tobago** | 1962 | | 1964 | S | 1970[1] SA[14] | 1986 SA | | | | | |
| **Tunisia** | 1967 | | 1965 | 1968 | | 1970 SA | 1971 | 1973 | 1978 | 1987 | |

| State | Geneva Protocol | Antarctic Treaty | Partial Test Ban Treaty | Outer Space Treaty | Treaty of Tlatelolco | Non-Proliferation Treaty | Seabed Treaty | BW Convention | Enmod Convention | 'Inhumane Weapons' Convention | Treaty of Rarotonga | CFE Treaty |
|---|---|---|---|---|---|---|---|---|---|---|---|---|
| **Turkey** | 1929 | | 1965 | 1968 | | 1980[13]<br>SA | 1972 | 1974 | S[3] | S | | 1992 |
| **Tuvalu** | | | | | | 1979<br>SA | | | | | 1986 | |
| **Uganda** | 1965 | | 1964 | 1968 | | 1982 | | 1991 | S | | | |
| **UK** | 1930[1] | 1960<br>CP | 1963 | 1967 | PI: 1969[11]<br>PII: 1969[11] | 1968<br>SA[14] | 1972 | 1975 | 1978 | S | | 1991 |
| **Ukraine** | | 1992 | 1963 | 1967 | | | 1971 | 1975 | 1978 | 1982 | | 1992 |
| **United Arab Emirates** | | | | | | | | S | | | | |
| **Uruguay** | 1977 | 1980[2]<br>CP | 1969 | 1970 | 1968[1]<br>SA[14] | 1970<br>SA | S | 1981 | | | | |
| **USA** | 1975[18] | 1960<br>CP | 1963 | 1967 | PI: 1981[12]<br>PII: 1971[13]<br>SA[15] | 1970<br>SA[15] | 1972 | 1975 | 1980 | S[8] | | 1992 |
| **Uzbekistan** | | | | | | 1992 | | | | | | |
| **Venezuela** | 1928 | | 1965 | 1970 | 1970[1]<br>SA[14] | 1975<br>SA | | 1978 | | | | |

| | | | | | | | | | | | |
|---|---|---|---|---|---|---|---|---|---|---|---|
| **Viet Nam** | 1980[1] | | | 1980 | | 1982 SA | 1980[8] | 1980 | 1980 | S | |
| **Yemen** | 1971[19] | | 1979 | 1979 | | 1979 | 1979 | 1979 | 1977 | | |
| **Yugoslavia** | 1929[20] | | 1964 | S | | 1970[16] SA | 1973[9] | 1973 | | 1983 | |
| **Zaire** | | | 1965 | S | | 1970 SA | | 1977 | S | | |
| **Zambia** | | | 1965 | 1973 | | 1991 | 1972 | | | | |
| **Zimbabwe** | | | | | | 1991 | | 1990 | | | |

*The 1925 Geneva Protocol*

[1] The Protocol is binding on this state only as regards states which have signed and ratified or acceded to it. The Protocol will cease to be binding on this state in regard to any enemy state whose armed forces or whose allies fail to respect the prohibitions laid down in it.

Australia withdrew its reservation to the Protocol in 1986, New Zealand in 1989, Romania, Bulgaria and Chile in 1991. In 1991, Canada and the UK withdrew their reservations only with regard to the right to retaliate in case of an attack by bacteriological weapons.

[2] In notifying its succession to the obligations contracted in 1930 by the UK, Barbados stated that as far as it was concerned the reservation made by the UK was to be considered as withdrawn.

[3] In a note of 2 Mar. 1970, submitted at the UN, Byelorussia stated that 'it recognizes itself to be a party' to the Protocol. However, it has not notified the depositary.

[4] The accession was made on behalf of the exiled coalition government of Democratic Kampuchea with a statement that the Protocol will cease to be binding on it in regard to any enemy state whose armed forces or whose allies fail to respect the prohibitions laid down in the Protocol. In Feb. 1990 the country was officially renamed Cambodia.

[5] On 13 July 1952 the People's Republic of China issued a statement recognizing as binding upon it the 1929 accession to the Protocol in the name of China. China considers itself bound by the Protocol on condition of reciprocity on the part of all the other contracting and acceding powers.

[6] Czechoslovakia shall cease to be bound by this Protocol towards any state whose armed forces, or the armed forces of whose allies, fail to respect the prohibitions laid down in the Protocol. This reservation was withdrawn in 1990.

[7] Ireland does not intend to assume, by this accession, any obligation except towards the states having signed and ratified this Protocol or which shall have finally acceded thereto, and should the armed forces or the allies of an enemy state fail to respect the Protocol, the government of Ireland would cease to be bound by the said Protocol in regard to such state. In 1972, Ireland declared that it had decided to withdraw the above reservations made at the time of accession to the Protocol.

[8] The Protocol is binding on Israel only as regards states which have signed and ratified or acceded to it. The Protocol shall cease to be binding on Israel as regards any enemy state whose armed forces, or the armed forces of whose allies, or the regular or irregular forces, or groups or individuals operating from its territory, fail to respect the prohibitions which are the object of the Protocol.

[9] Jordan undertakes to respect the obligations contained in the Protocol with regard to states which have undertaken similar commitments. It is not bound by the Protocol as regards states whose armed forces, regular or irregular, do not respect the provisions of the Protocol.

[10] The Dem. People's Rep. of Korea does not exclude the right to exercise its sovereignty *vis-à-vis* a contracting party which violates the Protocol in its implementation.

[11] In case of breach of the prohibition laid down in this Protocol by any of the parties, Kuwait will not be bound, with regard to the party committing the breach, to apply the provisions of this Protocol.

[12] The Protocol is binding on Libya only as regards states which are effectively bound by it and will cease to be binding on Libya as regards states whose armed forces, or the armed forces of whose allies, fail to respect the prohibitions which are the object of this Protocol.

[13] In the case of violation of this prohibition by any state in relation to Mongolia or its allies, Mongolia shall not consider itself bound by the obligations of the Protocol towards that state. This reservation was withdrawn in 1990.

[14] As regards the use in war of asphyxiating, poisonous or other gases and of all analogous liquids, materials or devices, this Protocol shall cease to be binding on the Netherlands with regard to any enemy state whose armed forces or whose allies fail to respect the prohibitions laid down in the Protocol.

[15] This is the date of receipt of Paraguay's instrument of accession. The date of the notification by the depositary government 'for the purpose of regularization' is 1969.

[16] The Protocol only binds the USSR in relation to the states which have signed and ratified or which have definitely acceded to the Protocol. The Protocol shall cease to be binding on the USSR in regard to any enemy state whose armed forces or whose allies *de jure* or *de facto* do not respect the prohibitions which are the object of this Protocol. On 29 Jan. 1992 the Russian President stated that Russia withdrew its reservation concerning the possibility of using biological weapons.

[17] For Spain the Protocol is binding *ipso facto*, without special agreement with respect to any other state accepting and observing the same obligation, that is, on condition of reciprocity. This reservation was withdrawn in 1992.

[18] The Protocol shall cease to be binding on the USA with respect to use in war of asphyxiating, poisonous or other gases, and of all analogous liquids, materials, or devices, in regard to an enemy state if such state or any of its allies fail to respect the prohibitions laid down in the Protocol.

[19] In case any party fails to observe the prohibition under the Protocol, the People's Democratic Republic of Yemen will consider itself free of its obligation. This reservation appears to be valid for the united state of Yemen, unless stated otherwise by the Government of Yemen.

[20] The Protocol shall cease to be binding on Yugoslavia in regard to any enemy state whose armed forces or whose allies fail to respect the prohibitions which are the object of the Protocol.

*The Antarctic Treaty*

[1] Romania stated that the provisions of Article XIII, para. 1 of the Treaty were not in accordance with the principle according to which multilateral treaties whose object and purposes concern the international community, as a whole, should be open for universal participation.

[2] In acceding to the Treaty, Uruguay proposed the establishment of a general and definitive statute on Antarctica in which the interests of all states involved and of the international community as a whole would be considered equitably. It also declared that it reserved its rights in Antarctica in accordance with international law.

*The Outer Space Treaty*

[1] Brazil interprets Article X of the Treaty as a specific recognition that the granting of tracking facilities by the parties to the Treaty shall be subject to agreement between the states concerned.

[2] Madagascar acceded to the Treaty with the understanding that under Article X of the Treaty the state shall retain its freedom of decision with respect to the possible installation of foreign observation bases in its territory and shall continue to possess the right to fix, in each case, the conditions for such installation.

*The Treaty of Tlatelolco*

[1] The Treaty is in force for this country due to a declaration, annexed to the instrument of ratification in accordance with Article 28, para. 2, which waived the requirements for the entry into force of the Treaty, specified in para. 1 of that Article. (Colombia made this declaration subsequent to the deposit of ratification, as did Nicaragua and Trinidad and Tobago.)

[2] On signing the Treaty, Argentina stated that it understands Article 18 as recognizing the rights of parties to carry out, by their own means or in association with third parties, explosions of nuclear devices for peaceful purposes, including explosions which involve devices similar to those used in nuclear weapons.

[3] On signing the Treaty, Brazil stated that, according to its interpretation, Article 18 of the Treaty gives the signatories the right to carry out, by their own means or in association with third parties, nuclear explosions for peaceful purposes, including explosions which involve devices similar to those used in nuclear weapons. This statement was reiterated at the ratification. Brazil has not waived the requirements for the entry into force of the Treaty laid down in Article 28. The Treaty is therefore not yet in force for Brazil.

[4] Chile has not waived the requirements for the entry into force of the Treaty laid down in Article 28. The Treaty is therefore not yet in force for Chile.

[5] On signing Protocol II, China stated, *inter alia*: China will never use or threaten to use nuclear weapons against non-nuclear Latin American countries and the Latin American nuclear weapon-free zone; nor will China test, manufacture, produce, stockpile, install or deploy nuclear weapons in these countries or in this zone, or send its means of transportation and delivery carrying nuclear weapons to cross the territory, territorial sea or airspace of Latin American countries.

China holds that, in order that Latin America may truly become a nuclear weapon-free zone, all nuclear countries, and particularly the superpowers, must undertake not to use or threaten to use nuclear weapons against the Latin American countries and the Latin American nuclear weapon-free zone, and implement the following undertakings: (1) dismantle all foreign military bases in Latin America and refrain from establishing new bases there, and (2) prohibit the passage of any means of transportation and delivery carrying nuclear weapons through Latin American territory, territorial sea or airspace.

[6] On signing Protocol I, France made the following reservations and interpretative statements: The Protocol, as well as the provisions of the Treaty to which it refers, will not affect the right of self-defence under Article 51 of the UN Charter; the application of the legislation referred to in Article 3 of the Treaty relates to legislation which is consistent with international law; the obligations under the Protocol shall not apply to transit across the territories of the French Republic situated in the zone of the Treaty, and destined to other territories of the French Republic; the Protocol shall not limit, in any way, the participation of the populations of the French territories in the activities mentioned in Article 1 of the Treaty, and in efforts connected with the national defence of France; the provisions of Articles 1 and 2 of the Protocol apply to the text of the Treaty as it stands at the time when the Protocol is signed by France, and consequently no amendment to the Treaty that might come into force under Article 29

thereof would be binding on the government of France without the latter's express consent. On ratifying Protocol I, France reiterated its statement made upon signature, and added that it did not consider the zone described in Article 4, paragraph 2, of the Treaty as established in accordance with international law; it could not, therefore, agree that the Treaty should apply to that zone.

[7] On signing Protocol II, France stated that it interprets the undertaking contained in Article 3 of the Protocol to mean that it presents no obstacle to the full exercise of the right of self-defence enshrined in Article 51 of the UN Charter; it takes note of the interpretation of the Treaty given by the Preparatory Commission for the Denuclearization of Latin America and reproduced in the Final Act, according to which the Treaty does not apply to transit, the granting or denying of which lies within the exclusive competence of each state party in accordance with the pertinent principles and rules of international law; it considers that the application of the legislation referred to in Article 3 of the Treaty relates to legislation which is consistent with international law. The provisions of Articles 1 and 2 of the Protocol apply to the text of the Treaty as it stands at the time when the Protocol is signed by France. Consequently, no amendment to the Treaty that might come into force under the provision of Article 29 would be binding on the government of France without the latter's express consent. If this declaration of interpretation is contested in part or in whole by one or more contracting parties to the Treaty or to Protocol II, these instruments would be null and void as far as relations between France and the contesting state or states are concerned. On depositing its instrument of ratification of Protocol II, France stated that it did so subject to the statement made on signing the Protocol. On 15 Apr. 1974, France made a supplementary statement to the effect that it was prepared to consider its obligations under Protocol II as applying not only to the signatories of the Treaty, but also to the territories for which the statute of denuclearization was in force in conformity with Article 1 of Protocol I.

[8] On signing the Treaty, Mexico said that if technological progress makes it possible to differentiate between nuclear weapons and nuclear devices for peaceful purposes, it will be necessary to amend the relevant provisions of the Treaty, according to the procedures established therein.

[9] Nicaragua stated that it reserved the right to use nuclear energy for peaceful purposes such as the removal of earth for the construction of canals, irrigation works, power plants, and so on, as well as to allow the transit of atomic material through its territory.

[10] The USSR signed and ratified Protocol II with the following statement:

The USSR proceeds from the assumption that the effect of Article 1 of the Treaty extends, as specified in Article 5 of the Treaty, to any nuclear explosive device and that, accordingly, the carrying out by any party to the Treaty of explosions of nuclear devices for peaceful purposes would be a violation of its obligations under Article 1 and would be incompatible with its non-nuclear status. For states parties to the Treaty, a solution to the problem of peaceful nuclear explosions can be found in accordance with the provisions of Article V of the Non-Proliferation Treaty and within the framework of the international procedures of the IAEA. The signing of the Protocol by the USSR does not in any way signify recognition of the possibility of the force of the Treaty being extended beyond the territories of the states parties to the Treaty, including airspace and territorial waters as defined in accordance with international law. With regard to the reference in Article 3 of the Treaty to 'its own legislation' in connection with the territorial waters, airspace and any other space over which the states parties to the Treaty exercise sovereignty, the signing of the Protocol by the USSR does not signify recognition of their claims to the exercise of sovereignty which are contrary to generally accepted standards of international law. The USSR takes note of the interpretation of the Treaty given in the Final Act of the Preparatory Commission for the Denuclearization of Latin America to the effect that the transport of nuclear weapons by the parties to the Treaty is covered by the prohibitions in Article 1 of the Treaty. The USSR reaffirms its position that authorizing the transit of nuclear weapons in any form would be contrary to the objectives of the Treaty, according to which, as specially mentioned in the preamble, Latin America must be completely free from nuclear weapons, and that it would be incompatible with the non-nuclear status of the states parties to the Treaty and with their obligations as laid down in Article 1 thereof.

Any actions undertaken by a state or states parties to the Treaty which are not compatible with their non-nuclear status, and also the commission by one or more states parties to the Treaty of an act of aggression with the support of a state which is in possession of nuclear weapons or together with such a state, will be regarded by the USSR as incompatible with the obligations of those countries under the Treaty. In such cases the USSR reserves the right to reconsider its obligations under Protocol II. It further reserves the right to reconsider its attitude to this Protocol in the event of any actions on the part of other states possessing nuclear weapons which are incompatible with their obligations under the said Protocol. The provisions of the articles of Protocol II are applicable to the text of the Treaty of Tlatelolco in the wording of the Treaty at the time of the signing of the Protocol by the Soviet Union, due account being taken of the position of the USSR as set out in the present statement. Any amendment to the Treaty entering into force in accordance with the provisions of Articles 29 and 6 of the Treaty without the clearly expressed approval of the USSR shall have no force as far as the USSR is concerned.

In addition, the USSR proceeds from the assumption that the obligations under Protocol II also apply to the territories for which the status of the denuclearized zone is in force in conformity with Protocol I of the Treaty.

[11] When signing and ratifying Protocol I and Protocol II, the UK made the following declarations of understanding: In connection with Article 3 of the Treaty, defining the term 'territory' as including the territorial sea, airspace and any other space over which the state exercises sovereignty in accordance with 'its own legislation', the UK does not regard its signing or ratification of the Protocols as implying recognition of any legislation which does not, in its view, comply with the relevant rules of international law.

The Treaty does not permit the parties to carry out explosions of nuclear devices for peaceful purposes unless and until advances in technology have made possible the development of devices for such explosions which are not capable of being used for weapon purposes.

The signing and ratification by the UK could not be regarded as affecting in any way the legal status of any territory for the international relations of which the UK is responsible, lying within the limits of the geographical zone established by the Treaty.

Should any party to the Treaty carry out any act of aggression with the support of a nuclear weapon state, the UK would be free to reconsider the extent to which it could be regarded as committed by the provisions of Protocol II.

In addition, the UK declared that its undertaking under Article 3 of Protocol II not to use or threaten to use nuclear weapons against the parties to the Treaty extends also to territories in respect of which the undertaking under Article I of Protocol I becomes effective.

[12] The USA ratified Protocol I with the following understandings: The provisions of the Treaty made applicable by this Protocol do not affect the exclusive power and legal competence under international law of a state adhering to this Protocol to grant or deny transit and transport privileges to its own or any other vessels or aircraft irrespective of cargo or armaments; the provisions of the Treaty made applicable by this Protocol do not affect rights under international law of a state adhering to this Protocol regarding the exercise of the freedom of the seas, or regarding passage through or over waters subject to the sovereignty of a state, and the declarations attached by the United States to its ratification of Protocol II apply also to its ratification of Protocol I.

[13] The USA signed and ratified Protocol II with the following declarations and understandings: In connection with Article 3 of the Treaty, defining the term 'territory' as including the territorial sea, airspace and any other space over which the state exercises sovereignty in accordance with 'its own legislation', the ratification of the Protocol could not be regarded as implying recognition of any legislation which does not, in the view of the USA, comply with the relevant rules of international law.

Each of the parties retains exclusive power and legal competence, unaffected by the terms of the Treaty, to grant or deny non-parties transit and transport privileges.

As regards the undertaking not to use or threaten to use nuclear weapons against the parties, the USA would consider that an armed attack by a party, in which it was assisted by a nuclear weapon state, would be incompatible with the party's obligations under Article 1 of the Treaty.

The definition contained in Article 5 of the Treaty is understood as encompassing all nuclear explosive devices; Articles 1 and 5 of the Treaty restrict accordingly the activities of the parties under para. 1 of Article 18.

Article 18, para. 4 permits, and US adherence to Protocol II will not prevent, collaboration by the USA with the parties to the Treaty for the purpose of carrying out explosions of nuclear devices for peaceful purposes in a manner consistent with a policy of not contributing to the proliferation of nuclear weapon capabilities.

The USA will act with respect to such territories of Protocol I adherents, as are within the geographical area defined in Article 4, para. 2 of the Treaty, in the same manner as Protocol II requires it to act with respect to the territories of the Parties.

[14] Safeguards agreements under the Non-Proliferation Treaty cover the Treaty of Tlatelolco.

[15] Safeguards agreements under Protocol I.

*The Non-Proliferation Treaty*

[1] When acceding to the Treaty China stated that the nuclear weapon states should: (*a*) undertake not to be the first to use nuclear weapons at any time and under any circumstances; (*b*) not to use or threaten to use nuclear weapons against non-nuclear weapon countries or nuclear-free zones; and (*c*) support the establishment of nuclear weapon-free zones, respect the status of such zones and undertake corresponding obligations. All states that have nuclear weapons deployed outside of their boundaries should withdraw all those weapons back to their own territories. China declared that it regards the signing and ratification of the NPT by Taiwan in the name of China as illegal and null and void.

[2] On the occasion of the deposit of the instrument of ratification, Egypt stated that since it was embarking on the construction of nuclear power reactors, it expected assistance and support from industrialized nations with a developed nuclear industry. It called upon nuclear weapon states to promote research and development of peaceful applications of nuclear explosions in order to overcome all the difficulties at present involved therein. Egypt also appealed to these states to exert their efforts to conclude an agreement prohibiting the use or threat of use of nuclear weapons against any state, and expressed the view that the Middle East should remain completely free of nuclear weapons.

[3] An agreement between France, the European Atomic Energy Community (Euratom) and the IAEA for the application of safeguards in France had entered into force in 1981. The agreement covers nuclear material and facilities notified to the IAEA by France.

[4] On depositing the instrument of ratification, FR Germany reiterated the declaration made at the time of signing: it reaffirmed its expectation that the nuclear weapon states would intensify their efforts in accordance with the undertakings under Article VI of the Treaty, as well as its understanding that the security of FR Germany continued to be ensured by NATO; it stated that no provision of the Treaty may be interpreted in such a way as to hamper further development of European unification; that research, development and use of nuclear energy for peaceful purposes, as well as international and multinational co-operation in this field, must not be prejudiced by the Treaty; that the application of the Treaty, including the implementation of safeguards, must not lead to discrimination of the nuclear industry of FR Germany in international competition; and that it attached vital importance to the undertaking given by the USA and the UK concerning the application of safeguards to their peaceful nuclear facilities, hoping that other nuclear weapon states would assume similar obligations.

[5] On acceding to the Treaty, the Holy See stated, *inter alia*, that the Treaty will attain in full the objectives of security and peace and justify the limitations to which the states party to the Treaty submit, only if it is fully executed in every clause and with all its implications.

[6] On signing the Treaty, Indonesia stated, *inter alia*, that it attaches great importance to the declarations of the USA, the UK and the USSR affirming their intention to provide immediate assistance to any non-nuclear weapon state party to the Treaty that is a victim of an act of aggression in which nuclear weapons are used. Of utmost importance, however, is not the action *after* a nuclear attack has been committed but the guarantees to prevent such an attack. Indonesia trusts that the nuclear weapon states will study further this question of effective measures to ensure the security of the non-nuclear weapon states. On depositing the instrument of ratification, Indonesia expressed the hope that the nuclear countries would be prepared to co-operate with non-nuclear countries in the use of nuclear energy for peaceful purposes and implement the provisions of Article IV of the Treaty without discrimination. It also stated the view that the nuclear weapon states would observe the provisions of Article VI of the Treaty relating to the cessation of the nuclear arms race.

[7] Italy stated that in its belief nothing in the Treaty was an obstacle to the unification of the countries of Western Europe; it noted full compatibility of the Treaty with the existing security agreements; it noted further that when technological progress would allow the development of peaceful explosive devices different from nuclear weapons, the prohibition relating to their manufacture and use shall no longer apply; it interpreted the provisions of Article IX, para. 3 of the Treaty, concerning the definition of a nuclear weapon state, in the sense that it referred exclusively to the five countries which had manufactured and exploded a nuclear weapon or other nuclear explosive device prior to 1 Jan. 1967, and stressed that under no circumstance would a claim of pertaining to such category be recognized by Italy for any other state.

[8] On depositing the instrument of ratification, Japan urged a reduction of nuclear armaments and a comprehensive ban on nuclear testing; appealed to all states to refrain from the threat or use of force involving either nuclear or non-nuclear weapons; expressed the view that peaceful nuclear activities in non-nuclear weapon states party to the Treaty should not be hampered and that Japan should not be discriminated against in favour of other parties in any aspect of such activities. It also urged all nuclear weapon states to accept IAEA safeguards on their peaceful nuclear activities.

[9] On depositing the instrument of ratification, the Republic of Korea took note of the fact that the depositary governments of the three nuclear weapon states had made declarations in June 1968 to take immediate and effective measures to safeguard any non-nuclear weapon state which is a victim of an act or an object of a threat of aggression in which nuclear weapons are used.

[10] On depositing the instruments of accession and ratification, Liechtenstein and Switzerland stated that activities not prohibited under Articles I and II of the Treaty include, in particular, the whole field of energy production and related operations, research and technology concerning future generations of nuclear reactors based on fission or fusion, as well as production of isotopes. Liechtenstein and Switzerland define the term 'source or special fissionable material' in Article III of the Treaty as being in accordance with Article XX of the IAEA Statute, and a modification of this interpretation requires their formal consent; they will accept only such interpretations and definitions of the terms 'equipment or material especially designed or prepared for the processing, use or production of special fissionable

material', as mentioned in Article III of the Treaty, that they will expressly approve; and they understand that the application of the Treaty, especially of the control measures, will not lead to discrimination of their industry in international competition.

[11] On signing the Treaty, Mexico stated, *inter alia*, that none of the provisions of the Treaty shall be interpreted as affecting in any way whatsoever the rights and obligations of Mexico as a state party to the Treaty of Tlatelolco.

It is the understanding of Mexico that at the present time any nuclear explosive device is capable of being used as a nuclear weapon and that there is no indication that in the near future it will be possible to manufacture nuclear explosive devices that are not potentially nuclear weapons. However, if technological advances modify this situation, it will be necessary to amend the relevant provisions of the Treaty in accordance with the procedure established therein.

[12] The agreement provides for the application of IAEA safeguards in Soviet peaceful nuclear facilities designated by the USSR.

[13] The ratification was accompanied by a statement in which Turkey underlined the non-proliferation obligations of the nuclear weapon states, adding that measures must be taken to meet adequately the security requirements of non-nuclear weapon states.

[14] This agreement, signed by the UK, Euratom and the IAEA, provides for the submission of British non-military nuclear installations to safeguards under IAEA supervision.

[15] This agreement provides for safeguards on fissionable material in all facilities within the USA, excluding those associated with activities of direct national security significance.

[16] In connection with the ratification of the Treaty, Yugoslavia stated, *inter alia*, that it considered a ban on the development, manufacture and use of nuclear weapons and the destruction of all stockpiles of these weapons to be indispensable for the maintenance of a stable peace and international security; it held the view that the chief responsibility for progress in this direction rested with the nuclear weapon powers, and expected these powers to undertake not to use nuclear weapons against the countries which have renounced them as well as against non-nuclear weapon states in general, and to refrain from the threat to use them.

*The Seabed Treaty*

[1] On signing and ratifying the Treaty, Argentina stated that it interprets the references to the freedom of the high seas as in no way implying a pronouncement of judgement on the different positions relating to questions connected with international maritime law. It understands that the reference to the rights of exploration and exploitation by coastal states over their continental shelves was included solely because those could be the rights most frequently affected by verification procedures. Argentina precludes any possibility of strengthening, through this Treaty, certain positions concerning continental shelves to the detriment of others based on different criteria.

[2] On signing the Treaty, Brazil stated that nothing in the Treaty shall be interpreted as prejudicing in any way the sovereign rights of Brazil in the area of the sea, the sea-bed and the subsoil thereof adjacent to its coasts. It is the understanding of Brazil that the word 'observation', as it appears in para. 1 of Article III of the Treaty, refers only to observation that is incidental to the normal course of navigation in accordance with international law. This statement was repeated at the time of ratification.

[3] In depositing the instrument of ratification, Canada declared: Article I, para. 1, cannot be interpreted as indicating that any state has a right to implant or emplace any weapons not prohibited under Article I, para. 1, on the sea-bed and ocean floor, and in the subsoil thereof, beyond the limits of national jurisdiction, or as constituting any limitation on the principle that this area of the sea-bed and ocean floor and the subsoil thereof shall be reserved for exclusively peaceful purposes. Articles I, II and III cannot be interpreted as indicating that any state but the coastal state has any right to implant or emplace any weapon not prohibited under Article I, para. 1 on the continental shelf, or the subsoil thereof, appertaining to that coastal state, beyond the outer limit of the sea-bed zone referred to in Article I and defined in Article II. Article III cannot be interpreted as indicating any restrictions or limitation upon the rights of the coastal state, consistent with its exclusive sovereign rights with respect to the continental shelf, to verify, inspect or effect the removal of any weapon, structure, installation, facility or device implanted or emplaced on the continental shelf, or the subsoil thereof, appertaining to that coastal state, beyond the outer limit of the sea-bed zone referred to in Article I and defined in Article II.

[4] The Chinese Government reaffirms that nothing in this Treaty shall be interpreted as prejudicing in any way the sovereign rights and the other rights of the People's Republic of China over its territorial sea, as well as the sea area, the seabed and subsoil thereof adjacent to its territorial sea.

[5] On the occasion of its accession to the Treaty, the government of India stated that as a coastal state, India has, and always has had, full and exclusive rights over the continental shelf adjoining its territory and beyond its territorial waters and the subsoil thereof. It is the considered view of India that other countries cannot use its continental shelf for military purposes. There cannot, therefore, be any restriction on, or limitation of, the sovereign right of India as a coastal state to verify, inspect, remove or

destroy any weapon, device, structure, installation or facility, which might be implanted or emplaced on or beneath its continental shelf by any other country, or to take such other steps as may be considered necessary to safeguard its security. The accession by the government of India to the Treaty is based on this position.

[6] On signing the Treaty, Italy stated, *inter alia*, that in the case of agreements on further measures in the field of disarmament to prevent an arms race on the sea-bed and ocean floor and in their subsoil, the question of the delimitation of the area within which these measures would find application shall have to be examined and solved in each instance in accordance with the nature of the measures to be adopted. The statement was repeated at the time of ratification.

[7] Mexico declared that in its view no provision of the Treaty can be interpreted to mean that a state has the right to emplace nuclear weapons or other weapons of mass destruction, or arms or military equipment of any type, on the continental shelf of Mexico. It reserves the right to verify, inspect, remove or destroy any weapon, structure, installation, device or equipment placed on its continental shelf, including nuclear weapons or other weapons of mass destruction.

[8] Viet Nam stated that no provision of the Treaty should be interpreted in a way that would contradict the rights of the coastal states with regard to their continental shelf, including the right to take measures to ensure their security.

[9] On 25 Feb. 1974, the Ambassador of Yugoslavia transmitted to the US Secretary of State a note stating that in the view of the Yugoslav Government, Article III, para. 1, of the Treaty should be interpreted in such a way that a state exercising its right under this Article shall be obliged to notify in advance the coastal state, in so far as its observations are to be carried out 'within the stretch of the sea extending above the continental shelf of the said state'.

*The BW Convention*

[1] On depositing its instrument of ratification, FR Germany stated that a major shortcoming of the BW Convention is that it does not contain any provisions for verifying compliance with its essential obligations. The Federal Government considers the right to lodge a complaint with the UN Security Council to be an inadequate arrangement. It would welcome the establishment of an independent international committee of experts able to carry out impartial investigations when doubts arise as to whether the Convention is being complied with.

[2] In a statement made on the occasion of the signature of the Convention, India reiterated its understanding that the objective of the Convention is to eliminate biological and toxin weapons, thereby excluding completely the possibility of their use, and that the exemption with regard to biological agents or toxins, which would be permitted for prophylactic, protective or other peaceful purposes, would not in any way create a loophole in regard to the production or retention of biological and toxin weapons. Also any assistance which might be furnished under the terms of the Convention would be of a medical or humanitarian nature and in conformity with the UN Charter. The statement was repeated at the time of the deposit of the instrument of ratification.

[3] Ireland considers that the Convention could be undermined if the reservations made by the parties to the 1925 Geneva Protocol were allowed to stand, as the prohibition of possession is incompatible with the right to retaliate, and that there should be an absolute and universal prohibition of the use of the weapons in question. Ireland notified the depositary government for the Geneva Protocol of the withdrawal of its reservations to the Protocol, made at the time of accession in 1930. The withdrawal applies to chemical as well as to bacteriological (biological) and toxin agents of warfare.

[4] Mexico considers that the Convention is only a first step towards an agreement prohibiting also the development, production and stockpiling of all chemical weapons, and notes the fact that the Convention contains an express commitment to continue negotiations in good faith with the aim of arriving at such an agreement.

[5] The ratification by Switzerland contains the following reservation: Owing to the fact that the Convention also applies to weapons, equipment or means of delivery designed to use biological agents or toxins, the delimitation of its scope of application can cause difficulties since there are scarcely any weapons, equipment or means of delivery peculiar to such use; therefore, Switzerland reserves the right to decide for itself what auxiliary means fall within that definition.

*The Enmod Convention*

[1] It is the understanding of the Republic of Korea that any technique for deliberately changing the natural state of rivers falls within the meaning of the term 'environmental modification techniques' as defined in Article II of the Convention. It is further understood that military or any other hostile use of such techniques, which could cause flooding, inundation, reduction in the water-level, drying up, destruction of hydrotechnical installations or other harmful consequences, comes within the scope of the Convention, provided it meets the criteria set out in Article I thereof.

[2] The Netherlands accepts the obligation laid down in Article I of the Enmod Convention as extending to states which are not party to the Convention and which act in conformity with Article I of this Convention.

[3] On signing the Convention, Turkey declared that the terms 'widespread', 'long-lasting' and 'severe effects' contained in the Convention need to be more clearly defined, and that so long as this clarification was not made, Turkey would be compelled to interpret for itself the terms in question and, consequently, reserved the right to do so as and when required. Turkey also stated its belief that the difference between 'military or any other hostile purposes' and 'peaceful purposes' should be more clearly defined so as to prevent subjective evaluations.

*The 'Inhumane Weapons' Convention*

[1] The accession of Benin refers only to Protocols I and III of the Convention.

[2] Upon signature, China stated that the Convention fails to provide for supervision or verification of any violation of its clauses, thus weakening its binding force. The Protocol on mines, booby traps and other devices fails to lay down strict restrictions on the use of such weapons by the aggressor on the territory of the victim and to provide adequately for the right of a state victim of an aggression to defend itself by all necessary means. The Protocol on incendiary weapons does not stipulate restrictions on the use of such weapons against combat personnel.

[3] Cyprus declared that the provisions of Article 7, para. 3b, and Article 8 of Protocol II of the Convention will be interpreted in such a way that neither the status of peace-keeping forces or missions of the UN in Cyprus will be affected nor will additional rights be, *ipso jure*, granted to them.

[4] France ratified only Protocols I and II. On signing the Convention France stated that it regretted that it had not been possible to reach agreement on the provisions concerning the verification of facts which might be alleged and which might constitute violations of the undertakings subscribed to. It therefore reserved the right to submit, possibly in association with other states, proposals aimed at filling that gap at the first conference to be held pursuant to Article 8 of the Convention and to utilize, as appropriate, procedures that would make it possible to bring before the international community facts and information which, if verified, could constitute violations of the provisions of the Convention and the Protocols annexed thereto. Reservation: Not being bound by the 1977 Additional Protocol I to the Geneva Conventions of 1949, France considers that para. 4 of the preamble to the Convention on prohibitions or restrictions on the use of certain conventional weapons, which reproduces the provisions of Article 35, para. 3, of Additional Protocol I, applies only to states parties to that Protocol. France will apply the provisions of the Convention and its three Protocols to all the armed conflicts referred to in Articles 2 and 3 common to the Geneva Conventions of 1949.

[5] Italy stated its regret that no agreement had been reached on provisions that would ensure respect for the obligations under the Convention. Italy intends to undertake efforts to ensure that the problem of the establishment of a mechanism that would make it possible to fill this gap in the Convention is taken up again at the earliest opportunity in every competent forum.

[6] The Netherlands made the following statements of understanding: A specific area of land may also be a military objective if, because of its location or other reasons specified in Article 2, para. 4, of Protocol II and in Article I, para. 3, of Protocol III, its total or partial destruction, capture, or neutralization in the prevailing circumstances offers a definitive military advantage; military advantage mentioned in Article 3, para. 3 under c, of Protocol II, refers to the advantage anticipated from the attack considered as a whole and not only from isolated or particular parts of the attack; in Article 8, para. 1, of Protocol II, the words 'as far as it is able' mean 'as far as it is technically able'.

[7] Romania stated that the provisions of the Convention and its Protocols have a restricted character and do not ensure adequate protection either to the civilian population or to the combatants as the fundamental principles of international humanitarian law require.

[8] The USA stated that it had strongly supported proposals by other countries to include special procedures for dealing with compliance matters, and reserved the right to propose at a later date additional procedures and remedies, should this prove necessary, to deal with such problems.

# III. UN member states and year of membership, as of 31 December 1992

Afghanistan, 1946
Albania, 1955
Algeria, 1962
Angola, 1976
Antigua and Barbuda, 1981
Argentina, 1945
Armenia, 1992
Australia, 1945
Austria, 1955
Azerbaijan, 1992
Bahamas, 1973
Bahrain, 1971
Bangladesh, 1974
Barbados, 1966
Belarus, 1945
Belgium, 1945
Belize, 1981
Benin, 1960
Bhutan, 1971
Bolivia, 1945
Botswana, 1966
Brazil, 1945
Brunei Darussalam, 1984
Bulgaria, 1955
Burkina Faso (*formerly Upper Volta*), 1960
*Burma (see Myanmar)*
Burundi, 1962
*Byelorussia (see Belarus)*
Cambodia (Kampuchea), 1955
Cameroon, 1960
Canada, 1945
Cape Verde, 1975
Central African Republic, 1960
Chad, 1960
Chile, 1945
China, 1945
Colombia, 1945
Comoros, 1975
Congo, 1960
Costa Rica, 1945
Côte d'Ivoire, 1960
Cuba, 1945
Cyprus, 1960
Czechoslovakia, 1945[a]
Denmark, 1945
Djibouti, 1977
Dominica, 1978
Dominican Republic, 1945
Ecuador, 1945
Egypt, 1945
El Salvador, 1945
Equatorial Guinea, 1968
Estonia, 1991
Ethiopia, 1945
Fiji, 1970
Finland, 1955
France, 1945
Gabon, 1960
Gambia, 1965
Germany, 1973
Ghana, 1957
Greece, 1945
Grenada, 1974
Guatemala, 1945
Guinea, 1958
Guinea-Bissau, 1974
Guyana, 1966
Haiti, 1945
Honduras, 1945
Hungary, 1955
Iceland, 1946
India, 1945
Indonesia, 1950
Iran, 1945
Iraq, 1945
Ireland, 1955
Israel, 1949
Italy, 1955
*Ivory Coast (see Côte d'Ivoire)*
Jamaica, 1962
Japan, 1956
Jordan, 1955
Kazakhstan, 1992
Kenya, 1963
Korea, Dem. People's Rep. of (North Korea), 1991
Korea, Rep. of (South Korea), 1991
Kuwait, 1963
Kyrgyzstan, 1992
Lao People's Democratic Republic, 1955
Latvia, 1991
Lebanon, 1945
Lesotho, 1966
Liberia, 1945
Libya, 1955
Liechtenstein, 1990
Lithuania, 1991
Luxembourg, 1945
Macedonia, Former Yugoslav Rep. of, 1993
Madagascar, 1960
Malawi, 1964
Malaysia, 1957
Maldives, 1965
Mali, 1960
Malta, 1964
Marshall Islands, 1991
Mauritania, 1961
Mauritius, 1968
Mexico, 1945
Micronesia, 1991
Moldova, 1992
Mongolia, 1961
Morocco, 1956
Mozambique, 1975
Myanmar (*formerly Burma*), 1948
Namibia, 1990
Nepal, 1955
Netherlands, 1945
New Zealand, 1945
Nicaragua, 1945
Niger, 1960
Nigeria, 1960
Norway, 1945
Oman, 1971
Pakistan, 1947
Panama, 1945
Papua New Guinea, 1975
Paraguay, 1945
Peru, 1945
Philippines, 1945
Poland, 1945
Portugal, 1955
Qatar, 1971
Romania, 1955
Rwanda, 1962
Saint Kitts (Christopher) and Nevis, 1983
Saint Lucia, 1979
Saint Vincent and the Grenadines, 1980
Samoa, Western, 1976
San Marino, 1992
Sao Tome and Principe, 1975
Saudi Arabia, 1945
Senegal, 1960
Seychelles, 1976
Sierra Leone, 1961
Singapore, 1965
Solomon Islands, 1978
Somalia, 1960
South Africa, 1945
Spain, 1955

Sri Lanka, 1955
Sudan, 1956
Suriname, 1975
Swaziland, 1968
Sweden, 1946
Syria, 1945
Tajikistan, 1992
Tanzania, 1961
Thailand, 1946
Togo, 1960
Trinidad and Tobago, 1962
Tunisia, 1956
Turkey, 1945
Turkmenistan, 1992
Uganda, 1962
UK, 1945
Ukraine, 1945
United Arab Emirates, 1971
*Upper Volta (see Burkina Faso)*
Uruguay, 1945
USA, 1945
USSR, 1945
Uzbekistan, 1992
Vanuatu, 1981
Venezuela, 1945
Viet Nam, 1977
Yemen, 1947
Yugoslavia, 1945[b]
Zaire, 1960
Zambia, 1964
Zimbabwe, 1980

[a] On 31 Dec. 1992 Czechoslovakia ceased to exist. It split into two independent states: the Czech Republic and the Slovak Republic. On 19 Jan. 1993 the two new states were admitted to the United Nations.

[b] A claim by Yugoslavia (i.e., Serbia and Montenegro) to continue automatically the membership of the former Yugoslavia was not accepted. Therefore, it is barred from participating in the work of UN bodies.

# **Annexe B.** Chronology 1992

RAGNHILD FERM

For the convenience of the reader, key words are indicated in the right-hand column, opposite each entry. They refer to the subject-areas covered in the entry. Definitions of the acronyms can be found on page xvi.

| | | |
|---|---|---|
| *10 Jan.* | An informal High Level Working Group (HLWG) (mandated by the Ministers of the North Atlantic Cooperation Council, NACC, on 20 Dec. 1991), meeting in Brussels, agrees that CFE Treaty obligations assumed by the former USSR should be wholly accounted for by all the newly independent states in the area of application. | CFE |
| *12 Jan.* | In a note delivered to the governments of the states accredited in Moscow, the Foreign Ministry of the Russian Federation declares that Russia continues to exercise the rights and discharge the obligations of the international treaties concluded by the USSR. | Russia |
| *16 Jan.* | The Salvadoran Government and the FMLN (Farabundo Marti National Liberation Front) sign, in Mexico City, a peace agreement to officially end the 11-year civil war. The UN Observer Mission in El Salvador (ONUSAL) is to monitor the disarmament of the FMLN forces and supervise a 50% reduction of the Salvadoran Army. | El Salvador; UN |
| *20 Jan.* | The Prime Ministers of North and South Korea, meeting in Pyongyang, formally issue a Joint Declaration on the denuclearization of the Korean Peninsula. The two states pledge to refrain from testing, manufacturing and possessing nuclear weapons and to use nuclear energy only for peaceful purposes. An Agreement on reconciliation, non-aggression, and exchanges and co-operation is also signed. (The two accords enter into force on 19 Feb. 1992.) | North Korea/ South Korea; Nuclear weapons |
| *28 Jan.* | In his State of the Union Address, President Bush announces deeper cuts in the US strategic weapons arsenals than those required by the 1991 START Treaty. Further production of the B-2 bomber as well as the programme for the MX missile and the small ICBM will be stopped, provided the CIS eliminates all its MIRVed ICBMs. Warheads for SLBMs will be reduced by about one-third. | USA; Nuclear weapons |

| | | |
|---|---|---|
| *29 Jan.* | President Yeltsin's Statement 'Russia's policy in the field of arms limitation and reduction' (dated 27 Jan.) is released. New deep cuts in strategic arms are proposed, reducing each side's nuclear warheads to 2000–2500. Russia is prepared to negotiate with the USA a ban on anti-satellite weapons and a global defence system to replace the US Strategic Defense Initiative (SDI). He suggests that the bilateral talks with the USA on limitation of nuclear tests are resumed. | Russia; Nuclear weapons; SDI; Nuclear tests |
| *30–31 Jan.* | The CSCE Council of Foreign Ministers, meeting in Prague, admits all the CIS republics as member states. (In its capacity as the successor of the USSR, Russia was already a member.) The membership is thereby extended to Asian states. A decision was made that the Council or the CSCE Committee of Senior Officers (CSO) should take appropriate action, if necessary, in the absence of the consent of the state concerned ('consensus minus one') in case of grave violations of CSCE commitments. | CSCE |
| *30 Jan.* | North Korea signs a safeguards agreement with the IAEA, allowing inspection of all its nuclear facilities. (The agreement enters into force on 10 Apr.) | IAEA/ North Korea |
| *31 Jan.* | The heads of state and government of the member states of the UN Security Council adopt a joint statement, asking the UN Secretary-General to make recommendations on how the Council can take a more active and positive role in preserving peace and averting crises. (This is the first UN Security Council summit meeting since the UN was created.) | UN |
| *7 Feb.* | The member states of the European Community (EC) formally sign the Treaty on a European Union agreed on at an EC summit meeting held in Maastricht, the Netherlands, on 9–11 Dec. 1991. | EC |
| *14 Feb.* | At a CIS meeting, held in Minsk, an agreement is signed stating that all CIS non-nuclear forces should be under a joint central command. (Azerbaijan, Moldova and Ukraine refuse to sign the agreement.) | CIS |
| *21 Feb.* | The UN Security Council unanimously adopts Resolution 743, establishing a UN Protection Force (UNPROFOR) to create the conditions of peace and security required for settlement of the Yugoslavian crises. | UN; Yugoslavia |
| *24 Feb.* | Canada announces that it will withdraw all 6500 of its troops from Europe by the end of 1994. | Canada; Withdrawal |
| *26 Feb.* | The IAEA Board of Governors agrees to strengthen the Agency's safeguards system to improve its ability to detect clandestine nuclear weapon programmes. | IAEA |

| | | |
|---|---|---|
| *27 Feb.* | President Yeltsin issues a decree that preparations for two to four underground nuclear tests per year will continue at the test site on Novaya Zemlya (renamed the Central Test Site of the Russian Federation), if the existing Russian test moratorium is terminated. | Russia; Nuclear tests |
| *28 Feb.* | The UN Security Council unanimously adopts Resolution 745, authorizing the establishment of a UN Transitional Authority in Cambodia (UNTAC), to oversee the country's transition to a new administration after multiparty elections. | UN; Cambodia |
| *29 Feb.–1 Mar.* | A referendum on independence is held in Bosnia and Herzegovina. Over 99% of the voters vote in favour. (A great majority of the Serb population boycotts the referendum.) (On 3 Mar. Bosnia and Herzegovina declares independence.) | Bosnia and Herzegovina/ Yugoslavia<br><br>UN |
| *4 Mar.* | The Vienna Document 1992, integrating and building upon the provisions of the Vienna Document 1990 and containing a new set of confidence- and security-building measures, is signed by all CSCE states. (It enters into force on 1 May.) | |
| *9 Mar.* | China accedes to the 1968 Non-Proliferation Treaty. | China; NPT |
| *20 Mar.* | At a CIS summit meeting, held in Kiev, agreements are signed on the non-use or threat of force between the CIS member states and on the creation of corps of military observers and collective peace-keeping forces in the CIS. | CIS |
| *24 Mar.* | The Treaty on Open Skies, providing for a system of unarmed reconnaissance flights over the entire territory of each of the parties, is signed in Helsinki by the NATO member states, Belarus, Bulgaria, Czechoslovakia, Georgia, Hungary, Poland, Romania, Russia and Ukraine. | Open Skies |
| *24 Mar.–8 July* | The fourth CSCE Follow-up Meeting is held in Helsinki. | CSCE |
| *25 Mar.* | The IAEA announces its decision to destroy Iraqi facilities and equipment designed for producing nuclear devices. (The demolition takes place on 13 Apr.) | Iraq; Nuclear weapons |
| *3 Apr.* | The Nuclear Suppliers Group (NSG), meeting in Warsaw, agrees on a new set of Guidelines for nuclear transfers (the revised 1977 London Guidelines) and on Guidelines for transfers of nuclear-related dual-use equipment, material and related technology (the Warsaw Guidelines). | NSG |
| *8 Apr.* | The French Prime Minister announces that France will suspend nuclear testing until the end of 1992. | France; Nuclear tests |
| *8 Apr.* | The President of Bosnia and Herzegovina declares a state of emergency. | Bosnia and Herzegovina |

| | | |
|---|---|---|
| *11 Apr.* | President Yeltsin issues a decree on the implementation of the 1972 Biological Weapons Convention. He states that all Russian development, production and stockpiling of biological weapons are prohibited. | BW; Russia |
| *24 Apr.* | The UN Security Council unanimously adopts Resolution 751, establishing a UN Operation in Somalia (UNOSOM). | UN; Somalia |
| *27 Apr.* | Serbia and Montenegro announce the formation of a new state: the Federal Republic of Yugoslavia (FRY). | Yugoslavia |
| *6 May* | The CIS Command announces that all tactical nuclear weapons in Ukraine have been withdrawn, thereby completing the withdrawal of all short-range nuclear weapons from former non-Russian Soviet republics. | Ukraine; SNF |
| *15 May* | At a summit meeting in Tashkent, Armenia, Azerbaijan, Belarus, Georgia, Kazakhstan, Moldova, Russia and Ukraine sign an Agreement on the principles and procedures of implementation of the CFE Treaty, with four protocols, confirming the allocation of treaty-limited equipment on their territories. The CIS Treaty on collective security is signed by Armenia, Kazakhstan, Kyrgyzstan, Russia, Tajikistan and Uzbekistan. An agreement on chemical weapons, reaffirming the obligations under the 1925 Geneva Protocol assumed by the former USSR, is signed by Armenia, Azerbaijan, Kazakhstan, Kyrgyzstan, Moldova, Russia, Tajikistan, Turkmenistan and Uzbekistan. It requires the parties to conclude a CW Convention and to control export of 'dual-purpose' chemicals. All the CIS member states adopt a statement on the reduction of the armed forces of the former USSR. They confirm the adherence of the CIS as the legal heir of the USSR to international commitments in the field of disarmament and arms control. | CIS; CFE; CW |
| *22 May* | At a summit meeting at La Rochelle, France, the French and German Presidents formalize the inauguration of the Franco-German army corps. A joint command will be established on 1 July and the corps of at least 35 000 soldiers will be operative by 1 Oct. 1995. | France/ Germany |

| | | |
|---|---|---|
| *23 May* | The Secretary of State of the USA and the Foreign Ministers of Russia, Belarus, Kazakhstan and Ukraine, meeting in Lisbon, sign a Protocol (the Lisbon Protocol) to facilitate the implementation of the 1991 START Treaty. Belarus, Kazakhstan, Russia and Ukraine, as successor states of the former USSR, will assume the obligations of the USSR under the START Treaty and exchange instruments of ratification with the USA. Belarus, Kazakhstan and Ukraine shall adhere to the NPT as non-nuclear weapon states in the shortest possible time. Associated letters from Belarus, Kazakhstan and Ukraine to President Bush, containing a guarantee to eliminate all nuclear weapons on their territories and a pledge to adhere to the NPT, are regarded as legally binding documents. | START; USA/Russia, Belarus, Kazakhstan, Ukraine |
| *29 May* | Representatives of the five permanent members of the UN Security Council, meeting in Washington, adopt guidelines on restrictions related to weapons of mass destruction. | UN; NBC |
| *30 May* | The UN Security Council adopts Resolution 757 imposing sanctions, including a trade embargo, on Serbia and Montenegro. China and Zimbabwe abstain from voting. | UN; Yugoslavia |
| *3–14 June* | The UN Conference on Environment and Development (UNCED, the 'Earth Summit') is held in Rio de Janeiro. The Rio Declaration on environment and development is adopted by consensus. It includes a principle stating that in times of armed conflict states should respect international law providing protection for the environment. | UN |
| *4 June* | The North Atlantic Council, meeting in Oslo, expresses willingness to support CSCE peace-keeping activities, including making available Alliance resources and expertise. | NATO; CSCE |
| *5 June* | At a NACC meeting held in Oslo, the Final Document of the Extraordinary Conference of the states parties to the CFE Treaty (the Oslo Document) is signed by the NATO states and Armenia, Azerbaijan, Belarus, Bulgaria, Czechoslovakia, Georgia, Hungary, Kazakhstan, Moldova, Poland, Romania, Russia and Ukraine. The Document modifies the language of the Treaty to make the countries of the former USSR with territories within the ATTU zone parties to the CFE Treaty. | CFE |
| *9 June* | President Mitterrand states that the level of alert in peacetime of the French nuclear forces will be reduced. | France; Nuclear weapons |
| *12 June* | France announces that the production of its short-range nuclear missile (Hadès) has been cancelled. However, 30 missiles of that kind will be kept in storage. | France; Nuclear weapons |

| | | |
|---|---|---|
| *15 June* | The British Defence Secretary states in Parliament that Royal Navy ships and aircraft and Royal Air Force maritime patrol aircraft will no longer have the capability to deploy tactical nuclear weapons. Weapons previously earmarked for this role will be destroyed. | UK; Nuclear weapons |
| *16–17 June* | President Bush and President Yeltsin meet in Washington. They sign the Joint Understanding on further reductions in strategic offensive arms (the De-MIRVing Agreement). Following the entry into force of the 1991 START Treaty, the two sides shall within a seven-year period reduce their nuclear forces to an overall total of 3800–4250 warheads each. By the year 2003 all MIRVed ICBMs will be eliminated and SLBM warheads reduced to no more than 1750. Each side would then possess a total of no more than 3000–3500 nuclear warheads. Joint US–Russian statements are made on co-operation in a Global Protection System against ballistic missiles and on the prohibition of chemical weapons. An Agreement is signed between the two states concerning the safe and secure transportation, storage and destruction of weapons and the prevention of weapons proliferation (the Weapons Destruction and Non-proliferation Agreement). | USA/Russia; Nuclear weapons; GPS; SSD |
| *19 June* | The WEU Council of Ministers, meeting in Petersburg, Germany, decides to empower the WEU with military forces drawn from its members to engage in activities ranging from rescue missions to actual combat missions. The planning and execution of these tasks will be fully compatible with the dispositions necessary to ensure the collective defence of all NATO member states. The Council also decides to set up a forum of consultation between the WEU and the Central and East European states plus the Baltic states. | WEU |
| *24 June* | Russia and Georgia agree on a cease-fire in the region of South Ossetia. | Russia/Georgia |
| *2 July* | The members of the Missile Technology Control Regime (MTCR), meeting in Oslo, agree to amend the MTCR Guidelines to extend the scope of the regime to missiles capable of delivering not only nuclear but also biological and chemical weapons. (The new Guidelines take effect on 7 Jan. 1993.) | MTCR |
| *2 July* | President Bush releases a statement that the withdrawal from abroad of all US tactical nuclear weapons (announced on 27 Sep. 1991) is completed. | USA; Nuclear weapons |
| *2 July* | Kazakhstan ratifies the 1991 START Treaty. | Kazakhstan; START |

| | | |
|---|---|---|
| *6 July* | A CIS summit meeting, held in Moscow, agrees to establish joint peace-keeping forces to intervene in CIS conflicts. | CIS |
| *6–8 July* | The heads of state and government of the Group of Seven Western industrialized countries (the G7), meeting in Munich, issue a Declaration supporting the MTCR regime. They state that the transparency and consultation in the transfer of conventional weapons should be improved and restraint in arms trade be encouraged. | G7; Nuclear weapons |
| *7 July* | The CSCE Committee of Senior Officials (CSO) decides not to allow the attendance of Yugoslavia (i.e., Serbia and Montenegro) at CSCE meetings until 14 Oct. 1992. This suspension was later extended. | CSCE; Yugoslavia |
| *9–10 July* | The heads of state or government of the CSCE, meeting in Helsinki, adopt the Helsinki Document 1992, 'The Challenges of Change', containing the Helsinki Summit Declaration and the Helsinki Decisions, including rules for CSCE institutions and structures. It is stated that CSCE peace-keeping operations will be undertaken with due regard to the responsibilities of the UN and that the UN Security Council will be kept informed about the operations. | CSCE |
| *10 July* | The Concluding Act of the negotiation on personnel strength of conventional armed forces in Europe, the CFE-1A Agreement, placing politically binding limits on military manpower in the ATTU zone, is signed in Helsinki by the NATO states and Armenia, Azerbaijan, Belarus, Bulgaria, Czechoslovakia, Georgia, Hungary, Kazakhstan, Moldova, Poland, Romania, Russia and Ukraine. (The CFE-1A Agreement enters into force on 17 July, simultaneously with the provisional application of the CFE Treaty.) | CFE |
| *13 July* | In announcing his 'Non-proliferation initiative', President Bush states that the USA will stop its production of plutonium and highly enriched uranium (HEU) for nuclear explosive purposes. (The USA has not produced plutonium since 1988 or HEU since 1964.) | USA; NPT |
| *14 July* | A joint US–Russian statement is released confirming the two Presidents' agreement to explore the potential for a joint early-warning centre and co-operation in developing ballistic missile defence capabilities. | USA/Russia; BMD |

| | | |
|---|---|---|
| *30 July* | The US Department of Defense and the Russian President's Committee on conventional problems of chemical and biological weapons sign, in Washington, the Agreement concerning the safe, secure and ecologically sound destruction of chemical weapons under which US CW destruction assistance will be provided to Russia at no cost. | USA/Russia; CW |
| *3 Aug.* | France accedes to the 1968 Non-Proliferation Treaty. | France; NPT |
| *3 Aug.* | The US Senate votes to suspend nuclear weapon testing for nine months, limit tests for the next three years and permanently ban all tests by 1996. | USA; Nuclear tests |
| *13 Aug.* | The UN Security Council adopts Resolution 770, authorizing the use of force, if necessary, to ensure the delivery of humanitarian aid to Bosnia and Herzegovina. (China, India and Zimbabwe abstain from voting.) | Bosnia and Herzegovina |
| *14 Aug.* | The UN Secretary-General introduces the report that sets the technical parameters of the UN Register of Conventional Arms. | UN |
| *17 Aug.* | The UN Security Council simultaneously adopts Resolution 772, authorizing the Secretary-General to deploy a UN observer mission in South Africa (UNOMSA). | UN; South Africa |
| *19 Aug.* | In a Joint Declaration on the complete prohibition of chemical weapons, signed in New Delhi, India and Pakistan undertake not to use, develop, produce or otherwise acquire chemical weapons or induce anyone to engage in such activities. | India/Pakistan; CW |
| *24 Aug.* | France ratifies Protocol I Additional to the 1967 Treaty for the prohibition of nuclear weapons in Latin America (the Treaty of Tlatelolco), committing it to apply the Treaty to its territorial possessions within the Treaty's zone of application. | France; Treaty of Tlatelolco |
| *27 Aug.* | An international conference on the former Yugoslavia, held in London under the co-chairmanship of the UN Secretary-General and the British Prime Minister, issues a document containing a plan for action aimed at ending the violence in the conflict area. The Conference will continue its work until a final settlement of the conflict in the former Yugoslavia has been reached. | Yugoslavia |
| *27 Aug.* | The UN General Assembly adopts Resolution 46/242, urging the Security Council to consider taking further measures to end the fighting and restore the territorial integrity of Bosnia and Herzegovina. (Vote: 136 to 1; five member states abstain from voting.) | UN; Bosnia and Herzegovina |

| | | |
|---|---|---|
| *27 Aug.* | The General Conference of the Agency for the Prohibition of Nuclear Weapons in Latin America (OPANAL), meeting in Mexico City, approves amendments to the Treaty of Tlatelolco dealing with verification of compliance. | Treaty of Tlatelolco |
| *27 Aug.* | The Czech and Slovak Prime Ministers, meeting in Prague, agree that Czechoslovakia should split into two independent states on 1 Jan. 1993. (The Slovak Republic adopts its new constitution on 3 Sep., the Czech Republic on 16 Dec.) | Czechoslovakia |
| *28 Aug.* | The UN Security Council unanimously adopts Resolution 775, endorsing the Secretary-General's request for 3500 security troops to be sent to Somalia. | UN; Somalia |
| *31 Aug.* | President Bush announces that the USA will buy Russian highly enriched uranium from dismantled nuclear weapons to convert it into fuel for commercial nuclear plants. | USA/Russia; Nuclear weapons |
| *3 Sep.* | The Conference on Dismarmament adopts by consensus the Draft Convention on the prohibition of the development, production, stockpiling and use of chemical weapons and on their destruction. | CW; CD |
| *8 Sep.* | Russia and Lithuania reach agreement on the withdrawal of all Russian troops from Lithuania by 31 Aug. 1993. (However, a suspension of the withdrawal of Russian troops from all three Baltic states is ordered by President Yeltsin on 29 Oct.) | Russia/ Lithuania; Withdrawal |
| *14 Sep.* | After a meeting in Moscow, US, British and Russian senior officials confirm their commitment to full compliance with the 1972 BW Convention. The three governments agree that all biological sites should be open for inspections by experts from all three states. | BW; USA; UK; Russia |
| *14–18 Sep.* | The Second Review Conference of the Parties to the 1977 Convention on the prohibition of military and other hostile use of environmental modification techniques (the Enmod Convention) is held in Geneva. Its Final Declaration states that the military or any other hostile use of herbicides as an environmental modification technique is prohibited by the Convention if such use upsets the ecological balance of a region, causing widespread or long-lasting destruction, damage or injury to any other state party. | Enmod |
| *19 Sep.* | The UN Security Council adopts Resolution 777, recommending that Yugoslavia (i.e., Serbia and Montenegro) shall not participate in the work of the General Assembly and should apply for membership in the UN. (China, India and Zimbabwe abstain from voting.) (On 22 Sep. the UN General Assembly adopts Resolution 47/1 to the same effect.) | UN; Yugoslavia |

| | | |
|---|---|---|
| *22 Sep.* | The Forum for Security Co-operation (FSC), the CSCE permanent body for harmonization of arms control obligations, enhancement of stability, co-operation in the strengthening of non-proliferaton regimes, conversion and verification, opens in Vienna. | CSCE |
| *29–30 Sep.* | The first multi-party general election is held in Angola. 54% of the voters vote for the Popular Movement for the Liberation of Angola–Workers' Party (MPLA–PT), 34% vote for the National Union for the Total Independence of Angola (UNITA). | Angola |
| *1 Oct.* | The US Senate approves ratification of the START Treaty (vote: 93 to 6) with the understanding that the Lisbon Protocol and its associated letters (see *23 May*) carry the same obligation as the Treaty. | USA; START |
| *2 Oct.* | President Bush signs the Energy and Water Development Appropriations bill, containing provisions for a nine-month moratorium on nuclear testing and a three-year period of regulated testing, culminating in a complete ban on all US nuclear testing from 1 Oct. 1996, provided no other state tests after that date. | USA; Nuclear tests |
| *4 Oct.* | The President of Mozambique and the leader of the MNR (Mozambique National Resistance) sign, in Rome, a General Peace Agreement, ending the country's 16-year civil war. | Mozambique |
| *9 Oct.* | The UN Security Council adopts Resolution 781, establishing an air-zone over Bosnia and Herzegovina from which all military flights are banned. (China abstains from voting.) | UN; Bosnia and Herzegovina |
| *19 Oct.* | President Yeltsin signs a decree stating that the Russian moratorium on nuclear tests, announced by President Gorbachev in Oct. 1991, will be extended until 1 July 1993. | Russia; Nuclear tests |
| *28 Oct.* | The last former Soviet combat unit stationed in Poland leaves Polish territory. | Russia/Poland; Withdrawal |
| *3 Nov.* | The French Foreign Minister proposes that the five nuclear weapon states should discuss the question of nuclear testing. | France; Nuclear tests |
| *4 Nov.* | The Russian Parliament approves ratification of the START Treaty (vote: 157 to 1) with the stipulation that the exchange of the instruments of ratification will be postponed until Belarus, Kazakhstan and Ukraine have acceded to the NPT as non-nuclear states and agreed on how to implement the 1991 START Treaty. | Russia; START |

| | | |
|---|---|---|
| *5 Nov.* | All NATO members agree on a Directive, containing basic principles of co-operation between the Alliance and the WEU. It is stated that co-operation is necessary to avoid parallel action. | NATO/ WEU |
| *9 Nov.* | The CFE Treaty enters into force. | CFE |
| *12 Nov.* | In response to the statement of the French Foreign Minister (see *3 Nov.*), a Chinese Foreign Military spokesman says that the Chinese Government is willing to discuss nuclear test issues with all the members of the Conference on Disarmament (CD), including France. | China; CD; Nuclear tests |
| *16 Nov.* | The UN Security Council adopts Resolution 787, authorizing necessary measures to ensure that the embargo on shipping to Serbia and Montenegro is observed. (China and Zimbabwe abstain from voting.) | UN; Yugoslavia |
| *20 Nov.* | At a meeting held in Rome, the WEU decides to extend full membership to Greece. Observer status is granted to Denmark and Ireland and associate membership to Iceland, Norway and Turkey. | WEU |
| *30 Nov.* | The UN General Assembly adopts, without a vote, Resolution 47/39, urging the Secretary-General to open the CW Convention for signature in Paris on 13 Jan. 1993. | UN; CW |
| *30 Nov.* | In a letter to the NATO Secretary-General, France and Germany state that the Franco-German corps (see *22 May*) will be placed under NATO's operational command in the event of an attack against a NATO member state. | NATO; France; Germany |
| *3 Dec.* | The UN Security Council unanimously adopts Resolution 794, authorizing a major multinational military force to use all necessary means to secure the delivery of humanitarian aid to Somalia. | UN; Somalia |
| *4 Dec.* | The US State Department announces that the USA has offered to lead a coalition of forces under UN auspices in Somalia. | UN; Somalia |
| *11 Dec.* | The UN Security Council adopts Resolution 743, authorizing the UN Secretary-General to establish a presence of the UN Protection Force (UNPROFOR) in the Former Yugoslav Republic of Macedonia. | UN; Macedonia |
| *14–15 Dec.* | The CSCE Council of Foreign Ministers meets in Stockholm. The Stockholm Convention on conciliation and arbitration, establishing the CSCE Court, is signed by 29 of the participating states. | CSCE |
| *16 Dec.* | The UN Security Council unanimously adopts Resolution 797, establishing a UN Operation in Mozambique (ONUMOZ) of up to 8000 UN troops and *c.* 350 military observers. | UN; Mozambique |

| | | |
|---|---|---|
| *16 Dec.* | The Israeli Cabinet approves to deport over 400 Palestinians (alleged Hamas activists) to Lebanon in retaliation for the killing of Israeli border police and military personnel. On 18 Dec. the UN Security Council unanimously adopts Resolution 799, condemning the Israeli action. | Israel/Palestine; UN |
| *29 Dec.* | The US Secretary of State and the Russian Foreign Minister agree on the text of the Treaty between the United States and the Russian Federation on further reduction and limitation of strategic offensive arms, the START II Treaty. (The Treaty is signed by the two Presidents on 3 Jan. 1993.) | START |
| *30 Dec.* | Mongolia announces that the last former Soviet troops have left Mongolian territory. (The withdrawal began in Apr. 1987.) | Russia/ Mongolia; Withdrawal |
| *31 Dec.* | Czechoslovakia dissolves. The Czech Republic and the Slovak Republic become sovereign states on 1 Jan. 1993. On 22 Dec. Czechoslovakia had notified the IAEA that the two new states intend to be considered parties, by virtue of succession, to agreements and conventions concluded with the IAEA, including NPT safeguards agreements. At the CSCE Council meeting in Stockholm (see *14–15 Dec.*), the two states were admitted as members as of 1 Jan. 1993. | Czechoslovakia |

# About the contributors

***Dr Ramses Amer*** (Sweden) is a Research Fellow at the Department of Peace and Conflict Research, Uppsala University. His recent publications include *The United Nations and Foreign Military Interventions: A Comparative Study of the Application of the Charter* (1992) and *The Ethnic Chinese in Vietnam and Sino-Vietnamese Relations* (1991). He has contributed articles to international journals and reports on issues of Asian security.

***Dr Ian Anthony*** (United Kingdom) is a Senior Researcher and Leader of the SIPRI arms transfers and arms production project. He is the editor of the SIPRI volume *Arms Export Regulations* (1991), author of the monograph *The Arms Trade and Medium Powers: Case Studies of India and Pakistan 1947–90* (1991), and the author of *The Naval Arms Trade* (SIPRI, 1990). He has written or co-authored chapters in the *SIPRI Yearbook* since 1988.

***Dr Eric Arnett*** (United States) is Co-leader of the SIPRI project on military technology and international security. In 1988–92 he was Senior Programme Associate of the American Association for the Advancement of Science Program on Science and International Security and Director of the programme's Project on Advanced Weaponry in the Developing World. His previous publications include *Sea-Launched Cruise Missiles and U.S. Security* (1991) and *Gunboat Diplomacy and the Bomb: Nuclear Proliferation and the U.S. Navy* (1989).

***Dr Vladimir Baranovsky*** (Russia) is a Senior Researcher and Leader of the SIPRI project on Russia's security agenda. He holds the position of Senior Researcher at the Institute of World Economy and International Relations (Moscow) where he was Head of the International Security section (1986–88) and Head of the European Studies Department (1988–92). He is the author of *The Countries of Southern Europe in the Modern World* (1990, in Russian), *1992: New Horizons of Western Europe* (1992, in Russian), *In from the Cold: Germany, Russia and the Future of Europe* (1992, in Russian), and *Western Europe: Military and Political Integration* (1988, in Russian). He co-edited *West European Integration: Political Aspects* (1985, in Russian) and has contributed to a number of joint volumes and journals.

***Professor Tamas Bartfai*** (Sweden), a biochemist, is Chairman of the Department of Neurochemistry and Neurotoxicology, Stockholm University. He is Secretary of the National Committee for Biochemistry and Molecular Biology of the Swedish Royal Academy of Sciences. He has published some 200 scientific articles on molecular neurobiology.

***Paul Claesson*** (United States) is a Research Assistant on the SIPRI arms transfers and arms production project. He was an editor at SIPRI in 1988–92. He is the editor of *Grønland, Middelhavets Perle: Et Indblik i Amerikansk Atomkrigsforberedelse* [Greenland, The Pearl of the Mediterranean: A View of American Preparation for Nuclear War] (1982, in Danish).

***Dr Saadet Deger*** (Turkey), an economist, is Senior Researcher and Leader of the SIPRI world military expenditure project. She received her PhD from the University of London and in 1977–87 was Research Fellow at Birkbeck College, University of London. She has been an expert adviser to the United Nations, a member of the UN Institute of Disarmament Research Working Group of Experts on the Economics of Disarmament, and consultant to the World Bank, the United Nations Children's Fund and the United Nations University. She is on the Executive Board of the International Defence Economics Association and the Editorial Board of the journal *Defence Economics.* She is the author of *Military Expenditure in Third World Countries: The Economic Effects* (1986), co-editor of *Defence, Security and Development* (1987), and co-author of *Military Expenditure: The Political Economy of International Security* (SIPRI, 1990). She has published many articles in international journals on economic aid and defence conditionality, the developmental effects of military expenditure, the economics of security, disarmament and development and the arms trade, and has co-authored chapters in the *SIPRI Yearbook* since 1988.

***Ambassador Rolf Ekéus*** (Sweden) is Executive Chairman of the United Nations Special Commission for overseeing the elimination of weapons of mass destruction in Iraq. Since 1989 he has been the chief of the Swedish Delegation to the Conference on Security and Co-operation in Europe in Vienna. In 1983–88 he was the Permanent Representative of Sweden to the Conference on Disarmament and *inter alia* served in 1984 and 1987 as Chairman of the negotiations on the Chemical Weapons Convention.

***Ragnhild Ferm*** (Sweden) is Researcher on the SIPRI arms control and disarmament project. She has published chapters on nuclear explosions, a comprehensive test ban and arms control agreements, and the annual chronologies of arms control and political events in the *SIPRI Yearbook* since 1982. She is the author of brochures and leaflets on SIPRI research topics in Swedish.

***Gülay Günlük-Senesen*** (Turkey) is an Associate Professor in Quantitative Economics at Istanbul Technical University and the vice-president of the ITÛ Technological and Economic Research Center (TEDRC), Turkey. Apart from her works on econometrics and applied economics, she has published articles and papers on the probable economic impacts of the installation of a domestic arms industry in Turkey, on the macroeconomic and microeconomic levels, since 1989. She is the author of a chapter in the SIPRI volume *Arms Industry Limited* (1993).

***Gerd Hagmeyer-Gaverus*** (Germany) is a Researcher on the SIPRI arms transfers and arms production project. He was formerly a Researcher at the Centre for Social Science Research at the Free University of Berlin, where he co-authored several research reports. He is a co-author of chapters on world military expenditure in the *SIPRI Yearbooks 1985, 1987* and *1988* and on arms production in the *SIPRI Yearbooks 1990* and *1991.*

***Gro Harlem Brundtland*** (Norway) who delivered the Olof Palme Memorial Lecture, 1992, is the Prime Minister of Norway. Since entering politics in 1974, she has been Minister of the Environment, a member of the parliamentary Standing Committee on Foreign Affairs, and three times, in 1981, 1986–89 and from November 1990 to the present, Prime Minister. She was a member of the Palme Commission on Security and Disarmament and chaired the World Commission on Environment and Development which published *Our Common Future* in 1987.

***Birger Heldt*** (Sweden) is a Research Associate and responsible for the armed conflicts data project at the Department of Peace and Conflict Research, Uppsala University. He is co-author of a chapter in *States in Armed Conflict 1989* (1991) and a chapter in the *SIPRI Yearbooks 1991* and *1992,* and the author of *States in Armed Conflict 1990–91* (1992).

***Roger Hill*** (Canada) is Senior Consultant to the Cooperative Security Competition Program, Ottawa. In 1987–91 he was Director of Research and Senior Research Fellow at the Canadian Institute for International Peace and Security, Ottawa, and in 1992 he was Professor, Université canadienne en France, Villefrance-sur-mer, France. His recent publications include reports, monographs and chapters and articles on NORAD renewal, Canadian defence policy, and European arms control and global security issues.

***Shannon Kile*** (United States) is a Research Assistant on the SIPRI project on Russia's security agenda. He is a doctoral candidate in the Department of Political Science at the Massachusetts Institute of Technology, USA.

***Dr Richard Kokoski*** (Canada), a physicist, is Co-leader of the SIPRI military technology and international security project. He is a co-author of the SIPRI volume *Conventional Arms Control in Europe: Perspectives on Verification* (forthcoming, 1993), co-editor of *Verification of Conventional Arms Control in Europe: Technological Constraints and Opportunities* (SIPRI, 1990) and the author of a chapter in the *SIPRI Yearbook 1990.*

***Dr Zdzislaw Lachowski*** (Poland) is a Researcher on the SIPRI project on building a co-operative security system in and for Europe. He has been a Researcher at the Polish Institute of International Affairs since 1980, where he dealt with problems of European security and the CSCE process in particular, as well as West European political integration. He has published extensively on these subjects. He is the co-editor of *Visions of Europe* (in Polish, 1989).

***Signe Landgren*** (Sweden) is Leader of the SIPRI project on the post-Soviet nations and conflicts, and the author of chapters in the *SIPRI Yearbooks 1972, 1974–81* and *1992*. She is a co-author of SIPRI's first arms trade study *The Arms Trade with the Third World* (1971), and was Leader of the SIPRI arms trade project in 1974–81. She is the author of the SIPRI volumes *Southern Africa: The Escalation of a Conflict* (1976) and *Embargo Disimplemented: South Africa's Military Industry* (1989), and author of a forthcoming SIPRI Research Report *Sufficient Defence? The Post-Soviet Nations and the Superpower Legacy* and the SIPRI monograph *New Thinking in Arms Control: Soviet Initiatives and US Responses* (both 1993).

***Dunbar Lockwood*** (United States) is a Senior Analyst at the Arms Control Association (ACA), Washington, DC, where he is responsible for monitoring nuclear arms control negotiations and agreements between the USA and the nations of the Commonwealth of Independent States. Before starting at the ACA in 1990, he worked at the Center for Defense Information (CDI) for four years as a senior analyst. He has written monthly news articles for the ACA journal *Arms Control Today* since 1990. He wrote six issues of the CDI *Defense Monitor* in 1986–90 and is the author of a chapter in *Satanfaust: Das nukleare Erbe der Sowjetunion* [The Fist of the Devil: The Nuclear Heritage of the Soviet Union] (1992).

***Evamaria Loose-Weintraub*** (Germany) is Research Assistant on the SIPRI world military expenditure project. She was previously Research Assistant on the SIPRI arms trade project. She has published chapters in the *SIPRI Yearbooks 1984–88,* is a co-author of the tables of world military expenditure in the *SIPRI Yearbook 1992,* and is the author of a chapter in the SIPRI volume *Arms Export Regulations* (1991). Her current research fields are military expenditure in the Central and East European countries and Latin American economic development.

***Dr S. J. Lundin*** (Sweden), retired, is a former Senior Director of Research at the Swedish National Defence Research Establishment (FOA) and in 1987–92 was Senior Researcher and Head of the SIPRI chemical and biological warfare programme and editor of the series *SIPRI Chemical & Biological Warfare Studies.* He was formerly Scientific Adviser to the Swedish Disarmament Delegation (1969–85) and adviser on NBC questions to the Swedish Supreme Commander (1985–87). He is the author of a number of chapters in the *SIPRI Yearbooks 1988–92* and has written extensively in the field of chemical and biological weapons.

***Erik Melander*** (Sweden) is a Research Assistant on the armed conflicts data project at the Department of Peace and Conflict Research, Uppsala University.

***Dr Richard H. Moss*** (United States) is Deputy Executive Director of the Human Dimensions of Global Environmental Change Programme and Programme Officer of the International Geosphere–Biosphere Programme, based at the Royal Swedish Academy of Sciences. Previously he worked at Princeton University and the Social Science Research Council in New York. He is currently planning and co-ordinating collaborative natural science–social science research activities for the IGBP and HDP on human land use and global land-cover change and the impacts of environmental changes on coastal areas, with special emphasis on South East Asia, Equatorial South

Ameria, and Africa north of the equator. He has written articles on international environmental issues and relations between developing and industrialized countries during the cold war.

***Dr Kjell-Åke Nordquist*** (Sweden) is Assistant Professor at the Department of Peace and Conflict Research, Uppsala University. He is co-author (with C. Ahlström) of *Casualties of Conflict* (1991), prepared for the Red Cross/Red Crescent Movements' World Campaign for the Protection of Victims of War, and author of *Peace After War: On Conditions for Durable Inter-State Boundary Agreements* (1992) and *Conflicting Peace Proposals: Four Peace Proposals in the Palestine Conflict Appraised* (1985). He is a co-author of chapters on major armed conflicts in the *SIPRI Yearbooks 1990–92*.

***Thomas Ohlson*** (Sweden) is a political scientist and economist, working with the Department for Peace and Conflict Research, Uppsala University since 1990. In 1982–87 he was Leader of the SIPRI arms trade research project. In 1987–90 he was a researcher at the Centre for African Studies in Maputo, Mozambique. He edited the SIPRI volume *Arms Production in the Third World* (with M. Brzoska, 1986) and is co-author of *Arms Transfers to the Third World, 1971–85* (SIPRI, 1987). He edited the SIPRI publication *Arms Transfer Limitations and Third World Security,* and co-authored chapters in the *SIPRI Yearbooks 1982–87.* His recent publications include *The New Is Not Yet Born: Conflict and Conflict Resolution In Southern Africa* (forthcoming, 1993) and several articles on South Africa.

***J. P. Perry Robinson*** (United Kingdom), a chemist and lawyer, is a Senior Fellow of the Science Policy Research Unit, University of Sussex. He has held research appointments at the Free University of Berlin and at Harvard University. He has been studying issues pertaining to chemical and biological weapons for the past 25 years, and has published extensively on the subject. As a consultant to SIPRI, where he worked in 1968–71, he was the founding editor of the series *SIPRI Chemical & Biological Warfare Studies.* With Professor Matthew Meselson of Harvard University, he now edits *Chemical Weapons Convention Bulletin,* published in Washington, DC. He has served as consultant to a variety of national and international organizations including the WHO and the UN Secretariat.

***John Pike*** (United States) is Director of the Space Policy Project at the Federation of American Scientists (FAS) in Washington, DC. He directs FAS research and public education on military and civilian space policy. A former political consultant and science writer, he is the author of over 170 studies and articles on space and national security and of chapters on the military uses of outer space in the *SIPRI Yearbook* since 1989.

***Dr Adam Daniel Rotfeld*** (Poland) is the Director of SIPRI and Leader of the project on building a co-operative security system in and for Europe. He has been head of the European Security Department in the Polish Institute of International Affairs, Warsaw since 1978. He was a member of the Polish Delegation to the Conference on Security and Co-operation in Europe (Geneva, 1973–75) and to CSCE Follow-up Meetings (Belgrade 1977–78, Madrid 1980–83 and Vienna, 1986–88). He is the author or editor of over 20 studies on European security and the CSCE process. His recent publications include the SIPRI volume *Germany and Europe in Transition* (1991, co-edited with W. Stützle) and *European Security System in Statu Nascendi* (1990, in Polish). He has contributed chapters to the *SIPRI Yearbook* since 1991.

***Dr Bo Rybeck*** (Sweden) has held the position of Director General of the Swedish National Defence Research Establishment (FOA) since 1985. In 1979–85 he was Director of the Medical Board of the Swedish Armed forces and Head of the Medical Corps of the Swedish Armed Forces with the rank of major general. Earlier engagements include work and positions at the University of Gothenburg and the Karolinska Institute (orthopedic surgery), Swedish National Defence College and various assignments as a surgeon in the Armed Forces. He has conducted research in fields of orthopedic surgery, disaster medicine and wound ballistics.

***Jane M. O. Sharp*** (United Kingdom) is a Senior Research Fellow at King's College and Director of the Defence and Security Programme at the Institute for Public Policy Research (IPPR), both in London. In 1987–91 she was Senior Researcher at SIPRI and before that held teaching and research fellowships at Harvard and Cornell Universities. She has written extensively on arms control and the security dilemmas of military alliances, with a current focus on security options available to the former members of the Warsaw Treaty Organization following the collapse of the WTO and dissolution of the USSR. She is the editor of *Europe After an American Withdrawal* (SIPRI, 1990), the author of *Bankrupt in the Balkans: British Policy in Bosnia* (IPPR, 1993) and *Negotiations on Conventional Armed Forces in Europe, 1989–91* (SIPRI, forthcoming), and has contributed to the *SIPRI Yearbook* since 1988.

***Elisabeth Sköns*** (Sweden) is a Researcher on the SIPRI arms transfers and arms production project. She was previously a Researcher at the Research Service of the Swedish Parliament (1987–91) and Research Assistant on the SIPRI world military expenditure and arms trade research projects (1978–87). She is the author of a chapter in the SIPRI volume *Arms Industry Limited* (1993). She is a co-author of chapters in the *SIPRI Yearbooks 1984, 1985*, *1987* and *1992*, and author of chapters in the *SIPRI Yearbooks 1983* and *1986*.

***Dr Thomas Stock*** (Germany) is Leader of the SIPRI chemical and biological warfare research project. He is an analytical chemist and was formerly a researcher at the Research Unit for Chemical Toxicology of the former Academy of Sciences of the GDR. He is a co-author of chapters in the SIPRI volumes *Verification of Conventional Arms Control in Europe* (1990) and *Non-Production by Industry of Chemical-Warfare Agents: Technical Verification under a Chemical Weapons Convention* (1988) and co-editor of *National Implementation of the Future Chemical Weapons Convention* (1990). He is also co-author of the CBW chapter in the *SIPRI Yearbooks 1991* and *1992* and has written several articles in the field of verification of chemical disarmament.

***Professor Ronald G. Sutherland*** (Canada) is Associate Dean, College of Arts and Science, University of Saskatchewan, Saskatoon, technical adviser to the Canadian Delegation to the Conference on Disarmament (Geneva), and consultant to the Arms

Control and Disarmament Division, Department of External Affairs and International Trade, Ottawa. He is the co-author of the SIPRI volume *National Implementation of the Future Chemical Weapons Convention* (1990) and of *The Phenoxyalkanoic Herbicides* (1981). He has written extensively on organic, organometallic, photochemical and sonochemistry and conducted research on verification related to chemical and biological weapons.

***Professor Peter Wallensteen*** (Sweden) holds the Dag Hammarskjöld Chair in Peace and Conflict Research and is the Head of the Department of Peace and Conflict Research, Uppsala University. He is the author of a chapter in *Behaviour, Society and Nuclear War, vol. II* (in ed. Tetlock *et al.*, 1991), and co-author of *Environmental Destruction and Serious Social Conflict* (1991). He is a co-author of chapters in the *SIPRI Yearbooks 1988–92* and the author of *Conflict Resolution in the Global System* (forthcoming).

***Siemon T. Wezeman*** (The Netherlands) is Research Assistant on the SIPRI arms transfers and arms production research project.

***Jon Brook Wolfsthal*** (United States) is a Senior Fesearch Analyst at the Arms Control Association (ACA) and is responsible for issues related to nuclear weapon and ballistic missile proliferation. He has written extensively for the ACA journal *Arms Control Today*. He worked previously on non-proliferation and arms control issues at the Science Applications International Corporation and has served as an intern at the Carter Center at Emory University.

***Dr Herbert Wulf*** (Germany) is a Senior Researcher at the University of Hamburg and an adviser to the United Nations Office of Disarmament Affairs. In 1989–92 he was Leader of the SIPRI project on arms transfers and arms production. He is the editor of the SIPRI volume *Arms Industry Limited* (1993) and co-author of the SIPRI Research Report *West European Arms Production: Structural Changes in the New Political Environment* (1990), author of chapters in *Arms Export Regulations* (SIPRI, 1991) and *Alternative Produktion statt Rüstung* [Alternative Production to Military Prodution] 1987), and the author of *Rüstungsexport aus Deutschland* [German Arms Exports] (1989). He has published extensively on security policy, arms production, arms transfers, conversion, and disarmament and development. He has contributed to the *SIPRI Yearbooks 1985* and *1990–92*.

*SIPRI Yearbook 1993: World Armaments and Disarmament*
Oxford University Press, Oxford, 1993, 834 pp.
(Stockholm International Peace Research Institute)
ISBN 0-19-829166-3

# ABSTRACTS

ROTFELD, A. D., 'Introduction: parameters of change', in *SIPRI Yearbook 1993,* pp. 1–12.

The dramatic end of the cold war was accompanied by important decisions in arms limitation. Arms control negotiations which reached or neared conclusion in 1992 led to the Open Skies Treaty, the CFE-1A Agreement, the 1993 Chemical Weapons Convention and the 1993 START II Treaty. As these important agreements were reached, unprecedented transformations in the world gave rise to new and formerly unknown threats and challenges in the sphere of international security. The former security system was tailored to solve conflicts between states and not inside them, as in Bosnia and Herzegovina, the post-Soviet armed conflicts in Nagorno-Karabakh, Georgia, Northern Caucasus, Moldova and Tajikistan, and the conflicts in Southern Africa and Cambodia. Shaping a new, effective system of international security will be a long and difficult process. It is encouraging that in 1992 global military spending fell steeply and arms sales and arms industry employment declined in the OECD and developing countries.

HARLEM BRUNDTLAND, G., 'The environment, security and development', in *SIPRI Yearbook 1993*, pp. 15–26.

A lasting new concept of of global security is needed now that the world has entered a period of genuine disarmament while also facing shortages of renewable resources and the imperatives of alleviating poverty and environmental degradation. A concept of common security, based on an acknowledgement of states' interdependence, multilateralism, the pooling of sovereignty and a better-organised world community, is replacing security through mutual deterrence. A juster international economic order, social justice and ecological stability are conditions of common security and are inextricably linked. Regional co-operation offers important and promising new opportunities.

MOSS, R. H., 'Resource scarcity and environmental security', in *SIPRI Yearbook 1993*, pp. 27–36.

Resource scarcity can be a source of conflict within and between nations, leading for example to movements of 'environmental refugees', support for insurgencies, heightened ethnic and religious tensions and direct conflict between states. Environmental security in future will depend on levels of resource exploitation, the social and political impacts, and the response capabilities of governments, and thus to a great extent on internal social, political and economic structures and conditions. Current analyses often conclude that violent conflicts are likely to become more common and severe. Pressures may, however, increase the likelihood of co-operation. Development of quantitative indicators of environmental security could help in assessing the probability of conflict and in the management of scarce renewable resources.

HILL, R., 'Preventive diplomacy, peace-making and peace-keeping', in *SIPRI Yearbook 1993*, pp. 45–80.

Efforts to control conflicts assumed a new importance. The need for action to deal with war in the former Yugoslavia, starvation in Somalia and other instances of inhumanity was accompanied by an opportunity for progress opened by the end of the cold war. The UN report *An Agenda for Peace*, claimed a central role for the UN and called for new approaches in preventive diplomacy, peace-making, peace-keeping and related efforts. UN fact-finding, mediation, peace enforcement, peace-keeping and similar activities also expanded, including the establishment of new or broadened operations in El Salvador, the former Yugoslavia, Cambodia, Somalia and Mozambique. Regional initiatives were also significant. The experience was mixed, with disappointments and frustrations as well as achievements. The approach to conflicts remained piecemeal. UN members need new, comprehensive policies.

AMER, R., HELDT, B., LANDGREN, S., MAGNUSSON, K., MELANDER, E., NORDQUIST, K-Å., OHLSON, T. and WALLENSTEEN, P., 'Major armed conflicts', in *SIPRI Yearbook 1993*, pp. 81–130

Major armed conflicts were waged in 30 locations in 1992. No major armed conflicts were recorded in 5 locations (El Salvador, Ethiopia, Iraq–Kuwait, Morocco/Western Sahara and Uganda) compared to 1991. Also, the major armed conflicts in the Aceh region of Indonesia and the Kurdish region in Iran were inactive. Major armed conflicts emerged in 5 new locations (Azerbaijan, India–Pakistan, Laos, Tajikistan, and Bosnia and Herzegovina. The number of locations with at least one major armed conflict has decreased from 32 in 1989 to 30 in 1992. The number of contested incompatibilities with at least one major armed conflict decreased from 36 in 1989, 37 in 1990 and 1991, to 33 in 1992. The end of the cold war is associated with a decrease of locations and contested incompatibilities with major armed conflicts.

BARANOVSKY, V., 'The post-Soviet conflict heritage and risks', in *SIPRI Yearbook 1993*, pp. 131–70.

The conflicts in the post-Soviet geopolitical space have developed as a dramatic by-product of fundamental transformations taking place in the territory of the former superpower. The conflict-generating trends are related to the extremely difficult transition from a totalitarian regime to a new type of society and from an empire to 15 independent states. The pattern of relations among the successors remains uncertain; they experience a painful process of state formation which often undermines domestic stability and erodes effectiveness and the very existence of political power is under way; the complex issue of territorial integrity and frontier claims is a source of serious conflicts which are also stimulated by the increasing role of ethnocentrism in political development and the search of national minorities for self-identification. Conflicts have also arisen over the military heritage of the former superpower (primarily on issues such as the status of nuclear weapons and withdrawal of Russian armed forces from the other new independent states' territory).

ROTFELD, A. D., 'The CSCE: towards a security organization', in *SIPRI Yearbook 1993*, pp. 171–218.

The multilateral institutions established by the Conference on Security and Co-operation in Europe (CSCE) underwent an evolutionary transformation in 1992 with the progress towards the construction of a security system across Europe, North America and the vast expanse of Asia. The new challenges and tasks facing the CSCE are examined in the light of the series of meetings held during the year. The CSCE summit meeting in Helsinki developed the new political strategy and character of CSCE activities, spelled out in the Helsinki Document 1992 which details the goals of the new Forum for Security Co-operation. The new structures and organs evolving within the CSCE constitute a framework for a pan-European security organization. Key CSCE documents from 1992 are appended to the chapter, notably extensive excerpts from the Helsinki Document 1992.

LOCKWOOD, D. and WOLFSTHAL, J. B., 'Nuclear weapon developments and proliferation', in *SIPRI Yearbook 1993*, pp. 221–53.

In 1992 the USA and Russia agreed to make deep reductions in their strategic nuclear forces and made commitments to scale down their strategic modernization programmes. France and the UK made unilateral commitments to curb their nuclear forces but did not abandon their principal modernization programmes. China continued to modernize its nuclear weapons slowly. Developments in the former Soviet Union and in Iraq and other developing nations made it clear in 1992 that the end of the East–West military confrontation increased rather than decreased both the dangers and incentives for the proliferation of nuclear weapons and ballistic missiles.

FERM, R., 'Nuclear explosions, 1945–92', in *SIPRI Yearbook 1993*, pp. 254–57.

The annual number of nuclear explosions has consistently declined since 1988. Only the USA and China carried out nuclear tests in 1992: the USA 6 and China 2 tests. France and Russia observed a test moratorium for the entire year. The USA announced a one-year moratorium on 1 October 1992. The tables provide data for all nuclear explosions carried out in 1945–92. For the tests in 1992, the date, exact time, location and body wave magnitude are given.

STOCK, T., 'Chemical and biological weapons: developments and proliferation', in *SIPRI Yearbook 1993*, pp. 259–92.

Negotiations on the Chemical Weapons Convention (CWC) were concluded in 1992. Proliferation of chemical weapons (CW) and to a lesser extent biological weapons (BW) continues as a concern of the 1990s, although many countries are strengthening national measures to prevent proliferation. Allegations of CW use in areas of tension remain difficult to confirm. CW destruction presents a challenge to several countries, especially Russia and the USA. The issue of dumped ammunition, particularly CW, received greater attention in 1992. The number of countries participating in the information exchange under the Biological Weapons Convention (BWC) did not increase significantly. New information confirmed that Russian BW R&D was carried out as late as 1992 and that CW activities were conducted in the 1980s.

BARTFAI, T., LUNDIN, S. J. and RYBECK, B., 'Benefits and threats of developments in biotechnology and genetic engineering', in *SIPRI Yearbook 1993*, pp. 293–305.

Developments in molecular genetics and biotechnology (genetic engineering) have proceeded more rapidly than expected, particularly as regards the mapping of the human genome (the HUGO project). While the biotechnology developments are clearly beneficial, some may deserve monitoring to avoid their misuse for the possible development of biological weapons which focus, for example, on differences in genetic characteristics. Such activities are prohibited by the 1972 Biological Weapons Convention and the 1948 Genocide Convention.

ARNETT, E. and KOKOSKI, R., 'Military technology and international security: the special case of the USA', in *SIPRI Yearbook 1993*, pp. 307–334.

There is a new consensus in the United States that military research and development (R&D) must be maintained at its cold war level, if not accelerated. This consensus assures that the USA will maintain technological supremacy for the foreseeable future but does not rule out deeper cuts in defence expenditure. In the two areas of R&D opposed by President Clinton, nuclear weapons and war in space, existing US advantages are likely to remain. Given these advantages, the necessity of US efforts to deny developing countries access to dual-use technologies may have been overstated.

DEGER, S., 'World military expenditure', in *SIPRI Yearbook 1993*, pp. 337–97.

World military expenditure in 1992 accelerated a downward trend begun in 1989, and fell by over 15 per cent per annum. The main reason was the halving of defence spending in one year in the Commonwealth of Independent States countries and drastic reductions in Eastern Europe forced by economic circumstances. NATO and Western European countries made slight cuts although many are planning drastic cuts in their future budgets. Procurement spending in these countries fell significantly while personnel costs were relatively protected and R&D increased its share. Military spending rose in the Middle East and slightly in the Far East but fell in all other regions.

LOOSE-WEINTRAUB, E., 'Military expenditure in the Central and East European countries', in *SIPRI Yearbook 1993*, pp. 398–414.

Economic difficulties and chaotic information-gathering in the former centrally-planned economies mean that estimates of their military expenditure must be very tentative. A broad picture emerges, however, of a strategic and military vacuum with the end of the WTO and the pull-out of Soviet troops and of drastic reductions in military expenditure enforced by economic crisis. The basic analysis of defence needs is seriously deficient in the present period of change. The necessary military restructuring has in most cases barely begun, and developments have been driven by economic factors. The share of procurement in military budgets is falling rapidly, with personnel costs accounting for an increasingly high share. A problem for the future is that military production may not be falling as fast as expenditure; it is unclear whether or not surplus weapon stocks are building up.

ANTHONY, I., CLAESSON, P., SKÖNS, E. and WEZEMAN, S. T., 'Arms production and arms trade', in *SIPRI Yearbook 1993*, pp. 417–68.

The global arms industry is undergoing major structural adjustments. Production cuts, closures, rationalization, mergers and acquisitions, diversification and privatization are major strategies. Arms industries in Central and Eastern Europe (Russia and Ukraine included) face the most serious crisis. Combined arms sales by the top 100 companies of the OECD and developing countries fell by 7% in real terms to $178.8 billion in 1991. Nearly 80% of these companies had reduced their employment levels by 1991. The global value of foreign deliveries of major conventional weapons in 1992 is estimated at $18 405 m. (in 1990 US$)—c. 25% less than in 1991. As Russian arms exports fell to 11% of all deliveries, the USA remained the dominant arms exporter, accounting for 46% in 1992. EC countries account for 26%, with Germany the predominant West European arms exporter for the first time. The Middle East accounted for 22%of deliveries in 1992; Asia 30%; and Europe 36%. This regional pattern may change since large orders have been placed by Egypt, Israel and Persian Gulf states, and the Middle East may re-emerge as the most important single international market for conventional weapons.

GÜNLÜK-SENESEN, G., 'An overview of the arms industry modernization programme in Turkey', in *SIPRI Yearbook 1993*, pp. 521–32.

In a period of falling military investment in NATO countries, Turkey is a significant exception. It ranks highest among NATO arms importers and is beginning to play a part as a producer of military equipment under its 10-year, $10 b. modernization programme. Investment started in late 1989 and initial projects led to Turkey's emergence as a significant arms importer and aircraft producer. It is expected that the Turkish F-16 plant might become a major supplier of F-16 aircraft once the USA stops production. The flow of armaments through the NATO CFE cascade and supplies during and after the 1991 Gulf War have served to upgrade Turkey's inventory. This trend stands as a challenge to domestic production and to the disarmament process advocated by the West, which is ironically supporting the upgrading process.

WULF, H., 'The United Nations Register of Conventional Arms', in *SIPRI Yearbook 1993*, pp. 533–44.

In Resolution 46/36 L, the UN General Assembly established the Register of Conventional Arms, for which UN member countries are requested to report, voluntarily and on an annual basis, their arms imports and exports, including battle tanks, armoured combat vehicles, large-calibre artillery systems, combat aircraft, attach helicopters, warships, and missiles and missile launchers. If the Register is adequately implemented, it will constitute a significant step forward towards transparency and confidence building. While it is not designed to control arms transfers, the information could proovide early warning of a country's 'exceessive and destabilizing accumulation of arms'. In the first stage, countries will report on arms transfers in 1992 and 1993; in the second stage, in 1994, the UN will consider an expansion of the Register to include data on stocks of military equipment, arms produc tion, technology transfers and weapons of mass destruction. The reports for 1992 are due by 30 April 1993. If the 10 leading arms-exporting countries all report, over 95% of the weapon systems in the seven categories will be recorded.

LOCKWOOD, D., 'Nuclear arms control', in *SIPRI Yearbook 1993*, pp. 549–89.

1992 was highlighted by the completion of the negotiations on the START II Treaty, signed 3 January 1993, and progress towards a comprehensive test ban (CTB). The USA and the former Soviet Union began to move towards co-operative denuclearization, discussing ways to improve the physical security and safety of nuclear weapons and technical co-operation on the safe transportation, storage and dismantlement of nuclear weapons. Despite this progress, at the end of 1992 the 1991 START Treaty had not yet entered into force and CTB negotiations and discussions to establish a warhead dismantlement regime with strict verification measures were still not under way.

SHARP, J. M. O., 'Conventional arms control', in *SIPRI Yearbook 1993*, pp. 591–617.

Several of the arms control negotiations begun at the end of the cold war were completed in 1992. The 1990 CFE Treaty came into force on 9 November. This chapter traces the ratification process, one made complex by the dissolution of the USSR and the difficulties of reallocating treaty-limited equipment ceilings among the former Soviet republics in the area of application. In the first phases of implementation of the Treaty, which started from the provisional entry into force on 17 July, data were exchanged in August and December and the programme of inspections was started. The CFE-1A Agreement was signed on 10 July, setting ceilings on military personnel in the area of application. Key documents on conventional arms control in Europe in 1992 are appended.

LACHOWSKI, Z., 'Implementation of the Vienna Document 1990 in 1992', in *SIPRI Yearbook 1993,* pp. 618–31.

On 4 March 1992, the CSCE participants adopted the Vienna Document 1992, which developed and built upon the existing confidence- and security-building measures (CSBMs) and supplemented them with more detailed parameters and some additional measures. However CSBMs in their traditional shape are losing ground and increasingly being overshadowed by other co-operative security arrangements in the new environment. The newly established Forum for Security Co-operation has been entrusted, *inter alia*, with the task of further developing and improving CSBMs, elaborating new measures with regional and sub-regional application, and harmonizing them with other security regimes (CFE, CFE-1A) to set all CSCE states on a common footing in military matters.

KOKOSKI, R., 'The Treaty on Open Skies', in *SIPRI Yearbook 1993*, pp. 632–44.

The ground-breaking Open Skies Treaty was signed on 24 March 1992. While the Treaty has not yet entered into force, the Open Skies Consultative Commission has already demonstrated that it is capable of operating effectively by eliminating some of the issues involving costs and sensors which were left for it to resolve at the time the Treaty was signed. In addition many successful trial over-flights took place during 1992 to test procedures as well as sensor calibration. Although 27 countries had signed the Treaty by the end of the year, only three had ratifies. Entry into force is expected during 1993.

EKÉUS, R., 'The United Nations Special Commission on Iraq: activities in 1992', in *SIPRI Yearbook 1993,* pp. 691–703.

The implementation of UN Security Council Resolution 687, the so-called cease-fire resolution, continued in 1992. For chemical weapon inspections both emphasis and resources shifted to destruction activities. There was an increase in the number of ballistic missile inspections and of missiles declared by Iraq. Missile production facilities and related equipment were destroyed by UN Special Commission on Iraq (UNSCOM) inspectors. Nuclear inspections, carried out by the International Atomic Energy Agency (IAEA), assisted and supported by UNSCOM, continued and provided insight into Iraq's nuclear programme. Some key nuclear facilities and equipment were destroyed. UNSCOM utilized U-2 aircraft and helicopters for aerial surveillance. Iraq continued to refuse to accept the implementation of compliance monitoring activities and attempted to deny access to one site. In January 1993 Iraq's obstruction of the continuation of UNSCOM operations ended after a missile strike carried out by the USA against an industrial facility in the vicinity of Baghdad.

STOCK, T., PERRY ROBINSON, J. P. and SUTHERLAND, R. G., 'The Chemical Weapons Convention: the success of chemical disarmament negotiations', in *SIPRI Yearbook 1993,* pp. 705–56.

The Conference on Disarmament in Geneva completed negotiations on the Chemical Weapons Convention (CWC) on 3 September 1992, and it was opened for signature in Paris on 13 January 1993. The CWC is a major multilateral disarmament and arms control agreement based on an interplay between national implementation obligations and international compliance measures; it balances the rights and obligations of States Parties to the CWC against the costs and benefits of adhering to the Convention. The history of the negotiations reveals how it was finally possible to achieve the Convention, which is reproduced in an appendix. The plans for implementation of the CWC and its signatory status as of 8 February are also presented.

# Errata

*SIPRI Yearbook 1992: World Armaments and Disarmament*

*Page 229, table 7.16, col. for 1985, row for Turkey:* The figure should read 336 (not 3 336). The figures for the totals in this column are correct.

*Page 429, in table 11A:* The heading for the region was not printed: 'Middle East' should appear at the top of the page, before the conflict in Iran.

*Page 470, table 12.4, cols 3 and 4:* The columns for Artillery and ACVs were incorrectly labelled. The entire table should read:

**Table 12.4.** WTO agreement on TLE entitlements, 3 November 1990

| State | Battle tanks | Artillery | ACVs | Combat aircaft | Attack helicopters |
|---|---|---|---|---|---|
| WTO 'group of six' | 20 000 | 20 000 | 30 000 | 6 800 | 2 000 |
| USSR | 13 150[a] | 13 175[b] | 20 000[c] | 5 150 | 1 500 |
| Bulgaria | 1 457 | 1 750 | 2 000 | 235 | 67 |
| Czechoslovakia | 1 435 | 1 150 | 2 050 | 345 | 75 |
| Hungary | 835 | 840 | 1 700 | 180 | 108 |
| Poland | 1 730 | 1 610 | 2 150 | 460 | 130 |
| Romania | 1 376 | 1 475 | 2 100 | 430 | 120 |

As agreed on 14 June 1991:

[a] 933 tanks deployed with Soviet naval infantry (120) and coastal defence forces (813) must count against this allowance of 13 150.

[b] 1080 artillery pieces deployed with Soviet naval infantry (234) and coastal defence forces (846) must count against this allowance of 13 175.

[c] 1725 ACVs deployed with Soviet naval infantry (753) and coastal defence forces (972) must count against this allowance of 20 000.

*Page 563, first line of section I:* Should read: '. . . Paris summit meeting of 19–21 November 1990'.

# INDEX

*Material on the independent republics of the former Soviet Union as a group will be found under* CIS


Abdulajanov, Prime Minister 106
Abkhazia:
  cease-fire 137
  chemical weapons and 262
  conflict in 94, 95, 99–101, 138
  Russia and 58, 134fn., 153, 154, 156, 606
ABM (Anti-Ballistic Missile) Treaty (1972) 317, 318, 319
Aérospatiale 230, 232, 429, 470
Afghanistan 86, 106, 123
Africa:
  aid for 21
  arms exports 477
  arms imports 418, 449, 453, 476
  conflicts in 86, 87, 88, 112–20
  economy 20
  military expenditure 393
A/F-X bomber 327
Agenda 21, 24, 39–42
Agni missile 249
Agusta 430, 431, 435, 474, 531
aircraft: R&D for 315
aircraft-carriers 349, 375
Akashi, Yasushi 108
Akayev, President Askar 105
Albania 92
Al Hussein missile 700
Alliant Techsystems 434, 472
Allied Command Europe 378, 380
Allison Transmission 439
America *see* United States of America
ANC (African National Congress) 112, 113, 116, 117
Anders, William 424
Andò, Salvo 426
Angola:
  cease-fire supervision 114
  conflict in 83, 88, 112, 116, 127
  MPLA 114
  peace accords 112, 114, 116, 118
  UN and 50, 52, 112, 114
  UNITA 112, 114, 115, 118
Antarctic Treaty (1959) 759, 763, 765–81
Antonov, Victor 362
Apache-MAW 442
Ardzinba, Vladislav 100
Argentina 571, 572
Armenia:
  Azerbaijan, conflict with 95–98, 134, 139, 143
  CFE and 597, 600
  Russian troops in 605–6
  *see also* Nagorno-Karabakh
arms control, normative 325–27
Armscor 429, 431, 437–48, 473
arms industry:
  employment 416, 429–32
  structures 416, 427–43
  trends 415–18
  *see also under names of countries*
arms trade:
  control of 459–68
  exporters 444
  importers 445
  SIPRI's sources and methods 519–20
  trends 415–18, 427–29, 443–68
artillery shells 225, 228
ASEAN (Association of South-East Asian Nations) 112, 476, 477
Asia:
  arms exports 477
  arms imports 417, 449, 452, 553–56, 476
  conflicts in 86, 87, 88
  military expenditure 454–55
  nuclear weapon developments 244–49
ASMP (Air Sol Moyenne Portée) missile 229
Aspin, Les 310, 316, 324, 328, 424, 434
Australia:
  chemical weapons and 715, 718
  Somalia and 51
Australia Group 268–69, 717
Austria 284
Aven, Peter 448
AWACS (Airborne Warning and Control System) 315, 418
Ayaz, Nevzat 523
Azerbaijan:
  Armenia, conflict with 95–98, 134, 139, 143
  CFE and 597
  CIS and 602
  conflict in 93, 95–98, 121
  military expenditure 356
  Russia and 147

B-1 bomber 234, 558
B-2 bomber 221, 323, 344, 556, 558, 559
B-52 bomber 223–24, 234
B-57 depth bomb 225, 231
B-61 bomb 225
Backfire bomber 228
Baker, James 96, 249, 550, 555, 593

Baltic Fleet 282, 283
Baltic Sea 282–83
Baltic states 135, 136, 139, 143, 151, 157, 593, 604–5 *see also under names of*
Bangladesh 124, 452
BMD International 433
Bear bomber 227, 235, 558
Beg, Mirza Aslam 452
Belarus:
  CFE and 597
  CIS and 133
  military expenditure 356
  nuclear weapons 4, 148, 149, 228, 601–2
  SSD talks and 566, 569
  START and 553
Belgium:
  military expenditure 368, 369, 370, 372, 373
  military personnel 375, 376, 377
  Somalia and 51
Bessarabia 101fn.
Biden, Joe 551
biological weapons:
  biotechnology and 300–4
  developments 286–88, 293
  proliferation 259, 268–72
  use allegations 260
  *see also following entries and* chemical weapons
Biological Weapons Convention (1972) 260, 271, 291, 761, 763, 765–81
biotechnology 293–305
Blackjack bomber 227, 235, 558
Black Sea 139, 156fn.
Black Sea Economic Co-operation Pact 522
Black Sea Fleet 151, 356, 358, 359, 602
Blix, Hans 241, 244
Boeing 424, 470
Bofors 436, 441, 472
bombers 221, 223, 227–28, 229, 236, 554–55, 558–59 *see also under designations of*
Bosnia and Herzegovina:
  casualties 91–92
  chemical weapons and 263, 267
  conflict in 88–93, 121
  Croats of 89, 90
  CSCE and 421
  EC and 90
  'ethnic cleansing' 91
  Geneva peace talks 56
  humanitarian aid 90
  independence 90
  Islamic dimension 92
  London Conference on 90–91
  major-power intervention in 92
  Muslims of 89, 90, 91
  no-fly zone 50, 55, 92
  refugees 92
  Russia and 92
  Sarajevo 90
  Serbs of 89, 90
  UN and 55, 90
  USA and 90
Botha, Pik 115
Botswana 51
Boutros-Ghali, Boutros:
  arms control after cold war 5, 6, 7
  Georgia and 100
  Somalia and 51
Brandt Commission 18, 26
Brazauskas, Algirdas 168
Brazil 436, 571, 572
Brezhnev, Leonid 713
Brilliant Pebbles 318, 320, 324
British Aerospace 422, 431, 439, 470
Brussels Declaration 708
Buccaneer aircraft 231, 238
Budapest Agreement 598
Bulgaria:
  armed forces 411
  economy 398, 400, 410–11
  future problems 92
  military expenditure 412
Burma *see* Myanmar
Bush, George:
  arms exports and 446
  chemical weapons and 273, 715, 716
  CTB and 561–62
  forces in Europe and 225
  Georgia and 100
  Iran and 251
  Middle East and 84
  military expenditure and 339, 342, 351
  'new world order' and 7
  NPT and 561
  nuclear weapons and 222
  R&D and 312, 313, 314
  SDI and 317
  SSD and 566
  START and 550, 553, 555, 557, 574
Butkevicius, Audrius 605

C-160 aircraft 695, 696, 697
Cambodia:
  conflict in 124
  elections 108, 109, 110, 111, 112
  major-power interest in 111–12
  name changes 107
  Thailand and 110, 111, 112
  UN and 52, 54–55, 56, 57
  Viet Nam and 107, 109, 111, 112
Cameroon 395
Canada:
  arms exports 446, 468
  chemical weapons and 281

military expenditure 368, 369, 370, 372, 373
military personnel 376, 377
OAS and 59
Somalia and 51
Caribbean Common Market 59
Carlsson, Ingvar 25
Carlyle Group 433, 434
Carter, Jimmy 713
Carter Center 59
Carus, Seth 324
CASA 431, 531
cease-fires, enforcement of 46
Celsius 429, 436, 472
Central America:
arms exports 477
arms imports 476
conflicts in 86, 87
Ceylon *see* Sri Lanka
CFE (Treaty on Conventional Armed Forces in Europe) (1990):
assessment 17
cascade 418, 451, 532
CIS and 150
data exchange 607
former Soviet republics ratify 593–98
implementation 606–13
inspections 607–13
Oslo Declaration 596–98, 677–82
ratification 591, 592–606
status 763, 765–81
summary 762
Tashkent Document 596, 598, 600, 601, 607, 671–77
TLE holdings 358, 361–62, 401, 414, 608–9
WTO states and 598
CFE-1A Agreement (1992) 591, 613–15, 617, 683–89, 762
Chad 88, 127
Chechnia 100, 103, 104, 146, 157
chemical weapons:
destruction 260, 273–82, 290
dumped 260, 282–85
possession alleged 264–68
proliferation 259, 268–72
security aspects 708, 710–11, 712, 717
use allegations 259, 260–63
Chemical Weapons Convention (CWC) (1993):
achievements of 706–7
assessment 732–34
challenge inspection 728–29, 733
Conference of States Parties 721
definitions 726
destruction and 707, 724–26, 732
entry into force 732
Executive Council 722
historical overview 708–18
Inspectorate 723
National Implementation provision 719–20
opened for signature 259, 705
Organization for the Prohibition of Chemical Weapons 707, 721–23
Preparatory Commission 730–32
signatory status 709–10
Technical Secretariat 722–23
text 735–56
use, investigation of 729–30
verification 707, 716, 719, 726–30, 733
Cheney, Dick 223
Chetek Corporation 280
Chile 572
China:
aid to 395
arms exports 417, 452–53, 479–82
arms imports 389, 449, 450, 455
chemical weapons and 283–84
conversion 388–89
CTB and 564–65
military expenditure 308, 386–90, 454
military expenditure on R&D 388
missile exports 249
nuclear weapons 221, 222, 232–33, 239
nuclear weapon tests 221, 254, 255, 256, 557
Russia and 140, 389
technology and 325
Tiananmen Square 386
Ukraine and 389
Viet Nam and 386
Chu, David 318
CIA (Central Intelligence Agency) 247, 250, 357
CIS (Commonwealth of Independent States):
armed forces of 133, 148, 150–55, 356, 466, 596
borders in 138, 140
CFE Treaty and 593–98, 599–601
Charter 132
chemical weapons and 261, 265, 266, 281
conflict containment forces 58
conflicts and 94, 134
corruption in 136
Cossack movement 136–37
crisis in 135–38
disintegration trends 144–48
economics 144, 157
ethno-nationalism in 140–48
foreign troops in states of 94
mafia in 137
military dimension 148–55
military expenditure 337, 338, 339, 351, 353, 354, 356, 360
military expenditure on R&D 338

minorities in, problems of 143–44
Minsk meetings 132, 552
*nomenklatura* system and 135, 136, 141, 142
nuclear weapons 148–49, 235
peace-keeping forces 58, 134, 603
political élites, new 141–42
problems of 93–94
Russian troops in 94, 96, 356, 603
START and 550
status of 132–34
structures lacking 135–38
Treaty on Collective Security 134, 600, 603
US disarmament aid 366
*for relations with individual countries see under names of countries*
Clapper, James 226
Clarke, Richard 272
Clinton, President Bill:
CTB and 562
military expenditure 327, 339, 342–44, 350–51
missile defences 318
nuclear weapon tests and 254
R&D and 307, 308, 309–11, 314, 317, 327
SDI and 317
START and 557
Yugoslavian conflict and 419
Club of Rome 18
cold war, end of:
arms control and 4
effects of 86–88
nuclear weapons and 10
research agenda and 7–9
Colombia 129
Committee on Disarmament 714
*Common Crisis* 18
*Common Security* 18
Confederation of Mountainous Peoples of the Caucasus 99, 137, 147
Conference on Disarmament (CD) 564, 705
conflicts: changes in, 1992 82–83
Congo 52
Convention on Conciliation and Arbitration 180, 181
Conyers, John 320
Cooper, Henry 318
Coordinating Committee on Multilateral Export Controls (COCOM) 270, 462, 464–65
Council for Mutual Economic Assistance (CMEA) 398, 404, 408, 410, 412
Council of Europe 22, 23, 176, 177
Cranborne, Viscount 565
Crimea, the 602
Croatia:
Bosnia and 89, 90
conflict in 121
independence 88
refugees 92
UN and 55, 624
CSBMs (confidence- and security-building measures) *see* Vienna Document 1990 and Vienna Document 1992
CSCE (Conference on Security and Co-operation in Europe):
arms imports 476
arms trade and 545
arms trade control and 462–63
Berlin meeting (1991) 173, 175, 176, 180
changes 171, 173, 175, 180, 182, 186, 187
Committee of Senior Officials (CSO) 174, 176, 178, 180, 184, 186
Conflict Prevention Centre 58, 174–75, 179, 625
Decision on Peaceful Settlement of Disputes 209–11
democracy and 171, 172, 174
Dispute Settlement Mechanism 180
Executive Committee 187
failures of 173
Forum for Security Co-operation 176, 182–84, 186, 202–6, 463, 591, 617, 619, 630
Helsinki Document 1992, 1, 190–209
Helsinki Final Act (1975) 172, 178, 180, 620
High Commission on National Minorities 179
human rights and 174
Madrid meeting 617
Japan and 185
organizations, relations with 176–78, 201–2
Paris summit meeting (1990) 171, 172–73, 181
participating states, new 184–85
peace-keeping 179, 180
Prague meeting (1992) 171, 174, 175, 176, 178, 184, 463
Programme for Immediate Action 183, 203, 205
Secretariat 58
Serbian suspension 171fn.
*Shaping a New Europe* 212–18
Stockholm Meetings (1992) 171, 180–85
transformations 171
'Troika' 178, 195
Vienna meetings 175, 183, 184
*see also under names of conflicts and of other organizations*
CTB (comprehensive test ban):
likelihood of 16
progress in 561–65

Cuba 572
Cyprus:
  Turkey and 523
  UN and 50, 52, 53
Czechoslovakia:
  arms exports 446, 479–82
  armed forces 403
  division of 398, 401–2
  economy 398, 400, 402–3
  military expenditure 403–4
  *see also following entry and* Slovak Republic
Czech Republic:
  armed forces 401, 403
  CFE and 598
  economy 402–3
  military expenditure 403–5

Dagestan 148
Daimler Benz 470
DASA 440–41, 442, 470, 472, 473
Dassault 431, 471, 473
DCN 431, 470
debts 20, 393
Declaration of Human Rights (1948) 172
de Klerk, President F. W. 113
De La Plata River Agreement 29
democracy 16, 19, 21, 23, 25
Deng Xiao Ping 390
Denmark:
  military expenditure 368, 369, 370, 372, 373
  military personnel 376, 377
Desert Storm, Operation 348, 718
developing world:
  arms exports 477
  arms imports 476
  military expenditure 340, 392–96
development 18–19, 21, 24
development aid 21, 24, 394–96
DF missile 232, 233, 239
Dhlakama, Afonso 116
Dienstbier, Jiri 97, 176
diplomacy, preventive 46–47, 49–50
disabling weaponry 326–27
DNA fingerprinting 300, 304
Dornier 431
drought 31
Dudayev, Dzhokhar 146, 157fn.
Dumas, Roland 564

Eagleburger, Lawrence 556, 557
EC (European Community):
  arms exports 477
  arms trade control and 461–62
  economic indicators 384–85
  enlargement of 382
  membership applications 382, 384–85
  military expenditure 369, 373, 381, 383–85, 386
  military matters and 381–86
  Nordic countries and 22
ECOMOG (Economic Community of West African States' Monitoring Group) 84
EFA (European Fighter Aircraft) 371, 422, 424–26, 427
EFIM 429, 471
EFTA (European Free Trade Association) 22
Egypt:
  arms imports 418
  chemical weapons 711
  Somalia and 51
  UN and 52
Eidgenössische Rüstrungsbetriebe 431, 473
Eisenhower, Dwight 632
Elciby, President Ebulfez 141fn., 168
El Salvador 52, 53, 56, 82
Embraer 436
EN Bazan 431
Enhanced Proliferation Controls Initiative 272
Enmod Convention (1977) 761–62, 765–81
environment:
  development and 17–19, 20, 24
  indicators 32, 33–35
  protecting 23–24, 27
  security 17–20, 26, 27–29
environmental concerns 18–19, 23, 24, 274, 282–85
Ericsson 436
Eritrea 57, 82
Estonia 151, 593, 604, 605
Ethiopia 57, 82
Euphrates River 522
Eurofighter 2000, 425 *see* EFA
Euroflag 442
Europatrol 442
Europe:
  arms imports 417, 449, 476
  conflicts in 86, 172 *see also* Bosnia and Herzegovina
  military expenditure 339, 340, 368, 369, 370, 372
  military personnel 376, 377
  Rapid Reaction Force 378, 380
  troop withdrawals 260, 617
European Bank for Reconstruction and Development 176, 177, 450
European Coal and Steel Community 22
European Economic Area 22
European Investment Bank 176

F-10 aircraft 389
F-15 aircraft 225
F-16 aircraft 421, 422, 446, 523, 531

F-22 aircraft 315, 423, 424
F-111 aircraft 225
F/A-18 aircraft 344, 423
Falklands/Malvinas War 326, 379
Far East, military expenditure 337
Ferranti International 441
FIAT 431, 472
Finmeccanica 435
FMC 433, 472
Fokker 440, 531
Ford 429
France:
  arms exports 417, 428, 446, 479–82
  arms production 422
  chemical weapons 710
  CTB and 563–64
  Germany, joint corps with 382
  military expenditure 308, 368, 369, 370, 372, 373, 374–75
  military expenditure on R&D 308, 374
  military personnel 375, 376, 377
  nuclear weapons 221, 222, 228–30, 233, 237
  nuclear weapon tests 255, 256, 257
  Somalia and 51
FROG missile 228
Fur, Lajar 407

G-7 countries 251, 463, 534
Gaddis, John Lewis 3
Gagauz region 95, 102
Gaidar, Yegor 169, 340, 358, 359, 449
Gamsakhurdia, President Zviad 98, 141, 157fn., 159
Garel-Jones, Tristan 565
Gates, Robert 247, 250, 568
GATT (General Agreement on Tariffs and Trade) 21
GDR (German Democratic Republic) 282
GEC 441, 470
General Dynamics 276, 421, 424, 429, 431, 432, 433, 439, 470, 522
General Electric 432, 433, 470, 531
General Motors 432, 439, 470
genetic engineering 294, 295–97, 298–99, 300, 303, 305
'genetic weapons' 300, 303–4
Geneva Protocol (1925) 271, 301, 304, 708, 710, 759, 763, 765–81
Genocide Convention (1948) 304
Genome Diversity Project 299, 304
genome mapping 294–95
Genscher, Hans-Dietrich 594
Georgia:
  Black Sea Fleet 356
  CFE and 597
  CIS and 602
  conflict in 83fn., 93, 94, 95, 98–100, 145
  history 98fn.
  NATO and 100
  Russia and 98–99, 100, 134fn., 139, 147, 153–54, 606
  UN and 100
  USA and 100
  *see also* Abkhazia
Germany:
  aid 394–95
  arms exports 417, 451–52, 479–82
  chemical weapons and 270–71, 281, 283, 284
  France, joint corps with 382
  military expenditure 308, 368, 369, 370, 371, 372–73, 374
  military expenditure on R&D 308
  military personnel 376, 377, 378
  troops withdrawn from 284–85
Gershwin, Lawrence 568
GIAT 430, 431, 471
Glavkosmos 248
Global Protection System 319
Glukhikh, Viktor 449
Glybin, Yuriy 361
Golan Heights 85
Goldas, Emran Shah 106
Gorbachev, Mikhail:
  CFE and 592
  chemical weapons and 710
  military expenditure and 355
  nuclear weapons and 228
  nuclear weapon tests 254
  SSD and 566
  Union treaty and 95
Gorbunovs, President Anatolijs 142fn.
Gorno-Badakhshan 145
Grachev, Pavel 156fn., 170, 557, 563, 600, 605, 606
Granby, Operation 379
Great Britain *see* United Kingdom
Greece:
  arms imports 418, 452
  military expenditure 368, 369, 370, 372, 373
  military personnel 376, 377
Gromov, Boris 604
Grumman 431, 471
GTAR 441–42
GTE 471
Guatemala 129

H-7 aircraft 232
Hadès missile 230, 237
Hague Gas Projectile Declaration 708
Hague Regulations 708
Haiti 49, 59
Hansen, Lynn 596
Harsco 433

Hawker Siddeley 474
Hawk missile 457
Helsinki Document 1992 *see under* CSCE
High-Altitude Endoatmospheric Defense Interceptor (HEDI) 321
Hindustan Aeronautics 431
Hollandse Signaalapparaten 431, 441, 474
Hughes Electronics 431, 432, 433, 434, 442, 470
*Human Development Report* 20
Human Genome Organization (HUGO) 294–95, 303
human rights 19, 23
Hungary:
  armed forces 406
  arms imports 465
  economy 398, 400, 405–6
  military expenditure 406–7
Hunting 431
Hurd, Douglas 8
Hussein, President Saddam 58

IAEA (International Atomic Energy Agency):
  safeguards 244
  special inspections 240–41, 561
  strengthening 240, 241
  *for relations with particular countries, see under names of countries*
IBM 471
Iceland 381, 615
Iliescu, President Ion 102
IMF (International Monetary Fund) 353, 360, 394, 405, 408
Imhausen 271
Independent Commission on Disarmament and Security Issues *see* Palme Commission
Independent Commission on International Development Issue *see* Brandt Commission
India:
  aid to 395
  arms imports 418, 448, 449, 450, 451
  China and 248
  conflict in 124
  military expenditure 308
  missiles 248–49
  nuclear explosion 243, 247, 256
  nuclear weapons 247–49
  Pakistan, conflict with 83, 125, 247
  Somalia and 51
  Space Research Organization 248
  technology and 325
Indian United Phosphorus 264
Indonesia 53, 82, 125
Indus Water Treaty (1960) 29
Ingush Republic 58, 94, 95, 103, 146–47
Inhumane Weapons Convention (1981) 762, 763, 765–81
INI 471, 473
International Court of Justice 181
International Institute for Strategic Studies (IISS) 355, 359
International Peace Academy 59
Iran:
  arms imports 450, 451
  chemical weapons and 264–65
  conflict in 122
  Kurds in 82
  nuclear weapons 250–51
Iraq:
  aerial surveillance of 694
  Al Atheer 693, 699
  Al Fallujah 700
  Al Muthanna 700
  Al Sharqat 699
  Baghdad 693, 696, 702
  biological weapons 259, 691, 692, 698, 701
  chemical weapons 259, 263, 264, 270, 271, 282, 691, 698, 700–1, 714, 716
  coalition against 50
  conflict in 122, 701
  foreign companies and 702
  IAEA and 241, 691, 692, 696, 697–99, 701–2, 703
  Kurds in 50, 83–84, 86, 701
  Kuwait and 82
  military expenditure 308, 309
  missiles 252, 691–93, 696, 698, 699–700
  no-fly zone in 50, 695
  nuclear materials 697
  nuclear weapons and 240, 251–52, 691, 693, 697–99
  Shiites in 50
  Tarmiya 697, 699
  technology 325
  UN and:
    resistance to 692–94, 702
    Resolutions on 691, 692, 695, 697, 702, 703
  *see also* Persian Gulf War
Iraq–Iran War 732
IRI 432, 471
Iskandarov, President Akbarsho 169
Islamic fundamentalism 184
Israel:
  arms imports 418
  arms industry 437
  missiles 250
  NBC protection and 285–86
  nuclear weapons 250
  Palestine conflict 84–86, 123
  Syria and 85
  technology 325

Israel Aircraft Industries 431, 437, 471
Israel Military Industries 431, 437, 473
Italy:
  arms exports 479–82
  arms industry 435
  military expenditure 368, 369, 370, 372, 373
  Somalia and 51
ITT 429, 471
Izetbegovic, Alija 89, 263

Japan:
  aid and 394–95
  arms imports 418
  chemical weapons and 271, 283–84
  Kurile Islands and 140
  military expenditure 308, 390–92
  military expenditure on R&D 391
JAS-39 Gripen aircraft 422, 436
Jennekens, Jon 251
Jericho missile 250
JL missile 232, 233, 239
Jordan 85
Joxe, Pierre 230

Kara-Kalpak 145
Karaoglanov, Sergey 389
Karimov, President Islam 105
Kashmir 83, 86
Kawasaki 472
Kazakhstan:
  CIS and 133
  military expenditure 356
  nuclear weapons 4, 148, 149, 228, 601–2
  SSD talks 566, 570
  START and 550
  Tajikistan and 105
KH-11 satellite 321
Khan, Shahryar 247
Khasbulatov, Ruslan 99
Khmer Rouge 54
Khudonazarov, Counsellor 106
Kim Yong Nam 244
Kiss, Csaba 407
Kissinger, Henry 712
Kitovani, Tenguiz 98, 100
Klaus, Vaclav 401
Kobets, Konstantin 595, 599
Kockums Group 436
Kohl, Chancellor Helmut 382, 395, 426
Kokoshin, Andrey 226
Korea, North:
  armed forces 392
  denuclearization agreement 245
  IAEA and 244, 246
  missile exports 246
  NPT withdrawal 245, 561
  nuclear suspicions 244–46
  research reactor 245
Korea, South:
  armed forces 392
  CBW activities alleged 265
  satellites and 322
Korean War 708
Kosovo 93
Kozyrev, Andrey 96, 138, 277, 552, 555, 556, 557, 573, 605
Krauss-Maffei 431, 474
Kravchuk, President Leonid 142fn., 159, 161, 164, 165, 168, 550, 553
Kuchma, Leonid 157fn., 450–51
Kuntsevich, Anatoly 267, 277, 280, 288
Kuwait:
  Iraqi raids into 50
  oil fires 260, 289
Kyrgyzstan 105, 356

Lacrosse satellite 321
Lance missile 225, 381
land cover 32
land resources 31, 35–36
Landsbergis, Vytautas 141fn., 157fn., 168
Laos 83, 125
Latvia 151, 593, 604, 605
Laurance, Edward J. 543, 544
League of Arab States 47, 717
Lebanon 85
Lewis and Hua 232
Lezghistan 148fn.
Liberia 59, 84, 88, 127
Libya:
  chemical weapons and 265, 271
  missiles sought by 246
Lithuania 593, 604, 605
Litton Industries 431, 470
Lockheed 421, 424, 429, 431, 432, 433, 434, 470
London Club *see* Nuclear Suppliers Group
Loral 429, 430, 431, 434, 471
LTV 416, 433, 438, 471
Lugar, Richard 566, 567, 569
Luxembourg:
  military expenditure 368, 369, 370, 372, 373
  military personnel 376, 377
Lynx helicopter 231

M-4 missile 229, 237
M-5 missile 228-29, 564
M-11 missile 249, 452
M-45 missile 229
Maastricht Treaty (1992) 381, 425, 462
McDonnell Douglas 421, 431, 439, 470
Macedonia 55, 89, 92, 93, 419
McNamara, Robert S. 560

McPeak, Merrill 423
Major, Prime Minister John 422, 426, 565
Malaysia 454
Malei, Mikhail 365
Malta 534
Mandela, Nelson 112
Mariam, Mengistu Haile 82
Martin Marietta 232, 431, 432, 433, 434, 470
Matra Group 472
MBB 471
Meciar, Vladimir 401
Menem, President Carlos 465
Middle East:
arms exports 477
arms imports 417–18, 421, 422, 449, 453, 476
conflict in 86, 87, 88
Madrid peace talks 84, 85
military expenditure 337
nuclear weapon-free zone in 250
nuclear weapons and 249
peace process 84–86, 250
UN and 50
Midgetman missile 221, 222
MiG-29 aircraft 365
MiG-31 aircraft 360
military exercises 620–21
military expenditure: world 16, 337–41, 396, 398–401 *see also under names of individual countries*
military expenditure on R&D 308, 309–10, 313
Milstar satellites 323–24
Ministrel, François le 432
MINURSO (UN Mission for the Referendum in Western Sahara) 52, 61, 65
Minuteman missile 222–23, 234, 556fn.
Mirage aircraft 229, 237, 375, 422
MIRVs (multiple independently targetable re-entry vehicles) 226, 229, 232, 554, 555, 575–76
missiles:
proliferation 10, 240, 452
smarter 314–15
technology, spread of 270, 452
trade in, control of 452, 460
*see also following entry and under names and designations of*
missiles, cruise:
Russia 221, 227–28
USA 221, 224, 225
Missile Technology Control Regime 248, 324, 464, 465
Mitsubishi 429, 471, 473
Mitterrand, President François 254, 382, 564
Moldova:
Bendery 101, 102, 156
CFE and 597, 601
CIS and 102
conflict in 83fn., 93, 94, 96, 101–3 *see also* Trans-Dniester dispute
CSCE and 102
military expenditure 356
Russia and 102, 152, 606
UN and 50
USA and 102
*see also* Trans-Dniester dispute
Montenegro 58, 89, 90, 92, 419, 420
Morocco:
Somalia and 51
Western Sahara and 82
Mozambique:
aid to 395
chemical weapons and 262
conflict in 84, 116, 128
drought 115
FRELIMO 116
peace accords 112, 115–16, 118
RENAMO 112, 115, 116, 117, 118, 128, 262
UN and 50, 52, 56, 57, 116
Murtha, John 327
Musawi, Secretary-General 85
mustard gas 264, 265, 278, 279, 280, 281, 282, 700
Mutalibov, President Ayaz 96, 97, 160, 162
MX missile 221, 222, 233, 234
Myanmar 126, 452, 453

Nabiyev, President Rakhmon 105, 106, 159, 162, 166
Nagorno-Karabakh dispute:
cause of 94
cease-fire 137
chemical weapons and 261–62, 283
CIS and 96, 97
CSCE and 97, 176
history of 95–98, 156
independence declaration 97
Iran and 97
Kazakhstan and 97
NACC and 97
peace talks 96
Russia and 58, 605
Turkey and 96, 907
UN and 50, 58, 96, 97
Namibia 53, 112
nation-state, concept of 25
NATO (North Atlantic Treaty Organization):
arms imports 476
CSCE and 176, 177, 375
European Union and 381–82
Franco-German Corps and 382
future of 3
military expenditure 337, 340, 367–80
military personnel 375–78

security and 16
TLE holdings 608
WTO's collapse and 16
natural resources:
access to 20
availability of 31–32
conflicts over 20, 27, 28, 32
degradation 28
demands for 30–31, 33–35
exploitation 29–32
international co-operation and 19, 24–25, 29
scarcity of 19, 27–28
technology and 30
Nazarbayev, President Nursultan 97, 105, 142fn., 159, 161, 550
NBC protection 260, 285–86
Nechaev, Andrey 357
nerve agents 264, 265, 280, 281, 282, 700, 708
Netherlands:
arms exports 479–82
military expenditure 368, 369, 370, 372, 373
military personnel 375, 376, 377
'new world order' 3, 7, 8
Niger 31, 83fn.
Niger River 29
Nikolayev, Andrey 449
Nimrod aircraft 231
Nixon, Richard 712, 713
NobelTech 436, 438
No Dong I missile 246
Nolan, Janne 324
non-lethal warfare 267–68, 326–27
Non-Proliferation Treaty (NPT) (1968) 7, 233, 241, 247, 248, 251, 252, 549, 560–61, 711, 760–61, 763, 765–81
Extension Conference 7, 561
North Atlantic Co-operation Council (NACC) 177, 182, 189, 594, 595, 605
Northern Caucasus *see under* Russia
North Ossetia 58, 103–4
Northrop 421, 431, 432, 434, 470
North–South relations 711, 717, 718
Norway 23, 24, 368, 369, 370, 372, 373, 376, 377, 381
Novichok 266, 267
nuclear exports, register of 241
Nuclear Suppliers Group 242–44, 565–66
nuclear trade: dual-use items 243–44
nuclear war 15
nuclear weapons:
proliferation 10, 240–44 *see also* Non-Proliferation Treaty
reduction in 221
*see also under names of possessor countries*
nuclear weapon tests 221, 254–57:
moratoria on 17, 254, 563
Nunn, Sam 327, 566, 567, 569

OAS (Organization of American States) 22, 47, 59
OAU (Organization of African Unity) 22, 47
Oberon-Export organization 389
Oceania 477
OECD (Organization for Economic Co-operation and Development):
arms exports 477
arms imports 476
CSCE and 176
developing countries and 394
economy 21
Oerlikon-Bührle 471
ONUC (UN Operation in the Congo) 52
ONUCA (UN Observer Group in Central America) 52
'Onumoz', Operation 116
ONUSAL (UN Observer Mission in El Salvador) 52, 53, 61–65
OPANAL 572
OPEC (Organization of Petroleum Exporting Countries):
arms exports 477
arms imports 476
Open Skies Treaty (1992) 591, 620, 632–35:
text 653–71
Ordnance Factories 431, 472
Organization for the Prohibition of Chemical Weapons 707, 721–23
Oslo Declaration 596–98, 677–82
Oto Melara 431, 435, 473
*Our Common Future* 18–19
Outer Space Treaty (1967) 760, 763, 765–81
Owen, Lord 91
ozone layer 19

Pakistan:
aid to 395
arms imports 452, 453
CBW allegations concerning 265
India and 83, 125, 247
missiles 249
nuclear weapons 247
Palestinians 84–86, 123
Palme Commission 15, 16, 18, 22, 24
Paris, Charter of (1990) 8, 178, 180
Paris Agreements/Conference 54, 108, 109, 110
Paris Club 410
Partial Test Ban Treaty (1963) 759, 763, 765–81
Patriot missile 315, 316
Pavlov, Valentin 354
peace-building 47, 57

'peace dividend' 81, 86, 397
peace enforcement 46, 47, 48, 50–51
peace-keeping 47, 49, 52–57
peace-making 47, 50
*perestroika* 95, 103
Persian Gulf War:
  arms trade and 533–34
  cease-fire 691
  chemical weapons and 718, 732
  electronic warfare 326
  environmental implications 289
  satellites and 32, 323
  UN Resolutions 691 *see also under* Iraq
Peru 130
Philippines 126
Philips 441
Plessey 441, 442
Pluton missile 230, 237
Poland:
  arms exports 446
  armed forces 408–9
  economy 398, 400, 407–8
  military expenditure 409–10
Polaris missile 230, 238
Polaris submarine 380
pollution 18, 19, 23, 24
population: world trends 18, 19, 30, 32
Portugal:
  military expenditure 368, 369, 370, 372, 373
  military personnel 376, 377
Poseidon submarine 223
poverty 20, 30
Powell, Colin 225, 423, 555
Pratt & Whitney 471
protein engineering 297, 302
public opinion 7

Q-5 aircraft 232
Qadhafi, Muammar 265

Rabin, Yitzak 84, 437
Rafale aircraft 229–30, 375, 422
Rakhmonov, Imamali 106, 169, 170
Ramphal, Sir Shridath 25
Rapid Reaction Force 378, 380
Rarotonga Treaty (1985) 762, 763, 765–81
Raytheon 429, 431, 470
Reagan, Ronald 715
Red Cross 106
Reed, Thomas 312
regional co-operation 22–26
regional organizations 58–59
Renwick, Sir Robin 565
research centres, international nuclear 17
resources *see* natural resources
retrofits 456–59
Rice, Donald 423
Rifkind, Malcolm 425, 426
Rio Conference *see* United Nations Conference on Environment and Development
Rio Declaration 37–39
riot control agents 260, 265, 279, 733
Ritter von Wagner, Adolf 706
Rock, Stephen 3
Rockwell 470
Rolls Royce 471
Romania:
  armed forces 412–13
  economy 398, 400, 412
  military expenditure 413–14
  Moldova and 101fn., 102, 143
Rühe, Volker 425, 426
Russia:
  ABM system 231
  aid to 366 *see also* US disarmament aid to
  armed forces, independent 351, 356
  arms exports 351, 362, 389, 416–17, 421–22, 443, 446, 448–51, 466–67, 479–82
  arms industry 338, 360–66, 415
  Azerbaijan and 147
  Baltic states and 143
  biological weapons and 271–72, 288, 290
  CFE and 593, 595, 597, 601–3, 606–7, 613, 616
  chemical weapons and 265–67, 271–72, 290, 712:
    destruction 260, 273, 277–81, 290
  China and 140
  CIS and 94, 133
  conversion 340, 354, 362–66
  CTB and 563
  CWC and 716
  disintegration trends 144
  economy 352, 354–56, 360, 365
  ethnic problems in 146
  ethno-nationalism 142
  Federal Treaty 146
  Georgia and 98–99, 100, 139, 147
  *glasnost* 354–59
  Japan and 140
  military expenditure 340, 351–67
  military expenditure on R&D 351, 355–56, 359, 364, 367
  minority populations abroad 94, 143
  missile defence 318–20
  missile exports 248–49
  Moldova and 102, 139
  'near abroad' 133fn., 143, 144
  Northern Caucasus 93, 103–4, 146–47, 156
  nuclear disarmament and 16
  nuclear weapons 4, 148, 226–28, 601–2:
    numbers 235

reductions 221, 236
withdrawals of 228
nuclear weapon tests 254
peace-keeping in other republics 58
plutonium disposal 573
President–Parliament conflict 103, 135
SSD talks 566, 567–69
START and 550, 551, 552, 554
Transcaucasus 147, 152–53
troops abroad 94, 96, 101, 152, 153, 356, 603–6
Ukraine and 138, 151
uranium sales 570–71
US disarmament aid to 273, 281, 290, 366, 554, 569
USA, co-operation with 6
USSR's UN Security Council seat 594
warhead dismantlement facilities 568
Rutskoy, Alexander 101, 366
Rwanda 88, 128

S3 missile 228, 229, 237
S45 missile 228
SA-2 missile 249
Saab 422, 431, 436, 473
Safety, Security and Dismantlement (SSD) Talks 6fn., 551, 566–71
Sann, Son 107
Santos, President José Eduardo 114–15
satellites:
*general references*
launches, 1992 330–34
trade in 322, 447
*individual countries*
China 331
CIS 330, 331–32, 332–33, 334
France 332
India 250
USA 320, 321–24, 331, 332, 333, 334
Saudi Arabia:
arms imports 421, 422
military expenditure 308
Savimbi, Jonas 115
Scud missile 228, 246, 700:
Iraqi 699, 700
SDI (Strategic Defense Initiative) 311, 324, 329, 345
SDIO (Strategic Defense Initiative Organization) 313, 317–21
Seabed Treaty (1971) 761, 765–81
Sea King helicopter 231
Seawolf submarine 316, 323, 327, 344
security:
common 15–17
comprehensive 17–21
concept of 9
economic 20–21
environmental *see* environment: security
Senegal River 29
Serbia:
Bosnia and 89, 90
CSCE expulsion 620
Kosovo and 93
non-lethal warfare agents and 267
refugees 92
sanctions 58, 90
UN trade embargo 419–21
Yugoslav unity and 88
Shakhrai, Sergey 104, 137fn.
Shanibov, Musa 167
Shaposhnikov, Marshal Yevgeniy 96, 102, 105, 160, 168, 553, 563, 595
Shavit space launch vehicle 250
Shevardnadze, Eduard 98, 99, 100, 142fn., 160, 165, 466, 592, 593, 600, 715
ships: R&D 315–16
Shushkevich, Stanislav 142fn., 550, 599
Siemens 441, 442, 472
Sierra Leone 83fn.
Sigua, Prime Minister Tengiz 170
Sihanouk, Prince Norodam 107
'silver bullet' technologies 325, 328
SIPRI (Stockholm International Peace Research Institute) 9
Sistemi 435
Slovak Republic:
armed forces 401, 403
arms exports 405
CFE and 598
economy 402–3
military expenditure 403–5
Slovenia 88, 89, 92, 419
Small ICBM 221, 222
SNECMA 471, 472
Snegur, President Mircea 101, 102, 164, 165, 168
Sobchak, Anatoly 355
Somalia:
aid to 349
conflict in 84, 128
Operation Restore Hope 50–51
starvation in 45
UN and 52, 56
South:
problems of 23
resource transfer to 23–24
South Africa:
Angola and 112
apartheid 112, 113
arms industry 437–48
Boipatong 113
CODESA (Convention for a Democratic South Africa) 113
conflict in 49, 113–14, 128
elections 116
Inkhata Freedom Party 117, 118

Mozambique and 112, 115
nuclear weapons 252–53
Pan African Congress 117
referendum 113
UN and 113–14
South America:
arms exports 477
arms imports 418, 476
conflicts in 86, 87
military expenditure 393
South Asian nuclear weapon-free zone 247
South Commission 26
South–North problems 18
South Ossetia 58, 94, 95, 98–99, 104, 134, 138, 156
Spain:
military expenditure 368, 369, 370, 372, 373
military personnel 376, 377
satellites 322
Sri Lanka 126
SS-11 missile 226
SS-17 missile 226, 235
SS-18 missile 226, 233, 235, 556, 557–58, 559, 570
SS-19 missile 235, 552, 556, 557, 558, 559, 570
SS-24 missile 226, 233, 235
SS-25 missile 226, 235, 553, 556
SS-N-8 missile 235
SS-N-18 missile 235
SS-N-20 missile 227, 235
SS-N-21 missile 228
SS-N-23 missile 235
START I (1991) 236, 549–54:
Lisbon Protocol 550, 551, 553, 574–75
START II 549, 554–60, 572, 573, 576–89:
limitations under 22, 223, 226, 227, 231, 236, 348, 366
Star Wars *see* SDI
Stepankov, Valentin 144
Stockholm Conference (1972) 18
Stoltenberg, Gerhard 418, 426, 451
Su-27 aircraft 389
submarines:
China 232
France 229, 375
Russia 221, 226, 227
UK 380
Sudan 129
summit meetings:
Moscow 713
Washington 555, 556
Super Etendard aircraft 237
Sweden:
arms exports 446, 467–68, 479–82
arms industry 435–36
dumped chemical ammunition and 283
Swedish Initiative 1991, 26
Switzerland 446
Syria 85, 264

TACSATS 322
Taiwan: arms imports 421, 422, 446, 447, 461
Taiwan Aerospace Corporation 439
Tajikistan:
CIS and 105, 106, 134
conflict in 93, 94, 95, 104–7, 137, 155
military expenditure 356
R&D 315
Russia and 58, 106, 107
Tashkent Document 596, 598, 600, 601, 607, 671–77
Tatarstan 137, 146
technology spread, controlling 324
Ter-Petrosyan, President Levon 97, 141fn., 166, 600
Texas Instruments 471
Textron 434, 471
Thailand 110, 111, 112, 453, 454
Thiokol 431, 473
Third World *see* developing countries
Thirsty Sabre project 314
Thomson 416, 429, 431, 439, 440, 441, 470, 474
Thorn EMI 441, 473
Tiger attack helicopter 375
Tigris River 522
Tlatelolco Treaty (1967) 571–72, 760, 763, 765–81
Tomahawk missile 225
Topaz reactors 320
Tornado aircraft 231, 230, 238, 379, 422
Transcaucasus 147, 152–53
Trans-Dniester dispute 58, 94, 95, 96, 101–3, 134, 138, 143, 152, 156, 606
Travkin, Nikolai 365
Trident missile 221, 222, 223, 230, 234, 560
Trident submarine 223, 230–31, 380
TUCAS Engine Industries 526, 531
Turkey:
aerospace industry 531
arms embargo on 523
arms imports 418, 450, 452, 530
arms industry:
administrative basis 524–25
financing 525–26
modernization 521–32
producers 526
projects 529
Azerbaijan and 96, 522
conflict in 84, 123, 418
Cyprus and 523
EC and 381, 522
Greece and 523

Kurds in 418, 451, 522
military aid to 521
military expenditure 368, 369, 370, 372, 373, 521
military personnel 376, 377
NATO's cascade and 532
PKK 522
strategic context 522
USA and 523, 531
Turkistan 522
Turkmenistan:
military expenditure 356
TUCAS 526, 531

U-2 aircraft 694
Uganda 82, 395
Ukraine:
arms exports 389, 417, 448–51
arms industry 362, 415
CFE and 358 594, 597, 599
chemical weapons 266
military expenditure 338, 340, 351–552, 356, 358–59
nuclear weapons 4, 148, 149, 228, 601–2
refugees 102
Russia and 139, 151
SSD talks 566, 569–70
START and 550, 552, 553–54, 566, 569
US disarmament aid 569
UNAVEM (UN Angola Verification Missions) 52, 61–65, 114
UNDOF (UN Disengagement Observer Force) 61–65
UNEF (UN Emergency Force) 52
UNFICYP 61–65
UNIDO (UN Industrial Development Organization) 355
UNIFIL (UN Interim Force in Lebanon) 61–65, 85
UNIIMOG (UN Iran–Iraq Military Observer Group) 52
UNIKOM (UN Iraq–Kuwait Observer Mission) 52, 57, 61–65
UNIMOGIP (UN Military Observer Group in India and Pakistan) 61–65
Union of Soviet Socialist Republics (USSR):
arms exports 449, 450
biological weapons 260, 287–88
breakup:
borders and 94
conflicts after 9, 49, 93–107, 131–58
nuclear weapons and 6
transition period 132–40
CFE treaty-limited equipment and 594–96
chemical weapons 260, 267, 283, 710, 712
*coup d'etat* 131, 136
*glasnost* 354, 715
military expenditure 338, 351–52, 353, 354–56, 359–60
nuclear weapon tests 255, 256, 257
*perestroika* 359–62
Sverdlovsk 287, 288, 290
troop withdrawals 398
*see also under names of independent republics*
United Arab Emirates 447
United Kingdom:
ammunition dumping 284
arms exports 417, 428, 446, 479–82
BAOR 380
chemical weapons 710
conflict in 122
CTB and 565
Lakenheath AFB 225
military expenditure 368, 369, 370, 372, 373, 374, 378–80
military expenditure on R&D 308
military personnel 376, 377, 378, 379, 380
nuclear weapons 221, 222–25, 230–32, 233, 238
nuclear weapon tests 254, 255, 256, 257, 265
'Options for Change' 379–80
United Nations:
*Agenda for Peace* 17, 45, 46–48, 56, 66–80
arms trade control and 459–61
Charter 17, 48, 172, 177, 186, 697
co-ordination in 24–25
CSCE and 177
diplomacy, preventive 46–47, 49–50
disarmament post-cold war 4–5
Economic Commission for Europe 176, 177
finances 48
member states 790–91
peace-building 47, 57
peace enforcement 46, 47, 48, 50–51
peace-keeping 47, 49, 59–65
peace-making 47, 50, 52–57
Regional Commissions 22
Register of Conventional Arms 463, 533–44
*Report, September 1992* 45, 48–49, 54
Weapons of Mass Destruction guidelines 545–46
*see also under names of peace-keeping forces; for relations with individual countries, see under names of countries*
United Nations Conference on Environment and Development (UNCED) (Rio Conference) 23, 24, 25, 32, 37–42
United Nations Development Programme 20, 28

United States of America (USA):
Aberdeen Proving Ground 275, 276
Anniston Army Depot 275, 276
arms exports 417, 422, 428, 446, 461, 467, 479–82
arms imports 438–39, 443–48
arms industry 416, 430, 433–35
ASAT programme 324
biological weapons and 712
chemical weapons and 272, 710, 712
chemical weapon destruction 260, 273–77, 290
CWC and 705, 714, 715, 716
Clayton Antitrust Act 434
Congress:
chemical weapons and 274
SDI and 317–19
CTB and 561–63
DARPA 313–17
Defense Production Act 439
economy 327, 341
elections 84
Energy and Water Development Appropriations Act 562
fissile material production cut-off 571
Former Soviet Union Demilitarization Act 566
Iran–Iraq Non-Proliferation Act 467
Johnston Atoll (JACADS) 273, 274–76
Los Alamos National Laboratory 313
Middle East and 84–85
military expenditure 307, 308, 337, 339, 340, 341–51, 368, 369, 370, 372, 373
military expenditure on R&D 308, 309–10, 313, 344–46, 347–48
military personnel 343, 346, 347–50, 376, 377
missile defence 318, 319–20
Nevada Test Site 562, 565
non-lethal technology 268
non-proliferation research 272
nuclear weapons:
numbers of 234
reductions 221, 236
withdrawals of 224–25
nuclear weapon tests 221, 254, 255, 256, 257
R&D 307–21
Rocky Flats plant 223
satellite exports 447
Somalia and 50–51
Soviet Nuclear Threat Reduction Act 566
space programmes 307, 321–24
START and 551
technical supremacy 307, 308, 328
Tooele Army Depot 275, 276, 277, 281
troops withdrawn 285
uranium purchases 570–51
United Technologies 470
UNOMSA (UN Observer Mission in South Africa) 114
UNOSOM (UN Operation in Somalia) 56, 61–65
UNPROFOR (UN Protection Force) 55, 61–65
UNSCOM (UN Special Commission on Iraq) 252, 263, 691–703:
inspection schedule 698–99
UNTAC (UN Transitional Authority in Cambodia) 54, 61–65, 107, 109, 110, 111
UNTAG (UN Transition Assistance Group in Namibia) 52
UNTEA (UN Temporary Executive Authority) 53
UNTSO (UN Truce Supervision Organization) 61–65
uranium, highly enriched 247, 253, 554, 570–71
Uruguay Round 21
Uzbekistan 106, 145

Valetta Experts Meeting 180
Vance, Cyrus 49, 91, 97, 114
Vanguard Class submarine 380
Vasilev, Sergey 363
Vienna Document 1990, 619
Vienna Document 1992:
communications 628
data bank 629–30
data exchange 622
future 630–41
implementation 620–30
military activities 620–21, 625–27
military contacts 624–25
participants 619, 628–29
risk reduction 624
signing 591
text 635–53
verification 627
Viet Nam:
chemical weapons in 712
*see also* UNTAC *and under* Cambodia
Visegrad Group 187, 410
Volski, Arkadi 365
Volvo 436
VSEL 431, 472
VX gas 266, 273–75, 278, 279

W-76 warheads 223
W-88 warheads 221, 223
Walesa, President Lech 163
Warbreaker programme 314
war crimes 23
Warsaw Guidelines 243–44
water resources 29, 31, 34
WE-177 bombs 231

Western European Union (WEU) 176, 177, 381, 382, 425
Western Sahara 50, 82
Westinghouse 471
West Irian 53
Westland Group 431, 473
Wolfowitz, Paul 326
Woolsey, James 251
World Bank 21, 395
World Commission on Environment and Development 18–19
World War I 716
World War II 710
WTO (Warsaw Treaty Organization):
  demise of 16, 398–99, 403, 406
  military expenditure 399
  TLE holdings 609
Wyoming Joint Memorandum (1989) 273

Yanpolsky, Gennadiy 448
Yasin, Yevgeni 363
Yeltsin, President Boris:
  arms exports 466
  biological weapons 287, 288
  chemical weapons and 273, 277
  fissile material and 571
  military expenditure 358
  Moldova and 102
  nuclear weapons and 228
  nuclear weapon tests 254
  Parliament, conflict with 103, 135
  Russia as peace-keeper 156
  South Ossetia 98
  START and 554, 555, 557
  Tajikistan and 105
  troops abroad and 606
Yockey, Donald 423
Yoselyani, Dzhaba 98
Yugoslavia:
  CSBMs and 618
  CSCE and 22–23, 173, 420
  EC and 58, 420
  NATO and 58
  refugees 92
  UN and 22–23, 48, 55–56
  UN arms embargo 418–21
  WEU and 58
  *see also* Bosnia and Herzegovina; Croatia; Macedonia; Montenegro; Serbia; Slovenia

Zaire, aid to 395
Zlenko, Anatoliy 553